Grand Challenges in Biology and Biotechnology

Series editor
Pabulo H. Rampelotto

More information about this series at http://www.springer.com/series/13485

Pabulo H. Rampelotto
Editor

Biotechnology of Extremophiles

Advances and Challenges

Springer

Editor
Pabulo H. Rampelotto
Federal University of Rio Grande do Sul
Porto Alegre, Rio Grande do Sul, Brazil

ISSN 2367-1017 ISSN 2367-1025 (electronic)
Grand Challenges in Biology and Biotechnology
ISBN 978-3-319-13520-5 ISBN 978-3-319-13521-2 (eBook)
DOI 10.1007/978-3-319-13521-2

Library of Congress Control Number: 2016934051

Springer Cham Heidelberg New York Dordrecht London

Printed on acid-free paper

Springer International Publishing AG Switzerland is part of Springer Science+Business Media (www.springer.com)

This first book of my series, which I am personally editing, is dedicated to the first person who inspired me in life, my dearest mother Veronete. I am also grateful to those who taught me the real meaning of science and its beauty, in special, Stephen J. Gould, Ernst Mayr, and Carl E. Sagan.

Foreword

"*Living organisms cannot colonize environmental territories on our planet that were thought to be extreme from an anthropogenic point of view.*" Such a statement was well accepted in the scientific community for a long time. However, the last century taught us that microorganisms are even capable of thriving in the harshest places on earth including the deep sea, hot springs, glaciers, deserts, and solfataric fields. Microorganisms belonging to the domains of *Bacteria* and *Archaea* exclusively populate such extremely demanding environments, and we now start to understand that the so-called extremophiles must be adequately adapted with regard to metabolic processes, biological functions, genomes, and transcriptomes to overcome the challenges of life. Nowadays, fantastic new species have been discovered in both natural and artificial extreme environments. Moreover, these kinds of microorganisms are characterized by their unique and highly adapted enzymes, the so-called extremozymes, which can be applied in various industrial processes.

The purpose of the book *Biotechnology of Extremophiles: Advances and Challenges* is to give an extensive overview about several contemporary fields of research on extremophilic microbes and their potential application in industry. Well-renowned scientists have contributed outstanding book chapters covering their respective fields of expertise. The topics of these chapters illustrate recent and remarkable progress and feature the best of the latest in academic and industrial research on extremophiles.

The first part of the book is focused on basic aspects on different types of extremophilic microbes, environmental conditions, and the potential for biotechnological applications. A comprehensive overview on growth and metabolism of extremophiles with a special focus on carbohydrates and amino acids is provided (Chap. 1). Other authors explore cold environments including Antarctica, Arctic, and Himalayan glaciers to discuss the biodiversity and biotechnological potential of psychrophiles (Chaps. 2 and 3). Thermophilic species are presented as valuable sources to produce bioactive exopolysaccharides (Chap. 4). Three chapters focus on the lifestyle of acidophilic microbes including acidophilic sulfur reducers and biofilm-forming acidophilic metal oxidizers (Chaps. 5–7), while alkaliphilic prokaryotes are described as producers of highly relevant enzymes for versatile industrial applications (Chap. 8). Finally, the nitrogen metabolism in halophiles is the topic of another interesting article (Chap. 9).

All these organisms represent a "treasure chest" of biocatalysts with a tremendous potential for versatile industrial applications (especially in "white industrial biotechnology"), which is the major topic of the second part of this book dealing with novel screening techniques including proteomics, biochemical characterization, and supply of important groups of extremozymes (Chaps. 10 and 11). Extensive efforts have been undertaken to understand their roles in metabolism, biochemical functionality, and evolution. These biocatalysts are well known to cope with extremes of temperatures, pressure, and high concentrations of toxic metal ions or organic solvents. In this book, experts in the field illustrate the potential of lipolytic enzymes and proteases to be used in different industrial applications (Chaps. 12–14) and demonstrate the great potential of cold-active beta-galactosidases that are relevant in cosmetics, pharmaceutical, and food industry (Chap. 15). Further authors shed light on the evolution and biotechnological application of pesticide-hydrolyzing lactonases and alpha-amylases (Chaps. 16 and 17), while additional chapters report on enzymes from thermophiles, including DNA-replication proteins (Chap. 18) and lignocellulolytic hydrolases for the degradation of plant waste materials (Chap. 19).

In the third part, a bioinformatics approach to reconstruct ancestral protein sequences that can be used for the creation of heat-stable proteins (Chap. 20) and a systems biology framework that uses high-throughput "omics" technologies to investigate cell function in response to temperature shifts are presented (Chap. 21). Moreover, the potential of experimental microbial evolution is critically evaluated using state-of-the-art methodologies (Chap. 22), and immobilization strategies to isolate, purify, and reuse thermozymes based on the utilization of solid-binding peptides are described (Chap. 23). Molecular dynamics simulations are portrayed as an efficient tool to investigate structure-function relationships in extremozymes from psychrophiles (Chap. 24), and finally, the establishment of an expression system to produce malaria vaccines in a halophilic heterologous host is discussed (Chap. 25).

In summary, this book sheds light on various aspects of extremophilic microorganisms and their enzymes. It is a unique read and greatly covers some of the most exciting and innovative areas in the wide research field of "biotechnology of extremophiles." My congratulations to the authors and the editor!

Hamburg, Germany Garabed Antranikian

Preface

Over the last decades, the study of extremophiles has been providing ground breaking discoveries that challenge our understanding of biochemistry and molecular biology. In the applied side, extremophiles and their enzymes have spawned a multibillion dollar biotechnology industry, with applications spanning biomedical, pharmaceutical, industrial, environmental, and agricultural sectors. Taq DNA polymerase (which was isolated from *Thermus aquaticus* from a geothermal spring in Yellowstone National Park) is the most well-known example of the potential biotechnological application of extremophiles and their biomolecules. Indeed, the application of extremophiles and their biologically active compounds has opened a new era in biotechnology. However, despite the latest advances, we are just in the beginning of exploring the biotechnological potentials of extremophiles.

For this reason, I have chosen this topic to be the *premier* book of my series. In addition, I am motivated by the success of my previous special issues involving extremophiles and biotechnology. Another compelling reason to develop this project is because the successful commercial application of extremophiles is not well documented in the scientific literature, a consequence of the highly competitive nature of industrial R&D.

Aimed at research scientists and biotechnologists, this book is an essential reading for those working with extremophiles and their potential biotechnological application. Here, we provide a comprehensive and reliable source of information on the recent advances and challenges in different aspects of the theme. Written in an accessible language to the general public, the book is also a recommended reference text for anyone interested in this thriving field of research.

Porto Alegre, Rio Grande do Sul, Brazil — Pabulo H. Rampelotto

About the Editor

Pabulo Henrique Rampelotto is editor in chief of the book series *Grand Challenges in Biology and Biotechnology* (Springer). He is also editor in chief, associate editor, guest editor, and member of the editorial board of several scientific journals in the field of life sciences and biotechnology. Most of his recent work has been dedicated to the editorial process of several scientific journals as well as on the organization of books and special issues in his fields of expertise. In his books and special issues, some of the most distinguished team leaders in the field have published their work, ideas, and findings, including Nobel laureates and several of the highly cited scientists according to the ISI. His research interests encompass different fields of science, from basic to applied research, including biotechnology, metagenomics, next-generation sequencing, molecular biology and biochemistry of microorganisms, extremophiles, and astrobiology.

When he is not working, Pabulo enjoys spending time walking in the woods, in the mountains, and near the sea…thinking, always thinking.

Contents

Contributors

Garabed Antranikian Institute of Technical Microbiology, Hamburg University of Technology (TUHH), Hamburg, Germany

Satoshi Akanuma Department of Applied Life Sciences, Tokyo University of Pharmacy and Life Sciences, Hachioji, Tokyo, Japan

Faculty of Human Sciences, Waseda University, Shinjuku, Japan

Vanesa Bautista Department of Agrochemistry and Biochemistry, University of Alicante, Alicante, Spain

Sören Bellenberg Aquatische Biotechnologie, Biofilm Centre, Universität Duisburg – Essen, Essen, Germany

Peter L. Bergquist Department of Molecular Medicine and Pathology, Medical School, University of Auckland, Auckland, New Zealand

Department of Chemistry and Biomolecular Sciences, Macquarie University, Sydney, NSW, Australia

Biomolecular Frontiers Research Centre, Macquarie University, Sydney, NSW, Australia

Aneta Białkowska Institute of Technical Biochemistry, Lodz University of Technology, Lodz, Poland

Jenny M. Blamey Fundación Científica y Cultural Biociencia, José Domingo Cañas, Santiago, Chile

Paul Blum School of Biological Science, University of Nebraska, Lincoln, NE, USA

Freddy Boehmwald Fundación Científica y Cultural Biociencia, José Domingo Cañas, Santiago, Chile

Ivanka Bojadjieva Institute of Microbiology, Bulgarian Academy of Sciences, Sofia, Bulgaria

María José Bonete Department of Agrochemistry and Biochemistry, University of Alicante, Alicante, Spain

Emanuele Bosi Department of Biology, University of Florence, Florence, Italy

Mónica Camacho Department of Agrochemistry and Biochemistry, University of Alicante, Alicante, Spain

Andrew Care ARC Centre of Excellence for Nanoscale BioPhotonics (CNBP), Macquarie University, Sydney, NSW, Australia

Madhab Kumar Chattopadhyay CSIR-Centre for Cellular and Molecular Biology, Hyderabad, India

Chi-Won Choi Division of Bioconvergence Analysis, Daejeon, Republic of Korea

Hubert Cieśliński Faculty of Chemistry, Department of Molecular Biotechnology and Microbiology, Gdańsk University of Technology, Gdańsk, Poland

Priya DasSarma Department of Microbiology and Immunology, University of Maryland School of Medicine, Baltimore, MD, USA

Institute of Marine and Environmental Technology, University System of Maryland, Baltimore, MD, USA

Shiladitya DasSarma Department of Microbiology and Immunology, University of Maryland School of Medicine, Baltimore, MD, USA

Institute of Marine and Environmental Technology, University System of Maryland, Baltimore, MD, USA

Angelina Lo Giudice Institute for Coastal Marine Environment, National Research Council (IAMC-CNR), Messina, Italy

Department of Biological and Environmental Sciences, University of Messina, Messina, Italy

Immacolata Del Giudice Institute of Protein Biochemistry, CNR, Naples, Italy

Paola Di Donato Institute of Biomolecular Chemistry ICB, CNR, Pozzuoli, NA, Italy

Department of Sciences and Technologies, Centro Direzionale, University of Naples 'Parthenope', Naples, Italy

Skander Elleuche Institute of Technical Microbiology, Hamburg University of Technology (TUHH), Hamburg, Germany

Julia Esclapez Department of Agrochemistry and Biochemistry, University of Alicante, Alicante, Spain

Renato Fani Department of Biology, University of Florence, Florence, Italy

Ilaria Finore Institute of Biomolecular Chemistry ICB, CNR, Pozzuoli, NA, Italy

Tomasz Florczak Institute of Technical Biochemistry, Lodz University of Technology, Lodz, Poland

Anna P. Florentino Laboratory of Microbiology, Wageningen University, Wageningen, The Netherlands

Sub-Department of Environmental Technology, Wageningen University, Wageningen, The Netherlands

Patricio Flores Fundación Científica y Cultural Biociencia, José Domingo Cañas, Santiago, Chile

Marco Fondi Department of Biology, University of Florence, Florence, Italy

María Teresa García Faculty of Pharmacy, Department of Microbiology and Parasitology, University of Sevilla, Sevilla, Spain

Dennis W. Grogan Department of Biological Sciences, University of Cincinnati, Cincinnati, OH, USA

Ewa Gromek Institute of Technical Biochemistry, Lodz University of Technology, Lodz, Poland

Gaetano Invernizzi Structural Biology and NMR Laboratory, Department of Biology, University of Copenhagen, Copenhagen, Denmark

Štefan Janeček Laboratory of Protein Evolution, Institute of Molecular Biology, Slovak Academy of Sciences, Bratislava, Slovakia

Faculty of Natural Sciences, Department of Biology, University of SS Cyril and Methodius, Trnava, Slovakia

Margarita Kambourova Institute of Microbiology, Bulgarian Academy of Sciences, Sofia, Bulgaria

Ram Karan Department of Microbiology and Immunology, University of Maryland School of Medicine, Baltimore, MD, USA

Institute of Marine and Environmental Technology, University System of Maryland, Baltimore, MD, USA

Jong-Myoung Kim Department of Microbiology and Immunology, University of Maryland School of Medicine, Baltimore, MD, USA

Institute of Marine and Environmental Technology, University System of Maryland, Baltimore, MD, USA

Department of Marine Bio-Materials and Aquaculture, PuKyong National University, Pusan, Republic of Korea

Seung Il Kim Division of Bioconvergence Analysis, Daejeon, Republic of Korea

Ewelina Krajewska Faculty of Chemistry, Department of Molecular Biotechnology and Microbiology, Gdańsk University of Technology, Gdańsk, Poland

Joanna Krysiak Institute of Technical Biochemistry, Lodz University of Technology, Lodz, Poland

Józef Kur Faculty of Chemistry, Department of Molecular Biotechnology and Microbiology, Gdańsk University of Technology, Gdańsk, Poland

Licia Lama Institute of Biomolecular Chemistry ICB, CNR, Pozzuoli, NA, Italy

Sang-Yeop Lee Division of Bioconvergence Analysis, Daejeon, Republic of Korea

Kesen Ma Department of Biology, University of Waterloo, Waterloo, ON, Canada

Gashaw Mamo Department of Biotechnology, Lund University, Lund, Sweden

Indienz AB, Billeberga, Sweden

Giuseppe Manco Institute of Protein Biochemistry, CNR, Naples, Italy

M. Carmen Márquez Faculty of Pharmacy, Department of Microbiology and Parasitology, University of Sevilla, Sevilla, Spain

Rosa María Martínez-Espinosa Department of Agrochemistry and Biochemistry, University of Alicante, Alicante, Spain

Bo Mattiasson Department of Biotechnology, Lund University, Lund, Sweden

Indienz AB, Billeberga, Sweden

Samuel McCarthy School of Biological Science, University of Nebraska, Lincoln, NE, USA

Encarnación Mellado Faculty of Pharmacy, Department of Microbiology and Parasitology, University of Sevilla, Sevilla, Spain

María de Lourdes Moreno Faculty of Pharmacy, Department of Microbiology and Parasitology, University of Sevilla, Sevilla, Spain

Patricio Muñoz Fundación Científica y Cultural Biociencia, José Domingo Cañas, Santiago, Chile

Thomas R. Neu Department of River Ecology, Helmholtz Centre for Environmental Research – UFZ, Magdeburg, Germany

Barbara Nicolaus Institute of Biomolecular Chemistry ICB, CNR, Pozzuoli, NA, Italy

Elena Papaleo Structural Biology and NMR Laboratory, Department of Biology, University of Copenhagen, Copenhagen, Denmark

Edmond Changkyun Park Division of Bioconvergence Analysis, Daejeon, Republic of Korea

Deepak Parashar Department of Microbiology, University of Delhi, New Delhi, India

Benjamin Pavlik Department of Chemical and Biomolecular Engineering, University of Nebraska, Lincoln, NE, USA

Anna Pawlak-Szukalska Faculty of Chemistry, Department of Molecular Biotechnology and Microbiology, Gdańsk University of Technology, Gdańsk, Poland

Wolf T. Pecher Department of Microbiology and Immunology, University of Maryland School of Medicine, Baltimore, MD, USA

Institute of Marine and Environmental Technology, University System of Maryland, Baltimore, MD, USA

College of Arts and Sciences, University of Baltimore, Baltimore, MD, USA

Laia Pedro-Roig Department of Agrochemistry and Biochemistry, University of Alicante, Alicante, Spain

Carmen Pire Department of Agrochemistry and Biochemistry, University of Alicante, Alicante, Spain

Francisco Pérez-Pomares Department of Agrochemistry and Biochemistry, University of Alicante, Alicante, Spain

Annarita Poli Institute of Biomolecular Chemistry ICB, CNR, Pozzuoli, NA, Italy

Elena Porzio Institute of Protein Biochemistry, CNR, Naples, Italy

Nadja Radchenkova Institute of Microbiology, Bulgarian Academy of Sciences, Sofia, Bulgaria

Deepak Rudrappa School of Biological Science, University of Nebraska, Lincoln, NE, USA

Irene Sánchez-Andrea Laboratory of Microbiology, Wageningen University, Wageningen, The Netherlands

Wolfgang Sand Aquatische Biotechnologie, Biofilm Centre, Universität Duisburg – Essen, Essen, Germany

Gundlapally Sathyanarayana Reddy CSIR-Centre for Cellular and Molecular Biology, Hyderabad, India

Tulasi Satyanarayana Department of Microbiology, University of Delhi, New Delhi, India

Carola Schröder Institute of Technical Microbiology, Hamburg University of Technology (TUHH), Hamburg, Germany

Archana Sharma Department of Biophysics, University of Delhi, New Delhi, India

Sisinthy Shivaji CSIR-Centre for Cellular and Molecular Biology, Hyderabad, India

Jhaveri Microbiology Centre, L V Prasad Eye Institute, Hyderabad, India

Raghuveer Singh School of Biological Science, University of Nebraska, Lincoln, NE, USA

Alfons J.M. Stams Laboratory of Microbiology, Wageningen University, Wageningen, The Netherlands

CEB-Centre of Biological Engineering, University of Minho, Braga, Portugal

Photini Sinnis Department of Molecular Microbiology and Immunology, Bloomberg School of Public Health, Johns Hopkins University, Baltimore, MD, USA

Anwar Sunna Department of Chemistry and Biomolecular Sciences, Macquarie University, Sydney, NSW, Australia

Biomolecular Frontiers Research Centre, Macquarie University, Sydney, NSW, Australia

Katarzyna Szulczewska Institute of Technical Biochemistry, Lodz University of Technology, Lodz, Poland

Matteo Tiberti Department of Biotechnology and Biosciences, University of Milano-Bicocca, Milan, Italy

Iva Tomova Institute of Microbiology, Bulgarian Academy of Sciences, Sofia, Bulgaria

Annabella Tramice Institute of Biomolecular Chemistry ICB, CNR, Pozzuoli, NA, Italy

Marianna Turkiewicz Institute of Technical Biochemistry, Lodz University of Technology, Lodz, Poland

Ching Tse Department of Biology, University of Waterloo, Waterloo, ON, Canada

Anna Vegara Department of Agrochemistry and Biochemistry, University of Alicante, Alicante, Spain

Mario Vera Institute for Biological and Medical Engineering, Schools of Engineering, Biological Sciences and Medicine, Pontificia Universidad Católica de Chile, Santiago, Chile

Department of Hydraulic and Environmental Engineering, School of Engineering, Pontificia Universidad Católica de Chile, Santiago, Chile

Marta Wanarska Faculty of Chemistry, Department of Molecular Biotechnology and Microbiology, Gdańsk University of Technology, Gdańsk, Poland

Jan Weijma Sub-Department of Environmental Technology, Wageningen University, Wageningen, The Netherlands

Monika Wicka Faculty of Chemistry, Department of Molecular Biotechnology and Microbiology, Gdańsk University of Technology, Gdańsk, Poland

Akihiko Yamagishi Department of Applied Life Sciences, Tokyo University of Pharmacy and Life Sciences, Hachioji, Tokyo, Japan

Sung Ho Yun Division of Bioconvergence Analysis, Daejeon, Republic of Korea

Ruiyong Zhang Aquatische Biotechnologie, Biofilm Centre, Universität Duisburg – Essen, Essen, Germany

Chapter 1
Growth and Metabolism of Extremophilic Microorganisms

Ching Tse and Kesen Ma

1 Introduction

Extremophiles are organisms growing optimally in extreme environments (Rothschild and Mancinelli 2001). Some of them can do so in more than one extreme environment. These environments include physical extremes of temperatures, radiation, and pressure or geochemical extremes such as desiccation, salinity, pH, oxygen species or redox potential (Rothschild and Mancinelli 2001). Extremophiles are present in three domains of life. In this review paper, the focus will be on microorganisms. Microorganisms are defined as a collection of organisms that can be visible only under the microscope (Rybicki 1990). These include bacteria, protozoa, algae and fungi with a size smaller than 55 μm, which is the smallest size of an object recognizable to the naked eyes (Keller and Friedli 1992). Those microorganisms that can tolerate or resist extreme conditions such as low water activity and high level of radiation and heavy metals, but cannot grow optimally under such conditions are considered to be extremotolerants or extremoresistants.

Extremophiles can be classified as the following groups: (a) acidophiles with a pH optimum for growth at ≤3.0 (Johnson et al. 2006), (b) alkaliphiles with a pH optimum for growth above 9 (often between 10 and 12) but cannot grow or grow slowly at near-neutral pH of 6.5 (Zhilina et al. 2004), (c) psychrophiles with an optimal temperature for growth of ≤15 °C and a maximum temperature for growth of <20 °C (Morita 1975), (d) thermophiles with optimal growth temperature of 45 to <80 °C (Kristjansson 1991), (e) hyperthermophiles with optimal growth temperature of ≥80 °C or capable of growing at ≥90 °C (Blumer-Schuette et al. 2008), (f) halophiles with optimal growth in the presence of high concentrations (>0.3 M)

C. Tse • K. Ma (✉)
Department of Biology, University of Waterloo,
200 University Avenue West, Waterloo, Canada, ON, N2L 3G1
e-mail: kma@uwaterloo.ca

P.H. Rampelotto (ed.), *Biotechnology of Extremophiles*, Grand Challenges in Biology and Biotechnology 1, DOI 10.1007/978-3-319-13521-2_1

of NaCl (DasSarma and DasSarma 2012), (g) piezophiles with optimal growth under high-pressure condition and can also grow at 50 MPa at a rate that is above 30 % of their growth rate at atmospheric pressure with other optimal growth conditions (Abe 2007), (h) xerophiles with optimal growth under at least one set of environmental conditions at a water activity (a_w) below 0.85 (Pitt and Hocking 2009), (i) osmophiles with optimal growth in the presence of high concentrations of organic solute, particularly sugars (Baross and Lenovich 1992), (j) oligotroph are organisms isolated using media containing 1–15 mg organic carbon/liter and sub-cultured with the same media (Poindexter 1981).

Extremotolerants and extremoresistants are listed as the following groups: (a) metallotolerants that can tolerate high concentrations (>1 mM) of heavy metals in their environments (Carapito et al. 2006), (b) radioresistants that can protect their cytosolic proteins from oxidation without the formation of spores and tolerate many DNA double-strand breakages after exposure to high, acute ionizing radiation (dose greater than 1 kilogray for 90 % reduction in Colony Forming Units), and can also resist prolonged desiccation (Sghaier et al. 2008).

Extremophiles exist in a wide range of environmental conditions and have some unique metabolic capacities and/or physical structures for thriving. Extremotolerants and extremoresistants also have some unique metabolic capacities and/or physical structures for surviving. Carbon and nitrogen sources are two major essential nutrients for all organisms, and they may be available in different forms and concentrations in their environments. It is very intriguing to understand how these microorganisms use various substrates and metabolic pathways for their thriving and survival. However, there are two modes for their growth, heterotrophy and autotrophy. In this review, the focus will be on heterotrophs and their catabolism of carbohydrates and peptides. Specifically, key enzymes, modified and alternate versions of the central metabolic pathways for the utilization of carbohydrates and peptides will be discussed in details, and a view about the metabolic connections between the metabolism of carbohydrates and peptides will be provided.

2 Growth, Substrates and Transporters

Extremophiles grow on many types of substrates including monosaccharides, disaccharides, polysaccharides, celluloses and hemicelluloses, peptides and amino acids (Kazak et al. 2010). Various substrates are tested using pure culture to determine which ones can be used for supporting their growth (Table 1.1). Different solutes are transported into the cell for utilization by varies transporters (Pflüger and Müller 2004; Albers and Driessen 2007; Siebold et al. 2001; Albers et al. 2004).

ATP-binding cassette transporters (ABC)-transporters, belonging to the class of primary transporter, is a major class of transporter in archaea and bacteria (Holland and Blight 1999). In both domains, the transporters have consensus sequences of a walker sequence, a Q-loop and an H-region in the ATPase subunit (Higgins 1995). In the system, there are two permeases and cytoplasmic ATPase domains are complemented by an extracellular binding protein (Albers et al. 2004).

Table 1.1 Growth of heterotrophic microorganisms under extreme conditions

Environment	Extremophiles	Growth conditions			Generation time (h)[d]	References
		Substrates[a]	pH[b]	Temperature (°C)[c]		
	Acidophile ($pH_{opt} \leq 3.0$)	*Bacteria* Glucose, sucrose, xylose, galatose, fructose, glycerate, glycerol, ethanol, malate, starch, yeast extract, proline, glutamic acid, leucine, casamino acids, tryptone	1.3–3.0	32–52	2.0–3.0	[1]
pH		*Archaea* Glucose, galactose, starch, yeast extract, peptone, casamino acid, tryptone, beef extract	1.0–2.2	53–96	4.0–5.0	[2]
	Alkaliphile ($pH_{opt} \geq 9$)	*Bacteria* Glucose, arabinose, galactose, lactose, maltose, xylose, cellobiose, maltose, D-trehalose, xylan, starch, glycogen, mannose, fructose, agar, saccharose, formate, trehalose, pectin, methionine, yeast extract, glutamate, peptone	9.0–13.0	45–60	3.0–4.0	[3]
	Psychrophile ($T_{opt} \leq 15$ °C and $T_{max} \leq 20$ °C)	*Bacteria* Glucose, fructose, sucrose, maltose, xylose, galactose, cellobiose, lactose, ribose, glycerol, mannitol, N-acetyl glucosamine, histidine, proline, alanine, tryptophan	5.0–10.0	5.0–15.0	1.0–6.0	[4]
		Archaea Glucose, xylose, yeast extract, tryptophan	6.0–9.0	1.0–15.0	5.0–20.0	[5]
Temperature	Thermophiles ($T_{opt} \geq 45$ °C but <80 °C)	*Bacteria* Glucose, xylose, arabinose, galactose, cellobiose, fructose, lactose, mannose starch, methionine, leucine, phenylalanine, yeast extract, tryptone	4.0–9.8	45.0–78.0	1.7–5.0	[6]
		Archaea glucose, sucrose, dextrose, lactose, galactose, starch, maltose, starch, yeast extract, trypticase, casein, peptone	2.7–8.0	50.0–75.0	1.65–4.0	[7]
	Hyperthermophiles ($T_{opt} \geq 80$ °C or $T_{max} \geq 90$ °C)	*Bacteria* Glucose, ribose, xylose, galactose, sucrose, maltose, starch, glycogen, yeast extract	6.5–7.5	80.0–85.0	1.5–2.0	[8]

(continued)

Table 1.1 (continued)

Environment	Extremophiles	Growth conditions			Generation time (h)[d]	References
		Substrates[a]	pH[b]	Temperature (°C)[c]		
		Archaea Starch, dextrin, xyloglucan, maltose, sucrose, lactose, glucose, xylose, galactose, arabinose, glycogen, raffinose, pullulanan, cellobiose, melibiose, lactate, pyruvate yeast extract, tryptone, peptone	5.5–7.0	80.0–106.0	0.5–3.2	[9]
Salt Concentration	Halophiles (≥0.3 M NaCl)	*Bacteria* Glucose, xylose, arabinose, raffinose, sucrose, galactose, lactose, rhamnose, trehalose, yeast extract, gelatin, starch, biotin, glutamic acid, proline, arginine, tyrosine, casamino acid	6.5–8.0	31.0–50.0	3.0–18.0	[10]
		Archaea Glucose, xylose, fructose, arabinose, lactose, xylulose, sucrose, starch, arginine, asparate, glutamate, glycine, glutamine, tryptophan, yeast extract	6.5–9.5	30.0–45.0	1.5–3.0	[11]
Water activity	Osmophiles (high concentrations of organic solute, sugars)	Eucarya Glucose, sucrose, fructose	3.5–5.5	28.0–35.0	1.80–12.0	[12]
		Bacteria Glucose, fructose, sucrose, glutamate	5.0–7.0	25.0–30.0	2.0–3.0	[13]
	Xerophiles (water activity $(a_w) \leq 0.85$.)	*Eucarya* Glucose, fructose, acetate, gluconate, D-galacturonate, sorbitol	5.5–6.8	20.0–41.0	2.0–4.0	[14]
Pressure	Piezophile ($P_{opt} \geq 50$ MPa)	*Bacteria* Glucose, cellobiose, fructose, maltose, mannose, salicin	6.5–8.0	10.0–98.0	0.3–6.2	[15]

Low nutrient concentration	Oligotroph (1–15 mg organic Carbon/liter)	*Bacteria* Glucose, xylose, sucrose, galactose, mannose, arabinose, methionine, glycine, serine, glutamine, galactose, mannose, mannitol, and ribose, lactose, L-proline	6.0–8.5	20.0–37.0	20.0–24.0	[16]
Heavy Metal	Metallotolerant (tolerate heavy metals of >1 mM)	*Bacteria* Glucose, maltose, sucrose, acetate, citrate, malate, phenylacetate, adipate, gluconate, alanine, peptone	6.0–8.5	25.0–55.0	6.0	[17]
Radiation	Radioresistant (resist ≥ 1 kilogray [kGy])	*Bacteria* Glucose, maltose, fructose, arabinose, glycerol 17 amino acid including cysteine, lysine, serine, histidine, leucine, lysine, methionine, phenylalanine, threonine, tryptophan, valine, asparagines, arginine, alanine, glutamic acid, glutamine, serine	6.4–9.5	25.0–55.0	2.0–10.0	[18]
		Archaea All 20 amino acids, yeast extract, tryptone, peptone, pyruvate, starch, maltodetrins, maltose, trehalose, glucose and lactose	6.0–7.5	75.0–88.0	4.0–5.0	[19]

[1] (Hallberg and Lindström 1994; Hallberg et al. 2010; Harrison Jr 1984); [2] (Dopson et al. 2004); [3] (Takai et al. 2001; Blotevogel et al. 1985; Pikuta et al. 2000; Horikoshi 1996; Zhilina et al. 1997); [4] (Margesin et al. 2003; Yumoto et al. 2003; Novitsky and Morita 1976; Ulitzur 1974); [5] (Saunders et al. 2003; Preston et al. 1996; Sheridan et al. 2003); [6] (Rainey et al. 1994; Degryse et al. 1978; Wiegel 1992; Ben-Bassat et al. 1981; Lovitt et al. 1984; Saiki et al. 1985; Huber and Stetter 1991); [7] (Amend and Shock 2001; Egorova and Antranikian 2005; Huber and Stetter 1991; González et al. 1995); [8] (Huber et al. 1986); [9] (Blöchl et al. 1997; González et al. 1998; Huber et al. 2000; Erauso et al. 1993; Sako et al. 1996; Hafenbradl et al. 1996); [10] (Franzmann et al. 1988; Antón et al. 2002; Gonzalez et al. 1978; Rodriguez-Valera et al. 1983; Oren et al. 1990; Bouchotroch et al. 2001); [11] (Mormile et al. 2003; Xu et al. 1999; Burns et al. 2007; Robinson et al. 2005); [12] (Restaino et al. 1983); [13] (Jojima et al. 2004; Osman et al. 1987); [14] (Gock et al. 2003; Marco et al. 2009; Mari et al. 2003; Palacios-Cabrera et al. 2005); [15] (Nogi et al. 2002, 2004; Simonato et al. 2006; Alain et al. 2002; Fang et al. 2010; Takai et al. 2009; Kato 1999); [16] (Merchant et al. 2007; Reddy et al. 2007; Poindexter 1981; Fegatella et al. 1998; Chrzanowski et al. 1996); [17] (Muller et al. 2006; Margaryan et al. 2010; Brim et al. 1999; Suresh et al. 2004); [18] (Yoshinaka et al. 1973; Ferreira et al. 1997; Suresh et al. 2004; Rainey et al. 2007; De Groot et al. 2005; Yoo et al. 2010); [19] (Jolivet et al. 2003, 2004)

[a]Major growth substrates determined in laboratories

[b]Optimal pH for growing most microorganisms of the corresponding extremophiles

[c]Optimal growth temperatures of most microorganisms in the corresponding extremophiles

[d]Generation time of most microorganisms in the corresponding extremophiles

In bacteria, the ABC transporters are classified into two groups, the carbohydrates uptake transporters and the di/oligopeptide transporters (Schneider 2001). The translocation of carbohydrates is also mediated by phosphoenolpyruvate (PEP): sugar phosphotransferase system (PTS), which is not found in archaea (Siebold et al. 2001). PTS uses energy from PEP and the phosphoryl group of PEP is transferred to the transported sugar through several proteins (Postma et al. 1993). PTS consists of two cytoplasmic proteins, enzyme I, histidine protein and a variable number of sugar-specific transport complexes also known as enzymes II (Postma et al. 1993). In brief, enzyme I transfers phosphoryl groups from PEP to histidine protein then it transfers the phosphoryl group to different transport complexes (Postma et al. 1993). As an example, for the transportation of glucose, an enzyme II protein phosphorylates glucose as it transports across the plasma membrane to form glucose-6-phosphate which subsequently enters the EM pathway (Bettenbrock et al. 2007).

In archaea, secondary transporters play important roles in solutes transport in addition to ABC-transporters (Paulsen et al. 2000; Horlacher et al. 1998). Secondary transporters usually depend on the electrochemical gradient of sodium ion or protons across the cytoplasmic membrane, whereas ABC-transporter does not have this limitation and can accumulate substrates to much higher concentrations inside the cells (Albers et al. 2004). The major difference between ABC-transporters in archaea and bacteria are the substrate-binding proteins (Albers et al. 2004). In bacteria, the extracellular protein captures the substrate, which will be released through permeases and translocated across the membrane (Quiocho and Ledvina 1996). In contrast, archaeal binding proteins are glycosylated with potential function of stabilizing the extracellular proteins against proteolytic degradation and the binding proteins are well-studied in hyperthermophilic archaea (Erra-Pujada et al. 2001; Hettmann et al. 1998; Xavier et al. 1996; Evdokimov et al. 2001; Albers et al. 1999). In both classes, the binding-proteins bind to a wide range of substrates and vary in length and composition from each other (Elferink et al. 2001; Koning et al. 2002). ABC-transporters for monosaccharides and most disaccharides in archaea are similar to those of carbohydrate uptake transporters in bacteria (Albers et al. 2004). However, the transporters for some disaccharides and oligosaccharides, such as cellobiose, β-glucoside and cello-oligomer from hyperthermophiles show sequence homologies to transporters from that of di/oligopeptide class in bacteria (Elferink et al. 2001; Koning 2001). It is also found that some transporters operon assigned as di/oligopeptide transporters are near those for sugar-degrading enzymes, suggesting the possibility of their involvement in catalyzing the uptake of di/oligosaccharides (Nelson et al. 1999).

The utilization of polysaccharides such as starch, glycogen, cellulose, hemicellulose, and xyloglucan is dependent on the presence of extracellular glycosyl hydrolases for hydrolyzing them into mono- or di-saccharides, which is subsequently transported into the cell by either ATP-binding cassette-type or secondary transporters (Elferink et al. 2001; Saurin et al. 1999). Among six-carbon sugars, glucose serves as a model substrate for each group of extremophiles although some species are unable to use it (Table 1.1). In five-carbon sugars, xylose is commonly utilized but many of the extremophiles may only use a few of such sugars due to the lack of corresponding transporters.

Single amino acid as the only substrate can't support the growth of many extremophiles (Table 1.1). However, extremophiles that lack the biosynthetic pathway for a specific amino acid often can't survive without that amino acid, for instance, *Pyrococcus furiosus* DSM3638 requires isoleucine and valine for growth (Hoaki et al. 1994). Some extremophiles, such as hyperthermophilic archaea *Hyperthermus butylicus* (Zillig et al. 1990) and *Staphylothermus marinus* (Hao and Ma 2003) preferentially grow on proteinaceous compounds and utilize them as both carbon and nitrogen sources (Table 1.1) (Schönheit and Schäfer 1995). In general, extremophiles that prefer to grow on poly-, oligo- and mono-saccharides can utilize proteinaceous compounds but often not the other way round. This is probably due to the lack of sugar transporters, such as PTS in archaea (Verhees et al. 2003).

Most extremophiles have optimal pH in the neutral range except acidophiles and alkaliphiles, and with a wider range of temperatures for optimal growth. One plausible explanation may be that their habitats are very diverse and with different temperatures. Piezophiles are with the widest range of optimal temperatures from 10.0 to 98.0 °C to cope with high pressures in deep sea and volcano areas (Nogi et al. 2002, 2004; Simonato et al. 2006; Alain et al. 2002; Fang et al. 2010; Takai et al. 2009; Kato 1999). Under optimal growth conditions, the generation time of all extremophiles except oligotrophs are relatively comparable, indicating that extremophiles have evolved to grow their best under such extreme conditions and nutrient is an indispensable factor (Merchant et al. 2007; Reddy et al. 2007; Poindexter 1981; Fegatella et al. 1998; Chrzanowski et al. 1996).

The utilization of different substrates is dependent on their specific enzymes and central metabolic pathways present in different extremophiles, which will be described in detail in the next section. The major mechanisms used by each group of extremophile to cope with environmental stress will also be discussed.

3 Utilization of Carbohydrates, Central Metabolic Pathways and Adaptation to Extreme Environments

A wide range of substrates can be used by extremophiles to support their growth. Polysaccharides are hydrolyzed by hydrolases secreted by the microorganisms before they can be transported into the cells (Warren 1996). Their major central metabolic pathways for metabolizing carbohydrates are summarized here.

3.1 Acidophiles and Alkaliphiles

Microorganisms that grow at extreme pHs, the acidophiles and alkaliphiles, are particularly interesting as pH affects the state of ionization of acidic or basic amino acids and enzymes and it is intriguing to understand how these species cope with such extreme conditions (Jaenicke 1981). It is shown that the cell membranes and intracellular environment of those extremophiles have homeostasis mechanisms to

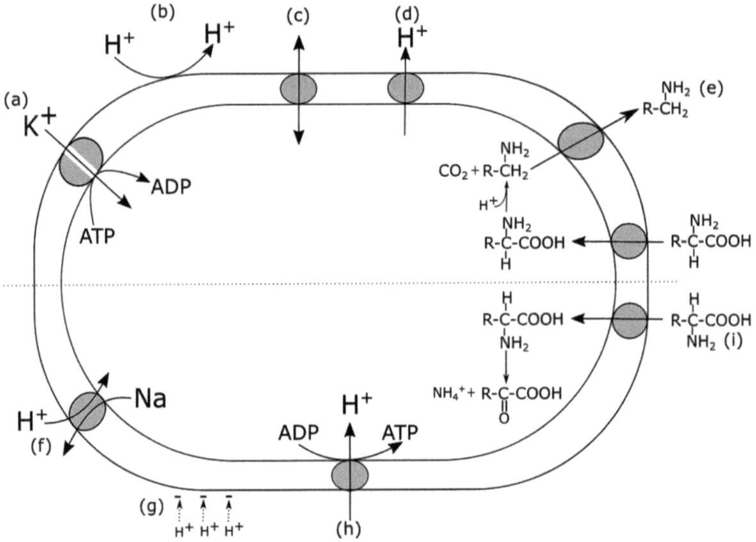

Fig. 1.1 pH homeostasis mechanisms in acidophiles and alkaliphiles. *The top* represents acidophiles: (*a*) Potassium transport by potassium-transporting ATPases to generate a reverse membrane potential. (*b*) Highly impermeable cell membrane to prevent the influx of protons. (*c*) Abundant secondary transporters to reduce the energy demand for transporting solutes and nutrient into the cells. (*d*) Active protons export by varies transporters. (*e*) Increased rate of decarboxylation of basic amino acids and decreased pH by the consumption of protons. *The bottom* represents alkaliphiles: (*f*) Sodium-hydrogen antiporters for the influx of protons. (*g*) Acidified cell membrane to attach and accumulate protons for transport into the cells. (*h*) Increased numbers of ATP synthase that couples the entry of protons and ATP synthesis. (*i*) Increased transporters and enzymes for deamination of amino acids for the production of corresponding α-keto acids

keep the internal pH to be constant (near neutral), which provides a stable environment inside the cell for enzymes to function properly (Jolivet et al. 2004).

In acidophiles, there are several proposed mechanisms in pH homeostasis (Fig. 1.1) (Baker-Austin and Dopson 2007). One potential mechanism is to generate a reverse membrane potential, an inside positive membrane potential, by the influx of potassium ion with the reverse reaction of ATPase to build a chemiosmotic barrier against the influx of proton (Baker-Austin and Dopson 2007). In the genomes of acidophiles, there is a predominance of potassium-transporting ATPases (Fütterer et al. 2004; She et al. 2001). Secondly, they have different cell membrane structures that are impermeable to protons (Konings et al. 2002). In acidophilic archaea, the cell membranes are composed of tetraether lipids instead of ester linkages found in neutrophiles (Shimada et al. 2002; Batrakov et al. 2002; van de Vossenberg et al. 1998). Moreover, excess protons are effectively pumped out of the system by proton efflux systems including H^+-ATPases, antiporters and symporters (Michels and Bakker 1985; Tyson et al. 2004). There is also a cytoplasmic buffering system (Castanie-Cornet et al. 1999). This is achieved primarily by decarboxylation of basic amino acids, lysine, histidine and arginine with the consumption of protons

(Castanie-Cornet et al. 1999; Álvarez-Ordóñez et al. 2010). In the example of lysine, once it enters the cell, it will be converted into cadaverine by lysine decarboxylase with the consumption of a proton and later transport out in exchange of a new lysine molecule (Álvarez-Ordóñez et al. 2010). The cytoplasmic pH is increased by proton consumption in the reaction (Álvarez-Ordóñez et al. 2010). In addition to these mechanisms, genome sequence analysis has suggested that the process of organic acid degradation, large number of genes for DNA and protein repair and maintenance, and a small genome size contribute to adaptation to low pH growth (Baker-Austin and Dopson 2007; Ciaramella et al. 2005).

In alkaliphiles, there are mainly four mechanisms for maintaining the pH homeostasis (Fig. 1.1) (Padan et al. 2005). In many alkaliphiles, an essential strategy is by increasing the expression and activity of monovalent cation/proton antiporters, most of them with the Na^+/H^+ antiporters for pH regulation (Kitada et al. 2000; Swartz et al. 2005). They also increase intracellular pH by enhancing the production of metabolic acids through amino acid deamination and sugar fermentation (Blankenhorn et al. 1999; Richard and Foster 2004). The activities of ATP synthase that couples H^+ entries are increased as well (Rozen and Belkin 2001; Krulwich et al. 1998). They also change their cell surfaces to be more acidic for binding cations and increased the availabilities of H^+ and Na^+ for pH homeostasis (Wang et al. 2004). The mechanisms in both acidophiles and alkaliphiles maintain pH homeostasis for the proper functions of metabolic enzymes and their metabolic pathways.

Embden-Meyerhof (EM), Entner-Doudoroff (ED) and Pentose Phosphate (PP) pathways are present in both acidophiles and alkaliphiles, however, archaeal ones have modified EM and ED pathways. The Tricarboxylic Acid (TCA) cycle in most microorganisms including both acidophiles and alkaliphiles are complete with all enzymes involved but some have incomplete one with gene(s) missing from the pathway.

In acidophilic bacteria, the main pathway used is EM pathway with exceptions such as *Acidiphilium cryptum* strain Lhet2, which does not have detectable key enzyme 6-phosphofructokinase in the EM pathway (Shuttleworth et al. 1985). It is shown that the use of EM, ED and PP pathways is dependent on the cultivation conditions (Barrie Johnson and Hallberg 2008). Under the mixotrophic conditions (presence of ferrous iron, thiosulfate, and organic compounds in the growth media), all three pathways are used for carbohydrate metabolism, which is examined by assaying the enzyme activities (Barrie Johnson and Hallberg 2008). Under the heterotrophic condition, both EM and ED pathways are operational (Barrie Johnson and Hallberg 2008). Under the autotrophic condition, only EM pathway is functional and there is a much higher activity of fructose-bisphosphase, suggesting its involvement in gluconeogenesis (Barrie Johnson and Hallberg 2008). The ability of acidophiles to utilize xylan due to the presence of xylanase makes it a very promising candidate in industrial applications (Kumar et al. 2013). Several acidophiles, such as *Acidithiobacillus ferrooxidans* and *Acidithiobacillus thiooxidans* are commonly used for bleaching of metals (Mishra et al. 2008; Wang et al. 2009). In *Acidithiobacillus caldus*, genes encoding succinate dehydrogenase and the 2-oxoglutarate dehydrogenase complex were lacking, suggesting an incomplete TCA cycle (You et al. 2011). The incomplete

TCA cycle is proposed to be an ancient biosynthetic pathway instead of being an energy generation cycle as in a complete TCA cycle (Wood et al. 2004).

Archaeal acidophiles also use EM, ED and PP pathways for carbohydrate metabolism (Kengen et al. 1996; Moracci et al. 2000; Budgen and Danson 1986). However, they have a modified EM pathway and two modified ED pathways, a semi-phosphorylative and a non-phosphorylative one. This might be consistent with the facts that acidophilic archaea are also thermophiles or hyperthermophiles. *Thermoproteus tenax*, a hyperthermophilic acidophilic archaeon, uses three pathways for sugar metabolism, and the EM pathway is modified with special enzymes and enzymes with modifications (Kengen et al. 1996). Firstly, the hexokinase is modified to have no regulatory properties (Kengen et al. 1996). Secondly, the replacement of the antagonistic enzyme couples ATP-dependent phosphofructokinase and fructose bisphosphatase by a bidirectional non-allosteric regulated PP_i-dependent enzyme (Kengen et al. 1996). It also has two different glyceraldehyde-3-phosphate dehydrogenases (GAPDH), differing in phosphate dependence, reversibility of the catalyzed reaction, and allosteric properties (Kengen et al. 1996; Brunner et al. 2001; Hensel et al. 1987). The NAD^+-dependent GAPDH is for the conversion of glyceraldehyde-3-phosphate (GA3P) into 3-phosphoglycerate, while the $NADP^+$-dependent one is for anabolism and catalyzing the conversion of 1,3-biphosphoglycerate into GA3P (Brunner et al. 2001). The NAD^+-dependent GAPDH is the one with allosteric regulation (Brunner et al. 1998, 2001). It is inhibited but also activated by a series of metabolites, which is rare among characterized aldehydes dehydrogenases. It is activated by AMP, ADP, and glucose-1-phosphate and fructose-6-phosphate but inhibited by NADP(H) and NADH (Brunner et al. 2001). The complex regulation suggests it may play a central role in controlling the catabolic carbon flux in the EM pathway and it compensates the lacking regulatory potential of the reversible non-allosteric PP_i-dependent phosphofructokinase (Brunner et al. 2001).

In the modified ED pathways, there are either semi-phosphorylated or non-phosphorylated (Fig. 1.2). In semi-phosphorylative ED pathway, which is used by *T. tenax* and thermophilic *Thermoplasma acidophilum*, phosphorylation only takes place at the level of glycerate and glyceraldehyde generated by the KDG aldolase (KDGA), which has the characteristic intermediate of the non-phosphorylative one (Kengen et al. 1996; Budgen and Danson 1986; van der Oost et al. 2005). In contrast, non-phosphorylative ED pathway is also found in the hyperthermophilic acidophilic archaea of *Sulfolobales*, where the phosphorylation step is omitted altogether in the formation of the first molecule of pyruvate, and the 2-keto-3-deoxygluconate is cleaved directly to pyruvate and glyceraldehyde (Budgen and Danson 1986; Kengen et al. 1996; Moracci et al. 2000).

All alkaliphilic bacteria show high activities of key enzymes in the EM pathway including hexokinase, glucose-6-phosphate dehydrogenase, phosphofructokinase, fructose-bisphosphate aldolase and glyceraldehydes-3-phosphate dehydrogenase, suggesting EM to be the major pathway used (Garnova and Krasil'nikova 2003; Zhilina et al. 2004; Paavilainen et al. 1999). A comprehensive study shows that saccharolytic alkaliphiles including *Amphibacillus*, *Halonatronum*, *Alkaliflexus* also

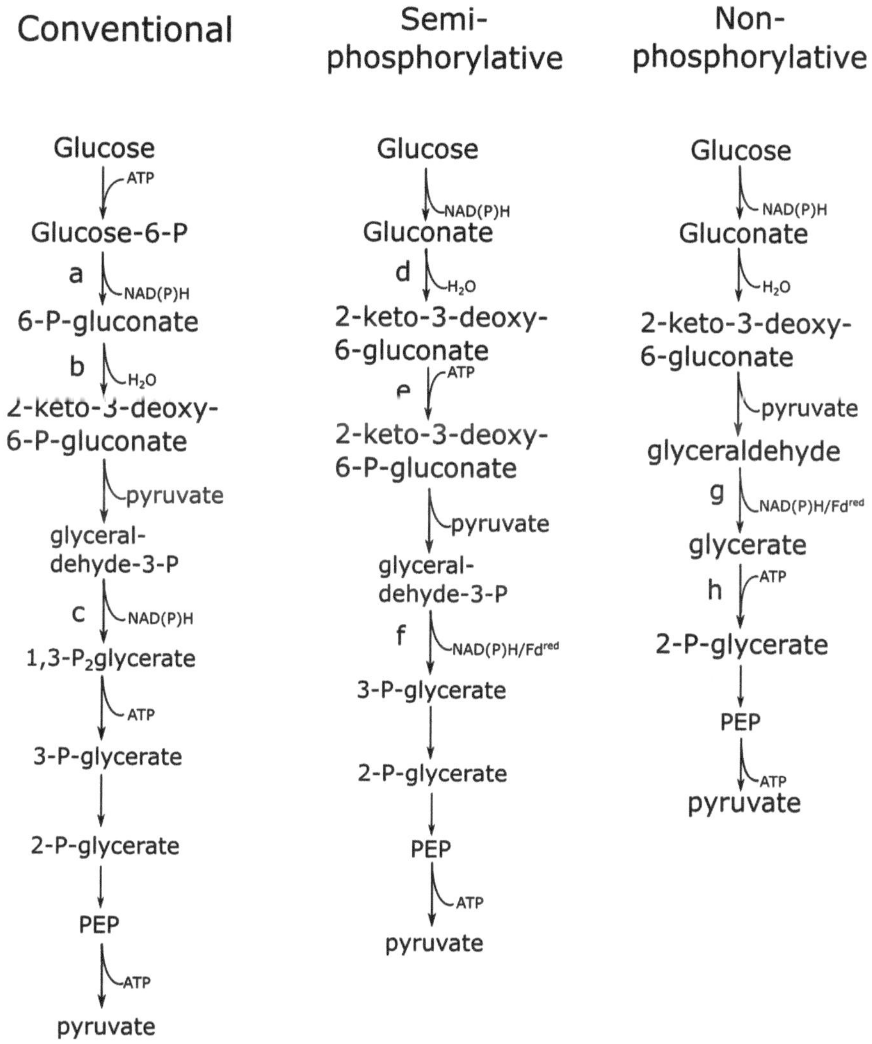

Fig. 1.2 Comparison of ED and modified ED pathways in extremophiles. Enzymes that are different in the three pathways are labeled from *a* to *h*. Enzymes in conventional ED pathway (*a*) Glucose-6-phosphate dehydrogenase. (*b*) 6-Phosphogluconate dehydratase (*c*) Glyceraldehyde-3-phosphate dehydrogenase are replaced by (*d*) gluconate dehydratase, (*e*) 2-keto-3-deoxygluconate kinase and (*f*) glyceraldehydes-3-phosphate dehydrogenase/glyceraldehyde-3-phosphate oxidoreductase in semi-phosphorylative ED pathway respectively, which are further modified as (*g*) aldehyde dehydrogenase/aldehyde oxidoreductase and (*h*) glycerate kinase in non-phosphorylated pathway respectively. *PEP* phosphoenolpyruvate, *P* phosphate group

have high activities of enzymes involved in other central metabolic pathways. *Amphibacillus* contains the key enzymes of the ED pathway, 6-phosphogluconate dehydratase and 2-keto-3-deoxy-6-phosphogluconate aldolase (Garnova and Krasil'nikova 2003). In *Halonatronum,* enzymes for the PP pathway are found, and in *A. fermentum* activities of the key enzymes of all three pathways of glucose metabolism are detected (Garnova and Krasil'nikova 2003). A special alkaliphilic bacterium, *Alkaliflexus imshenetskii* both a halophile and alkaliphile, metabolizes cellobiose through EM and ED pathway and has no enzymes for the PP pathway (Zhilina et al. 2004). *Bacillus circulans* var. *alkalophilus* metabolizes glucose through three pathways, but the majority is EM pathway (90–93 %), and less than 10 % is ED and PP pathways (Paavilainen et al. 1999). For alkaliphilic bacteria that are methanotrophs, for example *Methylobacter alcaliphilus*, α-ketoglutarate dehydrogenase in the TCA cycle is not found, resulting in an incomplete TCA cycle which is used by many Type-1 methanotrophs (Khmelenina et al. 1997). Other heterotrophic acidophiles and alkaliphiles show a complete TCA cycle from genome sequence analysis, and it suggests that the pathway of the TCA is influenced by the mode of nutrient as methanotrophs and chemolithoautotrophs often have an incomplete TCA cycle (Hanson and Hanson 1996).

Extremophiles growing at extreme temperatures not only have unique metabolic pathways, but also many novel and distinct enzymes are particularly found in archaea. Similar to acidophilic archaea that grow at high temperatures, some of the extremophiles that favor high temperatures have a non-phosphorylated ED pathway, suggesting such a modification is for growing at high temperatures instead of extreme pH. For coping with the stresses from the environment, extremophiles develop many strategies.

3.2 Psychrophiles

At low temperatures, two main physical challenges are low thermal energy and high viscosity (D'Amico et al. 2006). The target for modification is protein as it is involved in many processes important for cell survival (D'Amico et al. 2006).

In psychrophiles, cold-acclimation/cold-shock proteins and chaperones are abundant with aid in protein-folding, also there are antifreeze protein, trehalose and exopolysaccharides that have an important role in cryoprotection (Phadtare 2004; Jia and Davies 2002). Many enzymes in important processes such as translation and transcription have adapted to be optimally active at low temperatures (Lim et al. 2000), which is achieved by increasing the flexibility of their structures and decreasing the numbers of enthalpy-driven interactions that have to be broken during catalysis (Violot et al. 2005; Berger et al. 1996; Russell 2000). In the cell membranes of psychrophiles, they have a higher content of unsaturated, polyunsaturated and methyl-branched fatty acids and the fatty acids are with shorter acyl-chain length to increases the membrane fluidity to cope with low temperature (Chintalapati et al. 2004; Russell 1997). These mechanisms maintain the proper functions of metabolic enzymes at low temperatures for the utilization of nutrients and enable psychrophiles to grow optimally at low temperatures.

For psychrophilic bacteria and archaea, most utilize carbon sources through EM pathway, and the pathways for glycolysis and gluconeogenesis are complete (Xu et al. 2003; Bakermans et al. 2003; Médigue et al. 2005; Cavicchioli et al. 2000). One of the challenges faced in cold-environment is the decrease in enzymatic activities in physiological reactions including the generation of ATP and cofactors in the central metabolic pathways (Amato 2013). It is found using proteomic, transcriptomic methods and enzyme assays that psychrophiles overcome this problem by up-regulating or increasing activities of important enzymes in the central metabolic pathways (Amato 2013). In a cold-adapted strain of *Lactobacillus piscium*, glyceraldehyde-3-phosphate dehydrogenase is up-regulated to overcome the temperature-limited step of glycolysis (Garnier et al. 2010). Also, fructose biphosphate aldose and phosphoglycerate kinase are up-regulated (Garnier et al. 2010). In *Propionibacterium fredenreichii*, enzymes in the TCA cycle including aconitase, fumarase, succinate dehydrogenase and citrate synthase are up-regulated (Dalmasso et al. 2012). Similarly, in *Pseudoalteromonas haloplankis*, the enzyme activities in the TCA cycle are up-regulated to overcome cold-stress (Piette et al. 2010). The increases in TCA enzyme activities are to assure the catabolic process by providing substrates through intermediates from the TCA cycle and to increase the production of ATP. Psychrophilic archaea, *Methanococcoides burtonii* has a truncated oxidative TCA cycle in the reductive direction with genes almost completely missing from oxaloacetate to 2-oxoglutarate except for a heterodimeric fumarase present (Goodchild et al. 2004). In the oxidative direction, citrate synthase and aconitase are present, and isocitrate/isopropylmalate dehydrogenase is a likely candidate to catalyze the remaining step from isocitrate to 2-oxoglutarate (Goodchild et al. 2004). There are also pyruvate synthase and pyruvate carboxylase for catalyzing the synthesis of oxaloacetate from acetyl-CoA (Goodchild et al. 2004). *M. burtonii* is also a methylotroph (Goodchild et al. 2004).

3.3 Thermophiles

Thermophiles and hyperthermophiles have to cope with high temperatures that cause denature of proteins (Jaenicke and Böhm 1998). There are two mechanisms for stabilizing the proteins and prevent denature at high temperatures and it is interestingly found that the mechanism used is depended on the evolutionary origin of the extremophile (Berezovsky and Shakhnovich 2005). In brief, the first mechanism is through modification of protein structures to be significantly more compact compared to mesophilic homologues and this is through a high number of interactions (Berezovsky and Shakhnovich 2005). It is found in thermophiles and hyperthermophiles that originated from extreme environment of high temperatures (England and Shakhnovich 2003; England et al. 2003). The second mechanism is sequence-based and their structures are not significantly different from their mesophilic homologs (Dominy et al. 2004; Macedo-Ribeiro et al. 1996). The sequence modifications build up few strong interactions for high thermal stability; hence, the structures found to show no significant differences with mesophilic homologs (Li et al. 2005). This mechanism is found in extremophiles mainly thermophiles that was mesophiles but later recolonized a hot environment and it is an evolutionary strategy for adaptation (Berezovsky and Shakhnovich 2005). They also have chaperons, heat shock proteins, for helping fold proteins that cannot form the thermostable conformation by itself (Hendrick and Hartl 1993; Feder and Hofmann 1999). The chaperons also prevent the aggregation of unfolding proteins at high temperatures and direct misfolded and denatured proteins into the cellular protein degradation systems (Sterner and Liebl 2001). The common chaperon found is Hsp60 complex, which is also called thermosomes (Sterner and Liebl 2001).

At high temperatures, thermophilic bacteria utilize sugar mainly through the EM pathway and all the enzymes from the pathway are present (Selig et al. 1997). The activities of key enzymes phosphogluconate dehydratase and 2-keto-3-deoxy-6 phosphogluconate aldolase in the ED pathway have not been detected in several thermophilic bacteria *Geobacillus thermoglucosidasius* (Tang et al. 2009) and *Caldicellulosiruptor saccharolyticus* (De Vrije et al. 2007). It is also found that the utilization of PP pathway in *G. thermoglucosidasius* is different under different oxygen concentration (Tang et al. 2009). Lower activity of the oxidative part of the PP pathway is detected under glucose fermentation conditions but with a missing 6-phosphogluconolactonase gene (Tang et al. 2009). Under aerobic conditions, about two-third of the glucose go through glycolysis and the rest to PP pathway, and under microaerobic conditions, TCA cycle and PP pathway are reduced to half of the original level (Tang et al. 2009). At microaerobic condition, the growth rate was slower and reduced to 0.2 h^{-1}. In general, NADPH to support biomass synthesis is mainly from PP pathway (Christensen et al. 2002). Hyperthermophilic bacteria, such as *Thermotoga maritima* also metabolize glucose mainly through the conventional EM pathway (Schröder et al. 1994).

In thermophilic and hyperthermophilic archaea, modified ED and EM pathways are used and the oxidative PP pathway is replaced by non-oxidative PP pathway (Schönheit and Schäfer 1995; Van der Oost and Siebers 2007). The novel enzymes in each of the modified pathways will be described with examples.

In thermophilic archaea, the main pathway for carbohydrate metabolism is also through EM pathway but modified (Fig. 1.3). *Thermococcus zilligii* and *Thermococcus stetteri* have ADP-dependent phosphofructokinase and unique glyceraldehyde-3-phosphate: ferredoxin oxidoreductase (Ronimus et al. 1999; Xavier et al. 2000; Kengen et al. 1996). A novel glycolytic pathway with possible involvement of pentoses is also proposed in *T. zilligii* based on [^{13}C]-glucose-labelling experiments (Xavier et al. 2000). The pathway is proposed as the following:

Step 1 $\text{Glucose} + \text{ADP} \rightarrow \text{Glucose}\ \ 6\ \ \text{phosphate} + \text{AMP(hexokinase)}$

Step 2 $\text{Glucose}-6-\text{phosphate} + \text{NADP}^{+} \rightarrow 6-\text{phosphogluconate} + \text{NADPH} + \text{H}^{+}$ $\text{(glucose}-6-\text{phosphate dehydrogenase)}$

Step 3 $6-\text{phosphogluconate} \rightarrow \text{formate} + \text{xylulose}-5-\text{phosphate (novel lyase)}$

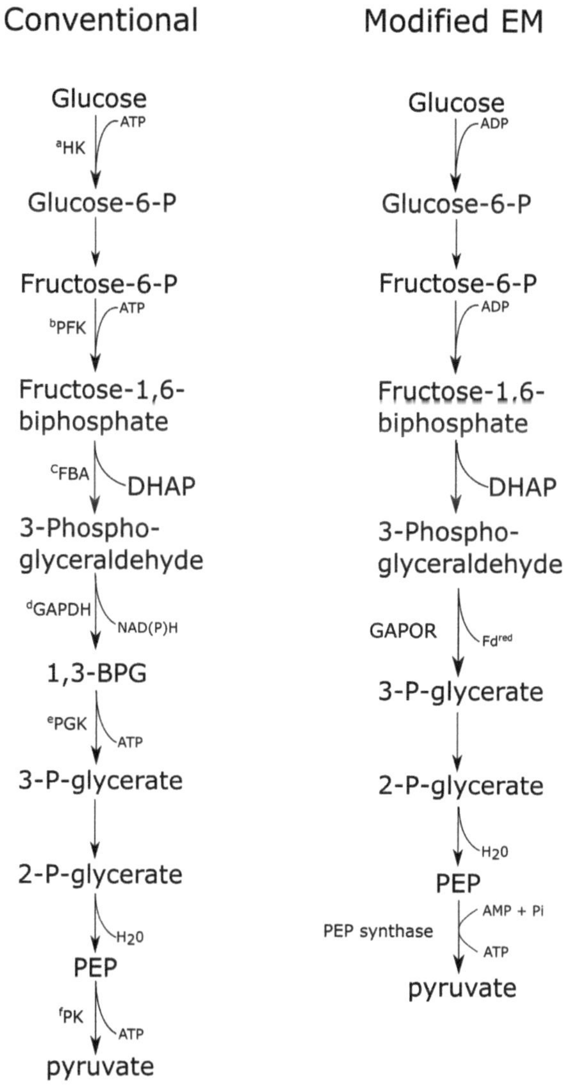

Fig. 1.3 Comparison of EM and Modified EM pathway in extremophiles. Enzymes from conventional EM pathway are either modified or replaced by novel enzymes in extremophiles and a modified EM pathway is present in hyperthermophilic archaea. Enzymes that are modified are labeled. (*a*) HK, hexokinase and (*b*) PFK, phosphofructose kinase are modified in hyper/thermophiles including hyper/thermophilic acidophiles. (*c*) FBA, fructose bisphosphatase aldolase and (*d*) GAPDH, glyceraldehyde-3-phosphate dehydrogenase are modified in hyperthermophilic acidophiles and upregulated in psychrophiles. (*e*) PGK, phosphoglycerate kinase is upregulated in psychrophiles. (*f*) PK, pyruvate kinase is replaced by an AMP-dependent PEP synthase in modified EM pathway in hyperthermophiles. GAPOR, glyceraldehyde-3-phosphate oxidoreductase replaces GAPDH and PGK in modified EM pathway in hyperthermophiles. *DHAP* Dihydroxyacetone phosphate, *1,3-BPG* 1,3-biphosphoglycerate, *PEP* phosphoenolpyruvate, *Fd(red)* reduced ferredoxin

$$\text{Xylulose}-5-\text{phosphate} \rightarrow \text{glyceraldehyde}-3-\text{phosphate}+\text{acetyl}$$
$$-\text{phosphate (Pentose phosphoketolase)}$$

Step 4
$$\text{Glyceraldehyde}-3-\text{phosphate} \rightarrow \text{pyruvate} \rightarrow \text{actetate}+CO_2$$
$$\text{Acetyl}-\text{phosphate} \rightarrow \text{acetate}$$
$$\text{(Pentose Phosphoketolase Pathway)}$$

In brief, it is suggested gluconate-6-phosphate is cleaved into xylulose-5-phosphate and formate with a novel type lyase (step 3) and the pentose phosphate is catabolized through the pentose phosphoketolase pathway (Xavier et al. 2000).

3.4 Hyperthermophiles

Hyperthermophilic archaea including *Desulfurococcales, Thermococcales, Pyrococcus and Archaeoglobus fulgidus* utilize the modified EM pathway with unique enzymes (Fig. 1.3) (Schönheit and Schäfer 1995; Huber and Stetter 2006; Bertoldo and Antranikian 2006; Kengen et al. 1994; Labes and Schönheit 2001; Siebers and Schönheit 2005). Hyperthermophilic bacterium *Thermotoga maritima* and some hyperthermophiles that are also acidophiles use both EM pathways and modified ED pathways (Siebers and Schönheit 2005). EM pathway is modified with unique enzymes, such as ADP-dependent kinases including ADP-dependent glucokinase and ADP-dependent phosphofructokinase, which in mesophilic one are ATP-dependent (Sakuraba et al. 2002; Koga et al. 2000; Dörr et al. 2003). From an energetic point of view, ADP is not inferior to ATP, the free energy change of ADP hydrolysis is about the same as that of ATP hydrolysis (Hongo et al. 2006). Also, ADP is envisaged to be more stable than ATP at higher temperatures, especially in the presence of some divalent metal ions (Hongo et al. 2006; Tetas and Lowenstein 1963). Another modification is the presence of glyceraldehyde-3-phosphate ferredoxin-linked oxidoreductase in replacement of classical glycolytic enzymes glyceraldehyde-3-phosphate dehydrogenase and phosphoglycerate kinase; it is a novel tungsten-containing iron-sulfur protein and was found in *Pyrococcus furiosus* (Mukund and Adams 1991, 1995; van der Oost et al. 1998), *Thermococcales* (Kletzin et al. 1995; Mukund and Adams 1993) and *Desulfurococcus amylolyticus* (Selig et al. 1997). It catalyzes the single step for the production of 3-phosphoglycerate and reduced ferredoxin from the oxidation of GA3P, which are two reactions in conventional one (de Vos et al. 1998). In this reaction, the ATP generation step from phosphoglycerate kinase is omitted, which reduces the overall substrate-level ATP yield to zero (Verhees et al. 2004). The need for such a modification might be that pyridine nucleotides are less thermostable than ferredoxins. Also, to avoid the production of a heat-liable intermediate 1,3 biphosphoglycerate (Brunner et al. 2001). To compensate for this loss in ATP production, it is found that in *Thermococcus kodakarensis*, the conversion of phosphoenolpyruvate (PEP) to pyruvate is replaced by PEP synthase (Sakuraba and Ohshima 2002; Sakuraba et al. 2004). The conventional enzyme pyruvate kinase uses ADP and produces one ATP, but

PEP synthase uses AMP and phosphate, which can produce an additional 2 ATP per glucose (Sakuraba and Ohshima 2002; Sakuraba et al. 2004; Imanaka et al. 2006). It is also demonstrated that the deletion of the gene for PEP synthetase in *T. kodakarensis* decreased its survival on carbohydrates medium suggesting it is essential for the growth on carbohydrates (Imanaka et al. 2006).

In aerobic hyperthermophiles, when they are also acidophiles, the non-phosphorylated ED pathway is used (Siebers and Schönheit 2005). In this pathway, glucose is oxidized to gluconate and 2-keto-3-deoxy-gluconate is formed, and a specific kinase phosphorylates the resulting glycerate (Fig. 1.3c) (Siebers and Schönheit 2005).

In hyperthermophilic archaea, many enzymes have sequences different from previous characterized and it is difficult to predict those enzymes solely based on genome sequence analysis. For example, in *T. kodakaraensis* KOD1, other than the functional 2-deoxyribose-5-phosphate aldolase predicted from genome sequence analysis, a structurally novel phosphopentomutase (PPM) with no significant sequence homology with the previously characterized PPM genes is discovered (Rashid et al. 2004). This provides insight into the presence of metabolic link between pentoses and central carbon metabolism, which is found in numerous bacteria and in hyperthermophiles (Rashid et al. 2004). More novel enzymes functioning in the central metabolic pathways in hyperthermophiles are potentially to be discovered.

3.5 Halophiles

High salts concentration causes high osmotic pressure and halophiles withstand the pressure by actively excluding salt in the cytoplasm and producing osmolytes inside the cells (Roberts 2000, 2004). All halophiles contain transport mechanisms like those of Na^+/H^+ antiporters to remove intracellular Na^+ ions (Oren et al. 1990; Obis et al. 1999). For achieving a high osmotic pressure in the cytoplasm, they have different strategies. The first strategy due to its limitation is only used by few archaea and bacteria. It builds up the KCl concentration in the cytoplasm at least as high as the surrounding NaCl concentration to cope with the stress (Pflüger and Müller 2004). A universal method is to accumulate organic solutes with the most common being glycerol, glycine betaine, ectoine, sucrose and trehalose (Imhoff and Rodriguez-Valera 1984; Brown 1976). This strategy needs more energy to build up the solute but the intracellular enzymatic system does not have to adapt to high KCl concentration as in the first strategy. Enzymes in halophiles are also modified to perform in high salt concentration.

In the central metabolic pathway of halophiles, enzymes are also modified to cope with the environmental conditions. For example, in the halophilic bacterium, *Salinibacter ruber*, glucose metabolism proceeds via a constitutive, salt-inhibited hexokinase and a constitutive salt-dependent NADP-linked glucose-6-phosphate dehydrogenase (Oren and Mana 2003). The pathway for metabolizing glucose is the conventional ED pathway, but not EM pathway as glucose dehydrogenase and fructose-1,6-bisphosphate aldolase activity are not detectable (Oren and Mana 2003).

In archaeal halophiles, *Halococcus saccharolytucus*, *Haloferax mediteranes*, and *Haloarcula valismurtis*, catabolism of glucose and galactose has been shown to proceed via semi-phosphorylated ED pathway found in other archaea (Johnsen et al. 2001; Falb et al. 2008). Also, they use a modified EM pathway for metabolizing fructose (Johnsen et al. 2001; Falb et al. 2008). They lack the gene encoding the key enzyme 6-phosphofructokinase, and the hexose part of the classical EM pathway from the glucose to fructose-6-phosphate are predicted to be missing (Johnsen et al. 2001; Falb et al. 2008).

3.6 Osmophiles and Xerophiles

Microorganisms that grow on high concentrations of solute in media have mechanisms for osmoregulation to enable cell survival and proper functions of metabolic enzymes for energy production (Pflüger and Müller 2004). Osmophilic yeasts are commonly found in the environment, and similarly, the central metabolic pathways in yeasts are through the EM, ED and PP pathways for the production of pyruvate that enters the TCA cycle (Rodrigues et al. 2006). Glycerol plays an essential role in osmoregulation (Nevoigt and Stahl 1997). The rate of glycerol production is significantly elevated by increasing cytoplasmic glycerol-3-phosphate dehydrogenase activity in response to decreased extracellular water activity (Nevoigt and Stahl 1997; Påhlman et al. 2001; Albertyn et al. 1994). Fps1p channel is also simultaneously closed to conserve glycerol produced within the cells to maintain an osmotic equilibrium with the external environment (Tamás et al. 1999; Luyten et al. 1995). Trehalose is another compatible solute for osmoregulation (De Smet et al. 2000; Lippert et al. 1993). The common pathway for synthesis of trehalose is from UDP-glucose and glucose-6-phosphate (De Smet et al. 2000; Vogel et al. 1998; Virgilio et al. 1993). The two compounds form trehalose-6-phosphate and free trehalose are formed through dephosphorylation (Wolf et al. 2003). The key enzymes in the pathway are trehalose-6-phosphate synthase and trehalose-6-phosphate phosphatase respectively (De Smet et al. 2000; Vogel et al. 1998; Virgilio et al. 1993). In osmophilic yeast, *Saccharomyces rouxii*, glycerol is produced in response to high concentrations of salt, and D-arabitol is produced in response to high glucose concentration (Moran and Witter 1979). Both glycerol and D-arabitol can be accumulated intracellularly (Moran and Witter 1979).

The metabolism of glucose is through EM (~80 %) and PP (~20 %) pathways (Moran and Witter 1979). As glucose concentration increases (60 %, w/v), the participation of PP pathway increases (31 %) while EM pathway decreases (69 %) (Moran and Witter 1979). An active PP pathway is involved in high sugar tolerance by providing a plentiful supply of NADPH which is necessary for cell survival and synthesis of structural lipid (Moran and Witter 1979; Tokuoka 1993). However, there is no similar observation when sucrose is used in the media (Moran and Witter 1979). This is because D-arabitol is formed from glucose metabolized through the PP pathway and D-arabitol is not involved in the sucrose-tolerant mechanism

(Moran and Witter 1979). *Zymomonas mobilis*, a mesophilic bacterium, metabolizes sugar through ED pathway and produces different by-products when different sugars are used as substrates (Sprenger 1996; Rogers et al. 1979; De Graaf et al. 1999). The high sugar concentration is not inhibitory to the enzymes in the ED pathway. The transportation of sugar is also unique to cope with the environmental stress, and glucose is transported by a high-velocity, carrier-mediated, facilitated diffusion system so that its intracellular concentration can quickly reach a plateau close to the external concentration and prevent damages from high sugar concentration (DiMarco and Romano 1985).

Under conditions of high concentrations of solutes resulting in low water activity, xerophilic fungi (*Penicillium janczewskii, Eurotium chevalieri, Xeromyces bisporus* and *Wallemia sebi*) produce glycerol as an important osmoregulatory solute and it accumulated inside the cell (Hocking 1986; Huang et al. 2009).

Xerophilic fungi metabolize sugar through the EM pathway (Hocking and Norton 1983). An extreme example, *Xeromyces bisporus*, loses all genes encoding enzymes for catalyzing the production of secondary metabolites (Leong et al. 2014).

3.7 *Oligotrophs*

Although nutrient is the fundamental source of energy and carbon for growth, a special group of extremophiles, the oligotroph can thrive in environments of low carbon (1–15 mg organic carbon/liter) (Poindexter 1981). The central metabolic pathways become a focus for investigation. Their metabolic pathways are rearranged and some genes are lost due to streamlining selection. It is found that one extreme oligotroph *Candidatus Pelagibacter* lacks the complete EM pathway and genes encoding pyruvate kinase and phosphofructokinase are absent (Smith et al. 2013). However, a putative operon encoding genes involved in a predicted modification of the ED pathway is present, so is a complete gluconeogenesis pathway (Smith et al. 2013). Similarly, *Caulobacter crescentus* lacks a homolog of phosphofructokinase (Hottes et al. 2004). From genomics analysis, its growth on glucose induces the gene expression of the ED pathway, and its utilization of xylose is not via PP pathway as important enzymes xylose isomerase and xylulose kinase are lacking and no homolog genes are found in the genome sequence (Hottes et al. 2004). It has the potential to degrade biopolymers from plants as it produces cellulases, endo-β-1,4-glucanases, β-glucosidases, xylanases and xylosidases, and polysaccharide deacetylases (Hottes et al. 2004).

3.8 *Piezophiles*

Among all extremophiles, piezophiles are found in the widest range of environments with diverse carbon source available (Simonato et al. 2006). High pressure environment alters membrane fluidity like that at low and high temperatures

(Simonato et al. 2006). Piezophiles overcome this by increasing the proportion of unsaturated fatty acids mainly monounsaturated in the lipid bilayers (Valentine and Valentine 2004; Bartlett and Bidle 1999). This also helps to maintain ion permeability for a bioenergetic purpose (Vossenberg et al. 1995) and adjust the membrane curvature to withstand more elastic stress at high pressure (Attard et al. 2000). The second challenge is the reduction of efficiency in transporters (Simonato et al. 2006). Pizeophiles often upregulate the transporters to compensate this. One well-studied example is the transporter Tat2, a high affinity tryptophan permease, which also helps the uptake of tryptophan (Abe and Horikoshi 2000; Abe 2003). As piezophiles are found in cold and hot environments, they produced numerous heat shock proteins and cold-shock proteins. They also produce intracellular solutes β-hydroxybutyrate and its oligomers for withstanding hydrostatic pressure (Martin et al. 2002). Many piezophiles live in deep sea environments, where there are many complex polymers, and piezophiles have many enzymes for breaking down these polymers, which have potentials for industrial applications (Simonato et al. 2006).

Most piezophiles use EM pathway as the major pathway for the metabolism of carbohydrates (Table 1.2). However, in one of the piezophilic bacteria, *Marinitoga piezophila* KA3, there is only a complete EM pathway (Lucas et al. 2012). The PP pathway is impaired as there is a lack of glucose-6-phosphate-dehydrogenase and 6-phosphogluconate dehydrogenase (Lucas et al. 2012). It also lacks enzymes necessary for the ED pathway, the 6-phosphogluconate dehydratase and 2-dehydro-3-deoxy-phosphogluconate aldolase enzymes (Lucas et al. 2012).

3.9 Radioresistants

Under conditions of a high level of radiation, which are harmful to most organisms, genetic recovery from DNA damage due to radiation is heavily dependent on energy metabolism and protein (Venkateswaran et al. 2000). *Deinococcus* species, radioresistant bacteria, metabolize glucose via both ED and PP pathways as EM pathway might affect the resistance toward UV radiation despite the key enzymes of EM pathway are found to be present (White et al. 1999; Zhang et al. 2000; Yuan et al. 2012). The EM pathway may deplete metabolites for DNA repair, and induce oxidative stress and lead to less radioresistance (White et al. 1999; Zhang et al. 2000; Yuan et al. 2012). It's shown that *Deinococcus* species grow on fructose, glucose and maltose without EM pathway (White et al. 1999; Zhang et al. 2000; Yuan et al. 2012). However, analysis of the genome of the archaeal *Thermococcus gammatolerans* shows it has a modified EM glycolytic pathway and a non-oxidative PP pathway (Zivanovic et al. 2009). Further study in addition to genome analysis is needed to test which pathway is mainly used for the utilization of sugars.

The central metabolic pathways of extremophilic bacteria are generally well conserved with little modification, but modified pathways including semi-phosphorylated, non-phosphorylated ED pathways and EM pathway with novel enzymes and distinct control are present in archaea, and the modified ED pathways are shared among extremophiles from different extreme environments. The TCA cycle in extremophiles

Table 1.2 Metabolic pathways and key enzyme activities of heterotrophic microorganisms under extreme conditions

Environment	Types of microorganisms	Metabolic pathways/Enzyme activities: Central metabolism[a]	Metabolic pathways/Enzyme activities: Peptides/amino acids[b]	References
pH	Acidophile ($pH_{opt} \leq 3.0$)	*Bacteria* EMP, ED and PP pathways present and the use of each depends on growth conditions. TCA cycle (some incomplete)	Most genes involved in the pathways for the synthesis of amino acids are identified	[1]
		Archaea EMP (modified and only in some) and ED (mainly used) pathways TCA cycle	Characterized proteases and peptidase and complete pathway for protein degradation	
	Alkaliphile ($pH_{opt} \geq 9$)	*Bacteria* EMP, ED and PP pathways present TCA cycle (some incomplete)	Characterized proteases and genes found for metabolism of amino acid except for the serine degradation pathway	[2]
	Psychrophile ($T_{opt} \leq 15$ °C and $T_{max} \leq 20$ °C)	*Bacteria* EMP (mainly used), ED and PP pathways TCA cycle	Characterized serine protease, peptidase and metalloprotease. Genes involve in degradation of amino acids are identified	[3]
		Archaea EMP (mainly used) ED and PP pathways present TCA cycle (some incomplete)	Characterized metallopeptidase. Genes in the pathway for degradation of all 20 amino acids are identified	
Temperature	Thermophiles ($T_{opt} \geq 45$ °C but <80 °C)	*Bacteria* EMP (mainly used), ED (only in some) and PP (lack of oxidative part) pathways TCA cycle	Characterized protease and peptidase, genes for typical enzymes in the amino acid degradation pathway are detected	[4]
		Archaea EMP (modified, mainly used), ED (only in some) and PP pathways TCA cycle	Characterized proteinases and gene for the catabolic pathways for amino acids are identified	

(continued)

Table 1.2 (continued)

Environment	Types of microorganisms	Metabolic pathways /Enzyme activities: Central metabolism[a]	Metabolic pathways /Enzyme activities: Peptides/amino acids[b]	References
	Hyperthermophiles ($T_{opt} \geq 80$ °C or $T_{max} \geq 90$ °C)	*Bacteria* EMP (modified, mainly used), ED (only in some) and PP pathways TCA cycle	Characterized proteases	[5]
		Archaea EMP (modified, mainly used), ED (only in some) and PP pathways TCA cycle	Characterized serine protease and pyrrolidone carboxyl peptidase, genes for breaking down of all amino acids are identified.	
Salt concentration	Halophiles (≥0.3 M NaCl)	*Bacteria* EMP (only in some), ED (mainly used) and PP pathways TCA cycle	Characterized protease, genes encoding transaminase are identified	[6]
		Archaea EMP (modified), ED (modified) and PP pathways TCA cycle	Characterized serine peptidase, genes for amino acid degradation pathways including mesaconate pathway are found but some cannot synthesize up to more than 8 amino acids	
Water activity	Osmophiles (high concentrations of organic solute, sugars)	*Eucarya* EMP (mainly used) and PP pathways TCA cycle	Characterized alkaline protease	[7]
		Bacteria EMP, ED (mainly used), PP (only in some) pathways TCA cycle (some incomplete)	Genes for the complete degradation pathway of amino acids including amino acid dehydrogenase and oxidase are identified	
	Xerophiles (water activity (a_w) ≤ 0.85.)	*Eucarya* EMP (mainly used), ED and PP pathways		[8]
Pressure	Piezophile ($P_{opt} \geq 50$ MPa)	*Bacteria* EMP pathway Incomplete TCA cycle	Many proteases and peptidase up to 46 are detected in the genome and suggest those species can degrade diverse substrates	[9]

Low nutrient concentration	Oligotroph (1–15 mg organic Carbon/liter)	*Bacteria* ED (modified) pathway TCA cycle	Abundant secreted peptidase for utilizing amino acid as carbon source and high expression of genes for the degradation pathway of amino acids	[10]
Heavy Metal	Metallotolerant (>1 mM heavy metals)	*Bacteria* Found in low carbon environment and are autotrophic		[11]
Radiation	Radioresistant (d≥1 kilogray [kGy])	*Bacteria* EMP (not used but key enzymes found) ED and PP pathways TCA cycle	Cannot utilize ammonia as a nitrogen source and is entirely dependent on exogenous amino acids as a nitrogen source, biosynthetic pathways of lysine, serine and cysteine are incomplete, but it can grow on these three amino acids. Degradation pathways of amino acids are complete	[12]
		Archaea EMP (mainly used) and PP pathway TCA cycle	Genes in biosynthetic pathway of isoleucine, proline, arginine, leucine, phenylalanine and valine are missing	

[1] (Dopson et al. 2004; Lobos et al. 1986; Kinoshita et al. 1995; Barrie Johnson and Hallberg 2008; Valdés et al. 2008; Karavaĭko et al. 2000; Zhang and Lovitt 2005; Barry et al. 2006; Angelov and Liebl 2006; Ruepp et al. 2000; Kurosawa et al. 1998; Fuchs et al. 1995; Voges et al. 1999); [2] (Zhilina et al. 1997, 2001, 2004; Garnova and Krasil'nikova 2003; Detkova and Kevbrin 2009; Paavilainen et al. 1999; Khmelenina et al. 1997; Habib et al. 2012; Rao et al. 1998); [3] (Herbert and Bell 1977; Zhu et al. 2003; Allen et al. 2009; Villeret et al. 1997; Aghajari et al. 2003; Irwin et al. 2001; Riley et al. 2008; Khan and Sylte 2009); [4] (Feng et al. 2009; Patel et al. 2006; van Niel et al. 2003; De Vrije et al. 2007; Verhees et al. 2003; Kengen et al. 1996; Budgen and Danson 1986; Guangrong et al. 2006; Gatti et al. 2004; Klingeberg et al. 1995); [5] (Schröder et al. 1994; Huber et al. 1986; Klingeberg et al. 1991; Morikawa et al. 1994; Eggen et al. 1990; Kaushik et al. 2002); [6] (Oren and Mana 2003; Gonzalez et al. 2008; Rawal et al. 1988; Baliga et al. 2004; Ng et al. 2000; Anithajothi et al. 2014; Sanchez-Porro et al. 2003; Kamekura et al. 1992; Fine et al. 2006); [7] (Moran and Witter 1979; Loos et al. 1994; Ogawa et al. 1990; de Montigny et al. 2000; Seo et al. 2005); [8] (Hocking and Norton 1983; Leong et al. 2014); [9] (Lucas et al. 2012; Tamegai et al. 2008; Aono et al. 2010); [10] (Carini et al. 2012; Hottes et al. 2004; Arellano et al. 2010); [11] (Carapito et al. 2006; Bertin et al. 2011; Bryan et al. 2009); [12] (Yuan et al. 2012; Sghaier et al. 2008; Zivanovic et al. 2009; Zhang et al. 2000; White et al. 1999)

[a]Major metabolic pathways of central metabolism, which are mainly obtained using enzyme activity assays and genome sequence analyses from representative examples of extremophiles

[b]Data mainly obtained using enzyme assays and gene identification from genome sequences, which are involved in the metabolism of amino acids in the corresponding extremophiles

is complete unless otherwise specified and it plays an important role in providing addition ATP and molecular building blocks (White et al. 2007). Each pathway has functions important for cell survival such as producing precursors for biosynthesis cofactors and energy. It is known that the ED pathway predates the EM pathway and is only present in prokaryotes (Stettner and Segrè 2013). Although the EM pathway produce one more ATP per glucose than ED pathway, ED pathway provides another advantage for its presence (Stettner and Segrè 2013). The ED pathway releases more free energy and has highly exergonic reactions, meaning that the forward reaction is highly favored (Flamholz et al. 2013; Bar-Even et al. 2012). In comparison, the EM pathway is close to an equilibrium reaction (Flamholz et al. 2013; Bar-Even et al. 2012). To compensate for this, the EM pathway has a higher protein cost (3.5 folds than ED pathway) for higher enzyme activities and protein synthesis (Flamholz et al. 2013; Bar-Even et al. 2012). In prokaryotes, if they are anaerobic, substrate-level phosphorylation represents primary means of energy conservation and most of them are dependent on the EM pathway (Stettner and Segrè 2013). Extremophiles that only use the ED pathway are thermoacidophilic archaea and some halophilic bacteria, osmophilic bacteria, and oligotroph. Those extremophiles do not have the key enzymes for the EM pathway, and it is thought that the evolution of an EM pathway is not critical for survival at those extreme environments. It remains a question to whether modified one found in archaea or the conventional one in most microorganisms is the origin of the metabolic pathway. Further study may provide insight into a better understanding of the central metabolic pathways commonly shared in extremophiles and the evolution lineage.

4 Peptides and Amino Acids Metabolism

Proteins, peptides, and amino acids are used by many organisms including extremophilic microorganisms. In general, proteins are hydrolyzed by proteases into peptides and amino acids, and peptides are further hydrolyzed into amino acids (Fig. 1.4) (White et al. 2012). After transporting amino acids inside the cell, they will either be used as building blocks for protein synthesis and/or undergoing deamination mainly by oxidative deamination and/or aminotransferase catalyzed reactions, producing corresponding α-keto acids such as pyruvate, oxaloacetic acid and α-ketoglutarate (White et al. 2012).

As there are different proteinaceous sources available from the environment, the decision to which type to utilize is dependent on the nitrogen availability in the environments (Leigh and Dodsworth 2007). The pathway for conversion of different nitrogen sources into ammonia requires different sources of energy and electrons. Hence, the pathway for utilizing a low concentration of nitrogen sources ie requiring more energy is usually repressed under conditions of nitrogen sufficiency but induced if the cell signals nitrogen deficiency. Also, nitrogen deficiency will induce the assimilatory pathways of ammonia by the interconversion of glutamate and α-ketoglutarate (Fig. 1.5). In bacteria and archaea, most of them use the metabolic

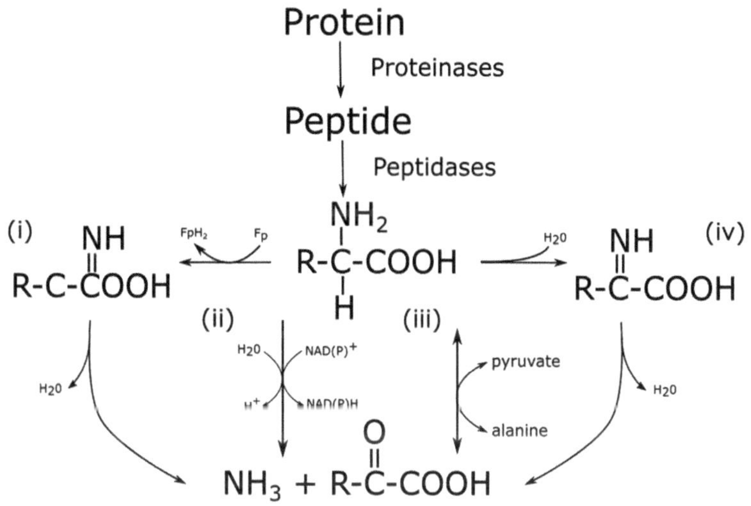

Fig. 1.4 Catabolism of protein, peptides and amino acids in extremophiles. Catabolism of amino acids into α-keto acids by using enzyme (*i*) Amino acid oxidases; (*ii*) Amino acid dehydrogenase; (*iii*) Transaminase and (*iv*) Serine/Threonine dehydratase in different pathways. *Fp* flavoprotein

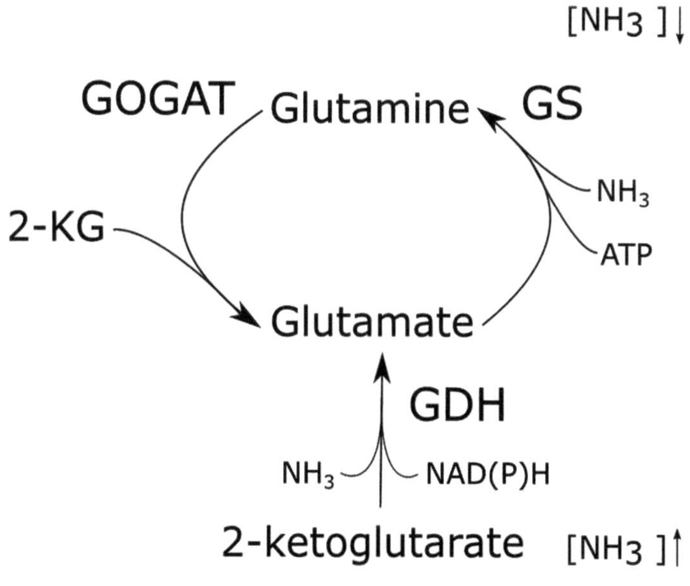

Fig. 1.5 Synthesis of glutamate using GDH and GOGAT system. At high ammonia concentrations, glutamate is synthesized by glutamate dehydrogenase and at low ammonia concentrations, glutamate is synthesized by glutamine and 2-ketoglutarate with enzyme glutamine oxoglutarate aminotransferase (GOGAT). *2-KG* 2-ketoglutarate, *GS* glutamine synthetase, *GDH* glutamate dehydrogenase

intermediates α-ketoglutarate for sensing nitrogen deficiency and glutamine for signaling nitrogen sufficiency (Doucette et al. 2011; Magasanik and Kaiser 2002; Kessler et al. 2001; Ehlers et al. 2005; Leigh and Dodsworth 2007). In most extremophiles they have a complete TCA cycle and α-ketoglutarate is an important intermediate in carbon and nitrogen metabolism (Ninfa and Jiang 2005; Fendt et al. 2013) but in organisms with an incomplete TCA cycle like some of the extremophiles, they usually have a partial pathway for generating α-ketoglutarate, which is used only as a precursor for nitrogen assimilation (Leigh and Dodsworth 2007).

Many proteases and peptidases have been well characterized in extremophiles (Table 1.2), and the pathways for both synthesis and catabolism of amino acids in different extremophiles are predicted mostly by genomics studies of their representative species. The carbon skeletons of amino acids after deamination will be fed into central metabolic pathways, but mostly incorporated into different intermediates in the TCA cycle and those intermediates are also precursors for the synthesis of amino acids (White et al. 2007). Generally, no significant differences are found in the synthesis and catabolic pathways of amino acids in extremophilic bacteria and archaea. In the following paragraphs, only those missing genes encoding enzymes functional in these pathways identified by genome sequence analyses are described.

In acidophilic bacteria, genes involved in the synthesis of amino acids pathways are all found except ornithine cyclodeaminase involved in proline biosynthesis, aromatic-amino-acid transaminase and arogenate dehydrogenase involved in tyrosine biosynthesis (Valdés et al. 2008).

In *Methanococcoidei burtonii*, a psychrophilic archaeon, glutamate is mainly synthesized by glutamine synthetase and glutamine oxoglutarate aminotransferase system (Allen et al. 2009). Moreover, the serine pathway enzymes including hydroxypyruvate reductase and serine glyoxylate aminotransferase are not found (Allen et al. 2009). For extremophiles that grow at high temperatures, the thermophiles, special enzymes and pathway are present. An alternate isoleucine synthesis pathway via the citramalate pathway which uses pyruvate and acetyl-CoA as precursors and key enzyme citramalate synthase to regulate the citramalate pathway is determined in thermophilic bacterium *Thermoanaerobacter* sp. strain X514 (Feng et al. 2009).

In halophiles, some archaea cannot synthesize more than 8 amino acids, hence amino acid is required as a substrate for growth (Ng et al. 2000). Amino acids can also play a role in the survival of osmophiles by osmoregulating the osmotic pressure, which is achieved by accumulation or synthesis of certain amino acids including proline, glycine, and glutamate (Galinski 1995).

One of the oligotrophs, *Caulobacter crescentus*, abundantly secreted peptidase for utilizing amino acids as a carbon source (Hottes et al. 2004). It also has different pathways for the degradation of different amino acids into pyruvate and is subsequently used for energy conservation and growth (Hottes et al. 2004).

One of the representative examples of radioresistant bacteria, *Deinococcus*, cannot utilize ammonia as a nitrogen source and is entirely dependent on exogenous amino acids as nitrogen source. Also, the biosynthetic pathways of lysine, serine and cysteine are incomplete, but it can grow on these three amino acids (White et al. 1999; Yuan et al. 2012; Zhang et al. 2000).

Extremophiles without any of the three central metabolic pathways may not have the precursor for the biosynthesis of those amino acids. This decreases the survival capability of those extremophiles in environment without the amino acid. This also might be the reason for the missing pathways of biosynthesis of amino acids from the genome analysis as they are not required.

5 Connections Between Central Metabolism and Amino Acids Metabolism

The connections between the central metabolic pathways are through the common intermediate phosphoglyceraldehyde that is oxidized to pyruvate (White et al. 2007). Each pathway has a unique metabolic purpose. The EM pathway is the pathway producing two ATP from one glucose while the ED pathway produces only one ATP and its oxidizing glucose into gluconate can provide a competitive advantage for survival by removing glucose from the living environment, and the PP pathway is for generating pentose and NADPH (White et al. 2007). Those extremophiles without one or more of the central metabolic pathways often have unique enzymes to compensate the functions of the missing pathway(s). For example, extremophiles without the oxidative part of the PP pathway will alternately use pyruvate:ferredoxin oxidoreductase and ferredoxin-NADP$^+$ reductase for the generation of additional NADPH needed (Blamey and Adams 1993; Hoek and Rydstrom 1988).

Besides the connections of the central metabolic pathways, some microorganisms utilize either carbohydrates or peptides as carbon and nitrogen sources, so the reversible conversion of amino acids and α-keto acids becomes essential (Fig. 1.6). Several intermediates from the central metabolic pathways and TCA cycle serve as precursors for the amino acid biosynthesis and the carbon skeletons of amino acids are also fed into intermediates of the TCA cycle and central metabolic pathways. From the TCA cycle, the intermediate oxaloacetate can be used for synthesizing six amino acids (Meister 2012), α-ketoglutarate can be used for two, glutamine and glutamate (Peng et al. 1993), the two most important amino acids used for the synthesis of other amino acids. Succinyl-CoA donates a succinyl group in the formation of intermediates for the synthesis of lysine and methionine (D'Mello and D'Mello 2003). From the central metabolic pathways, three glycolytic intermediates, pyruvate, phosphoglyceraldehyde, and phosphoenolpyruvate are precursors to nine amino acids (White et al. 2007). The PP pathway generates erythrose-4-P, which is for making aromatic amino acids (White et al. 2007). Glutamate is the amino donor for most other amino acids and important for the incorporation of inorganic nitrogen into cell material. Glutamate is synthesized by two different routes depending on ammonia concentration (Tempest et al. 1973). In environments of high concentrations of ammonia (>1 mM), glutamate dehydrogenase is used for catalyzing the reductive amination of α-ketoglutarate (Tempest et al. 1973). If the concentration of ammonia is low, which is the situation in many natural environments, two enzymes glutamate synthase and glutamine synthetase are involved and ATP is required (Tempest et al. 1973). Ammonia will be first incorporated into glutamine catalyzed by glutamine synthetase, and the ammonia from glutamine is transferred to α-ketoglutarate by glutamate synthase, producing glutamate [102]. In the catabolism of amino acids, all 20 amino acids are degraded to seven intermediates including pyruvate, acetyl-CoA, acetoacetyl-CoA, α-ketoglutarate, succinyl-CoA, fumarate, and oxaloacetate, and they enter the TCA cycle (Fig. 1.6) (Massey et al. 1976; Pederson et al. 1999; Barker 1981; Cunin et al. 1986). Amino acids are deaminated either oxidatively or by aminotransferase reaction to corresponding α-keto acids (Massey et al. 1976; Pederson et al. 1999; Barker 1981; Cunin et al. 1986). Firstly, the oxidation can be carried out by a flavooxidase such as D-amino oxidase and L-amino oxidase (D'aniello et al. 1993). Secondly, a more specific enzyme NAD(P)$^+$-linked amino acid dehydrogenase (Miflin and Lea 1977). Thirdly, by specific deaminases for serine and threonine into pyruvate, aspirate into oxaloacetate and histidine (Alföldi et al. 1968).

For those extremophiles that use only carbohydrates for their growth, intermediates of the central metabolic pathways including TCA cycle are used for the synthesis of amino acids (Fig. 1.6). However, some of them may not have all enzymes/pathways

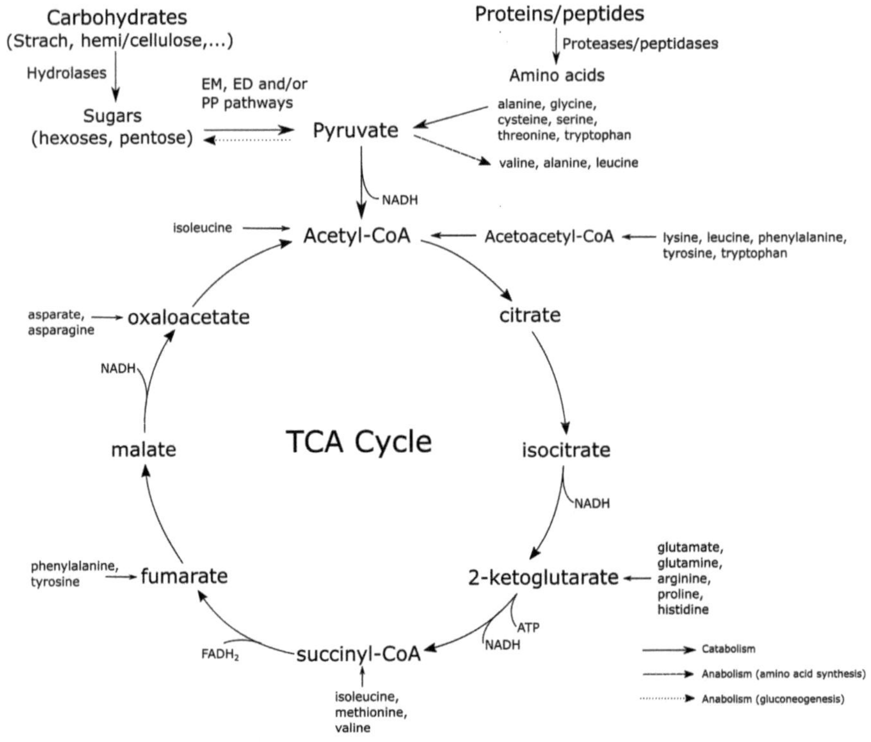

Fig. 1.6 Interconnection of metabolic pathways for the utilization of different substrates by extremophilic microorganisms. Amino acids entry points into the TCA cycle are labeled

for the synthesis of all amino acids, so they will require the addition of either peptides and/or amino acids into the growth media for their growth. There are a few extremophiles that are unable to use any carbohydrate but only peptides/amino acids for their growth, so they must have pathways for gluconeogenesis with a connection to TCA cycle and central metabolic pathways (Fig. 1.6).

6 Conclusions and Perspectives

Carbohydrates and/or peptides are essential substrates for growth of all microorganisms and extremophiles are not an exception. Although extremophiles adapt to the extreme environments, they still operate one or more of the EM, ED, PP pathways and TCA cycle or a modified version of these pathways with the common goals of producing ATP and recycle of cofactor NADP(H)/NAD(H). They have comparative growth rates when the same substrates are used. The pathways of either for the utilization of carbohydrates or peptides have the common intermediate of pyruvate (α-keto acid) to enter the subsequent reactions and pathways for the production of ATP, building block molecules and cofactors. The difference in substrate specificity is mainly dependent on the presence of specific transporters as they have a very similar metabolic capacity for utilizing carbohydrates and peptides/amino acids. It's intriguing and also challenging for further understanding the adaptation mechanisms of extremophiles and extremotolerants or extremoresistants. The molecular basis of their capability of growing under various extreme conditions would be a great resource for exploring bio-catalysis under extreme conditions. The unique properties of enzymes from the metabolic pathways have attracted attention for many applications, and the characterization of those enzymes from extremophiles serves as a very initial step for future exploitation.

Many enzymes discovered from different groups of extremophiles have successful applications in industries with DNA polymerases from hyperthermophiles as the most widely used one. Other examples including xylanases from thermophiles for paper bleaching, cellulases from acidophiles and amylases from psychrophiles in detergents (Van Den Burg 2003; Karan et al. 2012), and proteases for baking, brewing and dairy production (Bressan et al. 2008). The discovery of novel enzymes has been accelerated due to recent advances in studies using genomics, proteomics, and metabolomics, which facilitate the development of new industrial processes. Various genes involved in coping with extreme environmental stresses, for instance, proline synthase and glyceraldehydes-3-phosphate dehydrogenase related to salt stress, have also been introduced to agricultural crops. As newly discovered enzymes for applications in industries often require modification, the advances in protein engineering will enable the utilization of extremophilic enzymes in a much larger scale.

Conflict of Interest Ching Tse and Kesen Ma declare that they have no conflict of interest.

References

Abe F (2003) The Role of Tryptophan Permease Tat2 in Cell Growth of Yeast under High-Pressure Condition. In: Winter R (ed) Advances in High Pressure Bioscience and Biotechnology II, Springer, Berlin, pp 271–274

Abe F (2007) Exploration of the effects of high hydrostatic pressure on microbial growth, physiology and survival: perspectives from piezophysiology. Biosci Biotechnol Biochem 71(10):2347–2357

Abe F, Horikoshi K (2000) Tryptophan permease gene TAT2 confers high-pressure growth in *Saccharomyces cerevisiae*. Mol Cell Biol 20(21):8093–8102

Aghajari N, Van Petegem F, Villeret V, Chessa JP, Gerday C, Haser R, Van Beeumen J (2003) Crystal structures of a psychrophilic metalloprotease reveal new insights into catalysis by cold-adapted proteases. Proteins Struct Funct Bioinf 50(4):636–647

Alain K, Marteinsson VT, Miroshnichenko ML, Bonch-Osmolovskaya EA, Prieur D, Birrien JL (2002) *Marinitoga piezophila* sp. nov., a rod-shaped, thermo-piezophilic bacterium isolated under high hydrostatic pressure from a deep-sea hydrothermal vent. Int J Syst Evol Microbiol 52(4):1331–1339

Albers SV, Driessen AJ (2007) Membranes and Transport Proteins of Thermophilic Microorganisms. In: Robb F, Antranikian G (eds) Thermophiles: Biology and Technology at High Temperatures, CRC Press, New York, pp 39–54

Albers SV, Elferink MG, Charlebois RL, Sensen CW, Driessen AJ, Konings WN (1999) Glucose transport in the extremely thermoacidophilic *Sulfolobus solfataricus* involves a high-affinity membrane-integrated binding protein. J Bacteriol 181(14):4285–4291

Albers SV, Koning SM, Konings WN, Driessen AJ (2004) Insights into ABC transport in archaea. J Bioenerg Biomembr 36(1):5–15

Albertyn J, Hohmann S, Thevelein JM, Prior BA (1994) GPD1, which encodes glycerol-3-phosphate dehydrogenase, is essential for growth under osmotic stress in *Saccharomyces cerevisiae*, and its expression is regulated by the high-osmolarity glycerol response pathway. Mol Cell Biol 14(6):4135–4144

Alföldi L, Raskó I, Kerekes E (1968) L-Serine deaminase of *Escherichia coli*. J Bacteriol 96(5):1512–1518

Allen MA, Lauro FM, Williams TJ, Burg D, Siddiqui KS, De Francisci D, Chong KW, Pilak O, Chew HH, De Maere MZ (2009) The genome sequence of the psychrophilic archaeon, *Methanococcoides burtonii*: the role of genome evolution in cold adaptation. ISME J 3(9):1012–1035

Álvarez-Ordóñez A, Fernández A, Bernardo A, López M (2010) Arginine and lysine decarboxylases and the acid tolerance response of *Salmonella Typhimurium*. Int J Food Microbiol 136(3):278–282

Amato P (2013) Energy Metabolism in Low-temperature and Frozen Conditions in Cold-adapted Microorganism. In: Yumoto I (ed) Cold-adapted microorganisms, Caister Academic Press, UK, pp 71–96

Amend JP, Shock EL (2001) Energetics of overall metabolic reactions of thermophilic and hyperthermophilic Archaea and Bacteria. FEMS Microbiol Rev 25(2):175–243

Angelov A, Liebl W (2006) Insights into extreme thermoacidophily based on genome analysis of *Picrophilus torridus* and other thermoacidophilic archaea. J Biotechnol 126(1):3–10

Anithajothi R, Nagarani N, Umagowsalya G, Duraikannu K, Ramakritinan C (2014) Screening, isolation and characterization of protease producing moderately halophilic microorganism *Halomonas meridiana* associated with coral mucus. Toxicol Environ Chem 96(2):296–306

Antón J, Oren A, Benlloch S, Rodríguez-Valera F, Amann R, Rosselló-Mora R (2002) *Salinibacter ruber* gen. nov., sp. nov., a novel, extremely halophilic member of the Bacteria from saltern crystallizer ponds. Int J Syst Evol Microbiol 52(2):485–491

Aono E, Baba T, Ara T, Nishi T, Nakamichi T, Inamoto E, Toyonaga H, Hasegawa M, Takai Y, Okumura Y (2010) Complete genome sequence and comparative analysis of *Shewanella violacea*, a psychrophilic and piezophilic bacterium from deep sea floor sediments. Mol BioSyst 6(7):1216–1226

Arellano BH, Ortiz JD, Manzano J, Chen JC (2010) Identification of a dehydrogenase required for lactose metabolism in *Caulobacter crescentus*. Appl Environ Microbiol 76(9): 3004–3014

Attard GS, Templer RH, Smith WS, Hunt AN, Jackowski S (2000) Modulation of CTP: phosphocholine cytidylyltransferase by membrane curvature elastic stress. Proc Natl Acad Sci 97(16):9032–9036

Baker-Austin C, Dopson M (2007) Life in acid: pH homeostasis in acidophiles. Trends Microbiol 15(4):165–171

Bakermans C, Tsapin AI, Souza-Egipsy V, Gilichinsky DA, Nealson KH (2003) Reproduction and metabolism at −10 C of bacteria isolated from Siberian permafrost. Environ Microbiol 5(4):321–326

Baliga NS, Bonneau R, Facciotti MT, Pan M, Glusman G, Deutsch EW, Shannon P, Chiu Y, Weng RS, Gan RR (2004) Genome sequence of *Haloarcula marismortui*: a halophilic archaeon from the Dead Sea. Genome Res 14(11):2221–2234

Bar-Even A, Flamholz A, Noor E, Milo R (2012) Rethinking glycolysis: on the biochemical logic of metabolic pathways. Nat Chem Biol 8(6):509–517

Barker H (1981) Amino acid degradation by anaerobic bacteria. Annu Rev Biochem 50(1):23–40

Baross JA, Lenovich L (1992) Halophilic and osmophilic microorganisms. Compendium of methods for the microbiological examination of foods. American Public Health Association, Washington, DC, pp 199–212

Barrie Johnson D, Hallberg KB (2008) Carbon, iron and sulfur metabolism in acidophilic microorganisms. Adv Microb Physiol 54:201–255

Barry RC, Young MJ, Stedman KM, Dratz EA (2006) Proteomic mapping of the hyperthermophilic and acidophilic archaeon *Sulfolobus solfataricus* P2. Electrophoresis 27(14):2970–2983

Bartlett D, Bidle K (1999) Membrane-based adaptations of deep-sea piezophiles. In: Seckbach J (ed) Enigmatic Microorganisms and Life in Extreme Environments, Springer, Netherlands, pp 501–512

Batrakov SG, Pivovarova TA, Esipov SE, Sheichenko VI, Karavaiko GI (2002) β-D-Glucopyranosyl caldarchaetidylglycerol is the main lipid of the acidophilic, mesophilic, ferrous iron-oxidising archaeon *Ferroplasma acidiphilum*. BBA-Mol Cell Biol Lipids 1581(1):29–35

Ben-Bassat A, Lamed R, Zeikus J (1981) Ethanol production by thermophilic bacteria: metabolic control of end product formation in *Thermoanaerobium brockii*. J Bacteriol 146(1):192–199

Berezovsky IN, Shakhnovich EI (2005) Physics and evolution of thermophilic adaptation. Proc Natl Acad Sci 102(36):12742–12747

Berger F, Morellet N, Menu F, Potier P (1996) Cold shock and cold acclimation proteins in the psychotrophic bacterium *Arthrobacter globiformis* SI55. J Bacteriol 178(11):2999–3007

Bertin PN, Heinrich-Salmeron A, Pelletier E, Goulhen-Chollet F, Arsène-Ploetze F, Gallien S, Lauga B, Casiot C, Calteau A, Vallenet D (2011) Metabolic diversity among main microorganisms inside an arsenic-rich ecosystem revealed by meta-and proteo-genomics. ISME J 5(11):1735–1747

Bertoldo C, Antranikian G (2006) The order *Thermococcales*. In: Dworkin M, Falkow S, Rosenberg E, Schleifer K, Stackebrandt E (eds) The Prokaryotes, Springer, New York, pp 69–81

Bettenbrock K, Sauter T, Jahreis K, Kremling A, Lengeler JW, Gilles E-D (2007) Correlation between growth rates, EIIACrr phosphorylation, and intracellular cyclic AMP levels in *Escherichia coli* K-12. J Bacteriol 189(19):6891–6900

Blamey JM, Adams MW (1993) Purification and characterization of pyruvate ferredoxin oxidoreductase from the hyperthermophilic archaeon *Pyrococcus furiosus*. Biochim Biophys Acta 1161(1):19–27

Blankenhorn D, Phillips J, Slonczewski JL (1999) Acid-and base-induced proteins during aerobic and anaerobic growth of *Escherichia coli* revealed by two-dimensional gel electrophoresis. J Bacteriol 181(7):2209–2216

Blöchl E, Rachel R, Burggraf S, Hafenbradl D, Jannasch HW, Stetter KO (1997) *Pyrolobus fumarii*, gen. and sp. nov., represents a novel group of archaea, extending the upper temperature limit for life to 113 °C. Extremophiles 1(1):14–21

Blotevogel KH, Fischer U, Mocha M, Jannsen S (1985) *Methanobacterium thermoalcaliphilum* spec. nov., a new moderately alkaliphilic and thermophilic autotrophic methanogen. Arch Microbiol 142(3):211–217

Blumer-Schuette SE, Kataeva I, Westpheling J, Adams MW, Kelly RM (2008) Extremely thermophilic microorganisms for biomass conversion: status and prospects. Curr Opin Biotechnol 19(3):210–217

Bouchotroch S, Quesada E, del Moral A, Llamas I, Béjar V (2001) *Halomonas maura* sp. nov., a novel moderately halophilic, exopolysaccharide-producing bacterium. Int J Syst Evol Microbiol 51(5):1625–1632

Bressan RA, Bohnert HJ, Hasegawa PM (2008) Genetic engineering for salinity stress tolerance. Adv Plant Biochem Mol Biol 1:347–384

Brim H, Heuer H, Krögerrecklenfort E, Mergeay M, Smalla K (1999) Characterization of the bacterial community of a zinc-polluted soil. Can J Microbiol 45(4):326–338

Brown A (1976) Microbial water stress. Bacteriol Rev 40(4):803

Brunner NA, Brinkmann H, Siebers B, Hensel R (1998) NAD^+-dependent glyceraldehyde-3-phosphate dehydrogenase from *Thermoproteus tenax* the first identified archaeal membrane of the aldehyde dehydrogenase superfamily is a glycolytic enzyme with unusual regulatory properties. J Biol Chem 273(11):6149–6156

Brunner NA, Siebers B, Hensel R (2001) Role of two different glyceraldehyde-3-phosphate dehydrogenases in controlling the reversible Embden–Meyerhof–Parnas pathway in *Thermoproteus tenax*: regulation on protein and transcript level. Extremophiles 5(2):101–109

Bryan CG, Marchal M, Battaglia-Brunet F, Kugler V, Lemaitre-Guillier C, Lièvremont D, Bertin PN, Arsène-Ploetze F (2009) Carbon and arsenic metabolism in *Thiomonas* strains: differences revealed diverse adaptation processes. BMC Microbiol 9(1):127

Budgen N, Danson MJ (1986) Metabolism of glucose via a modified Entner-Doudoroff pathway in the thermoacidophilic archaebacterium *Thermoplasma acidophilum*. FEBS Lett 196(2):207–210

Burns DG, Janssen PH, Itoh T, Kamekura M, Li Z, Jensen G, Rodríguez-Valera F, Bolhuis H, Dyall-Smith ML (2007) *Haloquadratum walsbyi* gen. nov., sp. nov., the square haloarchaeon of Walsby, isolated from saltern crystallizers in Australia and Spain. Int J Syst Evol Microbiol 57(2):387–392

Carapito C, Muller D, Turlin E, Koechler S, Danchin A, Van Dorsselaer A, Leize-Wagner E, Bertin PN, Lett M-C (2006) Identification of genes and proteins involved in the pleiotropic response to arsenic stress in *Caenibacter arsenoxydans*, a metalloresistant beta-proteobacterium with an unsequenced genome. Biochimie 88(6):595–606

Carini P, Steindler L, Beszteri S, Giovannoni SJ (2012) Nutrient requirements for growth of the extreme oligotroph *Candidatus Pelagibacter ubique* HTCC1062 on a defined medium. ISME J 7(3):592–602

Castanie-Cornet MP, Penfound TA, Smith D, Elliott JF, Foster JW (1999) Control of acid resistance in *Escherichia coli*. J Bacteriol 181(11):3525–3535

Cavicchioli R, Thomas T, Curmi PM (2000) Cold stress response in Archaea. Extremophiles 4(6):321–331

Chintalapati S, Kiran M, Shivaji S (2004) Role of membrane lipid fatty acids in cold adaptation. Cell Mol Biol 50(5):631–642

Christensen HR, Frokiaer H, Pestka JJ (2002) Lactobacilli differentially modulate expression of cytokines and maturation surface markers in murine dendritic cells. J Immunol 168(1):171–178

Chrzanowski T, Kyle M, Elser J, Sterner R (1996) Element ratios and growth dynamics of bacteria in an oligotrophic Canadian shield lake. Aquat Microb Ecol 11(2):119–125

Ciaramella M, Napoli A, Rossi M (2005) Another extreme genome: how to live at pH 0. Trends Microbiol 13(2):49–51

Cunin R, Glansdorff N, Pierard A, Stalon V (1986) Biosynthesis and metabolism of arginine in bacteria. Microbiol Rev 50(3):314

D'Amico S, Collins T, Marx JC, Feller G, Gerday C (2006) Psychrophilic microorganisms: challenges for life. EMBO Rep 7(4):385–389

D'aniello A, D'onofrio G, Pischetola M, D'aniello G, Vetere A, Petrucelli L, Fisher G (1993) Biological role of D-amino acid oxidase and D-aspartate oxidase. Effects of D-amino acids. J Biol Chem 268(36):26941–26949

D'Mello J, D'Mello J (2003) An outline of pathways in amino acid metabolism. Amino acids in animal nutrition. CABI Publishing, Wallingford, UK, pp 71–86

Dalmasso M, Aubert J, Briard-Bion V, Chuat V, Deutsch SM, Even S, Falentin H, Jan G, Jardin J, Maillard M-B (2012) A Temporal-omic study of *Propionibacterium freudenreichii* CIRM-BIA1T adaptation strategies in conditions mimicking cheese ripening in the cold. PLoS One 7(1), e29083

DasSarma S, DasSarma P (2012) Halophiles. Wiley Online Library

De Graaf AA, Striegel K, Wittig RM, Laufer B, Schmitz G, Wiechert W, Sprenger GA, Sahm H (1999) Metabolic state of *Zymomonas mobilis* in glucose, fructose, and xylose-fed continuous cultures as analysed by 13C-and 31P-NMR spectroscopy. Arch Microbiol 171(6):371–385

De Groot A, Chapon V, Servant P, Christen R, Fischer-Le Saux M, Sommer S, Heulin T (2005) *Deinococcus deserti* sp. nov., a gamma-radiation-tolerant bacterium isolated from the Sahara Desert. Int J Syst Evol Microbiol 55(6):2441–2446

de Montigny J, Straub ML, Potier S, Tekaia F, Dujon B, Wincker P, Artiguenave F, Souciet JL (2000) Genomic exploration of the hemiascomycetous yeasts: *Zygosaccharomyces rouxii*. FEBS Lett 487(1):52–55

De Smet KA, Weston A, Brown IN, Young DB, Robertson BD (2000) Three pathways for trehalose biosynthesis in mycobacteria. Microbiology 146(1):199–208

de Vos WM, Kengen SW, Voorhorst WG, van der Oost J (1998) Sugar utilization and its control in hyperthermophiles. Extremophiles 2(3):201–205

De Vrije T, Mars A, Budde M, Lai M, Dijkema C, De Waard P, Claassen P (2007) Glycolytic pathway and hydrogen yield studies of the extreme thermophile *Caldicellulosiruptor saccharolyticus*. App Microbiol Biotechnol 74(6):1358–1367

Degryse E, Glansdorff N, Piérard A (1978) A comparative analysis of extreme thermophilic bacteria belonging to the genus *Thermus*. Arch Microbiol 117(2):189–196

Detkova E, Kevbrin V (2009) Cellobiose catabolism in the haloalkaliphilic hydrolytic bacterium *Alkaliflexus imshenetskii*. Microbiology 78(3):267–272

DiMarco AA, Romano AH (1985) D-Glucose transport system of *Zymomonas mobilis*. Appl Environ Microbiol 49(1):151–157

Dominy BN, Minoux H, Brooks CL (2004) An electrostatic basis for the stability of thermophilic proteins. Protein Struct Funct Bioinf 57(1):128–141

Dopson M, Baker-Austin C, Hind A, Bowman JP, Bond PL (2004) Characterization of *Ferroplasma* isolates and *Ferroplasma acidarmanus* sp. nov., extreme acidophiles from acid mine drainage and industrial bioleaching environments. Appl Environ Microbiol 70(4):2079–2088

Dörr C, Zaparty M, Tjaden B, Brinkmann H, Siebers B (2003) The Hexokinase of the Hyperthermophile *Thermoproteus tenax* ATP-Dependent Hexokinases and ADP-Dependent glucokinases, two alternatives for glucose phosphorylation in archaea. J Biol Chem 278(21):18744–18753

Doucette CD, Schwab DJ, Wingreen NS, Rabinowitz JD (2011) α-Ketoglutarate coordinates carbon and nitrogen utilization via enzyme I inhibition. Nature Chem Biol 7(12):894–901

Eggen R, Geerling A, Watts J, de Vos WM (1990) Characterization of pyrolysin, a hyperthermoactive serine protease from the archaebacterium *Pyrococcus furiosus*. FEMS Microbiol Lett 71(1–2):17–20

Egorova K, Antranikian G (2005) Industrial relevance of thermophilic Archaea. Current Opin Microbiol 8(6):649–655

Ehlers C, Weidenbach K, Veit K, Forchhammer K, Schmitz RA (2005) Unique mechanistic features of post-translational regulation of glutamine synthetase activity in *Methanosarcina mazei* strain Gö1 in response to nitrogen availability. Mol Microbiol 55(6):1841–1854
Elferink MG, Albers SV, Konings WN, Driessen AJ (2001) Sugar transport in *Sulfolobus solfataricus* is mediated by two families of binding protein-dependent ABC transporters. Mol Microbiol 39(6):1494–1503
England JL, Shakhnovich EI (2003) Structural determinant of protein designability. Phys Rev Lett 90(21):218101
England JL, Shakhnovich BE, Shakhnovich EI (2003) Natural selection of more designable folds: a mechanism for thermophilic adaptation. Proc Natl Acad Sci 100(15):8727–8731
Erauso G, Reysenbach A-L, Godfroy A, Meunier J-R, Crump B, Partensky F, Baross JA, Marteinsson V, Barbier G, Pace NR (1993) *Pyrococcus abyssi* sp. nov., a new hyperthermophilic archaeon isolated from a deep-sea hydrothermal vent. Arch Microbiol 160(5):338–349
Erra-Pujada M, Chang-Pi-Hin F, Debeire P, Duchiron F, O'Donohue MJ (2001) Purification and properties of the catalytic domain of the thermostable pullulanase type II from *Thermococcus hydrothermalis*. Biotech Lett 23(16):1273–1277
Evdokimov AG, Anderson DE, Routzahn KM, Waugh DS (2001) Structural basis for oligosaccharide recognition by *Pyrococcus furiosus* maltodextrin-binding protein. J Mol Biol 305(4):891–904
Falb M, Müller K, Königsmaier L, Oberwinkler T, Horn P, von Gronau S, Gonzalez O, Pfeiffer F, Bornberg-Bauer E, Oesterhelt D (2008) Metabolism of halophilic archaea. Extremophiles 12(2):177–196
Fang J, Zhang L, Bazylinski DA (2010) Deep-sea piezosphere and piezophiles: geomicrobiology and biogeochemistry. Trends Microbiol 18(9):413–422
Feder ME, Hofmann GE (1999) Heat-shock proteins, molecular chaperones, and the stress response: evolutionary and ecological physiology. Ann Rev Physiol 61(1):243–282
Fegatella F, Lim J, Kjelleberg S, Cavicchioli R (1998) Implications of rRNA operon copy number and ribosome content in the marine oligotrophic Ultramicrobacterium Sphingomonassp. Strain RB2256. Appl Environ Microbiol 64(11):4433–4438
Fendt SM, Bell EL, Keibler MA, Olenchock BA, Mayers JR, Wasylenko TM, Vokes NI, Guarente L, Vander Heiden MG, Stephanopoulos G (2013) Reductive glutamine metabolism is a function of the α-ketoglutarate to citrate ratio in cells. Nat Commun 4:2236
Feng X, Mouttaki H, Lin L, Huang R, Wu B, Hemme CL, He Z, Zhang B, Hicks LM, Xu J (2009) Characterization of the central metabolic pathways in *Thermoanaerobacter* sp. strain X514 via isotopomer-assisted metabolite analysis. Appl Environ Microbiol 75(15):5001–5008
Ferreira AC, Nobre MF, Rainey FA, Silva MT, Wait R, Burghardt J, Chung AP, Da Costa MS (1997) *Deinococcus geothermalis* sp. nov. and *Deinococcus murrayi* sp. nov., two extremely radiation-resistant and slightly thermophilic species from hot springs. Int J Syst Bacteriol 47(4):939–947
Fine A, Irihimovitch V, Dahan I, Konrad Z, Eichler J (2006) Cloning, expression, and purification of functional Sec11a and Sec11b, type I signal peptidases of the archaeon *Haloferax volcanii*. J Bacteriol 188(5):1911–1919
Flamholz A, Noor E, Bar-Even A, Liebermeister W, Milo R (2013) Glycolytic strategy as a tradeoff between energy yield and protein cost. Proc Natl Acad Sci 110(24):10039–10044
Franzmann P, Stackebrandt E, Sanderson K, Volkman J, Cameron D, Stevenson P, McMeekin T, Burton H (1988) *Halobacterium lacusprofundi* sp. nov., a halophilic bacterium isolated from Deep Lake, Antarctica. Syst Appl Microbiol 11(1):20–27
Fuchs T, Huber H, Teiner K, Burggraf S, Stetter KO (1995) *Metallosphaera prunae*, sp. nov., a novel metal-mobilizing, *Thermoacidophilic Archaeum*, isolated from a uranium mine in Germany. Syst Appl Microbiol 18(4):560–566
Fütterer O, Angelov A, Liesegang H, Gottschalk G, Schleper C, Schepers B, Dock C, Antranikian G, Liebl W (2004) Genome sequence of *Picrophilus torridus* and its implications for life around pH 0. Proc Natl Acad Sci 101(24):9091–9096
Galinski E (1995) Osmoadaptation in bacteria. Adv Microb Physiol 37:273

Garnier M, Matamoros S, Chevret D, Pilet MF, Leroi F, Tresse O (2010) Adaptation to cold and proteomic responses of the psychrotrophic biopreservative *Lactococcus piscium strain* CNCM I-4031. Appl Environ Microbiol 76(24):8011–8018

Garnova E, Krasil'nikova E (2003) Carbohydrate metabolism of the saccharolytic alkaliphilic anaerobes *Halonatronum saccharophilum*, *Amphibacillus fermentum*, and *Amphibacillus tropicus*. Microbiology 72(5):558–563

Gatti M, Fornasari M, Lazzi C, Mucchetti G, Neviani E (2004) Peptidase activity in various species of dairy thermophilic lactobacilli. J Appl Microbiol 96(2):223–229

Gock MA, Hocking AD, Pitt JI, Poulos PG (2003) Influence of temperature, water activity and pH on growth of some xerophilic fungi. Int J Food Microbiol 81(1):11–19

Gonzalez C, Gutierrez C, Ramirez C (1978) *Halobacterium vallismortis* sp. nov. An amylolytic and carbohydrate-metabolizing, extremely halophilic bacterium. Can J Microbiol 24(6): 710–715

González JM, Kato C, Horikoshi K (1995) *Thermococcus peptonophilus* sp. nov., a fast-growing, extremely thermophilic archaebacterium isolated from deep-sea hydrothermal vents. Arch Microbiol 164(3):159–164

González JM, Masuchi Y, Robb FT, Ammerman JW, Maeder DL, Yanagibayashi M, Tamaoka J, Kato C (1998) *Pyrococcus horikoshii* sp. nov., a hyperthermophilic archaeon isolated from a hydrothermal vent at the Okinawa Trough. Extremophiles 2(2):123–130

Gonzalez O, Gronau S, Falb M, Pfeiffer F, Mendoza E, Zimmer R, Oesterhelt D (2008) Reconstruction, modeling & analysis of *Halobacterium salinarum* R-1 metabolism. Mol BioSyst 4(2):148–159

Goodchild A, Saunders NF, Ertan H, Raftery M, Guilhaus M, Curmi PM, Cavicchioli R (2004) A proteomic determination of cold adaptation in the Antarctic archaeon, *Methanococcoides burtonii*. Mol Microbiol 53(1):309–321

Guangrong H, Tiejing Y, Po H, Jiaxing J (2006) Purification and characterization of a protease from thermophilic *Bacillus strain* HS08. Afr J Biotechnol 5(24):2433–2438

Habib S, Fakhruddin A, Begum S, Ahmed M (2012) Isolation and screening of thermo stable extracellular alkaline protease producing bacteria from *Tannery Effluents*. J Sci Res 4(2):515

Hafenbradl D, Keller M, Dirmeier R, Rachel R, Roßnagel P, Burggraf S, Huber H, Stetter KO (1996) *Ferroglobus placidus* gen. nov., sp. nov., a novel hyperthermophilic archaeum that oxidizes Fe^{2+} at neutral pH under anoxic conditions. Arch Microbiol 166(5):308–314

Hallberg KB, Lindström EB (1994) Characterization of *Thiobacillus caldus* sp. nov., a moderately thermophilic acidophile. Microbiology 140(12):3451–3456

Hallberg KB, González-Toril E, Johnson DB (2010) *Acidithiobacillus ferrivorans*, sp. nov.; facultatively anaerobic, psychrotolerant iron-, and sulfur-oxidizing acidophiles isolated from metal mine-impacted environments. Extremophiles 14(1):9–19

Hanson RS, Hanson TE (1996) Methanotrophic bacteria. Microbiol Rev 60(2):439–471

Hao X, Ma K (2003) Minimal sulfur requirement for growth and sulfur-dependent metabolism of the hyperthermophilic archaeon *Staphylothermus marinus*. Archaea 1(3):191–197

Harrison AP Jr (1984) The acidophilic *thiobacilli* and other acidophilic bacteria that share their habitat. Annu Rev Microbiol 38(1):265–292

Hendrick JP, Hartl F (1993) Molecular chaperone functions of heat-shock proteins. Annu Rev Biochem 62(1):349–384

Hensel R, Laumann S, Lang J, Heumann H, Lottspeich F (1987) Characterization of two D-glyceraldehyde-3-phosphate dehydrogenases from the extremely thermophilic archaebacterium *Thermoproteus tenax*. European J Biochem 170(1-2):325–333

Herbert R, Bell C (1977) Growth characteristics of an obligately psychrophilic *Vibrio* sp. Arch Microbiol 113(3):215–220

Hettmann T, Schmidt CL, Anemüller S, Zähringer U, Moll H, Petersen A, Schäfer G (1998) Cytochrome b 558/566 from the Archaeon *Sulfolobus acidocaldarius* a novel highly glycosylated, membrane-bound b-type hemoprotein. J Biol Chem 273(20):12032–12040

Higgins CF (1995) The ABC of channel regulation. Cell 82(5):693–696

Hoaki T, Nishijima M, Kato M, Adachi K, Mizobuchi S, Hanzawa N, Maruyama T (1994) Growth requirements of hyperthermophilic sulfur-dependent heterotrophic archaea isolated from a shallow submarine geothermal system with reference to their essential amino acids. Appl Environ Microbiol 60(8):2898–2904

Hocking AD (1986) Effects of water activity and culture age on the glycerol accumulation patterns of five fungi. J Gen Microbiol 132(2):269–275

Hocking AD, Norton RS (1983) Natural-abundance 13C nuclear magnetic resonance studies on the internal solutes of xerophilic fungi. J Gen Microbiol 129(9):2915–2925

Hoek JB, Rydstrom J (1988) Physiological roles of nicotinamide nucleotide transhydrogenase. Biochem J 254:1–10

Holland IB, Blight MA (1999) ABC-ATPases, adaptable energy generators fuelling transmembrane movement of a variety of molecules in organisms from bacteria to humans. J Mol Biol 293(2):381–399

Hongo K, Hirai H, Uemura C, Ono S, Tsunemi J, Higurashi T, Mizobata T, Kawata Y (2006) A novel ATP/ADP hydrolysis activity of hyperthermostable group II chaperonin in the presence of cobalt or manganese ion. FEBS Lett 580(1):34–40

Horikoshi K (1996) Alkaliphiles—from an industrial point of view. FEMS Microbiol Rev 18(2–3):259–270

Horlacher R, Xavier KB, Santos H, DiRuggiero J, Kossmann M, Boos W (1998) Archaeal binding protein-dependent ABC transporter: molecular and biochemical analysis of the trehalose/maltose transport system of the hyperthermophilic archaeon *Thermococcus litoralis*. J Bacteriol 180(3):680–689

Hottes AK, Meewan M, Yang D, Arana N, Romero P, McAdams HH, Stephens C (2004) Transcriptional profiling of *Caulobacter crescentus* during growth on complex and minimal media. J Bacteriol 186(5):1448–1461

Huang Y, Chapman B, Wilson M, Hocking AD (2009) Effect of agar concentration on the matric potential of glycerol agar media and the germination and growth of xerophilic and non-xerophilic fungi. Int J Food Microbiol 133(1):179–185

Huber G, Stetter KO (1991) *Sulfolobus metallicus*, sp. nov., a novel strictly chemolithoautotrophic thermophilic archaeal species of metal-mobilizers. Syst App Microbiol 14(4):372–378

Huber H, Stetter KO (2006) *Desulfurococcales*. In: Dworkin M, Falkow S, Rosenberg E, Schleifer K, Stackebrandt E (eds) The Prokaryotes, Springer, New York, pp 52–68

Huber R, Langworthy TA, König H, Thomm M, Woese CR, Sleytr UB, Stetter KO (1986) *Thermotoga maritima* sp. nov. represents a new genus of unique extremely thermophilic eubacteria growing up to 90 °C. Arch Microbiol 144(4):324–333

Huber H, Burggraf S, Mayer T, Wyschkony I, Rachel R, Stetter KO (2000) *Ignicoccus* gen. nov., a novel genus of hyperthermophilic, chemolithoautotrophic Archaea, represented by two new species, *Ignicoccus islandicus* sp nov and Ignicoccus pacificus sp nov. and *Ignicoccus pacificus* sp. nov. Int J Syst Evol Microbiol 50(6):2093–2100

Imanaka H, Yamatsu A, Fukui T, Atomi H, Imanaka T (2006) Phosphoenolpyruvate synthase plays an essential role for glycolysis in the modified Embden-Meyerhof pathway in *Thermococcus kodakarensis*. Mol Microbiol 61(4):898–909

Imhoff JF, Rodriguez-Valera F (1984) Betaine is the main compatible solute of halophilic eubacteria. J Bacteriol 160(1):478–479

Irwin JA, Alfredsson GA, Lanzetti AJ, Gudmundsson HM, Engel PC (2001) Purification and characterisation of a serine peptidase from the marine psychrophile strain PA-43. FEMS Microbiol Lett 201(2):285–290

Jaenicke R (1981) Enzymes under extremes of physical conditions. Annu Rev Biophys Bioeng 10(1):1–67

Jaenicke R, Böhm G (1998) The stability of proteins in extreme environments. Curr Opin Struct Biol 8(6):738–748

Jia Z, Davies PL (2002) Antifreeze proteins: an unusual receptor–ligand interaction. Trend Biochem Sci 27(2):101–106

Johnsen U, Selig M, Xavier KB, Santos H, Schönheit P (2001) Different glycolytic pathways for glucose and fructose in the halophilic archaeon *Halococcus saccharolyticus*. Arch Microbiol 175(1):52–61

Johnson DB, Stallwood B, Kimura S, Hallberg KB (2006) Isolation and characterization of *Acidicaldus organivorus*, gen. nov., sp. nov.: a novel sulfur-oxidizing, ferric iron-reducing thermo-acidophilic heterotrophic *Proteobacterium*. Arch Microbiol 185(3):212–221

Jojima Y, Mihara Y, Suzuki S, Yokozeki K, Yamanaka S, Fudou R (2004) *Saccharibacter floricola* gen. nov., sp. nov., a novel osmophilic acetic acid bacterium isolated from pollen. Int J Syst Evol Microbiol 54(6):2263–2267

Jolivet E, L'Haridon S, Corre E, Forterre P, Prieur D (2003) *Thermococcus gammatolerans* sp. nov., a hyperthermophilic archaeon from a deep-sea hydrothermal vent that resists ionizing radiation. Int J Syst Evol Microbiol 53(3):847–851

Jolivet E, Corre E, L'Haridon S, Forterre P, Prieur D (2004) *Thermococcus marinus* sp. nov. and *Thermococcus radiotolerans* sp. nov., two hyperthermophilic archaea from deep-sea hydrothermal vents that resist ionizing radiation. Extremophiles 8(3):219–227

Kamekura M, Seno Y, Holmes M, Dyall-Smith M (1992) Molecular cloning and sequencing of the gene for a halophilic alkaline serine protease (halolysin) from an unidentified halophilic archaea strain (172P1) and expression of the gene in *Haloferax volcanii*. J Bacteriol 174(3):736–742

Karan R, Capes MD, DasSarma S (2012) Function and biotechnology of extremophilic enzymes in low water activity. Aquat Biosyst 8(4):1–15

Karavaĭko G, Krasil'nikova E, Tsaplina I, Bogdanova T, Zakharchuk L (2000) Growth and carbohydrate metabolism of *sulfobacilli*. Microbiology 70(3):293–299

Kato C (1999) Barophiles (Piezophiles). In: Horikoshi K, Tsuji K (eds) Extremophiles in deep-sea environments. Springer, Japan, pp 91–111

Kaushik JK, Ogasahara K, Yutani K (2002) The unusually slow relaxation kinetics of the folding-unfolding of pyrrolidone carboxyl peptidase from a hyperthermophile, *Pyrococcus furiosus*. J Mol Biol 316(4):991–1003

Kazak H, Oner E, Dekker RF (2010) Extremophiles as sources of exopolysaccharides. In: Ito R, Matsuo Y (eds) Handbook of Carbohydrate Polymers Development and Properties, Nova Science Publishers, New York, pp 605–619

Keller H, Friedli T (1992) Visibility limit of naked-eye sunspots. Q J R Astron Soc 33:83–89

Kengen S, De Bok F, Van Loo N, Dijkema C, Stams A, De Vos W (1994) Evidence for the operation of a novel Embden-Meyerhof pathway that involves ADP-dependent kinases during sugar fermentation by *Pyrococcus furiosus*. J Biol Chem 269(26):17537–17541

Kengen S, Stams AJ, Vos WM (1996) Sugar metabolism of hyperthermophiles. FEMS Microbiol Rev 18(2-3):119–137

Kessler PS, Daniel C, Leigh JA (2001) Ammonia switch-off of nitrogen fixation in the methanogenic archaeon *Methanococcus maripaludis*: mechanistic features and requirement for the novel GlnB homologues, NifI1 and NifI2. J Bacteriol 183(3):882–889

Khan MTH, Sylte I (2009) Determinants for psychrophilic and thermophilic features of metallopeptidases of the M4 family. Silico Biol 9(3):105–124

Khmelenina VN, Kalyuzhnaya MG, Starostina NG, Suzina NE, Trotsenko YA (1997) Isolation and characterization of halotolerant alkaliphilic methanotrophic bacteria from Tuva soda lakes. Curr Microbiol 35(5):257–261

Kinoshita K, Takano M, Koseki T, Ito K, Iwano K (1995) Cloning of the xynNB gene encoding xylanase B from *Aspergillus niger* and its expression in *Aspergillus kawachii*. J Ferment Bioeng 79(5):422–428

Kitada M, Kosono S, Kudo T (2000) The Na^+/H^+ antiporter of alkaliphilic *Bacillus* sp. Extremophiles 4(5):253–258

Kletzin A, Mukund S, Kelley-Crouse TL, Chan MK, Rees DC, Adams M (1995) Molecular characterization of the genes encoding the tungsten-containing aldehyde ferredoxin oxidoreductase from *Pyrococcus furiosus* and formaldehyde ferredoxin oxidoreductase from *Thermococcus litoralis*. J Bacteriol 177(16):4817–4819

Klingeberg M, Hashwa F, Antranikian G (1991) Properties of extremely thermostable proteases from anaerobic hyperthermophilic bacteria. App Microbiol Biotechnol 34(6):715–719

Klingeberg M, Galunsky B, Sjoholm C, Kasche V, Antranikian G (1995) Purification and properties of a highly thermostable, sodium dodecyl sulfate-resistant and stereospecific proteinase from the extremely thermophilic archaeon *Thermococcus stetteri*. Appl Environ Microbiol 61(8):3098–3104

Koga S, Yoshioka I, Sakuraba H, Takahashi M, Sakasegawa S, Shimizu S, Ohshima T (2000) Biochemical characterization, cloning, and sequencing of ADP-dependent (AMP-forming) glucokinase from two hyperthermophilic archaea, *Pyrococcus furiosus* and *Thermococcus litoralis*. J Biochem 128(6):1079–1085

Koning HP (2001) Uptake of pentamidine in *Trypanosoma brucei* is mediated by three distinct transporters: implications for cross-resistance with arsenicals. Mol Pharmacol 59:586–592

Koning SM, Albers SV, Konings WN, Driessen AJ (2002) Sugar transport in (hyper) thermophilic archaea. Res Microbiol 153(2):61–67

Konings WN, Albers SV, Koning S, Driessen AJ (2002) The cell membrane plays a crucial role in survival of bacteria and archaea in extreme environments. Springer 81(1–4):61–72

Kristjansson JK (1991) Thermophilic bacteria. CRC Press, Boca Raton, FL

Krulwich TA, Ito M, Hicks DB, Gilmour R, Guffanti AA (1998) pH homeostasis and ATP synthesis: studies of two processes that necessitate inward proton translocation in extremely alkaliphilic Bacillus species. Extremophiles 2(3):217–222

Kumar V, Verma D, Satyanarayana T (2013) Extremophilic bacterial xylanases: production, characteristics and applications. Curr Biotechnol 2(4):380–399

Kurosawa N, Itoh YH, Iwai T, Sugai A, Uda I, Kimura N, Horiuchi T, Itoh T (1998) *Sulfurisphaera ohwakuensis* gen. nov., sp. nov., a novel extremely thermophilic acidophile of the order *Sulfolobales*. Int J Syst Bacteriol 48(2):451–456

Labes A, Schönheit P (2001) Sugar utilization in the hyperthermophilic, sulfate-reducing archaeon *Archaeoglobus fulgidus* strain 7324: starch degradation to acetate and CO2 via a modified Embden-Meyerhof pathway and acetyl-CoA synthetase (ADP-forming). Arch Microbiol 176(5):329–338

Leigh JA, Dodsworth JA (2007) Nitrogen regulation in bacteria and archaea. Annu Rev Microbiol 61:349–377

Leong SL, Lantz H, Pettersson OV, Frisvad JC, Thrane U, Heipieper HJ, Dijksterhuis J, Grabherr M, Pettersson M, Tellgren-Roth C (2014) Genome and physiology of the ascomycete filamentous fungus *Xeromyces bisporus*, the most xerophilic organism isolated to date. Environ Microbiol 17(2):496–513

Li W, Zhou X, Lu P (2005) Structural features of thermozymes. Biotechnol Adv 23(4):271–281

Lim J, Thomas T, Cavicchioli R (2000) Low temperature regulated DEAD-box RNA helicase from the Antarctic archaeon, *Methanococcoides burtonii*. J Mol Biol 297(3):553–567

Lippert K, Galinski EA, Trüper HG (1993) Biosynthesis and function of trehalose in *Ectothiorhodospira halochloris*. Antonie Van Leeuwenhoek 63(1):85–91

Lobos JH, Chisolm TE, Bopp LH, Holmes DS (1986) *Acidiphilium organovorum* sp. nov., an acidophilic heterotroph isolated from a *Thiobacillus ferrooxidans* culture. Int J Syst Bacteriol 36(2):139–144

Loos H, Krämer R, Sahm H, Sprenger GA (1994) Sorbitol promotes growth of *Zymomonas mobilis* in environments with high concentrations of sugar: evidence for a physiological function of glucose-fructose oxidoreductase in osmoprotection. J Bacteriol 176(24):7688–7693

Lovitt R, Longin R, Zeikus J (1984) Ethanol production by thermophilic bacteria: physiological comparison of solvent effects on parent and alcohol-tolerant strains of *Clostridium thermohydrosulfuricum*. Appl Environ Microbiol 48(1):171–177

Lucas S, Han J, Lapidus A, Cheng J-F, Goodwin LA, Pitluck S, Peters L, Mikhailova N, Teshima H, Detter JC (2012) Complete genome sequence of the thermophilic, piezophilic, heterotrophic bacterium *Marinitoga piezophila* KA3. Jo Bacteriol 194(21):5974–5975

Luyten K, Albertyn J, Skibbe WF, Prior BA, Ramos J, Thevelein JM, Hohmann S (1995) Fps1, a yeast member of the MIP family of channel proteins, is a facilitator for glycerol uptake and efflux and is inactive under osmotic stress. EMBO J 14:1360–1371

Macedo-Ribeiro S, Darimont B, Sterner R, Huber R (1996) Small structural changes account for the high thermostability of 1 [4Fe–4S] ferredoxin from the hyperthermophilic bacterium *Thermotoga maritima*. Structure 4(11):1291–1301

Magasanik B, Kaiser CA (2002) Nitrogen regulation in *Saccharomyces cerevisiae*. Gene 290(1):1–18

Marco MG, Rodríguez LV, Ramos EL, Renovato J, Cruz-Hernández MA, Rodríguez R, Contreras J, Aguilar CN (2009) A novel tannase from the xerophilic fungus *Aspergillus niger* GH1. J Microbiol Biotechnol 1:1–10

Margaryan A, Panosyan H, Popov Y (2010) Isolation and characterization of new metallotolernat *Bacilli* Strains. Biotech Biotechnol Equip 24(Suppl 1):450–454

Margesin R, Spröer C, Schumann P, Schinner F (2003) *Pedobacter cryoconitis* sp. nov., a facultative psychrophile from alpine glacier cryoconite. Int J Syst Evol Microbiol 53(5): 1291–1296

Marı S, Abellana M, Rubinat M, Sanchis V, Ramos A (2003) Efficacy of sorbates on the control of the growth of *Eurotium* species in bakery products with near neutral pH. Int J Food Microbiol 87 (3):251–258

Martin D, Bartlett DH, Roberts MF (2002) Solute accumulation in the deep-sea bacterium *Photobacterium profundum*. Extremophiles 6(6):507–514

Massey LK, Sokatch JR, Conrad RS (1976) Branched-chain amino acid catabolism in bacteria. Bacteriol Rev 40(1):42

Médigue C, Krin E, Pascal G, Barbe V, Bernsel A, Bertin PN, Cheung F, Cruveiller S, D'Amico S, Duilio A (2005) Coping with cold: the genome of the versatile marine Antarctica bacterium *Pseudoalteromonas haloplanktis* TAC125. Genome Res 15(10):1325–1335

Meister A (2012) Biochemistry of the amino acids. Elsevier, New York.

Merchant MM, Welsh AK, McLean RJ (2007) *Rheinheimera texasensis* sp. nov., a halointolerant freshwater oligotroph. Int J Syst Evol Microbiol 57(10):2376–2380

Michels M, Bakker EP (1985) Generation of a large, protonophore-sensitive proton motive force and pH difference in the acidophilic bacteria *Thermoplasma acidophilum* and *Bacillus acidocaldarius*. J Bacteriol 161(1):231–237

Miflin B, Lea P (1977) Amino acid metabolism. Annu Rev Plant Physiol 28(1):299–329

Mishra D, Kim DJ, Ralph D, Ahn JG, Rhee YH (2008) Bioleaching of metals from spent lithium ion secondary batteries using *Acidithiobacillus ferrooxidans*. Waste Manage 28(2): 333–338

Moracci M, Ponzano BC, Trincone A, Fusco S, De Rosa M, van der Oost J, Sensen CW, Charlebois RL (2000) Identification and molecular characterization of the first α-xylosidase from an archaeon. J Biol Chem 275(29):22082–22089

Moran JW, Witter LD (1979) Effect of sugars on D-arabitol production and glucose metabolism in *Saccharomyces rouxii*. J Bacteriol 138(3):823–831

Morikawa M, Izawa Y, Rashid N, Hoaki T, Imanaka T (1994) Purification and characterization of a thermostable thiol protease from a newly isolated hyperthermophilic *Pyrococcus* sp. Appl Environ Microbiol 60(12):4559–4566

Morita RY (1975) Psychrophilic bacteria. Bacteriol Rev 39(2):144

Mormile MR, Biesen MA, Gutierrez MC, Ventosa A, Pavlovich JB, Onstott TC, Fredrickson JK (2003) Isolation of *Halobacterium salinarum* retrieved directly from halite brine inclusions. Environ Microbiol 5(11):1094–1102

Mukund S, Adams M (1991) The novel tungsten-iron-sulfur protein of the hyperthermophilic archaebacterium, *Pyrococcus furiosus*, is an aldehyde ferredoxin oxidoreductase. Evidence for its participation in a unique glycolytic pathway. J Biol Chem 266(22):14208–14216

Mukund S, Adams M (1993) Characterization of a novel tungsten-containing formaldehyde ferredoxin oxidoreductase from the hyperthermophilic archaeon, *Thermococcus litoralis*. A role for tungsten in peptide catabolism. J Biol Chem 268(18):13592–13600

Mukund S, Adams MW (1995) Glyceraldehyde-3-phosphate ferredoxin oxidoreductase, a novel tungsten-containing enzyme with a potential glycolytic role in the hyperthermophilic archaeon *Pyrococcus furiosus*. J Biol Chem 270(15):8389–8392

Muller D, Simeonova DD, Riegel P, Mangenot S, Koechler S, Lièvremont D, Bertin PN, Lett MC (2006) *Herminiimonas arsenicoxydans* sp. nov., a metalloresistant bacterium. Int J Syst Evol Microbiol 56(8):1765–1769

Nelson KE, Clayton RA, Gill SR, Gwinn ML, Dodson RJ, Haft DH, Hickey EK, Peterson JD, Nelson WC, Ketchum KA (1999) Evidence for lateral gene transfer between Archaea and bacteria from genome sequence of *Thermotoga maritima*. Nature 399(6734):323–329

Nevoigt E, Stahl U (1997) Osmoregulation and glycerol metabolism in the yeast *Saccharomyces cerevisiae*. FEMS Microbiol Rev 21(3):231–241

Ng WV, Kennedy SP, Mahairas GG, Berquist B, Pan M, Shukla HD, Lasky SR, Baliga NS, Thorsson V, Sbrogna J (2000) Genome sequence of Halobacterium species NRC-1. Proc Natl Acad Sci 97(22):12176–12181

Ninfa AJ, Jiang P (2005) PII signal transduction proteins: sensors of α-ketoglutarate that regulate nitrogen metabolism. Curr Opin Microbiol 8(2):168–173

Nogi Y, Kato C, Horikoshi K (2002) *Psychromonas kaikoae* sp. nov., a novel from the deepest piezophilic bacterium cold-seep sediments in the Japan Trench. Int J Syst Evol Microbiol 52(5):1527–1532

Nogi Y, Hosoya S, Kato C, Horikoshi K (2004) *Colwellia piezophila* sp. nov., a novel piezophilic species from deep-sea sediments of the Japan Trench. Int J Syst Evol Microbiol 54(5): 1627–1631

Novitsky JA, Morita RY (1976) Morphological characterization of small cells resulting from nutrient starvation of a psychrophilic marine *vibrio*. Appl Environ Microbiol 32(4):617–622

Obis D, Guillot A, Gripon JC, Renault P, Bolotin A, Mistou MY (1999) Genetic and biochemical characterization of a high-affinity betaine uptake system (BusA) in *Lactococcus lactis* reveals a new functional organization within bacterial ABC transporters. J Bacteriol 181(20):6238–6246

Ogawa Y, Tatsumi H, Murakami S, Ishida Y, Murakami K, Masaki A, Kawabe H, Arimura H, Nakano E, Motai H (1990) Secretion of *Aspergillus oryzae* alkaline protease in an osmophilic yeast, *Zygosaccharomyces rouxii*. Agric Biol Chem 54(10):2521–2529

Oren A, Mana L (2003) Sugar metabolism in the extremely halophilic bacterium *Salinibacter ruber*. FEMS Microbiol Lett 223(1):83–87

Oren A, Ginzburg M, Ginzburg B, Hochstein L, Volcani B (1990) *Haloarcula marismortui* (*Volcani*) sp. nov., nom. rev., an extremely halophilic bacterium from the Dead Sea. Int J Syst Bacteriol 40(2):209–210

Osman Y, Conway T, Bonetti S, Ingram L (1987) Glycolytic flux in *Zymomonas mobilis*: enzyme and metabolite levels during batch fermentation. J Bacteriol 169(8):3726–3736

Paavilainen S, Oinonen S, Korpela T (1999) Catabolic pathways of glucose in *Bacillus circulans var. alkalophilus*. Extremophiles 3(4):269–276

Padan E, Bibi E, Ito M, Krulwich TA (2005) Alkaline pH homeostasis in bacteria: new insights. BBA Biomem 1717(2):67–88

Påhlman A-K, Granath K, Ansell R, Hohmann S, Adler L (2001) The yeast glycerol 3-phosphatases Gpp1p and Gpp2p are required for glycerol biosynthesis and differentially involved in the cellular responses to osmotic, anaerobic, and oxidative stress. J Biol Chem 276(5):3555–3563

Palacios-Cabrera H, Taniwaki MH, Hashimoto JM, Menezes HCd (2005) Growth of *Aspergillus ochraceus*, *A. carbonarius* and *A. niger* on culture media at different water activities and temperatures. Braz J Microbiol 36(1):24–28

Patel MA, Ou MS, Harbrucker R, Aldrich HC, Buszko ML, Ingram LO, Shanmugam K (2006) Isolation and characterization of acid-tolerant, thermophilic bacteria for effective fermentation of biomass-derived sugars to lactic acid. Appl Environ Microbiol 72(5):3228–3235

Paulsen IT, Nguyen L, Sliwinski MK, Rabus R, Saier MH (2000) Microbial genome analyses: comparative transport capabilities in eighteen prokaryotes. J Mol Biol 301(1):75–100

Pederson JA, Steele JL, Christensen JE, Dudley EG (1999) Peptidases and amino acid catabolism in lactic acid bacteria. In: Konings WN, Kuipers OP, Huis In't Veld JHJ (eds) Lactic acid bacteria: genetics, metabolism and applications. Springer, Netherlands, pp 217–246

Peng L, Hertz L, Huang R, Sonnewald U, Petersen S, Westergaard N, Larsson O, Schousboe A (1993) Utilization of glutamine and of TCA cycle constituents as precursors for transmitter glutamate and GABA. Dev Neurosci 15(3–5):367–377

Pflüger K, Müller V (2004) Transport of compatible solutes in extremophiles. J Bioenerg Biomem 36(1):17–24

Phadtare S (2004) Recent developments in bacterial cold-shock response. Curr Issues Mol Biol 6(2):125–136

Piette F, D'Amico S, Struvay C, Mazzucchelli G, Renaut J, Tutino ML, Danchin A, Leprince P, Feller G (2010) Proteomics of life at low temperatures: trigger factor is the primary chaperone in the Antarctic bacterium *Pseudoalteromonas haloplanktis* TAC125. Mol Microbiol 76(1):120–132

Pikuta E, Lysenko A, Chuvilskaya N, Mendrock U, Hippe H, Suzina N, Nikitin D, Osipov G, Laurinavichius K (2000) *Anoxybacillus pushchinensis* gen. nov., sp. nov., a novel anaerobic, alkaliphilic, moderately thermophilic bacterium from manure, and description of *Anoxybacillus flavitherms comb.* nov. Int J Syst Evol Microbiol 50(6):2109–2117

Pitt JI, Hocking AD (2009) Xerophiles. In: Fungi and food spoilage. Springer, US, pp 339–355

Poindexter JS (1981) Oligotrophy. In: Alexander M (ed) Advances in microbial ecology. Springer, US, pp 63–89

Postma P, Lengeler J, Jacobson G (1993) Phosphoenolpyruvate: carbohydrate phosphotransferase systems of bacteria. Microbiol Rev 57(3):543

Preston CM, Wu KY, Molinski TF, DeLong EF (1996) A psychrophilic crenarchaeon inhabits a marine sponge: *Cenarchaeum symbiosum* gen. nov., sp. nov. Proc Natl Acad Sci 93(13):6241–6246

Quiocho FA, Ledvina PS (1996) Atomic structure and specificity of bacterial periplasmic receptors for active transport and chemotaxis: variation of common themes. Mol Microbiol 20(1):17–25

Rainey F, Donnison A, Janssen P, Saul D, Rodrigo A, Bergquist PL, Daniel RM, Stackebrandt E, Morgan HW (1994) Description of *Caldicellulosiruptor saccharolyticus* gen. nov., sp. nov: an obligately anaerobic, extremely thermophilic, cellulolytic bacterium. FEMS Microbiol Lett 120(3):263–266

Rainey FA, Ferreira M, Nobre MF, Ray K, Bagaley D, Earl AM, Battista JR, Gómez-Silva B, McKay CP, da Costa MS (2007) *Deinococcus peraridilitoris* sp. nov., isolated from a coastal desert. Int J Syst Evol Microbiol 57(7):1408–1412

Rao MB, Tanksale AM, Ghatge MS, Deshpande VV (1998) Molecular and biotechnological aspects of microbial proteases. Microbiol Mol Biol Rev 62(3):597–635

Rashid N, Imanaka H, Fukui T, Atomi H, Imanaka T (2004) Presence of a novel phosphopentomutase and a 2-deoxyribose 5-phosphate aldolase reveals a metabolic link between pentoses and central carbon metabolism in the hyperthermophilic archaeon *Thermococcus kodakaraensis*. J Bacteriol 186(13):4185–4191

Rawal N, Kelkar S, Altekar W (1988) Alternative routes of carbohydrate metabolism in halophilic archaebacteria. Indian J Biochem Biophys 25(6):674–686

Reddy GS, Potrafka RM, Garcia-Pichel F (2007) *Modestobacter versicolor* sp. nov., an actinobacterium from biological soil crusts that produces melanins under oligotrophy, with emended descriptions of the genus *Modestobacter* and *Modestobacter multiseptatus*. Int J Syst Evol Biol 57(9):2014–2020

Restaino L, Bills S, Tscherneff K, Lenovich LM (1983) Growth characteristics of *Saccharomyces rouxii* isolated from chocolate syrup. Appl Environ Microbiol 45(5):1614–1621

Richard H, Foster JW (2004) *Escherichia coli* glutamate-and arginine-dependent acid resistance systems increase internal pH and reverse transmembrane potential. J Bacteriol 186(18):6032–6041

Riley M, Staley JT, Danchin A, Wang TZ, Brettin TS, Hauser LJ, Land ML, Thompson LS (2008) Genomics of an extreme psychrophile, *Psychromonas ingrahamii*. BMC Genomics 9(1):210

Roberts MF (2000) Osmoadaptation and osmoregulation in archaea. Front Biosci 5(362):796–812
Roberts MF (2004) Osmoadaptation and osmoregulation in archaea: update 2004. Front Biosci 9:1999–2019
Robinson JL, Pyzyna B, Atrasz RG, Henderson CA, Morrill KL, Burd AM, DeSoucy E, Fogleman RE, Naylor JB, Steele SM (2005) Growth kinetics of extremely halophilic Archaea (family Halobacteriaceae) as revealed by Arrhenius plots. J Bacteriol 187(3):923–929
Rodrigues F, Ludovico P, Leão C (2006) Sugar metabolism in yeasts: an overview of aerobic and anaerobic glucose catabolism. In: Peter G, Rosa C (eds) Biodiversity and ecophysiology of yeasts. Springer, Berlin, pp 101–121
Rodriguez-Valera F, Juez G, Kushner D (1983) *Halobacterium mediterranei* spec, nov., a new carbohydrate-utilizing extreme halophile. Syst App Microbiol 4(3):369–381
Rogers P, Lee KJ, Tribe D (1979) Kinetics of alcohol production by *Zymomonas mobilis* at high sugar concentrations. Biotechnol Lett 1(4):165–170
Ronimus RS, Koning J, Morgan H (1999) Purification and characterization of an ADP-dependent phosphofructokinase from *Thermococcus zilligii*. Extremophiles 3(2):121–129
Rothschild LJ, Mancinelli RL (2001) Life in extreme environments. Nature 409(6823):1092–1101
Rozen Y, Belkin S (2001) Survival of enteric bacteria in seawater. FEMS Microbiol Rev 25(5):513–529
Ruepp A, Graml W, Santos-Martinez M-L, Koretke KK, Volker C, Mewes HW, Frishman D, Stocker S, Lupas AN, Baumeister W (2000) The genome sequence of the thermoacidophilic scavenger *Thermoplasma acidophilum*. Nature 407(6803):508–513
Russell NJ (1997) Psychrophilic bacteria-molecular adaptations of membrane lipids. Comp Biochem Physiol A Physiol 118(3):489–493
Russell NJ (2000) Toward a molecular understanding of cold activity of enzymes from psychrophiles. Extremophiles 4(2):83–90
Rybicki E (1990) The classification of organisms at the edge of life, or problems with virus systematics. S Afr J Sci 86:182–186
Saiki T, Kobayashi T, Kawagoe K, Beppu T (1985) *Dictyoglomus thermophilum* gen. nov., sp. nov., a chemoorganotrophic, anaerobic, thermophilic bacterium. Int J Syst Bacteriol 35(3):253–259
Sako Y, Nomura N, Uchida A, Ishida Y, Morii H, Koga Y, Hoaki T, Maruyama T (1996) *Aeropyrum pernix* gen. nov., sp. nov., a novel aerobic hyperthermophilic archaeon growing at temperatures up to 100 °C. Int J Syst Bacteriol 46(4):1070–1077
Sakuraba H, Ohshima T (2002) Novel energy metabolism in anaerobic hyperthermophilic archaea: a modified Embden-Meyerhof pathway. J Biosci Bioeng 93(5):441–448
Sakuraba H, Yoshioka I, Koga S, Takahashi M, Kitahama Y, Satomura T, Kawakami R, Ohshima T (2002) ADP-dependent glucokinase/phosphofructokinase, a novel bifunctional enzyme from the hyperthermophilic archaeon *Methanococcus jannaschii*. J Biol Chem 277(15):12495–12498
Sakuraba H, Goda S, Ohshima T (2004) Unique sugar metabolism and novel enzymes of hyperthermophilic archaea. Chem Rec 3(5):281–287
Sanchez-Porro C, Mellado E, Bertoldo C, Antranikian G, Ventosa A (2003) Screening and characterization of the protease CP1 produced by the moderately halophilic bacterium *Pseudoalteromonas* sp. strain CP76. Extremophiles 7(3):221–228
Saunders NF, Thomas T, Curmi PM, Mattick JS, Kuczek E, Slade R, Davis J, Franzmann PD, Boone D, Rusterholtz K (2003) Mechanisms of thermal adaptation revealed from the genomes of the Antarctic Archaea *Methanogenium frigidum* and *Methanococcoides burtonii*. Genome Res 13(7):1580–1588
Saurin W, Hofnung M, Dassa E (1999) Getting in or out: early segregation between importers and exporters in the evolution of ATP-binding cassette (ABC) transporters. J Mol Evol 48(1): 22–41
Schneider E (2001) ABC transporters catalyzing carbohydrate uptake. Res Microbiol 152(3):303–310
Schönheit P, Schäfer T (1995) Metabolism of hyperthermophiles. World J Microbiol Biotechnol 11(1):26–57

Schröder C, Selig M, Schönheit P (1994) Glucose fermentation to acetate, CO_2 and H_2 in the anaerobic hyperthermophilic eubacterium *Thermotoga maritima*: involvement of the Embden-Meyerhof pathway. Arch Microbiol 161(6):460–470

Selig M, Xavier KB, Santos H, Schonheit P (1997) Comparative analysis of Embden-Meyerhof and Entner-Doudoroff glycolytic pathways in hyperthermophilic archaea and the bacterium *Thermotoga*. Arch Microbiol 167(4):217–232

Seo JS, Chong H, Park HS, Yoon KO, Jung C, Kim JJ, Hong JH, Kim H, Kim JH, Kil JI (2005) The genome sequence of the ethanologenic bacterium *Zymomonas mobilis* ZM4. Nat Biotechnol 23(1):63–68

Sghaier H, Ghedira K, Benkahla A, Barkallah I (2008) Basal DNA repair machinery is subject to positive selection in ionizing-radiation-resistant bacteria. BMC Genomics 9(1):297

She Q, Singh RK, Confalonieri F, Zivanovic Y, Allard G, Awayez MJ, Christina C-Y, Clausen IG, Curtis BA, De Moors A (2001) The complete genome of the crenarchaeon *Sulfolobus solfataricus* P2. Proc Natl Acad Sci 98(14):7835–7840

Sheridan PP, Loveland-Curtze J, Miteva VI, Brenchley JE (2003) *Rhodoglobus vestalii* gen. nov., sp. nov., a novel psychrophilic organism isolated from an Antarctic Dry Valley lake. Int J Syst Evol Microbiol 53(4):985–994

Shimada H, Nemoto N, Shida Y, Oshima T, Yamagishi A (2002) Complete polar lipid composition of *Thermoplasma acidophilum* HO-62 determined by high-performance liquid chromatography with evaporative light-scattering detection. J Bacteriol 184(2):556–563

Shuttleworth KL, Unz RF, Wichlacz PL (1985) Glucose catabolism in strains of acidophilic, heterotrophic bacteria. Appl Environ Microbiol 50(3):573–579

Siebers B, Schönheit P (2005) Unusual pathways and enzymes of central carbohydrate metabolism in Archaea. Curr Opin Microbiol 8(6):695–705

Siebold C, Flükiger K, Beutler R, Erni B (2001) Carbohydrate transporters of the bacterial phosphoenolpyruvate: sugar phosphotransferase system (PTS). FEBS Lett 504(3):104–111

Simonato F, Campanaro S, Lauro FM, Vezzi A, D'Angelo M, Vitulo N, Valle G, Bartlett DH (2006) Piezophilic adaptation: a genomic point of view. J Biotechnol 126(1):11–25

Smith DP, Thrash JC, Nicora CD, Lipton MS, Burnum-Johnson KE, Carini P, Smith RD, Giovannoni SJ (2013) Proteomic and transcriptomic analyses of *Candidatus Pelagibacter ubique* describe the first pii-independent response to nitrogen limitation in a free-living Alphaproteobacterium. Mol Biol 6:e00133–00112

Sprenger GA (1996) Carbohydrate metabolism in *Zymomonas mobilis*: a catabolic highway with some scenic routes. FEMS Microbiol Lett 145(3):301–307

Sterner R, Liebl W (2001) Thermophilic adaptation of proteins. Crit Rev Biochem Mol Biol 36(1):39–106

Stettner AI, Segrè D (2013) The cost of efficiency in energy metabolism. Proc Natl Acad Sci 110(24):9629–9630

Suresh K, Reddy G, Sengupta S, Shivaji S (2004) *Deinococcus indicus* sp. nov., an arsenic-resistant bacterium from an aquifer in West Bengal, India. Int J Syst Evol Microbiol 54(2):457–461

Swartz TH, Ikewada S, Ishikawa O, Ito M, Krulwich TA (2005) The Mrp system: a giant among monovalent cation/proton antiporters? Extremophiles 9(5):345–354

Takai K, Moser DP, Onstott TC, Spoelstra N, Pfiffner SM, Dohnalkova A, Fredrickson JK (2001) *Alkaliphilus transvaalensis* gen. nov., sp. nov., an extremely alkaliphilic bacterium isolated from a deep South African gold mine. Int J Syst Evol Microbiol 51(4): 1245–1256

Takai K, Miyazaki M, Hirayama H, Nakagawa S, Querellou J, Godfroy A (2009) Isolation and physiological characterization of two novel, piezophilic, thermophilic chemolithoautotrophs from a deep-sea hydrothermal vent chimney. Environ Microbiol 11(8):1983–1997

Tamás MJ, Luyten K, Sutherland FCW, Hernandez A, Albertyn J, Valadi H, Li H, Prior BA, Kilian SG, Ramos J (1999) Fps1p controls the accumulation and release of the compatible solute glycerol in yeast osmoregulation. Mol Microbiol 31(4):1087–1104

Tamegai H, Nakasone K, Siddiqui KS (2008) Protein Adaptation to High-Pressure Environments. In: Siddiqui K, Thomas T (eds) Protein Adaptation in Extremophiles, Nova Publishers, New York, pp 167–191

Tang YJ, Sapra R, Joyner D, Hazen TC, Myers S, Reichmuth D, Blanch H, Keasling JD (2009) Analysis of metabolic pathways and fluxes in a newly discovered thermophilic and ethanol-tolerant *Geobacillus* strain. Biotechnol Bioeng 102(5):1377–1386

Tempest D, Meers J, Brown C (1973) Glutamate synthetase (GOGAT); a key enzyme in the assimilation of ammonia by prokaryotic organisms. In: Prusiner S, Stadtman ER (eds) The enzymes of glutamine metabolism, Academic Press, UK, pp 167–182

Tetas M, Lowenstein JM (1963) the effect of bivalent metal ions on the hydrolysis of adenosine Di-and triphosphate. Biochemistry 2(2):350–357

Tokuoka K (1993) Sugar and salt tolerant yeasts. J App Bacteriol 74(2):101–110

Tyson GW, Chapman J, Hugenholtz P, Allen EE, Ram RJ, Richardson PM, Solovyev VV, Rubin EM, Rokhsar DS, Banfield JF (2004) Community structure and metabolism through reconstruction of microbial genomes from the environment. Nature 428(6978):37–43

Ulitzur S (1974) *Vibrio parahaemolyticus* and *Vibrio alginolyticus*: short generation-time marine bacteria. Microb Ecol 1(1):127–135

Valdés J, Pedroso I, Quatrini R, Dodson RJ, Tettelin H, Blake R, Eisen JA, Holmes DS (2008) *Acidithiobacillus ferrooxidans* metabolism: from genome sequence to industrial applications. BMC Genomics 9(1):597

Valentine RC, Valentine DL (2004) Omega-3 fatty acids in cellular membranes: a unified concept. Prog Lipid Res 43(5):383–402

van de Vossenberg JL, Driessen AJ, Zillig W, Konings WN (1998) Bioenergetics and cytoplasmic membrane stability of the extremely acidophilic, thermophilic archaeon *Picrophilus oshimae*. Extremophiles 2(2):67–74

van den Burg B (2003) Extremophiles as a source for novel enzymes. Curr Opin Microbiol 6(3):213–218

van der Oost J, Siebers B (2007) The glycolytic pathways of Archaea: evolution by tinkering. Archaea: evolution, physiology and molecular biology 22:247–260

van der Oost J, Schut G, Kengen SM, Hagen WR, Thomm M, de Vos WM (1998) The ferredoxin-dependent conversion of glyceraldehyde-3-phosphate in the hyperthermophilic archaeon *Pyrococcus furiosus* represents a novel site of glycolytic regulation. J Biol Chem 273(43):28149–28154

van der Oost J, Ahmed H, Ettema T, Tjaden B, Geerling A, Siebers B (2005) The semi-phosphorylative Entner-Doudoroff pathway in hyperthermophilic archaea: a re-evaluation. Biochem J 390:529–540

van Niel EW, Claassen PA, Stams AJ (2003) Substrate and product inhibition of hydrogen production by the extreme thermophile, *Caldicellulosiruptor saccharolyticus*. Biotechnol Bioeng 81(3):255–262

Venkateswaran A, McFarlan SC, Ghosal D, Minton KW, Vasilenko A, Makarova K, Wackett LP, Daly MJ (2000) Physiologic determinants of radiation resistance in *Deinococcus radiodurans*. Appl Environ Microbiol 66(6):2620–2626

Verhees C, Kengen S, Tuininga J, Schut G, Adams M, de Vos W, Van Der Oost J (2003) The unique features of glycolytic pathways in archaea. Biochem J 375:231–246

Verhees C, Kengen S, Tuininga J, Schut G, Adams M, de Vos W, Van der Oost J (2004) The unique features of glycolytic pathways in archaea. Biochem J 377:819–822

Villeret V, Beeumen JV, Chessa JP, Gerday C (1997) Preliminary crystal structure determination of the alkaline protease from the Antarctic psychrophile *Pseudomonas aeruginosa*. Protein Sci 6(11):2462–2464

Violot S, Aghajari N, Czjzek M, Feller G, Sonan GK, Gouet P, Gerday C, Haser R, Receveur-Bréchot V (2005) Structure of a full length psychrophilic cellulase from *Pseudoalteromonas haloplanktis* revealed by X-ray diffraction and small angle X-ray scattering. J Mol Biol 348(5):1211–1224

Virgilio C, Burckert N, Bell W, Jeno P, Boller T, Wiemken A (1993) Disruption of TPS2, the gene encoding the 100 kDa subunit of the trehalose-6-phosphate synthase/phosphatase complex in *Saccharomyces cerevisiae*, causes accumulation of trehalose-6-phosphate and loss of trehalose-6-phosphate phosphatase activity. Eur J Biochem 212(2):315–323

Vogel G, Aeschbacher RA, Müller J, Boller T, Wiemken A (1998) Trehalose-6-phosphate phosphatases from *Arabidopsis thaliana*: identification by functional complementation of the yeast TPS2 mutant. Plant J 13(5):673–683
Voges D, Zwickl P, Baumeister W (1999) The 26S proteasome: a molecular machine designed for controlled proteolysis. Ann Rev Biochem 68(1):1015–1068
Vossenberg JL, Ubbink-Kok T, Elferink MG, Driessen AJ, Konings WN (1995) Ion permeability of the cytoplasmic membrane limits the maximum growth temperature of bacteria and archaea. Mol Microbiol 18(5):925–932
Wang Z, Hicks DB, Guffanti AA, Baldwin K, Krulwich TA (2004) Replacement of amino acid sequence features of a-and c-subunits of ATP synthases of alkaliphilic *Bacillus* with the *Bacillus* consensus sequence results in defective oxidative phosphorylation and non-fermentative growth at pH 10.5. J Biol Chem 279(25):26546–26554
Wang J, Bai J, Xu J, Liang B (2009) Bioleaching of metals from printed wire boards by *Acidithiobacillus ferrooxidans* and *Acidithiobacillus thiooxidans* and their mixture. J Hazard Mater 172(2):1100–1105
Warren R (1996) Microbial hydrolysis of polysaccharides. Ann Rev Microbiol 50(1):183–212
White O, Eisen JA, Heidelberg JF, Hickey EK, Peterson JD, Dodson RJ, Haft DH, Gwinn ML, Nelson WC, Richardson DL (1999) Genome sequence of the radioresistant bacterium *Deinococcus radiodurans* R1. Science 286(5444):1571–1577
White D, Drummond JT, Fuqua C (2007) The physiology and biochemistry of prokaryotes. Oxford University Press, New York
White D, Drummond JT, Fuqua C (2012) The physiology and biochemistry of prokaryotes, 4th edn. Oxford University Press, New York
Wiegel J (1992) The obligately anaerobic thermophilic bacteria. In: Kristjansson J (ed) Thermophilic Bacteria, CRC Press, Boca Raton, pp 105–184
Wolf A, Krämer R, Morbach S (2003) Three pathways for trehalose metabolism in *Corynebacterium glutamicum* ATCC13032 and their significance in response to osmotic stress. Mol Microbiol 49(4):1119–1134
Wood AP, Aurikko JP, Kelly DP (2004) A challenge for 21st century molecular biology and biochemistry: what are the causes of obligate autotrophy and methanotrophy? FEMS Microbiol Rev 28:335–352
Xavier KB, Martins LO, Peist R, Kossmann M, Boos W, Santos H (1996) High-affinity maltose/trehalose transport system in the hyperthermophilic archaeon *Thermococcus litoralis*. J Bacteriol 178(16):4773–4777
Xavier KB, da Costa MS, Santos H (2000) Demonstration of a novel glycolytic pathway in the hyperthermophilic archaeon *Thermococcus zilligii* by 13C-labeling experiments and nuclear magnetic resonance analysis. J Bacteriol 182(16):4632–4636
Xu Y, Zhou P, Tian X (1999) Characterization of two novel haloalkaliphilic archaea *Natronorubrum bangense* gen. nov., sp. nov. and *Natronorubrum tibetense* gen. nov., sp. nov. Int J Syst Bacteriol 49(1):261–266
Xu Y, Feller G, Gerday C, Glansdorff N (2003) Metabolic enzymes from psychrophilic bacteria: challenge of adaptation to low temperatures in ornithine carbamoyltransferase from *Moritella abyssi*. J Bacteriol 185(7):2161–2168
Yoo SH, Weon HY, Kim SJ, Kim YS, Kim BY, Kwon SW (2010) *Deinococcus aerolatus* sp. nov. and *Deinococcus aerophilus* sp. nov., isolated from air samples. Int J Syst Evol Microbiol 60(5):1191–1195
Yoshinaka T, Yano K, Yamaguchi H (1973) Isolation of highly radioresistant bacterium, *Arthrobacter radiotolerans* nov sp. Agric Biol Chem 37(10):2269–2275
You XY, Guo X, Zheng HJ, Zhang MJ, Liu LJ, Zhu YQ, Zhu B, Wang SY, Zhao GP, Poetsch A (2011) Unraveling the *Acidithiobacillus caldus* complete genome and its central metabolisms for carbon assimilation. J Genet Genomics 38(6):243–252
Yuan M, Chen M, Zhang W, Lu W, Wang J, Yang M, Zhao P, Tang R, Li X, Hao Y (2012) Genome sequence and transcriptome analysis of the radioresistant bacterium *Deinococcus gobiensis*: insights into the extreme environmental adaptations. PLoS One 7(3), e34458

Yumoto I, Hirota K, Sogabe Y, Nodasaka Y, Yokota Y, Hoshino T (2003) *Psychrobacter okhotskensis* sp. nov., a lipase-producing facultative psychrophile isolated from the coast of the Okhotsk Sea. Int J Syst Evol Microbiol 53(6):1985–1989

Zhang DS, Lovitt R (2005) Studies on growth and metabolism of *Oenococcus oeni* on sugars and sugar mixtures. J Appl Microbiol 99(3):565–572

Zhang YM, Wong TY, Chen LY, Lin CS, Liu JK (2000) Induction of a futile Embden-Meyerhof-Parnas pathway in *Deinococcus radiodurans* by Mn: possible role of the pentose phosphate pathway in cell survival. Appl Environ Microbiol 66(1):105–112

Zhilina T, Zavarzin G, Rainey F, Pikuta E, Osipov G, Kostrikina N (1997) *Desulfonatronovibrio hydrogenovorans* gen. nov., sp. nov., an alkaliphilic, sulfate-reducing bacterium. Int J Syst Bacteriol 47(1):144–149

Zhilina T, Garnova E, Tourova T, Kostrikina N, Zavarzin G (2001) *Halonatronum saccharophilum* gen. nov. sp. nov.: a new haloalkaliphilic bacterium of the order *Haloanaerobiales* from Lake Magadi. Microbiology 70(1):64–72

Zhilina TN, Appel R, Probian C, Brossa EL, Harder J, Widdel F, Zavarzin GA (2004) *Alkaliflexus imshenetskii* gen. nov. sp. nov., a new alkaliphilic gliding carbohydrate-fermenting bacterium with propionate formation from a soda lake. Arch Microbiol 182(2–3):244–253

Zhu F, Wang S, Zhou P (2003) *Flavobacterium xinjiangense* sp. nov. and *Flavobacterium omnivorum* sp. nov., novel psychrophiles from the China No. 1 glacier. Int J Syst Evol Microbiol 53(3):853–857

Zillig W, Holz I, Janekovic D, Klenk H, Imsel E, Trent J, Wunderl S, Forjaz V, Coutinho R, Ferreira T (1990) *Hyperthermus butylicus*, a hyperthermophilic sulfur-reducing archaebacterium that ferments peptides. J Bacteriol 172(7):3959–3965

Zivanovic Y, Armengaud J, Lagorce A, Leplat C, Guérin P, Dutertre M, Anthouard V, Forterre P, Wincker P, Confalonieri F (2009) Genome analysis and genome-wide proteomics of *Thermococcus gammatolerans*, the most radioresistant organism known amongst the archaea. Genome Biol 10(6):R70

Chapter 2
Biodiversity, Adaptation and Biotechnological Importance of Bacteria Occurring in Cold Climates

Gundlapally Sathyanarayana Reddy, Madhab Kumar Chattopadhyay, and Sisinthy Shivaji

1 Biodiversity of Psychrophilic Bacteria

The cold habitats span from the Arctic to the Antarctic and include the high-mountains such as Himalayas, the alpine glaciers, the deep ocean, snow, permafrost, sea ice, lakes, cold soils (especially sub-soils), cold deserts, and caves. All these permanently cold habitats have been successfully colonized by a class of microorganisms known as psychrophiles (grow from subzero to 30 °C) (Morita 1975; Helmke and Weyland 2004; Laucks et al. 2005). The ability of psychrophiles to survive and proliferate at low temperatures is intriguing and the exact mechanism is puzzling (see Sect. 2). The first part of this review summarizes the prokaryotic diversity from three major habitats, the Antarctica, Arctic and Himalayan regions which was reviewed recently by Chattopadhyay et al. (2014).

1.1 Diversity of Prokaryotes from Antarctica

Antarctica, the coldest, driest, windiest and iciest (with an average thickness of 1.9 km) of all known habitats is considered as one of the most extreme habitats of the world (Vincent 1988; Claridge and Campbell 1977; Campbell and Claridge 2000; Smith et al. 1992). Despite the harsh climatic conditions, diverse life forms (such as mites, ticks, seals, penguins, mosses, lichens, bacteria, yeasts, algae etc)(http://

G. Sathyanarayana Reddy • M.K. Chattopadhyay
CSIR-Centre for Cellular and Molecular Biology, Hyderabad, India

S. Shivaji (✉)
CSIR-Centre for Cellular and Molecular Biology, Hyderabad, India

Jhaveri Microbiology Centre, L V Prasad Eye Institute, Hyderabad, India
e-mail: shivas@ccmb.res.in; shivas@lvpei.org

P.H. Rampelotto (ed.), *Biotechnology of Extremophiles*, Grand Challenges in Biology and Biotechnology 1, DOI 10.1007/978-3-319-13521-2_2

www.globalclassroom.org/antarct6.html) survive and multiply in this continent. The psychrophilic bacteria are the most dominant community in the continent and contribute to the Antarctic ecosystem especially with respect to nutrient recycling (Stokes and Reymond 1966; Herbert and Bell 1977; Tanner and Herbert 1981; Delille and Lagarde 1974; Tanner 1985; Voytek and Ward 1995; Chessa et al. 2000; Kelly et al. 1978; Cavanagh et al. 1996; Denner et al. 2001; Cavicchioli and Thomas 2000). Ekelof (1908a, b) was the first to demonstrate the occurrence of bacteria, yeasts and fungi in the soil and air of Antarctica and this observation was confirmed by Pirie (1904, 1912) and Tsiklinsky (1908). Five decades later these unique microorganisms from the continent of Antarctica were identified at the genus and species level (Boyd 1962; Boyd and Boyd (1962a, b); Friedman 1980; Margini and Castrelos 1963; Marshall and Ohye 1966; Meyer et al. 1962; Pfiser and Burkholder 1965; Tsyganov 1970). Subsequently, with the advent of culture independent identification of microorganisms based on 16S rRNA gene metagenomics a greater diversity of microbes was discovered from various habitats of Antarctica. The NCBI (http://ncbi.nlm.nih.gov) database has about 6000 16S rRNA gene sequences of bacteria from Antarctica. Of these about 500 sequences correspond to the viable isolates of bacteria from Antarctica (Shivaji and Reddy 2009; Chattopadhyay et al. 2014).

1.1.1 Bacterial Diversity of Antarctica as Studied by Culture Independent Methods

In an attempt to unravel the total bacterial diversity of the various habitats of Antarctica the 16S rRNA approach is used to identify the dominating and the not so dominant bacterial taxa from various habitats that include soil, cyanobacterial mats, water, subglacial out flow, sediments, crypto-endolithic sand stone communities, ice cores, geothermal vents, ornithogenic soils, penguins, sponges and intestinal microbiota of Antarctic fish (Shivaji and Reddy 2009; Chattopadhyay et al. 2014; Bottos et al. 2014). In the Antarctic soils bacterial diversity varied with respect to soil type (Bottos et al. 2014; Shivaji et al. 2004). For instance, Proteobacteria (47.0 %) is dominant in Antarctic Peninsula soils where as Actinobacteria and Bacteroidetes (42.0 % and 31 % respectively) are dominant in West Antarctic soils and Transantarctic Mountain soils (Bottos et al. 2014), In the Transantarctic Mountain soils *Deinococcus-Thermus* bacteria are the third most abundant community (18 %). Shivaji et al. (2011a) observed stratification of bacteria with respect to oxygen requirement in Antarctic soils and also identified quite a few clones belonging to *Caldiserica* group, a thermophilic candidate phylum from Antarctica. In addition, it is generally observed that up to 20.0 % of the bacteria in the Antarctic soils do not affiliate with any known taxa (Shivaji et al. 2004; Aislabie et al. 2006, 2009, 2013; Yergeau et al. 2007a, b; Niederberger et al. 2008; Lee et al. 2012; Tiao et al. 2012; Bajerski and Wagner 2013). These results would imply that the bacterial communities across Antarctic soils vary significantly with geography, climate, soil physicochemical parameters and local biological influences.

Bacterial diversity studies of Antarctic sediments both from fresh water sediments (Bratina island, Sjöling and Cowan 2003; Ardley island, Li et al. 2006a, b), marine sediments (Vestfold Hills, Bowman et al. 2000; Ross ice shelf, Carr et al. 2013) and glacial sediments (Wright Glacier, Stibal et al. 2012) indicated that Alphaproteobacteria, Betaproteobacteria, Gammaproteobacteria, Bacteroidetes, Actinobacteria, Acidobacteria Gemmatimonadetes, Firmicutes and Cyanobacteria, are common to soils and sediments. In addition taxa affiliated to Spirochaetaceae, Prochlorococcus, Deltaproteobacteria (specially the unique groups *Desulfosarcina*, *Syntrophus* and *Geobacter/Pelobacter/Desulphuromonas* group) (Bowman et al. 2000), order Chlamydiales (Parachlamydiaceae), *Planctomycetes* and members of *Spirochaetales* (Sjöling and Cowan 2003; Li et al. 2006a, b) are also present. Carr et al. (2013) also observed that the Betaproteobacteria contained predominantly (over 45 %) two genera *Thiobacillus* and *Teptidiphilus* and the occurrence of unusually high proportion of *Chloroflexi* group. All the above bacteria are uniformly distributed except the Gammaproteobacteria which is very high at a depth of 120–125 cm below sea surface. Bowman et al. (2000) detected more than 200 distinct phylotypes in anoxic marine sediment with close to 31.0 % belonging to a novel deep branch within the low GC Gram-positives.

Microbial diversity of water in Antarctica is limited as only 0.4 % of the total ice covered area of Antarctica is seasonally ice free (Wilkins et al. 2013). The aquatic habitat includes fresh to hypersaline water bodies, permanently ice covered to perennially ice free, and mixed to stratified lakes. Further the age of water within Antarctic lakes varies considerably from 1.5 million years in case of sub-glacial outflow from Blood Falls (Mikucki et al. 2009) to 300 years old in case of Lake Miers (Green et al. 1988) and as a result organisms inhabiting the lakes could be recent or ancient in origin (Gibson 2006; Cavicchioli 2007). Bacterial diversity in the water is generally reduced compared to soils and sediments (Laybourn-Parry 1997) and the majority of bacteria belong to the Proteobacteria (47.6 %, representing the Alphaproteobacteria, Betaproteobacteria and Deltaproteobacteria) with dominating genera being *Rhodobacter* and *Sphingomonas* (Huang et al. (2013). Besides, Bacteroidetes (15.1 %), Actinobacteria (14.8 %), Chloroflexi (10.2 %), Acidobacteria (6.7 %), Firmicutes (3.6 %), Gemmatimonadetes (1.3 %) and Verrucomicrobia (0.2 %) are the minor components and Deinococcus-Thermus, Nitrospira, Planctomycetes, and Fusobacteria are also present. The most common genera are *Sphingomonas Caulobacter, Brevundimonas, Janthinobacterium, Duganella, Polaromonas, Variovorax, Rhodoferax, Flavobacterium, Pedobacter, Prevotella, Hymenobacter,* and *Arcicella* (Wilkins et al. 2013). However, a shift in diversity is observed with depth wherein the dominant Bacteroidetes, Betaproteobacteria and Actinobacteria in the surface layers are replaced by Gammaproteobacteria and unclassified bacteria (10–20 %) in deeper zones of McMurdo ice shelf, Victoria land and in the Amundsen sea polynya (Archer et al. 2014; Kim et al. 2014). Two major clades representing *Polaribacter* (20–64 %) and uncultivated Oceanospirillaceae (7–34 %) are abundant in the surface water while the abundance of *Pelagibacter* increased with depth (7–42 %) (Kim et al. 2014).

Response of bacterial community to hydrocarbon contamination in sea water revealed that genera *Psychrobacter, Arcobacter, Formosa, Polaribacter, Ulvibacter* and *Tenacibaculum* are detected only in hydrocarbon contaminated water and the abundance of *Sulfitobacter* group was high in sea water and decreased in contaminated seawater (Prabagaran et al. 2007). Distinct variation in diversity was obvious in marine and fresh water aquatic systems in Antarctica. The fresh water ecosystems are dominated by *Flavobacterium, Pseudomonas* and *Polaromonas* (up to about 56 % of total sequences) (Michaud et al. 2012) while *Sulfitobacter, Thalassospira,* members of *Roseobacter, Gelidibacter, Polaribacter, Psychroflexus,* and *Pseudoalteromonas* were common to marine waters (Prabagaran et al. 2007; Guibert et al. 2012; Lo Giudice et al. 2012).

Sea ice harbors a network of brine pores and channels (Junge et al. 2004; Søgaard et al. 2010) and supports a diverse (Brown and Bowman 2001; Brinkmeyer et al. 2003; Maas et al. 2012) community of bacteria represented by photosynthetic, chemoautotrophic and heterotrophic bacteria besides Archaea and several eukaryotes. Gammaproteobacteria, Alphaproteobacteria and Bacteroidetes are the predominant groups. Actinobacteria are very rarely present. Among the Gammaproteobacteria species of the genera *Colwellia* and *Glaciecola* are the most abundant and *Marinobacter* spp. were rare. The Alphaproteobacteria are represented by members of the *Roseobacter* lineage while Bacteroidetes group included *Polaribacter* genera (Bowman et al. 1997; Gosink et al. 1998; Staley and Gosink 1999; Brown and Bowman 2001; Junge et al. 2002; Brinkmeyer et al. 2003; Kuhn et al. 2014; Lanoil et al. 2009). Other genera in sea ice include *Shewanella, Planococcus, Alteromonas, Pseudoalteromonas, Psychrobacter, Halomonas, Pseudomonas, Hyphomonas, Sphingomonas, Arthrobacter, Planococcus* and *Halobacillus* (Bowman et al. 1997). Temporal community analyses over a period of 2 years indicated a shift in community wherein *Paenisporosarcina* was reduced by fivefold while *Bacillus* increased by fourfold. In addition, *Acinetobacter* and *Cohnella* (Firmicutes) replaced *Paenibacillus* and *Jeotgalibacillus* in a span of 2 years (Doyle et al. 2013). In conclusion Fig. 2.1 summarizes the contribution of each community to the total diversity of Antarctica.

1.1.2 Bacterial Diversity of Antarctica as Studied by Using Functional Genes

Characterization of functional genes of a physiological group of bacteria has been used to identify diverse groups of bacteria like the sulphate reducing bacteria (Karr et al. 2005), photosynthetic bacteria (Karr et al. 2003; Stibal et al. 2012; Kong et al. 2014), hydrocarbon degrading bacteria (Muangchinda et al. 2014; Marcos et al. 2009) chitinase and ketosynthase producing bacteria (Xiao et al. 2005) and the ammonium oxidizing bacteria (Magalhães et al. 2014). Such studies are normally attempted with the assumption that the primer set would identify the specific gene in all bacteria. But in the absence of a consensus sequence or known conserved sequence for a specific gene this approach could be biased and the diversity reported will be a function of the coverage of the primers used.

Antarctic hyper saline lakes in the Vestfold Hills have high concentration of dimethylsulfide (Yau et al. 2013). Therefore, it is not surprising that by targeting the

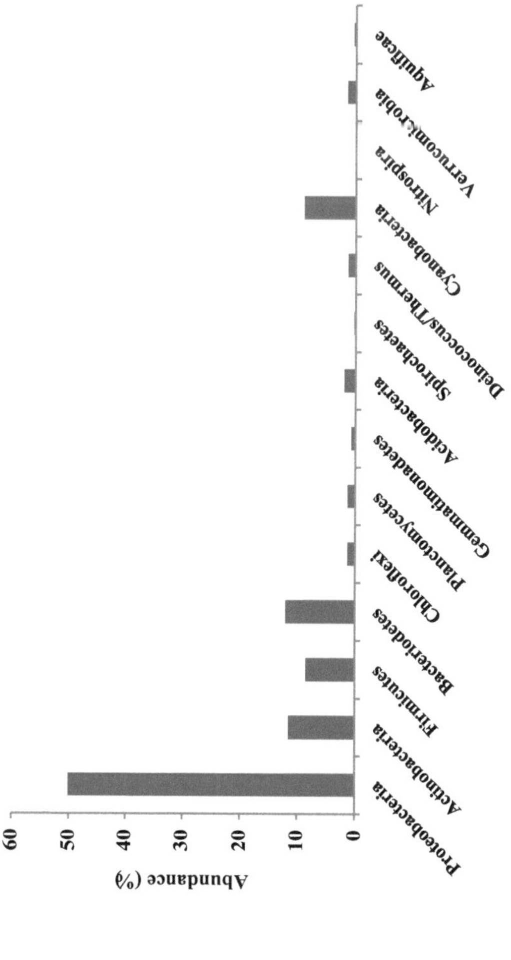

Fig. 2.1 Diversity and relative abundance (%) of bacteria from various samples from Antarctica based on culture independent methods. This figure was reconstructed from Shivaji and Reddy (2009)

gene *dsrA* (dissimilatory sulfite reductase), diverse group of sulfate-reducing bacteria were identified in Lake Fryxell, a permanently frozen freshwater lake, in Antarctica (Karr et al. 2005). It was also demonstrated that more than 70 % of sulfur transformation in subglacial lake sediments was contributed by two genera *Sideroxydans* and *Thiobacillus* (Purcell et al. 2014). Recently, Watanabe et al. (2013) based on *aprA* (adenosine-5′-phosphosulfate reductase alpha subunit) identified bacteria belonging to the genera *Thiocapsa, Sulfuricella, Desulfobacterium, Desulfofaba* and *Desulfotomaculum* from Antarctic fresh water lakes.

Scientific operations in Antarctica have led to the contamination of the pristine Antarctic environment with hydrocarbons in the form of fuel oils (Aislabie et al. 2004). Thus it was deemed relevant to identify bacteria with abilities to degrade polycyclic aromatic hydrocarbons (PAH). PAH degrading genes were detected from both Gram-positive and Gram-negative bacteria and were affiliated to the phyla Proteobacteria, Actinobacteria, Verrucomicrobia, Bacteroidetes, Firmicutes, Chloroflexi, Gemmatimonadetes, Cyanobacteria, Chlorobium, and Acidobacteria. *Sphingomonas*, one of the major PAH degraders in the environment, was observed in some locations. Thus indigenous bacteria have the potential to degrade PAHs and could thus facilitate bioremediation in Antarctica (Muangchinda et al. 2014). Based on the presence of the PAH degrading gene, bacteria belonging to the genera *Terrabacter, Mycobacterium, Diaphorobacter, Sphingomonas, Burkholderia, Rhodococcus*, *Mycobacterium*, *Nocardia*, *Terrabacter* and *Bacillus* were found to be predominant in Antarctic sediments (Muangchinda et al. 2014; Marcos et al. 2009) where as in soils *Pseudomonas* plays an important role (Ma et al. 2006). Diversity studies based on alkane monooxygenases genes (*alkB* and *alkM*), key enzymes in alkane degradation, from the Admiralty Bay and King George Island of Peninsula Antarctica indicated that the gene sequences were similar to the *alk*B gene reported in *Silicibacter pomeroyi, Gordonia sp., Prauserella rugosa, Nocardioides sp., Rhodococcus sp., Nocardia sp., Pseudomonas putida, Acidisphaera sp., Alcanivorax borkumensis*, and *alk*M described in *Acinetobacter* sp.(Kuhn et al. 2009). The penguins in Antarctica excrete huge amounts of chitin-containing droppings whose degradation is carried out by soil microbes. Studies based on chitinase genes (*chi67, chi69, chiA, chiB, chiF*) revealed the occurrence of *Janthinobacterium, Stenotrophomonas, Cytophaga* of Gram-negatives and *Streptomyces* and *Norcardiopsis* of Gram-positives as the abundant genera from a 1600 year old sediment obtained from Ardley Island, Antarctica (Xiao et al. 2005). It is well known that the microorganisms, particularly the members of Actinobacteria, are the reservoirs of natural compounds such as polyketides, alkaloids, peptides etc, and Antarctica prokaryotes are no exception. Owing to the significance of the community involved in polyketide synthesis, an attempt was made to investigate the diversity of bacteria based on ketosynthase (KS) gene. Phylogenetic analysis based on ketosynthase (KS) gene identified diverse bacterial groups, including Proteobacteria, Firmicutes, Planctomycetes, Cyanobacteria, Actinobacteria, some uncultured symbiotic bacteria and five independent clades. Most of the identified isolates showed below 80 % identity at the AA level to their closest match revealing great diversity and novelty of ketosynthase genes in Antarctic sediments (Zhao et al. 2008). In addition, Antarctic lakes are known to support a photosynthetic bacterial community though most of the lakes are covered with thick sheet of ice. Based on gene *pufM* (coding for photosynthetic pigment-binding protein) 33 unique phylotypes related to *Rubrivivax, Acidiphilum, Rhodoferax* and *Roseateles* were detected from Antarctic sediments (Karr et al. 2003; Stibal et al. 2012). While studying the diversity of *psbA* gene from Lake Bonney, unique phylotypes of genera *Nannochloropsis*, *Ochromonas* and *Isochrysis* were identified (Kong et al. 2014).

1.1.3 Bacterial Diversity of Antarctica as Studied by Cultivation Based Methods

The bacterial abundance from various habitats of Antarctica ranged from 0.2×10^7 to 0.6×10^{12} cells/g of ice (Carpenter et al. 2000; Priscu et al. 1999; Karl et al. 1999; Delille and Gleizon 2003), 0.2×10^2 to 10^7 cells/mL of water (Takii et al. 1986; Lo Giudice et al. 2012) and 8×10^6 to 2.4×10^7 cells/g of sediment (Lanoil et al. 2009; Stibal et al. 2012) and 10^5 to 10^{10} cells/g of soil (Aislabie et al. 2009). The culture dependent bacterial diversity was explored extensively from soil (Miwa 1975; Yi and Chun 2006; Ruckert 1985; Shivaji et al. 1988, 1989a, b; Wery et al. 2003; Bozal et al. 2007), water (Lo Giudice et al. 2012; Michaud et al. 2012; Söller et al. 2000; Labrenz et al. 2000; Cristóbal et al. 2011), cyanobacterial mats (Reddy et al. 2000, 2002a, b, 2003a, b, c, d, 2004; Van Trappen et al. 2002; Spring et al. 2003; Peeters et al. 2011), ice (Shivaji et al. 2004, 2013c; Antibus et al. 2012; Bowman et al. 1998) and sediments (Shivaji et al. 2011a; Yu et al. 2011) and have been reviewed by Shivaji and Reddy (2009) and Chattopadhyay et al. (2014) from Antarctica. Among the habitats, water, soil and mats are more diverse compared to ice and sediments (Fig. 2.1). In all the habitats communities representing Proteobacteria, Bacteroidetes, Actinobacteria and Firmicutes are wide spread while Deinococcus-thermus and *Spirochaetes* were detected only in soil (Hirsch et al. 2004; Antibus et al. 2012; Franzmann and Dobson 1992; Chattopadhyay et al. 2014). Among the Proteobacteria, Gammaproteobacteria were more commonly observed compared to Alphaproteobacteria, Betaproteobacteria and Deltaproteobacteria while members of Epsilonproteobacteria were absent. Alpha-, Beta- and Gamma-proteobacteria, Actinobacteria and Flavobacteria were also found to occur in Andean glacial ice, which is less impacted by anthropogenic interventions than the Arctic and Antarctic ice because of the difficulty in accessing it (Ball et al. 2014).

Relative percentage occurrence of cultivable bacteria from Antarctica indicated that the soils are dominated by Firmicutes (35 %), Actinobacteria (25 %) and Proteobacteria (18 %), water by Proteobacteria (42 %), Bacteroidetes (31 %) and Actinobacteria (23 %), algal mats by Proteobacteria (40 %), Firmicutes (28 %) and Bacteroidetes (24 %), ice by Proteobacteria (48 %), Bacteroidetes (21 %) and Firmicutes (19 %) and finally the sediment by Proteobacteria (50 %) and Actinobacteria (29 %) (Fig. 2.2). The above statistics clearly indicate that Proteobacteria is ubiquitously present abundantly in all the Antarctic habitats. The culturable bacteria isolated from Antarctica include 50 novel genera and 170 novel species. Only one novel family

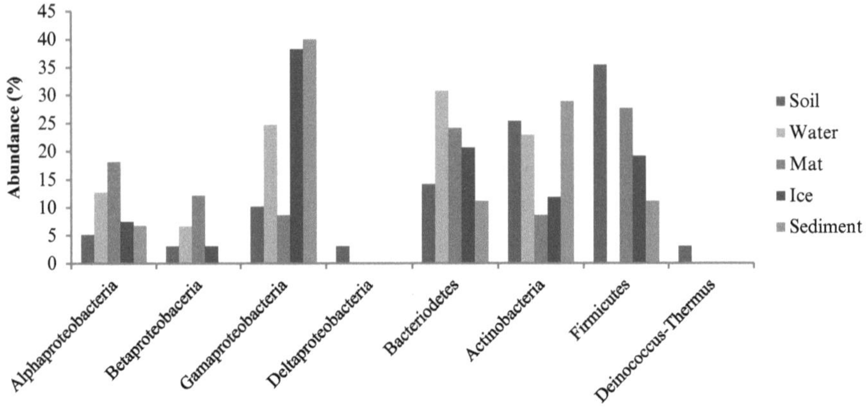

Fig. 2.2 Bacterial diversity and relative abundance (%) of culturable bacteria from soil, water, mat, ice and sediment samples from (communities contributing less than 3.0 % are omitted) Antarctica. Data compiled from references given in Sect. A-I-3

Granulosicoccaceae of Gammaproteobacteria has been described from Antarctica and it was isolated from seawater. It includes a single species *Granulosicoccus antarcticus* of the genus *Granulosicoccus* (Lee et al. 2007). In addition, some of the cultures attracted more interest, due to their thermophilic nature, such as *Alicyclobacillus pohliae, Aneurinibacillus terranovensis, Anoxybacillus amylolyticus, Bacillus thermantarcticus* and *Brevibacillus levickii* which were isolated from geothermal soils of Mount Melbourne and Mount Rittmann (Lama et al. 1996; Allan et al. 2005; Poli et al. 2006; Imperio et al. 2008). Nevertheless, it may be mentioned that many of the phyla are also found in other cold habitats.

1.2 Diversity of Prokaryotes from Arctic

The Arctic is the Earth region that lies between 66.5°N and the North Pole and geographically spans the Arctic Ocean and covers land areas in parts of Canada, Finland, Greenland, Iceland, Norway, Russia, Sweden and the United States (Alaska). The Arctic is very cold and harsh for most of the year with scanty solar radiation. One of the unique characteristics of Arctic is the year round permafrost and thus the lack of lush vegetation. Arctic microbiology started much later compared to Antarctica and the first evidence of microorganisms was reported by Kriss (1945), Pady et al. (1948), Polunin and Kelly (1952), Kelly and Layne (1957) and Boyd and Boyd (1962a, b). However, the first culture independent study of Arctic tundra soil was carried out by Zhou et al. (1997). Currently the NCBI (http://ncbi.nlm.nih.gov) database contains about 3429 16S rRNA gene sequences of bacterial clones from Arctic. Of these only 79 sequences correspond to novel isolates from Arctic.

1.2.1 Bacterial Diversity of Arctic as Studied by Culture Independent Methods

The bacterial abundance and diversity of various ecological niches such as soil (van Dorst et al. 2014; Chu et al. 2010; Edwards et al. 2011; Zhou et al. 1997; Neufeld and Mohn 2005; Campbell et al. 2010; Schütte et al. 2010), ice (Junge et al. 2002; Brinkmeyer et al. 2003; Yu et al. 2006; Bottos et al. 2008; Collins et al. 2010), permafrost soil (Wilhelm et al. 2011; Steven et al. 2007; Hansen et al. 2007) and sediments (Zhang et al. 2014; Forschner et al. 2009; Li et al. 2006a, b, 2009; Lysnes et al. 2004; Perreault et al. 2007; Bienhold et al. 2012) from Arctic were explored by both culture dependent and independent methods (reviewed by Chattopadhyay et al. 2014). These studies indicate that soil microbial communities are very similar to that observed in Antarctica and Acidobacteria, Alphaproteobacteria, Actinobacteria, Betaproteobacteria and Bacteroidetes, account for more than 83 % of the diversity (Chu et al. 2010; Edwards et al. 2011; Zhou et al. 1997; Neufeld and Mohn 2005; Campbell et al. 2010; Schütte et al. 2010). In addition, Gammaproteobacteria, Verrucomicrobia, Gemmatimonadetes and Deltaproteobacteria were also reported. When the soil diversity was compared with that of the diversity in permafrost, as anticipated, a subset of the taxa were common. For instance in the permafrost soils Proteobacteria, Actinobacteria and Acidobacteria (Steven et al. 2007; Wilhelm et al. 2011) were dominant whereas Bacteroidetes, Gemmatimonas, Verrucomicrobia, and Planctomyces were less dominant. Studies have also indicated that the bacterial communities of these high latitude polar biocrusts did not show a consensus response to intermittent flow in water tracks over high Arctic permafrost (Steven et al. 2013). It was also observed that low pH supported the Proteobacteria at the expense of Acidobacteria and high pH supported Bacteroidetes (Ganzert et al. 2014). In case of sediments Proteobacteria was exuberantly present contributing 50.0–85.0 % of the diversity (Li et al. 2009; Bienhold et al. 2012; Lysnes et al. 2004; Steven et al. 2008a). Zhang et al. (2014) identified potential novel genera by both culture dependent and independent methods from Arctic marine sediments.

Sea-ice contained bacterial lineages that belonged to Proteobacteria, Bacteroidetes, Planctomycetes and Chloroflexi (Junge et al. 2002; Brinkmeyer et al. 2003; Bottos et al. 2008) and sea water harbored significant proportions of Proteobacteria (>50 %) and Bacteroidetes (25 %) and Actinobacteria and Verrucomicrobia were present below 25 % (Kirchman et al. 2010). So far, close to 284 genera (Schütte et al. 2010) belonging to Proteobacteria (131 genera), Actinobacteria (57 genera), Firmicutes (22 genera), Bacteroidetes (22), Chloroflexi (5 genera), Plancotomyces (5 genera), Gemmatimonas (4 genera), Fibrobacter (2 genera), Acidobacteria (2 genera), Spirochaetes (1 genus), and other classes (34) have been detected from the various ecological niches of Arctic (Fig. 2.3). Relative percentage abundance in four extensively explored habitats soil, sediment, permafrost and ice indicated that the genera belonging to the class Proteobacteria are most dominating and members of the class Actinobacteria, Firmicutes and Bacteroidetes contribute 3–25 % (Collins et al. 2010).

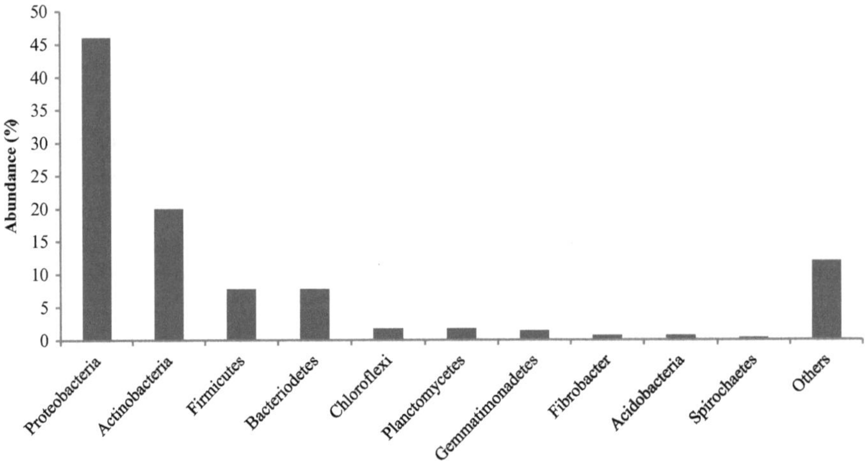

Fig. 2.3 Diversity and relative abundance (%) of bacteria from sea ice from Arctic based on a culture independent method. This figure was constructed from data obtained from Junge et al. (2002), Brinkmeyer et al. (2003) and Bottos et al. (2008)

1.2.2 Bacterial Diversity of Arctic as Studied by Functional Genes

Diversity of several physiological groups of bacteria such as nitrogen fixing communities (Deslippe and Egger 2006), ammonium oxidizers (Díez et al. 2012) and cyanobacterial diazotrophs (Díez et al. 2012 have been studied from various habitats of Arctic. Thus attempts have also been made to identify the diversity associated with nitrogen fixing communities based on the *nifH* gene. A good proportion (50–65 %) of the nitrogen fixing communities were affiliated to Alphaproteobacteria, Gammaproteobacteria, Deltaproteobacteria and Firmicutes and the genera were identified as *Rhodopseudomonas, Azospirillum, Azorhizobium, Rhizobium, Bradyrhizobium* of Alphaproteobacteria, *Herbaspirillum* and *Oryza* of Betaproteobacteria, *Pseudomonas, Azoarcus* of Gammaproteobacteria and *Paenibacillus* of Firmicutes. Using the same *nifH* gene the dominating cyanobacteria were identified as affiliated to Chroococcales, Nostocales, Oscillatoriales and *Phormidium* in brine and sea water (Díez et al. 2012). Further using *nitAB* gene sequences, *Nitrosospira* and *Nitrosomonas* were identified as the ammonium oxidizers present in the Arctic Ocean water (Bano and Hollibaugh 2000). In addition, thawing of permafrost due to climate change can result in mineralization of organic carbon in to carbon dioxide and methane through microbial activity. Exploration of methanogenic community that contribute to the green house gasses from Arctic would be interesting. It was also observed that *Methylobacter* (Type I methanotroph) and *Methylosinus* and *Methylocystis* (Type II methanotroph) were the methanotrophic members (Wartiainen et al. 2003; Liebner et al. 2009) from Arctic habitats and the community changed depending on the pH (Martineau et al. 2014). Arctic fresh water, marine water, ice shelf and coastal water receive solar radiation and thus support the photosynthetic bacterial community. Boeuf et al. (2013)

determined the Aerobic anoxygenic phototrophic (AAP) community in western Arctic Ocean using *puf*M gene and Alphaproteobacteria and Betaproteobacteria represented 49 % and 48 % respectively.

1.2.3 Bacterial Diversity of Arctic as Studied by Culture Dependent Methods

The bacterial abundance in major habitats from Arctic region was comparable to that observed in Antarctica and it ranged from 1.2 to 2.3×10^7 cells/g of sediment (Forschner et al. 2009), 5.2×10^4–1.7×10^9 cells/g of permafrost (Hansen et al. 2007), 0.9×10^3–2.4×10^6 cells/mL of ice (Møller et al. 2011; Brinkmeyer et al. 2003) and in case of water the abundance was 5×10^5 cells/mL (Møller et al. 2011). Cultivable bacterial diversity from major Arctic environments such as ice (Zhang et al. 2008; Auman et al. 2006, 2010; Gosink et al. 1998; Brinkmeyer et al. 2003; Groudieva et al. 2004; Bottos et al. 2008; Steven et al. 2008b), sediments (Prasad et al. 2014; Srinivas et al. 2009; Reddy et al. 2009b; Kim et al. 2009, 2012; Knittel et al. 2005; Knoblauch et al. 1999; Vandieken et al. 2006; Shivaji et al. 2012a; Begum et al. 2013), permafrost (Hansen et al. 2007; Steven et al. 2007, 2008a), mats (Dong et al. 2012), tundra soil (Jiang et al. 2012; Jiang et al. 2013; Männistö et al. 2012), water (Prasad et al. 2014; Yu et al. 2011; Van Trappen et al. 2004; Zhang et al. 2006; Kim et al. 2008; Jang et al. 2011; Al Khudary et al. 2008) and cryoconite holes (Singh et al. 2014) was explored and close to 675 strains belonging to 135 genera were identified and the bacterial diversity of the permafrost appeared to be more diverse compared to mats and water (Fig. 2.4). Relative percentage abundance indicated that Gammaproteobacteria, Bacteroidetes and Actinobacteria were abundant in ice and sediments and Bacteroidetes, Firmicutes and Actinobacteria were predominant in permafrost and soil (Fig. 2.2). The 675 bacteria isolated from Arctic included 12 novel genera (Ren et al. 2015; Storesund and Øvreås 2013; Qu et al. 2014; Prasad et al. 2013; Jiang et al. 2013; Steinsbu et al. 2011; Tamura et al. 2010; Steven et al. 2008b; Lee et al. 2007; Knoblauch et al. 1999; Gosink et al. 1998) and 70 novel species. The genera, *Parablastomonas* (Ren et al. 2015), *Psychroglaciecola* (Qu et al. 2014), *Arcticibacter* (Prasad et al. 2013), *Huanghella* (Jiang et al. 2013), *Dasania* (Lee et al. 2007) and *Augustibacter* (Tamura et al. 2010) have so far been reported only from the habitats in Arctic. Further, members of Deltaproteobacteria were detected in sediments (Knoblauch et al. 1999) and the class contains three novel genera *Desulfofrigus, Desulfofaba* and *Desulfotalea,* represented by nine novel species (Knoblauch et al. 1999). In addition, a novel species of the class *Verrucomicrobium* (*Luteolibacter luojiensis*) (Jiang et al. 2012), four novel species (*Granulicella arctica, Granulicella mallensis, Granulicella tundricola, Granulicella sapmiensis*) of the phylum *Acidobacteria* (Männistö et al. 2012) were described from Arctic. Most of the bacteria described from Arctic are psychrophilic except *Rhabdothermus arcticus* (37–65 °C), a novel thermophile isolated from a hydrothermal vent (Steinsbu et al. 2011). Comparison of the diversity between Antarctica and Arctic indicated similarity at the class level which was reduced at the genera level.

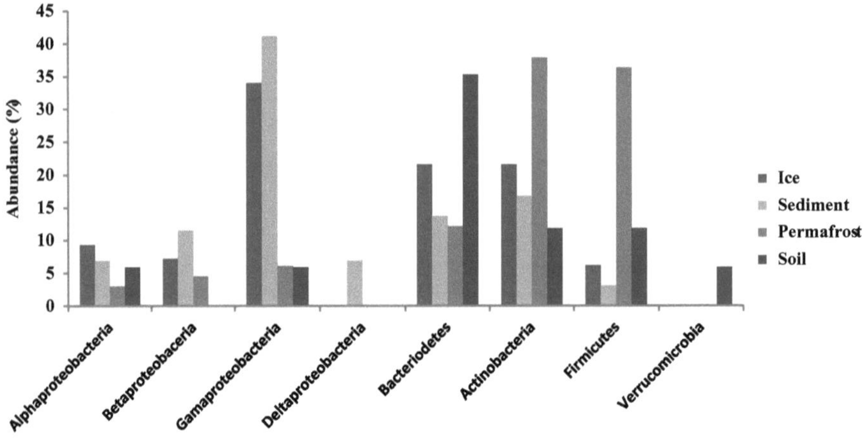

Fig. 2.4 Bacterial diversity and relative abundance (%) of culturable bacteria from ice, sediment, permafrost and soil samples from Arctic (communities contributing less than 3.0 % are omitted). Data compiled from references given in Sect. 1.2.3

1.3 Diversity of Prokaryotes from Himalayas

Seventy percent of the world's freshwater is frozen in glaciers, and the Himalayas have the largest concentration of glaciers outside the Polar Regions, covering about 33,000 km^2. The region has been referred to as the 'Water Tower of Asia', as it provides around 8.66106 m^3 of water annually (Dyurgerov and Meier 1997). Further, the Himalayan environments are distinct from those in Polar Regions and are characterized by dramatic seasonal shifts in physical and biochemical properties in addition to the altitude gradient of climatic changes. In spite of the recent spurt of research activities in cold habitats, the bacterial diversity of the Himalayas has remained largely unexplored.

1.3.1 Bacterial Diversity of Himalayas as Studied by Culture Independent Methods

Culture dependent and culture independent based diversity of soils from three major glaciers of Himalayas, Pindari (Shivaji et al. 2011b), Kafni (Srinivas et al. 2011) and Roopkund (Pradhan et al. 2010), soil from western region of Himalayas (Gangwar et al. 2009) and snow and water (Liu et al. 2006, 2009, 2011) was reported recently. The total bacterial abundance was low in case of water (1.1 to 25×10^4 cells per mL) (Liu et al. 2011) and higher in case of sediments wherein it ranged from 0.9×10^7 to 30.7×10^8 cell per gram sediment (Shivaji et al. 2011a, b; Pradhan et al. 2010; Srinivas et al. 2011) and the Shannon diversity indices ranged from 2.7 to 3.4 in case of water to 0.9 to 4.89 for sediments (Srinivas et al. 2011). The three habitats, soil, water and sediments from three glaciers shared common taxa affiliated to Acidobacteria,

Actinobacteria, Bacteroidetes, Chloroflexi, Chlamydiae, Firmicutes, Nitrospirae, Proteobacteria, and Verrucomicrobia. However, sediments from the glaciers differed in their diversity in that sediments from Kafni glacier contained clones affiliated to Spirochaetae and Tenericutes (Srinivas et al. 2011; Shivaji et al. 2011a, b) where as samples from Pindari and Roopkund had sequences affiliated to Gemmatimonadetes and Planctomycetes. Gangwar et al. (2009) concluded the contribution of Proteobacteria as the highest in soil wherein it contributed 45–70 % to their niche followed by Actinobacteria (10–86 %) and Bacteroidetes (6.4–28 %) (Liu et al. 2006; Gangwar et al. 2009; Srinivas et al. 2011). The difference in community composition in the three habitats studied so far can be attributed to the nature of the soil. So far 135 diverse genera were identified from various environments of Himalayas and the number of genera followed the trend Actinobacteria (28.0 %), Betaproteobacteria (21.5 %), Alphaproteobacteria (9.6 %), Gammaproteobacteria (8.9 %), Deltaproteobacteria (11.1 %), Bacteroidetes (11.1 %), Firmicutes (3.7 %), Acidobacteria (1.4 %), Verrucomicrobia (3.0 %) and Epsilonproteobacteria (1.5 %) (Fig. 2.5).

1.3.2 Bacterial Diversity of Himalayas as Studied by Functional Genes

Compared to Antarctica and Arctic, very limited functional diversity studies have been carried out in the Himalayan region. The prokaryotic community in the Himalayas has been explored with respect to communities contributing to nitrogen fixation or production of green house gasses. Studies based on *nifH* gene indicated that majority of the diazotrophic community in rhizospheric soils of Himalayas included members of *Rhizobium* (53 %), *Dechloromonas* (7 %), *Bradyrhizobium* (7 %) and non-culturable bacteria (29 %) (Suyal et al. 2014). Diversity of methanogenic communities using *mcrA* gene and 16S rRNA gene sequence analysis indicated that the community was represented by the genera *Methanosarcina*, *Methanocella* and *Methanobacterium*

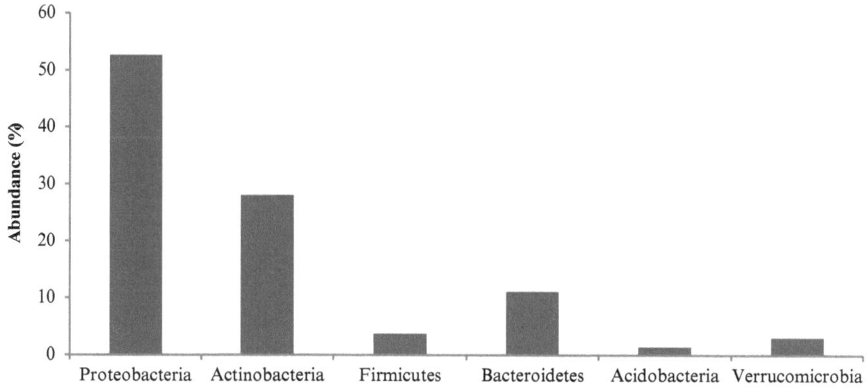

Fig. 2.5 Diversity and relative abundance (%) of bacteria from soil samples from Kafni Pindari and Roopkund glaciers of Himalayas based on culture independent methods. This figure was constructed from data compiled from references given in Sect. 1.3.1

(Aschenbach et al. 2013). More than 95 % of the methanotrophs as studied using *pmoA* gene belonged to the genera *Methylohalobius, Methylothermus, Methylococcus, Methylocaldum* and *Methylosoma* sp. (Zheng et al. 2012).

1.3.3 Bacterial Diversity of Himalayas as Studied by Culture Based Methods

Close to 466 strains have been isolated from Roopkund (Shivaji et al. 2008; Reddy et al. 2009a; Pradhan et al. 2010), Pindari (Shivaji et al. 2011a, b; Reddy et al. 2008, 2013a, b; Kishore et al. 2009; Pindi et al. 2009), Kafni (Kishore et al. 2009; Pindi et al. 2009; Srinivas et al. 2011), Hamta (Shivaji et al. 2005; Chaturvedi et al. 2005; Chaturvedi and Shivaji 2007) and Western Himalayan glaciers (Gangwar et al. 2009; Sahay et al. 2013; Yadav et al. 2015) and they represented the major taxa Proteobacteria, Bacteroidetes, Firmicutes and Actinobacteria. NCBI database revealed that so far only 21 novel species have been identified from the Himalayan mountain ranges. A novel genus, *Glaciihabitans* was described from Himalayan region (Li et al. 2013). Bioprospecting of these microbes indicated that a good proportion of the isolates from Pindari (Shivaji et al. 2011a, b) and Kafni (Srinivas et al. 2011) exhibited amylase, urease and or lipase activities at 4 and 20 °C. Gangwar et al. (2009) also demonstrated that cultures from high altitude cold habitats of Himalayan region exhibit lipase activity (Gangwar et al. 2009). In addition, culture based studies led to the identification of novel sulfate reducing and a phototrophic bacteria *Desulfovibrio psychrotolerans*(Jyothsna et al. 2008) and *Rhodobacter megalophilus* (Arunasri et al. 2008) from Himalayas (Fig. 2.6).

1.4 *Some Unique Features of Bacteria from Extremely Cold Environments*

Interestingly, bacterial isolates from Antarctica, Arctic and Himalayas share many common features with respect to growth wherein the maximum growth temperature is below 30 °C and growth is possible even at 4 °C with very few exceptions such as *Alicyclobacillus pohliae* (42–60 °C), *Aneurinibacillus terranovensis* (15–55 °C), *Anoxybacillus amylolyticus* (45–65 °C), *Bacillus thermantarcticus* (30–60 °C) and *Brevibacillus levickii* (20–55 °C) and *Rhabdothermus arcticus* (37–75 °C) (Lama et al. 1996; Allan et al. 2005; Poli et al. 2006; Imperio et al. 2008; Steinsbu et al. 2011). Earlier studies also revealed that the G+C content of 16S rRNA gene correlated well with the growth temperature in prokaryotes wherein the G-C content is concentrated in the stem regions of the molecule and offers structural stability and integrity to rRNA while its single stranded regions are enriched with purines (A and G). An elongation of helix 17 was observed in five out of eight true psychrophilic species of the genus *Cryobacterium* (unpublished) whose growth temperature was below 20 °C. Occurrence of 16S rDNA operons with psychrotolerant signatures, the transitions from C and G to T and A, higher uracil content and G:U mismatches from

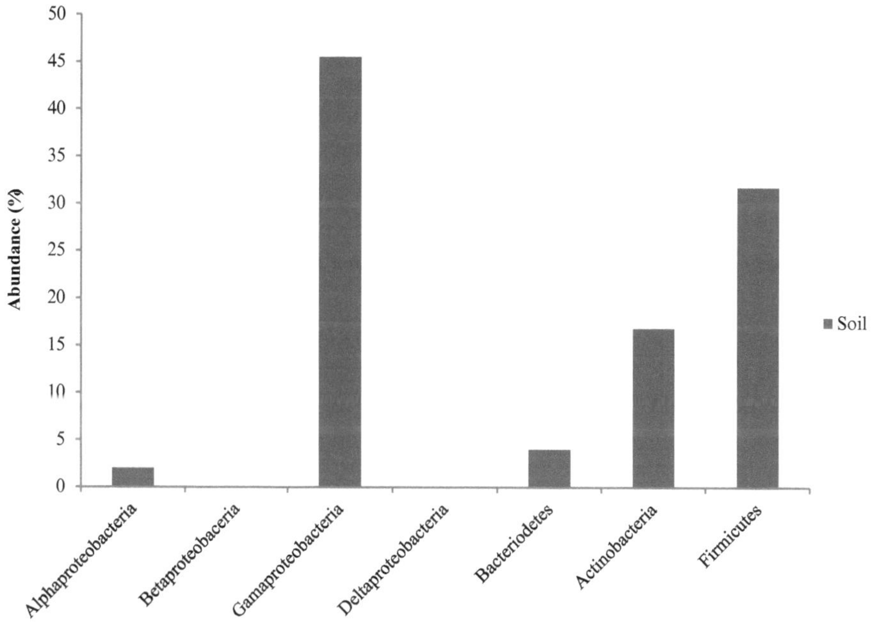

Fig. 2.6 Bacterial diversity and relative abundance (%) of culturable bacteria from soil samples collected from of Pindari, Hamta, Roopkund and Kafni glaciers of Himalayas. Data compiled from references given in Sect. 1.3.3

mesophilic to psychrotolerant, was reported in the strains of *Bacillus* (Prüss et al. 1999; Lauro et al. 2007). Further swollen sporangia were observed in psychrophilic strains of the genus *Bacillus* compared to their nearest mesophilic counterparts (Reddy et al. 2008; Abd El-Rahman et al. 2002; Larkin and Stokes 1967; Priest et al. 1988). *Bacillus cecembensis,* a psychrophilic species from Himalayas contained $C_{16:1}\omega7c$ alcohol in its membrane while it is absent in its nearest mesophilic neighbor (Reddy et al. 2008). In fact all the psychrophiles have a preponderance of unsaturated fatty acids compared to mesophilic counterparts. These bacteria also have cold active and heat labile enzymes (Ray et al. 1992; Reddy et al. 1994; Chattopadhyay et al. 1995) and have genes that are essential for their survival at low temperature (Singh et al. 2009; Singh and Shivaji 2010; Sundareswaran et al. 2010). Some other aspects of adaptation to low temperature are highlighted in the following section of this article.

2 Adaptation and Biotechnological Importance of Cold-Tolerant Bacteria

Bacterial isolates obtained from extreme cold habitats offer a challenging field of investigations. During the past few decades a vast body of information has been generated on various types of bacteria obtained from the Antarctic, Arctic and

Himalayan regions and also on the biochemical and genetic basis of their cold-tolerance (Chattopadhyay et al. 2014). A sizeable fraction of these cold-tolerant isolates are oligotrophic and eurythermal in nature. This section highlights some of the major clues obtained so far to the mechanism of cold-tolerance and biotechnological applications of cold-tolerant bacteria or their enzymes.

2.1 Mechanism of Cold-Tolerance: An Overview

Available evidences in the literature provide some idea as to the mechanism of bacterial cold-tolerance. Miniaturization and fragmentation of cells was a strategy used by an Arctic strain of *Psychrobacter* for increasing the chance of survival during lowering of environmental temperature (Ewert and Deming 2014). Further, as seen in fishes, insects, plants and fungi, some psychrophilic bacteria synthesize anti-freeze proteins to overcome the effect of sub-zero temperatures in the Antarctic environment. A report highlighting high level of thermal hysteresis and recrystallization inhibition activities in the cell-free extract of an Antarctic isolate added a new dimension to the present state of knowledge on bacterial anti-freeze proteins (Kawahara et al. 2007). The cold-adapted bacteria are known to produce some cold-tolerant and/or thermolabile enzymes (Feller and Gerday 2003; Singh et al. 2014), which help them sustain metabolic activities of the cell at non-permissible temperatures. It is also crucially important for them to sense the lowering of environmental temperature, which is accomplished by sensor molecules associated with the membrane (Ray et al. 1994). Temperature-dependent changes in the DNA topology is also postulated to play an important role in sensing the temperature of the outside environment (Shivaji and Prakash 2010). A two-component signal transduction mechanism has been demonstrated to be involved in sensing and responding to various stress conditions in the environment. Stress is normally detected by a membrane associated histidine kinase (HK) which acts as a sensor, gets phosphorylated under stress conditions and transduces the signal to a response regulator in the cytoplasm which upregulates genes essential for adaptation to the stress. Recently, the involvement of two HK-encoding genes viz, *yyc G* and *lis K* in the growth of the food-borne pathogen *Listeria monocytogenes* at low temperature is demonstrated (Pöntinen et al. 2015). Maintenance of the fluidity of cell membrane, which tends to get frigid at low temperature, is another challenge they are confronted with in the cold environment. Preferential synthesis of membrane-fluidizing fatty acids (unsaturated, short-chain and anteiso-fatty acids) coupled with involvement of some carotenoid pigments appear to help them in the homeoviscous adaptation of membrane fluidity (Chintalapati et al. 2004; Chattopadhyay and Jagannadham 2001). Most of the cold-adapted bacteria do not form endospores. So they are likely to continue metabolic activities at low temperature. It was believed earlier that at extreme low temperature, bacteria could continue only catabolic activities. Recent evidences of both catabolic and anabolic activities in bacteria at sub-zero temperatures have revealed flexibility of the metabolic machinery of cold-tolerant bacteria (Sengupta and

Chattopadhyay 2013). Transcription and translation were found to be continued at low temperature in the cell-free system of an Antarctic isolate while the same processes in the cell-free system of *Escherichia coli* did not continue at low temperature (Ray et al. 1998). Substantial increase in the intracellular concentration of ATP and ADP in *Psychrobacter cryohalolentis* K5, a eurythermal psychrophile, obtained from the Siberian permafrost, following lowering of the temperature indicated that it might be a strategy adopted by the organism to compensate for the lowering of the kinetic rate of metabolic reactions at low temperature (Amato and Christner 2009). Occurrence of the homologs of the gene *csp A*, which encodes the major cold shock protein in *E. coli*, was demonstrated in an Antarctic isolate (Ray et al. 1998). Hence cold shock proteins also might have a role in continuation of metabolism at low temperature. Growth of an Antarctic bacterium at low temperature was found to be enhanced in presence of some small molecular weight compounds (proline, betaine) in the culture (unpublished). So like some amphibians and insects, bacteria might get the help of some cellular cryoprotectants for adaptation to low temperature. Frequent arrest of the replication fork and fork-reversal during replication of DNA at low temperature was postulated as a reason behind inability of bacteria to grow in cold environments. The Rec BCD protein of an Antarctic lake isolate *Pseudomonas syringae* Lz 4 W, a psychrophile, appeared to degrade the reversed replication fork and facilitate re-initiation of replication (Sinha et al. 2013). The role of some other proteins (aspartate aminotransferase and t-RNA modification GTPase) in bacterial cold-tolerance was demonstrated using cold-sensitive mutants of an Antarctic bacterium as a tool (Singh et al. 2009; Sundareswaran et al. 2010). Degradation of RNA is a crucial step in the regulation of gene expression. The components of RNA-degradosome, a protein complex involved in RNA degradation, were found to be substantially different in an Antarctic bacterium compared to those in *E. coli* (Purusharth et al. 2007). Genomic and transcriptomic analysis of two cultures of a psychrophilic methanogen, *Methanolobus psychrophilus* R15, one incubated at 18 °C and the other at 4 °C, indicated acceleration in the exosome-mediated RNA degradation at low temperature (Chen et al. 2012). In natural environments, stress conditions (extremities of temperature and pH, high salt, oxidative stress, ultraviolet radiation) seldom occur alone and most often they occur in various combinations. Consequently, cold-adapted bacteria might have to be adapted to more than one stressor in order to survive. Role of some proteins in adaptation of bacteria to low temperature and other environmental stress factors indicates stress response in bacteria might be interlinked (Chattopadhyay 2008). Increase in the intracellular production of free radicals (that contributes to the oxidative stress) was observed in an Antarctic bacterium when it was grown at low temperature (Chattopadhyay et al. 2011). The use of Entner-Duodoroff pathway (which contributes to the increase in the intracellular level of NADPH) by cold-adapted bacteria may help them in coping with increase in oxidative stress, associated with growth at low temperature (Moreno and Rojo 2014). Metabolism of polyhydroxyalkanoates, the carbon-storage polymers, also appears to relieve the cells of the elevated oxidative stress at low temperature (Ayub et al. 2009). A recent meta-analysis of the sequence of 66 plasmids obtained from various cold-tolerant and psychrophilic bacteria belonging to various genera (e.g., *Allivibrio, Bacillus, Carnobacterium, Desulfotalea, Exiguobacterium, Flavobacterium, Pseudomonas, Psychrobacter, Runella, Spingopyxis*) revealed the presence of genes responsible for protection from low temperature and ultraviolet radiation and also resistance to heavy metals, metalloids and antibiotics (Dziewit and Bartosik 2014). Continuing investigations in this area are offering new clues to the mechanism of bacterial cryotolerance.

DNA sequencing has always provided insight into the regulation and functioning of genes and in the process has had a major impact on our understanding of processes involving human health and disease, agriculture and many other environmental processes involving living organisms. Thus any technology, which enhances the efficacy of deciphering genome sequences at an affordable cost and simultaneously provides whole genome sequence data within a short period of time, promises substantial addition to the present state of our knowledge on genes and their function. Next-generation sequencing (NGS) is such a technology which has facilitated whole-genome sequencing of multiple organisms from different and distinct habitats, helped generate genomic information and understand and appreciate unique features of organisms such as their adaptation to extreme environments. A recent study (Raiger Iustman et al. 2015) based on the genome sequence analysis of *Pseudomonas extremaustralis* provided new insights into environmental adaptability and adaptation to extreme conditions. In this psychrophilic bacterium from Antarctica, genes required for osmoregulation, cold adaptation, exopolysaccharide production and degradation of complex compounds were detected. Presence of genes acquired by horizontal transfer was also implied. Based on comparative genome analysis it was observed that genes coding for cold-shock proteins (Csps) and cold acclimation proteins (Caps) were present only in *Pseudomonas extremaustralis* which is a psychrophile. In contrast, in *P. Syringae* pv. *syringae* B278a and *P. aeruginosa* PAO1, which are mesophiles, the lowest number of Csps was present but no Cap was present. Thus, it was suggested that these proteins contribute to cold adaptation of *P. extremaustralis.* Draft genome sequence of a lake-isolate of *Hymenobacter*, tolerant to ultraviolet radiation and oxidative stress besides low temperature, was published some time back (Koo et al. 2014). Our group has recently reported the genome sequence of a number of psychrophiles from different cold habitats like Antarctica (Sreenivas et al. 2014; Reddy et al. 2013a, b, 2014; Kumar et al. 2013a, b), Arctic (Shivaji et al. 2013a, b; Kumar et al. 2013b), Himalayan glaciers (Reddy et al. 2014) and Stratosphere (Shivaji et al. 2012b).

A comparative genome analysis of Csps in the psychrophilic *Sphingobacterium antarcticum, Oceanisphaera arctica* and *Exiguobacterium indicum* indicated that CspA was present in all of them, Csp C, D, E and G were present in at least one of the three isolates whereas Csp B and I were absent (unpublished). The implications of this observation with respect to cold adaptation warrant further studies. Comparative genome analysis of psychrophiles besides providing information on the role of specific genes in cold adaptation would promise insight into the adaptive response of bacteria to other stressors (viz high salt, high and low pH).

2.2 Biotechnological Potential

Cold-active and /or thermolabile enzymes obtained from cold-tolerant organisms are of immense biotechnological importance. A thermolabile phosphatase, obtained from an Antarctic bacterium and patented (HK TM) some time back is active in neutral and alkaline pH. It is highly useful in performing sequential steps in a single tube viz, restriction enzyme digestion, dephosphorylation, enzyme inactivation, and ligation or end-labelling. Thus it simplifies the experimental procedure and minimizes the amount of nucleic acid required. Cold-adapted bacteria capable of degrading hydrocarbons are useful in waste water treatment (Margesin and Schinner 2001). Some strains isolated from the cryoconite holes in the Arctic glaciers were found to produce cellulase having higher activity at low temperature (Singh et al. 2014). Cold-active proteases isolated from cold-tolerant organisms are very suitable for food processing and in leather industries since they remain active at the temperature of tap-water and therefore eliminate the necessity of heat-activation. They are also useful in laundries for removal of stains. A cold-active lipase, obtained by cloning a lipolytic gene from an Arctic bacterium (*Rhodococcus* sp. AW25M09) in *E. coli*, has been found to retain activity at high pH, organic solvents (acetonitrile, diethyl ether) and salt (1 M NaCl). Thus it appears to be an interesting candidate for industrial applications (De Santi et al. 2014). A number of cold-adapted strains of Pseudomonads are known to produce lipases, which have high activity at moderate or low temperature. Such strains could be useful in lipase-mediated biodiesel production (Moreno and Rojo 2014). In view of the widespread resistance of pathogenic microorganisms to the presently used antibiotics, attention is focussed on alternative source of antibiotics. Bacteria occurring in Polar Regions are one of the potential candidates for this purpose. Notwithstanding substantial evidence of antibiosis in polar bacteria (Liu et al. 2013; Maida et al. 2014; Fondi et al. 2014) this aspect remains relatively unexplored.

Human activities in cold climate are leading to the accumulation of a lot of garbage, which poses a serious threat to the environment. Petroleum products are widely used for heating, which is unavoidable during camping in such extreme harsh climate. Consumption of the petroleum products generates pollutants. Moreover, spillage of petroleum products from the aircrafts and reservoirs leads to serious environmental pollution. In spite of precautionary measures, accidental spillage is likely to occur because of increase in human intervention in these pristine environments. Airlifting of the contaminated soil for removal of the pollutants to some other place being prohibitively costly, in situ bioremediation using indigenous bacteria appears to be the most preferred option. Bioremediation involves transformation of organic pollutants by microorganisms (or microbial enzymes) into less hazardous products which are integrated into the biogeochemical cycle. Biodegradation in high temperature environments is associated with some advantages (e.g., enhanced solubility, bioavailability), not available to the bacteria in the cold environments. However, isolation of autochthonous bacteria having capacities to degrade hydrocarbons from Arctic and Antarctic soil has been reported from time to time. They are found not only in oil contaminated soil but also in soil that has no history of oil contamination. An Antarctic strain of *Pseudomonas* was found to grow optimally in presence of ammonium sulphate, nitrate, nitrite and 3.5 % diesel at pH 7.0 within 10 to 20 °C (Shukor et al. 2009). In general, the degraders adapt rapidly to the contaminants and increase in number quickly following an episode of contamination. Hydrocarbon compounds associated with oil are widely diverse in nature (i.e., iso-, cyclo, and linear alkanes, monoaromatic compounds and polycyclic aromatic hydrocarbons). So microorganisms with specific biochemical mechanisms to metabolize different types of hydrocarbons are required for biodegradation of the pollutants (Timmis et al. 2010). In one study, organisms having previous exposure to oil-contamination were observed to degrade the contaminants at a faster rate indicating the importance of adaptation. So pre-exposure appears to be essential for quick degradation. Successful application of real-time PCR during the past decade has revealed the presence of a number of genes responsible for hydrocarbon biodegradation (e.g. monooxygenases, dioxygenases) in different oil-contaminated sites (Powell et al. 2006). However, despite their capacity to degrade a wide range of hydrocarbons, all the microbial communities occurring in cold climate, cannot degrade all sorts of petroleum-derived pollutants. Moreover, some of the pollutants (e.g. highly substituted and high molecular weight compounds) are recalcitrant in nature and resist microbial degradation. They are immune to microbial degradation. For example, asphalt (or bitumen) is a highly viscous liquid or semi-solid form of petroleum. Asphalt pavements (mineral aggregate bound with asphalt) were detected years after a spillage in Chile and another spillage in Canada. Following the Deep water Horizon blow out in August 2010, a substantial amount of hydrocarbons were found to be left though another large portion was degraded by microbial metabolism and removed by human intervention (Kimes et al. 2014). The rate of degradation is slow at low temperature. Significant improvement in the rate of oil-degradation could be achieved in one case by spraying surfactant-foam containing oil-degrading microbe. The surfactant facilitates dissolution of oil from the soil. Its insulating effect maintains a warmer environment in the soil (Jeong et al. 2015). Besides petroleum products, other types of environmental contaminants (e.g., heavy metals) also occur in Antarctic soil because of anthropogenic activities (Claridge et al. 1995; Lohan et al. 2001; Chaparro et al. 2007). A bacterial isolate obtained from the South Shetlands Islands, Antarctica, tentatively identified as a *Pseudomonas* sp. was found to convert sodium molybdate or Mo^{6+} to molybdenum blue. The optimum temperature range for the reaction was 15–20 °C. Thus, the strain was suitable for bioremediation in both cold and temperate regions (Ahmad et al. 2013). Polychlorinated biphenyls (PCBs) widely used for industrial and commercial purposes also occur in the polar regions because of aerial transportation, inputs by the rivers (in the Arctic regions) and improper disposal of PCB-containing articles (e.g., electrical equipment) at the research site. Microorganisms having the capacity to degrade PCBs have been isolated from both the Arctic and Antarctic regions from time to time (Master and Mohn 1998; De Domenico et al. 2004).

Besides low temperature, deficiency of nutrients in the soil is another limiting factor for growth of bacteria used for bioremediation. Hence, sometimes it may be needed to stimulate the activities of the soil bacteria by addition of nutrients (Ruberto et al. 2009). Sometimes it appears realistic to use bioremediation in combination with some physical process of cleaning (use of boomers and skim). It has been possible to improve the cold-tolerance of the mesophilic organism *E. coli* by cloning the genes encoding two chaperonin proteins obtained from an Antarctic organism (Ferrer et al. 2003) and also by cloning a gene encoding an antifreeze protein occurring in a cold-tolerant plant (Deng et al. 2014). The use of such genetically engineered microorganisms with high capacity to degrade pollutants appears to be an attractive option. But introduction of foreign microorganism in the environment is prohibited in Antarctica.

Acknowledgments Authors acknowledge Council for Scientific and industrial research (CSIR), Department of Biotechnology (DBT) and Department of science and technology, Govt of India for funding various projects.

Conflict of Interest Authors declare that they have no conflict of interest.

References

Abd El-Rahman HA, Fritze D, Spröer C, Claus D (2002) Two novel psychrotolerant species, *Bacillus psychrotolerans* sp. nov. and *Bacillus psychrodurans* sp. nov., which contain ornithine in their cell walls. Int J Syst Evol Microbiol 52:2127–2133

Ahmad SA, Shukor MY, Shamaan NA, Mac Cormack WP, Syed MA (2013) Molybdate reduction to molybdenum blue by an Antarctic bacterium. Biomed Res Int 2013:871–941

Aislabie J, Balks M, Foght J, Waterhouse E (2004) Hydrocarbon spills on Antarctic soils: Effects and management. Environ Sci Technol 38:1265–1274

Aislabie J, Saul DJ, Foght JM (2006) Bioremediation of hydrocarbon-contaminated polar soils. Extremophiles 10:171–179

Aislabie J, Jordan S, Ayton J, Klassen JL, Barker GM, Turner S (2009) Bacterial diversity associated with ornithogenic soil of the Ross Sea region, Antarctica. Can J Microbiol 55:21–36

Aislabie JM, Lau A, Dsouza M, Shepherd C, Rhodes P, Turner SJ (2013) Bacterial composition of soils of the Lake Wellman area, Darwin Mountains, Antarctica. Extremophiles 17:775–786

Al Khudary R, Stösser NI, Qoura F, Antranikian G (2008) *Pseudoalteromonas arctica* sp. nov., an aerobic, psychrotolerant, marine bacterium isolated from Spitzbergen. Int J Syst Evol Microbiol 58:2018–2024

Allan RN, Lebbe L, Heyrman J, De Vos P, Buchanan CJ, Logan NA (2005) *Brevibacillus levickii* sp. nov. and *Aneurinibacillus terranovensis* sp. nov., two novel thermoacidophiles isolated from geothermal soils of northern Victoria Land, Antarctica. Int J Syst Evol Microbiol 55:1039–1050

Amato P, Christner BC (2009) Energy metabolism response to low-temperature and frozen conditions in *Psychrobacter cryohalolentis*. Appl Environ Microbiol 75:711–718

Antibus DE, Leff LG, Hall BL, Baeseman JL, Blackwood CB (2012) Cultivable bacteria from ancient algal mats from the McMurdo Dry alleys, Antarctica. Extremophiles 16:105–114

Archer SD, McDonald IR, Herbold CW, Cary SC (2014) Characterisation of bacterioplankton communities in the meltwater ponds of Bratina Island, Victoria Land, Antarctica. FEMS Microbiol Ecol 89:451–464

Arunasri K, Venkata Ramana V, Spröer C, Sasikala C, Ramana CV (2008) *Rhodobacter megalophilus* sp. nov., a phototroph from the Indian Himalayas possessing a wide temperature range for growth. Int J Syst Evol Microbiol 58:1792–1796

Aschenbach K, Conrad R, Reháková K, Doležal J, Janatková K, Angel R (2013) Methanogens at the top of the world: occurrence and potential activity of methanogens in newly deglaciated soils in high-altitude cold deserts in the Western Himalayas. Front Microbiol 4:359

Auman AJ, Breezee JL, Gosink JJ, Kämpfer P, Staley JT (2006) *Psychromonas ingrahamii* sp. nov., a novel gas vacuolate, psychrophilic bacterium isolated from Arctic polar sea ice. Int J Syst Evol Microbiol 56:1001–1007

Auman AJ, Breezee JL, Gosink JJ, Schumann P, Barnes CR, Kämpfer P, Staley JT (2010) *Psychromonas boydii* sp. nov., a gas-vacuolate, psychrophilic bacterium isolated from an Arctic sea-ice core. Int J Syst Evol Microbiol 60:84–92

Ayub ND, Tribelli PM, Lopez NI (2009) Polyhydroxyalkanoates are essential for maintenance of redox state in the Antarctic bacterium *Pseudomonas* sp. 14-3 during low temperature adaptation. Extremophiles 13:59–66

Bajerski F, Wagner D (2013) Bacterial succession in Antarctic soils of two glacier forefields on Larsemann Hills, East Antarctica. FEMS Microbiol Ecol 85:128–142

Ball MM, Gómez W, Magallanes X, Rosales R, Melfo A, Yarzábal LA (2014) Bacteria recovered from a high-altitude, tropical glacier in Venezuelan Andes. World J Microbiol Biotechnol 30:931–941

Bano N, Hollibaugh JT (2000) Diversity and distribution of DNA sequences with affinity to ammonia-oxidizing bacteria of the beta subdivision of the class Proteobacteria in the Arctic Ocean. Appl Environ Microbiol 66:1960–1969

Begum Z, Srinivas TN, Manasa P, Sailaja B, Sunil B, Prasad S, Shivaji S (2013) *Winogradskyella psychrotolerans* sp. nov., a marine bacterium of the family Flavobacteriaceae isolated from Arctic sediment. Int J Syst Evol Microbiol 63:1646–1652

Bienhold C, Boetius A, Ramette A (2012) The energy-diversity relationship of complex bacterial communities in Arctic deep-sea sediments. ISME J 6:724–773

Boeuf D, Cottrell MT, Kirchman DL, Lebaron P, Jeanthon C (2013) Summer community structure of aerobic anoxygenic phototrophic bacteria in the western Arctic Ocean. FEMS Microbiol Ecol 85:417–432

Bottos EM, Vincent WF, Greer CW, Whyte LG (2008) Prokaryotic diversity of arctic ice shelf microbial mats. Environ Microbiol 10:950–966

Bottos EM, Woo AC, Zawar-Reza P, Pointing SB, Cary SC (2014) Airborne bacterial populations above desert soils of the McMurdo Dry Valleys, Antarctica. Microb Ecol 67:120–128

Bowman JP, McCammon SA, Brown MV, Nichols DS, McMeekin TA (1997) Diversity and association of psychrophilic bacteria in Antarctic sea ice. Appl Environ Microbiol 63:3068–3078

Bowman JP, Gosink JJ, McCammon SA, Lewis TE, Nichols DS, Nichols PD, Skerratt JH, Staley JT, McMeekin TA (1998) *Colwellia demingiae* sp. nov., *Colwellia hornerae* sp. nov., *Colwellia rossensis* sp. nov. and *Colwellia psychrotropica* sp. nov.: psychrophilic Antarctic species with the ability to synthesize docosahexanoic acid (22:6 omega 3). Int J Syst Bacterial 48:1171–1180

Bowman JP, Rea SM, McCammon SA, McMeekin TA (2000) Diversity and community structure within anoxic sediment from marine salinity meromictic lakes and a coastal meromictic marine basin, Vestfold Hills, Eastern Antarctica. Environ Microbiol 2:227–237

Boyd WL (1962) Comparison of soil bacteria and their metabolic activities in Arctic and Antarctic regions. Polar Rec 11:319

Boyd WL, Boyd JW (1962a) Presence of *Azotobacter* species in polar regions. J Bacterial 85:1121–1123

Boyd WL, Boyd JW (1962b) Viability of thermophiles and coliform bacteria in arctic soils and water. Can J Microbiol 8:189–192

Bozal N, Montes MJ, Mercadé E (2007) *Pseudomonas guinea* sp. nov., a novel psychrotolerant bacterium from an Antarctic environment. Int J Syst Evol Microbiol 57:2609–2612

Brinkmeyer R, Knittel K, Jurgens J, Weyland H, Amann R, Helmke E (2003) Diversity and structure of bacterial communities in Arctic versus Antarctic pack ice. Appl Environ Microbiol 69:6610–6619

Brown MV, Bowman JP (2001) A molecular phylogenetic survey of sea-ice microbial communities (SIMCO). FEMS Microbiol Ecol 35:267–275

Campbell IB, Claridge GGC (2000) Soil temperature, moisture and salinity patterns in Transantarctic Mountain cold desert ecosystems. In: Davidson W, Howard-Williams C, Broady P (eds) Antarctic ecosystems: models for wider ecological understanding. New Zealand Natural Sciences, Canterbury University, Christchurch, New Zealand, pp 233–240

Campbell BJ, Polson SW, Hanson TE, Mack MC, Schuur EA (2010) The effect of nutrient deposition on bacterial communities in Arctic tundra soil. Environ Microbiol 12:1842–1854

Carpenter EJ, Lin S, Capone DG (2000) Bacterial activity in South Pole snow. Appl Environ Microbiol 66:4514–4517

Carr SA, Vogel SW, Dunbar RB, Brandes J, Spear JR, Levy R, Naish TR, Powell RD, Wakeham SG, Mandernack KW (2013) Bacterial abundance and composition in marine sediments beneath the Ross Ice Shelf, Antarctica. Geobiology 11:377–395

Cavanagh J, Austin JJ, Sanderson K (1996) Novel *Psychrobacter* species from Antarctic ornithogenic soils. Int J Syst Bacterial 46:841–848

Cavicchioli R (2007) Antarctic metagenomics. Microbiol Aust 28:98–103

Cavicchioli R, Thomas T (2000) Extremophiles. In: Lederberg J (ed) Encyclopedia of microbiology, 2nd edn. vol 2. Academic Press, San Diego, pp 317–337

Chaparro MAE, Nunez H, Lirio JM, Gogorza CSG, Sinito AM (2007) Magnetic screening and heavy metal pollution studies in soils from Marambio Station, Antarctica. Antarctic Sci 19:379–393

Chattopadhyay MK (2008) Cryotolerance in bacteria: Interlink with adaptation to other stress factors. Trends Microbiol 16:455

Chattopadhyay MK, Jagannadham MV (2001) Maintenance of membrane fluidity in Antarctic bacteria. Polar Biol 24:386–388

Chattopadhyay MK, Uma Devi K, Gopisankar Y, Shivaji S (1995) Thermolabile alkaline phosphatase from *Sphingobacterium antarcticus*, a psychrotrophic bacterium from Antarctica. Polar Biol 15:215–219

Chattopadhyay MK, Raghu G, Sharma YVRK, Biju AR, Rajasekharan MVR, Shivaji S (2011) Increase in oxidative stress at low temperature in an Antarctic bacterium. Curr Microbiol 62:544–546

Chattopadhyay MK, Reddy GSN, Shivaji S (2014) Psychrophilic bacteria: Biodiversity, Molecular basis of cold adaptation and biotechnological implications. Curr Biotech 3:100–116

Chaturvedi P, Shivaji S (2007) *Exiguobacterium indicum* sp. nov., a psychrophilic bacterium from the Hamta glacier of the Himalayan mountain ranges in India. Int J Syst Evol Microbiol 56:2765–2770

Chaturvedi P, Reddy GSN, Shivaji S (2005) *Dyadobacter hamtensis* sp. nov., from Hamta Glacier, Himalayas, India. Int J Syst Evol Microbiol 55:2113–2117

Chen Z, Yu H, Li L, Hu S, Dong X (2012) The genome and transcriptome of a newly described psychrophilic archaeon, *Methanolobus psychrophilus* R15, reveal its cold adaptive characteristics. Environ Microbiol Rep 4:633–641

Chessa JP, Petrescu I, Bentahir M, Van Beeumen J, Gerday C (2000) Purification, physicochemical characterization and sequence of a heat labile alkaline metalloprotease isolated from a psychrophilic *Pseudomonas* species. Biochim Biophys Acta 1479:265–274

Chintalapati S, Kiran MD, Shivaji S (2004) Role of membrane lipid fatty acids in cold adaptation. Cell Mol Biol (Noisy-le-Grand) 50:631–642

Chu HY, Fierer N, Lauber CL, Caporaso JG, Knight R, Grogan P (2010) Soil bacterial diversity in the Arctic is not fundamentally different from that found in other biomes. Environ Microbiol 12:2998–3006

Claridge GGC, Campbell IB (1977) The salts in Antarctic soils, their distribution and relationship to soil processes. Soil Sci 123:377–384

Claridge GGC, Campbell IB, Powell HKJ, Amin ZH, Balks MR (1995) Heavy metal contamination in some soils of the McMurdo Sound region, Antarctica. Antarctic Sci 7:9–14

Collins RE, Rocap G, Deming JW (2010) Persistence of bacterial and archaeal communities in sea ice through an Arctic winter. Environ Microbiol 12:1828–1841

Cristóbal HA, López MA, Kothe E, Abate CM (2011) Diversity of protease-producing marine bacteria from sub-antarctic environments. J Basic Microbiol 51:590–600

De Domenico M, Lo Giudice A, Michaud L, Saitta M, Bruni V (2004) Diesel oil and PCB-degrading psychrotrophic bacteria isolated from Antarctic seawaters (Terra Nova Bay, Ross Sea). Polar Res 23:141–146

De Santi C, Tedesco P, Ambrosino L, Altermark B, Willassen NP, de Pascale D (2014) A new alkaliphilic cold-active esterase from the psychrophilic marine bacterium *Rhodococcus* sp.: functional and structural studies and biotechnological potential. Appl Biochem Biotechnol 172:3054–3068

Delille D, Gleizon F (2003) Distribution of enteric bacteria in Antarctic seawater surrounding the Port-aux-Français permanent station (Kerguelen Island). Mar Pollut Bull 46:1179–1183

Delille D, Lagarde E (1974) Contribution al'etude ecologique des Milieux subantarctiques. V Rev lnst Pasteur Lyon 7:149–165

Deng LQ, Yu HQ, Liu YP, Jiao PP, Zhou SF, Zhang SZ, Li WC, Fu FL (2014) Heterologous expression of antifreeze protein gene AnAFP from *Ammopiptanthusnanus* enhances cold tolerance in *Escherichia coli* and tobacco. Gene 539:132–140

Denner EBM, Mark B, Busse HJ, Turkiewicz M, Lubitz W (2001) *Psychrobacter proteolyticus* sp. nov., a psychrotrophic, halotolerant bacterium isolated from the Antarctic krill *Euphausia superba* Dana, excreting a cold-adapted metalloprotease. Syst Appl Microbiol 24:44–53

Deslippe JR, Egger KN (2006) Molecular diversity of nifH genes from bacteria associated with high arctic dwarf shrubs. Microb Ecol 51:516–525

Díez B, Bergman B, Pedrós-Alió C, Antó M, Snoeijs P (2012) High cyanobacterial nifH gene diversity in Arctic seawater and sea ice brine. Environ Microbiol Rep 4:360–366

Dong S, Yang J, Zhang XY, Shi M, Song XY, Chen XL, Zhang YZ (2012) Cultivable alginate lyase-excreting bacteria associated with the Arctic brown alga Laminaria. Mar Drugs 10:2481–2491

Doyle SM, Montross SN, Skidmore ML, Christner BC (2013) Characterizing microbial diversity and the potential for metabolic function at -15 °c in the Basal ice of Taylor Glacier, antarctica. Biology (Basel) 2:1034–1053

Dyurgerov MB, Meier MF (1997) Mass balance of mountain and subpolar glaciers: a new global assessment for 1961–1990. Arctic Alpine Res 29:379–391

Dziewit L, Bartosik D (2014) Plasmids of psychrophilic and psychrotolerant bacteria and their role in adaptation to cold environments. Front Microbiol 5:596

Edwards A, Anesio AM, Rassner SM, Sattler B, Hubbard B, Perkins WT, Young M, Griffith GW (2011) Possible interactions between bacterial diversity, microbial activity and supraglacial hydrology of cryoconite holes in Svalbard. ISME J 5:150–160

Ekelof E (1908a) Bakteriologische studien wahrend der Schwedishen Sudpolar expedition 1901–1903. In: Wissenschaftiche Ergebnisse der Schwedischen sudpolar Expedition 1901–1903. 0. Nordenskjold (ed) Lithogr. lnst. Generalstabs, Stockholm, p 210

Ekelof E (1908b) Studien uber den Bakteriengehalt der luft und des Erdbodeus der antarktishen Gegenden, ausgefiihrt wahrend der schwedischen sudpolar expedition 1901–1903. Zhyg lnfekt 56:344–370

Friedman E (1980) Endolithic microbial life in hot and cold deserts. Orig Life 10:223–235

Ewert M, Deming JW (2014) Bacterial responses to fluctuations and extremes in temperature and brine salinity at the surface of Arctic winter sea ice. FEMS Microbiol Ecol 89:476–489

Feller G, Gerday C (2003) Psychrophilic enzymes: hot topics in cold adaptation. Nat Rev Microbiol 1:200–208

Ferrer M, Chernikova TN, Yakimov MM, Golyshin PN, Timmis KN (2003) Chaperonins govern growth of *Escherichia coli* at low temperatures. Nat Biotechnol 21:1266–1267

Fondi M, Orlandini V, Perrin E, Maida I, Bosi E, Papaleo MC, Michaud L, Lo Giudice A, de Pascale D, Tutino ML, Liò P, Fani R (2014) Draft genomes of three Antarctic *Psychrobacter* strains producing antimicrobial compounds against *Burkholderiacepacia* complex, opportunistic human pathogens. Mar Genomics 13:37–38

Forschner SR, Sheffer R, Rowley DC, Smith DC (2009) Microbial diversity in Cenozoic sediments recovered from the Lomonosov Ridge in the Central Arctic basin. Environ Microbiol 11:630–639

Franzmann PD, Dobson SJ (1992) Cell wall-less, free-living spirochetes in Antarctica. FEMS Microbiol Lett 76:289–292

Gangwar P, Alam SI, Bansod S, Singh L (2009) Bacterial diversity of soil samples from the western Himalayas, India. Can J Microbiol 55:564–577

Ganzert L, Bajerski F, Wagner D (2014) Bacterial community composition and diversity of five different permafrost-affected soils of Northeast Greenland. FEMS Microbiol Ecol 89:426–441

Gibson JAE (2006) Limnology of Epiglacial Lakes of the Framnes Mountains, Antarctica: insights into a widespread but poorly Studied Lake Type. Scientific Committee on Antarctic Research (SCAR), vol XXIX, Hobart, 11–15 July, p 422

Gosink JJ, Woese CR, Staley JT (1998) *Polaribacter* gen. nov., with three new species, *P. irgensii* sp. nov., *P. franzmannii* sp. nov., and *P. filamentus* sp. nov., gas vacuolate polar marine bacteria of the Cytophaga-Flavobacterium-Bacteroides group and reclassification of *'Flectobacillus glomeratus'* as *Polaribacter glomeratus* comb. nov. Int J Syst Bacteriol 48:223–235

Green WJ, Angle MP, Chave KE (1988) The geochemistry of Antarctic streams and their role in the evolution of four lakes of the McMurdo Dry Valleys. Geochim Cosmochim Acta 52:1265–1274

Groudieva T, Kambourova M, Yusef H, Royter M, Grote R, Trinks H, Antranikian G (2004) Diversity and cold-active hydrolytic enzymes of culturable bacteria associated with Arctic sea ice, Spitzbergen. Extremophiles 8:475–488

Guibert LM, Loviso CL, Marcos MS, Commendatore MG, Dionisi HM, Lozada M (2012) Alkane biodegradation genes from chronically polluted subantarctic coastal sediments and their shifts in response to oil exposure. Microb Ecol 64:605–616

Hansen AA, Herbert RA, Mikkelsen K et al (2007) Viability, diversity and composition of the bacterial community in a high Arctic permafrost soil from Spitsbergen, Northern Norway. Environ Microbiol 9:2870–2884

Helmke E, Weyland H (2004) Psychrophilic versus psychrotolerant bacteria-occurrence and significance in polar and temperate marine habitats. Cell Mol Biol 50:553–561

Herbert RA, Bell CR (1977) Growth characteristics of an obligately psychrophilic Vibrio sp. Arch Microbiol 113:215–220

Hirsch P, Mevs U, Kroppenstedt RM, Schumann P, Stackebrandt E (2004) Cryptoendolithic actinomycetes from antarctic sandstone rock samples: *Micromonospora endolithica* sp. nov. and two isolates related to *Micromonospora coerulea* Jensen 1932. Syst Appl Microbiol 27:166–174

Huang J, Hoover RB, Swain A, Murdock C, Bej AK (2013) Comparison of the microbial diversity and abundance between the freshwater land-locked lakes of Schirmacher Oasis and the perennially ice-covered Lake Untersee in East Antarctica. In: Proceedings of International Society for Optical Engineering, Instruments, Methods and Mission for Astrobiology (SPIE) 7819, 78190W

Imperio T, Viti C, Marri L (2008) *Alicyclobacillus pohliae* sp. nov., a thermophilic, endospore-forming bacterium isolated from geothermal soil of the north-west slope of Mount Melbourne (Antarctica). Int J Syst Evol Microbiol 58:221–225

Jang GI, Hwang CY, Choi HG, Kang SH, Cho BC (2011) Description of *Spongiibacter borealis* sp. nov., isolated from Arctic seawater, and reclassification of *Melitea salexigens* Urios et al. 2008 as a later heterotypic synonym of *Spongiibacter marinus* Graeber et al. 2008 with emended descriptions of the genus *Spongiibacter* and *Spongiibacter marinus*. Int J Syst Evol Microbiol 61:2895–2900

Jeong SW, Jeong J, Kim J (2015) Simple surface foam application enhances bioremediation of oil-contaminated soil in cold conditions. J Hazard Mater 286C:164–170

Jiang F, Li W, Xiao M, Dai J, Kan W, Chen L, Li W, Fang C, Peng F (2012) *Luteolibacter luojiensis* sp. nov., isolated from Arctic tundra soil, and emended description of the genus *Luteolibacter*. Int J Syst Evol Microbiol 62:2259–2263

Jiang F, Xiao M, Chen L, Kan W, Peng F, Dai J, Chang X, Li W, Fang C (2013) *Huanghella arctica* gen. nov., sp. nov., a bacterium of the family *Cytophagaceae* isolated from Arctic tundra soil. Int J Syst Evol Microbiol 63:696–702

Junge K, Imhoff F, Staley T, Deming JW (2002) Phylogenetic diversity of numerically important Arctic sea-ice bacteria cultured at subzero temperature. Microb Ecol 43:315–328

Junge K, Eicken H, Deming JW (2004) Bacterial activity at –2 to –20 °C in Arctic wintertime sea ice. Appl Environ Microbiol 70:550–557

Jyothsna TSS, Sasikala C, Ramana CV (2008) *Desulfovibrio psychrotolerans* sp. nov., a psychrotolerant and moderately alkaliphilic sulfate-reducing deltaproteobacterium from the Himalayas. Int J Syst Evol Microbiol 58:821–825

Karl DM, Bird DF, Björkman K, Houlihan T, Shackelford R, Tupas L (1999) Microorganisms in the accreted ice of Lake Vostok, Antarctica. Science 286:2144–2147

Karr EA, Sattley WM, Jung DO, Madigan MT, Achenbach LA (2003) Remarkable diversity of phototrophic purple bacteria in a permanently frozen Antarctic lake. Appl Environ Microbiol 69:4910–4914

Karr EA, Sattley WM, Rice MR, Jung DO, Madigan MT, Achenbach LA (2005) Diversity and distribution of sulfate-reducing bacteria in permanently frozen Lake Fryxell, McMurdo Dry Valleys, Antarctica. Appl Environ Microbiol 71:6353–6359

Kawahara H, Iwanaka Y, Higa S, Muryoi N, Sato M, Honda M, Omura H, Obata H (2007) A novel, intracellular antifreeze protein in an antarctic bacterium, *Flavobacterium xanthum*. Cryo Lett 28:39–49

Kelly CD, Layne S (1957) Bacteria found in the air over Canada and the American Arctic. Can J Microbiol 3:447–455

Kelly MD, Lukaschewsky S, Anderson CG (1978) Bacterial flora *of* Antarctic krill and some of their enzymatic properties. J Food Sci 43:1196–1197

Kim HJ, Park S, Lee JM, Park S, Jung W, Kang JS, Joo HM, Seo KW, Kang SH (2008) *Moritella dasanensis* sp. nov., a psychrophilic bacterium isolated from the Arctic ocean. Int J Syst Evol Microbiol 58:817–820

Kim SJ, Park SJ, Yoon DN, Park BJ, Choi BR, Lee DH, Roh Y, Rhee SK (2009) *Marinobacterium maritimum* sp. nov., a marine bacterium isolated from Arctic sediment. Int J Syst Evol Microbiol 59:3030–3034

Kim SJ, Park SJ, Oh YS, Lee SA, Shin KS, Roh DH, Rhee SK (2012) *Shewanella arctica* sp. nov., an iron-reducing bacterium isolated from Arctic marine sediment. Int J Syst Evol Microbiol 62:1128–1133

Kim JG, Park SJ, Quan ZX, Jung MY, Cha IT, Kim SJ, Kim KH, Yang EJ, Kim YN, Lee SH, Rhee SK (2014) Unveiling abundance and distribution of planktonic Bacteria and Archaea in a polynya in Amundsen Sea, Antarctica. Environ Microbiol 16:566–578

Kimes NE, Callaghan AV, Suflita JM, Morris PJ (2014) Microbial transformation of the Deepwater Horizon oil spill-past, present, and future perspectives. Front Microbiol 5:603

Kirchman DL, Cottrell MT, Lovejoy C (2010) The structure of bacterial communities in the western Arctic Ocean as revealed by pyrosequencing of 16S rRNA genes. Environ Microbiol 12:1132–1143

Kishore KH, Begum Z, Pathan AAK, Shivaji S (2009) *Paenibacillus glacialis* sp. nov. isolated from Kafni glacier of Himalayas, India. Int J Syst Evol Microbiol 60:1909–1913

Knittel K, Kuever J, Meyerdierks A, Meinke R, Amann R, Brinkhoff T (2005) *Thiomicrospira arctica* sp. nov. and *Thiomicrospira psychrophila* sp. nov., psychrophilic, obligately chemolithoautotrophic, sulfur-oxidizing bacteria isolated from marine Arctic sediments. Int J Syst Evol Microbiol 55:781–786

Knoblauch C, Sahm K, Jørgensen BB (1999) Psychrophilic sulfate-reducing bacteria isolated from permanently cold arctic marine sediments: description of *Desulfofrigus oceanense* gen. nov., sp. nov., *Desulfofrigus fragile* sp. nov., *Desulfofaba gelida* gen. nov., sp. nov., *Desulfotalea psychrophila* gen. nov., sp. nov. and *Desulfotalea arctica* sp. nov. Int J Syst Bacteriol 49:1631–1643

Kong W, Li W, Romancova I, Prášil O, Morgan-Kiss RM (2014) An integrated study of photochemical function and expression of a key photochemical gene (psbA) in photosynthetic communities of Lake Bonney (McMurdo Dry Valleys, Antarctica). FEMS Microbiol Ecol 89:293–302

Koo H, Ptacek T, Crowley M, Swain AK, Osborne JD, Bej AK, Andersen DT (2014) Draft genome sequence of *Hymenobacter* sp. Strain IS2118, Isolated from a freshwater lake in Schirmacher Oasis, Antarctica, reveals diverse genes for adaptation to cold ecosystems. Genome Announc 2:e00739–14

Kriss AE (1945) Microorganisms of the eastern part of the Arctic Ocean. Mikrobiologiia 14:268–276

Kuhn E, Bellicanta GS, Pellizari VH (2009) New alk genes detected in Antarctic marine sediments. Environ Microbiol 11:669–673

Kuhn E, Ichimura AS, Peng V, Fritsen CH, Trubl G, Doran PT, Murray AE (2014) Brine assemblages of ultrasmall microbial cells within the ice cover of Lake Vida, Antarctica. Appl Environ Microbiol 80:3687–3698

Kumar PA, Sreenivas A, Singh A, Shivaji S (2013a) Draft genome sequence of *Winogradskyella psychrotolerans* RS-3 T, isolated from the marine transect of Kongsfjorden, Ny-Ålesund, Svalbard, Arctic. Genome Announc 1:e00630–13

Kumar PA, Singh A, Sreenivas A, Begum Z, Reddy GSN, Shivaji S (2013b) Draft genome sequence of *Leifsoniarubra* CMS 76R T, isolated from a cyanobacterial mat sample from a pond in Wright Valley, McMurdo, Antarctica. Genome Announc 1:e00633–13

Labrenz M, Tindall BJ, Lawson PA, Collins MD, Schumann P, Hirsch P (2000) *Staleya guttiformis* gen. nov., sp. nov. and *Sulfitobacter brevis* sp. nov., alpha-3-Proteobacteria from hypersaline, heliothermal and meromictic Antarctic Ekho Lake. Int J Syst Evol Microbial 50:303–313

Lama L, Nicolaus B, Calandrelli V, Esposito E, Gambacorta A (1996) Xylanase produced by *Bacillus thermoantarcticus,* a new thermophilic bacillus. Enzyme Eng XIII:284–286

Lanoil B, Skidmore M, Priscu JC, Han S, Foo W, Vogel SW, Tulaczyk S, Engelhardt H (2009) Bacteria beneath the West Antarctic ice sheet. Environ Microbiol 11:609–615

Larkin JM, Stokes JL (1967) Taxonomy of psychrophilic strains of Bacillus. J Bacteriol 94:889–895

Laucks ML, Sengupta A, Junge K, Davis EJ, Swan-son BD (2005) Comparison of psychro-active Arctic marine bacteria and common mesophilic bacteria using surface-enhanced Raman spectroscopy. Appl Spectrosc 10:1222–1228

Lauro FM, Chastain RA, Blankenship LE, Yayanos AA, Bartlett DH (2007) The unique 16S rRNA genes of piezophiles reflect both phylogeny and adaptation. Appl Environ Microbiol 73:838–845

Laybourn-Parry J (1997) The microbial loop in Antarctic Lakes. In: Howards-Williams C, Lyons WB, Hawes I (eds) Ecosystem processes in Antarctic ice-free landscapes. Balkema, Rotterdam/Brookfield, pp 231–240

Lee K, Lee HK, Choi TH, Kim KM, Cho JC (2007) *Granulosicoccaceae* fam. nov., to include *Granulosicoccus antarcticus* gen. nov., sp. nov., a non-phototrophic, obligately aerobic chemoheterotroph in the order Chromatiales, isolated from Antarctic seawater. J Microbiol Biotechnol 17:1483–1490

Lee CK, Barbier BA, Bottos EM, McDonald IR, Cary SC (2012) The inter-valley soil comparative survey: the ecology of dry valley edaphic microbial communities. ISME J 6:1046–1057

Li S, Xiao X, Yin X, Wang F (2006a) Bacterial community along a historic lake sediment core of Ardley Island, west Antarctica. Extremophiles 10:461–467

Li HR, Yu Y, Zeng YX, Chen B, Ren DM (2006b) Phylogenetic analysis of bacterial diversity in Pacific Arctic sediments. Wei Sheng Wu Xue Bao 46:177–183

Li H, Yu Y, Luo W, Zeng Y, Chen B (2009) Bacterial diversity in surface sediments from the Pacific Arctic Ocean. Extremophiles 13:233–246

Li AH, Liu HC, Xin YH, Kim SG, Zhou YG (2013) *Glaciihabitans tibetensis* gen. nov., sp. nov., a psychrotolerant bacterium of the family *Microbacteriaceae*, isolated from glacier ice water. Int J Syst Evol Microbiol 64:579–587

Liebner S, Rublack K, Stuehrmann T, Wagner D (2009) Diversity of aerobic methanotrophic bacteria in a permafrost active layer soil of the Lena Delta, Siberia. Microbiol Ecol 57:25–35

Liu Y, Yao T, Jiao N, Kang S, Zeng Y, Huang S (2006) Microbial community structure in moraine lakes and glacial meltwaters, Mount Everest. FEMS Microbiol Lett 265:98–105

Liu YQ, Yao TD, Jiao NZ, Kang SC, Xu BQ, Zeng YH, Huang SJ, Liu XB (2009) Bacterial diversity in the snow over Tibetan Plateau Glaciers. Extremophiles 13:411–423

Liu Y, Yao T, Jiao N, Tian L, Hu A, Yu W, Li S (2011) Microbial diversity in the snow, a moraine lake and a stream in Himalayan glacier. Extremophiles 15:411–421

Liu JT, Lu XL, Liu XY, Gao Y, Hu B, Jiao BH, Zheng H (2013) Bioactive natural products from the antarctic and arctic organisms. Mini Rev Med Chem 13:617–626

Lo Giudice A, Caruso C, Mangano S, Bruni V, De Domenico M, Michaud L (2012) Marine bacterioplankton diversity and community composition in an antarctic coastal environment. Microb Ecol 63:210–223

Lohan MC, Statham PJ, Peck L (2001) Trace metals in the Antarctic soft-shelled clam *Laternulaelliptica*: implications for metal pollution from Antarctic research stations. Polar Biol 24:808–817

Lysnes K, Thorseth IH, Steinsbu BO, Øvreås L, Torsvik T, Pedersen RB (2004) Microbial community diversity in seafloor basalt from the Arctic spreading ridges. FEMS Microbiol Ecol 50:213–230

Ma Y, Wang L, Shao Z (2006) Pseudomonas, the dominant polycyclic aromatic hydrocarbon-degrading bacteria isolated from Antarctic soils and the role of large plasmids in horizontal gene transfer. Environ Microbiol 8:455–465

Maas EW, Simpson AM, Martin A, Thompson S, Koh EY, Davy SK, Ryan KG, O'Toole RF (2012) Phylogenetic analyses of bacteria in sea ice at Cape Hallett, Antarctica. N Z J Mar Freshw Res 46:3–12

Magalhães CM, Machado A, Frank-Fahle B, Lee CK, Cary SC (2014) The ecological dichotomy of ammonia-oxidizing archaea and bacteria in the hyper-arid soils of the Antarctic Dry Valleys. Front Microbiol 5:515

Maida I, Fondi M, Papaleo MC, Perrin E, Orlandini V, Emiliani G, de Pascale D, Parrilli E, Tutino ML, Michaud L, Lo Giudice A, Romoli R, Bartolucci G, Fani R (2014) Phenotypic and genomic characterization of the Antarctic bacterium *Gillisia* sp. CAL575, a producer of antimicrobial compounds. Extremophiles 18:35–49

Männistö MK, Rawat S, Starovoytov V, Häggblom MM (2012) *Granulicella arctica* sp. nov., *Granulicella mallensis* sp. nov., *Granulicella tundricola* sp. nov. and *Granulicella sapmiensis* sp. nov., novel acidobacteria from tundrasoil. Int J Syst Evol Microbiol 62:2097–2106

Marcos MS, Lozada M, Dionisi HM (2009) Aromatic hydrocarbon degradation genes from chronically polluted subantarctic marine sediments. Lett Appl Microbiol 49:602–608

Margesin R, Schinner F (2001) Biodegradation and bioremediation of hydrocarbons in extreme environments. Appl Microbiol Biotechnol 56:650–663

Margini RA, Castrelos OD (1963) Examenes bacteriologicos de aire, rieve ysuelo de carbo primaruera y Estacion cientifica Ellsworth. Inst Antarcti Argent Pub 76:1–15

Marshall BJ, Ohye DF (1966) *Bacillus macquariensis* sp. nov, a psychrotrophic bacterium from sub-antarctic soil. J Gen Microbiol 44:41–46

Martineau C, Pan Y, Bodrossy L, Yergeau E, Whyte LG, Greer CW (2014) Atmospheric methane oxidizers are present and active in Canadian high Arctic soils. FEMS Microbiol Ecol 89:257–269

Master ER, Mohn WW (1998) Psychrotolerant bacteria Isolated from Arctic Soil that degrade polychlorinated biphenyls at low temperatures. Appl Environ Microbiol 64:4823–4829

Meyer GH, Morrow MB, Wyss O, Berg TE, Littlepage JL (1962) Antarctica: the microbiology of an unfrozen saline pond. Science 138:1103–1104

Michaud L, Caruso C, Mangano S, Interdonato F, Bruni V, Lo Giudice A (2012) Predominance of *Flavobacterium*, *Pseudomonas*, and *Polaromonas* within the prokaryotic community of freshwater shallow lakes in the northern Victoria Land, East Antarctica. FEMS Microbiol Ecol 82:391–404

Mikucki JA, Pearson A, Johnston DT, Turchyn AV, Farquhar J, Schrag DP, Anbar AD, Priscu JC, Lee PA (2009) A contemporary microbially maintained subglacial ferrous "ocean". Science 324:397–400

Miwa T (1975) Clostridia in soil of the Antarctica. Jpn J Med Sci Biol 28:201–213

Møller AK, Barkay T, Abu Al-Soud W, Sørensen SJ, Skov H, Kroer N (2011) Diversity and characterization of mercury-resistant bacteria in snow, freshwater and sea-ice brine from the High Arctic. FEMS Microbiol Ecol 75:390–401

Moreno R, Rojo F (2014) Features of pseudomonads growing at low temperatures: another facet of their versatility. Environ Microbiol Rep 6:417–426

Morita RY (1975) Psychrophilic bacteria. Bacteriol Rev 39:144–167

Muangchinda C, Chavanich S, Viyakarn V, Watanabe K, Imura S, Vangnai AS, Pinyakong O (2014) Abundance and diversity of functional genes involved in the degradation of aromatic hydrocarbons in Antarctic soils and sediments around Syowa Station. Environ Sci Pollut Res 22:4725–4735

Neufeld JD, Mohn WW (2005) Unexpectedly high bacterial diversity in Arctic tundra relative to boreal forest soils, revealed by serial analysis of ribosomal sequence tags. Appl Environ Microbiol 71:5710–5718

Niederberger TD, McDonald IR, Hacker AL, Soo RM, Barrett JE, Wall DH, Cary SC (2008) Microbial community composition in soils of Northern Victoria Land, Antarctica. Environ Microbiol 10:1713–1724

Pady SM, Kelly CD, Polunin N (1948) Arctic aerobiology; preliminary report on fungi and bacteria isolated from the air in 1947. Nature 162:379–381

Peeters K, Hodgson DA, Convey P, Willems A (2011) Culturable diversity of heterotrophic bacteria in Forlidas Pond (Pensacola Mountains) and Lundström Lake (Shackleton Range), Antarctica. Microb Ecol 62:399–413

Perreault NN, Andersen DT, Pollard WH, Greer CW, Whyte LG (2007) Characterization of the prokaryotic diversity in cold saline perennial springs of the Canadian high Arctic. Appl Environ Microbiol 73:1532–1543

Pfiser RM, Burkholder PR (1965) Numerical taxonomy of some bacteria isolated from Antarctic and tropical sea waters. J Bacterial 90:863–872

Pindi PK, Hara Kishore K, Reddy GSN, Shivaji S (2009) Description of *Leifsonia kafniensis* sp. nov. and *Leifsonia antarctica* sp. nov. Int J Syst Evol Microbiol 59:1348–1352

Pirie JHH (1904) First Antarctic voyage of the Scotia. J Bacterial Scot Geogr Mag 20:129–132

Pirie JHH (1912) Notes on Antarctic bacteriology. Rep Sci Res SY Scotia 3:157–168

Poli A, Esposito E, Lama L, Orlando P, Nicolaus G, de Appolonia F, Gambacorta A, Nicolaus B (2006) *Anoxybacillus amylolyticus* sp. nov., a thermophilic amylase producing bacterium isolated from Mount Rittmann (Antarctica). Syst Appl Microbiol 29:300–307

Polunin N, Kelly CD (1952) Arctic aerobiology; fungi and bacteria, etc., caught in the air during flights over the geographical North Pole. Nature 170:314–316

Pöntinen A, Markkula A, Lindström M, Korkeala H (2015) Two-component system histidine kinases involved in growth of *Listeria monocytogenes* EGD-e at low temperatures. Appl Environ Microbiol 81:3994–4004

Powell SM, Ferguson SH, Bowman JP, Snape I (2006) Using real-time PCR to assess changes in the hydrocarbon-degrading microbial community in Antarctic soil during bioremediation. Microb Ecol 52:523–532

Prabagaran SR, Manorama R, Delille D, Shivaji S (2007) Predominance of *Roseobacter, Sulfitobacter, Glaciecola* and *Psychrobacter* in seawater collected off Ushuaia, Argentina, Sub-Antarctica. FEMS Microbiol Ecol 59:342–355

Pradhan S, Srinivas TNR, Pindi PK, Hara Kishore K, Begum Z, Singh PK, Singh AK, Pratibha MS, Yasala AK, Reddy GSN, Shivaji S (2010) Bacterial biodiversity from Roopkund Glacier, Himalayan mountain ranges, India. Extremophiles 14:377–395

Prasad S, Manasa BP, Buddhi S, Pratibha MS, Begum Z, Bandi S, Tirunagari P, Shivaji S (2013) *Arcticibacter svalbardensis* gen. nov., sp. nov., of the family *Sphingobacteriaceae* in the phylum *Bacteroidetes*, isolated from Arctic soil. Int J Syst Evol Microbiol 63:1627–1632

Prasad S, Manasa P, Buddhi S, Tirunagari P, Begum Z, Rajan S, Shivaji S (2014) Diversity and bioprospective potential (cold-active enzymes) of cultivable marine bacteria from the subarctic glacial Fjord, Kongsfjorden. Curr Microbiol 68:233–238

Priest FG, Goodfellow M, Todd C (1988) A numerical classification of the genus Bacillus. J Gen Microbiol 134:1847–1882

Priscu JC, Adams EE, Lyons WB, Voytek MA, Mogk DW, Brown RL, McKay CP, Takacs CD, Welch KA, Wolf CF, Kirshtein JD, Avci R (1999) Geomicrobiology of subglacial ice above Lake Vostok, Antarctica. Science 286:2141–2144

Prüss BM, Francis KP, von Stetten F, Scherer S (1999) Correlation of 16S ribosomal DNA signature sequences with temperature-dependent growth rates of mesophilic and psychrotolerant strains of the *Bacillus cereus* group. J Bacteriol 181:2624–2630

Purcell AM, Mikucki JA, Achberger AM, Alekhina IA, Barbante C, Christner BC, Ghosh D, Michaud AB, Mitchell AC, Priscu JC, Scherer R, Skidmore ML, Vick-Majors TJ, The Wissard Science Team (2014) Microbial sulfur transformations in sediments from Subglacial Lake Whillans. Front Microbiol 5:594

Purusharth RI, Madhuri B, Ray MK (2007) Exoribonuclease R in Pseudomonas syringae is essential for growth at low temperature and plays a novel role in the 3′- end processing of 16 and 5 S ribosomal RNA. J Biol Chem 282:16267–16277

Qu Z, Jiang F, Chang X, Qiu X, Ren L, Fang C, Peng F (2014) *Psychroglaciecola arctica* gen. nov., sp. nov., isolated from Arctic glacial foreland soil. Int J Syst Evol Microbiol 64:1817–1824

Raiger Iustman LJ, Tribelli PM, Ibarra JG, Catone MV, Solar Venero EC, López NI (2015) Genome sequence analysis of *Pseudomonas extremaustralis* provides new insights into environmental adaptability and extreme conditions resistance. Extremophiles 19:207–220

Ray MK, Kumar GS, Shivaji S (1994) Phosphorylation of membrane proteins in response to temperature in an Antarctic *Pseudomonas syringae*. Microbiology 140:3217–3723

Ray MK, Devi KU, Kumar GS, Shivaji S (1992) Extracellular protease from the antarctic yeast Candida humicola. Appl Environ Microbiol 58:1918–1923

Ray MK, Kumar GS, Janiyani K, Kannan K, Jagtap P, Basu MK, Shivaji S (1998) Adaptation to low temperature and regulation of gene expression in Antarctic psychrotrophic bacteria. J Biosci 23:423–435

Reddy GSN, Rajagopalan G, Shivaji S (1994) Thermolabile ribonuclease from Antarctic psychrotropic bacteria: detection of the enzyme in various bacteria and purification from *Pseudomonas fluorescens*. FEMS Microbiol Lett 122:211–216

Reddy GSN, Agarwal RK, MatsumotoGI SS (2000) *Arthrobacter flavus*sp. nov., a psychrotropic bacterium isolated from a pond in Mc Murdo dry valley, Antarctica. Int J Syst Evol Microbiol 50:1553–1561

Reddy GSN, Prakash JSS, Matsumoto GI, Stackebrandt E, Shivaji S (2002a) *Arthrobacter roseus* sp. nov., a psychrotropic bacterium isolated from an Antarctic cyanobacterial mat sample. Int J Syst Evol Microbiol 52:1017–1021

Reddy GSN, Prakash JSS, Vairamani M, Prabhakar S, Matsumoto GI, Shivaji S (2002b) *Planococcus antarcticus* and *Planococcus psychrophilus* spp. nov. isolated from cyanobacterial mat samples collected from ponds in Antarctica. Extremophiles 6:253–261

Reddy GSN, Matsumoto GI, Shivaji S (2003a) *Sporosarcina macmurdoensis* sp. nov. from a cyanobacterial mat samples from a pond in the McMurdo dry valley, Antarctica. Int J Syst Evol Microbiol 53:1363–1367

Reddy GSN, Prakash JSS, Srinivas R, Matsumoto GI, Shivaji S (2003b) *Leifsonia rubra* sp. nov. and *Leifsonia aurea* sp. nov. psychrophiles from a Pond in Antarctica. Int J Syst Evol Microbiol 53:977–984

Reddy GSN, Raghavan PMU, Sarita NB, Prakash JSS, Nagesh N, Delille D, Shivaji S (2003c) *Halomonas glacies* sp. nov. isolated from fast ice of Adelie Land, Antarctica. Extremophiles 7:55–61

Reddy GSN, Prakash JSS, Prabahar V, Matsumoto GI, Stackebrandt E, Shivaji S (2003d) *Kocuria polaris* sp. nov., an orange pigmented psychrotrophic bacterium isolated from an Antarctic cyanobacterial mat sample. Int J Syst Evol Microbiol 53:183–187

Reddy GSN, Matsumoto GI, Shuman P, Stackebrandt E, Shivaji S (2004) Psychrophilic *Pseudomonas* from Antarctica: *Pseudomonasantarctica* sp. nov., *Pseudomonas meridianae* sp. nov. and *Pseudomonas proteolytica* sp. nov. Int J Syst Evol Microbiol 54:713–719

Reddy GSN, Uttam A, Shivaji S (2008) *Bacillus cecembensis* sp. nov., a bacterium isolated from the Pindari glacier of the Himalayan mountain ranges, India. Int J Syst Evol Microbiol 58:2330–2335

Reddy GSN, Pradhan S, Manorama R, Shivaji S (2009a) *Cryobacterium roopkundensis* sp. nov., a psychrophilic bacterium from Roopkund Glacier of the Himalayan mountain ranges. Int J Syst Evol Microbiol 60:866–870

Reddy PVV, Shiva Nageswara Rao SS, Pratibha MS, Sailaja B, Kavya B, Manorama RR, Singh SM, Radha Srinivas TN, Shivaji S (2009b) Bacterial diversity and bioprospecting for cold-active enzymes from culturable bacteria associated with sediment from a melt water stream of Midtre Lovenbreen glacier, an Arctic glacier. Res Microbiol 160:538–546

Reddy GSN, Poorna Manasa B, Singh SK, Shivaji S (2013a) *Paenisporosarcina indica* sp. nov., a psychrophilic bacterium from Pindari Glacier of the Himalayan mountain ranges and reclassification of *Sporosarcina antarctica* Yu et al., 2008 as *Paenisporosarcina antarctica* comb. nov. and emended description of the genus *Paenisporosarcina*. Int J Syst Evol Microbiol IJS/2012/047514

Reddy GSN, Ara S, Singh A, Kumar Pinnaka A, Shivaji S (2013b) Draft genome sequence of *Psychrobacter aquaticus* CMS 56^T, isolated from a cyanobacterial mat sample collected from water bodies in the McMurdo Dry Valley region of Antarctica. Genome Announc 1:e00918–13

Reddy GSN, Sreenivas A, Shivaji S (2014) Draft genome sequence *Cryobacterium roopkundensis* strain RuGl7, isolated from a soil sample in the vicinity of Roopkund Lake, Himalayas, India. Genome Announc 2:e01206–e01214

Ren L, Chang X, Jiang F, Kan W, Qu Z, Qiu X, Fang C, Peng F (2015) *Parablastomonas arctica* gen. nov., sp. nov., isolated from high Arctic glacial till. Int J Syst Evol Microbiol 65:260–266

Ruberto L, Dias R, Lo Balbo A, Vazquez SC, Hernandez EA, Mac Cormack WP (2009) Influence of nutrients addition and bioaugmentation on the hydrocarbon biodegradation of a chronically contaminated Antarctic soil. J Appl Microbiol 106:1101–1110

Ruckert G (1985) *Myxobacteria* from Antarctic soils. Biol Fert Soil 1:215–216

Sahay H, Babu BK, Singh S, Kaushik R, Saxena AK, Arora DK (2013) Cold-active hydrolases producing bacteria from two different sub-glacial Himalayan lakes. J Basic Microbiol 53:703–714

Schütte UM, Abdo Z, Foster J, Ravel J, Bunge J, Solheim B, Forney LJ (2010) Bacterial diversity in a glacier foreland of the high Arctic. Mol Ecol 1:54–66

Sengupta D, Chattopadhyay MK (2013) Metabolism in bacteria at low temperature: a recent report. J Biosci 38:409–12

Shivaji S, Prakash JS (2010) How do bacteria sense and respond to low temperature? Arch Microbiol 192:85–95

Shivaji S, Reddy GSN (2009) Chapter 2. Bacterial biodiversity of Antarctica: conventional polyphasic and rRNA approaches. In: Bej AK, Aislabie J, Atlas RM (eds) Polar microbiology: the ecology, biodiversity and bioremediation potential of microorganisms in extremely cold environments. Taylor & Francis, Boca Raton

Shivaji S, Rao NS, Saisree L, Sheth V, Reddy GSN, Bhargava PM (1988) Isolation and identification of *Micrococcus* roseus and *Planococcus* sp. from Schirmacher Oasis, Antarctica. J Biosci 113:409–414

Shivaji S, Rao NS, Saisree L, Sheth V, Reddy GSN, Bhargava PM (1989a) Isolation and identification of *Pseudomonas* sp. from Schirmacher Oasis, Antarctica. Appl Environ Microbiol 55:767–770

Shivaji S, Rao NS, Saisree L, Reddy GSN, Seshu Kumar G, Bhargava PM (1989b) Isolates of *Arthrobacter* from the soils of Schirmacher Oasis, Antarctica. Polar Biol 10:225–229

Shivaji S, Reddy GSN, Prasad RA, Kutty R, Ravenschlag K (2004) Bacterial diversity of a soil sample from Schirmacher Oasis, Antarctica. Cell Mol Biol 50:525–536

Shivaji S, Chaturvedi P, Reddy GSN, Suresh K (2005) *Pedobacter himalayensis* sp. nov. from Hamta glacier located in the Himalayan mountain ranges of India. Int J Syst Evol Microbiol 55:1083–1088

Shivaji S, Bhadra B, Rao RS, Pradhan S (2008) *Rhodotorula himalayensis* sp. nov., a novel psychrophilic yeast isolated from Roopkund lake of the Himalayan mountain ranges, India. Extremophiles 12:375–381

Shivaji S, Kumari K, Kishore KH, Pindi PK, Rao PS, Radha Srinivas TN, Asthana R, Ravindra R (2011a) Vertical distribution of bacteria in a lake sediment from Antarctica by culture-independent and culture-dependent approaches. Res Microbiol 162:191–203

Shivaji S, Pratibha MS, Sailaja B, Hara Kishore K, Singh AK, Begum Z, Anarasi U, Prabagaran SR, Reddy GSN, Srinivas TN (2011b) Bacterial diversity of soil in the vicinity of Pindari glacier, Himalayan mountain ranges, India, using culturable bacteria and soil 16S rRNA gene clones. Extremophiles 15:1–22

Shivaji S, Reddy PV, Rao SS, Begum Z, Manasa P, Srinivas TN (2012a) *Cyclobacterium qasimii* sp. nov., a psychrotolerant bacterium isolated from Arctic marine sediment. Int J Syst Evol Microbiol 62:2133–2139

Shivaji S, Ara S, Singh SK, Bandi S, Singh A, Pinnaka AK (2012b) Draft genome sequence of *Bacillus isronensis* strain B3W22 isolated from the upper atmosphere. J Bacteriol 194:6624–6625

Shivaji S, Ara S, Singh A, Kumar Pinnaka A (2013a) Draft genome sequence of *Cyclobacterium qasimii* M12-11B T, isolated from an Arctic marine sediment. Genome Announc 1:e00642–13

Shivaji S, Ara S, Prasad S, Manasa BP, Begum Z, Singh A, Kumar Pinnaka A (2013b) Draft genome sequence of *Arcticibacter svalbardensis* MN12-7 T, a member of the family Sphingobacteriaceae isolated from an Arctic soil sample. Genome Announc 1:e00484–13

Shivaji S, Begum Z, Shiva Nageswara Rao SS, Vishnu Vardhan Reddy PV, Manasa P, Sailaja B, Prathiba MS, Thamban M, Krishnan KP, Singh SM, Srinivas TN (2013c) Antarctic ice core samples: culturable bacterial diversity. Res Microbiol 164:70–82

Shukor MY, Hassan NA, Jusoh AZ, Perumal N, Shamaan NA, MacCormack WP, Syed MA (2009) Isolation and characterization of a *Pseudomonas* diesel-degrading strain from Antarctica. J Environ Biol 30:1–6

Singh AK, Shivaji S (2010) A cold-active and a heat-labile t-RNA modification GTPase from a psychrophilic bacterium *Pseudomonas syringae* (Lz4W). Res Microbiol 161:46–50

Singh AK, Pindi PK, Dube S, Sundareswaran VR, Shivaji S (2009) In the psychrophilic *Pseudomonas syringae*, trmE is important for low temperature growth. Appl Environ Microbiol 75:4419–4426

Singh P, Singh SM, Dhakephalkar P (2014) Diversity, cold active enzymes and adaptation strategies of bacteria inhabiting glacier cryoconite holes of High Arctic. Extremophiles 18:229–242

Sinha AK, Pavankumar TL, Kamisetty S, Mittal P, Ray MK (2013) Replication arrest is a major threat to growth at low temperature in Antarctic *Pseudomonas syringae* Lz4W. Mol Microbiol 89:792–810

Sjöling S, Cowan DA (2003) High 16S rDNA bacterial diversity in glacial meltwater lake sediment, Bratina Island, Antarctica. Extremophiles 7:275–282

Smith RC, Prezelin, BB, Baker KS, Bidigare RR, Boucher NP, Coley T, Karentz D, MacIntyre S, Matlick HA, Menzies D, Ondrusek M, Wan Z, Waters KJ (1992) Ozone depletion: ultraviolet radiation and phytoplankton biology in Antarctic waters. Science 255:952–959

Søgaard DH, Kristensen M, Rysgaard S, Glud RN, Hansen PJ, Hilligsøe KM (2010) Autotrophic and heterotrophic activity in Arctic first-year sea ice: Seasonal study from Malene Bight, SW Greenland. Mar Ecol Prog Ser 419:31–45

Söller R, Hirsch P, Blohm D, Labrenz M (2000) Differentiation of newly described antarctic bacterial isolates related to Roseobacter species based on 16S-23S rDNA internal transcribed spacer sequences. Int J Syst Evol Microbiol 50:909–915

Spring S, Merkhoffer B, Weiss N, Kroppenstedt RM, Hippe H, Stackebrandt E (2003) Characterization of novel psychrophilic clostridia from an Antarctic microbial mat: description of *Clostridium frigoris* sp. nov., *Clostridium lacusfryxellense* sp. nov., *Clostridium bowmanii* sp. nov., and *Clostridium psychrophilum* sp. nov., and reclassification of *Clostridium laramiense* as *Clostridium estertheticum* subsp. *laramiense* subsp. nov. Int J Syst Evol Microbiol 53:1019–1029

Sreenivas A, Reddy GSN, Shivaji S (2014) Draft genome sequence of a psychrophilic bacterium *Sphingomonas antarcticum* 4BY, isolated from the soils of Schirmacher Oasis, Antarctica. Genome Announc 2:e00696–14

Srinivas TN, Nageswara Rao SS, Vishnu Vardhan Reddy P, Pratibha MS, Sailaja B, Kavya B, Hara Kishore K, Begum Z, Singh SM, Shivaji S (2009) Bacterial diversity and bioprospecting for cold-active lipases, amylases and proteases, from culturable bacteria of kongsfjorden and Ny-alesund, Svalbard, Arctic. Curr Microbiol 59:537–547

Srinivas TN, Singh SM, Pradhan S, Pratibha MS, Kishore KH, Singh AK, Begum Z, Prabagaran SR, Reddy GSN, Shivaji S (2011) Comparison of bacterial diversity in proglacial soil from Kafni Glacier, Himalayan Mountain ranges, India, with the bacterial diversity of other glaciers in the world. Extremophiles 15:673–690

Staley JT, Gosink JJ (1999) Poles apart: biodiversity and biogeography of sea ice bacteria. Annu Rev Microbiol 53:189–215

Steinsbu BO, Tindall BJ, Torsvik VL, Thorseth IH, Daae FL, Pedersen RB (2011) *Rhabdothermus arcticus* gen. nov., sp. nov., a member of the family Thermaceae isolated from a hydrothermal vent chimney in the Soria Moria vent field on the Arctic Mid-Ocean Ridge. Int J Syst Evol Microbiol 61:2197–2204

Steven B, Briggs G, McKay CP, Pollard WH, Greer CW, Whyte LG (2007) Characterization of the microbial diversity in a permafrost sample from the Canadian high Arctic using culture-dependent and culture-independent methods. FEMS Microbiol Ecol 59:513–523

Steven B, Pollard WH, Greer CW, Whyte LG (2008a) Microbial diversity and activity through a permafrost/ground ice core profile from the Canadian high Arctic. Environ Microbiol 10:3388–3403

Steven B, Chen MQ, Greer CW, Whyte LG, Niederberger TD (2008b) *Tumebacillus permanentifrigoris* gen. nov., sp. nov., an aerobic, spore-forming bacterium isolated from Canadian high Arctic permafrost. Int J Syst Evol Microbiol 58:1497–1501

Steven B, Lionard M, Kuske CR, Vincent WF (2013) High bacterial diversity of biological soil crusts in water tracks over permafrost in the high arctic polar desert. PLoS One 8, e71489

Stibal M, Hasan F, Wadham JL, Sharp MJ, Anesio AM (2012) Prokaryotic diversity in sediments beneath two polar glaciers with contrasting organic carbon substrates. Extremophiles 16:255–265

Stokes JL, Reymond ML (1966) Quantitative ecology of psychrophilic microorganisms. Appl Microbiol 14:74–78

Storesund JE, Øvreås L (2013) Diversity of Planctomycetes in iron-hydroxide deposits from the Arctic Mid Ocean Ridge (AMOR) and description of *Bythopirellula goksoyri* gen. nov., sp. nov., a novel Planctomycete from deep sea iron-hydroxide deposits. Antonie Van Leeuwenhoek 104:569–584

Sundareswaran VR, Singh AK, Dube S, Shivaji S (2010) Aspartate aminotransferase is involved in cold adaptation in psychrophilic *Pseudomonas syringae*. Arch Microbiol 192:663–672

Suyal DC, Yadav A, Shouche Y, Goel R (2014) Differential proteomics in response to low temperature diazotrophy of Himalayan psychrophilic nitrogen fixing Pseudomonas migulae S10724 strain. Curr Microbiol 68:543–550

Takii S, Kondal T, Hiraishi A, Matsumoto GI, Kawano T, Torii T (1986) Vertical distribution in and isolation of bacteria from Lake Vanda: an Antarctic lake. Hydrobiologia 135:15–21

Tamura T, Ishida Y, Otoguro M, Yamamura H, Hayakawa M, Suzuki K (2010) *Angustibacter luteus* gen. nov., sp. nov., isolated from subarctic forest soil. Int J Syst Evol Microbiol 60:2441–2445

Tanner AC (1985) The role of bacteria in the cycling of nutrients within the maritime Antarctic environment. In: Siegfried WR, Condy PR, Laws RM (eds) Antarctic nutrient cycles and food webs. Springer, Berlin, pp 123–127

Tanner AC, Herbert RA (1981) Nutrient regeneration in Antarctic marine sediments. Kiel Meeresforsch Sanderh 5:390–395

Tiao G, Lee CK, McDonald IR, Cowan DA, Cary SC (2012) Rapid microbial response to the presence of an ancient relic in the Antarctic Dry Valleys. Nat Commun 3:660

Timmis KN, McGenity TJ, Meer JR, deLorenzo V (eds) (2010) Handbook of hydrocarbon and lipid microbiology. Springer, Berlin. doi:10.1007/978-3-540-77587-4

Tsiklinsky M (1908) La flore microbieene dans les regions due pole Sud. In: Expedition Antarctique Francais 1903–1905. 3:1–33

Tsyganov VA (1970) Detection and morphological cultural characteristics of *Actinomycetes* from the antarctic. Mikrobiologia 39:821–826

van Dorst J, Bissett A, Palmer AS, Brown M, Snape I, Stark JS, Raymond B, McKinlay J, Ji M, Winsley T, Ferrari BC (2014) Community fingerprinting in a sequencing world. FEMS Microbiol Ecol 89:316–330

Van Trappen S, Mergaert J, Van Eygen S, Dawyndt P, Cnockaert MC, Swings J (2002) Diversity of 746 heterotrophic bacteria isolated from microbial mats from ten Antarctic lakes. Syst Appl Microbiol 25:603–610

Van Trappen S, Tan TL, Yang J, Mergaert J, Swings J (2004) *Glaciecola polaris* sp. nov., a novel budding and prosthecate bacterium from the Arctic Ocean, and emended description of the genus Glaciecola. Int J Syst Evol Microbiol 54:1765–1771

Vandieken V, Mussmann M, Niemann H, Jørgensen BB (2006) *Desulfuromonas svalbardensis* sp. nov. and *Desulfuromusa ferrireducens* sp. nov., psychrophilic, Fe(III)-reducing bacteria isolated from Arctic sediments, Svalbard. Int J Syst Evol Microbiol 56:1133–1139

Vincent CF (1988) Microbial ecosystems of Antarctica. Cambridge University Press, Cambridge, p 303

Voytek MA, Ward BB (1995) Detection of ammonium-oxidizing bacteria of the beta-subclass of the class Proteobacteria in aquatic samples with the PCR. Appl Environ Microbiol 61:1444–1450

Wartiainen I, Hestnes AG, Svenning MM (2003) Methanotrophic diversity in high Arctic wetlands on the islands of Svalbard (Norway) – denaturing gradient gel electrophoresis analysis of soil DNA and enrichment cultures. Can J Microbiol 49:602–612

Watanabe T, Kojima H, Takano Y, Fukui M (2013) Diversity of sulfur-cycle prokaryotes in freshwater lake sediments investigated using aprA as the functional marker gene. Syst Appl Microbiol 36:436–443

Wery N, Gerike U, Sharman A, Chaudhuri JB, Hough DW, Danson MJ (2003) Use of a packed-column bioreactor for isolation of diverse protease-producing bacteria from antarctic soil. Appl Environ Microbiol 69:1457–1464

Wilhelm RC, Niederberger TD, Greer C, Whyte LG (2011) Microbial diversity of active layer and permafrost in an acidic wetland from the Canadian High Arctic. Can J Microbiol 57:303–315

Wilkins D, Yau S, Williams TJ, Allen MA, Brown MV, DeMaere MZ, Lauro FM, Cavicchioli R (2013) Key microbial drivers in Antarctic aquatic environments. FEMS Microbiol Rev 37:303–335

Xiao X, Yin X, Lin J, Sun L, You Z, Wang P, Wang F (2005) Chitinase genes in lake sediments of Ardley Island, Antarctica. Appl Environ Microbiol 71:7904–7909

Yadav AN, Sachan SG, Verma P, Tyagi SP, Kaushik R, Saxena AK (2015) Culturable diversity and functional annotation of psychrotrophic bacteria from cold desert of Leh Ladakh (India). World J Microbiol Biotechnol 31:95–108

Yau S, Lauro FM, Williams TJ, Demaere MZ, Brown MV, Rich J, Gibson JA, Cavicchioli R (2013) Metagenomic insights into strategies of carbon conservation and unusual sulfur biogeochemistry in a hypersaline Antarctic lake. ISME J 7:1944–1961

Yergeau E, Newsham KK, Pearce DA, Kowalchuk GA (2007a) Patterns of bacterial diversity across a range of Antarctic terrestrial habitats. Environ Microbiol 9:2670–2682

Yergeau E, Bokhorst S, Huiskes AH, Boschker HT, Aerts R, Kowalchuk GA (2007b) Size and structure of bacterial, fungal and nematode communities along an Antarctic environmental gradient. FEMS Microbiol Ecol 59:436–451

Yi H, Chun J (2006) *Flavobacterium weaverense* sp. nov. and *Flavobacterium segetis* sp. nov., novel psychrophiles isolated from the Antarctic. Int J Syst Evol Microbiol 56:1239–1244

Yu Y, Li HR, Chen B, Zeng YX, He JF (2006) Phylogenetic diversity and cold-adaptive hydrolytic enzymes of culturable psychrophilic bacteria associated with sea ice from high latitude ocean, Arctic. Wei Sheng Wu Xue Bao 46:184–190

Yu Y, Li HR, Zeng YX, Chen B (2011) Bacterial diversity and bioprospecting for cold-active hydrolytic enzymes from culturable bacteria associated with sediment from Nella Fjord, Eastern Antarctica. Mar Drugs 9:184–195

Zhang DC, Yu Y, Chen B, Wang HX, Liu HC, Dong XZ, Zhou PJ (2006) *Glaciecola psychrophila* sp. nov., a novel psychrophilic bacterium isolated from the Arctic. Int J Syst Evol Microbiol 56:2867–2869

Zhang DC, Li HR, Xin YH, Liu HC, Chen B, Chi ZM, Zhou PJ, Yu Y (2008) *Marinomonas arctica* sp. nov., a psychrotolerant bacterium isolated from the Arctic. Int J Syst Evol Microbiol 58:1715–1718

Zhang G, Cao T, Ying J, Yang Y, Ma L (2014) Diversity and novelty of actinobacteria in Arctic marine sediments. A Van Leeuw 105:743–754

Zhao J, Yang N, Zeng R (2008) Phylogenetic analysis of type I polyketide synthase and nonribosomal peptide synthetase genes in Antarctic sediment. Extremophiles 12:97–105

Zheng Y, Yang W, Sun X, Wang SP, Rui YC, Luo CY, Guo LD (2012) Methanotrophic community structure and activity under warming and grazing of alpine meadow on the Tibetan Plateau Appl Microbiol Biotechnol. doi:10.1007/s00253-011-3535-5

Zhou J, Davey ME, Figueras JB, Rivkina E, Gilichinsky D, Tiedje JM (1997) Phylogenetic diversity of a bacterial community determined from Siberian tundra soil DNA. Microbiology 143:3913–3919

Chapter 3
Antimicrobial Potential of Cold-Adapted Bacteria and Fungi from Polar Regions

Angelina Lo Giudice and Renato Fani

1 Introduction

The Earth's biosphere is predominantly cold in time (about 85 % of the year) being exposed to temperatures below 5 °C. Polar Regions constitute about 14 % of cold habitats on Earth. Such low-temperature environments often undergo a combination of environmental stresses including desiccation, nutrients limitation, high salinity, adverse solar radiation and low biochemical activity (Pearce 2012). Even though such harsh conditions preclude life in most of its forms, cold habitats have been successfully colonized by numerous organisms, especially by microorganisms that predominate over other organisms in terms of both biodiversity and biomass (Feller and Gerday 2003; Margesin 2007; Pearce 2012). On the basis of their cardinal temperatures, cold-adapted microorganisms are frequently distinguished in psychrophilic and psychrotolerant (or cold-loving and cold-tolerant, respectively) (Morita 1975). By definition, the optimal growth temperature of psychrophiles is ≤ 15 °C, whereas they are not able to grow above 20 °C. On the other side, psychrotolerants can grow over a wide temperature range with the fastest growth rates being above 20 °C. Accordingly, the heat-sensitive true psychrophilic microorganisms inhabit permanently cold habitats, whereas psychrotolerant are overrepresented in

A. Lo Giudice (✉)
Institute for Coastal Marine Environment, National Research Council (IAMC-CNR), Spianata San Raineri 86, I-98124 Messina, Italy

Department of Biological and Environmental Sciences, University of Messina, Viale F. Stagno d'Alcontrès 31, I-98166 Messina, Italy
e-mail: angelina.logiudice@iamc.cnr.it

R. Fani
Department of Biology, University of Florence, Via Madonna del Piano 6, 50019 Sesto Fiorentino, Firenze, Italy

P.H. Rampelotto (ed.), *Biotechnology of Extremophiles*, Grand Challenges in Biology and Biotechnology 1, DOI 10.1007/978-3-319-13521-2_3

environments undergoing seasonal or diurnal thermal fluctuations (Margesin 2007). In global terms, psychrotolerants exhibit a much wider distribution than psychrophiles (Pearce 2012).

Since cold-adapted microorganisms have been subjected to a number of environmental stresses on a long timescale, they have evolved a variety of structural and physiological modifications to ensure survival in restrictive environmental conditions (Pearce 2012). These include the production of cold-active enzymes with a more flexible 3D structure at low temperature, cold-acclimation (CAPs) and cold-shock proteins (CSPs), the incorporation of high amounts of unsaturated fatty acids and carotenoids in cell membranes to maintain optimum fluidity and permeability, and the synthesis of cryoprotective substances (Margesin et al. 2007; Russell 2008). In addition to cellular modifications, the recently assessed cold-active antagonistic properties of cold-adapted microorganisms may reduce the presence of competitive microorganisms thus contributing to the microbial adaptation to permanently low temperatures (Lo Giudice et al. 2007a; Mangano et al. 2009; Prasad et al. 2011; Bell et al. 2013). On the other side, such capability has highlighted the possibility to use cold-adapted microorganisms as a novel source of industrial exploitable antimicrobial compounds. Thus, many different, complex and sophisticated survival strategies, which are quite relevant for the ecology of cold-adapted microorganisms, might render them valuable resources also for biotechnological purposes (Cavicchioli et al. 2002).

2 Antibiotic Potential of Cold-Adapted Microorganisms from Polar Regions

Several factors such as the increasing global resistance to existing antibiotics, the development of multiple-drug-resistant pathogens and the emergence of new infections constitute a pressing public health problem. Moreover, in recent decades there has been a dearth of new classes of discovered antibiotics. In attempts to overcome this emergency, the discovery of new and efficient antimicrobials has become of great interest for natural product chemistry. Historically, most bioactive microbial products have been obtained from actinomycetes and filamentous fungi and mainly from terrestrial habitats (Biondi et al. 2008). Recently, the exploration of unusual and underexplored sources of medically useful substances and the screening of less exploited microbial groups (e.g. cyanobacteria) have been recognized as promising tools for the discovery of new natural drugs. Particularly interesting from this viewpoint are microorganisms from extreme environments that produce biomolecules under unusual conditions, thus representing a valuable source of novel metabolites, including antimicrobial compounds with unique structures and specific biological activity (Hemala et al. 2014). For these reasons, cold-adapted microbial producers (heterotrophic and autotrophic bacteria and fungi) of antimicrobial compounds (mainly antibacterial and antifungal) have been isolated from various aquatic and terrestrial environments in the Arctic and Antarctica. Several studies have been based on screening tests targeting human pathogens. In some cases, the microbial inhibitor has been extracted and (partially o fully) characterized. The versatile antimicrobial potential of microorganisms from Polar Regions will be discussed in the following sections.

2.1 Terrestrial Environments

Soil has been and is still the most exploited ecological niche for the discovery of useful natural bioproducts. Researches have been mainly addressed to the Actinobacteria and fungi, which are widespread in such habitat and able to produce several and different useful secondary metabolites and compounds exhibiting different biological properties. An overview of bacteria and fungi with antimicrobial potential from Polar soils is reported in this section.

2.1.1 Actinobacteria

Several bioactivity screenings on bacteria from Polar soils have been focused on the Actinobacteria (Moncheva et al. 2002; Nedialkova and Naidenova 2005; Gesheva 2010; Lee et al. 2012a; Pan et al. 2013). Such class of bacteria has represented the most fruitful source of antibiotics for decades; the largest fraction (about 80 %) of the new discovered antibiotics derived from *Streptomyces* species (order Actinomycetales), which possess a biosynthetic capacity that remains (at least so far) without rivals in the microbial world (Bérdy 2005; Bull and Stach 2007; Manivasagan et al. 2014).

However, nowadays, the discover of new commercially useful secondary metabolites from common streptomycetes is becoming increasingly difficult, thus emphasizing the need for isolating and testing novel members in this genus (Lyutskanova et al. 2009). Streptomycetes from underexplored habitats, including Polar soils, represent a rich source of novel bioactive compounds, as highlighted by results discussed below. A list of active actinobacterial isolates, including *Streptomyces* spp., is shown in Table 3.1.

Pan et al. (2013) selected 46 Antarctic actinobacterial isolates closely related to the psychrotolerant *Streptomyces beijiangensis* (Signy Island, South Orkney Islands) for their antibacterial activity against *Proteus vulgaris* and *Staphylococcus aureus*, whereas no activity was observed against *Escherichia coli* (only *Streptomyces* sp. PSY097 is reported in Table 3.1 as representative of the 46 *Streptomyces beijiangensis* related isolates). Interestingly, all strains were found to contain the nonribosomal peptide synthetase (NRPS) genes. As reported by Gesheva (2010), among actinomycetes that were able to synthesize extracellular biologically active products the methanol extract from *Streptomyces* sp. 5 biomass showed an antibiotic activity versus both Gram-positive bacteria and phytopathogenic fungi. Another psychrophilic

Table 3.1 Active actinobacterial isolates from Polar soils

	Isolate(s)	Location	Active against[a]	Main features of bioactive molecules	Reference
Antarctic	*Arthrobacter* sp. HPG8	East Antarctica	**7, 14, 15, 21, 25**	Proteinaceous nature	O'Brien et al. (2004)
	Arthrobacter sp. HPH17	East Antarctica	**7, 14, 25**	Proteinaceous nature	O'Brien et al. (2004)
	Arthrobacter sp. 1	East Antarctica	**5, 9, 11, 23, 25**		Gesheva (2010)
	Arthrobacter sp. 9	East Antarctica	**5, 23, 25**		Gesheva (2010)
	Brevibacterium spp. BV2, BV34 and BV35	Barrientos Island	**8**		Lee et al. (2012a)
	Brevibacterium sp. BV37	Barrientos Island	**17, 25**		Lee et al. (2012a)
	Demetria spp. DT40 and DT41	Barrientos Island	**25**		Lee et al. (2012a)
	Gordonia spp. G3 and G48	Barrientos Island	**8, 25**		Lee et al. (2012a)
	Janibacter sp. JB26	Barrientos Island	**17**		Lee et al. (2012a)
	Kocuria sp. KC21	Barrientos Island	**17, 25**		Lee et al. (2012a)
	Lapillicoccus sp. LC31	Barrientos Island	**8**		Lee et al. (2012a)
	Micromonospora sp. 18	East Antarctica	**5, 23, 25**		Gesheva (2010)
	Micromonospora sp. MM6	Barrientos Island	**17**		Lee et al. (2012a)
	Micromonospora sp. MM32	Barrientos Island	**20**		Lee et al. (2012a)
	Nocardioides sp. ND52	Barrientos Island	**8**		Lee et al. (2012a)
	Nocardioides sp. A-1	East Antarctica	**5, 16, 24, 27**	Glycolipidic and/or lipopeptidic nature	Gesheva and Vasileva-Tonkova (2012)

	Rhodococcus sp. 2	East Antarctica	**23**		Gesheva (2010)
	Rhodococcus sp. RC56	Barrientos Island	**25**		Lee et al. (2012a)
	Streptomyces sp. PSY097	Signy Island	**19, 25**		Pan et al. (2013)
	Streptomyces sp. NTK 97	Terra Nova Bay	**5, 25**	Frigocyclinone	Bruntner et al. (2005)
	Streptomyces flavovirens 6[7]	Livingston Island			Ivanova et al. (2002)
	Streptomyces sp. 5	East Antarctica	**5, 9, 11, 23, 25**	Actinomycin	Gesheva (2010)
	Streptomyces sp. 8	East Antarctica	**2, 5, 6, 9, 10, 11, 13, 18, 22, 23, 25**	Non-polyenic macrolide antibiotic, azalomycin B, and nigericin	Gesheva (2009)
Arctic	*Streptomyces* spp. SB9, SB72 and SB81	Svalbard Islands	**1, 3, 4, 5, 9, 12, 18, 22, 23, 25, 26**		Lyutskanova et al. (2009)
	Streptomyces spp. SB33 and SB47	Svalbard Islands	**1, 3, 4, 5, 9, 18, 22, 23, 25, 26**		Lyutskanova et al. (2009)

[a]Target strains: **1**, *Acinetobacter johnsonii*; **2**, *Aspergillus niger*; **3**, *Bacillus megatherium*; **4**, *B. mycoides*; **5**, *B. subtilis*; **6**, *Botrytis cinerea*; **7**, *Brochothrix thermosphacta*; **8**, *Candida albicans*; **9**, *C. tropicalis*; **10**, *C. utilis*; **11**, *Cladosporium cladosporioides*; **12**, *Escherichia coli*; **13**, *Fusarium oxysporum*; **14**, *Listeria innocua*; **15**, *L. monocytogenes*; **16**, *Micrococcus* sp.; **17**, MRSA; **18**, *Penicillium chrysogenum*; **19**, *Proteus vulgaris*; **20**, *Pseudomonas aeruginosa*; **21**, *P. fragi*; **22**, *Saccharomyces cerevisiae*; **23**, *Sarcina lutea*; **24**, *Staphylococcus* sp.; **25**, *S. aureus*; **26**, *Trichosporon cutaneum*; **27**, *Xanthomonas oryzae*

Streptomyces (strain 8) was able to inhibit the growth of Gram-positive bacteria, yeasts, and phytopathogenic fungi (Gesheva 2009). Antibacterial or/and antifungal activity (against *Bacillus subtilis*, *Candida tropicalis* and *Cladosporium cladosporioides*) of *Streptomyces* spp. were further reported by Gesheva and Negoita (2012) for *Streptomyces* spp. 10 and 21 isolated from soils of Haswell Island, Antarctica, but no details on their individual antimicrobial spectrum were reported.

From a panel of 91 psychrotolerant streptomycete strains isolated from the permafrost soils in Spitsbergen (Arctic Ocean) five ones exhibited a strong antimicrobial activity toward Gram-positive and Gram-negative bacteria, yeasts and fungi (Lyutskanova et al. 2009). The thin layer chromatography (TLC) profiles of their cell-free supernatants showed antibiotic complexes consisting of three major compounds, whose highest concentration was found in the supernatant obtained from cultures of *Streptomyces* sp. SB9. No biochemical characterization was reported.

Active actinobacterial isolates other than *Streptomyces* spp. have been also isolated from Polar soils. The methanol extract from the *Micromonospora* sp. 18 biomass showed a good antibacterial activity (Gesheva 2010). Actinobacterial strains from different locations of Barrientos Island, Antarctica, were selected by Lee and coworkers (2012) for their ability to produce secondary metabolites with antimicrobial and antifungal activities. Four screening models were used allowing identifying seven isolates that were active against *Candida albicans*, seven isolates versus *S. aureus*, four isolates toward methicillin-resistant *S. aureus* (MRSA) and one isolates against *Pseudomonas aeruginosa*. The most bioactive genus was *Brevibacterium* with four bioactive isolates. The relationship between taxonomic and metabolic diversity of bacteria was highlighted. Five isolates (within the genera *Demetria*, *Nocardioides*, *Lapillicoccus* and *Rhodococcus*) might be assigned to new genus or species, as they were separated from their respective type strains phylogenetic neighbors by sequence similarities of their 16S rRNA genes, strongly suggesting the possibility of discovering novel strains with antimicrobial activity.

Gesheva and Vasileva-Tonkova (2012) tested the antimicrobial activity of cell-free supernatants of *Nocardioides* sp. A-1 cultures in mineral salts medium supplemented with different carbon sources. These latter favored the production of a broad spectrum compounds with antimicrobial activity against Gram-positive and Gram-negative bacteria, especially *S. aureus* and *Xanthomonas oryzae*. A preliminary analysis by TLC showed that compounds with antimicrobial activity were mainly glycolipids and/or lipopeptides depending on the used carbon source.

2.1.2 Other Bacteria

Reports on antimicrobial producing heterotrophic bacteria, other than the Actinobacteria, from Polar soils are reported in Table 3.2. A first study was carried out by O'Brien et al. (2004) by screening 4496 bacterial isolates from East Antarctica soils for the production of cold-active antimicrobials compounds potentially useful in chilled-food preservation. Four of the inhibitor–producers (0.29 % of total isolates) were affiliated to the genera *Arthrobacter*, *Planococcus* and *Pseudomonas*.

Table 3.2 Active cyanobacteria and heterotrophic bacteria (other than Actinobacteria) from Antarctic soils

Phylum or class	Isolate	Location	Active against[a]	Notes	Reference
Alphaproteobacteria	*Bradyrhizobium* sp. BR45	Barrientos Island	**5**		Lee et al. (2012b)
	Bradyrhizobium sp. BR42	Barrientos Island	**5**		Lee et al. (2012b)
	Bradyrhizobium sp. BR62	Barrientos Island	**5**		Lee et al. (2012b)
	Bradyrhizobium sp. BR65	Barrientos Island	**5**		Lee et al. (2012b)
	Bradyrhizobium sp. BR82	Barrientos Island	**5**		Lee et al. (2012b)
	Bradyrhizobium sp. BR88	Barrientos Island	**5**		Lee et al. (2012b)
	Bradyrhizobium sp. BR96	Barrientos Island	**5**		Lee et al. (2012b)
	Bradyrhizobium sp. BR100	Barrientos Island	**5**		Lee et al. (2012b)
	Bradyrhizobium sp. BR105	Barrientos Island	**5**		Lee et al. (2012b)
	Methylobacterium sp. MB63	Barrientos Island	**18**		Lee et al. (2012b)
	Methylobacterium sp. MB104	Barrientos Island	**5**		Lee et al. (2012b)
	Methylobacterium sp. MB20	Barrientos Island	**5**		Lee et al. (2012b)
	Paracoccus sp. PC101	Barrientos Island	**5**		Lee et al. (2012b)
	Sphingomonas sp. SM14	Barrientos Island	**5**		Lee et al. (2012b)
Betaproteobacteria	*Janthinobacterium* sp. SMN33.6	Fildes Peninsula	**1, 10, 11, 14, 18**		Asencio et al. (2009)
Gammaproteobacteria	*Pseudomonas* sp. CrCD21	East Antarctica	**4, 12, 19**	Proteinaceous	O'Brien et al. (2004)
	Pseudomonas sp. CG21	King George Island	**9, 10**	Proteinaceous	Wong et al. (2011)
	Pseudomonas sp. MTC3	King George Island	**9, 10, 11, 16**	Proteinaceous	Wong et al. (2011)
	Pseudomonas sp. WEK1	King George Island	**20**		Wong et al. (2011)
	Pseudomonas sp. WEA1	King George Island	**20**		Wong et al. (2011)
	Pseudomonas sp. MA2	King George Island	**20**		Wong et al. (2011)
Bacteroidetes	*Pedobacter* sp. BG5	King George Island	**2, 9, 10, 11, 17**	Proteinaceous	Wong et al. (2011)
Firmicutes	*Planococcus* sp. CHF8	East Antarctica	**4, 12, 13**	Proteinaceous	O'Brien et al. (2004)
	Enterococcus sp. APR 210	Schirmacher Oasis	**5, 6, 11, 12, 13**	Bacteriocin-like	Shekh et al. (2011)
Cyanobacteria	*Fischerella* sp.	Haswell Island	**3, 7, 8**		Gesheva and Negoita (2012)

[a]Target strains: **1**, *Acinetobacter baumannii*; **2**, *Bacillus cereus*; **3**, *B. subtilis*; **4**, *Brochothrix thermosphacta*; **5**, *Candida albicans*; **6**, *C. krusei*; **7**, *C. tropicalis*; **8**, *Cladosporium cladosporioides*; **9**, *Enterobacter cloacae*; **10**, *Escherichia coli*; **11**, *Klebsiella pneumoniae*; **12**, *Listeria innocua*; **13**, *L. monocytogenes*; **14**, *Pseudomonas. aeruginosa*; **15**, *P. putida*; **16**, *Salmonella enterica*; **17**, *Salmonella* spp.; **18**, *Serratia marcescens*; **19**, *Staphylococcus aureus*; **20**, *Vibrio parahaemolyticus*

The proteinaceous nature of inhibitors synthesized by these strains was revealed by their sensitivity to protease. The same was true for antimicrobials from *Pedobacter* sp. BG5, and *Pseudomonas* spp. CG21 and MTC3 that were selected among 2465 bacterial isolates from Antarctic soils as they demonstrated inhibitory effects on the growth of one or more indicator foodborne pathogens (Wong et al. 2011). The activity of inhibitors from three additional *Pseudomonas* isolates (i.e. WEK1, WEA1 and MA2) was insensitive to catalase, lipase, α-amylase, and protease enzymes.

Shekh et al. (2011) selected *Enterococcus* sp. APR 210 from an Antarctic penguin rookery for its ability to inhibit the growth of the multidrug-resistant fungal pathogenic yeast strains *C. albicans* NCIM 3471 and *Candida krusei*, in addition to antibacterial activity. The effect of enzymes, heat and pH on the antifungal activity of cell-free supernatants of *Enterococcus* sp. APR 210 was investigated. The biological activity was completely lost at 100 and 121 °C, whereas it was maintained after repeated freezing and thawing or long-term storage at −20 and −80 °C. The activity was also completely lost at pH values 2, 4 and 10. However, no loss of activity at pH values 6, 6.9 and 8 was observed. The candidacidal principle was sensitive to proteinase K (which caused a complete loss of antimicrobial activity) and to pronase E (which led to partial loss of biological activity). Treatment with trypsin had no effect on the biological activity. . Results suggested that the antimicrobial compound(s) produced by *Enterococcus* sp. APR 210 belonged to a protein of class II bacteriocins or bacteriocin-like inhibitory substances.

Lee et al. (2012) tested the antibacterial and antifungal capability of 57 proteobacterium isolates from soils of Barrientos Island, Antarctica. After the screening, a total of 14 isolates producing bioactive metabolites were identified. Members of the genera *Bradyrhizobium*, *Paracoccus* and *Sphingomonas* were active against *C. albicans*, while three isolates from the genus *Methylobacterium* showed bioactivity against *S. aureus* and *C. albicans*. None of the isolates showed bioactivity against *P. aeruginosa* or MRSA.

The ethanolic extract of *Janthinobacterium* sp. SMN 33.6 from Antarctic soil possessed antibacterial activity against multi-resistant nosocomial isolates, such as *Serratia marcescens* (MIC = 0.5–2 μg/mL), *P. aeruginosa* (MIC = 1 μg/mL), *Klebsiella pneumoniae* (MIC = 16 μg/mL), *Escherichia coli* (MIC = 0.5–1 μg/mL) and *Acinetobacter baumannii* (MIC = 1 μg/mL) (Asencio et al. 2009).

Among Bacteria, cyanobacteria from Polar soils have been rarely reported as antimicrobial producers. Gesheva and Negoita (2012) described a *Fischerella* sp. isolate (Haswell Island, Antarctica) that inhibits the growth of *B. subtilis*, *C. tropicalis* and the ascomycete *C. cladosporioides*.

2.1.3 Fungi

Even though fungi may represent an important source of new natural bioactive molecules, few studies have been carried out on this issue. Among fungi, members of the genera *Penicillium* and *Aspergillus* generally prevail in soils and their antimicrobial activity has been often reported, including for Polar soils (Table 3.3).

Table 3.3 Ascomycetes with antimicrobial activity isolates from Polar soils

Isolate	Location	Mainly active against[a]	Notes	Reference
Aspergillus nidulans	Haswell Island, Antarctica	Not shown		Gesheva and Negoita (2012)
Aspergillus sydowii 9541	Ellsworth Mountains, Antarctica	**4, 7**		Godinho et al. (2015)
Geomyces sp. 2481	King George Island, Antarctica	**1, 5, 8**	Geomycin A-C	Li et al. (2008)
Penicillium allii-sativi 9451	Ellsworth Mountains, Antarctica	**8**		Godinho et al. (2015)
P. allii-sativi 9458	Ellsworth Mountains, Antarctica	**4**		Godinho et al. (2015)
P. allii-sativi 9508	Ellsworth Mountains, Antarctica	**8**		Godinho et al. (2015)
P. allii-sativi 9524	Ellsworth Mountains, Antarctica	**8**		Godinho et al. (2015)
P, brevicompactum 9446	Ellsworth Mountains, Antarctica	**8**		Godinho et al. (2015)
P, brevicompactum 9448	Ellsworth Mountains, Antarctica	**4, 8**		Godinho et al. (2015)
P. chrysogenum 9466	Ellsworth Mountains, Antarctica	**4**		Godinho et al. (2015)
P. chrysogenum 9534	Ellsworth Mountains, Antarctica	**8**		Godinho et al. (2015)
P. griseofulvum	Greenland		Griseofulvin Fulvic acid Mycelianamide Roquefortine C and D Chanoclavine I Elymoclavine	Frisvad et al. (2004)
P. griseofulvum strain VKM FW-2251	Kolyma Lowland, Russia	Not shown	Griseofulvin	Kozlovsky et al. (2012)
P. nalgiovense Laxa	Antarctica	**3, 5, 8**	Amphotericin B	Svahn et al. (2015)
P. rubens 9496	Ellsworth Mountains, Antarctica	**4**		Godinho et al. (2015)
P. verrucosum	Haswell Island, Antarctica	Not shown		Gesheva and Negoita (2012)
Phoma herbarum CCFEE 5015	Dry Valleys, Antarctica	**2, 5, 6, 7**		Onofri et al. (2000)
Phoma herbarum CFEE 5007	Dry Valleys, Antarctica	**2, 5, 6, 7**		Onofri et al. (2000)
Phoma herbarum CCFEE 5020	Dry Valleys, Antarctica	**2, 5, 6, 7**		Onofri et al. (2000)
Phoma herbarum CCFEE 459	Dry Valleys, Antarctica	**2, 5, 6, 7**		Onofri et al. (2000)

[a]Target strains: **1**, *Aspergillus fumigatus*; **2**, *Bacillus subtilis*; **3**, *Candida albicans*; **4**, *Cladosporium sphaerospermum* **5**, *Escherichia coli*; **6**, *Pseudomonas putida*; **7**, *Sarcina* sp.; **8**, *Staphylococcus aureus*

The profiles of secondary metabolites in fungal strains of the subgenus *Penicillium* (genus *Penicillium*) isolated from Arctic and Antarctic permafrost were studied to clarify their taxonomic positions (Kozlovsky et al. 2012). The strain *Penicillum griseofulvum* VKM FW-2251, from Kolyma Lowland (Russia), synthesizes a metabolite identical in terms of physicochemical properties to griseofulvin, a polyketide metabolite that is known as a fungicidal antibiotic but rarely used today because of its high toxicity. Gesheva and Negoita (2012) reported on the production of antibiotics by *P. verrucosum* and *A. nidulans* strains from Antarctic soil, but no details on their individual antimicrobial spectrum were reported. Additional 14 fungal strains isolated from Antarctic soils and rocks were screened for the production of antibiotic substances (Onofri et al. 2000). Among them *Phoma herbarum* CCFEE 5015 and CCFEE 5007 from soil, and strains CCFEE 5020 and CCFEE 459 from rocks showed inhibitory activity. The antibiotic activity of strain CCFEE 5020, showing the highest degree of bioactivity, was further experimentally characterized, revealing that it was still present in the temperature range 5°–25 °C, whereas at 30 °C it was reduced. Agitation and aeration strongly influenced the growth and antibiotic production by the same strain, suggesting that it required high levels of oxygen.

Very recently, Godinho et al. (2015) reported on 17 fungal isolates from cold-arid oligotrophic Antarctic soils (Ellsworth Mountains), mainly affiliated to the genera *Aspergillus* and *Penicillium*, with antibacterial and antifungal activities (the most active isolates are listed in Table 3.3). Bioactive extracts generally contained fatty acid functional groups and triglycerides. The presence of highly functionalized secondary metabolites due to the presence of protons in the aromatic and olefinic regions also occurred.

2.2 *Polar Lakes and Ponds*

Investigations on the antimicrobial activity of microorganisms inhabiting lakes in Polar Regions are very few and mainly addressed to Antarctic benthic mats, which have accumulated for thousands of years virtually undisturbed, due to the extreme climatic conditions and the absence of higher metazoans. The synthesis of antibiotics and/or toxins by microbes belonging to these dense communities may confer a survival advantage (Biondi et al. 2008).

2.2.1 Heterotrophic Bacteria

Rojas et al. (2009) screened bacterial strains isolated from microbial mats growing in the benthic environment of Antarctic lakes (Table 3.4). Among them, 122 isolates showed antibacterial activity against the Gram-positives *S. aureus* and to a lower extent *E. faecium*, and *vs* the Gram-negative *E. coli*. Few of these strains were also active against *Cryptococcus neoformans*, *A. fumigatus* and to a lower extent *C. albicans*. The active strains were affiliated to several lineages of the α-, β- and

Table 3.4 Active bacteria, cyanobacteria and fungi from Polar lakes and ponds

	Phylum or class	Isolate	Location	Active against[a]	Notes	Reference
Bacteria	Bacteroidetes	*Flavobacterium* sp. Ant342	Schirmacher Oasis	**7**	Flexirub n	Mojib et al. (2010)
	Betaproteobacteria	*Janthinobacterium* sp. Ant5-2	Schirmacher Oasis	**7**	Violacin	Mojib et al. (2010)
		Janthinobacterium sp. R-7687	Dry Valleys	**5, 6, 10**		Rojas et al. (2009)
	Gammaproteobacteria	*Shewanella* sp. R-8990	Vestfold Hills	**6, 10**		Rojas et al. (2009)
		Pseudomonas sp. R-12565	Larsemann Hills	**5, 6, 10**		Rojas et al. (2009)
		Pseudomonas sp. R-12533	Larsemann Hills	**5, 6, 9, 10**		Rojas et al. (2009)
		Pseudomonas sp. R-12535	Larsemann Hills	**6, 9, 10**		Rojas et al. (2009)
		Psychrobacter sp. R-12597	Larsemann Hills	**6, 10**		Rojas et al. (2009)
	Actinobacteria	*Arthrobacter* spp. R-7513 and R-7941	Dry Valleys	**5, 10**	Cyclic thiazolyl peptide	Rojas et al. (2009)
	Cyanobacteria	*Leptolyngbya antarctica* ANT.LG2.3	Not specified	**10**		Biondi et al. (2008)
		Leptolyngbya antarctica ANT.LG2.5	Not specified	**10**		Biondi et al. (2008)
		Leptolyngbya antarctica ANT.L18.2	Not specified	**10**		Biondi et al. (2008)
		Nostoc sp. ANT.L34.1	Not specified	**10**		Biondi et al. (2008)
		Nostoc sp. ANT.LPR.1	Not specified	**4**		Biondi et al. (2008)
		Nostoc ANT.L52B.1	Not specified	**4, 10**		Biondi et al. (2008)
		Nostoc ANT.L36.1	Not specified	**4, 10**		Biondi et al. (2008)
		Nostoc ANT.LG2.6	Not specified	**4, 10**		Biondi et al. (2008)
		Phormidium murray ANT.PE.1	Not specified	**10**		Biondi et al. (2008)
		Phormidium priesteyi ANT.LPR.6	Not specified	**10**		Biondi et al. (2008)
		Phormidium priesteyi ANT.L61.2	Not specified	**10**		Biondi et al. (2008)
		Phormidium priesteyi ANT.L52.4	Not specified	**1, 4**		Biondi et al. (2008)
		Phormidium priesteyi ANT.L52.6	Not specified	**1, 4, 10**		Biondi et al. (2008)
		Pseudophormidium ANT.LG2.1	Not specified	**4**		Biondi et al. (2008)
		Pseudophormidium ANT.LPR.2	Not specified	**10**		Biondi et al. (2008)
		Pseudophormidium ANT.LPR.3	Not specified	**4, 10**		Biondi et al. (2008)
		Pseudophormidium ANT.LG2.2	Not specified	**4, 10**		Biondi et al. (2008)

(continued)

Table 3.4 (continued)

	Phylum or class	Isolate	Location	Active against[a]	Notes	Reference
Fungi	Ascomycota	*Aspergillus clavatus* IWW 447	Lake Sarah Tarn	**2, 3, 5, 6, 10**		Brunati et al. (2009)
		Aspergillus niger IWW 1026	Lake Pendant	**10**		Brunati et al. (2009)
		Beauveria sp. IWW 1017	Lake Pendant	**3**		Brunati et al. (2009)
		Cladosporium sp. IWW 1019	Lake Pendant	**6**		Brunati et al. (2009)
		Penicillium sp. IWW 1054	Lake Pendant	**2, 3, 6**		Brunati et al. (2009)
		Penicillium sp. IWW 1059	Lake Ace	**2, 3, 6**		Brunati et al. (2009)
		P. chrysogenum IWW 1053	Lake Highway	**5, 6, 10**		Brunati et al. (2009)
		P. chrysogenum IWW 1055	Lake Pendant	**1, 6, 10**		Brunati et al. (2009)
		P. chrysogenum TF 3/3	Tarn Flat	**6, 8, 10**	beta-lactam	Montemartini Corte et al. (2000)
		P. chrysogenum G 3/2	Gondwana	**8,10**		Montemartini Corte et al. (2000)
		P. citrinum S1/4R	Skua Lake	**8,10**		Montemartini Corte et al. (2000)
		P. citrinum S1/4bis	Skua Lake	**8,10**		Montemartini Corte et al. (2000)
		P. crustosum IWW 1023	Lake Watts	**6**		Brunati et al. (2009)
		P. roseopurpureum S1/3 bis	Skua Lake	**8,10**		Montemartini Corte et al. (2000)
		P. waksmanii S1/5	Skua Lake	**6,10**		Montemartini Corte et al. (2000)
		P. waksmanii II/3	Inexepress. Island	**6, 8, 10**		Montemartini Corte et al. (2000)
		P. waksmanii G3/17	Gondwana	**6,10**		Montemartini Corte et al. (2000)
		P. waksmanii G3/18	Gondwana	**6,10**		Montemartini Corte et al. (2000)

[a]Target strains: **1**, *Aspergillus fumigatus*; **2**, *Candida albicans*; **3**, *C. neoformans*; **4**, *Cryptococcus neoformans*; **5**, *Enterococcus faecium*; **6**, *Escherichia coli*; **7**, *Mycobacterium tubercolosis*; **8**, *Micrococcus luteus*; **9**, *Pseudomonas aeruginosa*; **10**, *Staphylococcus aureus*

γ-proteobacteria, the Bacteroidetes branch, and the high and low G+C Gram-positives. LC–MS fractionation of extracts from a subset of strains that exhibited relatively strong antibacterial activity evidenced a chemical novelty that was further investigated. Two of these strains, R-7513 (from Lake Fryxell) and R-7941 (from Lake Hoare), both belonging to the genus *Arthrobacter*, produced potent antibacterial compounds active against Gram-positives. Since these drugs were related to siomycin, cyclothiazomycin or thiopeptin (with similar antibacterial spectrum against Gram-positive bacteria, including methicillin resistant bacteria) it is possible that they represent new cyclic thiazolyl peptide class of antibiotics.. In the case of the bioactive extracts obtained from *Janthinobacterium* sp. R-7687 and *Pseudomonas* sp. R-12535, the preliminary MS spectra were not associated with any known compound, suggesting the activity of these Antarctic bacteria might rely on production of novel chemicals..

2.2.2 Cyanobacteria

Members of the Cyanobacteria are over-represented in the microbial communities from benthic mats. Taton et al. (2006) and Biondi et al. (2008) first reported on mass cultivation and pharmaceutical screening of a number of Antarctic cyanobacteria for the production of new lead molecules for drug development (Table 3.4). Strains were isolated from 27 benthic microbial mat samples collected from 23 lakes and ponds in the Larsemann Hills, Bølingen Islands, Vestfold Hills, Rauer Islands and the McMurdo Dry Valleys. Seventeen cyanobacteria showed antimicrobial activity against *S. aureus*, the filamentous fungus *A. fumigatus* or the yeast *C. neoformans*. The highest frequency of strains exhibiting antimicrobial activity was detected for the genera *Pseudophormidium* and *Nostoc*. Seven isolates showed only antibacterial activity against *S. aureus*, three only antifungal activity and seven both antibacterial and antifungal activity. *P. priestleyi* strains ANT.L52.4 and ANT.L52.6, isolated from the same lake, showed potent antifungal and/or antibacterial activity and were further analyzed. The fractions active against *A. fumigatus* from the two strains showed very similar chromatographic profiles and eluted with the same retention time. In the same LC-MS system, *Pseudophormidium* sp. ANT.LPR.2, *L. antarctica* ANT.LG2.3 and *Nostoc* sp. L34.1 showed that the fractions active against *S. aureus* eluted with similar retention times, indicating that the three strains produced chemically similar antibacterial compounds and that bioactivity was strain-specific rather than species-specific. No known antimicrobial metabolite was identified thus suggesting that the activity of these Antarctic cyanobacteria was due to novel biochemical compounds.

2.2.3 Fungi

Brunati et al. (2009) isolated 47 filamentous fungi with antimicrobial activity from Antarctic lake microbial mats (Table 3.4). Most of them inhibited the growth of *S. aureus* (14 %), *E. coli* (10 %), and yeasts *C. albicans* (11 %) and *C. neoformans* (8 %).

Both enterobacteria and filamentous fungi were less sensitive. The most bioactive fungi were the cold-tolerant cosmopolitan hyphomycetes such as *Penicillium*, *Aspergillus*, *Beauveria* and *Cladosporium*. Cold-adapted Antarctic endemic *Thelebolus* was least productive in terms of antimicrobial activities than more opportunistic representatives of the genera *Penicillium*, *Aspergillus* and *Cladosporium*. LC–MS fractionation of extracts from *A. clavatus* IWW 447 and *A. niger* IWW 1026 that exhibited relatively potent antimicrobial activity, evidenced a chemical novelty of the antibiotic molecules. Indeed, two bioactive bis-anthraquinones, rugulosin and skyrin (having various medical and insecticidal applications) were identified by LC–MS as the main products in the *P. chrysogenum* strain IWW 1056 isolated from the saline Highway lake in the Vestfold Hills. They inhibit the growth of *S. aureus* (MIC, 8 and >128 μg/mL, respectively), *Moraxella catharrhalis* (MIC, 0.25 and 8 μg/mL, respectively), *E. coli* (MIC, 32 and >128 μg/mL, respectively), *P. aeruginosa* (MIC, >128 and >128 μg/mL, respectively), and *C. albicans* (MIC, >128 and >128 μg/mL, respectively),

Montemartini Corte et al. (2000) screened *Penicillium* spp. from the sediments of ponds in continental Antarctica for antibacterial activity. The MIC of liquid cultures of *Penicillium* strains was determined allowing the selection of nine isolates. Among them, *P. chrysogenum* TF 3/3 differed from the type strain in that it produced a red-brown pigmentation, and also in the presence of activity toward gram-negative bacteria. These preliminary data suggested the synthesis of beta-lactam antibiotics.

2.3 Marine Environments

More than 70 % of the Earth's surface is covered by water, mainly oceans; hence, the marine environment represents some of the world's most unexplored extreme environments. Today, the success rate of drug discovery from the marine world is 1 out of the 3140 known molecular entities, which is roughly twofold to threefold better than the industry average (i.e., 1 out of 5000–10,000 tested compounds) (Gerwick and Moore 2012; Giddings and Newman 2015). Hence, marine environments represent an untapped source to find microorganisms with the potential to produce new, invaluable pharmacophores (Giddings and Newman 2015). Microorganisms able to synthesize bioactive compounds with antimicrobial properties have been isolated also from Arctic and Antarctic marine environments, as described in the following sections.

2.3.1 Bacteria

Researches have been mainly addressed to the Actinobacteria and, at a lesser extent, to other heterotrophic bacteria, from Polar seawater and sediment (Table 3.5). Lo Giudice et al. (2007b) isolated from Antarctic seawater and sediments different

Table 3.5 Active bacteria from marine Polar environments

Phylum or class	Isolate	Origin	Active against[a]	Notes	Reference
Gammaproteobacteria	*Pseudoalteromonas* sp. F26	Antarctic seawater	**6, 14**		Lo Giudice et al. (2007b)
	Pseudoalteromonas sp. G24	Antarctic seawater	**6, 14**		Lo Giudice et al. (2007b)
	Pseudoalteromonas sp. 59	Antarctic seawater	**6**		Lo Giudice et al. (2007b)
	Pseudoalteromonas sp. 129	Antarctic seawater	**6, 14**		Lo Giudice et al. (2007b)
	Pseudoalteromonas sp. 131	Antarctic seawater	**6, 14**		Lo Giudice et al. (2007b)
	Pseudomonas sp. 65/3	Antarctic fish	**14**		Lo Giudice et al. (2007b)
	Pseudoalteromonas sp. MB33	Arctic copepods	**19**		Wietz et al. (2012)
	Pseudoalteromonas sp. MB205	Arctic surface water	**19**		Wietz et al. (2012)
	Pseudoalteromonas sp. MB220	Arctic sea-ice	**19**		Wietz et al. (2012)
	Pseudoalteromonas sp. MB240	Arctic sea-ice	**19**		Wietz et al. (2012)
	Psychrobacter sp. XX5	Arctic sea-ice	**19**		Wietz et al. (2012)
	Psychrobacter sp. ST4	Arctic sea-ice	**19**		Wietz et al. (2012)
	Vibrio sp. RR12	Arctic amphipods	**19**		Wietz et al. (2012)
	Vibrio sp. EF14	Arctic deep-sea	**17, 19**		Wietz et al. (2012)
	Vibrio sp. RS9	Arctic sea-ice	**17, 19**		Wietz et al. (2012)
Bacteroidetes	*Salegentibacter* sp. T436	Arctic sea-ice	**2, 4, 5, 6, 8, 9, 10, 11, 12, 13, 15, 18**	Nitro compounds	Al-Zereini et al. (2007)

(continued)

Table 3.5 (continued)

Phylum or class	Isolate	Origin	Active against[a]	Notes	Reference
Actinobacteria	*Arthrobacter* sp. B20	Antarctic seawater	**9, 14**		Lo Giudice et al. (2007b)
	Arthrobacter sp. F40	Antarctic seawater	**4**		Lo Giudice et al. (2007b)
	Arthrobacter sp. G18	Antarctic seawater	**6, 9**		Lo Giudice et al. (2007b)
	Arthrobacter sp. G75	Antarctic seawater	**6, 9, 14**		Lo Giudice et al. (2007b)
	Arthrobacter sp. PP12	Arctic sea-ice	**17, 19**		Wietz et al. (2012)
	Arthrobacter sp. MB182	Arctic sea-ice	**17, 19**		Wietz et al. (2012)
	Arthrobacter sp. SS14	Arctic copepods	**17, 19**		Wietz et al. (2012)
	Arthrobacter sp. TT4	Arctic meltwater	**17, 19**		Wietz et al. (2012)
	Arthrobacter sp. ZZ3 18	Arctic sea-ice	**17, 19**		Wietz et al. (2012)
	Arthrobacter sp. LM7	Arctic surface water	**17, 19**		Wietz et al. (2012)
	Arthrobacter sp. WX11	Arctic deep-sea	**1, 3, 7, 16, 17, 19, 20, 21, 22, 23, 24**	Arthrobacilins	Wietz et al. (2012)
	Janibacter sp. B8	Antarctic seawater	**6, 9, 14**		Lo Giudice et al. (2007b)
	Janibacter sp. F21	Antarctic seawater	**14**		Lo Giudice et al. (2007b)
	Janibacter sp. F34	Antarctic seawater	**6, 9, 14**		Lo Giudice et al. (2007b)
	Janibacter sp. F39	Antarctic seawater	**14**		Lo Giudice et al. (2007b)
	Janibacter sp. G4	Antarctic seawater	**6, 14**		Lo Giudice et al. (2007b)
	Janibacter sp. G5	Antarctic seawater	**6, 14**		Lo Giudice et al. (2007b)
	Janibacter sp. I44	Antarctic seawater	**6, 9, 14**		Lo Giudice et al. (2007b)
	Microlunatus sp. y400	Antarctic sediment	**4**		Yuan et al. (2014)
	Nesterenkonia sp. S1-21	Antarctic sediment	**6**		Lo Giudice et al. (2007b)
	Nesterenkonia sp. S1-40	Antarctic sediment	**6**		Lo Giudice et al. (2007b)

	Nocardiopsis sp. y4	Antarctic sediment	**4**	Pradimicin, Tetarimycin A	Yuan et al. (2014)
	Nocardiopsis sp. y47	Antarctic sediment	**4**	Pradimicin, Tetarimycin A	Yuan et al. (2014)
	Nocardiopsis sp. y17	Antarctic sediment	**4, 5**	Pradimicin, Tetarimycin A	Yuan et al. (2014)
	Nocardiopsis sp. y18	Antarctic sediment	**4, 5**	Pradimicin, Tetarimycin A	Yuan et al. (2014)
	Nocardiopsis sp. y64	Antarctic sediment	**4, 5**	Pradimicin, Tetarimycin A	Yuan et al. (2014)
	Rhodococcus sp. B7	Antarctic seawater	**6, 9, 14**		Lo Giudice et al. (2007b)
	Rhodococcus sp. G77	Antarctic seawater	**6**		Lo Giudice et al. (2007b)
	Rhodococcus sp. W4-5	Antarctic seawater	**4, 6**		Lo Giudice et al. (2007b)
	Streptomyces sp. ART5	Arctic sediment	**5**	Arcticoside C-1027 chromophore V	Moon et al. (2014)
	Streptomyces sp. y146	Arctic sediment	**4, 5, 6**		Yuan et al. (2014)
	Streptomyces sp. y2	Arctic sediment	**4, 5, 17**		Yuan et al. (2014)
	Streptomyces sp. y23	Arctic sediment	**4, 5**		Yuan et al. (2014)
	Streptomyces sp. y222	Arctic sediment	**4,5**		Yuan et al. (2014)
	Streptomyces sp. y481	Arctic sediment	**4**		Yuan et al. (2014)

[a]Target strains: **1**, *Aeromonas salmonicida*; **2**, *Bacillus brevis*; **3**, *B. cereus*; **4**, *B. subtilis*; **5**, *Candida albicans*; **6**, *Escherichia coli*; **7**, *Listeria monocytogenes*; **8**, *Magnaporthe grisea*; **9**, *Micrococcus luteus*; **10**, *Mucor miehei*; **11**, *Nematospora coryli*; **12**, *Paecilomyces variotii*; **13**, *Penicillium notatum*; **14**, *Proteus mirabilis*; **15**, *P. vulgaris*; **16**, *Salmonella enterica*; **17**, *Staphylococcus aureus*; **18**, *Ustilago nuda*; **19**, *Vibrio anguillarum*; **20**, *V. parahaemolyticus*; **21**, *V. harveyi*; **22**, *V. vulnificus*; **23**, *Yersinia enterolitica*; **24**, *Y. ruckeri*

bacteria embedded with antimicrobial activity. Sixteen Actinobacteria phylogenetically affiliated to the genera *Janibacter*, *Nesterenkonia*, *Arthrobacter* and *Rhodococcus* were active toward *E. coli* and *P. mirabilis* strains, and at a lesser extent *vs M. luteus* and *B. subtilis*. Interestingly, the genera *Nesterenkonia* and *Rhodococcus* are not known as producers of bioactive molecules and they could represent a new source of antimicrobials. Among heterotrophic bacteria, other than Actinobacteria, isolates from seawater column (five *Pseudoalteromonas* isolates) and the intestinal content of the fish *Trematomus bernacchii* (one *Pseudomonas* isolate) from the Terra Nova Bay (Antarctica) exhibited a strong antibacterial activity against *E. coli* and *P. mirabilis* (Lo Giudice et al. 2007b). Moreover, the finding that different (even though closely related) isolates exhibit different inhibition pattern suggests that antibacterial activity was very likely strain-specific and that members of the same species might probably synthesize several compounds acting on multiple/different targets.

Of 511 randomly selected strains from different Arctic marine sources (e.g. sea-ice, surface seawater, zooplankton, deep sea, and meltwater), Wietz et al. (2012) selected 16 isolates showing considerable antibacterial activity against *Vibrio anguillarum* and *S. aureus* upon repeated testing. Antibiotic-producing isolates belonged to the genera *Arthrobacter* (7 strains), *Pseudoalteromonas* (4 strains), *Psychrobacter* (2 strains), and *Vibrio* (3 strains). Ethanolic extracts from *Arthrobacter* sp. WX11 inhibited the growth of *Aeromonas salmonicida*, *B. cereus*, *Listeria monocytogenes*, *S. aureus*, *Salmonella enterica*, *Vibrio vulnificus*, *V. parahaemolyticus*, *V. harveyi*, *V. anguillarum*, *Yersinia enterolitica*, and *Y. ruckeri*. Arthrobacilins A, B and C were detected in both ethanol and ethyl acetate extracts, but in this latter case it lacked antibacterial activity, indicating that antibiosis may rely on the synergistic action of different compounds.

Yuan et al. (2014) reported on 11 actinobacterial strains from Arctic deep-sea sediment samples showing antibacterial and/or antifungal activity. Among them, seven ones showed activity against both *B. subtilis* and *C. albicans*. However, *Streptomyces* sp. y146 and *Streptomyces* sp. y2 had activity against *E. coli* and *S. aureus*, respectively. Active strains belonged to the genera *Streptomyces*, *Nocardiopsis* and *Microlunatus*. The presence of genes coding for type I polyketide synthase (PKS I), PKS II, nonribosomal peptide synthase (NRPS), aminodeoxyisochorismate synthase (phzE), dTDP-glucose-4, 6-dehydratase (dTGD), halogenase (Halo) or cytochrome P450 hydroxylase (CYP) was checked through PCR amplification, revealing that the genome of all isolates harbored at least two gene clusters involved in the biosynthesis of secondary-metabolites. The PCR products of the ketosynthase (KS) domain of PKS II from five *Nocardiopsis* strain genomes were sequenced. The KS sequences from these strains displayed a 67 % and 69 % amino acid sequence identity with the KS domain of *fabF* encoded protein involved in the biosynthesis of pradimicin, an antifungal antibiotic from *Actinomadura hibisca*, and had also 68 % similarity with TamM, which is related to the biosynthesis of Tetarimycin A, a tetracyclic MRSA-active antibiotic (Yuan et al. 2014).

2.3.2 Fungi

Montemartini Corte et al. (2000) screened *Penicillium* spp. isolated from wooden baits sunk to a 50 m depth in the Ross Sea (Antarctica). The MIC of liquid cultures of *Penicillium* strains was determined allowing the selection of seven isolates, mainly *P. chrysogenum*, with activity against *S. aureus* and *M. luteus*. However, excellent activity was observed for *P. melinii* R55 might rely on the synthesis of patulin, as already demonstrated for the type strain (Table 3.6). Additional active fungi, listed in Table 3.6, were isolated also from Antarctic macroalgae and sponges (see Sects. 2.4.2 and 2.4.3).

Table 3.6 Marine fungi showing antimicrobial activity

	Isolate	Origin	Target strains[a]	Notes	Reference
Ascomycota	*Dipodascus australiensis* 6031	Antarctic algae	**1, 2**		Furbino et al. (2014)
	Cladosporium sp. F09-T13-2	Antarctic sponge	**7**		Henríquez et al. (2014)
	Epicoccum sp. F09-T15-1	Antarctic sponge	**7, 8**		Henríquez et al. (2014)
	Epicoccum sp. F09-T15-4	Antarctic sponge	**8, 9**		Henríquez et al. (2014)
	Geomyces sp. F09-T1-8	Antarctic sponge	**4, 7, 8, 9**		Henríquez et al. (2014)
	Geomyces sp. F09-T3-19	Antarctic sponge	**4, 7, 8, 9**		Henríquez et al. (2014)
	Geomyces sp. F09-T3-5	Antarctic sponge	**7, 8, 9**		Henríquez et al. (2014)
	Metschnikowia australis MH47.1.2	Antarctic algae	**1, 2**		Furbino et al. (2014)
	Penicillium chrysogenum R9	Sunk wooden baits	**6, 8**		Montemartini Corte et al. (2000)
	Penicillium chrysogenum R28	Sunk wooden baits	**6, 8**		Montemartini Corte et al. (2000)
	Penicillium chrysogenum R31	Sunk wooden baits	**6, 8**		Montemartini Corte et al. (2000)
	Penicillium chrysogenum R34	Sunk wooden baits	**6, 8**		Montemartini Corte et al. (2000)
	Penicillium chrysogenum R36	Sunk wooden baits	**6, 8**		Montemartini Corte et al. (2000)
	Penicillium chrysogenum R38	Sunk wooden baits	**6, 8**		Montemartini Corte et al. (2000)
	P. commune F09-T8-1	Antarctic sponge	**8, 9**		Henríquez et al. (2014)

(continued)

Table 3.6 (continued)

	Isolate	Origin	Target strains[a]	Notes	Reference
	P. commune F09-T10-1	Antarctic sponge	**4, 8**		Henríquez et al. (2014)
	P. melinii R55	Sunk wooden baits	**5, 6, 8**	Patulin	Montemartini Corte et al. (2000)
	P. polonicum F09-T7-2	Antarctic sponge	**4, 8, 9**		Henríquez et al. (2014)
	Penicillium steckii 6012	Antarctic algae	**1, 2**		Furbino et al. (2014)
	Penicillium sp. 6034	Antarctic algae	**3**		Godinho et al. (2013)
	Penicillium sp. 6120	Antarctic algae	**3**		Godinho et al. (2013)
	Pseudogymnoascus sp. 1	Antarctic algae	**1, 2**		Furbino et al. (2014)
	Pseudogymnoascus sp. 2	Antarctic algae	**1, 2, 3**		Furbino et al. (2014)
	Pseudeurotium F09-T5-4	Antarctic sponge	**8**		Henríquez et al. (2014)
	Thelebolus F09-T14-3	Antarctic sponge	**8**		Henríquez et al. (2014)
	Trichocladium sp. F09-T24-1	Antarctic sponge	**4, 8, 9**		Henríquez et al. (2014)
Basidiomycota	*Guehomyes pullulans* MH33.1	Antarctic algae	**1, 2, 3**		Furbino et al. (2014)

[a]Target strains: **1**, *Candida albicans*; **2**; *C. krusei*; **3**, *Cladosporium sphaerospermum*; **4**, *Clavibacter michiganensis*; **5** , *Escherichia coli*; **6**, *Micrococcus luteus*; **7**, *Pseudomonas aeruginosa*; **8**, *Staphylococcus aureus*; **9**, *Xanthomonas campestris*

2.4 Microorganisms Living in Association with Other Organisms

The exceptional genetic and metabolic plasticity of microorganisms allow them to interact with many, if not all, eukaryotic (micro)organisms. Indeed, they commonly colonize either plants and/or animals with physiological and structural features enabling them to survive and synthesize a wide range of compounds that are able to protect their eukaryotic counterpart against pathogenic and fouling organisms. Several organisms (mainly invertebrates) have been attracting the attention of many researchers because of their diverse pharmaceutical potentials (antiviral, anti-proliferative, anti-inflammatory, anti-tumor, and antimycobacterial activities). However, some recent reports suggest that some metabolites obtained from algae, invertebrates and lichens might be synthesized by their fungal and bacterial symbionts. With regard to the Polar Regions, antimicrobial activity has been reported for bacteria and/or fungi associated with lichens and, as marine counterparts, algae and sponges.

2.4.1 Microorganisms Associated with Lichens and Mosses

The antibacterial potential of bacterial strains associated with Arctic and Antarctic lichens has been evaluated by Kim et al. (2012, 2013, 2014a, b). Active bacteria mainly belonged to the genera *Sphingomonas*, *Burkholderia* and *Rhodanobacter* within the Alpha-, Beta- and Gammaproteobacteria, respectively (Table 3.7). Bacterial extracts, obtained using different solvents (i.e. acetone, water, chloroform, diethyl ether, ethanol, methanol and petroleum ether), showed antibacterial activities against both Gram-positive (*S. aureus*, *B. subtilis*, and *M. luteus*) and Gram-negative (*E. cloacae*, *P. aeruginosa*, and *E. coli*) target pathogens. With respect to solvents, ethyl acetate extraction was more effective than solvents with high (e.g. ethanol) or very low polarity (e.g. petroleum ether). All the aqueous extracts do not exhibit any antibacterial activity against target bacteria, probably depending on the lack of water solubility of active compounds.

Melo et al. (2014) reported on the antibacterial activity of the Antarctic endophytic fungus *Mortierella alpina* strain ITA1-CCMA 952 isolated from the moss *Schistidium antarctici* (Admiralty Bay, King George Island). The strain inhibited the growth of *E. coli* (MIC of 26.9 μg/mL), *P. aeruginosa* and *E. faecalis* (both with a MIC of 107 μg/mL). Antifungal activity was absent. As it was revealed by the GC–MS analysis, antibacterial metabolites were identified as pyrrolopyrazine alkaloids.

To the best of our knowledge, there are no studies about the antimicrobial activity of fungi isolated from lichens or bacteria from mosses so far.

2.4.2 Microorganisms Associated with Sponges

Marine sponges are animals that harbour various organisms, and in some cases, up to 35 % of their wet weight is represented by microorganisms. Furthermore, high- and low-microbial abundance sponges can harbor microbial concentrations of 10^8–10^{10} cells/g and 10^5–10^6 cells/g, respectively, demonstrating the vast potential for the discovery of numerous new secondary metabolites isolated from microbes that live in environments with(in) other organisms (Hentschel et al. 2002). It is known that microbes associated to sponges possess tremendous biological activities covering a wide range of biological functions. The sponge-microbial association is a potential chemical and ecological phenomenon, which provides sustainable resource for developing new pharmaceutical leads. Keeping in view the importance of antimicrobial potential of microbes associated with marine sponge, targeting sponge microsymbionts is an essential focus nowadays. On this basis, since Antarctic sponges represent a potentially rich, untapped source of new antimicrobial compounds, the study by Papaleo et al. (2012) was aimed at characterizing the cultivable bacterial communities (a total of 132 strains) from three different Antarctic sponge species (*Lissodendoryx nobilis*, *Anoxycalyx joubini* and *Haliclonissa verrucosa*) from the Terra Nova Bay. The study was performed to

Table 3.7 Bacterial strains with antimicrobial activity living in association with lichens

Phylum or class	Isolate	Location	Active against[a]	Reference
Alphaproteobacteria	*Sphingomonas* sp. PAM26605	Svalbard Islands (Arctic)	**1, 3, 5, 6**	Kim et al. (2014a)
	Sphingomonas sp. PAM26625	Svalbard Islands (Arctic)	**3, 5, 6**	Kim et al. (2014a)
	Sphingomonas sp. PAMC26556	King George Island (Antarctica)	**5**	Kim et al. (2014b)
	Sphingomonas sp. PAMC26561	King George Island (Antarctica)	**3, 5**	Kim et al. (2014b)
	Sphingomonas sp. KOPRI26645	Svalbard Islands (Arctic)	**1, 2, 3, 4, 5**	Kim et al. (2012)
Betaproteobacteria	*Burkholderia* sp. PAM26606	Svalbard Islands (Arctic)	**1, 3, 4, 5**	Kim et al. (2014a)
	Burkholderia sp. PAM26607	Svalbard Islands (Arctic)	**3, 6**	Kim et al. (2014a)
	Burkholderia sp. PAM26608	Svalbard Islands (Arctic)	**1, 3, 5, 6**	Kim et al. (2014a)
	Burkholderia sp. PAMC26507	King George Island (Antarctica)	**4**	Kim et al. (2014b)
	Burkholderia sp. PAMC26537	King George Island (Antarctica)	**4**	Kim et al. (2014b)
	Burkholderia sp. PAMC26633	King George Island (Antarctica)	**1, 3, 5, 6**	Kim et al. (2014b)
	Burkholderia sp. KOPRI26643	Svalbard Islands (Arctic)	**1, 2, 3, 4, 5, 6**	Kim et al. (2012)
	Burkholderia sp. KOPRI26644	Svalbard Islands (Arctic)	**1, 2, 3, 4, 5, 6**	Kim et al. (2012)
	Burkholderia sp. KOPRI26646	Svalbard Islands (Arctic)	**1, 2, 3, 4, 5, 6**	Kim et al. (2012)
	Burkholderia sp. KOPRI26647	Svalbard Islands (Arctic)	**1, 2, 4**	Kim et al. (2012)
Gammaproteobacteria	*Rhodanobacter* sp. PAMC26515	King George Island (Antarctica)	**4**	Kim et al. (2014b)
	Rhodanobacter sp. PAMC26518	King George Island (Antarctica)	**4**	Kim et al. (2014b)
	Rhodanobacter sp. PAMC26538	King George Island (Antarctica)	**1, 2, 3, 5, 6**	Kim et al. (2014b)
	Rhodanobacter sp. PAMC26551	King George Island (Antarctica)	**1, 3, 5, 6**	Kim et al. (2014b)
	Rhodanobacter sp. PAMC26552	King George Island (Antarctica)	**4**	Kim et al. (2014b)
	Rhodanobacter sp. PAMC26557	King George Island (Antarctica)	**3, 4**	Kim et al. (2014b)
Bacteroidetes	*Hymenobacter* sp. PAMC26554	King George Island (Antarctica)	**1, 3, 6**	Kim et al. (2014b)
Actinobacteria	*Streptomyces* sp. PAMC26508	King George Island (Antarctica)	**1, 2, 3, 4, 5, 6**	Kim et al. (2014b)
	Frigoribacterium sp. PAMC26555	King George Island (Antarctica)	**1, 4, 6**	Kim et al. (2014b)
Firmicutes	*Paenibacillus* sp. PAMC26517	King George Island (Antarctica)	**1, 2, 6**	Kim et al. (2014b)

[a]Target strains: **1**, *Bacillus subtilis*; **2**, *Enterobacter cloacae*; **3**, *Escherichia coli*; **4**, *Micrococcus luteus*; **5**, *Pseudomonas aeruginosa*; **6**, *Staphylococcus aureus*

check whether some of these strains might inhibit the growth of a panel of more than 70 opportunistic pathogens, including those affecting cystic fibrosis (CF) patients. Data obtained revealed that most of these sponge-associated Antarctic bacteria were able to completely inhibit the growth of members of the *Burkholderia cepacia* complex (Bcc), representing one of the most important CF pathogens. On the other hand, the same strains did not interfere with the growth of other pathogenic bacteria, such as *P. aeruginosa* or *S. aureus*, suggesting a specific inhibition activity toward Bcc bacteria. Further experiments carried out on the most active isolates (i.e. *Pseudoalteromonas* spp. TB41 and AC163, *Shewanella* sp. TB4, *Psychrobacter* spp. TB47 and TB67) revealed that at least some of the antimicrobial compounds were very likely mVOCs, which constitute an important regulatory factor in the interrelationships among different organisms in microbial ecosystems. This finding was confirmed by solid phase micro extraction gas chromatography mass-spectrometry (SPME-GC-MS) performed on, representative set of Antarctic bacteria, which revealed the production of a large set of mVOCs whose synthesis was very likely constitutive, in that it was not induced by the presence of target strains (Romoli et al. 2011). More interestingly, the activity of the VOCs seemed to be more effective than most of the commonly used antibiotics in inhibiting the growth of Bcc bacteria. A metabolomic approach applied to *Pseudoalteromonas* sp. TB41 allowed a selection of 30 compounds, some of which presumably responsible for the inhibition of *B. cenocepacia* (Bc) LMG16654 growth by the Antarctic isolate (Romoli et al. 2014). The mVOCs profiles obtained from strain TB41 when grown alone or in the presence of Bc-LMG16654 were not significantly different, suggesting that the production of such molecules may be constitutive and not modified by the presence of the target strain, as this latter neither interfered with the production nor induced the synthesis of different mVOCs. Recently, Papaleo et al. (2013) also reported on the volatile profiles of *Psychrobacter* spp. (isolated from *A. joubini*; Mangano et al. 2009) under aerobic conditions. Results suggested that Antarctic bacteria exhibited an antimicrobial ability that might rely on a complex mixture of mVOCs whose relative concentration was dependent on the growth conditions (presence/absence of oxygen and growth media used). Results also revealed that only *Pseudoalteromonas* sp. TB41 (Mangano et al. 2009) possessed some *nrps–pks* genes, similarly to *Arthrobacter* sp. TB23 (isolated from the Antarctic sponge *L. nobilis*; Mangano et al. 2009) whose genome sequence was recently reported (Fondi et al. 2012) (see Sect. 3).

With regard to fungi, only recently Henríquez et al. (2014) described for the first time the biodiversity and the metabolic potential of fungi associated with Antarctic marine sponges. Fungal culture extracts were assayed for their possible antimicrobial activity (in addition to antitumoral and antioxidant ones, not reported). Among them, 52 extracts, mainly obtained from the genus *Geomyces* and unidentified relatives, showed antimicrobial activity against some of the bacteria tested, a selection of which is shown in Table 3.6. In general, fungal extracts were more active against Gram positive (particularly against *S. aureus*) than Gram negative bacteria. In particular, 5 isolates (belonging to the genera *Geomyces* and *Epicoccum*) exhibited antibacterial activity against *P. aeruginosa*, 44 isolates showed inhibitory activity *versus S. aureus*, 11 isolates (belonging to the genera *Geomyces* and *Penicillium*, or not identified) exhibited antibacterial activity against *Clavibacter michiganensis*, and 22 isolates (belonging to the genera *Geomyces*, *Penicillium*, *Epicoccum* and *Cladosporium*, or not identified) exhibited antibacterial activity against *Xanthomonas campestris*. Several fungal isolates with the same ITS type showed different antimicrobial activity. *Geomyces* was the most prolific fungal genus, with 18 isolates (54.5 % of *Geomyces* sp.) showing antimicrobial activity. Among them, *Geomyces* spp. F09-T1-8, and F09-T3-19 resulted particularly active. The antimicrobial activity in the genus *Pseudeurotium* was first described by Henríquez et al. (2014). More interestingly, most of un-identified isolates (among the Leotiomycetes) also were prolific as producers of antimicrobial activities.

2.4.3 Fungi Associated with Macroalgae

Marine algae play a key role in organic matter mineral cycling, particularly in both littoral and infralittoral ecosystem in Antarctic shallow waters. The Antarctic macroalgae are highly endemic, play a fundamental role as primary producers, food for marine herbivores as well as in habitat structure. To date, only two reports exist on the antimicrobial activity of fungi associated with Antarctic algae (Table 3.6).

Godinho et al. (2013) analyzed the distribution and diversity of fungi associated with eight Antarctic macroalgae and their ability to synthesize bioactive compounds. *Penicillium* sp. strains 6034 and 6120, isolated from the endemic species *Palmaria decipiens* (Rhodophyta) and *Monostroma hariotii* (Chlorophyta), respectively, yielded extracts exhibiting high and selective antifungal (and/or trypanocidal) activity. Penicillium sp. 6120 displayed antifungal activity against the filamentous fungus *Cladosporium sphaerospermum*, producing 96 % inhibition and a MIC value of 250 mg/mL. The preliminary proton nuclear magnetic resonance spectroscopy analysis indicated the presence of highly functionalised aromatic compounds.

More recently, Furbino et al. (2014) characterized the fungal communities associated with the endemic Antarctic macroalgae *Pyropia endiviifolia* (Rhodophyta) and *Monostroma hariotii* (Chlorophyta), which are among the most abundant species across the Antarctic Peninsula. A total of six algicolous fungal taxa were able to produce compounds with biological activities. All extracts showed selective antifungal activities against *Candida albicans* and *C. krusei*, whereas those obtained by *Pseudogymnoascus* sp. 2 and *Guehomyes pullulans* MH33.1 also inhibited *Cladosporium sphaerosperum*. Additionally, the extract of *Penicillium steckii* 6012 inhibited 96 % of yellow fever virus. No extract displayed antibacterial or trypanocidal activities.

To the best of our knowledge, no study about the antimicrobial activity of fungi isolated from Arctic macroalgae has been carried out so far.

3 Genetic and Genomic Aspects of Sponge-Associated Antarctic Bacteria with Antibiotic Activity

Further insights in the metabolic potential of sponge-associated bacteria were obtained by the molecular and genomic analyses of Antarctic bacterial strains belonging to different genera whose draft genome sequence was recently obtained (Fondi et al. 2012; Papaleo et al. 2013; Maida et al. 2014; Orlandini et al. 2014; Bosi et al. 2015; Maida et al. 2015).

Among the isolates characterized by Papaleo et al. (2012), representatives of the genera *Gillisia* (i.e isolate CAL575), *Psychrobacter* (i.e. isolates TB2, TB15 and AC24) and *Arthrobacter* (i.e. isolates TB26, CAL618 and TB23) were further analyzed by a genomic approach in order to shed some light on their inhibitory activity against Bcc (Fondi et al. 2014; Maida et al. 2014; Orlandini et al. 2014). Maida et al. (2014) characterized *Gillisia* sp. CAL575 from the sponge *H. verrucosa* using a combination of different techniques, including genomics, phenotypic characterization and analysis of mVOCs. Fondi et al. (2014) reported the draft genomes of *Psychrobacter* spp. TB2, TB15 (both from the sponge *L. nobilis*) and AC24 (from the sponge *H. verrucosa*). In particular, *Psychrobacter* sp. AC24 efficiently inhibited the growth of almost all the Bcc strains tested, regardless of the growth medium. Conversely, TB2 and TB15 displayed a reduced inhibitory ability compared to AC24 and, in some cases, the effect on the growth of Bcc strains was influenced by the corresponding growth medium. Moreover, very similar inhibition pattern *vs* Bcc strains were exhibited by *Arthrobacter* spp. TB26 (from the sponge *L. nobilis*) and CAL 618 (from *H. verrucosa*), suggesting that the genetic determinants responsible for the biosynthesis of antimicrobial compounds belonged to the *core* genome (Orlandini et al. 2014). Additionally, the whole body of cross-streaking data suggested that also diffusible organic molecules (in combination with volatile compounds) might interfere with the growth of Bcc strains. This is also true for some *Pseudoalteromonas* strains that were shown to be able to interfere with the growth of Bcc strains (Maida et al. 2015).

The draft genome sequence of 38 Antarctic bacterial strains (including those mentioned above) were searched for genes involved in the biosynthesis of secondary metabolites, known to often possess antimicrobial activity. Data obtained revealed that, according to previous experimental data (Papaleo et al. 2012), no gene involved in secondary metabolite biosynthesis (*pks* or *nrps*), except for the presence of a *pks* type III gene and a terpene biosynthetic cluster, were identified in the *Gillisa* sp. CAL575 strain. Moreover, in the genome of the *Arthrobacter* sp. TB23 three gene clusters including a type III PKS, a NRPS gene, and terpene biosynthetic genes, respectively, responsible for the biosynthesis of antimicrobial compounds, which might be targeted towards Bcc bacteria were disclosed. The comparative analysis of the genome of these strains highlighted the presence of few genes belonging to the *core* genome involved in the secondary metabolites biosynthesis (Papaleo et al. 2013). Besides, it was also suggested that the biosynthesis of these compounds might be synthesized by still unknown metabolic routes.

Results suggested that the metabolic strategies exploited by the three *Psychrobacter* strains to inhibit the growth of *Burkholderia* representatives fall outside the range of already characterized biochemical systems.

4 Bioactive Metabolites with Antimicrobial Activities from Cold-Adapted Microorganisms

The number of new bioactive molecules from cold-adapted microorganisms has significantly increased in the last decade, with some of them that possess antimicrobial activity (for a recent review see: Bratchkova and Ivanova 2011; Liu et al. 2013). Several biomolecules have been cited in the Sect. 2. Additional information is reported below.

With regard to streptomycetes, Bruntner et al. (2005) isolated a new angucyclinone antibiotic, called frigocyclinone, from the Antarctic *Streptomyces griseus* strain NTK 97 (from soil), consisting of a tetrangomycin moiety attached through a C-glycosidic linkage with the aminodeoxysugar ossamine (Fig. 3.1). Frigocyclinone showed good inhibitory activity against Gram-positive bacteria (such as *B. subtilis*, MIC 4.6 μg/mL, and *S. aureus*, MIC 15 μg/mL), whereas Gram-negative bacteria (i.e. *E. coli*, *P. fluorescens* and *P. mirabilis*), filamentous fungi (i.e. *Botrytis cinerea*, *Aspergillus viridinutans*, *Penicillium notatum* and *Paecilomyces variotii*) and yeast (i.e. *S. cerevisiae* and *C. albicans*) were not sensitive.

Streptomyces flavovirens 6^7, isolated from soil samples in the region of Livingston Island, Antarctica, synthesized molecules embedded with antimicrobial and antitumour activity, which belonged to the actinomycins, within peptidic antibiotics (Ivanova et al. 2002) (Fig. 3.1). The new substance 2-amino-9,13-dimethyl heptadecanoic acid and phthalic acid diethyl ester, 1,3-bis(3-phenoxyphenoxy)benzene, hexanedioic acid dioctyl ester as natural products were found in the culture broth of the strain *Streptomyces* sp. 1010, isolated from water samples in the region Livingston Island, Antarctica. The phthalic acid diethyl ester possessed an antibacterial property against *M. luteus* (MIC, 3 μg/mL), *B. subtilis* (MIC, 12 μg/mL) and *S. aureus* (MIC, 25 μg/mL) (Ivanova et al. 2001).

Gesheva (2010), by comparison with different antibiotic standards, suggested that *Streptomyces* sp. eight produced three antibiotics: non-polyenic macrolide antibiotic (composed by two components), azalomycin B, and nigericin.

New benzoxazine secondary metabolites, articoside and C-1027 chromophore V, have also been obtained by Moon et al. (2014) from an Arctic marine *Streptomyces* strain ART5, isolated from surface sediment (East Siberian continental margin; depth of 354 m) (Fig. 3.1). Interestingly, articoside is a benzoxazine disaccharide, a structure type that has not been previously reported, whereas C-1027 chromophore-V possesses a chlorine atom, amino sugar, as well as cyclopenta[a]indene and 3′-chloro-5′-hydroxy-β-tyrosine moieties. Articoside and C-1027 chromophore-V inhibited *C. albicans* isocitrate lyase, an enzyme that plays a role in the pathogenicity of such yeast, with IC50 values of 30.4 and 37.9 μM, respectively.

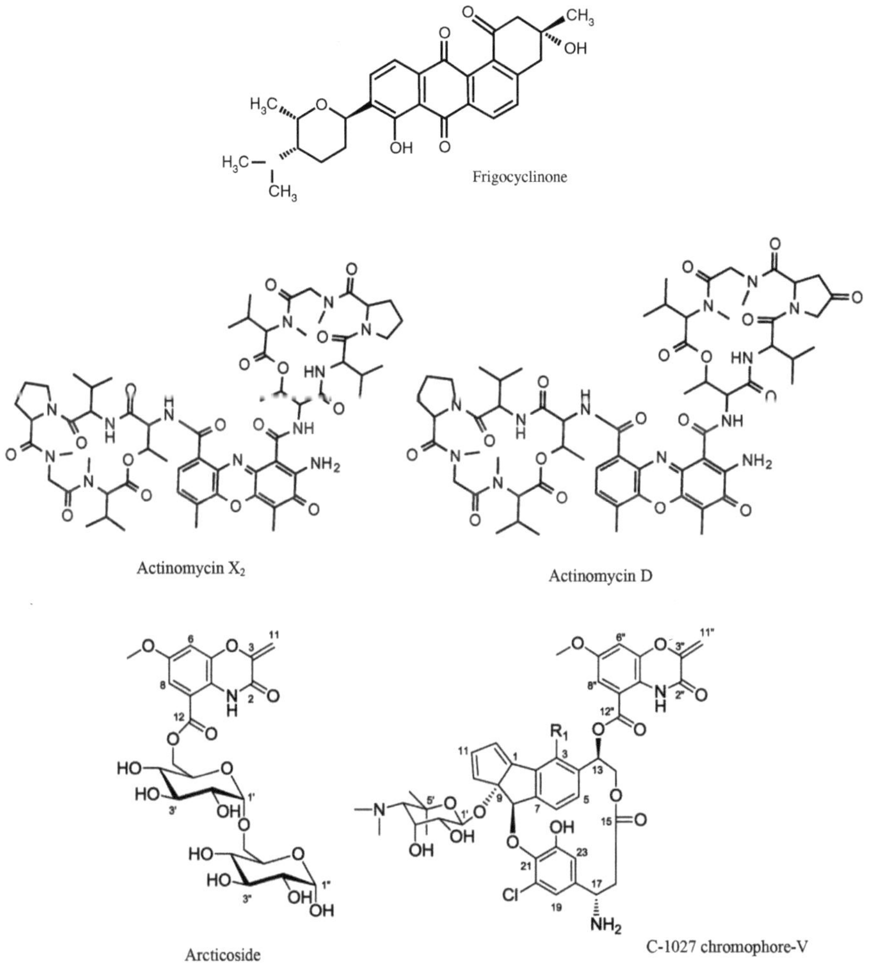

Fig. 3.1 Examples of bioactive compounds isolated from Actinobacteria from Polar sites. *Source*: Lo Giudice and Fani

Among bacteria, other than Actinobacteria, Mojib et al. (2010) tested for antimycobacterial activity two pigments, violacein, a purple violet pigment from *Janthinobacterium* sp. Ant5-2 (J-PVP), and flexirubin, a yellow-orange pigment from *Flavobacterium* sp. Ant342 (F-YOP). Both bacterial strains found were isolated from the land-locked freshwater lakes of Schirmacher Oasis, East Antarctica. Results indicated that the MICs of J-PVP and F-YOP were 8.6 and 3.6 μg/mL for avirulent *Mycobacterium smegmatis* mc^{2}155, 5 and 2.6 μg/mL for avirulent *Mycobacterium tuberculosis* mc^{2}6230, and 34.4 and 10.8 μg/mL for virulent *M. tuberculosis* H_{37}Rv, respectively. The effectiveness of J-PVP on Multiple drug resistant (MDR) and methicillin-resistant *Staphylococcus aureus* (MRSA) strains was further investigated by Huang et al. (2012). The structures J-PVP and F-YOP have not been elucidated.

Nineteen aromatic nitro compounds were isolated from *Salegentibacter* sp. strain T436 from Arctic pack ice, with four of them that were new and six that were never reported before from natural source (Al-Zereini et al. 2007; Schuhmann et al. 2009). The new natural products showed weak antifungal, antibacterial, and cytotoxic activities. The 2-nitro-4-(2′-nitroethenyl)-phenol was the most potent antimicrobial and cytotoxic substance, being active against all targets.

A bacterial strain of *Pseudomonas aeruginosa* from the Antarctic sponge *Isodictya setifera* was found to contain metabolites inhibiting the growth of *B. subtilis*, *S. aureus*, and *M. luteus* (Jayatilake et al. 1996). The culture broth of this bacterium contained a series of diketopiperazines, including a new natural product and two known phenazine alkaloid antibiotics.

Among cyanobacteria, a new lead antibacterial molecule with the proposed structure of 4-[(5-carboxy-2-hydroxy)-benzyl]-1,10-dihydroxy-3,4,7,11,11-pentamethyl-octahydrocyclopenta<a>naphthalene was reported by Asthana et al. (2009) to be produced by the Antarctic cyanobacterium *Nostoc* CCC 537. This intracellular biomolecule, structurally similar to noscomin, exhibited antibiotic activity against acid-fast *M. tuberculosis* H37Rv (MIC, 2.5 µg/mL), *S. aureus* (MIC, 0.5 µg/mL), *Enterobacter aerogenes* (MIC, 4.0 µg/mL), *Salmonella typhi* (MIC, 2.0 µg/mL), *P. aeruginosa* (MIC, 2.0 µg/mL), *E. coli* (MIC, 2.0 µg/mL), and multidrug-resistant strains of *E. coli* (MIC, 16 µg/mL).

Among fungi, the Antarctic ascomycete *Geomyces* sp. 2481, obtained from a soil sample (Fildes Peninsula, King George Island), was grown in solid-substrate fermentation culture (Li et al. 2008). Its organic solvent extract contained five new asterric acid derivatives, ethyl asterrate, n-butyl asterrate, and geomycins A-C. Geomycin B displayed significant antifungal activity against *A. fumigatus* with a MIC value of 20 µg/mL. Geomycin C exhibited antibacterial activities against *S. aureus* (MIC, 24 µg/mL) and *E. coli* (MIC, 20 µg/mL).

A number of metabolites with antimicrobial activity were isolated form *Penicillium griseofulvum* (Greenland) as follows: griseofulvin, fulvic acid, mycelianamide, roquefortine C and D, chanoclavine I and elymoclavine (Frisvad et al. 2004).

More recently, Svahn et al. (2015) isolated *Penicillium nalgiovense* Laxa from a soil sample of an abandoned penguin's nest. Amphotericin B, an antifungal agent used worldwide against fungal infections, was the only metabolite secreted from *P. nalgiovense* Laxa with noticeable inhibitory activity *versus C. albicans*, *S. aureus* and *E. coli*.

5 Conclusions

The review highlights the versatile antimicrobial potential of bacteria and fungi from Polar Regions, making such extreme and (often) underexplored habitats promising for the discovery of new natural compounds of pharmaceutical interest.

It is interesting to note that microorganisms belonging to the same species (isolated from the same or from different habitats) often show different antimicrobial spectra. This suggests that (1) inhibitory activity is more likely strain- rather than species-specific, (2) single species could probably synthesize a set compounds acting on multiple targets, and (3) the antimicrobial spectrum could depend on the habitat of origin. Thus, if secondary metabolic diversity is of interest, it becomes of great importance to keep different isolates of the same bacterial/fungal species in culture collections. To this purpose, our knowledge of the antimicrobial activity of microorganisms from Polar Regions derives from the screening of cultivable strains. This approach is the basis for most microbiological bioprospecting efforts as this both gives access to the totality of genomic information in microorganisms and makes the study of their phenotype possible in the laboratory (de Pascale et al. 2012). However, only a limited fraction of the total microbial community from a certain environment can be easily cultured under laboratory conditions, thus leaving the exceptional bioprospecting potential of the uncultured diversity unexplored (Vester et al. 2015). Recent developments in high-throughput molecular biology techniques have paved the way for employing cultivation-independent approaches (including metagenomics and single-cell genomic sequencing) for bioprospecting purposes. Coupling improved cultivation methods and metagenomics approaches for functional screening of cold-adapted microorganisms can lead to generate targeted information on communities enriched for antimicrobial activities. This emerging trend for microbial bioprospecting in Polar areas certainly merits to be taken into serious consideration in future research programs.

Conflict of Interest Angelina Lo Giudice and Renato Fani declare that they have no conflict of interest.

References

Al-Zereini W, Schuhmann I, Laatsch H, Helmke E, Anke H (2007) New aromatic nitro compounds from *Salegentibacter* sp. T436, an Arctic sea ice bacterium: taxonomy, fermentation, isolation and biological activities. J Antibiot 60:301–308

Asencio G, Lavina P, Alegría K, Domínguez M, Bello H, González-Rocha G, González-Aravena M (2009) Antibacterial activity of the Antarctic bacterium *Janthinobacterium* sp. SMN 33.6 against multi-resistant Gram-negative bacteria. Electron J Biotechnol 17:1–5

Asthana RK, Deepali TMK, Srivastava A, Singh AP, Singh SP, Nath G, Srivastava R, Srivastava BS (2009) Isolation and identification of a new antibacterial entity from the Antarctic cyanobacterium *Nostoc* CCC 537. J Phycol 21:81–88

Bell TH, Callender KL, Whyte LG, Greer CW (2013) Microbial competition in polar soils: a review of an understudied but potentially important control on productivity. Biology 2:533–554

Bérdy J (2005) Bioactive microbial metabolites. J Antibiot 58:1–26

Biondi N, Tredici MR, Taton A, Wilmotte A, Hodgson DA, Losi D, Marinelli F (2008) Cyanobacteria from benthic mats of Antarctic lakes as a source of new bioactivities. J Appl Microbiol 105:105–115

Bosi E, Fondi M, Maida I, Perrin E, de Pascale D, Tutino ML, Parrilli E, Lo Giudice A, Filloux A, Fani R (2015) Genome-scale phylogenetic and DNA composition analyses of Antarctic

Pseudoalteromonas bacteria reveal inconsistencies in current taxonomic affiliation. Hydrobiologia 761:85–95.
Bratchkova A, Ivanova V (2011) Bioactive metabolites produced by microorganisms collected in Antarctica and the Arctic. Biotechnol Biotechnol Equip 25:1–7
Brunati M, Rojas JL, Sponga F, Ciciliato I, Losi D, Göttlich E, de Hoog S, Genilloud O, Marinelli F (2009) Diversity and pharmaceutical screening of fungi from benthic mats of Antarctic lakes. Mar Genomics 2:43–50
Bruntner C, Binder T, Pathom-aree W, Goodfellow M, Bull AT, Potterat O, Puder C, Hörer S, Schmid A, Bolek W, Wagner K, Mihm G, Fiedler H-P (2005) Frigocyclinone, a novel angucyclinone antibiotic by a *Streptomyces griseus* strain from Antarctica. J Antibiot 58:346–349
Bull AT, Stach JEM (2007) Marine actinobacteria: new opportunities for natural product search and discovery. Trends Microbiol 15:491–499
Cavicchioli R, Siddiqui KS, Andrews D, Sowers KR (2002) Low-temperature extremophiles and their applications. Curr Opin Biotechnol 13:253–261
de Pascale D, de Santi C, Fu J, Landfald B (2012) The microbial diversity of Polar environments is a fertile ground for bioprospecting. Mar Genomics 8:15–22
Feller G, Gerday C (2003) Psychrophilic enzymes: hot topics in cold adaptation. Nat Rev Microbiol 1:200–208
Fondi M, Orlandini V, Maida I, Perrin E, Papaleo MC, Emiliani G, de Pascale D, Parrilli E, Tutino ML, Michaud L, Lo Giudice A, Fani R (2012) The draft genome of the VOCs-producing Antarctic bacterium *Arthrobacter* sp. TB23 able to inhibit Cystic Fibrosis pathogens belonging to the *Burkholderia cepacia* complex. J Bacteriol 194:6334–6335
Fondi M, Orlandini V, Perrin E, Maida I, Bosi E, Papaleo MC, Michaud L, Lo Giudice A, de Pascale D, Tutino ML, Liò P, Fani R (2014) Draft genomes of three Antarctic *Psychrobacter* strains known to have antimicrobial activity against *Burkholderia cepacia* complex opportunistic pathogens. Mar Genomics 13:37–38
Frisvad JC, Smedsgaard J, Larsen TO, Samson RA (2004) Mycotoxins, drugs and other extrolites produced by species in *Penicillium* subgenus. Stud Mycol 49:201–241
Furbino LE, Godinho VM, Santiago IF, Pellizari FM, Alves TM, Zani CL, Junior PA, Romanha AJ, Carvalho AG, Gil LH, Rosa CA, Minnis AM, Rosa LH (2014) Diversity patterns, ecology and biological activities of fungal communities associated with the endemic macroalgae across the Antarctic peninsula. Microb Ecol 67:775–787
Gerwick WH, Moore BS (2012) Lessons from the past and charting the future of marine natural products drug discovery and chemical biology. Chem Biol 19:85–98
Gesheva V (2009) Distribution of psychrophilic microorganisms in soils of Terra Nova Bay and Edmonson Point, Victoria Land and their biosynthetic capabilities. Polar Biol 32:1287–1291
Gesheva V (2010) Production of antibiotics and enzymes by soil microorganisms from the Windmill Islands Region, Wilkes Land, East Antarctica. Polar Biol 33:1351–1357
Gesheva V, Negoita T (2012) Psychrotrophic microorganism communities in soils of Haswell Island, Antarctica, and their biosynthetic potential. Polar Biol 35:291–297
Gesheva V, Vasileva-Tonkova E (2012) Production of enzymes and antimicrobial compounds by halophilic Antarctic *Nocardioides* sp. grown on different carbon sources. World J Microbiol Biotechnol 28:2069–2076
Giddings L-A, Newman DJ (2015) Bioactive compounds from marine extremophiles. Springer Briefs Microbiol 1–124
Godinho VM, Furbino LE, Santiago IF, Pellizzari FM, Yokoya NS, Pupo D, Alves TM, Junior PA, Romanha AJ, Zani CL, Cantrell CL, Rosa CA, Rosa LH (2013) Diversity and bioprospecting of fungal communities associated with endemic and cold-adapted macroalgae in Antarctica. ISME J 7:1434–1451
Godinho VM, Gonçalves VN, Santiago IF, Figueredo HM, Vitoreli GA, Schaefer CE, Barbosa EC, Oliveira JG, Alves TM, Zani CL, Junior PA, Murta SM, Romanha AJ, Kroon EG, Cantrell CL, Wedge DE, Duke SO, Ali A, Rosa CA, Rosa LH (2015) Diversity and bioprospection of fungal community present in oligotrophic soil of continental Antarctica. Extremophiles 19:585–596

Hemala L, Zhanga D, Margesin R (2014) Cold-active antibacterial and antifungal activities and antibiotic resistance of bacteria isolated from an alpine hydrocarbon-contaminated industrial site. Res Microbiol 165:447–456
Henríquez M, Vergara K, Norambuena J, Beiza A, Maza F, Ubilla P, Araya I, Chávez R, San-Martín A, Darias J, Darias MJ, Vaca I (2014) Diversity of cultivable fungi associated with Antarctic marine sponges and screening for their antimicrobial, antitumoral and antioxidant potential. World J Microbiol Biotechnol 30:65–76
Hentschel U, Hopke J, Horn M, Friedrich AB, Wagner M, Hacker J, Moore BS (2002) Molecular evidence for a uniform microbial community in sponges from different oceans. Appl Environ Microbiol 68:4431–4440
Huang JP, Mojib N, Rr G, Watkins S, Waites KB, Ravindra R, Andersen DT, Bej AK (2012) Antimicrobial activity of PVP from an Antarctic bacterium, *Janthinobacterium* sp. Ant5-2, on multi-drug and methicillin resistant *Staphylococcus aureus*. Nat Prod Bioprospect 2:104–110
Ivanova V, Oriol M, Montes M-J, García A, Guinea J (2001) Secondary metabolites from a *Streptomyces* strain isolated from Livingston Island, Antarctica Z Naturforsch C 56:1–5
Ivanova V, Yocheva L, Schlegel R, Graefe U, Kolarova M, Aleksieva K, Naidenova M (2002) Antibiotic complex from *Streptomyces flavovirens* 6[7], isolated from Livingston Island, Antarctica. Bulgarian Antarct Res Life Sci 3:35–42
Jayatilake GS, Thornton MP, Leonard AC, Grimwade JE, Baker BJ (1996) Metabolites from an Antarctic sponge-associated bacterium, *Pseudomonas aeruginosa*. J Nat Prod 59:293–296
Kim M-K, Park H, Oh T-J (2012) Antibacterial properties associated with microorganisms isolated from Arctic lichens. Korean J Microbiol Biotechnol 40:380–388
Kim M-K, Park H, Oh T-J (2013) Antibacterial properties of the bacterial associates of the Arctic lichen *Stereocaulon* sp. Afr J Microbiol Res 7:3651–3657
Kim M-K, Park H, Oh T-J (2014a) Antibacterial and antioxidant capacity of polar microorganisms isolated from Arctic lichen *Ochrolechia* sp. Pol J Microbiol 63:317–322
Kim M-K, Park H, Oh T-J (2014b) Antibacterial and antioxidant potential of polar microorganisms isolated from Antarctic lichen *Psoroma* sp. Afr J Microbiol Res 8:3529–3535
Kozlovsky AG, Zhelifonova VP, Antipova TV, Baskunov BP, Kochkina GA, Ozerskaya SM (2012) Secondary metabolite profiles of the *Penicillium* fungi isolated from the Arctic and Antarctic permafrost as elements of polyphase taxonomy. Microbiology 81:308–313
Lee L-H, Cheah Y-K, Sidik SM, Ab Mutalib N-S, Tang Y-L, Lin H-P, Hong K (2012a) Molecular characterization of Antarctic actinobacteria and screening for antimicrobial metabolite production. World J Microbiol Biotechnol 28:2125–2137
Lee L-H, Cheah Y-K, Syakima AMN, Shiran MS, Tang Y-L, Lin H-P, Hong K (2012b) Analysis of Antarctic proteobacteria by PCR fingerprinting and screening for antimicrobial secondary metabolites. Genet Mol Res 11:1627–1641
Li Y, Sun B, Liu S, Jiang L, Liu X, Zhang H, Che Y (2008) Bioactive asterric acid derivatives from the Antarctic ascomycete fungus *Geomyces* sp. J Nat Prod 1:1643–1646
Liu J-T, Lu X-L, Liu X-Y, Yun G, Bo H, Jiao B-H, Zheng H (2013) Bioactive natural products from the Antarctic and Arctic organisms source. Mini Rev Med Chem 13:617–626
Lo Giudice A, Brilli M, Bruni V, De Domenico M, Fani R, Michaud L (2007a) Bacterium-bacterium inhibitory interactions among psychrotrophic bacteria isolated from Antarctic seawaters (Terra Nova Bay, Ross Sea). FEMS Microbiol Ecol 60:383–396
Lo Giudice A, Bruni V, Michaud L (2007b) Characterization of Antarctic psychrotrophic bacteria with antibacterial activities against terrestrial microorganisms. J Basic Microbiol 47:496–505
Lyutskanova D, Ivanova V, Stoilova-Disheva M, Kolarova M, Aleksieva K, Raykovska V, Peltekovavska V, Peltekova V (2009) Isolation, characterization and screening for antimicrobial activities of psychrotolerant Streptomycetes isolated from polar permafrost soil. Biotechnol Biotechnol Eq 23.
Maida I, Fondi M, Papaleo MC, Perrin E, Orlandini V, Emiliani G, de Pascale D, Parrilli E, Tutino ML, Michaud L, Lo Giudice A, Romoli R, Bartolucci G, Fani R (2014) Phenotypic and genomic characterization of the Antarctic bacterium *Gillisia* sp. CAL575, a producer of antimicrobial compounds. Extremophiles 18:35–49

Maida I, Bosi E, Fondi M, Perrin E, Orlandini V, Papaleo MC, Mengoni A, de Pascale D, Tutino ML, Michaud L, Lo Giudice A, Fani R (2015. Antimicrobial activity of Pseudoalteromonas strains isolated from the Ross Sea (Antarctica) vs Cystic Fibrosis opportunistic pathogens. Hydrobiologia 761:443–457
Mangano S, Michaud L, Caruso C, Brilli M, Bruni V, Fani R, Lo Giudice A (2009) Antagonistic interactions among psychrotrophic cultivable bacteria isolated from Antarctic sponges: a preliminary analysis. Res Microbiol 160:27–37
Manivasagan P, Venkatesan J, Sivakumar K, Kim S-K (2014) Pharmaceutically active secondary metabolites of marine actinobacteria. Microbiol Res 169:262–278
Margesin R (2007) Alpine microorganism: useful tools for low-temperature bioremediation. J Microbiol 45:281–285
Melo IS, Santos SN, Rosa LH, Parma MM, Silva LJ, Queiroz SC, Pellizari VH (2014) Isolation and biological activities of an endophytic *Mortierella alpina* strain from the Antarctic moss *Schistidium antarctici*. Extremophiles 18:15–23
Mojib N, Philpott R, Huang JP, Niederweis M, Bej AK (2010) Antimycobacterial activity in vitro of pigments isolated from Antarctic bacteria. Antonie Van Leeuwenhoek 98:531–540
Moncheva P, Tishkov S, Dimitrova N, Chipeva V, Antonova-Nikolova S, Bogatzevska N (2002) Characteristics of soil Actinomycetes from Antarctica. J Cult Collect 3:3–14
Montemartini Corte A, Liotta M, Venturi CB, Calegari L (2000) Antibacterial activity of *Penicillium* spp. strains isolated in extreme environments. Polar Biol 23:294–297
Moon K, Ahn C-H, Shin Y, Won TH, Ko K, Lee SK, Oh K-B, Shin J, Nam SI, Oh DC (2014) New benzoxazine secondary metabolites from an Arctic actinomycetes. Mar Drugs 12:2526–2538
Morita RY (1975) Psychrophilic bacteria. Bacteriol Rev 39:144–167
Nedialkova D, Naidenova M (2005) Screening the antimicrobial activity of Actinomycetes strains isolated from Antarctica. J Cult Collect 4:29–35
O'Brien A, Sharp R, Russell NJ, Roller S (2004) Antarctic bacteria inhibit growth of food-borne microorganisms at low temperatures. FEMS Microbiol Ecol 48:157–167
Onofri S, Fenice M, Cicalini AR, Tosi S, Magrino A, Pagano S, Selbmann L, Zucconi L, Vishniac HS, Ocampo-Friedmann R, Friedmann EI (2000) Ecology and biology of microfungi from Antarctic rocks and soils. Ital J Zool 67:163–167
Orlandini V, Maida I, Fondi M, Perrin E, Papaleo MC, Bosi E, de Pascale D, Tutino ML, Michaud L, Lo Giudice A, Fani R (2014) Genomic analysis of three sponge-associated *Arthrobacter* Antarctic strains, inhibiting the growth of *Burkholderia cepacia* complex bacteria by synthesizing volatile organic compounds. Microbiol Res 169:593–601
Pan SY, Tan GYA, Convey P, Pearce DA, Tan IKP (2013) Diversity and bioactivity of actinomycetes from Signy Island terrestrial soils, maritime Antarctic. Adv Polar Sci 24:208–212
Papaleo MC, Fondi M, Maida I, Perrin E, Lo Giudice A, Michaud L, Mangano S, Bartolucci G, Romoli R, Fani R (2012) Sponge-associated microbial Antarctic communities exhibiting antimicrobial activity *against Burkholderia cepacia* complex bacteria. Biotechnol Adv 30:272–293
Papaleo MC, Romoli R, Bartolucci G, Maida I, Perrin E, Fondi M, Orlandini V, Mengoni A, Emiliani G, Tutino ML, Parrilli E, de Pascale D, Michaud L, Lo Giudice A, Fani R (2013) Bioactive volatile organic compounds from Antarctic (sponges) bacteria. New Biotechnol 30:824–838
Pearce DA (2012) Extremophiles in Antarctica: life at low temperatures. In: Stan-Lotter H, Fendrihan S (eds) Adaption of microbial life to environmental extremes. Springer, Vienna, pp 87–118
Prasad S, Manasa P, Buddhi S, Singh SM, Shivaji S (2011) Antagonistic interaction networks among bacteria from a cold soil environment. FEMS Microbiol Ecol 78:376–385
Rojas JL, Martín J, Tormo JR, Vicente F, Brunati M, Ciciliato I, Losi D, Van Trappen S, Mergaert J, Swings J, Marinelli F, Genilloud O (2009) Bacterial diversity from benthic mats of Antarctic lakes as a source of new bioactive metabolites. Mar Genomics 2:33–41

Romoli R, Papaleo MC, de Pascale D, Tutino ML, Michaud L, Lo Giudice A, Fani R, Bartolucci G (2011) Characterization of the volatile profile of Antarctic bacteria by using solid-phase microextraction – gas chromatography mass spectrometry. J Mass Spectrom 46:1051–1059

Romoli R, Papaleo MC, de Pascale D, Tutino ML, Michaud L, Lo Giudice A, Fani R, Bartolucci G (2014) GC-MS volatolomic approach to study the antimicrobial activity of the Antarctic bacterium *Pseudoalteromonas* sp. TB41. Metabolomics 10:42–51

Russell NJ (2008) Membrane components and cold sensing. In: Margesin R, Schinner F, Marx JC, Gerday C (eds) Psychrophiles: from biodiversity to biotechnology. Springer, Berlin, Heidelberg, pp 177–190

Schuhmann I, Yao CB, Al-Zereini W, Anke H, Helmke E, Laatsch H (2009) Nitro derivatives from the Arctic ice bacterium *Salegentibacter* sp. isolate T436. J Antibiot 62:453–460

Shekh RM, Singh P, Singh SM, Roy U (2011) Antifungal activity of Arctic and Antarctic bacteria isolates. Polar Biol 34:139–143

Svahn KS, Chryssanthou E, Olsen B, Bohlin L, Göransson U (2015) *Penicillium nalgiovense* Laxa isolated from Antarctica is a new source of the antifungal metabolite amphotericin B. Fungal Biol Biotechnol 2:1. doi:10.1186/s40694-014-0011-x

Taton A, Grubisic S, Ertz D, Hodgson DA, Piccardi R, Biondi N, Tredici MR, Mainini M, Losi D, Marinelli F, Wilmotte A (2006) Polyphasic study of antarctic cyanobacterial strains. J Phycol 42:1257–1270

Vester JK, Glaring MA, Stougaard P (2015) Improved cultivation and metagenomics as new tools for bioprospecting in cold environments. Extremophiles 19:17–29

Wietz M, Månsson M, Bowman JS, Blom N, Ng Y, Gram L (2012) Wide distribution of closely related, antibiotic-producing *Arthrobacter* strains throughout the Arctic Ocean. Appl Environ Microbiol 78:2039–2042

Wong CMV, Tam HK, Alias SA, González M, González–Rocha G, Domínguez–Yévenes M (2011) *Pseudomonas* and *Pedobacter* isolates from King George Island inhibited the growth of foodborne pathogens. Pol Polar Res 32:3–14

Yuan M, Yu Y, Li H-R, Dong N, Zhang X-H (2014) Phylogenetic diversity and biological activity of Actinobacteria isolated from the Chukchi Shelf marine sediments in the Arctic Ocean. Mar Drugs 12:1281–1297

Chapter 4
Thermophiles as a Promising Source of Exopolysaccharides with Interesting Properties

Margarita Kambourova, Nadja Radchenkova, Iva Tomova, and Ivanka Bojadjieva

1 Introduction

Thermophiles are a type of extremophilic microorganisms able to grow at comparatively high temperatures, between 45 and 122 °C (Takai et al. 2008). Extremophiles not only endure, but are functionally active in some of the harshest conditions of life found on Earth. "Extreme" is a relative term, which means the conditions too harsh for the existence of man (Satyanarayana et al. 2005). The conception for upper temperature for life has changed several times at the end of the last century as thermophilic representatives of the domains *Bacteria* and *Archaea* were isolated from geothermal and hydrothermal habitats at higher and higher temperature. Suggested groups of thermophiles are: facultative thermophiles growing up to 50 °C, obligate thermophiles able to grow between 50 and 70 °C and optimally at 55–65 °C, extreme thermophiles growing in the range 65–80 °C, and hyperthermophiles—optimum is higher than 80 °C (Wiegel and Canganella 2001).

Thermophilic niches include volcanic and geothermal areas (terrestrial, subterranean and marine hot springs), solfataric areas, sun heated refuse, oil reservoirs, and manmade habitats (Fig. 4.1). As high temperature in them limits growth of the representatives of the domain *Eukarya*, the representatives of domains *Bacteria*, *Archaea* and their viruses predominate in hot niches (López-López et al. 2013). Various arguments have been cited in support of the idea that the ancestors of the *Archaea* and *Bacteria* domains seem to be (hyper)thermophiles (Di Giulio 2003). Extensive global research efforts on the diversity and biotechnological potential in these extreme environments revealed unexpected number of taxa and presence of number of novel species. Extremophiles possess novel metabolic properties making

M. Kambourova (✉) • N. Radchenkova • I. Tomova • I. Bojadjieva
Institute of Microbiology, Bulgarian Academy of Sciences,
Acad. G. Bonchev str. 26, 1113 Sofia, Bulgaria
e-mail: margikam@microbio.bas.bg

P.H. Rampelotto (ed.), *Biotechnology of Extremophiles*, Grand Challenges in Biology and Biotechnology 1, DOI 10.1007/978-3-319-13521-2_4

Fig. 4.1 Geysera, Sapareva Banja, Bulgaria is characterized by water temperature of 103 °C in the hole drilling

them capable of physiologic activity in unfavorable conditions and production of unique metabolites (Rozanov et al. 2014). Such compounds significantly contribute to the development of biotechnology in the recent years.

Thermophilic microorganisms and their thermostable enzymes already have imperative biotechnological recognition because of their unique ability to function at high temperature and stability to a variety of harsh industrial conditions. Among compounds involved in thermophilic adaptation, exocellular polysaccharides are of a special interest due to their structural and functional diversity and their diverse physiological roles in cell. Exopolysaccharides are high-molecular-weight microbial polymers secreted into the surrounding environment. They are composed of identical or differing sugar residues arranged as repeated units within the polymer. Homopolysaccharides are composed of identical residues; different sugar residues form the molecule of heteropolysaccharides.

2 Physiological Role of EPSs in Cell

Accumulation of EPS on cell surface is a common adaptation strategy of extremophiles including participation in cell protection by stabilizing membrane structure. Polysaccharide layer over the cell surface is often several times ticker than the cell dimension. However, the physiological role of EPSs in bacteria is probably more diverse and complex than currently known. Some of these biopolymers perform the same function, whereas others fulfill distinct functions in dependence of taxonomic

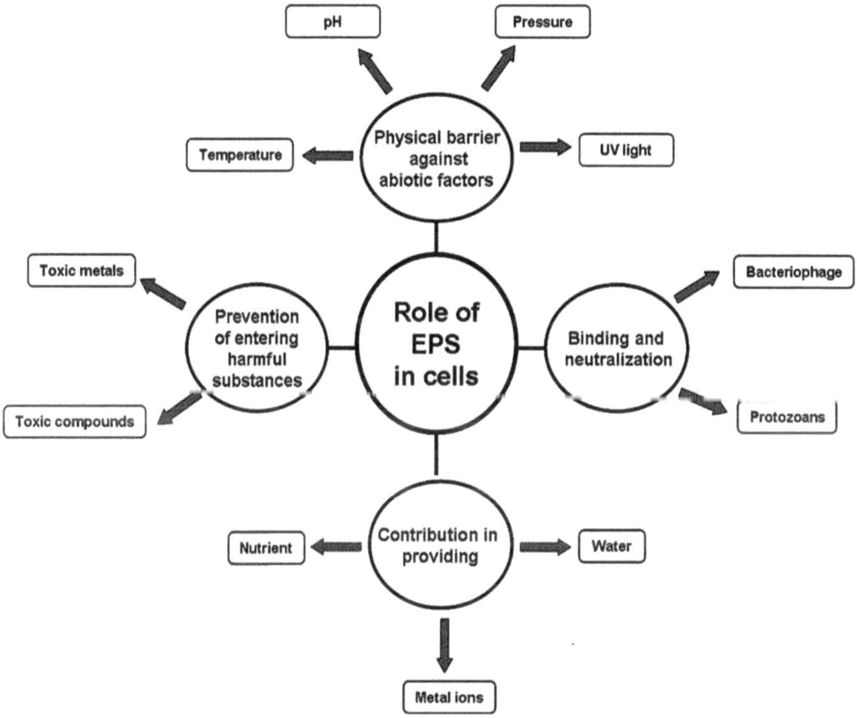

Fig. 4.2 Possible physiological roles of EPS in microbial cells that could contribute surviving in extreme conditions

affiliation and ecological niches (Poli et al. 2011; Nwodo et al. 2012). Despite the fact that the EPS production requires energy representing up to 70 % of total energy reserve, its impact in microbial growth is significantly higher and depends on the environment. Some of the functions are shown below (Fig. 4.2):

- EPSs protect microorganisms as a general physical barrier. Their production reflects selective environmental pressures including osmotic stress, temperature, pH, atmospheric pressure and light intensity and aid in adapting to extreme conditions (Otero and Vincenzini 2003).
- Surrounding of cell by EPS allows retention of water in water-deficient environments maintaining a hydrated microenvironment.
- At oligotrophic conditions like many of the extreme environments are EPSs can sequester nutrient materials from the surrounding environment EPS matrix binds and accumulates biodegradable compounds and cations from the bulk water phase. The anionic nature of the exterior polysaccharides due to the presence of sulphates and uronic acids residues can help interaction with cations such as metals and capture essential minerals and nutrients.
- EPSs may aid the organisms to adhere to surfaces and can serve as flocculants or emulsifiers. Many microorganisms use the synthesis of exopolysaccharides as a

strategy for growing and adhering to solid surfaces by covering the hydrophobic sites on the cell envelope.
- The presence of EPS layer around the cell may affect the diffusion of harmful substances, like antibiotics or toxic compounds (e.g. toxic metal ions, sulphur dioxide, and ethanol) into and out of the cell. Due to their anionic character, the polysaccharides can enhance the immobilization of ions such as Pb^{2+} and Cu^{2+} with a significant ecological impact. EPSs participate in the removal of heavy metals from the environment by flocculation and binding metal ions from solutions (Nicolaus et al. 2010). The polymer from a thermophilic bacterium *Geobacillus tepidamans* was proved to have an anti-cytotoxic activity against avarol (Kambourova et al. 2009).
- Participation of EPSs in cell protection is referred to binding and neutralizing bacteriophage (Vu et al. 2009).
- Although it is commonly accepted that EPSs are not used as energy reserves, and microorganisms are unable to degrade their own EPSs (Donot et al. 2012), EPSs produced by some hyper thermophilic species *(Sulfolobus*, *Thermococcus*, and *Thermotoga)* can act indirectly as extracellular storage polymers in extreme environments poor of other organic source (Nicolaus et al. 1993; Rinker and Kelly 2000).

2.1 Biofilms

In natural environment, matrix of extracellular polymeric substances synthesized by most bacteria comprises surface-associated communities and its integrity is essential for their survival (Moons et al. 2009). Biofilm is involved in the protective role to environmental stress, in adherence of biofilms to surfaces, in cell-cell interactions. Induction of biofilm formation by elevated pH, decreased and increased growth temperature, high salt, and exposure to UV light, oxygen, or antibiotic in *Archaeoglobus fulgidus* and ammonium chloride in *Thermococcus litoralis* demonstrated their protective role (Pysz et al. 2004). Such a way the biofilms create specific microenvironments, and could expand the limits of microbial growth creating more favorable conditions for life (Lowell et al. 2008). Multilayered biofilm structures consist of EPSs, extracellular DNAs (Flemming et al. 2007). Exopolysaccharides are the basis for biofilms, which have been observed in pure cultures or co-cultures. Although many thermophiles have been isolated from hot springs, still the reports of their biofilm EPSs formation are scared. The mechanism of biofilm formation and its importance for microbial survival in natural habitats has attracted increasing interest in recent years. Quorum sensing participation in interspecies interaction was demonstrated by participation of several GGDEF domain proteins in conjunction with a quorum sensing peptide TM0504 in co-cultivation of hyperthermophilic bacterium *Thermotoga maritima* with a methanogenic archaea *Methanococcus jannaschii* (Muralidharan et al. 1997).

3 Thermophilic EPS-Producing Prokaryotes

Synthesis of EPSs is in response to adaptation to an extreme environment, biotic stress (e.g., inter and intra species competition for substrate, water or growth factors), abiotic stress factors (e.g., temperature, light intensity, pH, salinity) (Donot et al. 2012). Thermophilic microorganisms could be found in almost every phylum of *Archaea* and *Bacteria.*

3.1 *Bacterial Producers*

Thermophilic bacteria belonging to obligate thermophilic genera *Bacillus, Geobacillus, Brevibacillus,* and *Aeribacillus,* extremely thermophilic genus *Thermus* and hyperthermophilic genus *Thermotoga* were reported as good thermophilic producers of EPSs. They were isolated from continental hot springs or shallow marine vents (Table 4.1). Hyperthermophilic bacterial species *Thermotoga maritima* is isolated originally from geothermally heated sea floor (Huber et al. 1986). This microorganism is strictly anaerobic heterotroph with optimal growth temperature of 80 °C. As microorganisms seldom behave in an isolated manner, mutually beneficial relationship probably expands the scope of natural hydrothermal environments (Kolter and Losick 1998). Accumulation of H_2 by methanogens is a preposition for often observed co-cultures of heterotrophs and methanogens in natural environments (Muralidharan et al. 1997). The co-cultivation of methanogenic archaea *Methanococcus jannaschii* and extremely thermophilic fermentative anaerobic bacteria *Thermotoga maritima* results in increasing the cell density. A novel extracellular polysaccharide from the biofilm of *Thermus aquaticus* YT-1 was isolated (Lin et al. 2011).

Despite of the enormous exploration of thermophilic bacilli as sources for thermostable enzymes, the knowledge on their ability to produce EPSs is still in its childhood. *Bacillus licheniformis* B3-15 and T14 (Maugeri et al. 2002; Spanò et al. 2013) and strains belonging to the genus *Geobacillus* isolated from shallow hydrothermal vents and terrestrial geothermal springs (Manca et al. 1996; Nicolaus et al. 2000, 2002, 2003, 2004; Kambourova et al. 2009) were described as producers of EPSs. Two different EPSs were produced by *Aeribacillus pallidus* 418 (Fig. 4.3) (Radchenkova et al. 2013).

3.2 *Archaea*

Discovery of extremophilic archaea has had a great significance to biocatalysis, as their enzymes allow improvements in multiple sectors of industry (Węgrzyn and Żukrowski 2014). Another type of biopolymer from *Archaea* with a potential impact for biotechnological industry is EPS.

Table 4.1 Exopolysaccharide production by thermophilic prokaryotes

Domain	Microorganism	Isolation source	Growth temperature (°C)	Carbon source	Maximum yield (mg/L)	Functional properties and potential application	Reference
Bacteria	*Thermus aquaticus* YT-1	Hot spring, Yellowstone, US	60	n.d.	n.d.	Immunomodulatory effects	Lin et al. (2011)
	Thermotoga maritima	Geothermal heated marine sediment at Vulcano, Italy	88	Maltose	120	n.d.	Rinker and Kelly (2000)
	Geobacillus thermoantarcticus	Crater of Mount Melbourne, Antarctica	65	Mannose	400	n.d.	Manca et al. (1996); Nicolaus et al. (2004)
	Bacillus licheniformis B3-15	Shallow marine hot spring, Vulcano, Italy	45	Glucose	165	Antiviral and immunoregulatory effects	Maugeri et al. (2002); Arena et al. (2006)
	Bacillus licheniformis T14	Shallow hydrothermal vent, Panarea Island, Italy	50	Sucrose	366	Antiviral and immunomodulatory effects	Spanò et al. (2013); Gugliandolo et al. (2014)
	Geobacillus sp. 4001	Shallow marine hydrothermal vents of Flegrean Ares, Italy	65	Galactose, sucrose	55	n.d.	Nicolaus et al. (2002); Nicolaus et al. (2004)
	Geobacillus sp. 4004	Shallow marine hydrothermal vents of Flegrean Ares, Italy	65	Trehalose, sucrose	65	n.d.	Nicolaus et al. (2002); Nicolaus et al. (2004)
	Geobacillus thermodenitrificans B3-72	Shallow hydrothermal vent, Vulcano island, Italy	65	Glucose, sucrose	70	EPS2: antiviral and immunomodulatory effects	Nicolaus et al. (2000); Arena et al. (2009)
	Geobacillus tepidamans V264	Hot spring, Velingrad	60	Maltose	111.4	High thermostability; anti-cytotoxic activity	Kambourova et al. (2009)
	Aeribacillus pallidus 418	Hot spring, Rupi basin, Bulgaria	55	Maltose	170	High thermostability; emulsifying properties	Radchenkova et al. (2013); Radchenkova et al. (2014)
	Brevibacillus thermoruber 423	Hot spring, Blagoevgrad region, Bulgaria	55	Maltose	897	Biomedical applications	Yasar Yildiz et al. (2014)

Archaea	*Thermococcus litoralis*	Shallow marine thermal spring, Naples, Italy	88	Maltose	180	Production of a mannan-like compound typically produced by eukaryotes	Rinker and Kelly (1996); Rinker and Kelly (2000)
	Sulfolobus solfataricus MT4	Hot acidic spring, Agnano, Italy	88	Glucose	8.4	n.d.	Nicolaus et al. (1993)
	Sulfolobus solfataricus MT3	Hot acidic spring, Agnano, Italy	75	Glucose	7	n.d.	Nicolaus et al. (1993)
	Sulfolobus tokodaii	Beppu hot springs, Japan	76	n.d.	n.d.	n.d.	Koerdt et al. (2010)
	Sulfolobus acidocaldarius	Hot spring, Yellowstone, US	76	n.d.	n.d.	n.d.	Koerdt et al. (2010)

n.d. no data

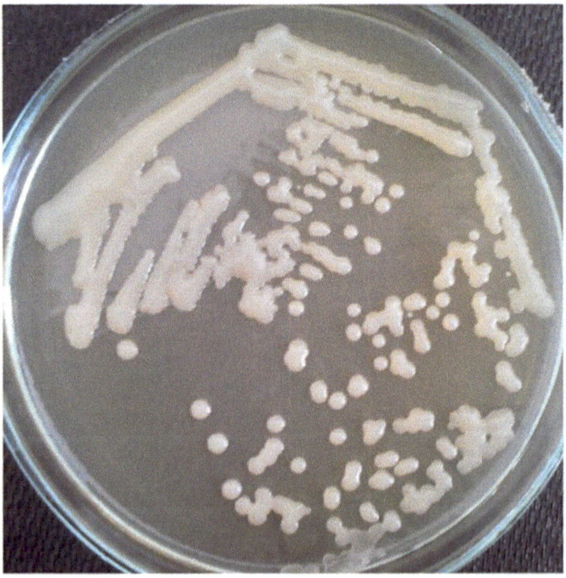

Fig. 4.3 Slime production by an obligate thermophile *Aeribacillus pallidus* 418 on peptone-yeast extract agar medium, 24 h culture

Archaea representatives are often observed component of biofilm communities from many different environments (Krüger et al. 2008; Zhang et al. 2008), but few studies report on biofilm formation by *Archaea*. The ability of *Sulfolobus solfataricus* to produce EPS was first studied more than 20 years ago (Nicolaus et al. 1993) and further *Sulfolobus acidocaldarius* and *Sulfolobus tokodaii* were reported to form biofilm (Koerdt et al. 2010). Since similar biofilms were observed for *Archaeoglobus profundus*, *Archaeoglobus fulgidus* (Lapaglia and Hartzell 1997), *Thermococcus litoralis* (Rinker and Kelly 1996), *Methanococcus jannaschii*, (Lapaglia and Hartzell 1997) and *Methanothermobacter thermoautotrophicus* (Thoma et al. 2008), the biofilm formation might be a common stress response mechanism among the *Archaea* (Hartzell et al. 1999).

Representatives of a genus *Sulfolobus* (phylum *Crenarchaeota*) exist all over the world in acidic, mostly muddy, hot springs in which the sharp variations in temperature, pH and geochemical conditions are often observed suggesting the need in a quick adaptation or survive undisturbed the changing conditions (Koerdt et al. 2010).

EPS production is not limited to *Crenarchaeota*; some species belonging to the phylum *Euryarchaeota* have this ability too. The cells of euryarchaeon *Pyrococcus furiosus* (notable for its optimum growth temperature of 100 °C and often used as a model organism for hyperthermophiles) are interconnected by flagella in microcolonies in biofilm-like structures (Näther et al. 2006). This microorganism was able to adhere also to biotic surfaces; it was proved to form biofilm in co-culture with another hyperthermophilic euryarchaeon, *Methanopyrus kandleri* habituating similar niches (temperature close to that of boiling water and anaerobic conditions). *Pyrococcus furiosus* and *Methanopyrus kandleri* cells were able to form archaeal bi-species biofilm under laboratory conditions in less than 24 h (Schopf et al. 2008). The heterotrophic facultative sulfur-dependent hyperthermophile *Thermococcus litoralis* (*Euryarchaeota*) isolated from shallow marine thermal spring also formed a biofilm on hydrophilic surfaces under a variety of conditions. Baker-Austin et al. (2010) observed two distinct biofilm morphologies in the extremely acidophilic *Euryarchaeote, Ferroplasma acidarmanus*. The ability of an anaerobic marine hyperthermophilic euryarchaeon *Archaeoglobus fulgidus* to colonize widely separated areas was related to biofilm formation as a stress mechanism for surviving variations in conditions like concentrations of nutrients, temperature, and potentially toxic compounds (Lapaglia and Hartzell 1997).

4 Chemical and Structural Composition of Thermophilic Exopolysaccharides

In contrast to plant polysaccharides, accumulated knowledge on the structural properties of bacterial exopolysaccharides and especially EPSs from thermophilic microorganisms is scarce. The EPSs synthesized by microbial cells are characterized by different composition and hence, different chemical and physical properties. Comparison the data for carbohydrate structures of mammals and bacteria accumulated in multiple databases showed diversity in bacteria that is 10 times higher for monosaccharides and 9 times higher for glycosidic bonds compared to those reported from mammals (Herget et al. 2008). Most EPSs are heteropolysaccharides consisting of three or four different monosaccharides forming groups of 10 or less to form the repeating units (Poli et al. 2011). The most commonly monosaccharides are linked in the backbones by strong 1,4-β-, 1,3-β-, or 2,6-β linkages and more flexible 1,2-α- or 1,6-α-linkages. Several linkages can occur at the same time in one polysaccharide. The repeating sugar units are mainly composed of glucose, galactose, mannose, uronic acids, N-acetyl glucosamine, N-acetyl galactosamine and rhamnose, in variable ratios. The composition and structure of the polysaccharides determine their primary conformation. Secondary configuration comprises aggregated helices; acyl substituents influence the transition from random coil to ordered helical aggregates (Sutherland 1994). The rigidity of glycosaminoglycans in EPS provides structural integrity of molecules and cells. Forming of random coils that tend to form helical aggregates by the acetyl groups of N-acetyl galactosamines in EPSs from *Thermus aquaticus* YT-1 was observed by Lin et al. (2011). Its presence contributes to a regular and stable structure. Often heteropolysaccharides contain non-sugar components like acetates, pyruvates, succinates, phosphates, sulphates, methyl esters, proteins, nucleic acids and lipids (Nicolaus et al. 2010) which components may play an important role, for example, the over-sulphation of EPSs can modify their biological activity (Courtois et al. 2014). Some EPSs are neutral macromolecules, polyanionic nature of many EPSs due to uronic acids or ketal-linked pyruvate or inorganic residues in the molecule (Nicolaus et al. 2010).

The polysaccharides from *Thermococcus litoralis* and *Geobacillus thermoantarcticus* are homopolymers composed by mannose (Rinker and Kelly 1996; Manca et al. 1996). Almost pure glucan is EPS synthesized by *Geobacillus tepidamans* (Kambourova et al. 2009).

The main sugar for most of the heteropolysaccharides synthesized by thermophilic microorganisms is glucose or mannose; however also fructose or galactose was reported (Table 4.2). Three *Geobacillus* strains isolated from shallow marine vent is reported to produce EPSs with different composition (Nicolaus et al. 2003). EPSs isolated from two *Geobacillus* sp. strains contained as main sugars glucose, galactose and mannose in different proportions and the third strain from the same genus contained glucosamine and arabinose together with galactose and mannose. EPS from the thermophilic bacterium *Brevibacillus thermoruber* 423 is composed by five different sugars (with glucose as a major monomer unit) (Yasar Yildiz et al. 2014). *Aeribacillus pallidus* 418 produced two high molecular weight EPSs consisting of high variety of sugars (six for EPS 1 and seven for EPS 2) with mannose as a major component (Radchenkova et al. 2013). Recently, *Aeribacillus pallidus* YM-1 was reported to produce a novel bioemulsifier consisting of lipids (47.6 %), carbohydrates (41.1 %), and proteins (11.3 %) (Zheng et al. 2012). Carbohydrate fraction consisted of glucose (36.6 %), altrose (30.9 %), mannose (24.4 %) and galactose (8.1 %).

A presence of pyruvate and sulphate for EPS produced by *Bacillus thermantarticus* was reported (Nicolaus et al. 2004). Lin et al. (2011) characterized the primary structure of TA-1, the major EPS secreted by *Thermus aquaticus*. The polymer consists of tetrasaccharide-repeating units of galactofuranose, galactopyranose, and N-acetylgalactosamine and lacks acidic sugars. Five per cent of the dry mass in the aggregates of *Thermotoga maritima* was polysaccharide consisting of 91.2 % glucose, 5.2 % ribose and 2.7 % mannose (Johnson et al. 2005). *Sulfolobus* strains produce extracellular polysaccharides containing mannose, glucose, galactose, and N-acetylglucosamine (Koerdt et al. 2010).

Another feature of EPSs from all reported thermophiles seems to be their high molecular weight (beginning from several hundred kDa). A molecular weight of 380, 400, 600 and 1000 kDa was reported for EPSs from thermophilic bacilli (Nicolaus et al. 2003). The molecular weight was determined to be about 700 and 1000 kDa for EPSs from *Aeribacillus pallidus* 418 (Radchenkova et al. 2013). EPS from another thermophile, *Geobacillus thermoantarcticus*, had a molecular weight approximately 300 kDa (Manca et al. 1996). The molecular weight of the polymer from *Geobacillus tepidamans* V264 was higher than 1000 kDa (Kambourova et al. 2009). According to Kumar et al. (2007) mesophilic bacteria synthesize polysaccharides with molecular weight of 10–30 kDa. These values are significantly lower than the established ones for thermophilic bacteria. High molecular weight was announced for some EPSs from mesophiles, like EPSs excreted by *Alteromonas macleodii* (300 and 1500 kDa) (Raguénès et al. 1996, 2003); alginates produced by *Pseudomonas* species (34–500 kDa) (Conti et al. 1994), gellan (250 and 490 kDa) (Milas et al. 1990). EPSs from lactobacilli are also characterized by a different molecular weight, from 10 to 1000 kDa according to Patel et al. (2010) and from 1000 to 5000 kDa according to Kralj et al. (2004).

Table 4.2 Properties of exopolysaccharide synthesized by thermophilic microorganisms

Domain	Microorganism	Growth temperature (°C)	Carbohydrate and protein content	Molecular weight (Da)	EPS composition	Reference
Bacteria	*Thermus aquaticus* YT-1	60	n.d.	500,000	Galactofuranose/galactopyranose/N-acetylgalactosamine (1:1:2)	Lin et al. (2011)
	Thermotoga maritima (Co-cultivation with *Methanococcus jannaschii*)	80	5 % carbohydrate	n.d.	Glucose/ribose/mannose (1:0.06:0.03)	Johnson et al. (2005)
	Geobacillus thermoantarcticus	65	96 % carbohydrate 0.2 % protein	EPS1: 300,000 EPS2: 300,000	EPS1: mannose/glucose (1.0:0.7); tEPS2: mannose/glucose (1:tr)	Manca et al. (1996; Nicolaus et al. (2004)
	Bacillus licheniformis B3-15	45	66 % carbohydrate 5 % protein	600,000	Mannose/glucose (1:3)	Maugeri et al. (2002)
	Bacillus licheniformis T14	50	99 % carbohydrate 1.2 % protein	1,000,000	Fructose/fucose/glucose/galactosamine/mannose (1.0 0.75:0.28:tr:tr)	Spanò et al. (2013);
	Geobacillus sp. 4001	65	81 % carbohydrate 7 % protein	380,000	Mannose/glucose/galactose/mannosamine (1:0.1:tr:tr)	Nicolaus et al. (2002); Nicolaus et al. (2004)
	Geobacillus sp. 4004	65	65 % carbohydrate 2.6 % protein	>1,000,000	Galactose/mannose/glucosamine/arabinose (1:0.8:0.4:0.2)	Nicolaus et al. (2002); Nicolaus et al. (2004)
	Geobacillus thermodenitrificans B3-72	65	80 % carbohydrate 3 % protein	EPS2: 400,000	EPS1: mannose/glucose (0.3:1) EPS2: mannose/glucose (1:0.2)	Nicolaus et al. (2000)
	Geobacillus tepidamans V264	60	100 % carbohydrate	>1,000,000	Glucose/galactose/fucose/fructose (1:0.07:0.04:0.02)	Kambourova et al. (2009)

(continued)

Table 4.2 (continued)

Domain	Microorganism	Growth temperature (°C)	Carbohydrate and protein content	Molecular weight (Da)	EPS composition	Reference
	Aeribacillus pallidus 418	55	80.5 % carbohydrate 19 % protein	EPS1: 700,000 EPS2: 1,000,000	EPS1: mannose/glucose/galactosamine/glucosamine/galactose/ribose (1:0.16:0.1:0.09:0.07:0.06:0.04) EPS2: mannose/galactose/glucose galactosamine/glucosamine/ribose/arabinose (1:0.5:0.46:0.35:0.24:0.16:0.14)	Radchenkova et al. (2013)
	Brevibacillus thermoruber 423	55	92 % carbohydrate 6 % protein	n.d.	Glucose galactose galactosamine mannose mannosamine (1:0.3:0.25:0.16:0.04)	Yasar Yildiz et al. (2014)
Archaea	*Thermococcus litoralis*	88	n.d.	41,000	Mannose	Rinker and Kelly (1996)
	Sulfolobus solfataricus MT4	88	75 % carbohydrate 0.1 % protein	n.d.	Glucose/mannose/glucosamine/galactose (1.2:1.0:0.18:0.13)	Nicolaus et al. (1993)
	Sulfolobus solfataricus MT3	75	75 % carbohydrate 0.1 % protein	n.d.	Glucose/mannose/glucosamine/galactose (1.2:1.0:0.77:0.73)	Nicolaus et al. (1993)
	Sulfolobus tokodaii	76	n.d.	n.d.	Mannose/glucose/galactose/N-acetylglucosamine	Koerdt et al. (2010)
	Sulfolobus acidocaldarius	76	n.d.	n.d.	Mannose/glucose/galactose/N-acetylglucosamine	Koerdt et al. (2010)

n.d. no data

EPSs from thermophilic bacteria were thermostable, with the highest thermostability reported for the polymers from *Geobacillus tepidamans* V264 (280 °C) (Kambourova et al. 2009), *Geobacillus thermodenitrificans* strain B3-72 (240 °C) (Arena et al. 2006) and *Bacillus licheniformis* (240 °C) (Spanò et al. 2013). The information concerning thermostability of exopolysaccharides from mesophilic bacteria is very scared. A moderately halophilic bacterium was described to produce highly thermostable exopolysaccharide (melting point at 207 °C) (Cojoc et al. 2009).

5 Specific Conditions for EPS Synthesis by Thermophiles

Increasing in the yield of bacterial EPSs traditionally is achieved by strain selection and/or optimization of cultivation conditions, however each given bacterium has physiological limits that could be difficult to overcome (Freitas et al. 2011a). Regulation of biosynthesis of bacterial EPSs is a complex process as a large number of enzymes and regulatory proteins are involved (Jaiswal et al. 2014). Despite the structural diversity of EPSs, four mechanisms are known in bacteria for the polymerization, namely, extracellular biosynthesis, synthase dependent biosynthesis, ABC-transporter dependent and the most commonly used wzx/wzy- dependent pathways. The mechanism of EPS biosynthesis involving sugar nucleotide synthesis, repeating unit synthesis, and polymerization of the repeating units was well studied in mesophilic microorganisms (De Vuyst et al. 2001; Freitas et al. 2011b). Recently a hypothetical mechanism for sugar uptake in EPS biosynthesis similar to those involved in the synthesis of polysaccharides from mesophiles was suggested for the thermophile *Brevibacillus thermoruber* 423 (Yildiz et al. 2015). Essential genes associated with EPS biosynthesis were detected by genome annotation and the biosynthesis of NDP-sugars was shown. Genome information revealed the presence of ABC-transporter dependent pathway (Bth.peg.2228, Bth.peg.4273, Bth.peg.3612, Bth.peg.3618, Bth.peg.4275) in EPS biosynthesis by *Brevibacillus thermoruber* 423. In addition, an exopolysaccharide synthesis pathway in the hyperthermophilic bacterium *Thermotoga maritima* was identified (Johnson et al. 2005). Transcriptional analysis of exopolysaccharide formation by this microorganism in a co-culture with *Methanococcus jannaschii* showed a strong upregulation of a gene, encoding a polypeptide. This polypeptide contains a motif found in peptide-signalling molecules in mesophilic bacteria. Characterization of the complete 15 kb St Sfi6 eps gene cluster of *Streptococcus thermophilus* revealed high degree of similarities among the products and known glycosyltransferases and their potential role in the synthesis of the repeating monomer was suggested (Delcour et al. 2000). The genes which encode the proteins or enzyme required for the biosynthesis of EPS are located on chromosomes in those which are thermopiles (Yildiz et al. 2015) and on plasmid in most lactic acid bacteria (Laws et al. 2001). In contrast, a lack of plasmids encoding components required for slime production was observed for thermophilic LAB *Streptococcus thermophilus* (Harutoshi 2013) and a location of the EPS gene cluster in chromosomal DNA was reported for *Lactobacillus fermentum* TDS030603 (Dan et al. 2009).

Despite of the fact that the composition and the amount of microbial EPSs are genetically determined, they also depends of several factors, such as type of strain, carbon and nitrogen sources, mineral salts, trace elements, the medium component ratio, fermentation conditions (temperature, pH, agitation and aeration (Nicolaus et al. 2010). The synthesis of EPS by microbial cells basically depends on the carbon and nitrogen availability in the culture medium. EPS producing microorganisms utilize sugars as their carbon and energy source; ammonium salts and amino acids are their source of nitrogen (Gandhi et al. 1997; Czaczyk and Wojciechowska 2003). Usually lower cost sugars like glucose, maltose or sucrose were used as a carbon source despite the fact that in some cases higher production was observed in a presence of other sugars. The polymer production was lower in a medium containing glucose in comparison with maltose in the case of several thermophilic producers (Rinker and Kelly 2000; Kambourova et al. 2009; Radchenkova et al. 2013; Yasar Yildiz et al. 2014). The increased production in abundance of carbon source and minimal nitrogen was reported by several authors (Radchenkova et al. 2013; Yasar Yildiz et al. 2014). The addition of extra nitrogen favors the biomass production but diminishes EPS production. Production may or may not be growth associated. EPS production is growth associated for *Aeribacillus pallidus* 418 (Radchenkova et al. 2013), *Brevibacillus thermoruber* 423 (Yasar Yildiz et al. 2014), *Geobacillus tepidamans* (Kambourova et al. 2009) and differed from the reports for some mesophilic producers synthesizing EPS during the whole stationary phase (Conti et al. 1994; Raguénès et al. 1997).

Oxygen is a key substrate in aerobic bioprocesses for EPS production, whose continuous supply should be ensured because of its low solubility in broths. Product formation by aerobic thermophiles could be increased by optimization of agitation and aeration rates. The size of gas bubbles and their dispersion throughout the reactor volume are critical for its performance. The smaller is the bubble size the larger is the surface area for gas contact that could improve oxygen transfer rate. Low solubility of oxygen in the medium especially at enhanced temperatures of thermophilic processes often determines the oxygen (air) transfer as a rate-limiting step in the aerobic bioprocess. High temperature sharply decreases oxygen solubility, one of the most important parameter in massive EPS production in bioreactors. Oxygen transfer in bioreactors is an object of investigations by many authors, especially in recent years (Garcia-Ochoa and Gomez 2009). However, the information concerning optimization of agitation and aeration conditions in thermophilic processes for EPS production is still very scarce. Aeration and agitation were proven to be crucial for attaining maximum productivity in microbial aerobic processes by *Aeribacillus pallidus* 418 (Radchenkova et al. 2014). According to some authors (Rau et al. 1992; Radchenkova et al. 2014), dissolved oxygen limitation is a desirable condition for enhanced polymer production. It is well known that one of the physiological roles of EPSs is cell adaptation to unfavorable conditions; for extremophiles it could serve as enhancer for bacterial survival at severe conditions of extreme niches (Nicolaus et al. 2010). A short oxygen limitation could provoke EPS synthesis as a cell response to this limitation. In stirred bioreactors, a high number of variables like stirrer speed, type and number of stirrers, gas flow rate influence mixing and mass transfer (Garcia-Ochoa and Gomez 2009). A number of impellers have been reported to ensure uniform distribution of substrates and high heat and mass transfer rates in polysaccharide-producing broths (Chhabra 2003). Radial flow turbines type Rushton are very popular in the equipment of the most laboratory scale stirred tank systems, changing a shear stress on the medium. Study on the influence of agitation and aeration revealed that both parameters influenced specific EPS production (Radchenkova et al. 2014) however agitation was much more effective than aeration.

Anaerobic processes for EPS production by hyperthermophiles often do not need even in agitation as EPS is accumulated as a biofilm. Sufficient biofilm formation on nylon mesh by continuous cultures of *Thermotoga maritima* was obtained in anaerobic chemostat at $D=0.25\ h^{-1}$ (Pysz et al. 2004). Batch and continuous cultures without agitation and at gas sparging were used to compare specific physiological features in EPS synthesis by the hyperthermophilic archaeon *Thermococcus litoralis* and hyperthermophilic bacterium *Thermotoga maritima* (Rinker and Kelly 2000) or to optimize growth of *Thermococcus hydrothermalis* (Postec et al. 2005). In other cases, low rate of agitation was reported as optimal, 100 rpm for *Thermotoga maritima* (Johnson et al. 2005), low mechanical agitation for two thermophilic archaea belonging to the genus *Sulfolobus* (Nicolaus et al. 1993).

Cultivation of thermophiles in bioreactors is characterized by some features. The need in good aeration and agitation is determined by the lower solubility of oxygen in higher temperature, although the oxygen transfer coefficient increases (Shih and Pan 2011; Kennes and Veiga 2013). Another disadvantage is the high investment cost required for the compressors and heat exchangers in thermo-reactor (Van Groenestijn et al. 2002). Higher temperature results in higher evaporation heat and this heat needs to be recovered. Sharp and Raven (1997) recommended bioreactors made of stainless steel with added nickel, molybdenum or chrome or with a teflon coating to prevent corrosion caused by high temperatures, salt concentrations and sulfide. In the case of anaerobic membrane bioreactor ceramic membranes are used at thermophilic conditions due to their thermal stability and long lifetime (Abeynayaka and Visvanathan 2011).

The observed levels of EPS production by thermophilic producers are lower than those reported for mesophilic producers and usually varied in the range 50–200 μg mL^{-1} (Kambourova et al. 2009; Manca et al. 1996; Nicolaus et al. 2003). A higher amount of EPS (366 μg/mL) has been reported for the facultative thermophile *Bacillus licheniformis* (Spanò et al. 2013) for 48 h cultivation at temperature of 50 °C in a complex medium. Highest production (863 μg/mL) in comparison with other thermophiles was reported for a thermophile *Brevibacillus thermoruber* 423 (Yasar Yildiz et al. 2014).

6 Biotechnological Potential of Exopolysaccharides from Thermophilic Microorganisms

Biotechnological applications of EPSs range from traditional areas as food, pharmaceutical and cosmetic industries to novel biomedicine areas. New microbial polysaccharides could conquer the traditional polysaccharide market in the case of using cheap substrates, development of low costly downstream processes, better functional or novel properties. According to Belsito et al. (2012), 19 microbial polysaccharides are currently used in cosmetic formulations. None of them is produced by a thermophilic microorganism.

About 12 reports on EPS production by thermophiles, polysaccharide composition and properties are currently known. EPS synthesis by thermophilic microorganisms suggests some advantages like:

- Short fermentation processes (often lasting several hours) (Kambourova et al. 2009; Radchenkova et al. 2013; Yasar Yildiz et al. 2014) due to the high growth rate at elevated temperature and lower concentration of nutrient components.
- Good mass transfer at high temperature for cultivation (Turner et al. 2007; Kumar et al. 2011).
- Viscosity of culture liquid is lower at high temperature that suggests lower energy consumption (Haki and Rakshit 2003).
- Performance of processes at high temperature reduces the risk of contamination (Turner et al. 2007; Kikani et al. 2010; Xiao et al. 2015).
- Non pathogenic products from thermophiles are applicable in food and cosmetic industry (Nicolaus et al. 2010).
- EPSs synthesized by thermophilic bacteria and archaea are suggested to keep their emulsifying and rheological properties at high temperature, in which many processes in food industry are performed (Sajna et al. 2013)
- Usually thermostable molecules can remain effective even at extreme conditions of pH, temperature, and salinity due to their more rigid molecule.
- They form stable oil/water emulsions needed for cosmetic industry (Radchenkova et al. 2014).

Among the disadvantages of exploring hot spring potential are the low biomass and respectively low EPS synthesis by thermophiles (Krebs et al. 2014). Additionally, the proportion of microorganisms reluctant to cultivation-based approaches is very high in extreme environments (Lorenz et al. 2002). These disadvantages could be overcome to some extent by exploration of better producers and optimizing of cultivation conditions. Exciting prospects for increasing production yield are found in genetic engineering. The genetic manipulation of bacteria is much easier than that for higher organisms (Morris and Harding 2014). Genetic manipulation could tailor chemical composition and structure of EPS, which further determine their specific explorations.

Another disadvantage is some cases could be the high cost of the received product resulting from the used substrate. Substrate utilization is lower in thermophilic processes than in mesophilic ones as the biomass yield is comparatively low. Use of agricultural waste or dairy waste could lead to reduction in EPS cost. The inherent costs of large-scale fermenters are significantly higher in comparison with simple extraction processes for plant polysaccharides however biotechnological advances in bacterial large-scale processes result in large EPS quantities and correspondingly lowering the cost of the product.

Comparatively low levels of EPS synthesis by thermophilic microorganisms and correspondingly higher production cost determine the interest for the development of microbial EPSs used in high-value market niches, where the desired properties or the degree of purity could not be suggested by the traditional polymers (Kumar et al. 2007). Novel biopolymers produced by thermophilic bacteria could suggest different chemical structure and correspondingly different physicochemical properties valuable for medical and pharmaceutical applications, especially for drug delivery, tissue engineering, as immunostimulatory, immunomodulatory, antitumor, antiviral, anti-inflammatory and antioxidant agents (Arena et al. 2009; Sam et al. 2011; Lee and Mooney 2012; Freitas et al. 2014).

Over the past several years, anti-tumour activity of microbial exopolysaccharides was reported by several authors (Khalikova et al. 2006; Nwodo et al. 2012). Marine bacterium isolated from hydrothermal vent was shown to produce new EPS (Courtois et al. 2014). In its native or over-sulphated form it modulated the complement system suggesting an effective treatment of diseases caused by deregulation of the immune system and overactivation of the complement system. A dose-dependent immunomodulatory and antiviral effects of extracellular polysaccharides, produced by *Bacillus licheniformis* strain T14 and *Geobacillus thermodenitrificans* strain B3-72 were proven and a partial restoration of immunological disorders after treatment with this EPS was observed (Arena et al. 2006, 2009). The novel EPS1-T14 was able to act as immunomodulator inhibiting herpes simplex virus type 2 (HSV-2) replication by triggering the production of Th1-type cytokines (Gugliandolo et al. 2014). Interesting chemical and rheological characteristics were reported for a new fucose containing EPS from *Bacillus licheniformis* strain T14 isolated from Panarea Island (Spanò et al. 2013).

A novel extracellular polysaccharide TA-1 from *Thermus aquaticus* YT-1, stimulated macrophage cells to produce the cytokines, which increases the immune response (Lin et al. 2011). D-galactofuranose residues in the novel TA-1 are probably responsible for TA-1 immunoregulatory activity within macrophages, the first line of host defense against bacterial infection. EPS from *Aeribacillus pallidus* 418 has a potential application in cosmetic industry due to its good emulsifying properties (Radchenkova et al. 2014). Properties of EPSs can be changed dramatically using mixtures with other biopolymers. A stable emulsion, especially valuable for cosmetic industry was received as a result of synergistic action of EPS from *Aeribacillus pallidus* 418 with xanthan (Radchenkova et al. 2014).

As many of the food production processes run at elevated temperature thermostability of exopolysaccharides synthesized by thermophiles is an important characteristic for their industrial applications (Sajna et al. 2013). High temperature of destruction of these EPSs suggests easy preparation of emulsions for different food and cosmetic creams at higher temperature where viscosity is lower and mixing is easier. It is also a preposition for long term preservation of the received products even at room temperature. The solutions of the thermostable EPSs are able to maintain high viscosity at high temperature of oil drilling fluids. They could have a great potential as flocculating agents in the thermophilic processes of municipal and wastewater treatment.

Investigations on exopolysaccharides from thermophilic microorganisms revealed their interesting properties like high molecular weight suggesting good viscosity; stability of the molecules in harsh industrial conditions, good emulsifying properties and good synergism, biological activity against cytotoxic compounds, antiviral and immunomodulating activities suggesting the potential of EPSs from thermophilic microorganisms in several biotechnological and biomedical processes.

Current review gathered the information on EPS synthesis by thermophilic microorganisms, industrially valuable properties of the novel biopolymers and their biological activities in trends for possible potential future applications.

Conflict of Interest Margarita Kambourova, Nadja Radchenkova, Iva Tomova, and Ivanka Bojadjieva declare that they have no conflict of interest.

References

Abeynayaka A, Visvanathan C (2011) Performance comparison of mesophilic and thermophilic aerobic sidestream membrane bioreactors treating high strength wastewater. Bioresour Technol 102:5345–5352

Arena A, Maugeri TL, Pavone B, Iannello D, Gugliandolo C, Bisignano G (2006) Antiviral and immunoregulatory effect of a novel exopolysaccharide from a marine thermotolerant *Bacillus licheniformis*. Int Immunopharmacol 6:8–13

Arena A, Gugliandolo C, Stassi G, Pavone B, Iannello D, Bisignano G, Maugeri TL (2009) An exopolysaccharide produced by *Geobacillus thermodenitrificans* strain B3-72: Antiviral activity on immunocompetent cells. Immunol Lett 123:132–137

Baker-Austin C, Potrykus J, Wexler M, Bond PL, Dopson M (2010) Biofilm development in the extremely acidophilic archaeon '*Ferroplasma acidarmanus*' Fer1. Extremophiles 14:485–491

Belsito MD, Hill RA, Klaassen CD, Liebler D, Marks Jr JG, Ronald C (2012) Safety Assessment of Microbial Polysaccharide Gums as Used in Cosmetics. http://www.cir-safety.org/sites/default/files/microb092012rep.pdf

Chhabra RP (2003) Fluid mechanics and heat transfer with non-Newtonian liquids in mechanically agitated vessels. Adv Heat Transfer 37:77–176

Cojoc R, Merciu S, Oancea P, Pincu E, Dumitru L, Enache M (2009) Highly thermostable exopolysaccharide produced by the moderately halophilic bacterium isolated from a man-made young salt lake in Romania. Pol J Microbiol 58:289–294

Conti E, Flaibani A, O'Regan M, Sutherland IW (1994) Alginate from *Pseudomonas fluorescence* and *P. putida*: production and properties. Microbiology 140:1125–1132

Courtois A, Berthou C, Guézennec J, Boisset C, Bordron A (2014) Exopolysaccharides isolated from hydrothermal vent bacteria can modulate the complement system. PLos One 9(4), e94965

Czaczyk K, Wojciechowska K (2003) Formation of bacterial biofilms—the essence of the matter and mechanisms of interactions. Biotechnologia 3:180–192

Dan T, Fukuda K, Sugai-Bannai M, Takakuwa N, Motoshima H, Urashima T (2009) Characterization and expression analysis of the exopolysaccharide gene cluster in *Lactobacillus fermentum* TDS030603. Biosci Biotechnol Biochem 73:2656–2664

De Vuyst L, De Vin F, Vaningelgem F, Degeest B (2001) Recent developments in the biosynthesis and applications of heteropolysaccharides from lactic acid bacteria. Int Dairy J 11:687–707

Delcour J, Ferain T, Hols P (2000) Advances in the genetics of thermophilic lactic acid bacteria. Curr Opin Biotech 11:497–504

Di Giulio M (2003) The universal ancestor was a thermophile or a hyperthermophile: tests and further evidence. J Theor Biol 221:425–436

Donot F, Fontana A, Baccou JC, Schorr-Galindo S (2012) Microbial exopolysaccharides: main examples of synthesis, excretion, genetics and extraction. Carbohydr Polym 87:951–962

Flemming HC, Neu TR, Wozniak DJ (2007) The EPS matrix: the "house of biofilm cells". J Bacteriol 189:7945–7947

Freitas F, Alves VD, Reis MA (2011a) Advances in bacterial exopolysaccharides: from production to biotechnological applications. Trends Biotechnol 29:388–398

Freitas F, Alves VD, Torres CAV, Cruz M, Sousa I, Melo MJ, Ramos AM, Reis MAM (2011b) Fucose-containing exopolysaccharide produced by the newly isolated *Enterobacter* strain A47 DSM23139. Carbohydr Polym 1:159–165

Freitas F, Alves VD, Reis M, Crespo J, Coelhoso I (2014) Microbial polysaccharide-based membranes: Current and future applications. J Appl Polym Sci 131: doi:10.1002/app.40047

Gandhi HP, Ray RM, Patel RM (1997) Exopolymer production by *Bacillus* species. Carbohydr Polym 34:323–327

Garcia-Ochoa F, Gomez E (2009) Bioreactor scale-up and oxygen transfer rate in microbial processes: an overview. Biotech Adv 27:153–176

Gugliandolo C, Spanò A, Lentini V, Arena A, Maugeri TL (2014) Antiviral and immunomodulatory effects of a novel bacterial exopolysaccharide of shallow marine vent origin. J Appl Microbiol 116:1028–1034

Haki GD, Rakshit SK (2003) Developments in industrially important thermostable enzymes: a review. Bioresour Technol 89:17–34

Hartzell PL, Millstein J, Lapaglia C (1999) Biofilm formation in hyperthermophilic archaea. Methods Enzymol 310:335–349

Harutoshi T (2013) Exopolysaccharides of lactic acid bacteria for food and colon health applications. In: Kongo M (ed) Exopolysaccharides of lactic acid bacteria for food and colon health applications. INTECH Open Access Publisher, pp 515–538

Herget S, Toukach P, Ranzinger R, Hull W, Knirel Y, Von Der Lieth C-W (2008) Statistical analysis of the bacterial carbohydrate structure data base (BCSDB): characteristics and diversity of bacterial carbohydrates in comparison with mammalian glycans. BMC Struct Biol 8:35

Huber R, Langworthy TA, König H, Thomm M, Woese CR, Sleytr UB, Stetter KO (1986) *Thermotoga maritima* sp. nov. represents a new genus of unique extremely thermophilic eubacteria growing up to 90 °C. Arch Microbiol 144:324–333

Jaiswal P, Sharma R, Sanodiya BS, Bisen PS (2014) Microbial exopolysaccharides: natural modulators of dairy products. J Appl Pharm Sci 4:105–109

Johnson MR, Montero CI, Conners SB, Shockley KR, Bridger SL, Kelly RM (2005) Population density-dependent regulation of exopolysaccharide formation in the hyperthermophilic bacterium *Thermotoga maritima*. Mol Microbiol 55:664–674

Kambourova M, Mandeva R, Dimova D, Poli A, Nicolaus B, Tommonaro G (2009) Production and characterization of a microbial glucan, synthesized by *Geobacillus tepidamans* V264 isolated from Bulgarian hot spring. Carbohydr Polym 77:338–343

Kennes C, Veiga MC (2013) In: Kennes C, Veiga MC (eds) Bioreactors for waste gas treatment, vol. 4: Springer Science & Business Media, Dordrecht, pp 47–98

Khalikova TA, Korolenko TA, Zhanaeva SY, Kaledin VI, Kogan G (2006) Enhancing effect of new biological response modifier sulfoethylated (1→3)- beta-D-glucan on antitumor activity of cyclophosphamide in the treatment of experimental murine leukoses. Exp Oncol 28:308–313

Kikani BA, Shukla RJ, Singh SP (2010) Biocatalytic potential of thermophilic bacteria and actinomycetes. In: Méndez-Vilas A (ed) Current research, technology and education. Topics in applied microbiology and microbial biotechnology, FORMATEX C/ Zurbarán 1, 2º - Oficina 1 06002 Badajoz Spain vol 2, pp 1000–1007

Koerdt A, Gödeke J, Berger J, Thormann KM, Albers S-V (2010) Crenarchaeal biofilm formation under extreme conditions. PLos One 5(11), e14104

Kolter R, Losick R (1998) One for all and all for one. Science 280:226–227

Kralj S, van Geel-Schutten GH, Dondorff MMG, Kirsanovs S, van der Maarel MJEC, Dijkhuizen L (2004) Glucan synthesis in the genus *Lactobacillus*: isolation and characterization of glucansucrase genes, enzymes and glucan products from six different strains. Microbiology 150:3681–3690

Krebs JE, Vaishampayan P, Probst AJ, Tom LM, Marteinsson VG, Andersen GL, Venkateswaran K (2014) Microbial community structures of novel icelandic hot spring systems revealed by PhyloChip G3 Analysis. Astrobiology 14:229–240

Krüger M, Blumenberg M, Kasten S, Wieland A, Känel L, Klock JH, Michaelis W, Seifert R (2008) A novel, multi-layered methanotrophic microbial mat system growing on the sediment of the Black Sea. Environ Microbiol 10:1934–1947

Kumar AS, Mody K, Jha B (2007) Bacterial exopolysaccharides: A perception. J Basic Microbiol 47:103–117

Kumar L, Awasthi G, Singh B (2011) Extremophiles: a novel source of industrially important enzymes. Biotechnology 10:121–135

Lapaglia C, Hartzell PL (1997) Stress-induced production of biofilm in the hyperthermophile *Archaeoglobus fulgidus*. Appl Environ Microbiol 63:3158–3163

Laws A, Gu Y, Marshall V (2001) Biosynthesis, characterisation, and design of bacterial exopolysaccharides from lactic acid bacteria. Biotechnol Adv 19:597–625

Lee KY, Mooney DJ (2012) Alginate: Properties and biomedical applications. Prog Polym Sci 37:106–126

Lin M-H, Yan Y-L, Chen Y-P, Hua K-F, Lu C-P, Sheu F, Lin G-H, Tsay S-S, Liang S-M, Wu S-H (2011) A novel exopolysaccharide from the biofilm of *Thermus aquaticus* YT-1 induces the immune response through toll-like receptor 2. J Biol Chem 286:17736–17745

López-López O, Cerdán ME, González-Siso MI (2013) Hot spring metagenomics. Life 3:308–320

Lorenz P, Liebeton K, Niehaus F, Eck J (2002) Screening for novel enzymes for biocatalytic processes: accessing the metagenome as a resource of novel functional sequence space. Curr Opin Biotechnol 13:572–577

Lowell RP, Seewald JS, Metaxas A, Perfit MR (2008) In: Lowell RP, Seewald JS, Metaxas A, Perfit, MR (eds) Modeling hydrothermal processes at ocean spreading centers: magma to microbe–an overview. Magma to Microbe. American Geophysical Union, Washington, DC. doi:10.1029/178GM02

Manca MC, Lama L, Improta R, Esposito E, Gambacorta A, Nicolaus B (1996) Chemical composition of two exopolysaccharides from *Bacillus thermoantarcticus*. Appl Environ Microbiol 62:3265–3269

Maugeri TL, Gugliandolo C, Caccamo D, Panico A, Lama L, Gambacorta A, Nicolaus B (2002) A halophilic thermotolerant *Bacillus* isolated from a marine hot spring able to produce a new exopolysaccharide. Biotechnol Lett 24:515–519

Milas M, Shi X, Rinaudo M (1990) On the physicochemical properties of gellan gum. Biopolymers 30:451–464

Moons P, Michiels CW, Aertsen A (2009) Bacterial interactions in biofilms. Crit Rev Microbiol 35:157–168

Morris GA, Harding SE (2014) Production of polysaccharides. In: Panesar PS, Marwaha SS (eds) Biotechnology in agriculture and food processing: opportunities and challenges. Taylor & Francis Group CRC Press, Boca Raton, Fl. pp 355–386

Muralidharan V, Rinker KD, Hirsh IS, Bouwer EJ, Kelly RM (1997) Hydrogen transfer between methanogens and fermentative heterotrophs in hyperthermophilic cocultures. Biotechnol Bioeng 56:268–278

Näther DJ, Rachel R, Wanner G, Wirth R (2006) Flagella of *Pyrococcus furiosus*: multifunctional organelles, made for swimming, adhesion to various surfaces, and cell-cell contacts. J Bacteriol 188:6915–6923

Nicolaus B, Manca MC, Romano I, Lama L (1993) Production of an exopolysaccharide from two thermophilic archaea belonging to the genus *Sulfolobus*. FEMS Microbiol Lett 109:203–206

Nicolaus B, Panico A, Manca MC, Lama L, Gambacorta A, Maugeri T, Gugliandolo C, Caccamo D (2000) A thermophilic *Bacillus* isolated from an Eolian shallow hydrothermal vent, able to produce exopolysaccharides. Syst Appl Microbiol 23:426–432

Nicolaus B, Lama L, Panico A, Gambacorta A (2002) Production and characterization of exopolysaccharides excreted by thermophilic bacteria from shallow, marine hydrothermal vents of Flegrean areas (Italy). Syst Appl Microbiol 25:319–325

Nicolaus B, Moriello V, Maugeri T, Gugliandolo C, Gambacorta A (2003) Bacilli from shallow mediterranean marine vents producers of exopolysaccharides. Recent Res Devel Microbiol 7:197–208

Nicolaus B, Schiano Moriello V, Lama L, Poli A, Gambacorta A (2004) Polysaccharides from extremophilic microorganisms. Orig Life Evol Biosph 34:159–169

Nicolaus B, Kambourova M, Oner ET (2010) Exopolysaccharides from extremophiles: from fundamentals to biotechnology. Environ Technol 31:1145–1158

Nwodo UU, Green E, Okoh AL (2012) Bacterial exopolysaccharides: functionality and prospects. Int J Mol Sci 13:14002–14015

Otero A, Vincenzini M (2003) Extracellular polysaccharide synthesis by *Nostoc* strains as affected by N source and light intensity. J Biotechnol 102:143–152

Patel AK, Michaud P, Singhania RR, Soccol CR, Pandey A (2010) Polysaccharides from probiotics: new developments as food additives. Food Technol Biotechnol 48:451–463

Poli A, Di Donato P, Abbamondi GR, Nicolaus B (2011) Synthesis, production, and biotechnological applications of exopolysaccharides and polyhydroxyalkanoates by *Archaea*. Archaea, Article ID 693253

Postec A, Pignet P, Cueff-Gauchard V, Schmitt A, Querellou J, Godfroy A (2005) Optimisation of growth conditions for continuous culture of the hyperthermophilic archaeon *Thermococcus hydrothermalis* and development of sulphur-free defined and minimal media. Res Microbiol 156:82–87

Pysz MA, Conners SB, Montero CI, Shockley KR, Johnson MR, Ward DE, Kelly RM (2004) Transcriptional analysis of biofilm formation processes in the anaerobic, hyperthermophilic bacterium *Thermotoga maritima*. Appl Environ Microbiol 70:6098–6112

Radchenkova N, Vassilev S, Panchev I, Anzelmo G, Tomova I, Nicolaus B, Kuncheva M, Petrov K, Kambourova M (2013) Production and properties of two novel exopolysaccharides synthesized by a thermophilic bacterium *Aeribacillus pallidus* 418. Appl Biochem Biotechnol 171:31–43

Radchenkova N, Vassilev S, Martinov M, Kuncheva M, Panchev I, Vlaev S, Kambourova M (2014) Optimization of the aeration and agitation speed of *Aeribacillus palidus* 418 exopolysaccharide production and the emulsifying properties of the product. Process Biochem 49:576–582

Raguénès G, Pignet P, Gauthier G, Peres A, Christen R, Rougeaux H, Guezennec J (1996) Description of a new polymer-secreting bacterium from a deep-sea hydrothermal vent, *Alteromonas macleodii* subsp. fijiensis, and preliminary characterization of the polymer. Appl Environ Microbiol 62:67–73

Raguénès G, Christen R, Guezennec J, Pignet P, Barbier G (1997) *Vibrio diabolicus* sp. nov., a new polysaccharide-secreting organism isolated from a deep-sea hydrothermal vent polychaete annelid, *Alvinella pompejana*. Int J Syst Bacteriol 47:989–995

Raguénès G, Cambon-Bonavita MA, Lohier JF, Boisset C, Guezennec J (2003) A novel, highly viscous polysaccharide excreted by an *Alteromonas* isolated from a deep-sea hydrothermal vent shrimp. Curr Microbiol 46:0448–0452

Rau U, Gure E, Olszewski E, Wagner F (1992) Enhanced glucan formation of filamentous fungi by effective mixing, oxygen limitation and fed-batch processing. J Ind Microbiol 9:19–26

Rinker KD, Kelly RM (1996) Growth physiology of the hyperthermophilic archaeon *Thermococcus litoralis*: development of a sulfur-free defined medium, characterization of an exopolysaccharide, and evidence of biofilm formation. Appl Environ Microbiol 62:4478–4485

Rinker KD, Kelly RM (2000) Effect of carbon and nitrogen sources on growth dynamics and exopolysaccharide production for the hyperthermophilic archaeon *Thermococcus litoralis* and bacterium *Thermotoga maritima*. Biotechnol Bioeng 69:537–547

Rozanov AS, Bryanskaya AV, Malup TK, Meshcheryakova IA, Lazareva EV, Taran OP, Ivanisenko TV, Ivanisenko VA, Zhmodik SM, Kolchanov NA, Peltek SE (2014) Molecular analysis of the benthos microbial community in Zavarzin thermal spring (Uzon Caldera, Kamchatka, Russia). BMC Genomics 15(Suppl 12):S12

Sajna KV, Sukumaran RK, Gottumukkala LD, Jayamurthy H, Dhar KS, Pandey A (2013) Studies on structural and physical characteristics of a novel exopolysaccharide from *Pseudozyma* sp. NII 08165. Int J Biol Macromol 59:84–89

Sam S, Kucukasik F, Yenigun O, Nicolaus B, Toksoy Öner E, Yukselen MA (2011) Flocculating performances of exopolysaccharides produced by a halophilic bacterial strain cultivated on agro-industrial waste. Biores Technol 102:1788–1794

Satyanarayana T, Raghukumar C, Shivaji S (2005) Extremophilic microbes: diversity and perspectives. Curr Sci 89:78–90

Schopf S, Wanner G, Rachel R, Wirth R (2008) An archaeal bi-species biofilm formed by *Pyrococcus furiosus* and *Methanopyrus kandleri*. Arch Microbiol 190:371–377

Sharp RJ, Raven NDH (1997) In: Rhodes PM, Stanbury PF (eds) Isolation and growth of hyperthermophiles. Applied microbial physiology. IRL Press, New York, pp 23–52

Shih TW, Pan TM (2011) Stress responses of thermophilic Geobacillus sp. NTU 03 caused by heat and heat-induced stress. Microbiol Res 166:346–359

Spanò A, Concetta Gugliandolo C, Lentini V, Maugeri TL, Anzelmo G, Poli A, Nicolaus B (2013) A novel EPS-producing strain of *Bacillus licheniformis* isolated from a shallow vent off Panarea Island (Italy). Curr Microbiol 67:21–29

Sutherland IW (1994) Structure-function relationships in microbial exopolysaccharides. Biotech Adv 12:393–448

Takai K, Nakamura K, Toki T, Tsunogai U, Miyazaki M, Miyazaki J, Hirayama H, Nakagawa S, Nunoura T, Horikoshi K (2008) Cell proliferation at 122 °C and isotopically heavy CH4 production by a hyperthermophilic methanogen under high-pressure cultivation. Proc Natl Acad Sci U S A 105:10949–10954

Thoma C, Frank M, Rachel R, Schmid S, Näther D, Wanner G, Wirth R (2008) The Mth60 fimbriae of *Methanothermobacter thermoautotrophicus* are functional adhesins. Environ Microbiol 10:2785–2795

Turner P, Mamo G, Karlsson EN (2007) Potential and utilization of thermophiles and thermostable enzymes in biorefining. Microb Cell Fact 6:1–23

Van Groenestijn JW, Hazewinkel JHO, Nienoord M, Bussmann PJT (2002) Energy aspects of biological hydrogen production in high rate bioreactors operated in the thermophilic temperature range. Int J Hydrogen Energy 27:1141–1147

Vu B, Chen M, Crawford RJ, Ivanova EP (2009) Bacterial extracellular polysaccharides involved in biofilm formation. Molecules 14:2535–2554

Węgrzyn A, Żukrowski K (2014) Biotechnological applications of archaeal extremozymes. Chemik 68:717–722

Wiegel J, Canganella F (2001) Extreme thermophiles. In: Encyclopedia of life sciences, article 392. Wiley, Chichester. http://www.els.net/WileyCDA/ElsArticle/refId-a0000392.html

Xiao Z, Zhang Y, Xi L, Huo F, Zhao JY, Li J (2015) Thermophilic production of polyhydroxyalkanoates by a novel *Aneurinibacillus* strain isolated from Gudao oilfield. China J Basic Microbiol doi:. doi:10.1002/jobm.201400843

Yasar Yildiz S, Anzelmo G, Ozer T, Radchenkova N, Genc S, Di Donato P, Nicolaus B, Toksoy Oner E, Kambourova M (2014) Brevibacillus themoruber: a promising microbial cell factory for exopolysaccharide production. J Appl Microbiol 116(2):314–324. ISSN: 1364–5072.

Yildiz SY, Radchenkova N, Arga KY, Kambourova M, Toksoy Oner E (2015) Genomic analysis of *Brevibacillus thermoruber* 423 reveals its biotechnological and industrial potential. Appl Microbiol Biotechnol 99:2277–2289

Zhang CL, Ye Q, Huang Z, Li W, Chen J, Song Z, Zhao W, Bagwell C, Inskeep WP, Ross C, Gao L, Wiegel J, Romanek CS, Shock EL, Hedlund BP (2008) Global occurrence of archaeal amoA genes in terrestrial hot springs. Appl Environ Microbiol 74:6417–6426

Zheng C, Li Z, Su J, Zhang R, Liu C, Zhao M (2012) Characterization and emulsifying property of a novel bioemulsifier by *Aeribacillus pallidus* YM-1. J Appl Microbiol 113:44–51

Chapter 5
Ecophysiology and Application of Acidophilic Sulfur-Reducing Microorganisms

Anna P. Florentino, Jan Weijma, Alfons J.M. Stams, and Irene Sánchez-Andrea

1 Sulfur Compounds in Nature

Sulfur is an important element in the Earth crust, representing about 0.05 % of the lithosphere weight (Steudel and Eckert 2003). However, it is highly concentrated in various continental rocks, such as metal sulfide ore deposits [e.g. pyrite (FeS_2), chalcopyrite ($CuFeS_2$), pyrrhotite (FeS)] or sulfate deposits [e.g. gypsum ($CaSO_4{\cdot}2H_2O$), barite ($BaSO_4$)]. Sulfur exhibits nine different oxidation states, however the most abundant in nature are −2 (sulfide and reduced organic sulfur), 0 (elemental sulfur) and +6 (sulfate) (Steudel 2000; Tang et al. 2009).

Electronic supplementary material: The online version of this chapter (doi:10.1007/978-3-319-13521-2_5) contains supplementary material, which is available to authorized users.

A.P. Florentino
Laboratory of Microbiology, Wageningen University,
Stippeneng 4, 6708 WE Wageningen, The Netherlands

Sub-Department of Environmental Technology, Wageningen University,
Bornse Weilanden 9, 6708 WG Wageningen, The Netherlands

J. Weijma
Sub-Department of Environmental Technology, Wageningen University,
Bornse Weilanden 9, 6708 WG Wageningen, The Netherlands

A.J.M. Stams
Laboratory of Microbiology, Wageningen University,
Stippeneng 4, 6708 WE Wageningen, The Netherlands

CEB-Centre of Biological Engineering, University of Minho,
Campus de Gualtar, 4710-057 Braga, Portugal

I. Sánchez-Andrea (✉)
Laboratory of Microbiology, Wageningen University,
Stippeneng 4, 6708 WE Wageningen, The Netherlands
e-mail: irene.sanchezandrea@wur.nl

P.H. Rampelotto (ed.), *Biotechnology of Extremophiles*, Grand Challenges in Biology and Biotechnology 1, DOI 10.1007/978-3-319-13521-2_5

The oxidation reaction of sulfide to sulfate implies the transference/loss of eight electrons and can be performed in different steps, in which elemental sulfur, thiosulfate, sulfite, and polysulfide (Hedderich et al. 1999) can appear as intermediates. The importance and stability of these intermediates in solution depends on pH, temperature, presence of chemical oxidizing and reducing agents, catalysts, and the species involved (Knickerbocker et al. 2000).

Transformations of sulfur forms in the environment are highly dependent on microbial activities (Steudel 2000). Transformation of organic and inorganic sulfur compounds performed by microorganisms greatly affects chemical, physical and biological properties of the biosphere.

The sulfur cycle can be analyzed through two points of view (Canfield and Farquhar 2012). From a geological perspective, the generation of oceanic crust is associated with the transfer of sulfur from the earth mantle to the earth surface and to the oceans (Canfield 2004), which occurs via volcanic outgassing of SO_2 and H_2S, release of H_2S during hydrothermal circulation, and the erosion of igneous sulfide minerals (Canfield and Farquhar 2012). From the biological perspective, sulfate and/or sulfur reduction may be either assimilatory, when the sulfide produced is used for anabolic reactions, or dissimilatory, when used for energy conservation and growth (Tang et al. 2009; Canfield and Farquhar 2012).

Sulfide is used as electron donor by several anoxygenic phototrophic bacteria which perform photosynthesis. They form elemental sulfur, sulfate (Ghosh and Dam 2009) or, sometimes, thiosulfate (Pfennig 1975) as products (Eqs. 5.1–5.3). Sulfide may be also oxidized by chemotrophic prokaryotes coupled to O_2, nitrate, manganese or iron reduction (Hedderich et al. 1999; Ohmura et al. 2002).

$$H_2S + ½O_2 \rightarrow S^0 + H_2O \tag{5.1}$$

$$H_2S + 2O_2 \rightarrow H_2SO_4 \tag{5.2}$$

$$4H_2S + 5O_2 \rightarrow 2S_2O_3 + 4H_2O \tag{5.3}$$

Elemental sulfur (S^0), thiosulfate ($S_2O_3^{2-}$) and sulfite (SO_3^{2-}), as products of sulfide oxidation, can be oxidized, reduced, or disproportionated to sulfate and sulfide by microorganisms. The disproportionation of elemental sulfur seems to be of great significance in the environment (Steudel 2000; Tang et al. 2009; Canfield and Farquhar 2012).

Biological reactions described in this section are summarized in Fig. 5.1. Chemical reactions are described in the next section.

2 Chemistry of Elemental Sulfur

Elemental sulfur ($S_8{}^0$) is the molecule with the largest number of solid structural forms that can be divided into ambient pressure and high-pressure allotropes. Although there exist over 180 different allotropes and polymorphs (Box 5.1), the only steady form of elemental sulfur at standard temperature and pressure conditions (273.15 K and 1 bar) is the orthorhombic α-$S_8{}^0$ modification (Steudel and Eckert 2003).

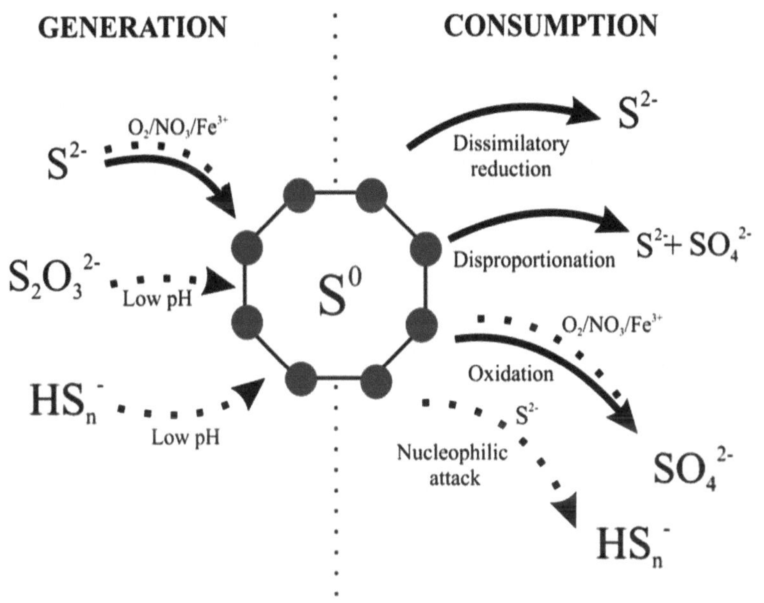

Fig. 5.1 Possible reactions with elemental sulfur as product or reagent. On the left side, reactions that lead to sulfur (by oxidation processes or acidification of the medium) are shown. On the right side, consuming reactions (sulfur reduction, disproportionation, oxidation and nucleophilic attack by sulfide) shown. Biological reactions are represented as *full lines* and chemical reactions as *dashed lines*

Box 5.1
Allotropy: ability of a material to have more than one structure under different conditions of temperature and pressure and to regain these structures when conditions are reversed. Hence, allotropy is a reversible polymorphism.

Polymorphism: ability of solid material to exist in more than one form or crystal structure. If there is change in temperature and pressure, and it is not accompanied by melting or vaporization of the solid, it will cause the solid to change its internal structure of atoms.

Oxo-compounds: compounds containing an oxygen atom doubly bound to carbon or another element (=O).

Sulfur is hardly soluble in water; the solubility of the α-S_8^0 at 20 °C is only 5 µg L^{-1} (Boulegue 1978). In general, the higher the molecular size of the sulfur allotropes, the lower is the solubility in organic solvents. Carbon disulfide, toluene and dichloromethane are the best sulfur solvents, while *cyclo*-alkanes are worthy only at ambient temperatures, dissolving smaller ring molecules (Steudel and Eckert 2003). At higher temperatures (65–140 °C), elemental sulfur is also soluble in compressed gases like nitrogen, methane, carbon dioxide, and hydrogen sulfide, which is of importance for the gas industry since many natural gas reservoirs also contain H_2S

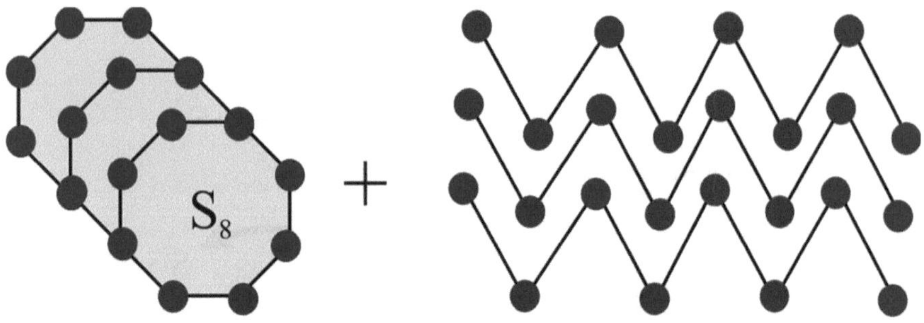

Fig. 5.2 Rings and chain-like macromolecules of polymeric sulfur that compose the commercialized sulfur flower

and elemental sulfur. For example, in a range of pressure from 10 to 30 MPa, solubility of elemental sulfur in hydrogen sulfide increases from 38.6 mg L^{-1} at 65 °C (Roof 1971) to 65.7 at 90 °C (Gu et al. 1993), 68.1 at 100 °C, 91.2 at 110 °C (Roof 1971) and 110.8 mg L^{-1} at 140 °C (Brunner and Woll 1980).

The customary form in which elemental sulfur is typically traded, also called sulfur flower, mainly consists of S_8^0-rings and some polymeric sulfur composed by chain-like macromolecules (Steudel and Eckert 2003) (Fig. 5.2). The heat of reaction from S_8^0 (ring) to S (chain) is 115.14 kJ mol^{-1} per sulfur atom, which is 2.3 kJ mol^{-1} stronger than the bond strength between S–S bonds in polymeric sulfur (Franz et al. 2007). Therefore, polymeric sulfur might be easier to access by sulfur-reducing or sulfur-oxidizing microorganisms.

When sulfide (S^{2-}) is present in the same environment as elemental sulfur, normally at high pH values, a nucleophilic attack of HS^- anion cleaves the S_8^0-ring of elemental sulfur, generating polysulfide (Blumentals et al. 1990). Polysulfide is considered to be preferred over elemental sulfur as electron acceptor by microbes at high temperature and neutral-high pH values due to its higher solubility at these conditions (Schauder and Müller 1993). The most important polysulfide species are tetrasulfide S_4^{2-} and pentasulfide S_5^{2-} (Rabus et al. 2006) which can interconvert rapidly at neutral environments, supporting the growth of neutrophilic sulfur-reducing microorganisms (Schauder and Müller 1993).

The equilibrium concentration of polysulfide (S_n^{2-}) in sulfide solution depends on pH, temperature and sulfide concentration. When the pH decreases, the pH equilibrium concentration of polysulfide immensely shrinks, due to the instability of S_n^{2-} at low pH, and the reaction goes towards elemental sulfur and sulfide, as represented in Eq. (5.4) (Schauder and Müller 1993).

$$2S_5^{2-} + 4H^+ \leftrightarrow \alpha S_8 + 2H_2S \qquad (5.4)$$

However, the equilibrium concentration increases with increasing temperatures. Thus, 0.1 mM will dissolve at pH 6.7 and at 30 °C, while at pH 5.5, the same amount will only dissolve at 90 °C. Due to the dissociation constant, the maximum amount of S_8^0 that can be converted into polysulfide in a sulfide solution at pH 8.0

and 37 °C is roughly comparable to the sulfide concentration (Klimmek et al. 1991). However, at pH values below the pK_a1 of H_2S, which is 7.0 at 25 °C, polysulfide is formed in much lower concentration (Hedderich et al. 1999).

Also thiosulfate is unstable under acidic pH conditions and decomposes into sulfur oxides, sulfide and colloidal/dissolved sulfur as nanocrystals (Eq. 5.5) (Wang et al. 1998), which turn the solutions into a milky suspension. In natural environments, an adhesion of organic polymers to colloidal sulfur particles occurs, which alters the surface properties of elemental sulfur and increases its hydrophilicity (Breher 2004). As sulfur particles are generated together with sulfide, they can react, producing an aqueous solution of polysulfide ions, which affects the mobility of sulfur in the environment, its availability for bio-oxidation, and the formation kinetics of polysulfide and sulfide (Breher 2004). However, colloidal sulfur is thermodynamically unstable and eventually precipitates as small settleable crystals (Kleinjan et al. 2005).

$$3S_2O_3^{2-} + 4H^+ \leftrightarrow 2H_2S + SO_3^{2-} + 3SO_2 \quad (5.5)$$

Another form of elemental sulfur, more hydrophilic than the orthorhombic form, is the so-called bio-sulfur (Steudel and Eckert 2003), which is formed when sulfide is biologically oxidized and can be stored inter- or extra-cellularly as sulfur globules (Kleinjan et al. 2005). It has been suggested that adsorbed organic polymers, such as proteins, or organic end groups are the responsible for the more hydrophilic nature of the bio-sulfur, and so, its structure may differ between species of sulfur bacteria (Steudel et al. 2003). Organic groups also stabilize the long sulfur chains that are produced by phototrophic bacteria. The chemotrophic bacteria, however, mainly form rings composed by eight sulfur atoms (Kleinjan et al. 2005).

3 Sulfur-Reducing Microorganisms

Many prokaryotes are able to colonize environments without any presence of oxygen, evolving not only fermentation pathways, but also respiration, conserving energy for anaerobic growth by coupling the oxidation of hydrogen or organic substrates with the reduction of organic or inorganic compounds (Hedderich et al. 1999; Rabus et al. 2006). Nitrate, manganese (IV), ferric iron, carbon dioxide, protons, selenite, uranium (VI), chromate (VI), arsenate, trymethylamine-N-oxide (TMAO), and sulfur compounds, such as sulfate, elemental sulfur, sulfite, thiosulfate, sulfoxides, dimethylsulfoxides (DMSO), and organic disulfides can be used as electron acceptors by prokaryotes under anoxic conditions (Rabus et al. 2006).

Dissimilatory reduction of Fe (III) and sulfur compounds are significant geobiochemical reactions that occur in soils, aquatic and subsurface environments (Lovley et al. 1995). Reduction of iron has a pronounced influence on the dispersal of iron and trace metals and nutrients. Additionally, it is involved in the degradation of organic matter and can be a promising agent for bioremediation of organic and metals contaminated environments (Lovley et al. 2004). Reduction of Fe (III) can be performed by several microorganisms in the presence of sulfur compounds as energy source.

Reduction of sulfur compounds by it turn attracts attention as it generates hydrogen sulfide as the main end product. Sulfide is known by its remarkable impact on the chemistry of the environment and, furthermore, can serve as electron donor for a considerable high diversity of microorganisms (Rabus et al. 2006). Due to the abundancy and thermodynamic stability, sulfate is the sulfur compound most studied as electron acceptor for anaerobic respiration.

Elemental sulfur reduction, however, is of great importance especially in deep-sea vents, hot springs and other extreme environments, from where microorganisms have most frequently been isolated and their diversity is equivalent to that of sulfate reducers (Stetter 1996).

3.1 Ecophysiology of Sulfur Reducers

Currently known sulfur reducers are spread over about 69 genera within 9 phyla in the *Bacteria* domain (Fig. 5.3a, b) and 37 genera within 2 phyla in the *Archaea* domain (Fig. 5.4). They use elemental sulfur as the main electron acceptor for the oxidation of organic compounds or H_2.

Although microbial sulfur reduction was already reported in early studies as for example by Beijerink (1895), Pelsh (1936) reported the first evidence of elemental sulfur reduction as the only source of energy for microbial growth in enrichments of a vibrioid bacterium from mud, with sulfur and H_2 as electron acceptor and donor, respectively. The first pure culture growing by sulfur reduction was *Desulfuromonas acetoxidans*, an obligatory anaerobic acetate-degrading mesophile and obligate sulfur reducer, not able to use sulfate (SO_4^{2-}) (Pfennig and Biebl 1976).

Afterwards, many sulfur reducers were isolated and showed the ability to reduce other compounds such as thiosulfate, iron (III), nitrate and even oxygen, though anoxic environments are more favorable (Rabus et al. 2006). The capability for sulfur reduction was also observed for microorganisms isolated with other electron acceptors, such as sulfate (Biebl and Pfennig 1977), iron (III) (Caccavo et al. 1994) and manganese (IV) (Myers and Nealson 1988). Only a few species of sulfate reducers are able to grow by sulfur reduction, and sometimes the growth can even be inhibited by elemental sulfur (Bak and Pfennig 1987; Burggraf et al. 1990).

Sulfur-reducing prokaryotes are able to grow at a broad range of temperature (from −2 to 110 °C) and pH (from 1 to 10.5) (Supplemental material—Table 1). Most of the sulfur reducers identified thrive at neutral environments. However, some hyperthermophilic *Archaea* isolated from solfatara fields are reported to grow at pH as low as 1, such as *Acidianus ambivalens*, *Acidianus brierleyi*, *Styogiolobus azoricus*, *Thermoplasma volcanium* and *Thermoplasma acidophilum* (Segerer et al. 1986, 1988, 1991). The lowest pH reported so far for sulfur-reducing bacteria growth is 1.3 for *Acidithiobacillus ferrooxidans* (Ohmura et al. 2002), but several

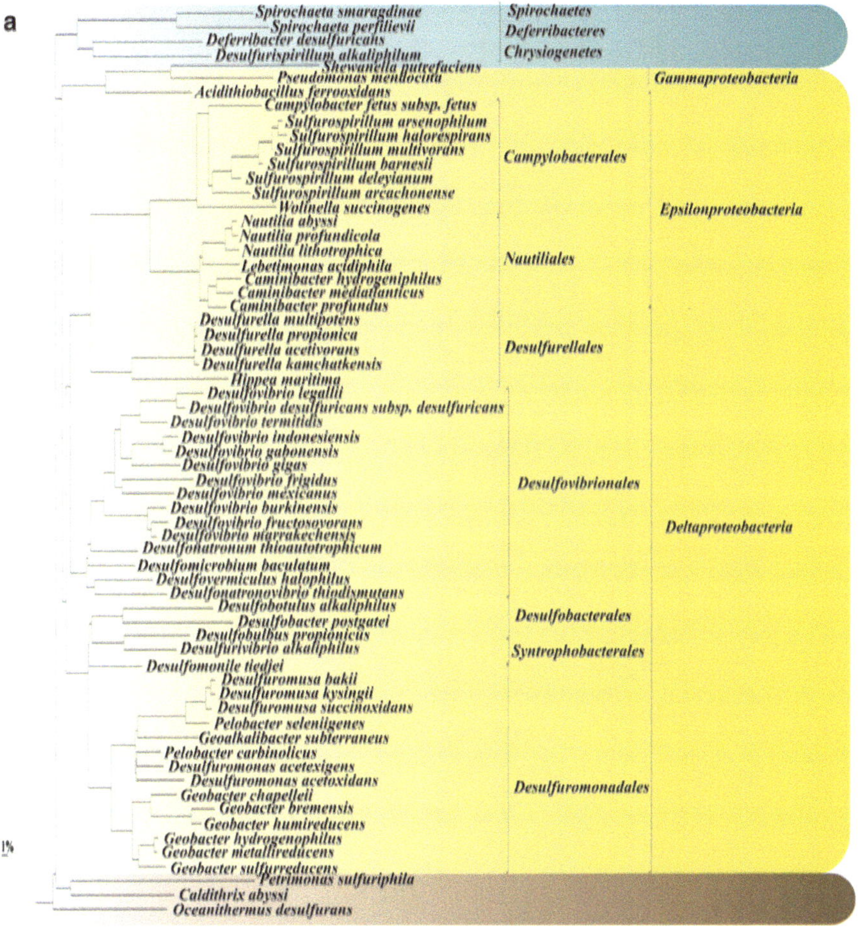

Fig. 5.3 Phylogenetic affiliation of 16S rRNA gene sequences of sulfur-reducing bacteria in The All-Species Living Tree Project (Yarza et al. 2008). In **a**, the sequences belonging to the phyla *Proteobacteria*, *Spirochaetes*, *Deferribacteres* and *Chrysiogenetes* are represented, and in **b** sequences belonging to the phyla *Firmicutes*, *Aquificae*, *Thermodesulfobacteria*, *Synergistetes* and *Thermotogae* are represented. 1 % estimated sequence divergence

acidophilic and acidotolerant species have been described within the bacterial domain, such as *Desulfosporosinus acididurans* (pH 3.8), *Desulfurobacterium thermolithotrophum*, *Marinitoga hydrogenitolerans* and *Thermanaerovibrio velox* (pH 4.5) (L'Haridon et al. 1998; Zavarzina et al. 2000; Postec et al. 2005; Sánchez-Andrea et al. 2015).

Even though several mesophilic microorganisms able to reduce elemental sulfur have been described such as *Desulfuromonas*, *Beggiatoa*, or *Sulfurospirillum* (Pfennig and Biebl 1976), sulfur respiration seems to be more widespread at higher temperature. Slightly thermophilic bacteria (T_{opt} = 40–60 °C) such as *Desulfurella*

b

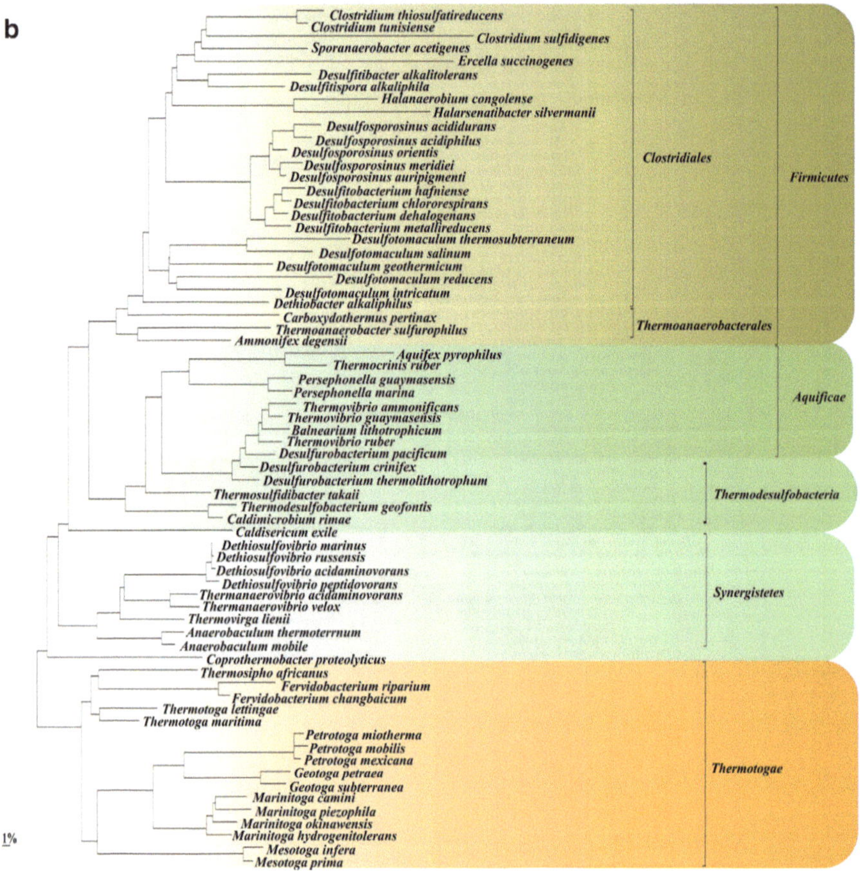

Fig. 5.3 (continued)

and *Thermoanaerobacter* (Bonch-Osmolovskaya et al. 1990b; Bonch-Osmolovskaya et al. 1997) and moderately thermophilic bacteria (T_{opt} = 60–80 °C), such as *Ammonifex* (Huber et al. 1996) and *Desulfurobacterium* (L'Haridon et al. 1998) have been described as well as some hyperthermophilic sulfur reducers, such as *Aquifex* (Huber et al. 1992).

Extreme habitats, such as hot water pools in solfataric fields, acidic hot springs, hydrothermal systems in shallow and deep sea, hypersaline lakes and anoxic mud sediments harbor sulfur reducers that grow at high temperature and low pH (Stetter 1996; Rabus et al. 2006). Due to their abundance and specialized metabolic activities sulfur-reducing prokaryotes are thought to play an important role in the sulfur biogeochemical cycle in deep-sea vents, hot springs and other extreme environments (Bonch-Osmolovskaya et al. 1990a; Alain et al. 2009; Birrien et al. 2011).

In anoxic mud sediment environments, sulfur-reducing microorganisms often form associations with sulfide oxidizers, which provide them with elemental sulfur. The sulfur reducers by their turn reduce the elemental sulfur back to sulfide that is used as electron donor by the sulfide oxidizers (Pfennig 1975). In hydrothermal vents, some sulfur reducers can be found as free-living organisms on vent chimneys or plumes, or as endosymbionts of animals such as tube worms and shrimps, in which they play the same role as their counterparts in the vents by reducing and oxidizing sulfur compounds (Alain et al. 2009).

Described sulfur-reducing bacteria are widespread within the phylogenetic tree of life. They belong to the phyla *Proteobacteria* (*Delta*, *Epsilon*- and *Gammaproteobacteria* classes), *Thermodesulfobacteria, Spirochaetes, Deferribacteres, Chrysiogenetes, Firmicutes, Aquificiae, Synergistetes and Thermotogae* (Fig. 5.3a, b). In the order *Clostridiales* and *Thermoanaerobacterales*, sulfur reduction seems to be a quite widespread metabolic trait (Hernandez-Eugenio et al. 2002; Sallam and Steinbüchel 2009). Within the *Archaea*, sulfur reduction occurs in the phyla *Euryarchaeota* (Fiala and Stetter 1986; Burggraf et al. 1990) and *Crenarchaeota* (Fig. 5.4) (Itoh et al. 1998; Prokofeva et al. 2000; Itoh et al. 2003).

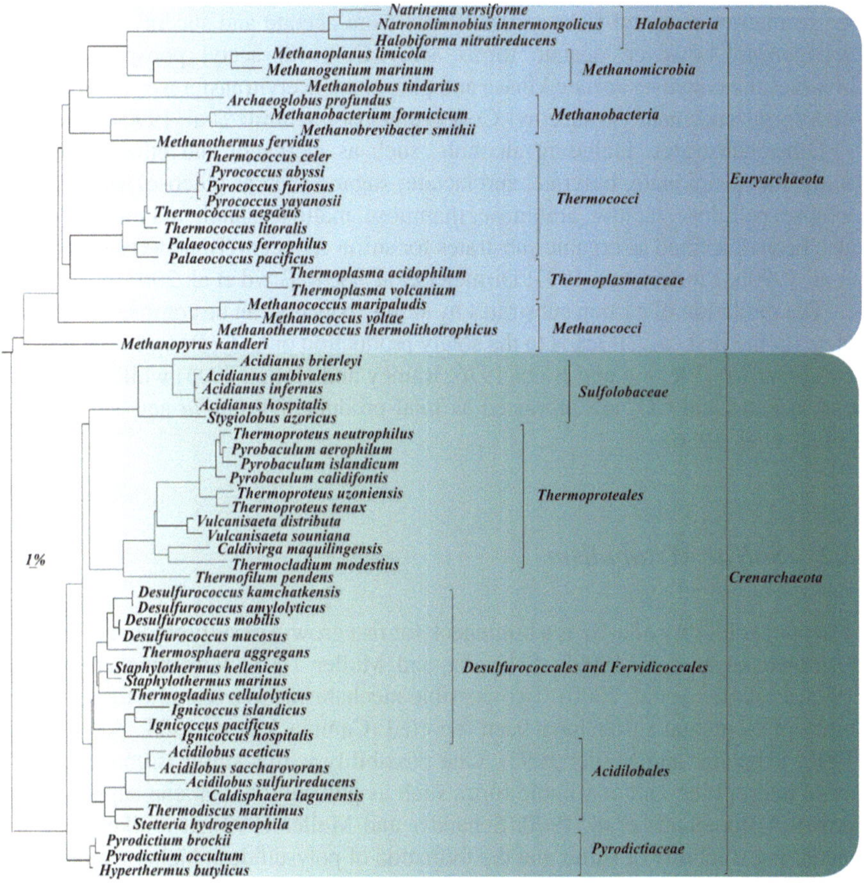

Fig. 5.4 Phylogenetic affiliation of 16S rRNA gene sequences of sulfur-reducing archaea in The All-Species Living Tree Project (Yarza et al. 2008). 1 % estimated sequence divergence

The metabolism of sulfur reducers has been poorly studied, with the exception of few microorganisms, such as the bacterium *Wolinella succinogenes* and the archaeon *Pyrococcus furiosus*. Besides to the biochemistry and bioenergetics of sulfur respiration, little attention has been paid to the conversion of the electron donors in sulfur reducers. Most of the literature related to metabolic pathways and energy conservation is focused on lithotrophic growth on hydrogen or formate as electron donors. Heterotrophic growth on acetate has been investigated only in a few bacteria (Schröder et al. 1988; Klimmek et al. 1991; Kreis-Kleinschmidt et al. 1995). For instance, oxidation of acetate with sulfur as electron acceptor was studied in *Desulfurella* and *Desulfuromonas* species, which occurs via the citric acid cycle. The electron transport is carried out by ferredoxin that might accept electrons from the 2-oxoglutarate via NADP in a 2-oxoglutarate dehydrogenase reaction and menaquinone mediates electron flow to sulfur reductase (Schmitz et al. 1990; Rosenberg et al. 2013). Acetate activation and succinate formation, however, are performed by different mechanisms in those bacteria. In *D. acetoxidans,* it is most likely that only one enzyme (succinyl-CoA:acetate CoA transferase) is involved in the formation of acetyl-CoA and succinate from acetate and succinyl-CoA. In *D. acetivorans,* however, acetate forms acetyl-CoA via acetyl phosphate, which involves the enzymes acetate kinase and phosphate acetyltransferase, and succinyl-CoA forms succinate via succinyl-CoA synthetase (Schmitz et al. 1990).

Other substrates, including alcohols, such as methanol and ethanol; organic acids, like propionate, butyrate, and lactate; sugars, such as glucose, fructose, cellobiose, cellulose, lactose, arabinose, rhamnose, maltose; starch and molasses have also been described as organic substrates for sulfur reducers (Bonch-Osmolovskaya et al. 1990b; Finster et al. 1997; Dirmeier et al. 1998; Boyd et al. 2007).

The oxidation of carbon substrates by sulfur reducers can be complete or incomplete. In the first case, it leads to the solely production of CO_2 (*Desulfuromonas* and *Desulfurella*) (Pfennig and Biebl 1976; Rainey and Hollen 2005) while in the second, acetate and CO_2 are produced as final products (*Wolinella* and *Shewanella*) (Macy et al. 1986).

3.2 Sulfur Metabolism

The poor solubility of α-S_8^0 is a bottleneck for fast growth of sulfur reducers (Bonch-Osmolovskaya et al. 1990b; Schauder and Müller 1993; Miroshnichenko et al. 1998; Prokofeva et al. 2000). Two possible mechanisms to overcome the low solubility of elemental sulfur have been reported (Cammack et al. 1984; Zöphel et al. 1991; Schauder and Müller 1993). One possibility is that sulfur is converted to a more hydrophilic and/or soluble form, such as polysulfide, that can support faster growth (Blumentals et al. 1990; Schauder and Müller 1993). It is likely that the increasing solubility of sulfur and the formation of polysulfide at higher temperatures and pH is beneficial for growth of thermophilic and hyperthermophilic microorganisms (Belkin et al. 1985).

However, as polysulfides are unstable at low pH, it can be that the binding proteins synthesized by sulfur reducers, such as polysulfide sulfur transferases, allow fast polysulfide respiration at low polysulfide concentration (Klimmek 2005), and thus polysulfide is still the substrate. Alternatively, it could be that acidophiles use nanocrystalline that is formed from polysulfide decomposition as electron acceptor. So far, there is still no agreement if polysulfides or nanocrystalline can serve as electron acceptor for acidophilic/acidotolerant microorganisms (Boyd and Druschel 2013). Besides polysulfide, hydrophilic sulfur formed by the association of elemental sulfur with small portions of oxo-compounds (Box 5.1), such as aldehydes, carboxylic acids, ketones, amides, and esters (Steudel et al. 1989) can serve as electron acceptor for microorganisms.

It is remarkable, however, that some bacteria are reported to grow with elemental sulfur when there is no possibility of solubilization in the form of polysulfide (Thamdrup et al. 1993; Finster et al. 1998). As an alternative mechanism, a direct conversion of sulfur into sulfide is suggested to occur due to a physical attachment of the microorganisms to the elemental sulfur.

Even though it is still not clear which mechanism of sulfur reduction is used by the different sulfur reducers, it is likely that hyperthermophilic chemolithoautotrophic archaea reduce elemental sulfur to sulfide via physical attachment (Pihl et al. 1989; Stetter et al. 1993). Moreover, since polysulfides are unstable at low pH and rapidly dissociate into sulfur and sulfide, it is reasonable to hypothesize that elemental sulfur can be the real substrate for the sulfur reductase identified in *A. ambivalens*, an extreme acidophile (Laska et al. 2003).

The reductases that mediate sulfur reduction (either via attachment or via polysulfide) have been purified and characterized from a few sulfur reducers (Schröder et al. 1988; Childers and Noll 1994; Ng et al. 2000; Laska et al. 2003), but sulfur reduction via polysulfide has only been confirmed in *W. succinogenes* (Klimmek et al. 1991), *P. furiosus* (Blumentals et al. 1990) and some *Clostridium* species (Takahashi et al. 2010).

3.3 Enzymes Involved in Sulfur Reduction

In general, the nomenclature of the enzymes involved in sulfur reduction is not well standardized in the published literature. Sometimes the enzymes receive one name related to specific characteristics when they are first isolated and, afterwards, due to more general properties, the name is changed. That was the case for the enzyme sulfhydrogenase. The two hydrogenases isolated from *P. furiosus* were formerly called sulfhydrogenases (Shy). However, as these enzymes seem to be regulated by metabolites other than sulfur, the name sulfhydrogenase became confusing and out of date; so, it was proposed to rename as hydrogenase from hyperthermophiles (Hyh) (Vignais et al. 2001). However, sulfhydrogenase is still present in the database as the main name of the enzyme and is therefore used in this manuscript.

In the genomes database, it is common to find enzymes in reported sulfur reducers named only as sulfur reductase, without specificity about the groups to which they are related. It is also possible to find the mentioned names as synonyms, when they actually refer to different enzymes. In some searches on the available databases, for example MetaCyc (http://metacyc.org/), sulfide dehydrogenase can be referred as sulfhydrogenase and vice-versa.

So far, three enzymes involved in reduction of elemental sulfur and polysulfide to hydrogen sulfide are characterized and described in literature: polysulfide reductase, isolated from *Wolinella succinogenes* (Hedderich et al. 1999), and sulfide dehydrogenase and sulfhydrogenase, both isolated from *P. furiosus* (Ma and Adams 1994).

3.3.1 Polysulfide Reductase

The membrane-bound enzyme is a molybdopterin-containing protein that consists of three subunits predicted by the operon *psrABC* (Krafft et al. 1995). The molybdopterein cofactor is located at the catalytic subunit PsrA, which has an [4Fe-4S] iron-sulfur center. The purified enzyme contains 20 mol of free iron and sulfide per mol of enzyme. Since the subunit PsrB contains four [4Fe-4S] iron-sulfur centers, the mentioned amount is consistent with the whole enzyme (Hedderich et al. 1999).

The hydrophobic subunit of polysulfide reductase (PsrC) anchors the enzyme in the membrane. The transference of electron from the membrane anchor and the catalytic subunit of the enzyme is most likely mediated by the subunit PsrB. The PsrB is probably bound to the other subunits at the periplasmic side of the membrane (Dietrich and Klimmek 2002). The purified enzyme contains menaquinone as cofactor. Due to its lipophilic nature, it is likely that the menaquinone is bound to the subunit PsrC of the enzyme.

The hypothetical mechanism of polysulfide reduction at the catalytic subunit PsrA indicates that the polysulfide chain is cleaved at the last sulfur atom, which is released and bound to the molybdenum cofactor that is further oxidized. The molybdenum cofactor in the PsrA is most likely coordinated by two molybdopterin guanine nucleotide molecules. Thus, after the uptake of a proton, probably via sulfide dehydrogenase, and two electrons, HS^- is released and the molybdenum returns to its reduced stage (Fig. 5.5) (Klimmek et al. 1991).

Sequences of the gene subunits deposited in the JGI genome database are available under accession numbers: PsrA: NP906381; PsrB: NP906382; PsrC: NP906383.

3.3.2 Sulfide Dehydrogenase

Sulfide dehydrogenase, also called flavocytochrome c sulfide dehydrogenase, is a bifunctional cytoplasmic enzyme that catalyzes the reduction of polysulfides to sulfide using NADPH as electron donor (Ma and Adams 1994), but it can also function as a ferredoxin:$NADP^+$ oxidoreductase (Ma and Adams 1994). Reduction of $NADP^+$ is thought to be a required step in the disposal of reducing equivalents as H_2.

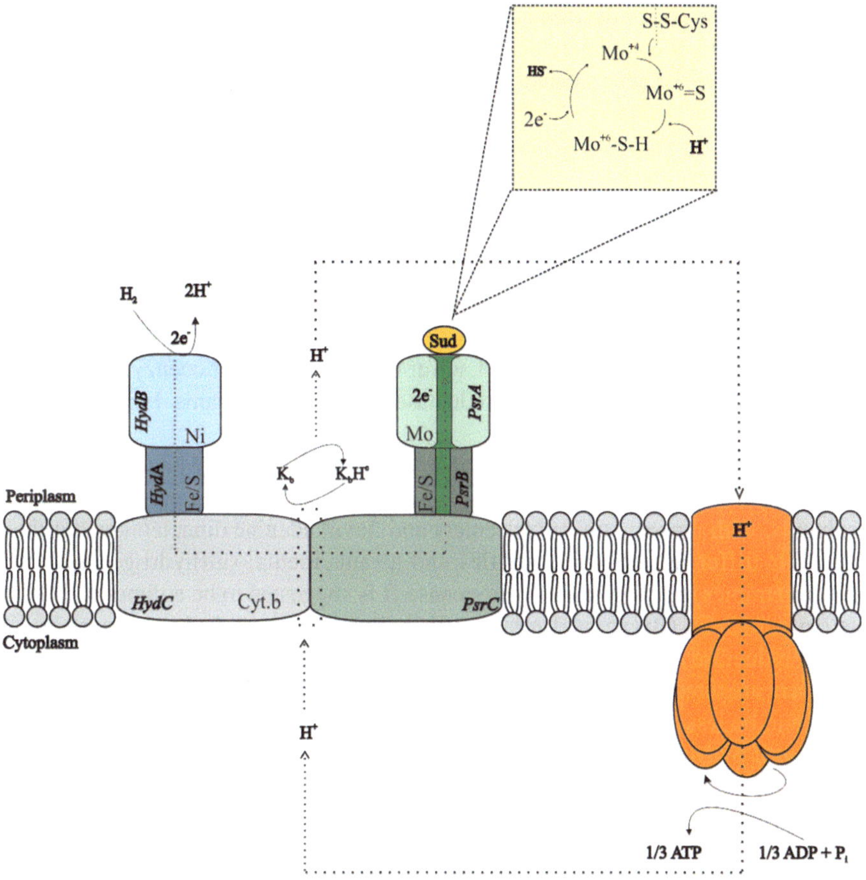

Fig. 5.5 Hypothetical view of elemental sulfur reduction (via polysulfide) and anaerobic electron transport chain in *W. succinogenes*. For the electron transfer to happen between the enzymes, collision of the enzymes is assumed to be required and menaquinone seems to be bound to the subunit C of the polysulfide reductase. Protons are also assumed to be translocated to the periplasm via menaquinone. Subunits of the hydrogenase are labeled HydA, HydB and HydC and subunits of the polysulfide reductase are labeled PsrA, PsrB and PsrC. K stands for quinone and Sud stands for a sulfur/polysulfide transferase. Model adapted from Hedderich et al. (1999) and Rosenberg et al. (2013)

The protein contains two flavins and three different [Fe-S] centers: a putative [2Fe-2S] cluster coordinated by a motif of an aspartate and three cysteine (Asp(Cys)$_3$) that combines physic-chemical properties known as exclusive from protein clusters coordinated by hystidine (Rieske-type), a regular [3Fe-4S] cluster with high reduction potential, and a [4Fe-4S] cluster also with unusual reduction properties (Hagen et al. 2000). The role of the high reduction potentials for the last two clusters is not yet clear, but the redox potential of the flavins is consistent with the function of sulfide dehydrogenase and ferredoxin: NADP$^+$ oxidoreductase.

As the properties of the iron-sulfur clusters in the subunits of the sulfide dehydrogenase are not yet completely understood, the mechanism of action is not clear.

Sequences of the gene subunits deposited in the JGI genome database are available under accession numbers: SudHA: AAL81451/AAL82034; SudHB: AAL81452/AAL82035.

3.3.3 Sulfhydrogenase

Two different cytoplasmic hydrogen-metabolizing enzymes were purified from *P. furiosus* and showed sulfur reductase activity. Both are referred as sulfhydrogenases, I and II, also called NAD(P)H:sulfur oxidoreductase, or coenzyme A (CoA)-dependent NADP(H) sulfur oxidoreductase (Bryant and Adams 1989; Ma et al. 1993, 2000).

Both, sulfhydrogenases I (Bryant and Adams 1989) and II (Ma et al. 2000) can reduce $S_8^0$and polysulfide to H_2S using H_2 as electron donor. Both proteins have four subunits, with nickel, iron-sulfur centers and flavin adenine dinucleotide, but their subunits differ in catalytic activities and arrangements; sulfhydrogenase I is a heterotetramer ($\alpha\beta\gamma\delta$) and sulfhydrogenase II is suggested to be a dimer of heterotetramer $(\alpha\beta\gamma\delta)_2$ (Bryant and Adams 1989). In both cases β and γ subunits play a sulfur reductase role, while α and δ function as hydrogenases.

There are three main differences between the enzymes: (1) sulfhydrogenase II was shown to be less active for hydrogen production, uptake and sulfur reduction assays developed by Ma et al. (2000). (2) The authors also showed that sulfhydrogenase II has higher affinity for elemental sulfur and polysulfide, suggesting a physiological relevance of this enzyme when the concentration of sulfur is low. (3) Sulfhydrogenase II shows greater affinity for NAD(H) and NADP(H) the sulfhydrogenase I, and potentially uses both nucleotides with equivalent efficiency.

Sequences of the gene subunits of the two complexes deposited in the JGI genome database are available under accession numbers: shyA: AAL81018/AAL81456; shyB: AAL81015/AAL81453; shyC: AAL81016/AAL81454; shyD: AAL81017/AAL81455.

A possible novel enzyme involved in elemental sulfur reduction was purified from the acidophilic archaeon *A. ambivalens* (Laska et al. 2003), which reduces elemental sulfur with H_2 or $NADPH_2$ as electron donors. The sulfur reductase is shown to be a membrane-bound protein related to the one from *W. succinogenes,* with subunits sharing similar structure and properties. At least three proteins likely compose the main structure of the core enzyme: a catalytic subunit, probably a molybdopterin (SreA), an iron-sulfur protein (SreB) and a membrane anchor (SreC). The membrane anchor, however, was shown to be phylogenetically unrelated to the analogous protein in *W. succinogenes*. As the enzyme was isolated in the absence of sulfide, it is most likely that it reduces elemental sulfur itself, instead of polysulfide. Deeper investigations on the sulfur reductase were not possible, as the enzyme could not be purified in the absence of hydrogenase (Laska et al. 2003). A complete characterization of the enzyme is still necessary to reveal if it is a true novel enzyme in sulfur-reducing microorganisms which will help in the elucidation of the mechanisms.

A similar enzyme is present in several microorganisms within *Archaea* and *Bacteria* domains, such as *Deferribacter desulfuricans, Desulfitobacterium dehalogenans, Pelobacter carbinolicus Desulfovibrio frigidus, Acidilobus sulfurireducens, Desulfurella acetivorans, Thermanaerovibrio acidaminovorans, Thermodesulfobacterium geofontis, Acidilobus sulfurireducens, Caldisphaera lagunensis, Vulcanisaeta distributa, Pyrobaculum islandicum, Methanococcus maripaludis* and *Natronolimnobius innermongolicus.*

A general overview of the enzymes present in reported sulfur reducers is given as supplemental material (Supplemental material—Table 2). A search on the online Joint Genome Institute database (http://img.jgi.doe.gov/) shows that the aforementioned enzymes are present in the genome of many microorganisms not reported so far as sulfur reducers. These potential sulfur-reducing prokaryotes are spread over the tree of life, including some phyla without reported species of sulfur-reducing bacteria, such as *Chloroflexi, Actinobacteria, Nitrospira, Chlorobi* or *Rikenellaceae* (Supplemental material—Figure 1). In *Archaea*, the potential sulfur reducers are spread only over the phyla *Crenarchaeota* and *Euryarchaeota* (Supplemental material—Figure 2), where the reported sulfur reducers are also distributed. Even though some of these microorganisms have been tested and did not show sulfur reduction activity, it is not known whether the conditions applied were optimal for growth and/or sulfur reduction. In some cases, e.g. *Desulfonatronovibrio thiodismutans, Desulfonatronum thioautotrophicum* and *Desulfobotulus alkaliphilus* elemental sulfur reduction occurred in resting cells, but sulfur did not support growth. It is suggested that the reaction between the sulfide produced and elemental sulfur generates polysulfide. Due to its toxicity, the polysulfide produced inhibits growth of some of those microorganisms (Sorokin et al. 2011).

3.4 Reduction of Sulfur via Polysulfide

Analyzing *Sulfurospirillum deleyianum*, formerly called *Spirillum* 5175, Zöphel et al. (1991) showed that the addition of thiols, such as glutathione and sulfide, to the medium facilitated elemental sulfur reduction by the membrane fractions of cell extract; and cleaving of S–S bonds by nucleophilic attack was enhanced, which increased the activity. It has also been suggested that polysulfide chains formed from sulfide and sulfur are intermediates in the reduction of sulfur by cytochrome c_3 of *Desulfovibrio desulfuricans* (Cammack et al. 1984). The sulfide (S^{2-}) formed by reduction of the polysulfide cleaves the $S_8{}^0$-ring by nucleophilic attack leading to the generation of new polysulfide molecules, which are quickly reduced to S^{2-} by cytochrome c_3 (Cammack et al. 1984).

Sulfur reduction via polysulfide has been extensively studied in *W. succinogenes.* Macy et al. (1986) reported growth of *W. succinogenes* on formate and elemental sulfur, with H_2S and CO_2 as products. Later, Klimmek et al. (1991) reported growth of *W. succinogenes* with formate and polysulfide.

Ringel et al. (1996) questioned the involvement of polysulfide as intermediate for sulfur respiration in *W. succinogenes* and added Fe^{2+} to the medium to precipitate all the sulfide produced by the bacterium as FeS. In that case, polysulfide formation was prevented. Under the mentioned conditions, anaerobic growth of *W. succinogenes* was observed with formate and elemental sulfur and it was concluded that elemental sulfur was the terminal electron acceptor for sulfur reduction in *W. succinogenes*. Three years later, Hedderich et al. (1999) isolated a soluble sulfur-containing fraction and a periplasmic sulfide dehydrogenase, so-called Sud protein, from the cultures to which Fe^{2+} was added. When they treated the isolated protein with CN^{-} and thiosulfate, no reaction was observed; but when polysulfide was added to the medium, thiocyanate was formed (Eq. 5.6).

$$S_n^{2-} + CN^- \rightarrow SCN^- + S_{n-1}^{2-} \tag{5.6}$$

The Sud protein was found to be involved in the transfer of sulfur from polysulfide in solution to the catalytic site of the polysulfide reductase (Psr) (Klimmek et al. (1991). The menaquinone present in the Psr is thought to serve as electron acceptor of the hydrogenase in polysulfide/sulfur reduction (Rosenberg et al. 2013). The electron transport chain of polysulfide reduction with hydrogen or formate is composed by polysulfide reductase (Psr) and hydrogenases or formate dehydrogenase. Hydrogenases and polysulfide reductase are assumed to be randomly dispersed in the membrane of *W. succinogenes* (Jankielewicz et al. 1995).

Later studies indicated that 8-methyl-menaquinone is essential for sulfur reduction in *W. succinogenes* (Jankielewicz et al. 1995; Hedderich et al. 1999). As most of the menaquinones are assumed to be dissolved in the lipid bilayer phase of the membrane and to play a role in the transference of electrons by diffusion, this was the first hypothesis for its involvement in the mechanisms of sulfur/polysulfide reduction by *W. succinogenes*. However, the redox potential of the menaquinone dissolved in the membrane is much more positive than that of polysulfide, which makes the electron transfer from formate dehydrogenase to polysulfide reductase mediated by diffusion improbable (Hedderich et al. 1999). Alternatively, the menaquinone is likely bound to polysulfide reductase and is the primary electron acceptor for the cytochrome b subunit of the hydrogenase (Hedderich et al. 1999). Therefore, it is possible that the electron transfer from hydrogenase to polysulfide reductase requires collision or aggregation of the two enzymes within the membrane (Fig. 5.5). As the menaquinone is intramembrane, it is assumed that its reduction is coupled to the uptake of protons from the cytoplasm by the hydrogenase and the oxidation is coupled to proton release at the periplasm, by the polysulfide reductase (Dietrich and Klimmek 2002).

Several genes were subcloned from genomic libraries of *W. succinogenes*, such as *frh* genes, encoding for formate dehydrogenase (Bokranz et al. 1991), psr genes encoding for polysulfide reductase (Krafft et al. 1995), and sud genes encoding for sulfide dehydrogenase (Kreis-Kleinschmidt et al. 1995).

Blumentals et al. (1990) investigated the mechanism of sulfur reduction in the archaeon *P. furiosus*. The authors observed sulfide and polysulfide formation in cultures in which elemental sulfur was physically separated from the microorganism, indicating that contact between the archaeon and elemental sulfur is not necessary for the metabolism and that soluble polysulfides serve as substrates for sulfur reduction. It is not yet clear whether sulfur reduction in *P. furiosus* is coupled to energy conservation. Sulfur can serve merely as electron sink allowing a more effective fermentation of organic compounds (Rosenberg et al. 2013).

P. furiosus can use protons as terminal electron acceptors, coupling directly the production of H_2 to the synthesis of ATP. The multiprotein membrane bound hydrogenase complex and the ferredoxin, which functions as an electron donor of low-potential, couple the electron transfer to proton reduction and proton translocation (Sapra et al. 2003).

3.5 *Reduction of Sulfur via Physical Attachment to Solid Phase*

Due to the low solubility of elemental sulfur in water, some microorganisms reduce it at the surface of the outer membrane. The mechanisms adopted by these microorganisms are poorly studied. As some prokaryotes are also able to reduce insoluble mineral-oxides outside the membrane (Lovley 1991; Lovley et al. 2004; Hartshorne et al. 2009), different strategies for electron transfer have been proposed, which can be related to sulfur reducers.

For example, in species of the iron-reducing genera Shewanella and Geobacter, in which some sulfur reducer members can be found, external insoluble iron oxides reduction is reported to happen by four different mechanisms: (1) cytochrome c extends the respiratory chain to the cell surface (Richardson 2000; Lovley et al. 2004; Richter et al. 2012); (2) extracellular redox mediators, such as humic acids, quinones, phenazines and cysteine, can shuttle electrons between the terminal electron donor of the electron transport chain and the insoluble acceptor (Lovley et al. 1998; Scott et al. 1998; Newman and Kolter 2000; Hernandez and Newman 2001); (3) in the absence of cytochrome c, microorganisms can produce modified pili, so-called nanowires, that can serve as an electrical connection between the cell and the surface of the oxides (Reguera et al. 2005); and, some strains can construct electrically conductive networks with nanoparticles of crystalline, conductive or semiconductive minerals, such as iron oxides (Kato et al. 2010).

Some microorganisms are reported to reduce elemental sulfur directly to sulfide, such as *A. ambivalens, A. ferrooxidans, Pyrodictium abyssi* and *Pyrodictium brockii*, from which several studies were performed and are here summarized.

Hydrogenase, quinone and cytochrome c were detected in membranes of *P. brockii* (Pihl and Maier 1991; Pihl et al. 1992). The purified hydrogenase is of the Ni-Fe type, with two subunits (Pihl and Maier 1991). Even though the quinone in this microorganism shows chromatographic properties of migration like ubiquinone-6, nuclear magnetic resonance analysis performed by Pihl et al. (1992) evidentiated a quinone different from all the compared quinones. When the quinone was inactivated by exposition to UV light, the electron transport activity was inactivated. The addition of quinone reactivated the process, implying that the electron transfer sequence is: hydrogenase → quinone → cytochrome c. With this, cytochrome c is supposed to be the electron donor for the not yet identified sulfur reductase.

Dirmeier et al. (1998) isolated a sulfur-oxidoreductase complex from the membrane fraction of *P. abyssi* isolate TAG11 and showed that the electron transport chain that catalyzes sulfur reduction by hydrogen is different from *P. brockii* in composition and organization of its constituents. The complete respiratory chain of the organism is suggested to be represented by an enzyme multi-complex, in which the components of the electron transport, the hydrogenase and the sulfur reductase are consistently arranged. The reductase is composed by at least nine subunits, with two b-type cytochromes and one c-type. No quinone has been detected in the membrane fraction complex enzyme of *P. abyssi.* The presence of nickel in the sulfur-oxidoreductase indicates that its hydrogenase is of the Ni-Fe type (Rosenberg et al. 2013), as for *P. brockii.*

A sulfur reductase purified from *A. ambivalens* was shown to reduce elemental sulfur with hydrogen as electron donor in the presence of a co-purified hydrogenase, with a quinone as electron carrier (Laska et al. 2003). The hydrogenase has similar main subunits as the hydrogenase purified from *W. succinogenes*, one homologous Ni-containing catalytic subunits (HynL/HydB), one homologous Fe-S containing electron transfer subunit (HynS/HydB) and one non-homologous membrane anchor (IspI/HydC) (Laska et al. 2003). Thus, the electron transport chain in this microorganism is most likely composed of the two enzymes connected by quinones (Fig. 5.6).

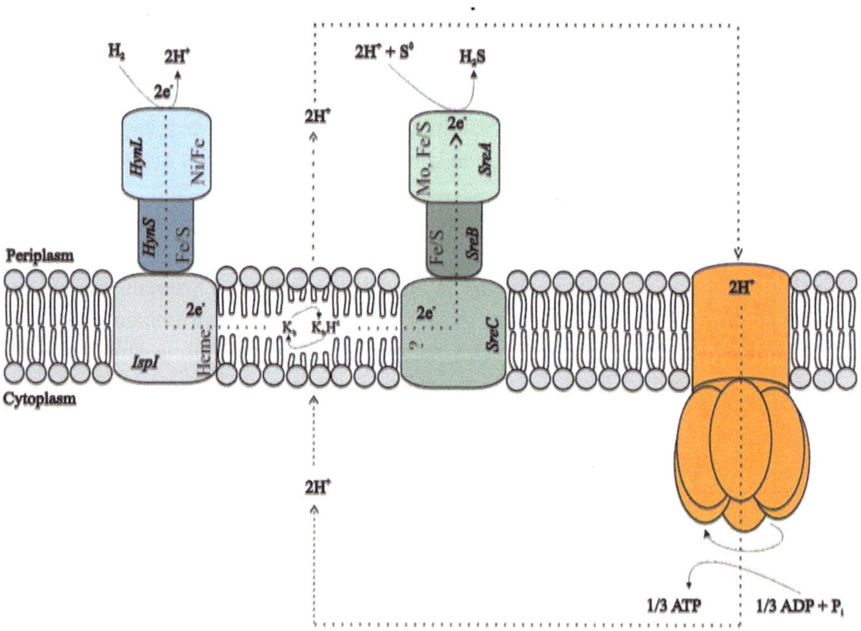

Fig. 5.6 Hypothetical view of elemental sulfur reduction and anaerobic electron transport chain in *A. ambivalens.* Protons are assumed to be translocated to the periplasm via quinone. Only major structural subunits are represented. Subunits of the hydrogenase are labeled HynL, HynS and IspI and subunits of the sulfur reductase are labeled SreA, SreB and SreC. K stands for quinone. Model adapted from Laska et al. (2003)

As the net balance of protons from the periplasmic reactions is zero, an electrochemical gradient is most likely generated with protons taken up by quinone from the cytoplasm and released at the periplasm.

4 Mechanisms of Adaptation to Acidic Conditions

Many sulfur-reducing microorganisms prefer neutral pH to grow. Nonetheless, several species that are capable to thrive in acidic environments have been identified (Stetter 1996; Hedderich et al. 1999; Yoneda et al. 2012) (Supplemental material—Table 1). Those species of acidophiles or acidotolerants tolerate larger pH gradients across the cytoplasmic membrane than neutrophilic organisms. These microorganisms normally face a proton motive force beyond the cell membrane which can drive energy dependent processes to promote pH homeostasis (Baker-Austin and Dopson 2007). To maintain a physiological pH despite the external acidic conditions, microorganisms adopt several strategies. Baker-Austin and Dopson (2007) presented an extremely valuable review on the pathways and mechanisms proposed that enable microorganisms to thrive at low pH, which are summarized in this section, such as utilization of specific transporters and enzymes for proton export, adoption of particular permeability properties, increment of buffer capacity and enhancement of positive surface charges.

In general, acidophiles and acidotolerants have a highly impermeable cell membrane or low membrane fluidity to restrict proton influx to the cytoplasm (Benjamin and Datta 1995; Dilworth and Glenn 1999; Konings et al. 2002). The membranes of some acidophilic archaea are composed of tetraether lipids which make them rather impermeable to protons. Additionally, ether linkages are less sensitive to acid hydrolysis than ester linkages, commonly found in bacterial and eukaryotic cell membranes (Macalady and Banfield 2003; Golyshina and Timmis 2005). Moreover, the lipids from the membranes are also characterized by a substantially higher content of glycolipids, in which one or more sugar units are exposed at the outer surface of the cell (De Rosa et al. 1983; Chong 2010). Although there is still a lack of direct evidence, it was suggested that the abundant modifications of sugar on the cell surface of archaea can provide a protection against proton influx (Shimada et al. 2008; Wang et al. 2012)

Reduction of the size and permeability of the membrane channels is another mechanism for pH homeostasis in acidophiles. The membrane pore reduces its size and the selection of ions to enter the porin occurs based on their charge and size (Amaro et al. 1991).

Another mechanism adopted by acidophiles to reduce the influx of protons is the maintenance of a difference in electrical potential between the intra and extracellular environment without current coursing through the membrane, developing an inside positive $\Delta\Psi$ against the inside negative $\Delta\Psi$ in neutrophiles, the so-called Donnan potential. This Donnan potential is probably generated by a greater influx of potassium ions. The importance of this mechanism is suggested by a very high number of putative cation transporters identified in the genomes of several acidophiles, including some related to sulfur cycle, such as *Acidithiobacillus thiooxidans* (Suzuki et al. 1999), *Acidithiobacillus caldus* (Dopson et al. 2002), *A. ferrooxidans* (Cox et al. 1979) and *Acidiphilium acidophilum* (Goulbourne et al. 1986).

Proton efflux pump systems, such as proton ATPases, antiporters and symporters (Box 5.2), are also used by some acidophiles to maintain the pH homeostasis (Tyson et al. 2004; Golyshina and Timmis 2005; Baker-Austin and Dopson 2007). Protons that enter the cell must be balanced by extrusion during electron transport and reduction of terminal electron acceptors.

Box 5.2

Antiporters: integral membrane proteins that actively transport a substance through the membrane, while transporting ions in the opposite direction. The ions, typically hydrogen (H^+) or sodium (Na^+) ions, flow down their concentration gradient, and in doing so provide the energy for the transport of another substance in the other direction.

Symporters: integral membrane proteins that simultaneously transports two substances across membrane in the same direction. Often, one molecule can move up an electrochemical gradient because the movement of the other molecule is more favourable.

The cytoplasm of all microbes presents a buffering capacity (Box 5.3) to sequester or release protons, according to the shifts in pH. Amino acids or other small organic molecules and ionizable groups in proteins and inorganic polymers, such as polyphosphates, have this buffering capacity (Slonczewski et al. 1982; Zychlinsky and Matin 1983; Krulwich et al. 1985; Leone et al. 2007). Zychlinsky and Matin (1983) compared the buffering capacity of *Acidiphilium acidophilum* and *Escherichia coli* and the result showed a slightly higher capacity for the acidophile, 97 and 85 mmol H^+ per pH unit, respectively. It was also found by Krulwich et al. (1985) that *Bacillus acidocaldarius* has a higher buffering capacity (around 600 mmol H^+ per pH unit) than other bacilli in neutrophilic conditions (around 400–550 mmol H^+ per pH unit). However, the results obtained in both studies show that the buffering capacity of the acidophiles is not necessarily higher than their counterpart of neutrophiles. This suggests that the buffering capacity can contribute to pH homeostasis only together with other mechanisms.

Box 5.3

Buffering capacity: It is the ability of a solution to resist to changes in pH by either absorbing or desorbing H^+ and OH^- ions. It is represented by the moles of an acid or base needed to change the pH of a solution by 1, divided by the pH change and the volume of buffer.

The low pH of the environments can damage biomolecules in the cell, which requires repair mechanisms. This can explain the great number of DNA and protein repair genes present in the genomes of several acidophiles (Crossman et al. 2004). At low pH, chaperones involved in protein refolding are highly expressed in a wide range of acidophiles, suggesting that they can play a role in the survival of microorganisms under acidic conditions.

Investigation on *Ferroplasma acidiphilum*, an obligate acidophile with an intracellular predicted pH of 5.6 during active growth, showed that several enzymes were functional at pH values in a range of 1.7–4.0, suggesting that they need to be functional to get the metabolism started when the cells grow at extreme low pH values. It has been detected a higher amount of iron proteins in the proteoms of many acidophiles which contributes to the pH stability of enzymes at low pH (Ferrer et al. 2007). The removal of iron from purified proteins of these acidophiles makes them to lose the secondary structure of the proteins and, therefore, their activity. Iron is then thought to play an important role on the maintenance of three dimensional structures of the proteins and then serves as an iron rivet—an early property that has a role in stabilizing proteins in acidic condition (Ferrer et al. 2007).

Most of the organic acids, such as acetic and lactic acid facilitate transfer of protons across the membrane at low pH. In this condition, there is a diffusion of acids in protonated form into the cell and consequently the protons dissociate in the cytoplasm, where the pH is higher (Baker-Austin and Dopson 2007). Therefore, the organic acid degradation ability in some acidophiles can play a detoxifying role.

5 Biotechnological Application

5.1 Industrial Wastes and Acid Mine Drainage

The biological oxidation of sulfidic minerals and formation of acidic metal-rich mine drainage waters have been described in several studies (Hoffert 1947; Johnson 1995, 2003). Briefly, due to their exposure to oxidants (O_2 or Fe^{3+}), the geobiochemical oxidation of metal sulfides such as pyrite is the root cause of acid mine drainage (AMD) (Johnson and Hallberg 2005). In most situations, ferric iron is the primary oxidant which chemically oxidize the ores (Eqs. 5.7 and 5.8) and its biological regeneration (Eq. 5.9) maintains the open-ended oxidation of the mineral (Schippers and Sand 1999; Johnson and Hallberg 2005; Vera et al. 2013) and the acidic environment formation, in which metals are commonly dissolved.

$$FeS_2 + 6Fe^{3+} + 3H_2O \rightarrow S_2O_3^{2-} + 7Fe^{2+} + 6H^+ \tag{5.7}$$

$$S_2O_3^{2-} + 8Fe^{3+} + 5H_2O \rightarrow 2SO_4^{2-} + 8Fe^{2+} + 10H^+ \tag{5.8}$$

$$4Fe^{2+} + O_2 + 4H^+ \rightarrow 4Fe^{3+} + 2H_2O \tag{5.9}$$

Copper, zinc, cadmium, arsenic, manganese, aluminium, lead, nickel, silver, mercury, chromium and iron are metals of remarkable interest in acid mine drainage and industrial wastewaters, as they can be present in a wide range of concentration, from 10^{-6} to 10^{2} g L^{-1} (Huisman et al. 2006). As examples, in Tinto River, a natural acidic rock drainage, iron can be detected up to 20.2 g L^{-1}, copper up to 0.7 g L^{-1}, and zinc up to 0.56 g L^{-1} (Lopez-Archilla et al. 2001); while in the effluent of a textile industry iron was detected up to 0.11 g L^{-1}, and copper and zinc up to 0.01 g L^{-1} (Joshi and Santani 2012).

5.2 *State of the Art Methods for Metal Removal and Recovery*

5.2.1 Chemical/Physical Methods

Many chemical/physical methods have been applied to remove heavy metals from contaminated wastewaters, such as absorption, ion exchange, complex formation and precipitation by addition of chemicals, which is the most widely applied chemical/physical approach for the treatment of acid mine drainage (AMD) and other metal-contaminated streams (Johnson and Hallberg 2005).

To raise the pH and consequently precipitate metals in a mitigation process, some neutralizing agents are added to the medium, such as calcium carbonate, calcium oxide, calcium hydroxide or sodium hydroxide (Weijma et al. 2002). Despite effective treatments, these methods are relatively expensive and produce large volumes of residual metal-contaminated sludge with no or low metal reuse potential (Gallegos-Garcia et al. 2009; Tekerlekopoulou et al. 2010).

5.2.2 Microbiological Methods

Microbial processes, such as methanogenesis, denitrification, and reduction of iron and manganese, generate alkalinity, which may result in metal precipitation as hydroxides (Johnson and Hallberg 2005). Even though hydroxides can be removed from the effluent, as all the metals precipitate together, the generated waste needs to be disposed, which results in extra costs of the process. Metals may also be recovered bioelectrochemical systems, where an organic substrate is biologically oxidized at the anode, thereby generating electrons which are used to reduce metals like Cu^{2+} at the cathode (Heijne et al. 2010). Much research in the past used the concept of metal biosorption, i.e. the adsorption of metal ions to the surface of biological matter such as bacterial cells and plants. This method is not widely applied, presumably due to the low metal loading capacity and the production of a residue from which metal recovery is hardly feasible.

Bioreactors systems to precipitate metals based on sulfidogenesis are as effective as the physical methods while operating at substantially lower costs and producing lower amounts of residual sludge (Johnson and Hallberg 2005). Sulfidogenesis is based on the oxidation of simple organic compounds or hydrogen by microorganisms under anaerobic conditions, generating sulfide from the reduction of sulfur compounds, such as sulfate, sulfite, thiosulfate, organic sulfoxides, elemental sulfur, polysulfide, and organic disulfides. The versatility of sulfidogenic microorganisms allows for many combinations of electron donor and sulfur sources, and also for a wide range of operational conditions for the process (temperature, salinity, pH).

5.3 Sulfidogenesis for Metal Removal and Recovery

In sulfidogenic processes for metal removal and recovery, the biologically produced sulfide binds to dissolved heavy metals, such as Cu^{2+}, Zn^{2+}, and Ni^{2+} precipitating as insoluble metal sulfides (Hulshof et al. 2006; Neculita et al. 2007). The theoretical solubility of most metal sulfides at neutral to alkaline pH is extremely low, much lower than that of the corresponding metal hydroxides. Thus, better effluent qualities can be reached and more metal can be recovered. Also the reactions rates are higher and the acid-stable metal sulfides, such as Co, Ni and Cu, present good settling properties and high potential for re-use (Tsukamoto et al. 2004; Gallegos-Garcia et al. 2009; Lewis 2010; Sánchez-Andrea et al. 2014). Smelter facilities for base metal production use ore concentrates that often contain the metal in their sulfidic mineral form, such as sphalerite in the case of ZnS. This facilitates the use of biologically precipitated metals sulfides as feedstock for smelters. For ZnS, this is practised at the zinc refinery of Nyrstar in The Netherlands (Weijma et al. 2002).

Sulfate reduction is the most used biological process for the treatment of mining and metallurgical streams. However, there are only a few described species of moderate acidophilic sulfate-reducing bacteria: *Thermodesulfobium narugense*, which can grow at pH 4 (Mori et al. 2003), *Desulfosporosinus acidiphilus,* which can grow at pH 3 (Jameson et al. 2010), and *Desulfosporosinus acididurans*, which can grow at pH 3.8 (Sánchez-Andrea et al. 2015). The use of biogenic sulfur is of particular interest for the treatment of acid mine and acid rock drainage (Hoffert 1947; Johnson 1995, 2003).

For treatment of metal-contaminated streams such as acid mine/rock drainage, two designs of sulfidogenic bioreactors have been proposed. One is based on a biological and a chemical compartment operating independently (Tabak et al. 2003). In the biological compartment, hydrogen sulfide is produced and transferred via a gas circulation to the chemical circuit, which receives the raw influent (Fig. 5.7a). Thus, the biological production of sulfide and the precipitation of metals are separated by stripping hydrogen sulfide from the biological solution with a carrier gas (nitrogen) and then the hydrogen sulfide gas dissolves in the metal-contaminated (waste)water. In this device, there is no contact between the sulfidogenic biomass and the metal-contaminated stream. This is the major advantage of this design because it prevents possible biomass toxicity effects due to high acidity and metal concentrations (Johnson and Hallberg 2005). The drawback is that the carrier gas recycle requires a high energy input. This technique has been studied with metals like Cu and Zn

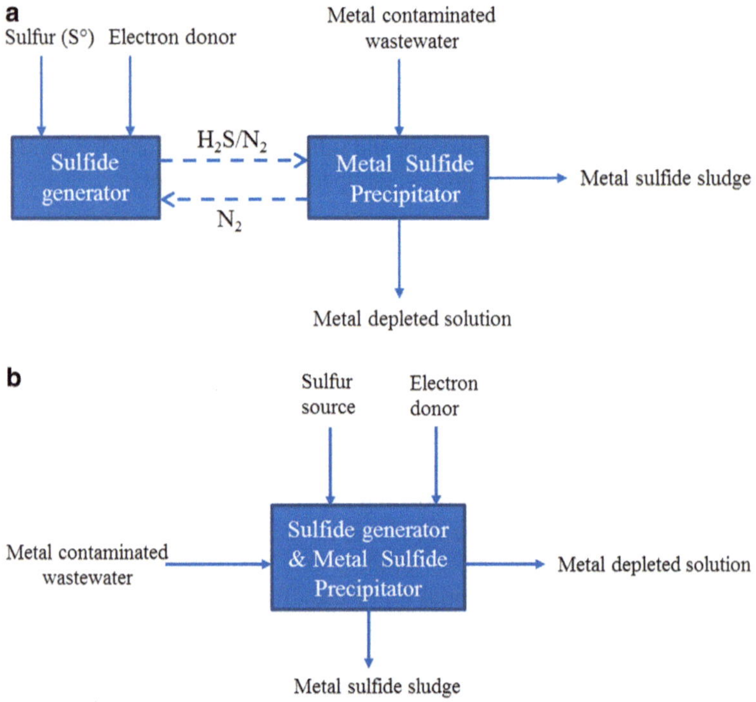

Fig. 5.7 Flowsheet for two-stage biological metal removal with no direct contact between the sulfidogenic microorganisms and the metal-contaminated wastewater (**a**). One-stage biological metal removal with direct contact between the sulfidogenic microorganisms and the metal-contaminated wastewater (**b**)

(Foucher et al. 2001; Al-Tarazi et al. 2005; Gramp et al. 2009). Because of the separate sulfide production and metal sulfide precipitation, both process parts can be controlled at their optimal conditions. For example, selective precipitation of individual heavy metals can be achieved by carefully controlling the pH and the pS (-log[S^{2-}]) in the precipitator (Veeken and Rulkens 2003; König et al. 2006; Sampaio et al. 2009). These results in relatively pure precipitates of metal sulfides that have a higher value as supplement to ore concentrate feedstock in the metallurgical industry (Grootscholten et al. 2008).

The other designed system has only one compartment, in which biological sulfide production and metal precipitation occur simultaneously (Fig. 5.7b).

In this configuration, since the sulfidogenic culture comes into contact with the dissolved metals from the influent, metal toxicity is a design and operation concern. By keeping some excess of hydrogen sulfide relative to the metals, a 'sulfide buffer' is created that can accommodate fluctuations in metal loading and biological activity. The advantage of this configuration is that sulfide generation and metal sulfide precipitation take place in a single unit, thereby eliminating the need for energy-intensive recirculation of a carrier gas. This flow scheme has been studied by amongst others (Labrenz et al. 2000; Steed et al. 2000; Kaksonen et al. 2003; Johnson and Hallberg 2005;

Sierra-Alvarez et al. 2006; Gallegos-Garcia et al. 2009; Sánchez-Andrea et al. 2012). Full-scale operations for biogenic sulfide production are described in Weijma et al. (2002) and Möbius et al. (2015).

5.4 Comparative Analysis of Cost Between Sulfate and Sulfur Reduction Processes

Wastewater from mining or metals industries contains, normally, low organic matter content. To completely reduce the sulfur compounds to sulfide, electron donors need to be added (Liamleam and Annachhatre 2007). Based on the stoichiometry of the reactions, elemental sulfur is more attractive as electron acceptor than sulfate, since only two electrons per mol of sulfide produced are needed in the process (Eq. 5.10), instead of eight needed for sulfate (Eq. 5.11). The sulfide produced determines the amount of metals to be recovered (Eq. 5.12), therefore with the same amount of metal precipitated, the process needs four times less of electron donor for sulfur reduction in comparison with sulfate reduction.

Hedrich and Johnson (2014) performed analysis of costs for modular reactors to oxidize iron and reduce sulfate to precipitate metals operating at low pH. The 42 m^3 sulfidogenic reactor needed to treat 1 m^3 mine water operated with glycerol as electron donor, would produce 3.96 mol of sulfide. As the stoichiometric reaction of glycerol with sulfate is 4–7 (Eq. 5.12), 2.26 mol (208.52 g) of glycerol would be required in the reactor. Assuming the market price of glycerol as 2400 $/ton, the cost of this reagent in the process result on 0.5 $, as described in the article.

If instead of sulfate, sulfur is applied as electron acceptor, to reach the same amount of sulfide in a 42 m^3 reactor, an input of 0.126 kg of sulfur is required. As an estimated market price of sulfur of 61 $/ton, an additional cost of 0.008 $ is needed in the process. However, as sulfur reduction requires four times less electron donors (Eq. 5.13), the same amount of sulfide is reached with only 52.13 g of glycerol, implying a global reduction in costs of $ 0.37 per m^3 of mine water treated.

In accordance with Eq. (5.11) and as expressed in Hedrich and Johnson (2014), with the amount of sulfide produced via sulfur or sulfate reduction, 0.46 kg can be recovered, which represents about 0.80 $ of return per m^3 of mine water treated. Considering copper, which is common in acid mine drainage, 0.46 kg Cu recovered would imply 2.71 $ of return per m^3 of mine treated, taking 5900 $/ton as an average market price of copper.

Another advantage of implementing elemental sulfur reduction for remediation of AMD streams is that sulfur reducers can generally reduce elemental sulfur at pH values lower than the so far described sulfate reducers. Sulfur reduction is reported in extremely acidophilic microorganisms, such as *A. ferrooxidans* (pH 1.8) (Osorio et al. 2013), *Acidilobus sulfurireducens* (pH 2) (Boyd et al. 2007), *Acidianus infernus* (pH 1.5) (Stetter 1996), *Stygiolobus azoricus* (pH 1) (Svetlichnyi et al. 1987; Stetter 1996), *Thermoplasma acidophilum* and *volcanicum* (pH 1) (Segerer et al. 1988). The lowest reported pH for sulfate reduction by isolates is 3.6–3.8 by members of *Desulfosporosinus* genus (Alazard et al. 2010;

Sánchez-Andrea et al. 2015) and Ňancucheo and Johnson (2012) reported activity at a pH as low as 2.5 in bioreactors.

$$\text{Half reaction}: H_2 \rightarrow 2e^- + 2H^+$$

$$\text{Half reaction}: S + 2e^- \rightarrow S^{2-}$$

$$\textit{Global reaction}: S + H_2 \rightarrow S^{2-} + 2H^+ \tag{5.10}$$

$$\text{Half reaction}: 4H_2 \rightarrow 8e^- + 8H^+$$

$$\text{Half reaction}: SO_4^{2-} + 8e^- + 8H^+ \rightarrow S^{2-} + 4H_2O$$

$$\textit{Global reaction}: SO_4^{2-} + 4H_2 \rightarrow S^{2-} + 4H_2O \tag{5.11}$$

$$H_2S + Me^{2+} \rightarrow MeS_{(s)} + 2H^+ \tag{5.12}$$

where Me^{2+} = metal, such as Zn^{2+}, Cu^{2+}, Pb^{2+} and Ni^{2+}

$$4C_3H_8O_3 + 7SO_4^{2-} \rightarrow 7H_2S + 12CO_2 + 16H_2O \tag{5.13}$$

$$C_3H_8O_3 + 7S + 3H_2O \rightarrow 7H_2S + 3CO_2 \tag{5.14}$$

Additionally, many sulfate reducers are incomplete oxidizers (e.g.: *Desulfotomaculum* sp., *Desulfobulbus* sp., *Archaeoglobus* sp. (Castro et al. 2002), *Desulfovibrio* sp., *Thermodesulfobacterium* sp. (Widdel and Pfennig 1981; Widdel 1988; Widdel and Pfennig 1991), *Desulfosporosinus* sp. (Sánchez-Andrea et al. 2015) which means that they contribute to the accumulation of acetic acid in the medium, with the consequent possible inhibition of the process. This is not the case for most of the sulfur reducers, especially the ones belonging to the *Deltaproteobacteria* class, which are able to oxidize organic substrates to CO_2, such as *Desulfuromonas* sp., *Geobacter* sp., *Pelobacter* sp. *and Desulfurella* sp. the latter ones are usually found in acid environments (Bonch-Osmolovskaya et al. 1990b; Miroshnichenko et al. 1998).

Sulfur reduction looks more promising for treatment of metal-laden streams in metallurgical processes, which are free of sulfate, often acidic and sometimes hot. However, for obvious reasons such as the natural presence of sulfate in AMD water, sulfate reduction might be still the easiest option for *in situ* systems such as permeable reactive barriers.

6 Concluding Remarks and Future Perspectives

Microorganisms involved in the sulfur cycle are of great importance from the industrial and environmental point of view, especially the ones that perform sulfidogenesis. Sulfur-reducing prokaryotes are ubiquitously distributed in marine and terrestrial environments and able to grow in a broad range of temperature and pH. Species able to thrive in acidic environments are of interest for selective metals precipitation and bioremediation processes.

Several acidophilic sulfur reducers were described but their physiology and specific mechanisms adopted to face extreme conditions are still poorly understood. Ongoing and future research on these microorganisms will provide more insight into the real substrate used by sulfur reducers, physiology and ecology of those microorganisms and their behavior in engineered ecosystems such as reactors for the selective precipitation and recovery of heavy metals from mining and metallurgical industries.

Acknowledgements The doctoral study program of A.P. Florentino is supported by the organization of the Brazilian Government for the development of Science and Technology CNPq (Conselho Nacional de Desenvolvimento Científico e Tecnológico). Research of I. Sánchez-Andrea and A.J.M. Stams is financed by ERC grant project 323009 and by Gravitation grant project 024.002.002 from the Netherlands Ministry of Education, Culture and Science.

Conflict of Interest Anna P. Florentino, Jan Weijma, Alfons J. M. Stams, and Irene Sánchez-Andrea declare that they have no conflict of interest.

References

Alain K, Callac N, Guégan M, Lesongeur F, Crassous P, Cambon-Bonavita MA, Querellou J, Prieur D (2009) *Nautilia abyssi* sp. nov., a thermophilic, chemolithoautotrophic, sulfur-reducing bacterium isolated from an East Pacific Rise hydrothermal vent. Int J Syst Evol Microbiol 59(6):1310–1315

Alazard D, Joseph M, Battaglia-Brunet F, Cayol JL, Ollivier B (2010) *Desulfosporosinus acidiphilus* sp. nov.: a moderately acidophilic sulfate-reducing bacterium isolated from acid mining drainage sediments: New taxa: Firmicutes (Class Clostridia, Order Clostridiales, Family Peptococcaceae). Extremophiles 14(3):305–312

Al-Tarazi M, Heesink ABM, Versteeg GF, Azzam MOJ, Azzam K (2005) Precipitation of CuS and ZnS in a bubble column reactor. AIChE J 51(1):235–246

Amaro AM, Chamorro D, Seeger M, Arredondo R, Peirano I, Jerez CA (1991) Effect of external pH perturbations on in vivo protein synthesis by the acidophilic bacterium *Thiobacillus ferrooxidans*. J Bacteriol 173(2):910–915

Bak F, Pfennig N (1987) Chemolithotrophic growth of *Desulfovibrio sulfodismutans* sp. nov. by disproportionation of inorganic sulfur compounds. Arch Microbiol 147(2):184–189

Baker-Austin C, Dopson M (2007) Life in acid: pH homeostasis in acidophiles. Trends Microbiol 15(4):165–171

Beijerink WM (1895) Über Spirillum desulfuricans als Ursache von Sulfatreduction. Centralb Bakteriol II 1:49–59

Belkin S, Wirsen CO, Jannasch HW (1985) Biological and abiological sulfur reduction at high temperatures. Appl Environ Microbiol 49:1057–1061

Benjamin MM, Datta AR (1995) Acid tolerance of enterohemorrhagic *Escherichia coli*. Appl Environ Microbiol 61(4):1669–1672

Biebl H, Pfennig N (1977) Growth of sulfate-reducing bacteria with sulfur as electron acceptor. Arch Microbiol 112(1):115–117

Birrien JL, Zeng X, Jebbar M, Cambon-Bonavita MA, Quérellou J, Oger P, Bienvenu N, Xiao X, Prieur D (2011) *Pyrococcus yayanosii* sp. nov., an obligate piezophilic hyperthermophilic archaeon isolated from a deep-sea hydrothermal vent. Int J Syst Evol Microbiol 61(12):2827–2881

Blumentals II, Itoh M, Olson GJ, Kelly RM (1990) Role of polysulfides in reduction of elemental sulfur by the hyperthermophilic archaebacterium *Pyrococcus furiosus*. Appl Environ Microbiol 56(5):1255–1262

Bokranz M, Gutmann M, Körtner C, Kojro E, Fahrenholz F, Lauterbach F, Kröger A (1991) Cloning and nucleotide sequence of the structural genes encoding the formate dehydrogenase of *Wolinella succinogenes*. Arch Microbiol 156(2):119–128

Bonch-Osmolovskaya EA, Miroshnichenko ML, Kostrikina NA, Chernych NA, Zavarzin GA (1990a) *Thermoproteus uzoniensis* sp. nov., a new extremely thermophilic archaebacterium from Kamchatka continental hot springs. J Bacteriol 154(6):556–559

Bonch-Osmolovskaya EA, Sokolova TG, Kostrikina NA, Zavarzin GA (1990b) *Desulfurella acetivorans* gen. nov. and sp. nov.–a new thermophilic sulfur-reducing eubacterium. Arch Microbiol 153:151–155

Bonch-Osmolovskaya EA, Miroshnichenko ML, Chernykh NA, Kostrikina NA, Pikuta EV, Rainey FA (1997) Reduction of elemental sulfur by moderately thermophilic organotrophic bacteria and the description of thermoanaerobacter *Sulfurophilus* sp. nov. Microbiol 66(5):581–587

Boulegue J (1978) Solubility of elemental sulfur in water at 298 K. Phosphorus Sulfur Silicon Relat Elem 5(1):127–128

Boyd ES, Druschel GK (2013) Involvement of intermediate sulfur species in biological reduction of elemental sulfur under acidic, hydrothermal conditions. Appl Environ Microbiol 79(6):2061–2068

Boyd ES, Jackson RA, Encarnacion G, Zahn JA, Beard T, Leavitt WD, Pi Y, Zhang CL, Pearson A, Geesey GG (2007) Isolation, characterization, and ecology of sulfur-respiring crenarchaea inhabiting acid-sulfate-chloride-containing geothermal springs in Yellowstone National Park. Appl Environ Microbiol 73(20):6669–6677

Breher F (2004) Elemental sulfur and sulfur-rich compounds, Angewandte Chemie. International ed.; WILEY-VCH Verlag, 43(28): 3629a– 3630

Brunner E, Woll W (1980) Solubility of sulfur in hydrogen sulfide and sour gases C1–SPE. Soc Petrol Eng J 8 doi:10.2118/8778-pa

Bryant FO, Adams MW (1989) Characterization of hydrogenase from the hyperthermophilic archaebacterium, *Pyrococcus furiosus*. J Biol Chem 264(9):5070–5079

Burggraf S, Jannasch HW, Nicolaus B, Stetter KO (1990) *Archaeoglobus profundus* sp. nov., represents a new species within the sulfate-reducing archaebacteria. Syst Appl Microbiol 13(1):24–28

Caccavo F, Lonergan DJ, Lovley DR, Davis M, Stolz JF, McInerney MJ (1994) *Geobacter sulfurreducens* sp. nov., a hydrogen- and acetate-oxidizing dissimilatory metal-reducing microorganism. Appl Environ Microbiol 60(10):3752–3759

Cammack R, Fauque G, Moura JJG, LeGall J (1984) ESR studies of cytochrome c3 from *Desulfovibrio desulfuricans* strain Norway 4: Midpoint potentials of the four haems, and interactions with ferredoxin and colloidal sulfur. Biochim Biophys Acta 784(1):68–74

Canfield DE (2004) The evolution of the Earth surface sulfur reservoir. Am J Sci 304(10):839–861

Canfield DE, Farquhar J (2012) The global sulfur cycle. Fundamentals of geobiology, John Wiley & Sons, Ltd, Chichester, UK, pp 49–64

Castro H, Reddy KR, Ogram A (2002) Composition and function of sulfate-reducing prokaryotes in eutrophic and pristine areas of the Florida everglades. Appl Environ Microbiol 68(12):6129–6137

Childers SE, Noll KM (1994) Characterization and regulation of sulfur reductase activity in *Thermotoga neapolitana*. Appl Environ Microbiol 60(7):2622–2626

Chong PLG (2010) Archaebacterial bipolar tetraether lipids: physico-chemical and membrane properties. Chem Phys Lipids 163(3):253–265

Cox JC, Nicholls DG, Ingledew WJ (1979) Transmembrane electrical potential and transmembrane pH gradient in the acidophile *Thiobacillus ferrooxidans*. Biochem J 178(1):195–200

Crossman L, Holden M, Pain A, Parkhill J (2004) Genomes beyond compare. Nat Rev Microbiol 2(8):616–617

De Rosa M, Gambacorta A, Nicolaus B (1983) A New type of cell membrane, in thermophilic archaebacteria, based on bipolar ether lipids. J Membr Sci 16:287–294

Dietrich W, Klimmek O (2002) The function of methyl-menaquinone-6 and polysulfide reductase membrane anchor (PsrC) in polysulfide respiration of *Wolinella succinogenes*. Eur J Biochem 269(4):1086–1095

Dilworth MJ, Glenn AR (1999) Problems of adverse pH and bacterial strategies to combat it. Novartis Found Symp 221:4–14, discussion 14–18

Dirmeier R, Keller M, Frey G, Huber H, Stetter KO (1998) Purification and properties of an extremely thermostable membrane-bound sulfur-reducing complex from the hyperthermophilic *Pyrodictium abyssi*. Eur J Biochem 252(3):486–491

Dopson M, Lindstrom EB, Hallberg KB (2002) ATP generation during reduced inorganic sulfur compound oxidation by *Acidithiobacillus caldus* is exclusively due to electron transport phosphorylation. Extremophiles 6(2):123–129

Ferrer M, Golyshina OV, Beloqui A, Golyshin PN, Timmis KN (2007) The cellular machinery of *Ferroplasma acidiphilum* is iron-protein-dominated. Nature 445(7123):91–94

Fiala G, Stetter K (1986) *Pyrococcus furiosus* sp. nov. represents a novel genus of marine heterotrophic archaebacteria growing optimally at 100 °C. Arch Microbiol 145(1):56–61

Finster K, Coates JD, Liesack W, Pfennig N (1997) *Desulfuromonas thiophila* sp. nov., a new obligately sulfur-reducing bacterium from anoxic freshwater sediment. Int J Syst Bacteriol 47(3):754–758

Finster K, Leiesack W, Thamdrup BO (1998) Elemental sulfur and thiosulfate disproportionation by *Desulfocapsa sulfoexigens* sp. nov., a new anaerobic bacterium isolated from marine surface sediment. Appl Environ Microbiol 64(1):119–125

Foucher S, Battaglia-Brunet F, Ignatiadis I, Morin D (2001) Treatment by sulfate-reducing bacteria of Chessy acid-mine drainage and metals recovery. Chem Eng Sci 56(4):1639–1645

Franz B, Lichtenberg H, Hormes J, Modrow H, Dahl C, Prange A (2007) Utilization of solid 'elemental' sulfur by the phototrophic purple sulfur bacterium *Allochromatium vinosum*: a sulfur K-edge X-ray absorption spectroscopy study. Microbiology 153(4):1268–1274

Gallegos-Garcia M, Celis LB, Rangel-Mendez R, Razo-Flores E (2009) Precipitation and recovery of metal sulfides from metal containing acidic wastewater in a sulfidogenic down-flow fluidized bed reactor. Biotechnol Bioeng 102(1):91–99

Ghosh W, Dam B (2009) Biochemistry and molecular biology of lithotrophic sulfur oxidation by taxonomically and ecologically diverse bacteria and archaea. FEMS Microbiol Rev 33(6):999–1043

Golyshina OV, Timmis KN (2005) *Ferroplasma* and relatives, recently discovered cell wall-lacking archaea making a living in extremely acid, heavy metal-rich environments. Environ Microbiol 7(9):1277–1288

Goulbourne E, Matin M, Zychlinsky E, Matin A (1986) Mechanism of delta pH maintenance in active and inactive cells of an obligately acidophilic bacterium. J Bacteriol 166(1):59–65

Gramp JP, Wang H, Bigham JM, Jones FS, Tuovinen OH (2009) Biogenic synthesis and reduction of Fe(III)-hydroxysulfates. Geomicrobiol J 26(4):275–280

Grootscholten T, Keesman K, Lens P (2008) Modelling and on-line estimation of zinc sulfide precipitation in a continuously stirred tank reactor. Sep Purif Technol 63(3):654–660

Gu MX, Li Q, Zhou SY, Chen WD, Guo TM (1993) Experimental and modeling studies on the phase behavior of high H_2S-content natural gas mixtures. Fluid Phase Equilib 82:173–182

Hagen WR, Silva PJ, Amorim MA, Hagedoorn PL, Wassink H, Haaker H, Robb FT (2000) Novel structure and redox chemistry of the prosthetic groups of the iron-sulfur flavoprotein sulfide dehydrogenase from *Pyrococcus furiosus*; evidence for a [2Fe-2S] cluster with $Asp(Cys)_3$ ligands. J Biol Inorg Chem 5(4):527–534

Hartshorne RS, Reardon CL, Ross D, Nuester J, Clarke TA, Gates AJ, Mills PC, Fredrickson JK, Zachara JM, Shi L, Beliaev AS, Marshall MJ, Tien M, Brantley S, Butt JN, Richardson DJ (2009) Characterization of an electron conduit between bacteria and the extracellular environment. Proc Natl Acad Sci USA 106(52):22169–22174

Hedderich R, Klimmek O, Kröger A, Dirmeier R, Keller M, Stetter KO (1999) Anaerobic respiration with elemental sulfur and with disulfides. FEMS Microbiol Rev 22:29

Hedrich S, Johnson DB (2014) Remediation and selective recovery of metals from acidic mine waters using novel modular bioreactors. Environ Sci Technol 48(20):12206–12212

Heijne AT, Liu F, Weijden RVD, Weijma J, Buisman CJN, Hamelers HVM (2010) Copper recovery combined with electricity production in a microbial fuel cell. Environ Sci Technol 44(11):4376–4381

Hernandez ME, Newman DK (2001) Extracellular electron transfer. Cell Mol Life Sci 58(11):1562–1571

Hernandez-Eugenio G, Fardeau ML, Cayol JL, Patel BKC, Thomas P, Macarie H, Garcia JL, Ollivier B (2002) *Sporanaerobacter acetigenes* gen. nov., sp. nov., a novel acetogenic, facultatively sulfur-reducing bacterium. Int J Syst Evol Microbiol 52(4):1217–1223

Hoffert JR (1947) Acid mine drainage. Ind Eng Chem 39(5):642–646

Huber R, Wilharm T, Huber D, Trincone A, Burggraf S, König H, Reinhard R, Rockinger I, Fricke H, Stetter KO (1992) *Aquifex pyrophilus* gen. nov. sp. nov., represents a novel group of marine hyperthermophilic hydrogen-oxidizing bacteria. Syst Appl Microbiol 15(3):340–351

Huber R, Rossnagel P, Woese CR, Rachel R, Langworthy TA, Stetter KO (1996) Formation of ammonium from nitrate during chemolithoautotrophic growth of the extremely thermophilic bacterium *Ammonifex degensii* gen. nov. sp. nov. Syst Appl Microbiol 19(1):40–49

Huisman JL, Schouten G, Schultz CE (2006) Biologically produced sulfide for purification of process streams, effluent treatment and recovery of metals in the metal and mining industry. Hydrometallurgy 83(1–4):106–113

Hulshof AH, Blowes DW, Gould WD (2006) Evaluation of in situ layers for treatment of acid mine drainage: a field comparison. Water Res 40(9):1816–1826

Itoh T, Suzuki K, Nakase T (1998) *Thermocladium modestius* gen. nov., sp. nov., a new genus of rod-shaped, extremely thermophilic crenarchaeote. Int J Syst Bacteriol 48(3):879–887

Itoh T, Suzuki K, Sanchez PC, Nakase T (2003) *Caldisphaera lagunensis* gen. nov., sp. nov., a novel thermoacidophilic crenarchaeote isolated from a hot spring at Mt Maquiling, Philippines. Int J Syst Evol Microbiol 53(4):1149–1154

Jameson E, Rowe OF, Hallberg KB, Johnson DB (2010) Sulfidogenesis and selective precipitation of metals at low pH mediated by *Acidithiobacillus* spp. and acidophilic sulfate-reducing bacteria. Hydrometallurgy 104(3–4):488–493

Jankielewicz A, Klimmek O, Kröger A (1995) The electron transfer from hydrogenase and formate dehydrogenase to polysulfide reductase in the membrane of *Wolinella succinogenes*. Biochim Biophys Acta 1231(2):157–162

Johnson DB (1995) Acidophilic microbial communities: candidates for bioremediation of acidic mine effluents. Int Biodeter Biodegr 35(1-3):41–58

Johnson DB (2003) Chemical and microbiological characteristics of mineral spoils and drainage waters at abandoned coal and metal mines. Water Air Soil Pollut 3(1):47–66

Johnson DB, Hallberg KB (2005) Acid mine drainage remediation options: a review. Sci Total Environ 338(1-2):3–14

Joshi VJ, Santani DD (2012) Physicochemical characterization and heavy meta concentration in effluent of textile industry. Univers J Environ Res Technol 2(2):93–96

Kaksonen AH, Riekkola-Vanhanen ML, Puhakka JA (2003) Optimization of metal sulfide precipitation in fluidized-bed treatment of acidic wastewater. Water Res 37(2):255–266

Kato S, Nakamura R, Kai F, Watanabe K, Hashimoto K (2010) Respiratory interactions of soil bacteria with (semi) conductive iron-oxide minerals. Environ Microbiol 12(12):3114–3123

Kleinjan WE, de Keizer A, Janssen AJH (2005) Equilibrium of the reaction between dissolved sodium sulfide and biologically produced sulfur. Colloids Surf B 43(3–4):228–237

Klimmek O (2005) The biological cycle of sulfur. Metal ions in biological systems, CRC Press. 43:105–130

Klimmek O, Kröger A, Steudel R, Holdt G (1991) Growth of *Wolinella succinogenes* with polysulfide as terminal acceptor of phosphorylative electron transport. Arch Microbiol 155(2):177–182

Knickerbocker C, Nordstrom DK, Southam G (2000) The role of "blebbing" in overcoming the hydrophobic barrier during biooxidation of elemental sulfur by Thiobacillus thiooxidans. Chem Geol 169(3-4):425–433

König J, Keesman KJ, Veeken AHM, Lens PNL (2006) Dynamic modelling and process control of ZnS precipitation. Separ Sci Technol 41(6):1025–1042

Konings WN, Albers SV, Koning S, Driessen AJM (2002) The cell membrane plays a crucial role in survival of bacteria and archaea in extreme environments. A Van Leeuw 81:61–72

Krafft T, Gross R, Kröger A (1995) The function of *Wolinella succinogenes* psr genes in electron transport with polysulfide as the terminal electron acceptor. Eur J Biochem 230(2):601–606

Kreis-Kleinschmidt V, Fahrenholz F, Kojro E, Kröger A (1995) Periplasmic sulfide dehydrogenase (sud) from *Wolinella succinogenes*: Isolation, nucleotide sequence of the sud gene and its expression in *Escherichia coli*. Eur J Biochem 227(1-2):137–142

Krulwich TA, Agus R, Schneier M, Guffanti AA (1985) Buffering capacity of bacilli that grow at different pH ranges. J Bacteriol 162(2):768–772

Labrenz M, Druschel GK, Thomsen-Ebert T, Gilbert B, Welch SA, Kemner KM, Logan GA, Summons RE, De Stasio G, Bond PL, Lai B, Kelly SD, Banfield JF (2000) Formation of sphalerite (ZnS) deposits in natural biofilms of sulfate-reducing bacteria. Science 290(5497):1744–1747

Laska S, Lottspeich F, Kletzin A (2003) Membrane-bound hydrogenase and sulfur reductase of the hyperthermophilic and acidophilic archaeon *Acidianus ambivalens*. Microbiology Pt 149(9):2357–2371

Leone L, Ferri D, Manfredi C, Persson P, Shchukarev A, Sjöberg S, Loring J (2007) Modeling the acid–base properties of bacterial surfaces: a combined spectroscopic and potentiometric study of the gram-positive bacterium *Bacillus subtilis*. Environ Sci Technol 41(18):6465–6471

Lewis AE (2010) Review of metal sulfide precipitation. Hydrometallurgy 104(2):222–234

L'Haridon S, Cilia V, Messner P, Raguénès G, Gambacorta A, Sleytr UB, Prieur D, Jeanthon C (1998) *Desulfurobacterium thermolithotrophum* gen. nov., sp. nov., a novel autotrophic, sulfur-reducing bacterium isolated from a deep-sea hydrothermal vent. Int J Syst Bacteriol 48(3):701–711

Liamleam W, Annachhatre AP (2007) Electron donors for biological sulfate reduction. Biotechnol Adv 25(5):452–463

Lopez-Archilla AI, Marin I, Amils R (2001) Microbial community composition and ecology of an acidic aquatic environment: The Tinto River, Spain. Microbiol Ecol 41(1):20–35

Lovley DR (1991) Dissimilatory Fe(III) and Mn(IV) reduction. Microbiol Rev 55(2):259–287

Lovley DR, Phillips EJP, Lonergan DJ, Widman PK (1995) Fe(III) and S0 reduction by *Pelobacter carbinolicus*. Appl Environ Microbiol 61(6):2132–2138

Lovley DR, Fraga JL, Blunt-Harris EL, Hayes LA, Phillips EJP, Coates JD (1998) Humic substances as a mediator for microbially catalyzed metal reduction. Acta Hydrochim Hydrobiol 26(3):152–157

Lovley DR, Holmes DE, Nevin KP (2004) Dissimilatory Fe(III) and Mn(IV) reduction. Adv Microb Physiol 49:219–286

Ma K, Adams MW (1994) Sulfide dehydrogenase from the hyperthermophilic archaeon *Pyrococcus furiosus*: a new multifunctional enzyme involved in the reduction of elemental sulfur. J Bacteriol 176(21):6509–6517

Ma K, Schicho RN, Kelly RM, Adams MW (1993) Hydrogenase of the hyperthermophile *Pyrococcus furiosus* is an elemental sulfur reductase or sulfhydrogenase: evidence for a sulfur-reducing hydrogenase ancestor. Proc Natl Acad Sci USA 90(11):5341–5344

Ma K, Weiss R, Adams MW (2000) Characterization of hydrogenase II from the hyperthermophilic archaeon *Pyrococcus furiosus* and assessment of its role in sulfur reduction. J Bacteriol 182(7):1864–1871

Macalady J, Banfield JF (2003) Molecular geomicrobiology: genes and geochemical cycling. Earth Planet Sci Lett 209(1–2):1–17

Macy JM, Schröder IE, Thauer RK, Kröger A (1986) Growth the *Wolinella succinogenes* on H_2S plus fumarate and on formate plus sulfur as energy sources. Arch Microbiol 144(2):147–150

Miroshnichenko ML, Rainey FA, Hippe H, Chernyh NA, Kostrikina NA, Bonch-Osmolovskaya EA (1998) *Desulfurella karnchatkensis* sp. nov. and *Desulfurella propionica* sp. nov., new sulfur respiring thermophilic bacteria from Kamchatka thermal environments. Int J Syst Bacteriol 48:475–479

Möbius CH, Demel I, Schmid A, Temper U, Carozzi A, Buisman C, Weijma J, Rijnaarts HHM, Hendrickx TLG, Brockmann M (2015) Anwendung für organisch und anorganisch belastete Abwässer anderer Industriebereiche. In: Rosenwinkel K-H, Kroiss H, Dichtl N, Seyfried C-F, Weiland P (eds) Anaerobtechnik. Springer, Berlin, pp 541–601

Mori K, Kim H, Kakegawa T, Hanada S (2003) A novel lineage of sulfate-reducing microorganisms: *Thermodesulfobiaceae* fam. nov., *Thermodesulfobium narugense*, gen. nov., sp. nov., a new thermophilic isolate from a hot spring. Extremophiles 7(4):283–290

Myers CR, Nealson KH (1988) Bacterial manganese reduction and growth with manganese oxide as the sole electron acceptor. Science 240(4857):1319–1321

Ňancucheo I, Johnson DB (2012) Selective removal of transition metals from acidic mine waters by novel consortia of acidophilic sulfidogenic bacteria. Microbial Biotech 5(1):34–44

Neculita CM, Zagury GJ, Bussière B (2007) Passive treatment of acid mine drainage in bioreactors using sulfate-reducing bacteria. J Environ Qual 36(1):1–16

Newman DK, Kolter R (2000) A role for excreted quinones in extracellular electron transfer. Nature 405(6782):94–97

Ng KY, Sawada R, Inoue S, Kamimura K, Sugio T (2000) Purification and some properties of sulfur reductase from the iron-oxidizing bacterium *Thiobacillus ferrooxidans* NASF-1. J Biosci Bioeng 90(2):199–203

Ohmura N, Sasaki K, Matsumoto N, Saiki H (2002) Anaerobic respiration using Fe^{3+}, S^0, and H_2 in the chemolithoautotrophic bacterium *Acidithiobacillus ferrooxidans*. J Bacteriol 184(8):2081–2087

Osorio H, Mangold S, Denis Y, Ñancucheo I, Esparza M, Johnson DB, Bonnefoy V, Dopson M, Holmes DS (2013) Anaerobic sulfur metabolism coupled to dissimilatory iron reduction in the extremophile *Acidithiobacillus ferrooxidans*. Appl Environ Microbiol 79(7):2172–2181

Pelsh AO (1936) Hydrobiology of Karabugaz bay of the Caspian sea. Trudy Solyan Laboratory Akademie Nauk 5(49):31

Pfennig N (1975) The phototrophic bacteria and their role in the sulfur cycle. Plant Soil 43(1–3):1–16

Pfennig N, Biebl H (1976) *Desulfuromonas acetoxidans* gen. nov. and sp. nov., a new anaerobic, sulfur-reducing, acetate-oxidizing bacterium. Arch Microbiol 110(1):3–12

Pihl TD, Maier RJ (1991) Purification and characterization of the hydrogen uptake hydrogenase from the hyperthermophilic archaebacterium *Pyrodictium brockii*. J Bacteriol 173(6): 1839–1844

Pihl TD, Schicho RN, Kelly RM, Maier RJ (1989) Characterization of hydrogen-uptake activity in the hyperthermophile *Pyrodictium brockii*. Proc Natl Acad Sci USA 86(1):138–141

Pihl TD, Black LK, Schulman BA, Maier RJ (1992) Hydrogen-oxidizing electron transport components in the hyperthermophilic archaebacterium *Pyrodictium brockii*. J Bacteriol 174(1):137–143

Postec A, Breton CL, Fardeau ML, Lesongeur F, Pignet P, Querellou J, Ollivier B, Godfroy A (2005) *Marinitoga hydrogenitolerans* sp. nov., a novel member of the order Thermotogales isolated from a black smoker chimney on the Mid-Atlantic Ridge. Int J Syst Evol Microbiol 55(3):1217–1221

Prokofeva MI, Miroshnichenko ML, Kostrikina NA, Chernyh NA, Kuznetsov BB, Tourova TP, Bonch-Osmolovskaya EA (2000) *Acidilobus aceticus* gen. nov., sp. nov., a novel anaerobic thermoacidophilic archaeon from continental hot vents in Kamchatka. Int J Syst Evol Microbiol 50(6):2001–2008

Rabus R, Hansen TA, Widdel F (2006) Dissimilatory sulfate- and sulfur-reducing prokaryotes. In: Dworkin M, Falkow S, Rosenberg E, Schleifer K-H, Stackebrandt E (eds) The prokaryotes, vol 2. Springer, New York, pp 659–768

Rainey FA, Hollen B (2005) *Desulfurella* Bonch-Osmolovskaya, Sokolova, Kostrikina, and Zavarzin 1993, 624 VP emend. Miroshnichenko, Rainey, Hippe, Chernyh, Kostrikina and Bonch-Osmolovskaya 1998, 478 (Effective publication: Bonch-Osmolovskaya, Sokolova, Kostrikina, and Zavarzin 1990, 155). Bergey's Manual® of Systematic Bacteriology, Springer 923–924

Reguera G, McCarthy KD, Mehta T, Nicoll JS, Tuominen MT, Lovley DR (2005) Extracellular electron transfer via microbial nanowires. Nature 435(7045):1098–1101

Richardson DJ (2000) Bacterial respiration: a flexible process for a changing environment. Microbiol 146(3):551–571

Richter K, Schicklberger M, Gescher J (2012) Dissimilatory reduction of extracellular electron acceptors in anaerobic respiration. Appl Environ Microbiol 78(4):913–921

Ringel M, Gross R, Krafft T, Kröger A, Schauder R (1996) Growth of *Wolinella succinogenes* with elemental sulfur in the absence of polysulfide. Arch Microbiol 165(1):62–64

Roof JG (1971) Solubility of sulfur in hydrogen sulfide and in carbon disulfide at elevated temperature and pressure C1–SPE. Soc Petrol Eng J 11(3):5

Rosenberg E, DeLong E, Lory S, Stackebrandt E, Thompson F, Rabus R, Hansen TA, Widdel F (2013) Dissimilatory sulfate- and sulfur-reducing prokaryotes. In: The prokaryotes. Springer, Berlin, pp 309–404

Sallam A, Steinbüchel A (2009) *Clostridium sulfidigenes* sp. nov., a mesophilic, proteolytic, thiosulfate- and sulfur-reducing bacterium isolated from pond sediment. Int J Syst Evol Microbiol 59(7):1661–1665

Sampaio RMM, Timmers RA, Xu Y, Keesman KJ, Lens PNL (2009) Selective precipitation of Cu from Zn in a pS controlled continuously stirred tank reactor. J Hazard Mater 165(1):256–265

Sánchez-Andrea I, Triana D, Sanz JL (2012) Bioremediation of acid mine drainage coupled with domestic wastewater treatment. Water Sci Technol 66(11):2425–2431

Sánchez-Andrea I, Sanz JL, Bijmans MFM, Stams AJM (2014) Sulfate reduction at low pH to remediate acid mine drainage. J Hazard Mater 269:98–109

Sánchez-Andrea I, Stams AJM, Hedrich S, Nancucheo I, Johnson DB (2015) *Desulfosporosinus acididurans* sp. nov.: an acidophilic sulfate-reducing bacterium isolated from acidic sediments. Extremophiles 19(1):39–47

Sapra R, Bagramyan K, Adams MWW (2003) A simple energy-conserving system: proton reduction coupled to proton translocation. Proc Natl Acad Sci USA 100(13):7545–7550

Schauder R, Müller E (1993) Polysulfide as a possible substrate for sulfur-reducing bacteria. Arch Microbiol 160(5):377–382

Schippers A, Sand W (1999) Bacterial leaching of metal sulfides proceeds by two indirect mechanisms via thiosulfate or via polysulfides and sulfur. Appl Environ Microbiol 65(1):319–321

Schmitz RA, Bonch-Osmolovskaya EA, Thauer RK (1990) Different mechanisms of acetate activation in *Desulfurella acetivorans* and *Desulfuromonas acetoxidans*. Arch Microbiol 154(3):274–279

Schröder I, Kröger A, Macy JM (1988) Isolation of the sulfur reductase and reconstitution of the sulfur respiration of *Wolinella succinogenes*. Arch Microbiol 149(6):572–579

Scott DT, McKnight DM, Blunt-Harris EL, Kolesar SE, Lovley DR (1998) Quinone moieties act as electron acceptors in the reduction of humic substances by humics-reducing microorganisms. Environ Sci Technol 32(19):2984–2989

Segerer A, Neuner A, Kristjansson JK, Stetter KO (1986) *Acidianus infernus* gen. nov., sp. nov., and *Acidianus brierleyi* comb. nov.: Facultatively aerobic, extremely acidophilic thermophilic sulfur-metabolizing archaebacteria. Int J Syst Bacteriol 36(4):559–564

Segerer A, Langworthy TA, Stetter KO (1988) *Thermoplasma acidophilum* and *Thermoplasma volcanium* sp. nov. from Solfatara Fields. Syst Appl Microbiol 10(2):161–171

Segerer AH, Trincone A, Gahrtz M, Stetter KO (1991) *Stygiolobus azoricus* gen. nov., sp. nov. represents a novel genus of anaerobic, extremely thermoacidophilic archaebacteria of the order *Sulfolobales*. Int J Syst Bacteriol 41(4):495–501

Shimada H, Nemoto N, Shida Y, Oshima T, Yamagishi A (2008) Effects of pH and temperature on the composition of polar lipids in *Thermoplasma acidophilum* HO-62. J Bacteriol 190(15):5404–5411

Sierra-Alvarez R, Karri S, Freeman S, Field JA (2006) Biological treatment of heavy metals in acid mine drainage using sulfate reducing bioreactors. Water Sci Technol 54(2):179–185

Slonczewski JL, Macnab RM, Alger JR, Castle AM (1982) Effects of pH and repellent tactic stimuli on protein methylation levels in *Escherichia coli*. J Bacteriol 152(1):384–399

Sorokin DY, Tourova TP, Kolganova TV, Detkova EN, Galinski EA, Muyzer G (2011) Culturable diversity of lithotrophic haloalkaliphilic sulfate-reducing bacteria in soda lakes and the description of *Desulfonatronum thioautotrophicum* sp. nov., *Desulfonatronum thiosulfatophilum* sp. nov., *Desulfonatronovibrio thiodismutans* sp. nov., and *Desulfonatronovibrio magnus* sp. nov. Extremophiles 15(3):391–401

Steed VS, Suidan MT, Gupta M, Miyahara T, Acheson CM, Sayles GD (2000) Development of a sulfate-reducing biological process to remove heavy metals from acid mine drainage. Water Environ Res 72(5):530–535

Stetter KO (1996) Hyperthermophilic prokaryotes. FEMS Microbiol Rev 18:149–158

Stetter KO, Huber R, Blochl E, Kurr M, Eden RD, Fielder M, Cash H, Vance I (1993) Hyperthermophilic archaea are thriving in deep North Sea and Alaskan oil reservoirs. Nature 365(6448):743–745

Steudel R (2000) The sulfur cycle. In: Lens PNL, Hulshoff Pol L (eds) Environmental technologies to treat sulfur pollution. IWA Publishing, London, pp 1–31

Steudel R, Eckert B (2003) Solid sulfur allotropes sulfur allotropes. Elemental sulfur and sulfur-rich compounds I. Springer, Berlin Heidelberg. 230:1–80

Steudel R, Göbel T, Holdt G (1989) The molecular nature of the hydrophilic sulfur prepared from aqueous sulfide and sulfite (selmi sulfur sol). Z Naturforsch B 44(5):526–530

Steudel R, Kleinjan WE, de Keizer A, Janssen AJH (2003) Biologically produced sulfur. Elemental sulfur and sulfur-rich compounds I. Springer, Berlin Heidelberg, 230:167–188

Suzuki I, Lee D, Mackay B, Harahuc L, Oh JK (1999) Effect of various ions, pH, and osmotic pressure on oxidation of elemental sulfur by *Thiobacillus thiooxidans*. Appl Environ Microbiol 65(11):5163–5168

Svetlichnyi VA, Slesarev AI, Svetlichnaya TP, Zavarzin GA (1987) *Caldococcus litoralis,* gen. nov. sp. nov.–a new marine, extremely thermophilic, sulfur-reducing archaebacterium. Mikrobiologiya 56:831–838

Tabak H, Scharp R, Burckle J, Kawahara F, Govind R (2003) Advances in biotreatment of acid mine drainage and biorecovery of metals: 1. Metal precipitation for recovery and recycle. Biodegradation 14(6):423–436

Takahashi Y, Suto K, Inoue C (2010) Polysulfide reduction by *Clostridium* relatives isolated from sulfate-reducing enrichment cultures. J Biosci Bioeng 109(4):372–380

Tang K, Baskaran V, Nemati M (2009) Bacteria of the sulfur cycle: An overview of microbiology, biokinetics and their role in petroleum and mining industries. Biochem Eng J 44(1):73–94

Tekerlekopoulou AG, Tsiamis G, Dermou E, Siozios S, Bourtzis K, Vayenas DV (2010) The effect of carbon source on microbial community structure and Cr(VI) reduction rate. Biotechnol Bioeng 107(3):478–487

Thamdrup B, Finster K, Hansen JW, Bak F (1993) Bacterial disproportionation of elemental sulfur coupled to chemical reduction of iron or manganese. Appl Environ Microbiol 59(1):101–108

Tsukamoto TK, Killion HA, Miller GC (2004) Column experiments for microbiological treatment of acid mine drainage: low-temperature, low-pH and matrix investigations. Water Res 38(6):1405–1418

Tyson GW, Chapman J, Hugenholtz P, Allen EE, Ram RJ, Richardson PM, Solovyev VV, Rubin EM, Rokhsar DS, Banfield JF (2004) Community structure and metabolism through reconstruction of microbial genomes from the environment. Nature 428(6978):37–43

Veeken AHM, Rulkens WH (2003) Innovative developments in the selective removal and reuse of heavy metals from wastewaters. Water Sci Technol 47(10):9–16

Vera M, Schippers A, Sand W (2013) Progress in bioleaching: fundamentals and mechanisms of bacterial metal sulfide oxidation–part A. Appl Microbiol Biotechnol 97(17):7529–7541

Vignais PM, Billoud B, Meyer J (2001) Classification and phylogeny of hydrogenases I. FEMS Microbiol Rev 25(4):455–501

Wang F, Tessier A, Buffle J (1998) Voltammetric determination of elemental sulfur in pore waters. Limnol Oceanogr 43(6):1353–1361

Wang X, Lv B, Cai G, Fu L, Wu Y, Wang X, Ren B, Ma H (2012) A proton shelter inspired by the sugar coating of acidophilic archaea. Sci Rep 2(892):1–5

Weijma J, Copini CFM, Buisman CJN, Schultz CE (2002) Biological recovery of metals, sulfur and water in the mining and metallurgical industry. In: Lens PNL, Hulshoff Pol LW, Wilderer P, Asano T (eds) Water recycling and recovery in Industry. IWA Publishing, London, pp 605–622

Widdel F (1988) Microbiology and ecology of sulfate–and sulfur–reducing bacteria. Willey–Interscience Publication, New York

Widdel F, Pfennig N (1981) Studies on dissimilatory sulfate-reducing bacteria that decompose fatty acids. I. Isolation of new sulfate-reducing bacteria enriched with acetate from saline environments. Description of *Desulfobacter postgatei* gen. nov., sp. nov. Arch Microbiol 129(5):395–400

Widdel F, Pfennig N (1991). The genus *Desulfuromonas* and other Gram-negative sulfur-reducing eubacteria. In: Balows A, Trüper HG, Dworkin M, Harder W, Schleifer KH (eds) The prokaryotes. Springer, Berlin

Yarza P, Richter M, Peplies J, Euzeby J, Amann R, Schleifer KH, Ludwig W, Glockner FO, Rossello-Mora R (2008) The all-species living tree project: a 16S rRNA-based phylogenetic tree of all sequenced type strains. Syst Appl Microbiol 31(4):241–250

Yoneda Y, Yoshida T, Kawaichi S, Daifuku T, Takabe K, Sako Y (2012) *Carboxydothermus pertinax* sp. nov., a thermophilic, hydrogenogenic, Fe(III)-reducing, sulfur-reducing carboxydotrophic bacterium from an acidic hot spring. Int J Syst Evol Microbiol 62(7):1692–1697

Zavarzina DG, Zhilina TN, Tourova TP, Kuznetsov BB, Kostrikina NA, Bonch-Osmolovskaya EA (2000) *Thermanaerovibrio velox* sp. nov., a new anaerobic, thermophilic, organotrophic bacterium that reduces elemental sulfur, and emended description of the genus *Thermanaerovibrio*. Int J Syst Evol Microbiol 50(3):1287–1295

Zöphel A, Kennedy MC, Beinert H, Kroneck PMH (1991) Investigations on microbial sulfur respiration. Eur J Biochem 195(3):849–856

Zychlinsky E, Matin A (1983) Cytoplasmic pH homeostasis in an acidophilic bacterium, *Thiobacillus acidophilus*. J Bacteriol 156(3):1352–1355

Chapter 6
The Biofilm Lifestyle of Acidophilic Metal/Sulfur-Oxidizing Microorganisms

Ruiyong Zhang, Sören Bellenberg, Thomas R. Neu, Wolfgang Sand, and Mario Vera

1 Introduction

The most common and widespread lifestyle of microbes on earth is in form of biofilms. These are communities of different species of microorganisms embedded in a matrix of extracellular polymeric substances (EPS), which mainly consist of polysaccharides, proteins, nucleic acids and lipophilic compounds. Biofilms can be found as surface-associated or "floating mats", occurring in air-water interfaces. The biofilm lifestyle protects cells from environmental stress like desiccation, nutrient starvation, radiation and/or oxidative stress (Flemming and Wingender 2010).

Bioleaching of metal sulfides (MS) like pyrite (FeS_2) or chalcopyrite ($CuFeS_2$) is accelerated by a diverse group of acidophilic metal/sulfur-oxidizing microorganisms (AMOM). These can effect MS dissolution through the oxidation of iron(II)-ions and/or reduced inorganic sulfur compounds (RISCs). Thereby iron(III)-ions

R. Zhang • S. Bellenberg • W. Sand
Aquatische Biotechnologie, Biofilm Centre, Universität Duisburg – Essen, Universitätsstraße 5, 45141 Essen, Germany
e-mail: ruiyong.zhang@uni-due.de

T.R. Neu
Department of River Ecology, Helmholtz Centre for Environmental Research – UFZ, Brückstrasse 3a, 39114 Magdeburg, Germany

M. Vera (✉)
Institute for Biological and Medical Engineering, Schools of Engineering, Biological Sciences and Medicine, Pontificia Universidad Católica de Chile, Vicuña Mackenna 4860, 7820436 Macul, Santiago, Chile

Department of Hydraulic and Environmental Engineering, School of Engineering, Pontificia Universidad Católica de Chile, Vicuña Mackenna 4860, 7820436 Macul, Santiago, Chile
e-mail: mariovera@uc.cl

P.H. Rampelotto (ed.), *Biotechnology of Extremophiles*, Grand Challenges in Biology and Biotechnology 1, DOI 10.1007/978-3-319-13521-2_6

and protons, the MS attacking agents, are available. As a consequence, acid mine/rock drainage (AMD/ARD) is generated. This is a serious problem due to water and soil contamination by heavy metals and sulfuric acid. Several species of acidophilic microorganisms from the genera *Acidithiobacillus*, *Leptospirillum*, *Acidiferrobacter*, *Acidiphilium*, *Ferroplasma*, *Acidimicrobium*, *Ferrimicrobium*, *Sulfobacillus*, *Metallosphaera*, *Sulfolobus* and *Acidianus*, among others, are commonly found at AMD sites as well as in biomining operations under mesophilic, moderately- and thermophilic conditions. In the "contact-leaching" mode microorganisms thrive attached to the MS surface and the electrochemical processes resulting in the dissolution of sulfide minerals, occur at the interface between bacterial cells and the MS surface. This space is filled with EPS, which in *Acidithiobacillus ferrooxidans* are mainly composed of polysaccharides, lipids and uronic acids (Gehrke et al. 1998).

The process of biofilm formation by AMOM on MS is not completely understood. Under laboratory conditions, most AMOM form monolayer biofilms on MS, which are also their energy substrate. Environmental acidophilic biofilms are characterized by a relative low species abundance, which is restricted by the acid pH, high concentration of heavy metals as well as the narrow range of electron donors and acceptors available. In general these biofilms are dominated by iron-oxidizing chemolithoautotrophs such as *Leptospirillum* Group II and sub-dominated by mixed populations of heterotrophic or mixotrophic bacteria, archaea and eukaryotes (Denef et al. 2010).

Biofilm formation and EPS biosynthesis in acidophilic metal/sulfur-oxidizing bacteria are regulated at different levels, comprising among others, energy substrates, inorganic phosphate (P_i) limitation, and cell-cell communication mechanisms of "quorum sensing" (QS). This chapter presents an overview about the biofilm lifestyle of AMOM. This includes surface science, microscopy, cell-cell communication, interspecies interactions as well as molecular and high-throughput studies. Future perspectives in this field include the elucidation of EPS biosynthesis pathways and comprehensive analysis of the chemical nature of the EPS polymers. Cell-cell communication and microbial interactions within multispecies biofilms of acidophiles are considered to be crucial determinants of metabolic activity of AMOM. A comprehensive knowledge of these aspects of AMOM biofilms will be beneficial for biotechnological applications such as biomining, in which an enhancement of biofilm formation and leaching rates is desired and for mitigation of AMD generation, in which inhibition of biofilm formation, or inactivation of biofilm cells might contribute significantly.

2 Biofilms of Acidophilic Metal/Sulfur-Oxidizing Microorganisms

The predominant MS dissolving microorganisms include acidophilic bacteria and archaea (i.e. microorganisms that thrive at pH values lower than 3). These oxidize RISCs and/or iron(II)-ions. Leaching bacteria are distributed among the *Proteobacteria* (*Acidithiobacillus*, *Acidiphilium*, *Acidiferrobacter*, *Ferrovum*), *Nitrospirae* (*Leptospirillum*), *Firmicutes* (*Alicyclobacillus*, *Sulfobacillus*) and

Actinobacteria (*Ferrimicrobium*, *Acidimicrobium*, *Ferrithrix*). Several recent reviews have given an overview of the microbial diversity in mining biotopes (Schippers 2007; Hedrich et al. 2011; Dopson and Johnson 2012; Johnson 2014). Within all bacterial groups, mesophilic and moderately-thermophilic microorganisms have been described (Clark and Norris 1996; Norris and Johnson 1998; Norris et al. 2000). Most leaching archaea belong to the *Sulfolobales*, which are thermophilic, sulfur- and/or iron(II)-oxidizers such as *Sulfolobus*, *Acidianus*, *Metallosphaera* and *Sulfurisphaera* (Norris et al. 2000; Wheaton et al. 2015). Also, within the *Thermoplasmales* the mesophilic iron(II)-oxidizing species, *Ferroplasma acidiphilum* (Golyshina et al. 2000), *Ferroplasma acidarmanus* (Edwards et al. 2000b; Dopson et al. 2004) and *Ferroplasma thermophilum* (Zhou et al. 2008) are known.

2.1 Mechanisms of Metal Sulfide Dissolution

Based on the acid solubility of MS, there are two different bioleaching pathways. The thiosulfate pathway is involved in the dissolution of acid-insoluble MS, such as pyrite, molybdenite (MoS_2), and tungstenite (WS_2). The polysulfide pathway is relevant for dissolution of acid-soluble MS, such as sphalerite (ZnS), galena (PbS) or chalcopyrite (Schippers et al. 1996; Schippers and Sand 1999; Sand et al. 2001). These two pathways are indirect mechanisms, since in acidic solution the acid-insoluble MS are solely oxidized chemically by iron(III)-ions, which are supplied by the microbial oxidation of iron(II)-ions (Schippers et al. 1996; Rodriguez et al. 2003; Gleisner et al. 2006). Acid-soluble MS are also dissolved by proton attack. Two leaching modes have been proposed: "contact" and "non-contact" leaching (Sand et al. 2001; Rawlings 2002). Non-contact leaching is carried out by planktonic cells, which oxidize iron(II)-ions. The resulting iron(III)-ions are reduced at the mineral surface, and the sulfide moiety is oxidized. Thus iron(II)-ions enter the cycle again. In a strict sense, this corresponds to the "indirect mechanism" (Sand et al. 1995). In contrast, contact leaching takes into account biofilm forming cells on the surface of MS. In this case the electrochemical processes resulting in MS dissolution are occurring in the EPS located at the interface between the bacterial cell and the MS surface.

The anodic and cathodic reactions have been reviewed for the case of pyrite. In the first step, electrons are transferred from the surface of the pyrite to the aqueous oxidant species, usually O_2 or iron(III)-ions (cathodic reaction). In the second step a charge is transported from the site of an anodic reaction to replace the electron lost from the cathodic site. Later, at an anodic site, the oxygen atom of a water molecule interacts with a sulfur atom to create a sulfoxy species. This step releases an electron into the solid and one or two hydrogen ions to solution (Rimstidt and Vaughan 2003). Probably cells are chemotactically attracted towards these local anodes. The anodes and cathodes may be resulting from imperfections in the crystal lattice where the iron to sulfur ratio is imbalanced due to inclusion of other metal atoms during the process of crystallization and/or from variations of temperature during crystallization (causing amorphous up to highly crystalline structures). Recently, a new proposal was made by

Crundwell (2013). The new hypothesis of mineral dissolution is based on an electrochemical mechanism which updates contents on the kinetics of mineral dissolution. It proposed that: (1) no separation of the surface into anodic sites and cathodic sites exist, (2) no flow of electrons occurs across the bulk of the mineral, (3) the first step of the dissolution reaction is not by acid, (4) the charge exchange across the mineral-solution interface is the rate-controlling step, and (5) bacterial leaching does not change the mechanism of reaction. This means that biological effects are limited to reaction steps which do not affect the rate-determining step.

2.2 Microbial Attachment to Metal Sulfides, EPS

Microbial attachment to ores and subsequent biofilm formation increase leaching activities, since a unique microenvironment is formed between the bacterium and the MS surface (Fig. 6.1). In case of *At. ferrooxidans*, iron(III)-ions, each one probably complexed by two uronic acid residues, are present in the EPS. Thus, the first function of complexed iron(III)-ions is the mediation of cell attachment, in which cells are electrostatically attached to the negatively charged pyrite. The second function of complexed iron(III)-ions is the oxidative dissolution of the MS, similar to the role of free iron(III)-ions in "non-contact leaching". In this species, the electrons extracted from the MS reduce molecular oxygen via a redox chain forming a supercomplex spanning the periplasmic space and connecting both outer and inner membranes (Castelle et al. 2008).

In general, the majority of AMOM can form biofilms on MS. Assuming of a non-limiting surface, up to 80–90 % of a certain inoculum may attach to MS within 24 h (Dispirito et al. 1983; Bagdigian and Myerson 1986; Gehrke et al. 1998; Harneit et al. 2006). It is important to remark that the initial attachment ratios strongly depend on the nature of the species as well as their pre-cultivation conditions. In general *Leptospirillum* spp. show higher attachment rates to pyrite than iron-oxidizing *Acidithiobacillus* spp. and, among the latter ones, iron(II)- or pyrite-grown cells show higher attachment rates to pyrite than sulfur-grown cells due to a considerable chemical modification of their EPS (Gehrke et al. 1998). Nevertheless, some cells always remain in planktonic state, even though the surface area for attachment is not the limiting factor. For example, when *At. ferrooxidans* ATCC 23270^T or *Acidithiobacillus ferrivorans* SS3 is incubated with pyrite for 24 h, about 50 % of the cells attached to the ore (Bellenberg et al. 2015). The reason(s) for this remains unknown.

In case of *At. ferrooxidans* strain R1 it was demonstrated that pyrite-grown cells had a similar EPS composition compared to iron(II)-grown cells. These EPS consist of the monosaccharides glucose, rhamnose, fucose, xylose, mannose, C12–C20 saturated fatty acids, glucuronic acid, and iron(III)-ions (Gehrke et al. 1998; Gehrke et al. 2001). However, pyrite-grown cells possessed more than tenfold amount of EPS. As already mentioned, the initial attachment is mainly driven by electrostatic interactions (in which most likely 2 mol negatively charged glucuronic acid residues complex 1 mol positively charged iron(III)-ions resulting in a net positive charge) with the negatively charged

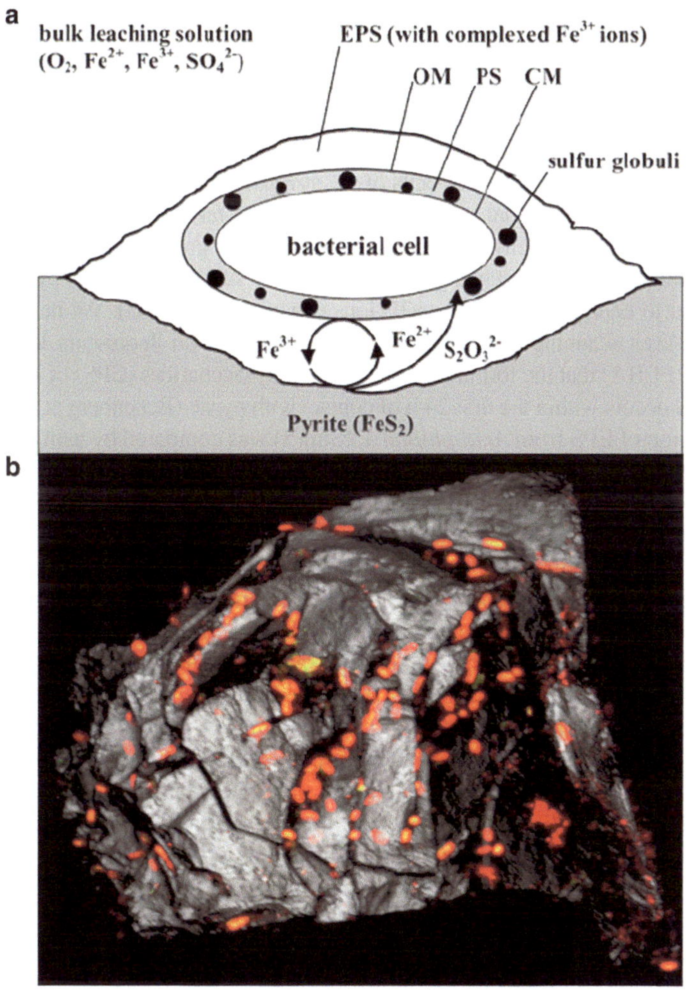

Fig. 6.1 Model for contact leaching catalyzed by a bacterial cell (from Vera et al. 2013b). A: Overview showing a biofilm cell embedded in the EPS layer attached to pyrite. Compounds like Iron(II)/(III)-ions, thiosulfate present during MS dissolution are shown. CM, cytoplasmic membrane; PS, periplasmic space; OM, outer membrane. B: CLSM image showing a 3-D projection of a pyrite grain (50–100 µm) colonized with cells of *At. ferrooxidans*^{T} after 1 week of incubation. Cells were double stained with Syto 9 (*green*) for nucleic acids and Con A. Color allocation: *green*=Syto 9, *red*=Con A-tetramethyl rhodamine isothiocyanate (TRITC), *grey*=reflection. The merged image from all three channels is shown. The bacterial colonization pattern strongly correlates with surface imperfections

pyrite surface (at pH 2) in sulfuric acid solution (Solari et al. 1992; Blake et al. 1994). Also hydrophobic interactions contribute to the attachment to MS surfaces (Gehrke et al. 1998; Sampson et al. 2000b). This applies especially to very hydrophobic surfaces e.g. of elemental sulfur (S^0). Hydrophobic interactions as well as covalent bonds seem to mediate the secondary (tight) surface attachment. Cells grown on S^0 do not attach well to pyrite, since their EPS composition is different compared to pyrite-grown ones. Their EPS contain considerably less monosaccharides and lack uronic acids, resulting in a complete absence of complexed iron(III)-ions or other positively charged ions. In contrast, EPS from sulfur-grown cells possess much more fatty acids than EPS extracted from pyrite-grown cells. Consequently, it seems that hydrophobic interactions are exclusively relevant for attachment of cells of *At. ferrooxidans* to S^0 (Gehrke et al. 1998). A FTIR analysis indicated that cell surface charge of *At. ferrooxidans* was different from soluble iron(II)-grown cells compared to solid substrate grown ones. Cells possessed a higher amount of protein when grown on insoluble substrates such as pyrite or sulfur compared to cells grown with iron(II)-ions (Sharma et al. 2003). We have shown by confocal laser scanning microscopy (CLSM) together with fluorescent lectin binding analysis (FLBA) that the formation of capsular polysaccharides (CPS) of *At. ferrooxidans*^{T} EPS occurs within the first 24 h of contact with pyrite (Bellenberg et al. 2012).

Isolation of EPS from *Acidiphilum* 3.2Sup(5) was compared by using five extraction methods: EDTA, NaOH, ion exchange resin, heating and centrifugation. The extracted EPS mainly contained carbohydrates and proteins by all methods. However, higher EPS amounts as well as a less degree of cell lysis were obtained by using EDTA. This study confirmed that both, the amount and the chemical composition of EPS strongly depended on the extraction method (Tapia et al. 2009). EPS from a mixed culture predominated by *Acidithiobacillus caldus* and *Leptospirillum ferriphilum* grown on chalcopyrite concentrate were characterized to contain proteins, lipids, carbohydrates and iron(III)-ions (Zeng et al. 2010). Recently, tightly bound EPS of the crenarchaeote *Sulfolobus metallicus* DSM 6482^T were shown to be mainly composed of proteins and carbohydrates. By contrast, the loosely bound EPS were mainly containing carbohydrates. Extracellular DNA and proteins were detected in *S. metallicus*^{T} EPS from biofilms grown on S^0, whereas proteins were characterized as the main components (Zhang et al. 2015b).

EPS production in stirred reactors dominated by *L. ferriphilum* BRGM1 was studied in order to correlate factors for the optimization of bioleaching processes. Under nitrogen limitation, bacterial attachment and leaching efficiency were both decreased in accordance with reduced levels of EPS production. CO_2 limitation caused a significant decrease of exopolysaccharide production (d'Hugues et al. 2008). Another study on EPS composition was done for mixed cultures containing mesophilic, moderate thermophilic and thermophilic leaching strains grown on several MS such as pyrite, sphalerite and chalcopyrite in several continuously operated bioleaching systems. It was found that 70 % of the EPS extracted were mainly composed of carbohydrates and smaller amount of proteins with trace levels of humic and uronic acids (Govender and Gericke 2011). Characterization of two acidophilic microbial biofilms from Iron Mountain, California showed that their EPS were composed of carbohydrates, metals, proteins and minor quantities of DNA and lipids (Jiao et al. 2010). Studies on acidophiles grown on organic compounds were also reported. Two strains of *Sulfolobus* grown with glucose excreted an extracellular polysaccharide containing monosaccharides like of glucose, mannose, glucosamine and galactose (Nicolaus et al. 1993). Proteins and carbohydrates were major components in EPS of *Sulfolobus solfataricus* P2 grown on tryptone, N-Z-amine or glucose (Koerdt et al. 2012).

The site of attachment of AMOM on metal sulfides and how the detection/sensing of this specific location(s) occurs are still open questions. Several evidences from the literature (Andrews 1988; Ohmura et al. 1993; Shrihari et al. 1995; Dziurla et al. 1998; Sanhueza et al. 1999; Edwards et al. 2001) and our own work suggest that microbial attachment to MS does not occur randomly (Gehrke et al. 1998; Sand et al. 1998; Gehrke et al. 2001; Noël et al. 2010; Zhang et al. 2014, 2015a). It has been observed by atomic force microscopy (AFM) as well as CLSM that *At. ferrooxidans* and other AMOM preferentially (>90 %) attach to sites with visible surface imperfections e.g. pores, scratches, etc. (see also Sect. 4). Cell attachment to areas with a low degree of crystallization seems to be favored, resulting in cell orientation along crystallographic axes, in whose direction oxidation fronts propagate (Sanhueza et al. 1999). Cell adhesion to pores and scratches may be explained by contact area enhancement and protection from weak shear forces. In contrast cell attachment to areas with low crystallization and crystallographic axis is not often related to changes in surface topography. Therefore, the presence of some attractants may explain attachment to specific sites on the mineral surface. This is most likely caused by charge imbalances on the surface as caused e.g. by oxidation processes. Several strains of *At. ferrooxidans* and *Leptospirillum ferrooxidans* possess chemosensory systems (Acuña et al. 1992; Meyer et al. 2002). Chemotaxis and attraction to gradients of iron(II)/(III)-ions, thiosulfate and other compounds which may occur compulsorily during MS leaching (Edwards et al. 2000a). Dissolution, occurring at local anodes, brings iron(II)-ions and thiosulfate in solution. Recent work using AFM equipped with a Kelvin probe for force mapping indicates that the cells of *L. ferrooxidans* attached to a pyrite surface are more negatively charged (about 100–200 mV) than the surrounding surface (i.e. extracting electrons from the pyrite) (Vera et al. 2013b). In a similar case, Little and co-workers showed that sulfate-reducing bacteria attached in the immediate vicinity (nanometer range) of the anode on steel surfaces. The latter is negatively charged until a release of iron(II)-ions occurs. As a consequence of bacterial attachment, the anode and the cathode become permanent (manifest), and steel dissolution commences (Little et al. 2000). This observation may be also relevant for bioleaching of MS. In summary, cells seem to be transiently attracted (electrically charged) to dissolution sites by their chemotaxis systems and cause the anodes and cathodes on the MS surface to become permanent. In case of biofilm forming cells the dissolution process occurs within the EPS layer (Fig. 6.1), which can be considered a reaction space filling the void volume between the outer cell membrane and the surface of the MS. Tributsch and co-workers demonstrated that this distance is 10–100 nm wide (Rodriguez-Leiva and Tributsch 1988). The thickness of *At. ferrooxidans* EPS was estimated for iron(II)-grown cells by *in vivo* AFM to be 28.7 nm (±13.5) (Taylor and Lower 2008). The EPS thickness values for S^0- or pyrite-grown cells *in vivo* remains to be elucidated. Presumably these values will be higher than the above mentioned ones since it is already known that EPS levels increase when the bacteria are grown with these solid substrates (Sand et al. 1998).

Newly developed AFM probes coated with acidophilic bacteria allowed to measure interactive forces between AMOM and mineral surfaces (Diao et al. 2014a, b). It was found that *At. ferrooxidans* grown on chalcopyrite showed the strongest interaction forces with the same substrate. Those compared with the ones of cells grown with iron(II)-ions or S^0 were considerably lower. Also, chalcopyrite leaching rate and cell attachment capacity were found to correlate positively with the cell adhesion force (Zhu et al. 2015). Among the three species *At. ferrooxidans*, *Acidithiobacillus thiooxidans* and *L. ferrooxidans*, the later one showed the highest adhesion force to chalcopyrite. EPS-deficient cells exhibited reduced adhesion forces and initial cellular attachment to MS surfaces (Zhu et al. 2012).

Still several questions remain open: Are some strains benefiting from the presence of primary colonizers in order to attach to these previously existing biofilms? How is the biochemical/molecular nature of antagonistic or synergistic interactions known to occur among certain species? We have recently shown in binary species biofilms that the presence of active biofilms of iron-oxidizers may influence subsequent cell attachment by other species. *At. thiooxidans* cells attached 40 % more to pyrite precolonized with biofilms of *At. ferrooxidans* or *L. ferrooxidans*. Interestingly, its cell attachment was faster to pyrite precolonized with *L. ferrooxidans* than with *At. ferrooxidans*. As *L. ferrooxidans* leach pyrite more efficiently than *At. ferrooxidans*, the faster attachment observed for *At. thiooxidans* may be related to a chemotactic response towards RISCs like thiosulfate which are known to be released after pyrite leaching (Bellenberg et al. 2014). The analysis of the complete *At. thiooxidans* genome sequence revealed a complete suite of genes for flagellar formation and chemotaxis (Valdes et al. 2011). In contrast, the cell attachment of *At. ferrooxidans* to pyrite grains precolonized with *L. ferrooxidans* was strongly dependent on its pre-cultivation. Thiosulfate-grown cells were influenced by the presence of *L. ferrooxidans*, while iron(II)-grown cells were not (Bellenberg et al. 2014). Summarizing, the presence of iron oxidizers, which have been described as primary colonizers in natural AMD biofilms (Wilmes et al. 2009) may be a relevant factor for sulfur-oxidizers to efficiently attach to MS.

3 Physiological Adaptations of Iron-Oxidizing Acidithiobacilli to the Biofilm Lifestyle

3.1 Responses Against Oxidative Stress

AMOM are confronted to high levels of extracellular reactive oxygen species (ROS), which are formed at the mineral/liquid interface. Hydrogen peroxide (H_2O_2) is formed in reactions of oxygen (in acidic aqueous solutions) with mineral lattice-bound iron (Nooshabadi and Rao 2014) and it is generated by pyrite, among other MS, especially when ores are crushed before biohydrometallurgical processing (Jones et al. 2011). When pyrite is present, H_2O_2 reacts with iron(II)-ions and highly reactive hydroxyl radicals are formed by the Fenton reaction. In addition, ROS such as O_2^- and H_2O_2 are easily penetrating cell membranes at acid pH values at diffusion limited rates, likely causing damage to DNA and enzymes when elevated concentrations of these compounds occur (Imlay 2013). We have recently shown that pyrite-grown *At.*

ferrooxidans or *At. ferrivorans* cells, in contrast to iron(II)-grown cells, were able to oxidize iron(II)-ions or pyrite after 24 h iron starvation and incubation with 1 mM H_2O_2. This indicates that these cells were already adapted to enhanced levels of ROS, which are generated on MS surfaces. We have also shown that the mere presence of pyrite has a clear inhibitory effect on the iron oxidation activity of both, the planktonic sub-population (PP) and the mineral-attached, biofilm cell sub-population (BP), after transfer of iron(II)-grown cells to a medium containing pyrite. The inhibition measured was especially higher in the BP compared to PP (Bellenberg et al. 2015). The reasons for this phenomenon may be manifold. Among these (1) the metabolic transition of BP cells to a biofilm lifestyle and (2) the release of inhibitory ROS, which cause a metabolic inhibition or a reduction of the amount of actively metabolizing cells (Jones et al. 2011), can be included. In this context, another function of biofilm formation and EPS production in AMOM might be related to confer an enhanced resistance against the damages caused by ROS. The enhanced EPS levels of biofilm cells might act by blocking reactive sites on the mineral where generation of ROS might occur as well by acting as a ROS scavenger (Bellenberg et al. 2015).

3.2 Inorganic Phosphate and Polyphosphate Metabolism

Every microbial biofilm is a consequence of physical interactions of microorganisms with a surface via EPS or other attachment factors. EPS biosynthesis is mediated by environmental signals, such as nutrient availability and stress influence (Janczarek 2011). P_i is an essential nutrient for all living cells. Acidophilic leaching bacteria must be able to deal with the problem of P_i scarcity due to its precipitation in environments, where iron(III)-ions are present (Tuovinen 1990). For survival under P_i starvation, *E. coli* and many other bacteria induce the "Pho regulon", which is a genetic system controlled by the PhoB/PhoR two-component regulatory system and plays a key role in P_i homeostasis (Wanner 1996a, b). The *At. ferrooxidans* P_i starvation response and the presence of a Pho regulon have been described (Seeger and Jerez 1993; Vera et al. 2003; Alvarez and Jerez 2004). Besides its role in bacterial P_i homeostasis, this system is part of a complex regulatory network connecting stationary phase responses, virulence, QS and especially biofilm formation, which is known to be induced in several bacterial species as a response against P_i starvation (Lamarche et al. 2008). E.g. in *Sinorizobium melliloti* the DNA-binding transcriptional factor PhoB positively regulates the synthesis of exopolysaccharides involved in bacterial attachment and infection of root nodules, when P_i is limiting (Rüberg et al. 1999). Also in *Agrobacterium tumefaciens* an enhancement of biofilm formation under P_i limited growth conditions has been described (Danhorn et al. 2004). It has been observed that *At. ferrooxidans* cells, which were sub-cultured in iron(II) medium lacking P_i attach more efficiently to MS and S^0 surfaces (Amaro et al. 1993). We have shown by epifluorescence microscopy (EFM) and FLBA that *At. ferrooxidans* cells grown under P_i starvation showed an enhanced amount of CPS compared with cells grown with sufficient P_i (Bellenberg et al. 2012).

AMOM accumulate high levels of inorganic polyphosphates (polyP), which are polymers of phosphoanhydride-linked phosphate residues, found as chains up to 1000 residues long in cells from all three domains of life. Among its several functions, polyP can serve as P_i reservoir, modulator of stress responses as well as a contributing factor to the high copper resistance shown by AMOM (Alvarez and Jerez 2004; Remonsellez et al. 2006; Kanao et al. 2007; Orell et al. 2010). However, probably the most important of its functions is to participate in the physiological adjustments of bacteria to environmental changes and stress conditions. Mutant strains lacking polyP are more sensitive in the stationary phase and show decreased resistance to heat, oxidants, osmotic challenge, antibiotics and UV radiation (Rao and Kornberg 1996; Kim et al. 2002). Recent evidence has pointed out new roles of polyP in protection against oxidative stress. These can be achieved by direct and indirect mechanisms such as (1) protein chaperone, since unlike proteins, polyP does not react with ROS, (2) to participate in the formation of Mn^{2+}/polyP complexes, which can detoxify O_2^- after their hydrolysis or through a non-catalytic mechanism involving the formation of Mn^{3+}-polyP, (3) to provide defense against the Fenton reaction, in which the most probable mechanism involved is facilitating the chelation and export of copper(II)-ions (Grillo-Puertas et al. 2014), (4) to play an essential role in the molecular mechanism by which bacteria enter to a persistent state, in which they become highly stress-resistant and (5) to participate directly in the regulation of general stress response networks (Gray and Jakob 2015). Future studies of changes in polyP levels of leaching bacteria upon biofilm formation as well as the construction of polyP deficient strains will contribute to a better understanding of the role of its versatile polymer in the biofilm lifestyle of AMOM.

3.3 *Other Factors Affecting Biofilm Formation of Iron Oxidizing Acidithiobacilli*

A broad range of other factors influencing the biofilm formation on pyrite of iron-oxidizing acidithiobacilli have been identified. In general, cultivation at non-optimum growth temperatures or under increased ionic strengths led to a decreased colonization of pyrite. *At. ferrivorans* SS3 showed enhanced biofilm formation on pyrite if compared to *At. ferrooxidans*^{T}. The presence of iron(III)-ions increased pyrite colonization, especially when pyrite-grown cells were used, while the addition of 20 mM copper(II)-ions in case of both above-mentioned species resulted in reduced biofilm formation on pyrite. This observation correlated with a different EPS composition of copper-exposed cells in pyrite cultures. Some attachment effectors were also identified. The addition of 1 mM glucuronic acid, especially in combination with 1 mM iron(III)-ions, had strong enhancing effects on attachment of all tested strains. These positive effects regarding pyrite colonization caused by the addition of 1 mM glucuronic acid to *At. ferrooxidans* correlated with a pyrite dissolution enhancement of around 25 % and significantly elevated planktonic cell numbers (Bellenberg et al. 2015).

Several questions remain open: Which other abiotic and biotic factors influence biofilm formation? Are these factors different among species of iron-oxidizing acidithiobacilli? Does the presence of the PP influence the start of the MS dissolution? Are there differences in the iron oxidation activity between BP and PP in pyrite cultures? Does the increase of the initial amount of attached cells correlate with improved pyrite dissolution? In order to answer these questions, we have compared the contribution of the BP and PP from pyrite cultures to the iron oxidation activity in *At. ferrooxidans*^{T}. It was shown that within the first 4–5 days, only the BP was responsible for pyrite dissolution (Bellenberg et al. 2015).

The molecular mechanisms used by leaching bacteria to adapt the composition and amount of EPS according to the energy source (planktonic cells grown with iron(II)-sulfate, produce almost no EPS), as well as the physiological adjustments to a sessile lifestyle remain to be elucidated in detail. Recent results have highlighted that some similarities with biofilm formation in other bacterial systems might occur in acidophilic metal/sulfur-oxidizing bacteria, e.g. involvement of P*i*/polyP metabolism, EPS biosynthesis pathways, coordination of stress responses, altered metabolic rates and cell-cell signaling networks. Our current knowledge in this area is summarized in Sect. 5.

4 Visualization of Biofilms by Acidophilic Metal/Sulfur-Oxidizing Microorganisms

The microbial biofilm lifestyle generally includes attachment, biofilm development, maturation and dispersal (Stoodley et al. 2002; Hall-Stoodley et al. 2004). The EPS components enable formation of 3-dimensional biofilm structures. Several advanced microscopy techniques have been applied to reveal biofilm formation and dynamics. CLSM in combination with fluorescent probes provide detailed 3-dimensional structure and compositional information (Lawrence et al. 2003; Neu and Lawrence 2014a). EPS represent a crucial part of microbial biofilms and a key element in terms of biofilm functionality (Neu and Lawrence 2009). Due to the complexity of EPS, an *in situ* approach to analyze the EPS glycoconjugates by means of FLBA has been developed (Staudt et al. 2003; Peltola et al. 2008; Zippel and Neu 2011; Bennke et al. 2013; Castro et al. 2014). FLBA allows simultaneous visualization and characterization of EPS glycoconjugates. If combined with other stains specific for proteins, nucleic acids or lipids, among others, additional EPS components may be visually characterized (Neu and Lawrence 2014a, b). Transmission electron microscopy (TEM) as well as AFM, offer the highest resolution for biofilm structural inspection (Dufrêne 2003; Lawrence et al. 2003). In addition, scanning electron microscopy (SEM) has been widely used in the field for biofilm visualization and monitoring (Baldensperger et al. 1974; Bryant et al. 1984; Mikkelsen et al. 2007). Also, environmental scanning electron microscopy (ESEM) can minimize biofilm dehydration and preserve native biofilm morphology and support surfaces (Priester et al. 2007). These techniques, as well as magnetic resonance imaging

(MRI), scanning transmission X-ray microscopy (STXM), Raman microscopy (RM) and surface-enhanced Raman scattering (SERS) allow *in situ* and non-destructive analysis of the structure, species and EPS composition as well as dynamic processes within microbial biofilms (Lawrence et al. 2003; Manz et al. 2003; Ivleva et al. 2009; Wagner et al. 2009). RM provides single-organism fingerprints for biological samples with spatial resolution in the nanoscale range and enables correlations between optical and chemical images to be made. Since water is the major component of the biofilm matrix, RM is ideal for *in situ* studies. In contrast, STXM allows visualization of biofilm structures by biochemical and elemental imaging. Especially, these techniques do not involve the use of additional probes. Nevertheless, although STXM reveals unique data sets, it requires synchrotron beam time and has limitations in sample mounting.

As mentioned above, AMOM live in an extremely special niche, usually characterized by a high content of metal ions, low pH and high temperature gradients. Although biofilm occurrence and architecture have been studied for decades, detailed knowledge on acidophilic biofilms is still rather limited if compared with microbial biofilms related to medical and industrial fields. Nevertheless, in recent years progress has been made for visualization and characterization of biofilms produced by acidophilic microorganisms.

4.1 Biofilm Formation by Acidophilic Metal/Sulfur-Oxidizing Microorganisms at Laboratory Scale

4.1.1 Attachment to and Biofilm Formation on Elemental Sulfur

S^0 has a very low solubility in water (<5 μg/L). Hence, a direct contact is prerequisite for microbial oxidation (Vogler and Umbreit 1941). Waksman first noticed that S^0 particles in a growing culture of *At. thiooxidans* were surrounded by bacteria using light microscopy (Waksman 1932). Visualization of biofilms formed by acidophiles on S^0 revealed that cells are forming monolayers and erosion sites associated with cells were evenly distributed on S^0 surface. In general, acidophilic bacteria and archaea showed preferential attachment to defect sites present on S^0 surface. This was observed for *At. thiooxidans* (Schaeffer et al. 1963), *Sulfolobus* sp. (Weiss 1973), *Thiobacillus denitrificans* (Baldensperger et al. 1974) and *At. ferrooxidans* (Espejo and Romero 1987). In addition, pili and cell wall components like a glycocalyx were found to be involved in connecting cells of *Sulfolobus* sp. and *Thiobacillus albertis* and S^0 (Weiss 1973; Bryant et al. 1983; Bryant et al. 1984; Laishley et al. 1986). The presence of mucous polysaccharides in EPS was visualized by ruthenium red staining (Bryant et al. 1983). Additionally, membrane blebs were visualized and these were hypothesized to aid the cells overcoming the hydrophobic barrier necessary for their growth on S^0 (Knickerbocker et al. 2000; Crescenzi et al.

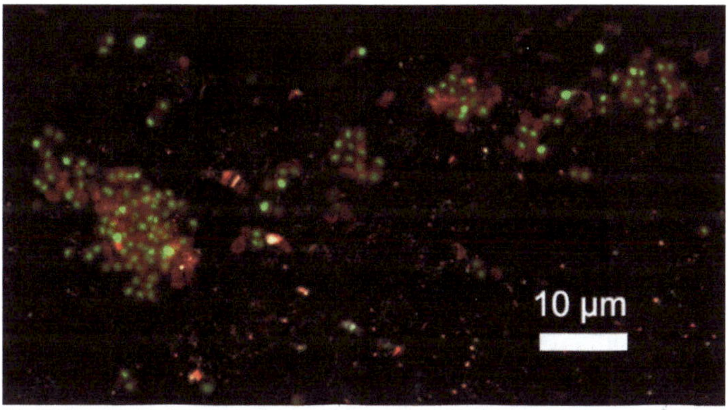

Fig. 6.2 Maximum intensity projection of biofilms by *Acidianus* sp. DSM 29099 on S^0. Cells and biofilm matrix were double stained by Con A and SybrGreen. Color allocation: *green* = SybrGreen, *red*=Con A-tetramethyl rhodamine isothiocyanate (TRITC), *grey* = reflection. Biofilm cells in the form of microcolonies were embedded in EPS matrix containing monosaccharides mannose and glucose

2006). Recently, we studied the biofilm formation of the thermophilic archaeon *Acidianus* sp. DSM 29099 on S^0 by means of FLBA. It was found that biofilm cells were heterogeneously distributed as individual groups of cell clusters and micro-colonies. 21 lectins were shown to be useful for the study of EPS glycoconjugates produced by this archaeon on S^0 surfaces. In addition, various glycoconjugates, containing monosaccharides such as fucose, glucose, galactose, mannose, N-acetyl glucosamine (GlcNAc) and N-acetyl galactosamine (GalNAc) were detected in these biofilms (Zhang et al. 2015a). Figure 6.2 shows an example of biofilm matrix of *Acidianus* sp. DSM 29099 on S^0 visualized by CLSM.

Apart from labelled lectins, other fluorochromes may be simultaneously applied to visualize and characterize additional EPS compounds. For instance, the Syto and Sypro series are used to detect cells via their nucleic acids and cellular proteins, respectively. FM dyes (FM1-43 and FM4-64) and Nile red are specific to stain membranes and lipophilic compounds. These stains can be used also to stain extra-cellular compounds in biofilms (Lawrence et al. 2007; Neu and Lawrence 2014a). DDAO (7-hydroxy-9H-1,3-dichloro-9,9-dimethylacridin-2-one) stains nucleic acids and normally does not penetrate cell membranes. Thus, it has been selected as the preferred fluorochrome for staining extracellular DNA (eDNA) (Koerdt et al. 2010). Combination of several types of stains allowed us to get detailed information about biofilms of acidophiles on different energetic substrates. For instance, biofilms of *S. metallicus*^{T} were shown to be embedded in an EPS matrix containing proteins and eDNA by combination of Sypro, DDAO and several lectins (Zhang et al. 2015b). In Fig. 6.3 the presence of eDNA in biofilms of *S. metallicus* grown on S^0 is shown.

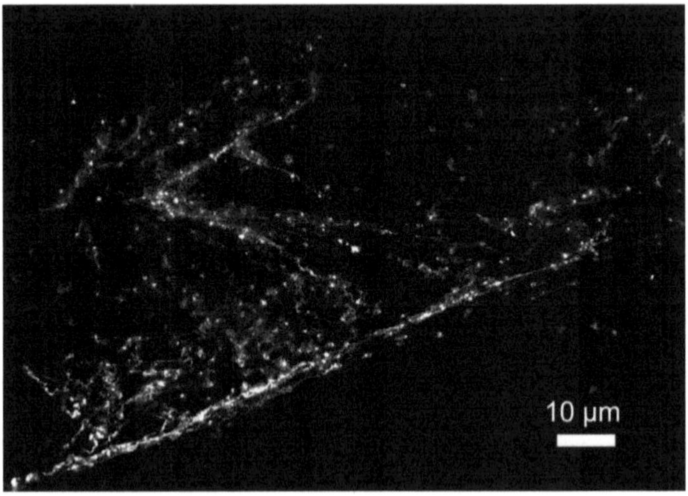

Fig. 6.3 CLSM image of *S. metallicus*^{T} biofilms on S^0. Cells and biofilm matrix were stained by SybrGreen. Diffuse and thread DNA signals around/connecting cells are visible

4.1.2 Biofilm Formation on Mineral Sulfides

In general, cells are forming monolayer biofilms and show selective attachment, preferentially to sites with crystal defects, fractures and pores. Examples of these observations include visualization of cells of *Caldariella* (a thermo-acidophilic archaeon) on pyrite or chalcopyrite phases of a low-grade copper ore (Murr and Berry 1976), *At. ferrooxidans* on pyrite or chalcopyrite (Wakao et al. 1984; Gehrke et al. 1998; Sanhueza et al. 1999; Sampson et al. 2000a; Tributsch and Rojas-Chapana 2000; Lei et al. 2009; Noël et al. 2010), *A. caldus* on pyrite (Edwards et al. 2000a) and enrichment cultures obtained from the Iron Mountain in California on pyrite surfaces (Edwards et al. 1998, 1999b). In contrast, an *in situ* AFM analysis of *Sulfobacillus thermosulfidooxidans* attached to pyrite indicated that cells on the surface were distributed in small clusters instead of forming a continuous biofilm. No evidence was found to suggest a preferential attachment to certain sites or a preferred orientation (Becker et al. 2011). Also, cell attachment of *Metallosphaera sedula* and *S. metallicus* show no preferential orientation. However, pyrite oxidation and pit etching were influenced by surface symmetries (Etzel et al. 2008). Interestingly, two distinct biofilm morphologies were described for a moderate thermophilic archaeon *F. acidarmanus* Fer1. A multilayer biofilm was developed on pyrite surfaces, and up to 5 mm-long filaments were found on sintered glass spargers taken from gas lift bioreactors (Baker-Austin et al. 2010). Cells of *M. sedula* were found to be wiggling along the metal ore by EFM, suggesting that cell appendages were involved in cell attachment to the ore (Huber et al. 1989). Production of EPS (slime or an organic capsule) was often observed. Such phenomenon has been reported for cells of *Sb. thermosulfidooxidans* (Golovacheva 1978), *At. ferrooxidans*, *At. thiooxidans* and/or *L. ferrooxidans* (Rojas-Chapana et al. 1996; Telegdi et al. 1998; Tributsch and Rojas-Chapana 2000; Bellenberg et al. 2012; González et al. 2012). The production of EPS can be induced by several factors like direct contact with solid substrates or with its dissolution products (Bellenberg et al. 2012; Zhang et al. 2015a). Interactions of three axenic cultures of thermophiles *Acidianus brierleyi*, *M. sedula* and *S. metallicus* with pyrite were first documented using SEM and TEM. Several of deposited structures were formed on the pyrite surface, including sub-micron precipitates and disc-shaped structures (Mikkelsen et al. 2007).

The combination of AFM and EFM allows for confirmation of biological or chemical origins of structures at a high resolution (Mangold et al. 2008). Surface properties of minerals and their modification due to microbial activities can be recorded. In addition, the probes used (e.g. fluorescently labelled lectins) can provide additional biochemical information of the biofilm cells. The formation of monolayer biofilms as well as EPS production were found for cells of *F. acidiphilum* on pyrite (Zhang et al. 2014), *Metallosphaera hakonensis* on chalcopyrite (Africa et al. 2013), *At. ferrooxidans* and mixed cultures of *At. thiooxidans* and *At. ferrooxidans* on pyrite (Harneit et al. 2006; Florian et al. 2010; Noël et al. 2010; Florian et al. 2011; Gonzalez et al. 2013). The biofilm formation was accompanied with the production of EPS containing mannose or glucose. By combination with other techniques, e.g. a Kelvin probe, relative surface potential differences and charge distributions on the surface can be measured (Vera et al. 2013b). We have applied FLBA to analyze biofilms of axenic cultures of *F. acidiphilum*, *S. metallicus* and *Acidianus* sp. DSM 29099 on pyrite. Glycoconjugates containing monosaccharides such as fucose, glucose, galactose, mannose, sialic acids, GlcNAc and GalNAc were detected in these biofilms. The two main binding patterns, e.g. tightly or loosely bound EPS were detected (Zhang et al. 2015a). Cells showed preferential attachment to specific sites of the pyrite surfaces (Fig. 6.4).

The molecular mechanisms controlling biofilm formation in acidophilic metal/sulfur-oxidizing archaea are far less explored (Orell et al. 2013). Preliminary work on crenarchaeal biofilms on other surfaces with respect to their morphology, architecture and chemical components has been done. The first crenarchaeal biofilm analysis was described for three closely related *Sulfolobus* sp. Biofilms with "carpet-like" structures by *S. solfataricus* and *S. tokodaii* and high-density "tower-like" structures by *S. acidocaldarius* were detected (Koerdt et al. 2010). Cell appendages such as pili or flagella were shown to be involved in initial attachment of *S. solfataricus* to various surfaces, including glass, mica, pyrite and carbon-coated gold grids (Zolghadr et al. 2010). Taking advantage of the expression of fluorescent proteins in archaea, three type IV pili-like cell appendages of *S. acidocaldarius* were found to be involved, but possessed different functions in cell colonization and biofilm formation on glass surfaces (Henche et al. 2012). By using comparative fluorescence microscopy and CSLM, the enzyme mannosidase in *S. solfataricus* was found to be important in archaeal biofilm formation and modulation of EPS composition (Koerdt et al. 2012).

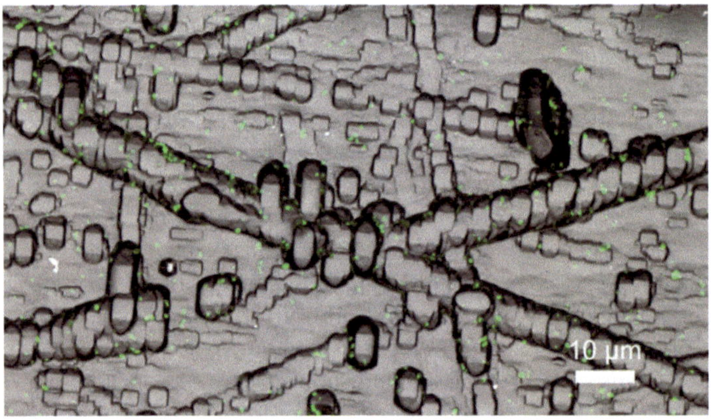

Fig. 6.4 CLSM image of biofilms by *Acidianus* sp. DSM 29099 on pyrite. Cells were stained by the lectin AAL. Color allocation: *green*=AAL-fluorescein isothiocyanate (FITC). The pyrite surface is shown in reflection mode (=*grey*). Cells preferentially attach to pyrite surfaces along surface defective patterns

4.2 Biofilms of Acidophiles in Natural Environments

Microbial biofilms associated with low pH and high metal concentration are often found in natural environments, like AMD sites at Rio Tinto, Spain or the Iron Mountain (Richmond, California). In contrast to laboratory biofilms, environmental ones usually form extended, macroscopic structures of different morphology. In these biofilms, bacteria, archaea and eukaryotic microorganisms have been detected e.g. in the slimes at the Iron mountain AMD site (Edwards et al. 1999a, 2000b). Macroscopic biofilm streamers of around 1 cm in diameter and dominated by *Leptospirillum* sp. have been found (Bond et al. 2000b). Several studies have focused on the microbial communities in biofilm streamers and snottites (Tyson et al. 2004; Wilmes et al. 2009; Jones et al. 2012; Yelton et al. 2013; Aliaga Goltsman et al. 2015). A benthic community in an acidic (pH $\leq$2) stream was analyzed by CLSM. The dominating microbes *Gloeochrysis* were observed as a brown, mucilaginous biofilm growing on stones within a dense matrix containing also inorganic particles and fungal hyphae, all held together by EPS (Baffico et al. 2004). Macroscopic filaments studied at Rio Tinto identified *At. ferrooxidans*, *L. ferrooxidans* and *Acidiphilum* spp. as the dominant species (García-Moyano et al. 2007). Biolims (streamers and snottites) obtained from a subsurface mine in Königstein (Saxony, Germany) were visualized after staining with four acid stable fluorescent dyes. These dyes were proven to be suitable for staining acidophilic biofilms of acidophilic microorganisms under ambient pH conditions (Brockmann et al. 2010).

4.3 Detachment of Acidophilic Biofilms

Biofilm detachment as well as cell dispersal from acidophilic biofilms is still an unexplored field and these processes are not well understood. However, there is evidence that AMOM show such a behavior since microbial footprints on MS have been observed. These are mainly composed of organic substances which are left behind by microbes during transient contact with mineral surfaces or due to a cell programmed detachment as part of the biofilm lifestyle cycle. This ability of microbes to label relevant surfaces was first reported as 'footprints', for polymeric substances left on a glass surface after cells of *Pseudomonas* detached by a shear force (Marshall et al. 1971). Microbial footprints are mainly composed of EPS remaining on surfaces after cell detachment or mechanical removal (Neu and Marshall 1990, 1991). Footprints have been described for *T. intermedius* on iron (Telegdi et al. 1998), *At. ferrooxidans* (Rojas-Chapana et al. 1996; Mangold et al. 2008), *St. thermosulfidooxidans* (Becker et al. 2011) and a mixed culture of mesophilic acidophilic chemolithotrophs where *L. ferriphilum* was predominant (Ghorbani et al. 2012). We detected that cells of *S. metallicus*^{T} and *Acidianus* sp. DSM 29099 left behind mannose or glucose containing materials on surfaces after cell detachment (Zhang & Vera, unpublished data). We have also observed different dynamics of colonization, biofilm formation and cell detachment of iron-oxidizing acidithiobacilli and leptospirilli. In this case the biofilms formed by the latter species are stable for longer incubation periods than the ones formed by iron-oxidizing acidithiobacilli. Probably these detachment processes form part of a response to oxidative stress caused by the presence of ROS, the increase of the ionic strength and the decrease of the pH. It has been shown that EPS compounds like lipopolysaccharides (LPS) might detach from biofilms during their maturation (Jiao et al. 2010).

Several extracellular enzymes have been found to be present in EPS from environmental biofilms. Many of these are probably involved in the degradation of biopolymers, resulting in cell detachment and biofilm dispersal (Flemming and Wingender 2010). Currently there are no identified enzymes which could be involved in detachment of acidophilic biofilms. However, transcriptomic studies have suggested that several genes could be involved in maintenance and detachment/dispersal of acidophilic biofilms (Moreno-Paz et al. 2010).

4.4 Chemical Mapping of Extracellular Components

Recently, synchrotron radiation based STXM imaging and μ-XRF mapping have been applied for an *in situ* comparative analysis of the extracellular thiol groups (SH) of *At. ferrooxidans* cells. It was found that the SH content of *At. ferrooxidans* grown on S^0 is roughly four times higher than those contents of iron(II)-grown cells. These data suggest that extracellular SH play an important role in sulfur activation prior to its oxidation (Xia et al. 2013). STXM has been used to visualize biofilms and EPS formed by *At. ferrooxidans* on pyrite. The distribution of polysaccharides and proteins in these biofilms was visually correlated with the optical overview. Polysaccharide-rich biomolecules were abundant at the pyrite-cell boundary, while lipid- and protein-rich regions were detected in the center region of the cells (Mitsunobu et al. 2015).

5 Molecular Studies of Biofilm Lifestyle of Acidophilic Metal/Sulfur-Oxidizing Microorganisms

5.1 Proteomics of Biofilm Formation in At. ferrooxidans^T

As mentioned earlier, the different EPS amount and composition between cells grown with different energy substrates, adaptation to enhanced levels of ROS and the obvious physiological adaptations necessary for the transition to a biofilm lifestyle strongly suggest that planktonic and sessile cells differ in their gene expression and, consequently, in their proteomic patterns. *At. ferrooxidans*^{T} possess homologous systems for CPS biosynthesis and export to the ones described in Gram-negative bacteria (Barreto et al. 2005). The genes *Afe_1339* and *Afe_2975* encode for two outer membrane polysaccharide export proteins, homologous to *E. coli* Wzy and KpsD (Whitfield 2006). The expression levels of *Afe_2975* were found to be enhanced in PP and BP after transfer of iron (II)-grown cells to pyrite cultures (Bellenberg et al. 2011). A bioinformatic analysis of putative CPS synthesis genes in *At. ferrivorans* SS3 complete genome sequence revealed the presence of homologous genes of the Wz CPS biosynthesis and export system, while homologous genes for the Kps system are absent. Most of putative genes including *wza* (polysaccharide biosynthesis export domain; *Acife_1181*), *wzc* (tyrosine-protein-kinase; *Acife_1180*), *wzb* (tyrosine-protein-phosphatase; *Acife_1179*), and *wzx* (polysaccharide biosynthesis protein; *Acife_1172*), were clustered in a possible operon in this bacterium. The *wzy* gene (polysaccharide biosynthesis protein; *Acife_0130*) and two copies of *wbaP* gene (undecapyrenyl-galactose-phosphate transferase; *Acife_1166* and *Acife_1194*) are present but not clustered in operons. This finding allows us to speculate that if these gene clusters are the main responsible ones for the biosynthesis of EPS polysaccharide polymers, presumably these would be different among both species. However, the functionality of the Wz system in *At. ferrivorans* remains to be determined.

The first high-throughput proteomic study of the biofilm formation in *At. ferrooxidans*^{T} was done in order to map the changes during its early biofilm formation process. It compared proteomes from PP and BP after 24 h of biofilm formation on pyrite by using quantitative shotgun proteomics. In total 1319 proteins, (42 % of the predicted *At. ferrooxidans*^{T} proteome) were identified. At least 16 % of the total amount of detected proteins were found to have increased or decreased levels among both cell subpopulations. Functions such as biosynthesis of EPS and electron carrier containing molecules, transport of reduced inorganic sulfur compounds as well as osmotic stress response were found to be enhanced in the BP.

The presence of three transcriptional factors, which were found to have increased (MerR, IclR) or decreased levels (AbrB) in the BP, suggests their involvement on the regulation of *At. ferrooxidans* biofilm formation process. In addition, membrane and outer membrane transport functions, including increased levels of proteins for CPS biosynthesis, efflux pumps, lipoproteins, ABC transporters and several proteins related to stress responses had increased levels in BP. It is well accepted that bacteria show stress responses during biofilm formation (Otto and Silhavy 2002).

As mentioned in subchapter 3, especially the BP must deal with changes in the osmolarity as well as with the presence of enhanced levels of ROS, which are generated in presence of pyrite. The genome of *At. ferrooxidans*^{T} possesses several genes encoding proteins related to responses against oxidative stress. These include a Mn-dependent superoxide dismutase (SodA), as well as members of the alkylhydroperoxidase family (AhpC and AhpD). Also, several components for non-enzymatic neutralization of ROS are encoded. These include thioredoxins, genes for biosynthesis and export of glutathione (GSH), and disulfide reductases, which maintain the thiol/disulfide balance of other molecules (Valdes et al. 2008). In this context, changes in proteins related to responses against osmotic and oxidative stress were detected. Apart from the osmolarity sensor protein EnvZ, a protein encoding an iron and 2-oxoglutarate (2-OG) dependent dioxygenase (AFE_3138) was also found to have increased levels in biofilm cells. Members of this protein family might be involved in ectoine biosynthesis as a response to high osmolarity (Reuter et al. 2010). Glutathione (GSH) is a major biological antioxidant, which contributes to maintain redox balance in prokaryotes (Smirnova and Oktyabrsky 2005). It is especially important for re-establishing the redox balance of proteins and lipids. In addition, GSH possesses a dual role in *At. ferrooxidans,* since it has been shown that low molecular weight thiol-containing compounds like GSH are relevant for the oxidation of elemental sulfur to sulfite in a reaction catalyzed by a periplasmic sulfur dioxygenase (SDO) (Rohwerder and Sand 2003). Several proteins involved in GSH metabolism were found to be induced in BP. These include AFE_1390 (CydC), a GSH transporter, AFE_0366 and AFE_2773, a GSH reductase (GR), and a protein of the YghU family of GSH S-transferases. This suggests an increase in periplasmic GSH levels, as part of a GSH-driven response against oxidative stress or related sulfur and RISCs oxidation (Vera et al. 2013a). In this context, it has been shown that the GR encoding gene is induced after *At. ferrooxidans* is exposed to copper, suggesting its involvement in recovering GSH pools as a response against oxidative stress (Xia et al. 2011).

Certain molecules can modulate biofilm formation. Polyamines are linear organic polycations widespread in bacteria. A common function of polyamines does not seem to exist but recent evidence has suggested their involvement in biofilm enhancement or inhibition in different bacterial species, acting extra- or intracellularly (Karatan and Michael 2013). The asymmetrical polyamine spermidine is widespread in bacteria. In *E. coli*, GSH is mostly present under aerobic growth and glutathionyl-spermidine (GspSH) is predominating under anaerobic growth conditions and stationary phase (Smith et al. 1995). A protein of the GSH S-transferase YghU family probably related with the synthesis of GspSH (AFE_2773) was found to be enhanced in *At. ferrooxidans*^{T} BP. It has been suggested that YghU belongs to a new family of GSH S-transferases (Stourman et al. 2011). Spermidine has been found ubiquitously in *At. thiooxidans* cells grown with different energy sources such as iron(II)-sulfate, sulfur and chalcopyrite (Martínez et al. 2013).

Biofilm cells are characterized to have "altered metabolic rates" (Seneviratne et al. 2012). In *At. ferrooxidans* BP there was an increased amount of proteins involved in recycling of cellular components and decreased levels of respiratory functions. These include several nucleotidases and ATPases, which may enhance the pools of building blocks for *de novo* macromolecule biosynthesis. Also an enhanced biosynthesis of cofactors and coenzymes was suggested to occur in BP, including proteins containing iron-sulfur clusters, heme, pyrroloquinoline quinone (PQQ) and functions related to ubiquinone metabolism (Vera et al. 2013a). Around one fifth of the total proteins detected were hypothetical proteins (231 proteins, from which 12 % were found with altered levels among BP and PP). Therefore, it is highly likely that still unknown metabolic pathways of *At. ferrooxidans* are relevant during biofilm formation. Among these ones, some "biofilm specific molecules" with potential roles as antibiotics or inhibitors may be considered. The latter ones have been shown to occur in some bacterial species, being probably involved in survival against competitors and predators (Beloin et al. 2004; Rendueles and Ghigo 2012). The increased presence of hypothetical proteins induced in biofilm proteomes, including those from AMD biofilms will require further studies of heterologous expression and subsequent biochemical screening in order to address their functions (Denef et al. 2010).

It is well accepted that biofilm development is an organized series of sequential events: in which planktonic microorganisms adhere to a surface; later on these adhered organisms produce a discrete set of microbial colonies in which EPS production starts together with the appearance of a three-dimensional community structure. At mature stages, some of the biofilm cells leave to seek for new surfaces (Flemming and Wingender 2010). As mentioned before, in acidophilic biofilms we have microscopic evidence pointing out the existence of cell detachment processes (see Sect. 4). However, to the best of our knowledge, there is no detailed knowledge of the physical factors as well as the physiological responses and molecular events probably involved in cell detachment processes.

5.2 *Molecular Diversity Studies in Environmental Acid Mine Drainage Biofilms*

A major contribution to the ecology and molecular study of AMD biofilms has been published by Jillian Banfield & co-workers. These studies include extensive metagenomic, metaproteomic and proteogenomic analyses of natural AMD communities and their biofilms (Ram et al. 2005; Denef et al. 2009; Wilmes et al. 2009).

Sequencing of metagenomes obtained from the Richmond Mine at the Iron Mountain in northern California allowed the reconstruction of near complete genomes of *Leptospirillum* Group II (which is the dominating microorganism), later characterized as *Leptospirillum rubarum* (Goltsman et al. 2009) and a *Ferroplasma* type II (Tyson et al. 2004). Subsequent studies have allowed to perform genomic reconstructions of some other microorganisms present within these communities such as *Leptospirillum* group III, later characterized as *L. ferrodiazotrophum* (Goltsman et al. 2009) and *Leptospirillum* group IV UBA BS (Goltsman et al. 2013), as well as new uncultivated species of archaea (Yelton et al. 2013).

AMD biofilms grow at the air-liquid interfaces of sites like streams and pools that overlay pyrite containing sediments. Biofilm growth (up to hundred microns thick) generates iron(III), which subsequently catalyzes pyrite oxidation, coupling inorganic and biological oxidation processes. The main characteristics of this system is low species diversity and reproducible stages of biofilm development and succession over time (Wilmes et al. 2009; Denef et al. 2010). In the earliest developmental stages of AMD biofilms *Leptospirillum* group II dominates in presence of a small proportion of *Leptospirillum* group III and some archaea. When the biofilms enter into a mature stage, the proportion of *Leptospirillum* group III and archaea increases and some eukaryotes appear. Interestingly, although *Leptospirillum* groups II and III are present through all the biofilm development, they strongly differ in their spatial arrangement. *Leptospirillum* group II was observed to form tight agglomerations of cells while *Leptospirillum* group III was mostly observed as microcolonies or single cells (Wilmes et al. 2009).

5.3 Proteomics in Acid Mine Drainage Biofilms

The genomic types of natural biofilm populations in a community sample were inferred from proteomics data. Samples from two locations (UBA and 5-way GC) within the Richmond mine were used, each one dominated by a distinct *Leptospirillum* group II type, with 0.3 % differences at the 16S rRNA sequence level. Both genomes possessed 80 % identity at the gene level with an average amino acid identity of 95.2 %. A proteomics inferred genome typing (PIGT) provided evidence of recombination among two closely related *Leptospirillum* group II populations (Lo et al. 2007). Later on, PIGT was used to genotype the dominant *Leptospirillum* group II population by analysis of 27 biofilm samples taken from the Richmond mine over a 4-year period. Six distinct phenotypes, which are recombinants derived from two parental genotypes were identified, confirming that homologous recombination is the main strategy used for fine scale environmental adaptation within those biofilms (Denef et al. 2009).

Comparative proteomic analyses of AMD biofilms have been also performed with laboratory-cultivated biofilms, inoculated from environmental samples (Belnap et al. 2010, 2011). Interestingly, laboratory-grown biofilms were also dominated by *Leptospirillum* group II, with a lower abundance of *Leptospirillum* group III. Biofilm field samples showed a higher abundance of proteins related to functional categories such as energy production and conversion, cell motility, cell wall, membrane and envelope biogenesis, intracellular and secretion and vesicular transport functions. In contrast, transcriptional proteins as well as proteins probably involved in defence mechanisms were more abundant in laboratory-grown biofilms (Belnap et al. 2010). Further studies determined that *Leptospirillum* group II proteins involved in amino acid and nucleotide metabolism, together with proteins involved in cell membrane and envelope biogenesis were overrepresented at high pH. In addition, a pH-specific niche partitioning was shown to occur for some low abundance bacteria and archaea, since *Leptospirillum* group III was more abundant in biofilms grown at higher pH values, whereas archaeal species were more abundant at lower pH values (Belnap et al. 2011).

By high-throughput proteomics and metabolomics it has been shown that the identified proteins and metabolites from *Leptospirillum* groups II and III exhibit organism related correlation patterns, which suggest a restructuration of their metabolic and/or regulatory networks reducing their competition and allowing them to occupy distinct niches (Wilmes et al. 2010). After evaluation of the reproducibility of AMD community proteomes, a set of reliable classifier proteins was identified which may be used to predict growth stages of biofilm communities. During early stages of biofilm growth, *Leptospirillum* group II cell population responded to some abiotic stresses by reorganizing their metabolism, since functions such as protein biosynthesis, cell division as well as metabolism of 1 or 2 carbon compounds were more abundant. Stress responses were abundant in early stage biofilms, since proteins involved in metal efflux were found to be increased. Among the increased proteins in this growth stage, a cytochrome 572, a cytochrome oxidase with high O_2 affinity as well as a cytochrome c-553 were found. Cell division functions were also more abundant in early stage biofilms and these seem to further decrease due to accumulation of DNA mutations. This is in agreement with a 26-fold increase in the abundance of ribosomal proteins detected in early stage biofilms, compared to the mature ones. Several enzymes involved in the metabolism of more complex carbohydrates, amino acid biosynthesis, amino- and nucleotide-sugar metabolism, lipopolysaccharide biosynthesis, starch and trehalose metabolism were more abundant in mature biofilms. Proteomes from late stage biofilms also showed an increased level of proteins involved in P*i* and molybdenum transport, suggesting that essential nutrients such as oxygen, phosphorous, nitrogen and molybdenum become limiting in those biofilms (Mueller et al. 2011).

The response of AMD microbial biofilm communities to temperature gradients has been recently studied. Cultivation at elevated temperatures e.g. increase from 40 °C to 46 °C repressed carbon fixation in two *Leptospirillum* genotypes while a third one was probably subjected to a viral stress, which would increase the carbon turnover through the release of the viral lysate (Mosier et al. 2015). Several enzymes upregulated could probably be involved in changes in the EPS composition in these AMD biofilms, which have been shown to possess carbohydrates such as glucose, galactose, rhamnose, heptose and mannose (Jiao et al. 2010). Recently it has been shown that approximately 29 % of the proteins reliably detected in the dominant *Leptospirillum* type II biofilm population carry post-translational modifications (PTMs), among these, 43 % carry more than one PTM. The PTMs profile strongly differs between early and mature biofilms, as well as between orthologous of two ecologically differentiated *Leptospirillum* group II bacteria (Li et al. 2014). By metabolomic analysis it has been found that natural acidophilic microbial biofilms dominated by bacteria of the genus *Leptospirillum* contained unusual lysophosphatidylethanolamine (PE) lipids in high abundance. The unusual polar head group structure of these lipids may be related to their affinity for iron and calcium ions (Fischer et al. 2012). Around 3500 metabolic features were identified in biofilms grown at AMD solutions at pH 0.9. By stable isotope labeling, several unrepresented metabolites were found in these biofilm communities, which may represent novel compounds. Among the known metabolites found, taurine, ectoine and hydroxyectoine were suggested to provide protection against osmotic stress. As taurine is not known to be synthesized by bacteria or archaea (but probably assimilated by some bacterial members of the AMD community), it was speculated that *Acydomyces richmondensis*, the dominant fungus in the Richmond mine AMD community, could be responsible for its biosynthesis (Mosier et al. 2013). The genomes of *Leptospirillum* group II possess genes for ectoine or hydroxyectoine biosynthesis. The production of ectoine and hydroxyectoine decreased in mature biofilms, suggesting that during the early biofilm development ectoine and hydroxyectoine are more abundant in order to help bacteria to the exposure to the a high ionic strength AMD solution. In later stages, the biofilm structure may provide some protection against high metal and proton concentrations (Mosier et al. 2013).

6 Cell-Cell Communication in Acidophilic Metal/Sulfur-Oxidizing Bacteria

Microbial biofilm development is a complex process which is regulated at different levels by a diverse set of mechanisms. Microbes do not only exist as single cells and they are able to coordinate their activities in a concerted manner, in a similar way to multicellular organisms. Several bacterial species employ sophisticated intercellular communication systems, which rely on the secretion and sensing of small signal molecules which allow to control the expression of multiple target genes. In many Gram-negative bacteria biofilm formation and EPS production are modulated by cell-cell communication mechanisms of Quorum Sensing (QS) (Marketon et al. 2003). QS cell-cell signalling is mediated by diffusible molecules named autoinducers (AIs), which facilitate the regulation of cellular processes affected in a cell-density dependent way (Labbate et al. 2007).

6.1 QS Studies in At. ferrooxidans^T

At. ferrooxidans^{T} has one AI-1-type QS system, involving the production of several AIs of the acyl-homoserine lactone (AHL) type (Farah et al. 2005; Rivas et al. 2005). The system is composed of an AHL synthase (AfeI), a transcriptional regulator (AfeR) that binds AHLs, an *afe*-box which is the target of the binary complex [AfeR-AHL] and different AHL signalling molecules. It has been shown that at least 9 different AHLs with diverse C-3 substitutions (oxo or hydroxy) and a range of carbon atoms in the acyl chain oscillating between 8 and 16 are synthesized in *At. ferrooxidans*^{T}. P_i starvation increased the transcription of *afeI* and AHL levels (Farah et al. 2005; Valenzuela et al. 2007). Moreover, compared to iron(II)-grown cells, *afeI*-transcript levels were also increased in cells grown with S^0. Altogether, these information allowed to propose that the *At. ferrooxidans*^{T} QS system (*afeI/afeR*) could be involved in the regulation of EPS production and consequently in the regulation of biofilm formation, as it has been demonstrated in other Gram-negative bacteria (Marketon et al. 2003; Parsek and Greenberg 2005; Labbate et al. 2007; Decho et al. 2010). If so, synthetic AHLs and AHL-analogues (Choudhary and Schmidt-Dannert 2010; Stevens et al. 2010; Galloway et al. 2012) could be used in cells of *At. ferrooxidans*^{T} to enhance or inhibit QS-regulated phenotypes, such as EPS production and biofilm formation. This has been demonstrated, since biofilm formation on pyrite or S^0 surfaces was stimulated by the external addition of synthetic long-chain AHLs (Gonzalez et al. 2013). Concominantly with this, increased levels of CPS were observed in biofilms grown on polycarbonate filters in presence of long-chain AHLs. In agreement with these findings, a bioinformatic study predicted that at least 75 genes possess *afe*-boxes, suggesting their expression to be regulated by AfeR-AHL complexes in *At. ferrooxidans*^{T}. Among these, genes encoding glycosyltransferases, metallo-beta lactamases, proteins probably related to RNA metabolism and active transport-related proteins were found, suggesting these to be directly related in EPS biosynthesis and export (Banderas and Guiliani 2013).

Cell attachment to MS during bioleaching is a selective process in which microorganisms attach preferentially to certain sites on the pyrite surface. In consequence, increased cell densities may occur at these interfaces. Such kind of cell trapping may consequently result in QS auto-induction in attached cells but not in planktonic ones. This activation presumably contributes to the establishment of the EPS/CPS matrix and the biofilm phenotype which can be observed in solid substrate cultures. In contrast, in cultures with iron(II)-ions as soluble substrate AHL detection was either unsuccessful or it indicates levels below auto-induction threshold. A similar mechanism of regulation during the initiation of biofilm formation has been postulated to exist in *Vibrio cholerae* (Waters et al. 2008).

6.2 Cell-Cell Communication in Other Acidophilic Leaching Bacteria

Interaction(s) of bacterial species in natural environments represents a complex research field. In AMOM this field is poorly understood. However, microbial interactions, such as competitive and cooperative traits were demonstrated with laboratory strains and are hypothesized to have strong influences on the propagation of single species in laboratory cultures or microbial community compositions in natural ecological niches (Johnson 1998). These interactions are supposed to act in addition to specific adaptation or strain specific fitness towards abiotic factors, such as ranges of suitable temperatures, nutrients, electron acceptors, concentrations of heavy metals or enhanced osmotic pressure. As previously mentioned, natural habitats of AMOM are characterized by a low microbial diversity compared to neutrophilic environments due to their extremes regarding pH, heavy metal concentration and electron acceptors (Bond et al. 2000a). However, microbial intra- and interspecies interactions are complex due to the amount of environmental variables which are undoubtedly involved. These include temperature, hydration, pH, presence of diverse inorganic ions, organic compounds, such as secreted metabolites, EPS, proteins, secondary metabolites with antibiotic activity or other compounds which may originate from cell lysis or allochthonous biomass. Niche specialization of AMOM is of particular relevance, since in mixed species cultures these are supposed to compete for common nutrients, electron acceptors and carbon sources. The array of possibilities also extends when considering that acidophilic heterotrophs, fungi and algae as well as bacteriophages are also involved in ecological processes such as synergy, mutualism and predation in nature, which may affect the bacterial community composition.

In order to gain insights about cell communication between different species coexisting in bioleaching habitats, the presence of AHLs by mass spectrometry and *Agrobacterium tumefaciens* NTL4 bioreporter assays (Cha et al. 1998) was analyzed. Cultures of *At. ferrivorans* SS3, *At. thiooxidans* DSM 14887, *L. ferrooxidans* DSM 2391 and two strains of *Acidiferrobacter* spp. were assayed. Interestingly, similar as described for *At. ferrooxidans*^{T}, *Acidiferrobacter* strains produced a great diversity of AHLs: C10-AHL, C12-AHL, 3-hydroxy-C12-AHL, C14-AHL, 3-hydroxy-C14-AHL and C16-AHL. In *At. thiooxidans* two kinds of unsubstituted AHLs were detected (C10 and C12-AHL), whereas no AHLs were detected in cultures of *L. ferrooxidans* DSM 2391 and *At. ferrivorans* SS3 (Bellenberg et al. 2014).

The influence of defined mixtures of synthetic AHLs added to pyrite leaching assays in pure and binary mixed cultures containing *At. ferrooxidans*^{T}, *At. ferrivorans* SS3, *Acidiferrobacter* sp. SPIII/3 and *L. ferrooxidans* DSM 2391 was tested. Reduced cell attachment to pyrite upon AHL mixture additions correlated with lowered mineral dissolution. Consequently AHL-based interspecies interactions are likely to occur in natural habitats and they were also demonstrated in binary mixed cultures of *L. ferrooxidans* DSM 2391 and *Acidiferrobacter* sp. SPIII/3, in which an interdependent inhibition was observed.

A search for the presence of QS systems in the complete genomes of leaching bacteria revealed the presence genes encoding for LuxR-like proteins in *At. ferrivorans* SS3 (Acife_1471, 56 % identity) and *L. ferrooxidans* C2-3 (LFE_1606, 29 % identity). Interestingly, no AHLs were detected in cultures of *L. ferrooxidans* or *At. ferrivorans* SS3. The fact that these strains possess LuxR-like receptors may explain in part some of the inhibitory effects observed after addition of certain AHL mixtures. Interestingly, *At. thiooxidans* strains produced different AHLs, in a range of C8 to C10, and could also sense them since the addition of C8, oxo-C8 and C10-AHLs resulted in an enhanced biofilm formation on S^0. However no homologous genes to *luxI/R* are encoded in the genome of *At. thiooxidans* DSM 14887^T (Valdes et al. 2008). At the moment it is unknown how *At. thiooxidans* may sense AHLs and respond differentially after their addition without any LuxR-like receptor. This strongly suggests the presence of novel pathways probably involved in AHL biosynthesis and sensing. Probably these pathways may be widespread among some species of *Acidithiobacillus*. Taken together, these results clearly suggest that cell-cell communication mechanisms and their connection with biofilm formation phenotypes must be thoroughly understood in order to further influence biotechnological processes and develop countermeasures against unwanted natural leaching of MS.

6.3 The c-di-GMP Pathway in Acidithiobacilli

In many bacterial species the transition to a biofilm lifestyle is controlled by the second messenger c-di-GMP. It is synthesized from two GTP molecules by diguanylate cyclases (DGCs) and phosphodiesterases (PDEs). Several proteins are known as c-di-GMP effectors, being the ones possessing a PilZ domain one the most characterized to date (Amikam and Galperin 2006; Hengge 2009). These regulate phenotypes such flagellar and twitching motility and EPS biosynthesis. In *At. ferrooxidans*^{T}, c-di-GMP levels were increased in biofilm cells on S^0 or pyrite, strongly suggesting its involvement in *At. ferrooxidans* biofilm formation (Ruiz et al. 2011). Very recently it has been shown that the c-di-GMP pathway is also functional in *At. caldus* ATCC 51756^T. Several genes encoding DGC and PDE effector proteins were identified in its genome sequence and the presence of c-di-GMP by mass spectrometry was confirmed. In addition, genes for several of the enzymes with DGC domains were functional as shown by heterologous genetic complementation in *Salmonella enterica* serovar Typhimurium mutants. A deletion mutant lacking the DGC encoded by the gene ACAty_C1319, which is presumably the main DGC enzyme in *At. caldus*, was recently constructed. This mutant strain had almost 14-fold less c-di-GMP levels than the wild type strain. Interestingly, it showed an increased motility and a reduced capacity to attach to S^0, indicating that the c-di-GMP pathway is involved in the regulation of swarming motility and cell attachment to S^0 surfaces by *At. caldus* (Castro et al. 2015).

7 Future Perspectives

Our current knowledge on the chemical compositions and microbial dynamics of acidophilic biofilms is still limited. The presence and functional roles of macromolecules and metabolites within EPS from these biofilms remain to be clarified. High-throughput FLBA lectin studies may allow a deeper understanding of detailed biofilm formation and interfacial interactions. Acid-stable fluorescent stains, e.g. isolation of lectins from metal/sulfur-oxidizing acidophiles shall be developed. A combination of novel physical and chemical microscopic techniques, e.g. Raman microscopy and nanoscopy techniques such as STED and blink microscopy may allow a detailed investigation. Together with molecular techniques it may greatly extend the understanding of acidophilic biofilm function, chemical structure and dynamics at different scales (Neu and Lawrence 2015). Future molecular studies shall include polymer chemistry, metabolomics as well as and molecular cloning and protein chemistry of proteins remaining as "unknown". These may have important roles in controlling the dynamics of acidophilic biofilm phenotypes and interactions with other microbial populations. The detailed studies of QS, cell signaling and development of QS analogs chemically stable and with enhanced activities could also contribute to develop future methods to influence bioleaching of metal sulfides. The detailed understanding of biofilm lifestyle and cell-cell interactions will be helpful to develop strategies for bioleaching manipulation both in metal recovery where enhancement of leaching rates is necessary and mitigation of AMD, where reduction or inactivation of biofilms might reduce environmental problems.

Acknowledgements We would like to acknowledge the excellent technical assistance of Ute Kuhlicke (Department of River Ecology, Helmholtz Centre for Environmental Research-UFZ, Magdeburg) in CLSM and image processing. Ruiyong Zhang appreciates China Scholarship Council (CSC) for financial support (No. 2010637124).

Conflict of Interest Ruiyong Zhang, Sören Bellenberg, Thomas R. Neu, Wolfgang Sand, and Mario Vera declare that they have no conflict of interest.

References

Acuña J, Rojas J, Amaro AM, Toledo H, Jerez CA (1992) Chemotaxis of *Leptospirillum ferrooxidans* and other acidophilic chemolithotrophs: comparison with the *Escherichia coli* chemosensory system. FEMS Microbiol Lett 75:37–42

Africa C-J, van Hille RP, Sand W, Harrison ST (2013) Investigation and in situ visualisation of interfacial interactions of thermophilic microorganisms with metal-sulphides in a simulated heap environment. Miner Eng 48:100–107

Aliaga Goltsman DS, Comolli LR, Thomas BC, Banfield JF (2015) Community transcriptomics reveals unexpected high microbial diversity in acidophilic biofilm communities. ISME J 9:1014–1023

Alvarez S, Jerez CA (2004) Copper ions stimulate polyphosphate degradation and phosphate efflux in *Acidithiobacillus ferrooxidans*. Appl Environ Microbiol 70:5177–5182

Amaro AM, Seeger M, Arredondo R, Moreno M, Jerez CA (1993) The growth conditions affect *Thiobacillus ferrooxidans* attachment to solids. In: Torma AE, Apel ML, Brierley CL (eds) Biohydrometallurgical technologies. The Minerals, Metals & Materials Society, Warrendale, PA pp 577–585

Amikam D, Galperin MY (2006) PilZ domain is part of the bacterial c-di-GMP binding protein. Bioinformatics 22:3–6

Andrews GF (1988) The selective adsorption of *Thiobacilli* to dislocation sites on pyrite surfaces. Biotechnol Bioeng 31:378–381

Baffico GD, Diaz MM, Wenzel MT, Koschorreck M, Schimmele M, Neu TR, Pedrozo F (2004) Community structure and photosynthetic activity of epilithon from a highly acidic (pH ≤ 2) mountain stream in Patagonia, Argentina. Extremophiles 8:463–473

Bagdigian RM, Myerson AS (1986) The adsorption of *Thiobacillus ferrooxidans* on coal surfaces. Biotechnol Bioeng 28:467–479

Baker-Austin C, Potrykus J, Wexler M, Bond P, Dopson M (2010) Biofilm development in the extremely acidophilic archaeon '*Ferroplasma acidarmanus*' Fer1. Extremophiles 14:485–491

Baldensperger J, Guarraia L, Humphreys W (1974) Scanning electron microscopy of thiobacilli grown on colloidal sulfur. Arch Microbiol 99:323–329

Banderas A, Guiliani N (2013) Bioinformatic prediction of gene functions regulated by quorum sensing in the bioleaching bacterium *Acidithiobacillus ferrooxidans*. Int J Mol Sci 14:16901–16916

Barreto M, Jedlicki E, Holmes DS (2005) Identification of a gene cluster for the formation of extracellular polysaccharide precursors in the chemolithoautotroph *Acidithiobacillus ferrooxidans*. Appl Environ Microbiol 71:2902–2909

Becker T, Gorham N, Shiers D, Watling H (2011) *In situ* imaging of *Sulfobacillus thermosulfidooxidans* on pyrite under conditions of variable pH using tapping mode atomic force microscopy. Process Biochem 46:966–976

Bellenberg S, Vera M, Sand W (2011) Transcriptomic studies of capsular polysaccharide export systems involved in biofilm formation by *Acidithiobacillus ferrooxidans*. In: Qiu GZ, Jiang T, Qin WQ, Liu XD, Yang Y, Wang HD (eds) Biohydrometallurgy: Biotech Key to Unlock Mineral Resources Value: Proceedings of the 19th International Biohydrometallurgy Symposium. Central South University Press, Changsha, China, pp 460–464

Bellenberg S, Leon-Morales C-F, Sand W, Vera M (2012) Visualization of capsular polysaccharide induction in *Acidithiobacillus ferrooxidans*. Hydrometallurgy 129–130:82–89

Bellenberg S, Diaz M, Noel N, Sand W, Poetsch A, Guiliani N, Vera M (2014) Biofilm formation, communication and interactions of leaching bacteria during colonization of pyrite and sulfur surfaces. Res Microbiol 165:773–781

Bellenberg S, Barthen R, Boretska M, Zhang R, Sand W, Vera M (2015) Manipulation of pyrite colonization and leaching by iron-oxidizing *Acidithiobacillus* species. Appl Microbiol Biotechnol 99:1435–1449

Belnap CP, Pan C, VerBerkmoes NC, Power ME, Samatova NF, Carver RL, Hettich RL, Banfield JF (2010) Cultivation and quantitative proteomic analyses of acidophilic microbial communities. ISME J 4:520–530

Belnap CP, Pan C, Denef VJ, Samatova NF, Hettich RL, Banfield JF (2011) Quantitative proteomic analyses of the response of acidophilic microbial communities to different pH conditions. ISME J 5:1152–1161

Beloin C, Valle J, Latour-Lambert P, Faure P, Kzreminski M, Balestrino D, Haagensen JA, Molin S, Prensier G, Arbeille B, Ghigo JM (2004) Global impact of mature biofilm lifestyle on *Escherichia coli* K-12 gene expression. Mol Microbiol 51:659–674

Bennke CM, Neu TR, Fuchs BM, Amann R (2013) Mapping glycoconjugate-mediated interactions of marine *Bacteroidetes* with diatoms. Syst Appl Microbiol 36:417–425

Blake RC, Howard GT, McGinness S (1994) Enhanced yields of iron-oxidizing bacteria by in situ electrochemical reduction of soluble iron in the growth medium. Appl Environ Microbiol 60:2704–2710

Bond PL, Druschel GK, Banfield JF (2000a) Comparison of acid mine drainage microbial communities in physically and geochemically distinct ecosystems. Appl Environ Microbiol 66:4962–4971

Bond PL, Smriga SP, Banfield JF (2000b) Phylogeny of microorganisms populating a thick, subaerial, predominantly lithotrophic biofilm at an extreme acid mine drainage site. Appl Environ Microbiol 66:3842–3849

Brockmann S, Arnold T, Schweder B, Bernhard G (2010) Visualizing acidophilic microorganisms in biofilm communities using acid stable fluorescence dyes. J Fluoresc 20:943–951

Bryant R, McGroarty K, Costerton J, Laishley E (1983) Isolation and characterization of a new acidophilic *Thiobacillus* species (*T. albertis*). Can J Microbiol 29:1159–1170

Bryant R, Costerton J, Laishley E (1984) The role of *Thiobacillus albertis* glycocalyx in the adhesion of cells to elemental sulfur. Can J Microbiol 30:81–90

Castelle C, Guiral M, Malarte G, Ledgham F, Leroy G, Brugna M, Giudici-Orticoni M-T (2008) A new iron-oxidizing/O_2-reducing supercomplex spanning both inner and outer membranes, isolated from the extreme acidophile *Acidithiobacillus ferrooxidans*. J Biol Chem 283:25803–25811

Castro L, Zhang R, Muñoz JA, González F, Blázquez ML, Sand W, Ballester A (2014) Characterization of exopolymeric substances (EPS) produced by *Aeromonas hydrophila* under reducing conditions. Biofouling 30:501–511

Castro M, Deane SM, Ruiz L, Rawlings DE, Guiliani N (2015) Diguanylate cyclase null mutant reveals that C-Di-GMP pathway regulates the motility and adherence of the extremophile bacterium *Acidithiobacillus caldus*. PLoS One 10, e0116399

Cha C, Gao P, Chen Y-C, Shaw PD, Farrand SK (1998) Production of Acyl-homoserine lactone Quorum-sensing signals by Gram-negative plant-associated bacteria. Mol Plant-Microbe Interact 11:1119–1129

Choudhary S, Schmidt-Dannert C (2010) Applications of quorum sensing in biotechnology. Appl Microbiol Biotechnol 86:1267–1279

Clark DA, Norris PR (1996) *Acidimicrobium ferrooxidans* gen. nov., sp. nov.: mixed-culture ferrous iron oxidation with *Sulfobacillus* species. Microbiology 142:785–790

Crescenzi F, Crisari A, D'Angel E, Nardella A (2006) Control of acidity development on solid sulfur due to bacterial action. Environ Sci Technol 40:6782–6786

Crundwell F (2013) The dissolution and leaching of minerals: mechanisms, myths and misunderstandings. Hydrometallurgy 139:132–148

d'Hugues P, Joulian C, Spolaore P, Michel C, Garrido F, Morin D (2008) Continuous bioleaching of a pyrite concentrate in stirred reactors: Population dynamics and exopolysaccharide production vs. bioleaching performance. Hydrometallurgy 94:34–41

Danhorn T, Hentzer M, Givskov M, Parsek MR, Fuqua C (2004) Phosphorus limitation enhances biofilm formation of the plant pathogen *Agrobacterium tumefaciens* through the PhoR-PhoB regulatory system. J Bacteriol 186:4492–4501

Decho AW, Frey RL, Ferry JL (2010) Chemical challenges to bacterial AHL signaling in the environment. Chem Rev 111:86–99

Denef VJ, VerBerkmoes NC, Shah MB, Abraham P, Lefsrud M, Hettich RL, Banfield JF (2009) Proteomics-inferred genome typing (PIGT) demonstrates inter-population recombination as a strategy for environmental adaptation. Environ Microbiol 11:313–325

Denef VJ, Mueller RS, Banfield JF (2010) AMD biofilms: using model communities to study microbial evolution and ecological complexity in nature. ISME J 4:599–610

Diao M, Taran E, Mahler S, Nguyen AV (2014a) A concise review of nanoscopic aspects of bioleaching bacteria–mineral interactions. Adv Colloid Interface Sci 212:45–63

Diao M, Taran E, Mahler SM, Nguyen AV (2014b) Comparison and evaluation of immobilization methods for preparing bacterial probes using acidophilic bioleaching bacteria *Acidithiobacillus thiooxidans* for AFM studies. J Microbiol Methods 102:12–14

Dispirito AA, Dugan PR, Tuovinen OH (1983) Sorption of *Thiobacillus ferrooxidans* to particulate material. Biotechnol Bioeng 25:1163–1168

Dopson M, Johnson DB (2012) Biodiversity, metabolism and applications of acidophilic sulfur-metabolizing microorganisms. Environ Microbiol 14:2620–2631

Dopson M, Baker-Austin C, Hind A, Bowman JP, Bond PL (2004) Characterization of *Ferroplasma* isolates and *Ferroplasma acidarmanus* sp. nov., extreme acidophiles from acid mine drainage and industrial bioleaching environments. Appl Environ Microbiol 70:2079–2088

Dufrêne YF (2003) Recent progress in the application of atomic force microscopy imaging and force spectroscopy to microbiology. Curr Opin Microbiol 6:317–323

Dziurla MA, Achouak W, Lam BT, Heulin T, Berthelin J (1998) Enzyme-linked immunofiltration assay to estimate attachment of thiobacilli to pyrite. Appl Environ Microbiol 64:2937–2942

Edwards KJ, Schrenk MO, Hamers R, Banfield JF (1998) Microbial oxidation of pyrite; experiments using microorganisms from an extreme acidic environment. Am Mineral 83: 1444–1453

Edwards KJ, Gihring TM, Banfield JF (1999a) Seasonal variations in microbial populations and environmental conditions in an extreme acid mine drainage environment. Appl Environ Microbiol 65:3627–3632

Edwards KJ, Goebel BM, Rodgers TM, Schrenk MO, Gihring TM, Cardona MM, Mcguire MM, Hamers RJ, Pace NR, Banfield JF (1999b) Geomicrobiology of pyrite (FeS_2) dissolution: case study at Iron Mountain, California. Geomicrobiol J 16:155–179

Edwards KJ, Bond PL, Banfield JF (2000a) Characteristics of attachment and growth of *Thiobacillus caldus* on sulphide minerals: a chemotactic response to sulphur minerals? Environ Microbiol 2:324–332

Edwards KJ, Bond PL, Gihring TM, Banfield JF (2000b) An archaeal iron-oxidizing extreme acidophile important in acid mine drainage. Science 287:1796–1799

Edwards KJ, Hu B, Hamers RJ, Banfield JF (2001) A new look at microbial leaching patterns on sulfide minerals. FEMS Microbiol Ecol 34:197–206

Espejo RT, Romero P (1987) Growth of *Thiobacillus ferrooxidans* on elemental sulfur. Appl Environ Microbiol 53:1907–1912

Etzel K, Klingl A, Huber H, Rachel R, Schmalz G, Thomm M, Depmeier W (2008) Etching of {111} and {210} synthetic pyrite surfaces by two archaeal strains, *Metallosphaera sedula* and *Sulfolobus metallicus*. Hydrometallurgy 94:116–120

Farah C, Vera M, Morin D, Haras D, Jerez CA, Guiliani N (2005) Evidence for a functional quorum-sensing type AI-1 system in the extremophilic bacterium *Acidithiobacillus ferrooxidans*. Appl Environ Microbiol 71:7033–7040

Fischer CR, Wilmes P, Bowen BP, Northen TR, Banfield JF (2012) Deuterium-exchange metabolomics identifies N-methyl lyso phosphatidylethanolamines as abundant lipids in acidophilic mixed microbial communities. Metabolomics 8:566–578

Flemming H-C, Wingender J (2010) The biofilm matrix. Nat Rev Microbiol 8:623–633

Florian B, Noël N, Sand W (2010) Visualization of initial attachment of bioleaching bacteria using combined atomic force and epifluorescence microscopy. Miner Eng 23:532–535

Florian B, Noël N, Thyssen C, Felschau I, Sand W (2011) Some quantitative data on bacterial attachment to pyrite. Miner Eng 24:1132–1138

Galloway WR, Hodgkinson JT, Bowden S, Welch M, Spring DR (2012) Applications of small molecule activators and inhibitors of quorum sensing in Gram-negative bacteria. Trends Microbiol 20:449–458

García-Moyano A, González-Toril E, Aguilera A, Amils R (2007) Prokaryotic community composition and ecology of floating macroscopic filaments from an extreme acidic environment, Rio Tinto (SW, Spain). Syst Appl Microbiol 30:601–614

Gehrke T, Telegdi J, Thierry D, Sand W (1998) Importance of extracellular polymeric substances from *Thiobacillus ferrooxidans* for bioleaching. Appl Environ Microbiol 64:2743–2747

Gehrke T, Hallmann R, Kinzler K, Sand W (2001) The EPS of *Acidithiobacillus ferrooxidans*-a model for structure-function relationships of attached bacteria and their physiology. Water Sci Technol 43:159–167

Ghorbani Y, Petersen J, Harrison ST, Tupikina OV, Becker M, Mainza AN, Franzidis J-P (2012) An experimental study of the long-term bioleaching of large sphalerite ore particles in a circulating fluid fixed-bed reactor. Hydrometallurgy 129:161–171

Gleisner M, Herbert RB, Kockum PCF (2006) Pyrite oxidation by *Acidithiobacillus ferrooxidans* at various concentrations of dissolved oxygen. Chem Geol 225:16–29

Golovacheva R (1978) Attachment of *Sulfobacillus thermosulfidooxidans* cells to the surface of sulfide minerals. Mikrobiologiia 48:528–533

Goltsman DSA, Denef VJ, Singer SW, VerBerkmoes NC, Lefsrud M, Mueller RS, Dick GJ, Sun CL, Wheeler KE, Zemla A (2009) Community genomic and proteomic analyses of chemoautotrophic iron-oxidizing "*Leptospirillum rubarum*"(Group II) and "*Leptospirillum ferrodiazotrophum*"(Group III) bacteria in acid mine drainage biofilms. Appl Environ Microbiol 75:4599–4615

Goltsman DS, Dasari M, Thomas BC, Shah MB, VerBerkmoes NC, Hettich RL, Banfield JF (2013) New group in the *Leptospirillum* clade: cultivation-independent community genomics, proteomics, and transcriptomics of the new species "*Leptospirillum* group IV UBA BS". Appl Environ Microbiol 79:5384–5393

Golyshina OV, Pivovarova TA, Karavaiko GI, Kondrateva TF, Moore ER, Abraham WR, Lunsdorf H, Timmis KN, Yakimov MM, Golyshin PN (2000) *Ferroplasma acidiphilum* gen. nov., sp. nov., an acidophilic, autotrophic, ferrous-iron-oxidizing, cell-wall-lacking, mesophilic member of the *Ferroplasmaceae* fam. nov., comprising a distinct lineage of the Archaea. Int J Syst Evol Microbiol 50:997–1006

González DM, Lara RH, Alvarado KN, Valdez-Pérez D, Navarro-Contreras HR, Cruz R, García-Meza JV (2012) Evolution of biofilms during the colonization process of pyrite by *Acidithiobacillus thiooxidans*. Appl Microbiol Biotechnol 93:763–775

Gonzalez A, Bellenberg S, Mamani S, Ruiz L, Echeverria A, Soulere L, Doutheau A, Demergasso C, Sand W, Queneau Y, Vera M, Guiliani N (2013) AHL signaling molecules with a large acyl chain enhance biofilm formation on sulfur and metal sulfides by the bioleaching bacterium *Acidithiobacillus ferrooxidans*. Appl Microbiol Biotechnol 97:3729–3737

Govender Y, Gericke M (2011) Extracellular polymeric substances (EPS) from bioleaching systems and its application in bioflotation. Miner Eng 24:1122–1127

Gray MJ, Jakob U (2015) Oxidative stress protection by polyphosphate—new roles for an old player. Curr Opin Microbiol 24:1–6

Grillo-Puertas M, Schurig-Briccio LA, Rodríguez-Montelongo L, Rintoul MR, Rapisarda VA (2014) Copper tolerance mediated by polyphosphate degradation and low-affinity inorganic phosphate transport system in *Escherichia coli*. BMC Microbiol 14:72

Hall-Stoodley L, Costerton JW, Stoodley P (2004) Bacterial biofilms: from the natural environment to infectious diseases. Nat Rev Microbiol 2:95–108

Harneit K, Göksel A, Kock D, Klock J-H, Gehrke T, Sand W (2006) Adhesion to metal sulfide surfaces by cells of *Acidithiobacillus ferrooxidans*, *Acidithiobacillus thiooxidans* and *Leptospirillum ferrooxidans*. Hydrometallurgy 83:245–254

Hedrich S, Schlomann M, Johnson DB (2011) The iron-oxidizing proteobacteria. Microbiology 157:1551–1564

Henche AL, Koerdt A, Ghosh A, Albers SV (2012) Influence of cell surface structures on crenarchaeal biofilm formation using a thermostable green fluorescent protein. Environ Microbiol 14:779–793

Hengge R (2009) Principles of c-di-GMP signalling in bacteria. Nat Rev Microbiol 7:263–273

Huber G, Spinnler C, Gambacorta A, Stetter KO (1989) *Metallosphaera sedula* gen, and sp. nov. represents a new genus of aerobic, metal-mobilizing, thermoacidophilic archaebacteria. Syst Appl Microbiol 12:38–47

Imlay JA (2013) The molecular mechanisms and physiological consequences of oxidative stress: lessons from a model bacterium. Nat Rev Microbiol 11:443–454

Ivleva N, Wagner M, Horn H, Niessner R, Haisch C (2009) Towards a nondestructive chemical characterization of biofilm matrix by Raman microscopy. Anal Bioanal Chem 393:197–206

Janczarek M (2011) Environmental signals and regulatory pathways that influence exopolysaccharide production in rhizobia. Int J Mol Sci 12:7898–7933

Jiao Y, Cody GD, Harding AK, Wilmes P, Schrenk M, Wheeler KE, Banfield JF, Thelen MP (2010) Characterization of extracellular polymeric substances from acidophilic microbial biofilms. Appl Environ Microbiol 76:2916–2922

Johnson DB (1998) Biodiversity and ecology of acidophilic microorganisms. FEMS Microbiol Ecol 27:307–317

Johnson DB (2014) Biomining—biotechnologies for extracting and recovering metals from ores and waste materials. Curr Opin Biotechnol 30:24–31

Jones GC, Corin KC, van Hille RP, Harrison STL (2011) The generation of toxic reactive oxygen species (ROS) from mechanically activated sulphide concentrates and its effect on thermophilic bioleaching. Miner Eng 24:1198–1208

Jones DS, Albrecht HL, Dawson KS, Schaperdoth I, Freeman KH, Pi Y, Pearson A, Macalady JL (2012) Community genomic analysis of an extremely acidophilic sulfur-oxidizing biofilm. ISME J 6:158–170

Kanao T, Kamimura K, Sugio T (2007) Identification of a gene encoding a tetrathionate hydrolase in *Acidithiobacillus ferrooxidans*. J Biotechnol 132:16–22

Karatan E, Michael AJ (2013) A wider role for polyamines in biofilm formation. Biotechnol Lett 35:1715–1717

Kim KS, Rao NN, Fraley CD, Kornberg A (2002) Inorganic polyphosphate is essential for long-term survival and virulence factors in *Shigella* and *Salmonella* spp. Proc Natl Acad Sci U S A 99:7675–7680

Knickerbocker C, Nordstrom D, Southam G (2000) The role of "blebbing" in overcoming the hydrophobic barrier during biooxidation of elemental sulfur by *Thiobacillus thiooxidans*. Chem Geol 169:425–433

Koerdt A, Gödeke J, Berger J, Thormann KM, Albers S-V (2010) Crenarchaeal biofilm formation under extreme conditions. PLoS One 5, e14104

Koerdt A, Jachlewski S, Ghosh A, Wingender J, Siebers B, Albers S-V (2012) Complementation of *Sulfolobus solfataricus* PBL2025 with an α-mannosidase: effects on surface attachment and biofilm formation. Extremophiles 16:115–125

Labbate M, Zhu H, Thung L, Bandara R, Larsen MR, Willcox MD, Givskov M, Rice SA, Kjelleberg S (2007) Quorum-sensing regulation of adhesion in *Serratia marcescens* MG1 is surface dependent. J Bacteriol 189:2702–2711

Laishley E, Bryant R, Kobryn B, Hyne J (1986) Microcrystalline structure and surface area of elemental sulphur as factors influencing its oxidation by *Thiobacillus albertis*. Can J Microbiol 32:237–242

Lamarche MG, Wanner BL, Crepin S, Harel J (2008) The phosphate regulon and bacterial virulence: a regulatory network connecting phosphate homeostasis and pathogenesis. FEMS Microbiol Rev 32:461–473

Lawrence J, Swerhone G, Leppard G, Araki T, Zhang X, West M, Hitchcock A (2003) Scanning transmission X-ray, laser scanning, and transmission electron microscopy mapping of the exopolymeric matrix of microbial biofilms. Appl Environ Microbiol 69:5543–5554

Lawrence JR, Korber DR, Neu TR (2007) Analytical imaging and microscopy techniques. In: Hurst CJ, Crawford RL, Garland JL, Lipson DA, Mills AL, Stetzenbach LD (eds) Manual of environmental microbiology. ASM Press, Washington, DC, pp 40–68

Lei J, Huaiyang Z, Xiaotong P, Zhonghao D (2009) The use of microscopy techniques to analyze microbial biofilm of the bio-oxidized chalcopyrite surface. Miner Eng 22:37–42

Li Z, Wang Y, Yao Q, Justice NB, Ahn TH, Xu D, Hettich RL, Banfield JF, Pan C (2014) Diverse and divergent protein post-translational modifications in two growth stages of a natural microbial community. Nat Commun 5:4405

Little B, Ray B, Pope R, Franklin M, White DC (2000) Spatial and temporal relationships between localised corrosion and bacterial activity on iron-containing substrata. In: Sequeira CAC (ed)

Microbial corrosion. European Federation of Corrosion Publications; Institute of Materials, London, pp 21–35

Lo I, Denef VJ, VerBerkmoes NC, Shah MB, Goltsman D, DiBartolo G, Tyson GW, Allen EE, Ram RJ, Detter JC (2007) Strain-resolved community proteomics reveals recombining genomes of acidophilic bacteria. Nature 446:537–541

Mangold S, Harneit K, Rohwerder T, Claus G, Sand W (2008) Novel combination of atomic force microscopy and epifluorescence microscopy for visualization of leaching bacteria on pyrite. Appl Environ Microbiol 74:410–415

Manz B, Volke F, Goll D, Horn H (2003) Measuring local flow velocities and biofilm structure in biofilm systems with Magnetic Resonance Imaging (MRI). Biotechnol Bioeng 84:424–432

Marketon MM, Glenn SA, Eberhard A, González JE (2003) Quorum sensing controls exopolysaccharide production in *Sinorhizobium meliloti*. J Bacteriol 185:325–331

Marshall K, Stout R, Mitchell R (1971) Mechanism of the initial events in the sorption of marine bacteria to surfaces. J Gen Microbiol 68:337–348

Martínez P, Gálvez S, Ohtsuka N, Budinich M, Cortés MP, Serpell C, Nakahigashi K, Hirayama A, Tomita M, Soga T (2013) Metabolomic study of Chilean biomining bacteria *Acidithiobacillus ferrooxidans* strain Wenelen and *Acidithiobacillus thiooxidans* strain Licanantay. Metabolomics 9:247–257

Meyer G, Schneider-Merck T, Böhme S, Sand W (2002) A simple method for investigations on the chemotaxis of *A. ferrooxidans* and *D. vulgaris*. Acta Biotechnol 22:391–399

Mikkelsen D, Kappler U, Webb R, Rasch R, McEwan A, Sly L (2007) Visualisation of pyrite leaching by selected thermophilic archaea: nature of microorganism–ore interactions during bioleaching. Hydrometallurgy 88:143–153

Mitsunobu S, Zhu M, Takeichi Y, Ohigashi T, Suga H, Makita H, Sakata M, Ono K, Mase K, Takahashi Y (2015) Nanoscale identification of extracellular organic substances at the microbe–mineral interface by scanning transmission X-ray microscopy. Chem Lett 44:91–93

Moreno-Paz M, Gomez M, Arcas A, Parro V (2010). Environmental transcriptome analysis reveals physiological differences between biofilm and planktonic modes of life of the iron oxidizing bacteria *Leptospirillum* spp. in their natural microbial community. BMC Genomics 11:404.

Mosier AC, Justice NB, Bowen BP, Baran R, Thomas BC, Northen TR, Banfield JF (2013) Metabolites associated with adaptation of microorganisms to an acidophilic, metal-rich environment identified by stable-isotope-enabled metabolomics. MBio 4:e00484–12

Mosier AC, Li Z, Thomas BC, Hettich RL, Pan C, Banfield JF (2015) Elevated temperature alters proteomic responses of individual organisms within a biofilm community. ISME J 9:180–194

Mueller RS, Dill BD, Pan C, Belnap CP, Thomas BC, Verberkmoes NC, Hettich RL, Banfield JF (2011) Proteome changes in the initial bacterial colonist during ecological succession in an acid mine drainage biofilm community. Environ Microbiol 13:2279–2292

Murr L, Berry V (1976) Direct observations of selective attachment of bacteria on low-grade sulfide ores and other mineral surfaces. Hydrometallurgy 2:11–24

Neu T, Lawrence J (2009) Extracellular polymeric substances in microbial biofilms. In: Moran AP, Holst O, Brennan PJ, von Itzstein M (eds) Microbial glycobiology: structures, relevance and applications. Elsevier, San Diego, pp 735–758

Neu TR, Lawrence JR (2014a) Advanced techniques for in situ analysis of the biofilm matrix (structure, composition, dynamics) by means of laser scanning microscopy. In: Donelli G (ed) Microbial biofilms: methods and protocols, methods in molecular biology. Springer, New York, pp 43–64

Neu TR, Lawrence JR (2014b) Investigation of microbial biofilm structure by laser scanning microscopy. Adv Biochem Eng Biotechnol 146:1–51

Neu TR, Lawrence JR (2015) Innovative techniques, sensors, and approaches for imaging biofilms at different scales. Trends Microbiol 23:233–242

Neu TR, Marshall KC (1990) Bacterial polymers: physicochemical aspects of their interactions at interfaces. J Biomater Appl 5:107–133

Neu TR, Marshall KC (1991) Microbial "footprints"—a new approach to adhesive polymers. Biofouling 3:101–112

Nicolaus B, Manca MC, Romano I, Lama L (1993) Production of an exopolysaccharide from two thermophilic archaea belonging to the genus *Sulfolobus*. FEMS Microbiol Lett 109:203–206

Noël N, Florian B, Sand W (2010) AFM & EFM study on attachment of acidophilic leaching organisms. Hydrometallurgy 104:370–375

Nooshabadi AJ, Rao KH (2014) Formation of hydrogen peroxide by sulphide minerals. Hydrometallurgy 141:82–88

Norris PR, Johnson DB (1998) Acidophilic microorganisms. In: Horikoshi K, Grant WD (eds) Extremophiles: microbial life in extreme environments. Wiley-Liss, New York, pp 133–153

Norris PR, Burton NP, Foulis NA (2000) Acidophiles in bioreactor mineral processing. Extremophiles 4:71–76

Ohmura N, Kitamura K, Saiki H (1993) Selective adhesion of *Thiobacillus ferrooxidans* to pyrite. Appl Environ Microbiol 59:4044–4050

Orell A, Navarro CA, Arancibia R, Mobarec JC, Jerez CA (2010) Life in blue: copper resistance mechanisms of bacteria and archaea used in industrial biomining of minerals. Biotechnol Adv 28:839–848

Orell A, Fröls S, Albers S-V (2013) Archaeal biofilms: the great unexplored. Annu Rev Microbiol 67:337–354

Otto K, Silhavy TJ (2002) Surface sensing and adhesion of *Escherichia coli* controlled by the Cpx-signaling pathway. Proc Natl Acad Sci U S A 99:2287–2292

Parsek MR, Greenberg E (2005) Sociomicrobiology: the connections between quorum sensing and biofilms. Trends Microbiol 13:27–33

Peltola M, Neu TR, Raulio M, Kolari M, Salkinoja-Salonen MS (2008) Architecture of *Deinococcus geothermalis* biofilms on glass and steel: a lectin study. Environ Microbiol 10:1752–1759

Priester JH, Horst AM, Van De Werfhorst LC, Saleta JL, Mertes LA, Holden PA (2007) Enhanced visualization of microbial biofilms by staining and environmental scanning electron microscopy. J Microbiol Methods 68:577–587

Ram RJ, VerBerkmoes NC, Thelen MP, Tyson GW, Baker BJ, Blake RC, Shah M, Hettich RL, Banfield JF (2005) Community proteomics of a natural microbial biofilm. Science 308:1915–1920

Rao NN, Kornberg A (1996) Inorganic polyphosphate supports resistance and survival of stationary-phase *Escherichia coli*. J Bacteriol 178:1394–1400

Rawlings DE (2002) Heavy metal mining using microbes. Annu Rev Microbiol 56:65–91

Remonsellez F, Orell A, Jerez CA (2006) Copper tolerance of the thermoacidophilic archaeon *Sulfolobus metallicus*: possible role of polyphosphate metabolism. Microbiology 152:59–66

Rendueles O, Ghigo J-M (2012) Multi-species biofilms: how to avoid unfriendly neighbors. FEMS Microbiol Rev 36:972–989

Reuter K, Pittelkow M, Bursy J, Heine A, Craan T, Bremer E (2010) Synthesis of 5-hydroxyectoine from ectoine: crystal structure of the non-heme iron(II) and 2-oxoglutarate-dependent dioxygenase EctD. PLoS One 5, e10647

Rimstidt JD, Vaughan DJ (2003) Pyrite oxidation: A state-of-the-art assessment of the reaction mechanism. Geochim Cosmochim Acta 67:873–880

Rivas M, Seeger M, Holmes DS, Jedlicki E (2005) A Lux-like quorum sensing system in the extreme acidophile *Acidithiobacillus ferrooxidans*. Biol Res 38:283–297

Rodriguez Y, Ballester A, Blazquez ML, Gonzalez F, Munoz JA (2003) New information on the pyrite bioleaching mechanism at low and high temperature. Hydrometallurgy 71:37–46

Rodriguez-Leiva M, Tributsch H (1988) Morphology of bacterial leaching patterns by *Thiobacillus ferrooxidans* on synthetic pyrite. Arch Microbiol 149:401–405

Rohwerder T, Sand W (2003) The sulfane sulfur of persulfides is the actual substrate of the sulfur-oxidizing enzymes from *Acidithiobacillus* and *Acidiphilium* spp. Microbiology 149:1699–1710

Rojas-Chapana JA, Giersig M, Tributsch H (1996) The path of sulfur during the bio-oxidation of pyrite by *Thiobacillus ferrooxidans*. Fuel 75:923–930

Rüberg S, Pühler A, Becker A (1999) Biosynthesis of the exopolysaccharide galactoglucan in *Sinorhizobium meliloti* is subject to a complex control by the phosphate-dependent regulator PhoB and the proteins ExpG and MucR. Microbiology 145:603–611
Ruiz LM, Castro M, Barriga A, Jerez CA, Guiliani N (2011) The extremophile *Acidithiobacillus ferrooxidans* possesses a c-di-GMP signalling pathway that could play a significant role during bioleaching of minerals. Lett Appl Microbiol 54:133–139
Sampson M, Phillips C, Ball A (2000a) Investigation of the attachment of *Thiobacillus ferrooxidans* to mineral sulfides using scanning electron microscopy analysis. Miner Eng 13:643–656
Sampson MI, Phillips CV, Blake RCI (2000b) Influence of the attachment of acidophilic bacteria during the oxidation of mineral sulfides. Miner Eng 13:373–389
Sand W, Gerke T, Hallmann R, Schippers A (1995) Sulfur chemistry, biofilm, and the (in) direct attack mechanism—a critical evaluation of bacterial leaching. Appl Microbiol Biotechnol 43:961–966
Sand W, Gehrke T, Hallmann R, Schippers A (1998) Towards a novel bioleaching mechanism. Miner Process Extract Metall Rev 19:97–106
Sand W, Gehrke T, Jozsa PG, Schippers A (2001) (Bio) chemistry of bacterial leaching—direct vs. indirect bioleaching. Hydrometallurgy 59:159–175
Sanhueza A, Ferrer I, Vargas T, Amils R, Sánchez C (1999) Attachment of *Thiobacillus ferrooxidans* on synthetic pyrite of varying structural and electronic properties. Hydrometallurgy 51:115–129
Schaeffer W, Holbert P, Umbreit W (1963) Attachment of *Thiobacillus thiooxidans* to sulfur crystals. J Bacteriol 85:137–140
Schippers A (2007) Microorganisms involved in bioleaching and nucleic acid-based molecular methods for their identification and quantification. In: Donati ER, Sand W (eds) Microbial processing of metal sulfides. Springer, Dordrecht, pp 3–33
Schippers A, Sand W (1999) Bacterial leaching of metal sulfides proceeds by two indirect mechanisms via thiosulfate or via polysulfides and sulfur. Appl Environ Microbiol 65:319–321
Schippers A, Jozsa P, Sand W (1996) Sulfur chemistry in bacterial leaching of pyrite. Appl Environ Microbiol 62:3424–3431
Seeger M, Jerez CA (1993) Phosphate-starvation induced changes in *Thiobacillus ferrooxidans*. FEMS Microbiol Lett 108:35–41
Seneviratne CJ, Wang Y, Jin L, Wong SS, Herath TD, Samaranayake LP (2012) Unraveling the resistance of microbial biofilms: has proteomics been helpful? Proteomics 12:651–665
Sharma P, Das A, Rao KH, Forssberg K (2003) Surface characterization of *Acidithiobacillus ferrooxidans* cells grown under different conditions. Hydrometallurgy 71:285–292
Shrihari RK, Modak JM, Kumar R, Gandhi KS (1995) Dissolution of particles of pyrite mineral by direct attachment of *Thiobacillus ferrooxidans*. Hydrometallurgy 38:175–187
Smirnova GV, Oktyabrsky ON (2005) Glutathione in bacteria. Biochem (Mosc) 70:1199–1211
Smith K, Borges A, Ariyanayagam MR, Fairlamb AH (1995) Glutathionylspermidine metabolism in *Escherichia coli*. Biochem J 312:465–469
Solari JA, Huerta G, Escobar B, Vargas T, Badilla-Ohlbaum R, Rubio J (1992) Interfacial phenomena affecting the adhesion of *Thiobacillus ferrooxidans* to sulphide mineral surfaces. Colloid Surf 69:159–166
Staudt C, Horn H, Hempel D, Neu T (2003) Screening of lectins for staining lectin-specific glycoconjugates in the EPS of biofilms. In: Lens P, O'Flaherty V, Moran AP, Stoodley P, Mahony T (eds) Biofilms in medicine, industry and environmental technology. IWA Publishing, London, pp 308–327
Stevens AM, Queneau Y, Soulere L, Sv B, Doutheau A (2010) Mechanisms and synthetic modulators of AHL-dependent gene regulation. Chem Rev 111:4–27
Stoodley P, Sauer K, Davies D, Costerton JW (2002) Biofilms as complex differentiated communities. Annu Rev Microbiol 56:187–209
Stourman NV, Branch MC, Schaab MR, Harp JM, Ladner JE, Armstrong RN (2011) Structure and function of YghU, a nu-class glutathione transferase related to YfcG from *Escherichia coli*. Biochemistry 50:1274–1281

Tapia J, Munoz J, Gonzalez F, Blazquez M, Malki M, Ballester A (2009) Extraction of extracellular polymeric substances from the acidophilic bacterium *Acidiphilium* 3.2 Sup (5). Water Sci Technol 59:1959–1967

Taylor ES, Lower SK (2008) Thickness and surface density of extracellular polymers on *Acidithiobacillus ferrooxidans*. Appl Environ Microbiol 74:309–311

Telegdi J, Keresztes Z, Pálinkás G, Kálmán E, Sand W (1998) Microbially influenced corrosion visualized by atomic force microscopy. Appl Phys A Mater Sci Process 66:S639–S642

Tributsch H, Rojas-Chapana JA (2000) Metal sulfide semiconductor electrochemical mechanisms induced by bacterial activity. Electrochim Acta 45:4705–4716

Tuovinen O (1990) Biological fundamentals of mineral leaching processes. In: Ehrich HL, Brierley CL (eds) Microbial mineral recovery. McGraw-Hill, New York, pp 55–77

Tyson GW, Chapman J, Hugenholtz P, Allen EE, Ram RJ, Richardson PM, Solovyev VV, Rubin EM, Rokhsar DS, Banfield JF (2004) Community structure and metabolism through reconstruction of microbial genomes from the environment. Nature 428:37–43

Valdes J, Pedroso I, Quatrini R, Holmes DS (2008) Comparative genome analysis of *Acidithiobacillus ferrooxidans*, *A. thiooxidans* and *A. caldus*: insights into their metabolism and ecophysiology. Hydrometallurgy 94:180–184

Valdes J, Ossandon F, Quatrini R, Dopson M, Holmes DS (2011) Draft genome sequence of the extremely acidophilic biomining bacterium *Acidithiobacillus thiooxidans* ATCC 19377 provides insights into the evolution of the *Acidithiobacillus* genus. J Bacteriol 193:7003–7004

Valenzuela S, Banderas A, Jerez CA, Guiliani N (2007) Cell-cell communication in bacteria. A promising new approach to improve bioleaching efficiency? In: Donati ER, Sand W (eds) Microbial processing of metal sulfides. Springer, Dordrecht, pp 253–264

Vera M, Guiliani N, Jerez CA (2003) Proteomic and genomic analysis of the phosphate starvation response of *Acidithiobacillus ferrooxidans*. Hydrometallurgy 71:125–132

Vera M, Krok B, Bellenberg S, Sand W, Poetsch A (2013a) Shotgun proteomics study of early biofilm formation process of *Acidithiobacillus ferrooxidans* ATCC 23270 on pyrite. Proteomics 13:1133–1144

Vera M, Schippers A, Sand W (2013b) Progress in bioleaching: fundamentals and mechanisms of bacterial metal sulfide oxidation-part A. Appl Microbiol Biotechnol 97:7529–7541

Vogler K, Umbreit W (1941) The necessity for direct contact in sulfur oxidation by *Thiobacillus thiooxidans*. Soil Sci 51:331–338

Wagner M, Ivleva NP, Haisch C, Niessner R, Horn H (2009) Combined use of confocal laser scanning microscopy (CLSM) and Raman microscopy (RM): Investigations on EPS–matrix. Water Res 43:63–76

Wakao N, Mishina M, Sakurai Y, Shiota H (1984) Bacterial pyrite oxidation III. Adsorption of *Thiobacillus ferrooxidans* cells on solid surfaces and its effect on iron release from pyrite. J Gen Appl Microbiol 30:63–77

Waksman SA (1932) Principles of soil microbiology. Tindall & Cox, Bailliere, London

Wanner BL (1996a) Phosphorus assimilation and control of phosphate regulon. In: Neidhardt FC, Ingraham JL, Low KB, Magasanik B, Schaechter M, Umbarger HE (eds) *Escherichia coli* and *Salmonella typhimurium*: cellular and molecular biology. ASM Press, Washington, DC, pp 1357–1381

Wanner BL (1996b) Signal transduction in the control of phosphate-regulated genes of *Escherichia coli*. Kidney Int 49:964–967

Waters CM, Lu W, Rabinowitz JD, Bassler BL (2008) Quorum sensing controls biofilm formation in *Vibrio cholerae* through modulation of cyclic di-GMP levels and repression of *vpsT*. J Bacteriol 190:2527–2536

Weiss R (1973) Attachment of bacteria to sulphur in extreme environments. J Gen Microbiol 77:501–507

Wheaton G, Counts J, Mukherjee A, Kruh J, Kelly R (2015) The confluence of heavy metal biooxidation and heavy metal resistance: Implications for bioleaching by extreme thermoacidophiles. Minerals 5:397–451

Whitfield C (2006) Biosynthesis and assembly of capsular polysaccharides in *Escherichia coli*. Annu Rev Biochem 75:39–68

Wilmes P, Remis JP, Hwang M, Auer M, Thelen MP, Banfield JF (2009) Natural acidophilic biofilm communities reflect distinct organismal and functional organization. ISME J 3:266–270

Wilmes P, Bowen BP, Thomas BC, Mueller RS, Denef VJ, Verberkmoes NC, Hettich RL, Northen TR, Banfield JF (2010) Metabolome-proteome differentiation coupled to microbial divergence. MBio 1:e00246–10

Xia JL, Wu S, Zhang RY, Zhang CG, He H, Jiang HC, Nie ZY, Qiu GZ (2011) Effects of copper exposure on expression of glutathione-related genes in *Acidithiobacillus ferrooxidans*. Curr Microbiol 62:1460–1466

Xia JL, Liu HC, Nie ZY, Peng AA, Zhen XJ, Yang Y, Zhang XL (2013) Synchrotron radiation based STXM analysis and micro-XRF mapping of differential expression of extracellular thiol groups by *Acidithiobacillus ferrooxidans* grown on Fe^{2+} and S^0. J Microbiol Methods 94:257–261

Yelton AP, Comolli LR, Justice NB, Castelle C, Denef VJ, Thomas BC, Banfield JF (2013) Comparative genomics in acid mine drainage biofilm communities reveals metabolic and structural differentiation of co-occurring archaea. BMC Genomics 14:485

Zeng W, Qiu G, Zhou H, Liu X, Chen M, Chao W, Zhang C, Peng J (2010) Characterization of extracellular polymeric substances extracted during the bioleaching of chalcopyrite concentrate. Hydrometallurgy 100:177–180

Zhang R, Bellenberg S, Castro L, Neu TR, Sand W, Vera M (2014) Colonization and biofilm formation of the extremely acidophilic archaeon *Ferroplasma acidiphilum*. Hydrometallurgy 150:245–252

Zhang R, Neu T, Bellenberg S, Kuhlicke U, Sand W, Vera M (2015a) Use of lectins to in situ visualize glycoconjugates of extracellular polymeric substances in acidophilic archaeal biofilms. Microb Biotechnol 8:448–461

Zhang R, Neu T, Zhang Y, Bellenberg S, Kuhlicke U, Li Q, Sand W, Vera M (2015b) Visualization and analysis of EPS glycoconjugates of the thermoacidophilic archaeon *Sulfolobus metallicus*. Appl Microbiol Biotechnol 99:7343–7356

Zhou H, Zhang R, Hu P, Zeng W, Xie Y, Wu C, Qiu G (2008) Isolation and characterization of *Ferroplasma thermophilum* sp. nov., a novel extremely acidophilic, moderately thermophilic archaeon and its role in bioleaching of chalcopyrite. J Appl Microbiol 105:591–601

Zhu J, Li Q, Jiao W, Jiang H, Sand W, Xia J, Liu X, Qin W, Qiu G, Hu Y (2012) Adhesion forces between cells of *Acidithiobacillus ferrooxidans*, *Acidithiobacillus thiooxidans* or *Leptospirillum ferrooxidans* and chalcopyrite. Colloids Surf B 94:95–100

Zhu J, Wang Q, Zhou S, Li Q, Gan M, Jiang H, Qin W, Liu X, Hu Y, Qiu G (2015) Insights into the relation between adhesion force and chalcopyrite-bioleaching by *Acidithiobacillus ferrooxidans*. Colloids Surf B 126:351–357

Zippel B, Neu T (2011) Characterization of glycoconjugates of extracellular polymeric substances in tufa-associated biofilms by using fluorescence lectin-binding analysis. Appl Environ Microbiol 77:505–516

Zolghadr B, Klingl A, Koerdt A, Driessen AJ, Rachel R, Albers S-V (2010) Appendage-mediated surface adherence of *Sulfolobus solfataricus*. J Bacteriol 192:104–110

Chapter 7
Acidophilic Microbes: Biology and Applications

Archana Sharma, Deepak Parashar, and Tulasi Satyanarayana

1 Introduction

Microorganisms account for most of the diversity of life on the planet Earth. A variety of microbes colonize extreme environments that exist on the Earth such as geothermal areas, polar regions, acid and alkaline springs and the cold pressurized depths of the oceans. The exact definition of extreme environment is debatable. It is an environment where a restricted range of microbes survive and grow. The extreme environments vary in some aspects from those which humans consider as 'normal', where moderate conditions exist with pH around neutral, temperatures ranges between 20 °C and 40 °C, pressures around 0.1 MPA (1 atm), and adequate levels of nutrients and salts (Satyanarayana et al. 2005; Thiel 2011). Extreme environments harbor well adapted organisms, called extremophiles, which can survive in the presence of heavy metals, acids, petroleum and natural gas; these are toxic to human beings. Microorganisms capable of growth in harsh environments increased scientific curiosity to understand the mechanisms that allow their growth in extreme environments. The extremophiles as well as their novel products could be useful in biotechnological processes.

Natural and man-made environments with varying degrees of acidity are present on the Earth. The majority of areas with pH lower than 3.0 are those where comparatively huge amounts of sulphur or pyrite are exposed to oxygen. Both sulphur and pyrite are oxidized abiotically through an exothermic reaction where the former is oxidized to sulphuric acid, and the ferrous iron in the latter to ferric iron.

A. Sharma
Department of Biophysics, University of Delhi, New Delhi, India

D. Parashar • T. Satyanarayana (✉)
Department of Microbiology, University of Delhi,
South Campus, Benito Juarez Road, New Delhi 110021, India
e-mail: tsnarayana@gmail.com

P.H. Rampelotto (ed.), *Biotechnology of Extremophiles*, Grand Challenges in Biology and Biotechnology 1, DOI 10.1007/978-3-319-13521-2_7

Both processes take place abiotically, and are accelerated 10^6 times by the activity of acidophiles. Mostly the acidic pyrite areas have been generated by mining and are usually formed around coal and sulphur mines. These areas have very high sulphide concentrations and pH values as low as 1.0, and are very low in organic matter and quite toxic due to high concentrations of heavy metals. In all acidic niches, the acidity is mainly due to sulphuric acid. The spontaneous combustion leads refuse piles to undergo self-heating and offers the high temperature environment needed to sustain thermophiles. The illuminated regions, such as mining outflows and tailings, support the growth of phototrophic algae (Satyanarayana et al. 2005). In this spectrum of acidic environments, the clear boundary between extreme and moderate acidophily is open to debate. Johnson (1998, 2008) defined extreme acidophiles as those organisms that grow optimally at pH 3 or less. This definition allows inclusion of a number of autotrophic and heterotrophic organisms present in three major biological lineages, *Archaea*, *Bacteria*, and *Eukarya*. These are known to thrive in natural acidic environments like sulfuric pools, solfataric fields and artificial environments such as areas associated with mining of coal and metal ores. Most known thermoacidophiles are archaea, which normally thrive in the most extreme acidic niches.

Acidophiles play an important role in biomining of metals from the low grade ores and the enzymes produced by them have found several applications in food and feed industries. The other applications of acidophiles include bioremediation and electricity generation. The pictorial representation of the applications of acidophiles is shown in Fig. 7.1. The metagenomic and metatranscriptomic analysis of AMD microorganisms provide hints for the microbial reaction and adaptation mechanisms in the oligotrophic and extremely acidic environments (Chen et al. 2014). The recent developments on the biology and applications of acidophilic microbes are reviewed in this chapter.

2 Sources of Acidophilic Microorganisms

There are many environmental niches on the Earth which are naturally acidic. These sites are home to a variety of acidophilic microbes with unique adaptations for survival in hostile low pH environments. Solfataric fields and geothermal sulfur rich acidic sites have been found to be rich in acidophiles such as *Sulfolobus solfataricus* (Yellowstone National Park, USA) (Jeffries et al. 2001), *Sulfolobus hakonensis* (geothermal area in Hakone, Japan) (Takayanagi et al. 1996).

Acidianus sulfidivorans (Lihir Island, Papua New Guinea) (Plumb et al. 2007), *Vulcanisaeta thermophila* (Mayon volcano in the Republic of the Philippines) (Yim et al. 2015). *Picrophilus torridus* and *Picrophilus oshimae* isolated from solfataric locations in Northern Japan are moderately thermophilic heterotrophic archaea reported to have the lowest pH optima for growth (pH 0.7) among all known non-mineral oxidizing acidophilic microbes (Schleper et al. 1996). Among the acidophilic archaea, *Picrophilus*, *Sulfolobus*, *Metallosphaera* and *Sulfurococcus* are

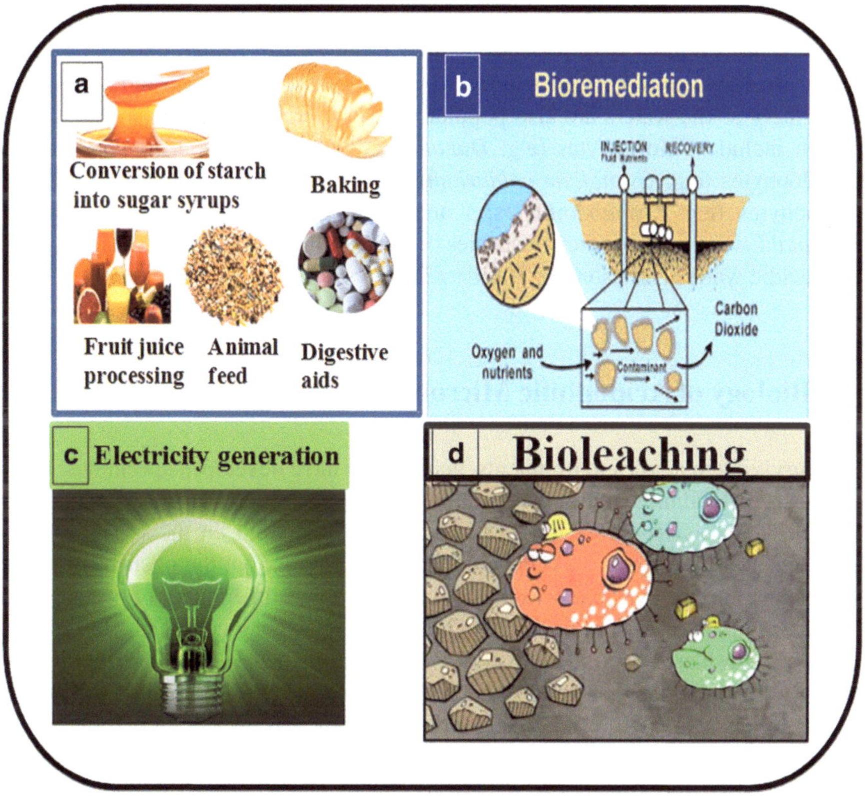

Fig. 7.1 Potential applications of acidophiles and their enzymes

obligate aerobes, while *Thermoplasma* and *Acidianus* are facultative anaerobes, and *Stygiolobus azoricus* is an obligate anaerobe (Johnson 1998).

There are many reports of mesophilic and thermophilic acidophiles, but very few investigations have been focused on psychrophilic and psychrotolerant acidophiles, although a number of extremely acidic, low temperature sites such as subterranean mine waters in the mid-high latitudes are present. Many iron oxidizers and heterotrophic isolates are known to be psychrotolerant, but none of them are truly psychrophilic (Berthelot et al. 1994).

Acid mine drainages (AMD) are areas associated with mining of metals and coal. The well studied AMD are the Rio Tinto River in Spain, Killingdal mine dump and King's mine stream in Norway, Parys mine in Wales, and Iron Mountain in California. Bacteria belonging to the genera *Leptospirillum, Acidithiobacillus, Ferroplasma* and *Alicyclobacillus* are the predominant microbes of AMD. Many types of yeast such as *Rhodotorula, Candida, Cryptococcus,* and *Trichosporon*are dominant in AMD. *Acontium velatum* (pH range 0.2–0.7) *and Scytalidium acidophilum* (pH 0) are copper-tolerant acidophilic fungi reported from AMD (Schleper et al. 1995).

Among protozoa, *Urotricha, Vorticella* and *Oxytricha* are the residents of acid rich mine water. These acidophilic/acid-tolerant protozoa use chemolithotrophic iron/sulfur-oxidizers and heterotrophic acidophilic bacteria act as source of food. Extremely acidic waters are also populated with microalgae as primary producers which include Chlorophytes (e.g. *Dunaliella acidophila, Chlamydomonas* spp.), Rhodophytes (e.g. *Cyanidium caldarium*), Chrysophytes (e.g. *Ochromonas* spp.), Dinophytes (e.g. *Gymnodinium* sp.) and Euglenoids (e.g. *Euglena mutabilis*). *E. mutabilis* is a useful indicator species of AMD pollution and is dominant in metal rich acidic waters (Johnson and Hallberg 2003).

3 Biology of Acidophilic Microbes

For surviving in acidic environments, acidophiles have developed specialized strategies for maintaining neutral intracellular pH, which implies that a gradient of several pH units exists across the cell membrane. This could, in theory, be used to generate a large amount of ATP using the F0F1 ATPase, but the unchecked influx of protons into the cytoplasm would impair the normal functioning of proteins and nucleic acids due to rapid acidification, eventually causing cell death. In order to prevent disturbances in vital intracellular processes such as DNA transcription, protein synthesis and enzyme activities, a mechanism is required to balance the proton influx by an equivalent efflux. Acidophiles are believed to use a combination of mechanisms such as a proton impermeable cell membrane, reversed membrane potential and cytoplasmic buffering (Baker-Austin and Dopson 2007) (Fig. 7.2).

The cell membrane acts as the primary defense against low pH for acidophiles thriving in acidic environments. The cell membranes of acidophiles differ greatly from neutrophiles in fatty acid and lipid composition. The relatively high proportion of archaea among acidophiles is partially attributable to the low permeability of archaeal membranes to protons. Archaeal membranes are characterized by the presence of tetraether lipids (Batrakov et al. 2002; Pivovarova et al. 2002; Shimada et al. 2002; Macalady et al. 2004). These ether linkages are less susceptible to acid hydrolysis as compared to the ester linkages found in bacterial and eukaryotic membrane lipids (Golyshina and Timmis 2005). Studies on liposomes derived from *P. oshimae* membrane lipids indicate that the impermeability to protons might also be due to the presence of rigid monolayer preventing splitting of the membrane and the bulky isoprenoid core (Van de Vossenberg et al. 1998a, b). Some archaea such as *Ferroplasma* and *Thermoplasma* are cell wall less microbes (Golyshina et al. 2000). The cell membrane of *Thermoplasma* spp. does not have sterols, glycoproteins and lipoglycans. The detailed analysis of membrane structure of archaea revealed the important role of membrane lipids in maintaining constant pH inside the cell, membrane proton gradient and proper functioning of membrane ATP synthase.

Acidophiles use a reverse membrane potential (positive $\Delta\Psi$) as a pH homeostasis mechanism, as opposed to the negative $\Delta\Psi$ seen in neutrophiles. This positive $\Delta\Psi$ is generated in acidophiles by the Donnan potential of positively charged ions,

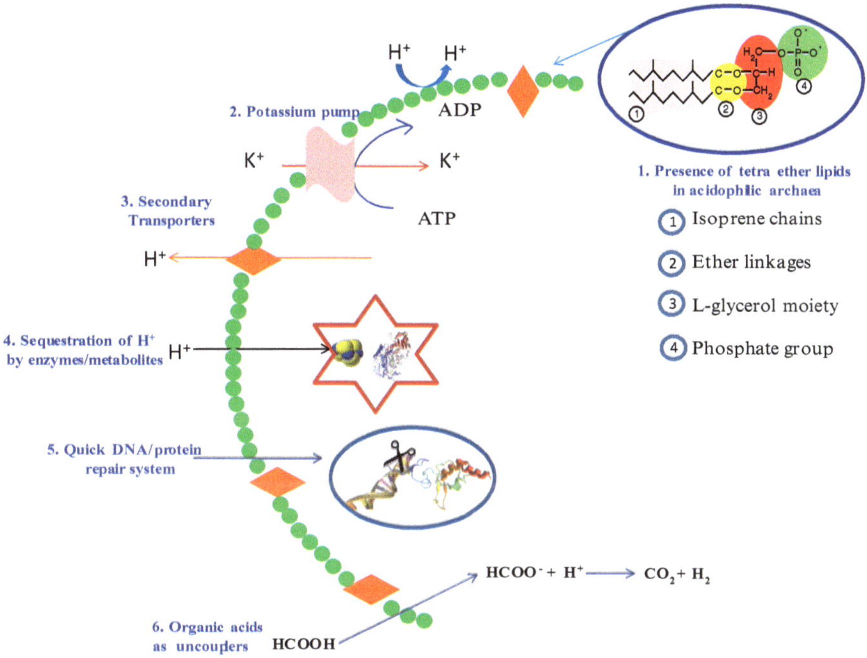

Fig. 7.2 Adaptations of acidophiles in acidic environments

usually potassium, which inhibits the entry of protons into the cell. The genome analysis of acidophiles such as *P. torridus* (Futterer et al. 2004), *F. acidarmanus, S. solfataricus* (Jeffries et al. 2001) and *Leptospirillum* (Tyson et al. 2004) suggested the presence of disproportionately high number of putative cation transporters, possibly involved in the generation of Donnan potential (Dopson et al. 2004).

In spite of highly impermeable cell membrane of acidophiles, if there is any sudden influx of protons, the buffering capacity of the cytoplasm can sequester those protons and prevent ensuing damage. All acidophiles contain cytoplasmic buffering molecules which have basic amino acids such as lysine, histidine, arginine and others that are capable of capturing protons. A comparison between the cytoplasmic buffering capacity of the extremophile *A. acidophilum* and neutrophile *E. coli* suggested that the latter has more buffering capacity (Zychlinsky and Matin 1983). This suggests that pH homeostasis in acidophiles through cytoplasmic buffering does not imply higher buffering capacity than neutrophiles. Other buffering molecules include dihydrogen phosphate ion and potassium (Spijkerman et al. 2007).

Genome sequence analysis of most of the acidophiles like *Ferroplasma, Leptospirillum, Acidithiobacillus ferrooxidans, A. thioxidans, A. caldus* confirmed the presence of putative proton efflux system that includes H^+ATPases, antiporters and symporters (Tyson et al. 2004). Abundance of secondary transporters has been reported from acidophiles. Active secondary transporters are membrane proteins that uses transmembrane electrochemical gradient of protons or sodium ions to drive transport. In *P. torridus* and *T. acidophilum*, these secondary transporters are present in huge numbers, representing the adaptation of these acidophiles to low pH (Futterer et al. 2004).

Generally heterotrophic acidophiles are capable of degrading organic acids such as acetic and lactic acid (Alexander et al. 1987; Ciaramella et al. 2005). These acids are deleterious to acidophiles as they act as uncouplers of respiratory chain at acidic pH. These are protonated acids/conjugated bases which have dissociable protons that can easily pass across the cell membrane (Baker-Austin and Dopson 2007). Genes encoding the enzymes of organic acid degradation pathways are present in the genomes of extreme acidophiles; its association with low pH is, however, unclear. Interestingly, all acidophiles capable of growing at extreme acidic pH are heterotrophs and are efficient in degrading organic acids.

Chaperones are proteins that are involved in the proper refolding of other proteins (Crossman et al. 2004). Interestingly, in acidophiles, a high expression of heat shock proteins/chaperones has been reported. These chaperones enable the rapid and efficient repair of damaged proteins (Laksanalamai and Robb 2004).

4 Molecular Adaptations of Acidstable Proteins

Acidophilic enzymes possess properly folded structure and stability in acidic environments and are catalytically active at pH as low as 1.0. The adaptation is necessary for proteins to function at low pH, since acid hinders with the charge on amino acid residues, which may destabilize the native structure of proteins. The exact adaptation of acidstable proteins has not been clearly understood, but the presence of acidic amino acids (negatively charged at a neutral pH) on the surface of these enzymes appear to enable them to function at low pH. The presence of numerous glutamic and aspartic surface residues on the modeled endo-β-glucanase from *S. solfataricus* generates a high negative surface charge at neutral pH, a significant adaptation of acidstable enzymes at low pH. Numerous acidic residues also correspond to a lower isoelectric point (pI) for the endo-β-glucanase. However neutral β-glucanases from *S. solfataricus* also have a similar isoelectric point to that of acidstable β-glucanases, although the former exhibits optimal activity at neutral to slightly acidic pH. This signifies that the presence of large number of acidic surface residues cannot be the only factor in determining the acid stability of endoβ-glucanase (Huang et al. 2005).

Acidstable α-amylase from *A. acidocaldarius* possesses a reduced density of both positive and negative charges on the surface of the protein; this avoids the electrostatic repulsion of charged groups at acidic pH and may be considered an adaptation for acid stability (Schwermann et al. 1994).

The analysis of several proteins from *F. acidiphilum* suggested the presence of enzymes active at pH lower than its cytoplasmic pH. This may be due to the intracellular compartmentalization of these enzymes and the pH gradient that exists within the cytoplasm. Another possible reason for this is that these enzymes form multienzyme complexes which increase the pH optima closer to that of cytoplasm. The proteome analysis of *F. acidiphilum* indicated the presence of a high proportion of iron proteins that contributes to the pH stability of enzymes. This iron functions as an 'iron rivet' that stabilizes the 3D structure of the protein (Golyshina and Timmis 2005).

5 Metagenomic and Metatranscriptomic Analysis of Acidophiles

A total of 56 draft or completely sequenced genomes of acidophiles are reported that includes 30 bacteria and 26 archaea. Availability of multiple genome sequences permits the prediction of metabolic and genetic interactions among the members of the bioleaching microbial community and the analysis of main evolutionary developments that shape genome architecture and evolution. Chen et al. (2014) studied the comparative metagenomics and metatranscriptomics analysis of microbial collections from geochemically distinct AMD sites. The species of *Acidithiobacillus*, *Leptospirillum* and *Acidiphilium* are predominantly present in the microbial communities and exhibit high transcriptional activities.

The comparative analyses of microbial community of AMD showed that the microorganisms are adapted to various environmental conditions through regulating the expression of genes playing roles in multiple *in situ* functional activities such as low pH adaptation, assimilation of carbon, nitrogen and phosphate, energy generation, environmental stress resistance and many other functions. The comparative analysis of acidophiles revealed diverse strategies employed by *Acidithiobacillus ferrivorans* and *Leptospirillum ferrodiazotrophum* in nutrient assimilation and energy generation for survival under different conditions. Ram et al. (2005) analyzed the microbial biofilm community of AMD sites and reported the dominance of proteins involved in protein refolding and response to oxidative stress. This suggests that damage to biomolecules is a main challenge in the survival of microorganisms in the extreme environments.

6 Acidophilic Bacteria and Archaea in Microbiohydrometallurgy

Most of the extremely acidic environments are the result of human activity such as mining of metals and coal. The microbial dissimilatory oxidation of elemental sulfur, reduced sulfur compounds (RSCs) and ferrous iron generates acidity.

Elemental sulfur is mainly found in geothermal areas where it is formed by the condensation of sulfur dioxide and hydrogen sulfide

$$SO_2 + 2H_2S \rightarrow 2H_2O + 3S^o$$

This elemental sulfur is oxidized by autotrophic and heterotrophic microorganisms to sulphuric acid. If this sulphuric acid is not further neutralized by basic minerals, it results in the generation of acidity.

$$S^o + H_2O + 1.5O_2 \rightarrow H_2SO_4$$

Iron disulfide mineral pyrite is the most common and abundant sulphidic mineral on the earth and is also associated with other metal sulfide ores. Ferric iron, a potent oxidizing agent, oxidizes the sulfur associated with minerals to thiosulphate and is reduced to ferrous iron.

$$FeS_2 + 6Fe^{3+} + 3H_2O \rightarrow 7Fe^{2+} + S_2O_3^{\;2-} + 6H^+$$

This ferrous iron is oxidized by a variety of acidophilic bacteria and archaea.

$$4Fe^{2+} + O_2 + 4H^+ \rightarrow 4Fe^{3+} + 2H_2O$$

Thiosulphate is unstable in acidic liquors in the presence of ferric iron, and is further oxidized to other reduced inorganic sulfur compounds (RISCs) such as trithionate ($S_3O_6^{2-}$), tetrathionate ($S_4O_6^{2-}$) and elemental sulphur (S^0). These RISCs are used as substrates by sulfur oxidizing bacteria and archaea.

Most of the iron sulfur oxidizing acidophiles is autotrophs. The two most well studied acidophilic microorganisms that oxidize iron/sulfur are *A. ferrooxidans* and *A. thiooxidans*. These are autotrophic chemolithotrophs, use inorganic electron donors and fix CO_2. They are generally isolated from rivers, canals, and acidified sulfate soils apart from acid mine drainage (AMD) sites.

Prokaryotes that carry out the dissimilatory oxidation of iron and/or RSCs are either mixotrophic or obligately heterotrophic. A number of heterotrophic microorganisms are reported from most extremely acidic environments. Obligately acidophilic heterotrophs are distributed among archaea, bacteria, fungi, yeasts and protozoa. Some mesophilic prokaryotes like *Ferromicrobium acidophilum* and *L. ferrooxidans* are iron oxidizers and play a direct role in the dissimilatory oxidoreduction of iron (Pronk and Johnson 1992; Johnson 1998). The extremely thermophilic iron oxidizing acidophiles include *Acidianus brierleyi*, *A. infernus*, *A. ambivalens*, *Metallosphaera sedula* and *Sulfurococcus yellowstonii*. Among sulfur oxidizers *Sulfolobus shibatae* (mixotrophic), *S. Solfataricus* (mixotrophic), *S. hakonensis* (mixotrophic), *S. metallicus* (autotrophic), *Metallosphaera prunae* (mixotrophic) and *Sulfurococcus mirabilis* (mixotrophic) are grouped within the thermophilic acidophiles (Johnson 1998).

Microorganisms have had a significant impact on the extraction and recovery of metals from ores and wastes for a long time, but their roles were not recognized. During eighteenth–nineteenth century, 'precipitation ponds' at the Rio Tinto mine (southern Spain) and the Parys mine (Anglesey, north Wales) were constructed to recover copper from leached rocks. In the middle of the twentieth century, bacterially mediated dissolution of metal-containing sulfide minerals was discovered, leading to the concept of 'biomining', i.e. the biotechnological application of microbes to mining processes (Rawlings and Johnson 2007; Johnson 2008).

The bioprocessing of ores and concentrates has several advantages over conventional approaches such as pyrometallurgy. The major benefits include the ability to process low grade ores and the much lower energy inputs required. Environmental benefits of bioprocessing are the significantly lower harmful wastes, recovery of metals from metallurgical waste and reduced production of chemically active tailings (Johnson 2008).

Sulfide minerals can be processed through bioleaching that results in the solubilization of target metals (e.g. copper from chalcopyrite and covellite) or through biooxidation, which is used for dissolution of pyrite and arsenopyrite associated with fine-grain gold, allowing extraction of the precious metal by cyanidation (Johnson 2008). Many commercial methods are known for bioleaching which includes *in situ* dump, heap and vat leaching. The *in situ* leaching involves pumping of solution and air under pressure into the mine or ore bodies made permeable by explosive charging. The metal enriched solutions are recovered through wells drilled below the ore body (Bampton et al. 1983; Brierley and Brierley 2000). In dump leaching, the uncrushed waste rocks are used. These dumps contain very low amount of copper (0.1–0.5 %) which is very difficult to recover by the conventional methods. Heap leaching requires preparation of the ore such as size reduction so that the mineral-lixiviant interaction increases and formation of impermeable base to prevent lixiviant loss and pollution of water bodies (Rawlings 1997). A lixiviant is a liquid medium used in hydrometallurgy to selectively extract the desired metal from the ore or mineral. It assists in rapid and complete leaching. The metal is then recovered after leaching in the concentrated form. In both dump and heap leaching, lixiviant is applied at the top of the dump and on the surface of the heap and metal rich solution is recovered. On the top of the dump, dilute sulphuric acid is sprinkled which percolates through the dump. This decreases the pH and promotes the growth of acidophiles. The acid run off is collected at the bottom of the dump and sent to recovery stations. From these acid run offs, metal is extracted by various methods such as cementation, solvent extraction and electrowinning. Vat leaching is applied to oxide ores that involves dissolution of crushed material in a tank or bioreactors (Siddiqui et al. 2009) (Fig. 7.3).

Biomining has been harnessed to extract copper, gold, uranium and cobalt, and other metals, including nickel and zinc. The detailed biomining methodology of some important metals like copper, gold and uranium are discussed below.

6.1 Copper

Copper ores such as chalcopyrite (Cu_2S) or covellite (CuS) are crushed, acidified with sulfuric acid, agglomerated in rotated drums to bind fine material to coarse particles before piling in heaps (Schnell 1997). When iron containing solution is passed through the heap, acidophilic microbes that grow on the surface of the ore

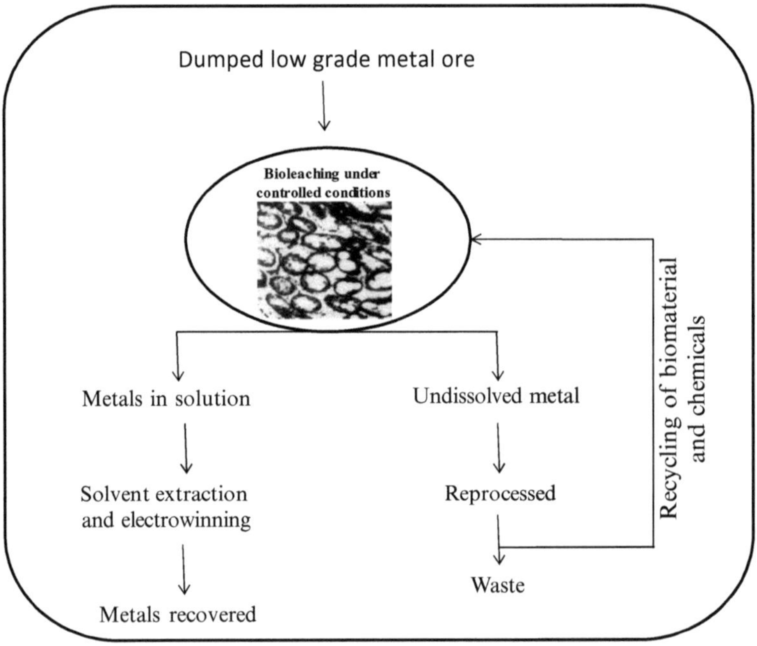

Fig. 7.3 Steps involved in bioleaching of metals from low grade ores

and in solution generate ferric iron that plays an important role in the production of copper sulfate. The soluble copper and iron is collected and pumped to recovery plant where copper is finally recovered.

6.2 Gold

In ores known as refractory, gold particles are covered by insoluble sulphides. Gold is recovered from ores by solubilization with the cyanide solution. In biooxidation process, bacteria partly oxidize the sulfur covered gold microparticles in the ores and concentrates (Dew et al. 1997). Initially bacteria catalyze breakdown of the mineral arsenopyrite (FeAsS) by oxidizing the sulphur and metal to higher oxidation states, while reducing dioxygen by H_2 and Fe^{3+}. This permits the dissolution of soluble products.

$$FeAsS_{(s)} \rightarrow Fe^{2+}{}_{(aq)} + As^{3+}{}_{(aq)} + S^{6+}{}_{(aq)}$$

This process is carried out at the bacterial cell membrane, electrons pass into the cell are utilized in biochemical processes to generate energy to reduce oxygen molecule to water. In second stage, bacteria oxidize ferrous (Fe^{2+}) to ferric ions (Fe^{3+}). Further it oxidizes metal to a higher positive oxidation state. With the gain of electrons, Fe^{3+} is further reduced to Fe^{2+} in a continuous cycle. Finally the gold gets separated from ore and it is recovered. The gold recovery prócess is higher following biooxidation (Siddiqui et al. 2009).

6.3 Uranium

The recovery of uranium is similar to that of copper. Uranium is recovered by the conversion of insoluble uranium oxides to soluble sulfates by the production of ferric iron and sulphuric acid by the microbes (Siddiqui et al. 2009).

$$UO_2 + Fe_2(SO_4)_3 \rightarrow UO_2SO_4 + 2FeSO_4$$

$$UO_3 + H_2SO_4 \rightarrow UO_2SO_4 + H_2O$$

Metals that are present in an insoluble reduced sulphur form such as NiS, ZnS and cobalt containing pyrite turns soluble when oxidized to a sulfate, which can be recovered by biomining. Lead is recovered from lead acetate containing solution and the solution is then recycled to further leaching of lead sulphidic minerals or lead sulfide containing particles (Geisler and Pudington 1996). Biomining helps in the recovering of metals from many low-grade ores that would be regarded as waste, and its application mainly depends on the significance of the metal to be recovered (Rawlings 2002). A main challenge is to discover appropriate match between an ore body and biomining technology and to recognize correct concentration and size that results in economic recovery.

7 Need for Acidstable Enzymes

Many enzymes are explored from acidophilic bacteria and archaea. The properties of enzymes derived from thermoacidophiles show activity at low pH and high temperature, which are of potential significance in many industrial applications such as starch, fruit juice, feed and baking industries. Amylolytic, xylanolytic and proteolytic enzymes, cellulases, acid phosphatases and maltose binding proteins have been reported from acidophiles (Sharma et al. 2012). Some of these important enzymes and the scope of application are discussed below.

7.1 Starch Industry

Starch is a primary source of various sugar syrups that provides basis for several pharmaceutical and confectionary industries. Amylases are one of the most important enzymes with wide applications in starch saccharification, baking, paper and textile industries. Of these, starch saccharification and baking can benefit greatly from the use of acidic α-amylases. The α-amylases presently used in starch processing are optimally active at 95 °C and pH 6.8 and are stabilized by Ca^{2+}. Therefore, the industrial processes by using these enzymes cannot be carried out at low pH (3.2–4.5), the pH of native starch (Shivaramakrishnan et al. 2006; Sharma et al. 2012). In order to be well-suited with the optimal pH of the enzyme required for liquefaction, the pH of the starch slurry is increased from its native pH 3.2–4.5 to 5.8–6.2, and further, Ca^{2+} is supplement to increase the activity and/or stability of enzyme. The next saccharification step again needs pH adjustment to pH 4.2–4.5. Both of these steps (pH adjustment and salts removal) should be excluded, as they are time consuming and increase the cost of the products. Extremozymes from extremophiles are, therefore, needed that are naturally possess properties required for specific industrial applications (Sharma et al. 2012). The α-amylase from *A. acidocaldarius* is the first example of heat and acid stable protein with a pH and temperature optima of 3.0 and 75 °C, respectively (Matzke et al. 1997; Bertoldo et al. 2004). There are very few reports of thermo-acidstable α-amylases (Bai et al. 2012; Sharma and Satyanarayana 2010; Liu and Xu 2008). Acidstable α-amylases from *A. acidocaldarius* and *Bacillus acidicola* are useful in starch industry (Sharma and Satyanarayana 2012; Bai et al. 2012). Another category of amylolytic enzymes used in starch industry are glucoamylases. These are known from thermoacidophilic archaea such as *P. torridus*, *P. oshimae*, and *T. acidophilum*. Archaeal glucoamylases are reported to be optimally active at pH 2.0 and 90 °C, on the other hand glucoamylases produced by fungi, yeast and bacteria are optimally active at 70 °C in the pH range between 3.5 and 6.

7.2 Baking Industry

While the baking industry also uses amylase, it requires different properties than those required by the starch industry. Maltogenic amylases with intermediate thermostability are desired in baking. The maltogenic nature has antistaling effect while intermediate thermostability leads to inactivation of the enzyme at the end of baking, preventing residual enzyme activity and product deterioration. Acidic α-amylase of *B. acidicola,* an acidophilic bacterium, is optimally active at 4.5, and thus, useful in baking.

In the recent years, there are several applications for xylanases. The key role of xylanases in baking is the breakdown of hemicellulose present in wheat flour and redistribution of water, leaving the dough soft and easy to knead. Its supplementation in dough helps in absorption of water, resistance to fermentation and increase in the volume of bread. Since the pH of dough is acidic, acidstable xylanases are needed in baking. Shah et al. (2006) reported the use of acid stable xylanases (optimum pH 5.3) from acidophilic fungus *Aspergillus foetidus* as bread improver for making whole wheat bread.

7.3 Fruit Juice Industry

The production of fruit and vegetable juice needs methods of extraction, clarification and stabilization. In the past, when the production of citrus fruit juices started, the yields were low due to filtration problems and turbidity. Nowadays with the use of enzymes such as pectinases, xylanases, α-amylases, cellulases and others, the yields of fruit juice have improved due to the reduction in viscosity and turbidity, along with enhanced recovery of aroma, essential oils, vitamins and mineral salts.

Among all enzymes used in fruit juice processing, the most important is pectinases. It causes the degradation of pectin, which is a structural polysaccharide present in the middle lamella and the primary cell walls of young plant cells. Acidic pectinases are used commercially in the production of clear juices of apple, pear and grapes. Pectic enzymes with high levels of polygalacturonase activity are used tin stabilizing the cloud of citrus juices, prune juice, tomato juice, purees, nectars and unicellular products. Unicellular products are produced by the transformation of organized tissues into a suspension of whole cells, and the product formed is used as the base material for pulpy juices and nectars, baby foods, components for dairy products such as pudding and yogurt. This process is known as maceration, and the enzymes used for this are known as 'macerases', usually a combination of cellulases, hemicellulases and pectic enzymes. Acidic pectinases used in fruit juice processing industry and wine making mainly come from fungal sources, especially from *A. niger*. The other sources of acidstable pectinases are listed in Table 7.1.

7.4 Animal Feed

Animal feed supplementation with enzymes such as xylanases, amylases, cellulases, pectinases, phytases and proteasescauses reduction in unwanted residues such as phosphorus, nitrogen, copper and zinc in the excreta which plays important role in reducing environmental contamination. Among these enzymes, acidic xylanases and phytases are significantly used in animal feeds. Xylanases are added in animal feeds to hydrolyze arabinoxylans present in the feed. This arabinoxylan is found in the cell walls of grains and shows anti-nutrient effect in the poultry. Phytases are another group of enzymes which are used in animal feeds. Microbial phytases are mainly added to animal (swine and poultry) and human feed and foodstuffs to improve mineral bioavailability and food processing. Xylanases and phytases are generally reported from fungi and yeast. There are very few reports of xylanases and phytases from acidophilic bacteria and archaea. Some of the reports of acid-stable xylanases include production of enzyme from *S. solfataricus* that displayed activity on carboxymethylcellulose with optimum activity at pH 3.5 and 95 °C. Another report of xylanases is from *Acidobacterium capsulatum* which exhibits optimum activity and stability in the acidic range (Inagaki et al. 1998). To the best of our knowledge, there is no report of phytases from acidophilic bacteria and

Table 7.1 Sources of industrially important acidstable enzymes

Enzymes	Organisms	pH	References
α-Amylases			
	Alicyclobacillus acidocaldarius	3.0	Matzke et al. (1997)
	Bacillus acidicola	4.0	Sharma and Satyanarayana (2010)
	Bacillus sp. YX1	5.0	Liu and Xu (2008)
Glucoamylases			
	Thermoplasma acidophilum	2.0	Serour and Antranikian (2002)
	Picrophilus torridus	2.0	Serour and Antranikian (2002)
	P. oshimae	2.0	Serour and Antranikian (2002)
Proteases			
	Xanthomonas sp.	2.7	Oda et al. (1987a)
	Pseudomonas sp.	3.0	Oda et al. (1987b)
	Sulfolobus acidocaldarius	2.0	Murao et al. (1988)
	Thermoplasma volcanium	3.0	Fusek et al. (1990)
Endo-glucanases			
	A. acidocaldarius	4.0	Eckert and Schneider (2003)
Pectinases			
	Aspergillus niger CH4	4.5–6.0	Acuna-Arguelles et al. (1995)
	Penicillium frequentans	4.5–4.7	Borin et al. (1996)
	Sclerotium rolfsii	3.5	Channe and Shewal (1995)
	Rhizoctonia solani	4.8	Marcus et al. (1986)
	Mucor pusillus	5.0	Al-Obaidi et al. (1987)
Xylanases			
	A. foetidus	5.3	Shah et al. (2006)
	A. awamori	5.0	Do et al. (2012)
Phytases			
	A. niger	5.0	Soni et al. (2010)

archaea. There is a need to explore bacterial acidic phytases that can be used in animal feed, as they have higher substrate specificity and better catalytic efficiency than the fungal phytases (Rodriguez et al. 1999; Kim et al. 2003).

7.5 Pharmaceutical Industry

Aspartic proteases, also known as carboxyl group proteases, corresponds to the group of proteolytic enzymes that digest proteins and peptides in acidic solutions. These are reported from various organisms such as mammals, fungi, plants, and retroviruses and recently in archaea and bacteria. Acidic proteases have significant applications in food, beverage and pharmaceutical industries. The existence of two aspartate residues at the active site (Asp32 and Asp215, according to pepsin numbering) shifts the optimum pH of these enzymes in the low pH range (Davies 1990). Thermopsin is an acid protease from *S. acidocaldarius*, which lacks aspartyl residue in the active site, is optimally active at pH 2.0 and 90 °C.

Collagenase is another class of proteases involved in proteolytic degradation of collagen. This collagen is used as non-allergic preservative for medicine and cosmetics (Gaffney et al. 1996; Honda 1998). Collagenase with pH optimum in the acidic range has been reported from *Bacillus* strain NTAP-1 and *Alicyclobacillus sendaiensis* NTAP-1, both are acidophiles (Nakayama et al. 2000).

8 Commercialized Acidstable Enzymes

The history of enzymes started in 1811 when the first starch hydrolyzing enzyme was discovered by Kirchhoff (Gupta et al. 2003). The story of commercialisation of enzymes began in 1830 when the first enzyme diastase was available in market for the production of dextrins in bakeries, beer and wine from fruits in France in 1830. Later Christian Hansen in 1874 in Denmark started the first company (Christian Hansen's Laboratory) which produced the rennet for cheese making (Chandel et al. 2007). In 1894, amylase from a fungal source was made available for use as a digestive aid (Pandey et al. 2000). But the enzymes gained the status of a household commodity when microbial proteases were introduced as detergent additives. The first bacterial protease was marketed in 1959, and its market value rose when Novozyme company began manufacturing it (Leisola et al. 2002).

Survey suggests that global market for enzymes in 2013 was USD 4.4 billion and is expected to rise to USD 7.65 billion by 2020 (http://www.bccresearch.com). At present, among all commercial enzymes available, carbohydrases dominates the market because of their diverse applications in the food and beverage industry, followed by proteases that accounted for 27 % of global market in 2013 (http://www.grandviewresearch.com). The enzymes are gaining importance because they not only reduce the cost of the products but benefit the environment.

Most enzymes available commercially for industrial applications work best around neutral pH and moderate temperatures. The extreme conditions such as low pH make them lose enzyme activity rapidly, and therefore, it is necessary to look for acidstable enzymes. In starch based industries, there is a demand for acid stable enzymes (e.g. amylase, glucose isomerase) because the pH of native starch is 3.0–4.5. There is a tremendous potential for acid stable enzymes to revolutionize existing industrial processes and to make many novel applications possible. The efforts to make cellulosic biofuel cost competitive with gasoline is the most important trend in the world today. Major enzyme producers like Novozyme and Danisco are investing heavily in this area. With the increase in the investments in the area of enzymology from other biotech giants such as Verenium Corporation and Dyadic International (USA) and Quest International (Irish Republic), Genzyme, DSM and CHR-Hansen (Danish company) are expected to bring changes in the market for acidstable enzymes.

In order to develop cost-effective, environment friendly acid stable enzymes, it is imperative that the broad scientific knowledge and process technology should exist in tandem with each other. Table 7.2 presents the major acidstable commercially available enzymes. Today the market has become excessively competitive with technological advancements; the profit margins have plummeted and production of profitable enzymes has become a challenge. Despite a promising future, the global enzyme market faces certain obstacles because of regulations, especially in developed countries (http://www.transparency market-research.com). There is, however, a hope for growth of acid stable enzymes in future.

9 Electricity Generation Using Acidophilic Microbes

Microbial fuel cell (MFC) is being viewed as a promising bio-electrochemical device that can produce energy in the form of bioelectricity from biodegradable compounds present in the waste water using the catalytic reactions that occur in microbes (Habermann and Pommer 1991; Logan et al. 2006; Wen et al. 2009; Raghavulu et al. 2009). In this way, they not only generate electricity, but also can be used in the treatment of wastewater simultaneously (Fig. 7.4). In a microbial fuel cell (MFC), bacteria are kept at the anode which is separated from a terminal electron acceptor at the cathode so that bacteria can respire only by transferring electrons to the anode. The organic or inorganic matter present in waste water is oxidized in the anode chamber by bacteria that produce carbon dioxide, protons and electrons.

The protons and electrons generated move towards cathode via proton exchange membrane and external electrical circuit, respectively (Oh et al. 2010; Rabaey and Verstraete 2005). At the cathode, an oxidant (normally oxygen) is being reduced. Equations given below illustrate the basic reactions (Jadhav and Ghangrekar 2009). However, MFCs have not yet been commercialized because power generated by these systems is limited due to high internal resistance (Cheng and Logan 2011; Feng et al. 2008). Improvements in the system design are anticipated to generate power from microbial metabolic reactions.

$$\text{Anode}: C_{12}H_{22}O_{11} + 13H_2O \rightarrow 12CO_2 + 48H^+ + 48e^-$$

$$\text{Cathode}: 48H^+ + 48e^- + 12O_2 \rightarrow 24H_2O$$

$$\text{Overall reaction}: C_{12}H_{22}O_{11} + 12O_2 \rightarrow 12CO_2 + 11H_2O$$

$$\Delta G = -5792.2\,\text{kJ/mol}$$

More recently, the generation of bioelectricity employing microbial fuel cell (MFC) seems to be gaining prominence. The use of acidophiles could play a major role in MFC because these microorganisms can work more efficiently as bioelectrocatalysts in a system that operates under acidic conditions than their neutrophilic microbial counterparts (Jadhav and Ghangrekar 2009; Raghavulu et al. 2009).

Table 7.2 The details of commercially available acidstable enzymes

Source	Enzyme	Commercial name	Optimum pH	Application	Company
Trichoderma longibrachiatum	β-Glucanase	BrewZyme LP	4.2	Baking, starch-gluten separation, alcohol fermentation and animal feeds	Danisco
NA[a]	Xylanase	CeluStar XL	3.5–7.5		Danisco
NA	Cellulase	HTec3	4.8–5.2	Biomass degradation in cellulosic biofuels applications	Novozymes
NA	Amylase	SAN™ Extra	3.0–4.5	Starch hydrolysis	Novozymes
NA	Pectinase	Pectinex® Ultra Mash	3.0–4.2	For mashing of apples and pears, and grapes for juice production	Novozymes
Aspergillus niger	Glucoamylase	Boli GA-150	3.5–4.5	Industries such as ethanol, brewage, glutamate and antibiotic fermentation, etc.	Boli bioproducts
Bacillus deramificans	Pullulanase	OPTIMAX L - 1000.	5	High fructose corn syrups (HFCS), and in the production of beer and potable alcohol	Genencor
A. niger	Amylase	Hazyme	4.5–6	Juice clarification	DSM
	Protease	Enzyclean	2.5–8.0	Cleaning of in plant machinery and floor of tea manufacturing	Nivshakti Bioenergy Pvt Ltd.
NA	Cellulase	ROCKSOFT™ACL CONC	4.5	Use on denim garments stone washing	Dyadic International, Inc.

[a]Details not available

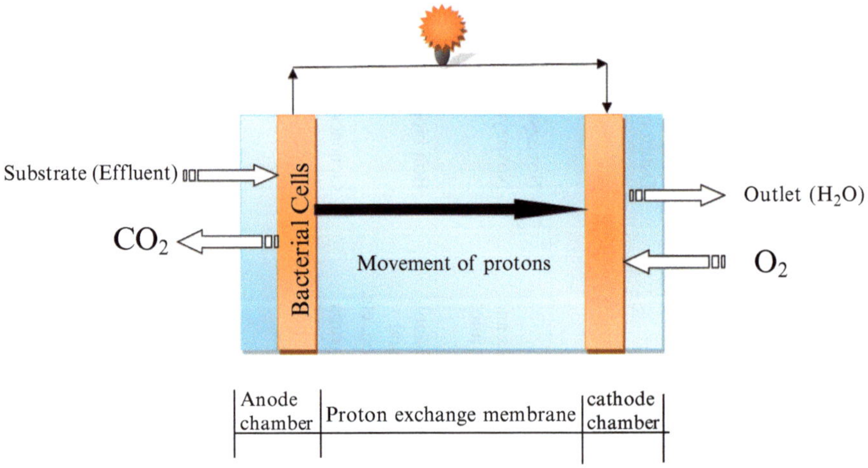

Fig. 7.4 Schematic diagram of microbial fuel cell

The effect of electrolyte pH directly correlates with the efficiency of MFC. Higher the pH difference, between the anodic and cathodic solutions, more will be the electricity generated because of change in internal resistance of MFC. Internal resistance of a cell is the summation of resistance of anode electrolyte, resistance of cathode electrolyte and resistance due to proton exchange membrane (PEM). Internal resistance of MFC decreases with increase in pH difference between anode and cathode solutions because higher pH difference increases the proton flux rate through the PEM. Hence, the power output of the MFC is more, when the difference of pH between anode and cathode is more (Jadhav and Ghangrekar 2009). Borole et al. (2008) have reported electricity generation in acidic condition (below pH 4). They used an acidophilic bacterium, *Acidiphilium cryptum*, as biocatalyst. Similarly, Jadhav and Ghangrekar (2009) showed that high pH difference is correlated with maximum power density.

Most MFCs are operated at neutral pH in order to facilitate bacterial growth. This process, however, faces a problem due to low concentration of protons at this pH, resulting in high internal resistance of the cell. Thus, if low pH electrolyte is used in combination with acidophiles, the performance of MFC can be enhanced significantly (Ieropoulos et al. 2005; Daniel et al. 2009). Moreover, the acidophilic ecosystems can help us to overcome the limitation of bacterial efficiency, when acidic waste water is used in MFC (Biffinger et al. 2008; Erable et al. 2009). One of the natural acidic ecosystems is the Río Tinto (Huelva, Spain), where the average pH is 2.3 ± 0.6. The abundance of ferric ions in the water acts as buffer, maintaining the acidic pH of the river. Microbial ecology studies have confirmed that 80 % of the prokaryotic diversity in the water column corresponds to three bacterial genera *A. ferrooxidans*, *Acidiphilium* spp., and *Leptospirillum* spp., which are conspicuous members of the iron cycle (García-Muñoz et al. 2011; González-Toril et al. 2003; Amaral-Zettler et al. 2002). The behaviour of single chambermicrobial fuel cells operating under different anodic pH [acidophilic (pH 6), neutral (pH 7) and alkaline (pH 8)] using anaerobic mixed cultures as anodic biocatalyst at room temperature (29 °C) during chemical wastewater treatment was evaluated (Raghavulu et al. 2009). The study showed that the acidic pH generates higher power relative to the neutral and alkaline operations irrespective of the nature of the catholyte used. Recently, Sulonen et al. (2015) provided the proof of model that electricity can be generated in MFC in the pH range of 1.2–2.5 using tetrathionate as a substrate (electron donor) for biological electricity production using the species of *Acidithiobacillus* and *Ferroplasma*. Furthermore, the use of acidophiles becomes more important, when pH of waste water is in the acidic range (e.g. starch, chocolate and brewery industry effluents), so that the addition of buffer to maintain pH can be avoided (Lu et al. 2009; Patil et al. 2009). The efforts made towards optimal design have paid the dividends. The reported power outputs in the laboratory scale MFCs have increased from 0.001 to several 6.9 Wm^{-2} (Fan et al. 2008; Oh et al. 2010) in less than a decade. Material costs have been reduced, but need more affordable process in order to make MFCs attractive alternatives to other forms of wastewater treatment.

10 Acidophiles in Bioconversions and Bioremediation

Bioremediation is a process in which microorganisms transform or mineralize organic contaminants into the non-hazardous substances, which then become part of natural biogeochemical cycles. Efforts have been made to accelerate the naturally occurring biodegradation process of toxic compounds through the optimization of several factors such as nutrients, oxygen, pH, composition and concentration of the contaminants (Allard and Neilson 1997; Margesin and Schinner 2001). Many environments like acid mine drainage or effluent of some industries are characterized by low pH. Acidophilic microorganisms are adapted to grow under extreme conditions. Hydrocarbon degrading or heavy metal accumulating acidophiles (e.g. *Acidiphilium rubrum*) have potential to remove contaminants from polluted extreme habitats (Johnson 1995; Stapleton et al. 1998; Roling et al. 2006).

AMD is one of the most serious forms of water pollution in the world and causes environmental hazards. AMD is a major environmental challenge that the mining industry is facing globally. AMD is not only associated with surface and groundwater pollution, but is also responsible for the dispersion of heavy metals into the environment. Furthermore, toxic substances such as cyanides and heavy metals present in effluents from metal mining have serious health hazard issues and ecological implications (Sheoran and Sheoran 2006; Hallberg 2010). Besides this, the high acidity of AMD further increases the mineral dissolution by solubilising other metals and metalloids to higher level as compared to neutral environments. Aluminium, copper, lead, zinc, cadmium, nickel and arsenic are among the elements that generally found in high concentrations (Sullivan and Yelton 1988; Johnson 1995).

AMD increases toxicity to other water bodies, since it leads to a reduction in pH of the recipient water making it unsuitable for aquatic life. Furthermore, oxidation and precipitation of metals in AMD will also lead to the reduction of neutralization capacity, and thus lowering the pH of recipient waters. As the pH of recipient waters is lowered, the solubility of the toxic metal increases causing toxic effects on aquatic life. Besides the toxicity of metals, the precipitation of metals, especially iron and aluminium, leads to aggregation at the bottom of recipient waters, where they break food chains of aquatic organisms and disturb the life cycle of aquatic organisms by inhibiting the reproduction of benthic organisms (Hallberg 2010).

The conventional method used for treatment of AMD is the addition of a source of alkalinity to raise the pH above the certain level required by iron oxidizing bacteria, thereby decreases the rate of acid generation. However, the operating cost of the conventional treatment technologies used in the treatment of acid mine drainage are not economical (Sheoran and Sheoran 2006). Nevertheless, the removal of toxic metals from contaminated soils can be achieved by using acidophilic microorganisms that can interact with these elements. The bacteria have specific property that they can tolerate high levels of metals (e.g. active efflux or metal ion trapping by metal chaperones). Moreover, gene duplications, the presence of genomic islands and an inorganic polyphosphate-driven metal resistance mechanism makes acidophiles suitable candidates for bioremediation (Dopson et al. 2003; Franke and Rensing 2007; Navarro et al. 2009; Krulwich et al. 2011).

In many cases, environments get contaminated from acid mine drainage and oil spills from industries containing polycyclic aromatic hydrocarbons. These compounds are generally considered to be hazardous for living beings as well as for the environment (Sutherland 1992; Pothuluri and Cerniglia 1994; Stapleton et al. 1998). As a result, there has been significant interest in the potential acidophiles that can help in bioremediation of polluted extreme environments.

Acidophiles have been reported to degrade various hydrocarbons, including aliphatic, aromatic, halogenated and nitrated compounds. Hydrocarbon-degrading acidophiles, adapted to grow in these environments, play a major role in the eco-friendly treatment of polluted habitats. The biodegradation (transformation or mineralization) of a wide range of hydrocarbons has been shown to occur in various extreme habitats. The biodegradation of many components of petroleum hydrocarbons by acidophiles have been reported (Stapleton et al. 1998; Christen et al. 2012).

A number of heavy metal tolerant acidophiles that can metabolize a range of aliphatic hydrocarbons have been isolated under acidic conditions. Moreover, aliphatic organic acids, which are generally toxic to acidophiles (Alexander et al. 1987), were utilized as substrates for energy and growth. A potential acidophilic bacterium, closely related to the genus *Acidocella*, has been isolated that tolerates high concentrations of acetic acid, if provided in sequentially small doses (1 mM, about 0.006 % v/v) (Gemmell and Knowles 2000). Furthermore, Roling et al. (2006) extracted DNA from a natural, surface petroleum seep and subjected to culture independent analysis that suggested the dominance of acidophilic bacteria, especially α-Proteobacteria group (mainly *Acidiphilium* and *Acidocella*). The presence of archaea was not confirmed, but fungi were present and the pH of the sample ranged between 3.0 and 5.0.

Stapleton et al. (1998) showed the potential of acidophiles isolated from soil samples of long term coal pile storage basin of pH 2.0 in degrading aromatic hydrocarbons. Even at such a low pH, more than 40 % biodegradation of parent hydrocarbons, naphthalene and toluene, to carbon dioxide and water was recorded. The DNA hybridization analysis suggested that the nucleic acids isolated from the whole community of these samples did not hybridize with genes (*nahA*, *nahG*, *nahH*, *todC1C2*, and *tomA*) which belong to neutrophilic bacteria. These data suggested that the degradation of aromatic hydrocarbons can occur in environments with extremely low pH values. Similarly, Hamamura et al. (2005) reported that the bacterial communities grow in the presence of acyclic alkanes (e.g., *n* alkanes with chain lengths of C15 to C30, as well as branched alkanes), predominately pristane and phytane at low pH values (pH 2.8–3.8), which are characteristics of acid-sulfate geothermal activity. The bacterial community was characterized through 16S rRNA gene clone library which showed that sequences were related to heterotrophic acidophilic bacteria of the species of *Acidisphaera*, *Acidiphilium* of *Proteobacteria* and chemolithotroph *Acidithiobacillus* spp. The denaturing gradient gel electrophoresis (DGGE) of 16S rRNA gene fragments of hydrocarbon-amended soil-sand mixtures showed the heterotrophic acidophile-related sequences as dominant DGGE bands. Besides this, an alkane-degrading isolate was cultivated, which confirmed the alkane degradation capability of one population indigenous to acidic hydrocarbon seep soil. Recently, Christen et al. (2012) reported biodegradation of phenol by a well-acclimatized strain of *S. solfataricus*, a thermoacidophilic archaeon, at 80 °C and pH 3.2. Phenol is an organic pollutant present in wastewater from various industries such as refining, coking, coal processing and petrochemicals production. In some industrial effluents, it can reach concentrations up to 6.8 g L^{-1}. The potential of acidophiles to grow and degrade phenol at high temperature and low pH is significant for bioremediation from hot acidic sites contaminated by phenol.

The significance of acidophilic bacteria in the treatment of acidic effluents has been already established because of their capability of the reduction as well as the oxidation of iron and sulphur. The hydrocarbon degrading acidophiles are very important for the bioremediation of oil polluted acidic effluents.

11 Acidophilic Microbes in Foods

Generally high acid foods are resistant to spoilage. A few microorganisms can, however, grow at acidic pH and are important spoilage microorganisms. For example, the species of *Alicyclobacillus* are the major contaminants in the fruit juice industry. In 1984, a significant case of apple juice spoilage in Germany was attributed to *Alicyclobacillus*, followed by many reports of *Alicyclobacillus* mediated spoilage of juices, juice blends, juice concentrates, carbonated fruit drinks and shelf stable ice tea. *Alicyclobacillus* spores are highly acid and heat resistant (Walker and Phillips 2008). They can easily survive common pasteurization treatments (92 °C for 10 s), and cause spoilage. *A. acidoterrestris* is used as the reference organism in designing pasteurization for high acidic foods (Silva and Gibbs 2001).

12 Conclusions and Future Perspectives

Acidophiles are a diverse group of microorganisms. There is an urgent need to explore their diversity in view of their multifarious applications. Despite several attempts to explain the strategies evolved by acidophiles to survive and thrive in acidic environments, further research is called for understanding their adaptations satisfactorily. Acid stable enzymes/proteins of acidophilic microbes have been shown to be useful in industrial processes. Acidophiles have been extensively used in bioleaching of metals from low grade ores. The possibility of using acidophiles in microbial fuel cells for generating electricity is an exciting application.

Conflict of Interest Archana Sharma, Deepak Parashar, and Tulasi Satyanarayana declare that they have no conflict of interest.

References

Acuna-Arguelles ME, Gutierrez-Rajas M, Viniegra-Gonzalez G, Favela-Toress E (1995) Production and properties of three pectinolytic activities produced by *A. niger* in submerged and solid state fermentation. Appl Microbiol Biotechnol 43:808–814

Alexander B, Leach S, Ingledew WJ (1987) The relationship between chemiosmotic parameters and sensitivity to anions and organic acids in the acidophile *Thiobacillus ferrooxidans*. J Gen Microbiol 133:1171–1179

Allard AS, Neilson AH (1997) Bioremediation of organic waste sites: a critical review of microbiological aspects. Int Biodeter Biodegr 39:253–285

Al-Obaidi ZS, Aziz GM, Al-Bakir AY (1987) Screening of fungal strains for polygalacturonase production. J Agric Water Resour Res 6:125–182

Amaral-Zettler LA, Gómez F, Zettler E, Keenan BG, Amils R, Sogin ML (2002) Eukaryotic diversity in Spain's river of fire: this ancient and hostile ecosystem hosts a surprising variety of microbial organisms. Nature 417:37

Bai Y, Huang H, Meng K, Shi P, Yang P, Luo H, Luo C, Feng Y, Zhang W, Yao B (2012) Identification of an acidic α-amylase from *Alicyclobacillus* sp. A4 and assessment of its application in the starch industry. Food Chem 131:1473–1478

Baker-Austin C, Dopson M (2007) Life in acid: pH homeostasis in acidophiles. Trends Microbiol 15(4):165–171

Bampton KF, Bologiannis F, Canterford JH, Smith AN (1983) Development of experimental in-situ leaching at the Mutooroo copper mine, South Australia. In: Australasian IMM Annual Conference, Broken Hill, NSW, pp 371–379

Batrakov SG, Pivovarova TA, Esipov SE, Sheichenko VI, Karavaiko GI (2002) Beta-D-glycopyranosyl caldarchaetidylglycerol is the main lipid of the acidophilic, mesophilic, ferrous iron-oxidizing archaeon *Ferroplasm acidiphilum*. Biochim Biophys Acta 1581:29–35

Berthelot D, Leduc LG, Ferroni GD (1994) The absence of psychrophilic *Thiobacillus ferrooxidans* and acidophilic heterotrophic bacteria in cold, tailings effluents from a uranium mine. Can J Microbiol 40:60–63

Bertoldo C, Dock C, Antranikian G (2004) Thermoacidophilic microorganisms and their novel biocatalysts. Eng Life Sci 4:521–531

Biffinger JC, Pietron J, Bretschger O, Nadeau LJ, Johnson GR, Williams CC, Nealson KH, Ringeisen BR (2008) The influence of acidity on microbial fuel cells containing *Shewanella oneidensis*. Biosens Bioelectron 24:900–905

Borin MDF, Said S, Fonseca MJV (1996) Purification and biochemical characterization of an extracellular endopolygalacturonase from *Penicillium frequentans*. J Agric Food Chem 44:1616–1620

Borole AP, Neill HO, Tsouris C, Cesar S (2008) A microbial fuel cell operating at low pH using the acidophile *Acidiphilium cryptum*. Biotechnol Lett 30:1367–1372

Brierley CL, Brierley JA (2000) Bioheap processes: operational requirements and techniques. In: Proceedings of Randol Copper Hydromet Roundtable, Tucson, Ariz. pp 95–103

Chandel AK, Rudravaram R, Rao LV, Ravindra P, Narasu ML (2007) Industrial enzymes in bio-industrial sector development: An Indian perspective. J Commer Biotechnol 13:283–291

Channe PS, Shewal JG (1995) Pectinase production by *Sclerotium rolfsii*: Effect of culture conditions. Folia Microbiol 40:111–117

Chen L-X, Hu M, Huang L-N, Hua Z-S, Kuang J-L, Li S-J, Shu W-S (2014) Comparative metagenomic and metatranscriptomic analyses of microbial communities in acid mine drainage. Int Soc Microb Ecol 9:1–14

Cheng SA, Logan BE (2011) Increasing power generation for scaling up single-chamber air cathode microbial fuel cells. Biores Technol 102:4468–4473

Christen P, Vega A, Casalot L, Simon G, Auria R (2012) Kinetics of aerobic phenol biodegradation by the acidophilic and hyperthermophilic archaeon *Sulfolobus solfataricus* 98/2. Biochem Eng J 62:56–61

Ciaramella M, Napoli A, Rossi M (2005) Another extreme genome: how to live at pH 0. Trends Microbiol 13:49–51

Crossman L, Holden M, Pain A, Parkhill J (2004) Genomes beyond compare. Nat Rev Microbiol 2:616–617

Daniel DK, Mankidy BD, Ambarish K, Manogari R (2009) Construction and operation of a microbial fuel cell for electricity generation from wastewater. Int J Hydrogen Energy 34:7555–7560

Davies DR (1990) The structure and function of the aspartic proteinases. Annu Rev Biophys Chem 19:189–215

Dew DW, Lawson EN, Broadhurst JL (1997) The Biox® process for biooxidation of gold-bearing ores or concentrates. In: Rawlings DE (ed) Biomining: theory microbes and Industrial processes. Springer, Berlin, pp 45–80

Do TT, Quyen DT, Dam TH (2012) Purification and characterization of an acid-stable and organic solvent-tolerant xylanase from *Aspergillus awamori* VTCC-F312. Sci Asia 38:157–165

Dopson M, Baker-Austin C, Koppineedi PR, Bond PL (2003) Growth in sulfidic mineral environments: metal resistance mechanisms in acidophilic micro-organisms. Microbiology 149:1959–1970

Dopson M, Baker-Austin C, Hind A, Bowman JP, Bond PL (2004) Characterization of *Ferroplasma* isolates and *Ferroplasma acidarmanus* sp. nov., extreme acidophiles from acid mine drainage and industrial bioleaching environments. Appl Environ Microbiol 70:2079–2088

Eckert K, Schneider E (2003) A thermoacidophilic endoglucanase (CelB) from *Alicyclobacillusacidocaldarius* displays high sequence similarity to arabinofuranosidases belonging to family 51 of glycoside hydrolases. Eur J Biochem 270:3593–3602

Erable B, Etcheverry L, Bergel A (2009) Increased power from a two chamber microbial fuel cell with a low pH air-cathode compartment. Electrochem Commun 11:619–622

Fan Y, Sharbrough E, Liu H (2008) Quantification of the internal resistance distribution of microbial fuel cells. Environ Sci Technol 42:8101–8107

Feng Y, Wang X, Logan BE, Lee H (2008) Brewery wastewater treatment using air cathode microbial fuel cells. Appl Microbiol Biotechnol 78:873–880

Franke S, Rensing C (2007) Acidophiles. Mechanisms to tolerate metal and acid toxicity. In: Gerday C, Glansdorff N (eds) Physiology and biochemistry of extremophiles. ASM Press, Washington, DC, pp 271–278

Fusek M, Lin XL, Tang J (1990) Enzymic properties of thermopsin. J Biol Chem 265:1496–1501

Futterer O, Angelov A, Liesegang H, Gottschalk G, Schleper C, Schepers B, Dock C, Antranikian G, Liebl W (2004) Genome sequence of *Picrophilus torridus* and its implications for life around pH 0. Proc Natl Acad Sci USA 101:9091–9096

Gaffney PJ, Edgell TA, Dawson PA, Ford AW, Stocker E (1996) A pig collagen peptide fraction. A unique material for maintaining biological activity during lyophilization and during storage in the liquid state. J Pharm Pharmacol 48:896–898

García-Muñoz J, Amils R, Fernández VM, Lacey ALD, Malki M (2011) Electricity generation by microorganisms in the sediment-water interface of an extreme acidic microorganism. Int Microbiol 14:73–81

Geisler RA, Pudington IE (1996) Treatment of lead sulfide bearing minerals. US Patent 5:523-066

Gemmell RT, Knowles CJ (2000) Utilisation of aliphatic compounds by acidophilic heterotrophic bacteria: the potential for bioremediation of acidic wastewaters contaminated with toxic organic compounds and heavy metals. FEMS Microbiol Lett 192:185–190

Golyshina OV, Timmis KN (2005) *Ferroplasma* and relatives, recently discovered cell wall-lacking archaea making a living in extremely acid, heavy metal-rich environments. Environ Microbiol 7:1277–1288

Golyshina OV, Pivovarova TA, Karavaiko GI, Moore ER, Abracham WR, Lunsdorf H, Timmis KN, Yakimov MM, Golyshin PN (2000) *Ferroplasma acidiphilum gen. Nov., sp. Nov*., an acidophilic autotrophic, ferrous-iron oxidizing, cell wall lacking, mesophilic member of the Ferroplasmaceae fam. Nov., comprising distinct lineage of Archaea. Int J Syst Evol Microbiol 50:997–1006

González-Toril E, Llobet-Brossa E, Casamayor EO, Amann R, Amils R (2003) Microbial ecology of an extreme acidic environment, the Tinto River. Appl Environ Microbiol 69:4853–4865

Gupta R, Gigras P, Mohapatra H, Goswami VK, Chauhan B (2003) Microbial α-amylase: a biotechnological perspective. Process Biochem 38:1599–1616

Habermann W, Pommer EH (1991) Biological fuel cells with sulphide storage capacity. Appl Microbiol Biotechnol 35:128–133

Hallberg KB (2010) New perspectives in acid mine drainage microbiology. Hydrometallurgy 104:448–453

Hamamura N, Olson SH, Ward DM, Inskeep WP (2005) Diversity and functional analysis of bacterial communities associated with natural hydrocarbon seeps in acidic soils at Rainbow Springs, Yellowstone National Park. Appl Environ Microbiol 71:5943–5950

Honda S (1998) Dietary use of collagen and collagen peptides for cosmetics. Food Style 21:54–60

Huang Y, Krauss G, Cottaz S, Driguez H, Lipps G (2005) A highly acid-stable and thermostable endo-b-glucanase from the thermoacidophilic archaeon *Sulfolobus solfataricus*. Biochem J 385:581–588

Ieropoulos IA, Greenman J, Melhuish C, Hart J (2005) Comparative study of three types of microbial fuel cell. Enzyme Microb Tech 37:238–245

Inagaki K, Nakahira K, Mukai K, Tamura T, Tanaka H (1998) Gene cloning and characterization of an acidic xylanase from *Acidobacterium capsulatum.* Biosci Biotechnol Biochem 62:1061–1067

Jadhav GS, Ghangrekar MM (2009) Performance of microbial fuel cell subjected to variation in pH, temperature, external load and substrate concentration. Biores Technol 100:717–723

Jeffries AC, Kozera CJ, Medina N, Peng X, Thi-Ngoc HP, Redder P, Schenk ME, Theriault C, Tolstrup N, Charlebois RL, Doolittle WF, Duguet M, Gaasterland T, Garrett RA, Ragan MA, Sensen CW, Vander Oost J (2001) The complete genome of the Crenarchaeon *Sulfolobus solfataricus* P2. Proc Natl Acad Sci USA 98:7835–7840

Johnson DB (1995) Acidophilic microbial communities: candidates for bioremediation of acidic mine effluents. Int Biodeter Biodegrad 35:41–58

Johnson D (1998) Biodiversity and ecology of acidophilic microorganism. FEMS Microbiol Ecol 27:307

Johnson D (2008) Biodiversity and interactions of acidophiles: Key to understanding and optimizing microbial processing of ores and concentrates. Trans of Nonferr Metal Soc China 18: 1367–1373

Johnson D, Hallberg K (2003) The microbiology of acidic mine waters. Res Microbiol 154:466–473

Kim HW, Kim YO, Lee JH, Kim KK, Kim YJ (2003) Isolation and characterization of a phytase with improved properties from *Citrobacter braakii*. Biotechnol Lett 25:1231–1234

Krulwich TA, Sachs G, Padan E (2011) Molecular aspects of bacterial pH sensing and homeostasis. Nat Rev Microbiol 9:330–343

Laksanalamai P, Robb FT (2004) Small heat shock proteins from extremophiles: a review. Extremophiles 8:1–11

Leisola M, Jokela J, Pastinen O, Turunen O, Schoemaker H (2002) Industrial use of enzymes. In: Encyclopedia of life support systems (EOLSS), EOLSS Publishers, Oxford

Liu XD, Xu Y (2008) A novel raw starch digesting a-amylase from a newly isolated *Bacillus* sp. YX-1: purification and characterization. Biores Technol 99:4315–4320

Logan BE, Hamelers B, Rozendal R, Schrorder U, Keller J, Freguia S, Aelterman P, Verstraete W, Rabaey K (2006) Microbial fuel cells: methodology and technology. Environ Sci Technol 40:5181

Lu N, Zhou SG, Zhuang L, Zhnag JT, Ni JR (2009) Electricity generation from starch processing wastewater using microbial fuel cell technology. Biochem Eng J 43:246–251

Macalady JL, Vestling MM, Baumler D, Boekelheide N, Kaspar CW, Banfield JF (2004) Tetraether-linked membrane monolayers in *Ferroplasma* spp: a key to survival in acid. Extremophiles 8:411–419

Marcus L, Barash I, Sneh B, Koltin Y, Finker A (1986) Purification and characterization of pectolytic enzymes produced by virulent and hypovirulent isolates of *Rhizoctonia solani* Kuhn. Physiol Mol Plant Pathol 29:325–336

Margesin R, Schinner F (2001) Biodegradation and bioremediation of hydrocarbons in extreme environments. Appl Microbiol Biotechnol 56:650–663

Matzke J, Schwermann B, Baker EP (1997) Acidstable and acidophilic proteins: the example of the alpha amylase from *Alicyclobacillus acidocaldarius*. Comp Biochem Physiol A Mol Integr Physiol 118:411–419

Murao S, Okhuni K, Naganao M (1988) A novel thermostable S-PI (pepstatin Ac)-insensitive acid proteinase from thermophilic *Bacillus* novo sp. strain Mn-32. Agric Biol Chem 52:1029–1031

Nakayama T, Tsuruoka N, Akai M, Nishino T (2000) Thermostable collagenolytic activity of a novel thermophilic isolate, *Bacillus* sp. Strain NTAP-1. J Biosci Bioeng 89:612–614

Navarro CA, Orellana LH, Mauriaca C, Jerez CA (2009) Transcriptional and functional studies of *Acidithiobacillus ferrooxidans* genes related to survival in the presence of copper. Appl Environ Microbiol 75:6102–6109

Oda K, Nakazima T, Terashita T, Suziki KA, Murao S (1987a) Purification and properties of an S-PI (Pepstatin Ac) insensitive carboxyl proteinase from a *Xanthomonas* sp. bacterium. Agric Biol Chem 51:3073–3080

Oda K, Sugitani M, Fukuhara K, Murao S (1987b) Purification and properties of a pepstatin-insensitive carboxyl proteinase from a gram negative bacterium. Biochim Biophys Acta 923:463–469

Oh JR, Kim GC, Premier TH, Lee C, Kim WT (2010) Sloan Sustainable wastewater treatment: how might microbial fuel cells contribute. Biotechnol Adv 28:871–881

Pandey A, Nigam P, Soccol CR, Soccol VT, Singh D, Mohan R (2000) Advances in microbial amylases. Biotechnol Appl Biochem 31:135–152

Patil SA, Surakasi VP, Koul S, Ijmulwar S, Vivek A, Shouche YS, Kapadnis BP (2009) Electricity generation using chocolate industry wastewater and its treatment in activated sludge based microbial fuel cell and analysis of developed microbial community in the anode chamber. Biores Technol 100:5132–5139

Pivovarova TA, Kondrateva TF, Batrakov SG, Esipov SE, Sheichenko VI, Bykova SA (2002) Phenotypic features of *Ferroplasma acidiphilum* strains Y^{t} and Y-2. Microbiol (Moscow) 71:809–818

Plumb JJ, Haddad CM, Gibson JA, Franzmann PD (2007) *Acidianus sulfidivorans* sp. nov., an extremely acidophilic, thermophilic archaeon isolated from a solfatara on Lihir Island, Papua New Guinea, and emendation of the genus description. Int J Syst Evol Microbiol 57:1418–1423

Pothuluri JV, Cerniglia CE (1994) Microbial metabolism of polycyclic aromatic hydrocarbons. In: Chaundry GR (ed) Biological degradation and bioremediation of toxic chemicals. Dioscorides Press, Portland, OR, pp 92–124

Pronk JT, Johnson DB (1992) Oxidation and reduction of iron by acidophilic bacteria. Geomicrobiol J 10(3–4):153–171

Rabaey K, Verstraete W (2005) Microbial fuel cells: novel biotechnology for energy generation. Trends Biotechnol 23:291–298

Raghavulu SV, Mohan SV, Goud RK, Sarma PN (2009) Effect of anodic pH microenvironment on microbial fuel cell (MFC) performance in concurrence with aerated and ferricyanide catholytes. Electrochem Commun 11:371–375

Ram RJ, VerBerkmoes NC, Thelen MP, Tyson GW, Baker BJ, Blake RC II, Shah M, Hettich RL, Banfield JF (2005) Community proteomics of a natural microbial biofilm. Science 308:1915–1999

Rawlings DE (ed) (1997) Biomining: theory, microbes and industrial processes. Springer, Berlin

Rawlings DE (2002) Heavy metal mining using microbes. Annu Rev Microbiol 56:65–91

Rawlings DE, Johnson DB (2007) The microbiology of biomining: development and optimization of mineral–oxidizing microbial consortia. Microbiology 153:315–324

Rodriguez E, Han Y, Lei XG (1999) Cloning, sequencing and expression of an *Escherichia coli* acid phosphatase, phytase gene *(app*A2*)* isolated from pig colon. Biochem Biophys Res Commun 257:117–123

Roling WFM, Ortega-Lucach S, Larter SR, Head IM (2006) Acidophilic microbial communities associated with a natural, biodegraded hydrocarbon seepage. J Appl Microbiol 101:290–299

Satyanarayana T, Raghukumar C, Shivaji S (2005) Extremophilic microbes: diversity and perspectives. Curr Sci 89:78–90

Schleper C, Püehler G, Kuhlmorgen B, Ziling W (1995) Life at extremely low pH. Nature 375:741–742

Schleper C, Puehler G, Klenk HP, Zillig W (1996) *Picrophilus oshimae Picrophilus torridus fam. nov., gen. nov., sp. nov.*, two species of hyperacidophilic, thermophilic, heterotrophic aerobic archaea. Int J Syst Bacteriol 46:814–816

Schnell HA (1997) Bioleaching of copper. In: Rawlings DE (ed) Biomining: theory microbes and industrial processes. Springer-Verlag and Landes Bioscience, Berlin, Germany, pp 21–43

Schwermann B, Pfau K, Liliensiek B, Schleyer M, Fischer T, Bakker EP (1994) Purification, properties and structural aspects of the thermoacidophilic α-amylase from *Alicyclobacillusacidocaldarius* ATCC 27009. Insight into acidostability of proteins. Eur J Biochem 226:981–991

Serour E, Antranikian G (2002) Novel thermoactive glucamylases from the thermoacidophilic archaea *Thermoplasma acidophilum, Picrophilus torridus* and *Picrophilus oshimae*. Antonie van Leewenhoek 81:73–83

Shah AR, Shah RK, Madamwar D (2006) Improvement of the quality of whole wheat bread by supplementation of xylanase from *Aspergillus foetidus*. Biores Technol 97:2047–2053

Sharma A, Satyanarayana T (2010) High maltose-forming, Ca^{2+}- independent and acid stable α-amylase from a novel acidophilic bacterium *Bacillus acidicola* TSAS1. Biotechnol Lett 32:1503–1507

Sharma A, Satyanarayana T (2012) Production of acid-stable and high-maltose-forming α-amylase of *Bacillus acidicola* by solid-state fermentation and immobilized cells and its applicability in baking. Appl Biochem Biotechnol 168:1025–1034

Sharma A, Kawarabayasi Y, Satyanarayana T (2012) Acidophilic bacteria and archaea: acid stable biocatalysts and their potential applications. Extremophiles 16:1–19

Sheoran AS, Sheoran V (2006) Heavy metal removal mechanism of acid mine drainage in wetlands: A critical review. Miner Eng 19:105–116

Shimada H, Nemoto N, Shida Y, Oshima T, Yamagishi A (2002) Complete polar lipid composition of *Thermoplasma acidophilum* HO-62 determined by high performance liquid chromatography with evaporative light scattering detection. J Bacteriol 184:556–563

Shivaramakrishnan S, Gangadharan D, Nampoothiri KM, Soccol CR, Pandey A (2006) α-Amylases from microbial sources–an overview on recent developments. Food Technol Biotechnol 44:173–184
Siddiqui MH, Kumar A, Kesari KK, Arif JM (2009) Biomining–A useful approach toward metal extraction. American-Eurasian J Agron 2:84–88
Silva FVM, Gibbs P (2001) *Alicyclobacillus acidoterrestris* spores in fruit products and design of pasteurization processes. Trends Food Sci Technol 12:68–74
Soni SK, Magdum A, Khire JM (2010) Purification and characterization of two distinct acidic phytases with broad pH stability from *Aspergillus niger* NCIM 563. World J Microbiol Biotechnol 26:2009–2018
Spijkerman E, Bissinger V, Meister A, Gaedke U (2007) Low potassium and inorganic carbon concentrations influence a possible phosphorus limitation in *Chlamydomonas acidophila* (Chlorophyceae). Eur J Phycol 42:327–339
Stapleton RD, Savage DC, Sayler GS, Stacey G (1998) Biodegradation of aromatic hydrocarbons in an extremely acidic environment. Appl Environ Microbiol 64:4180–4184
Sullivan PJ, Yelton JL (1988) An evaluation of trace element release associated with acid mine drainage. Environ Geol Water Sci 12:181–186
Sulonen ML, Kokko ME, Lakaniemi AM, Puhakka JA (2015) Electricity generation from tetrathionate in microbial fuel cells by acidophiles. J Hazard Mater 284:182–189
Sutherland JB (1992) Detoxification of polycyclic aromatic hydrocarbons by fungi. J Ind Microbiol 9:53–62
Takayanagi S, Kawasaki H, Sugimori K, Yamada T, Sugai A, Ito T, Yamasato K, Shioda M (1996) *Sulfolobus hakonensis* sp. nov., a novel species of acidothermophilic archaeon. Int J Syst Bacteriol 46:377–382
Thiel V (2011) Extreme environments. Encyclopedia of Geobiology 362–366
Tyson GW, Chapman J, Hugenholtz P, Allen EE, Ram JR, Richardson MP, Solovyev VV, Rubin ME, Rokhsar SD, Banfield FJ (2004) Community structure and metabolism through reconstruction of microbial genomes from the environment. Nature 428:37–43
Van de Vossenberg JL, Driessen AJ, Konings WN (1998a) The essence of being extremophilic: the role of the unique archaeal membrane lipids. Extremophiles 2:163–170
Van de Vossenberg JL, Driessen AJ, Zillig W, Konings WN (1998b) Bioenergetics and cytoplasmic membrane stability of the extremely acidophilic, thermophilic archaeon *Picrophilus oshimae*. Extremophiles 2:67–74
Walker M, Phillips CA (2008) *Alicyclobacillus acidoterrestris*: an increasing threat to the fruit juice industry. Int J Food Sci Technol 43:250–260
Wen Q, Wu Y, Cao D, Zhao L, Sun Q (2009) Electricity generation and modeling of microbial fuel cell from continuous beer brewery wastewater. Biores Technol 100:4171–4175
Yim KJ, Cha IT, Rhee JK, Song HS, Hyun DW, Lee HW, Kim D, Kim KN, Nam YD, Seo MJ, Bae JW, Roh SW (2015) *Vulcanisaeta thermophila* sp. nov., a hyperthermophilic and acidophilic crenarchaeon isolated from solfataric soil. Int J Syst Evol Microbiol 65:201–205
Zychlinsky E, Matin A (1983) Effect of starvation on cytoplasmic pH, proton motive force, and viability of an acidophilic bacterium *Thiobacillus acidophilus*. J Bacteriol 153:371–374

Web addresses

http://www.bccresearch.com/market-research/biotechnology/enzymes-industrial-applications-bio030h.html
http://www.transparencymarketresearch.com/enzymes-market.html
http://www.grandviewresearch.com/industry-analysis/enzymes-industry

Chapter 8
Alkaliphilic Microorganisms in Biotechnology

Gashaw Mamo and Bo Mattiasson

1 Introduction

Microorganisms that grow at alkaline conditions are often referred to alkaliphiles. Some of these organisms grow optimally at high pH while others can lavishly grow at high pH but not optimally. Those organisms that optimally grow at or above pH 9 are widely recognized as alkaliphiles, which are further categorized as obligate alkaliphiles (which do not grow at neutral pH) and facultative alkaliphiles (which are able to grow at neutral condition). The group of microbes that grow optimally below pH 9 but thrives well in high pH environment (> pH 9) are known as alkali tolerant. Several factors such as type and concentration of nutrients, and cultivation temperature can affect the optimum pH for microbial growth (Horikoshi 1999). Thus, the optimum pH for growth of an organism can vary depending on the growth conditions. It seems that the only distinguishing feature of alkaliphiles is their ability to grow optimally above pH 9 at least in one set of cultivation conditions (such as medium and temperature).

1.1 Habitats and Diversity of Alkaliphiles

Alkaliphiles have been isolated from different natural and man-made alkaline environments which are 'extreme' to other groups of organisms. Soda lakes and soda deserts are the most stable naturally occurring alkaline environments, where the pH values are usually in excess of 10. These habitats are the major sources of alkaliphiles

G. Mamo (✉) • B. Mattiasson (✉)
Department of Biotechnology, Lund University, P.O. Box 124, Lund, Sweden

Indienz AB, Billeberga, Sweden
e-mail: gashaw.mamo@biotek.lu.se; bo.mattiasson@biotek.lu.se

P.H. Rampelotto (ed.), *Biotechnology of Extremophiles*, Grand Challenges in Biology and Biotechnology 1, DOI 10.1007/978-3-319-13521-2_8

from which several novel organisms have been isolated and described (Jones et al. 1999; Horikoshi 1999; Delgado et al. 2006; Canganella and Wiegel 2011; Grant and Sorokin 2011; Sorokin et al. 2014; Glaring et al. 2015). Pockets of alkaline environments with high pH can arise from human activities such as cement manufacturing, paper and pulp production, mining activities, indigo dye preparation, food and textile processing, electroplating, etc. (Grant and Tindall 1986). Thus, from the waste streams and areas of these industrial activities it is possible to enrich alkaliphiles. Alkaliphiles have also been found in the guts of termites (Taksawan et al. 2005), isolated from fecal material (Horikoshi 1999), and even from neutral soils (Horikoshi 1991).

Alkaline environments are known to harbor diverse groups of organisms that belong to eubacteria, archaea and eukarya (Hayashi et al. 1996; Horikoshi 1999; Duckworth et al. 1996; Canganella and Wiegel 2011). The great majority of alkaliphiles identified so far are members of the eubacteria, and a good portion of them are isolated from soda lake samples. These extremophiles are not only taxonomically diverse but they are also physiologically heterogeneous and belong to different groups such as aerobes, facultative anaerobes, obligate anaerobes, halophilies, thermophilies, psychrophilies, mesophiles, etc. (Grant and Sorokin 2011; Sorokin et al. 2015).

1.2 Adaptation to High pH Environment

Organisms evolve adaptive strategies to live and thrive in an environment they colonize. Alkaliphiles in the course of their evolution developed adaptive mechanisms that allow them to flourish in extremely high pH environment. Some of these organisms can grow impressively even above pH 13 (Roadcap et al. 2006). A sizable number of studies have been committed to unravel the secret how these organisms thrive in such extremely high pH environments. Most of these studies have been focused on mechanisms of lowering the cytoplasmic pH by means of solute transport mainly by Na^+/H^+ antiporter (Kitada et al. 1994; Krulwieh et al. 1997), cell wall components that protect the cells from the effect of high pH (Aono et al. 1999) and bioenergetic problems across the membrane (Sturr et al. 1994; Krulwieh et al. 1998).

Results from transcriptome and proteome analyses complemented the existing physiological and genetic information on the high pH adaptation of this fascinating group of microorganisms. Based on the findings, growth at alkaline environment is associated with metabolic shift that promote increased acid production, enhanced expression of transporters and enzymes that facilitate the capture and retention of protons (such as ATP synthase and monovalent cation/proton antiporters), and change in cell surface composition that helps to retain the cytoplasmic proton (Krulwich et al. 2001a, b; Padan et al. 2005). These studies indicated that alkaliphiles use combination of different strategies to thrive in their high pH environment. The ability of producing enzymes that remains active and stable at alkaline conditions is another important strategy that has often been less pronounced when it comes to adaptive mechanism of alkaliphiles.

1.3 *Why are Alkaliphilic Microorganisms of Interest to Biotechnology?*

These extremophiles developed adaptive mechanisms for their survival and we humans make use of these mechanisms to our own advantages (Fig. 8.1). Among the adaptive mechanisms of alkaliphiles, their extracellular enzymes that are operationally stable at alkaline conditions are very useful in several biotechnological applications and attracted a great deal of attention (Fujinami and Fujisawa 2010). These organisms produce extracellular enzymes that are highly active in the environmental conditions prevalent where the organisms are growing (Fig. 8.2). There are many industrial and environmental applications that require enzymes which are active and stable at elevated pH. On the other hand, the pH-activity profile of the intracellular enzymes of alkaliphiles is often similar to their counterparts from non-alkaliphilic microbes. This is mainly because of the alkaliphilic adaptation machinery that keeps the cytoplasmic pH close to neutral, and hence the enzymes are adapted to work under conditions found in the cytoplasm. Observations that intracellular enzymes from extremophiles do not have properties that fully correlate to the conditions at which the cells are capable of growing is seen, not only in alkaliphiles but also for acidophiles (Baker-Austin and Dopson 2007) and to some extent in thermostability of enzymes from thermophiles (Turner et al. 2007).

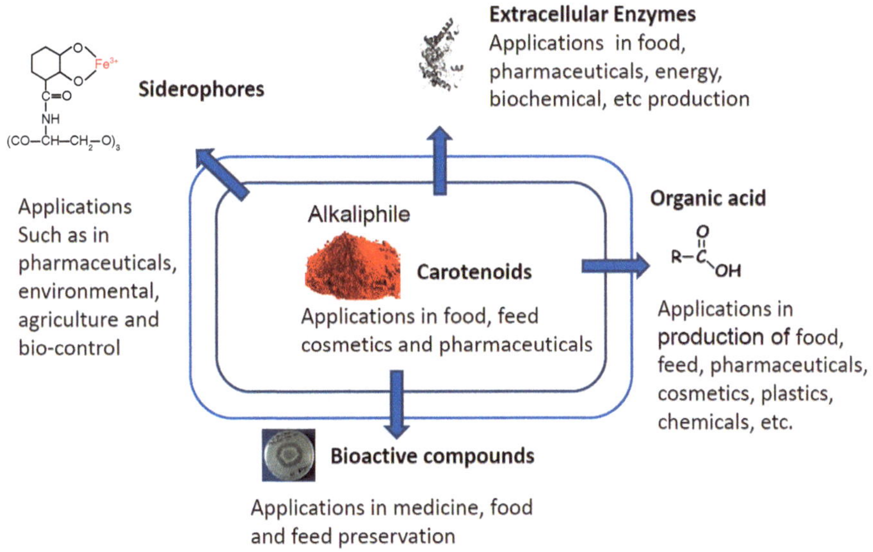

Fig. 8.1 Some of the adaptive mechanisms of alkaliphiles and their biotechnological importance

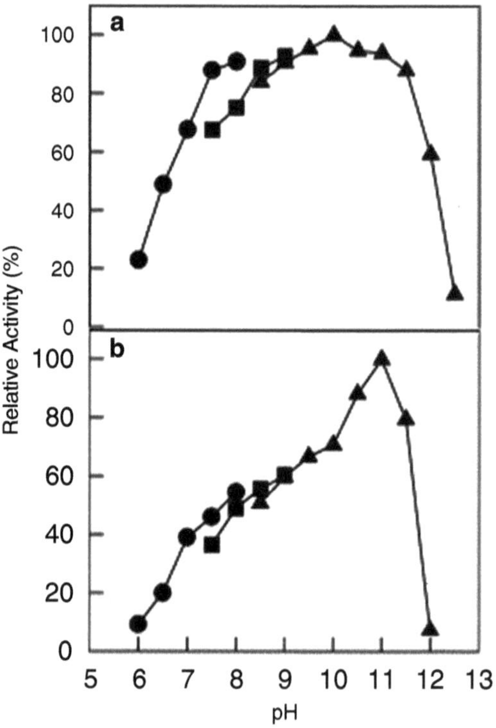

Fig. 8.2 The effect of pH on the activity of AL20 (a) and AL89 (b) proteases produced by two alkaliphilic strains. The activity was assayed at 50 °C. The buffers used were phosphate, pH 6–8 (*filled circle*); Tris–HCl, pH 7.5–9 (*filled square*); and glycine–NaOH, pH 8.5–12 (*filled triangle*), each at a concentration of 50 mM (Reproduced with permission from Gessesse et al. 2003)

Unlike most other extremophiles, alkaliphiles are known as very good producers of extracellular enzymes. Moreover, these organisms grow fast, often even faster than e.g. neutralophiles (Niehaus et al. 1999; Horikoshi 2006), and at relatively higher density. These are desirable properties when organisms are considered for biotechnological applications.

In addition to their enzymes, alkaliphiles are of interest to biotechnology because of their remarkable ability of organic acid production which is an input in an array of industrial processes. Some of the alkaliphiles are also known producing siderophores and carotenoids that have immense potential in wide areas of applications. These organisms have been less explored as sources of bioactive compounds; however, the limited studies made so far clearly indicate their potential as sources of novel bioactive compounds.

In this chapter, some of the real and potential applications of the major extracellular alkaline active enzymes are discussed. The potential of alkaliphiles in production of organic acids, carotenoids, bioactive compounds, and siderophores is assessed. However, these are not the only products of alkaliphiles that are of interest to biotechnology, they are just selected examples.

2 Alkaline Active Enzymes of Alkaliphiles and Their Industrial Applications

Application of enzymes is one of the most important components of industrial biotechnology. A large number of enzymes are being used in various sectors and expanding even more and faster with emergence of Green Chemistry. Some of these applications require enzymes that are operationally stable at alkaline condition (Fig. 8.3). The enzymes of alkaliphiles, especially the extracellular enzymes are catalytically adapted to perform at alkaline environment. Although not all enzymes used in alkaline applications originated from alkaliphiles, alkaliphiles have been considered as reliable sources of alkaline active enzymes. Hence, a large number of alkaline active enzymes from alkaliphiles have been isolated, characterized, evaluated and some of them are applied in different industries. The majority of commercially important alkaline active enzymes are biopolymer degrading enzymes and in this chapter, an effort is made to summarize the applications of these enzymes in some selected areas.

2.1 *Application of Alkaline Active Proteases*

Proteases, enzymes that degrade proteins are used in various areas including in food processing, medicine, cosmetic, synthetic chemistry, waste management, etc. The annual global sale of industrial enzymes in 2013 was estimated to be around 3.7 Billion US dollars and nearly half of this market was accounted for proteases. Alkaline active proteases have been used in detergent and leather tanning industries, and they are among the important components of the commercial proteases.

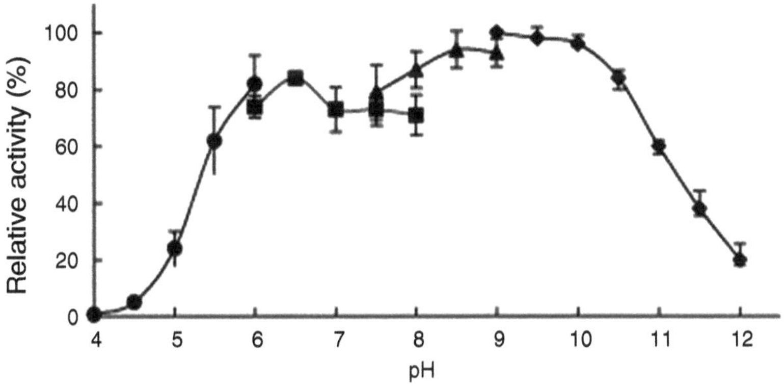

Fig. 8.3 The effect of pH on the activity of *Bacillus halodurans* S7 xylanse. The enzyme was incubated at 70 °C with 1 % (w/v) birchwood xylan dissolved in 50 mM sodium acetate (*filled circle*), sodium phosphate (*filled square*), Tris–HCl (*filled triangle*) and glycine–NaOH (*filled diamond*). About 0.5 U/mL of the xylanase activity at pH 9 was taken as 100 % (Reproduced with permission from Mamo et al. 2006)

2.1.1 Detergent

Detergent preparations containing enzymes have been in use for more than five decades. The proven benefits of using enzymes in cleaning made these biological catalysts important ingredient in the formulations of common detergents, stain removers, automatic dishwashing detergents, laundry pre-spotters as well as in industrial and medical cleaning agents. Today, in developed countries more than half of all detergents available in the market contain at least one enzyme, and the detergent industry accounts for about 30 % of the total global enzyme sale. The success story of using enzymes in detergents is mainly due to the washing performance as well as its eco-friendly nature. For instance, according to the information retrieved from the Novozymes webpage (http://www.novozymes.com), laundry washing is one of the most energy consuming activities in common household. The mere reduction of the washing temperature from 40–60 °C to 30 °C in Europe alone can reduce the emission of 12 million tons of carbon dioxide which is comparable to the annual emission of 3 million cars. One of the benefits of using enzymes in laundry washing is reduction of washing temperature without compromising the cleaning efficiency. This suggests that the use of enzymes in this application has clear economical and environmental benefits.

Detergents are of different kinds that are used in a variety of cleaning. Among cleaning agents, alkaline detergents (which are in the pH range of 9–12.5) are the most common detergents used in dish and laundry washing. The alkalinity enhances dirt removal and results in better washing performance. Alkaline detergents contain sodium derivatives such as tripolyphosphate, carbonate, phosphate, bicarbonate and sulphate which contribute to the alkalinity of the detergent. Although it is known that protein dirt treated with proteases can be removed easily and effectively, not all proteases are important to formulate detergents. It is vital to use proteases that are compatible with the detergent ingredients and its high pH.

Ideal proteases for alkaline detergent formulation should: (a) be operationally stable in the alkaline conditions, (b) be resistant to chelating and bleaching agents, (c) have reasonably high activity and stability at 30–60 °C, and (d) be stored at room temperature for months without losing activity. However, the great majority of proteases studied so far do not have all these properties and the search for more amenable alkaline proteases has continued through screening of new proteases and also by engineering of existing proteases to fine-tune their properties for detergent application. A large number of alkaline active proteases have been studied so far and in Table 8.1 some examples of alkaline active enzymes produced by alkaliphiles is given.

Most of the commercially available detergent proteases are from *Bacillus* spp. The known major detergent protease suppliers such as Novozymes, GistBrocades, Nagase Chemtex, Showa Denko K.K., Kao and Genencor developed the first and second generation of *Bacillus* proteases for detergent application. Novozymes supply the market with Alcalase, Esparase and Savinase. Alcalase is from *B. licheniformis.* Esperase and Savinase are from alkaliphilic strains of *B. licheniformis* and *B. amyloliquefaciens,* respectively. The other well-known protease brand Maxatase is from *B. licheniformis* and the patent for it is owned by GistBrocades. The *B.*

Table 8.1 Some alkaline active proteases

Organism	Optimum pH for protease activity	References
Paenibacillus tezpurensis AS-S24-II	9.5	Rai et al. (2010)
Nocardiopsis prasina HA-4	7	Ningthoujam et al. (2009)
Bacillus sp. GUS1	6–12	Seifzadeh et al. (2008)
Stenotrophomonas maltophilia-MTCC 7528	10	Kuddus and Ramteke (2009)
Nesterenkornia sp.	7–11.5	Gessesse et al. (2003)
Bacillus circulans		Rao et al. (2009)
Bacillus sp. KSM-K 16	12.3	Kobayashi et al. (1995)
Bacillus sp.	8–9	Haile and Gessesse (2012)
Bacillus patagoniensis	9–12	Olivera et al. (2006)
Bacillus sp. L21	11	Genckal and Tari (2006)
Bacillus sp. JB-99	11	Johnvesly and Naik (2001)
Streptomyces sp.	11	Tatineni et al. (2008)
Bacillus cereus MCM B-326	9	Nilegaonkar et al. (2007)
Bacillus halodurans JB 99	11	Shrinivas and Naik (2011)

licheniformis enzymes, Alcalase and Maxatase can be used effectively in the pH range of 7–10.5 and at 10–65 °C. Savinase and Esperase may be used even at more alkaline condition of pH 11 and 12, respectively. An alkaline active protease from *Nesterencornia* sp. shows interesting properties as an additive to detergents due to high pH-stability (Fig. 8.4) and resistance to chelating agents found in washing medium (Bakhtiar et al. 2003). In recent years, the third generation (engineered *Bacillus* proteases) of detergent proteases from Novozymes (Kannase®, Durazyme®) and Genencor (Properase®, Purafect OxP®) have been available in the market.

2.1.2 Dehairing

Leather tanning involves several steps and dehairing is one of the essential pre-tanning steps of leather processing. Dehairing refers to removal of hair together with epidermis and non-collagenous proteins. The conventional dehairing processes use lime and sulfide which is one of the major sources of pollution emanating from leather tanning industries. The use of sulfide may cause formation of hydrogen sulfide (H_2S) which is toxic with obnoxious odor and if left untreated poses a serious environmental threat. The growing concern about environment and the strict legislation concerning pollution, have forced leather tanning industries to be looking for alternative, greener processes. In this line, the use of alkaline active proteases in hide dehairing is becoming attractive. These proteases are known as keratinases (Anbu et al. 2005; Brandelli 2008; Cai et al. 2008). A number of enzymes have been studied with regard to their dehairing efficiency (Shrinivas and Naik 2011; Senthilvelan et al. 2012; Haile and Gessesse 2012). Enzymatic dehairing is often performed in the pH range of 8 to 10 and currently there are some commercially available dehairing enzymes.

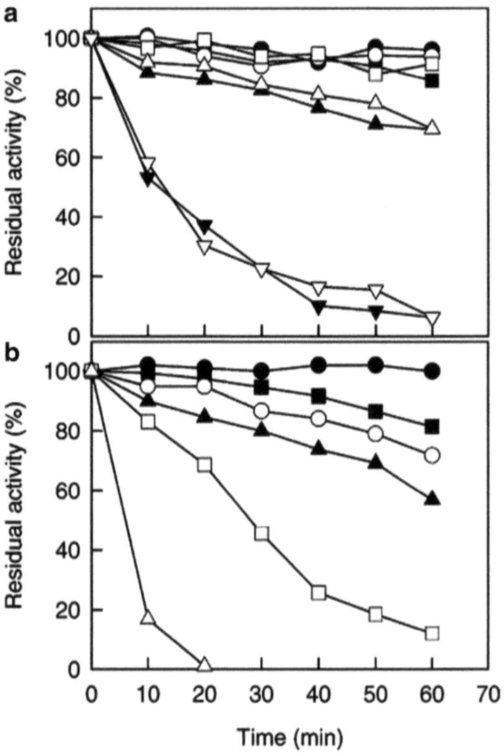

Fig. 8.4 Thermal stability of (**a**) protease AL-20 and (**b**) protease AL-89 determined in the absence (open symbols) or presence (closed symbols) of 5 mM Ca^{2+}. The enzymes were incubated at 55 °C (*open circle, filled circle*), 60 °C (*open square, filled square*), 65 °C (*open triangle, filled triangle*) and 70 °C (*open inverted triangle, filled inverted triangle*) in 50 mM glycine-NaOH buffer, pH 10. Samples were withdrawn at time interval and residual activity was measured following standard assay procedure (Reproduced with permission from Gessesse et al. 2003)

The technical and economic advantages of enzymatic dehairing have been demonstrated using enzyme/sodium metasilicate instead of the lime/sulfide system. The use of enzymes resulted in reduced chemical oxygen demand (COD) and total solid loads by 55 % and 25 %, respectively (Bhavan et al. 2008). The use of enzymes in the dehairing step reduces the consumption of lime by half, and also shortens the process time (Soerensen et al. 2011; de Souza and Gutterres 2012). This in turn will lead to cut down the amount of deliming agent such as ammonium chloride or sulphate needed to remove lime used at the dehairing step and further contribute to reduce the process cost and environmental burden.

2.1.3 Bating

The process of softening hides by removing non-collagenous proteins such as elastin, albumin, globulin, and mucoids is called bating. This allows rupturing of the collagen fibers and enhances the absorption of tanning chemicals which gives the finished leather the desired properties. Among the leather processing steps, bating has been the first step where enzymes were used. Studies revealed that the use of enzymes significantly reduces the amount of tanning agents (Kamini et al. 1999) which is attractive from environmental viewpoint.

Proteolytic enzymes extracted from bovine and pig pancreas have been used in bating. The extract contains chymotrypsinogen, trypsinogen, procarboxy-peptidase which later are activated to chymotrypsin, trypsin and carboxypeptidase, respectively. Over the years bacterial and fungal enzymes have been replacing the pancreas preparations, mainly due to their easy production and desirable properties. Leathers bated with alkaline active proteases have shown good physicochemical properties (Thanikaivelan et al. 2004; Choudhary et al. 2004). Moreover, the microbial enzymes do not attack the collagen. Therefore, there has been interest in using the alkaline active microbial proteases and today a number of these enzymes are available in the market.

2.2 Application of Alkaline Active Cellulases

Cellulases, enzymes that brake down cellulosic materials have got applications in industries producing food, feed, bioenergy carriers, biochemicals, textiles, etc. Most of these applications use acid- or neutral-active cellulases which are often obtained from fungi. The revelation of alkaline active cellulases (Horikoshi et al. 1984) has opened new application windows. In the last three decades, a number of alkaline active cellulases have been isolated and characterized, the great majority happened to be from strains of *Bacillus* as shown in Table 8.2.

Table 8.2 Examples of carbohydrate degrading alkaline active hydrolases

Organism	Enzyme	Optimum pH for activity	References
Bacillus sp. KSM-635	Cellulase	9.5	Ito et al. (1989a, b)
Cephalosporium sp. Rym-202	Cellulase	7–9.5	Kang and Rhe (1995)
Bacillus sp.	Cellulase	8–10.9	Sashihara et al. (1984)
Bacillus licheniformis C108	Cellulase	10	Aygan et al. (2011)
Bacillus sp. HSH-810	Cellulase	10	Kim et al. (2005)
B. circulans	Cellulase	8.5	Hakamada et al. (2002)
Bacillus sp. KSM-S237	Cellulase	8.6–9	Hakamada et al. (1997)
B. halodurans S7	Xylanase	9–10	Mamo et al. (2006)
Bacillus sp. AR-009	Xylanase	9–10	Gessesse (1998)
Micrococcus sp. AR-135	Xylanase	7.5–9	Gessesse and Mamo (1998)
B. pumilus 13a	Xylanase	9	Duarte et al. (2000)
Bacillus sp.	Xylanase	9–10	Balakrishnan et al. (1992)
Bacillus sp.	Xylanase	9–0	Nakamura et al. (1994)
Bacillus sp. NT-33	Pectinase	10.5	Cao et al. (1992)
Bacillus sp. MG-cp-2	Pectinase	10	Kapoor et al. (2001)
Amycolata sp.	Pectinase	10.25	Bruhlmann et al. (1994)
B. licheniformis	Pectinase	11	Singh et al. (1999)
B.halodurans LBK 34	Amylase	10.5–11.5	Hashim et al. (2005)
Bacillus sp. KSM-K38	Amylase	8–9.5	Hagihara et al. (2001)
Bacillus sp. TS-23	Amylase	9	Lin et al. (1998)
B. agaradhaerens LS-3C	CGTase	9	Martins and Hatti-Kaul (2002)
B. pseudalcaliphilus 20RF	CGTase	5–10	Atanasova et al. (2011)

2.2.1 Detergent

The discovery of alkaline cellulases allowed formulation of cellulase containing detergents. Unlike other detergent enzymes, cellulases do not degrade or modify the dirt or stains, instead, these enzymes modify the surface of the garment cellulosic fiber to improve the over-all washing performance. Cellulases intended for alkaline detergent formulation must be active and stable in alkaline condition and modify the amorphous cellulosic structures (microfibrils) but not the main crystalline cellulose fibers of the fabric. Alkaline cellulases which are known so far either do not degrade crystalline cellulose or degrade it only slightly (Horikoshi 2011). These enzymes act on the amorphous interfiber celluloses and boost the detergent ability of removing dirt including those which are oily in nature (Ito et al. 1989a, b; Murata et al. 1991). The removal of the non-crystalline cellulose from the surface of the fabric discharges dirt attached to it and hence results in superior cleaning. In addition to dirt removal, the use of alkaline active cellulases in detergents has colour brightening effect. Garments made of native or blended cellulose fibers, when used and washed many times, become 'fluffy' and its color fades. This is due to the defibrillation of the crystalline cellulose to microfibrils which are loose from the main fibers surface. When light shine on this surface, a portion of the light will be reflected back before it reaches the main fiber and that leaves the observer with an impression that the garment color is dull. The degradation of the microfibrils by the detergent cellulase removes the surface fuzziness and gives back the used garment a smooth and glossy appearance, and restores its original color brightness. Moreover, it has been observed that the removal of the microfibrils improves the hydrophilicity and moisture absorbance of the fabric (Bhat 2000) which may positively contribute to the washing performance. The use of alkaline active cellulases in detergents offers another desirable effect on cellulose based garments—greater softness.

The alkaline active cellulase EglK of the alkaliphilic *Bacillus* sp. KSM-635 was the first cellulase used in detergent formulation (Ito et al. 1989a, b). Since then a sizable number of alkaline active cellulases have been reported and evaluated for potential detergent application (Ito et al. 1998). Several companies including Kao Corporation and Novozymes have been developing and marketing alkaline active cellulases for detergent application (Olsen and Falholt 1998).

2.2.2 Biostoning

Cellulases have another interesting application in textile industry. These enzymes have been used in the finishing step of cellulose based textiles which aims to improve the textile hand-feel and appearance (Hebeish and Ibrahim 2007; Karmakar and Ray 2011). The use of cellulases in processing cotton based textiles has technical and economic advantages. For instance, jeans are traditionally stonewashed in washing machines with pumice stone (1–2 kg/pair of jeans). This stonewashing affects the machine as well as the garment quality. To alleviate these problems, cellulases have been used instead of pumice stone in a process known as biostoning. The cellulases used in this process modify the cotton fabric and degrade the small fiber on the yarn surface which facilitates the removal of the dye during the washing cycle. This process has proven to be more effective in solving the washing machine problem. Moreover, biostoning improves the fabric quality by minimizing wear; and it is cheaper, faster, and has a low environmental impact (Sukumaran et al. 2005; Singh et al. 2007; Araújo et al. 2008; Andreaus et al. 2014).

Although neutral and acid active cellulases are used in biostoning process, due to backstaining problems use of alkaline active cellulases has been suggested for biostoning (Cavaco-Paulo et al. 1998). However, despite its potential there is no available report so far on the use of alkaline cellulases in biostoning.

2.2.3 Toner Removal

A significant amount of paper consumed in the world is recycled. Compared to glass and plastic, more paper is recovered for recycling. According to the Environmental Protection Agency (EPA, http://www.epa.gov/osw/education/quest/pdfs/sections/u2_chap2.pdf), recycling a single ton of paper saves about 17 trees, 26,500 L of water and 1750 L of oil. These advantages drive more paper to recycling and there has been a continuous effort to improve the technical and economic aspects of the recycling process.

Enzymes have been used to improve the efficiency of paper recycling. Over 70 % of the paper is uncoated and printed with toners (Thakur 2006), removal of this toner is an important component of the recycling process. However, it is often difficult to remove toners by the conventional alkaline deinking process which consumes large amount of chemicals and water (Shrinath et al. 1991; Thomas 1994). Paper recycling is done at alkaline conditions which swell the cellulose fibers and increase their flexibility. These alkali-induced changes enhance the removal of the ink from the fibers' surface (Wielen et al. 1999). In addition to swelling of the fibers, the alkali can also interact directly with the dye which may weaken its interaction to the fibers and facilitate its removal (Shrinath et al. 1991). Therefore, if enzymes are considered for deinking process, it is necessary that the enzymes should be active and stable in the alkaline recycling process.

The use of alkaline active cellulases in toner removal have been studied and the results show that the application of these enzymes greatly improve toner removal from waste paper and facilitate recycling process (Gubitz et al. 1998; Bajpai and Bajpai 1998; Lee et al. 2007; Vyas and Lachke 2003; Ibarra et al. 2012). Thus, the use of enzymes is desirable from economic, technical as well as environmental viewpoint. Although it is believed that the hydrolysis of the microfibrils by the cellulase helps the separation of the dye during flotation/washing step (Ibarra et al. 2012), the exact mechanism of dye removal by these enzymes is not yet precisely known.

2.3 Alkaline Active Xylanases and Their Application in Delignification of Kraft Pulp

The first xylan degrading enzyme from alkaliphilic microorganism was reported in 1973 (Horikoshi and Atsukawa 1973). This xylanase from the alkaliphilic *Bacillus* sp. C-59-2 was optimally active in a pH range of 6–8. Since then several xylanases from a wide variety of alkaliphiles have been purified and characterized. The great majority of these xylanases have pH optima around neutral condition but with relatively good activity in the alkaline range. However, few xylanases with even more alkaline pH optima for activity have also been reported. The most alkaline active xylanases reported to date include XylB from *Bacillus* sp. AR-009 (Gessesse 1998); xylanases from *Bacillus* sp. TAR-1 (Nakamura et al. 1994), *Bacillus halodurans* C-125 (Honda et al. 1985), *Bacillus halodurans* S7 (Mamo et al. 2006) and *Bacillus* sp. (NCL-86-6-10) (Balakrishnan et al. 1992) which are optimally active at pH 9-10. Other highly alkaliphilic xylanases include xylanase J from *Bacillus* sp. strain 41 M-1 (Nakamura et al. 1993) and a xylanase from *Bacillus pumilus* 13a (Duarte et al. 2000) both of which have a pH optima at pH 9.

Bacillus spp. are fascinating sources of several alkaline active and stable enzymes. Members of this genus are so far the promising sources of xylanases which are operationally stable at elevated pH. So far, very few xylanases are described which are optimally active up to pH 10 and at or above 70 °C (Honda et al. 1985; Gessesse 1998; Chang et al. 2004; Mamo et al. 2006). Moreover, these xylanases are quite stable at temperatures as high as 65 °C when incubated in substrate free buffer. A better stability is expected in the presence of substrate.

The most important application of alkaline active xylanases is in the delignification of pulp. The modern pulp and paper industry uses different methods of recovering cellulose fibers from lignocellulosic biomass, most of which is from wood. Chemical pulping which includes Kraft, sulfite and semi-chemical processes is the dominant method among the pulping processes. Kraft pulping accounts for over 90 % of the total chemical pulping. The Kraft process which is also called the sulfate process involves an alkaline treatment with solutions of sodium sulfide and sodium hydroxide at about 170 °C for 2 h resulting in the degradation and solubilization of lignin. The resulting pulp from this alkaline cooking has brownish color primarily due to the residual lignin and lignin derivatives which are covalently attached to the hemicellulose fraction. Removal of this lignin requires bleaching process, which often uses chlorine. Although the use of chlorine effectively bleaches the pulp, it leads to formation of undesirable chlorinated organic by-products which are known to be highly toxic and mutagenic (Nagarathnamma and Bajpai 1999; Pokhrel and Viraraghavan 2004). Due to the growing public awareness about environment and the appearance of strong legislations concerning pollution, it was necessary to look for alternative ways to minimize or avoid the use of chlorine in bleaching pulp. Currently, different environmental friendly bleaching chemicals like H_2O_2, O_2 and O_3 are substituting the toxic chlorine based bleaching chemicals (Abad et al. 2001).

Enzymatic treatment of Kraft pulp prior to bleaching hydrolyses the xylan that exists in the pulp. The xylan hydrolysis facilitates lignin removal and hence the enzymatic treatment allows to substantially minimize the use of chlorine (Vicuna et al. 1997; Ximenes et al. 1999; Haarhoff et al. 1999) and has emerged as one of the most promising alternative approaches. The use of xylanases has also proved important in enhancing the bleaching efficiency and decrease the consumption of H_2O_2, O_2 and O_3 (Srinivasan and Rele 1999; Abad et al. 2001). Several studies have been conducted to understand the mechanism(s) by which xylanases enhance pulp bleaching. The findings of these studies center around three major mechanisms. Xylanases enhancing bleaching because xylanase treatment can: (1) improve lignin extraction, (2) alter carbohydrate and lignin associations, or (3) cleave re-deposited xylan (Kenealy and Jeffries 2003).

So far, the use of xylanases for pre-bleaching of Kraft pulp became one of the greatest success stories of enzyme applications in the pulp and paper industry. Since the Kraft pulp is alkaline and hot, it is necessary to cool and neutralize the pulp before adding enzymes. However, this is time consuming and expensive. If thermostable alkaline xylanases are used, it significantly downsizes the cooling and the pH-readjustment cost. This is the main driving force behind the aggressive search for alkaline active thermostable xylanases.

2.4 *Alkaline Active Pectinases*

Pectins are macromolecular acidic polysaccharides built with a backbone of α-(1-4) linked D-galacturonic acid residues together with some rhamnose residues, and the side chains are rich in arabinose, galactose and xylose. Pectic substances are among the more complex polymeric carbohydrates. A number of processes that modify plant biomass involve some degree of pectin removal, which can be done either by chemical or/and enzymatic treatments. However, the use of enzymes known as pectinases has several advantages over the chemical based alternative. Acid- and neutral-active pectinases are used in industries processing food, juice, wine, etc (Alkorta et al. 1998). The discovery of alkaline active pectinases paved the way for new application. Today, these enzymes are among the most important groups of industrial enzymes and used in different applications such as textile processing, plant bast fibers degumming, waste management and paper production. Some of these applications are discussed below.

2.4.1 Bioscouring of Cotton

The textile industry heavily depends on cotton fibers. Processing of cotton to yarn that can be used in textile production involves several steps. One of the important steps is the removal of pectin which binds waxes and proteins together to form the water insoluble cellulose fiber protective structure. The removal of pectin from cotton is known as scouring—a conventional process that dissolves pectin using sodium hydroxide solution and high temperature. The alkaline condition solubilises pectin, hemicellulose and lignin fraction of the plant material and leaves behind the cellulose fibers. The high pH, in addition to facilitating the removal of the non-cellulosic components of the cotton, prevents growth of microorganisms (especially fungi) that are capable of degrading the cellulose fibers. However, this process is expensive and not environmentally benign as it uses excess water for rinsing and requires high energy input to heat the suspension. The alternative process often referred to as bioscouring uses enzymes that selectively hydrolyse the pectin fraction of the cotton at low temperature and significantly reduces the cost and makes the process environmentally benign (Bruhlmann et al. 1994). Moreover, the use of alkaline active pectinases reduces the consumption of sodium hydroxide substantially and results in better quality fibers than the conventional approach (Hoondal et al. 2002).

2.4.2 Degumming of Bast Fibers

The removal of gummy substances that covers the cellulosic part of plant fibers is known as degumming, a step necessary for production of fibers for different applications (Said et al. 1991). Traditionally, bast fibers are degummed in 2–20 % (w/v) sodium hydroxide solution supplemented with wetting and reducing agents (Cao et al. 1992). After 24 h of soaking, the suspension will be boiled for up to 4 h, rinsed, neutralized and washed extensively. Thus, this process is not only expensive but it also has a negative impact on the environment.

Like in bioscouring that recovers fibers from cotton, fibers from other botanical sources can also be processed using pectinases and benefit from the economic and environment advantages. So far, pectinases have been tried to degumme fibers of flax, ramie, jute and hemp (Cao et al. 1992; Bruhlmann et al. 1994; Henriksson et al. 1997; Kapoor et al. 2001). Alkaline active pectinases are desirable for degumming of plant fibers since a high pH prevents contamination by cellulose fiber degrading fungi and allows an open fermentation system to be adopted which is expected to substantially reduce the processing cost. Moreover, it removes lignin and hence may minimize the amount of bleaching agents needed in subsequent steps.

2.5 Alkaline Active Starch Degrading Enzymes

Starch degrading enzymes are one of the most important groups of enzymes used in processing applications such as food, textile, fuel ethanol, etc. Amylases are of different types and produced by diverse group of organisms. Despite they are important in several applications; alkaline active amylases are relatively less studied. Two important application examples of alkaline active starch modifying enzymes are given below.

2.5.1 Alkaline Active α-Amylases

Alkaline active α-amylases can potentially be used to hydrolyze starch under alkaline condition in starch and textile industries. Moreover, it can be used as detergent ingredient for automatic dishwashers and laundries. Although not from alkaliphilic organism and as such not alkaline active, Novozymes developed amylases, Termamyl® and its mutant Duramyl® have been used to formulate detergents to remove food based starch stain (Olsen and Falholt 1998). There has been interest to develop ideal amylases that are optimally active and stable in alkaline detergents. This interest has contributed to the discovery of a number of alkaline active amylases (Hashim et al. 2004; Burhan et al. 2003; Hashim et al. 2005; Saxena et al. 2007).

Amylases are known to be metallozymes which require calcium ions at their active centre and detergents contain chelating agents that easily strip the calcium ions from amylases and inactivate them. Thus, one of the targets in developing amylases for detergent application has been to look for alkaline active amylases that are functional in the presence of the detergent chelating agents. In this regard, a unique enzyme, AmyK38, from *Bacillus* sp. KSMK38 has been reported that exhibit at highly alkaline condition a remarkable resistance to the effect of not only chelating agents but also to oxidants and surfactant reagents (Hagihara et al. 2001). Detailed studies revealed that this amylase was completely free of calcium ions (Nonaka et al. 2003) and its activity is calcium independent.

2.5.2 Cyclodextrin Glycosyl Transferase (CGTase)

Cyclodextrins are widely used in foods, cosmetics, pharmaceuticals, toiletries, etc. for stabilization of volatile compounds, as molecular shields against the effects of oxidants and UV radiation, as deodorising agents and solubilizers of water-insoluble substances (Szente and Szejtli 2004; Kurkov and Loftsson 2013; Crini 2014). Cyclodextrins are produced from starch using CGTases, enzymes with multifunctional activity. CGTases catalyse the formation of cyclic oligosaccharides commonly comprising 6, 7 or 8 glucose units which are known as α-, β- and γ-cyclodextrins, respectively. The ring closure reaction in the formation of these molecules is catalyzed by the enzyme which uses the non-reducing end of a linear oligosaccharide bound to the enzyme as an acceptor of the intramolecular transglycosylation reaction. CGTases from different microorganisms often produce different levels of α-, β- and γ-cyclodextrins.

Although several CGTases have been studied, CGTases from alkaliphilic *Bacillus* strains are found to be the most promising enzymes for production of cyclodextrins (Matzuzawa et al. 1975; Martins et al. 2001; Niehaus et al. 1999). The majority of alkaline active CGTases convert starch into β-cyclodextrin as main product which contains small amount of other cyclodextrins (Horikoshi 1999; Atanasova et al. 2011; Martins and Hatti-Kaul 2002). This ability of CGTases from alkaliphiles is attractive as it simplifies the subsequent downstream processing of the major product. Currently, the major commercial enzymes used in industrial production of cyclodextrins are from alkaliphilic *Bacillus* strains.

3 Other Products of Biotechnological Importance from Alkaliphiles

In addition to serving as sources of enzymes for industrial and environmental applications, alkaliphiles can also provide an impressively wide range of products for various applications. Since these organisms grow easily, faster and at higher density with minimized risk of contamination they are attractive for a variety production processes. Moreover, the high physiological diversity of these organisms may allow choosing an ideal product and suitable organism to develop production process. However, in this regard, there has been only a limited effort so far which aims to exploit the potential of alkaliphiles. Some selected examples of products and the potentials of using alkaliphiles to make these products are discussed below.

3.1 Organic Acid Production by Alkaliphiles

Organic acids are very useful ingredients in production of vast array of products ranging from food to medicine. Organic acids can be produced through chemical or biological (microbiological) routes; however, there is a growing trend that microbiological production is becoming an important alternative. These valuable substances account a significant amount of the global fermentation-based market. Filamentous fungi, lactic acid bacteria and some genetically modified bacteria such as *E. coli* and yeast have been used for production of different organic acids. However, there has been little attention given to tap the potential of other groups of microorganisms. For instance, it is not uncommon to detect organic acids in the fermentation broths of alkaliphilic organisms but studies made on production of organic acids by alkaliphiles are scarce. A summary on the studies made previously on alkaliphiles in relation to acid production is provided below to highlight their remarkable potential in this sector.

One of the fascinating features of alkaliphiles when grown in media composed of sugars is their ability to dramatically shift the pH of the media. This pH shift is mainly due to acid formation. Also this has not been properly studied; production of organic acids by alkaliphiles may have a relevance to adaptation of their high pH environment. From biotechnology point of view, production of organic acids by alkalipiles seems very attractive. Studies on sugar catabolism revealed that some alkaliphiles produce organic acids such as acetic acid, formic acid, succinic, propionic acid, butyric acid and lactic acid from different carbohydrates (Paavilainen et al. 1994; Garnova and Krasilnikova 2003; Zhilina et al. 2004; Kulshreshtha et al. 2012; Yokaryo and Tokiwa 2014). This fermentation performance may allow using cheap mixed carbohydrate substrates such as plant biomass hydrolysate. Moreover, most of the alkaliphiles are salt tolerant, which in fact cannot grow in the absence of sodium or other monovalent ions, and hence bases such as NaOH can be used to maintain the pH of the production media instead of $CaCO_3$. The use of $CaCO_3$

makes mixing during the ferementation process difficult. The other advantage of using alkaliphiles for organic acid production is the low risk of contamination due to media high pH and salt content that keeps away contaminants (Calabia et al. 2011; Jiang et al. 2013). Hence, it can be speculated that alkaliphiles may be promising candidates for production of organic acids (Paavilainen et al. 1994).

Halophilic and alkaliphilic lactic acid bacteria which are able to produce organic acids have been isolated from cheese, decaying marine algae, fish, and salted and fermented shrimp paste (Ishikawa et al. 2007, 2009, 2013). An alkaliphilic microorganism isolated from a marine environment, produced about 66 g/L L-lactic acid at pH 9 (Calabia et al. 2011). An alkaliphilic *Exiguobacterium* strain has been reported producing 125 g/L 100 % optically pure L-lactic acid (Jiang et al. 2013). Alkaliphilic *Bacillus* strains have also been reported producing different organic acids in carbohydrate containing media, and it has also been observed that formation of acids improves with increasing pH and/or buffer concentration (Paavilainen et al. 1994). Among alkaliphiles reported so far, *Bacillus* sp. WL-S20 achieved a high level of final acid accumulation (Meng et al. 2012). Using sodium hydroxide as neutralizing agent, this organism was able to produce 225 g/L L-lactic acid with of 99.3 % yield. It is also interesting that peanut meal is used as cheap nitrogen source for this *Bacillus* and there was no detectable D-lactic acid in the fermentation broth. The results achieved in this work is comparable to what has been reported for *Lactobacillus rhamnosus* (Berry et al. 1999), *L. casei* (Ding and Tan 2006), and *L. lactis* (Bai et al. 2003). None of these *Lactobacillus* strains achieved the impressive optical purity achieved by this alkaliphilic *Bacillus*.

Although a number of alkaliphiles are known to produce organic acids, except lactic acid production, detailed studies on production of other organic acids like formic acid, succinic acid or acetic acid is very limited. The available reports simply describe the presence of different organic acids in the fermentation broth of alkaliphiles. Under unoptimized condition, accumulation of upto 5 g/L acetic acid (Paavilainen et al. 1994) and 2 g/L formic acid (Kulshreshtha et al. 2012) have been reported. Further studies on screening and production optimization of different organic acids are expected to result in higher productivity.

3.2 Carotenoids

Carotenoids are pigment molecules produced by plants and a large number of microorganisms. These compounds are tetraterpenoids and comprised of 40 carbon atoms. So far, over 700 carotenoid structures are known which arise from the modification of the carbon chain through oxidation, hydroxylation and other reactions. Carotenoids are yellow, orange, or red in color and are known to protect organisms from oxidative damage. These properties led to the use of carotenoids as food colorants, source of vitamin A, feed supplements, and nutraceuticals for cosmetic and pharmaceutical applications. The global annual sell of carotenoids is projected to reach $1.4 billion in 2018 which is higher than its 2010 sell estimated to be $1.2 billion (BCC Research 2011). Studies have shown that carotenoids have remarkable anti-carcinogenic effect (Palozza et al. 2009; Miyashita 2009) and this may potentially lead to higher demand.

Due to the expanding market (from existing and new applications) there has been an interest to look for an alternative method to the traditional production of carotenoids by synthetic reactions and extraction from plants. Production using microorganisms is becoming the new trend due to the low cost, high yield and safety benefits (Lee and Schmidt-Dannert 2002; Mata-Gómez et al. 2014).

Several alkaliphilic bacteria have been known producing carotenoids, and it has been even suggested that it may have survival importance to alkaliphiles (Aono and Horikoshi 1991); however, this needs further study to confirm. The ability of producing carotenoids among alkaliphiles seems widely distributed among different genera. The carotenoid producing alkaliphiles isolated by Aono and Horikoshi were *Bacillus* strains. A strain of *Microbacterium arborescens* has been reported producing lycopene-like carotenoid when grown at pH 10.5 (Godinho and Bhosle 2008). The sulphur oxidizing *Roseinatronobacter thiooxidans* produce carotenoids in alkaline medium containing organic nitrogen (Sorokin et al. 2000). *Paracoccus bogoriensis* cells which synthesizes carotenoid have been able to secret 0.4 mg astaxanthin per gram of wet cells (Osanjo et al. 2009). Astaxanthin is known for its anti-cancer effect (Palozza et al. 2009; Miyashita 2009), and significant anti-oxidant and immunomodulatory activity (Jyonouchi et al. 1995). It has been reported that different alkaliphilic strains of *Heliorestis* produce novel carotenoid glucoside esters (Takaichi et al. 2003) which needs further study for evaluating their application potential.

It seems that the ability of producing these valuable products is widely distributed among alkaliphiles and further studies may reveal novel carotenoids from this group of organisms. In addition, since alkaliphiles are tolerant to salt and cultivation is done at high pH, contamination is expected to be lower when alkaliphiles are used as production strains.

3.3 Antimicrobial Agents

In today's world the dramatic appearance of drug resistance among pathogens is becoming a serious threat to human society. Contrary to the needs, the rate of new antibiotics discovery has been declining which may be partly due to the focus of antibiotic production screening studies only on certain groups of microorganisms. It may be rewarding to look at less explored groups of microorganisms, especially extremophiles for novel antimicrobial agents. In fact, screening of extremophiles may also yield other bioactive compounds that can be used in treatment of non-infectious disease including cancer and cardiovascular problems.

Alkaliphilic strains of *Streptomyces* are known producing different antimicrobial agents. For instance, an unidentified isolate, *Streptomyces* sp. AK produces pyrocoll which is an active agent against fungi, protozoa, and different human cancer cell lines (Dietera et al. 2003). The alkaliphilic strains of *S. sannanensis* and *S. aburavi-*

ensis produce unidentified compounds that are effective against Gram positive bacteria (Vasavada et al. 2006; Thumar et al. 2010). Furthermore, it has been reported that *S. tanashiensis* secret an antifungal and antibacterial metabolite (Singh et al. 2009). Production of antimicrobial agents has also been observed in other alkaliphilic genera. *Nocardiopsis dassonvillei* synthesizes and accumulates intracellular phenazine antibiotics (Tsujibo et al. 1988) and a strain of *Paecilomyces lilacinus* produces two antifungal peptides (Sato et al. 1980). Unidentified member of genus *Nocardiopsis* produces a unique highly functionalized polycyclic metabolite, Naphthospironone which exhibited antibacterial activity as well as cytotoxicity (Ding et al. 2010). One of the most studied alkaliphilic *Bacillus*, *B. halodurans* secrets a novel two component bacteriocin known as haloduracin (Lawton et al. 2007; Danesh et al. 2011).

Screening studies on production of antimicrobial agents by alkaliphilic bacteria are limited and so far, there is no commercially available antibiotic developed from this group of microorganisms. However, the few available studies indicated that alkaliphiles can potentially serve as sources of novel antibiotics if well-designed extensive screening is undertaken. This may contribute to meet the growing need for novel antimicrobial agents.

3.4 Siderophores

Organisms require iron for their cellular processes including the vital electron transport chain that generates ATP and as cofactor for enzymes that mediate an array of biochemical reactions. Thus, organisms have evolved mechanisms that allow them to acquire this vital resource. One of the strategies discovered among bacteria is the production of low molecular weight chelating compounds known as siderophores. Siderophores are produced in large quantity when producer strains are grown often under iron limited conditions. Currently, about 500 types of siderophores are known.

There is an impressive application potential of microbial siderophores in agriculture, medicine, environment, etc. In agriculture, it helps to improve soil fertility, reduce excessive accumulation of heavy metals, serve as biocontrol agent that inhibits the growth of many phytopathogenic fungi, etc. In medicine, these fascinating compounds can be used in treatment of various diseases related to excess iron in the human body. For instance hemochromatosis, hemosiderosis, and accidental iron poisoning needs the removal of excess iron from the body and siderophore based drug can be used to treat this kind of diseases (Pietrangelo 2002). Another interesting application of these compounds is in the delivery of antibiotics to drug resistant pathogens. In nature there are many siderophore-antibiotic conjugates such as salimycins (Vértesy et al. 1995) and albomycins (Benz et al. 1982). If we successfully mimic this natural trick and prepare specific siderophore-antibiotic conjugates, we may have chance to overcome some of the drug resistance problems that the world is facing. Furthermore, siderophores can be used in magnetic resonance imaging (MRI) (Doble et al. 2003), and treatment of cancer (Chua et al. 2003;

Miethke and Marahiel 2007) and malaria (Gysin et al. 1991). In area of environmental applications, siderophores can be used to remove or immobilize heavy metal contaminants such as in drinking water and soil. Similarly it can be used to decontaminate soils and water from radioactive contaminants (Ruggiero et al. 2000; Von Gunten and Benes 1995).

Inorganic iron and other metal ions are extremely insoluble in alkaline environment. In soda lakes, in addition to high pH, the abundant oxygen released by photosynthetic cyanobacteria facilitates the precipitation of iron and hence the concentration of iron ions is lower than what is required for the optimal growth of bacteria. To ensure acquiring iron and maybe other metal ions as well, many alkaliphiles are known to produce siderophores (Gascoyne et al. 1991; McMillan et al. 2010). Thus, alkaliphiles can potentially be attractive sources of siderophores. With further studies, a wide range of novel siderophores can be identified from alkaliphiles which could be amenable for several applications.

4 Alkaliphiles in Environmental Applications

Microorganisms are very vital components of every environment in the biosphere and regulate virtually most of the biogeochemical systems of the planet. Microbes have been used directly or indirectly in environmental protection. The biosynthetic capabilities of microbial enzymes or whole cells used in many industries have been replacing toxic catalysts and process, and positively contribute to the environment. Microorganisms have been used to clean and rehabilitate contaminated soil and water. They have been used to substitute chemical fertilizers, recalcitrant chemical and toxic insecticides, etc. Microbes have been used to produce chemicals from renewable resources that have been produced from petroleum. There are many more examples that these organisms have enabled feasible and cost-effective processes that protect our environment which would have been impossible without their involvement. Below the use of alkaliphiles in selected environmental applications has been summarized.

4.1 Neutralizing Alkaline Effluents

A number of human activities result in alkaline waste which has to be neutralized before released to the environment. For instance industrial cleaning, electroplating, pulping, mining, textile processing, etc often have alkaline waste effluents. This waste should be neutralized to release the effluents, and often this is done with mineral acids. However, alkaliphiles are known in producing organic acids and lower the pH of their media even when it is strongly buffered. Thus, the alkaline effluents from these industrial activities can be neutralized using alkaliphiles and avoid the cost and handling problem of concentrated mineral acid. An interesting example could be the use of *Exiguobacterium* sp. which lowers the pH of an industrial waste from pH 12 to 7.5 (Kulshreshtha et al. 2012). The most important thing to be assessed if alkaliphiles are considered for neutralization of alkaline waste is to make sure that the waste contains carbohydrate, if not supplement it with some cheap carbohydrate sources such as plant biomass hydrolysate.

4.2 *Removal of Hydrogen Peroxide from Textile Process Media*

Bleaching of textile fabrics is often done using hydrogen peroxide under alkaline conditions ($\geq$ pH 9) (Spiro and Griffith 1997). The unreacted hydrogen peroxide needs to be removed after the bleaching process which otherwise can interfere in the subsequent processing steps of the textile material. One of the methods is to wash the fabric extensively and remove the hydrogen peroxide, which is time consuming and requires copious amounts of water. Moreover, the waste-stream will be alkaline and contain peroxide (Weck 1991), which means for a proper discharge needs further treatment with extra cost. The other alternative is degrading the hydrogen peroxide which has earlier been carried out by adding e.g. sodium bisulphite (Fruhwirth et al. 2002; Paar et al. 2001). However, this results in formation of salt which increases the ionic strength of the solution that might negatively influence the fiber processing if the water is used in subsequent steps. Thus, there has been a desire to decompose hydrogen peroxide to water and oxygen using enzyme, catalase. The great majority of catalases known are active in acidic to neutral condition, but bleaching using hydrogen peroxide is done at high pH and hence alkaline active catalases are desirable as it avoids pH re-adjustment and associated salt formation. Few alkaline active catalases have been reported from alkaliphiles and alkali-tolerant thermophiles (Gudelj et al. 2001; Kagawa et al. 1999; Michaud-Soret et al. 1998; Thompson et al. 2003). This shows the possibility of developing alkaline active catalases from alkaliphiles that are suitable and effective to remove hydrogen peroxide from textile fabric bleaching media.

Use of free catalase is not recommended due to interactions between the protein and the dye molecules (Tzanov et al. 2001). Thus, it is important to immobilize the enzyme and this has been proven useful (Costa et al. 2002). An alternative approach is to use the whole cell where the intracellular catalase is protected from interacting with the dye molecules, but still exposed to the hydrogen peroxide which diffuses to cell (Oluoch et al. 2006). The result of this study (Oluoch et al. 2006) was very encouraging and shows the potential of whole-cell alkaliphiles in removing hydrogen peroxide and possibly other pollutants as well.

5 Concluding Remarks

Alkaliphiles are widely distributed in natural and manmade alkaline environments. They are highly diverse and relatively easy to cultivate, which make them attractive for biotechnological applications. Although there have been a number of applications that use alkaliphile whole-cells or their products, there is still an immense potential of biotechnological applications that should be realized. However, even some of the examples described in this chapter that exhibit enormous application potential need further studies to come to commercial level. In general, this group of organisms is relatively less studied, and this is one of the main reasons that they are not so far widely used in biotechnology. However, the trickling results from the existing research, the emergence and availability of new technologies and the growing need for microbial based products and process will undoubtedly expand the biotechnological use of these remarkable organisms.

Acknowledgements Part of this work was supported by the Swedish Research Council via a regular research grant and a Research link grant.

Conflict of Interest Gashaw Mamo and Bo Mattiasson declare that they have no conflict of interest.

References

Abad S, Santos V, Parajó JC (2001) Totally chlorine-free bleaching of Acetosolv pulps: a clean approach to dissolving pulp manufacture. J Chem Technol Biotechnol 76:1117–1123

Alkorta I, Garbisu C, Llama M, Serra J (1998) Industrial application of pectic enzymes. Process Biochem 33:21–28

Anbu P, Gopinath SCB, Hilda A, Priya TL, Annadurai G (2005) Purification of keratinase from poultry farm isolate – *Scopulariopsis brevicaulis* and statistical optimization of enzyme activity. Enzyme Microb Technol 36:639–647

Andreaus J, Olekszyszen DN, Silveira MHL (2014) Processing of cellulosic textile materials with cellulases in cellulose and other naturally occurring polymers. In: Fontana JD, Tiboni M, Grzybowski A (eds) Cellulose and other naturally occurring polymers. Research Signpost, Kerala, India, pp 11–19 ISBN: 978-81-308-0543-6

Aono R, Horikoshi K (1991) Carotenes produced by alkaliphilic yellow pigmented strains of *Bacillus*. Agric Biol Chem 55:2643–2645

Aono R, Ito M, Maehida T (1999) Contribution of the cell wall component teichuronopeptide to pH homeostasis and alkaliphily in the alkaliphile *Bacillus lentus* C-125. J Bacteriol 181:6600–6606

Araújo R, Casal M, Cavaco-Paulo A (2008) Application of enzymes for textile fibres processing. Biocatal Biotransform 26:332–349

Atanasova N, Kitayska T, Bojadjieva I, Yankov D, Tonkova A (2011) A novel cyclodextrin glucanotransferase from alkaliphilic *Bacillus pseudalcaliphilus* 20RF: purification and properties. Process Biochem 46:116–122

Aygan A, Karcioglu L, Arikan B (2011) Alkaline thermostable and halophilic endoglucanase from *Bacillus licheniformis* C108. Afr J Biotechnol 10:789–796

Bai DM, Wei Q, Yan ZH, Zhao XM, Li XG, Xu SM (2003) Fed-batch fermentation of *Lactobacillus lactis* for hyper-production of L-lactic acid. Biotechnol Lett 25:1833–1835

Bajpai P, Bajpai PK (1998) Deinking with enzymes: a review. Tappi J 81:111–117

Baker-Austin C, Dopson M (2007) Life in acid: pH homeostasis in acidophiles. Trends Microbiol 15:165–171

Bakhtiar S, Andersson MM, Gessesse A, Mattiasson B, Hatti-Kaul R (2003) Stability characteristics of a calcium-independent alkaline protease from *Nesterenkornia* sp. Enzyme Microb Technol 32:525–531

Balakrishnan H, Dutta-Choudhary N, Srinivasan MC, Rele MV (1992) Cellulase-free xylanase production from an alkalophilic *Bacillus* sp. NCL-87-6-10. World J Microbiol Biotechnol 8:627–631

BCC Research: The global market for carotenoids (2011) http://www.bccresearch.com/market-research/food-and-beverage/carotenoids-global- market-fod025d.html.

Benz G, Schroder T, Kurz J, Wunsche C, Karl W, Steffens G, Pfitzner J, Schmidt D (1982) Konstitution der desferriform der albomycine d1, d2 and e. Angew Chem 94:552–553 and Suppl 1322–1335.

Berry AR, Franco CMM, Zhang W, Middelberg APJ (1999) Growth and lactic acid production in batch culture of *Lactobacillus rhamnosus* in a defined medium. Biotechnol Lett 21:163–167

Bhat MK (2000) Cellulases and related enzymes in biotechnology. Biotechnol Adv 18:355–383

Bhavan S, Rao JR, Nair BU (2008) A potential new commercial method for processing leather to reduce environmental impact. Environ Sci Pollut Res Int 15:293–295

Brandelli A (2008) Bacterial keratinases: useful enzymes for bioprocessing agroindustrial wastes and beyond. Food Bioprocess Technol 1:105–116

Bruhlmann F, Kim KS, Zimmerman W, Fletcher A (1994) Pectinolytic enzymes from *Actinomycetes* for degumming of ramie bast fibers. Appl Environ Microbiol 60:2107–2112

Burhan A, Nisa U, Gökhan C, Ömer C, Ashabil A, Osman G (2003) Enzymatic properties of a novel thermostable, thermophilic, alkaline and chelator resistant amylase from an alkaliphilic *Bacillus* sp. isolate ANT-6. Process Biochem 38:1397–1403

Cai C-G, Chen J-S, Ql J-J, Yin Y, Zheng X-d (2008) Purification and characterization of keratinase from a new *Bacillus subtilis* strain. J Zhejiang Univ Sci B 9:713–720

Calabia BP, Tokiwa Y, Aiba S (2011) Fermentative production of L-(+)-lactic acid by an alkaliphilic marine microorganism. Biotechnol Lett 33:1429–1433

Canganella F, Wiegel J (2011) Extremophiles: from abyssal to terrestrial ecosystems and possibly beyond. Naturwissenschaften 98:253–279

Cao J, Zheng L, Chen S (1992) Screening of pectinase producer from alkalophilic bacteria and study on its potential application in degumming of ramie. Enzyme Microbiol Technol 14:1013–1016

Cavaco-Paulo A, Morgado J, Almeida L, Kilburn D (1998) Indigo backstaining during cellulase washing. Text Res J 68:398–401

Chang P, Tsai WS, Tsai CL, Tseng MJ (2004) Cloning and characterization of two thermostable xylanases from an alkaliphilic *Bacillus firmus*. Biochem Biophys Res Commun 319:1017–1025

Choudhary RB, Jana AK, Jha MK (2004) Enzyme technology applications in leather processing. Indian J Chem Technol 11:659–671

Chua AC, Ingram HA, Raymond KN, Baker E (2003) Multidentate pyridinones inhibit the metabolism of nontransferrin-bound iron by hepatocytes and hepatoma cells. Eur J Biochem 270:1689–1698

Costa SA, Tzanov T, Carneiro F, Gübitz GM, Cavaco-Paulo A (2002) Recycling of textile bleaching effluents for dyeing using immobilized catalase. Biotechnol Lett 24:173–176

Crini G (2014) A history of cyclodextrins. Chem Rev 114:10940–10975

Danesh A, Mamo G, Mattiasson B (2011) Production of haloduracin by *Bacillus halodurans* using solid-state fermentation. Biotechnol Lett 33:1339–1344

de Souza FR, Gutterres M (2012) Application of enzymes in leather processing: a comparison between chemical and coenzymatic processes. Braz J Chem Eng 29:473–481

Delgado O, Quillaguamán J, Bakhtiar S, Mattiasson B, Gessesse A, Hatti-Kaul R (2006) *Nesterenkonia aethiopica* sp. nov., an alkaliphilic, moderate halophile isolated from an Ethiopian soda lake. Int J Syst Evol Microbiol 56:1229–1232

Dietera A, Hamm A, Fiedler HP, Goodfellow M, Muller WE, Brun R, Bringmann G (2003) Pyrocoll, an antibiotic, antiparasitic and antitumor compound produced by a novel alkaliphilic *Streptomyces* strain. J Antibiot 56:639–646

Ding S, Tan T (2006) L-Lactic acid production by *Lactobacillus casei* fermentation using different fed-batch feeding strategies. Process Biochem 41:1451–1454

Ding ZG, Li MG, Zhao JY, Ren J, Huang R, Xie MJ, Cui XL, Zhu HJ, Wen ML (2010) Naphthospironone a: an unprecedented and highly functionalized polycyclic metabolite from an alkaline mine waste extremophile. Chemistry 16:3902–3905

Doble DMJ, Melchior M, OSullivan B, Siering C, Xu J, Pierre VC, Raymond KN (2003) Toward optimized high-relaxivity MRI agents: the effect of ligand basicity on the thermodynamic stability of hexadentate hydroxypyridonate/catecholate gadolinium(III) complexes. Inorg Chem 42:4930–4937

Duarte MC, Pellegrino AC, Portugal EP, Ponezi AN, Franco TT (2000) Characterization of alkaline xylanases from *Bacillus pumilus*. Braz J Microbiol 31:90–94

Duckworth AW, Grant WD, Jones BE, van Steenbergen R (1996) Phylogenetic diversity of soda lake alkaliphiles. FEMS Microbiol Lett 19:181–191

Fruhwirth GO, Paar A, Gudelj M, Cavaco-Paulo A, Robra K-H, Gübitz GM (2002) An immobilized catalase peroxidase from the alkalothermophile *Bacillus* SF for the treatment of textile-bleaching effluents. Appl Microbiol Biotechnol 60:313–319

Fujinami S, Fujisawa M (2010) Industrial applications of alkaliphiles and their enzymes – past, present and future. Environ Technol 31:845–856

Garnova ES, Krasilnikova EN (2003) Carbohydrate metabolism of the saccharolytic alkaliphilic anaerobes *Halonatronum saccharophilum*, *Amphibacillus fermentum*, and *Amphibacillus tropicus*. Mikrobiologiia 72:558–563

Gascoyne DJ, Connor JA, Bull AT (1991) Capacity of siderophore – producing alkalophilic bacteria to accumulate iron, gallium and aluminium. Appl Microbiol Biotechnol 36:136–141

Genckal H, Tari C (2006) Alkaline protease production from alkalophilic *Bacillus* sp. isolated from natural habitats. Enzyme Microb Technol 39:703–710

Gessesse A, Mamo G (1998) Purification and characterization of an alkaline xylanase from alkaliphilic Micrococcus sp AR-135. J Ind Microbiol Biotechnol 20:210–214

Gessesse A, Hatti-Kaul R, Gashe BA, Mattiasson B (2003) Novel alkaline proteases from alkaliphilic bacteria grown on chicken feather. Enzyme Microb Technol 32:519–524

Glaring MA, Vester JK, Lylloff1 JE, Al-Soud WA, Sørensen SJ, Stougaard P (2015) Microbial diversity in a permanently cold and alkaline environment in Greenland. PLoS One. doi:10.1371/journal.pone.0124863

Godinho A, Bhosle S (2008) Carotenes produced by alkaliphilic orange-pigmented strain of *Microbacterium arborescens*—AGSB isolated from coastal sand dunes. Indian J Mar Sci 37:207–312

Grant WD, Sorokin DY (2011) Distribution and diversity of soda lake alkaliphiles In: Horikoshi K (ed) Extremophiles handbook. Springer, Tokyo, pp 28–54

Grant WD, Tindall BJ (1986) The alkaline saline environment. In: Herbert RA, Codd GA (eds) Microbes in extreme environments. Academic Press, London, pp 25–54

Gubitz GM, Mansfield SD, Bo¨hm D, Saddler JN (1998) Effect of endoglucanases and hemicellulases in magnetic and flotation deinking of xerographic and laser-printed papers. J Biotechnol 65:209–215

Gudelj M, Fruiwirth GO, Paar A, Lottspeich F, Robra K-H, Cavaco-Paulo A, Gübitz GM (2001) A catalase-peroxidase from a newly isolated thermoalkaliphilic *Bacillus* sp. with potential for the treatment of textile bleaching effluents. Extremophiles 5:423–429

Gysin J, Crenn Y, Pereira da Silva L, Breton C (1991) Siderophores as anti parasitic agents. US Patent 5:192–807

Haarhoff J, Moes CJ, Cerff C, Wyk WJV, Gerischer G, Janse BJH (1999) Characterization and biobleaching effect of hemicellulases produced by thermophilic fungi. Biotechnol Lett 21:415–420

Hagihara H, Igarashi K, Hayashi Y, Endo K, Ikawa-Kitayama K, Ozaki K, Kawai S, Ito S (2001) Novel alpha-amylase that is highly resistant to chelating reagents and chemical oxidants from the alkaliphilic *Bacillus* isolate KSM-K38. Appl Environ Microbiol 67:1744–1750

Haile G, Gessesse A (2012) Properties of alkaline protease C45 Produced by alkaliphilic *Bacillus* sp. isolated from Chitu, Ethiopian Soda Lake. J Biotechnol Biomater 2:136

Hakamada Y, Koike K, Yoshimatsu T, Mori H, Kobayashi T, Ito S (1997) Thermostable alkaline cellulase from an alkaliphilic isolate, *Bacillus* sp. KSM-S237. Extremophiles 1:151–156

Hakamada Y, Endo K, Takizawa S, Kobayashi T, Shirai T, Yamane T, Ito S (2002) Enzymatic properties, crystallization and deduced aminoacid sequence of an alkaline endoglucanase from *Bacillus circulans*. Biochim Biophys Acta 1570:174–180

Hashim SO, Delgado O, Hatti-Kaul R, Mulaa FJ, Mattiasson B (2004) Starch hydrolysing *Bacillus halodurans* isolates from a Kenyan soda lake. Biotechnol Lett 26:823–828

Hashim SO, Delgado OD, Martínez MA, Kaul RH, Mulaa FJ, Mattiasson B (2005) Alkaline active maltohexaose-forming α-amylase from *Bacillus halodurans* LBK 34. Enzyme Microb Technol 36:139–146

Hayashi K, Youichi N, Ohara N, Uichimura T, Suzuki H, Komagata K, Kozaki M (1996) Low-temperature-active cellulase produced by *Acremonium alcalophilum* JCM 7366. J Ferment Bioeng 81:185–190

Hebeish A, Ibrahim NA (2007) The impact of frontier sciences on textile industry. Colourage 54:41–55

Henriksson G, Akin DE, Hanlin RT, Rodriguez C, Archibald DD, Rigsby LL, Eriksson KEL (1997) Identification and retting efficiencies of fungi isolated from dew-retted flax in the United States and Europe. Appl Environ Microbiol 63:3950–3956

Honda H, Kudo T, Ikura Y, Horikoshi K (1985) Two types of xylanases of alkalophilic *Bacillus* sp. No. C-125. Can J Microbiol 31:538–542

Hoondal GS, Tiwari RP, Tewari R, Dahiya N, Beg QK (2002) Microbial alkaline pectinases and their industrial applications: a review. Appl Microbiol Biotechnol 59:409–418

Horikoshi K (1991) Microorganisms in alkaline environments. Kodansha, Tokyo, VCH, Weinheim, New York

Horikoshi K (1999) Alkaliphilies: some applications of their products for biotechnology. Microbiol. Molecul Biol Rev 63:735–750

Horikoshi K (2006) Alkaliphiles: genetic properties and applications of enzymes. Springer, Berlin

Horikoshi K (2011) Enzymes isolated from alkaliphiles. In: Extremophiles handbook. Springer, Japan, pp 164–177

Horikoshi K, Atsukawa Y (1973) Xylanase produced by alkalophilic *Bacillus* no C-59-2. Agric Biol Chem 37:2097–2103

Horikoshi K, Nakao M, Kurono Y, Saschihara N (1984) Cellulases of an alkalophilic *Bacillus* strain isolated from soil. Can J Microbiol 30:774–779

Ibarra D, Monte MC, Blanco A, Martinez AT, Martinez MJ (2012) Enzymatic deinking of secondary fibers: cellulases/hemicellulases versus laccase-mediator system. J Ind Microbiol Biotechnol 39:1–9

Ishikawa M, Kodama K, Yasuda H, Okamoto-Kainuma A, Koizumi K, Yamasato K (2007) Presence of halophilic and alkaliphilic lactic acid bacteria in various cheese. Lett Appl Microbiol 44:308–313

Ishikawa M, Tanasupawat S, Nakajima K, Kanamori H, Ishizaki S, Kodama K, Okamoto-Kainuma A, Koizumi Y, Yamamoto Y, Yamasato K (2009) *Alkalibacterium thalassium* sp. nov., *Alkalibacterium pelagium* sp. nov., *Alkalibacterium putridalgicola* sp. nov. and *Alkalibacterium kapii* sp. nov., slightly halophilic and alkaliphilic marine lactic acid bacteria isolated from marine organisms and salted foods collected in Japan and Thailand. Int J Syst Evol Microbiol 59:1215–1226

Ishikawa M, Yamasato K, Kodama K, Yasuda H, Matsuyama M, Okamoto-Kainuma A, Koizumi Y (2013) *Alkalibacterium gilvum* sp. nov., slightly halophilic and alkaliphilic lactic acid bacterium isolated from soft and semi-hard cheeses. Int J Syst Evol Microbiol 63:1471–1478

Ito S, Shdcata S, Ozaki K, Kawai S, Okamoto K, Inoue S, Takei A, Ohta Y, Satoh T (1989a) Cellulase for laundry detergents: production by *Bacillus* sp. KSM-635 and enzymatic properties. Agric Biol Chem 53:1275–1281

Ito S, Shikata S, Ozaki K, Kawai S, Okamoto K, Inoue S, Takei A, Ohta Y, Satoh T (1989b) Alkaline cellulase for laundry detergents: production by *Bacillus* sp. KSM-635 and enzymatic properties. Agric Biol Chem 53:1275–1281

Ito S, Kobayashi T, Ara K, Ozaki K, Kawai S, Hatada Y (1998) Alkaline detergent enzymes from alkaliphiles: enzymatic properties, genetics, and structures. Extremophiles 2:185–190

Jiang X, Xue Y, Wang A, Wang L, Zhang G, Zeng Q, Yu B, Ma Y (2013) Efficient production of polymer-grade L-lactate by an alkaliphilic *Exiguobacterium* sp. strain under nonsterile open fermentation conditions. Bioresour Technol 143:665–668

Johnvesly B, Naik GR (2001) Studies on production of thermostable alkaline protease from thermophilic and alkaliphilic *Bacillus* sp. JB-99 in a chemically defined medium. Process Biochem 37:139–144

Jones BE, Grant WD, Duckworth AW, Owenson GG (1999) Microbial Diversity of soda lakes. Extremophiles 2:191–200

Jyonouchi H, Sun S, Gross M (1995) Effect of carotenoids on in vitro immunoglobulin production by human peripheral blood mononuclear cells: Astaxanthin, a carotenoid without vitamin A activity, enhances in vitro immunoglobulin production in response to a T-dependent stimulant and antigen. Nutr Cancer 23:171–183

Kagawa M, Murakoshi N, Nishikawa Y, Matsumoto G, Kurata Y, Mizobata T, Kawata Y, Nagai J (1999) Purification and cloning of a thermostable manganese catalase from thermophilic bacterium. Arch Biochem Biophys 362:346–355

Kamini NR, Hemachander C, Mala JGS, Puvanakrishnan R (1999) Microbial enzyme technology as an alternative to conventional chemicals in leather industry. Curr Sci 77:80–86

Kang MK, Rhe YH (1995) Carboxymethyl cellulases active and stable at alkaline pH from alkalophilic *Cephalosporium* sp. Rym-202. Biotechnol Lett 17:507–512

Kapoor M, Beg QK, Bhushan B, Singh K, Dadhich KS, Hoondal GS (2001) Application of an alkaline and thermostable polygalacturonase from *Bacillus* sp. MG-cp-2 in degumming of ramie (*Boehmeria nivea*) and sunn hemp (*Crotalaria juncea*) bast fibers. Process Biochem 36:803–807

Karmakar M, Ray RR (2011) Current trends in research and application of microbial cellulases. Res J Microbiol 6:41–53

Kenealy WR, TW Jeffries (2003) Enzyme processes for pulp and paper: a review of recent developments. In: Goodell B, Nicholas DD, Schultz TP (eds) Wood deterioration and preservation: advances in our changing world. Oxford University Press, Oxford, pp 210–239

Kim JY, Hur SH, Hong JH (2005) Purification and characterization of an alkaline cellulase from a newly isolated alkalophilic *Bacillus sp.* HSH-810. Biotechnol Lett 27:313–316

Kitada M, Hashimoto M, Kudo T, Horikoshi K (1994) Properties of two different Na^+/H^+ antiport systems in alkaliphilic *Bacillus* sp. strain C-125. J Bacteriol 176:6464–6469

Kobayashi T, Hakamada Y, Adachi S, Hitomi J, Yoshimatsu T, Koike K, Kawai S, Ito S (1995) Purification and properties of an alkaline protease from alkalophilic *Bacillus* sp. KSM-K16. Appl Microbiol Biotechnol 43:473–481

Krulwich TA, Ito M, Gilmour R, Hicks DB, Guffanti AA (1998) Energetics of alkaliphilic *Bacillus* species: physiology and molecules. Adv Microb Physiol 40:401–438

Krulwich TA, Ito M, Guffani AA (2001a) The Na-dependence of alkaliphily in *Bacillus*. Biochim Biophys Acta 1501:158–168

Krulwich TA, Ito M, Guffanti AA (2001b) The Na^+-dependency of alkaliphiliy in *Bacillus*. Biochem Biophys Acta 1505:158–168

Krulwieh TA, Ito M, Gilmour R, Guffanti AA (1997) Mechanisms of cytoplasmic pH regulation in alkaliphilic strains of *Bacillus*. Extremophiles 1:163–169

Kuddus M, Ramteke PW (2009) Cold-active extracellular alkaline protease from an alkaliphilic *Stenotrophomonas maltophilia*: production of enzyme and its industrial applications. Can J Microbiol 55:1294–1301

Kulshreshtha NM, Kumar A, Bisht G, Pasha S and Kumar R (2012) Usefulness of organic acid produced by *Exiguobacterium* sp. 12/1 on neutralization of alkaline wastewater. Sci World J 2012:345101 doi:10.1100/2012/345101

Kurkov SV, Loftsson T (2013) Cyclodextrins. Int J Pharma 453:167–180

Lawton EM, Cotter PD, Hill C, Ross RP (2007) Identification of a novel two-peptide lantibiotic, haloduracin, produced by the alkaliphile *Bacillus halodurans* C-125. FEMS Microbiol Lett 267:64–71

Lee PC, Schmidt-Dannert C (2002) Metabolic engineering towards biotechnological production of carotenoids in microorganisms. Appl Microbiol Biotechnol 60:1–11

Lee CK, Darah I, Ibrahim CO (2007) Enzymatic deinking of laser printed office waste papers: some governing parameters on deinking efficiency. Bioresour Technol 98:1684–1689

Lin LL, Chyau CC, Hsu WH (1998) Production and properties of a raw-starch-degrading amylase from the thermophilic and alkaliphilic *Bacillus* sp. TS-23. Biotechnol Appl Biochem 28:61–68

Mamo G, Hatti-Kaul R, Mattiasson B (2006) A thermostable alkaline active endo- β-1-4-xylanase from *Bacillus halodurans* S7: Purification and characterization. Enzyme Microb Technol 39:1492–1498

Martins RF, Hatti-Kaul R (2002) A new cyclodextrin glycosyltransferase from an alkaliphilic *Bacillus agaradhaerens* isolate: purification and characterization. Enzyme Microb Technol 30:116–124

Martins RF, Davids W, Abu Al-Soud W, Levander F, Rådström P, Hatti-Kaul R (2001) Starch-hydrolyzing bacteria from Ethiopian soda lakes. Extremophiles 5:135–144

Mata-Gómez LC, Montañez JC, Méndez-Zavala A, Aguilar CN (2014) Biotechnological production of carotenoids by yeasts: an overview. Microb Cell Fact 13:12

Matzuzawa M, Kawano M, Nakamura N and Horikoshi K (1975) An improved method for the preparation of Schardinger-Dextrin on an industrial scale by cyclodextrin glycosyl transferase of an alkalophilic *Bacillus* sp. (ATCC 21783). Starch–Stärke 27:410–413

McMillan DGG, Velasquez I, Nunn BL, Goodlett DR, Hunter KA, Lamont I, Sander SG, Cook GM (2010) Acquisition of iron by alkaliphilic *Bacillus* species. Appl Environ Microbiol 76:6955–6961

Meng Y, Xue Y, Yu B, Gao C, Ma Y (2012) Efficient production of L-lactic acid with high optical purity by alkaliphilic *Bacillus* sp. WL-S20. Bioresour Technol 116:334–339

Michaud-Soret I, Jacquamet L, Debaecker-Petit N, Le Pape L, Barynin VV, Latour J-M (1998) The existence of two oxidized Mn(III) forms from *Thermus thermophilus* manganese catalase. Inorg Chem 37:3874–3876

Miethke M, Marahiel MA (2007) Siderophore-based iron acquisition and pathogen control. Microbiol. Mol Biol Rev 71:413–451

Miyashita K (2009) Function of marine carotenoids. Forum Nutr 61:136–146

Murata M, Hoshino E, Yokosuka M, Suzuki A (1991) New detergent mechanism with use of novel alkaline cellulose. J Am Oil Chem Soc 68:553–558

Nagarathnamma R, Bajpai P (1999) Decolorization and detoxification of extraction-stage effluent from chlorine bleaching of Kraft pulp by *Rhizopus oryzae*. Appl Environ Microbiol 65:1078–1082

Nakamura S, Wakabayashi K, Nakai R, Aono R, Horikoshi K (1993) Purification and some properties of an alkaline xylanase from alkaliphilic *Bacillus* sp. strain 41M-1. Appl Environ Microbiol 59:2311–2316

Nakamura S, Nakai R, Wajabatacgu K, Ishiguro Y, Aono R, Horikoshi K (1994) Thermophilic alkaline xylanase from newly isolated alkaliphilic and thermophilic *Bacillus* sp. strain TAR-1. Biosci Biotechnol Biochem 58:78–81

Niehaus F, Bertoldo C, Kähler M, Antranikian G (1999) Extremophiles as a source of novel enzymes for industrial applications. Appl Mircrobiol Biotechnol 51:711–729

Nilegaonkar SS, Zambare VP, Kanekar PP, Dharephalkar PK, Sarnaik SS (2007) Production and partial characterization of dehairing protease from *Bacillus cereus* MCM B-326. Bioresour Technol 98:1238–1245

Ningthoujam DS, Kshetri P, Sanasam S, Nimaichand S (2009) Screening, identification of best producers and optimization of extracellular proteases from moderately halophilic alkalithermotolerant indigenous actinomycetes. World Appl Sci J 7:907–916

Nonaka T, Fujihashi M, Kita A, Hagihara H, Ozaki K, Ito S, Miki K (2003) Crystal structure of calcium-free alpha-amylase from *Bacillus* sp. strain KSM-K38 (AmyK38) and its sodium ion binding sites. J Biol Chem 278:24818–24824

Olivera N, Sequeiros C, Sineriz F, Breccia J (2006) Characterization of alkaline proteases from novel alkali-tolerant bacterium *Bacillus patagoniensis*. World J Microbiol Biotechnol 22:737–743

Olsen HS, Falholt P (1998) The role of enzymes in modern detergency. J Surfactants Deterg 1:555–566

Oluoch KR, Welander U, Andersson MM, Mulaa FJ, Mattiasson B, Hatti-Kaul R (2006) Hydrogen peroxide degradation by immobilized cells of alkaliphilic *Bacillus halodurans*. Biocat Biotransfor 24:215–222

Osanjo GO, Muthike EW, Tsuma L, Okoth MW, Bulimo WD, Lünsdorf H, Abraham WR, Dion M, Timmis KN, Golyshin PN, Mulaa FJ (2009) A salt lake extremophile, *Paracoccus bogoriensis* sp. nov., efficiently produces xanthophyll carotenoids. Afr J Microbiol Res 3:426–433

Paar A, Costa S, Tzanov T, Gudelj M, Robra K-H, Cavaco-Paulo A, Gübitz GM (2001) Thermo-alkali-stable catalases from newly isolated *Bacillus* sp. for treatment and recycling of textile bleaching effluents. J Biotechnol 89:147–153

Paavilainen S, Helistö P, Korpela T (1994) Conversion of carbohydrates to organic acids by alkaliphilic bacilli. J Ferment Bioeng 78:217–222

Padan E, Bibi E, Ito M, Krulwich TA (2005) Alkaline pH homeostasis in bacteria: new insights. Biochim Biophys Acta 1717:67–88

Palozza P, Torelli C, Boninsegna A, Simone R, Catąlano A, Mele MC, Picci N (2009) Growth-inhibitory effects of the astaxanthin- rich alga *Haematococcus pluvialis* in human colon cancer cells. Cancer Lett 283:108–117

Pietrangelo A (2002) Mechanism of iron toxicity. In: Hershko C (ed) Iron chelation theraphy, vol. 509. Kluwer Academic/Plenum Publishers, New York, pp 19–43

Pokhrel D, Viraraghavan T (2004) Treatment of pulp and paper mill wastewater-a review. Sci Total Environ 333:37–58

Rai SK, Roy JK, Mukherjee AK (2010) Characterisation of a detergent-stable alkaline protease from a novel thermophilic strain *Paenibacillus tezpurensis* sp. nov. AS-S24-II. Appl Microbiol Biotechnol 85:1437–1450

Rao CS, Sathish T, Ravichandra P, Prakasham RS (2009) Characterization of thermo- and detergent stable serine protease from isolated *Bacillus circulans* and evaluation of eco-friendly applications. Process Biochem 44:262–268

Roadcap GS, Sanford RA, Jin Q, Pardinas JR, Bethke CM (2006) Extremely alkaline (pH>12) ground water hosts diverse microbial community. Groundwater 44:511–517

Ruggiero CE, Neu MP, Matonic JH, Reilly SD (2000) Interactions of Pu with desferrioxamine siderophores can affect bioavailability and mobility. Actinide Res Q 2000:16–18

Said S, Fonseca MJV, Siessere V (1991) Pectinase production by *Penicillium frequentans*. World J Microbiol Biotechnol 7:607–608

Sashihara N, Kudo T, Horikoshi K (1984) Molecular cloning and expression of cellulase genes of alkalophilic *Bacillus* sp. strain N-4 in *Escherichia coli*. J Bacteriol 158:503–506

Sato M, Beppu T, Arima K (1980) Properties and structure of a novel peptide antibiotic no. 1970. Agric Biol Chem 44:3037–3040

Saxena RK, Dutt K, Agarwal L, Nayyar P (2007) A highly thermostable and alkaline amylase from a *Bacillus* sp. PN5. Bioresour Technol 98:260–265

Seifzadeh S, Sajedi RH, Sariri R (2008) Isolation and characterization of thermophilic alkaline proteases resistant to sodium dodecyl sulfate and ethylene diamine tetraacetic acid from *Bacillus* sp. GUS1. Iranian J Biotechnol 6:214–221

Senthilvelan T, Kanagaraj J, Mandal AB (2012) Application of enzymes for dehairing of skins: cleaner leather processing. Clean Technol Envir 14:889–897

Shrinath A, Szewczak JT, Bowen IJ (1991) A review of ink removal techniques in current deinking technology. Tappi J 74:85–93

Shrinivas D, Naik GR (2011) Characterization of alkaline thermostable keratinolytic protease from thermoallkalophilic *Bacillus halodurans* JB 99 exhibiting dehairing activity. Int Biodeterior Biodegrad 65:29–35
Singh SA, Plattner H, Diekmann H (1999) Exopolygalacturonate lyase from a thermophilic *Bacillus* sp. Enzyme Microb Technol 25:420–425
Singh A, Kuhad RC, Ward OP (2007) Industrial application of microbial cellulases. In: Kuhad RC, Singh A (eds) Lignocellulose biotechnology: future prospects. I.K. International Publishing House, New Delhi, pp 345–358
Singh LS, Mazumder S, Bora TC (2009) Optimisation of process parameters for growth and bioactive metabolite produced by a salt-tolerant and alkaliphilic actinomycete, *Streptomyces tanashiensis* strain A2D. J Mycol Med 19:225–233
Soerensen NH, Hoff T, Oestergaard PR, Cassland P (2011) Enzyme dehairing of skins and hides. WO 2011161135 A1
Sorokin DY, Turova TP, Kuznetsov BB, Briantseva IA, Gorlenko VM (2000) *Roseinatronobacter thiooxidans* gen. nov., sp. nov., a new alkaliphilic aerobic bacteriochlorophyll a- containing bacterium isolated from a soda lake. Mikrobiologia 69:89–97
Sorokin DY, Berben T, Melton ED, Overmars L, Vavourakis CD, Muyzer G (2014) Microbial diversity and biogeochemical cycling in soda lakes. Extremophiles 18:791–809
Sorokin DY, Banciu HL, Muyzer G (2015) Functional microbiology of soda lakes. Curr Opin Microbiol 25:88–96
Spiro MC, Griffith WP (1997) The mechanism of hydrogen peroxide bleaching. Text Chem Color 29:12–13
Srinivasan MC, Rele MV (1999) Microbial xylanases for paper industry. Curr Sci 77:137–142
Sturr MG, Guffanti AA, Krulwieh TA (1994) Growth and bioenergetics of alkaliphilic *Bacillus firmus* OF4 in continuous culture at high pH. J Bacteriol 176:3111–3116
Sukumaran RK, Singhania RR, Pandey A (2005) Microbial cellulases – production, applications and challenges. J Sci Ind Res 64:832–844
Szente L, Szejtli J (2004) Cyclodextrins as food ingredients. Trends Food Sci Tech 15:137–142
Takaichi S, Oh-Oka H, Maoka T, Jung DO, Madigan MT (2003) Novel carotenoid glucoside esters from alkaliphilic heliobacteria. Arch Microbiol 179:95–100
Taksawan T, Yuichi H, Saori K, Moriya O, Savitr T, Napavarn N, Toshiaki K (2005) Comparison of bacterial communities in the alkaline gut segment among various species of higher termites. Extremophiles 9:229–238
Tatineni R, Doddapaneni KK, Potumarthi RC, Vellanki RN, Kandathil MT, Kolli N, Mangamoori LN (2008) Purification and characterization of an alkaline keratinase from *Streptomyces* sp. Bioresour Technol 99:1596–1602
Thakur IS (2006) Industrial biotechnology: problems and remedies. I.K. International, New Delhi, p 36
Thanikaivelan P, Rao JR, Nair BU, Ramasami T (2004) Progress and recent trends in biotechnological methods for leather processing. Trends Biotechnol 22:181–188
Thomas WJ (1994) Comparison of enzyme-enhanced with conventional deinking of xerographic and laser-printed paper. Tappi J 77:173–179
Thompson VS, Schaller KD, Apel WA (2003) Purification and characterization of a novel thermo-alkali stable catalase from *Thermus brokianus*. Biotechnol Prog 19:1292–1299
Thumar JT, Dhulia K, Singh SP (2010) Isolation and partial purification of an antimicrobial agent from halotolerant alkaliphilic *Streptomyces aburaviensis* strain Kut-8. World J Microbiol Biotechnol 26:2081–2087
Tsujibo H, Sato T, Inui M, Yamamoto H, Inamori Y (1988) Intracellular accumulation of phenazine antibiotics production by an alkalophilic actinomycete. Agric Biol Chem 52:301–306
Turner P, Mamo G, Karlsson EN (2007) Potential and utilization of thermophiles and thermostable enzymes in biorefining. Microb Cell Fact 6:9–10
Tzanov T, Costa S, Gübitz GM, Cavaco-Paulo A (2001) Dyeing in catalase treated bleaching baths. Color Technol 117:1–5
Vasavada SH, Thumar JT, Singh SP (2006) Secretion of a potent antibiotic by salt-tolerant and alkaliphilic actinomycete *Streptomyces sannanensis* strain RJT-1. Curr Sci 91:1393–1397

Vértesy L, Aretz W, Fehlhaber HW, Kogler H (1995) Salimycin A-D, antibiotoka aus *Streptomyces violaveus*, DSM 8286, mit siderophor-aminoglycosid-struktur. Helv Chim Acta 78:46–60
Vicuna R, Escobar F, Osses M, Jara A (1997) Bleaching of Eucalyptus Kraft pulp with commercial xylanases. Biotechnol Lett 19:575–578
Von Gunten HR, Benes P (1995) Speciation of radionuclides in the environment. Radiochim Acta 69:1–29
Vyas S, Lachke A (2003) Biodeinking of mixed office waste paper by alkaline active cellulases from alkalotolerant *Fusarium* sp. Enzyme Microb Technol 32:236–245
Weck M (1991) Hydrogen peroxide – an environmentally safe textile bleaching agent. Text Prax Int 46:144–147
Wielen LCV, Panek JC, Pfromm PH (1999) Fracture of toner due to paper swelling. Tappi J 82:115–121
Ximenes FA, Sousa MV, Puls J, Silva FG, Filho EXF (1999) Purification and characterization of a low-molecular-weight xylanase produced by *Acrophialophora nainiana*. Curr Microbiol 38:18–21
Yokaryo H, Tokiwa Y (2014) Isolation of alkaliphilic bacteria for production of high optically pure L-(+)-lactic acid. J Gen Appl Microbiol 60:270–275
Zhilina TA, Appel R, Probian C, Brossa EL, Harder J, Widdel F, Zavarzin GA (2004) *Alkaliflexus imshenetskii* gen. nov. sp. nov., a new alkaliphilic gliding carbohydrate-fermenting bacterium with propionate formation from a soda lake. Arch Microbiol 182:244–253

Chapter 9
Recent Advances in the Nitrogen Metabolism in Haloarchaea and Its Biotechnological Applications

Julia Esclapez, Mónica Camacho, Carmen Pire, Vanesa Bautista, Anna Vegara, Laia Pedro-Roig, Francisco Pérez-Pomares, Rosa María Martínez-Espinosa, and María José Bonete

1 Introduction

In the biosphere, nitrate assimilation is the significant pathway for turning inorganic nitrogen into organic forms. This pathway requires the presence of two assimilative enzymes, nitrate reductase and nitrite reductase. First, NO_3^- is transported into the cells by high-affinity transporters and next reduced to NH_4^+ by two sequential reactions catalysed by assimilative nitrate reductase (Nas; EC 1.6.6.2) and assimilative nitrite reductase (Nir; EC 1.7.7.1). The resulting ammonium is combined with carbon skeletons by the glutamine synthetase/glutamate synthase pathway (GS-GOGAT; EC 6.3.1.2, EC 1.4.7.1, respectively) or via glutamate dehydrogenase (GDH; EC 1.4.1.2). The GS/GOGAT pathway is especially important as it allows ammonia assimilation into L-glutamate (Glu) at low intracellular ammonia concentrations and it efficiently substitutes the other glutamate biosynthetic reaction (GDH) in these conditions (Bonete et al. 2008).

Genes encoding the main proteins involved in nitrate and ammonium assimilation have been found in the genomes of the two major Archaea subgroups: Crenarchaeota and Euryarchaeota (Feng et al. 2012). On the other hand, studies of haloarchaea communities from soil revealed that the haloarchaeal assimilatory nitrate-reducing community seems to be important in salty and/or alkaline environments (Alcántara-Hernández et al. 2009). However, biochemical characterisation of haloarchaeal enzymes involved in NO_3^-, NO_2^- and NH_4^+ assimilation is insufficient and it has been performed mainly in *Haloferax mediterranei* at the time of writing this

J. Esclapez • M. Camacho • C. Pire • V. Bautista • A. Vegara • L. Pedro-Roig • F. Pérez-Pomares • R.M. Martínez-Espinosa • M.J. Bonete (✉)
Department of Agrochemistry and Biochemistry, University of Alicante, Carretera San Vicente del Raspeig s/n - 03690 San Vicente del Raspeig, Alicante, Spain
e-mail: mjbonete@ua.es

P.H. Rampelotto (ed.), *Biotechnology of Extremophiles*, Grand Challenges in Biology and Biotechnology 1, DOI 10.1007/978-3-319-13521-2_9

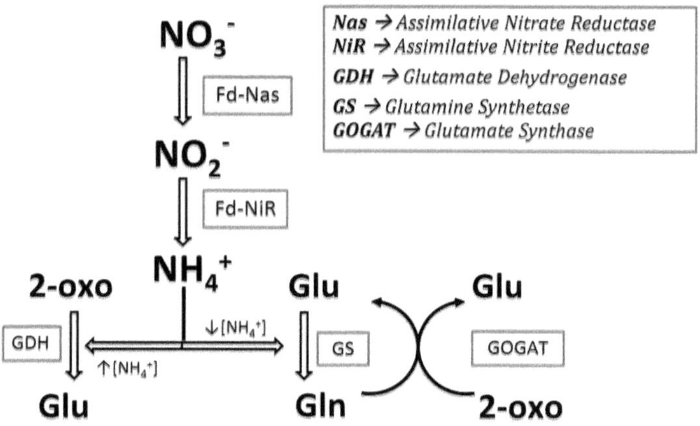

Fig. 9.1 Nitrate and ammonium assimilation pathways in *Haloferax mediterranei*. *2-oxo* 2-oxoglutarate, *Glu* L-glutamate, *Gln* L-glutamine

chapter (Fig. 9.1). All the known biochemical parameters already described from haloarchaeal protein transporters and enzymes involved in assimilatory nitrate reduction are summarized in the following sections.

2 Enzymes Involved in Nitrate/Nitrite and Ammonium Assimilation

2.1 NO_3^- and NO_2^- Transporters

The uptake of nitrogen compounds such as nitrate or nitrite is the first step to use these compounds as nitrogen sources. In prokarya, the genes encoding proteins required for NO_3^- uptake and its subsequent reduction are, in general, clustered (Moreno-Viviàn and Flores 2007). Furthermore, it is common that NO_3^- is transported into the cells by an active system. At the time of writing, two kind of proteins involved in nitrate transport are described in prokarya: ATP-dependent ABC transporters (constitute by (i) an integral membrane subunit, (ii) a cytoplasmic ATP-binding component and (iii) a periplasmic substrate binding protein) and the monomeric NarK-type transporters belonging to the major facilitator superfamily (MFS-type permease). The last type of transporter depends on proton-motive force. ABC transporters can be found in Archaea, Bacteria and Eukarya (Wanner and Soppa 1999). Bacterial NarK-like transporters can be divided into two subgroups: NarK1 (proton: nitrate symporter that allows initiation of nitrate respiration) and NarK2 (nitrate: nitrite antiporter required for maintenance of a steady state rate) (Wood et al. 2002). Some of these proteins are involved in NO_3^-/ NO_2^- exchange rather than simply in the uptake of one of the mentioned ions. However, the transport mechanism by means of these MFS importers has not been elucidated yet. It has been stated that bacterial nitrate assimilation requires ATP-dependent ABC nitrate transporters whereas nitrate respiration is mainly associated with proton-motive-force driven NarK transporters (Wood et al. 2002). Database comparisons of the genes involved in haloarchaea nitrate assimilation, in particular in *Hfx. mediterranei*, revealed that *nasB* gene (Q703N4) encodes a NO_3^- transporter, which is a membrane protein (Mr around 46.1 kDa) with 12 potential α helices and is most closely related with the NarK1 type of transporters. The best fits of *nasB* gene were with bacterial homologues belonging to the genera *Thermus*, *Paracoccus* and *Pseudomonas* (Lledó et al. 2005). *Hfx. mediterranei* NarK is the first Archaea NarK transporter reported to date, not being exclusive to Bacteria and Eukarya as it was initially suggested. This data suggest that NarK transporters could be involved in the nitrate transport not only in the assimilatory process but also in the denitrification pathway. This transporter could be in charge of the nitrate/nitrite exchange. However, more biochemical studies on that kind of transporters are required to properly understand how nitrate and nitrite are up taken in haloarchaea.

2.2 Ammonium Transporters

If ammonium is the nitrogen source for growth, the first step is the ammonium uptake facilitated by the ammonium transporter (Amt). Amt transporters are characterised by: high affinity and selectivity for ammonium, non-permeability to alkaline cations, permeability to methylammonium and quick saturation at low millimolar concentrations. This transporters are trimers containing 11-crossing membrane fragments. Each monomer contains a hydrophobic channel that conducts NH_3 but not any water or ions. The re-protonation of NH_3 on the receiving side raises the pH on that side in the absence of metabolism of NH_3, and there is no transfer of protons through the protein (Khademi and Stroud 2006). Crystallization of homologue proteins from the three domains pointed out that the functional entity corresponds to a trimer, with each monomer maintaining a conductive pore (Pantoja 2012). Although, Amt transporter is well characterised in Bacteria and Eukarya domains from a biochemical point of view, it still remains quite unknown in Archaea. In fact only the Amt from *Archaeoglobus fulgidus* (Andrade et al. 2005; Andrade and Einsle 2007) and *Hfx. mediterranei* (Pedro-Roig et al. 2013) have been studied to date from this domain.

2.3 Assimilatory Nitrate Reductase

Once nitrate enters into cells, it is reduced to nitrite by assimilatory nitrate reductase (Nas). Nas are cytoplasmic enzymes that catalyse the following reactions which implies the mobilization of two-electrons:

$$NO_3^- + NADH\ (or\ reduced\ ferredoxin) + H^+ \rightarrow$$
$$NO_2^- + NAD^+\ (or\ oxidized\ ferredoxin) + H_2O$$

Most of the characterised Nas use either NADH or ferredoxin as physiological electron donors, although some of them use flavodoxin instead of ferredoxin. Nas activity is only detected under aerobic conditions and it is induced by nitrate and repressed by ammonium, as described for bacterial Nas (Richardson et al. 2001; Luque-Almagro et al. 2011).

Ferredoxin-dependent Nas are usually monomeric enzymes while NADH-dependent Nas are heterodimers (Richardson et al. 2001). Fd-Nas and NADH-Nas are structurally and functionally different from the dissimilatory periplasmic nitrate reductases (Nap; EC 1.7.99.4) and the respiratory membrane-bound nitrate reductases (Nar; EC 1.7.99.4) present in many prokaryotes. Gene sequence analysis has revealed that *Hfx. mediterranei* ferredoxin-dependent Nas (Q703N5) is a molybdoenzyme and the synthesis of Mo-bis Pyranopterin guanine dinucleotide cofactor (Mo-bis-PGD) is mediated by MobA. This enzyme catalyses the binding of GTP to molybdopterin, resulting the active cofactor PGD. On the other hand, spectroscopic studies have revealed the presence of one [4Fe-4S] cluster. This monomeric enzyme (molecular mass around 75 kDa) is the only Nas isolated and characterised from haloarchaea up to know. In this case, the electrons could flux as follow: from the [2Fe-2S] cluster-containing ferredoxin (negative redox potential electron donor) to the [4Fe-4S] cluster and from this centre to the Mo-cofactor for the reduction of NO_3^- (Martínez-Espinosa et al. 2001a; Lledó et al. 2005; Bonete et al. 2008).

The best *Hfx. mediterranei* Nas protein fits were to Nas proteins of the genera *Pseudomonas*, *Xanthomonas* and *Synechococcus*. The comparison with the products of the putative archaeal genes coding for Nas showed that there was only a low overall similarity between these and assimilatory nitrate reductase from *Hfx. mediterranei*, with conserved residues predominantly those related to the cofactor binding sites. Nas kinetic parameters were determined by varying the concentration of one substrate (MV) at several concentrations of the other substrate (nitrate). The apparent K_m values for nitrate and MV were 0.95 ± 0.12 and 0.07 ± 0.12 mM, respectively. This halophilic Nas showed a strong dependence on the temperature and NaCl (maximum activity detected at 80 °C for NaCl concentrations between 3.1 and 2.2 M). Nas from *Hfx. mediterranei* can receive electrons from methylviologen and benzylviologen but not NAD(P)H. Nas was present during aerobic growth, and the activity decreases and even disappeared when the culture was near the stationary phase. In the presence of ammonium ion (or an alternative nitrogen source to nitrate) Nas activity is not detected (Martínez-Espinosa et al. 2001a).

2.4 Assimilatory Nitrite Reductases

Nitrite produced in the previous reactions is reduced to ammonium by ferredoxin-dependent assimilatory nitrite reductase (Q703N2) that catalyses the following reaction involving the mobilisation of six-electrons:

$$NO_2^- + 3\,NADH\,(or\,3\,reduced\,ferredoxin) + 5\,H^+ \rightarrow$$
$$NH_4^+ + 3\,NAD^+\,(or\,3\,oxidized\,ferredoxin) + 2\,H_2O$$

Assimilatory nitrite reductases are classified on the basis of the electron donor specificity: Fd-dependent Nir (described from eukaryotic and prokaryotic photosynthetic organisms) and NAD(P)H-dependent Nir (present in fungi and most heterotrophic bacteria). Fd-Nirs are cytoplasmic monomeric proteins containing a siroheme and a [4Fe-4S] cluster as redox centres. The electrons flow the [2Fe-2S] cluster-containing ferredoxin to the [4Fe-4S] cluster and from this centre to the siroheme for the reduction of nitrite. NADH-Nirs dimers and contain non covalently bound FAD, a [4Fe-4S] cluster, and siroheme as prosthetic groups (Fernández et al. 1998).

The Nir purified from *Hfx. mediterranei*, the first one purified from this kind of microorganisms, is a 66 kDa monomer that shows a strong sequence similarity with known ferredoxin-dependent nitrite reductases, such as those purified from cyanobacteria and bacteria. Some of its biochemical properties have been analysed in detail: K_m for nitrite around 8.6 mM and maximal activity at 60 °C in presence of 3.3 M NaCl. Like most of the bacterial assimilatory nitrite reductases described so far, the *Hfx. mediterranei* Nir contains siroheme and Fe-S centres as redox centres (Martínez-Espinosa et al. 2001b; Lledó et al. 2005).

2.5 Glutamine Synthetase-Glutamate Synthase (GS-GOGAT) Cycle and Glutamate Dehydrogenase (GDH)

In most bacteria and archaea, the formation of L-glutamate is the main way for ammonia and nitrogen assimilation. Three enzymes are involved in this process in *Hfx. mediterranei*. These are: glutamine synthetase (GS), which incorporates ammonium into glutamine, glutamate synthase (GOGAT), which converts the glutamine to glutamate, and glutamate dehydrogenase (GDH), which catalyzes the reversible reductive amination of 2-oxoglutarate to produce glutamate (Ferrer et al. 1996; Díaz et al. 2006; Martínez-Espinosa et al. 2006; Pire et al. 2014). The GS-GOGAT cycle requires ATP but has high affinity for ammonium, whereas GDH does not consume ATP but is less effective in cells growing in N-limited conditions. These enzymes are present in all three domains of life.

2.5.1 Glutamine Synthetase

Glutamine synthetase (GS) enzyme, which belongs to the GS- GOGAT route. It has been performed biochemical and molecular studies of GS protein from some hyperthermophilic and methanogenic archaea, prokaryotes and mammals (Adul Rahman et al. 1997; Reitzer 2003). This protein has a double function, producing glutamine and assimilating ammonium in collaboration with glutamate synthase. In brief, the GS can be classified into three types according to their molecular size and the number of subunits contained. GSI contains 12 subunits with a range of 44–60 kDa, and it is in archaea and bacteria groups (Brown et al. 1994; Robertson and Alberte 1996). GSII, is an octameric enzyme with subunit size of 35–50 kDa, and it is found in a small number of soil-dwelling bacteria and eukaryotes (Kumada et al. 1993). At last, GSIII is a hexameric protein with 75 kDa subunits, which was found in some anaerobic bacteria and cyanobacteria (Reyes and Florencio 1994).

Few analyses have been carried out with GS from members of Archaea, and just two are from GS from haloarchaea; these enzymes, purified from *Hfx. mediterranei* and *Hbt. salinarium* are octamers belonging to the GS type II (Manitz and Holldorf 1993; Martínez-Espinosa et al. 2006). However, a few GSs described from methanogenic or hyperthermophilic archaea are dodecamers of about 600 kDa. So, GS from haloarchaea exhibits typical properties of GS from eukaryotes and soil bacteria species. The results obtained from *Hfx. mediterranei* suggest that GS from this haloarchaeon could allow the assimilation of the ammonium produced by assimilatory nitrite reductase (Martínez-Espinosa et al. 2006, 2007), while GDH would allow the assimilation of ammonium when this nitrogen source is present in the culture media at high concentrations. The biochemical characterization of *Hfx. mediterranei* GS shows an optimum pH value for activity around 8. Either in the presence of NaCl or KCl, the maximum stability was found at the highest salt concentration described by us (3.5 M and 2.5 M, respectively). The K_m value for ADP could not be calculated for the transferase activity, as all concentrations analysed produced similar results (K_m around 3.10 ± 0.5 mM). The apparent K_m for NH_2OH, Gln, ATP and Glu were 10.5 ± 3.5, 25 ± 1.8, 0.30 ± 0.08 and 4.9 ± 1.5 mM, respectively.

These values are similar to those described in *Hbt salinarium* GS and are in the range described for GS from different cyanobacteria and Archaea (Manitz and Holldorf 1993). Studies about the effectiveness of different metal ions on the activity assay revealed that Mn^{2+} was the most effective ion for transferase activity as it was expected due to its essential role as a cofactor.

2.5.2 Glutamate Synthase

Glutamate synthase (GOGAT or GltS), the second enzyme of the GS/GOGAT cycle catalyses the reductive transfer of the amide nitrogen from L-glutamine to 2-oxoglutarate to form two molecules of L-glutamate. There are different classes of GOGAT which differ for distribution among different organisms and tissues, subunit composition, cofactor content and physiological reductant (Vanoni and Curti 2005). The main GOGAT classification was (a) Bacterial NADPH-dependent GOGAT (NADPH-GOGAT), formed by two dissimilar enzyme subunits (α, the large subunit, and β, the small subunit, products of *gltB* and *gltD* genes, respectively). The α subunit consists of four domains, including the N-terminal amidotransferase domain, the FMN-binding domain and the C-terminal domain. The β subunit contains the NADPH binding site. (b) Ferredoxin-dependent GOGAT (Fd-GOGAT), from photosynthetic cells, consists of a single polypeptide chain similar to the α subunit of bacterial NADPH-GOGAT. (c) NADH-dependent GOGAT (NADH-GOGAT), from yeast, fungi and lower animals, consists of a single polypeptide chain with an N-terminal region similar to the bacterial alpha subunit, linked to a C-terminal region similar to the bacterial beta subunit. Many different types of GOGAT are found in archaeal organisms.

In archaea there are different types of GOGAT that do not belong to any of the enzymes mentioned above. In *Archaeoglobus fulgidus, Methanococcus jannaschii,* and *Methanobacterium thermoautotrophicum* there are truncated *gltB* genes; in the archaeon *Pyrococcus sp.* KOD1, a homolog of the bacterial small subunit (*gltD* gene product) has been shown to function alone, and the genomes of other *Pyrococcus* strains contain only *gltD* genes (Nesbo et al. 2001). The enzyme from *Hfx. mediterranei* is ferredoxin-dependent as the enzyme from plant and cyanobacteria, and it is similar to the α-type subunit found in bacterial NADPH-GOGAT. The activity was only observed with one of the two different 2Fe-2S ferredoxins chromatographically isolated from *Hfx. mediterranei.* The enzyme also displayed typical halophilic behaviour, being fully stable, and producing maximal activity, at salt concentrations from 3 to 4 M NaCl, pH 7.5 and a temperature of 50 °C (Pire et al. 2014). Similar genes can be found in the available genomes of other halophilic archaea.

The only previous Fd-GOGAT characterised from a non phototrophic organism is that from *Hydrogenobacter thermophilus,* a hydrogen-oxidizing chemoautotrophic bacterium (Kameya et al. 2007). Among known Fd-GOGATs, there is a conserved insert region that has been designated the Fd loop, this insertion is not present in NADPH-GOGATs and is presumed to be involved in the interaction with ferredoxin, but Fd-GOGAT from *H. thermophilus* and from *Hfx. mediterranei* does not have such a conserved insertion (Kameya et al. 2007; Pire et al. 2014), suggesting that the insertion may be dispensable for the interaction with ferredoxin in these Fd-GOGATs. The enzyme from *H. thermophilus* is activated by some organic acids that are metabolized in the TCA cycle as succinate, oxaloacetate, malate, or citrate.

2.5.3 Glutamate Dehydrogenase (GDH)

Glutamate dehydrogenase, that converts 2-oxoglutarate into L-glutamate, and the inverse reaction, may be considered an important connection of both, carbon and nitrogen metabolism, of which these metabolites are key components. Regulating the concentration of glutamate and 2-oxoglutarate, cell may control the adequate function of the pathways responsible for nitrogen and carbon incorporation. The assimilation of nitrogen from ammonium by *Hfx. mediterranei* was reported to be mediated by glutamate dehydrogenase (GDH) (Ferrer et al. 1996). The demonstrated presence in this organism of the other via that determines the concentrations of glutamate and 2-oxoglutarate, the GS-GOGAT pathway, requires the study of the conditions that determine which via is preferred, or the contribution of each of them to assure the correct function of the metabolic pathways working at the same time at the particular conditions cell has to endure. The availability of nitrogen has been demonstrated to be determinant in bacteria. The via mediated by glutamate dehydrogenase, that directly converts 2-oxoglutarate and ammonium into L-glutamate, mainly functions when ammonium is available at enough high concentrations, meanwhile, under ammonium starvation is the GS/GOGAT the main responsible of the production of

L-glutamate. The higher affinity of GS for ammonium, accordingly to the low K_m values this enzyme displays for ammonium, makes this route the preferred when ammonium is not readily available in the required concentrations. However, exceptions may be found, and in some bacteria, GDH has been reported to be highly active under low ammonium concentration (Hochman et al. 1988).

GDHs of different sources display high coenzyme specificity. This feature determines its classification in three groups: NAD-dependent glutamate dehydrogenases, NADP-specific glutamate dehydrogenases, and those that are not NAD or NADP-specific, and so may function with both coenzymes. The halophilic archaeon *Hfx. mediterranei* has been reported to have at least two different GDHs: NADP-GDH and NAD-GDH, both of them isolated and fully characterized (Ferrer et al. 1996; Díaz et al. 2006). The NADP-dependent GDH from *Hfx. mediterranei* was probed to be a hexameric enzyme of 320 kDa, composed of six monomers of approximately 55 kDa. For the amination reaction at pH 8.5, the K_m values for its substrates were 0.18 mM for the coenzyme $NADP^+$, 0.34 mM for 2-oxoglutarate and 4.2 mM for ammonium.

The regulation of the activity of glutamate dehydrogenases from different sources by a variety of metabolites is another feature commonly reported. The effect of the metabolites on the activity depends on the particular glutamate dehydrogenase tested. For example, compounds such as GTP, ATP, ADP and AMP that were known to be allosteric modifiers of GDHs from mammals, displayed no effect on the activity of the NAD-GDH from *Hbt. halobium* (Bonete et al. 1996) and so indicated that they had no role in the regulation of this enzyme. However, this purified halophilic enzyme was highly affected by TCA intermediates, such as fumarate, oxalacetate, succinate and malate, that caused strong inhibition on both amination and deamination reactions, pointing out its close relation to this important metabolic pathway.

Initial rate studies carried out on this halophilic enzyme for the oxidative deamination reaction, also showed the regulatory effect of a variety of metabolites such as $NADP^+$, D-glutamate and glutarate, analogues of the reaction substrates, as well as dicarboxylic compounds such as adipate, besides the previously mentioned effect of TCA metabolites. On the other hand, amino acids were activators of the enzyme, except D-glutamate, that was competitive inhibitor, due to its similarity with the substrate, the L-isomer glutamate.

2.6 Ferredoxin as Physiological Electron Donors in Nitrate/Nitrite Assimilation and in GS/GOGAT Cycle

Ferredoxins are small proteins containing iron and sulphur atoms organized as iron-sulphur clusters. These metallic clusters can accept or release electrons, changing the oxidation states (+2 or +3) of the iron atoms. As a consequence of these redox reactions, ferredoxin acts as electron transfer agents in biological redox reactions (Beinert 2000). The *Hfx. mediterranei* ferredoxin involved in the electron transfer during nitrate and nitrite assimilation is a small protein (Mr ≈ 21 kDa on SDS-PAGE) (Martínez-Espinosa et al. 2003; Zafrilla et al. 2011) that shows UV-visible, nuclear magnetic resonance (NMR) and electron paramagnetic resonance (EPR) spectra similar to those described from plant and bacteria [2Fe-2S] ferredoxins. UV-visible spectropotentiometric analysis determined a midpoint redox potential for the [2Fe-2S]2+/1+ transition of around −285 mV vs. SHE that was independent of salt concentration. This ferredoxin was a highly thermostable protein; the highest salt concentration present within the media (4 M) the highest thermal stability (80 °C). Ferredoxins from other halophilic archaea such as *Halobacterium halobium* (Geiger et al. 1978), *Haloarcula marismortui* (Frolow et al. 1996), and *Haloarcula japonica* (Sugimori et al. 2000) have been studied and in all cases they contain a cluster constituted by 2Fe and 2S.

3 Regulation of Enzymes Involved in Nitrate/Nitrite and Ammonium Assimilation

3.1 Regulation of Nitrate and Nitrite Reductases

Related to molecular biology analysis of the nitrate assimilation genes in the halophilic archaeon *Hfx. mediterranei*, Lledó et al. 2005 carried out the first studies about its identification and transcriptional analysis. This study presented the sequence determination and analysis of a segment of the *Hfx. mediterranei* genome. The bioinformatic analysis of the sequence revealed the presence of four open reading frames (ORF): *nasA* (2144 bp), *nasB* (1509 bp), *nasC* (633 bp) and *nasD* (1761 bp). These genes showed high similarity with their counterpart in bacteria and eukaryotes. The previous ORFs correspond to:

- *nasA*: protein with 707 amino acid residues identified as assimilatory nitrate reductase. This sequence also presents the typical residues involved in Fe-S clusters and Mo-MGD (Campbell and Kinghorn 1990). Based on the protein sequence analysis the halophilic nitrate reductase was classified into the monomeric ferredoxin Nas group.
- *nasB*: protein with 503 amino acid residues. It corresponds to nitrate transporter which belongs to the NarK family, a group of proteins not identified previously in Archaea. Specifically, the halophilic NarK matched with type I (Moir and Wood 2001), which are involved in the nitrate uptake in its assimilation pathway.
- *nasC*: protein with 211 amino acid residues wich corresponds to molybdopterin guanine dinucleotide biosynthesis protein. This data implied the presence of Mo cofactor in the halophilic nitrate reductase.
- *nasD*: protein with 587 residues identified as monomeric ferredoxin-dependent nitrite reductase. It has also identified residues involved in the Fe-S cluster formation.

The sequence of this 6720 bp segment of the *Hfx. mediterranei* genome represents a novel organization of nitrate assimilation genes. Moreover, unlike what goes on in bacteria (Lin and Stewart 1998), the genes are transcribed as two independent messengers, the first one as a polycistronic (*nasABC*) and the second one as monocistronic (*nasD*) (Lledó et al. 2005). This atypical arrangement raises different questions in relation to regulation of these two operons, because each one is under the control of different promoters. An interesting characteristic of both promoters was the presence of palindromic sequences, which could be a suitable candidates for the binding of transcriptional regulators as in other microorganisms belonging to the Archaea *Domain*.

The preliminary analysis related to the nitrate assimilation regulation was carried out by RT-PCR with the object to determine the effect of nitrogen source on the expression of *nasABC* and *nasD* (Lledó et al. 2005; Martínez-Espinosa et al. 2007). When *Hfx. mediterranei* was grown in the presence of ammonium as nitrogen source, no transcription of *nasABC* or *nasD* was detected. However, the expression of these operons was detected when *Hfx. mediterranei* was grown with nitrate (Fig. 9.2), indicating that these genes were needed for nitrate assimilation. The expression of *nasA* was higher when the optical density (OD) of the culture rose above 0.4, supporting previous studies related to Nas activity assays where the maximum Nas activity was detected at OD = 0.9. Otherwise, the expression of *nasD* was higher when *Hfx. mediterranei* was grown in the presence of nitrite than in the presence of nitrate, suggesting that nitrite could have a positive effect on *nasD*.

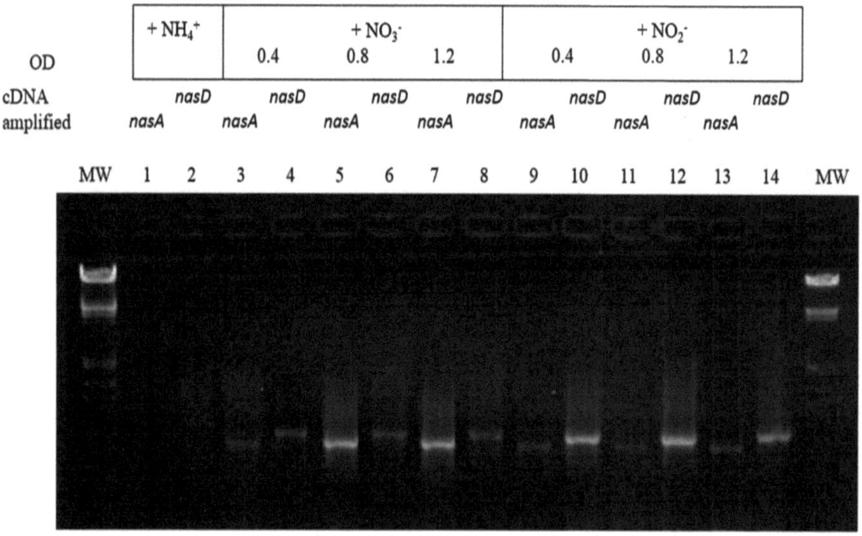

Fig. 9.2 Analysis of the expression of *nasA* and *nasD* genes by agarose gel electrophoresis of double-stranded DNA fragments generated in RT-PCR reactions. Reactions had been performed with total RNA isolated from ammonium medium (Lanes 1 and 2), nitrate (Lanes 3, 4, 5, 6, 7 and 8) or nitrite (9, 10, 11, 12, 13, and 14) supplemented medium from different OD and a pair of primers that amplify *nasA* or *nasD* cDNA. Lane MW: molecular size markers (Lledó et al. 2005)

Therefore, these results concluded that both operons, *nasABC* and *nasD,* were regulated at transcriptional level, being the ammonium the key element which repressed the transcription of nitrate and nitrite reductases genes. Moreover, nitrate and nitrite was involved in a specific control of the assimilation pathway because their presence increased the expression level of *nasABC* and *nasD.*

Martínez-Espinosa et al. (2009) carried out physiological and biochemical studies with *Hfx. mediterranei* in the presence of different concentrations of nitrate (0.5 mM–100 mM) and nitrite (0.5–2 mM). These studies revealed that in the presence of nitrate concentrations higher than 5 mM, the activity of nitrate reductase appeared in first place followed by the activity of nitrite reductase. Under these conditions, the intracellular nitrite concentrations increased shortly before nitrite reductase activity was detected, suggesting that the nitrite accumulation could act as a signal to increase the activity of this enzyme. The expression of *nasA* and *nasD* under the described conditions was analyzed by real-time quantitative PCR experiments (Fig. 9.3a and b), showing that their mRNA concentrations increased during exponential phase whereas they decreased at the beginning of the stationary phase. The maximum *nasA* and *nasD* mRNA concentrations were detected before the maximum enzyme activity in all conditions. Specifically, the espression level of *nasA* mRNA not showed significant differences in cultures with a range of 5–25 mM nitrate, whereas from 25 to 100 mM nitrate the mRNA content increased reaching the maximum expression at 100 mM nitrate (Fig. 9.3a). Moreover, the *nasA* mRNA expression decreased faster after the maximum expression in those media containing 100 mM NO_3^-. This fact allowed to propose that the product of nitrate assimilation could repress the expression of *nasA* mRNA. On the other hand, the *nasD* mRNA level showed a clear dependence on nitrate concentration (Fig. 9.3b) and, always, the maxima *nasD* expressions were detected after the maximum Nas activity. Therefore, the accumulation of nitrite inside the cells could act as inductor of *nasD* expression.

In relation to the experiments carried out when *Hfx. mediterranei* was grown in the presence of nitrite as sole nitrogen source, it was observed that nitrite was consumed by cells under all concentrations assayed and stationary phase was reached at a relatively low optical density (Martínez-Espinosa et al. 2009). Surprisingly, *nasA* mRNA was detected in the cultures with nitrite as nitrogen source independently of its concentration. The concentration of *nasA* mRNA under this conditions was lower than in presence of nitrate, therefore the nitrate has a positive effect on the expression of this gene. This basal *nasA* expression, in media with a different nitrogen source of nitrate, is not frequent as it has described by Chai and Stewart (1998). Consequently, this data suggests that other kind of regulation could be working in this halophilic microorganism. Regarding *nasD,* its mRNA expression was logically detected in the cultures with nitrite as nitrogen source. The Fig. 9.4 shows like its expression depends on nitrite concentration as well as on the optical density of the culture.

More recently, a transcriptome analysis of the operon *nasABC* and *nasD* has been carried out in the presence of inorganic nitrogen salt (nitrate and ammonium) as well as in the presence of four amino acids (glutamate, glutamine, aspartate, asparagine) as the sole nitrogen source (Esclapez et al. 2014). Dot-blot analysis of

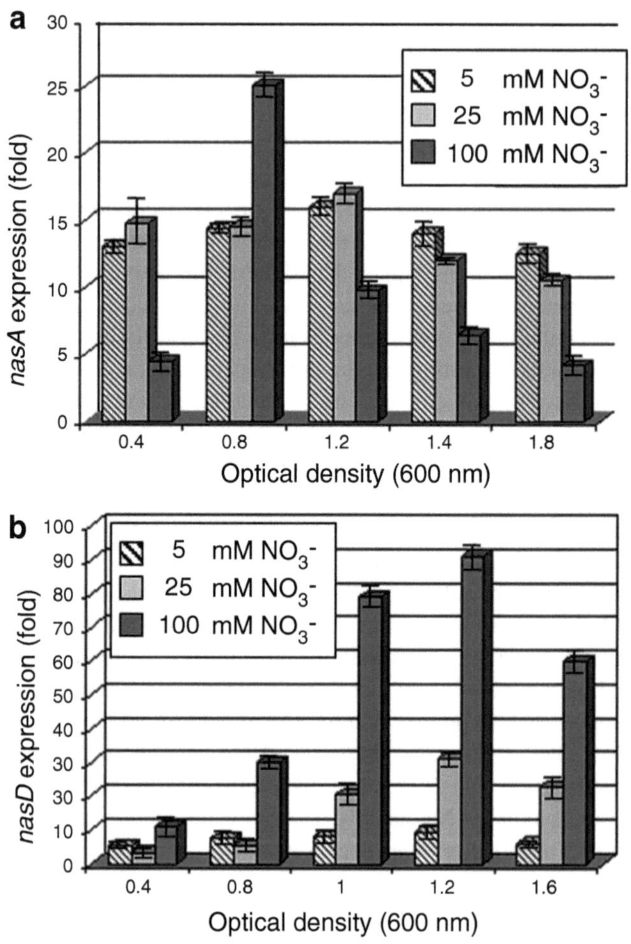

Fig. 9.3 *nasA* and *nasD* mRNA expression levels under different growth conditions during different optical density OD. Comparisons between results obtained from cultures grown in 5 mM NO_3^- (*hatched bar*), 25 mM NO_3^- (*light shaded bar*) and 100 mM NO_3^- (*dark shaded bar*) as sole nitrogen source. Data plotted are the average of the results obtained from quadruplicate experiments. Bars represent the standard deviations (Martínez-Espinosa et al. 2009)

mRNA isolated from different cultures showed hybridization with specific probes in the samples corresponding to the exponential growth phase, obtaining the stronger signals when the higher nitrate concentration was (Fig. 9.5a). Northern analysis showed hybridization with *nasA* and *nasD* probes in media with nitrate as nitrogen source, but any hybridization was observed when the samples came from complex medium (Fig. 9.5b). Therefore, nitrate and nitrite reductases are expressed in the presence of nitrate in the exponential growth phase. These data were confirmed by RT-PCR experiments, which determined that the *nasA* and *nasD* expression was 10-fold and 15-fold higher in the nitrate culture compared to the ammonium culture. These data are in agreement with other previous studies (McCarty and Bremner 1992;

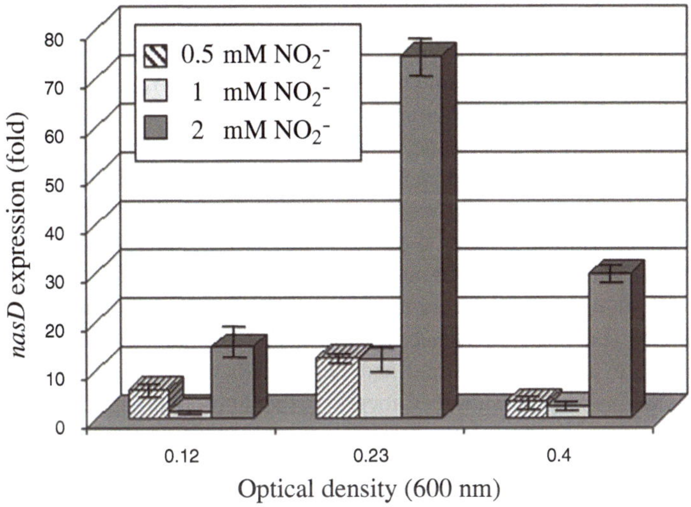

Fig. 9.4 *nasD* mRNA expression level under different growth conditions during different optical density OD. Comparisons between results obtained from cultures grown in 0.5 mM NO_2^- (*hatched bar*), 1 mM NO_2^- (*light shaded bar*) and 2 mM NO_2^- (*dark shaded bar*) as sole nitrogen source. Data plotted are the average of the results obtained from quadruplicate experiments. Bars represent the standard deviations (Martínez-Espinosa et al. 2009)

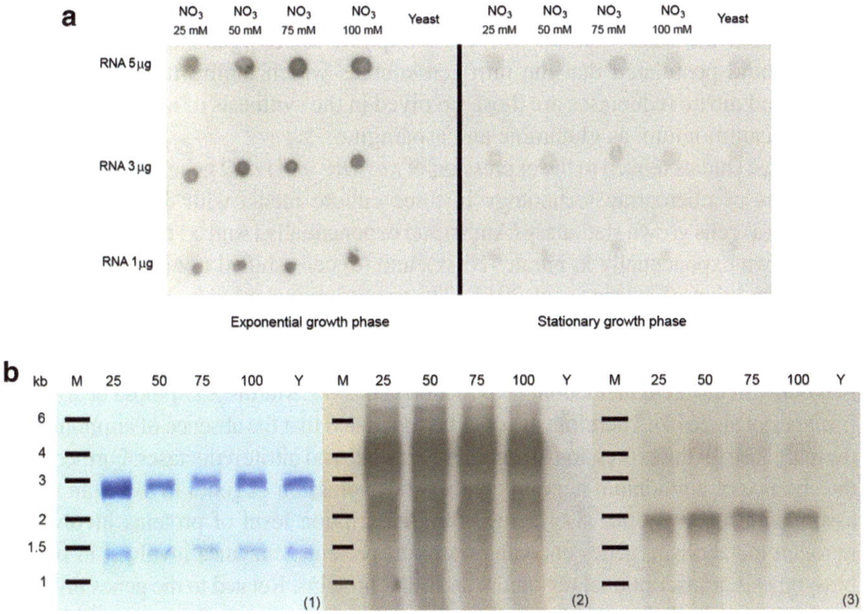

Fig. 9.5 (**a**) Dot-blot using *nasA*-specific probe. The samples of mRNA were obtained in exponential and stationary phase from cultures with different nitrate concentration. (**b**) Northern blot of mRNA obtained in exponential growth phase from cultures with nitrate in a range of 25–100 mM: (1) Methylene blue-stained membrane indicates the 23S and 16S rRNA genes and was used as control for the quality and quantity of total RNA, (2) Northern analysis using *nasA* probe, (3) Northern analysis using *nasD* probe (Esclapez et al. 2014)

Lledó et al. 2005; Bonete et al. 2008; Martínez-Espinosa et al. 2009), but the effect of organic nitrogen source have not been studied in halophilic microorganisms.

To analyse the effect of organic nitrogen sources on *nasABC* and *nasD* transcription, dot-blot and Northern blot analysis were carried out in cultures with glutamate, glutamine, aspartate or asparagine as sole nitrogen source. The mRNA from these cultures was isolated in the exponential phase, because of that is the phase in which nitrate and nitrite reductases are expressed. Dot-blot as well as Northern-blot analysis revealed hybridization with specific probes of *nasA* and *nasD* in the samples from cultures with nitrate, glutamate or aspartate as nitrogen source. However, any signal was obtained in the samples from cultures with ammonium, glutamine, asparagine or yeast extract. Hence, the presence of glutamate, aspartate or nitrate produces the expression of nitrate and nitrite reductases, obtaining the strongest signals in the presence of nitrate. Otherwise, the ammonium, glutamine or asparagine acts as repressor of the transcription of genes related to nitrate assimilation pathway. Previous studies carried out with plants postulated that glycine, glutamine and asparagine inhibit the stimulation performed by nitrate over nitrate reductase enzyme, whereas the amino acids with only one amine group, as glutamate and aspartate, induce its expression (Radin 1977). Later, it has also described the inhibitor effect of glutamine, ammonium or analog compounds over assimilatory nitrate reductase activity in soil (McCarty and Bremner 1992). The last study related with the use of amino acids as nitrogen source suggested that is the internal amino acids pools which acts as intracellular signal of nitrogen level (Miller et al. 2008). Taking into account the data obtained with the halophilic microorganism, *Hfx.mediterranei*, it could postulated that the nitrogen sources which inhibit the expression of nitrate and nitrite reductases are those involved in the synthesis of amino acids coming from ammonium, as glutamine and asparagine.

The last studies related to the expression of *nasABC* and *nasD* genes were performed by means of microarray technology in three culture media with different nitrogen sources: (a) cells grown stationary (AmSt) and exponentially (AmEx) in ammonium, (b) cells grown exponentially in nitrate (NiEx), and (c) cells shifted to nitrogen starvation (NSta) conditions (Esclapez et al. 2015). The transcriptional profiles showed significant differences between the cultures with ammonium as nitrogen source and those with nitrate or nitrogen starvation. These results are in agreement with previous analysis carried out with different methodologies (Lledó et al. 2005; Martínez-Espinosa et al. 2009; Esclapez et al. 2014). Therefore, it could be concluded that the absence of ammonium is the real cause for the expression of assimilative nitrate and nitrite reductases. Furthermore, the absence of ammonium not only modify the expression of genes involved in nitrate assimilation pathway but also change the transcription level of proteins involved in nitrogen metabolism, genes encoding transport proteins, enzymes involved in protein biosynthesis, transcriptional regulators and other proteins. Related to the genes involved in nitrogen assimilation, *nasA, nasB, nasC* and *nasD* genes are over-expressed under nitrogen starvation and in cultures with nitrate as sole nitrogen source in relation to ammonium in exponential growth phase (Table 9.1). Therefore, it may be concluded that nitrate is not a real inductor since *nas* gene expression with nitrate (like under nitrogen starvation) is a consequence of the absence of ammonium rather than an induction by nitrate (Martínez-Espinosa et al. 2007; Esclapez et al. 2014, 2015).

Table 9.1 Expression level of *nasABC* and *nasD* genes in NiEx-AmEx and NSta-AmEx contrasts

Annotation	Gene name	NSta-AmEx		NiEx-AmEx		NCBI accession number
		Log_2FC	SD	Log_2FC	SD	
Ferredoxin-nitrite reductase	*nasD*	5.54	0.03	4.12	0.02	AHZ23086.1
Ferredoxin-nitrate reductase	*nasA*	5.46	0.50	4.70	0.50	AHZ23083.1
Nitrate/nitrite transporter	*nasB*	4.22	0.07	3.60	0.11	AHZ23084.1
Molybdopterin guanine biosynthesis protein A	*nasC*	4.12	0.23	2.50	0.30	AHZ23085.1

3.2 Regulation of Ammonium Assimilation: Glutamate Dehydrogenases, Glutamine Synthetase and Glutamate Synthase

There are two alternative routes to assimilate ammonium which allow the great variety of living organisms cope with changing environments where they have to survive. So, when ammonium is readily available, is glutamate dehydrogenase (GDH), and the route it represents, the preferred to ensure the appropriate uptake of ammonium, meanwhile, when ammonium is not available, or its concentration is not enough high, the GS-GOGAT pathway is the working way to incorporate nitrogen into the cell.

Meanwhile the amination reaction catalyzed by GDH provides the nitrogen, as amino-group, that will be used in many biosynthetic pathways, the oxidative deamination reaction of GDH, on the other hand, converting L-glutamate in 2-oxoglutarate, will provide carbon to the tricarboxylic acid cycle (TCA), producing the adequate concentrations of L-glutamate to balance the glutamine to glutamate ratio and also regulating the necessary levels or concentrations of 2-oxoglutarate, pivotal and essential metabolite, since its belonging to the TCA, as well as its participation in the transamination reactions. These two different roles, let us name them catabolic and anabolic role, respectively, are commonly carried out by GDHs activities depending of NADH or NADPH as cofactors, that also differ in their properties and regulation (Bonete et al. 1986, 1987; Ferrer et al. 1996; Ingoldsby et al. 2005; Díaz et al. 2006; Tomita et al. 2010). As stated before, amino acids display an activator effect on the NAD-GDH from *Hbt salinarum*, meanwhile TCA intermediates and other metabolites inhibit its activity (Bonete et al. 1996; Pérez-Pomares et al. 1999)

Although it is reasonable that the presence of these two activities may be related to the essential roles previously discussed, in the metabolism of organisms (Smith et al. 1975; Santero et al. 2012), including extremophiles such as halophilic Archaea, the exact role of the different GDH proteins still need to be further investigated, studies that becomes even more interesting when, not only two, but, at least four putative *gdh* genes have been found in *Hbt salinarum* genome (Ingoldsby et al. 2005).

Recently Tomita et al. (2010) has shown that glutamate dehydrogenase from *Thermus thermophilus* displays interesting features regarding enzyme regulation. Studies on the genome of this extreme thermophile bacteria shows two genes, putatively forming an operon, *gdhA* and *gdhB*, which correspond to two different glutamate dehydrogenases. The products of these two genes were purified and characterized by Tomita et al. (2010) in order to assign them its real function. One of the products, GdhA, showed no GDH activity and the other, GdhB, displayed GDH activity, being 1.3 fold higher the rate for the reductive amination than that found for the oxidative deamination. The co- expression and purification of GdhA with His-tag-fused GdhB also displayed glutamate dehydrogenase activity, let us name it GdhA–GdhB. The GdhA–GdhB activity for the reductive amination was lower than that for GdhB alone, meanwhile for the oxidative deamination was higher. GdhA–GdhB had decreased reductive amination activity and increased oxidative deamination activity, in a ratio of 3.1-fold oxidative deamination versus reductive amination.

Just as reported for the halophilic archaeon *Hbt salinarum* NAD-glutamate dehydrogenase (Bonete et al. 1996), and previously discussed, amino acids had an activating effect on the hetero-complex GdhA–GdhB, and also it was higher for hydrophobic amino acids. The most effective activating amino acid for GdhA–GdhB was leucine, which, at 1 mM leaded to an activity 9.74 fold higher than that without leucine for the reductive amination, and 2.45 fold for the oxidative deamination. The activating effect of leucine was much lower for GdhB alone. The kinetic analysis of this activating effect showed that leucine enhanced the turnover number of glutamate dehydrogenase. This product acted as a hetero-oligomeric GDH system, where the role for GdhA subunit was regulatory, and GdhB subunit was catalytic. In this hetero-complex, GdhA acts modulating the activity of GdhB through the formation of the hetero-complex that depends on the concentration of the activator hydrophobic amino acid. This study of Tomita et al. illustrated the allosteric regulation of a hetero-oligomeric glutamate dehydrogenase (Tomita et al. 2011).

The regulation of glutamate synthase is not well known and there are different approaches depending on the organism. In plants there are two forms of GOGAT enzyme: Fd-GOGAT, which is usually present in high activities in the chloroplasts of photosynthetic tissues, and NADH-dependent enzyme, which is also present in plastids of non-photosynthesizing cells (Bowsher et al. 2007). Both enzymes supply glutamate during the nitrogen assimilation, and glutamate dehydrogenase does not represent a significant alternate route for glutamate formation. GDH may be expected to function mainly in the deaminating direction providing 2-oxoglutarate and ammonium that can be used in respiration and amide formation respectively (Miflin 2002; Forde and Lea 2007) and also would act with NADH-GOGAT controlling the homeostasis of Glu in the plant (Labboun et al. 2009).

In *Escherichia coli*, the GS-GOGAT pathway is used when the cell is not under energy limitation. At low ammonium concentration is the main way for glutamate synthesis and it is essential for regulation of the glutamine pool. In ammonium and phosphate abundance and when the cell is limited for energy and carbon, the GDH pathway is used in glutamate synthesis (Helling 1994).

There are a number of transcriptional regulators that control the GOGAT synthesis in bacteria. The *gltB* and *gltD* genes code for the large and small subunits, respectively, of the GOGAT. In *E. coli* expression of *gltBD* is influenced by regulators including the leucine-responsive regulatory protein, LRP and CRP (Reitzer 2003).

Synechocystis sp. strain PCC 6803 utilizes the GS-GOGAT pathway as the primary pathway of ammonia assimilation, but the presence of GDH appears to offer a selective advantage for the cyanobacterium under no exponential growth conditions (Chávez et al. 1999). In *Corynebacterium glutamicum*, transcription of the operon formed by the genes *gltB* and *gltD* is under control of AmtR, a repressor of nitrogen-regulated genes in this organism (Beckers et al. 2001), which regulates the transcription of other nitrogen genes including *amtB* and *glnK*. Schulz et al. (2001) found that disruption of the *hkm* gene, encoding a putative histidine kinase upstream of *gltBD* in *Corynebacterium glutamicum*, reduced the levels of GOGAT activity two-fold under nitrogen-rich and nitrogen-limiting conditions. Transcription of *hkm* was moderately induced by nitrogen starvation, indicating that the Hkm protein may play a role in signal transduction of the nutritional status of the growth medium (Schulz et al. 2001).

In *B. subtilis,* the two enzymes responsible for the unique pathway of ammonium assimilation are glutamine synthetase and glutamate synthase since GDH is absent in this organism. The *gltAB* operon requires a specific positive regulator, GltC, for its expression and is repressed by TnrA, a regulator of several other genes of nitrogen metabolism (Belitsky et al. 2000).

In haloarchaea, studies at transcriptional level have been done to elucidate the role of the different enzymes implied in the different pathways. Figure 9.6a shows the comparison of the expression profiles of *gdh-1*, *glnA* and *gltS*, tested at exponential phase, from different media (each containing ammonium, nitrate, glutamate or glutamine as nitrogen source). The results revealed that *gltS* expression was 70-fold higher in the nitrate culture compared with the ammonium culture. The presence of nitrate also enhanced the expression of *glnA*, but repressed the expression of *gdh-1*. This is in agreement with the hypothesis that the GS/GOGAT pathway operates under conditions of ammonium restriction, whereas ammonium assimilation by GDH would occur preferentially at high ammonium concentrations, although under these conditions the levels of transcription of *gdh-1, gltS* and *glnA* were very similar (Fig. 9.6b). In medium containing glutamate as the only nitrogen source, *gdh-1, glnA* and *gltS* expression was lower than that seen in the ammonium medium. However, in the medium with glutamine, *glnA* expression was repressed, whereas the expression of *gltS* and *gdh-1* was enhanced, when compared with levels in media containing ammonium. This profile suggests that high levels of glutamate in the cell can be a signal to repress ammonium assimilation, but high levels of glutamine are a signal of ammonium abundance; indeed, expression of both *gltS* and *gdh-1* was higher than in media containing ammonium. The RT-qPCR carried out by Pire et al. (2014) strongly suggests that the GS/ GOGAT pathway could be the preferred route for ammonium assimilation under conditions of ammonium starvation or deficiency. In the medium with ammonium as nitrogen source, *gdhA-1* and *gdh1* were transcribed at similar level, with a relative change of 1.1 ± 0.3 for the expression of *gdhA-1* compared with the expression of *gdh1*. Both enzymes work in the amination reaction (Ferrer et al. 1996; Díaz et al. 2006), but whereas the gene expression of *gdhA-1* did not change when the nitrogen source was nitrate instead of ammonium, the expression of gdh1 was highly repressed. Although the expression

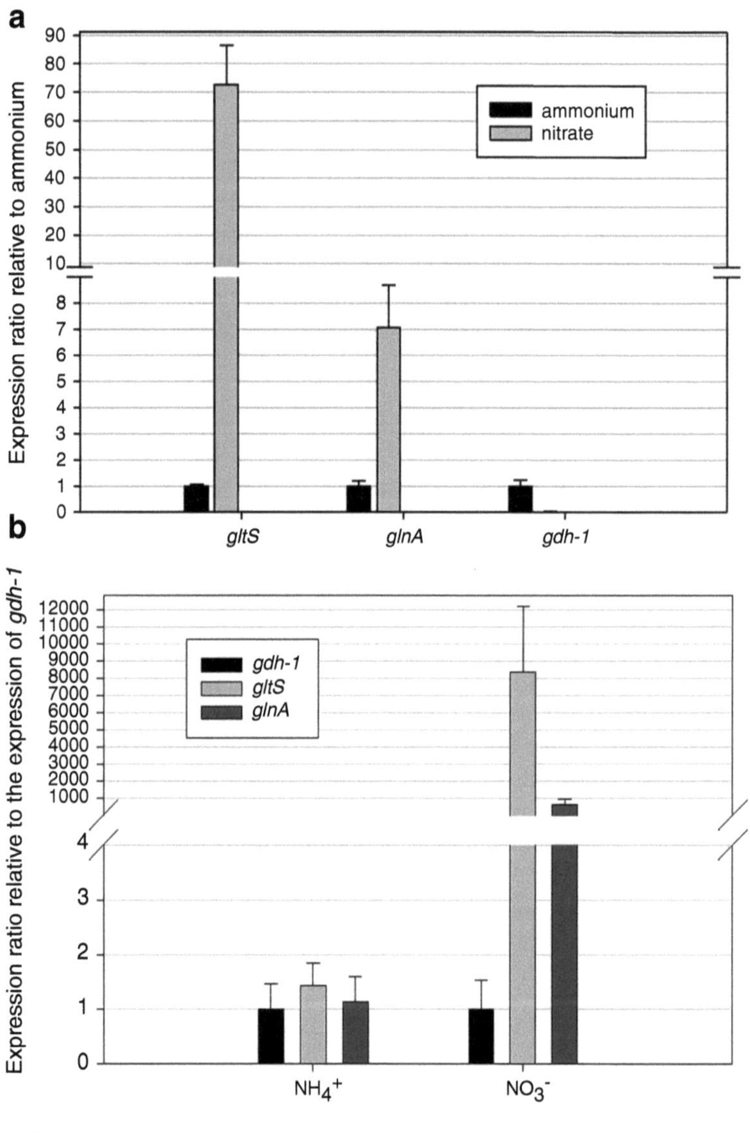

Fig. 9.6 Transcription level of *gltS*, *glnA*, and *gdh-1* genes from cultures at exponential phase with ammonium and nitrate as a nitrogen source. (**a**) The results from ammonium culture were used for calibrating. (**b**) The results for *gdh-1* expression were used for calibrating (Pire et al. 2014)

of *gdhA-1* seems to be constitutive, previous studies of the enzyme showed that the specific activity of NADPH-GDH was higher in ammonium than in nitrate, suggesting an inhibition of the enzyme activity (Ferrer et al. 1996).

Not only the transcription of *gdh-1* was highly repressed in a media with nitrate, but also the transcription of *gltS* and *glnA* was enhanced in the same media.

The GDH reaction would assume ammonium uptake when this is available in the medium, since ammonium is the preferred nitrogen source over nitrate or other nitrogen- containing compounds (Bonete et al. 2008). However, the Northern blot analysis, as well as the RT-qPCR results, showed a basal level of Fd-GOGAT transcription even when ammonium is available.

As shown in Fig. 9.6b, the transcription level of *gltS* in media containing ammonium is similar to that of *gdh-1*. Under these conditions, GDH activity predominates due to its higher affinity for 2-oxoglutarate and ammonium. These results would indicate that the level of 2-oxoglutarate is essential. It has been proposed that the "in vivo" rate of GOGAT reaction is controlled by the production of glutamine by GS, but whereas GS activity is determined by the cellular levels of glutamine and 2-oxoglutarate, GS would also be modulated by GOGAT through its influence on the 2-oxoglutarate concentration (Vanoni and Curti 1999). In our case, according to the K_m values, 2-oxoglutarate would be a key metabolite in determining whether assimilation takes place via the GS/GOGAT route (Pire et al. 2014).

All these results indicated that in *Hfx. mediterranei* the ammonia produced by the nitrate and nitrite reductase reactions is incorporated into the carbon skeletons mainly by the GS/GOGAT cycle. The global gene expression in *Hfx. mediterranei* in three culture media with different nitrogen sources (ammonium, nitrate, and in cells shifted to nitrogen starvation), has been analyzed (Esclapez et al. 2015). The main differences in the transcriptional profiles have been identified between the cultures with ammonium as nitrogen source and the cultures with nitrate or nitrogen starvation. Some transcriptional regulators have been identified, two ArsR type and other putative transcriptional regulators. Analysis of these transcriptional regulators, would allow identifying nitrogen regulators involved in the ammonium deficiency adaptation in haloarchaea.

4 Nitrogen Regulatory Proteins (PII) in Haloarchaea

In 1969, the PII signal transduction proteins were discovered when posttranslational modification studies were carried out with *Escherichia coli* glutamine synthetase (Shapiro 1969). Currently it is known that PII proteins are involved in regulatory processes involved in the assimilation of nitrogen (Ninfa and Jiang 2005; Leigh and Dodsworth 2007; Forchhammer 2008). Apparently, they can act as devices that collect the level of 2-oxoglutarate and adenylylate energy charge (Forchhammer 2010). Proteins regulated by PII are: AmtB ammonium transporter, which is able to retain GlnK to complex with it when the cellular nitrogen level is sufficiently high (Javelle et al. 2004); glutamine synthetase adenylyltransferase (GlnE), which modificates covalently the activity of glutamine synthetase (Jiang et al. 2007); DraG/DraT, which regulates the nitrogenase enzyme in nitrogen fixing bacteria (Dixon and Kahn 2004) and NAGK, important enzyme involved in arginine biosynthesis in cyanobacteria (Heinrich et al. 2004), among others.

Previous articles have been shown with the GlnB-K protein family in Archaea have been carried out, specifically with *Methanosarcina mazei* (Ehlers et al. 2002, 2005) and *Archaeoglobus fulgidus* (Helfmann et al. 2010; Litz et al. 2011). With halophilic Archaea, the study of these proteins began in 2011, when the existence of two homologues PII gene copies in the genome of *Hfx. mediterranei* (Pedro-Roig et al. 2011) was observed, with 84 % of identical amino acids, classified as GlnKs ($GlnK_1$ and $GlnK_2$), and both linked to genes encoding for ammonium transporters (*amt*). The two pairs of genes are located consecutively in the genome, but separated by a 300 nt gap of non-coding sequence. The C-terminal region is highly conserved in both GlnKs (PS00638, PROSITE), but only a fragment of the uridylylation site of PII proteins is conserved (PS00496, PROSITE). The amino acid sequence WRGEEY is also present in both GlnK proteins of *Hfx. mediterranei*, with two changes with regard to other proteins non halophilic, the first residue, W, is changed to Y, and the fourth residue, E, is changed to A or S. In the latter case, the substitution may be related to the halophilic adaptation. There is a tyrosine residue which can participate in the uridylylation, like in other PII proteins, at position 60 for $GlnK_1$ and 61 for $GlnK_2$. Studies carried out by the technique of cross-linking using glutaraldehyde concluded that GlnKs proteins have a trimeric quaternary structure (45 kDa) (Pedro-Roig et al. 2013b). The crystal structure has also been determined to $GlnK_2$ protein, being the first structural study of haloarchaea PII protein, and confirming the high degree of structural conservation in all PII proteins of all organisms in which they are located (Palanca et al. 2014).

4.1 GlnK Proteins in Different Nitrogen Sources and Their Role in the Regulation of Nitrogen Metabolism in Hfx. mediterranei

Hfx. mediterranei was grown in several different conditions: with yeast extract as a source of organic nitrogen, or either ammonium or nitrate as inorganic nitrogen sources (glucose was added in excess as a carbon source for these synthetic media). In the presence of nitrate both GlnKs were observed, while that in the presence of ammonium or yeast no signal could be detected (Pedro-Roig et al. 2011). With GS, the same thing happens (Martínez-Espinosa et al. 2006), so it can deduce that there is a relationship between PII and GS, activating the last one. Growth of *Hfx. mediterranei* in 75 mM nitrate, in exponential phase neither GS activity nor GlnK intense signal were observed. While at the same concentration of nitrate in stationary phase, high activity of GS and GlnK intense signal (Fig. 9.7b, lane 9) were observed. Therefore, it has been suggested an activating function for GlnK in ammonium assimilation in *Hfx. mediterranei*, through the GS/GOGAT via (Pedro-Roig et al. 2011). This fact was proved *in vitro*. Biosynthetic activity measurements were realized in the presence and absence of GlnK, and an increase of GS enzymatic activity up to 50 % as a result of purified recombinant GlnK presence in the assays was obtained. Moreover, in SDS-PAGE analysis from cells grown in nitrate, a sole band was seen, corresponding to $GlnK_1$, present under these conditions, so you could think they have different electrophoretic mobility (Pedro-Roig et al. 2013b).

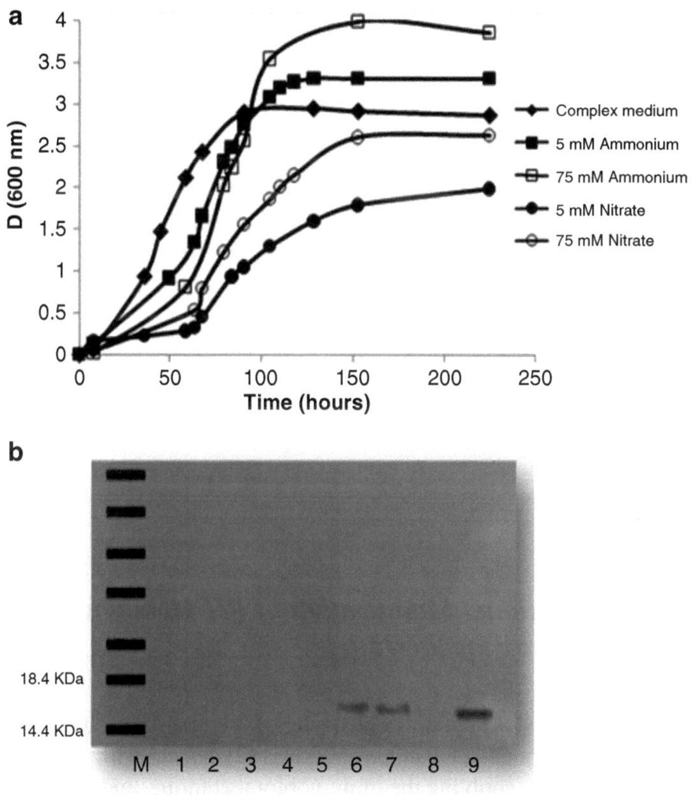

Fig. 9.7 *Hfx. mediterranei* growth curves with different nitrogen sources (**a**) and analysis of GlnK expression by SDS-PAGE and Western blot (**b**): Lane M, molecular-mass markers; lane 1, complex medium, exponential phase; lane 2, complex medium, stationary phase; lane 3, 5 mM ammonium, exponential phase; lane 4, 5 mM ammonium, stationary phase; lane 5, 75 mM ammonium, exponential phase; lane 6, 5 mM nitrate, exponential phase; lane 7, 5 mM nitrate, stationary phase; lane 8, 75 mM nitrate, exponential phase; lane 9, 75 mM nitrate, stationary phase (Pedro-Roig et al. 2011)

Biosynthetic activity of GS was increased in presence of 10 mM 2-oxoglutarate and in presence and absence of GlnK, showing an increase of 12-fold with the metabolite alone, and 18-fold in the presence of PII protein. 2-oxoglutarate is a metabolite that promotes assimilation of ammonium by GS when there is nitrogen deficiency. GS activity increases in the presence of GlnK because a complex between both two proteins is formed, as it has been demonstrated by gel filtration chromatography followed by SDS-PAGE and immunoblotting with specific antibodies for the regulatory protein (data not shown) (Pedro-Roig et al. 2013b). The presence of 2-oxoglutarate is important to form GS-GlnK while the GS activity must be increased, and ammonium assimilation should be produced by the route GS/GOGAT when the cell has low nitrogen. In the GlnK–GS complex from *Hfx. mediterranei*, the stoichiometry remains unclear, but one dodecamer of GS to four trimers of PII is calculated.

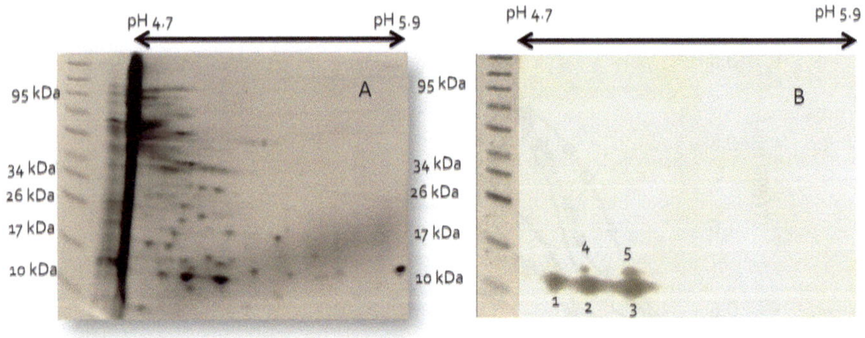

Fig. 9.8 (**a**) 2DE of 500 g of total cell protein isolated from 20-h nitrogen starved cells using an IPG strip pH 4.7–5.9 and an *Any kD TGX* gel, both from Bio-Rad. The gel was stained with Coomassie Blue. (**b**) Anti-GlnK immunoblot of a 50 g replicate of the 2DE in (**a**). Specific *Hfx. mediterranei* anti-GlnK antibodies were used, and ECL detection was performed using HRP-conjugated secondary antibodies and luminol as substrate (Pedro-Roig et al. 2013a)

4.2 *Post-Translational Modification of PII Homologues from Haloferax mediterranei*

To know if PII proteins of haloarchaeal species were post-translationally modified it has been used two-dimensional electrophoresis (2DE) and with specific detection using antibodies for PII, applying the immunoblot technique, and subsequent analysis of MALDI-TOF-MS. With this type of study should determine PTM for these proteins as they undergo a small displacement in the isoelectric point. The theoretical pI for $GlnK_1$ protein from *Hfx. mediterranei* is 5.15 and for $GlnK_2$ is 5.11. Specific antibodies were used for GlnK from *Hfx. mediterranei*, using as key the T-loop area of both proteins, and applying immunoblotting technique, five different signals which coincided with the pI values and expected size were obtained (Fig. 9.8b) (Pedro-Roig et al. 2013a). Spots 1, 2 and 3 (Fig. 9.8b) belong to $GlnK_1$, and spots 4 and 5 belong to $GlnK_2$. The high primary structure identity (84 %) between the two GlnK from *Hfx. mediterranei* could mask differentiation between both proteins with this technique; however, through the mass spectra, which analyzed peptides with different masses for each protein, allowed specific identification.

Previously, it has been proposed that there was no presence of $GlnK_2$ in *Hfx. mediterranei* grown with nitrate (Pedro-Roig et al. 2013b). When the cell is nitrogen starved, with the immunoblotting technique the second GlnK (Pedro-Roig et al. 2013a) also appears. This is important physiologically, since when the cell is in the presence of a nitrogen source, only is necessary one of the two GlnKs to metabolic regulation, but in total absence of nitrogen, both GlnKs are present. This indicates that GlnKs are activators of nitrogen assimilation in shortage of this component, which coincides with that both of them activate the GS *in vitro* (Pedro-Roig et al. 2013b).

In the spectra from spots 2 and 4 in Fig. 9.8b an increase in mass of 306 Da was observed, with respect to the theoretical mass of the peptide, which did not occur in spots 3 and 5. This mass matches with an UMP group (monoisotopic mass of 306.03), so it could be concluded that in the spots 2 and 4, the $GlnK_1$ and $GlnK_2$ proteins are uridylylated, respectively, while in the spots 3 and 5, the proteins are not modified. It could be said that the modification occurs at the conserved tyrosine residue (Y60 for $GlnK_1$, Y61 for $GlnK_2$), since, as previously mentioned, only a fragment of the uridylylation site of the PII protein is conserved (PS00496, PROSITE). This could be an important conclusion: PII proteins from the Archaea Domain undergo post-translational modifications. However, in the *Hfx. mediterranei* genome, a halophilic homologue of the uridylyltransferase/uridylyl-removing enzyme (encoded by *glnD*) has not found (Pedro-Roig et al. 2013a). Therefore, there should be some other similar halophilic Archaea protein that can perform this function.

4.3 Cotranscription of glnK and amtB Genes From Haloferax mediterranei Under Conditions of Ammonium Limitation

A transcriptional level, the two *amtB-glnK* genes pairs from *Hfx. mediterranei* have been studied (Pedro-Roig et al. 2013). When the medium was rich in ammonia or yeast extract, there was no evidence of *amtB-glnK* transcript, so that the pairs of genes are not transcribed or transcripts suffered degradation due to ammonia excess. Conversely, when the medium was rich in nitrate, the *amtB-glnK* transcripts of the two pairs of genes in RNA samples (data not shown) were obtained. With the Northern blot was found that *amtB-glnK* genes are cotranscribed in pairs, and the pair *amtB1-glnK1* is expressed primarily in conditions of nitrate in the medium, rather than *amtB2-glnK2*. Furthermore, transcription levels suffer regulation by the availability of nitrogen in the medium, with no expression in ammonium source in the medium and maximum when the medium lacks ammonium.

5 Biotechnological Applications

Human activity has led to an increase in the concentration of salts, such as nitrates and nitrites, in soils and groundwater. Companies engaged in the manufacture of pesticides, herbicides, explosives and dyes, contribute to this problem. Nitrate and nitrite have a significant involvement in agriculture, environment and public health. Most organisms are usually affected by even low nitrate and nitrite concentrations. *Hfx. mediterranei* is the first reported haloarchaea having the ability to assimilate nitrate and nitrite, due to the presence of nitrate (Nas) and nitrite (NiR) reductases in a high salt concentration environment. Physiological characterization carried out in *Hfx. mediterranei*, has shown that this microorganism is able to grow aerobically on minimal culture with nitrate (up to 2 M concentration) or nitrite (up to 50 mM concentration) as unique nitrogen sources (Martínez-Espinosa et al. 2009). The high nitrite concentration tested in the assay cited above is one of the highest yet described for a microorganism belonging to Prokarya. The growth of *Hfx. mediterranei* in these toxic conditions for most microorganisms studied up to now, involves removing nitrate and nitrite in the culture media. The concentrations of nitrate and nitrite in wastewater are usually lower than the concentrations studied in *Hfx. mediterranei*. So it is possible to think that this microorganism could be employed in bioremediation processes for salted wastewater or brines. In the same way, Nas and NiR enzymes could be immobilized to be used in bioremediation techniques or like sensors to detect the presence of nitrate and nitrite.

On the other hand, carotenoids have been studied because of their potential functions on nutrition and human health in the last years. Numerous studies have shown that these pigments have some physiological roles in the retardation of cancer and heart affections by quenching of singlet oxygen or free radicals, increasing the production of antibodies *in vitro*, etc. (Edge et al. 1997; Carpenter et al. 1997; Palozza et al. 1998; Chew et al. 1999). The carotenoids use in the field of nutrition as processed food coloring agents, is also studied (Bauernfeind 1981). Traditionally there have been studies with carotenoids producing organisms such as *Haematococcus pluvialis, Blaskeslea trispora* and *Dunaliella salina* (Olaizola 2000; Mehta et al. 2003; Raja et al. 2007). However there are very few studies in extreme halophilic microorganisms that are excellent candidates because of their characteristics for the production of such pigments. In *Hfx. alexandrinus* has conducted a study on the production of canthaxanthin (Asker and Ohta 2002) and another in *Hfx. mediterranei* in which the influence of the composition of the culture media for the production of C_{50} carotenoids is studied (Fang et al. 2010). Based on these studies, there have been realized various assays in *Hfx. mediterranei* in our research group in which it has been found that according to the carbon/nitrogen rate in the culture media different carotenes are produced (unpublished data).

References

Adul Rahman RN, Jongsareejit B, Fujiwara S, Imanaka T (1997) Characterization of recombinant glutamine synthetase from the hyperthermophilic archaeon *Pyrococcus* sp. strain KOD1. Appl Environ Microbiol 63:2472–2476

Alcántara-Hernández RJ, Valenzuela-Encinas C, Zavala-Díaz de la Serna FJ, Rodriguez-Revilla J, Dendooven L, Marsch R (2009) Haloarchaeal assimilatory nitrate-reducing communities from a saline alkaline soil. FEMS Microbiol Lett 298(1):56–66

Andrade SL, Einsle O (2007) The Amt/Mep/Rh family of ammonium transport proteins. Mol Membr Biol 24:357–365

Andrade SL, Dickmanns A, Ficner R, Einsle O (2005) Crystal structure of the archaeal ammonium transporter Amt-1 from *Archaeoglobus fulgidus*. Proc Natl Acad Sci U S A 102:14994–14999

Asker D, Ohta Y (2002) Production of canthaxanthin by *Haloferax alexandrines* under non-aseptic conditions and a simple, rapid method for its extraction. Appl Microbiol Biotechnol 58:743–750

Bauernfeind JC (1981) Natural food colors. In: Bauernfeind JC, Stewart GF, Schweigert BS, Hawthorn J (eds) Carotenoids as colorants and vitamin A precursors. Technological and nutritional applications. Academic Press, New York, pp 1–45. ISBN:978-0-12-082850-0

Beckers G, Nolden L, Burkovski A (2001) Glutamate synthase of *Corynebacterium glutamicum* is not essential for glutamate synthesis and is regulated by the nitrogen status. Microbiology 147(11):2961–2970

Beinert H (2000) Iron–sulfur proteins: ancient structures, still full of surprises. J Biol Inorg Chem 5:2–15

Belitsky BR, Wray LV, Fisher SH, Bohannon DE, Sonenshein AL (2000) Role of TnrA in nitrogen source-dependent repression of bacillus subtilis glutamate synthase gene expression. J Bacteriol 182(21):5939–5947

Bonete MJ, Camacho ML, Cadenas E (1986) Purification and some properties of NAD+-dependent glutamate dehydrogenase from *Halobacterium halobium*. Int J Biochem 18:785–789

Bonete MJ, Camacho ML, Cadenas E (1987) A new glutamate dehydrogenase from *Halobacterium halobium* with different coenzyme specificity. Int J Biochem 19:1149–1155

Bonete MJ, Perez-Pomares F, Ferrer J, Camacho ML (1996) NAD-glutamate dehydrogenase from *Halobacterium halobium*: inhibition and activation by TCA intermediates and amino acids. Biochim Biophys Acta 1289(1):14–24

Bonete MJ, Martínez-Espinosa RM, Pire C, Zafrilla B, Richardson DJ (2008) Nitrogen metabolism in haloarchaea. Saline Syst 4:9. doi:10.1186/1746-1448-4-9

Bowsher CG, Lacey AE, Hanke GT, Clarkson DT, Saker LR, Stulen I, Emes MJ (2007) The effect of Glc6P uptake and its subsequent oxidation within pea root plastids on nitrite reduction and glutamate synthesis. J Exp Bot 58(5):1109–1118

Brown JR, Masuchi Y, Robb FT, Doolittle WF (1994) Evolutionary relationships of bacterial and archaeal glutamine synthetase genes. J Mol Evol 38(6):566–576

Campbell WH, Kinghorn JR (1990) Functional domains of assimilatory nitrate reductase and nitrite reductases. Trends Biochem Sci 15:315–319

Carpenter KLH, van der Veen C, Hird R, Dennis IF, Ding T, Mitchinson MJ (1997) The carotenoids: β-carotene, canthaxanthin and zeaxanthin inhibit macrophage-mediated LDL oxidation. FEBS Lett 401:262–266

Chai W, Stewart V (1998) NasR, a novel RNA-binding protein, mediates nitrate-responsive transcription antitermination of the *Klebsiella oxytoca* M5al *nasF* operon leader *in vitro*. J Mol Biol 283:339–351

Chávez S, Lucena JM, Reyes JC, Florencio FJ, Candau P (1999) The presence of glutamate dehydrogenase is a selective advantage for the cyanobacterium *Synechocystis* sp. strain PCC 6803 under nonexponential growth conditions. J Bacteriol 181(3):808–813

Chew BP, Park JS, Wong MW, Wong TSA (1999) Comparison of the anticancer activities of dietary β-carotene, canthaxanthin and astaxanthin in mice in vivo. Anticancer Res 19:1849–1853

Díaz S, Pérez-Pomares F, Pire C, Ferrer J, Bonete MJ (2006) Gene cloning, heterologous overexpression and optimized refolding of the NAD-glutamate dehydrogenase from *Haloferax mediterranei*. Extremophiles 10(2):105–115

Dixon R, Kahn D (2004) Genetic regulation of biological nitrogen fixation. Nat Rev Microbiol 2(8):621–631

Edge R, McGarvey DJ, Truscott TG (1997) The carotenoids as antioxidants – a review. J Photochem Photobiol B Biol 41:189–200

Ehlers C, Grabbe R, Veit K, Schmitz RA (2002) Characterization of GlnK1 from *Methanosarcina mazei* Strain Gö1: complementation of an *Escherichia coli glnK* Mutant Strain by $GlnK_1$. J Bacteriol 184:1028–1040

Ehlers C, Weidenbach K, Veit K, Forchhammer K, Schmitz RA (2005) Unique mechanistic features of post-translational regulation of glutamine synthetase activity in *Methanosarcina mazei* strain Gö1 in response to nitrogen availability. Mol Microbiol 55:1841–1854

Esclapez J, Bravo-Barrales G, Bautista V, Pire C, Camacho M, Bonete MJ (2014) Effects of nitrogen sources on the nitrate assimilation in *Haloferax mediterranei*: growth kinetics and transcriptomic analysis. FEMS Microbiol Lett 350:168–174

Esclapez J, Pire C, Camacho M, Bautista V, Martínez-Espinosa RM, Zafrilla B, Vegara A, Alcaraz LA, Bonete MJ (2015) Transcriptional profiles of *Haloferax mediterranei* based on nitrogen availability. J Biotechnol 193:100–107

Fang CJ, Ku KL, Lee MH, Su NW (2010) Influence of nutritive factors on C_{50} carotenoids production by *Haloferax mediterranei* ATCC 33500 with two-stage cultivation. Bioresour Technol 101(16):6487–6493

Feng J, Liu B, Zhang Z, Ren Y, Li Y, Gan F, Huang Y, Chen X, Shen P, Wang L, Tang B, Tang XF (2012) The complete genome sequence of *Natrinema* sp. J7-2, a haloarchaeon capable of growth on synthetic media without amino acid supplements. PLoS One 7(7), e41621

Fernández E, Galvan A, Quesada A (1998) Nitrogen assimilation and its regulation. In: The molecular biology of chloroplast and mitochondria in *Chlamydomonas*. Kluwer Academic Publishers, The Netherlands, pp 637–659

Ferrer J, Pérez-Pomares F, Bonete MJ (1996) NADP-glutamate dehydrogenase from the halophilic archaeon *Haloferax mediterranei*: enzyme purification, N-terminal sequence and stability. FEMS Microbiol Lett 141(1):59–63

Forchhammer K (2008) P(II) signal transducers: novel functional and structural insights. Trends Microbiol 16(2):65–72

Forchhammer K (2010) The network of PII signalling protein interactions in unicellular cyanobacteria. Adv Exp Med Biol 675:71–90. Chap 5. doi:10.1007/978-1-4419-1528-3_5

Forde BG, Lea PJ (2007) Glutamate in plants: metabolism, regulation, and signalling. J Exp Bot 58(9):2339–2358

Frolow F, Harel M, Sussman JL, Mevarech M, Shoham M (1996) Insights into protein adaptation to a saturated salt environment from the crystal structure of a halophilic 2Fe-2S ferredoxin. Nat Struct Biol 3(5):452–458

Geiger B, Mevarech M, Werber MM (1978) Immunochemical characterization of ferredoxin from *Halobacterium* of the Dead Sea. Eur J Biochem 84(2):449–455

Heinrich A, Maheswaran M, Ruppert U, Forchhammer K (2004) The *Synechococcus elongates* PII signal transduction protein controls arginine synthesis by complex formation with N-acetyl-L-glutamate kinase. Mol Microbiol 52:1303–1314

Helfmann S, Lü W, Litz C, Andrade SLA (2010) Cooperative binding of MgATP and MgADP in the trimeric PII protein GlnK2 from *Archaeoglobus fulgidus*. J Mol Biol 402:165–177

Helling RB (1994) Why does Escherichia coli have two primary pathways for synthesis of glutamate? J Bacteriol 176:4664–4668

Hochman A, Nissany A, Amizur M (1988) Nitrate reduction and assimilation by a moderately halophilic, halotolerant bacterium Ba1. Biochim Biophys Acta 965(1):82–89

Ingoldsby LM, Geoghegan KF, Hayden BM, Engel PC (2005) The discovery of four distinct glutamate dehydrogenase genes in a strain of *Halobacterium salinarum*. Gene 349:237–244

Javelle A, Severi E, Thornton J, Merrick M (2004) Ammonium sensing in *Escherichia coli*. Role of the ammonium transporter AmtB and AmtB–GlnK complex formation. J Biol Chem 279:8530–8538

Jiang P, Mayo AE, Ninfa AJ (2007) *Escherichia coli* glutamine synthetase adenylyltransferase (ATase, EC 2.7.7.49): kinetic characterization of regulation by PII, PII-UMP, glutamine, and α-ketoglutarate. Biochemistry 46:4133–4146

Kameya M, Ikeda T, Nakamura M, Arai H, Ishii M, Igarashi Y (2007) A novel ferredoxin-dependent glutamate synthase from the hydrogen-oxidizing chemoautotrophic bacterium *Hydrogenobacter thermophilus* TK-6. J Bacteriol 189(7):2805–2812

Khademi S, Stroud RM (2006) The Amt/MEP/Rh family: structure of AmtB and the mechanism of ammonia gas conduction. Physiology (Bethesda) 21:419–429

Kumada Y, Benson DR, Hillemann D, Hosted TJ, Rochefort DA, Thompson CJ, Wohlleben W, Tateno Y (1993) Evolution of the glutamine synthetase gene, one of the oldest existing and functioning genes. Proc Natl Acad Sci USA 90:3009–3013

Labboun S, Tercé-Laforgue T, Roscher A, Bedu M, Restivo FM, Velanis CN et al (2009) Resolving the role of plant glutamate dehydrogenase. I. In vivo real time nuclear magnetic resonance spectroscopy experiments. Plant Cell Physiol 50:1761–1773

Leigh JA, Dodsworth JA (2007) Nitrogen regulation in bacteria and archaea. Annu Rev Microbiol 61:349–377

Lin JT, Stewart V (1998) Nitrate assimilation by bacteria. Adv Microb Physiol 39:330–379

Litz C, Helfmann S, Gerhardt S, Andrade SLA (2011) Structure of GlnK1, a signaling protein from *Archaeoglobus fulgidus*. Acta Crystallogr 67:178–181
Lledó B, Marhuenda-Egea FC, Martínez-Espinosa RM, Bonete MJ (2005) Identification and transcriptional analysis of nitrate assimilation genes in the halophilic archaeon *Haloferax mediterranei*. Gene 361:80–88
Luque-Almagro VM, Gates AJ, Moreno-Vivián C, Ferguson SJ, Richardson DJ, Roldán MD (2011) Bacterial nitrate assimilation: gene distribution and regulation. Biochem Soc Trans 39:1838–1843
Manitz B, Holldorf AW (1993) Purification and properties of glutamine synthetase from the archaebacterium *Halobacterium salinarium*. Arch Microbiol 159:90–97
Martínez-Espinosa RM, Marhuenda-Egea FC, Bonete MJ (2001a) Assimilatory nitrate reductase from the haloarchaeon *Haloferax mediterranei*: purification and characterisation. FEMS Microbiol Lett 204(2):381–385
Martínez-Espinosa RM, Marhuenda-Egea FC, Bonete MJ (2001b) Purification and characterisation of a possible assimilatory nitrite reductase from the halophile archaeon *Haloferax mediterranei*. FEMS Microbiol Lett 196(2):113–118
Martínez-Espinosa RM, Marhuenda-Egea FC, Donaire A, Bonete MJ (2003) NMR studies of a ferredoxin from *Haloferax mediterranei* and its physiological role in nitrate assimilatory pathway. Biochim Biophys Acta 1623(1):47–51
Martínez-Espinosa RM, Esclapez J, Bautista V, Bonete MJ (2006) An octameric prokaryotic glutamine synthetase from the haloarchaeon *Haloferax mediterranei*. FEMS Microbiol Lett 264(1):110–116
Martínez-Espinosa RM, Lledó B, Marhuenda-Egea FC, Bonete MJ (2007) The effect of ammonium on assimilatory nitrate reduction in the haloarchaeon *Haloferax mediterranei*. Extremophiles 11(6):759–767
Martínez-Espinosa RM, Lledó B, Marhuenda-Egea FC, Díaz S, Bonete MJ (2009) NO_3^-/NO_2^- assimilation in halophilic archaea: physiological analysis, *nasA* and *nasD* expressions. Extremophiles 13:785–792
McCarty GW, Bremner J (1992) Regulation of assimilatory nitrate reductase activity in soil by microbial assimilation of ammonium. Proc Natl Acad Sci USA 89:453–456
Mehta BJ, Obraztsova IN, Cerda-Olmedo E (2003) Mutants and intersexual heterokaryons of *Blakeslea trispora* for production of beta-carotene and lycopene. Appl Environ Microbiol 69:4043–4048
Miflin BJ (2002) The role of glutamine synthetase and glutamate dehydrogenase in nitrogen assimilation and possibilities for improvement in the nitrogen utilization of crops. J Exp Bot 53(370):979–987
Miller A, Fan X, Shen Q, Smith SJ (2008) Amino acids and nitrate as signals for the regulation of nitrogen acquisition. J Exp Bot 59:111–119
Moir W, Wood NJ (2001) Nitrate and nitrite transport in bacteria. Cell Mol Life Sci 58:215–224
Moreno-Vivián C, Flores E (2007) Nitrate assimilation in Bacteria. In: Bothe H, Ferguson SF, Newton WE (eds) Biology of the nitrogen cycle. Elsevier BV, Amsterdam, pp 263–282
Nesbo CL, L'Haridon S, Stetter KO, Doolittle WF (2001) Phylogenetic analyses of two 'archaeal' genes in thermotoga maritima reveal multiple transfers between archaea and bacteria. Mol Biol Evol 18(3):362–375
Ninfa AJ, Jiang P (2005) PII signal transduction proteins: sensors of [alpha]-ketoglutarate that regulate nitrogen metabolism. Curr Opin Microbiol 8:168–173
Olaizola M (2000) Commercial production of astaxanthin from *Haematococcus pluvialis* using 25,000-liter outdoor photobioreactors. J Appl Phycol 12:499–506
Palanca C, Pedro-Roig L, Llácer JL, Camacho M, Bonete MJ, Rubio V (2014) The structure of a PII signaling protein from a halophilic archaeon reveals novel traits and high-salt adaptations. FEBS J 281:3299–3314
Palozza P, Maggiano N, Calviello G, Lanza P, Piccioni E, Ranelletti FO, Bartoli GM (1998) Canthaxanthin induces apoptosis in human cancer cell lines. Carcinogenesis 19:373–376
Pantoja O (2012) High affinity ammonium transporters: molecular mechanism of action. Front Plant Sci 3:34. doi:10.3389/fpls.2012.00034

Pedro-Roig L, Camacho M, Bonete MJ (2011) In vitro proof of direct regulation of glutamine synthetase by GlnK proteins in the extreme halophilic archaeon *Haloferax mediterranei*. Biochem Soc Trans 39:259–262

Pedro-Roig L, Camacho M, Bonete MJ (2013a) *Haloferax mediterranei* GlnK proteins are post-translationally modified by uridylylation. Proteomics 13:1371–1374

Pedro-Roig L, Camacho M, Bonete MJ (2013b) Regulation of ammonium assimilation in *Haloferax mediterranei*: Interaction between glutamine synthetase and two GlnK proteins. Biochim Biophys Acta 1834:16–23

Pedro-Roig L, Lange C, Bonete MJ, Soppa J, Maupin-Furlow J (2013c) Nitrogen regulation of protein-protein interactions and transcript levels of GlnK PII regulator and AmtB ammonium transporter homologs in archaea. Microbiologyopen 2(5):826–840

Pérez-Pomares F, Ferrer J, Camacho M, Pire C, Llorca F, Bonete MJ (1999) Amino acid residues implied in the catalytic mechanism of NAD-dependent glutamate dehydrogenase from *Halobacterium salinarum*. Biochim Biophys Acta 1426:513–525

Pire C, Martínez-Espinosa RM, Pérez-Pomares F, Esclapez J, Bonete MJ (2014) Ferredoxin-dependent glutamate synthase: involvement in ammonium assimilation in *Haloferax mediterranei*. Extremophiles 18(1):147–159

Radin JW (1977) Amino acid interactions in the regulation of nitrate reductase induction in cotton root tips. Plant Physiol 60:467–469

Raja R, Hemaiswarya S, Rengasamy R (2007) Exploitation of *Dunaliella* for betacarotene production. Appl Microbiol Biotechnol 74:517–523

Reitzer L (2003) Nitrogen assimilation and global regulation in *Escherichia coli*. Annu Rev Microbiol 57:155–176

Reyes JC, Florencio FJ (1994) A mutant lacking the glutamine synthetase gene (*glnA*) is impaired in the regulation of the nitrate assimilation system in the cyanobacterium *Synechocystis* sp. strain PCC 6803. J Bacteriol 176:7516–7523

Richardson DJ, Berks BC, Russell DA, Spiro S, Taylor CJ (2001) Functional, biochemical and genetic diversity of prokaryotic nitrate reductases. Cell Mol Life Sci 58(2):165–178

Robertson DL, Alberte RS (1996) Isolation and characterization of glutamine synthetase from the marine diatom *Skeletonema costatum*. Plant Physiol 111(4):1169–1175

Santero E, Hervás AB, Canosa I, Govantes F (2012) Glutamate dehydrogenases: enzymology, physiological role and biotechnological relevance; Chapter 12. In: Canuto RA (ed) Dehydrogenases. InTech, under CC BY 3.0 license. doi:10.5772/2903. ISBN 978-953-307-019-3

Schulz AA, Collett HJ, Reid SJ (2001) Nitrogen and carbon regulation of glutamine synthetase and glutamate synthase in *Corynebacterium glutamicum* ATCC 13032. FEMS Microbiol Lett 205(2):361–367

Shapiro BM (1969) The glutamine synthetase deadenylylating enzyme system from *Escherichia coli*. Resolution into two components, specific nucleotide stimulation, and cofactor requirements. Biochemistry 8(2):659–670

Smith EL, Austen BM, Blumenthal KM, Nyc JF (1975) Glutamate dehydrogenase. In: Boyer PD (ed) The enzymes, vol 11, 3rd edn. Academic Press, New York, pp 293–367

Sugimori D, Ichimata T, Ikeda A, Nakamura S (2000) Purification and characterization of a ferredoxin from *Haloarcula japonica* strain TR-1. Biometals 13(1):23–28

Tomita T, Miyazaki T, Miyazaki J, Kuzuyama T, Nishiyama M (2010) Hetero-oligomeric glutamate dehydrogenase from *Thermus thermophilus*. Microbiology 156:3801–3813

Tomita T, Kuzuyama T, Nishiyama M (2011) Structural basis for leucine-induced allosteric activation of glutamate dehydrogenase. J Biol Chem 286:37406–37413

Vanoni MA, Curti B (1999) Glutamate synthase: a complex iron-sulfur flavoprotein. Cell Mol Life Sci 55:617–638

Vanoni MA, Curti B (2005) Structure and function studies on the ironsulfur flavoenzyme glutamate synthase: an unexpectedly complex self-regulated enzyme. Arch Biochem Biophys 433(1):193–211

Wanner C, Soppa J (1999) Genetic Identification of three ABC transporters as essential elements for nitrate respiration in *Haloferax volcanii*. Genetics 152(4):1417–1428

Wood NJ, Alizadeh T, Richardson DJ, Ferguson SJ, Moir JW (2002) Two domains of a dual-function NarK protein are required for nitrate uptake, the first step of denitrification in *Paracoccus pantotrophus*. Mol Microbiol 44(1):157–170

Zafrilla B, Martínez-Espinosa RM, Bonete MJ, Butt JN, Richardson DJ, Gates AJ (2011) A haloarchaeal ferredoxin electron donor that plays an essential role in nitrate assimilation. Biochem Soc Trans 39(6):1844–1848

Chapter 10
A Proteomics Approach for the Identification of Novel Proteins in Extremophiles

Sung Ho Yun, Chi-Won Choi, Sang-Yeop Lee, Edmond Changkyun Park, and Seung Il Kim

1 Introduction

Extremophiles are organisms that live in extreme environments, including those with extremely high or low temperatures, strong acidity or salinity, strong radiation, high atmospheric pressure, and limited nutrient supply (Burg et al. 2011). Identifying novel genes or proteins in extremophiles is currently a hot research topic because a lot of genes in extremophiles are predicted to have unique properties that permit adaptation to the extreme environments and to be absent from most bacteria growing in natural environments. *Tag* DNA polymerase of thermophilic *Thermus aquaticus* (Chien et al. 1976) and antifreeze proteins (AFPs) of Antarctic lake bacteria (Gilbert et al. 2004) are the representative unique proteins from extremophiles.

Genomic sequencing technology and functional genomics have been successful in amassing large amount of genomic data on extremophiles, including information about genes at the DNA sequence level (Lee et al. 2008; Vezzi et al. 2005). Recently, proteomic technology has emerged as a new functional genomics tool for the elucidation of novel proteins in extremophiles. Because protein expression and post-translational modifications are dynamic and have profound effects on cell function, proteomics is necessary part of functional genomics. Until now, difficulties in culturing cells under native growth conditions and in obtaining pure cultures have limited applying proteomics to extremophiles. However, recent developments in proteomic methods and cultivation techniques have overcome these limitations. Basic proteomic technologies used for bacterial proteomic analysis can be generally applied for extremophiles, but specific procedures for optimizing sample preparation still should be developed (Burg et al. 2011; Ellen et al. 2009; Yun et al. 2011a).

S.H. Yun • C.-W. Choi • S.-Y. Lee • E.C. Park • S.I. Kim (✉)
Division of Bioconvergence Analysis, Daejeon 305-806, Republic of Korea
e-mail: ksi@kbsi.re.kr

P.H. Rampelotto (ed.), *Biotechnology of Extremophiles*, Grand Challenges in Biology and Biotechnology 1, DOI 10.1007/978-3-319-13521-2_10

Proteomic studies on extremophiles can provide information about the number of proteins induced under specific conditions (Burg et al. 2011). Proteomics data can then be used to elucidate the physiological characteristics of extremophiles, and categorize proteins according to their biological roles with the help of bioinformatics. Because extremophiles have unique metabolic activities and survival mechanisms, they provide a valuable resource for discovering new enzymes and bio-compounds, especially those adapted to working under extreme conditions, and thus their enzymes, such as psychrophilic lipases, alkaliphilic cellulases, thermostable DNA polymerases, have broad applications in biotechnology industries (Antranikian et al. 2005; Podar and Reysenbach 2006). Therefore, improving screening methods using proteomic technique is essential for identifying novel proteins in extremophiles.

In the first part of this chapter, we introduce the basic principles of proteomics, which include protein separation methods, protein identification methods, application of bioinformatics, and high-throughput screening. In the second part, we summarize screening methods for identifying novel proteins in extremophiles.

2 Development of Proteomic Technology

This section describes the basic principles and concepts of protein separation and identification. Mass spectrometry and bioinformatic programs, which are essential tools for proteomic analysis, are also described.

2.1 Separation and Preparation of Protein Mixture

The first step in sample preparation for proteomic analysis is cell disruption to obtain soluble samples. Several disruption methods can be used, such as repeated freezing and thawing, sonication, and homogenization under high pressure. The most appropriate method can be selected after considering the fragility or sturdiness of the organisms under study.

Protein samples in soluble fractions are separated and enriched according to their physicochemical properties, such as solubility, net charge (pI), size (molecular weight), and hydrophobicity. In gel-based method, the proteins are separated using various polyacrylamide gel electrophoresis (PAGE). The three most popular gel-based separation techniques for intact proteins are one dimensional-sodium dodecyl sulfate polyacrylamide gel electrophoresis (1D SDS-PAGE), two-dimensional electrophoresis (2-DE), and preparative isoelectric focusing (IEF). 1D SDS-PAGE is simple and the most efficient method for protein separation and preparation for proteomic analysis. The protein sample is typically dissolved in loading buffer containing SDS and a thiol reducing agent, such as β-mercaptoethanol or DTT. 2-DE is the best gel-based method, and is commonly used for global analysis of complex samples. In this technique, protein samples are separated on the basis of their isoelectric points (pI) in an isoelectric focusing gel, and are then separated in a second dimension by their molecular size by SDS-PAGE (Fig. 10.1a).

An alternative protein separation method is liquid chromatography (LC), in which samples are separated on LC columns according to their physicochemical properties.

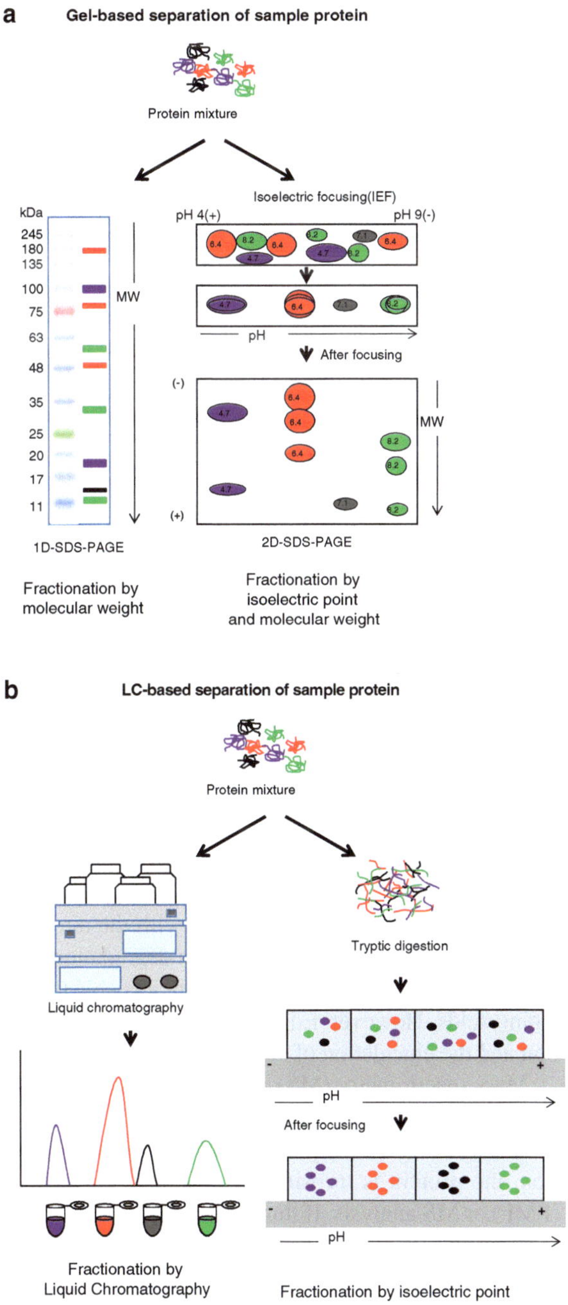

Fig. 10.1 Two protein separation techniques. (**a**) Gel-based: Protein mixtures are separated according to the molecular weight and isoelectric point on the electrophoresis. (**b**) Gel-free: Protein mixtures (peptide mixtures) are fractionated according to their physicochemical properties (size, charge or overall hydrophobicity, and isoelectric point) on LC columns

Table 10.1 Types of column for LC

Type of column	Separation property	Features
Ion exchange	Charge	High resolution High capacity High speed
Gel filtration/size exclusion	Size	High resolution
Normal/reversed phase	Hydrophobicity	High resolution
Hydrophobic interaction	Hydrophobicity	Good resolution Good capacity Good speed
Affinity	Ligand specificity	High resolution High capacity High speed

Reverse phase (C-4, C-8, and C-18), size exclusion, ion exchange, and affinity chromatography columns are frequently used for sample preparation and purification before proteomic analysis (Table 10.1). High-performance liquid chromatography (HPLC) is used for protein or peptide separation (Fig. 10.1b).

2.2 *Protein Identification*

In general, two types of method, protein sequencing using Edman degradation and tandem mass spectrometry (MS/MS) analysis, are commonly used for identifying unknown proteins (Choi et al. 2012). Edman degradation identifies peptide sequence by labeling and cleaving the amino-terminal residue of proteins. In this process, isothiocyanates is used for labeling the N-terminus of a protein or peptide. Under suitable conditions, a single round reaction of Edman degradation cleaves the N-terminal amino acid residue to produce a derivative of the amino acid and a free amino terminus corresponding to the next amino acid in the polypeptide chain. The limitation of Edman degradation is that it is not effective on proteins with N-terminal modifications. If the N-terminal amino acid has been chemically modified, the cleavage chemical (isothiocyanates) cannot interact with the amino group of the N-terminal amino acid, leading to termination of Edman degradation, in a process known as N-terminal blocking. Also, by Edman degradation, it is able to accurately sequence only up to 30 amino acids. However, Edman sequencing is still a useful tool for protein identification and the elucidation of sulfur bridges in proteins (Fig. 10.2a).

Mass spectrometry (MS) can accurately measure the mass/z of peptide fragments, and thus it is a highly sensitive and accurate method for determining the precise molecular weight of a protein and for identifying proteins. For protein identification, the protein is trypsinized and then subjected to peptide mass fingerprinting (PMF) or MS analysis. If this information does not provide clear-cut data for protein identification, the tryptic peptides of proteins may be subjected to a *de novo* sequencing using MS/MS (Fig. 10.2b).

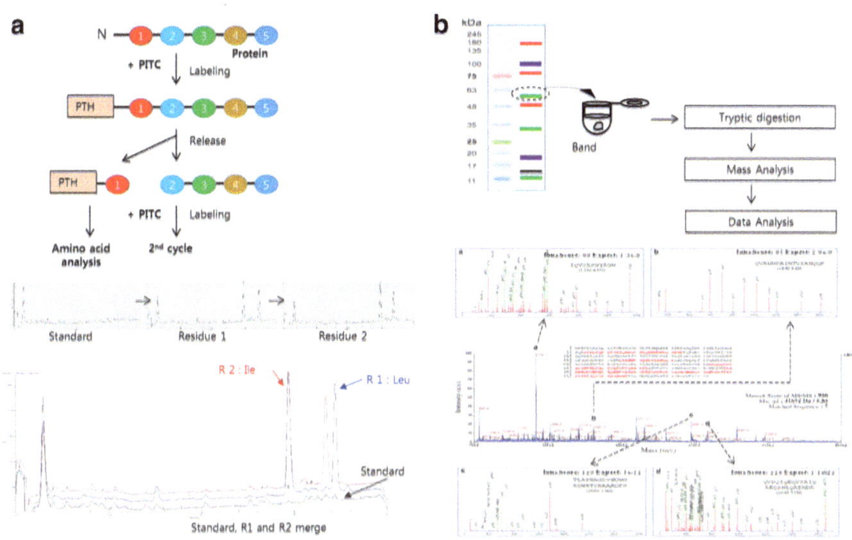

Fig. 10.2 Diagram of N-terminal sequencing. (**a**) Edman degradation: N-terminal amino acid group of protein is labeled with phenylisothiocyanate (PITC). The labeled N-terminal amino acid is selectively cleaved, and the extracted phenylthiohydantoin (PTH)-amino acid is identified by MS. This procedure can be repeated again to identify the next amino acid. (**b**) Peptide mass fingerprinting (PMF) and MS/MS analysis: In PMF analysis, the unknown protein of interest is cleaved into smaller peptides by restriction enzymes like trypsin, and their masses are accurately measured by MS. Then the peptide masses are compared to protein sequence database. In MS/MS analysis, a peptide is further fragmented using collision induced dissociation (CDI), then determined its sequence from the list of masses in the MS/MS spectrum. Part of the figure is adapted from a previous report (Choi et al. 2012)

2.3 *Mass Spectrometer*

In 2002, the Nobel Prize in Chemistry was awarded to John Bennett Fenn for the development of electrospray ionization (ESI) and to Koichi Tanaka for the development of matrix-assisted laser desorption ionization (MALDI) and the application of these two methods to the ionization of biological macromolecules, especially proteins. For MS analysis these two ionization methods are combined with various mass analyzers. MALDI method is typically combined with time of flight (TOF) mass analyzer. MALDI-TOF MS has been used for measuring the molecular weight of proteins and peptides and peptide mass fingerprinting (PMF). MALDI-TOF/TOF MS is used for MS/MS analysis of tryptic peptides and *de novo* amino acid sequencing. ESI method is normally combined with quadrupole and TOF. ESI-Q TOF MS is also used for MS/MS analysis. Recently, due to its ease of use, ESI ion trap (LTQ/Obitrap) MS has become a popular MS/MS analyzer for high-throughput proteomic analysis (Fig. 10.3).

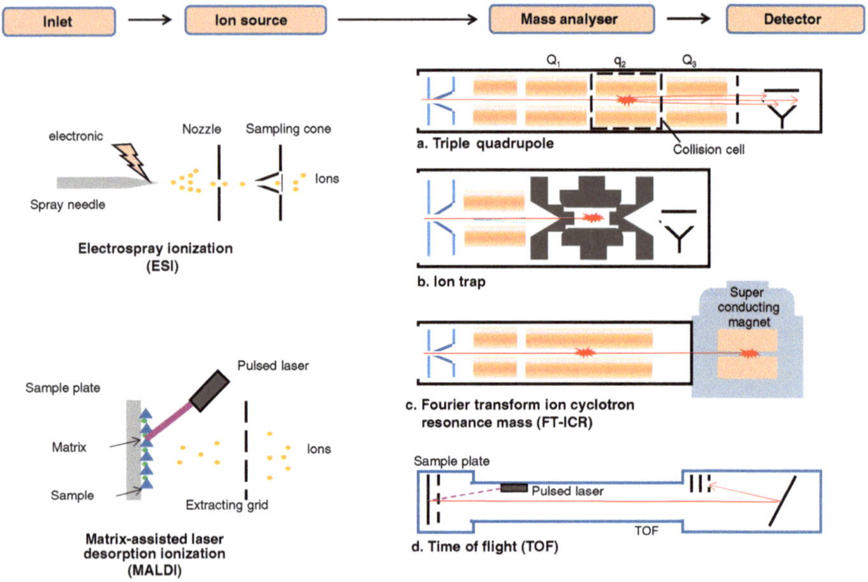

Fig. 10.3 Diagram showing the basic MS configuration. MS is composed of four parts: inlet, ion source, mass analyzer, and detector. ESI and MALDI are representative ion sources. Triple quadrupole, ion trap, FT-ICR, and TOF are mass analyzers used for proteomic analysis

2.4 Protein Identification Software and Bioinformatics Tools

Several protein identification software programs have been developed to analyze MS/MS spectra (Table 10.2). Because each program has been designed and optimized for specific mass spectrometers or specific MS analysis methods, it is important to understand the merits and limitations of each method to obtain the best proteomic results. APEX, Census, and MaxQuant are freely accessible on the internet, but MASCOT and Proteome Discoverer are only commercially available (Braisted et al. 2008; Cox and Mann 2008; Park et al. 2008). High-throughput proteomic analysis is possible by using these programs. After completion of quantitative proteomic analysis, different types of bioinformatic tools are needed to elucidate the specific functions of the identified proteins. Several bioinformatic tools have been developed for this purpose (Table 10.2).

Cello and PsortDB are popular programs used to identify the localization of proteins in the cell (Yu et al. 2006, 2011). Cello predicts protein localization based on sequence information supplied in FASTA format. However, Cello cannot utilize accessible genome databases, which usually include information of cell localization. On the other hand, when information about protein localization in bacteria or archaea from genome databases is required, PsortDB is a good choice because it can be

Table 10.2 Representative bioinformatic programs used for proteins identification and characterization

Program	Web site	Function
TMHMM Server v. 2.0	http://www.cbs.dtu.dk/services/TMHMM	Prediction of transmembrane topology
Phobius	http://phobius.sbc.su.se/	Prediction of transmembrane topology
SignalP 4.0	http://www.cbs.dtu.dk/services/SignalP/	Prediction of presence and location of signal peptide cleavage sites
MASCOT	http://www.matrixscience.com/	Protein identification
SEQUEST	http://fields.scripps.edu/sequest/	Protein identification
PSORTdb 3.0	http://www.psort.org/psortb	Prediction of subcellular localization
CELLO v 2.5	http://cello.life.nctu.edu.tw	Prediction of subcellular localization
KEGG PATHWAY Database	http://www.genome.jp/kegg/pathway.html	Pathway analysis
FIGfams	http://www.nmpdr.org/FIG/wiki/view.cgi/FIG/FigFam	Functional categories
COGs	http://www.ncbi.nlm.nih.gov/COG/	Functional categories
The Gene Ontology	http://www.geneontology.org/	Functional categories
UM-BBD	https://umbbd.ethz.ch/	Biocatalysis/Biodegradation Database

used with BlastP to predict cellular localizations of novel proteins. Another well-known bioinformatic tool is the THMHH server (Moller et al. 2001), which is available for the identification and prediction of the transmembrane position of proteins. It also predicts the range of protein sequences of the inside and outside of membrane, as well as the transmembrane helix region. An alternative program, SignalP, predicts the signal peptide cleavage site in the protein sequence (Petersen et al. 2011). Both of the TMHMM server and SignalP use FASTA format as the input format for protein sequences.

The aforementioned programs are useful for the analysis and characterization of individual proteins. However, comprehensive proteomic studies need more advanced bioinformatic tools for elucidation of protein complex and its biological functions. The Cluster of Orthologous Groups (COGs) database and Gene Ontology (GO) databases were designed for these purposes (Ashburner et al. 2000; Tatusov et al. 2003). The identified proteins can be classified according to their biological functions using these programs. The STRING database is designed for the analysis of protein-protein interactions (Franceschini et al. 2013). Particularly, the STRING includes visualization tools that show protein-protein interactions. The KEGG database is the most popular network analysis database for determining protein functions (Kanehisa and Goto 2000).

3 Screening for Novel Proteins

This section provides several case studies of screenings for novel proteins in extremophiles. Screening is usually performed using two high-throughput screening methods: 2DE-MALDI-TOF/TOF and 1D-LC-MS/MS.

3.1 High-Throughput Screening of Extremophiles

2DE-MALDI-TOF/TOF and 1D-LC-MS/MS are commonly used proteomic tools for high-throughput screening. 2DE-MALDI-TOF/TOF is composed of two techniques: 2-DE and MALDI-TOF/TOF. The advantage of 2-DE is the visualization of protein samples on the gels. Each protein can be visualized according to its molecular weight, isoelectric point (pI), and abundance. Information on post-translational modifications and protein degradation can also be obtained from the gels. However, 2-DE is limited because it cannot detect low abundance proteins. In addition, this method has difficulties in separating extremely acidic or basic proteins and hydrophobic proteins such as membrane proteins. However, 2DE-MALDI-TOF/TOF is a very useful tool for proteomic analysis in bacteria and extremophiles. Separated protein spots on 2D gels are normally identified by peptide mapping, MS/MS analysis, or *de novo* amino acid sequencing using MALDI-TOF/TOF MS (Chong and Wright 2005; Yun et al. 2014). MALDI-TOF based protein identification method is relatively quick and inexpensive in compared with LC-MS/MS. However, relatively pure samples are required. Halophilic archaeon *Halobacterium salinarum* (strain R1, DSM 671) is predicted to contain 2784 protein coding genes on the basis of its genome sequence. Cytosolic proteins from *H. salinarum* were separated on 2-D gels between pH 3.5 and 5.5 (Fig. 10.4). Each silver-stained spot was analyzed, resulting in the identification of 661 proteins out of about 1800 different protein spots. Ninety-four proteins were found in multiple spots, indicating that they were post-translational modification. From this analysis, about 40 % of the cytosolic proteome *H. salinarum* was identified (Tebbe et al. 2005).

The alternative to 2DE-MALDI-TOF/TOF is 1D-LC-MS/MS, which combines SDS-PAGE and LC-MS/MS for protein separation and identification. In this method, protein samples are separated on SDS-PAGE gels. Normally, proteins separated on these gels are divided into 7–20 fractions according to molecular size. Each gel fragment is then treated with a proteolytic enzyme such as trypsin, and the tryptic peptide mixtures are subjected to reverse phase (C-18 or C-8) chromatography before MS/MS. SDS-PAGE is particularly useful for the separation of membrane proteins and hydrophobic proteins (Yun et al. 2014). LC-MS/MS based protein identification method offers the highest confidence identifications since it is highly sensitive and reliable. However, this method is relatively slow and difficult to perform in compared with MALDI-TOF/TOF MS. LC-MS/MS has been used for the proteome analysis of the hyperthermophilic archaeon *Thermococcus onnurineus* NA1 cultured under carboxydotrophic conditions (Fig. 10.5). Out of the 1976 predicted genes of *T. onnurineus*

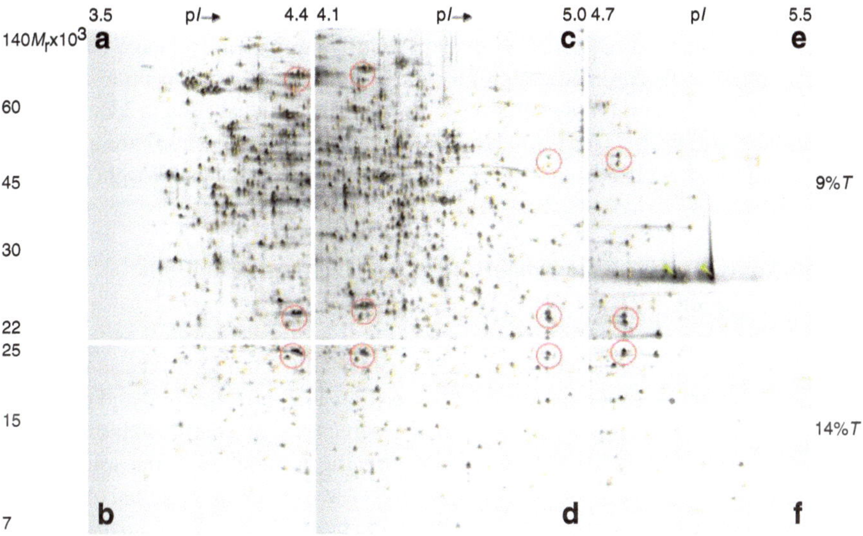

Fig. 10.4 2-D reference map of *H. salinarum*. The cytosolic proteins were separated on overlapping zoom gels covering the pI range 3.5–5.5 and stained with silver nitrate. Every pI range was covered with two different acrylamide concentrations (9 % and 14 % T). The different reference map sectors are marked (**a**–**f**) (taken from Tebbe et al. 2005)

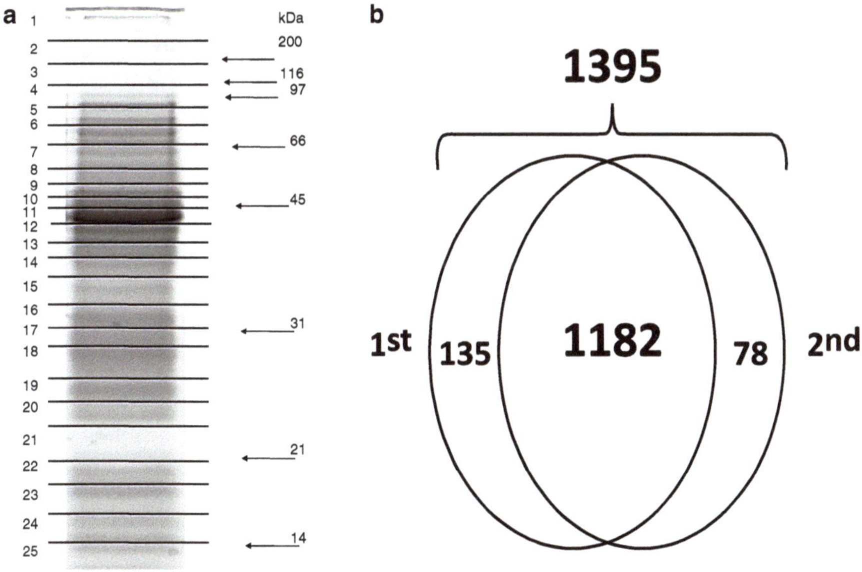

Fig. 10.5 (**a**) 12 % SDS-PAGE of 100 μg proteins from *Thermococcus onnurineus* NA1 cultured in yeast extract/peptone/sulfur medium (YPS) and CO condition. The gels were stained with Coomassie Brilliant Blue R-250 and separated into 25 pieces according to molecular weight. Each gel fragment was digested with trypsin. (**b**) Venn diagram of the identified NA1 proteins as determined by 1D-LC-MS/MS analysis (taken from Yun et al. 2011b)

NA1, 70.6 % (1395) were identified by twice LC-MS/MS analysis (Yun et al. 2011b). An alternative method of 1D-LC is tandem LC (strong cation exchange (SCX) and reversed-phase (RP) chromatography) for the separation of tryptic peptide mixtures. This technology is commonly referred to as multidimensional protein identification technology (MudPIT). Protein mixtures (before proteolytic digestion) are also separated by liquid-phase IEF and LC as well as SDS-PAGE (Chong and Wright 2005).

3.1.1 Analysis of Thermostable Protein from Hyperthermophilic Archaeon

Proteomic approaches had not been extensively used for the identification of thermostable proteins of hyperthermophilic archaeons. Recently, thermal and chemical perturbation methods have been used for screening of thermostable proteins (Prosinecki et al. 2006). These methods lowered the complexity of the soluble proteins and allow enrichment of cytosolic hyperthermostable proteins. A similar approach was used for screening of thermostable proteins from *T. onnurineus* NA1, of which optimum growth temperature is above 85 °C. Hyperthermostable proteins of *T. onnurineus* NA1 were selected and identified by a combined approach. Heat treatment at 100 °C for 2 h was used to enrich hyperthermostable proteins, which were then identified by 2-DE and 1-DE/MS–MS (Yun et al. 2011a) (Fig. 10.6). This resulted in the identification of thermostable DNA polymerase, aminopeptidase, and α-amylase from *T. onnurineus* NA1.

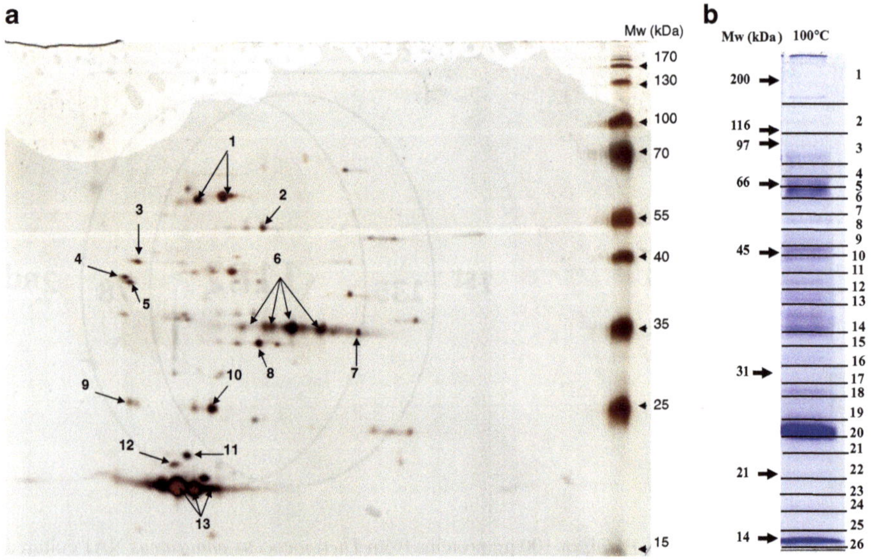

Fig. 10.6 2-DE of thermostable proteins of from *T. onnurineus* NA1. Protein samples were heat-treated at 100 °C for 10 min. After removal of coagulated proteins by centrifugation, enriched thermostable soluble proteins were used for 2-DE (**a**) and 12 % SDS-polyacrylamide gel (**b**) (taken from Yun et al. 2011a)

3.1.2 Proteomic Analysis of Radioresistant Bacterium

Radioresistant bacterium, *Deinococcus deserti* and *Deinococcus radiodurans*, can repair broken DNA in a few hours. Recently, their radiation resident proteins were discovered by proteomic analysis. 2DE-MALDI-TOF/TOF analysis identified that proteins related with DNA damage response were accumulated by drastic irradiation in *D. deserti* (Dedieu et al. 2013). In *D. radiodurans*, proteins involved in DNA repair, oxidative stress alleviation, and protein translation/folding are identified under irradiation condition (Basu and Apte 2012).

3.2 Membrane Proteins of Extremophiles

Membrane proteins play important functions in many processes, including nutrient transport, signal transduction, and energy conversion. Although they are functionally important in living cells, quantitative proteomic studies are still rare, mainly due to technical difficulties. In general, membrane proteins are fractionated by ultracentrifugation and sucrose density gradient centrifugation. Cell disruption is performed using a French presser or homogenizer, or by osmotic shock. Cell lysates are then fractionated on sucrose density gradient (20–50 %). This procedure was used for preparation of membrane proteins of extremophiles such as *Thermococcus* stains. Using this procedure, most membrane proteins were enriched in the heavy density layers (Fig. 10.7). Membrane proteins were analyzed by 2DE-MALDI-TOF/TOF or 1D-LC-MS/MS. Membrane proteins of the halophilic archaeon, *Halobacterium,* were also purified by ultracentrifugation and sucrose density gradient centrifugation. From this study, 165 proteins were identified in the membrane fraction although they were not all membrane proteins (Klein et al. 2005). Membrane protein prediction programs, such as TMHMM (www.cbs.dtu.dk/services/TMHMM), are necessary for categorizing and characterizing the identified proteins.

3.3 Exoproteome and Membrane Vesicle of Extremophiles

Ultracentrifugation is a simple but well-known way to enrich secretion proteins from culture medium. Precipitation procedures (ammonium sulfate or acetone up to 80 %) are often applied for the efficient preparation of secreted proteins. Precipitated pellet are re-suspended and dialyzed in appropriate buffers. Then the secreted protein samples are concentrated by micro-filtration tool such as viva spin or centricon before proteomic analysis. The secreted proteins of hyperthermophilic archaeon, *Pyrococcus furiosus* have been also purified by ultrafiltration (Sartorius AG, Germany, filter cut off 5 kDa) and concentrated using a Vivaspin 20 centrifugal concentrator (Sartorius AG, Germany, filter cut off 5 kDa). In total, 58 proteins were identified by 1DE-LC-MS/MS, including major enzymes for starch degradation (Schmid et al. 2013).

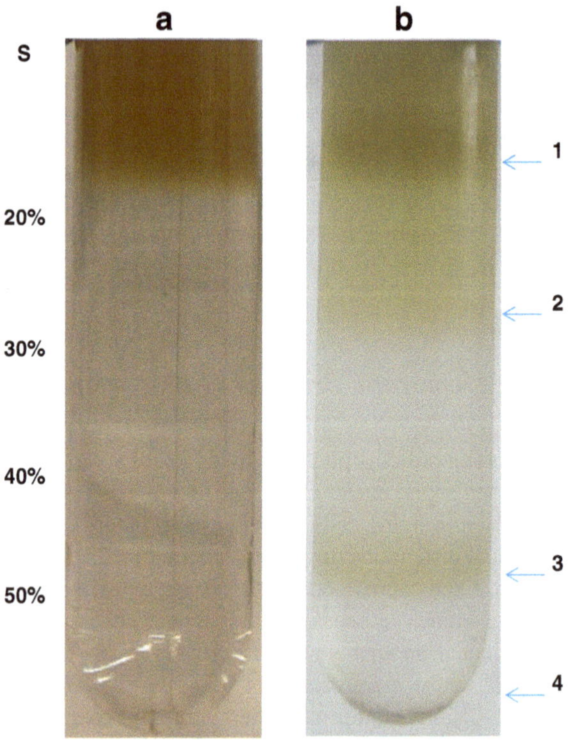

Fig. 10.7 Sucrose density gradient (20–50 %) centrifugation. Cell lysates were loaded on top of gradient (**a**). After ultracentrifugation, membrane proteins were concentrated in 3rd and 4th fractions (**b**)

Membrane vesicles are considered as novel secretion process occurred in gram-negative and gram-positive bacteria. Since membrane vesicles are frequently found in extremophiles, identification and functional studies of novel proteins in membrane vesicles are important. Because membrane vesicles are mixed with secreted proteins, they should be separated from secreted proteins. Filtration and concentration using filtration system such as Quix-stand system is used to separate membrane vesicles from other contaminated proteins (Fig. 10.8). Protein components of purified membrane vesicles are analyzed by shotgun proteomics (1D-LC-MS/MS) or 2DE-MALDI-TOF/TOF. Membrane vesicles purified from *Sulfolobus acidocaldarius*, *Sulfolobus solfataricus*, and *Sulfolobus tokodaii* were analyzed by MS/MS analysis using a MALDI-TOF MS spectrometer (Ellen et al. 2009). Electron microscopy of membrane vesicles revealed that the diameter of the vesicles ranged from about 90 nm to 230 nm in three *Sulfolobus* species, and that they were surrounded by a protein layer, presumably formed by the surface-layer protein. LC-MS analysis identified proteins of three *Sulfolobus* membrane vesicles.

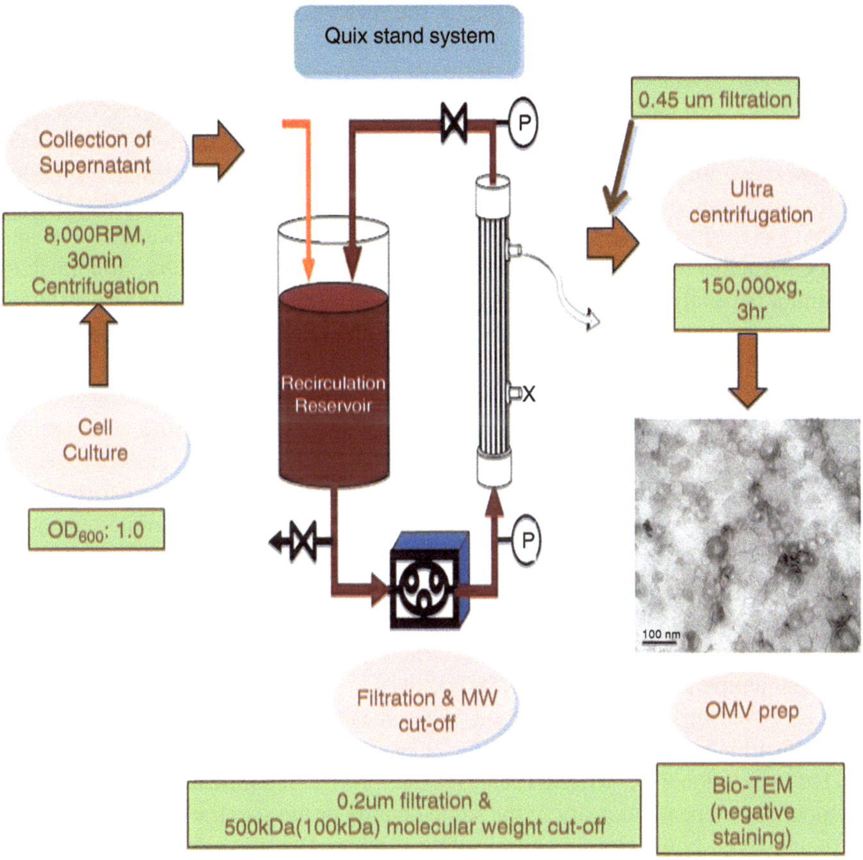

Fig. 10.8 Purification scheme of membrane vesicles by the Quick stand system. Membrane vesicles were concentrated by a filter (0.2 μm) and a molecular weight cut off cartridge (500 kDa or 100 kDa). Membrane vesicles purified by ultracentrifugation (150,000 g, 3 h) were confirmed by TEM analysis. The purified membrane vesicles are stored under deep freezing until proteome analysis

3.4 *Quantitative Proteomic Analysis*

Recently, *in vitro* labeling methods were designed and applied for quantitative proteomic analysis. The common principal is the use of isotope labeling for comparative analysis. In isotope-coded affinity tag (ICAT) method, the cysteine residues of sample proteins are isotopically labeled with heavy (C^{13}) or light (C^{12}) isobaric tags before trypsin digestion. Isobaric tag for relative and absolute quantitative (iTRAQ) method involves the covalent labeling of the N-terminus and the side chain amines of tryptic peptides. Labeled peptide mixtures are then analyzed by LC/MS-MS. Several proteomic studies have been performed using iTRAQ to screen for novel proteins of extremophiles. One example is *Methanococcoides burtonii*. This stain is a cold-adapted methanogenic archaeon (Goodchild et al. 2005). Two quantitative proteomic studies were performed to assess the effects of growth conditions at low temperature on this organism. In total, 14 of the 163 proteins identified by iTRAQ analysis showed significantly difference in expression between at 4 °C and 23 °C. Another example is hyperthermophilic archaeon *Sulfolobus solfataricus*. Quantitative analysis of proteins in *Sulfolobus solfataricus*, grown at three different culture temperature (65 °C, 70 °C, and 80 °C) identified 246 proteins as membrane protein (Pham et al. 2010).

Two-dimensional difference in gel electrophoresis (2D-DIGE) is a form of gel electrophoresis that allows of comparing protein quantity between different samples. Different protein samples are labeled with different fluorescent dyes, mixed together, and then separated by identical 2-DE. The gel is scanned with the excitation wavelength of each dye one after the other, so we are able to see each sample separately (Unlu et al. 1997). This technique is used to see changes in protein abundance, post-translational modifications, truncations and any modification that might change the size or isoelectric point of proteins. 2D-DIGE method was applied to compare proteomic expression of Antarctic bacterium *Pseudoalteromonas haloplanktis* in different temperatures (Piette et al. 2011).

Label-free quantification method is rapid and low-cost alternatives to other quantification proteomic methods. For label-free quantification, samples are subjected to individual LC-MS/MS analysis. The protein quantification is based on the comparison of peak intensity of the same peptide or the spectral count of the same protein. Thus, both relative and absolute protein quantification can be obtained (Zhu et al. 2010).

4 Conclusions

Proteomics is an important omics technology that is finding broad application extremophiles. The quality of proteomic data obtained relies largely on excellent sample preparation and applying the best proteomic methods. Therefore, further development of efficient sample preparation methods and proteomic methods is essential. Improvements in proteomic technology will expand our knowledge of how extremophiles adapt to extreme environments (Fig. 10.9).

Conflict of Interest Sung Ho Yun, Chi-Won Choi, Sang-Yeop Lee, Edmond Changkyun Park, and Seung Il Kim declare that they have no conflict of interest.

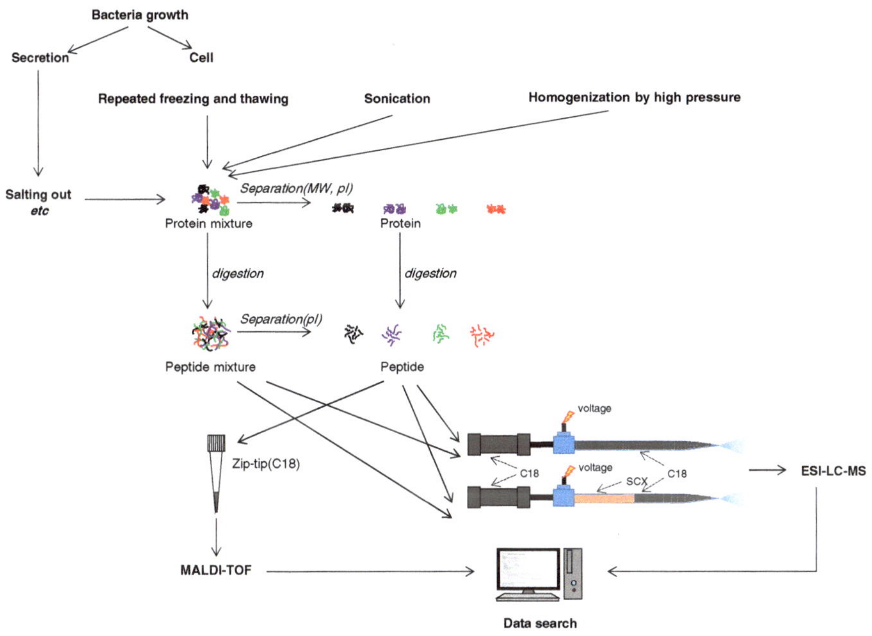

Fig. 10.9 Summarized diagram of the proteomic approach for searching novel proteins in extremophiles

References

Antranikian G, Vorgias CE, Bertoldo C (2005) Extreme environments as a resource for microorganisms and novel biocatalysts. Adv Biochem Eng Biotechnol 96:219–262

Ashburner M, Ball CA, Blake JA, Botstein D, Butler H, Cherry JM, Davis AP, Dolinski K, Dwight SS, Eppig JT, Harris MA, Hill DP, Issel-Tarver L, Kasarskis A, Lewis S, Matese JC, Richardson JE, Ringwald M, Rubin GM, Sherlock G (2000) Gene ontology: tool for the unification of biology. The gene ontology consortium. Nat Genet 25:25–29

Basu B, Apte SK (2012) Gamma radiation-induced proteome of Deinococcus radiodurans primarily targets DNA repair and oxidative stress alleviation. Mol Cell Proteomics 11(M111):011734

Braisted JC, Kuntumalla S, Vogel C, Marcotte EM, Rodrigues AR, Wang R, Huang ST, Ferlanti ES, Saeed AI, Fleischmann RD, Peterson SN, Pieper R (2008) The APEX quantitative proteomics tool: generating protein quantitation estimates from LC-MS/MS proteomics results. BMC Bioinformatics 9:529

Burg D, Ng C, Ting L, Cavicchioli R (2011) Proteomics of extremophiles. Environ Microbiol 13:1934–1955

Chien A, Edgar DB, Trela JM (1976) Deoxyribonucleic acid polymerase from the extreme thermophile Thermus aquaticus. J Bacteriol 127:1550–1557

Choi CW, Lee YG, Kwon SO, Kim HY, Lee JC, Chung YH, Yun CY, Kim SI (2012) Analysis of Streptococcus pneumoniae secreted antigens by immuno-proteomic approach. Diagn Microbiol Infect Dis 72:318–327

Chong PK, Wright PC (2005) Identification and characterization of the Sulfolobus solfataricus P2 proteome. J Proteome Res 4:1789–1798
Cox J, Mann M (2008) MaxQuant enables high peptide identification rates, individualized p.p.b.-range mass accuracies and proteome-wide protein quantification. Nat Biotechnol 26: 1367–1372
Dedieu A, Sahinovic E, Guerin P, Blanchard L, Fochesato S, Meunier B, de Groot A, Armengaud J (2013) Major soluble proteome changes in Deinococcus deserti over the earliest stages following gamma-ray irradiation. Proteome Sci 11:3
Ellen AF, Albers SV, Huibers W, Pitcher A, Hobel CF, Schwarz H, Folea M, Schouten S, Boekema EJ, Poolman B, Driessen AJ (2009) Proteomic analysis of secreted membrane vesicles of archaeal Sulfolobus species reveals the presence of endosome sorting complex components. Extremophiles 13:67–79
Franceschini A, Szklarczyk D, Frankild S, Kuhn M, Simonovic M, Roth A, Lin J, Minguez P, Bork P, von Mering C, Jensen LJ (2013) STRING v9.1: protein-protein interaction networks, with increased coverage and integration. Nucleic Acids Res 41:D808–D815
Gilbert JA, Hill PJ, Dodd CE, Laybourn-Parry J (2004) Demonstration of antifreeze protein activity in Antarctic lake bacteria. Microbiology 150:171–180
Goodchild A, Raftery M, Saunders NF, Guilhaus M, Cavicchioli R (2005) Cold adaptation of the Antarctic archaeon, Methanococcoides burtonii assessed by proteomics using ICAT. J Proteome Res 4:473–480
Kanehisa M, Goto S (2000) KEGG: Kyoto encyclopedia of genes and genomes. Nucleic Acids Res 28:27–30
Klein C, Garcia-Rizo C, Bisle B, Scheffer B, Zischka H, Pfeiffer F, Siedler F, Oesterhelt D (2005) The membrane proteome of Halobacterium salinarum. Proteomics 5:180–197
Lee HS, Kang SG, Bae SS, Lim JK, Cho Y, Kim YJ, Jeon JH, Cha SS, Kwon KK, Kim HT, Park CJ, Lee HW, Kim SI, Chun J, Colwell RR, Kim SJ, Lee JH (2008) The complete genome sequence of Thermococcus onnurineus NA1 reveals a mixed heterotrophic and carboxydotrophic metabolism. J Bacteriol 190:7491–7499
Moller S, Croning MD, Apweiler R (2001) Evaluation of methods for the prediction of membrane spanning regions. Bioinformatics 17:646–653
Park SK, Venable JD, Xu T, Yates JR 3rd (2008) A quantitative analysis software tool for mass spectrometry-based proteomics. Nat Methods 5:319–322
Petersen TN, Brunak S, von Heijne G, Nielsen H (2011) SignalP 4.0: discriminating signal peptides from transmembrane regions. Nat Methods 8:785–786
Pham TK, Sierocinski P, van der Oost J, Wright PC (2010) Quantitative proteomic analysis of Sulfolobus solfataricus membrane proteins. J Proteome Res 9:1165–1172
Piette F, D'Amico S, Mazzucchelli G, Danchin A, Leprince P, Feller G (2011) Life in the cold: a proteomic study of cold-repressed proteins in the antarctic bacterium pseudoalteromonas haloplanktis TAC125. Appl Environ Microbiol 77:3881–3883
Podar M, Reysenbach AL (2006) New opportunities revealed by biotechnological explorations of extremophiles. Curr Opin Biotechnol 17:250–255
Prosinecki V, Botelho HM, Francese S, Mastrobuoni G, Moneti G, Urich T, Kletzin A, Gomes CM (2006) A proteomic approach toward the selection of proteins with enhanced intrinsic conformational stability. J Proteome Res 5:2720–2726
Schmid G, Mathiesen G, Arntzen MO, Eijsink VG, Thomm M (2013) Experimental and computational analysis of the secretome of the hyperthermophilic archaeon Pyrococcus furiosus. Extremophiles 17:921–930
Tatusov RL, Fedorova ND, Jackson JD, Jacobs AR, Kiryutin B, Koonin EV, Krylov DM, Mazumder R, Mekhedov SL, Nikolskaya AN, Rao BS, Smirnov S, Sverdlov AV, Vasudevan S, Wolf YI, Yin JJ, Natale DA (2003) The COG database: an updated version includes eukaryotes. BMC Bioinformatics 4:41
Tebbe A, Klein C, Bisle B, Siedler F, Scheffer B, Garcia-Rizo C, Wolfertz J, Hickmann V, Pfeiffer F, Oesterhelt D (2005) Analysis of the cytosolic proteome of Halobacterium salinarum and its implication for genome annotation. Proteomics 5:168–179

Unlu M, Morgan ME, Minden JS (1997) Difference gel electrophoresis: a single gel method for detecting changes in protein extracts. Electrophoresis 18:2071–2077
Vezzi A, Campanaro S, D'Angelo M, Simonato F, Vitulo N, Lauro FM, Cestaro A, Malacrida G, Simionati B, Cannata N, Romualdi C, Bartlett DH, Valle G (2005) Life at depth: Photobacterium profundum genome sequence and expression analysis. Science 307:1459–1461
Yu CS, Chen YC, Lu CH, Hwang JK (2006) Prediction of protein subcellular localization. Proteins 64:643–651
Yu NY, Laird MR, Spencer C, Brinkman FS (2011) PSORTdb—an expanded, auto-updated, user-friendly protein subcellular localization database for Bacteria and Archaea. Nucleic Acids Res 39:D241–D244
Yun SH, Choi CW, Kwon SO, Lee YG, Chung YH, Jung HJ, Kim YJ, Lee JH, Choi JS, Kim S, Kim SI (2011a) Enrichment and proteome analysis of a hyperthermostable protein set of archaeon Thermococcus onnurineus NA1. Extremophiles 15:451–461
Yun SH, Kwon SO, Park GW, Kim JY, Kang SG, Lee JH, Chung YH, Kim S, Choi JS, Kim SI (2011b) Proteome analysis of Thermococcus onnurineus NA1 reveals the expression of hydrogen gene cluster under carboxydotrophic growth. J Proteomics 74:1926–1933
Yun SH, Lee YG, Choi CW, Lee SY, Kim SI (2014) Proteomic exploration of extremophiles. Curr Biotechnol 3:87–99
Zhu W, Smith JW, Huang CM (2010) Mass spectrometry-based label-free quantitative proteomics. J Biomed Biotechnol 2010:840518

Chapter 11
Functional Screening for the Discovery of New Extremophilic Enzymes

Freddy Boehmwald, Patricio Muñoz, Patricio Flores, and Jenny M. Blamey

1 Introduction

Worldwide, over $400 billion in goods are currently produced from biomass in conventional manufacturing every year (USDA 2008). These products include inorganic and organic chemicals, pharmaceuticals, soaps, detergents, pulp and paper, lumber, fuels, lubricants, greases, paints, among others. The emerging biobased economy is shifting from manufacturing petroleum-based feedstock to one based on biomass. Biocatalysis has proven to be an essential tool to convert biomass into more valuable bioproducts. Enzymes allow the efficient use of raw material and energy, minimizing waste generation by reducing the number of synthesis steps in industrial processes.

Enzymes for manufacturing of biobased products can be classified in: (1) bulk enzymes (sometimes referred as "industrial") and (2) specialty enzymes. So far as the manufacturing process of a biobased product requires more specialized enzymes, with outstanding performance and more sophisticated features, the added value of the final product is greater and the required amount of enzyme is less (Fig. 11.1). Thus, the discovery of new enzymes, development of new processes for enzyme production, and microbial systems with specific enzymatic steps will drive future innovation to a more efficient production of a big variety of new greener biobased products, specifically in the field of specialty and fine chemicals.

In the chemical manufacture context, the use of enzymes presents additional benefits: higher selectivity, increased sustainability, increased safety for operators and a decrease of toxic production of compounds derived from the synthetic procedures. These benefits make the trend towards cleaner production processes and

F. Boehmwald • P. Muñoz • P. Flores • J.M. Blamey (✉)
Fundación Científica y Cultural Biociencia,
José Domingo Cañas 2280, Ñuñoa, Santiago, Chile
e-mail: jblamey@bioscience.cl

P.H. Rampelotto (ed.), *Biotechnology of Extremophiles*, Grand Challenges in Biology and Biotechnology 1, DOI 10.1007/978-3-319-13521-2_11

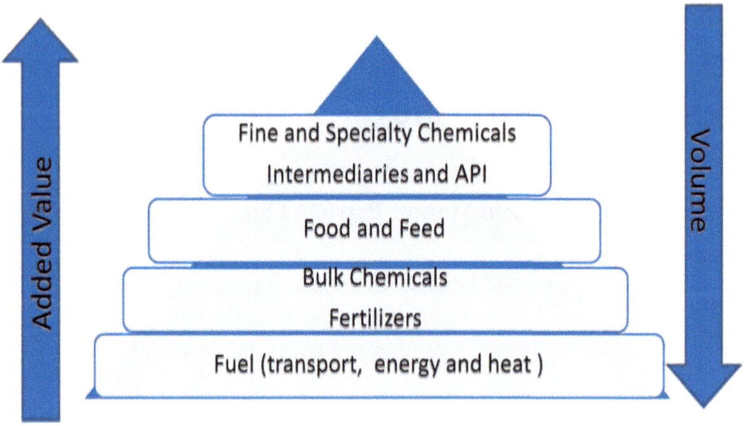

Fig. 11.1 Products and fields related with biomass bioprocessing. Enzymes participate in the manufacturing process of a variety of products. So far as the manufacturing process of a bioproduct requires specialized enzymes and more sophisticated features, the added value in the final product is greater and the required amount of enzyme is less

lower environmental impact. Indeed, the increasing social awareness about the environment and the implementation of new environmental regulations will transform this "trend" into a requirement in coming years. Under this future scenario, the use of enzymes and biocatalytic technologies will become a cornerstone for the transition to a greener and more sustainable economy.

Despite the paramount importance of biobased products, its utilization is still limited mainly to the lack of enzymes that can be integrated into existing production processes. In this regard, the main technological challenges related to the development of enzymes are:

- Development of more efficient and robust enzymes (WEF 2010). The low stability and efficiency of available enzymes hinders its integration into full-scale industrial processes. Currently, the repertoire of enzymes for industrial applications is limited by properties such as substrate concentration, sheering forces, temperature, pH and organic solvents, all common features in industrial settings. Since enzymes are typically the most expensive reagent in a biocatalytic process, more stable enzymes are required in order to create cost effective bioprocesses.
- Development of novel enzymes. The number and types of industrial enzymes available do not cover the number of chemical transformations carried out in the industry that could benefit from biocatalysis. Nowadays, only about 20 microbial enzymes are produced and sold at industrial scale (Li et al. 2012). Even more, this low number of industrial enzymes does not satisfy the demand of the specialized segments for enzymes capable to produce optically pure compounds.

The specialty and fine chemicals industry seeks to replace traditional chemical transformation processes for biocatalysis. However, for this to be economically feasible it will be necessary to increase the number of available industrial enzymes in order to develop highly stable enzymes (capable of withstand extreme industrial conditions), highly active, enantioselective and capable to catalyze the transformation of the right range of substrates (Andexer et al. 2009).

Biocatalysts derived from extremophilic microorganisms (particularly thermophilic and hyperthermophilic enzymes) offer an attractive solution to industrial biocatalysis since unlike most industrial enzymes; several extremozymes are much more stable and robust, therefore ideal candidates for the development of new industrial enzymes.

Extremophiles are organisms that thrive naturally in extreme temperatures, pH, pressure, salinity, etc. The key of extremophiles for survival in their extreme environments lies in the adaptation processes that enable them to develop highly specific molecular solutions. Moreover, extremophiles clearly possess some completely new metabolic pathways, which provide of a high number of enzymes with novel activities for current and new applications. Thus, extremophilic microorganisms represent an unexplored source of enzymes for the development of industrial biotechnology solutions of great importance.

2 Screening for Extremozymes: A Functional Approach

Nature has developed through evolution a vast source of biocatalysts. However, the probability to find the right enzymatic activity for a particular industrial application depends on the technical capability to efficiently assess this large biodiversity. Technologies such as metagenomic screening, genome mining and direct exploration of extremophilic enzymes (Adrio and Demain 2014; Bachmann 2014; Leis et al. 2013) are currently used.

The importance of metagenomic screening for searching new enzymes relies in the fact that less than 1 % of the microorganisms in a determined environment can be cultivated and this number is even lower when referred to extremophilic environments. This independent culture technique is performed through the preparation of a large genomic library from the environmental samples and searching for open reading frames (ORF) that can potentially encode for an enzyme. The results of this technique are based on two approaches, functional detection of activity for a heterologous expressed protein or the sequence approach to determine the presence of a homolog gene (Adrio and Demain 2014; Schmitz et al. 2008).

On the other hand, genome mining consists in finding the right ORF coding for putative enzymes through the alignment and comparison with annotated sequences that are deposited in databases and then cloned and expressed in a heterologous system. Beside the difficulties previously mentioned for this technique, the fact of trusting in previously annotated sequences provide more uncertainty to the expected results (Bachmann 2014).

Metagenomic screening and genome mining are based in the presence of DNA/RNA sequences codifying for a determined enzyme. This implies that the search for a novel enzyme is based on the genetic sequence homology to already described enzymes. This creates a bias and hinders the discovery of truly new enzymes. Additionally, the discovery of new enzymes based on genetic sequences does not always give accurate information about substrate specificity and/or efficiency under the industrial conditions of pH, temperature and others (Fernández-Arrojo et al. 2010). Often, due to the low efficiency of the recombinant enzymes discovered, this approach requires a follow up and further development through methodologies such as directed evolution and protein engineering to improve their features (Adrio and Demain 2014; Leis et al. 2013).

Direct exploration of extremophilic enzymes is based on functional screening of enzymatic activities in large collections of microorganisms. The functional detection of an enzymatic activity determines the existing bio-catalytic transformations. These tests verify the real existence of the biocatalytic reaction for a determined industrial setting. The functional screening approach has several competitive advantages over the molecular and metagenomic methodologies commonly used. First, it is important to note that for the industry to confirm the functionality of a biocatalyst under the actual conditions of a particular process makes the difference between obtaining or not the required product. Although, molecular approaches allow access to the genes of interest to clone and obtain large amount of proteins, it is not possible by genomic methodologies to discern specific characteristics of a enzyme such as: specific activity, pH and thermal stability, among others. These can be detected and measured through a functional approach.

With this approach it is possible to solve problems such as inhibition by the substrate or the product, stability, substrate specificity or enantiomeric selectivity by searching the right enzyme for a specific industrial requirement. However, although it can sound ideal to solve all industrial problems, it is not simple, since in order to find the right enzyme able to catalyze a specific transformation of industrial interest a large number of microorganisms/crude extracts or samples need to be screened and new robust enzymatic assays should be developed every time a novel enzymatic activity is screened. Not only that, in order for this approach to become industrially useful it is needed to set it in a miniaturized manner for allowing a massive and rapid discovery of the most efficient bio-catalyst for a specific industrial application.

For many years, this functional enzyme bio-discovery process was performed manually at a speed, volume and accuracy of analysis highly influenced by human capabilities and limitations. This technological barrier increment the developing time for a particular enzyme. However, deep knowledge in physiology of microorganism along with the growth of recombinant DNA, DNA sequencing and other technologies have been able to accelerate and improve the development of better and new enzymatic activities (Hallam et al. 2014; Schallmey et al. 2014). At the same time, the rapid growth of new extremophilic taxon descriptions, and advanced molecular techniques, for example metagenomics, have helped to discover novel microbial enzymes with improved/modified activities based on rational, semi-rational and random directed evolution strategies. For example, several anti-*Bacillus* lysins have been cloned from bacteriophage genomes and an aerolysin was cloned using metagenomic approaches (Adrio and Demain 2014; Schmitz et al 2008). Additionally, several extremophilic lipases and esterases with industrial and biotechnological applications have been obtained using metagenomic approaches (López-López et al. 2014).

3 High Throughput Functional Screening for Biocatalysts

As previously mentioned, functional screening have many advantages but those advantages are linked with the fact that big number of samples needs to be tested. So, it is no longer possible or practical to manually test a great number of reagents to develop an enzymatic assay, where testing will be error-prone and longer period of human work hours would be needed.

For the last 10 years High Throughput Screening (HTS) has been used for discovery of different compounds of industrial interest. The implementation of this technology can also be used for screening new biocatalyst. In HTS, large libraries of chemical or biological samples are tested for activity using automated assays (Glaser and Venus 2014; Inglese et al. 2006). In practice, HTS is an experimental instrument that allows the analysis of many biological processes at the same time. The experiments are conducted using plates of 96, 384 or 1536 wells. In these plates, all the components are incorporated following a specific protocol. The current goal of most companies working with this technology is to go beyond those formats for higher density and lower volume. This will result in lower costs. Reagents are stored in micro plates or 2D-barcoded mini-tubes. HTS is an automated methodology to carry out thousands of chemical and/or biological detection tests daily using liquid dispensers, handling devices, sensitive detectors, robotic devices, data processing, and control software (Klumpp et al. 2006; Omtaa et al. 2012; Zhu et al. 2010).

Currently, most HTS technology is centered on pharmaceutical companies searching for new drugs or chemicals (Zhu et al. 2010). However, this technology can be implemented in different research areas, including the search for novel biocatalysts (Donadio et al. 2009; Glaser and Venus 2014; Inglese et al. 2006; King et al. 2010). HTS technology for the screening of enzyme activities from extreme microorganisms is poorly developed and further effort will be needed to properly adapt this technology to functionally search for new enzymes.

The goal of HTS technology is to provide useful information to allow the correct identification of a hit among millions of samples in a highly reproducible manner. In the case of screening large numbers of extracts from extreme microorganisms, HTS would be able to select the right microorganisms with the specific enzymatic activity we search for. So, stringent requirement that surpass those used for bench-top assays are needed (Acker and Auldn 2014; Brooks et al. 2012).

In general, to screen enzymatic activities by HTS, three main determinations are performed: detection of substrate depletion, detection of product formation or direct binding of a ligand to the enzyme (Acker and Auldn 2014).

To develop an experiment under HTS standards is not an easy task. Extreme care is necessary in every stage of the design. The general process to implement HTS for the screening of extremophilic biocatalyst has been described as an association of several stages. However, the core of HTS relies on reagent settings, assay development and validation. Optimal enzyme/substrate ratios, cofactors, buffers, detection methods, compounds interferences and validation are crucial components and factors in the implementation (Liauda et al. 2014).

Crude extract libraries from extremophilic microorganisms do not require extensive preparation, but the steps for the identification of a biocatalyst are slow and complex processes. The robustness of screening assays in biological samples is challenged by chemical complexity of the extracts (Zhu et al. 2010). Commercially available enzymes are often used to set the initial test conditions for the reactions in order to detect a specific enzymatic activity. This allows obtaining the first insight on the main requirements of buffer, pH, salt concentration among other factors (Acker and Auldn 2014; Liauda et al. 2014). This is the starting point to create a more robust and sensitive assay. However, the purity of an enzyme in the extract used could alter its stability, affecting the decision of putative candidates and final hit selection. Additionally, the substrate concentration is directly related with the signal intensity of the assay. Higher concentrations of substrates increase turnover of the assay until the enzyme becomes saturated. Although it is recommended to use a substrate concentration over the Michaelis-Menten constant of the enzyme, this could lead to fewer probabilities to detect competitive inhibitors of the enzyme. Some enzymes need cofactors in order to perform their function. For those enzymes, cofactor availability needs the same consideration than substrates, even when by definition cofactors are not consumed in the reactions. Characteristics of buffers and pH are important not only because they affect the signals of the assays depending on the optimum pH of the enzyme, but also because it is important not to affect the behavior of other components of the assay. For instance, phosphate buffer is incompatible with reactions requiring Mg^{2+} or Ca^{2+} cofactors because the phosphate salts of these metals are poorly soluble in an aqueous buffer (Acker and Auldn 2014; Liauda et al. 2014). Compound interference is a term commonly utilized in HTS technology and it refers to the properties of some compounds that may affect the performance of an assay, producing false-positives. For instance, colored or fluorescent compounds can produce interference in a wide range of assays; weakly or insoluble substrates can aggregate themselves and interfere with optical detection system (Johnston et al. 2008; Moger et al. 2006). If the compounds present in an assay for a HTS campaign interfere significantly with the bioassay that gives a genuine biological activity, it will cause difficulties in the analysis of the results obtained for the identification of a hit (Noah 2010).

The ability to identify quickly and accurately active compounds in large chemical and/or biological libraries has been the ultimate goal in the development of HTS assays. However, the success of a particular HTS assay depends greatly on the specialty or fitness of the specific assay employed in this technology. For this, it was defined the Z factor (Zhang et al. 1999). This ratio is a reflection for both the test signal and the variation associated with the measurement data, being suitable for the evaluation of test quality. The Z factor is a dimensionless statistic value, simple characteristic for each HTS assay. This provides an advantageous tool for comparison and quality assessment of trials, optimization of critical parameters and subsequent statistical validation. For Z values between 0 and 0.5 are defined as indicative of a dual test or the existence of a low signal separation between positive (enzyme activity) and negative control, indicating that it is not a good test. Values between 0.5 and 1 define a good screening tool by HTS, with optimal value 1.

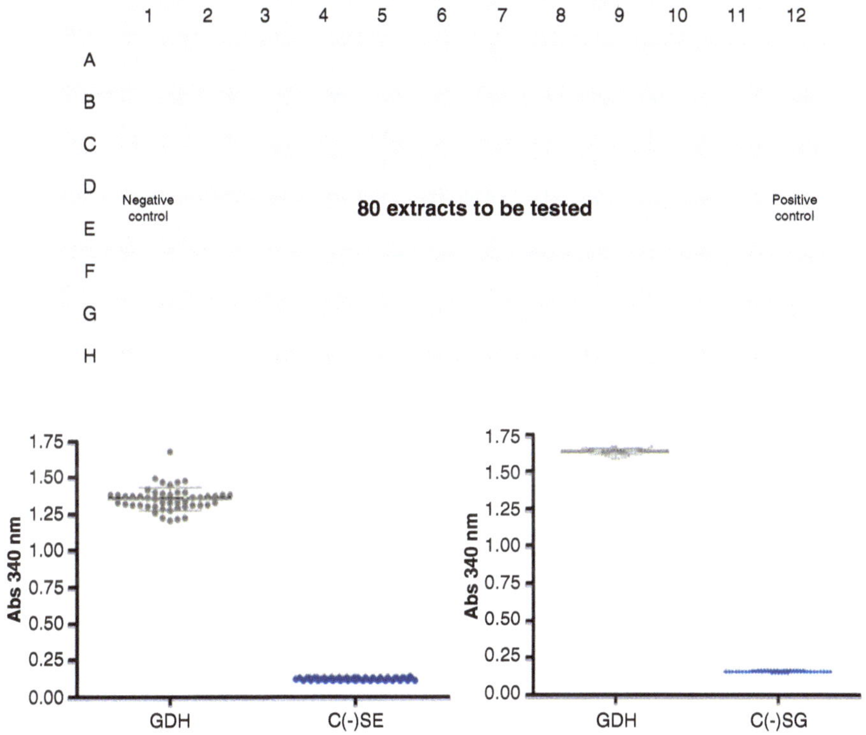

Fig. 11.2 HTS conformation and validation for a screening of glutamate dehydrogenase enzyme (GDH). (**a**) 96 wells plate typical conformation for HTS. (**b**) HTS Validation when manual process was utilized (Z factor: =0.78; Window: 10.5). (**c**) HTS Validation when an automatized system was utilized (Z factor: =0.94; Window: 9.9

Miniaturization of assays is the applied strategy in current HTS development. This reduction of volumes leads to a decrease of costs, materials and time, and such decrement can be achieved through the reduction of the size of microplate wells (Brooks et al. 2012). We designed a classic disposition for a screening under HTS conformation in a 96 wells plate for a glutamate dehydrogenase enzyme (GDH) (Fig. 11.2a). In this plate, 1 and 12 columns are used for positive and negative controls, respectively, leaving 80 spaces for different thermophilic extracts (unpublished results). The number of plates used for a specific assay will determine the number of different extracts and replicates studied. However, less number of plates can be utilized if the plate format is increased at 384–1536 wells, but more accurate and sensitive liquid dispenser will be required in order to set the initial low volumes. In the same figure, Z factor is calculated for a GDH assay manually performed (b) and an automatized assay (c). At first, it is possible to note that both assay versions can be validated under HTS standards, since both obtained Z factor values over 0.5 and similar signal windows (Zhang et al. 1999). However it is also possible to note that the dispersion is higher under the manual assay, affecting considerably the value of Z factor in comparison with the automatized assay. The time spent for only one plate is substantially higher than the automatized assay. It would not be cost effective to perform thousands of measures and replicates under a manual conformation. So, there are benefits in using HTS technology for a screening of large number of extracts from different types of microorganisms including the reduction in time of the screening process and reagent consumption.

The ultimate goal within HTS is that all the information obtained from the output of a particular screening should be promptly put in the correct biological or chemical context to facilitate decision-making and hits identification. Humans mainly do these analyses, but it is also possible to automatize it. In order to do this it is important to have a good Laboratory Information Management System (LIMS) since large amount of data are generated and even if all the previous requirements were perfectly set, it is frequent to have problems recovering the data and managing it (Omtaa et al. 2012).

Finally, the validation of HTS assays requires a statistical analysis for which several statistical methodologies are accepted. On the other hand one of the most difficult problems associated with manual techniques is the low number of performed replicates, which involves the use of conventional and less diligent statistical approaches. HTS assays include a large number of repeated measurements of controlled data. This means that there is a very rigorous measurement error associated to the samples and controls.

4 Extremophilic Enzymes with Potential in Biotechnology

Although the advantages of using biocatalysts as substitutes in chemical processes are often compelling from an environmental perspective, their industrial applications require some degree of cost-effectiveness to justify their cost competitiveness with conventional chemical catalysts (Jaeger 2004). For a given biocatalytic process, however, native enzymes do not often meet the requirements for large-scale application, and its physical–chemical properties need to be modulated. In fact, industrial processes often operate under harsh conditions that can inactivate the enzyme, including elevated pressure and temperature, extreme pH and non-aqueous solutions and oxidative conditions. The successful implementation of an enzyme as an industrial biocatalyst would require the availability of an economical and suitable enzyme with high activity, specificity and stability that can improve the performance and cost-effectiveness of the process under the required operational conditions (Jemli et al. 2014).

A great part of the current enzymes are obtained from mesophiles. In spite of their many advantages, their application is restricted owing to their limited stability at extreme temperatures, pH and ionic strength. In contrast to this, extremophiles are a natural source of extremozymes that possess an exceptional stability under these harsh conditions. Thus, the use of extremophiles and their extremozymes is being transformed from a basic science into an industrially viable technology. The unique features of each group of extremophiles can be exploited to provide enzymes with specific applications. Currently, many resources have been invested worldwide in order to develop industrial and biomedical applications of extremozymes (Bhattacharya and Pletschke 2014).

According to the type of reaction that enzymes from extremophiles and mesophiles are able to catalyze, they can be classified in six groups (Table 11.1) (MacDonald and Tipton 2014).

To the date, most attempts in the search of new extremophilic enzymes are focused on the screening of many cultured extremophiles that possess a particular biocatalyst of commercial interest. Additionally, metagenomic procedures allow obtaining numerous isolates from cultivable and non-cultivable microorganisms. Both approaches are time consuming and demand high reagent consumption for a single screening experiment. To overcome these limitations is necessary to implement HTS technology in order to rapidly identify new and novel enzymes from large libraries of extremophiles. In the following section several extremophilic enzymes with a high industrial potential are presented and how HTS technologies could be employed for specific enzymes.

4.1 Extremophilic Oxidoreductases

Oxidoreductases catalyze reactions of oxidation and reduction, transferring electrons from an electron donor (reductant) to an electron acceptor (oxidant). This group of enzymes uses cofactor such as NAD/NADH or FAD/$FADH_2$.

From this group, laccases, catalases, peroxidases and dehydrogenases are highlighted due to their applications in several processes. It was demonstrated that the improvement of their stability increased its biotechnological value (Guzik et al. 2014).

Dehydrogenases remove and transfer of hydrogen from a substrate in an oxidation-reduction reaction and can be employed to reduce carbonyl groups from aldehydes or ketones and carbon/carbon double bonds. Within this class, alcohol dehydrogenases (ADHs) represent a useful group of biocatalysts because they can synthesize optically active alcohols, which are key building blocks for the construction of pharmaceutical compounds. The physiological function of ADH is to catalyze the transformation between alcohols and aldehydes or ketones compounds occurs with the reduction of $NAD(P)^+$ to NAD(P)H. ADHs are widely distributed among microorganisms, including extremophiles. The ADH from *Sulfolobus solfataricus* uses NAD^+ as cofactor and contains Zn ions. In the other hand, the ADH from *Thermococcus litoralis* does not requires metallic ions and uses $NADP^+$ for the oxidation of primary alcohols. This enzyme is thermostable and its half-life is 15 min at 98 °C and 2 h at 85 °C (Egorova and Antranikian 2005). Another thermostable ADHs are the $NADP^+$-dependent ADH was purified from *Thermococcus hydrothermalis* (stereospecific for monoterpenes), the NAD^+-dependent alcohol-aldehyde oxidoreductase from *Alicyclobacillus acidocaldarius* (optimally active at 60 °C), the NADP-dependent Zn-ADH from *Thermoanaerobacter brockii*, among

Table 11.1 Biochemical classification of extremophilic enzymes

Classification	Example of extremozymes	Applications	Examples (source/obtention)	Reference
1. Oxidoreductase	Thermophilic and psychrophilic alcohol dehydrogenases	Chiral synthesis	*Sulfolobus solfataricus*/not specified	Egorova and Antranikian (2005)
		Chemicals production	*Thermococcus litoralis*/not specified	Egorova and Antranikian (2005)
			Thermococcus hydrothermalis/Recombinant	Antoine et al. (1999)
			Flavobacterium frigimaris KUC-1/Native	Kazuoka et al. (2007)
	Thermophilic and psychrophilic superoxide dismutases	Medical treatment	*Thermus thermophilus* HB27/Recombinant	Liu et al. (2011)
		Cosmetic industry	*Alicyclobacillus* sp. CC2/Native	Correa-Llantén et al. (2014)
		Food industry	*Exiguobacterium* sp. OS-77/Native	Nonaka et al. (2014)
		Chemicals production	*Pseudoalteromonas haloplanktis*/Native	Castellano et al. (2006)
	Psychrophilic catalases	Cosmetic industry	*Vibrio salmonicida*/Native	Lorentzen et al. (2006)
			Serratia sp. I1P/Commercial, recombinant	Unpublished results
	Halophilic catalases	Food industry	*Halobacterium halobium*/Native	Brown-Pterson and Salin (1995)
	Thermophilic laccases	Textile industry	*Thermus thermophilus* HB27/Native and recombinant	Miyazaki (2005)
		Pulp and paper industry	*Bacillus* sp. FNT/Native	Unpublished results
		Food industry		
		Bioremediation		
2. Transferases	Thermophilic and hyperthermophilic DNA polymerases	Molecular biology	*Thermus aquaticus* (Tag polymerase)/ Commercial, recombinant	Chien et al. (1976), Peake (1989)

			Pyrococcus furiosus (Pfu polymerase)/ Commercial, native and recombinant	Lundberg et al. (1991)
			Thermotoga maritima (ULTIMA DNA polymerase)/Commercial, discontinued	Diaz and Sabino (1998)
			Sulfolobus solfataricus/HTS of a mutant library	Kardashliev et al. (2014)
	Thermophilic aminotransferases	Chiral synthesis	*Sulfolobus solfataricus*/Native	Marino et al. (1988)
3. Hydrolases	Thermophilic and psychrophilic alkaline lipases	Detergent	*Geobacillus* sp. ID17/Native	Muñoz et al. (2013)
		Chiral synthesis	*Thermosyntropha lipolytica*/Native	Salameh and Wiegel (2007)
		Chemicals production	*Pseudomonas fluorescens* B68/Recombinant	Luo et al. (2006)
		Biodiesel production	*Thermus* sp. P1074/HTS of a recombinant library	Lagarde et al. (2002)
	Thermophilic nitrilases	Chiral synthesis	*Pyrococcus abyssi*/Recombinant	Mueller et al. (2006)
			Pyrococcus sp. M24/Native	Unpublished results
	Thermophilic xylanases	Bleaching pulp and paper	*Pyrodictium abyssi*/Native	Egorova and Antranikian (2005)
	Thermophilic and psychrophilic proteases	Detergent	*Aeropyrum pernix* K1/Native and recombinant	Bouzas et al. (2006)
		Chiral synthesis	*Serratia rubidaea*/Native	Cavicchioli et al. (2011)
	Thermophilic amylases	Starch processing	*Bacillus licheniformis*/Native	Bouzas et al. (2006)
			Pyrococcus woesei/Recombinant	Bouzas et al. (2006)
	Thermophilic and hyperthermophilic glycoside hydrolases	Lignocellulose deconstruction	Thermophilic Clostridia/Native	Blumer-Schuette et al. (2014)

(continued)

Table 11.1 (continued)

Classification	Example of extremozymes	Applications	Examples (source/obtention)	Reference
			Caldicellulosiruptor saccharolyticus/ Recombinant	Blumer-Schuette et al. (2014)
4. Isomerases	Halophilic xylose isomerase	High fructose syrups	*Halothermothrix orenii*/Not produced	Bhattacharya and Pletschke (2014)
	Thermophilic racemase	D-amino acids synthesis	*Geobacillus stearothermophilus*/Native	Soda et al. (1988)
5. Lyases	Alkaline alginate lyase	Pharmaceutical	*Agarivorans* sp. JAM-A1m/Native	Kobayashi et al. (2009)
	Psychrophilic and alkaline pectate lyase	Wastewater treatment	*Pseudoalteromonas haloplanktis*/ Recombinant	van Truong et al. (2001)
			Mrakia frigida/Native	Margesin et al. (2005)
6. Ligases	Thermophilic and psychrophilic DNA ligases	Molecular biology	*Aquifex pyrophilus*/Recombinant	Lim et al. (2001)
			Pseudoalteromonas haloplanktis/ Recombinant	Georlette et al. (2000)

others (Radianingtyas and Wright 2003). Thermophilic and hyperthermophilic ADHs play a considerable role in several processes and production, for example in the generation of alcohol, solvents and acetic acid. ADHs from thermophiles are of interest for industrial alcohol and enzyme production, because these microbes are uniquely appropriate for direct biomass fermentation to produce ethanol via reduced-pressure distillation, and to yield active thermostable enzymes of commercial interest (Andrade et al. 2001). There is also considerable interest in the employment of such enzymes for the synthesis of chemical compounds particularly pharmaceuticals, where the production of chiral synthons is one of the most important steps in the synthesis of chirally pure agents (Radianingtyas and Wright 2003).

On the other hand, psychrophilic ADHs have been described from Antarctic microorganisms including members of genera *Moraxella* and *Flavobacterium* (Tsigos et al. 1998; Kazuoka et al. 2007). The psychrotolerant bacterium *Flavobacterium frigidimaris* KUC-1 produces a cold-active, thermostable and NAD^+-dependent ADH. This enzyme is active in the range of temperatures from 0 °C to 85 °C and its optimum temperature is 70 °C. However, it shows a high catalytic efficiency at temperatures below 20 °C, similar to other psychrophilic enzymes. Its half-life at 60 °C in the presence of NAD^+ is 143 min, about three times longer than this enzyme in the absence of its cofactor (50 min) (Kazuoka et al. 2007). These properties suggest its potential for application in chiral synthesis of pharmaceutical compounds, a process that can be carried out at low temperatures using a very stable enzyme, reducing the formation of by-products.

The majority of ADHs requires NAD^+ or $NADP^+$ as cofactor and its reduction can be monitored through the increase in the absorbance at 340 nm. Unfortunately, this approach is generally not suitable for HTS methods due to the background noise created by cell lysates and the 96-well plate itself. It is necessary the use of special plates that do not absorb in UV range, but they are very expensive when considering a large-scale effort. Colorimetric assays resolve most of these difficulties and are quite amenable to HTS methods. For example, tetrazolium salts such as nitroblue tetrazolium (NBT) are suitable because they are reduced to formazan dyes, and this compound absorbs visible-light. These reactions can be easily monitored visually on filter discs or on a standard 96-well plate reader. A number of reactions leads to the formation of a colored formazan linking the production of NAD(P)H to the catalytic activity of a ADH (Johannes et al. 2006).

An NBT/phenazine methosulfate (PMS) assay that works well has been developed for different HTS formats, for the identification of novel ADH without prior enzyme purification. This new screening methods employs bioinformatics, molecular biology techniques, and the direct *in vitro* expression of enzymes in order to rapidly detect and characterize novel ADHs. A pool of 18 novel thermoactive ADHs with a broad substrate range was discovered and characterized (Ravot et al. 2003). This procedure can be extrapolated to any dehydrogenase that uses $NAD(P)^+$ as cofactor and could offer a new alternative for different extremophilic dehydrogenases.

Other type of oxidoreductase with biotechnological potential is superoxide dismutase (SOD). This enzyme catalyzes the dismutation of superoxide anion into hydrogen peroxide and molecular oxygen and is the first line of defense against oxidative stress. All SODs are metalloenzymes with a redox metal, such as Cu^{+2}, Zn^{+2}, Mn^{+2}, and Fe^{+2} at the active site. This enzyme has antioxidant effects and is widely applied in medical applications, and in cosmetic, food, agricultural, and chemical industries (Liu et al. 2011).

Thermal stability is a major requirement for commercial SOD, because thermal denaturation is a common cause of enzyme inactivation. Currently, there is an increasing interest in SOD from thermophilic and hyperthermophilic microorganisms, which probably produce thermostable SOD. A Mn-SOD was purified from *Thermus thermophilus* HB27. This enzyme retains 57 % of its activity after a heat treatment at 100 °C and is considered a highly stable enzyme (Liu et al. 2011). Other thermostable SOD was purified from the bacterium *Alicyclobacillus sp.* CC2. This enzyme retains 80 % of its initial activity when it was incubated for 6 h at 50 °C, and it has been proposed as part of the antioxidant defense system of CC2, allowing to this bacterium to survive under extreme conditions, avoiding the effects of oxidative stress in Antarctica (Correa-Llantén et al. 2014).

A Mn/Fe-SOD was purified from the UV-resistant bacterium *Deinococcus radiophilus*. The enzyme was stable at pH range from 5.0 to 11.0, but quite unstable under acidic conditions, and thermostable up to 40 °C (Yun and Lee 2004). Other thermostable Mn-SOD was obtained from the psychrophilic bacterium *Exiguobacterium sp.* OS-77. In this case, the optimal reaction pH and temperature of the enzyme were pH 9.0 and 5 °C, respectively. Notably, the purified Mn-SOD was thermostable up to 45 °C and retained almost 50 % of its activity after 21.2 min at 60 °C (Nonaka et al. 2014). The thermostable SOD from the psychrophile *Pseudoalteromonas haloplanktis* called PhSOD showed an inactivation profile that suggests a greater heat resistance with a half-life inactivation at temperature of 54.2 °C (Castellano et al. 2006). Its biochemical properties suggest that this enzyme possesses potential applications in medical and cosmetic industries, reducing the free radical damage to the skin, by reducing fibrosis during radiation for breast cancer (Shafey et al. 2010). In this way, it would be possible to reduce lipid peroxidation in the skin using a topical application of these thermostable SODs.

Currently, there is no method to screen SOD activity for microbial cells by HTS technologies. However, the company Trevigen® has developed a system that allows measure SOD activity from mammalian tissues and cell lysates using a 96 well format for HTS (HT Superoxide Dismutase Assay Kit). In this assay, superoxide radical ions are generated from the conversion of xanthine to uric acid, and hydrogen peroxide by xanthine oxidase generates a formazan salt that absorbs at 450 nm. SODs reduce superoxide ion concentrations and lower the formation of the formazan salt. The extent of reduction in the appearance of formazan salt is a measure of SOD activity present in experimental samples and could be very useful to screen this enzymatic activity in extremophilic samples.

Catalase (CAT) is a part of the detoxification system inside living cells against reactive oxygen species formed during the metabolism. This enzyme scavenges hydrogen peroxide generating oxygen and water. It is very useful in analytical and diagnostic methods acting as a biosensor and biomarker in addition to its other applications in textile, paper, food and pharmaceutical industries (Sooch et al. 2014).

CATs have been described from psychrophilic members of the genus *Vibrio*. From this genus, a CAT from *Vibrio salmonicida* was described to be optimally active at 0–10 °C and pH 8.8. This enzyme is functional until 60 °C. However, its thermostability is reduced with a half-life of 34 min at 50 °C, being rapidly inactivated at 70 °C, but at 37 °C it retained the 50 % of its initial activity after 6.5 h of incubation at this temperature (Lorentzen et al. 2006). On the other hand, a more thermostable CAT was purified in Bioscience Foundation from an Antarctic psychrotolerant bacterium member of genus *Serratia* named I1P. The thermostability profile of purified CAT from I1P showed retention of 60 % of its initial activity after 8 h of incubation at 50 °C. Additionally, when the enzyme was heterologously expressed in *E. coli*, optimal parameters of this CAT were pH 6.0–7.0 and 50 °C. The recombinant version of I1P CAT retains 50 % of its activity after six of incubation at 50 °C. Moreover, this enzyme is active in the temperature range from 20 °C to 70 °C. Higher stability and functional temperature range of CAT from I1P offer the possibility to use this CAT in cosmetic industry in anti-age creams reducing the oxidation of the skin by hydrogen peroxide.

The halophilic bacterium *Halobacterium halobium* produces a CAT that needs the presence of NaCl to be optimally functional. The lack of this salt in the buffer results in an inactive enzyme (Brown-Pterson and Salin 1995). This kind of enzymes is useful in food industry in combination with a glucose oxidase as a preserving system. Moreover, new applications for catalases are emerging owing to their high turnover rate, and its relatively simple and well-defined reaction mechanisms.

There is a lack of a HTS format for CAT activity for the different groups of extremophiles. However, Li and Schellhorn (2007) have proposed a rapid kinetic microassay for this enzyme. This method is a modification of the Beers and Sizer assay (1952) and can be used to screen large numbers of samples in parallel in a short time and is thus well-suited to HTS for extremophilic CAT.

Laccase is a multicopper blue oxidase that couples the four-electron reduction of oxygen with the oxidation of different organic substrates, such as phenols, polyphenols, anilines, and even certain inorganic compounds by a one-electron transfer mechanism. Laccases have great biotechnological potential due to their wide reaction capabilities as well as the broad substrate specificity. Promising application includes textile-dye bleaching, pulp bleaching, food improvement, bioremediation of soils and water, polymer synthesis, and the development of biosensors and biofuel cells (Kunamneni et al. 2008).

A few laccases are used in textile, food and other industries, and it is necessary to develop new candidates for future commercialization. To be more accurate, an energy saving, and biodegradable laccase adjusts well with the development of highly efficient, sustainable, and environmentally friendly industries. However, the main limitation is their low operational stability under harsh industrial conditions and its rapid degradation during storage. Laccases from thermophilic microorganisms will probably overcome these limitations.

A hyperthermophilic laccase was purified from the thermophilic bacterium *T. thermophilus* HB27. The enzyme was most notable for its striking thermophilicity; the optimum temperature was 92 °C and its half-life at 80 °C was >14 h, ranking it as the most thermophilic laccase reported, even more than CotA from *Bacillus subtilis* (Miyazaki 2005). These properties allow this enzyme to be useful in food industry to improve the flavor quality of vegetable oils by eliminating dissolved oxygen. Additionally, its stability is important to catalyze reactions in organic synthesis, where the presence of solvents inactivates mesophilic enzymes but not the thermophilic ones. However, its optimal pH was 5.0, limiting its application in pulp and paper and textile industries, where the processes are carry out with an alkaline pH. To overcome this limitation, Bioscience Foundation has developed an alkaline laccase obtained from a thermophilic *Bacillus* called FNT. This laccase has an optimum temperature at 70 °C, but it is still active at 100 °C. The optimum pH ranged is 7.0–8.0, but it is active even at pH 9.5, being more adequate for pulp and paper industry.

The ABTS assay is a very flexible assay applicable to laccases. The production of the green radical cation by a laccase is observed spectrophotometrically (Johannes et al. 2006). This assay was used to quantify the functional expression of a fungal laccase in *Saccharomyces cerevisiae* using directed evolution. Improvement of protein expression by eight-fold to the highest level yet reported with an additional increase of 22-fold in the turnover number (Bulter et al. 2003). This assay is used for the development of functional selection and enhanced stability screening, and should prove helpful in the future for thermophilic laccases.

4.2 *Extremophilic Transferases*

Transferases are a group of enzymes that catalyze the transfer of specific functional groups from one molecule (donor) to another (acceptor). They are involved in different types of metabolic reactions including transference of acyl chains, nucleotides (DNA and RNA polymerases), monosaccharides, ketones, aldehydes, and amines, among others.

Taq polymerase (DNA polymerase I from *Thermus aquaticus*) is the most famous representative enzyme among the thermostable DNA polymerases. This enzyme was purified from *T. aquaticus*, a thermophilic bacterium isolated from Yellowstone National Park, USA (Chien et al. 1976). Taq polymerase is very useful for the amplification of specific DNA fragments at high temperatures in the Polymerase Chain Reaction or PCR (Peake 1989).

Thermophilic microorganisms in general have thermostable DNA polymerases, and therefore, they became more popular as source of this kind of enzyme. Generally, hyperthermophiles can provide more heat-stable enzymes than thermophiles. For example, the Pfu polymerase from *Pyrococcus furiosus*, which is more stable and possesses a higher fidelity than Taq polymerase (Lundberg et al. 1991).

The recombinant KOD1 DNA polymerase from *Thermococcus kodakaraensis* KOD1 possesses low error rates, high processivity and high extension rates. This results in the accurate amplification of target DNA sequences of up to 6 kb (Egorova and Antranikian 2005). Other example of DNA polymerase is Tma DNA polymerase obtained from *Thermococcus marinus*. The enzyme was suitable to perform PCR reactions (Bae et al. 2009).

The first commercial enzyme obtained from the hyperthermophilic bacterium *Thermotoga maritima* is ULTIMA DNA polymerase. This enzyme has 3′–5′ exonuclease activity and thus was expected to perform PCR more accurately with its proofreading activity. However, it was not a commercial success, since no significant differences in fidelity were determined in sequencing (Diaz and Sabino 1998).

A screening system for directed evolution of DNA polymerases employing a fluorescent Scorpion probe as a reporter in a 96 well plate format has been developed and validated in a directed evolution experiment of a DNA polymerase from *Sulfolobus solfataricus* that was improved in elongation efficiency of consecutive mismatches. The Scorpion probe screening system enables to reengineer polymerases with low processivity and fidelity, and no secondary activities, such as exonuclease activity or strand displacement activity to match demands in diversity generation for directed protein evolution (Kardashliev et al. 2014). This method could be very useful to screen new DNA polymerases from hyperthermophilic microorganisms that are helpful in molecular biology experiments.

Transaminases (TAs) or aminotransferases are ubiquitous enzymes. They catalyze the synthesis and degradation of amino acids and amines in nature. The amino group of an amino donor is transfer to a carbonyl carbon atom of an α-keto acid, ketone or aldehyde, the amino acceptor (Höhne and Bornscheuer 2012).

Based on their substrate specificity, TAs are classified in 3 groups: α-TAs convert substrates with a carboxyl group at the α-position and they can synthesize non-natural amino acids (non-proteinogenic and/or D-configuration); ω-TAs transfer a terminal amino group attached to a primary carbon atom at least one carbon atom away from a carboxyl group; and amino-TAs or ATAs act on substrates lacking any carboxyl group, such as ketones and amines, using pyruvate as universal amino acceptor. This last offer the possibility for the synthesis of chiral primary amines and these ATAs are useful for the synthesis of chiral organic compounds (Höhne and Bornscheuer 2012).

Especially α-TAs and ATAs show high enantioselectivity for a given enantiomer and can be useful for the synthesis of different amino acids and synthetic compounds of pharmaceutical interest. However, due to the inactivating effect of the majority of the substrates used for the synthesis of chiral compounds, and to the presence of organic solvents that inactivate the current TAs employed in chiral synthesis, it is necessary to search for new thermostable TAs.

Currently, two pyridoxal phosphate-containing TAs have been described from *S. solfataricus*: an aspartate TA (Marino et al. 1988) and a serine TA (Littlechild 2011). Aspartate TA reversibly transfer an α-amino group between aspartate and glutamate and is an important enzyme in amino acid metabolism, but its biotechnological properties have not been exploited yet. Serine TA is involved in the synthesis of serine and it is not found in bacteria. The transaminase reaction that the enzyme carries out is the conversion of L-serine and pyruvate into 3-hydroxypyruvate and alanine. The enzyme shows activity towards methionine, asparagine, glutamine, phenylalanine, histidine and tryptophan. The enzyme can be combined with transketolase for the synthesis of chiral compounds (Chen et al. 2007). This enzyme is an example of a thermophilic archaeal α-TA with potential commercial application.

Walton and Chica (2013) reported the development of a high-throughput assay in a 96 wells plate format to screen α-ketoglutarate-dependent TA mutant libraries. They adapted the L-glutamate dehydrogenase coupled assay to allow screening mutant libraries of either L- or D-amino acid specific TA in a continuous fashion. This assay is reproducible, rapid, and sensitive. This HTS assay has the potential to discover new TA from thermophilic and hyperthermophilic microorganisms.

4.3 Extremophilic Hydrolases

Hydrolases catalyze the breakage of chemical bonds through the addition of a water molecule therein. They hydrolyze the cleavage of complex macromolecules like proteins, starch, fat, and nucleic acid into simpler molecules.

These enzymes have been extensively studied for a variety of applications. Within this group of enzymes are lipases, enzymes that catalyze the hydrolysis and synthesis of ester bonds of long chain acyl-glycerols. They are versatile biocatalysts with application in several industries such as food, dairy, pharmaceutical, detergent, textile, pulp and paper, animal feed, leather and cosmetic. Lipases are gaining increasing attention not only for their catalytic properties to hydrolase ester bonds, but also for their ability to synthetize optically pure compounds. Many lipases are commercially available, being some of them useful for one specific or several applications (Houde et al. 2004). However, despite the relatively high number of commercial lipases, industrial applications are limited because to their low stability under harsh conditions.

Thermophilic lipases are generally stable and active at high temperatures and in organic solvents. Examples of this are the lipases LipA and LipB secreted by *Thermosyntropha lipolytica*. These enzymes are highly thermoactive (optimum temperature 96 °C) and thermostable. Both enzymes could be useful as additives in laundry detergents for high temperature washing; for the production of fatty acids, diacylglycerols and different ester compounds at elevated temperature (Salameh and Wiegel 2007).

Lip1 is one of the four lipases secreted by the antarctic *Geobacillus* sp. ID17. This enzyme is optimally active at 65 °C and pH 9.0, offering the opportunity to be used in washing detergents. Moreover, it retains 70 % of its activity after 8 h of incubation at 70 °C (Muñoz et al. 2013). Its catalytic properties can be modulated through the presence of ionic liquids, improving its catalytic activity by the liquids salt 1-butyl-3-methylimidazolium hexafluorophosphate (Muñoz et al. 2015) and 1-butyl-3-methylimidazolium methylsulfate. Additionally, Lip1 is able to synthesize selectively ester compounds derived from menthol isomers, being its enantioselective properties modulated through the engineering of the reaction medium. All these offer the opportunity to exploit the biotechnological potential of Lip1.

In order to find novel enzymes for industrial bioconversions, Lagarde et al. (2002) have used a new generation of thermostable reporter substrates CLIPS-O (CataLyst Identification ProcesS per Oxidization) esterified with acyl chain of different lengths for the screening of esterases and lipases in microorganisms belonging to thermophilic genus. CLIPS-O is a versatile assay for enzymes that uses chromogenic or fluorogenic enzyme-labile functional group, which are not activated due to a chemical modification. *Thermus* sp. P1074 exhibited a high level of constitutive intracellular esterase activity on short and long chain ester CLIPS-O substrates. The genomic library of this strain was screened using a HTS technology. Three positive clones expressed thermophilic esterases and the enzymes were characterized during screening for chain length specificity. Using CLIPS-O substrates is possible to implement a HTS technology that allow to screen lipases and esterases from thermophilic and hyperthermophilic microorganisms. Moreover, it will be possible to determine substrate especificity using CLIPS-O esterified with acyl chains of several lengths.

Nitrile-degrading enzymes, specifically nitrilases are gaining more importance, especially in chiral biosynthesis of drugs in pharmaceutical industry. One thermostable nitrilase has been obtained from the hyperthermophilic archaeon *Pyrococcus abyssi*. The enzyme is active in a temperature range from 60 to 90 °C and its half-life was 9 h at 70 °C (Mueller et al. 2006). These characteristics reveal a great industrial potential. However, this enzyme is inhibited in the presence of acetone, a common organic solvent used in industry. Recently, in our laboratory, a thermostable nitrilase was purified and characterized from a new member of genus *Pyrococcus*. This nitrilase has an optimum temperature at 85 °C and show enantiopreference for the (S)-mandelonitrile isomer, indicating its potential property to synthesize (S)-mandelic acid, a compound with multiple pharmacological properties (unpublished).

Various HTS methods have been developed for nitriles hydrolyzing enzymes. For example, He et al. (2011) reported a fast and easy HTS method for assaying nitrile-hydrolyzing enzymes based on ferric hydroxymate spectrophotometry. More recently, Coady et al. (2013) have developed a HTS method for the production of enantiopure β-hydroxy acids using nitrile-hydrolyzing enzymes. The strategy includes toxicity, starvation and induction studies with subsequent colorimetric screening for activity using the substrates of interest. This high-throughput strategy uses a 96 well plate system, and enabled the rapid biocatalytic screening of 256 novel bacterial isolates towards β-hydroxynitriles. Moreover, *Rhodococcus erythropolis* SET1 was identified and found to catalyse the hydrolysis of 3-hydroxybutyronitrile with a high enantioselectivity under mild conditions. Therefore, it will be possible to screen enantioselective nitrile-hydrolyzing enzymes using this HTS method in order to look for novel thermophilic and hyperthermophilic enzymes with high enantioselectivity.

Another hydrolase used in the bleaching of pulp and paper is the cellulose-free thermostable xylanase. These enzymes degrade xylan, the most abundant form of hemicellulose. Among thermophilic archaea, xylanases are produced by *Pyrodictium abyssi* and *Thermococcus zilliggi* AN1. Both enzymes are active at temperatures higher than 100 °C and they are suitable for their utilization in pulp and paper industry (Egorova and Antranikian 2005).

The production of commercial proteases is greater than that of any other enzyme group of biotechnological relevance. They are the most important industrial enzymes and constitute around 65 % of the global market. The biotechnological interest has further driven the search for thermostable proteases able to withstand the harsh conditions of industrial processes. Proteases can be classified in two groups: endopeptidases—proteases that cleave peptide bonds within the protein—and exopeptidases, which cleave off amino acids from the ends of the protein. Many proteolytic enzymes from hyperthermophilic archaea and bacteria have been identified. Most of them are active at high temperatures, and in the presence of detergents and other denaturing substances, such as urea, guanidine-HCl, dithiotreitol or 2-mercaptoethanol. Hyperthermophilic proteases have been identified in *T. maritima, T. aggregans, T. celer, T. litoralis* and *Pyrococcus* sp. KODI. The thiol protease from *Pyrococcus* sp. KODI has an optima temperature of 110 °C and is the most hyperthermophilic protease. A protease from the aerobic hyperthermophile *Aeropyrum pernix* K1 was isolated. This microorganism grows optimally at 90 °C and its aerobic growth pattern and nutritional requirements are amenable to large-scale culture techniques. These characteristics make this hyperthermophile an attractive source of thermostable proteases (Bouzas et al. 2006).

The industrial enzymatic processing of starch is based on (partial) hydrolysis to maltodextrins, maltose and glucose syrups. This process requires enzymes stable and active at temperatures above 65–70 °C. Well-known starch-acting enzymes are α-amylases and pullulanases, which degrade it to maltooligosaccharides and glucose (Kaper et al. 2003).

Enzymatic processing of starch involves three steps: gelatinization (dissolution of starch granules, generating a viscous suspension through the heating the starch with water at high temperatures); liquefaction (partial hydrolysis of starch, with a loss of viscosity); and saccharification (production of glucose and maltose via further hydrolysis) (Bouzas et al. 2006).

The first thermostable α-amylases were isolated from *Bacillus subtilis*, *Bacillus amyloliquefaciens* and *Bacillus licheniformis*. However, their industrial use is restricted due to the majority of α-amylases require calcium for activity, and the calcium added to the reaction precipitates as calcium oxalate (Saha et al. 2014). To reduce the concentration of calcium oxalate is necessary to lower the reaction pH. For this reason is important the search for acid-stable α-amylases. Hyperthermophilic and acidic α-amylases have been obtained from *Pyrococcus woesei*, *Pyrococcus furiosus*, *Thermococcus profundus* and *Thermococcus hydrothermalis*. The optimal temperature for the activity of these enzymes is 100 °C, and the optimal pH is lower than that found for *Bacillus* α-amylases (Bouzas et al. 2006).

α-Glucosidases are generally used in the last step of starch degradation. A hyperthermophilic α-glucosidase has been isolated from *T. maritime*, which required NAD^+ and Mn^{+2} for its activity. Another hyperthermophilic α-glucosidases have been detected in *P. furiosus*, *P. woesei*, *Sulfolobus shibatae*, *Sulfolobus solfataricus*, *Thermococcus* strain AN1 and *Thermococcus hydrothermalis*, being the intracellular α-glucosidase from *P. furiosus* the most hyperthermophilic, with an optimal activity in the range of temperatures from 105 °C to 115 °C and a pH range from 5.0 to 6.0 (Bouzas et al. 2006).

Pullulan-degrading enzymes which also belong to the α-amylase family are ubiquitous produced by different species, including hyperthermophilic archaea. These enzymes have been isolated from *Thermococcusceler*, *Desulfurococcus mucosus*, *Staphylothermus marinus*, *Thermococcus agregans*, *P. furiosus*, *Thermococcus litoralis*, *T. hydrothermalis*, *Pyrococcus* strain ES4 *and T. maritima*. They exhibit great stability at temperatures above 100 °C, even in the absence of substrate or calcium ions (Bouzas et al. 2006).

There is a need for lignocellulolytic enzymes working under extreme conditions, like high salt concentrations, organic solvents, extreme pH values and temperatures. These harsh conditions are not optimal for the current used enzymes, lowering the turnover rates and lifetime of them. Thermophilic microorganisms are attractive candidates for conversion of lignocellulose into biofuels and they produce robust and effective carbohydrate degrading enzymes. Thermophilic members of genera *Clostridium* and *Caldicellulosiruptor* possess cellulosomal and noncellulosomal enzymes, respectively, capable of converting plant biomass into fermentable sugars. This offers the opportunity that little or no pretreatment may be required for the production of biofuels (Blumer-Schuette et al. 2014).

A random mutagenesis library of the cellulase Cel8H from the *Halomonas sp.* was generated and a RoboLector was used to select and characterize promising candidates from this library. This HTS system was developed by Kunze et al. (2012). This consists of an on-line monitoring system for cultivations/reactions in micro-scale (BioLector) and an automated liquid-handling robot. This interaction between the robot and the BioLector allows high-content data generation and systematic high-throughput experimentation. RoboLector can provide extensive kinetic data from the whole enzyme production process including the cultivation and expression step in a certain host and the final characterization of the enzymes. This example of automated screening systems can support the enzyme development and production process with minimal manual effort in the searching of novel extremozymes for lignocellulose treatment.

High demand of new thermostable hydrolases with appropriate performance and novel functionalities could provide huge savings in time, money and energy for industrial processes. The HotZyme project from The University of Copenhagen (http://hotzyme.com/project/) aims to identify such enzymes, using metagenomic approaches. A range of HTS technologies are employed to identify novel hydrolases from microorganisms isolated from hot terrestrial environments. However, no publications associated to this project are available.

In the other extreme of temperatures are psychrophilic hydrolases. Cold-adapted enzymes have a high biotechnological value due to their high k_{cat} at low temperatures. Their low thermostability and high capacity to function in organic solvents provide economic benefits, including higher productivity than mesophilic homologues at low temperature, thereby providing energy savings to the processes where they are used. As a result, cold-adapted hydrolases have found application in industries as household detergents.

Their ability to hydrolyze substrates at low temperatures is very useful for cleaning applications in a wide range of industries, including laundry and soaps, food, dairy and brewing medical devices and water treatment. Life cycle assessments of cleaning methods have been reported for dairy, water treatment, detergent, and brewing industries. Detergent manufacturers recognize the link between a reduced wash temperature and improved energy conservation. A decrease in wash temperature from 40 °C to 30 °C produces a 30 % reduction in electricity used, equating to a reduction of 100 g of CO_2 per wash. Proteases, amylases, lipases and cellulases, such as Alcalase (protease), Natalase (amylase) and Lipolase Ultra (lipase) from Novozymes have been used to low temperature washing. The ability of enzymes to clean at low temperatures has generated a reduction in temperature used for washing procedures in a range of industries; including automated dishwashers, the cleaning of membranes for water treatment, and cleaning of equipment in brewing and dairy. Enzymes from psychrophiles, such as proteases from *Serratia rubidaea* and *Stenotrophomonas maltophilia* are the kind of enzymes with a high potential to extend the effectiveness of enzyme-based, low-temperature cleaning formulations (Cavicchioli et al. 2011).

Surfaces that are at ambient temperatures, such as buildings, carpets and benches, cannot easily be heated or immersed in cleaning solutions and tend to be cleaned using sprays or wipes, providing good avenues for the use of cold adapted enzymes. Illustrating the value of enzymes, a lipase and glycoside hydrolase have been used in a cleaning solution in a building conservation project to improve the removal of mould from stone and reduce the damage normally associated with the use of standard cleaning agents (Cavicchioli et al. 2011).

4.4 Extremophilic Isomerases

Isomerases catalyze the conversion of a molecule from one isomer to another, facilitating the intramolecular rearrangement. These enzymes use one substrate that yields one product. Isomerases catalyze reactions across many metabolic pathways, such as glycolysis. This group of enzymes includes racemases, epimerases, *cis–trans* isomerases, intramolecular oxidoreductases, intramolecular lyases and intramolecular transferases.

Alanine racemase is an enzyme dependent of pyridoxal 5′-phosphate and catalyzes the racemization of L- and D-alanine. Bacteria utilize alanine racemase for the synthesis of peptidoglycan of cell walls providing the D-enantiomer (Okubo et al. 1999).

Soda et al. (1988) developed a procedure for the synthesis of various D-amino acids by coupling the product generated by alanine racemase of *Geobacillus stearothermophilus* to different enzymes. However, the yield of the procedure is very low to generate an industrial method of chemical synthesis.

Fructose is used commercially in food and beverages industry. This sugar is found in many foods and has a high relative sweetness. It can be obtained by isomerization of glucose to fructose, a reversible reaction catalyzed by glucose/xylose isomerase. This process is very important in the production of high fructose syrup. Industrially, starch is used for production of fructose in a process that depends on hydrolysing the starch into highly concentrated glucose syrup, which is transformed in the presence of the isomerase to fructose. The classical process for production of fructose syrups employs alpha-amylase to liquefy starch and then glucoamylase to saccharify the hydrolyzed starch. This resulting product will be directed for isomerization (Gaily et al. 2013).

Many xylose isomerases have been isolated from mesophilic microorganisms, with an optimal pH in the range of 7.5–9.0. Higher temperatures and alkaline pHs leads to the formation of by-products in the resulting fructose solution. Therefore, industrial process for fructose is limited to 60 °C and requires neutral or slightly acidic pHs.

Geobacillus thermodenitrificans TH2 produces an extremely stable xylose isomerase retaining its activity after 96 h of incubation at 4 °C and 50 °C. However, the optimum temperature of this enzyme was 80 °C and optimal pH was 7.5, properties that are not compatible with the production of high-fructose corn syrups (Konak et al. 2014).

The genome of *Halothermothrix orenii* codes for a high thermostable xylose isomerase and the evaluation of its activity and stability under pH range 6.0–7.0 could offer a candidate enzyme to be used in the production of high-fructose corn syrups (Bhattacharya and Pletschke 2014). However, more studies are needed in order to determine its biotechnological application in this procedure.

Currently, there is no method for HTS format that allow us to screen any type of isomerase even for mesophilic organisms.

4.5 Extremophilic Lyases

Lyases catalyze the cleavage of C–C, C–O, C–N bonds by neither hydrolysis nor oxidation. The process of elimination cleaves these bonds, resulting in the formation of a double bond or a ring. This class of enzymes uses two substrates in one reaction direction, but only one substrate is involved in the other direction.

Alginate lyases degrade alginate through β-elimination of the glycosidic bond. They produce oligosaccharides with unsaturated uronic acid at the non-reducing terminus and unsaturated uronic acid monomers. Alginate oligosaccharides have interesting biological activities, because they stimulate the cytokines production. For these reason alginate lyases have gained much attention as biocatalysts for the production of functional oligosaccharides. Alginate lyases could be used as pharmaceuticals in cystic fibrosis patients due to they increase antibiotic killing of mucoid *Pseudomonas aeruginosa* (Kim et al. 2011). A high-alkaline, salt-activated alginate lyase was purified from the deep-sea bacterium *Agarivorans* sp. JAM-A1m. This enzyme has a high activity at alkaline pH (10.0), and its maximum can be shifted to 9.0 when 200 mM NaCl is added to the reaction buffer. This alginate lyase can produce mannuronate–guluronate and guluronate-rich fragments from alginate (Kobayashi et al. 2009). These fragments stimulate the growth of human endothelial cells and the secretion of cytotoxic cytokines from human macrophage.

Pectinolytic enzymes can be used in food and beverage industries and on retting processes of plant fibers. Pectin is an abundant structure component in plant cell walls and functions as a matrix holding cellulose and hemicellulose fibers. It is composed of a main chain of methyl-esterified α-1,4-D-polygalacturonate and highly branched rhamnogalacturonan, in which the latter can serve as a binding site for 1,4-linked side chains like arabinans, galactans or arabinogalactans (Kluskens et al. 2003). From these enzymes, pectate lyase is the pectin depolymerizing enzyme that cleaves α-1,4-galacturonosidic linkages of polygalacturonic acid (PGA) by a *trans*-elimination mechanism requiring Ca^{+2} ions for its activity (Kobayashi et al. 2000).

Thermostable pectate lyases have been described from the alkaliphilic *Bacillus sp*. P-4-N. This enzyme was very stable in a pH range from 5.0 to 11.5 (60 h at 50 °C) in the presence of 100 mM NaCl (Kobayashi et al. 2000). The thermostable pectate lyase purified from *T. maritima* is a highly thermoactive and thermostable enzyme, which optimal parameters are 90 °C and pH 9.0. It possesses a half-life of almost 2 h at 95 °C (Kluskens et al. 2003).

The use of pectate lyases for the removal of pectic substances from the wastewater could represent a cost-effective and environmentally friendly alternative treatment method. The use of psychrophilic enzymes is especially desirable for such degradation processes and would enable low energy treatment. Pectate lyases have been purified from the psychrophilic microorganisms *Pseudoalteromonas haloplanktis* strain ANT/505 (van Truong et al. 2001) and *Mrakia frigida* (Margesin et al. 2005). These enzymes were optimally active at 30 °C under alkaline conditions, being suitable to remove compounds derived from pectin.

There are not methods for the screening of extremophilic lyases using HTS technology. It is important to develop this kind of technology to screen new lyases and isomerases that have a great industrial potential.

4.6 Extremophilic Ligases

Ligases join two large molecules by forming a chemical bond. These enzymes are widely distributed in nature and their application has been restricted to analytical methods.

DNA ligases are enzymes required for cellular processes, such as DNA replication, DNA recombination and DNA repair. They catalyze the formation of a phosphodiester bond between adjacent 5′-phosphoryl and 3′-hydroxyl groups in double stranded DNA (Georlette et al. 2000).

DNA ligases are employed in molecular biology in standard protocols, for example, for the cloning of genes on expression vectors. Thermostable DNA ligases could be useful for the construction of sequencing primers by high temperature ligation of hexameric primers, the detection of trinucleotide repeats through repeat expansion detection, or DNA detection by circularization of oligonucleotides (Egorova and Antranikian 2005).

The DNA ligase gene from *Aquifex pyrophilus* was successfully expressed in *Escherichia coli* and purified to homogeneity. The activity of this DNA was higher at 65 °C and pH 8.0–8.6 for nick-closing activity. The enzyme retains more than 75 % of its activity after incubation at 95 °C for 60 min, whereas the half-lives at 95 °C for *T. aquaticus* and *E. coli* DNA ligases were 15 min and 5 min, respectively. The thermostable DNA ligase from *A. pyrophilus* was applied to repeat expansion detection and could be a useful enzyme in DNA diagnostics (Lim et al. 2001).

On the other hand, a psychrophilic DNA ligase has been reported *from P. haloplanktis*. This cold-adapted enzyme showed a higher catalytic efficiency at low and moderate temperatures than *E. coli* DNA ligase. *P. haloplanktis* DNA ligase is useful for investigation of the adaptation of enzymes at low temperatures, and offers a novel tool for biotechnology. Commercial DNA ligases have low activities at temperatures below 15 °C and require long incubation times. These conditions give the opportunity to the action of residual nucleases, interfering with the ligation reaction. An additional advantage of the cold-active ligase is also the fact that a relatively low inactivation temperature can be used, avoiding DNA denaturation (Georlette et al. 2000).

5 Concluding Remarks

Despite the inherent advantages of extremozymes the actual number of available extremophilic biocatalytic tools is low; hence several scientific challenges are waiting to be taken to fully display the potential of extremozymes. Although, molecular approaches allow access to the genes of interest in a rapid manner to clone and obtain large amount of proteins it is not possible by the sole use of genomic methodologies. Discern specific enzymatic properties such as: thermal stability, specific activity, among others, is only possible through a functional approach. Even more, it is important to stress the relevance of integrating the functional approach in what is call large-scale biotechnology, in order to potentiate the availability of biocatalytic tools able to perform under extreme industrial conditions.

Conflict of Interest Freddy Boehmwald, Patricio Muñoz, Patricio Flores, and Jenny M. Blamey declare that they have no conflict of interest.

References

Acker M, Auldn D (2014) Considerations for the design and reporting of enzyme assays in high-throughput screening applications. Perspect Sci 1:56–73

Adrio J, Demain A (2014) Microbial enzymes: tools for biotechnological processes. Biomolecules 4:117–139

Andexer J, Langermann J, Kragl U, Pohl M (2009) How to overcome limitations in biotechnological processes examples from hydroxynitrile lyase applications. Trends Biotechnol 27:599–607

Andrade C, Aguiar W, Antranikian G (2001) Physiological aspects involved in production of xylanolytic enzymes by deep-sea hyperthermophilic archaeon *Pyrodictium abyssi*. Appl Biochem Biotechnol 91–93:655–669

Antoine E, Rolland J, Raffin J, Dietrich J (1999) Cloning and over-expression in *Escherichia coli* of the gene encoding NADPH group III alcohol dehydrogenase from *Thermococcus hydrothermalis*. Characterization and comparison of the native and the recombinant enzymes. Eur J Biochem 264:880–889

Bachmann B (2014) Microbial genome mining for accelerated natural products discovery: is a renaissance in the making? J Ind Microbiol Biotechnol 41:175–184

Bae H, Kim K, Lee J, Song J, Kil E, Kim J, Kwon S (2009) Characterization of DNA polymerase from the hyperthermophilic archaeon *Thermococcus marinus* and its application to PCR. Extremophiles 13:657–667

Beers R, Sizer I (1952) A spectrophotometric method for measuring the breakdown of hydrogen peroxide by catalase. J Biol Chem 195:133–140

Bhattacharya A, Pletschke B (2014) Review of the enzymatic machinery of *Halothermothrix orenii* with special reference to industrial applications. Enzyme Microb Technol 55:159–169

Blumer-Schuette S, Brown S, Sander K, Bayer E, Kataeva I, Zurawski J, Conway J, Adams M, Kelly R (2014) Thermophilic lignocellulose deconstruction. FEMS Microbiol Rev 38:393–448

Bouzas T, Barros-Velázquez J, González T (2006) Industrial applications of hyperthermophilic enzymes: a review. Protein Pept Lett 13:645–651

Brooks H, Geeganage S, Kahl S, Montrose C, Sittampalam S, Smith M, Weidner J (2012) Basics of enzymatic assays for HTS. In: Sittampalam G, Coussens N, Nelson H et al (eds) Assay guidance manual. Eli Lilly & Company and The National Center for Advancing Translational Sciences, Indianapolis, Bethesda, pp 1–13

Brown-Pterson N, Salin M (1995) Purification and characterization of a mesohalic catalase from the halophilic bacterium *Halobacterium halobium*. J Bacteriol 177:378–384

Bulter T, Alcalde M, Sieber V, Meinhold P, Schlachtbauer C, Arnold F (2003) Functional expression of a fungal laccase in *Saccharomyces cerevisiae* by directed evolution. Appl Environ Microbiol 69:987–995

Castellano I, Di Maro A, Ruocco M, Chambery A, Parente A, Di Martino M, Parlato G, Masullo M, De Vendittis E (2006) Psychrophilic superoxide dismutase from *Pseudoalteromonas haloplanktis*: biochemical characterization and identification of a highly reactive cysteine residue. Biochimie 88:1377–1389

Cavicchioli R, Charlton T, Ertan H, Mohd Omar S, Siddiqui K, Williams T (2011) Biotechnological used of enzymes from psychrophiles. Microb Biotechnol 4:449–460

Chen B, Sayar A, Kaulmann U, Dalby P, Ward J, Woodley J (2007) Reaction modelling and simulation to assess the integrated use of transketolase and ω-transaminase for the synthesis of an aminotriol. Biocatal Biotransform 24:449–457

Chien A, Edgar D, Trela J (1976) Deoxyribonucleic acid polymerase from the extreme thermophile *Thermus aquaticus*. J Bacteriol 127:1550–1557

Coady T, Coffey L, O'Reilly C, Owens E, Lennon C (2013) A high throughput screening strategy for the assessment of nitrile-hydrolyzing activity towards the production of enantiopure β-hydroxy acids. J Mol Catal B Enzym 97:150–155

Correa-Llantén D, Amenábar M, Muñoz P, Monsalves M, Castro M, Blamey J (2014) *Alicyclobacillus* sp. strain CC2, a thermo-acidophilic bacterium isolated from Deception Island (Antarctica) containing a thermostable superoxide dismutase enzyme. Adv Pol Sci 25:92–96

Diaz R, Sabino E (1998) Accuracy of replication in the polymerase chain reaction. Comparison between *Thermotoga maritima* DNA polymerase and *Thermus aquaticus* DNA polymerase. Braz J Med Biol Res 31:1239–1242

Donadio S, Monciardini P, Sosio M (2009) Approaches to discovering novel antibacterial and antifungal agents. Methods Enzymol 458:3–28

Egorova K, Antranikian G (2005) Industrial relevance of thermophilic Archaea. Curr Opin Microbiol 8:649–655

Fernández-Arrojo L, Guazzaroni M, López-Cortés N, Beloqui A, Ferrer M (2010) Metagenomic era for biocatalyst identification. Curr Opin Biotechnol 21:725–733

Gaily M, Sulieman A, Abasaeed A (2013) Kinetics of a three-step isomerization of glucose to fructose using immobilized enzyme. Int J Chem Eng Appl 4:31–34

Georlette D, Jónsson Z, van Petegem F, Chessa J, van Beeumen J, Hübscher U, Gerday C (2000) A DNA ligase from the psychrophile *Pseudoalteromonas haloplanktis* gives insights into the adaptation of proteins to low temperatures. Eur J Biochem 267:3502–3512

Glaser R, Venus J (2014) Screening of *Bacillus coagulans* strains in lignin supplemented minimal medium with high throughput turbidity measurements. Biotechnol Rep 4:60–65

Guzik U, Hupert-Kocurek K, Wojcieszyńska D (2014) Immobilization as a strategy for improving enzyme properties – application to oxidoreductases. Molecules 19:8995–9018

Hallam S, Nelson H, Greger V, Perreault-Micale C, Davie J, Faulkner N, Neitzel D, Casey K, Umbarger M, Chennagiri N, Kramer A, Porreca G, Kennedy C (2014) Validation for clinical use of and initial clinical experience with a novel approach to population-based carrier screening using high-throughput, next-generation DNA sequencing. J Mol Diagn 2:180–189

He Y, Ma C, Xu J, Zhou L (2011) A high-throughput screening strategy for nitrile-hydrolyzing enzymes based on ferric hydroxamate spectrophotometry. Appl Microbiol Biotechnol 89: 817–823

Höhne M, Bornscheuer U (2012) Application of transaminases. In: Drauz K, Gröger H, May O (eds) Enzyme catalysis in organic synthesis, 3rd edn. Wiley-VCH Verlag GmbH & Co, Weinheim, pp 779–820

Hotzyme Project. The University of Copenhagen. http://hotzyme.com/project/

Houde A, Kademi A, Leblanc D (2004) Lipases and their industrial applications. Appl Biochem Biotechnol 118:155–170

Inglese J, Auld D, Jadhav A, Johnson R, Simeonov A, Yasgar A, Zheng W, Austin C (2006) Quantitative high-throughput screening: a titration-based approach that efficiently identifies biological activities in large chemical libraries. Proc Natl Acad Sci USA 31:11473–11478

Jaeger K (2004) Protein technologies and commercial enzymes: white is the hype-biocatalyst on the move. Curr Opin Biotechnol 15:269–271

Jemli S, Ayadi-Zouari D, Hlima H, Bejar S (2014) Biocatalyst: application and engineering for industrial purposes. Crit Rev Biotechnol 6:1–13

Johannes T, Woodyer R, Zhao H (2006) High-throughput screening methods developed for oxidoreductases. In: Reymond J (ed) Enzyme assays: high-throughput screening, genetic selection and fingerprinting. John Wiley & Sons, New York, pp 77–93

Johnston P, Soares K, Shinde S, Foster C, Shun T, Takyi H, Wipf P, Lazo J (2008) Development of a 384-well colorimetric assay to quantify hydrogen peroxide generated by the redox cycling of compounds in the presence of reducing agents. Assay Drug Dev Technol 4:505–518

Kaper T, van der Marel M, Euverink G, Dijkhuizen L (2003) Exploring and exploiting starch-modifying amylomaltases from thermophiles. Biochem Soc Trans 32:279–282

Kardashliev T, Ruff A, Zhao J, Schwaneberg U (2014) A high-throughput screening method to reengineer DNA polymerases for random mutagenesis. Mol Biotechnol 56:274–283

Kazuoka T, Oikawa T, Maraoka I, Kuroda S, Soda K (2007) A cold-active and thermostable alcohol dehydrogenase of a psychrotorelant from Antarctic seawater, *Flavobacterium frigidimaris* KUC-1. Extremophiles 11:257–267

Kim H, Lee C, Lee E (2011) Alginate lyase: structure, property, and application. Biotechnol Bioprocess Eng 16:843–851

King O, Li X, Sakurai M, Kawamura A, Rose N, Quinn A, Rai G, Mott B, Beswick P, Klose R, Oppermann U, Jadhav A, Heightman T, Maloney D, Schofield C, Simeonov A (2010) Quantitative high-throughput screening identifies 8-hydroxyquinolines as cell-active histone demethylase inhibitors. PLoS One 5:1–12

Klumpp M, Boettcher A, Becker D, Meder G, Blank J, Leder L, Forstner M, Ottl J, Mayr L (2006) Readout technologies for highly miniaturized kinase assays applicable to high-throughput screening in a 1536-well format. J Biomol Screen 11:617–633

Kluskens L, van Alebeek G, Voragen A, de Vos W, van der Oost J (2003) Molecular and biochemical characterization of the thermoactive family 1 pectate lyase from the hyperthermophilic bacterium *Thermotoga maritima*. Biochem J 370:651–659

Kobayashi T, Hatada Y, Suzumatsu A, Saeki K, Hakamada Y, Ito S (2000) Highly alkaline pectate lyase Pel-4A from alkaliphilic *Bacillus* sp. strain P-4-N: its catalytic properties and deduced amino acid sequence. Extremophiles 4:377–383

Kobayashi T, Uchimura K, Miyazaki M, Nogi Y, Horikoshi K (2009) A new high-alkaline alginate lyase from a deep-sea bacterium *Agarivorans* sp. Extremophiles 13:121–129

Konak L, Kolcuoglu Y, Ozbek E, Colak A, Ergenoglu B (2014) Purification and characterization of an extremely stable glucose isomerase from *Geobacillus thermodenitrificans* TH2. Appl Biochem Microbiol 50:25–29

Kunamneni A, Plou F, Ballesteros A, Alcalde M (2008) Laccases and their applications: a patent review. Recent Pat Biotechnol 2:10–24

Kunze M, Schmidt T, Fischer R, Commandeur U, Büchs (2012) Application of high-throughput screening methods for the development of improved enzymes for cellulose degradation. In: Abstracts of the ACHEMA conference 2012, ACHEMA, Frankfurt am Main, 18–22 June 2012

Lagarde D, Nguyen H, Ravot G, Wahler D, Reymond J, Hills G, Veit T, Lefevre F (2002) High-throughput screening of thermostable esterases for industrial bioconversions. Org Process Res Dev 6:441–445

Leis B, Angelov A, Liebl W (2013) Screening and expression of genes from metagenomes. Adv Appl Microbiol 83:1–68

Li Y, Schellhorn H (2007) Rapid kinetic microassay for catalase activity. J Biomol Tech 18:185–187

Li S, Yang X, Yang S, Zhu M, Wang X (2012) Technology prospecting on enzymes: applications, marketing and engineering. Comput Struct Biotechnol J 2:1–11

Liauda N, Navarroa D, Vidald N, Sigoillota JC, Raouche S (2014) High throughput automated colorimetric method for the screening of L-lactic acid producing microorganisms. MethodsX 1:254–257

Lim J, Choi J, Han S, Kim S, Hwang H, Jin D, Ahn B, Han Y (2001) Molecular cloning and characterization of thermostable DNA ligase from *Aquifex pyrophilus*, a hyperthermophilic bacterium. Extremophiles 5:161–168

Littlechild J (2011) Thermophilic archaeal enzymes and applications in biocatalysis. Biochem Soc Trans 39:155–158

Liu J, Yin M, Zhu H, Lu J, Cui Z (2011) Purification and characterization of a hyperthermostable Mn-superoxide dismutase from *Thermus thermophiles* HB27. Extremophiles 15:221–226

López-López O, Cerdán M, González M (2014) New extremophilic lipases and esterases from metagenomics. Curr Protein Pept Sci 15:445–455

Lorentzen M, Moe E, Jouve M, Willanssen N (2006) Cold adapted features of *Vibrio salmonicida* catalase: characterization and comparison to the mesophilic counterpart from *Proteus mirabilis*. Extremophiles 10:427–440

Lundberg K, Shoemaker D, Adams M, Short J, Sorge J, Mathur E (1991) High-fidelity amplification using a thermostable DNA polymerase isolated from *Pyrococcus furiosus*. Gene 198:1–6

Luo Y, Zheng Y, Jiang Z, Ma Y, Wei D (2006) A novel psychrophilic lipase from *Pseudomonas fluorescens* with unique property in chiral resolution and biodiesel production via transesterification. Appl Microbiol Biotechnol 73:349–355

MacDonald A, Tipton K (2014) Fifty-five years of enzyme classification: advances and difficulties. FEBS J 281:583–592

Margesin R, Fauster V, Fonteyne P (2005) Characterization of cold-active pectate lyases from psychrophilic *Mrakia frigida*. Lett Appl Microbiol 40:453–459

Marino G, Nitti G, Arcone M, Sannia G, Gambacorta A, De Rosa M (1988) Purification and characterization of aspartate aminotransferase from thermoacidophilic archaebacterium *Sulfolobus solfataricus*. J Biol Chem 263:12305–12309

Miyazaki K (2005) A hyperthermophilic laccase from *Thermus thermophiles* HB27. Extremophiles 9:415–425

Moger J, Gribbon P, Sewing A, Winlove P (2006) The application of fluorescence lifetime readouts in high-throughput screening. J Biomol Screen 7:765–772

Mueller P, Egorova K, Vorgias C, Boutou E, Trauthwein H, Verseck S, Antranikian G (2006) Cloning, overexpression, and characterization of a thermoactive nitrilase from the hyperthermophilic archaeon *Pyrococcus abyssi*. Protein Expr Purif 47:672–681

Muñoz P, Correa-Llantén D, Blamey J (2013) Production, purification and partial characterization of four lipases from a thermophile isolated from Deception Island. Lipids 48:527–533

Muñoz P, Correa-Llantén D, Blamey J (2015) Ionic liquids increase the catalytic efficiency of a lipase (Lip1) from an Antarctic thermophilic bacterium. Lipids 50:49–55

Noah J (2010) New developments and emerging trends in high-throughput screening methods for lead compound identification. Int J High Throughput Screen 1:141–149

Nonaka K, Yoon K, Ogo S (2014) Biochemical characterization of psychrophilic Mn-superoxide dismutase from newly isolated *Exiguobacterium* sp. OS-77. Extremophiles 18:363–373

Okubo Y, Yokoigawa K, Esaki N, Soda K, Kawai H (1999) Characterization of psychrophilic alanine racemase from *Bacillus psychrosaccharolyticus*. Biochem Biophys Res Commun 256:333–340

Omtaa W, Egana D, Spruitb M, Brinkkemperb S (2012) Information architecture in high throughput screening. Procedia Technol 5:696–705

Peake I (1989) The polymerase chain reaction. J Clin Pathol 42:673–676

Radianingtyas H, Wright P (2003) Alcohol dehydrogenases from thermophilic and hyperthermophilic archaea and bacteria. FEMS Microbiol Rev 27:593–616

Ravot G, Wahler D, Favre-Bulle O, Cilia V, Lefevre F (2003) High throughput discovery of alcohol dehydrogenases for industrial biocatalysis. Adv Synth Catal 345:691–694

Saha K, Maity S, Roy S, Pahan K, Pathak R, Majumdar S, Gupta S (2014) Optimization of amylase production from *Bacillus amyloliquefaciens* (MTCC 1270) using solid state fermentation. Int J Microbiol 2014:7640–7646

Salameh M, Wiegel J (2007) Lipases from extremophiles and potential for industrial applications. Adv Appl Microbiol 61:253–283

Schallmey M, Frunzke J, Eggeling L, Marienhagen J (2014) Looking for the pick of the bunch: high-throughput screening of producing microorganisms with biosensors. Curr Opin Biotechnol 26:148–154

Schmitz J, Daniel A, Collin M, Schuch R, Fischetti V (2008) Rapid DNA library construction for functional genomic and metagenomic screening. Appl Environ Microbiol 74:1649–1652

Shafey H, Bahashwan S, Alghaithy A, Ghanem S (2010) Microbial superoxide dismutase enzyme as therapeutic agent and future gene therapy. Cur Res Technol Edu Top Appl Microbiol Microbial Biotech 435–440

Soda K, Tanaka H, Tanizawa K (1988) Thermostable alanine racemase and its application to D-amino acid synthesis. Ann N Y Acad Sci 542:375–382

Sooch B, Kauldhar B, Puri M (2014) Recent insights into microbial catalases: isolation, production and purification. Biotechnol Adv 32:1429–1447

Tsigos I, Velonia K, Smonou I, Bouriotis V (1998) Purification and characterization of an alcohol dehydrogenase from the Antarctic psychrophile *Moraxella* sp. TAE123. Eur J Biochem 254:356–362

USDA (2008) U.S. Biobased Products: Market Potential and Projection Through 2025. United States Department of Agriculture (OCE-2008-1) www.usda.gov/oce/reports/energy/index.htm

van Truong L, Tuyen H, Helmke E, Binh L, Schweder T (2001) Cloning of two pectate lyase genes from the marine Antarctic bacterium *Pseudoalteromonas haloplanktis* strain ANT/505 and characterization of the enzymes. Extremophiles 5:35–44

Walton C, Chica R (2013) A high-throughput assay for screening L- or D-amino acid specific aminotransferase mutant libraries. Anal Biochem 441:190–198
WEF (2010) The Future of Industrial Biorefineries. World Economic Forum. REF: 210610. http://www3.weforum.org/docs/IP/2012/CH/WEF_CollaborativeInnovation_FutureofIndustrialBiorefineries.pdf
Yun Y, Lee Y (2004) Purification and some properties of superoxide dismutase from *Deinococcus radiophilus*, the UV-resistant bacterium. Extremophiles 8:237–242
Zhang J, Chung T, Oldenburg K (1999) A simple statistical parameter for use in evaluation and validation of high throughput screening assays. J Biomol Screen 4:67–73
Zhu Y, Zhang Z, Zhang M, Mais D, Wang M (2010) High throughput screening for bioactive components from traditional Chinese medicine. Comb Chem High T Screen 13:837–848

Chapter 12
Lipolytic Extremozymes from Psychro- and (Hyper-)Thermophilic Prokaryotes and Their Potential for Industrial Applications

Skander Elleuche, Carola Schröder, and Garabed Antranikian

1 Introduction

Water insoluble lipids and fats compose a major part of earth's natural biomass. Lipolytic enzymes (lipases and esterases) are autocatalytic triacylglycerol acyl hydrolases catalyzing the release of free fatty acids and glycerol, monoacylglycerol and diacylglycerol from triglycerides (Fig. 12.1a). Lipases (EC 3.1.1.3) prefer water-insoluble long-chain acyl esters (>10 carbon atoms), while esterases (EC 3.1.1.1) catalyze the hydrolysis and synthesis of water-soluble short-chain fatty acids (<10 carbon atoms) (Fig. 12.2) (Bornscheuer 2002). In literature enzymes are mostly named "esterase" or "lipase" due to their preferred substrate. However, authors may refer to enzymes with preference for middle-chains length (C_{10}) as esterases or lipases, and there is no current agreement as to what chain length substrate preference delimits classification as a lipase versus esterase (Fuciños et al. 2011, 2014; Chow et al. 2012). Hence, regarding the differentiation based on substrate chain length, lipases and esterases can even be structurally engineered and transformed into one another.

Lipases are hydrophobic in nature and exhibit large hydrophobic surfaces at the active site region. Another feature of lipases is their interfacial activation, which is responsible for higher activity towards substrates at a water-micelle interface compared to substrate compounds dissolved in liquid solutions. Moreover, they do not only hydrolyze but also modify ester bonds including reverse esterification, and interesterification as well as aminolysis, alcoholysis (transesterification) and acidolysis (Wicka et al. 2013). Lipolytic enzymes usually exhibit enantio-, chemo- and regioselectivity making them superior over organic chemistry methods and valuable

S. Elleuche (✉) • C. Schröder • G. Antranikian
Institute of Technical Microbiology, Hamburg University of Technology (TUHH), Kasernenstr. 12, D-21073 Hamburg, Germany
e-mail: skander.elleuche@tuhh.de

P.H. Rampelotto (ed.), *Biotechnology of Extremophiles*, Grand Challenges in Biology and Biotechnology 1, DOI 10.1007/978-3-319-13521-2_12

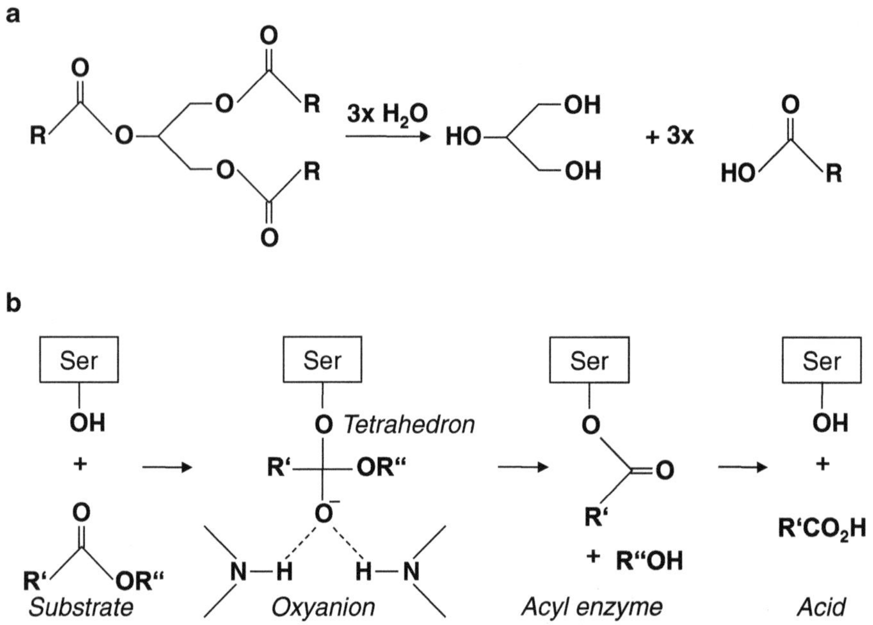

Fig. 12.1 Reaction scheme depicting lipolytic enzyme catalyzed hydrolysis of ester bonds to release three fatty acid residues and a glycerol (**a**) and illustration of reaction steps (**b**)

candidates for versatile applications. The physiological role of lipases in prokaryotes and eukaryotes is mainly the mobilization of lipids. These enzymes exhibit broad substrate specificity and solvent tolerance (organic solvents, ionic liquids, non-conventional solvents such as two-phase aqueous organic systems or microemulsion-based organogels) making them highly relevant for industrial applications including detergent, food and chemical industry as well as pharmaceutical, agrochemical and flavour sciences. Lipases also play a dominant role in the production of biofuels from lipids (Joseph et al. 2008; Hwang et al. 2014). Nowadays, lipolytic enzymes are considered to be one of the most important biocatalysts for biotechnological applications (Fig. 12.3) (Wicka et al. 2013). In particular, enzymes from extremophiles offer versatile ranges of applications including organic synthesis of chiral compounds, additives in food industries and in the biomedicine.

2 General Characteristics and Enzymatic Properties of Lipolytic Enzymes

Lipolytic enzymes belong to a large family of phylogenetically related biocatalysts. These enzymes have been identified in pro- and eukaryotic microorganisms, plants and animals. Most lipases and esterases consist of a compact and minimal domain

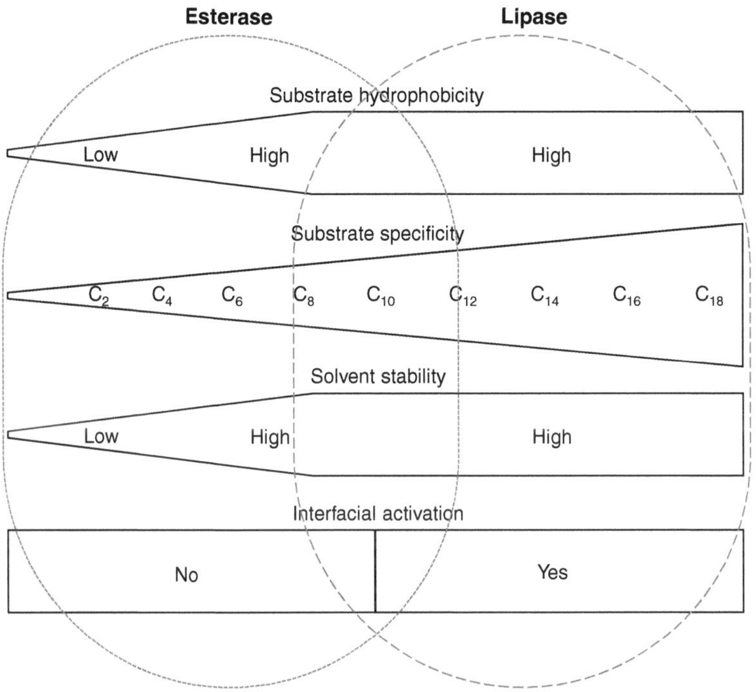

Fig. 12.2 Schematic representation of esterase and lipase properties with an overlap where features cannot be clearly assigned (Modified from Bornscheuer 2002)

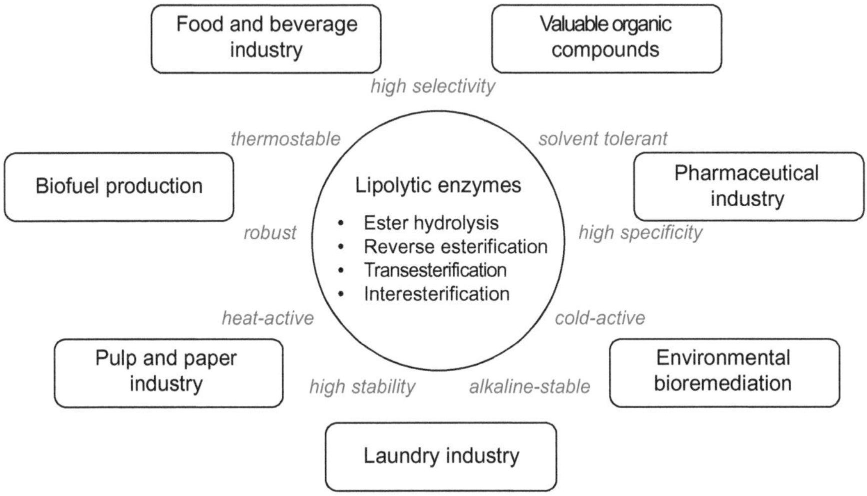

Fig. 12.3 Lipolytic enzyme catalyzed reactions, features and several applications

that can be clearly assigned to the α/β-hydrolase fold proteins often lacking additional structural domains. This typical fold consists of two layers of amphiphilic α-helices that enwrap a central hydrophobic core composed of eight β-sheets. However, there are some recent descriptions of variations including insertions of additional b-sheets in the conserved structural element (Arpigny and Jaeger 1999; Shaw et al. 2002; Siew et al. 2005). Their serine-protease like catalytic domain consists of a Ser-Asp-His triad, which is widespread in all serine hydrolases and is usually either solvent-exposed or closed by a flexible lid structure. The lid is composed of one or two α-helices that flips open in an activated conformation thereby guaranteeing access of a substrate to the active enzyme site. The lid was described as exclusive characteristic of lipases due to the movement of this domain in the presence of water-lipid interface (interfacial activation) (Bornscheuer 2002). The serine in the catalytic region acts as the nucleophile, histidine as the base and aspartate or glutamate as the acidic residue (Bornscheuer 2002). Moreover, this serine is embedded in a highly conserved pentapeptide Gly-X-Ser-X-Gly that is typical for the α/β-hydrolase superfamily (Bornscheuer 2002).

A classification for lipolytic enzymes based on amino acid sequence comparisons and some biological properties was proposed in 1999 (Arpigny and Jaeger 1999). Eight families (I–VIII) were specified with family I containing true lipases with the catalytic serine being embedded in the highly conserved motif Gly-His-Ser-X-Gly. Further six subgroups (I.1–I.6) were determined. Subfamily I.3 comprises enzymes from *Pseudomonas* sp. and others, while lipolytic enzymes from *Geobacillus* sp. belong to subfamily I.5 (Fig. 12.4). In family II (GDSL-family)

```
      Family I
                  ●                     ●                          ●
Pfl   GLSGKDVVVSGHSLGGLAVNSMAD 218   YENDPVFRALDGSSF 266   VNLPTWVSHLPT 316  | Family I.3
Psp   GLSGKDVLVSGHSLGGLAVNSLAD 218   YENDPVFRALDGSSF 266   VNLPTWVSHLPT 316  |
Pfr   QVGAQRVNLIGHSQGALTARYVAA 94    RENDGMVGRFSSHLG 214   IRSDYPLDHLDT 236
Gza   LKRGGRIHIIAHSQGGQTARMLVS 152   LENDGIVNTVSMNGP 356   DMGTYNVDHLEI 389  | Family I.5
Gth   LKRGGRIHIIAHSQGGQTARMLVS 152   LENDGIVNTISMNGP 356   DMGTYNVDHLEI 389  |
                 ** *                  ***                          **

      Family IV
          ----                         ●                       ●                  ●
Mse   LVYYHGGG-FVFGS 84   AVAGDSAGGNLSAVV 159   LVITAEYDPLRDQGE 251   QGMIHGFLSFY 280
Pca   VVYYHGGG-FVLGS 91   AVAGDSAGGNLAAVT 166   LVITAEYDPLRDEGE 261   NGVIHGFVNFY 290
Psp   LVFFHGGG-FVMGN 89   ALAGDSAGGNLALAV 164   TLITAEFDPLRDEGE 257   EGMIHGFISMA 286
Sac   LVFYHGGG-FVFGD 89   VVAGDSAGGNLAAVV 163   LVITAEYDPLRDEGE 255   DGMIHGFMTMP 284
Uba   VVYIHGGGPFVYGQ 70   QKTGDSAGGNLAAVV 175   VIITAELDPLRDQGE 248   NGMIHGADVIF 285
       *  **** ** *          ********             **** ***** *           ***
```

Fig. 12.4 Partial amino acid sequence alignments of conserved motifs of lipolytic enzymes assigned to family I, with two sequences grouped into subfamily I.3 and I.5 respectively, and family IV. Family I enzymes are depicted from *Pseudomonas fluorescens* (*Pfl,* AY694785), *Pseudomonas* sp. YY31 (*P*sp, AB642679), *Pseudomonas fragi* (*Pfr*, AJ250176), *Geobacillus zalihae* T1 (*Gza*, AY260764) and *Geobacillus thermoleovorans* YN (*Gth*, DQ298518). Family IV enzymes are depicted from *Metallosphaera sedula* DSM5348 (*Mse*, YP001191160), *Pyrobaculum calidifontis* (*Pca*, AB078331), *Pseudomonas* sp. B11-1 (*P*sp, AF034088), *Sulfolobus acidophilus* DSM10332 (*Sac*, AEW03609) and an uncultured bacterium (Uba, EF563989). Residues participating in the catalytic triad (•) and in oxyanion hole formation are highlighted (–)

Gly-Asp-Ser-(Leu) replaces the pentapeptide, while family III contains extracellular lipases from species of the genera *Streptomyces* and *Moraxella*. Family IV comprises the typical serine encompassing motif Gly-Asp-Ser-Ala-Gly-(Gly). This family is also described as the HSL (hormone-sensitive lipase) family due to high similarities to mammalian HSL. Lipolytic enzymes from psychrophiles (growth at 0–30 °C), mesophiles (30–45 °C) and thermophiles (50–80 °C) as well as from hyperthermophilic archaea (80–110 °C) could be assigned to HSL (Fig. 12.4). The pentapeptide is modified in family V to give Gly-X-Ser-(Met)-Gly-Gly, in family VI to Gly-Phe-Ser-Gln-Gly, in family VII to Gly-Phe-Ser-Gln-Gly-Gly and in family VIII to Ser-X-X-Lys (Arpigny and Jaeger 1999). In addition, other conserved amino acid arrangements were described for some of the different families. Since 1999 this classification has been extended by extensive elucidation of metagenomic sequences coding for novel lipolytic hydrolases showing low similarities to known enzymes (Kim et al. 2008; Nacke et al. 2011). Newly proposed families were rarely specific with regard to motifs or family names. Thereupon a continuation of the established family classification was proposed with family IX (PhaZ7), X (EstD), XI (LipG), XII (LipEH166), XIII (Est30) and XIV (EstA3). The conserved pentapeptides of these families were reported to be composed of Ala-His-Ser-Met-Gly, Gly-His-Ser-Leu-Gly, Gly-His-Ser-Leu-Gly-Gly, Gly-His-Ser-Leu-Gly, Gly-Leu-Ser-Leu-Gly-Gly and Cys-His-Ser-Met-Gly, respectively (Rao et al. 2011). Simultaneously, family X was identified as the LipR-cluster and consists of enzymes with the pentapeptide Gly-Tyr-Ser-Gly-Gly (Bassegoda et al. 2012b). Another new family (XV) was proposed for lipolytic enzymes exhibiting the serine-embedding motif Ser-His-Ser-Gln-Gly (Bayer et al. 2010). Subsequently, family XVI (Bacterial_Est97) further expanded the classification (Lenfant et al. 2013). Moreover, an alignment of proteins from different *Thermus* species (growth at 50–70 °C) resulted in the discovery of the putative pentapeptide Gly-Cys-Ser-Ala-Gly probably representing yet another family (Fuciños et al. 2011). These publications reflect the large number and the diversity of lipolytic enzymes from mesophiles and extremophiles.

The catalytic mechanism of α/β-hydrolases is composed of five reaction steps, beginning with the binding of an ester substrate and formation of a tetrahedral intermediate formed by catalytic serine mediated nucleophilic attack. An oxyanion hole is stabilized by two to three hydrogen bonds before the ester bond is cleaved. Afterwards, the alcohol moiety is released from the enzyme and finally the acyl enzyme is hydrolyzed (Joseph et al. 2008) (Fig. 12.1b). Among the most important properties of lipolytic enzymes for industrial applications are their thermostability and temperature range for optimal activity. Moreover, different lipolytic enzymes exhibit versatile properties with regard to their pH optima, tolerance towards detergents and metal ions, kinetics, fatty acid specificity and positional specificity. Structural modifications that are relevant for enzyme flexibility at certain conditions were investigated in cold- and heat-active lipolytic enzymes (extremozymes). In this context, it was hypothesized that lysine residues are replaced by arginine in enzymes from psychrophiles. Moreover, the number of prolines, loop structures, disulphide and salt bridges, aromatic-aromatic and hydrophobic interactions is reduced to adapt for cold environments and thermolability. In addition, an increased number of glycine residues was speculated to be important for local mobility of peptide regions (Joseph et al. 2008). It is also well accepted that the flexible catalytic site of cold-active enzymes is more susceptible to heat-inactivation compared to the complete enzyme structure, while biocatalysts from thermo- and most mesophiles usually display highest activity near the protein denaturation temperature (Feller and Gerday 2003; Wi et al. 2014).

3 Lipolytic Enzymes from Psychrophiles

Psychrophiles are microorganisms that populate cold environments such as sea ice, deep sea, and snowfields. Their enzymes function more effectively at cold temperatures compared to their counterparts from meso- and thermophilic microorganisms (Elleuche et al. 2014). The most studied lipase from a psychrophilic microorganism is the highly enantioselective enzyme CalB, which has been isolated from the hemiascomycetous yeast *Candida antarctica*. CalB is only mentioned for completeness due to its eukaryotic origin and because of its importance as a model lipase (Joseph et al. 2008).

3.1 Diversity of Cold-Active Esterases and Lipases

Although most of earth's biosphere is cold and inhabited by psychrophilic microorganisms, only a few cold-active esterases and lipases have been isolated from bacterial strains thriving in permanently cold sea ice, soil or glaciers, in the deep sea or mountain regions (Al Khudary et al. 2010; Wu et al. 2013; Wi et al. 2014; Parra et al. 2015; Tchigvintsev et al. 2015). Other cold-active bacterial lipolytic enzymes were isolated from food samples e.g. *Serratia marcescens* lipase (Abdou 2003).

Most of the lipolytic enzymes that are considered to be "cold-active", display optimal activity around 30 °C and are rapidly inactivated at temperatures above 45 °C, but there are some exceptions, e.g. a lipase from the Alaskan psychrothrophic bacterium *Pseudomonas* sp. B11-1 that is optimally active at 45 °C (Choo et al. 1998). Moreover, such biocatalysts often retain activity around the freezing point of water. A cold-active lipase from a metagenomic library, derived from oil contaminated soil, retained activity even at −5 °C (Elend et al. 2007). While most cold-active lipolytic enzymes were purified and characterized from their natural hosts, there are sparse descriptions of heterologously expressed genes encoding lipolytic enzymes (Table 12.1). To provide cold-active lipolytic enzymes in quantitative amounts, recombinant enzymes have been produced in *E. coli* and other mesophilic heterologous hosts (Feller et al. 1996; Trincone 2011). Modern recombinant DNA-technology provides a tool to increase enzyme production of isozymes that can only be produced and purified in low yield. Cloning of lipase-encoding genes from psychrotrophic microorganisms to be expressed in *E. coli* had already been reported at the end of the 1980s and beginning of 1990s, when lipases from *P. fragi* and *Moraxella* TA144 were cloned and sequenced (Aoyama et al. 1988; Feller et al. 1991). Comparison

Table 12.1 Recombinant lipolytic enzymes from psychrophilic bacteria

Sources	T_{opt} (°C)	pH_{opt}	MW (kDa)	Preferred substrate	References
Esterases[a]					
Uncultured bacterium (Lipo1)	10	7.5	35.6	*p*NP-butyrate (C_4)	(Roh and Villatte 2008)
Bacillus pumilus ArcL5 (BpL5)	20	9.0	19.2	*p*NP-caprylate (C_8)	(Wi et al. 2014)
Uncultured bacterium (CHA2)	20	11	34.7	*p*NP-propionate (C_3)	(Hu et al. 2012)
Pseudoalteromonas sp. 643A (EstA)	20	7.5	23.0	*p*NP-butyrate (C_4)	(Dlugolecka et al. 2009)
Pseudoalteromonas arctica (EstO)	25	7.5	44.1	*p*NP-butyrate (C_4)	(Al Khudary et al. 2010)
Psychrobacter pacificensis (Est10)	25	7.5	24.6	*p*NP-butyrate (C_4)	(Wu et al. 2013)
Pseudomonas fragi (rPFL)	29	8.0	32.0	Tributyrin (C_4)	(Alquati et al. 2002)
Pseudomonas fluorescens (LipA)	30	9.5	42.0	(S)-ketoprofen ethyl ester (C_3)	(Choi et al. 2003)
Rhodococcus sp. (RhLip)	30	11.0	38.0	*p*NP-butyrate (C_4)	(De Santi et al. 2014)
Uncultured bacterium (LipA)	35	8.0	32.2	*p*NP-butyrate (C_4)	(Couto et al. 2010)
Pseudomonas sp. B11-1 (LipP)	45	8.0	33.7	*p*NP-butyrate (C_4)	(Choo et al. 1998)
Lipases[a]					
Pseudomonas fluorescens (LipB68)	20	8.0	50.2	*p*NP-caprate (C_{10})	(Luo et al. 2006)
Photobacterium lipolyticum M37 (M37 lipase)	25	8.0	38.0	*p*NP-caprate (C_{10})	(Ryu et al. 2006)
Uncultured bacterium (rEML1)	25	8.0	33.6	Trilaurin (C_{12})	(Jeon et al. 2009)
Colwellia psychrerythraea 34H (CpsLip)	25	7.0	35.0	*p*NP-laurate (C_{12})	(Do et al. 2013)
Pseudomonas sp. YY31 (LipYY31)	25–30	8.0	49.5	*p*NP-caprate (C_{10})	(Yamashiro et al. 2013)
Pseudomonas sp. TK-3 (LipTK-3)	25–30	8.0	50.1	*p*NP-caprate (C_{10})	(Tanaka et al. 2012)
Psychrobacter sp. 7195 (LipA1)	30	9.0	35.2	*p*NP-myristate (C_{14})	(Zhang et al. 2007)
Pseudomonas sp. 7323 (LipA)	30	9.0	64.4	*p*NP-caprate (C_{10})	(Zhang and Zeng 2008)
Uncultured bacterium (LipCE)	30	7.0	53.2	*p*NP-caprate (C_{10})	(Elend et al. 2007)
Pseudomonas sp. KB700A (KB-Lip)	35	8.0–8.5	49.9	*p*NP-caprate (C_{10})	(Rashid et al. 2001)

[a]Enzymes were assigned to the groups of esterases or lipases based on their substrate preferences

with wildtype enzymes revealed that such cold-active lipases conserved their main properties as recombinant proteins (Feller et al. 1991). Nowadays, cold environments have been shown to be a rich source of diverse cold-adapted enzymes. To identify novel lipolytic enzymes from cold environments, activity-based screenings were established starting with gene expression induced by a lipid-containing feedstock such as olive or soybean oil, milk, triolein, tributyrin or tricaprylin (Henne et al. 2000; Joseph et al. 2008; Wi et al. 2014). In a recent study, 23 unique active clones were isolated by screening 274,000 clones from gene libraries that were prepared from different seawater samples collected in the Barents Sea (Russia) and the Mediterranean Sea (Italy) with temperatures between 3 °C and 15 °C at the sampling sites. The catalytic properties of five highly active carboxylesterases were examined, displaying broad biochemical versatility including temperature optima at 15 °C or 30 °C and substrate preferences for *para*-nitrophenol- (*p*NP-) valerate (C_5), *p*NP-propionate (C_3) or α-naphthyl-propionate (C_3), and *p*NP-octanoate (C_8), respectively (Tchigvintsev et al. 2015). A different strategy was recently applied to identify lipase-encoding genes from Antarctic seawater. Consensus-degenerate hybrid oligonucleotide primers were designed to amplify novel genes from isolated genomic DNA, allowing the identification of a typical cold-active lipase that prefers *p*NP-caproate (C_{10}) at temperatures between 15 °C and 25 °C (Parra et al. 2015).

Among the members with the lowest temperature optimum is an esterase (Lipo1) from an uncultivated microorganism that is optimally active at 10 °C (Roh and Villatte 2008). Moreover, Lipo1 displayed a higher stability at 10 °C compared to 40 °C, making it a potential candidate for applications in organic chemistry. Screening of a genomic library of the deep sea psychrotrophic bacterium *Psychrobacter pacificensis*, which was isolated from sediments in the Gulf of Mexico, revealed the existence of a cold-adapted and salt-tolerant esterase. The enzyme Est10 was optimally active at 25 °C and retained 55 % of its activity at 0 °C. Moreover, the catalytic activity and stability was even increased in the presence of high salt concentrations (between 2 and 5 M NaCl). As a typical esterase, this enzyme preferred *p*NP-butyrate (C_4) as substrate (Wu et al. 2013). Besides lipolytic enzymes from "true" psychrophilic bacteria, there are a number of candidates from mesophiles that exhibit cold-active enzymes including several lipases from different species of the genus *Bacillus* e.g. *Bacillus* sp. HH-01 (Kamijo et al. 2011). A further example is the cold-active esterase EstC from *Streptomyces coelicolor* A3(2). Genome mining revealed that this bacterium has at least 50 genes coding for putative lipolytic enzymes. EstC preferred *p*NP-valerate (C_5) as substrate and was optimally active at 35 °C. Moreover, this hydrolase retained 25 % of its activity when incubated at 10 °C (Brault et al. 2012).

3.2 Improving Enzymatic Properties of Psychrophiles by Genetic Engineering

Cold-active enzymes represent a promising target to increase thermal robustness or efficiency by molecular biology techniques including directed evolution and rational protein design (Bassegoda et al. 2012a; Joshi and Satyanarayana 2015). The thermolabile nature of cold-active enzymes is highly amenable to sequence and structure modifications. Site-directed mutagenesis resulting in a single amino acid mutation (T103G) increased CalB's stability at high temperatures, but went along with 50 % reduced activity at the same time (Patkar et al. 1998). The lipolytic enzyme BpL5 from the arctic psychrophile *Bacillus pumilus* ArcL5 was optimally active at 20 °C and retained around 85 % of its activity at 5 °C. Site-directed mutagenesis of a serine based on structural model predictions significantly improved the catalytic activity for *p*NP-caprylate and tricaprylin (C_8), without affecting pH and temperature optima (Wi et al. 2014). In another approach, directed evolution was applied to enhance the organic solvent-stability of a cold-active lipase, Lip9, from *Pseudomonas aeruginosa* LST-03. This microorganism has tolerance of organic solvents itself and secretes a highly stable enzyme that has been produced in recombinant form in *E. coli*. In a first approach, Lip9 was purified from inclusion bodies and refolded by the assistance of a natural lipase-specific foldase from the same bacterium (Ogino et al. 2007). Enhancement of enzyme stability against organic solvents was evaluated by a developed screening assay using dimethyl sulfoxide (DMSO) in addition to the substrate tri-*n*-butyrin. Identified candidates were sequenced and tested with different organic solvents. Using this approach, the mutant enzyme LST-03-R65 was identified displaying a 9- to 11-fold increased stability in the presence of cyclohexane and *n*-decane. This enzyme accumulated amino acid mutations on the surface of the protein and it has been speculated that such amino acid substitutions might be beneficial to prevent organic solvents from penetrating the protein's interior (Kawata and Ogino 2009). Most efforts were undertaken to improve thermal stability of enzymes, because thermal denaturation is the major route for enzyme inactivation under industrial process conditions. In good agreement with mutations mediating stability towards organic solvents, amino acid residues on the protein surface were also shown to be relevant for thermal robustness. A lipase variant from *Bacillus subtilis* with nine amino acid substitutions displayed a 15 °C increased melting temperature in addition to a 20 °C shifted temperature optimum (Ahmad et al. 2008). The psychrophilic bacterium *Pseudomonas fragi* produced a cold-active lipase, PFL, with a temperature optimum at 29 °C (Alquati et al. 2002). PFL is even highly active and stable at 10 °C, but completely inactivated when incubated for short time periods at moderate temperatures, i.e. 19 min at 42 °C. Error-prone polymerase chain reaction led to the development of an enzyme variant with four amino acid substitutions that exhibited a fivefold increase in half-life at 42 °C and a 10 °C shift of the temperature optimum (Gatti-Lafranconi et al. 2008). Another strategy to predict stabilizing mutations would be to compare lipolytic enzymes from psychrophiles with their homologous meso- or thermophilic counterparts (Bassegoda et al. 2012a). In case of the multidomain carboxylesterase EstO from the arctic psychrophile *P. arctica*, it has been shown that the deletion of the non-catalytic OsmC domain resulted in an increased thermal stability of the functionally active α/β-hydrolase fold domain. Detailed investigation of the primary and secondary structure in combination with characterization experiments revealed that the OsmC domain is probably important for flexibility and adaptation to low temperatures (Al Khudary et al. 2010; Elleuche et al. 2011). Such genetically modified enzyme candidates are attractive for biotechnological applications in diverse industrial processes (Fig. 12.3).

3.3 Applications of Cold-Active Lipolytic Enzymes

Unusual specificities and high activity at low temperatures offer opportunities for cold-active lipolytic enzymes to be exploited and applied in versatile industrial fields. Lipases and esterases from psychrophilic microorganisms are mainly of interest for use in the organic synthesis of fragile chiral compounds, but they are also applied as additives in food industries (cheese manufacture, bakery) and as detergents (energy saving, cold washing) as well as in environmental biotransformations (bioremediations), biomedicine (pharmaceuticals) or molecular biology approaches (Joseph et al. 2008; Joshi and Satyanarayana 2015). Moreover, lipases were also shown to be of interest to prevent inclusion bodies formation in psychrophilic host organisms (Feller et al. 1996). Development of economically feasible industrial applications of cold-active lipolytic enzymes is currently hindered in some instances by their low specific activity, poor recyclability, and unavailability in purified form. However, there are several examples of cold-active lipolytic enzymes that are established in industrial processes, or that have been patented for diverse applications or that have been investigated with regard to their potential for industrial biotechnology. Such an application not only saves energy by using lower temperatures, but also allows for the reduction of the amount of chemicals and detergents that are harmful to the environment, i.e. enzymatic desizing of materials is environmentally favourable compared to the traditional utilization of acidic and oxidizing chemical compounds. Moreover, non-aqueous solvents offer opportunities for the biotechnological production of valuable fine chemicals catalyzed by lipolytic enzymes. This has been shown to be useful in the modification of oils and fats and for the production of sugar-based polymers or to produce surfactants, emulsifiers, structured lipids and wax esters or to synthesize fragrance compounds (Joseph et al. 2008). The cold-active prototype enzyme for industrial applications is CalB from the hemiascomycetous yeast *C. antarctica* that has been patented for applications in food, chemical and pharmaceutical applications. Interestingly, this enzyme is highly thermostable in non-aqueous solutions when immobilized on solid carriers (Koops et al. 1999).

In principle, lipases are most successful as ingredients in detergent solutions for laundry and in household dishwaters. Lipolytic enzymes applicable in detergents remain active and stable at alkaline conditions (pH 10–11) and at moderate temperatures (Wicka et al. 2013). Novozymes launched the first genetically engineered industrial enzyme – Lipolase® from the fungus *Thermomyces lanuginosa* produced in the filamentous ascomycete *Aspergillus oryzae* – in 1988. This enzyme is applicable in the laundry industry to remove clothing stains such as fats, butter, lipstick, sauces and salad oil. Furthermore, it has been engineered and Novozyme came up with second and third generation variants that have also found their way to the market in different detergent formulations. Another well-known example for a washing enzyme is Lipomax®, an engineered variant of *Pseudomonas alcaligenes* lipase that has been introduced by Gist-Brocades in 1995 (Joshi and Satyanarayana 2015). Moreover, lipolytic enzymes were used as additives in bleaching compositions, in contact lens cleaning or in liquid leather cleaning (Joseph et al. 2008).

It is also desirable in the food industry to carry out reactions at low temperatures. Thereby, contaminations and changes in the composition of ingredients and flavour can be avoided. Especially lipases from mould have been used to improve traditional chemical processes in food manufacture. The potential of lipolytic enzymes of prokaryotic origin has also been evaluated to improve the aroma or enhance the flavour of food, e.g. a lipase from psychrotrophic *Pseudomonas* strain P38 catalyzes ester synthesis of a flavouring compound with an organic solvent phase at low temperatures (Tan et al. 1996). Moreover, lipolytic enzymes are important to enrich manifold unsaturated fatty acids from animal and plant lipids or to change the position or replace the fatty acid attached to the glycerol backbone (Wicka et al. 2013).

Cold-active lipases have also great potential for the production of biodiesel via transesterification reactions. An enzyme from the psychrophilic bacterium *Pseudomonas fluorescens* B68, LipB68, was highly active at low temperatures and was capable of producing 92 % of biodiesel from soybean oil after 120 h of incubation at 20 °C. Running the process at lower temperatures allows substantial savings in energy consumption (Luo et al. 2006).

Lipolytic enzymes are also attractive for the pharmaceutical industry due to substrate promiscuity and the catalysis of regioselective reactions in various organic solvents. Compared to organic chemistry, enzyme catalyzed processes are environmentally favourable and offer a higher degree of selectivity in the production of optimally active chiral drugs (Fig. 12.5). Moreover, none-racemic chiral building block chemicals represent valuable compounds for the fine chemical industry. One of the most prevalent anti-inflammatory drugs is ketoprofen [(R,S)-2-(3-benzoylphenyl) propionic acid] whereof S-ketoprofen is pharmacologically active (de O. Carvalho et al. 2006). Further examples of non-steroidal anti-inflammatory drugs to combat human diseases that are processed by lipolytic enzymes are naproxen and ibuprofen (Hess et al. 2008). Moreover, a lipase from *S. marcescensis* is applied to produce the calcium channel blocking drug Diltiazem (Wicka et al. 2013). Although protein engineering of cold-active lipolytic enzymes has steadily improved enzyme properties, industrial applications have not been rapidly developed compared to heat-active enzymes from thermophiles (Joseph et al. 2008).

4 Lipolytic Enzymes from (Hyper-)Thermophiles

Thermophilic microorganisms grow at high temperatures between 50 °C and 80 °C and mainly include members of the *Bacteria* and *Archaea*, while only some eukaryotic moderate thermophiles (growth below 60 °C) have been identified. In addition, archaea dominate the group of hyperthermophiles that are even adapted to temperatures up to 110 °C. Typical environments that are inhabited by thermophilic microorganisms include hot springs, hydrothermal vents and volcanic islands. Although, thermostable enzymes (thermozymes) are found in mesophilic and thermophilic microorganisms, thermophiles produce enzymes that are comparatively robust and hence are preferably used for industrial applications (Hasan et al. 2005).

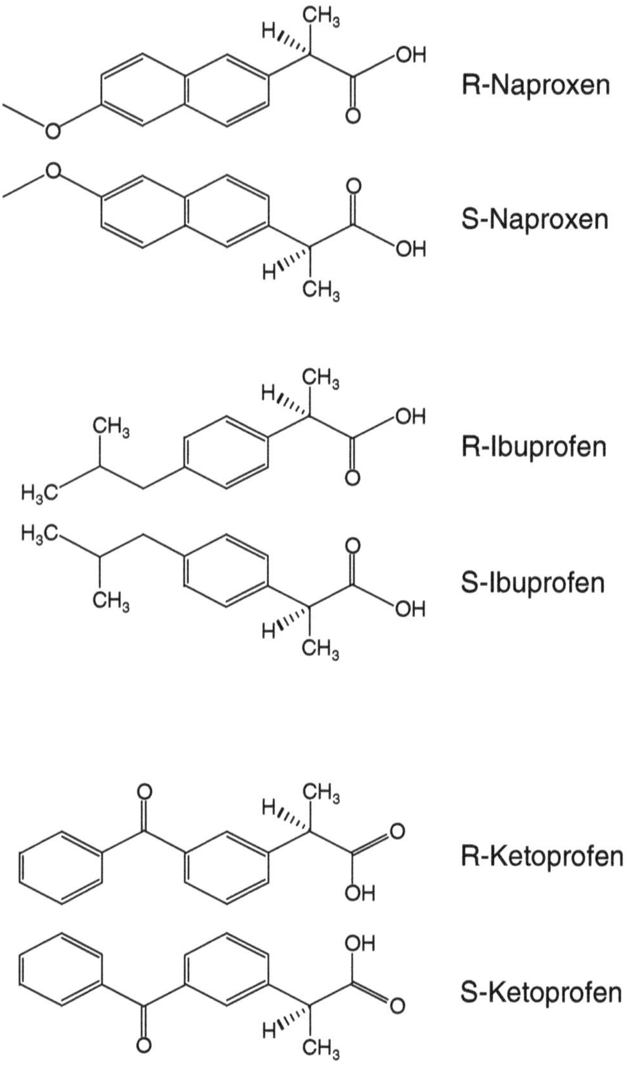

Fig. 12.5 Molecular structures of chiral drugs that can be produced by lipolytic enzyme catalyzed esterifications

4.1 Diversity of Heat-Active Lipases and Esterases

The recent growing demand and interest for enzymes from thermophilic microorganisms led to the identification and characterization of many heat-active lipases. Table 12.2 focuses on promising recombinant lipolytic enzymes from thermophilic bacteria. These enzymes showed the best catalytic performance between 55 °C and 80 °C and have molecular masses between 19 and 53 kDa. The highest temperature optimum of a characterized recombinant esterase EstTs1 from a thermophilic

Table 12.2 Recombinant lipolytic enzymes from thermophilic bacteria

Sources	T_{opt} (°C)	pH $_{opt}$	MW (kDa)	Preferred substrate	References
Esterases[a]					
Acidicaldus USBA-GBX-499	55	9.0	34	*p*NP-caproate (C_6)	(Lopez et al. 2014)
Thermus thermophiles HB27	58.2	6.3	37.7	*p*NP-caprate (C_{10})	(Fuciños et al. 2014)
Anoxybacillus sp. PDF1	60	8.0	26	*p*NP-butyrate (C_4)	(Ay et al. 2011)
Geobacillus thermoleovorans YN	60–65	9.5	29	*p*NP-acetate (C_2)	(Soliman et al. 2007)
Metagenome (activated sludge)	70	8.5	53	*p*NP-acetate (C_2)	(Shao et al. 2013)
Thermoanaerobacter tengcongensis MB4	70	9.5	50	*p*NP-caproate (C_6)	(Rao et al. 2011)
Thermosyntropha lipolytica	70	8.0	–	*p*NP-butyrate (C_4)	(Gumerov et al. 2012)
Thermotoga maritima	70	5.0–5.5	35.2	*p*NP-butyrate (C_4)	(Tao et al. 2013)
Thermus scotoductus SA-01	80	7.0	28.6	*p*NP-butyrate (C_4)	(du Plessis et al. 2010)
Lipases[a]					
Geobacillus sp. EPT9	55	8.5	44.8	*p*NP-palmitate (C_{16})	(Zhu et al. 2015)
Bacillus thermoleovorans ID-1	60	8.0–9.0	43	*p*NP-caprate (C_{10})	(Lee et al. 2001)
Bacillus thermoleovorans ID-1	60–65	9.0	19	*p*NP-caprylate (C_8)	(Lee et al. 2001)
Geobacillus sp. strain T1	70	9.0	43	Trilaurin (C_{12})	(Leow et al. 2007)
Geobacillus thermoleovorans YN	70	9.5	43	*p*NP-caprate (C_{10})	(Soliman et al. 2007)
Enrichment cultures (soil and water samples)	70	8.0	31.7	*p*NP-caprylate (C_8)	(Chow et al. 2012)
Enrichment cultures (soil and water samples)	75	8.0	36	*p*NP-caprate (C_{10})	(Chow et al. 2012)
Thermoanaerobacter thermohydrosulfuricus SOL1	75	8.0	34.2	*p*NP-caprate (C_{10})	(Royter et al. 2009)
Caldanaerobacter subterraneus subsp. *tengcongensis*	75	7.0	32.1	*p*NP-caprate (C_{10})	(Royter et al. 2009)
Fervidobacterium changbaicum	78	7.8	33	*p*NP-caprate (C_{10})	(Cai et al. 2011)

[a]Enzymes were assigned to the groups of esterases or lipases based on their substrate preferences

bacterium is 80 °C isolated from *Thermus scotoductus* (du Plessis et al. 2010). A recombinant lipase with a temperature optimum of 78 °C was derived from *Fervidobacterium changbaicum* (Cai et al. 2011). Two lipases purified directly from culture supernatant of *Thermosyntropha lipolytica* DSM1103 were described with highest activity at 96 °C (Salameh and Wiegel 2007). Within the scope of metagenomic approaches various novel lipolytic enzymes were discovered. Thermophilic microorganisms present in environmental samples can be enriched at elevated temperatures and used for metagenomic library construction. A metagenome can be isolated directly from suitable environments predominated by high temperatures to exploit the diversity and to gain access to non-cultivable organisms. Unique enzymes with low similarities to known proteins were found as shown for an esterase identified from a soil metagenome with 32–45 % identity to putative α/β-hydrolases (Choi et al. 2013). Another functional screening approach led to the identification of two genes encoding lipases that were optimally active at 70 °C and 75 °C (Chow et al. 2012).

The majority of characterized carboxylic ester hydrolases from thermophilic bacteria were described as esterases. Lipolytic enzymes are usually assigned to the class of esterases or lipases due to their substrate spectrum. Regarding the chain length of linked fatty acids within artificial substrates, some lipolytic enzymes display esterase and lipase activity (Fig. 12.2). TSLip1 from *Thermosynthropha lipolytica* exhibited the highest activity at 70 °C towards *p*NP-butyrate (C_4), minimal performance (<10 %) with the C_{10}-substrate, but activity increased again when *p*NP-palmitate (C_{16}) was used as substrate (Gumerov et al. 2012). Furthermore, various vegetable oils, such as soybean, olive, corn and sunflower oil were hydrolysed. A broad substrate spectrum was also reported at 70 °C for the recombinant lipase from *Geobacillus* sp. T1. Various natural oils and triacylglycerol substrates containing chain length between C_2 and C_{18} with an optimum towards C_{12} were hydrolyzed (Leow et al. 2007). Besides stability and maximal activity, substrate specificity could also be altered by single point mutations. A lipase mutant (N355K) exhibited highest activity toward *p*NP-palmitate (C_{16}) whereas the wildtype enzyme was most active toward *p*NP-laurate (C_{12}) with residual activity of approximately 15 % with palmitate (Sharma et al. 2014). A recombinant archaeal carboxylesterase displayed highest activity toward *p*NP-octanoate (C_8) with no residual activity towards *p*NP-butyrate (C_4). The C-terminal His-tagged protein showed similar results. When the gene was cloned with a N-terminal His-tag-encoding region, the enzyme exhibited a broader substrate spectrum with comparable activities towards *p*NP-butyrate (C_4) and *p*NP-octanoate (C_8). The substrate-binding pocket was located at the N-terminus, thus being somehow altered by the attached tag (Killens-Cade et al. 2014). Recombinant proteins may provide feature variations compared to the naturally occurring enzyme (Leow et al. 2007).

Generally, most bacterial lipolytic enzymes are optimally active at neutral or alkaline pH although they can act over a wide pH range (Gupta et al. 2004). Numerous thermostable lipolytic enzymes were reported to be active and stable exclusively at alkaline pH values. Esterases and lipases from metagenomes derived from not extreme habitats were often found to prefer alkaline conditions (Choi et al.

2013). Likewise, an esterase from the thermoacidophilic bacterium *Acidicaldus* sp. exhibited the highest activity at pH 9.0 and <20 % residual activity at pH 7.0. Only two heat-stable enzymes were described in the literature with slightly acidic pH optima including an esterase from *Thermus thermophilus* (pH optimum at 6.3) and an esterase from *Thermotoga maritima* with a pH optimum at 5.0–5.5 (Tao et al. 2013; Fuciños et al. 2014). The latter was described as first acid tolerant esterase from a thermophilic bacterium.

4.2 Hyperthermophilic Archaea as a Source for Lipolytic Enzymes

Thermophilic organisms grow at 50–80 °C, whereas hyperthermophiles grow between 80 °C and 113 °C. Archaea represent the group with the highest temperature optima (Egorova and Antranikian 2005). Accordingly, optimal temperature for heat-active biocatalysts from archaea often exceeds the one described for bacterial enzymes. Furthermore, they exhibit a high intrinsic thermal and chemical stability compared to bacterial proteins (Levisson et al. 2009). Some recombinantly produced and characterized archaeal lipolytic enzymes are shown in Table 12.3. Highest optimal temperature with 100 °C for activity was reported for the recombinant esterase from *Pyrococcus furiosus* with a half-life of 34 h at 100 °C (Ikeda and Clark 1998). The enzyme from *Sulfolobus solfataricus* P1 exhibited the highest activity at 80 °C and retained 41 % of its activity after 5 days of incubation at the same temperature (Park et al. 2006). The esterase from *Metallosphaera sedula* showed highest activity at 95 °C with C_8-substrates. This enzyme also showed a wide range for catalytic performance with residual activity even at 37 °C. Moreover, it exhibited more than 70 % residual activity after 6 h at 90 °C (Killens-Cade et al. 2014). *Sulfolobus acidophilum* esterase with an optimum at 70 °C appears comparatively low, but residual activity of >20 % was observed at 90 °C (Zhang et al. 2014). Enzymes from *Pyrobaculum* species and *Aeropyrum pernix* exhibited their highest activity at temperatures between 80 °C and 90 °C as well (Hotta et al. 2002; Gao et al. 2003; Shao et al. 2014). All esterases were active in a neutral or slightly alkaline pH range with one exception possessing activity at pH 10.0–11.0 (Table 12.3). This enzyme was active even at pH 12.0 being among the most alkaline pH range for hydrolases (Rusnak et al. 2005). No recombinant esterase from a hyperthermophilic organism was found with optimal activity below pH 7.0. An esterase from the extreme thermoacidophilic archaeon *Picrophilus torridus* (growth at 60 °C and pH 1.0–2.0) was reported to exhibit highest activity at pH 6.5 and 70 °C (Hess et al. 2008). An exceptional pH optimum of 2.0 at 50 °C with no residual activity at pH 5.0 was reported for an esterase from *Ferroplasma acidophilum* that grows at moderate temperatures and at pH 1.7 (Golyshina et al. 2006).

In the literature, no lipase with exclusive preference for long-chain fatty acid substrates (>C_8) was found to belong to the group of hyperthermophilic archaea.

Table 12.3 Recombinant lipolytic enzymes from hyperthermophilic archaea

Sources	T_{opt} (°C)	pH $_{opt}$	MW (kDa)	Preferred substrate	References
Archaeoglobus fulgidus DSM 4304	70	10.0–11.0	52.8	Methyl butyrate (C_4)/*p*NP-acetate (C_2)	(Rusnak et al. 2005)
Sulfolobus acidophilus DSM10332	70	8.0	34.1	*p*NP-butyrate (C_4)	(Zhang et al. 2014)
Pyrobaculum sp. 1860	80	9.0	23	*p*NP-acetate (C_2)	(Shao et al. 2014)
Pyrococcus furiosus	80	7.0	26.5	4-methylumbelliferyl heptanoate (C_7)	(Alqueres et al. 2011)
Archaeoglobus fulgidus DSM 4304	80	7.0–8.0	27.5	*p*NP-butyrate (C_4)	(Kim et al. 2008)
Thermococcus kodakarensis KOD1	85	8.0	29	*p*NP-butyrate (C_4)	(Cui et al. 2012)
Sulfolobus solfataricus P1	85	8.0	34	*p*NP-caprylate (C_8)	(Park et al. 2006)
Sulfolobus solfataricus	90	6.5–7.0	32	*p*NP-valerate (C_5)	(Morana et al. 2002)
Aeropyrum pernix K1	90	8.0	63	*p*NP-caprylate (C_8)	(Gao et al. 2003)
Pyrobaculum calidifontis VA1	90	7.0	34	*p*NP-caproate (C_6)	(Hotta et al. 2002)
Metallosphaera sedula DSM5348	95	7.0	33.2	*p*NP-caprylate (C_8)	(Killens-Cade et al. 2014)
Pyrococcus furiosus	100	7.6	–	4-methylumbelliferyl acetate (C_2)	(Ikeda and Clark 1998)

Nevertheless, some of the esterases shown in Table 12.3 were capable of hydrolyzing long-chain substrates to some extent. The enzyme from *Pyrococcus furiosus* exhibited 15 % residual activity toward *p*NP-palmitate (C_{16}) (Alqueres et al. 2011). The enzyme from *Aeropyrum pernix* hydrolyzed *p*NP-stearate (C_{18}) with 19 % activity compared to 100 % towards *p*NP-caprylate (C_8) (Gao et al. 2003). The lipolytic enzyme AFL from *Archaeglobus fulgidus* contained a small lid domain and an additional C-terminal lipid-binding domain (Chen et al. 2009). AFL was previously characterized and described as an esterase due to higher activity toward *p*NP-acetate (C_2) compared to the corresponding palmitate (Rusnak et al. 2005). Nevertheless, the extraordinary C-terminal domain of AFL was reported to be essential for binding long-chain substrates and thus, this enzyme was subsequently classified as true lipase due to structural findings (Chen et al. 2009). By unravelling structures of lipolytic enzymes, classification can be conducted more reliably than by investigation of artificial substrate preference. However, a lipase with preference for the long-chain substrate *p*NP-myristate (C_{14}) was recently identified from the mesophilic archaeon *Haloarcula* sp. G41 with optimal activity at 70 °C, pH 8.0 (Li and Yu 2014). This demonstrates the high potential of archaea as a source for novel lipases with unique features.

4.3 Enzyme Engineering

Thermal denaturation of proteins is reduced due to various features, such as higher rigidity or compact packing. However, higher rigidity is accompanied by less flexibility, which is thought to result in decreased activity. This assumption was recently disproved by identification of a lipase mutant with a more rigid active site than the wildtype but simultaneously enhanced activity and stability (Kamal et al. 2012). Different factors seem to contribute to the stability of thermoactive enzymes, such as higher amounts of disulphide bridges, hydrophobic interactions, hydrogen bonds or metal bindings (Vieille and Zeikus 2001). Exchange of a single amino acid (E315G) within a thermostable lipase showed that one residue close to the active site resulted in a decrease of thermal stability. In contrast to this substitution that enabled higher loop-flexibility, the N355K mutant enhanced the thermostability by formation of an additional hydrogen bond (Sharma et al. 2014). *Geobacillus* sp. EPT9 lipase contained a zinc-binding domain and higher percentages of proline and arginine compared to a lipase from a mesophilic *Bacillus* strain (Zhu et al. 2015). Proline was the most rigid amino acid and arginine was shown to participate in multiple non-covalent interactions. Disruption of the zinc-binding site of lipase L1 from *Geobacillus stearothermophilus* resulted in a shift of optimal temperature for enzyme activity from 60 °C to 45–50 °C with lower specific activity and in a decrease of thermal stability (Choi et al. 2005). Thus, the zinc-binding domain may participate in structural conformation of the active site. No ubiquitous generalized pattern for activity and stability at elevated temperatures can be determined. Nevertheless, it was shown that exchange of one residue might result in a small change of inner-protein interactions. This conformational deviation may lead to a significant decrease in stability and/or enzyme activity.

Several enzyme-engineering approaches were conducted to enhance thermal stability, resulting in mutants that exhibited maximal activity at higher temperatures. However, there are also some contrasting examples, e.g. one amino acid exchange resulted in the reduction of optimal temperature from 50 °C to 40 °C compared to a wildtype lipase, which was identified from a soil metagenome (Sharma et al. 2014). Hence, advantage can be taken of engineering robust enzymes from thermophiles or hyperthermophiles to create lipolytic enzymes suitable for processes that run at harsh conditions and moderate temperatures. For instance washing processes are favoured to function at lower temperatures nowadays to save energy. At the same time harsh conditions are provided under alkaline environment requiring stable fat cleaving enzymes (Jaeger and Reetz 1998). One of the latest achievements in developing new and improved lipolytic enzymes is the incorporation of non-canonical amino acids. Substitution of a high number of non-canonical amino acid residues in the heat-active lipase from *Thermoanaerobacter thermohydrosulfuricus* (TLL) resulted in a "cold-washing" enzyme that was activated without heat induction (Hoesl et al. 2011).

4.4 Applications of Heat-Stable Lipases and Esterases

Most industrial processes run at temperatures higher than 45 °C (Sharma et al. 2002). Higher operation temperature contributes to higher reaction rates of applied biocatalysts, increases solubility of substrates, and lowers viscosity and contamination risks. Furthermore, enzymes from thermophiles and hyperthermophiles are more stable under harsh denaturing conditions (Mozhaev 1993). High-level recombinant production of bacterial and archaeal enzymes has proven to yield sufficient amount of biocatalysts for industrial usage. Employing lipolytic enzymes for a conversion reaction implies milder conditions compared to treatment with chemicals, steam or pressure. The specificity of esterases and lipases contributes to reduction or elimination of unwanted side products and costs can be reduced due to unnecessary separation steps. It was shown that reaction temperature and incubation time were the major factors influencing the yield of produced menthyl butyrate by T1 lipase in a solvent-free system (Wahab et al. 2014). The high-yield synthesis was conducted with the recombinant enzyme without need of flammable or toxic solvents that have to be subsequently separated from the product. Nevertheless, when solvents are inevitable the lipolytic enzyme must show high tolerance. The lipolytic enzymes from *T. thermohydrosulfuricus* SOL1 and *Caldanaerobacter subterraneus* subsp. *tengcongensis* were resistant against several solvents or detergents up to a concentration of 99 % (Royter et al. 2009). Furthermore, the esterase from *T. thermohydrosulfuricus* SOL1 was reported to show high preference for esters of secondary alcohols and a high selectivity for R-enantiomers of pharmaceutically relevant substrates (Royter et al. 2009). This enzyme being active at high temperatures and resistant against organic solvents represents an attractive candidate for biotransformation in water-free media. The esterase Tm1160 from *Thermotoga maritima* showed high enantioselectivity hydrolyzing racemic ketoprofen ethyl ester (Tao et al. 2013). Likewise, the esterase from *Acidicaldus* sp. can be used to synthesize medically relevant S-enantiomers of naproxen and ibuprofen esters at harsh conditions (Fig. 12.5) (Lopez et al. 2014).

An example for industrial interesterification processes is the conversion of palm oil into cocoa butter fat substitute that exhibits a higher melting point and is therefore attractive as food, confection and cosmetic supplements. Cheap oils can be upgraded by enzymatic conversion into nutritionally important structured fats. High value products, such as human milk fat substitute or cocoa butter equivalents are commercially distributed (Hasan et al. 2005). Selective hydrolysis of fat triacylglycerides is also utilized for flavour development or flavour precursor formation (Jaeger and Reetz 1998). These compounds are difficult to isolate from parent sources and hence, industrial extraction procedures are often unprofitable. Terpene esters containing short-chain fatty acids, such as terpinyl esters, geraniol esters or citronellol esters, produced by lipolytic enzymes are applied in the beverage, food and pharmaceutical industry. Green synthetic routes for valuable compound production are favoured over chemical synthesis. Furthermore, new odour molecules are discovered by elucidating the wide range of action of lipolytic enzymes towards many substrates (Dhake et al. 2013).

Heat-stable lipases can also be applied within the scope of algae-based biofuel production. The amount of free fatty acids can be enhanced by cleavage of storage triacylglycerides. The use of thermostable lipolytic enzymes reduces time and energy consumption during the high temperature conversion of lipids to fuel (Killens-Cade et al. 2014). Furthermore, in the pulp and paper industry lipolytic enzymes are employed to remove hydrophobic components of wood, such as triacylglycerides and waxes, from the pulp at elevated temperatures (Jaeger and Reetz 1998).

5 Conclusion

Due to their versatile enzymatic properties, lipolytic extremozymes represent promising candidates to be used in several industries including organic synthesis of chiral compounds and pharmaceuticals, food and detergent industry and environmental bioremediations. A portfolio of cold- and heat-active esterases and lipases has been identified by modern molecular biology and microbiology approaches in extremophilic archaea and bacteria and are nowadays available for specific applications and engineering approaches. However, there are still a couple of barriers to overcome in the development of a tailor-made biocatalyst for certain industrial applications. Establishing novel and improved techniques to identify and engineer genes encoding biocatalysts and to produce proteins in recombinant and active form in high yield has made particular progress. Recent examples in the "omics" era are the development and allocation of appropriate expression hosts, directed evolution, gene shuffling, efficient cloning approaches, enzyme congeners containing non-canonical amino acids, fusion enzymes and site-directed mutagenesis (Sharma et al. 2002; Gatti-Lafranconi et al. 2008; Al Khudary et al. 2010; Merkel et al. 2010; Hoesl et al. 2011; Bassegoda et al. 2012a; Marquardt et al. 2014; Elleuche 2015; Joshi and Satyanarayana 2015).

References

Abdou AM (2003) Purification and partial characterization of psychrotrophic *Serratia marcescens* lipase. J Dairy Sci 86:127–132

Ahmad S, Kamal MZ, Sankaranarayanan R, Rao NM (2008) Thermostable *Bacillus subtilis* lipases: in vitro evolution and structural insight. J Mol Biol 381:324–340

Al Khudary R, Venkatachalam R, Katzer M, Elleuche S, Antranikian G (2010) A cold-adapted esterase of a novel marine isolate, *Pseudoalteromonas arctica*: gene cloning, enzyme purification and characterization. Extremophiles 14:273–285

Alquati C, De Gioia L, Santarossa G, Alberghina L, Fantucci P, Lotti M (2002) The cold-active lipase of *Pseudomonas fragi*. Heterologous expression, biochemical characterization and molecular modeling. Eur J Biochem 269:3321–3328

Alqueres SM, Branco RV, Freire DM, Alves TL, Martins OB, Almeida RV (2011) Characterization of the recombinant thermostable lipase (Pf2001) from *Pyrococcus furiosus*: effects of thioredoxin fusion tag and triton X-100. Enzyme Res 2011:316939

Aoyama S, Yoshida N, Inouye S (1988) Cloning, sequencing and expression of the lipase gene from *Pseudomonas fragi* IFO-12049 in E. coli. FEBS Lett 242:36–40

Arpigny JL, Jaeger KE (1999) Bacterial lipolytic enzymes: classification and properties. Biochem J 343(Pt 1):177–183

Ay F, Karaoglu H, Inan K, Canakci S, Belduz AO (2011) Cloning, purification and characterization of a thermostable carboxylesterase from *Anoxybacillus* sp. PDF1. Protein Expr Purif 80:74–79

Bassegoda A, Cesarini S, Diaz P (2012a) Lipase improvement: goals and strategies. Comput Struct Biotechnol J 2, e201209005

Bassegoda A, Pastor FI, Diaz P (2012b) *Rhodococcus* sp. strain CR-53 LipR, the first member of a new bacterial lipase family (family X) displaying an unusual Y-type oxyanion hole, similar to the *Candida antarctica* lipase clan. Appl Environ Microbiol 78:1724–1732

Bayer S, Kunert A, Ballschmiter M, Greiner-Stoeffele T (2010) Indication for a new lipolytic enzyme family: isolation and characterization of two esterases from a metagenomic library. J Mol Microbiol Biotechnol 18:181–187

Bornscheuer UT (2002) Microbial carboxyl esterases: classification, properties and application in biocatalysis. FEMS Microbiol Rev 26:73–81

Brault G, Shareck F, Hurtubise Y, Lepine F, Doucet N (2012) Isolation and characterization of EstC, a new cold-active esterase from *Streptomyces coelicolor* A3(2). PLoS ONE 7, e32041

Cai J, Xie Y, Song BK, Wang Y, Zhang Z, Feng Y (2011) *Fervidobacterium changbaicum* Lip1: identification, cloning, and characterization of the thermophilic lipase as a new member of bacterial lipase family V. Appl Microbiol Biotechnol 89:1463–1473

Chen CK, Lee GC, Ko TP, Guo RT, Huang LM, Liu HJ, Ho YF, Shaw JF, Wang AH (2009) Structure of the alkalohyperthermophilic *Archaeoglobus fulgidus* lipase contains a unique C-terminal domain essential for long-chain substrate binding. J Mol Biol 390:672–685

Choi GS, Kim JY, Kim JH, Ryu YW, Kim GJ (2003) Construction and characterization of a recombinant esterase with high activity and enantioselectivity to (S)-ketoprofen ethyl ester. Protein Expr Purif 29:85–93

Choi WC, Kim MH, Ro HS, Ryu SR, Oh TK, Lee JK (2005) Zinc in lipase L1 from *Geobacillus stearothermophilus* L1 and structural implications on thermal stability. FEBS Lett 579:3461–3466

Choi JE, Kwon MA, Na HY, Hahm DH, Song JK (2013) Isolation and characterization of a metagenome-derived thermoalkaliphilic esterase with high stability over a broad pH range. Extremophiles 17:1013–1021

Choo DW, Kurihara T, Suzuki T, Soda K, Esaki N (1998) A cold-adapted lipase of an Alaskan psychrotroph, *Pseudomonas* sp. strain B11-1: gene cloning and enzyme purification and characterization. Appl Environ Microbiol 64:486–491

Chow J, Kovacic F, Dall Antonia Y, Krauss U, Fersini F, Schmeisser C, Lauinger B, Bongen P, Pietruszka J, Schmidt M, Menyes I, Bornscheuer UT, Eckstein M, Thum O, Liese A, Mueller-Dieckmann J, Jaeger KE, Streit WR (2012) The metagenome-derived enzymes LipS and LipT increase the diversity of known lipases. PLoS ONE 7, e47665

Couto GH, Glogauer A, Faoro H, Chubatsu LS, Souza EM, Pedrosa FO (2010) Isolation of a novel lipase from a metagenomic library derived from mangrove sediment from the south Brazilian coast. Genet Mol Res 9:514–523

Cui Z, Wang Y, Pham BP, Ping F, Pan H, Cheong GW, Zhang S, Jia B (2012) High level expression and characterization of a thermostable lysophospholipase from *Thermococcus kodakarensis* KOD1. Extremophiles 16:619–625

de O. Carvalho P, Contesini FJ, Bizaco R, Calafatti SA, Macedo GA (2006) Optimization of enantioselective resolution of racemic ibuprofen by native lipase from *Aspergillus niger*. J Ind Microbiol Biotechnol 33:713–718

De Santi C, Tedesco P, Ambrosino L, Altermark B, Willassen NP, de Pascale D (2014) A new alkaliphilic cold-active esterase from the psychrophilic marine bacterium *Rhodococcus* sp.: functional and structural studies and biotechnological potential. Appl Biochem Biotechnol 172:3054–3068

Dhake KP, Thakare DD, Bhanage BM (2013) Lipase: a potential biocatalyst for the synthesis of valuable flavour and fragrance ester compounds. Falvour Fragr J 28:71–83

Dlugolecka A, Cieslinski H, Bruzdziak P, Gottfried K, Turkiewicz M, Kur J (2009) Purification and biochemical characteristic of a cold-active recombinant esterase from *Pseudoalteromonas* sp. 643A under denaturing conditions. Pol J Microbiol 58:211–218

Do H, Lee JH, Kwon MH, Song HE, An JY, Eom SH, Lee SG, Kim HJ (2013) Purification, characterization and preliminary X-ray diffraction analysis of a cold-active lipase (CpsLip) from the psychrophilic bacterium *Colwellia psychrerythraea* 34H. Acta Crystallogr Sect F Struct Biol Cryst Commun 69:920–924

du Plessis EM, Berger E, Stark T, Louw ME, Visser D (2010) Characterization of a novel thermostable esterase from *Thermus scotoductus* SA-01: evidence of a new family of lipolytic esterases. Curr Microbiol 60:248–253

Egorova K, Antranikian G (2005) Industrial relevance of thermophilic Archaea. Curr Opin Microbiol 8:649–655

Elend C, Schmeisser C, Hoebenreich H, Steele HL, Streit WR (2007) Isolation and characterization of a metagenome-derived and cold-active lipase with high stereospecificity for (R)-ibuprofen esters. J Biotechnol 130:370–377

Elleuche S (2015) Bringing functions together with fusion enzymes-from nature's inventions to biotechnological applications. Appl Microbiol Biotechnol 99:1545–1556

Elleuche S, Piascheck H, Antranikian G (2011) Fusion of the OsmC domain from esterase EstO confers thermolability to the cold-active xylanase Xyn8 from *Pseudoalteromonas arctica*. Extremophiles 15:311–317

Elleuche S, Schröder C, Sahm K, Antranikian G (2014) Extremozymes—biocatalysts with unique properties from extremophilic microorganisms. Curr Opin Biotechnol 29:116–123

Feller G, Gerday C (2003) Psychrophilic enzymes: hot topics in cold adaptation. Nat Rev Microbiol 1:200–208

Feller G, Thiry M, Arpigny JL, Gerday C (1991) Cloning and expression in *Escherichia coli* of three lipase-encoding genes from the psychrotrophic antarctic strain *Moraxella* TA144. Gene 102:111–115

Feller G, Narinx E, Arpigny JL, Aittaleb M, Baise E, Genicot S, Gerday C (1996) Enzymes from psychrophilic organisms. FEMS Microbiol Rev 18:189–202

Fuciños P, Pastrana L, Sanroman A, Longo MA, Hermoso JA, Rua ML (2011) An esterase from *Thermus thermophilus* HB27 with hyper-thermoalkalophilic properties: purification, characterisation and structural modelling. J Mol Catal B Enzym 70:127–137

Fuciños P, Atanes E, Lopez-lopez O, Solaroli M, Cerdan ME, Gonzalez-Siso MI, Pastrana L, Rua ML (2014) Cloning, expression, purification and characterization of an oligomeric His-tagged thermophilic esterase from *Thermus thermophilus* HB27. Process Biochem 49:927–935

Gao R, Feng Y, Ishikawa K, Ishida H, Ando S, Kosugi Y, Cao S (2003) Cloning, purification and properties of a hyperthermophilic esterase from archaeon *Aeropyrum pernix* K1. J Mol Catal B Enzym 24–25:1–8

Gatti-Lafranconi P, Caldarazzo SM, Villa A, Alberghina L, Lotti M (2008) Unscrambling thermal stability and temperature adaptation in evolved variants of a cold-active lipase. FEBS Lett 582:2313–2318

Golyshina OV, Golyshin PN, Timmis KN, Ferrer M (2006) The 'pH optimum anomaly' of intracellular enzymes of *Ferroplasma acidiphilum*. Environ Microbiol 8:416–425

Gumerov VM, Mardanov AV, Kolosov PM, Ravin NV (2012) Isolation and functional characterization of lipase from the thermophilic alkali-tolerant bacterium *Thermosyntropha lipolytica*. Prikl Biokhim Mikrobiol 48:376–382

Gupta R, Gupta N, Rathi P (2004) Bacterial lipases: an overview of production, purification and biochemical properties. Appl Microbiol Biotechnol 64:763–781

Hasan F, Shah AA, Hameed A (2005) Industrial applications of microbial lipases. Enzyme Microb Technol 39:235–251

Henne A, Schmitz RA, Bomeke M, Gottschalk G, Daniel R (2000) Screening of environmental DNA libraries for the presence of genes conferring lipolytic activity on *Escherichia coli*. Appl Environ Microbiol 66:3113–3116

Hess M, Katzer M, Antranikian G (2008) Extremely thermostable esterases from the thermoacidophilic euryarchaeon *Picrophilus torridus*. Extremophiles 12:351–364

Hoesl MG, Acevedo-Rocha CG, Nehring S, Royter M, Wolschner C, Wiltschi B, Budisa N, Antranikian G (2011) Lipase congeners designed by genetic code engineering. Chemcatchem 3:213–221

Hotta Y, Ezaki S, Atomi H, Imanaka T (2002) Extremely stable and versatile carboxylesterase from a hyperthermophilic archaeon. Appl Environ Microbiol 68:3925–3931

Hu XP, Heath C, Taylor MP, Tuffin M, Cowan D (2012) A novel, extremely alkaliphilic and cold-active esterase from Antarctic desert soil. Extremophiles 16:79–86

Hwang HT, Qi F, Yuan C, Zhao X, Ramkrishna D, Liu D, Varma A (2014) Lipase-catalyzed process for biodiesel production: protein engineering and lipase production. Biotechnol Bioeng 111:639–653

Ikeda M, Clark DS (1998) Molecular cloning of extremely thermostable esterase gene from hyperthermophilic archaeon *Pyrococcus furiosus* in *Escherichia coli*. Biotechnol Bioeng 57:624–629

Jaeger KE, Reetz MT (1998) Microbial lipases form versatile tools for biotechnology. Trends Biotechnol 16:396–403

Jeon JH, Kim JT, Kim YJ, Kim HK, Lee HS, Kang SG, Kim SJ, Lee JH (2009) Cloning and characterization of a new cold-active lipase from a deep-sea sediment metagenome. Appl Microbiol Biotechnol 81:865–874

Joseph B, Ramteke PW, Thomas G (2008) Cold active microbial lipases: some hot issues and recent developments. Biotechnol Adv 26:457–470

Joshi S, Satyanarayana T (2015) In vitro engineering of microbial enzymes with multifarious applications: prospects and perspectives. Bioresour Technol 176:273–283

Kamal MZ, Mohammad TA, Krishnamoorthy G, Rao NM (2012) Role of active site rigidity in activity: MD simulation and fluorescence study on a lipase mutant. PLoS One 7, e35188

Kamijo T, Saito A, Ema S, Yoh I, Hayashi H, Nagata R, Nagata Y, Ando A (2011) Molecular and enzymatic characterization of a subfamily I.4 lipase from an edible oil-degrader *Bacillus* sp. HH-01. Antonie Van Leeuwenhoek 99:179–187

Kawata T, Ogino H (2009) Enhancement of the organic solvent-stability of the LST-03 lipase by directed evolution. Biotechnol Prog 25:1605–1611

Killens-Cade R, Turner R, MacInnes C, Grunden A (2014) Characterization of a thermostable, recombinant carboxylesterase from the hyperthermophilic archaeon *Metallosphaera sedula* DSM5348. Adv Enzym Res 2:1–13

Kim SB, Lee W, Ryu YW (2008) Cloning and characterization of thermostable esterase from *Archaeoglobus fulgidus*. J Microbiol 46:100–107

Koops BC, Papadimou E, Verheij HM, Slotboom AJ, Egmond MR (1999) Activity and stability of chemically modified *Candida antarctica* lipase B adsorbed on solid supports. Appl Microbiol Biotechnol 52:791–796

Lee D-W, Kim H-W, Lee K-W, Kim B-C, Choe E-A, Lee H-S, Kim D-S, Pyun Y-R (2001) Purification and characterization of two distinct thermostable lipases from the gram-positive thermophilic bacterium *Bacillus thermoleovorans* ID-1. Enzyme Microb Technol 29:363–371

Lenfant N, Hotelier T, Velluet E, Bourne Y, Marchot P, Chatonnet A (2013) ESTHER, the database of the alpha/beta-hydrolase fold superfamily of proteins: tools to explore diversity of functions. Nucleic Acids Res 41:D423–D429

Leow TC, Rahman RN, Basri M, Salleh AB (2007) A thermoalkaliphilic lipase of *Geobacillus* sp. T1. Extremophiles 11:527–535

Levisson M, van der Oost J, Kengen SW (2009) Carboxylic ester hydrolases from hyperthermophiles. Extremophiles 13:567–581

Li X, Yu HY (2014) Characterization of an organic solvent-tolerant lipase from *Haloarcula* sp. G41 and its application for biodiesel production. Folia Microbiol (Praha) 59:455–463

Lopez G, Chow J, Bongen P, Lauinger B, Pietruszka J, Streit WR, Baena S (2014) A novel thermoalkalostable esterase from *Acidicaldus* sp. strain USBA-GBX-499 with enantioselectivity isolated from an acidic hot springs of Colombian Andes. Appl Microbiol Biotechnol 98:8603–8616

Luo Y, Zheng Y, Jiang Z, Ma Y, Wei D (2006) A novel psychrophilic lipase from *Pseudomonas fluorescens* with unique property in chiral resolution and biodiesel production via transesterification. Appl Microbiol Biotechnol 73:349–355

Marquardt T, von der Heyde A, Elleuche S (2014) Design and establishment of a vector system that enables production of multifusion proteins and easy purification by a two-step affinity chromatography approach. J Microbiol Methods 105:47–50

Merkel L, Schauer M, Antranikian G, Budisa N (2010) Parallel incorporation of different fluorinated amino acids: on the way to "teflon" proteins. Chembiochem 11:1505–1507

Morana A, Di Prizito N, Aurilia V, Rossi M, Cannio R (2002) A carboxylesterase from the hyperthermophilic archaeon Sulfolobus solfataricus: cloning of the gene, characterization of the protein. Gene 283:107–115

Mozhaev VV (1993) Mechanism-based strategies for protein thermostabilization. Trends Biotechnol 11:88–95

Nacke H, Will C, Herzog S, Nowka B, Engelhaupt M, Daniel R (2011) Identification of novel lipolytic genes and gene families by screening of metagenomic libraries derived from soil samples of the German Biodiversity Exploratories. FEMS Microbiol Ecol 78:188–201

Ogino H, Katou Y, Akagi R, Mimitsuka T, Hiroshima S, Gemba Y, Doukyu N, Yasuda M, Ishimi K, Ishikawa H (2007) Cloning and expression of gene, and activation of an organic solvent-stable lipase from *Pseudomonas aeruginosa* LST-03. Extremophiles 11:809–817

Park YJ, Choi SY, Lee HB (2006) A carboxylesterase from the thermoacidophilic archaeon *Sulfolobus solfataricus* P1; purification, characterization, and expression. Biochim Biophys Acta 1760:820–828

Parra LP, Espina G, Devia J, Salazar O, Andrews B, Asenjo JA (2015) Identification of lipase encoding genes from Antarctic seawater bacteria using degenerate primers: expression of a cold-active lipase with high specific activity. Enzyme Microb Technol 68:56–61

Patkar S, Vind J, Kelstrup E, Christensen MW, Svendsen A, Borch K, Kirk O (1998) Effect of mutations in *Candida antarctica* B lipase. Chem Phys Lipids 93:95–101

Rao L, Xue Y, Zhou C, Tao J, Li G, Lu JR, Ma Y (2011) A thermostable esterase from *Thermoanaerobacter tengcongensis* opening up a new family of bacterial lipolytic enzymes. Biochim Biophys Acta 1814:1695–1702

Rashid N, Shimada Y, Ezaki S, Atomi H, Imanaka T (2001) Low-temperature lipase from psychrotrophic *Pseudomonas* sp. strain KB700A. Appl Environ Microbiol 67:4064–4069

Roh C, Villatte F (2008) Isolation of a low-temperature adapted lipolytic enzyme from uncultivated microorganism. J Appl Microbiol 105:116–123

Royter M, Schmidt M, Elend C, Hobenreich H, Schafer T, Bornscheuer UT, Antranikian G (2009) Thermostable lipases from the extreme thermophilic anaerobic bacteria *Thermoanaerobacter thermohydrosulfuricus* SOL1 and *Caldanaerobacter subterraneus* subsp. *tengcongensis*. Extremophiles 13:769–783

Rusnak M, Nieveler J, Schmid RD, Petri R (2005) The putative lipase, AF1763, from *Archaeoglobus fulgidusis* is a carboxylesterase with a very high pH optimum. Biotechnol Lett 27:743–748

Ryu HS, Kim HK, Choi WC, Kim MH, Park SY, Han NS, Oh TK, Lee JK (2006) New cold-adapted lipase from *Photobacterium lipolyticum* sp. nov. that is closely related to filamentous fungal lipases. Appl Microbiol Biotechnol 70:321–326

Salameh MA, Wiegel J (2007) Purification and characterization of two highly thermophilic alkaline lipases from *Thermosyntropha lipolytica*. Appl Environ Microbiol 73:7725–7731

Shao H, Xu L, Yan Y (2013) Isolation and characterization of a thermostable esterase from a metagenomic library. J Ind Microbiol Biotechnol 40:1211–1222

Shao H, Xu L, Yan Y (2014) Biochemical characterization of a carboxylesterase from the archaeon *Pyrobaculum* sp. 1860 and a rational explanation of its substrate specificity and thermostability. Int J Mol Sci 15:16885–16910

Sharma R, Sona SK, Vohra RM, Gupta LK, Gupta JK (2002) Purification and characterisation of a thermostable alkaline lipase from a new thermophilic *Bacillus* sp. RSJ-1. Process Biochem 37:1075–1084

Sharma PK, Kumar R, Garg P, Kaur J (2014) Insights into controlling role of substitution mutation, E315G on thermostability of a lipase cloned from metagenome of hot spring soil. 3 Biotech 4:189–196

Shaw E, McCue LA, Lawrence CE, Dordick JS (2002) Identification of a novel class in the alpha/beta hydrolase fold superfamily: the N-myc differentiation-related proteins. Proteins 47:163–168

Siew N, Saini HK, Fischer D (2005) A putative novel alpha/beta hydrolase ORFan family in *Bacillus*. FEBS Lett 579:3175–3182

Soliman NA, Knoll M, Abdel-Fattah YR, Schmid RD, Lange S (2007) Molecular cloning and characterization of thermostable esterase and lipase from *Geobacillus thermoleovorans* YN isolated from desert soil in Egypt. Process Biochem 42:1090–1100

Tan S, Owusu Apenten RK, Knapp J (1996) Low temperature organic phase biocatalysis using cold-adapted lipase from psychrotrophic *Pseudomonas* P38. Food Chem 57:415–418

Tanaka D, Yoneda S, Yamashiro Y, Sakatoku A, Kayashima T, Yamakawa K, Nakamura S (2012) Characterization of a new cold-adapted lipase from *Pseudomonas* sp. TK-3. Appl Biochem Biotechnol 168:327–338

Tao W, Shengxue F, Duobin M, Xuan Y, Congcong D, Xihua W (2013) Characterization of a new thermophilic and acid tolerant esterase from *Thermotoga maritima* capable of hydrolytic resolution of racemic ketoprofen ethyl ester. J Mol Catal B Enzym 85–86:23–30

Tchigvintsev A, Tran H, Popovic A, Kovacic F, Brown G, Flick R, Hajighasemi M, Egorova O, Somody JC, Tchigvintsev D, Khusnutdinova A, Chernikova TN, Golyshina OV, Yakimov MM, Savchenko A, Golyshin PN, Jaeger KE, Yakunin AF (2015) The environment shapes microbial enzymes: five cold-active and salt-resistant carboxylesterases from marine metagenomes. Appl Microbiol Biotechnol 99:2165–2178

Trincone A (2011) Marine biocatalysts: enzymatic features and applications. Mar Drugs 9:478–499

Vieille C, Zeikus GJ (2001) Hyperthermophilic enzymes: sources, uses, and molecular mechanisms for thermostability. Microbiol Mol Biol Rev 65:1–43

Wahab RA, Basri M, Rahman RNZRA, Salleh AB, Rahman MBA, Chaibakhsh N, Leow TC (2014) Enzymatic production of a solvent-free menthyl butyrate via response surface methodology catalyzed by a novel thermostable lipase from *Geobacillus zalihae*. Biotechnol Biotechnol Equip 28:1065–1072

Wi AR, Jeon SJ, Kim S, Park HJ, Kim D, Han SJ, Yim JH, Kim HW (2014) Characterization and a point mutational approach of a psychrophilic lipase from an arctic bacterium, *Bacillus pumilus*. Biotechnol Lett 36:1295–1302

Wicka M, Krajewska E, Pawlak A (2013) Cold-adapted bacterial lipolytic enzymes and their applications. PhD Interdisp J 2:107–112

Wu G, Wu G, Zhan T, Shao Z, Liu Z (2013) Characterization of a cold-adapted and salt-tolerant esterase from a psychrotrophic bacterium *Psychrobacter pacificensis*. Extremophiles 17: 809–819

Yamashiro Y, Sakatoku A, Tanaka D, Nakamura S (2013) A cold-adapted and organic solvent-tolerant lipase from a psychrotrophic bacterium *Pseudomonas* sp. strain YY31: identification, cloning, and characterization. Appl Biochem Biotechnol 171:989–1000

Zhang JW, Zeng RY (2008) Molecular cloning and expression of a cold-adapted lipase gene from an Antarctic deep sea psychrotrophic bacterium *Pseudomonas* sp. 7323. Mar Biotechnol 10:612–621

Zhang J, Lin S, Zeng R (2007) Cloning, expression, and characterization of a cold-adapted lipase gene from an antarctic deep-sea psychrotrophic bacterium, *Psychrobacter* sp 7195. J Microbiol Biotechnol 17:604–610

Zhang XY, Fan X, Qiu YJ, Li CY, Xing S, Zheng YT, Xu JH (2014) Newly identified thermostable esterase from *Sulfobacillus acidophilus*: properties and performance in phthalate ester degradation. Appl Environ Microbiol 80:6870–6878

Zhu Y, Li H, Ni H, Xiao A, Li L, Cai H (2015) Molecular cloning and characterization of a thermostable lipase from deep-sea thermophile *Geobacillus* sp. EPT9. World J Microbiol Biotechnol 31:295–306

Chapter 13
Halophilic Bacteria and Archaea as Producers of Lipolytic Enzymes

María de Lourdes Moreno, M. Carmen Márquez, María Teresa García, and Encarnación Mellado

1 Introduction

Extremophiles are microorganism adapted to live in a wide range of extreme conditions, including salinity, pH, temperature and pressure, among others (Pikuta et al. 2007). Hypersaline environments such as solar marine salterns constitute the natural habitat for a group of microorganisms termed halophiles that are adapted to live and thrive in high salt concentrations (Grant et al. 1998; Ventosa 2006; Ventosa et al. 2014).

Halophilic microrganisms constitute a population considerably diverse phylogenetically, including representatives from within each of the following categories, e.g.: prokaryotes including archaea and bacteria, and eukaryotes including fungi and yeasts, protists, protozoa and algae, which have been studied concerning to their ecology, physiology and genetics (Ventosa et al. 1998; Uratani et al. 2014). This review will be related on the two main representative groups of microorganisms that live in these environments: the haloarchaea and the moderately halophilic bacteria. Haloarchaea are extremely halophilic aerobic members of the *Archaea* that require for growing at least 10 % NaCl and in most cases grow optimally in media containing 20–25 % NaCl. They are included in the family *Halobacteriaceae* and>70 species grouped in different genera are represented. Moderately halophilic bacteria have the ability to grow over a wider range of salt concentrations; they are considered as a very euryhaline group of microorganisms, growing optimally at 5–15 % NaCl (most at about 10 % NaCl) and are members of many different genera, including either only halophilic species or halophilic and non-halophilic representatives (Ventosa 2006).

M.d. L. Moreno • M.C. Márquez • M.T. García • E. Mellado (✉)
Faculty of Pharmacy, Department of Microbiology and Parasitology,
University of Sevilla, Profesor García González, 2, 41012 Sevilla, Spain
e-mail: emellado@us.es

P.H. Rampelotto (ed.), *Biotechnology of Extremophiles*, Grand Challenges in Biology and Biotechnology 1, DOI 10.1007/978-3-319-13521-2_13

Halophiles have evolved different strategies for the osmoadaptation to saline habitats: the "*salt-in*" strategy involves the accumulation of molar concentrations of inorganic cations that produce a saline cytoplasm and the "*salt-out*" strategy based on the synthesis and/or accumulation of low-molecular-weight organic compatible solutes that do not interfere with cell metabolism (Empadinhas and Da Costa 2008; Goh et al. 2011; Ceylan et al. 2012).

Because halophiles thrive under extreme environmental conditions naturally, they are viewed as a source of occurring enzymes with advantageous properties, known as extremozymes (Antranikian et al. 2005). The enzymes from halophiles in industrial applications are not limited to their activity at high salt concentrations (few industrial processes are running at saline conditions), since these extremozymes are also tolerant to organic solvents and high temperatures (Oren 2010).

Among the hydrolases, lipases (E.C.3.1.1.1) and carboxilesterases (E.C.3.1.1.3) constitute the more representative classes, which are widely distributed in nature from bacteria to higher eukaryotes. These enzymes catalyse different reactions such as esterification, transesterification and aminolysis (Jaeger et al. 1999). Lipolytic enzymes include both, lipases and esterases. While the lipases hydrolyze long-chain acylglycerols ($\geq C_{10}$), the esterases hydrolyze ester substrates with short-chain fatty esters ($\leq C_{10}$) (Jaeger et al. 1999). The difference between both enzymes is based on two criteria, the substrates able to recognize and the interface where these enzymes act. The lipases catalyze the hydrolysis of triglycerides at the oil–water interface; however, esterases catalyze reactions of water-soluble substrates. The lipases are activated at the interface of a biphasic system, containing the hydrophobic substrate in water (Brockerhoff and Jensen 1974; Jaeger et al. 1999). Lipases and esterases are included into the structural superfamily of α/β-hydrolases (Ollis et al. 1992), comprising enzymes which share the catalytic triad normally formed by Ser, His and Asp residues; the serine residue normally within the consensus sequence G-x-S-x-G (Arpigny and Jaeger 1999). This motif is usually located in a nucleophile elbow (Ollis et al. 1992). In contrast, most esterases present the motif S-x-x-K as the active site which is conserved also in the class C β-lactamases (Knox et al. 1996) and penicillin binding proteins (PBPs) (Joris et al. 1988).

The classification of lipolytic enzymes most commonly accepted was established by Arpigny and Jaeger (1999), grouping the microbial lipases into eight families on the basis of their conserved aminoacid residues and biological properties. Afterwards, other enzymes were characterized mainly using metagenomic approaches and six new families have been proposed (Handrick et al. 2001; Ewis et al. 2004; Lee et al. 2006, 2010; Kim et al. 2009; Bayer et al. 2010; Rao et al. 2011; Bassegoda et al. 2012; Fu et al. 2013).

The first lipases described were obtained from *Penicillium oxalicum* and *Aspergillus flavus* in 1935 (Kirsh 1935). Since them, lipases have acquired a great industrial relevance; thus, it is estimated these enzymes represent the third largest group of enzymes from the commercialization point of view (after proteases and carbohydrases) (Hasan et al. 2006). Lipolytic enzymes can be used in different industrial processes due to their wide-range versatility, being able to catalyze the hydrolysis of triglycerides and their synthesis from glycerol and fatty acids (Jaeger et al. 1999). Moreover, the quimio-regio selectivity showed by lipases on complex molecules is crucial for their use in industrial applications as biocatalysts, as these molecules display extreme difficulties to achieve selectivity. In addition, most industrial processes run under harsh conditions. Thus, it is important to isolate novel lipases exhibiting optimal activity at extreme of pH, temperature, different salt concentrations and in the presence of the usual organic compounds used as solvents in the industrial reactions (Mellado et al. 2005). In this sense, future trends relevant to the perspective of industrial biocatalyst include the characterization of lipolytic enzymes from halophiles as excellent alternative in industrial processes.

There have been a few reviews on lipases and esterases from extremophilic microorganism, including the halophiles, focused mainly on their biotechnological applications (Mellado et al. 2005; Ventosa et al. 2005; Salameh and Wiegel 2007; Litchfield 2011; Schreck and Grunden 2014; Dalmaso et al. 2015). In this chapter we will review the hypersaline sources of the main halophilic microorganisms producing lipolytic enzymes. We will focus on the diversity of these archaeal and bacterial producers, including a summary of the most important characteristics of the studied enzymes and highlighting their biotechnological potential.

2 Hypersaline Environments as a Source of Lypolytic Enzymes

Hypersaline environments constitute a large and diverse source of useful enzymes at industrial level. However, they have received little attention in comparison with other extreme habitats. In the last decades, different strategies have been used for discovering new lipases from hypersaline habitats, most of them based on culturing methodologies. More recently, functional metagenomic based approaches have also been successfully implemented (López-López et al. 2014). Among the conventional approaches, screening methods using saline solid media with the appropriate emulsified substrate for lipases have been the most thoroughly used up to now. In these media, the formation of halos surrounding colonies is a simple assay for detecting the presence of lipolytic activity. Tweens and tributiryn are the two alternative substrates most frequently used for lipase detection. In the case of Tweens, the appearance of halos around lipase-producing colonies is formed by released fatty acid salts; whereas the hydrolysis of tributiryn is detected by the presence of clear halos around the colonies. Screenings for specifically detecting true lipases have also been designed using longer substrates that are not hydrolyzed by esterases, such as emulsified olive oil in the presence of Victoria blue (blue color around the bacterial growth is observed) or the fluorescent dye rhodamine B (in this case an orange fluorescent color can be observed around the colonies after they are irradiated with UV at 350 nm) (Hasan et al. 2009; Babavalian et al. 2014).

Studies concerning the microbial diversity and extracellular lipases production in hypersaline habitats have been performed in a wide geographical distribution around the world. These studies have been focused mainly in solar salterns and saline lakes.

In 2003, Sánchez-Porro and colleagues studied the enzymatic diversity among halophiles of Spanish origin from water and sediment samples in various salterns located in south Spain (Sánchez-Porro et al. 2003). A group of 892 moderately halophilic bacteria producing different hydrolases were isolated. The hydrolytic activities obtained were quite similar when both sources (water and sediment) were compared. These authors also investigated a total of 21 moderately halophilic strains from culture collections but the majority of them did not have any of the hydrolytic activities tested. From the total of the fresh isolates, 217 were positive for lipase production. The characterization of 23 selected strains showed a wide phylogenetic diversity among them. Most of these isolates (15 strains) were Gram-negative rods belonging to the genera *Chromohalobacter*, *Halomonas* and *Salinibrivio*. These three genera include a large number of moderately halophilic *Gammaproteobacteria* which are frequently found in hypersaline habitats (Amoozegar et al. 2008b; de la Haba et al. 2012). On the other hand, the Gram-positive bacteria were assigned to the genera *Bacillus, Salibacillus, Salinicoccus* and *Marinococcus.* One of this strain, SM19, was proposed as a novel *Marinobacter* species (Martín et al. 2003) and an intracellular halophilic enzyme LipBL of this bacterium was further deeply characterized (Pérez et al. 2011, 2012). In addition, the draft genome sequence of strain SM19 was obtained and its analysis ratified the presence of genes not only for lipase, but also for other enzymes, such as amylase, protease, and DNase (Papke et al. 2013). On the other hand, in the course of a study performed in two salterns located also in the southwest region of Spain, Moreno and co-workers (2009) isolated 150 strains able to produce different hydrolases, such as protease, amylase, lipase and DNAse. From this study, 43 extremophilic halophilic strains were further identified by partial sequencing of their 16S rRNA gene. Only three isolates were bacteria related to the genera *Salicola, Salinibacter* and *Pseudomonas,* while the rest were halophilic archaea identified as members of the genera *Haloarcula, Halorubrum* and *Halobacterium.* From the 43 isolates selected, lipase activity was only detected in four strains. Furthermore, four strains presented combined hydrolytic activities, while three strains showed amylase and lipase activities and one strain displayed protease and lipase activities. Phylogenetically these six isolates were more closely related to the genus *Salicola* and to the species *Pseudomonas halophila, Halorubrum trapanicum* and *Halobacterium salinarum.* The isolate identified as *Salicola* sp. strain IC10, with protease and lipase activities, was selected for further studied due to its potential biotechnological interest (Moreno et al. 2009, 2013). These studies showed that the solar salterns of South Spain represent a valuable reservoir of byocatalyst useful in different economical areas.

In Turkey, Birbir et al. (2004) studied 12 extremely halophilic microorganisms isolated from a salt mine that exhibited activity against Tween-80. A polyphasic taxonomic approach showed that they belonged to the family *Halobacteriaceae*, a typical and dominant group in hypersaline habitats, and were related to the genera *Haloarcula, Halobacterium, Natrinema* and *Halorubrum* (Birbir et al. 2004). Three saline lakes in Turkey were also screened with the objective of detecting lipolytic enzymes in haloarchaea. On the basis of their high esterase activity, five strains were selected from an initial group of 118 isolates (Ozcan et al. 2009).

In a further characterization, the five isolates were taxonomically identified as member of the genera *Halovivax*, *Natrinema* and *Haloterrigena*, while one of them, named strain E49 was not closely related to any previously described species (Ozcan et al. 2012). This study indicated that lipase and esterase activities are often found in halophilic archaea.

Algeria is a country with many natural hypersaline habitats located generally in arid and semi-arid areas. In 2004, Hacěne and co-workers studied the microbial diversity in the Sebkha of El Golea, a saline lake located almost at the center of Algeria. These authors isolated 471 strains, which belonged to 31 genera within the domains *Archaea* and *Bacteria* (Hacěne et al. 2004). From this culture collection, Bhatnagar et al. (2005) selected 35 *Halobacteria* strains and screened them for lipolytic activity. The extremely halophilic archaeon *Natronococcus* strain TC6 was selected as it was the best lipase producer and it was tested for optimal lipolytic activity using different temperatures, pH values, substrates, and salt concentrations (Bhatnagar et al. 2005). Later, the enzyme produced by *Natronococcus* strain TC6 was characterized and described as the first true lipase identified in the *Archaea* domain (Boutaiba et al. 2006). On the other hand, Kharroub et al. (2014) performed a screening of extreme halophiles producing amylase, gelatinase and lipase activities, from three hypersaline Algerian lakes: Ezzemoul, Bethioua and Melghir. These lakes are subjected to high salinities (30–36 %, w/v) and high temperatures (up to 50 °C in summer) and thus, represent optimal environments for halophilic microbial growth, where diverse viable microbial communities with industrial applications may be available. These authors isolated 44 extremely halophilic archaea and found that the amylase activity was the predominant (22 strains) among them, whereas extracellular lipase activity was detected only in four strains phylogenetically related to the genera *Halorubrum*, *Halomicrobium* and *Haloferax* (Kharroub et al. 2014).

In the Bay of Bengal, Sana et al. (2007) isolated 400 strains from sediments of the Lothian Island. From this screening, five halotolerant bacteria with high extracellular esterase activity were selected and assayed for salt, acid, temperature and solvent tolerance. The best esterase producer of this group was identified as *Bacillus* sp. The isolation of two esterase enzymes obtained from other marine *Bacillus* bacterium had been firstly reported in 2005 by Karpushova and colleagues. Members of the genus *Bacillus* are characterized by their metabolic versatility, which gives them a high degree of ubiquity; thus, they have been isolated from many different sources, including saline habitats, and some of them have found application in different industries as enzyme producers (Karpushova et al. 2005; Ventosa et al. 2008; Liu et al. 2013).

Studies on the diversity of halophilic microorganisms showing hydrolase activities have also been carried out in Iran. In 2008 Amoozegar and coworkers reported the isolation of 55 moderately halophilic lipase producer bacteria from various salty environments (Amoozegar et al. 2008a). Strain SA-2 was selected for further study and placed in the genus *Salinivibrio* (Amoozegar et al. 2008b). On the other hand, Rohban et al. (2009) isolated and characterized 231 moderately halophilic bacterial strains from an Iranian hypersaline lake (Howz Soltan Lake) and assayed them for extracellular lipase activity. In this study, the greatest lypolitic activity was observed among members to the genera *Oceanobacillus, Halomonas* and *Gracilibacillus.* Two years later, 33 halophilic bacteria with lipase activity were isolated from the Maharlu Salt Lake and 13 moderately halophilic strains were selected for their activity and assigned to the Gram-positive genera *Bacillus* and *Staphylococcus* (Ghamesi et al. 2011). Recently, Babavalian et al. (2014) isolated 581 strains from three different salt lakes of Iran. Halotolerant, moderately and extremely halophilic bacteria able to produce nine different hydrolases (amylase, protease, lipase, DNase, inulinase, xylanase, carboxy methyl cellulase, pectinase and pullulanase) were isolated. Most lipase producers found in these habitats belonged to the phylum *Firmicutes* and were assigned to the genera *Piscibacillus, Bacillus, Oceanobacillus, Gracilibacillus, Halobacillus* and *Thalasobacillus,* followed by gamma *Proteobacteria* closely related to the genera *Idiomarina, Halomonas* and *Salicola.* However, a different distribution of these genera was found when the three lakes were compared.

The diversity among halophilic bacteria able to produce enzymes in saline habitats of India (west and south coast as well as Sambhar Salt Lake, located in Rajasthan region) was investigated by Kumar et al. (2012). These authors obtained 108 moderately halophilic bacteria with different hydrolase activities (amylases, lipases and proteases). Characterization of eight lipase producers by phenotypical and 16S rRNA sequence analysis related them closely to the genera *Marinobacter, Halomonas, Chromohalobacter* and *Geomicrobium.* As mentioned previously, *Marinobacter lipolyticus, Halomonas* sp. and *Chromohalobacter* sp. have been isolated from other hypersaline habitats and described for lipase production (Martín et al. 2003; Sánchez-Porro et al. 2003).

In 2012, Ardakani et al. described the screening for extracellular hydrolytic enzymes in Persan Gulf. Five moderately halophilic *Pseudoalteromonas* were isolated and tested for protease, lipase and amylase production using direct plating on agar media supplemented with skin milk, Tween-80 and starch as specific substrates of each enzyme. From the five isolates, two strains showed to have the three hydrolytic activities tested, having application for biotechnological purposes (Ardakini et al. 2012).

In conclusion, all these studies prove that the hypersaline ecosystems constitute optimal environments for halophilic bacterial and archaeal growth, where a diverse microbial community is available for discovering new lipolytic enzymes with novel features and potential use in industrial applications. Although among the hydrolases investigated, lipase producer microorganisms are apparently less abundant than protease and amylase microorganisms, further optimization of screening methodologies will contribute to a better understanding of this unexplored diversity.

3 Properties of Halophilic and Halotolerant Enzymes

As previously mentioned, halophiles require the presence of NaCl for growth. The enzymes isolated from this group of extremophiles are referred as halophilic enzymes, which commonly need high NaCl concentrations to survive, requiring salt for catalytic activity.

However, some proteins isolated from no halophilic sources tolerate high salt concentrations. These proteins are designated as halotolerant, and although they do not have salt dependence, they are active over a broad range of NaCl concentrations (Graziano and Merlino 2014).

Halophiles have developed two different basic mechanisms to achieve an osmotic equilibrium: the salt-in-cytoplasm and the organic-osmolyte accumulation (Goh et al. 2011; Ceylan et al. 2012).

Haloarchaea accumulate Na^+ and K^+ ions in the cytoplasm to maintain osmotic balance with the external medium. Therefore, halophilic adaptation includes different mechanisms at both structural and functional levels to overcome the effect of salt accumulation (Madern et al. 2000; Joo and Kim 2005). Different purification and crystallographic studies of halophilic proteins have led to understand their osmoregulation capacity and other unique features (Detkova and Boltyanskaya 2007). Concerning the surface of the halophilic enzymes, two important differences have been detected in the crystallographic studies. On one hand, the excess of acidic aminoacids in the enzyme surface, contributing to the formation of a hydration shell, responsible of the stability (Sinha and Khare 2014; Graziano and Merlino 2014) and on the other hand, the low number of hydrophobic residues found in the surface (Siglioccolo et al. 2011). The acidic residues conforms salt bridges with basic residues distributed among the enzymes (Lanyi 1974; Eisenberg et al. 1992; Danson and Hough 1997; Madern et al. 2000; Detkova and Boltyanskaya 2007; Bonete and Martínez-Espinosa 2011; Sinha and Khare 2013). Oren (2013) reviewed the proteomes of halophiles in order to clarify the different adaptation strategies detected in this group of microorganisms.

In general, proteins from haloarchaea require high salt concentrations to maintain their activity; these concentrations reach values as high as 2–4 M Na^+ or K^+, although there are some exceptions. These proteins not only require high salt concentrations to be active, but they also need high salt concentrations to maintain their stability.

Thus, most of these proteins are denatured in solutions with a lower content of 1.2 M NaCl. Moreover, these proteins present a high proportion of glutamic acid since this amino acid has a high ability to be hydrated. These characteristics have been found in several proteins of haloarchaea such as a dihydrofolate reductase and a isocitrate deshidrogenase from *Haloferax volcanii* (Pieper et al. 1998), a malate deshydrogenase and a 2Fe-2S ferredoxine from *Haloarcula marismortui* (Zaccai et al. 1989; Madern et al. 1995; Mevarech et al. 2000) and a glutamate deshidrogenase from *Halobacterium salinarum* (Britton et al. 1998). An esterase from *Haloarcula marismortui* has been studied. A 3D model was built confirming the abundance of acidic residues in its surface and the reduction in basic residues (Müller-Santos et al. 2009). Moreover, the analysis of the genome of some representative of the haloarchaea confirms the above-mentioned characteristics (Ng et al. 2000; Lee et al. 2014).

In spite of *Salinibacter ruber* (Antón et al. 2002) is an extreme halophile, this bacterium uses the same strategy than haloarchaea to cope the osmotic stress, by accumulation of K^+ and Na^+ in its cytoplasm. However, the study of the enzymes from *Salinibacter ruber* determined that these enzymes do not follow a general pattern and that individual behaviour explains their functionality at different salt concentrations. Thus, characterized enzymes as the isocitrate deshidrogenase, the fatty acid sintetase and a glucose-6-P- deshidrogenase exhibited behaviour similar to the halophilic enzymes, being highly dependent on salt for their activity (Oren and Mana 2002; Oren et al. 2003). However, other characterized enzymes from *S. ruber* are active both in the absence and presence of salt, such as an isocitrate dehydrogenase (Oren and Mana 2002) and a glycerol kinase (Sher et al. 2004).

Most moderate halophiles and certain extremely halophilic bacteria accumulate compatible solutes to cope the osmotic stress in saline environments. Thus, these bacteria do not present high ionic concentrations in the cytoplasm. Some authors have observed a slight increase in acidic residues in the characterized proteins of these organisms although it is not as high as in haloarchaea (Gandbhir et al. 1995; Coronado et al. 2000; Moreno et al. 2010).

The compatible solutes include organic compounds such as glycerol, sugars and their derivatives and quaternary amines (glycine betaine and ectoines). Moreover, it has been reported in several microorganisms, the accumulation of different compatible solutes (e.g. mannosylglycerate or di-myo-inositol phosphate) depending on the stress to which the organism is exposed (temperature or salinity) (Esteves et al. 2014) and even in the resistance against radiation (Webb and DiRuggiero 2012). *Halobacillus halophilus* synthesize different compatible solutes to cope the osmotic stress, switching its osmolyte strategy with the external salinity and growth phase (Saum et al. 2013).

In conclusion, the mechanisms of adaptation to salt exhibited by the enzymes from extreme halophiles are different to those of enzymes from moderate halophiles. While the enzymes from extreme halophiles have to withstand high salt concentrations both inside (cytoplasm) and in the extracellular medium, the enzymes from moderate halophiles found a low concentration of salt in the cytoplasm and an increase in the extracellular medium, so that they must be adapted to this changes in salinity.

The three-dimensional structures for a hundred of lipases have been determined (http://www.rcsb.org.pdb/home/), but only few structures of lipases from halophiles have been elucidated (Jiang et al. 2012). The resolution of the crystal structure of some of the promising enzymes described in this review will give some clues about the intrinsic structural factors that determine the strategy of haloadaptation of this group of proteins.

4 Halophilic and Halotolerant Lipolytic Enzymes of Potential Biotechnological Interest

4.1 Lipolytic Enzymes from Bacteria

There are numerous reports about lipase production from halophiles as previously described (Boutaiba et al. 2006; Amoozegar et al. 2008a; Niño de Guzman et al. 2008; Ozcan et al. 2009; Rohban et al. 2009; Ghamesi et al. 2011; Moreno et al. 2012);

however, scarce studies on their purification and characterization have been reported. Table 13.1 shows the described enzymes, the organism from which they originated, and their characteristics.

The regiospecificity, thermostability, substrate specificity, pH optimum and kinetics in non-aqueous system are especially varying properties in lipases from different sources (Gupta et al. 2004). Enantioselectivity, wide range of substrate utilization coupled with an enhanced efficiency in non-aqueous media are special characteristics desirable in new lipases (Doukyu and Ogino 2010).

The versatility of enzymes to exert its role both in the absence and presence of salts in the medium is one of the important features from the biotechnology point of view because the enzyme can be used in both situations in industrial processes, while other lipolytic enzymes fall off dramatically and irreversibly when the enzyme is exposed to low salt concentrations. As indicate in Table 13.1, extreme halotolerance has been reported in certain lipases from halophiles. An extracellular lipase from the genus *Salinivibrio*, SA-2 exhibited maximum activity with salt free enzyme solution although it showed high salt tolerance (0–3 M NaCl) (Amoozegar et al. 2008a). The extracellular esterase of *Halobacillus* sp. strain LY5 was also reported to retain its activity at high and low salt concentrations or even in its absence (Li et al. 2012). The lipase from *Idiomarina* sp. W33 was highly active and stable over a broad NaCl concentration range (2.5–25 %) and retained 70 % of its total activity in the absence of NaCl (Li et al. 2014). The characterization of the esterase PE10 from *Pelagibacterium halotolerans* $B2^T$ revealed its strong tolerance to NaCl, showing activity from 0 to 4 M NaCl although it was most active under 3 M NaCl. The three-dimensional modeling of PE10 allowed elucidating its tolerance to high salinity environments through the correlation with the high negative electrostatic potential on the surface of this enzyme (Jiang et al. 2012). In spite of the maximal activity of the halotolerant lipase LipBL from *Marinobacter lipolyticus* SM19 occurred in the absence of NaCl, the enzyme demonstrated 20 % activity at NaCl concentrations up to 4 M (Pérez et al. 2011).

The application of lipases in industrial processes operating under high temperatures is based on the desirable thermostability (Janssen et al. 1994). Thermal stability in proteins is attributed to the combination of factors like increased ionic interactions, decreased hydrophobic surface area, improved core packing, helix stabilization and reduced conformational strain (Sinha and Khare 2013). It has been proven that the presence of salt regulates the stability of halophilic proteins at high temperatures (Boutaiba et al. 2006). Most halophilic lipases were reported to show maximal activities in a temperature range from 45 °C to 55 °C (Table 13.1). However, excellent thermostability was observed in certain lipases. The extracellular lipase from *Chromohalobacter* sp. LY7-8 classified as moderately thermoactive showed optimal activity at 60 °C although the enzyme was highly active over a broad range of temperatures (30–90 °C) (Li and Yu 2012). The lipase from *Idiomarina* sp. W33 displayed excellent thermostability under temperatures ranging from 20 °C to 80 °C and the enzyme retained>80 % activity at 90 °C for 24 h although its optimal activity was found at 60 °C (Li et al. 2014). The lipase from *Marinobacter lipolyticus* SM19 could be used in numerous industrial processes as

Table 13.1 Characteristics of halophilic/halotolerant lipolytic enzymes and notable features industrially

Organism	Bacteria phylum	Enzyme	Optimal NaCl concentration (M)	Optimal temperature (°C)	Optimal pH	Substrate specificity (optimal substrate)	Interesting properties or comments from biotechnology point of view	References
Salinivibrio sp. strain SA-2	γ-Proteobacteria	Extracellular lipase	0	50	7.5	Short-chain *p*-nitrophenyl	Moderate thermoactivity (90 % activity at 60 °C for 4 h and 80 °C 30 min) High salt halotolerance (3 M NaCl, KCl, Na_2SO_4 and $NaNO_3$)	Amoozegar et al. (2008a)
Thalassobacillus sp. strain DF-E4	Firmicutes	Extracelular carboxylesterase	0.5	40	8.5	Short-chain *p*-nitrophenylesters (*p*-NPB)	Alkalistable (6–9.5)	Lv et al. (2011)
Halobacillus sp. strain LY5	Firmicutes	Extracellular esterase	1.7	50	10	Short-chain *p*-nitrophenylesters (*p*-NPB)	Thermoactive (30–90 °C) Alkalistable (pH 6–12) Halotolerant (0–20%NaCl) Remarkable stability towards surfactants (SDS and triton X-100)	Li et al. (2012)
Pelagibacterium halotolerans B2^T	α-Proteobacteria	Esterase	3	45	7.5	Short-chain *p*-nitrophenylesters (*p*-NPA)	Halotolerant (0–4 M NaCl)	Jiang et al. (2012)
Chromohalobacter sp. strain LY7-8	γ-Proteobacteria	Extracellular lipase	2.1	60	9.0	Long-chain *p*-nitrophenylesters (*p*-NPM)	Thermoactive (30–90 °C) Alkalistable (pH 6–12) Halotolerant (0–20%NaCl) High activity in presence of surfactants (SDS and triton X-100)	Li and Yu (2012)
Marinobacter lipolyticus strain SM19	γ-Proteobacteria	Intracellular lipase	0	80	8.0	Short-medium-chain *p*-nitrophenylesters (*p*-NPH)	Remarkable activity and stability in organic solvents Alkalistable (pH 6–10) Thermostable (5–90 °C) Halotolerant (0–3 M NaCl) High efficiency in enzymatic hydrolysis of oils (enrichment of PUFAs)	Pérez et al. (2011)

Salimicrobium sp. strain LY19	Bacteoidetes/ Chlorobi	Extracellular esterase	0.8	50	7.0	Short-chain *p*-nitrophenylesters (*p*-NPB)	High activity and stability in the presence of hydrophobic organic solvents (activity increased in isooctane) Halotolerant (2.5–25 % NaCl) Active and stable over broad ranges of temperature (20–70 °C), pH (7–10) and NaCl (2.5–25 %). Excellent thermostability (20–70 °C) Alkalistable (pH 7–11)	Xin and Hui-Ying (2013)
Staphylococcus sp. strain CJ3	Firmicutes	Extracelular lipase	0	45	8.0	Short-chain *p*-nitrophenylesters	Halotolerant (0–5 M NaCl) Good stability in organic solvents (long-chain alcohols)	Daoud et al. (2013)
Idiomarina sp. strain W33	γ-Proteobacteria	Extracellular lipase	1.7	60	7.0–9.0	Long-chain *p*-nitrophenylesters (*p*-NPM)	Excellent thermostable (20–90 °C) Alkali-stable (7–11) Halotolerant (2.5–25 %) Strong tolerance towards organic solvents Potentially useful for biodiesel production and in non-aqueous catalysis	Li et al. (2014)
Halobacillus sp. AP-MSU8	Firmicutes	Extracellular lipase	2.5	40	9.0	ND	ND	Esakkiraj et al. (2014)

ND not determined

it showed advantages over other halophilic and non-halophilic lipases. It was active over a wide range of temperature (from 5 °C to 90 °C) with maximum hydrolytic activity at 80 °C (Pérez et al. 2011). Excellent thermostability was reported by the extracellular esterase from *Salimicrobium* sp. LY19. The enzyme showed high stability under temperatures below 70 °C and after incubation at 80 °C for 2 h, the activity is still retained (Xin and Hui-Ying 2013).

Lipolytic enzymes to be used in organic synthesis require a high activity and stability (Doukyu and Ogino 2010). It is well known that organic solvents act as mild chaotropic agents disrupting hydrogen bonds between protein subunits and reducing the catalytic efficiency by affecting the critical water concentration at the active site. However, the stability in solvents is a general characteristic of halophilic enzymes (Zaccai 2004; Gupta and Khare 2009; Kumar and Khare 2012).

Moreover, in the hydrolytic extremozymes the role of the water is fundamental because of the thermodynamic equilibrium between the hydrolysis and synthesis reactions can move to the esters synthesis when working in organic media with low concentrations of water or to move to the hydrolysis controlling the degree of hydration of the enzyme (Ebel and Zaccai 2004). Saline environments have been reported as an excellent source of extremophiles producing hydrolases that are very stable in solutions containing organic solvents (Ventosa et al. 2005).

A possible explanation of the general behavior of most lipases can be based on the interaction of the hydrophobic amino acid residues of the lid covering the catalytic site of the enzyme with the organic solvent molecules, thereby the open conformation and catalysis property of the enzymes are maintained. Solvents with log P values lower than 2, distorting water-biocatalyst interactions and stripping essential water from enzymes cause more enzyme denaturation than solvents with high log P values (Laane et al. 1987). In contrast to the general tendency, the lipase LipBL from *Marinobacter lipolyticus* sp. SM19 was proved to be active in the presence of various organic solvents relevant in the industry both, hydrophilic and hydrophobic, increasing almost twofold in the presence of propan-2-ol or toluene among others (Pérez et al. 2012). Besides, the lipase from *Idiomarina* sp. W33 retained>80 % activity in the presence of glycerol, DMSO, toluene, cyclohexane or n-hexane and it retained >60 % activity in the presence of hydrophilic organic solvents (methanol and acetonitrile) (Li et al. 2014). However, the extracellular esterase from *Salimicrobium* sp. LY19 showed significant low instability in polar solvents as well as increased its activity in the presence of non-polar hydrophobic solvents (i.e. isooctane) (Xin and Hui-Ying 2013). The lipase from *Halobacillus* sp. AP-MSU 8 was stimulated by the presence of 10 % concentration of organic solvents such as benzene and acetone (Esakkiraj et al. 2014). The esterase from *Pelagibacterium halotolerans* PE10 showed stability in the presence of some detergents and organic solvents. This enzyme maintained >50 % of its activity in individual reaction mixes containing 1 % Triton X-100, 15 % DMSO or 15 % methanol (Jiang et al. 2012).

4.2 Lipolytic Enzymes from Archaea

The extreme conditions under the haloarchaea live, especially in solar salt evaporation ponds, imply reduced water activity due to the high salt content, high level of sunlight and often oligotrophic nutrient conditions (Litchfield 2011). Due to the unusual characteristics described by some enzymes produced by halophilic archaea, they are becoming the choice for various biocatalytic processes (Sellek and Chaudhuri 1999). An earlier review of biotechnological applications of haloarchaea summarized the possible uses in such diverse areas as bioplastics, wastewater treatment, agriculture, and the medical field (Margesin and Schinner 2001). Moreover, there have been a number of reviews on microbial lipases and esterases but the halophilic *Archaea* are generally not mentioned (Jaeger and Eggert 2002).

A preliminary screening for lipase activity in Algerian Sahara was performed and the alkaliphilic halophilic archaeon *Natronococcus* sp. was selected for its lipolytic activity (Bhatnagar et al. 2005). The discovery of the first true lipase detected in the Archaea domain caused the uncertainty of knowing its properties. Therefore, the thermostable lipase from *Natronococcus* sp. TC6 was partially characterized showing optimal activity at 4 M NaCl and no activity was detected in the absence of NaCl. This lipase was able to hydrolyze olive oil in the presence of high salt concentrations (Boutaiba et al. 2006).

The genome of the halophilic archaeon *Haloarcula marismortui* was sequenced and analyzed, finding genes coding for putative lipases and esterases (Baliga et al. 2004). Later, these genomic predictions were verified by Camacho et al. (2009) and the biochemical characterization of the enzyme was performed. Esterase activity was showed maximal at 45 °C and lipase activity at 50 °C however, previous studies using other haloarchaeal enzymes reported maximal activities between 50 °C and 60 °C, which was slightly higher (Studdert et al. 2001; Johnsen and Schonheit 2004; Hutcheon et al. 2005). Esterase and lipase activities were found using the crude extracts of *Haloarcula marismortui* at different NaCl concentrations (0.5 and 5 M), showing high activity in extremely saline conditions which imply that significant conformational changed occurred to maintain the activity (Camacho et al. 2009).

Esterases are intracellular, mainly involved in the hydrolysis of various types of carboxyl ester bonds, or extracellular as part of the general group of lipases. The carboxyl ester hydrolase of *Halobacterium* sp. NRC-1 has been studied by Camacho et al. (2010). Compared to other halophilic enzymes, the maximum intracellular esterase activity of *Halobacterium* sp. NRC-1 occurred at 80 °C.

The organic solvent-tolerant lipase from the haloarchaeon *Haloarcula* sp. G41 resulted to be halostable, thermostable, and alkali-stable. The lipase was highly active and stable over broad ranges of temperature (30–80 °C), pH (6–11) and NaCl concentrations (10–25 %), with an optimum at 70 °C, pH 8 and 15 % NaCl. This lipase showed activity and stability in the presence of various organic solvents. Interestingly, glycerol and n-hexane even increased the lipase activity. Contrary to the normal tendency of higher inactivation in the presence of the hydrophilic solvents, the lipase from strain G41 retained 50 % activity in their presence. Studies to improve the performance of the free and immobilized enzyme were carried out to catalyze the synthesis of biodiesel in *tert*-butanol. The immobilized lipase from strain G41 showed high efficiency (Li and Yu 2014).

5 Future Prospects

As previously described in this chapter, apart the ecological interest, halophiles constitute an interesting source of lipolytic enzymes with a great potential for biotechnological applications. The regioselectivity and versatility of lipolytic enzymes in aqueous and non-aqueous media make them excellent proteins to be used as biocatalysts in different reactions including the synthesis of biopolymers, the production of enantiopure pharmaceuticals and compounds used in the food industry (Hasan et al. 2006). The treatment of agricultural waste and wastes from food processing industries constitute other areas of interest for lipolytic enzymes (Jordan and Mullen 2007).

So far, most commercialized enzymes used in the above mentioned industrial applications have been isolated from mesophilic microorganisms (Jaeger and Eggert 2002). Notwithstanding, these enzymes do not function properly in industrial processes: lose activity rapidly during the reaction, low tolerance to organic solvents and they are not active at high temperatures. All of these factors increase the cost and lower the effectiveness of industrial use of enzymes. In fact, most characterized enzymes are highly sensitive and under certain conditions used in the chemical reactions enzymes denature, making difficult to handle. The stability of purified halophilic lipolytic enzymes summarized in this review opened new possibilities in industrial processes, being active in the presence of severe conditions.

Although the biotechnological industry has come to understand that halophiles represent a potential for biological prospecting of novel lipases with special characteristics, their biotechnological possibilities have not been extensively exploited as the commercialization of these enzymes has not been forthcoming. In fact, only a few lipases from halophiles are presently commercially exploited, as the lipase LipBL from *M. lipolyticus* (Pérez et al. 2011).

Halophiles have been found arduous to culture on a commercially significant scale. Therefore, enough scale to permit cost-effective isolation of the enzymes in sufficient quantity for marketing purposes is desirable. Moreover, in most cases, there is not an efficient scalable method, with expression levels typically poor; i.e. less than 5 % total cell protein (Pire et al. 2001). While different experimental methodologies have been used for the optimization of culture conditions of mesophilic lipase producers (Turki 2013), scarce studies concerning the improvement of halophilic microorganism for the production of lipolytic enzymes have been taken in a systematic way (Chauhan et al. 2013). Genetic engineering has been tried to express the commercial lipase-encoding genes in common prokaryotic (*Escherichia coli*) and eukaryotic (*Sacharomyces cerevisiae* and *Pichia pastoris*) expression host organisms in order to obtain high expression yields of the desired enzymes (Valero 2012). Mutagenic strategies have to be designed to develop "tailor-made" enzymes with improved stability and novel catabolic properties. Thus, the optimization of enzyme production and the economical overproduction of the characterized halophilic enzymes in novel heterologous hosts are great challenges for the future.

In recent years, environmental genome and whole genome sequencing have provided new opportunities for the biotechnological exploration of halophilic enzymes producers. As genes encoding hydrolases are particular interesting examples of nonessential genes in uncultured halophilic bacteria and archaea, it will be an important goal to compare the lipase genes of cultured and uncultured microorganisms to examine the relationship between the phylogeny of these genes.

Finally, we will highlight different future novel opportunities for lipases from halophiles.

5.1 Biodiesel Industry

In the last years, an increase in the production of biodiesel has been observed, mainly due to the increase in the petroleum price. Biodiesel (fatty acid alkyl esters) is conventionally produced in the industry by an alkali-catalysis method, characterized by high reaction rates. However, the presence of water and free fatty acids in the oils produces saponification of the ester, making difficult the downstream process (Uthoff et al. 2009). The use of lipases for biodiesel production has overcome most of the problems of usual chemical biodiesel production processes (Ribeiro et al. 2011; Hama and Kondo 2013). It is true that the enzymatic processes are still in the development stage and considerable efforts have to be made for the commercialization of novel lipases stable enough to not be inactivated during the transesterification reaction, the selection of cheaper oily substrates and the design of specific bioreactors for the easy recovery of the final product.

5.2 Pharmaceutical Industry

Concerning the use of halophilic lipases in the production of compounds useful in the pharmaceutical industries, the research will be focus on the development of new environmental friendly reactions, highly selective leading to the production of key compounds useful as intermediates in the synthesis of complex molecules of biological properties. The use of lipases as biocatalysts in the regio- and chemoselective preparation of partially acylated carbohydrates shows great interest as the products of these reactions are intermediate compounds of tailor-oligosaccharides and glycoconjugates of biomedical interest such as anticancer agents (Morris et al. 2011), antiviral (Howe et al. 2013) or vaccines (Peri 2013).

Thus, the complex glycoconjugates can be obtained from different disaccharide building blocks using glycal assembly methods. The challenge of appropriate protection and deprotection sequences play a determining role in achieving this goal, due to the high number of similar functional groups in these carbohydrates. A great advance in the synthesis of these interesting molecules would be the use of different and novel lipases capable to deacetylate and acetylate at different speeds and in different positions of the sugars so as to obtain a battery of regioisomers in each of the positions of a sugar, constituting building blocks to rapidly build-up the oligosaccharides of pharmaceutical interest.

5.3 Nutraceutical Industry

Excellent properties of natural polyphenols as potent antioxidants and their relation with prevention of human disease is well known (Díaz-Rubio et al. 2014). Some studies have highlighted the importance of the lipophilic character of the polyphenols, with a remarkable antioxidant capacity and with improved physicochemical properties. Therefore, new lipophilic antioxidants by modification of natural polyphenols are of great interest for the food industry. It is expected these modified analogues to have an improved bioavailability than either of the natural polyphenol used as starting material, being useful in the formulation of "beneficial foods" which exhibit beneficial health effects (Lucas et al. 2010; Gupta et al. 2013). The chemical methods described for the synthesis of these products are complex, presenting enormous technical difficulties due to the multifunctionality of the natural polyphenols, as has often to resort to lengthy protection-deprotection sequences to get a monoacylation at specific positions.

The enzyme-catalyzed approach could be an alternative to the chemical methods due to the regio and chemo selectivity of lipases. It would be of great interest to expand the repertoire of lipases robust enough able to perform chemo-regioselective acylation in natural polyphenols.

Acknowledgments This work was supported by grants from the Junta de Andalucía (P08-RMN-3515 and P11-CVI-7427).

References

Amoozegar MA, Salehghamari E, Khajeh K, Kabiri M, Naddaf S (2008a) Production of an extracellular thermohalophilic lipase from a moderately halophilic bacterium, *Salinivibrio* sp. strain SA-2. J Basic Microbiol 48:160–167

Amoozegar MA, Schumann P, Hajighasemi M, Fatemi AZ, Karbalaei-Heidari HR (2008b) *Salinivibrio proteolyticus* sp. nov. a moderately halophilic and proteolytic species from a hypersaline lake in Iran. Int J Syst Evol Microbiol 58:1159–1163

Antón J, Oren A, Benlloch S, Rodríguez-Valera F, Amann R, Rosselló-Mora R (2002) *Salinibacter ruber* gen. nov., sp. nov., a novel, extremely halophilic member of the Bacteria from saltern crystallizer ponds. Int J Syst Evol Microbiol 52:485–491

Antranikian G, Vorgias CE, Bertoldo C (2005) Extreme environments as a resource for microorganisms and novel biocatalysts. Adv Biochem Eng Biotechnol 96:219–262

Ardakini MR, Roayaie M, Poshtkouhian A, Amozeegar MA, Zolgharnein H (2012) Isolation of moderately halophilic *Pseudoalteromonas* producing extracellular hydrolytic enzymes from Persian Gulf. Indian J Microbiol 52:94–98

Arpigny LL, Jaeger K-E (1999) Bacterial lipolytic enzymes: classification and properties. Biochem J 343:177–183

Babavalian H, Amoozegar MA, Zahraei S, Rohban R, Shakeri F, Moghaddam MM (2014) Comparison of bacterial biodiversity and enzyme production in three hypersaline lakes; Urmia, Howz-Soltan and Aran-Bidgol. Indian J Microbiol 54:444–449

Baliga NS, Bonneau R, Facciotti MT, Pan M, Glusman G, Deutsch EW, Shannon P, Chiu Y, Weng RS, Gan RR, Hung P, Date SV, Marcotte E, Hood L, Ng WV (2004) Genome sequence of *Haloarcula marismortui*: a halophilic archaeon from the Dead Sea. Genome Res 14:2221–2234

Bassegoda A, Pastor FI, Diaz P (2012) *Rhodococcus* sp. strain CR-53 LipR, the first member of a new bacterial lipase family (family X) displaying an unusual Y-type oxyanion hole, similar to the *Candida antarctica* lipase clan. Appl Environ Microbiol 78:1724–1832

Bayer S, Kunert A, Ballschmiter M, Greiner-Stoeffele T (2010) Indication for a new lipolytic enzyme family: isolation and characterization of two esterases from a metagenomic library. J Mol Microbiol Biotechnol 18:181–187

Bhatnagar T, Boutaiba S, Hacène H, Cayol JL, Fardeau ML, Ollivier B, Baratti JC (2005) Lipolytic activity from *Halobacteria*: screening and hydrolase production. FEMS Microbiol Lett 248:133–140

Birbir M, Ogan A, Calli B, Mertoglu B (2004) Enzyme characteristics of extremely halophilic archaeal community in Tuzkoy salt mine, Turkey. World J Microbiol Biotechnol 20:613–621

Bonete MJ, Martínez-Espinosa RM (2011) Enzymes from halophilic archaea: open questions. In: Ventosa A, Oren A, Ma Y (eds) Halophiles hypersaline environments. Springer, Berlin/Heidelberg, pp 359–371

Boutaiba S, Bhatnagar T, Hacène H, Mitchell DA, Baratti JC (2006) Preliminary characterisation of a lipolytic activity from an extremely halophilic archaeon, *Natronococcus* sp. J Mol Catal B: Enzym 41:21–26

Britton KL, Stillman TJ, Yip KS, Forterre P, Engel PC, Rice DW (1998) Insights into the molecular basis of salt tolerance from the study of glutamate dehydrogenase from *Halobacterium salinarum*. J Biol Chem 273:9023–9030

Brockerhoff J, Jensen RG (1974) Lipolytic enzymes. In: Brockerhoff J, Jensen RG (eds) Lipases. Academic, New York, pp 25–175

Camacho RM, Mateos JC, González-Reynoso O, Prado LA, Córdova J (2009) Production and characterization of esterase and lipase from *Haloarcula marismortui*. J Ind Microbiol Biotechnol 36:901–909

Camacho RM, Mateos-Díaz JC, Diaz-Montaño DM, González-Reynoso O, Córdova J (2010) Carboxyl ester hydrolases production and growth of a halophilic archaeon, *Halobacterium* sp. NRC-1. Extremophiles 14:99–106

Ceylan S, Yilan G, Akbulut BS, Poli A, Kazan D (2012) Interplay of adaptive capabilities of *Halomonas* sp. AAD12 under salt stress. J Biosci Bioeng 114:45–52

Chauhan M, Chauhan RS, Garlapati VK (2013) Modelling and optimization studies on a novel lipase production by *Staphylococcus arlettae* through submerged fermentation. Enzyme Res 2013:353954, 8 p

Coronado MJ, Vargas C, Mellado E, Tegos G, Drainas C, Nieto JJ, Ventosa A (2000) The α-amylase gene *amy*H of the moderate halophile *Halomonas meridiana*: cloning and molecular characterization. Microbiology 146:861–868

Dalmaso GZL, Ferreira D, Vermelho AB (2015) Marine extremophiles: a source of hydrolases for biotechnological applications. Mar Drugs 13:1925–1965

Danson MJ, Hough DW (1997) The structural basis of protein halophilicity. Comp Biochem Physiol A Physiol 117:307–312

Daoud L, Kamoun J, Ali MB, Jallouli R, Bradai R, Mechichi T, Gargouri Y, Ali YB, Aloulou A (2013) Purification and biochemical characterization of a halotolerant *Staphylococcus* sp. extracellular lipase. Int J Biol Macromol 57:232–237

De la Haba RR, Márquez MC, Papke RT, Ventosa A (2012) Multilocus sequence analysis of the family *Halomonadaceae*. Int J Syst Evol Microbiol 62:520–538

Detkova EN, Boltyanskaya YV (2007) Osmoadaptation of haloalkaliphilic bacteria: role of osmoregulators and their possible practical application. Microbiology 76:511–522

Ribeiro BD, de Castro AM, Coelho MA, Freire DM (2011) Production and use of lipases in bioenergy: a review from the feedstocks to biodiesel production. Enzyme Res 2011:615803, 16 p

Díaz-Rubio ME, Pérez-Jiménez J, Martínez-Bartolomé MA, Álvarez I, Saura-Calixto F (2014) Regular consumption of an antioxidant-rich juice improves oxidative status and causes metabolome changes in healthy adults. Plant Foods Hum Nutr 70:9–14

Doukyu N, Ogino H (2010) Organic solvent-tolerant enzymes. Biochem Eng J 48:270–282

Ebel C, Zaccai G (2004) Crowding in extremophiles: linkage between solvation and weak protein-protein interactions, stability and dynamics, provides insight into molecular adaptation. J Mol Recognit 17:382–389

Eisenberg H, Mevarech M, Zaccai G (1992) Biochemical, structural, and molecular genetic aspects of halophilism. Adv Protein Chem 43:1–6

Empadinhas N, da Costa MS (2008) Osmoadaptation mechanisms in prokaryotes: distribution of compatible solutes. Int Microbiol 11:151–161

Esakkiraj P, Prabakaran G, Maruthiah T, Immanuel G, Palavesam A (2014) Purification and characterization of halophilic alkaline lipase from *Halobacillus* sp. Proc Natl Acad Sci India Sect B Biol Sci. doi:10.1007/s40011-014-0437-1

Esteves AM, Chandrayan SK, McTernan PM, Borges N, Adams MWW, Santos H (2014) Mannosylglycerate and Di-myo-Inositol phosphate have interchangeable roles during adaptation of *Pyrococcus furiosus* to heat stress. Appl Environ Microbiol 80:4226–4233

Ewis HE, Abdelal AT, Lu CD (2004) Molecular cloning and characterization of two thermostable carboxyl esterases from *Geobacillus stearothermophilus*. Gene 329:187–195

Fu J, Leiros HK, de Pascale D, Johnson KA, Blencke HM, Landfald B (2013) Functional and structural studies of a novel cold-adapted esterase from an Arctic intertidal metagenomic library. Appl Microbiol Biotechnol 97:3965–3978

Gandbhir M, Rasched I, Marliere P, Mutzel R (1995) Convergent evolution of amino acid usage in archaebacterial and eubacterial lineages adapted to high salt. Res Microbiol 146:113–120

Ghamesi Y, Rasoul-Amini S, Kazemi A, Zarrinic G, Morowvat MH, Kargar M (2011) Isolation and characterization of some moderately halophilic bacteria with lipase activity. Mikrobiologiia 80:477–481

Goh F, Jeon YJ, Barrow K, Neilan BA, Burns BP (2011) Osmoadaptive strategies of the archaeon *Halococcus hamelinensis* isolated from a hypersaline stromatolite environment. Astrobiology 6:529–536

Grant WD, Gemmel RT, GcGenity TJ (1998) Halophiles. In: Horikoshi K, Grant WD (eds) Extremophiles: microbial life in extreme environments. Wiley-Liss Inc., New York, pp 93–132

Graziano G, Merlino A (2014) Molecular bases of protein halotolerance. Biochim Biophys Acta 4:850–858

Gupta A, Khare SK (2009) Enzymes from solvent tolerant microbes: useful biocatalysts for non-aqueous enzymology. Crit Rev Biotechnol 29:44–54

Gupta R, Gupta N, Rathi P (2004) Bacterial lipases: an overview of production, purification and biochemical properties. Appl Microbiol Biotechnol 64:763–781

Gupta A, Kagliwal LD, Singhal RS (2013) Biotransformation of polyphenols for improved bioavailability and processing stability. Adv Food Nutr Res 69:183–217

Hacěne H, Rafaa F, Chebhounia N, Boutaibaa S, Bhatnagarb T, Baratti JC, Ollivier B (2004) Biodiversity of prokaryotic microflora in El Golea Salt Lake, Algerian Sahara. J Arid Environ 58:273–284

Hama S, Kondo A (2013) Enzymatic biodiesel production: an overview of potential feedstocks and process development. Bioresour Technol 135:386–395

Handrick R, Reinhardt S, Focarete ML, Scandola M, Adamus G, Kowalczuk M, Jendrossek D (2001) A new type of thermoalkalophilic hydrolase of *Paucimonas lemoignei* with high specificity for amorphous polyesters of short chain-length hydroxyalkanoic acids. J Biol Chem 276:36215–36224

Hasan F, Shah AA, Hameed A (2006) Industrial applications of microbial lipases. Enzyme Microbiol Technol 39:235–251

Hasan F, Shah AA, Hameed A (2009) Methods for detection and characterization of lipases: a comprehensive review. Biotechnol Adv 27:782–798

Howe JD, Smith N, Lee MJ, Ardes-Guisot N, Vauzeilles B, Désiré J, Baron A, Blériot Y, Sollogoub M, Alonzi DS, Butters TD (2013) Novel imino sugar α-glucosidase inhibitors as antiviral compounds. Bioorg Med Chem 21:4831–4838

Hutcheon GW, Vasisht N, Bolhuis A (2005) Characterisation of a highly stable alpha-amylase from the halophilic archaeon *Haloarcula hispanica*. Extremophiles 9:487–495

Jaeger KE, Eggert T (2002) Lipases for biotechnology. Curr Opin Biotechnol 13:390–397

Jaeger KE, Dijkstra BW, Reetz MT (1999) Bacterial biocatalysts: molecular biology, three-dimensional structures, and biotechnological applications of lipases. Annu Rev Microbiol 53:315–351

Janssen PH, Monk CR, Morgan HW (1994) A thermophilic, lipolytic *Bacillus* sp., and continuous assay of its p-nitro-phenyl-palmitate esterase activity. FEMS Microbiol Lett 120:195–200

Jiang X, Huo Y, Cheng H, Zhang X, Zhu X, Wu M (2012) Cloning, expression and characterization of a halotolerant esterase from a marine bacterium *Pelagibacterium halotolerans* B2^{T}. Extremophiles 16:427–435

Johnsen U, Schonheit P (2004) Novel xylose dehydrogenase in the halophilic archaeon *Haloarcula marismortui*. J Bacteriol 186:6198–6207

Joo WA, Kim CW (2005) Proteomics of halophilic archaea. J Chromatogr B Analyt Technol Biomed Life Sci 815:237–250

Jordan SN, Mullen GJ (2007) Enzymatic hydrolysis of organic waste materials in a solid-liquid system. Waste Manag 27:1820–1828

Joris B, Ghuysen JM, Dive G, Renard A, Dideberg O, Charlier P, Frère JM, Kelly JA, Boyington JC, Moews PC, Knox JR (1988) The active-site-serine penicillin-recognizing enzymes as members of the *Streptomyces* R61 DD-peptidase family. Biochem J 250:313–324

Karpushova A, Brümmer F, Barth S, Lange S, Schmid RD (2005) Cloning, recombinant expression and biochemical characterisation of novel esterases from *Bacillus* sp. associated with the marine sponge *Aplysina aerophoba*. Appl Microbiol Biotechnol 67:59–69

Kharroub K, Gomri MA, Aguilera M, Monteoliva-Sánchez M (2014) Diversity of hydrolytic enzymes in haloarchaea isolated from Algerian sabkhas. Afr J Microbiol Res 8:3992–4001

Kim EY, Oh KH, Lee MH, Kang CH, Oh TK, Yoon JH (2009) Novel cold-adapted alkaline lipase from an intertidal flat metagenome and proposal for a new family of bacterial lipases. Appl Environ Microbiol 75:257–260

Kirsh D (1935) Lipase production by *Penicillium oxalicum* and *Aspergillus flavus*. Bot Gaz 97:321–333

Knox JR, Moews PC, Frere JM (1996) Molecular evolution of bacterial β-lactam resistance. Chem Biol 3:937–947

Kumar S, Khare SK (2012) Purification and characterization of maltooligosaccharide-forming alpha-amylase from moderately halophilic *Marinobacter* sp. EMB8. Bioresour Technol 116:247–251

Kumar S, Karan R, Kapoor S, Singh SP, Khare SK (2012) Screening and isolation of halophilic bacteria producing industrially important enzymes. Braz J Microbiol 43:1595–1603

Laane C, Boeren S, Vos K, Veeger C (1987) Rules for optimization of biocatalysis in organic solvents. Biotechnol Bioeng 30:81–87

Lanyi JK (1974) Salt-dependent properties of protein from extremely halophilic bacteria. Bacteriol Rev 38:272–290

Lee MH, Lee CH, Oh TK, Song JK, Yoon JH (2006) Isolation and characterization of a novel lipase from a metagenomic library of tidal flat sediments: evidence for a new family of bacterial lipases. Appl Environ Microbiol 72:7406–7409

Lee MH, Hong KS, Malhotra S, Park JH, Hwang EC, Choi HK, Kim YS, Tao W, Lee SW (2010) A new esterase EstD2 isolated from plant rhizosphere soil metagenome. Appl Microbiol Biotechnol 88:1125–1134

Lee HW, Yim KJ, Song HS, Nam YD, Choi HJ, Seo MJ, Kim KN, Kim D, Roh SW, Rhee JK (2014) Draft genome sequence of *Halorubrum halophilum* B8^{T}, an extremely halophilic archaeon isolated from salt-fermented seafood. Mar Genomics 18:117–118

Li X, Yu H-Y (2012) Characterization of a novel extracellular lipase from a halophilic isolate, *Chromohalobacter* sp. LY7-8. Afr J Microbiol Res 14:3516–3522

Li X, Yu H-Y (2014) Characterization of an organic solvent-tolerant lipase from *Haloarcula* sp. G41 and its application for biodiesel production. Folia Microbiol 59:455–463

Li X, Yu H-U, Lin Y-F (2012) Purification and characterization of an extracellular esterase from a moderately halophilic bacterium, *Halobacillus* sp. strain LY5. Afr J Biotechnol 23:6327–6334

Li X, Qian P, Wu SG, Yu HY (2014) Characterization of an organic solvent-tolerant lipase from *Idiomarina* sp. W33 and its application for biodiesel production using Jatropha oil. Extremophiles 18:171–178

Litchfield CD (2011) Potential for industrial products from the halophilic *Archaea*. J Ind Microbiol Biotechnol 38:1635–1647

Liu L, Liu Y, Shin HD, Chen RR, Wang NS, Li J, Du G, Chen J (2013) Developing *Bacillus* spp. as a cell factory for production of microbial enzymes and industrially important biochemicals in the context of systems and synthetic biology. Appl Microbiol Biotechnol 97:6113–6127

López-López O, Cerdán ME, González Siso MI (2014) New extremophilic lipases and esterases from metagenomics. Curr Protein Pept Sci 15:445–455

Lucas R, Comelles F, Alcántara D, Maldonado OS, Curcuroze M, Parra JL, Morales JC (2010) Surface-active properties of lipophilic antioxidants tyrosol and hydroxytyrosol fatty acid esters: a potential explanation for the nonlinear hypothesis of the antioxidant activity in oil-in-water emulsions. J Agric Food Chem 14:8021–8026

Lv X-Y, Guo L-Z, Song L, Fu Q, Zhao K, Li A-X, Luo X-L, Lu W-D (2011) Purification and characterization of a novel extracellular carboxylesterase from the moderately halophilic bacterium *Thalassobacillus* sp. strain DF-E4. Ann Microbiol 61:281–290

Madern D, Pfister C, Zaccai G (1995) Mutation at behaviour of malate deshydrogenase from *Haloarcula marismortui* in physiological salts. Eur J Biochem 230:1088–1095

Madern D, Ebel C, Zaccai G (2000) Halophilic adaptation of enzymes. Extremophiles 4:91–98

Margesin R, Schinner F (2001) Potential of halotolerant and halophilic microorganisms for biotechnology. Extremophiles 5:73–83

Martín S, Márquez MC, Sánchez-Porro C, Mellado E, Arahal DR, Ventosa A (2003) *Marinobacter lipolyticus* sp. nov., a novel moderate halophile with lipolytic activity. Int J Syst Evol Microbiol 53:1383–1387

Mellado E, Martín S, Sánchez-Porro C, Ventosa A (2005) Lipolytic enzymes from extremophilic microorganisms. In: Mellado E, Barredo JL (eds) Microorganisms for industrial enzymes and biocontrol. Research Signpost, Kerala/India, pp 25–43

Mevarech M, Frolow F, Gloss LM (2000) Halophilic enzymes: proteins with a grain of salt. Biophys Chem 86:155–164

Moreno ML, Garcia MT, Ventosa A, Mellado E (2009) Characterization of *Salicola* sp. IC10, a lipase- and protease producing extreme halophile. FEMS Microbiol Ecol 68:59–71

Moreno ML, García MT, Ventosa A, Iglesias-Guerra F, Mellado E (2010) The extremely halophilic bacterium *Salicola marasensis* IC10 accumulates the compatible solute betaine. Syst Appl Microbiol 33:308–310

Moreno ML, Piubeli F, Bonfá MRL, García MT, Durrant LR, Mellado E (2012) Analysis and characterization of cultivable extremophilic hydrolytic bacterial community in heavy-metal-contaminated soils from the Atacama Desert and their biotechnological potentials. J Appl Microbiol 113:550–559

Moreno ML, Pérez D, Garcia MT, Mellado E (2013) Halophilic bacteria as a source of novel hydrolytic enzymes. Life 3:38–51

Morris JC, Chiche J, Grellier C, Lopez M, Bornaghi LF, Maresca A, Supuran CT, Pouyssegur J, Poulsen S-A (2011) Targeting hypoxic tumor cell viability with carbohydrate-based carbonic anhydrase IX and XII inhibitors. J Med Chem 54:6905–6918

Müller-Santos M, de Souza EM, Pedrosa Fde O, Mitchell DA, Longhi S, Carrière F, Canaan S, Krieger N (2009) First evidence for the salt- dependent folding and activity of an esterase from the halophilic archaea *Haloarcula marismortui*. Biochim Biophys Acta 1791:719–729

Ng WV, Kennedy SP, Mahairas GG, Berquist B, Pan M, Shukla HD, Lasky SR, Baliga NS, Thorsson V, Sbrogna J, Swartzell S, Weir D, May J, Dahl TA, Welti R, Goo YA, Leithauser B, Séller K, Cruz R, Danson MJ, Hough DW, Maddocks DG, Jablonski PE, Krebs MP, Angevine CM, Dale H, Isenbarger TA, Peck RF, Pohlschroder M, Spudich JL, Jung KW, Alam M, Freitas T, Hou S, Daniels CJ, Dennis PP, Omer AD, Ebhardt H, Lowe TM, Liang P, Riley M, Hood L, DasSarma S (2000) Genome sequence of *Halobacterium* species NRC-1. Proc Natl Acad Sci USA 97:12176–12181

Niño de Guzman M, Virginia A, Antezana H, Svoboda M (2008) Lipolytic enzyme production by halophilic/halotolerant microorganisms isolated from Laguna Verde, Bolivia. Rev Boliv Quim 25:14–23

Ollis DL, Cheah E, Cygler M, Dijkstra B, Frolow F, Franken SM, Harel M, Remington SJ, Silman I, Schrag J, Sussman JL, Verschueren KHG, Goldman A (1992) The alpha/beta hydrolase fold. Protein Eng 5:197–211

Oren A (2010) Industrial and environmental applications of halophilic microorganisms. Environ Technol 31:825–834

Oren A (2013) Life at high salt concentrations, intracellular KCl concentrations, and acidic proteomes. Front Microbiol 4:315

Oren A, Mana L (2002) Amino acid composition of bulk protein and salt relationships of selected enzymes of *Salinibacter ruber*, an extremely halophilic bacterium. Extremophiles 6:217–223

Oren A, Rodríguez-Valera F, Antón J, Benlloch S, Rosselló-Mora R, Amann R, Coleman J, Russell NJ (2003) Red, extremely halophilic, but not archaeal: the physiology and ecology of *Salinibacter ruber*, a bacterium isolated from saltern crystallizer ponds. In: Ventosa A (ed) Halophilic microorganisms. Springer, Berlin/Heidelberg, pp 63–76

Ozcan B, Ozyilmaz G, Cokmus C, Caliskan M (2009) Characterization of extracellular esterase and lipase activities from five halophilic archaeal strains. J Ind Microbiol Biotechnol 36:105–110

Ozcan B, Ozyilmaz G, Cihan A, Cokmus C, Caliskan M (2012) Phylogenetic analysis and characterization of lipolytic activity of halophilic archaeal isolates. Mikrobiologiia 81:205–213

Papke RT, de la Haba R, Infante-Domínguez C, Pérez D, Sánchez-Porro C, Lapierre P, Ventosa A (2013) Draft genome sequence of the moderately halophilic bacterium *Marinobacter lipolyticus* strain SM19. Genome Announc 1:e00379–13

Pérez D, Martin S, Fernández-Lorente G, Filice M, Guisán JM, Ventosa A, Garcia MT, Mellado E (2011) A novel halophilic lipase, LipBL, showing high efficiency in the production of eicosapentaenoic acid (EPA). PLoS One 6, e23325

Pérez D, Kovacic F, Wilhelm S, Jaeger KE, Garcia MT, Ventosa A, Mellado E (2012) Identification of amino acids involved in hydrolytic activity of lipase LipBL from *Marinobacter lipolyticus*. Microbiology 158:2192–2203

Peri F (2013) Clustered carbohydrates in synthetic vaccines. Chem Soc Rev 42:4543–4556

Pieper U, Kapadia G, Mevarech M, Herzberg O (1998) Structural features of halophilicity derived from the crystal structure of dihydrofolate reductase from the Dead Sea halophilic archaeon, *Haloferax volcanii*. Structure 6:75–88

Pikuta EV, Hoover RB, Tang J (2007) Microbial extremophiles at the limits of life. Crit Rev Microbiol 33:183–209

Pire C, Esclapez J, Ferrer J, Bonete MJ (2001) Heterologous overexpression of glucose dehydrogenase from the halophilic archaeon *Haloferax mediterranei*, an enzyme of the medium chain dehydrogenase/reductase family. FEMS Microbiol Lett 200:221–227

Rao L, Xue Y, Zhou C, Tao J, Li G, Lu JR, Ma Y (2011) A thermostable esterase from *Thermoanaerobacter tengcongensis* opening up a new family of bacterial lipolytic enzymes. Biochim Biophys Acta 1814:1695–1702

Rohban R, Amoozegar MA, Ventosa A (2009) Screening and isolation of halophilic bacteria producing extracellular hydrolyses from Howz Soltan Lake, Iran. J Ind Microbiol Biotechnol 36:333–340

Salameh M, Wiegel L (2007) Lipases from extremophiles and potential for industrial applications. Adv Appl Microbiol 61:253–283

Sana BGD, Saha M, Mukherjee J (2007) Purification and characterization of an extremely dimethylsulfoxide tolerant esterase from a salt-tolerant *Bacillus* species isolated from the marine environment of the Sundarbans. Process Biochem 42:1571–1578

Sánchez-Porro C, Martín S, Mellado E, Ventosa A (2003) Diversity of moderately halophilic bacteria producing extracellular hydrolytic enzymes. J Appl Microbiol 94:295–300

Saum SH, Pfeiffer F, Palm P, Rampp M, Schuster SC, Müller V, Oesterhelt D (2013) Chloride and organic osmolytes: a hybrid strategy to cope with elevated salinities by the moderately halophilic, chloride-dependent bacterium *Halobacillus halophilus*. Environ Microbiol 15:1619–1633

Schreck SD, Grunden AM (2014) Biotechnological applications of halophilic lipases and thioesterases. Appl Microbiol Biotechnol 98:1011–1021

Sellek GA, Chaudhuri JB (1999) Biocatalysis in organic media using enzymes from extramophiles. Enzyme Microbiol Technol 25:471–482

Sher J, Elevi R, Mana L, Oren A (2004) Glycerol metabolism in the extremely halophilic bacterium *Salinibacter ruber*. FEMS Microbiol Lett 232:211–215

Siglioccolo A, Paiardini A, Piscitelli M, Pascarella S (2011) Structural adaptation of extreme halophilic proteins through decrease of conserved hydrophobic contact surface. BMC Struct Biol 11:50

Sinha R, Khare SK (2013) Thermostable proteases. In: Satyanarayana T, Littlechild J, Kawarabayasi Y (eds) Thermophilic microbes in environmental and industrial biotechnology. Springer, Dordrecht, pp 859–880

Sinha R, Khare SK (2014) Protective role of salt in catalysis and maintaining structure of halophilic proteins against denaturation. Front Microbiol 5:165

Studdert CA, Herrera-Seitz MK, Plasencia-Gil MI, Sanchez JJ, De Castro RE (2001) Purification and biochemical characterization of the haloalkaliphilic archaeon *Natronococcus occultus* extracellular serine protease. J Basic Microbiol 41:375–383

Turki S (2013) Towards the development of systems for high-yield production of microbial lipases. Biotechnol Lett 35:1551–1560

Uratani JM, Kumaraswamy R, Rodriguez J (2014) A systematic strain selection approach for halotolerant and halophilic bioprocess development: a review. Extremophiles 18:629–639

Uthoff S, Bröker D, Steinbüchel A (2009) Current state and perspectives of producing biodiesel like compounds by biotechnology. Microb Biotechnol 2:551–565

Valero F (2012) Heterologous expression systems for lipases: a review. In: Sandoval G (ed) Lipases and phospholipases, Methods in molecular biology. Humana Press, New York, pp 161–178

Ventosa A (2006) Unusual micro-organisms from unusual habitats: hypersaline environments. In: Logan NA, Lappin-Scott HM, Oyston PCF (eds) Prokaryotic diversity: mechanisms and significance. Cambridge University Press, Cambridge, pp 223–253

Ventosa A, Nieto JJ, Oren A (1998) Biology of moderately halophilic aerobic bacteria. Microbiol Mol Biol Rev 62:504–544

Ventosa A, Sánchez-Porro C, Martin S, Mellado E (2005) Halophilic achaea and bacteria as a source of extracellular hydrolytic enzymes. In: Gunde-Cimerman N, Oren A, Plemenitas A (eds) Adaptation of life at high salt concentrations in *Archaea, Bacteria* and *Eukarya*. Springer, Berlin/Heidelberg, pp 337–355

Ventosa A, Mellado E, Sánchez-Porro C, Márquez MC (2008) Halophilic and halotolerant microorganisms from soils. In: Dion P, Nautiyal CS (eds) Microbiology of extreme soils. Soil biology. Springer, Berlin/Heidelberg, pp 87–115

Ventosa A, Fernández AB, León MJ, Sánchez-Porro C, Rodriguez-Valera F (2014) The Santa Pola saltern as a model for studying the microbiota of hypersaline environments. Extremophiles 18:811–824

Webb KM, DiRuggiero J (2012) Role of Mn2+ and compatible solutes in the radiation resistance of thermophilic bacteria and archaea. Archaea 2012:845756, 11 p

Xin L, Hui-Ying Y (2013) Purification and characterization of an extracellular esterase with organic solvent tolerance from a halotolerant isolate, *Salimicrobium* sp. LY19. BMC Biotechnol 13:108

Zaccai G (2004) The effect of water on protein dynamics. Philos Trans R Soc Lond B Biol Sci 359:1269–1275

Zaccai G, Cendrin F, Haik Y, Borochov N, Eisenberg H (1989) Stabilization of halophilic malate dehydrogenase. J Mol Biol 208:491–500

Chapter 14
Extremophilic Proteases: Developments of Their Special Functions, Potential Resources and Biotechnological Applications

Aneta Białkowska, Ewa Gromek, Tomasz Florczak, Joanna Krysiak, Katarzyna Szulczewska, and Marianna Turkiewicz

1 Introduction: General Characteristics and Classification of Proteases

Proteases (proteolytic enzymes, peptidases) were the first enzymes to be introduced into industrial practice a few decades ago. At first, they found application in the laundry detergent industry (Rao et al. 1998), which has remained the biggest recipient of protease preparations till this day (Kumar et al. 2008; Khan 2013). Currently, proteolytic enzymes, especially those of microbiological origin, are used on a wide scale in food, feed, textile, leather and pharmaceutical industries. They are also more and more widely applied in medical therapy and cosmetology. In terms of production, proteases represent the heart of the global market for enzymes which has been growing at a high rate for several decades now (Sarethy et al. 2011).

Proteases occur in various variants in all living organisms, from viruses and bacteria to a human being. The genes encoding proteases represent approx. 2 % of the total human genome (Li et al. 2013). They are involved not only in the process of dietary protein digestion and the cellular protein turnover, but also in numerous metabolic and regulatory processes that are significant for the proper functioning of an organism. Proteases are among the best-characterized group of enzymes and the studies devoted to them have handsomely contributed to the current state of knowledge on the relations between the structure and function of proteins.

In a formal sense, proteases were classified by the Nomenclature Committee of IUBMB (International Union of Biochemistry and Molecular Biology) as hydrolases (EC 3) that act on peptide bonds (subclass 4, EC 3.4), otherwise known as peptidases (Enzyme Nomenclature, http://www.chem.qmul.c.uk/iubmb/enzyme).

A. Białkowska (✉) • E. Gromek • T. Florczak • J. Krysiak • K. Szulczewska • M. Turkiewicz
Institute of Technical Biochemistry, Lodz University of Technology,
Stefanowskiego 4/10, 90-924 Lodz, Poland
e-mail: aneta.bialkowska@p.lodz.pl

P.H. Rampelotto (ed.), *Biotechnology of Extremophiles*, Grand Challenges in Biology and Biotechnology 1, DOI 10.1007/978-3-319-13521-2_14

The classification of this group of enzymes kept changing until the second half of the 1950s, when the first list of enzymes was officially presented. New proteins that hydrolyze peptide bonds were being described and molecular studies on highly diversified mechanisms, used by the enzymes in the process of catalysis, were being undertaken. Currently, the EC 3.4 class consists of 15 subclasses which include exo- (amino-, carboxy- and dipeptidases) and endopeptidases (aspartic, cysteine, serine, threonine, metallopeptidases and proteases of unknown catalytic mechanism; EC 3.4.21-25 and EC 3.4.99). At the same time, owing to the studies on the primary and secondary structure of proteolytic enzymes, we expand our knowledge on the evolutionary relatedness between these proteins, on the basis of which they were divided into aspartic, cysteine, glutamyl, serine, threonine proteases, metallopeptidases and proteases of mixed and unknown catalytic mechanism, including, along with hydrolases, asparagine peptide lyases. Each group is further divided into families, which include enzymes that display a great sequence homology, and clans, which include families with a similar 3D structure or the same catalytic amino acid sequence in a molecule. Currently, the MEROPS database (http://merops.sanger.ac.uk), which groups peptidases into families and clans and provides descriptions of thousands of individual peptidases, itemizes 353 families, 106 subfamilies and 49 clans of proteolytic enzymes, which shows how structurally diversified these proteins are. Therefore, despite the fact that they all hydrolyze peptide bonds, they can participate in various metabolic processes.

Our knowledge of proteolytic enzymes has considerably expanded owing to the studies on proteases derived from extremophilic microorganisms and, to a lesser extent, from extreme and moderate metagenomes, where genes of proteases with extremophilic properties might be present. In the following chapter we would like to present the current state of the research conducted in this field, including the actual use of extremophilic proteases in economy and their possible application in biotechnology.

2 Sources, Properties and Structural Adaptations of Extremophilic Proteases

2.1 Thermophilic Proteases

Thermophilic microorganisms constitute the greatest source of proteases, adapted to thrive at high temperatures. They can be classified into facultative thermophiles, which grow at temperatures of both 60–65 °C and 37 °C; obligate thermophiles, which grow at high temperatures of 65–70 °C, but do not grow at a temperature below 40 °C; extreme thermophiles, which thrive at temperatures of 40–70 °C and whose optimum growth temperature is 65 °C; and finally, hyperthermophiles, which require a temperature of 80–115 °C to grow (Kikani et al. 2010). Thermophilic microorganisms which produce proteolytic enzymes can be found in various biotopes, such as tropical soils (De Azeredo et al. 2004; Jaouadi et al. 2010a), marine geothermal springs (Klingeberg et al. 1995), including hydrothermal vents located 2500 m below the sea level at a pressure exceeding 250 atm., geysers (Matsuzawa et al. 1988; van den Burg et al. 1991), geothermal sediments, geothermal hot streams (Jang et al. 2002), solfataras (Morikawa et al. 1994; Chavez Croocker et al. 1999), volcanoes as well as fermenting compost (Hasbay Ifrij and Ogel 2002), boiling outflows of geothermal power plants, hot water pipeline systems (Dib et al. 1998), thermophilic digester fed with tannery waste and cattle manure (Majeed et al. 2013) and many other environments. Proteases adapted to high temperatures are isolated from hyperthermophilic and thermophilic archaea, bacteria and filamentous fungi. They are stable and active at temperatures above 60–70 °C or, in some cases, even above 100 °C (Sako et al. 1997; Morikawa et al. 1994). In comparison to their mesophilic counterparts they are more resistant to organic solvents, detergents (Jaouadi et al. 2010a; Lagzian and Asoodeh 2012), extreme pH values and other denaturing factors (Synowiecki 2010). They are used under high temperatures due to, for instance, a low solubility of substrates at a moderate temperature or endoergic character of the reaction catalyzed by enzymes. Higher temperature enhances the solubility of substrates as well as accelerates the diffusion thanks to a lower viscosity of the environment. It also decreases the possibility of contamination caused by meso- and psychrophilic microflora, which helps control the sterility of the process.

Increased thermoactivity and thermostability are determined by numerous structural adaptations. Even minor changes in the amino acid sequence can boost high temperature resistance. The difference in only three amino acid positions can cause that a neutral protease from *Bacillus caldolyticus* maintains 50 % of its activity after 30 min of incubation at a 8.2 °C higher temperature than Nrp protease from *Bacillus stearothermophilus* in a parallel experiment (van den Burg et al. 1991). Stability at higher temperatures depends on the level of molecular packing. A rise in temperature causes the structure to loose and increases the molecular flexibility, which can eventually lead to the loss of native properties as a result of irreversible conformational changes. Thermophilic proteases are characterized by a more closed-packed and rigid structure than their mesophilic and psychrophilic homologs, which helps them maintain their catalytic properties despite a high temperature. The molecular rigidity is influenced by additional ionic bonds, disulfide bridges, hydrogen bonds and hydrophobic interactions.

Furthermore, thermophilic enzymes that bind metal ions, have fewer empty inter-spaces as well as hydrophile regions on the surface and possess shorter loops and a lower number of residues, prone to deamination and oxidization (Li et al. 2005). In the case of most metals, especially Ca^{2+}, but also Mg^{2+}, Mn^{2+}, Zn^{2+} or Sr^{2+}, which bind either specifically or non-specifically with the molecule of an enzyme, the ionic bonding is essential to stabilize the structure and prevent unfolding at a higher temperature (De Azeredo et al. 2004; Li et al. 2007).

Electrostatic interactions in the molecules of thermophilic proteases are not only more numerous, but also more stable. In comparison with meso- and psychrophilic subtilisins, thermitase, a serine protease from *Thermoactinomyces vulgaris* (Kleine 1982), which displays the highest activity at a temperature of 85–90 °C and the highest stability at 70 °C, has a very complex and stable network of salt bridges, often formed by arginine residues (Tiberti and Papaleo 2011). If a salt bridge is removed from *Thermus aquaticus* YT-1 aqualysin I after an appropriate mutation, the thermostability of this protein is significantly reduced—there is a decrease in its half-life from 120 to 40 min at 80 °C (Matsuzawa et al. 1988).

Disulfide bridges are also significant when it comes to the stability and catalytic activity of aqualysin I at high temperatures. Their degradation through suitable mutations leads to a sharp decrease in thermostability. Moreover, it has been proved that the presence of proline residues in the surface loops of the protein contributes to its hyperstability (Sakaguchi et al. 2008). All these modifications ensure increased activity at high temperatures, but significantly reduce the activity at lower temperatures, which facilitates the inactivation of thermophilic proteases through cooling after the process is completed, so expensive inhibitors do not have to be used.

So far, it was possible to isolate and describe numerous thermophilic proteases (Table 14.1), active within the pH range of 2.0–12.0, whose optimum temperature of activity ranges from 45 to 115 °C. Those produced by Archaea proved to be the most thermophilic. They include serine proteases, cysteine proteases, metalloproteases and aspartic proteases. Strains of hyperthermophilic archaeon *Pyrococcus* sp. produce serine proteases with optimum temperature of activity above 100 °C. The S66 enzyme, isolated from *Pyrococcus furiosus,* shows maximum activity at 105 °C and pH 7.0. Its measured half-life at 98 °C is 33 h (Blumentals et al. 1990). Pyrolysin has an even higher optimum temperature of activity (115 °C), but it is less thermostable ($t_{1/2}$ only 9 h at 95 °C) than S66 (Eggen et al. 1990; Voorhorst et al. 1996). Serine proteases produced by *Aeropyrum pernix* K1 are equally active and thermostable. One of them, pernisine with an optimum temperature of 90 °C and optimum pH of 8.0–9.0 is the most thermostable serine protease of all the proteases described so far. It remains active for 4 h at 90 °C and its $t_{1/2}$ is 30 min at 120 °C. The presence of 1 mM Ca^{2+} further increases its thermostability to 4 h at 120 °C, without loss of its activity (Catara et al. 2003). *Sulfolobus solfataricus* produces SsMTP, a protease which shows optimum activity at 70 °C and pH 2.0 and is characterized by high thermostability (its half-life is 20 h at 80 °C).

Thermophilic bacterial proteases constitute an even more numerous group. So far, more than 40 proteases, active at temperatures between 50 and 95 °C, have been isolated and described. The highest optimum temperature of activity, 95 °C, has been reported in the case of a subtilisin-like protease from *Bacillus* sp. MLA64 (Lagzian and Asoodeh 2012). In comparison with other bacterial proteases, the subtilisin-like protease proves extremely thermostable as its measured half-life is 150 and 10 min at 100 °C and 120 °C, respectively. It remains active in the presence of Tween 80, SDS and Triton X-100, similarly to an alkaline serine protease from *Coprothermobacter proteolyticus*, which is additionally resistant to H_2O_2 and whose activity is further enhanced by SDS (Majeed et al. 2013). Another alkaline protease from *C. proteolyticus*, proteolysin, whose optimum temperature of protein hydrolysis is 85 °C at pH 9.5, tolerates the presence of organic solvents far better than commercially available subtilisin A (Toplak et al. 2013). A serine protease from *B. laterosporus* AK1 is another enzyme which remains stable in the presence of detergents. It reaches optimum activity at 75 °C and pH 9.0 and retains from 75 to 38 %

Table 14.1 Examples of characterized thermophilic proteases

Protease (PDB code)	Organism	T_{opt}	pH_{opt}	Thermostability	References
Archaea					
Pernisine, subtilisin-like serine protease	*Aeropyrum pernix* K1	90 °C	8.0	Stable at 90 °C, $t_{1/2}$ 60 min at 100 °C; in the presence of 1 mM $CaCl_2$ stable at 120 °C for 4 h	Catara et al. (2003)
Pernilase, serine protease (1VE6, 1VE7, 2QR5, 2QZP, 3O4G, 3O4H, 3O4I, 3O4J, 4RE5, 4RE6)	*Aeropyrum pernix* K1	90 °C	9.0	Stable for 4 h at 90 °C; $t_{1/2}$ 85 min at 100 °C and 12 min at 110 °C	Chavez Croocker et al. (1999)
Pyrolysin, serine protease	*Pyrococcus furiosus*	115 °C	6.5–10.5	Stable at 80 °C; $t_{1/2}$ 9 h at 95 °C, 4 h at 100 °C, 20 min at 105 °C and 3 min at 110 °C	Eggen et al. (1990), Voorhorst et al. (1996)
S66 Serine protease	*Pyrococcus furiosus*	105 °C	7.0	$t_{1/2}$ 33 h at 98 °C	Blumentals et al. (1990)
Aeropyrolysin, metalloproteinase	*Aeropyrum pernix* K1	110 °C in 1 mM $CaCl_2$	6.0–8.0	Stable at 100 °C with 1 mM $CaCl_2$, $t_{1/2}$ 2.5 h at 120 °C and 1.2 h at 125 °C with 1 mM $CaCl_2$	Sako et al. (1997)
Cysteine protease	*Pyrococcus* sp. KOD1	110 °C	7.0	Stable at 90 °C for 2 h; $t_{1/2}$ 60 min at 100 °C	Morikawa et al. (1994)
Thermopsin, acid protease	*Sulfolobus acidocaldarius*	90 °C	2.0	Stable at 80 °C for 48 h at pH 4.5	Lin and Tang (1990), Fusek et al. (1990)
SsMTP, thermopsin-like protease, acid protease	*Sulfolobus solfataricus* P2	70 °C (40–90 °C)	2.0	Stable at 50–80 °C, $t_{1/2}$ 20 days at 80 °C	Cannio et al. (2010)
Bacteria					
Subtilisin-like protease	*Bacillus* sp. MLA64	95 °C	9.5	Stable at 90 °C for 1 h, $t_{1/2}$ 25 min at 110 °C and 10 min at 120 °C	Lagzian and Asoodeh (2012)
Thermicin, subtilisin-like protease	*Thermoanaerobacter yonseiensis* KB-1	92.5 °C	9.0	$t_{1/2}$ 30 h at 80 °C	Jang et al. (2002)
Caldolysin (aqualysin I), serine protease	*Thermus aquaticus* T-351	75 °C	8.0	Stable at 80 °C, $t_{1/2}$ 1 h at 90 °C in 10 mM $CaCl_2$	Cowan and Daniel (1982)

(continued)

Table 14.1 (continued)

Protease (PDB code)	Organism	T_{opt}	pH_{opt}	Thermostability	References
Rt41A protease, serine protease	*Thermus* sp. Rt41A	90 °C in the presence of 5 mM $CaCl_2$	8.0	Stable at 70 °C for 24 h; $t_{1/2}$ 13.5 h at 80 °C in presence of 5 mM $CaCl_2$	Peek et al. (1992)
Proteolysin, serine protease	*Coprothermobacter proteolyticus* (expressed in *E. coli*)	85 °C	9.5	Stable at 70 °C for 22 h; $t_{1/2}$ 40 min at 80 °C	Toplak et al. (2013)
Serine protease	*Aquifex pyrophilus*	85–95 °C	7.0–9.0	$t_{1/2}$ 90 h at 85 °C; $t_{1/2}$ 6 h at 105 °C	Choi et al. (1999)
Alkaline serine protease	*Geobacillus stearothermophilus* B-1172 (expressed in *E. coli*)	60 °C	8.0	Stable at 90 °C for 1 h; $t_{1/2}$ 1 h at 100 °C	Iqbal et al. (2015)
Neutral protease	*Bacillus* sp. HUTBS62	80 °C	6.8	$t_{1/2}$ 6 h at 70 °C; $t_{1/2}$ 3 h at 90 °C	Aqel et al. (2012)
Actinomycetes					
TfpA, serine protease	*Thermomonospora fusca* YX (expressed in *Pichia pastoris*)	80 °C	8.5	Stable at 80 °C, 50 % activity after 15 min at 85 °C	Kim and Lei (2005)
Fungi					
Serine protease	*Chaetomium thermophilum*	60 °C	8.0	Stable at 60 °C, $t_{1/2}$ 60 min at 70 °C and 10 min at 90 °C	Li and Li (2009)
Keratinolytic alkaline proteinase	*Chrysosporium keratinophilum*	90 °C	9.0	$t_{1/2}$ 30 min at 90 °C	Dozie et al. (1994)

of its activity after 1 h of incubation with commercial detergents. It also remains stable at 80 °C, in the pH range of 7.0–11.0, which indicates its possible use as an ingredient of the new generation detergents (Arulmani et al. 2007). Some bacterial thermophilic proteases show high resistance to denaturing substances like urea and guanidine hydrochloride. Aqualysin I from *T. aquaticus* YT-1, with optimum pH of 10.0 at 80 °C, remains active in the presence of 7 M urea, 6 M guanidine hydrochloride and 1 % SDS (Matsuzawa et al. 1988), similarly to caldolysin from *T. aquaticus* T-351 (Cowan and Daniel 1982).

Proteases produced by thermophilic fungi and actinomycetes are not so numerous. In case of fungal enzymes, they include PRO33 and PRO66 from *Chaetomium thermophilum* (Li et al. 2007), thermomycolase from *Malbranchea pulchella* var. *sulfurea* (Ong and Gaucher 1976) and protease I from *Thermoascus aurantiacus* var. *levisporus* (Marcy et al. 1984). Among actinomycetes, we can differentiate between KERAB from *Streptomyces* sp. AB1 (Jaouadi et al. 2010a), keratinase from *Thermoactinomyces candidus* (Ignatova et al. 1999) and TfrA from *Thermomonospora fusca* YX (Kim and Lei 2005). Their optimum temperature of catalysis is lower than in the case of enzymes from archaea and bacteria, ranging from 45 to 70 °C. A keratinolytic alkaline proteinase from *Chrysosporium keratinophilum,* which is optimally active at 90 °C and pH 9.0, is one of the two exceptions (Dozie et al. 1994). The second one is a serine protease, TfpA from *T. fusca* YX, which is optimally active at 80 °C and pH 8.5. It remains stable under reaction conditions and its half-life at 85 °C is 15 min (Kim and Lei 2005).

2.2 *Psychrophilic Proteases*

Cold-adapted proteases are produced by psychrophilic and psychrotolerant microorganisms. Both these groups are capable of growth at 0 °C. In case of true psychrophiles, the optimum temperature of growth does not exceed 15 °C, whereas the maximum temperature is 20 °C. For psychrotolerant microorganisms, optimum growth temperatures are higher and range approx. from 20 °C to 35 °C (Morita 1975). Both of these groups inhabit permanently cold areas and environments with periodic temperature drops. They can be found in sea waters, arctic soils, glaciers, alpine soils and permafrost. They constitute a microflora of plants and animals living in cold areas as well as contaminations detected in cooled and frozen food.

One of the most important adaptations of cold-loving microorganisms, enabling them to sustain in the cold environment, is the production of enzymes that are kinetically and structurally adapted to thrive at low temperatures. Kinetic adaptations involve a relatively low optimum temperature of activity, which is usually 20–30 °C lower than their mesophilic counterparts and is in the range of 30–40 °C. Next characteristic features of this group of enzymes are higher values of catalytic constant k_{cat} and catalytic efficiency k_{cat}/K_m for a temperature range of 0–30 °C as well as low thermostability. The enzymes synthesized by psychrophiles exhibit much greater flexibility of the structure in comparison with their mesophilic and thermophilic counterparts, caused by weaker intramolecular interactions, especially in the active site, and the presence of surface loops that interact with the polar environment surrounding the enzyme. These proteins contain fewer proline residues, which increase the rigidity of a native conformation, and more glycine residues whose presence, especially in clusters, has a favorable influence on flexibility. A decrease in the concentration of basic amino acid residues, like arginine and lysine, may be observed. At appropriate pH of environment, these amino acids can form ionic bonds that add rigidity to protein molecules. An important feature of the structure is a decreased number of hydrophobic interactions, thanks to which the core of a molecule occupies less space than in the case of homologous mesophilic proteins. When it comes to cold-active enzymes, there is an increase in the number of hydrophilic interactions between a molecule and a solvent owing to the presence of an increased number of polar amino acid residues on the surface. Furthermore, in such a case, there is usually a better access to the active site, which offsets the decreased diffusivity of substrates (Feller 2003, 2013).

Among cold-active proteases-producing bacteria that have been described so far, the most dominant bacterial groups are *Pseudomonas, Pseudoalteromonas, Bacillus, Clostridium, Colwellia, Serratia, Shewanella, Sulfitobacter, Halomonas* and *Marinomonas.* Much less attention has been devoted to proteases from psychrophilic yeast and filamentous fungi (Table 14.2). One of the most interesting enzymes from this eukaryotic group is a psychropiezophilic *Aspergillus ustus* serine protease from deep-sea sediments of the Central Indian Basin, which maintains its activity even at 300 bar pressure (Damare et al. 2006). There are also other strains of proteolytic bacteria isolated from deep-sea sediments, for instance *Pseudoalteromonas* sp. SM9913 (Chen et al. 2003), *Halobacillus* sp. SCSIO 20089 (Yang et al. 2013), *Pseudomonas lundensis* (Yang et al. 2010) and *Colwellia psychrerythraea* (Huston et al. 2004). Often, these strains are also present in sea waters. Yuan et al. (2009), for instance, isolated a strain from the Sea of Japan known as *Enterococcus faecalis* TN-9, which produces psychrophilic metalloprotease. Another strain which secretes serine protease, *Pseudomonas aeruginosa* HY1215, was found in the Yellow Sea (Hao and Sun 2015). A similar enzyme is produced from a strain identified as *Penicillium chrysogenum* FS010, isolated from the Huanghai Sea (Zhu et al. 2009). Another example of a cold environment abundant in proteolytic strains is Antarctic soil from where yeast identified as *Cryptococcus humicola (*formerly *Candida humicola*) (Ray et al. 1992) has been isolated. It secretes the only one psychrophilic extracellular aspartyl protease from yeast that has been described in literature so far. Serine protease-producing bacteria such as *Stenotrophomonas maltophilia* (Vazquez et al. 2005) and *Clostridium* sp. (Alam et al. 2005) were also found there. Another important source of psychrophilic proteolytic microorganisms are highlands, where bacteria such as *Serratia marcescens* TS1 (Tariq et al. 2011), *S. maltophilia* MTCC 7528 (Kuddus and Ramteke 2009) and *Pedobacter cryoconitis* (Margesin et al. 2005) can be found.

The majority of psychrophilic proteases described in literature achieve a maximum activity at temperatures ranging from 30 to 40 °C (Table 14.2). Nevertheless, there are some enzymes that have a lower optimum temperature, for instance an

Table 14.2 Examples of extracellular proteases from psychrophilic microorganisms

Protease (PDB code)	Organism	T_{opt}	pH_{opt}	Thermal stability	References
Bacteria					
Serine protease	*Pseudoalteromonas* sp. SM9913	30 °C	7.0	Stable up to 30 °C for 20 min	Chen et al. (2003)
Serine protease	*Pseudoalteromonas* sp. NJ276	30 °C	8.0	Stable up to 40 °C for 400 min	Wang et al. (2008)
Serine protease	*Clostridium* sp.	37 °C	7.0	More than 40 % of activity after 1 h at 60 °C	Alam et al. (2005)
Serine protease	*Clostridium* sp. LP3	45 °C	7.0–8.0	Stable up to 50 °C for 30 min	Alam et al. (2006)
Serine protease	*Colwellia* sp. NJ341	35 °C	8.0	50 % of activity at 40 °C for 50 min	Wang et al. (2005)
Serine protease	*Pseudomonas aeruginosa* HY1215	25 °C	10.0	Stable up to 35 °C for 30 min	Hao and Sun (2015)
Serine protease	*Pseudomonas* sp. strain DY-A	40 °C	10.0	Stable up to 30 °C for 90 min	Zeng et al. (2003)
Serine protease	*Bacillus amyloliquefaciens* S94	45 °C	10.0—for protein substrate; 8.0—for synthetic substrate	n/d	Son and Kim (2003)
Serine protease	*Planomicrobium* sp. 547	35 °C	9.0	40 % of activity at 50 °C for 2 h	Yang et al. (2011)
Serine protease	*Bacillus* TA41	40 °C	9.0	Enzyme loss entire activity after 15 min exposure to 50 °C	Davail et al. (1994)
Serine protease	*Bacillus* TA39	n/d	9.0–10.0	$t_{1/2}$ 6 min at 50 °C	Narinx et al. (1997)
Metalloprotease	*Enterococcus faecalis* TN-9	30 °C	7.5–8.0	Stable up to 45 °C for 10 min	Yuan et al. (2009)
Metalloprotease	*Serratia marcescens* TS1	20 °C	8.5	n/d	Tariq et al. (2011)
Metalloprotease	*Pseudomonas lundensis*	30 °C	10.4	Stable at 25–40 °C for 2 h in the presence of Ca^{2+}	Yang et al. (2010)
Metalloprotease	*Stenotrophomonas maltophilia* MTCC 7528	20 °C	10.0	Stable up to 20 °C for 3 h	Kuddus and Ramteke (2009)
Metalloprotease	*Pedobacter cryoconitis*	40 °C	8.0	Stable up to 20–30 °C for 1 h	Margesin et al. (2005)
Metalloprotease (3CIA)	*Colwellia psychrerythraea* strain 34H	19 °C	7.0	$t_{1/2}$ at 30, 40, 45 and 50 °C, respectively 67, 38, 10 and 5 min	Huston et al. (2004)
Yeasts and filamentous fungi					
Serine protease	*Aspergillus ustus*	45 °C	9.0	80 % activity at 50 °C for 10 min; 55 % activity at 60 °C for 10 min	Damare et al. (2006)
Serine protease	*Leucosporidium antarcticum* (now *Glaciozyma antarctica*)	25 °C	8.0–8.5	Stable up to 30 °C	Turkiewicz et al. (2003)
Serine protease	*Penicillium chrysogenum* FS010	35 °C	9.0	10 % activity at 60 °C for 5 min	Zhu et al. (2009)
Aspartyl protease	*Candida humicola* (now *Cryptococcus humicola*)	37 °C	1.0–1.2	Stable up to 37 °C for 2 h	Ray et al. (1992)

alkaline protease from *Stenotrophomonas* sp. IIIM-ST045, isolated from a soil sample collected from the Thajiwas glacier of Kashmir, India, which displays significant activity in casein hydrolysis within a temperature range of 4–37 °C with maximum activity at 15 °C and at pH 10.0 (Saba et al. 2012). In turn, a metalloprotease from *S. maltophilia* MTCC 7528, isolated from the Gangotri glacier in the Western Himalaya, exhibits maximal activity at 20 °C and pH 10.0 during the hydrolysis of azocasein. The enzyme remains stable in commercially available detergents (maintains from 65 to 80 % of its activity after a 3 h incubation) and is an excellent remover of protein-containing stains at low temperatures. It is also resistant to repeated freezing and thawing which may be related with the habitat of *S. maltophilia* MTCC 7528 (Kuddus and Ramteke 2009). Similar features are showed by a metalloprotease from *P. cryoconitis,* isolated from an alpine glacier cryoconite (Margesin et al. 2005), and a metalloprotease produced by *P. lundensis* HW08 isolated from the Yellow Sea (Yang et al. 2010). Serine proteases from *P. aeruginosa* HY1215 (Hao and Sun 2015) and *Glaciozyma antarctica* (formerly *Leucosporidium antarcticum*) (Turkiewicz et al. 2003) are two other enzymes that can thrive at relatively low temperatures, with their optimum temperature of activity being 25 °C. They maintain 30 and 20 % of their maximum activity at 0 °C, respectively. Another important feature of cold-adapted enzymes is their thermolability, related with high catalytic efficiency at a temperature that is optimal for the existence of an enzyme producer. The majority of cold-adapted proteases described in literature possess a low thermostability, undergoing a partial or complete inactivation at temperatures ranging from 40 to 50 °C. In this context, it is worth mentioning three serine proteases produced by the following strains: *Clostridium* sp. (Alam et al. 2005), *Clostridium* sp. LP3 (Alam et al. 2006) and *Planomicrobium* sp. 547 (Yang et al. 2011). The first one maintains 40 % of its maximum activity after a 1 h preincubation at 60 °C, the second one—80 % of its activity after 30 min exposure to 50 °C, and the third one—40 % of its maximum activity after 2 h at 50 °C.

It is also important to mention the role of calcium ions, which stabilize the structure of particular proteases. The presence of these ions usually leads to an increase in thermostability, like in the case of metalloproteases from *P. lundensis* HW08 and *Flavobacterium psychrophilum.* In the absence of calcium ions, the first enzyme loses 20 % of its maximum activity, observed in the presence of 5 mM Ca^{2+} at temperatures between 20 and 35 °C, and 50 % at 40 °C (Yang et al. 2010). Regarding the second strain, the presence of 5 mM $CaCl_2$ causes the optimum temperature of metalloprotease Fpp2 to shift from 24 °C to 33–37 °C and leads to an increase in the thermostability of the enzyme, which after a 5 min exposure to 40 °C maintains 95 % of its activity that was otherwise completely lost in the absence of calcium ions (Secades et al. 2003).

A vast majority of psychrophilic proteases described in literature are enzymes showing maximum activity in a neutral or alkaline environment. So far, only one extracellular aspartic protease, with pH optimum of 1.0, has been described (Ray et al. 1992). In the context of detergent industry, proteases active at high pH values seem especially interesting. These include serine enzymes from such strains as *Pseudomonas* sp. strain DY-A (Zeng et al. 2003), *Bacillus amyloliquefaciens* S94

(Son and Kim 2003), *Bacillus* TA39 (Narinx et al. 1997) and metalloproteases from *P. lundensis* (Yang et al. 2010) and *S. maltophilia* MTCC 7528 (Kuddus and Ramteke 2009).

Metalloprotease Ps5 from *P. lundensis* (Yang et al. 2010) has a potential for the use in the laundry detergent industry as it maintains maximum activity during the hydrolysis of casein at 30 °C and pH 10.4. Moreover, it shows a significant stability in H_2O_2 solutions. In 1 % H_2O_2 solution Ps5 displays a 24 % higher stability than in a control sample, whereas in a 10 % solution its activity decreases only by 27 %. In addition, this enzyme remains stable in 1 % non-ionic detergent solutions (Tween 20 and Tween 80) and decreases its stability by half only in a 5 % solution. A serine protease from *P. aeruginosa* HY1215, a strain of bacteria isolated from the Yellow Sea (Hao and Sun 2015), has a similar stability. At concentrations from 0.2 to 1.2 % of H_2O_2, it has a 45 % higher activity than in a control sample, but in a 2 % H_2O_2 solution, the activity falls by 60 %. Furthermore, the enzyme exhibits a high stability in non-ionic surfactants, like Tween 40, Tween 80 and Triton X-100, retaining its total activity in solutions with a concentration of less than 1.8 % for 1 h.

2.3 *Alkaliphilic Proteases*

Alkaliphilic proteases show activity and stability in alkaline environments. They are produced by alkaliphiles (including thermo-, psychro- and haloalkaliphiles), which are organisms that grow optimally or very well at pH values above 9.0 but cannot grow or grow slowly at neutral pH values (Horikoshi 1999; Sarethy et al. 2011). These enzymes are also produced by other groups of extremophiles, which do not require high pH values to grow.

Alkaliphilic microorganisms secreting alkaliphilic proteases, were isolated from mud, soils from various environments, alkaline soda lakes, salt lakes, mine-water containment dam 3.2 km below the earth's surface in an ultra-deep gold mine, compost, tile-joint of a bathroom, feather samples collected at the shore of lakes and seawater (Gessesse et al. 2003; Saeki et al. 2002; Kobayashi et al. 2007; Karan et al. 2011; Dastager et al. 2008; Deng et al. 2010; Mitsuiki et al. 2002; Bakhtiar et al. 2002; Raval et al. 2014). These are mainly bacteria, including those classified as actinomycetes (Fig. 14.1) and haloalkaliphilic Archaea described in Sect. 2.4.

Alkaliphilic proteases are mainly extracellular serine enzymes (Table 14.3) inhibited by PMSF (phenylmethanesulfonyl fluoride) or 3,4-DCl (3,4-dichlorocoumarin), as in the case of PMSF-insensitive protease AL-20 from *Nesterenkonia* sp. (Gessesse et al. 2003). However, Kuddus and Ramteke (2009) described a unique metalloprotease which can be classified as alkaliphilic since it shows maximum activity in the hydrolysis of azocasein at pH 10.0, retains approx. 84 % of its maximum activity in pH range of 8.0–11.0 and remains stable in pH range of 8.0–10.0 for 1 h at 20 °C. Considering the fact that the producers of alkaliphilic proteases are isolated from various biotopes, the enzymes display a broad range of optimal temperatures, from 15 °C in the case of a protease from *Stenotrophomonas* sp. IIIM-ST045 (Saba et al. 2012)

Fig. 14.1 Phylogenetic tree showing relationship among the representatives of alkaliphilic microorganisms described in Sect. 2.3 prepared on the basis http://www.ncbi.nlm.nih.gov/Taxonomy/CommonTree/wwwcmt.cgi (Sayers et al. 2009; Benson et al. 2009)

to 70 °C in the case of a protease AL-20 from *Nesterenkonia* sp. AL-20 (Gessesse et al. 2003). The most important factor influencing the activity and stability of the enzymes is the alkalinity of the environment. Most often, the optimum pH value varies between 9.0 and 11.0, but Kobayashi et al. (2007) described a high-alkaline protease ALTP, isolated from strictly anaerobic and extremely alkaliphilic *Alkaliphilus transvaalensis*, which showed maximum activity in the hydrolysis of casein at pH 12.6 and 40 °C. The enzymes retain stability in a pH range of 5.0–12.0, as in the case of protease BCAP from *Bacillus clausii* I-52, which remained stable under such conditions for 72 h (Joo et al. 2003). It seems that the enzymes are able to thrive in high pH values thanks to some subtle structural differences that distinguish them from those with lower optimum pH values. Strongly alkaliphilic enzymes (opt. pH 10.0–11.0) showed increased number of alkaline Arg and His residues in a molecule and decreased number of acidic Asp and Glu residues. The increased concentration of arginine residues leads to the formation of more hydrogen bonds and ionic pairs (mainly Arg-Asp), which stabilize the structure of a protein in alkaline environments. Such amino acid composition results in higher isoelectric point values of these proteins (Fujinami and Fujisawa 2010).

Among alkaliphilic peptide hydrolases, there is a group identified as oxidatively stable serine proteases (OSPs) which is resistant to oxidizing agents (Saeki et al. 2002). It includes protease SAPB from *Bacillus pumilus* CBS, whose activity increased by 28 % in comparison to a control sample in the presence of 5 % H_2O_2 after a 24 h preincubation at 40 °C (Jaouadi et al. 2008), and a protease from *B. clausii* I-52, which retains its activity during a 72 h preincubation with 2.5 % sodium perborate at room temperature and increases its activity by 16 % in the presence of 5 % H_2O_2 in comparison with a control sample (Joo et al. 2003). Some enzymes remain stable also in the presence of detergents used under laboratory conditions (SDS, Triton X-100, Tween 20, cetrimonium bromide) and commercially available washing powders, a good example can be protease SABP. It remains active for 15 min at 65 °C and does not lose its stability for 24 h at 40 °C in the presence of 5 % Tween 20, Tween 80, Triton X-100 and 1 % SDS (Jaouadi et al. 2008) and for 1 h at 40 °C in the presence of commercially available liquid detergents, such as Dinol, Lav+, Nadhif, and solid detergents, like OMO, Dixan or Det (Jaouadi et al. 2009). A serine protease from *Bacillus subtilis* VSG-4 is another example of an

Table 14.3 Examples of extracellular alkaliphilic proteases from alkaliphilic microorganisms

Protease (PDB code)	Organism	T_{opt}	pH_{opt}	pH stability	References
Bacteria					
Serine protease AL-89	*Bacillus pseudofirmus* AL-89	60 °C	11.0	6.5–10.5 at 50 °C for 1 h	Gessesse et al. (2003)
Serine protease BCAP	*Bacillus clausii* I-52	60–65 °C	11.0	5.0–12.0 for 72 h	Joo et al. (2003), Joo and Chang (2005)
Serine protease KP-43	*Bacillus* sp. KSM-KP43	60 °C (without Ca^{2+}) 70 °C (with Ca^{2+})	11.0–12.0	6.0–12.0 for 24 h at 25 °C	Saeki et al. (2002)
Serine protease SABP	*Bacillus pumilus* CBS	65 °C	9.0–10.6	9.0–10.6 for 72 h at 40 °C	Jaouadi et al. (2008)
Serine protease ALTP	*Alkaliphilus transvaalensis*	70 °C	12.6	5.0–11.0 for 10 min at 50 °C	Kobayashi et al. (2007)
Alkaline protease	*Stenotrophomonas* sp. IIIM-ST045	15 °C	10.0	6.8–12.0 for 1 h	Saba et al. (2012)
Serine protease Ve_2-20-9_1	Strain Ve_2-20-9_1	50 °C	10.0	9.0–11.0 for 24 h at 50 °C	Raval et al. (2014)
Alkaline protease	*Stenotrophomonas maltophilia* strain SK	40 °C	9.0	n/d	Waghmare et al. (2015)
Serine protease AprB	*Bacillus* sp. B001	60 °C (without Ca^{2+}) 70 °C (with Ca^{2+})	10.0	5.0–12.0 for 6 h at 30 °C	Deng et al. (2010)
Serine protease AP-2	*Bacillus* spp. NCDC 180	55 °C	12.0	6.0–12.0 for 4 h at 30 °C (with and without 5 mM Ca^{2+}); 6.0–9.0 for 24 h at 30 °C with 5 mM Ca^{2+}	Kumar et al. (1999)
Serine protease	*Bacillus subtilis* VSG-4	50 °C	9.0	9.0 for 15 h at 37 °C	Giri et al. (2011)
Serine protease Protease M	*Bacillus* sp. KSM-K16	55 °C (without 5 mM Ca^{2+}) 70 °C (with 5 mM Ca^{2+})	12.3	6.0–11.0 (without 2 mM Ca^{2+}) and 5.0–12.0 (with 2 mM Ca^{2+}) for 10 min at 55 °C	Kobayashi et al. (1995)
Serine protease	*Bacillus* sp. SSR1	40 °C (without 5 mM Ca^{2+}) 45 °C (with 5 mM Ca^{2+})	10.0	8.0–11.0 for 1 h at 40 °C	Singh et al. (2001)

(continued)

Table 14.3 (continued)

Protease (PDB code)	Organism	T_{opt}	pH_{opt}	pH stability	References
Serine protease rBLAP	*Bacillus lehensis* (expressed in *E. coli*)	50 °C	12.8	10.0–12.8 for 26 h	Joshi and Satyanarayana (2013b)
Serine protease	Strain AH-6	37 °C	9.5–11.0	10.5–13.0 at 37 °C for 1 h	Dodia et al. (2008)
Actinomycetes					
Serine protease AL-20	*Nesterenkonia* sp. AL-20	70 °C	10.0	6.0–11.0 at 50 °C for 1 h	Gessesse et al. (2003)
Serine protease OM-6	*Brachystreptospora xinjiangensis* OM-6	60 °C	10.0–11.0	6.0–12.0 at 40 °C for 6 h	Gohel and Singh (2013)
Serine protease	*Nocardiopsis alba* strain OK-5	70 °C	10.0–11.0	6.0–12.0 at 40 °C for 6 h	Gohel and Singh (2012)
Serine protease protease Mit-1	*Streptomyces clavuligerus* strain Mit-1	70 °C	10.0–11.0	8.0–10.0 for 6 h	Thumar and Singh (2007)
Serine protease NAPase	*Nocardiopsis* sp. TOA-1	70–75 °C	11.0–11.5	1.5–12.0 for 24 h at 30 °C	Mitsuiki et al. (2002)
Serine protease Protease 2	*Nocardiopsis* sp. NCIM 5124	60 °C	10.0	4.0–10.0 for 1 h	Dixit and Pant (2000)

enzyme which did not lose its activity for 1 h at 37 °C in 1 % solution of ionic detergents (Tween 80 and Triton X-100), while in the presence of 5 mM SDS lost only 25 % of its activity. Under the same conditions of preincubation, the enzyme retained from 68 to 100 % of its activity in the presence of 1 % solid detergents (Surf excel, Rin, Tide and Ariel) (Giri et al. 2011).

Most usually, alkaliphilic proteases effectively hydrolyze casein, azocasein or hemoglobin (Gupta et al. 2005; Jaouadi et al. 2008; Kobayashi et al. 2007; Deng et al. 2010), but some of them also demonstrate keratinolytic properties, which are very much desired, like in the case of a protease from *Nocardiopsis* sp. SD5 that degraded native feather keratin at temperatures between 45 and 50 °C for 96–120 h (Saha et al. 2013). On the other hand, bacterial strains identified as *Nesterenkonia* sp. AL-20 and *Bacillus pseudofirmus* AL-89, isolated and described by Gessesse et al. (2003), utilize feathers, which did not undergo grinding or any other mechanical pretreatment. It was the only source of nitrogen in a medium, enabling the growth of these strains and production of extracellular proteases. In this context, the strain AL-89 demonstrated outstanding properties with a high-level production of proteases and a quick and efficient hydrolysis of feathers, with no other supplements added to the medium. Alkaliphilic proteases displaying keratinolytic activity are able to bind and hydrolyze keratin substrates in solid form. This is an important feature of enzymes used in detergents that should act upon protein substrates bound to a solid surface, such as keratin stains on shirt collars which constitute a huge problem in case of enzymes currently used in detergents (Gessesse et al. 2003).

2.4 Halophilic Proteases

Halophilic proteases are proteins produced by halophilic organisms that usually require the presence of NaCl for their catalytic activity. When it comes to halotolerant proteins, they do not always derive from halophiles, but they remain active in a broad range of NaCl concentrations, without any specific dependence on NaCl (Graziano and Merlino 2014).

Microorganisms producing salt-dependent extracellular proteases belong to archaea, bacteria and eukaryotes (Fig. 14.2) which are isolated from saline and hypersaline lakes, salt pans, salt marshes, playas, brine springs from underground salt deposits, solar salterns, salt maines, soda lakes, deep-sea sediments, sea water, fermented fish sauces, pastes and others (Yin et al. 2014; Setati 2010; Sinha and Khare 2013). In case of Archaea, these are usually extreme halophiles which, depending on a given genus, require 3.4–5.1 M of NaCl in a medium for growth (Graziano and Merlino 2014). Bacterial strains include halophiles, haloalkalophiles as well as bacteria that do not require an increased concentration of NaCl in a growth medium. Halophilic bacteria can be divided into slight halophiles, which show

Fig. 14.2 Phylogenetic tree showing relationship among the representatives of halophilic microorganisms described in Sect. 2.4 prepared on the basis http://www.ncbi.nlm.nih.gov/Taxonomy/CommonTree/wwwcmt.cgi (Sayers et al. 2009; Benson et al. 2009). *—strains belong to Archaea

optimum growth at 0.5 M NaCl, moderate halophiles, which grow best at 0.5–2.6 M NaCl, and few examples of extreme halophiles, which require 4.3 M NaCl to grow (Setati 2010).

All extracellular proteases from haloarchaea that have been described so far (Table 14.4) are serine enzymes, inhibited by PMSF. In case of bacterial enzymes, serine enzymes represent the largest group although several metalloproteases have also been described, e.g. proteases from *Pseudoalteromonas* sp. (Sánchez-Porro et al. 2003; Xiong et al. 2007). Both groups include enzymes whose activity is inhibited by β-mercaptoethanol and dithiothreitol or by *p*-chloromercuribenzoic acid (for instance, proteases from *Pseudoalteromonas* sp. (Sánchez-Porro et al. 2003), *Geomicrobium* sp. EMB2 (Karan and Khare 2010), *Natrialba magadii* (Giménez et al. 2000), *Haloferax lucentensis* VKMM 007 (Manikandan et al. 2009)), which indicates the significance of disulfide bridges and/or cysteine residues in maintaining biologically active conformations of the proteins. The optimum temperature for halophilic proteases at a proper NaCl concentration varies from 35 °C, as in the case of a cold-adapted metalloprotease from *Pseudoalteromonas* sp., isolated from sediment samples found in the area of the Aleutian margin in the Gulf of Alaska, Pacific Ocean (Xiong et al. 2007), to 75 °C, as in the case of a serine protease from *Chromohalobacter* sp. TVSP101 (Vidyasagar et al. 2009). The optimum pH range is quite broad, though alkaline (7.0–11.0) and it is even broader when it comes to stability—from 5.0 up to 12.0.

The most important factor influencing the activity and stability of these enzymes is the salinity of the environment in which a given reaction takes place. It is undoubtedly connected with habitats of halophilic microorganisms which, in order to survive in environments with a high osmotic pressure, must remain isosmotic. Therefore, they retain high concentrations of NaCl intracellularly (Manikandan et al. 2009). Halophilic proteins have an altered amino acid sequence when compared with other enzymes. Frequently, they have even 20 % more acidic amino acid residues (Asp and Glu), localized as clusters on the surface of a protein, in comparison with non-halophilic enzymes. At the same time, they have fewer Lys residues and an increased number of small hydrophobic residues, Ala and Gly, as well as polar residues, Ser and Thr, at the expense of large non-polar residues, Leu, Ile, Met and Phe. Owing to an increased negative charge, halophilic proteins may bind more hydrogen ions, which reduces the hydrophobicity of the molecular surface and reduces their tendency towards aggregation at high NaCl concentrations. This is a basic halophilic adaptation to a decreased amount of water that is bound by salts present in the environment at high concentrations (Graziano and Merlino 2014; Gomes and Steiner 2004). While proteases from *Halobacterium* sp. TuA4 (Schmitt et al. 1990) and unidentified haloalkaliphilic strain A2 (Yu 1991) need no more than 0.3 M NaCl (De Castro et al. 2005), the majority of enzymes show maximum activity at higher NaCl concentrations—from 1.0 to 5.1 M. Still, the number is even higher when synthetic substrate is being hydrolyzed, for instance in the case of protease Fll from *Natrialba asiatica* (Kamekura and Seno 1990), which breaks down N-succ-AAPF-pNa most quickly at 5.1 M NaCl concentration while azocasein at 1.7–2.4 M. It might be explained by the fact that at higher NaCl concentrations, azocasein becomes less hydrophobic and loses its proper conformation, which makes the hydrolysis more difficult. On the other hand, oligopeptides do not change their conformation under the same conditions and thus, can be effectively hydrolyzed by proteases. Several authors (Nordberg and von Hofsten 1969; Studdert et al. 2001; Shi et al. 2006) compared the efficiency of proteases in the presence of NaCl and KCl, receiving higher values of activity with NaCl, which is understandable considering the fact that producers of these enzymes live in environments rich in NaCl. A relatively high level of salinity, from 1.0 M for protease Nep (Giménez et al. 2000) up to 4.5 M for protease from *Haloferax mediterranei* strain 1583 (Stepanow et al. 1992), is also required to maintain the stability of these proteins. In case of several enzymes, such as halolysin R4 (Kamekura et al. 1996), a protease from *Halobacterium halobium* (Izotova et al. 1983) and a protease from *Hfx. mediterranei* strain 1583 (Stepanow et al. 1992), studies have shown that the removal of NaCl from the environment leads to the irreversible inactivation of these proteases, caused by their accelerated autodegradation in environments with low ionic strength. Therefore, it is worth noted that protease SptA (Shi et al. 2006) retains 20 % of its activity in the absence of NaCl, only to regain 60 % of its initial activity once the NaCl concentration is increased to 2.5 M. According to the authors, this indicates that in the absence of NaCl, the enzyme undergoes only a reversible denaturation. On the other hand, a protease from *Halogeometricum borinquense* TSS101 (Vidyasagar et al. 2006) maintains its total activity in the absence of NaCl, provided that the concentration of sucrose and betaine in the environment is 10 and 20 %, respectively. This shows that in order to maintain its activity, the enzyme needs an adequate osmotic pressure or reduced water activity (a_w) in the environment rather than the presence of Na^+ or Cl^- (Vidyasagar et al. 2006; Litchfield 2011).

Halophilic and halotolerant bacterial proteases described so far show a greater salinity tolerance than proteins from Archaea. The optimal salinity conditions for enzymes from *Bacillus* sp. EMB9 (Sinha and Khare 2013), *Pseudoalteromonas* sp. CP76 (Sánchez-Porro et al. 2003), *Geomicrobium* sp. EMB2 (Karan and Khare 2010) or *B. subtilis* FP-133 (Setyorini et al. 2006) vary from 0 to 1.0 M NaCl. In turn, a serine protease from *Virgibacillus* sp. SK33 shows maximum activity at NaCl concentration between 1.7 and 4.3 M (Sinsuwan et al. 2010). For a serine protease from *Chromohalobacter* sp. TVSP101, the optimal salinity is 4.5 M NaCl (Vidyasagar et al. 2009). Purohit and Singh (2011) described an interesting property of an enzyme isolated from *Oceanobacillus iheyensis* O.M.$A_1$8, which changes its optimal temperature depending on the salinity of the environment. At 0.25–0.5 M NaCl, its optimal temperature was 60 °C, but it increased by 20 °C, when the concentration of NaCl rose to 2.0–3.0 M. Although these enzymes can hydrolyze substrates without the presence of salt in the environment, they tend to require some small amounts of NaCl (e.g. 90 mM) in order to maintain stability. Still, the majority of proteins described so far remain stable in solutions with 2.0–4.5 M of NaCl (Table 14.4). In the case of a protease from *Chromohalobacter* sp. TVSP101, one of the few bacterial strains belonging to the group of extreme halophiles, the activity is completely and irreversibly lost at salinity below 1.0 M. However, the enzyme maintains its total activity if NaCl is replaced by glycine, sucrose (10 %) or glycerol

Table 14.4 Examples of extracellular halophilic and halotolerant proteases from halophilic microorganisms

Protease	Organism	T_{opt}	pH_{opt}	Opt. NaCl conc and stability in NaCl/KCl	References
Archaea					
Serine protease Nep (halolysin)	*Natrialba magadii*	60 °C	8.0–10.0	1.0–1.5 M NaCl; Stable in 1.0–3.0 M NaCl or 3.0 M KCl at 4 °C for 7 days (more stable in NaCl than in KCl)	Giménez et al. (2000), Ruiz and De Castro (2007)
Serine protease EP (chymotrypsin-like)	*Natronococcus occultus* NCBM 2192	60 °C	7.0–9.0	1.0 M NaCl or KCl; Stable in 3.0 M NaCl or KCl at 4 °C for at least 7 days	Studdert et al. (2001)
Serine endopeptidase	*Halobacterium halobium* S9	40 °C	8.7	4.0 M for N-succ-AAPF-pNA	Capiralla et al. (2002)
Serine protease SptA (halolysin)	*Natrinema* sp. J7	50 °C	8.0	2.5 M NaCl and 2.0 M KCl, higher activity for NaCl; Stable for at least 1 month at 4 °C in 2.5 M NaCl; Stable for 2 h at 37 °C in the 2.5 M NaCl	Shi et al. (2006)
Serine protease	*Halogeometricum borinquense* TSS101	60 °C	10.0	3.4–4.3 M; Stable in 3.4 M NaCl at 60 °C for 2 h and 30 days at 4 °C	Vidyasagar et al. (2006)
Serine protease (trypsin-like)	*Haloferax lucentensis* VKMM 007	60 °C	8.0	4.3–5.13 M	Manikandan et al. (2009)
Bacteria					
Metalloprotease (protease CP1)	*Pseudoalteromonas* sp. CP76	55 °C	8.5	0–1.0 M; Stable in 2.0 M NaCl at 4 °C for 24 h	Sánchez-Porro et al. (2003)
Metalloprotease	*Pseudoalteromonas* spp.	35 °C	7.5	2.0 M	Xiong et al. (2007)
Serine protease	*Bacillus* sp. EMB9	55 °C	9.0	0.17 M NaCl/0.13 M KCl; Stable in 0.86 M NaCl and 0.67 M KCl at 4 °C for 48 h	Sinha and Khare (2013)
Serine protease	*Chromohalobacter* sp. TVSP101	75 °C	8.0	4.5 M NaCl; Stable in 4.5 M NaCl at 60–70 °C for 2 h; Stable in 4.5 M NaCl at room temp. for 20 days and at –4 °C for 45 days	Vidyasagar et al. (2009)

Serine protease	*Filobacillus* sp. RF2-5	60 °C	10.0–11.0	0–0.86 M (casein as substrate), 2.6–4.3 M (N-succ-AAF-pNA as substrate); Stable in 4.3 M NaCl at 30 °C for 24 h	Hiraga et al. (2005)
Serine protease	*Halobacillus* sp. SR5-3	50 °C	10.0	4.3 M (N-succ-AAPF-MCA as a substrate); Stable in 3.4–6.0 M NaCl at 37 °C for 24 h	Namwong et al. (2006)
Serine protease	*Virgibacillus* sp. SK33	55 °C	7.5	1.7–3.4 M; Stable in 0–4.3 M NaCl at 55 °C for 60 min	Sinsuwan et al. (2010)
Serine protease	*Geomicrobium* sp. EMB2	50 °C	10.0	0.17 M	Karan and Khare (2010)
Serine protease	*Oceanobacillus iheyensis* O.M.$A_1$8	50 °C	11.0	0.25–0.5 M at 60 °C, 2.0–3.0 M at 80 °C	Purohit and Singh (2011)

(20 %) (Vidyasagar et al. 2009). Another protease from moderately haloalkaliphilic bacterium, *Geomicrobium* sp. EMB2, which requires only 0.17 M NaCl to maintain its activity, loses more than half of it once NaCl is removed from the environment. Nevertheless, when NaCl or other osmolytes (mannitol, sucrose, glycerol or betaine) are re-added, so that the final concentration is 5 %, the enzyme regains most of its initial activity (Karan and Khare 2010).

3 Application of Extremophilic Proteases

Currently, the global market for enzymes is estimated at approx. 5.1 billion USD. It is one of the best developing markets in the world, with an annual growth rate of 7 % (Sarethy et al. 2011). Proteases, as a dominating class of enzymes, occupy a pivotal position in the market, accounting for approx. 60 % of all commercially available enzymes. Their main sources are animals (e.g. calf stomach), plants (e.g. pineapple, fig and papaya) and microorganisms (e.g. *Bacillus* sp., *Pseudomonas* spp.). For several decades, proteases, especially microbiological proteases, have occupied a dominant position in the market and have been used for industrial purposes. When it comes to plant proteases, their production is to a large extent hindered by climate conditions, whereas the production of animal enzymes stirs up ethical debates. Nevertheless, regardless of their origin, proteases are used on an industrial scale mainly for their ability to hydrolyze peptide bonds. Numerous proteases can also efficiently catalyze synthesis reactions in micro-aqueous environments. This property is mainly used for pharmaceutical and nutritional purposes.

At the very beginning, the market for proteases was dominated by mesophilic enzymes (most often alkalitolerant), but an intensive development of research on extremophilic microorganisms proved that it is possible to improve the efficiency of biocatalytic technologies and lower their costs by replacing currently used enzymes by extremozymes. Depending on their origin, extremophilic proteases are naturally adapted to thrive at low or high temperatures and in acidic, alkaline or extremely saline environments. They can either show an exceptional resistance to thermal denaturation or, quite the contrary, be extremely thermolabile. Thus, they can be selectively inactivated under process conditions. Examples of proteases which achieved the biggest commercial success include subtilisin (for industrial purposes, it is produced mainly from alkaliphilic or alkalitolerant strains) and other alkaline proteases which constitute approx. 50 % of all proteolytic enzymes (Table 14.5). The enzymes have a wide range of applications thanks to their properties, analyzed in details in Sect. 2.3. The list of commercially available enzymes used in detergents, silk degumming, food and feed industry, leather dehairing, cosmetics and pharmaceuticals can be found in Table 14.5. The majority of alkaliphilic proteases were isolated from *Bacillus* strains. The main recipient of these enzymes is the laundry detergent industry which uses them as additives to washing powders and automatic dishwashing detergents. Their function is to degrade proteinaceous stains, which typically include blood, milk, egg, grass and sauces.

Table 14.5 Commercial bacterial alkaline proteases, sources, applications and their industrial suppliers (modified table from Gupta et al. 2002)

Supplier	Product trade name	Bacterial sources	Application
Novo Nordisk, Denmark	Alcalase	*B. licheniformis*	Detergent, silk degumming
	Savinase	*Bacillus* sp.	Detergent, textile industry
	Esperase	*B. lentus*	Detergent, food industry, silk degumming
	Everlase[a]	*B. clausii*	Detergent
	Kannase[a]	*B. clausii*	Detergent
	Durazym	*Bacillus* sp.	Detergent
	Novozyme 243	*B. licheniformis*	Denture cleaners
	Biofeed pro	*B. licheniformis*	Feed industry
	Nue	*Bacillus* sp.	Leather industry
Gist-Brocades, The Netherlands	Subtilisin	*B. alcalophilus*	Detergent
	Maxacal	*Bacillus* sp.	Detergent
	Maxatase	*Bacillus* sp.	Detergent
Genencor International, USA	Purafect	*B. lentus*	Detergent
	PurafectOxP[a]	*B. lentus*	Detergent
	Properase[a]	*B. alkalophilus* PB92	Detergent
	Primatan	Bacterial sources	Leather industry
	FNA[a]	*B. amyloliquefaciens*	Detergent
	FN4[a]	*B. lentus*	Detergent
Solvay Enzymes, Germany	Opticlean	*B. alcalophilus*	Detergent
	Optimase	*B. licheniformis*	Detergent
	Maxapem[a]	*Bacillus* sp.	Detergent
Amano Pharmaceuticals, Japan	PROTIN SD-NY10	*B. amyloliquefaciens*	Food industry
	PROTIN SD AY10F	*B. licheniformis*	Food industry
	Proleather	*Bacillus* sp.	Food industry
	Collagenase	*Clostridium* sp.	Technical use
Nagase Biochemicals, Japan	Bioprase SP-10	*B. subtilis*	Food industry
	Bioprase concentrate	*B. subtilis*	Cosmetic, pharmaceuticals industry
	Bioprase	*B. subtilis*	Chemical industry
Godo Shusei, Japan	Godo-Bap	*B. licheniformis*	Detergent, food
Enzyme Development, USA	Enzeco alkaline protease—L FG	*B. licheniformis*	Food industry
AB Enzymes, Germany	Corolase 7089	*B. subtilis*	Food industry
	Biotouch® ROC 250L	*Bacillus* sp.	Food industry
Wuxi Synder Bioproducts, China	Wuxi	*Bacillus* sp.	Detergent
Advance Biochemicals, India	Protosol	*Bacillus* sp.	Detergent
Henkel, Germany	BLAP S[a] or BLAP X[a]	*B. lentus*	Detergent

[a]Protein engineered

Once the success of detergent enzymes was recognized, numerous detergent proteases with specific uses were discovered. One of them was Alkazym (Novodan, Copenhagen, Denmark), an enzyme which plays a significant role in the process of membrane cleaning. Tergazyme (Alconox, New York, USA), Ultrasil (Henkel, Dusseldorf, Germany) and P3-paradigm (Henkel-Ecolab, Dusseldorf, Germany) are other enzymes used for this very purpose. Another example is Pronod 153L, an enzyme-based cleaner that is used to remove blood proteins from surgical instruments. A good example of other commercial protease is subtilopeptidase A used in optical cleaner in India (Gupta et al. 2002).

3.1 Detergent Industry

The first detergent containing a serine protease isolated from bacteria was introduced in 1956 as BIO-40. Seven years later, a Danish company Novo Industry A/S introduced Alcalase from alkalitolerant strain, identified as *Bacillus licheniformis,* under the trade name of Biotex (Maurer 2004). It is a serine endoprotease which is resistant to anionic or non-ionic surfactants and displays maximal activity at 60 °C and pH 8.3.

As previously mentioned, alkaline serine proteases, known as subtilisins and produced by *Bacillus* strains, account for almost half of all proteolytic enzymes produced in the market; in 2002, their production in EU amounted to 900 tons in terms of pure protein (Maurer 2004). The commercial success of this enzyme was guaranteed thanks to its properties, which include high alkalinity, stability in alkaline environment, broad specificity and its secretion by cells that make production of the enzyme preparation easier. Since 1984, numerous attempts have been undertaken to improve subtilisins and other proteases. Still, out of hundreds of different variants, only few entered industrial practice, the first one in 1990. These were solely subtilisin variants with improved characteristics. In 2004, eight such subtilisins, improved through site-directed mutagenesis or directed evolution, were available in the market. The improved characteristics included increased thermostability and activity in organic solvents, changed substrate specificity and pH optimum (Table 14.6). In the first engineered subtilisin Met222 (an easily oxidized amino acid, located in the neighborhood of catalytic Ser221) was replaced by non-oxidizable amino acids, Ala, Ser or Leu, in order to obtain a protein with increased oxidation tolerance.

The subtilisins from *Bacillus* strains, currently used in detergents are the subject of numerous patents with broadly-defined claims and therefore constantly classical screening is performed in order to isolate a new form of alkaline proteases from mesophilic and extremophilic microorganisms. For this purpose also genomes and metagenomic libraries are searched. For instance, Rai et al. (2010) described a potential alternative for *Bacillus* sp. proteases, widely used in the laundry detergent industry, which is a laundry-detergent stable serine protease, isolated from thermophilic bacterial strain, *Paenibacillus tezpurensis* sp. nov. The enzyme is Ca^{2+} inde-

Table 14.6 Subtilisins engineering for industrial applications

Protease and its microbial source	Strategies employed	Outcomes	References
Rational mutagenesis			
Subtilisin E, *B. subtilis*	Additional disulfide bridge C61/C98	Significantly enhanced thermal stability	Takagi et al. (1990)
Subtilisin BPN', *B. subtilis*	Additional disulfide bridge C22/C87	Improved stability, particularly in the absence of calcium ions	Pantoliano et al. (1987)
Subtilisin, *B. amyloliquefaciens*	Replacement of M222 in the neighbourhood of the active centre with non-oxidizable amino acids: Ala, Ser or Leu	Improved oxidative tolerance	Estell et al. (1985)
Subtilisin, *B. lentus*	M222C mutation coupled with chemical modification of the cysteinyl group with methyl methane thiosulfonate (Me-MTS)	Enhanced stability and catalysis under oxidative conditions	Grøn et al. (1990)
aprA-subtilisin, *B. amyloliquefaciens*	Introducing point mutations: N109S, N218S, N76D	Improved stability, with a 7 °C higher transition temperature	Narhi et al. (1991)
Subtilisin, *Bacillus* sp. AK.1	Introducing negatively charged Asp into the low-affinity calcium binding site via substitution at G131D and P172D	Improved stability via stronger electrostatic interactions between subtilisin and calcium ions	Pantoliano et al. (1988)
Subtilisin, *B. gibsonii* (alkaline protease BgAP)	I21V, S39E, N74D, D87E, M122L, N253D mutations	Increased activity at 15 °C (increased k_{cat} from 23.2 to 35.3 s^{-1}); increased half-life at 60 °C from 2 to 224 min	Martinez et al. (2013)
SABP, *B. pumilus*	Mutations: (1) single—N99Y; (2) double L31I/N99Y and T33S/N99Y; (3) Triple L31I/T33S/N99Y	Improved $t_{1/2}$ at 50 and 60 °C; shifted t_{opt} from 65 to 75 °C; 31-fold increase in k_{cat}/K_m for N-succ-AAPF-pNA	Jaouadi et al. (2010b)
Subtilisin BPN', *B. subtilis*	Mutations P129G and E156R	Double mutant tailored to cleave phosphotyrosine containing peptides (2500-fold enhanced selectivity)	Knight et al. (2007)
Subtilisin, *Bacillus* YaB	Mutations G124A, G124V and G151A in S1 pocket, affecting the substrate specificity of subtilisin	Increased elastolytic activity	Mei et al. (1998)
Subtilisin BPN', *B. subtilis*	Substitution of catalytic histidine 64 with alanine	Increased specificity against histidine-containing substrates	Carter et al. (1989)
Subtilisin BPN', *B. subtilis*	Mutations Y144A, I107G and L126A	Improved specificity for substrates with large hydrophobic P_4 side chains	Rheinnecker et al. (1993, 1994)

(continued)

Table 14.6 (continued)

Protease and its microbial source	Strategies employed	Outcomes	References
Random mutagenesis			
Subtilisin E, *B. subtilis*	Error-prone PCR, screening	Activity increased up to 170-fold in DMF	Chen and Arnold (1993)
Subtilisin BPN', *B. subtilis*	Loop removal, cassette mutagenesis, screening	$t_{1/2}$ increased up to 1000-fold	Strausberg et al. (1995)
Subtilisin, *B. lentus*	Error-prone PCR, enrichment in hollow fibers	50 % increase in enzyme secretion	Naki et al. (1998)
Subtilisin E, *B. subtilis*	Random-priming, screening	Up to eightfold increase in $t_{1/2}$ at 65 °C	Shao et al. (1998)
Subtilisin E, *B. subtilis*	Error-prone PCR, DNA shuffling, screening	Increased activity at different temperatures and 17 °C rise in T_m	Zhao and Arnold (1999)
Subtilisin, *Bacillus* sp. TA41	Saturation mutagenesis and in vitro recombination (DNA shuffling)	Increased activity at low temperatures and stability at high temperatures	Miyazaki et al. (2000)

pendent and displays maximal activity at pH 9.5 and in temperature range of 45–50 °C. It also exhibited a significant stability and compatibility with surfactants and the majority of tested commercial laundry detergents at room temperature. Proteases from halophilic strains, for instance *Geomicrobium* sp. EMB2 (Karan and Khare 2010) or *Virgibacillus* sp. SK33 (Sinsuwan et al. 2010), also have a great potential for application in the detergent industry. It should be emphasized that frequently, extremophilic proteases are obtained during the selection of mesophilic strains. For instance, an alkaliphilic protease from a mesophilic filamentous fungus, *Conidiobolus coronatus*, is an enzyme that is compatible with commercial detergents used in India and retains 43 % of its activity at 50 °C for 56 min in the presence of Ca^{2+} (25 mM) and glycine (1 M) (Phadatare et al. 1993). Also, as reported by Khan (2013), an alkaline serine protease from *Aspergillus clavatus* ES1 can be used in the detergent industry because this enzyme is salt, solvent, detergent and bleach tolerant.

During the metagenomic research, Neveu et al. (2011) isolated two serine proteases, DV1 and M30, from metagenomic libraries derived from samples of surface sand from the Gobi and Death Valley deserts, respectively. Both enzymes are active in alkaline environments (pH optimum for DV1—8.0, pH optimum for M30—11.0) and remain stable in the presence of non-ionic detergents and SDS, which renders them useful for household chemistry. Often, metagenomic screening of conventional environments results in the discovery of proteases that are capable of thriving in extreme conditions, for instance in extremely alkaline environments, which is one of the determining factors when it comes to the application of particular proteins in detergents. Examples of such enzymes include an alkaline protease KP-43, isolated from the A horizon of a Belgian deciduous forest (Biver et al. 2013), or a serine protease (AS-protease) isolated from the goat skin surface metagenome (Pushpam et al. 2011).

Another approach towards searching for subtilisin-like proteases, to be used mainly in detergents, was suggested by Toplak et al. (2013). They looked for new variants of subtilisin through genome mining, with regard to the presence of genes encoding these proteolytic enzymes. In this way, the authors selected genes that encode potential homologues of subtilisin E in the genomes of thermophilic bacteria and archaea. They looked for sequences that were at least 30 % identical in sequence to the gene of subtilisin. They found 24 putative homologous genes of subtilisin E and isolated sixteen of them from selected strains via PCR, using primers designed on the basis of preprosubtilisin-coding genes. Next, they cloned them into pBAD vectors and expressed them in *E. coli*. Through the selection of clones on agar plates with selective growth compounds (skim milk, antibiotic, inducer), six different functional clones with genes of proteases isolated from *Thermus aquaticus, Pseudomonas mendocina, Geobacillus thermodenitrificans, Deinococcus geothermalis* and *Coprothermobacter proteolyticus* were obtained. A clone that contained a gene of subtilase from the latter strain, *C. proteolyticus*, showed the highest activity with an ability to grow at 80 °C. This protease, named as proteolysin, was expressed in *E. coli* with efficiency of 20 mg/L. The enzyme proved thermostable and it shows optimum activity at 85 °C and pH 8.0. After 20 h of incubation at 70 °C, it retains 35 % of its activity. Another characteristic feature of the proteolysin is its great tolerance to DMSO, DMF, ethanol and guanidine hydrochloride. This thermophilic enzyme is also activated by a 10 % non-ionic detergent Tween 20 (80 % increase in activity).

The detergent industry is also the recipient of psychroalkaliphilic proteases which show stability in organic solvents and which are able to thrive at low temperatures. This is significant especially when it comes to cleaning applications in a wide range of industries, including laundry, dishwashing, food, dairy and brewing, medical devices and water treatment (Cavicchioli et al. 2011). For these purposes, thermolabile character of these proteins is used along with their high efficiency in catalyzing reactions at low temperatures. Their thermolability leads to a selective inactivation of proteins after the process is complete through a subtle increase in temperature. In this way, the risk of enzymatic degradation of the final product is reduced.

It is generally thought that detergents containing cold-adapted proteases are much more effective in comparison with non-enzyme detergents and they remove protein contamination from clothes stained with blood, milk, grass or sweat much more efficiently (Kuddus and Ramteke 2011). A cold-adapted enzyme, designed for washing clothes at lower temperatures, is produced by Novozymes under the trade name of Polarzyme. The estimations concerning potential energy savings, made on the assumption that in every household, the temperature of washing is lowered from 40 to 30 °C, show how important the availability of such products is. In this way, we could reduce the energy consumption by 30 % and decrease the CO_2 emission by 100 g per one laundry (Cavicchioli et al. 2011). Apart from economic and environmental advantages, the use of cold-adapted proteases in detergents may lead to the improvement in the quality of textile materials, especially those containing synthetic fibers, which do not tolerate temperatures above 50–60 °C (Joshi and Satyanarayana 2013a). Consequently, more and more attention is devoted to the production of such proteases. For instance, Baghel et al. (2005) described a cold-active protease from *B. subtilis*, which shows stability in SDS solutions as well as increased activity in Tween 80 and Wheel detergents. Recently, a cold-active serine protease (CP70) from *Flavobacterium balustinum* was patented. Its optimum temperature is lower by 20 °C compared to a protease Savinase that has been traditionally used in the detergent industry. The enzyme remains stable for 1 h at 30 °C and its optimum pH values range from 6.5 to 10.0. Moreover, surface-active components/bleaching agents do not affect its activity (Hasan and Tamiya 2001). Another alkaline cold-active protease, isolated from *Stenotrophomonas maltophilia*, shows maximum activity at 20 °C and pH 10.0. Its other properties include great stability and compatibility with commercial detergents, resulting in its high efficiency in removing various types of protein stains at low temperatures (Kuddus and Ramteke 2009). According to Kuddus and Ramteke (2011), a cold-adapted protease from *S. maltophilia* can completely remove blood and grass stains and increases their reflectance by 26 and 23 %, respectively. According to the authors, the protease can be used in mild detergents, designed for washing delicate clothes, and in the textile industry to remove gums from raw silk, as their proteinaceous content adds rigidity and dullness to materials. Doddapaneni et al. (2007) also see a potential in the use of low-temperature detergents. They mention two proteases from *Serratia rubidaea* which remain active at 30–40 °C and in the pH range of 8.0–10.0. Pawar et al. (2009) isolated and described a thermolabile protease from *Bacillus* sp. 158 (stable at 30–40 °C and pH 6.0–7.0) which could be used in contact lens clearing to increase the transmittance of lenses.

Halophilic enzymes also show some potential as laundry detergents. Apart from their ability to thrive in extremely saline environments, they remain stable in the presence of detergents and organic solvents. Examples of such enzymes are proteases isolated from halophilic strains: *Bacillus* sp. EMB9 (Sinha and Khare 2013), *Geomicrobium* sp. EMB2 (Karan and Khare 2010) or *Virgibacillus* sp. EMB13 (Sinha and Khare 2012). The latter was tested in terms of its compatibility with various detergents (after a thermal denaturation of enzymes present in commercial washing powders) and its ability to remove blood stains from fabrics in combination with these detergents. The results were extremely encouraging.

3.2 Food Industry

Microbiological proteases are broadly applied in the food industry, especially the extremophilic proteases, which are often used to increase the efficiency of a process and enhance the quality of the final product. Their main advantage is higher specificity of enzyme, lower risk of adverse reactions, lower risk of contamination with mesophilic microflora and, in the case of psychro- and thermophilic enzymes, the possibility of selective thermal inactivation. Without a doubt, in the context of food processing, alkaliphilic proteases come to the front, as it was the case with household chemicals. The primary function of alkaliphilic proteases is the hydrolysis of naturally-occurring protein substrates. For more than 40 years, protein hydrolysates of high nutritional value have been obtained thanks to alkaliphilic proteases. Commercial protein hydrolysates can be derived from the decomposition of such substances as casein (Miprodan; MD Foods, Germany), whey (Lacprodan; MD Foods, Germany), soy (Proup; Novo Nordisk, Denmark) and meat (Flavourzyme, Novo Nordisk, Denmark) (Gupta et al. 2002). They play a significant role in blood pressure regulation and are used in infant food formulations, specific therapeutic dietary products and as the fortification of fruit juices and soft drinks. Alkaline proteases also find their use in meat processing, e.g. a commercially available protease named SEB Tender 70. The enzyme is extensively used in meat tenderization for breaking down collagen and making the meat more tasty for consumers (Singhal et al. 2012). One of the alkaliphilic proteases, often discussed in the context of its application in the food industry, is Alcalase from *B. licheniformis*. It is used in the processing of soy meal to obtain a soluble, non-bitter hydrolysate, used as component of protein-fortified soft drinks and dietetic food. Alcalase is also helpful for the recovery of proteins from by-products of fish and meat industries and from crustacean shell waste during chitin production (Synowiecki 2010). Cold-active proteases could be used for similar purposes, but this would necessitate the use of highly thermolabile substrates or products. These enzymes are efficient and highly specific at low temperatures and thus, they reduce the amount of by-products, facilitating the selective inactivation of labile proteins throughout the process, or after its completion, without a huge input of energy. They are used for the treatment of beer, in bakeries and for the accelerated maturation of cheese. Similarly to alkalozymes, cold-adapted proteases, e.g. cold-adapted collagenolytic protease MCP-01, can be used to tenderize meat or improve the taste of refrigerated meat. This enzyme decreases the meat shear force by 23 % and enhances the relative myofibrillar fragment at ion index of the meat by 91.7 % at 4 °C. MCP-01 also helps maintain the fresh color of the meat as well as its moisture because it showed a unique tenderization mechanism and had a strong selectivity for the degradation of collagen at 4 °C in comparison to papain and bromelain as the commercially used tenderizers (Zhao et al. 2012). Otherwise, psychrophilic proteases facilitate the removal of the membrane from the fish roe, the evaporation of fish/meat stickwater and the rendering of fat. Cold-active proteases may be used in the production of digest, which is either coated onto or mixed into dry pet food to improve its tastiness (Kuddus and Ramteke 2012).

Halophilic and halotolerant bacteria also produce proteases that will be more suitable for the application in food industry, performed under saline conditions or in saline-free systems. Some examples may include saline fermentation processes, involved in the production of protein-rich food, such as processing of fish and meat-based products, and the production of soy sauce (Setati 2010). Enzymes from *Halobacillus* sp. SR5-3 (Namwong et al. 2006) and *Halobacterium* constitute good examples of halophilic proteases, used for the production of fish sauces (Akolkar et al. 2010).

3.3 Leather and Textile Industry

Alkaliphilic properties of proteases isolated from extremophiles have been used in the leather industry for a long time. These enzymes may serve an important role in the treatment of leather by substituting harmful substances used in soaking, dehairing and bating (e.g. sodium chloride). Currently, alkaline proteases with hydrated lime and sodium chloride are involved in a selective hydrolysis of non-collagenous components of the skin as well as for the removal of globular proteins, such as albumins and globulins. The increased usage of enzymes for dehairing and bating not only prevents pollution, but is also time-saving and enhances the quality of leather. At the moment, pancreatic proteases are used for the treatment of leather, but their alkaliphilic substitutes also constitute some alternative. For instance, Varela et al. (1997) used an alkaline protease from *B. subtilis* HQDB32, to unhair sheep skin. George et al. (1995) used a similar enzyme from *B. amyloliquefaciens* to unhair hides and skins, while Hameed et al. (1999) for bating and leather processing. Over the last few years, several new extremophilic proteolytic enzymes have been described as showing dehairing activity and, at the same time, displaying no collagen and keratin degrading properties that destroy the collagen structure of hide and hinder keratin recovery. These include proteases isolated from an alkaliphilic bacterium, *B. pumilus* CBS (Jaouadi et al. 2009), and from a mesophilic bacterium, *P. aeruginosa* PD 100 (Najafi et al. 2005). The latter enzyme, despite being isolated from a mesophilic strain is a unique protein since given its numerous properties, like alkali- and thermostability or resistance to organic solvents and detergents. It can be considered a polyextremozyme. Furthermore, it has a broad substrate specificity and therefore, can be used not only in dehairing, but also in the animal food industry, clearing of beverages, skin bating, production of amino acids and peptides and in molecular biology for the purification of DNA during its isolation. Another detergent-stable serine alkaline protease from *B. pumilus* CBS, which can efficiently remove hair from the skin with minimal damage of the collagen, effectively degraded feather-meal, chicken feather, goat hair and bovine hair (Jaouadi et al. 2009).

Alkaliphilic cold-adapted proteases can be also applied in the textile industry for improving production methods and fabric finishing. They prove effective in degumming threads of raw silk to remove sericin, a proteinaceous substance that covers silk fiber. Traditionally, degumming is done in an alkaline solution containing soap. This is a harsh process because not only sericin but also the fiber is attacked. On the contrary, proteolytic enzymes remove sericin without attacking silk fiber which is not damaged. Therefore, silk threads are stronger than in the case of traditional treatments. The surface of the wool and silk fibers treated by cold-active protease can provide new and unique finishes (Najafi et al. 2005). Example of this enzyme is an alkaline protease from *Bacillus* sp. RGR-14 described by Puri (2001). It is characterized by the silk-degumming efficiency. The scanning electron microscopy (SEM) of the fibers revealed that some clusters had broken apart, whereas the treated fiber maintained its smooth and compact structure.

3.4 Poultry Industry

Recently, much thought has been given to the possible application of thermostable proteases with keratinolytic activity in the decomposition and disposal of various types of waste (e.g. meat industry by-products or feathers from the poultry industry), whose present production amounts to approx. 10,000 tons per year (Suzuki et al. 2006). Currently, such raw materials are degraded through nonenzymatic alkaline hydrolysis and steam pressure cooking, during which particular amino acids are broken down and non-nutritive lysinoalanine and lantionine are produced. Enzymatic degradation of wastes from the poultry industry leads to the formation of hydrolysates, which can be used as fertilizers and dietary protein supplement for animals. Moreover, keratinases can be used as dehairing agents in the leather and cosmetic industry and as a component of detergents and edible films. The majority of keratinases described so far are produced by mesophilic microorganisms, despite their high thermostability. For instance, a mesophilic strain *B. licheniformis* K-19, cultivated at 37 °C, produces keratinase which is active in the temperature range of 30–90 °C. It shows optimum activity at 60 °C and at pH 7.5–8.0 (Xu et al. 2009). On the contrary, a protease isolated from a thermophilic bacterial strain identified as *Fervidobacterium islandicum* AW-1, shows keratinolytic activity at 100 °C and pH 9.0, whereas its half-life at optimum temperature is 90 min (Nam et al. 2002). Another keratinolytic protease from a thermoalkalitolerant halophilic strain, *Bacillus tequilensis* hsTKB2, which was described by Paul et al. (2014), is alkaliphilic (pH 9.0–10.5), thermostable (50–80 °C) and halophilic (0–30 % NaCl).

3.5 Medical and Pharmaceutical Industry

Currently, there is little information on the application of microbiological proteases from extremophiles in medicine. In the 1990s, there was some discussion concerning the use of elastolytic activity of *B. subtilis* 316 M for the preparation of elastoterase, which is applied in the treatment of burns, purulent wounds, carbuncles, furuncles and deep abscesses (Kudrya and Simonenko 1994). Another idea was to use an alkaline protease from *Bacillus* sp. strain CK 11-4, which shows fibrinolytic activity, as a thrombolytic agent (Kim et al. 1996). Still, if we share the opinion expressed by Feller (2013), who said that coldwater fish and crustaceans may be classified as extremophiles, we cannot forget about the trypsin from Atlantic cod (*Gadus morhua*) which is commercially available under the trade name of ColdZyme®. Trypsin is used in mouthwashes and it prevents the formation of dental plaque, which is a biofilm of caries-associated bacteria, mainly *Streptococcus mutans* and *Lactobacillus acidophilus*. Another product used to prevent tooth decay and periodontal diseases is known as Krillase®. It contains proteolytic enzymes isolated from the Antarctic krill, *Euphausia superba* Dana, with trypsin- and chymotrypsin-like proteases and carboxypeptidases A and B. The product is also used for wound dressing, in the treatment of hard-to-heal wounds and ulcerations. It not only removes necrotic tissue, but also accelerates the granulation of a healthy tissue (Fornbacke and Clarsund 2013).

In recent years, there have appeared new possibilities for the use of peptide hydrolases, especially thermophilic proteases with keratinolytic activity, in the degradation of infectious isoforms of prion proteins, PrPSc, which tend to aggregate and form anomalous structures known as amyloids. The deposition of amyloids may lead to such disorders as bovine spongiform encephalopathy (BSE) and Creutzfeldt-Jakob disease in a human organism. Numerous studies proved that prion aggregates are extremely resistant to conventional proteases and do not undergo degradation even during sterilization at 121 °C (Langeveld et al. 2003). First reports on the degradation of PrPSc without the use of detergents and in a non-alkaline environment were related with the application of keratinolytic proteases isolated from *Streptomyces, Thermoanaerobacter* and *Thermococcus* (Hui et al. 2004). Proteolytic potential of the last two extremophiles was used for the hydrolysis of a thermally pre-denatured prion aggregate; the possibility of using these two strains in the disinfection of animal wastes was also mentioned (Suzuki et al. 2006).

3.6 Protease Applications in Non-Conventional Media

Kinetic and thermodynamic studies on the reaction of proteolysis have already shown that it is possible to shift its equilibrium towards the formation of peptide bonds through synthesis (Glass 1981; Bongers and Heimer 1994; Elmore 2002). The kinetic control of the reaction consists in maintaining the concentration of a condensation product below the equilibrium concentration. Only then it is possible to conduct a protease-catalyzed peptide synthesis in the presence of water (Morihara and Oka 1981), but given the ionization constant of the substrates (thermodynamic control), the majority of enzymes require an environment with lower polarity. In general, three types of such environments are currently used: (1) water-organic solvent systems (water + water-miscible organic solvent), (2) two-phase systems (water + water-immiscible organic solvent system) (Doukyu and Ogino 2010) and (3) microaqueous solvent systems with fixed water activity (a_w) (Halling 1994). Nevertheless, the conditions which are favorable for the reaction of synthesis may negatively affect the activity of an enzyme. Therefore, in order to use a given protease for the catalysis in a water-restricted system, the enzyme should display high activity and stability in such environments (Deetz and Rozzell 1988). The analysis of these parameters is essential to properly assess the usefulness of an enzyme in the reaction of synthesis performed in non-conventional media.

3.6.1 Properties of Extremophilic Proteases in Low-Polarity Media

There are numerous reasons for which enzymes lose their activity in non-polar environments. It may be due to conformational changes in proteins, decreased conformational flexibility, loss of crucial water or interfacial inactivation (Doukyu and

Ogino 2010). These changes may be prevented via genetic and chemical enzyme engineering, immobilization of enzymes and optimization of the mixing during a two-phase reaction (Kumar and Bhalla 2005). Another approach is the use of extremophilic proteases, mainly psychrophilic and halophilic, which are naturally predisposed to catalysis in non-polar environments thanks to their structural adaptations (see Sects. 2.2 and 2.4).

Extremely valuable tools for syntheses in low-polarity media are extracellular proteases from halophilic microorganisms, since they are adapted to environments of low water activity due to hygroscopic properties of salts, which are usually present at high concentrations. A serine protease from γ-*Proteobacterium* is an example of such an enzyme as it maintains approx. 80 % of its initial activity after 18 h incubation at 35 % NaCl. When tested in a water-organic solvent system (ethylene glycol, ethanol, butanol, acetone, dimethyl sulfoxide, xylene, perchloroethylene), both one-phase and two-phase, the enzyme shows higher activity (with even a two-fold increase) and stability (up to 10 days) than without a solvent (Sana et al. 2006).

Unlike the protease from γ-*Proteobacterium*, some other enzymes, like a haloalkaliphilic protease from archaeon *Natrialba magadii*, require higher salt concentration to act in an organic medium. The enzyme was tested in media with various logP values (octanol/water partition coefficient; Laane et al. 1987) ranging from −1.76 (glycerol) to 0.5 (isopropanol), with 0.5 or 1.5 M NaCl. The protease showed higher activity in all the solvents with higher NaCl concentrations (Ruiz and De Castro 2007). Studies on the stability of an extremely halophilic protease from *Halobacterium halobium* (Kim and Dordick 1997) revealed it was 40-fold more stable in 40 % dimethyl sulfoxide (DMSO) containing 0.2 M NaCl than in 0.2 M NaCl. Its stability increased proportionally with NaCl concentration in 40 % DMSO and was 150-fold higher for 2 M NaCl. However, when DMSO was replaced by 40 % 1,4-dioxane, the stability of the protein significantly decreased, despite the fact that the concentration of NaCl was maintained. Also tetrahydrofuran (THF) containing 0.2 M NaCl increased the stability of this protease, however, only in the concentration range up to 40 %. When THF concentration reached 80 %, the stability was lower than in the absence of the solvent. These significant differences in the *Hbt. halobium* protease stability in various organic solvents could stem from their salting-in-out nature (Kim and Dordick 1997). Sellek and Chaudhuri (1999) who studied properties of the same enzyme found that it showed maximum synthetic activity in 32 % dimethylformamide (DMF), in which the efficiency of hydrolysis catalyzed by this protein, expressed as catalytic efficiency (k_{cat}/K_m), was threefold lower. For comparison, this constant was not changed in case of subtilisin Carlsberg. The differences between these two proteins suggest that halophilic enzymes are particularly adapted to carry out synthesis in organic media.

Despite limited usage, psychrophilic proteases are attractive catalysts of peptide synthesis in low-polarity media. Firstly, a decrease in temperature leads to a shift in the equilibrium from hydrolysis towards synthesis as the formation of acyl-enzyme intermediate is slowed down. Secondly, a decrease in temperature leads to lower energy consumption. Finally, psychrophilic enzymes are more stable in organic media as the number of ion-pairs that reduce the hydrophobic effect on protein folding increases. Moreover, owing to a greater flexibility of psychrophilic proteins, it is possible to use them with more hydrophobic solvents which as such, decrease the conformational mobility of a protein and thus, its catalytic efficiency (Sellek and Chaudhuri 1999).

Apart from halophiles and psychrophiles, many other thermophilic microorganisms produce proteases that are stable in organic solvents. A protease from *Thermus* sp. Rt4A2 is a good example of such an enzyme. It loses only 25 % of its activity at 4 °C, in the presence of 90 % acetonitrile. However, when acetonitrile is replaced by butanol at the same concentration, the decrease in activity is twice higher (59 %) (Freeman et al. 1993). Different stability of proteases from *P. aeruginosa* PST-01 and *Thermus* sp. Rt4A2, in the presence of butanol, prove that there is no universal rule which can be applied while choosing a proper solvent for the reaction of synthesis catalyzed by these enzymes. Subtle and very often hardly noticeable differences in the structure of enzymes may result in their different stability and suitable solvents for a particular reaction catalyzed by a given enzyme must be selected experimentally. One of thermostable proteases that might be useful in peptide synthesis is a proteolysin from *Coprothermobacter proteolyticus* (Toplak et al. 2013). The enzyme shows great tolerance to such solvents as DMSO, DMF and ethanol, although it is not stabilized by them. The research on the stability of immobilized thermolysin from *B. thermoproteolyticus*, which is one of the enzymes that are most frequently used to catalyze peptide synthesis, confirm that although thermophilic proteases are not stabilized by organic solvents, they retain much of their activity in the presence of such substances as acetonitrile (thermolysin—72 % of its activity after a 5 h incubation), tert-butyl alcohol (93 %) or tert-amyl alcohol (98 %).

Noteworthy, not only enzymatic proteins may be more stable in organic solvents. There are also solvent-tolerant strains which can be classified as another group of extremophilic microorganisms. They embrace not only halophiles, such as archaeon *Halobacterium* ssp. (Akolkar et al. 2008), but also some non-halophilic strains, like *P. aeruginosa* PST-01. PST-01 protease shows significantly greater stability (five to tenfold) in the presence of organic solvents (25 %), such as ethylene glycol, 1,4-butanediol, 1,5-pentanediol, ethanol, n-hexanol, methanol, butanol, DMSO and others, than in the aqueous environment. Its stability is much greater than that of subtilisin Carlsberg, which is commonly applied in biotransformations (Ogino et al. 1999a, b).

3.6.2 Synthesis Using Proteases

Only few proteases from extremophiles have been applied in peptide synthesis processes. Their examples are presented in Table 14.7. It is to note, that in hydrophobic environments with low water content, proteases synthesize mainly di- and tripeptides. Therefore, several proteases with different substrate specificities are needed to synthesize longer peptides. For example, Kimura et al. (1990b, Table 14.7) needed three steps to synthesize Z-L-Tyr-Gly-Gly-L-Phe-L-Leu-OEt. In the first step they received Z-Gly-Gly-OBut and Z-L-Tyr-Gly-Gly-OBut using papain and α-chymotrypsin. The latter peptide was converted to Z-L-Phe-L-Leu-OEt, and that

Table 14.7 Examples of peptide synthesis reactions catalyzed by extremophilic proteases

Protease	Synthetized peptide	References
Thermus sp. RT41A protease	Bz-Ala-Tyr-NH_2	Wilson et al. (1994)
Thermolysin	Asp-Phe (aspartame)	Kühn et al. (2002)
	α-L-Asp-Phe-OM	Rao et al. (1998)
	Z-L-Tyr-Gly-Gly-L-Phe-L-Leu-OEt[a]	Kimura et al. (1990a, b, c)
	Asp-Tyr-Met-Gly-Trp-Met-Asp-Phe[a] (Cholecystokinin)	Kullmann (1982)
	Asp-Tyr-Thr-Gly-Trp-Met-Asp-Phe[a] (Caerulein)	Takai et al. (1981)
Pst-01 Protease	Cbz-Lys-Leu-NH_2	Ogino et al. (1999a, b)
	Cbz-Ala-Leu-NH_2	
	Cbz-Ala-Phe-NH_2	
Natrialba magadii protease	Ac-Phe-Gly-Phe-NH_2	Ruiz et al. (2010)

[a]The peptide synthesis was carried out with the protease and an additional catalyst

to the target peptide using thermolysin. Kullman (1982, Table 14.7) used another method, in which the synthesis of cholecystokinin was conducted via chemical condensation of peptides, obtained using several enzymes: papain, α-chymotrypsin, arylsulfatase (EC 3.1.6.1; for desulfation of tyrosine-*O*-sulfate), thermolysin and aminopeptidase M.

In kinetically controlled peptide synthesis, an important role is played by subtilisins from alkaliphilic *Bacillus* strains that show optimum activity in alkaline pH. For these enzymes, the rate of peptide bond synthesis was found to be greater than the rate of its hydrolysis (Stepanov 1996). One of such enzymes, most commonly used in organic synthesis, is subtilisin Carlsberg, whose great potential was reported by Klein et al. (2000). In the medium with pH 9.5, containing 50 % (v/v) DMF and strongly basic piperidine (added to neutralize acid groups in the amine substrate) subtilisin catalyzed synthesis of peptide bonds between the acyl donor (Z-Val-Trp-OMe) and (1) amino acid amides as nucleophiles (with more than 70 % yield for Gly-NH_2) and (2) dipeptides Gly-Xaa (where Xaa = Gly, Ala, Phe, Gln, Ser, Val, Lys, Trp). The reaction of hydrolysis was not observed.

Subtilisins are known for their broad substrate specificity in the reaction of hydrolysis. In synthesis reactions, their specificity can be altered through a change in the medium. For instance, Klein et al. (2000) who studied subtilisin Carlsberg, observed that addition of acetonitrile (acetonitrile/DMF/H_2O = 8/1/1) increased the synthesis yield for single amino-acid nucleophiles (with the exception of Gly-NH_2) and the highest yield (at least 70 %) was achieved in the condensation reaction of Z-Val-Trp-OMe and Lys-NH_2 and Ala-NH_2. For dipeptide nucleophiles Xaa-Gly, the addition of acetonitrile resulted in a threefold increase in the yield of the synthesis for Xaa = Ser, Lys, Thr, Asn, Gln, Met, His, Val, Glu, Asp, and Arg. Still, for Gly-Xaa, the change in the yield was significantly lower. A marked increase in the yield was observed only for Xaa = Asp and Pro.

Some extremophilic proteases are capable of synthesizing non-protein compounds, such as esters via transesterification. An alkaline protease from *B. pseudofirmus* AL-89 (opt pH 10.0) is a good example. It catalyzes the synthesis of sucrose laurate (Pedersen et al. 2003) in an optimal solvent system (DMF-DMSO). Subtilisin Carlsberg is another example of a protease used in the synthesis of esters. It can be applied for the acylation of di- and oligosaccharides containing D-fructose moiety (Riva et al. 1998). Riva et al. (1998) received 1-*O*-butanoyl-derivatives of lactulose, maltulose, stachyose, and sucrose with 50 % yield in anhydrous DMF with activated ester—trifluoroethyl butanoate. Such sugar esters, produced via selective esterification, are extremely valuable surfactants.

4 Concluding Remarks and Challenges

Proteases, particularly these of microbial origin, play a leading role in industrial processes which use enzymatic catalysis for fabrication of a wide range of high value-added products. A unique diversity of these enzymes provides the basis, and undoubtedly will provide the basis also in future, for their newer and newer applications in various branches of industry and in medicine. Some of these uses have been currently even difficult to define owing to a small number of known proteases synthesized by microbial species. However, studies on proteolytic enzymes from extremophilic microorganisms caused that over the last decades the number of available commercial preparations of proteases has been increasing and will increase in future, as may be predicted based on the current tendency. Notwithstanding this progress, only a small part of microflora populating extreme biotopes has been known so far, and it means that the majority of catalytic proteins, synthesized by this microflora, including most valuable, unique enzymes, have not been available for industry. Analyses of metagenomes and metatranscriptomes derived from extreme environments, which already enabled isolation of genes encoding unique proteases are a promising approach to synthesis of a greater number of such enzymes, in particular those from microorganisms that have hitherto been unculturable under laboratory conditions. Such studies should be intensified. However, also classical prospecting for novel extremophilic microorganisms, synthesizing proteolytic enzymes should be continued. Application of improved, quick techniques of DNA sequencing caused a rapid growth in a number of available complete microbial genomes in databases. Mining of these genomes may give rise to finding new variants of proteolytic enzymes. Also the studies on subtle molecular adaptations of extremophilic proteases are very promising. Their results may facilitate rational engineering of already known proteases, and in further perspective, may be used to design novel variants of these enzymes that will be tailored to specific applications and industrial conditions. There is also an urgent need to establish novel expression systems, dedicated to enzymes from extremophilic microorganisms because expression of genes encoding these enzymes in mesophilic hosts is either insufficiently efficient or often, even impossible, that generally hinders broader application of many proteases with properties interesting for industry and medicine.

Conflict of Interest Aneta Białkowska, Ewa Gromek, Tomasz Florczak, Joanna Krysiak, Katarzyna Szulczewska, and Marianna Turkiewicz declare that they have no conflict of interest.

References

Akolkar AV, Desai AJ (2010) Catalytic and thermodynamic characterization of protease from *Halobacterium* sp. SP1(1). Res Microbiol 161(5):355–362

Akolkar AV, Deshpande GM, Raval KN, Durai D, Nerurkar AS, Desai AJ (2008) Organic solvent tolerance of *Halobacterium* sp. SP1(1) and its extracellular protease. J Basic Microbiol 48(5):421–425

Akolkar AV, Durai D, Desai AJ (2010) *Halobacterium* sp. SP1(1) as a starter culture for accelerating fish sauce fermentation. J Appl Microbiol 109(1):44–53

Alam SI, Dube S, Reddy GSN, Bhattacharya BK, Shivaji S, Singh L (2005) Purification and characterization of extracellular protease produced by *Clostridium* sp. from Schirmacher oasis, Antarctica. Enzyme Microb Technol 36(5–6):824–831

Alam S, Dube S, Agarwal M, Singh L (2006) Purification and characterization of an extracellular protease produced by psychrotolerant *Clostridium* sp. LP3 from lake sediment of Leh, India. Can J Microbiol 52(12):1238–1246

Amoozegar MA, Fatemi ZA, Karbalaei-Heidari HR, Razavi MR (2007) Production of an extracellular alkaline metalloprotease from a newly isolated, moderately halophile, *Salinivibrio* sp. strain AF-2004. Microbiol Res 162(4):369–377

Aqel H, Al-Quadan F, Yousef TK (2012) A novel neutral protease from thermophilic *Bacillus* strain HUTBS62. J Biosci Biotechnol 1(2):117–123

Arulmani M, Aparanjini K, Vasanthi K, Arumugam P, Arivuchelvi M, Kalaichelvan PT (2007) Purification and partial characterization of serine protease from thermostable alkalophilic *Bacillus laterosporus*-AK1. World J Microbiol Biotechnol 23(4):475–481

Baghel VS, Tripathi RD, Ramteke RW, Gopal K, Dwivedi S, Jain RK, Rai UN, Singh SN (2005) Psychrotrophic proteolytic bacteria from cold environments of Gangotri glacier, Western Himalaya, India. Enzyme Microb Technol 36(5–6):654–659

Bakhtiar S, Andersson MM, Gessesse A, Mattiasson B, Hatti-Kaul R (2002) Stability characteristics of a calcium-independent alkaline protease from *Nesterenkonia* sp. Enzyme Microb Technol 32(5):525–531

Benson DA, Karsch-Mizrachi I, Lipman DJ, Ostell J, Sayers EW (2009) GenBank. Nucleic Acids Res 37:D26–D31

Biver S, Portetelle D, Vandenbol M (2013) Characterization of a new oxidant-stable serine protease isolated by functional metagenomics. SpringerPlus 2:410

Blumentals II, Robinson AS, Kelly RM (1990) Characterization of sodium dodecyl sulfate-resistant proteolytic activity in the hyperthermophilic archaebacterium *Pyrococcus furiosus*. Appl Environ Microbiol 56(7):1992–1998

Bongers J, Heimer EP (1994) Recent applications of enzymatic peptide synthesis. Peptides 15(1):183–193

Cannio R, Catara G, Fiume I, Balestrieri M, Rossi M, Palmieri G (2010) Identification of a cell-bound extracellular protease overproduced by *Sulfolobus solfataricus* in peptide-rich media. Protein Pept Lett 17(1):78–85

Capiralla H, Hiroi T, Hirokawa T, Maeda S (2002) Purification and characterization of a hydrophobic amino acid-specific endopeptidase from *Halobacterium halobium* S9 with potential application in debittering of protein hydrolysates. Process Biochem 38(4):571–579

Carter P, Nilsson B, Burnier JP, Burdick D, Wells JA (1989) Engineering subtilisin BPN' for site-specific proteolysis. Proteins: Struct Funct Bioinf 6(3):240–248

Catara G, Ruggiero G, La Cara F, Digilio FA, Capasso A, Rossi M (2003) A novel extracellular subtilisin-like protease from the hyperthermophile *Aeropyrum pernix* K1: biochemical properties, cloning, and expression. Extremophiles 7(5):391–399

Cavicchioli R, Charlton T, Ertan H, Mohd Omar S, Siddiqui KS, Williams TJ (2011) Biotechnological uses of enzymes from psychrophiles. Microb Biotechnol 4(4):449–460

Chavez Croocker P, Sako Y, Uchida A (1999) Purification and characterization of an intracellular heat-stable proteinase (pernilase) from the marine hyperthermophilic archaeon *Aeropyrum pernix* K1. Extremophiles 3(1):3–9

Chen K, Arnold FH (1993) Tuning the activity of an enzyme for unusual environments: sequential random mutagenesis of subtilisin E for catalysis in dimethylformamide. Proc Natl Acad Sci U S A 90(12):5618–5622

Chen XL, Zhang YZ, Gao PJ, Luan XW (2003) Two different proteases produced by a deep-sea psychrotrophic bacterial strain, *Pseudoaltermonas* sp. SM9913. Mar Biol 143(5):989–993

Choi IG, Bang WG, Kim SH, Yu YG (1999) Extremely thermostable serine-type protease from *Aquifex pyrophilus*. Molecular cloning, expression, and characterization. J Biol Chem 274(2):881–888

Cowan DA, Daniel RM (1982) Purification and some properties of an extracellular protease (caldolysin) from an extreme thermophile. Biochim Biophys Acta 705(3):293–305

Damare S, Raghukumar C, Muraleedharan U, Raghukumar S (2006) Deep-sea fungi as a source of alkaline and cold-tolerant proteases. Enzyme Microb Technol 39(2):172–181

Dastager SG, Dayanand A, Li WJ, Kim CJ, Lee JC, Park DJ, Tian XP, Raziuddin QS (2008) Proteolytic activity from an alkali-thermotolerant *Streptomyces gulbargensis* sp. nov. Curr Microbiol 57(6):638–642

Davail S, Feller G, Narinx E, Gerday C (1994) Cold adaptation of proteins. Purification, characterization, and sequence of the heat-labile subtilisin from the Antarctic psychrophile *Bacillus* TA41. J Biol Chem 269(26):17448–17453

De Azeredo L, Freire D, Soares R, Leite S, Coelho R (2004) Production and partial characterization of thermophilic proteases from *Streptomyces* sp. isolated from Brazilian cerrado soil. Enzyme Microb Technol 34(3–4):354–358

De Castro RE, Maupin-Furlow JA, Giménez MI, Herrera Seitz MK, Sánchez JJ (2005) Haloarchaeal proteases and proteolytic systems. FEMS Microbiol Rev 30(1):17–35

Deetz JS, Rozzell JD (1988) Enzyme-catalysed reactions in non-aqueous media. Trends Biotechnol 6(1):15–19

Deng A, Wu J, Zhang Y, Zhang G, Wen T (2010) Purification and characterization of a surfactant-stable high-alkaline protease from *Bacillus* sp. B001. Bioresour Technol 101(18):7100–7106

Dib R, Chobert JM, Dalgalarrondo M, Barbier G, Haertlé T (1998) Purification, molecular properties and specificity of a thermoactive and thermostable proteinase from *Pyrococcus abyssi*, strain st 549, hyperthermophilic archaea from deep-sea hydrothermal ecosystem. FEBS Lett 431(2):279–284

Dixit VS, Pant A (2000) Comparative characterization of two serine endopeptidases from *Nocardiopsis* sp. NCIM 5124. Biochim Biophys Acta 1523(2–3):261–268

Doddapaneni KK, Tatineni R, Vellanki RV, Gandu B, Panyala NR, Chakali B, Mangamoori LN (2007) Purification and characterization of two novel extra cellular proteases from *Serratia rubidaea*. Process Biochem 42(8):1229–1236

Dodia MS, Rawal CM, Bhimani HG, Joshi RH, Khare SK, Singh SP (2008) Purification and stability characteristics of an alkaline serine protease from a newly isolated Haloalkaliphilic bacterium sp. AH-6. J Ind Microbiol Biotechnol 35(2):121–131

Doukyu N, Ogino H (2010) Organic solvent-tolerant enzymes. Biochem Eng J 48(13):270–282. doi:10.1016/j.bej.2009.09.009

Dozie IN, Okeke CN, Unaeze NC (1994) A thermostable, alkaline-active, keratinolytic proteinase from *Chrysosporium keratinophilum*. World J Microbiol Biotechnol 10(5):563–567

Eggen R, Geerling A, Watts J, de Vos WM (1990) Characterization of pyrolysin, a hyperthermoactive serine protease from the archaebacterium *Pyrococcus furiosus*. FEMS Microbiol Lett 71(1–2):17–20

Elmore DT (2002) Peptide synthesis. In: Barret GC, Davies JS (eds) Amino acids, peptides and proteins: Volume 33. RSC Publishing, UK, pp. 83–134

Estell DA, Graycar TP, Wells JA (1985) Engineering an enzyme by site-directed mutagenesis to be resistant to chemical oxidation. J Biol Chem 260(11):6518–6521

Feller G (2003) Molecular adaptations to cold in psychrophilic enzymes. Cell Mol Life Sci 60(4):648–662

Feller G (2013) Psychrophilic enzymes: from folding to function and biotechnology. Scientifica. Article ID 512840

Fornbacke M, Clarsund M (2013) Cold-adapted proteases as an emerging class of therapeutics. Infect Dis Ther 2(1):15–26

Freeman SA, Peek K, Prescott M, Daniel R (1993) Characterization of a chelator-resistant protease from *Thermus* strain Rt4A2. Biochem J 295(Pt 2):463–469

Fujinami S, Fujisawa M (2010) Industrial applications of alkaliphiles and their enzymes – past, present and future. Environ Technol 31(8–9):845–856. doi:10.1080/09593331003762807

Fusek M, Lin XL, Tang J (1990) Enzymic properties of thermopsin. J Biol Chem 265(3):1496–1501

George S, Raju V, Krishnan MRV, Subramanian TV, Jayaraman K (1995) Production of protease by *Bacillus amyloliquefaciens* in solid-state fermentation and its application in the unhairing of hides and skins. Process Biochem 30(5):457–462

Gessesse A, Hatti-Kaul R, Gashe BA, Mattiasson B (2003) Novel alkaline proteases from alkaliphilic bacteria grown on chicken feather. Enzym Microb Technol 32(5):519–524

Giménez MI, Studdert CA, Sánchez JJ, De Castro RE (2000) Extracellular protease of *Natrialba magadii*: purification and biochemical characterization. Extremophiles 4(3):181–188

Giri SS, Sukumaran V, Sen SS, Oviya M, Banu BN, Jena PK (2011) Purification and partial characterization of a detergent and oxidizing agent stable alkaline protease from a newly isolated *Bacillus subtilis* VSG-4 of tropical soil. J Microbiol 49(3):455–461

Glass JD (1981) Enzymes as reagent in the synthesis of peptides. Enzyme Microb Technol 3(1):2–8

Gohel SD, Singh SP (2012) Purification strategies, characteristics and thermodynamic analysis of a highly thermostable alkaline protease from a salt-tolerant alkaliphilic actinomycete, *Nocardiopsis alba* OK-5. J Chromatogr B Analyt Technol Biomed Life Sci 889–890:61–68

Gohel SD, Singh SP (2013) Characteristics and thermodynamics of a thermostable protease from a salt-tolerant alkaliphilic actinomycete. Int J Biol Macromol 56:20–27

Gomes J, Steiner W (2004) The biocatalytic potential of extremophiles and extremozymes. Food Technol Biotechnol 42(4):223–235

Graziano G, Merlino A (2014) Molecular bases of protein halotolerance. Biochim Biophys Acta 1844(4):850–858

Grøn H, Bech LM, Branner S, Breddam K (1990) A highly active and oxidation-resistant subtilisin-like enzyme produced by a combination of side-directed mutagenesis and chemical modification. Eur J Biochem 194(3):897–901

Gupta R, Beg QK, Lorenz P (2002) Bacterial alkaline proteases: molecular approaches and industrial application. Appl Microbiol Biotechnol 59(1):15–32

Gupta A, Roy I, Patel RK, Singh SP, Khare SK, Gupta MN (2005) One-step purification and characterization of an alkaline protease from haloalkaliphilic *Bacillus* sp. J Chromatogr A 1075(1–2):103–108

Halling PJ (1994) Thermodynamic prediction for biocatalysis in nonconventional media: theory, tests, and recommendations for experimental design and analysis. Enzyme Microb Technol 16(3):178–206

Hameed A, Keshavarz T, Evans CS (1999) Effect of dissolved oxygen tension and pH on the production of extracellular protease from a new isolate of *Bacillus subtilis* K2, for use in leather processing. J Chem Technol Biotechnol 74(1):5–8

Hao JH, Sun M (2015) Purification and characterization of a cold alkaline protease from a psychrophilic *Pseudomonas aeruginosa* HY1215. Appl Biochem Biotechnol 175(2):715–722

Hasan AKMQ, Tamiya E (2001) Cold-active protease CP70. Patent US 6200793

Hasbay Ifrij I, Ogel ZB (2002) Production of neutral and alkaline extracellular proteases by the thermophilic fungus, *Scytalidium thermophilum*, grown on microcrystalline cellulose. Biotechnol Lett 24(13):1107–1110

Hiraga K, Nishikata Y, Namwong S, Tanasupawat S, Takada K, Oda K (2005) Purification and characterization of serine proteinase from a halophilic bacterium, *Filobacillus* sp. RF2-5. Biosci Biotechnol Biochem 69(1):38–44

Horikoshi K (1999) Alkaliphiles: some applications of their products for biotechnology. Microbiol Mol Biol Rev 63(4):735–750

Hui Z, Doi H, Kanouchi H, Matsuura Y, Mohri S, Nonomura Y, Oka T (2004) Alkaline serine protease produced by *Streptomyces* sp. degrades PrP(Sc). Biochem Biophys Res Commun 321(1):45–50

Huston A, Methe B, Deming J (2004) Purification, characterization, and sequencing of an extracellular cold-active aminopeptidase produced by marine psychrophile *Colwellia psychrerythraea* strain 34H. Appl Environ Microbiol 70(6):3321–3328

Ignatova Z, Gousterova A, Spassov G, Nedkov P (1999) Isolation and partial characterization of extracellular keratinase from a wool degrading thermophilic actinomycete strain *Thermoactinomyces candidus*. Can J Microbiol 45(3):217–222

Iqbal I, Aftab MN, Afzal M, Ur-Rehman A, Aftab S, Zafat A, Ud-Din Z, Khuharo AR, Iqbal J, Ul-Haq I (2015) Purification and characterization of cloned alkaline protease gene of *Geobacillus stearothermophilus*. J Basic Microbiol 55(2):160–171

Izotova LS, Strongin AY, Chekulaeva LN, Sterkin VE, Ostoslavskaya VI, Lyublinskaya LA, Timokhina EA, Stepanov VM (1983) Purification and properties of serine protease from *Halobacterium halobium*. J Bacteriol 155(2):826–830

Jang HJ, Kim BC, Pyun YR, Kim YS (2002) A novel subtilisin-like serine protease from *Thermoanaerobacter yonseiensis* KB-1: its cloning, expression, and biochemical properties. Extremophiles 6(3):233–243

Jaouadi B, Ellouz-Chaabouni S, Rhimi M, Bejar S (2008) Biochemical and molecular characterization of a detergent stable serine alkaline protease from *Bacillus pumilus* CBS with high catalytic efficiency. Biochimie 90(9):1291–1305

Jaouadi B, Ellouz-Chaabouni S, Ben Ali M, Ben Messaoud E, Naili B, Dhouib A, Bejar S (2009) Excellent laundry detergent compatibility and high dehairing ability of the *Bacillus pumilus* CBS alkaline proteinase (SAPB). Biotechnol Bioprocess Eng 14:503–512

Jaouadi B, Abdelmalek B, Fodil D, Ferradji FZ, Rekik H, Zarai N, Bejar S (2010a) Purification and characterization of a thermostable keratinolytic serine alkaline proteinase from *Streptomyces* sp. strain AB1 with high stability in organic solvents. Bioresour Technol 101(21):8361–8369

Jaouadi B, Aghajari N, Haser R, Bejar S (2010b) Enhancement of the thermostability and the catalytic efficiency of *Bacillus pumilus* CBS protease by site-directed mutagenesis. Biochimie 92(4):360–369

Joo HS, Chang CS (2005) Oxidant and SDS-stable alkaline protease from a halo-tolerant *Bacillus clausii* I-52: enhanced production and simple purification. J Appl Microbiol 98(2):491–497

Joo HS, Kumar CG, Park GC, Paik SR, Chang CS (2003) Oxidant and SDS-stable alkaline protease from *Bacillus clausii* I-52: production and some properties. J Appl Microbiol 95(2):267–272

Joshi S, Satyanarayana T (2013a) Biotechnology of cold-active proteases. Biology 2(2):755–783

Joshi S, Satyanarayana T (2013b) Characteristics and applications of a recombinant alkaline serine protease from a novel bacterium *Bacillus lehensis*. Bioresour Technol 131:76–85

Kamekura M, Dyall-Smith ML (1995) Taxonomy of the family Halobacteriaceae and the description of two new genera *Halorubrobacterium* and *Natrialba*. J Gen Appl Microbiol 41(4):333–350

Kamekura M, Seno Y (1990) A halophilic extracellular protease from a halophilic archaebacterium strain 172 P1. Biochem Cell Biol 68(1):352–359

Kamekura M, Seno Y, Dyall-Smith ML (1996) Halolysin R4, a serine proteinase from the halophilic archaeon *Haloferax mediterranei*; gene cloning, expression and structural studies. Biochim Biophys Acta 1294(2):159–167

Karan R, Khare SK (2010) Purification and characterization of a solvent-stable protease from *Geomicrobium* sp. EMB2. Environ Technol 31(10):1061–1072

Karan R, Singh SP, Kapoor S, Khare SK (2011) A novel organic solvent tolerant protease from a newly isolated *Geomicrobium* sp. EMB2 (MTCC 10310): production optimization by response surface methodology. New Biotechnol 28(2):138–145

Karbalaei-Heidari HR, Shahbazi M, Absalan G (2013) Characterization of a novel organic solvent tolerant protease from a moderately halophilic bacterium and its behavior in ionic liquids. Appl Biochem Biotechnol 170(3):573–586

Khan F (2013) New microbial proteases in leather and detergent industries. Innov Res Chem 1(1):1–6

Kikani BA, Shukla RJ, Singh SP (2010) Biocatalytic potential of thermophilic bacteria and actinomycetes. In: Mendez-Vilas A (ed) Current research, technology and education topics in applied microbiology and microbial biotechnology. Formatex Research Center, Badajoz, Spain, pp 1000–1007

Kim J, Dordick JS (1997) Unusual salt and solvent dependence of a protease from an extreme halophile. Biotechnol Bioeng 55(3):471–479

Kim T, Lei XG (2005) Expression and characterization of a thermostable serine protease (TfpA) from *Thermomonospora fusca* YX in *Pichia pastoris*. Appl Microbiol Biotechnol 68(3):355–359

Kim W, Choi K, Kim Y, Park H, Choi J, Lee Y, Oh H, Kwon I, Lee S (1996) Purification and characterization of a fibrinolytic enzyme produced from *Bacillus* sp. strain CK 11-4 screened from Chungkook-Jang. Appl Environ Microbiol 62(7):2482–2488

Kimura Y, Muraya K, Araki Y, Matsuoka H, Nakanishi K, Matsuno R (1990a) Synthesis peptides consisting of essential amino acids by a reactor system using three proteinases and an organic solvent. Agric Biol Chem 54(12):3331–3333

Kimura Y, Nakanishi K, Matsuno R (1990b) Enzymatic synthesis of the precursor of Leu-enkephalin in water-immiscible organic solvent systems. Enzyme Microb Technol 12(4):273–280

Kimura Y, Yoshida T, Muraya K, Nakanishi K, Matsuno R (1990c) Continuous synthesis of a tripeptide by successive condensation and transesterification catalyzed by two immobilized proteinases in organic solvent. Agric Biol Chem 54(6):1433–1440

Klein UJ, Prykhodzka A, Cerovsky V (2000) The applicability of subtilisin Carlsberg in peptide synthesis. J Pept Sci 6(11):541–549

Kleine R (1982) Properties of thermitase, a thermostable serine protease from *Thermoactinomyces vulgaris*. Acta Biol Med Ger 41(1):89–102

Klingeberg M, Galunsky B, Sjoholm C, Kasche V, Antranikian G (1995) Purification and properties of a highly thermostable, sodium dodecyl sulfate-resistant and stereospecific proteinase from the extremely thermophilic archaeon *Thermococcus stetteri*. Appl Environ Microbiol 61(8):3098–3104

Knight ZA, Garrison JL, Chan K, King DS, Shokat KM (2007) A remodelled protease that cleaves phosphotyrosine substrates. J Am Chem Soc 129(38):11672–11673

Kobayashi T, Hakamada Y, Adachi S, Hitomi J, Yoshimatsu T, Koike K, Kawai S, Ito S (1995) Purification and properties of an alkaline protease from alkalophilic *Bacillus* sp. KSM-K16. Appl Microbiol Biotechnol 43(3):473–481

Kobayashi T, Lu J, Li Z, Hung VS, Kurata A, Hatada Y, Takai K, Ito S, Horikoshi K (2007) Extremely high alkaline protease from a deep-subsurface bacterium, *Alkaliphilus transvaalensis*. Appl Microbiol Biotechnol 75(1):71–80

Kuddus M, Ramteke PW (2009) Cold-active extracellular alkaline protease from an alkaliphilic *Stenotrophomonas maltophilia*: production of enzyme and its industrial applications. Can J Microbiol 55(11):1294–1301

Kuddus M, Ramteke PW (2011) Production optimization of an extracellular cold-active alkaline protease from *Stenotrophomonas maltophilia* MTCC 7528 and its application in detergent industry. Afr J Microbiol Res 5(7):809–816

Kuddus M, Ramteke PW (2012) Recent developments in production and biotechnological applications of cold-active microbial proteases. Crit Rev Microbiol 38(4):330–338

Kudrya VA, Simonenko IA (1994) Alkaline serine proteinase and lectin isolation from the culture fluid of *Bacillus subtilis*. Appl Microbiol Biotechnol 41(5):505–509

Kühn D, Dürrschmidt P, Mansfeld J, Ulbrich-Hofmann R (2002) Boilysin and thermolysin in dipeptide synthesis: a comparative study. Biotechnol Appl Biochem 36(Pt 1):71–76

Kullmann W (1982) Protease-catalyzed peptide bond formation: application to synthesis of the COOH-terminal octapeptide of cholecystokinin. Proc Natl Acad Sci U S A 79(9):2840–2844

Kumar D, Bhalla TC (2005) Microbial proteases in peptide synthesis: approaches and applications. Appl Microbiol Biotechnol 68(6):726–736

Kumar D, Savitri, Thakur N, Verma R, Bhalla TC (2008) Microbial proteases and application as laundry detergent additive. Res J Microbiol 3(12):661–672

Kumar CG, Tiwari MP, Jany KD (1999) Novel alkaline serine proteases from alkalophilic *Bacillus* spp.: purification and some properties. Process Biochem 34(5):441–449

Laane C, Boeren S, Vos K, Veeger C (1987) Rules for optimization of biocatalysis in organic solvents. Biotechnol Bioeng 30(1):81–87

Lagzian M, Asoodeh A (2012) An extremely thermotolerant, alkaliphilic subtilisin-like protease from hyperthermophilic *Bacillus* sp. MLA64. Int J Biol Macromol 51(5):960–967

Lama L, Romano I, Calandrelli V, Nicolaus B, Gambacorta A (2005) Purification and characterization of a protease produced by an aerobic haloalkaliphilic species belonging to the *Salinivibrio* genus. Res Microbiol 156(4):478–484

Langeveld JP, Wang JJ, Van de Wiel DF, Shih GC, Garssen GJ, Bossers A, Shih JC (2003) Enzymatic degradation of prion protein in brain team stem from infected cattle and sheep. J Infect Dis 188(11):1782–1789

Li AN, Li DC (2009) Cloning, expression and characterization of the serine protease gene from *Chaetomium thermophilum*. J Appl Microbiol 106(2):369–380

Li WF, Zhou XX, Lu P (2005) Structural features of thermozymes. Biotechnol Adv 23(4):271–281

Li AN, Ding AY, Chen J, Liu SA, Zhang M, Li DC (2007) Purification and characterization of two thermostable proteases from the thermophilic fungus *Chaetomium thermophilum*. J Microbiol Biotechnol 17(4):624–631

Li Q, Yi L, Marek P, Iverson BL (2013) Commercial proteases:present and future. FEBS Lett 587(8):1155–1163

Lin X, Tang J (1990) Purification, characterization, and gene cloning of thermopsin, a thermostable acid protease from *Sulfolobus acidocaldarius*. J Biol Chem 265(3):1490–1495

Litchfield CD (2011) Potential for industrial products from the halophilic Archaea. J Ind Microbiol Biotechnol 38(10):1635–1647

Lü J, Wu X, Jiang Y, Cai X, Huang L, Yang Y, Wang H, Zeng A, Li A (2014) An extremophile *Microbacterium* strain and its protease production under alkaline conditions. J Basic Microbiol 54(5):378–385

Majeed T, Tabassum R, Orts WJ, Lee CC (2013) Expression and characterization of *Coprothermobacter proteolyticus* alkaline serine protease. Sci World J. Article ID 396156

Manikandan M, Pašić L, Kannan V (2009) Purification and biological characterization of a halophilic thermostable protease from *Haloferax lucentensis* VKMM 007. World J Microbiol Biotechnol 25(12):2247–2256

Marcy RM, Engelhardt TC, Upadhyay JM (1984) Isolation, partial characterization, and some properties of protease I from a thermophilic mold *Thermoascus aurantiacus* var. *levisporus*. Mycopathologia 87(1–2):57–65

Margesin R, Dieplinger H, Hofmann J, Sarg B, Lindner H (2005) A cold-active extracellular metalloprotease from *Pedobacter cryoconitis* - production and properties. Res Microbiol 156(4):499–505

Martinez R, Jakob F, Tu R, Siegert P, Maurer KH, Schwaneberg U (2013) Increasing activity and thermal resistance of *Bacillus gibsonii* alkaline protease (BgAP) by directed evolution. Biotechnol Bioeng 110(3):711–720

Matsuzawa H, Tokugawa K, Hamaoki M, Mizoguchi M, Taguchi H, Terada I, Kwon ST, Ohta T (1988) Purification and characterization of aqualysin I (a thermophilic alkaline serine protease) produced by *Thermus aquaticus* YT-1. Eur J Biochem 171(3):441–447

Maurer KH (2004) Detergent proteases. Curr Opin Biotechnol 15(4):330–334

Mei HC, Liaw YC, Li YC, Wang DC, Takagi H, Tsai YC (1998) Engineering subtilisin YaB: restriction of substrate specificity by the substitution of Gly124 and Gly151 with Ala. Protein Eng 11(2):109–117

Mitsuiki S, Sakai M, Moriyama Y, Goto M, Furukawa K (2002) Purification and some properties of a keratinolytic enzyme from an alkaliphilic *Nocardiopsis* sp. TOA-1. Biosci Biotechnol Biochem 66(1):164–167

Miyazaki K, Wintrode PL, Grayling RA, Rubingh DN, Arnold FH (2000) Directed evolution study of temperature adaptation in a psychrophilic enzyme. J Mol Biol 297(4):1015–1026

Moreno ML, Garcia MT, Ventosa A, Mellado E (2009) Characterization of *Salicola* sp. IC10, a lipase- and protease-producing extreme halophile. FEMS Microbiol Ecol 68(1):59–71

Morihara K, Oka T (1981) Peptide bond synthesis catalyzed by subtilisin, papain, and pepsin. J Biochem 89(2):385–395

Morikawa M, Izawa Y, Rashid N, Hoaki T, Imanaka T (1994) Purification and characterization of a thermostable thiol protease from a newly isolated hyperthermophilic *Pyrococcus* sp. Appl Environ Microbiol 60(12):4559–4566

Morita RY (1975) Psychrophilic bacteria. Bacteriol Rev 39(2):144–167

Najafi MF, Deobagkar D, Deobagkar D (2005) Potential application of protease isolated from *Pseudomonas aeruginosa* PD 100. Electron J Biotechnol 8(2):197–203

Naki D, Paech C, Ganshaw G, Schellenberger V (1998) Selection of a subtilisin-hyperproducing *Bacillus* in a highly structured environment. Appl Microbiol Biotechnol 49(3):290–294

Nam GW, Lee DW, Lee HS, Lee NJ, Kim BC, Choe EA, Hwang JK, Suhartono MT, Pyun YR (2002) Native-feather degradation by *Fervidobacterium islandicum* AW-1, a newly isolated keratinase-producing thermophilic anaerobe. Arch Microbiol 178(6):538–547

Namwong S, Hiraga K, Takada K, Tsunemi M, Tanasupawat S, Oda K (2006) A halophilic serine proteinase from *Halobacillus* sp. SR5-3 isolated from fish sauce: purification and characterization. Biosci Biotechnol Biochem 70(6):1395–1401

Narhi LO, Stabinsky Y, Miller L, Sachdev R, Finley S, Park S, Kolvenbach C, Arakawa T, Zukowski M (1991) Enhanced stability of subtilisin by three point mutations. Biotechnol Appl Biochem 13(1):12–24

Narinx E, Baise E, Gerday C (1997) Subtilisin from psychrophilic antarctic bacteria: characterization and site-directed mutagenesis of residues possibly involved in the adaptation to cold. Protein Eng 10(11):1271–1279

Neveu J, Regeard C, DuBow MS (2011) Isolation and characterization of two serine proteases from metagenomic libraries of the Gobi and Death Valley deserts. Appl Microbiol Biotechnol 91(3):635–644

Nordberg P, von Hofsten B (1969) Proteolytic enzymes from extremely halophilic bacteria. J Gen Microbiol 55(2):251–256

Ogino H, Watanabe F, Yamada M, Nakagawa S, Hirose T, Noguchi A, Yasuda M, Ishikawa H (1999a) Purification and characterization of organic solvent-stable protease from organic solvent-tolerant *Pseudomonas aeruginosa* PST-01. J Biosci Bioeng 87(1):61–68

Ogino H, Yamada M, Watanabe F, Ichinose H, Yasuda M, Ishikawa H (1999b) Peptide synthesis catalyzed by organic-solvent-stable protease from *Pseudomonas aeruginosa* PST-01 in monophasic aqueous-organic solvent systems. J Biosci Bioeng 88(5):513–518

Ong PS, Gaucher GM (1976) Production, purification and characterization of thermomycolase, the extracellular serine protease of the thermophilic fungus *Malbranchea pulchella* var. *sulfurea*. Can J Microbiol 22(2):165–176

Pandey S, Rakholiya KD, Raval VH, Singh SP (2012) Catalysis and stability of an alkaline protease from a haloalkaliphilic bacterium under non-aqueous conditions as a function of pH, salt and temperature. J Biosci Bioeng 114(3):251–256

Pantoliano MW, Ladner RC, Bryan PN, Rollence ML, Wood JF, Poulos TL (1987) Protein engineering of subtilisin BPN': enhanced stabilization through the introduction of two cysteines to form a disulfide bond. Biochemistry 26(8):2077–2082

Pantoliano MW, Whitlow M, Wood JF, Rollence ML, Finzel BC, Gillialand GL, Poulos TL, Bryan PN (1988) The engineering of binding affinity at metal ion binding sites for the stabilization of proteins: subtilisin as a test case. Biochemistry 27(22):8311–8317

Paul T, Das A, Mandal A, Jana A, Halder SK, Das Mohapatra PK, Pati BR, Mondal KC (2014) Smart cleaning properties of a multi tolerance keratinolytic protease from an extremophilic *Bacillus tequilensis* hsTKB2: prediction of enzyme modification site. Waste Biomass Valoriz 5(6):931–945

Pawar R, Zambare V, Barve S, Paratkar G (2009) Application of protease isolated from *Bacillus* sp. 158 in enzymatic cleansing of contact lenses. Biotechnology 8(2):276–280

Pedersen NR, Wimmer R, Matthiesen R, Pedersen LH, Gessesse A (2003) Synthesis of sucrose laurate using a new alkaline protease. Tetrahedron Asymmetry 14(6):667–673

Peek K, Daniel RM, Monk C, Parker L, Coolbear T (1992) Purification and characterization of a thermostable proteinase isolated from *Thermus* sp. strain Rt41A. Eur J Biochem 207(3):1035–1044

Phadatare SU, Deshpande VV, Srinivasan MC (1993) High activity alkaline protease from *Conidiobolus coronatus* (NCL 86.8.20): enzyme production and compatibility with commercial detergents. Enzyme Microb Technol 15(1):72–76

Phrommao E, Yongsawatdigul J, Rodtong S, Yamabhai M (2011) A novel subtilase with NaCl-activated and oxidant-stable activity from *Virgibacillus* sp. SK37. BMC Biotechnol 11:65

Puri S (2001) An alkaline protease from a *Bacillus* sp.: production and potential applications in detergent formulation and degumming of silk. MSc thesis, University of Delhi, New Delhi

Purohit MK, Singh SP (2011) Comparative analysis of enzymatic stability and amino acid sequences of thermostable alkaline proteases from two haloalkaliphilic bacteria isolated from Coastal region of Gujarat, India. Int J Biol Macromol 49(1):103–112

Pushpam PL, Rajesh T, Gunasekaran P (2011) Identification and characterization of alkaline serine protease from goat skin surface metagenome. AMB Express 1(1):3

Qua D, Simidu U, Taga N (1981) Purification and some properties of halophilic protease produced by a moderately halophilic marine *Pseudomonas* sp. Can J Microbiol 27(5):505–510

Rai SK, Roy JK, Mukherjee AK (2010) Characterization of a detergent-stable alkaline protease from a novel thermophilic strain *Paenibacillus tezpurensis* sp. nov. AS-S24-II. Appl Microbiol Biotechnol 85(5):1437–1450

Rao MB, Tanksale AM, Ghatge MS, Deshpande VV (1998) Molecular and biotechnological aspects of microbial proteases. Microbiol Mol Biol Rev 62(3):597–635

Raval VH, Pillai S, Rawal CM, Singh SP (2014) Biochemical and structural characterization of a detergent-stable serine alkaline protease from seawater haloalkaliphilic bacteria. Process Biochem 49(6):955–962

Ray MK, Devi KU, Kumar GS, Shivaji S (1992) Extracellular protease from the Antarctic yeast *Candida humicola*. Appl Environ Microbiol 58(6):1918–1923

Rheinnecker M, Baker G, Eder J, Fersht AR (1993) Engineering a novel specificity in subtilisin BPN'. Biochemistry 32(5):1199–1203

Rheinnecker M, Eder J, Pandey PS, Fersht AR (1994) Variants of subtilisin BPN' with altered specificity profiles. Biochemistry 33(1):221–225

Riva S, Nonini M, Ottolina G, Danieli B (1998) Subtilisin-catalyzed esterification of di- and oligosaccharides containing a D-fructose moiety. Carbohydr Res 314(3–4):259–266

Ruiz DM, De Castro RE (2007) Effect of organic solvents on the activity and stability of an extracellular protease secreted by the haloalkaliphilic archaeon *Natrialba magadii*. J Ind Microbiol Biotechnol 34(2):111–115. doi:10.1007/s10295-006-0174-4

Ruiz DM, Iannuci NB, Cascone O, De Castro RE (2010) Peptide synthesis catalysed by a haloalkaliphilic serine protease from the archaeon *Natrialba magadii* (Nep). Lett Appl Microbiol 51(6):691–696

Saba I, Qazi PH, Rather SA, Dar RA, Qadri QA, Ahmad N, Johri S, Taneja SC, Shawl S (2012) Purification and characterization of a cold active alkaline protease from *Stenotrophomonas* sp., isolated from Kashmir, India. World J Microbiol Biotechnol 28(3):1071–1079

Saeki K, Hitomi J, Okuda M, Hatada Y, Kageyama Y, Takaiwa M, Kubota H, Hagihara H, Kobayashi T, Kawai S, Ito S (2002) A novel species of alkaliphilic *Bacillus* that produces an oxidatively stable alkaline serine protease. Extremophiles 6(1):65–72

Saha S, Dhanasekaran D, Shanmugapriya S, Latha S (2013) *Nocardiopsis* sp. SD5: a potent feather degrading rare actinobacterium isolated from feather waste in Tamil Nadu, India. J Basic Microbiol 53(7):608–616

Sakaguchi M, Takezawa M, Nakazawa R, Nozawa K, Kusakawa T, Nagasawa T, Sugahara Y, Kawakita M (2008) Role of disulphide bonds in a thermophilic serine protease Aqualysin I from *Thermus aquaticus* YT-1. J Biochem 143(5):625–632

Sako Y, Croocker PC, Ishida Y (1997) An extremely heat-stable extracellular proteinase (aeropyrolysin) from the hyperthermophilic archaeon *Aeropyrum pernix* K1. FEBS Lett 415(3):329–334

Sana B, Ghosh D, Saha M, Mukherjee J (2006) Purification and characterization of a salt, solvent, detergent and bleach tolerant protease from a new gamma-*Proteobacterium* isolated from marine environment of the *Sundarbans*. Process Biochem 41(1):208–215

Sánchez-Porro C, Mellado E, Bertoldo C, Antranikian G, Ventosa A (2003) Screening and characterization of the protease CP1 produced by the moderately halophilic bacterium *Pseudoalteromonas* sp. strain CP76. Extremophiles 7(3):221–228

Santos AF, Valle RS, Pacheco CA, Alvarez VM, Seldin L, Santos ALS (2013) Extracellular proteases of *Halobacillus blutaparonensis* strain M9, a new moderately halophilic bacterium. Braz J Microbiol 44(4):1299–1304

Sarethy IP, Saxena Y, Kapoor A, Sharma M, Sharma SK, Gupta V, Gupta S (2011) Alkaliphilic bacteria: applications in industrial biotechnology. J Ind Microbiol Biotechnol 38(7):769–790

Sayers EW, Barrett T, Benson DA, Bryant SH, Canese K, Chetvernin V, Church DM, DiCuccio M, Edgar R, Federhen S, Feolo M, Geer LY, Helmberg W, Kapustin Y, Landsman D, Lipman DJ, Madden TL, Maglott DR, Miller V, Mizrachi I, Ostell J, Pruitt KD, Schuler GD, Sequeira E, Sherry ST, Shumway M, Sirotkin K, Souvorov A, Starchenko G, Tatusova TA, Wagner L, Yaschenko E, Ye J (2009) Database resources of the National Center for Biotechnology Information. Nucleic Acids Res 37:D5–D15

Schmitt W, Rdest U, Goebel W (1990) Efficient high-performance liquid chromatographic system for the purification of halobacterial serine protease. J Chromatogr A 521(2):211–220

Secades P, Alvarez B, Guijarro J (2003) Purification and properties of a new psychrophilic metalloprotease (Fpp2) in the fish pathogen *Flavobacterium psychrophilum*. FEMS Microbiol Lett 226(2):273–279

Selim S, Hagagy N, Aziz MA, El-Meleigy ES, Pessione E (2014) Thermostable alkaline halophilic-protease production by *Natronolimnobius innermongolicus* WN18. Nat Prod Res 28(18):1476–1479

Sellek GA, Chaudhuri JB (1999) Biocatalysis in organic media using enzymes from extremophiles. Enzyme Microb Technol 25(6):471–482

Setati ME (2010) Diversity and industrial potential of hydrolase-producing halophilic/halotolerant eubacteria. Afr J Biotechnol 9(11):1555–1560

Setyorini E, Takenaka S, Murakami S, Aoki K (2006) Purification and characterization of two novel halotolerant extracellular protease from *Bacillus subtilis* strain FP-133. Biosci Biotechnol Biochem 70(2):433–444

Shao Z, Zhao H, Giver L, Arnold FH (1998) Random-priming in vitro recombination: an effective tool for directed evolution. Nucleic Acids Res 26(2):681–683

Shi W, Tang XF, Huang Y, Gan F, Tang B, Shen P (2006) An extracellular halophilic protease SptA from a halophilic archaeon *Natrinema* sp. J7: gene cloning, expression and characterization. Extremophiles 10(6):599–606

Shi WL, Zhong CQ, Tang B, Shen P (2007) Purification and characterization of extracellular halophilic protease from haloarchaea *Natrinema* sp. R6-5. Acta Microbiol Sin 47(1):161–163

Singh J, Batra N, Sobti RC (2001) Serine alkaline protease from a newly isolated *Bacillus* sp. SSR1. Process Biochem 36(8–9):781–785

Singhal P, Nigam VK, Vidyarthi AS (2012) Studies on production, characterization and applications of microbial alkaline proteases. Int J Adv Biotechnol Res 3(3):653–669
Sinha R, Khare SK (2012) Isolation of a halophilic *Virgibacillus* sp. EMB13: characterization of its protease for detergent application. Indian J Biotechnol 11:416–426
Sinha R, Khare SK (2013) Characterization of detergent compatible protease of a halophilic *Bacillus* sp. EMB9: differential role of metal ions in stability and activity. Bioresour Technol 145:357–361
Sinsuwan S, Rodtong S, Yongsawatdigul J (2010) A NaCl-stable serine proteinase from *Virgibacillus* sp. SK33 isolated from Thai fish sauce. Food Chem 119(2):573–579
Son E, Kim JI (2003) Multicatalytic alkaline serine protease from psychrotrophic *Bacillus amyloliquefaciens* S94. J Microbiol 41(1):58–62
Stepanov VM, Rudenskaya GN, Revina LP, Gryaznova YB, Lysogorskaya EN, Filippova IY, Ivanova II (1992) A serine proteinase of an archaebacterium, *Halobacterium mediterranei*. A homologue of eubacterial subtilisins. Biochem J 285(Pt 1):281–286
Stepanov VM (1996) Proteinases as catalysts in peptide synthesis. Pure Appl Chem 68(6):1335–1339
Strausberg SL, Alexander PA, Gallagher DT, Gillialand GL, Barnett BL, Bryan PN (1995) Directed evolution of a subtilisin with calcium-independent stability. Nat Biotechnol 13:669–673
Studdert CA, Herrera Seitz MK, Plasencia Gil MI, Sanchez JJ, De Castro RE (2001) Purification and biochemical characterization of the haloalkaliphilic archaeon *Natronococcus occultus* extracellular serine protease. J Basic Microbiol 41(6):375–383
Suzuki Y, Tsujimoto Y, Matsui H, Watanabe K (2006) Decomposition of extremely hard-to-degrade animal proteins by thermophilic bacteria. J Biosci Bioeng 102(2):73–81
Synowiecki J (2010) Some application of thermophiles and their enzymes for protein processing. Afr J Biotechnol 9(42):7020–7025
Takagi H, Takahashi T, Momose H, Inouye M, Maeda Y, Matsuzawa H, Ohta T (1990) Enhancement of the thermostability of subtilisin E by introduction of a disulfide bond engineered on the basis of structural comparison with thermophilic serine protease. J Biol Chem 265(12):6874–6878
Takai H, Sakato K, Nakamizo N, Isowa Y (1981) Enzymatic synthesis of caerulein peptide. In: Oyama K, Nikimura S (eds) Peptide chemistry. Protein Research Foundation, Osaka, Japan, pp 213–241
Tariq AL, Reyaz AL, Prabakaran J (2011) Purification and characterization of 56 kDa cold active protease from *Serratia marcescens*. Afr J Microbiol Res 5(32):5841–5847
Thangam EB, Rajkumar GS (2002) Purification and characterization of alkaline protease from *Alcaligenes faecalis*. Biotechnol Appl Biochem 35(2):149–154
Thumar J, Singh SP (2007) Two-step purification of a highly thermostable alkaline protease from salt-tolerant alkaliphilic *Streptomyces clavuligerus* strain Mit-1. J Chromatogr B 854(1–2):198–203
Thumar JT, Singh SP (2009) Organic solvent tolerance of an alkaline protease from salt-tolerant alkaliphilic *Streptomyces clavuligerus* strain Mit-1. J Ind Microbiol Biotechnol 26(2):211–218
Tiberti M, Papaleo E (2011) Dynamic properties of extremophilic subtilisin-like serine-proteases. J Struct Biol 174(1):69–83
Toplak A, Wu B, Fusetti F, Quaedflieg PJ, Janssen DB (2013) Proteolysin, a novel highly thermostable and cosolvent-compatible protease from the thermophilic bacterium *Coprothermobacter proteolyticus*. Appl Environ Microbiol 79(18):5625–5632
Turkiewicz M, Pazgier M, Kalinowska H, Bielecki S (2003) A cold-adapted extracellular serine proteinase of the yeast *Leucosporidium antarcticum*. Extremophiles 7(6):435–442
van den Burg B, Enequist HG, van den Haar ME, Eijsink VG, Stulp BK, Venema G (1991) A highly thermostable neutral protease from *Bacillus caldolyticus*: cloning and expression of the gene in *Bacillus subtilis* and characterization of the gene product. J Bacteriol 173(13):4107–4115
Varela H, Ferrari MD, Belobrajdic L, Vazquez A, Loperena ML (1997) Skin unhairing proteases of *Bacillus subtilis*: production and partial characterization. Biotechnol Lett 19(8):755–758

Vazquez S, Ruberto L, Mac Cormak W (2005) Properties of extracellular proteases from three psychrotolerant *Stenotrophomonas maltophilia* isolated from Antarctic soil. Polar Biol 28(4):319–325

Vidyasagar M, Prakash S, Litchfield C, Sreeramulu K (2006) Purification and characterization of a thermostable, haloalkaliphilic extracellular serine protease from the extreme halophilic archaeon *Halogeometricum borinquense* strain TSS101. Archaea 2:51–57

Vidyasagar M, Prakash S, Mahajan V, Shouche YS, Sreeramulu K (2009) Purification and characterization of an extreme halothermophilic protease from a halophilic bacterium *Chromohalobacter* sp. TVSP101. Braz J Microbiol 40(1):12–19

VijayAnand S, Hemaprita J, Selvin J, Kiran S (2010) Production and optimization of haloalkaliphilic protease by an extremophile – *Halobacterium* sp. Js1, isolated from thalassohaline environment. Glob J Biotechnol Biochem 5(1):44–49

Voorhorst WGB, Eggen RIL, Geerling ACM, Platteeuw C, Siezen RJ, de Vos WM (1996) Isolation and characterization of the hyperthermostable serine protease, pyrolysin, and its gene from the hyperthermophilic archaeon *Pyrococcus furiosus*. J Biol Chem 271(34):20426–20431

Waghmare SR, Gurav AA, Mali SA, Nadaf NH, Jadhav DB, Sonawane KD (2015) Purification and characterization of novel organic solvent tolerant 98 kDa alkaline protease from isolated *Stenotrophomonas maltophilia* strain SK. Protein Expr Purif 107:1–6

Wang QF, Miao JL, Hou YH, Ding Y, Wang GD, Li GY (2005) Purification and characterization of an extracellular cold-active serine protease from the psychrophilic bacterium *Colwellia sp.* NJ341. Biotechnol Lett 27(16):1195–1198

Wang QF, Hou HY, Xu Z, Miao JL, Li GY (2008) Purification and properties of an extracellular cold-active protease from the psychrophilic bacterium *Pseudoalteromonas* sp. NJ276. Biochem Eng J 38(3):362–368

Wilson SA, Daniel RM, Peek K (1994) Peptide synthesis with a proteinase from the extremely thermophilic organism *Thermus* Rt41A. Biotechnol Bioeng 44(3):337–346

Xiong H, Song L, Xu Y, Tsoi MY, Dobretsov S, Qian PY (2007) Characterization of proteolytic bacteria from the Aleutian deep-sea and their proteases. J Ind Microbiol Biotechnol 34(1):63–71

Xu B, Zhong Q, Tang X, Yang Y, Huang Z (2009) Isolation and characterization of a new keratinolytic bacterium that exhibits significant feather-degrading capability. Afr J Biotechnol 8(18):4590–4596

Yang C, Wang F, Hao J, Zhang K, Yuan N, Sun M (2010) Identification of a proteolytic bacterium, HW08, and characterization of its extracellular cold-active alkaline metalloprotease Ps5. Biosci Biotechnol Biochem 74(6):1220–1225

Yang XS, Chen XL, Xu XZ, Zeng RY (2011) Cold-adaptive alkaline protease from the psychrophilic *Planomicrobium* sp. 547: enzyme characterization and gene cloning. Adv Polar Sci 22(1):49–54

Yang J, Li J, Mai Z, Tian X, Zhang S (2013) Purification, characterization, and gene cloning of a cold-adapted thermolysin-like protease from *Halobacillus* sp. SCSIO 20089. J Biosci Bioeng 115(6):628–632

Yin J, Chen JC, Wu Q, Chen GQ (2014) Halophiles, coming stars for industrial biotechnology. Biotechnol Adv 33(7):1433–1442

Yu TX (1991) Proteases of haloalkaliphiles. In: Horikoshi K, Grand WD (eds) Superbugs: microorganisms in extreme environments. Japan Scientific Societies Press/Springer-Verlag, Tokyo/Berlin, pp 76–83

Yuan Q, Hayashi A, Kitamura Y, Shimada T, Na R, Jin X (2009) Purification and characterization of cold-adapted metalloprotease from deep sea water lactic acid bacteria *Enterococcus faecalis* TN-9. Int J Biol 1(2):12–21

Zeng R, Zhang R, Zhao J, Lin N (2003) Cold-active serine alkaline protease from the psychrophilic bacterium *Pseudomonas* strain DY-A: enzyme purification and characterization. Extremophiles 7(4):335–337

Zhao H, Arnold FH (1999) Directed evolution converts subtilisin E into a functional equivalent of thermitase. Protein Eng 12(1):47–53

Zhao GY, Zhou MY, Zhao HL, Chen XL, Xie BB, Zhang XY, He HL, Zhou BC, Zhang YZ (2012) Tenderization effect of cold-adapted collagenolytic protease MCP-01 on beef meat at low temperature and its mechanism. Food Chem 134(4):1738–1744
Zhu HY, Tian Y, Hou YH, Wang TH (2009) Purification and characterization of the cold-active alkaline protease from marine cold-adaptive *Penicillium chrysogenum* FS010. Mol Biol Rep 36(8):2169–2174

Internet Website

http://www.chem.qmul.c.uk/iubmb/enzyme
http://merops.sanger.ac.uk
http://www.ncbi.nlm.nih.gov/Taxonomy/CommonTree/wwwcmt.cgi

Chapter 15
Cold-Active β-Galactosidases: Sources, Biochemical Properties and Their Biotechnological Potential

Hubert Cieśliński, Marta Wanarska, Anna Pawlak-Szukalska, Ewelina Krajewska, Monika Wicka, and Józef Kur

1 β-D-Galactosidase: Basic Characteristic

β-D-Galactosidase (EC 3.2.1.23), which is also known as β-D-galactoside galactohydrolase (systematic name), is a highly specific exoglycosidase that catalyzes the hydrolysis of terminal non reducing β-D-galactoside residues in β-D-galactosides (Wanarska and Kur 2005). One of the most well known β-D-galactosides is the disaccharide lactose, which is found in milk and can be enzymatically hydrolyzed to D-glucose and D-galactose. In the presence of a high lactose concentration, some β-D-galactosidases can also exhibit galactosyltransferase activity (Cruz et al. 1999). Several β-D-galactosidases have also been reported to show other enzymatic activities. For example, the β-D-galactosidase isolated from *Rhodotorula minuta* IFO897 hydrolyzed some β-D-glucosides, β-D-fucosides and α-L-arabinosides (Onishi and Tanaka 1996).

β-D-Galactosidases can be found in a broad range of bacteria, fungi, plants and animals. β-D-Galactosidase enzymes isolated from different sources can vary considerably in terms of their properties, including their molecular weight, structure, metal ion requirements and substrate specificity, as well as the temperature and pH conditions required for their optimum enzymatic activity. The most abundant source of β-D-galactosidase enzymes are bacteria, with some bacteria producing more than one β-D-galactosidase. For example, *Bifidobacterium adolescentis* DSM 20083 (Van Laere et al. 2000) and *Bacillus circulans* (Vetere and Paoletti 1998) produces two and three different β-D-galactosidases, respectively. The most well known and widely characterized of the β-D-galactosidase enzymes isolated from bacteria is *LacZ* β-D-galactosidase, which was originally isolated from *E. coli*. The extensive

H. Cieśliński • M. Wanarska • A. Pawlak-Szukalska • E. Krajewska • M. Wicka • J. Kur (✉)
Faculty of Chemistry, Department of Molecular Biotechnology and Microbiology, Gdańsk University of Technology, 80-233 Gdańsk, Poland
e-mail: kur@pg.gda.pl

P.H. Rampelotto (ed.), *Biotechnology of Extremophiles*, Grand Challenges in Biology and Biotechnology 1, DOI 10.1007/978-3-319-13521-2_15

research on this particular enzyme led to the development of our current understanding of the mechanisms of the reactions catalyzed by β-D-galactosidases, and the mechanism of the regulation of gene expression in Jacob and Monod's operon model. To date, the ability of LacZ to produce an easily recognizable blue reaction product from the hydrolysis of 5-bromo-4-chloro-3-indolyl-β-D-galactopyranoside (X-Gal) has made it a principal component of blue/white screening processes in cloning and several other molecular biology procedures (Juers et al. 2012). Several other bacterial and fungal β-D-galactosidases have also been employed recently in a variety of other scientific and industrial applications, and the nature of these applications and the β-D-galactosidases themselves will be discussed in greater detail in the next chapter.

2 β-D-Galactosidases: Industrial Applications and Sources of Industrial Enzymes

β-D-Galactosidases have been widely used in a variety of different dairy products and foodstuffs. In terms of their role in dairy products, β-D-galactosidases have been used to achieve the hydrolysis of lactose in milk and whey. There are several technological, economical and medical reasons for wanting to achieve the enzymatic hydrolysis of lactose in milk. First, low lactose-containing milk (abr. LLM) is a dairy product dedicated to lactose-intolerant individuals who do not possess high enough levels of β-D-galactosidase in their own digestive system to achieve the hydrolysis of lactose. In these individuals the symptoms of lactose intolerance can be occurrence of abdominal pain, bloating borborygmi and loose stools after consuming milk and some milk products (Juers et al. 2012). It has been reported that large numbers of Asian (more than 90 %), African (80–100 %), Native American (more than 90 %) and Southern European (more than 80 %) populations are lactose intolerant (Mlichova and Rosenberg 2006). The high levels of lactose intolerance experienced in these populations could therefore be solved in part by the large-scale production of LLM as a commercial product to be sold in standard retail stores. For this purpose, in theory, LLM could be produced by the chemical or enzymatic hydrolysis of lactose present in milk. However, in practice, a comparison of these two processes reveals that there are several undeniable benefits to enzymatic hydrolysis, including the lack of any additional by-products, no degradation to any of the other valuable compounds contained in milk and the introduction of no additional nasty flavors, odors or colors. Therefore, the chemical hydrolysis of lactose (acid hydrolysis) is not used for production of LLM. Furthermore, milk treated with a β-D-galactosidase enzyme retains its original nutritional value. This is because that the products of enzymatic lactose hydrolysis, D-glucose and D-galactose are preserved in LLM (Ladero et al. 2003). Recently, the ways for production of LLM on industrial scale may base on enzymatic hydrolysis lactose in milk or combination of physical methods of lactose separation such as chromatographic separation process or membrane techniques combined with enzymatic hydrolysis of lactose (Harju et al. 2012).

Considerable interest has also been expressed in the development of new methods for the enzymatic hydrolysis of lactose by researchers working in food technology because of the poor solubility of lactose and the adverse impact that it can have on the quality of different foodstuffs. For example, high concentrations of lactose in milk used to produce of ice cream or condensed milk, can lead to lactose crystallization, which is responsible for an unpleasant sandy, gritty or mealy texture of these milk products. The hydrolysis of the lactose present in ice cream and other dairy products leads to significant improvements in the creaminess and scoopability of these products, while also making them more digestible. Furthermore, the monosaccharides formed by the hydrolysis of lactose can be fermented more readily than lactose itself by the starter cultures used for the production of several dairy products, including yoghurts and cottage cheese. The hydrolysis of lactose therefore shortens the time needed to obtain the desired low pH in these products. Moreover, the presence of glucose and galactose in LLM considerably increases its sweetness (about 50 %), which means that the amount of sweeteners added to low-calorie yoghurts can be reduced considerably.

Besides cow's milk, one of the other major sources of lactose is whey, which is formed during the production of cheese. The annual worldwide production of whey is 150 million tons, and the main components of this material are lactose (44–52 g L^{-1}), proteins (6–8 g L^{-1}) and minerals (4.3–9.5 g L^{-1}) (Johansen et al., 2002). Nowadays, cheese whey is used primarily in industry for the production of whey protein concentrate (WPC), which is a multifunctional food ingredient of high nutritional value. This process also creates a large amount of whey permeate, which still contains 4–5 % lactose (Mlichová and Rosenberg 2006). Lactose purified by crystallization is mainly used as a supplement in foods, as well as an excipient for the formulation of pharmaceutical products. However, despite its many uses, it is currently estimated that 47 % of the whey generated worldwide each year remains unused and represents a significant disposal problem (Guimaraes et al. 1992; Mlichová and Rosenberg 2006). For this reason, the presence of lactose in whey has been identified as a major problem in terms of its adverse impact on the utilization of this material. The lactose present in whey is responsible for more than 90 % of its biochemical oxygen demand (BOD), and the development of an effective method for the recovery of this disaccharide from whey is therefore crucial to allow for the effective utilization of this material. In this context, the enzymatic hydrolysis of lactose would present several benefits. In contrast to lactose, the products resulting from its hydrolysis, galactose and glucose, can be directly utilized as carbon sources by a large number of microorganisms, including recombinant systems, which could be used to produce valuable bioproducts, such as lactates, acetates, ethanol, butanediol and biopolymers (Guimaraes et al. 1992; Coté et al. 2004; Mehaia et al. 1993; Nath et al. 2014). Moreover, the same products of lactose hydrolysis in whey permeate could also be used for the production of galactose-glucose syrup (dairy syrup), which is used as a sweetener in the food industry. Furthermore, the hydrolysis of lactose would prevent it from crystallizing during the concentration of whey by evaporation, which is an extremely desirable process because it would allow for a significant reduction in the costs associated with the transport and storage of whey.

β-D-Galactosidase also exhibits galactosyltransferase activity, making it an especially attractive enzyme for food industry. The galactosyltransferase activity of β-D-galactosidase can be employed to achieve the synthesis of galactooligosaccharides (GOSs) containing two or more galactose units, starting from lactose. The conditions required for the efficient synthesis of GOSs include a high lactose concentration in the reaction medium. Moreover, the composition of GOSs bearing glycosidic linkages such as β(1,4), β(1,3), β(1,2) and β(1,6) formed during enzymatic process depends on the source of the enzyme and the overall reaction conditions. Nowdays, GOSs have attracted considerable attention from researchers working in the food industry because of their natural presence in human and cow's milk. GOSs present in milk stimulate the growth of bifidobacteria, which subsequently lead to a decrease in the concentration of putrefactive bacteria in the gut of humans and other animals. It has been suggested that the prebiotic properties of GOSs have several beneficial effects on their consumers, including (i) a decrease in the number of putrefactive products in their feces; (ii) a reduction in their blood cholesterol content; (iii) the absorption of larger quantities of Ca^{2+} from their diet; and (iv) a lower incidence of colon cancer amongst these individuals. Furthermore, GOSs are stable under the acidic and high temperature conditions used during food processing, and maintain excellent taste quality. For these reasons, GOSs are recognized as attractive carbohydrate-based functional food ingredients, and considerable research effort has been directed towards the development of methods for the efficient and economical production of these materials. Nowadays, GOSs are not only used as prebiotic food ingredients but also used as cosmetic ingredients and low calorie sweeteners (Mlichová and Rosenberg 2006).

In addition to being used for the synthesis of GOSs, β-D-galactosidases can also be used to produce heterooligosaccharides (HOSs) via the transfer of a galactosyl moiety to any other sugar, except for lactose, glucose or galactose. At present, 4-*O*-β-D-galactopyranosyl-β-D-fructofuranose, which is more commonly known as lactulose, is the most widely used HOS in food industry. Lactulose exhibits prebiotic properties and has consequently been used as a functional food ingredient in a similar manner to GOSs. On the other hand, lactulose in medicine, is applied to the treatment of constipation and hepatic encephalopathy. Moreover, this HOS can be also applied to the diagnosis of colonic disorders using a hydrogen breath test. The current route for the industrial production of lactulose involves the alkaline isomerization of lactose. Unfortunately, the harsh production conditions of this process (pH of 10.5–11.5 and temperatures of 70–100 °C) lead to the formation of several byproducts as a result of the degradation of lactose. In comparison to this, the enzymatic synthesis of lactulose could be conducted under much milder conditions and, therefore the crude lactose materials such as whey or whey permeate could be employed in this process. Moreover, it is noteworthy that the heterologous galactosyl transfer reaction catalyzed by β-D-galactosidase yields a virtually unlimited diversity of oligosaccharides (Pawlak-Szukalska et al. 2014).

The galactosyltransferase activity of β-D-galactosidase can not only be employed for the synthesis of GOSs and HOSs, but can also be employed for the synthesis of wide range of β-glycosides in which D-galactose residue is bond via β-1→4 linkage to such aglycons as various alcohols (Stevenson et al. 1993), antibiotics

(Scheckermann et al. 1997), ergot alkaloids (Kren et al. 1992) and also the flavonol glycoside myricitrin (Shimizu et al. 2006). These results therefore suggest that β-D-galactosidases could potentially be used for the industrial production of selected pharmaceuticals and a range of other biologically active compounds.

3 β-D-Galactosidases: Source for Industry

Although the LacZ enzyme from *E. coli* is the most well studied β-D-galactosidase, it is unlikely that this enzyme will ever be used in the food industry because of the inherent risk factors associated with the use of *E. coli* as a source of enzyme. The β-D-galactosidases used on an industrial scale for the manufacture of LLM and other dairy products are therefore mainly isolated from the yeast *Kluyveromyces lactis* (Maxilact® product line from DSM, Lactozym® Pure from Novozymes and GODO-YNL2 Lactase from DuPont) and the fungus *Aspergillus oryzae* (Tolerase™ L from DSM), which are generally recognized as safe microorganisms (GRAS status). The choice of a relevant β-D-galactosidase, which is used for the hydrolysis of lactose, depends on the reaction conditions. For example, the optimum pH for the hydrolysis of lactose in sweet whey or cow's milk using the β-D-galactosidase derived from yeast is in the range of 6.5–7.0. However, the optimum pH for the hydrolysis of the lactose in acidic whey using the fungal β-D-galactosidase is in the range of 3.0–5.0 (Boon et al. 2000; Harju 1987). The important feature for industrial enzymes is lack of feedback inhibition by reaction products. For this reason, the mechanism of β-D-galactosidase-catalyzed hydrolysis of lactose has been intensively studied. The most commonly proposed was a Michaelis-Menten kinetic model with competitive inhibition by D-galactose for both β-D-galactosidase from *Kluyveromyces* yeast and *Aspergillus* fungus (Santos et al. 1998 and references therein; Jurado et al. 2002). It is noteworthy that both of these industrial enzymes are produced by mesophilic microorganisms, and are therefore not well suited for the development of novel biotechnologies that require enzymatic reactions to be conducted at low or high temperatures. For this reason, there has been a significant growth during the last 15 years in the number of studies devoted to the assessment of the biotechnological potential of thermostable and cold-active β-D-galactosidases. In general, the use of thermostable and cold-active β-D-galactosidases in industrial processes provides a number of significant advantages, and the next section will focus specifically on the promising profits combined with the use of cold active β-D-galactosidases in the food industry.

4 Cold-Active β-D-Galactosidases

The distinguishing features of cold-active enzymes, such as their high catalytic efficiency at low and moderate temperatures and thermolability at moderate or high temperatures, offer a unique benefits that are important for the development of new biotechnology processes. The most important of these advantages include (i) reduced processing times; (ii) reduced energy costs; (iii) reduction in the loss of volatile compounds, including enzymatic reaction substrates and/or products; (iv) the ability to perform reactions involving thermo-sensitive compounds; and (v) the rapid inactivation of the enzyme at moderate temperatures upon completion of the enzymatic step of the industrial process. However, the poor thermostability of these enzymes can also be a disadvantage when good stability properties are required for medium- to long-term storage. For this reason, there have been a growing number of studies focused on the biotechnological potential of cold-active enzymes in science (molecular biology) from scientists working in a variety of different industries. In this respect, β-D-galactosidases have become one of the most heavily investigated groups of "cold-loving" enzymes because they could be used in the same way as their mesophilic counterparts in the dairy industry for the production of LLM or the hydrolysis of lactose in whey and whey permeate. However, in contrast to their mesophilic counterparts, cold-active enzymes can effectively carry out the hydrolysis of lactose at low temperatures (approximately at 10 °C). Under these conditions, the hydrolysis of lactose could be conducted during the shipping and storage of milk, which would allow for a significant reduction in the time required for the entire production process (i.e., energy saving). Furthermore, the risk of the final product being contaminated with any mesophilic microflora would be significantly decreased. The use of cold-active enzymes would also prevent the generation of nonenzymatic browning products, which are formed in milk at the higher temperatures required for the effective hydrolysis of lactose using the mesophilic β-D-galactosidases.

The main sources of cold-active β-D-galactosidases are psychrophilic and psychrotolerant microorganisms and the vast majority of these enzymes have been isolated from *Arthrobacter* species. The reason for this apparent preference can be explained by the fact that most of these strains were isolated from soil samples and *Arthrobacter* sp. are basic soil bacteria. As far as we know, only two cold-active β-D-galactosidases have been identified, isolated and biochemically characterized to date from eukaryotic microorganism, the yeast-like fungus *Guehomyces pullulans* (Song et al. 2010; Nakagawa et al. 2006b).

Most of the known bacterial strains producing cold-active β-D-galactosidases were isolated from low temperature environments. For example, *Arthrobacter* sp. C2-2 was isolated from fellfield soil (Karasová-Lipovová et al. 2003), *Planococcus* sp. SOS Orange from a hypersaline pond in Antarctica (Sheridan and Brenchley 2000) and *Carnobacterium piscicola* BA was isolated from a soil sample taken in late winter from a field treated with whey in Pennsylvania, USA (Coombs and Brenchley 1999). Several other interesting examples of bacterial strains producing cold-active β-D-galactosidases include *Pseudoalteromonas* sp. 22b, which was isolated from the digestive tract of Antarctic krill, *Thysanoessa macrura* (Cieśliński et al. 2005) and *Alkalilactibacillus ikkense,* which was isolated from the Ikka columns in South-West Greenland (Schmidt and Stougaard 2010). However, the living conditions of these microbes are generally impossible to recreate in the laboratory (using conventional methods) and culturable microorganisms represent less than 1 % of those occurring in nature (White et al. 2000). For this reason, Wang et al. investigated the direct cloning of the DNA found in environmental samples collected from an oil field in the Heilongjiang Province, China and constructed a series of metagenomic libraries. Using this method, the researchers successfully identified the *zd410* gene, which encoded a cold-active β-D-galactosidase (Wang et al. 2010).

Besides the origin of cold-active β-D-galactosidases, an even more important criterion for the comparison of their enzymatic properties is their affiliation to the glycoside hydrolase (GH) families. One hundred and thirty-three GH families have recently been distinguished, and proteins belonging to the same family have been reported to have an amino acid sequence homology of no less than 30 % (Poltorak et al. 2007). It is currently commonly accepted, that the GH1, GH2, GH35 and GH42 families contain enzymes that possess β-D-galactosidase activity. In terms of their ability to affect the removal of lactose from dairy products, enzymes belonging to the GH2 family perform more effectively than enzymes belonging to any of the other families because they demonstrate higher activity towards lactose. It is noteworthy, however, that β-D-galactosidases from the GH42 family possess a variety of other activities that are not exhibited by β-D-galactosidases in the GH2 family. Several proteins belonging to the GH42 family have also been reported to exhibit β-D-fucosidase (EC 3.2.1.38) and α-L-arabinosidase (EC 3.2.1.55) activities (Di Lauro et al. 2008; Lee et al. 2011; Kosugi et al. 2002; Saishin et al. 2010; Sheridan and Brenchley 2000).

Most of the cold-active β-D-galactosidases that have been identified, isolated and biochemically characterized to date are members of the GH2 and GH42 families, although a few cold-active enzymes with β-D-galactosidase activity have also been found belonging to the GH35 family. It has been also reported that some "cold-loving" microorganisms possess more genes that are capable of encoding enzymes with β-D-galactosidase activity. For example, the psychrotolerant bacterium *Arthrobacter* sp. B7 carries genes of three different isoenzymes with β-D-galactosidase activity, which can be classified into the GH2, GH35 and GH42 families (Trimbur et al. 1994; Gutshall et al. 1995, 1997). It has also been found that the bacterium *Carnobacterium piscicola* BA produces two β-D-galactosidases, which are members of the GH35 and GH42 families, respectively (Coombs and Brenchley 1999, 2001). Furthermore, *Arthrobacter* sp. ON14 has been reported to contain two cold-active β-D-galactosidases, which are members of the GH2 and GH42 families (Xu et al. 2011). Coombs and Brenchley (2001) proposed that the presence of two or more GH enzymes in one microorganism could be explained by the need for the synergistic activity of these enzymes to assist in the degradation of other polysaccharides with α- and β-galactoside linkages, which are abundant in the environment. This hypothesis was based on the fact that the genes of two cold-active β-D-galactosidases and one α–galactosidase are located on one operon in *Carnobacterium piscicola* BA. By cooperating in the degradation of certain sugars, these enzymes could provide good alternative sources of carbon for their microorganisms (Coombs and Brenchley 2001). Moreover, an investigation of the gene encoding the cold-active β-D-galactosidase from *A. ikkense* revealed that it was only separated by 16 bp from that of the α–galactosidase gene. It was also assumed in this case that the analyzed genes are under the control of the same promoter (Schmidt and Stougaard 2010), and they could therefore be contributing to the hydrolysis of sugar complexes in "cold" environments.

Based on the data presented above, the remainder of this chapter will focus on the enzymatic properties of the cold-active β-D-galactosidases in terms of their characterization into different GH families.

4.1 Cold-Active β-D-Galactosidases Belonging to the GH2 Family

Selected properties of cold-active β-D-galactosidases belonging to the GH2 family are presented in Table 15.1.

β-D-Galactosidases from the GH2 family show high specificity towards a variety of β-D-galactopyranosides, including lactose. Some cold-active β-D-galactosidases can be inhibited by the products resulting from the hydrolysis of lactose. D-Galactose is a competitive inhibitor of *Arthrobacter* sp. 32cB and *Arthrobacter* sp. SB β-D-galactosidases (Pawlak-Szukalska et al. 2014; Coker et al. 2003). In contrast, D-glucose has been reported to inhibit the activity of the β-D-galactosidase from *Pseudoalteromonas* sp. 22b (Makowski et al. 2009), while the β-D-galactosidase from *Paracoccus* sp. 32d is inhibited by both monosaccharides (Wierzbicka-Woś et al. 2011).

The molecular mass of subunit of cold-active β-D-galactosidases, members of the GH2 family, is around 110 kDa, although these enzymes mainly exist as tetramers in their native form. The exceptions to this rule, however, include the β-D-galactosidase from *Flavobacterium* sp. *4214* (Sørensen et al. 2006), which is active in its monomeric conformation, and the β-D-galactosidases from *Arthrobacter* sp. 32cB (Pawlak-Szukalska et al. 2014) and *Paracoccus* sp. 32d (Wierzbicka-Woś et al. 2011), which exist as dimers in their native form. As it is shown in Table 15.1, the apparent optimal temperature and pH values for enzymes belonging to the GH2 family ranged from 10 °C to 45 °C and pH 6.0–8.5, respectively. With the exception of the β-D-galactosidase enzymes derived from *Arthrobacter* sp. SB (Coker and Brenchley 2006) and *Arthrobacter* sp. 32cB (Pawlak-Szukalska et al. 2014), other cold-acive enzymes, which are presented in Table 15.1, show thermal inactivation at 50 °C. Furthermore, divalent metal ions such as Mg^{2+} and Mn^{2+} have a positive impact of the enzymatic activity of analyzed enzymes (Table 15.1), whereas heavy metal ions such as Cu^{2+}, Ni^{2+}, Pb^{2+} and Zn^{2+} behave as inhibitors towards these enzymes. Also, for some β-D-galactosidases belonging to GH2 family, the enzymatic activity was positively stimulated by presence of K^+ and Na^+.

Table 15.1 Biochemical characterization of cold-active β-D-galactosidases belonging to family GH2

Organism	Molecular mass [kDa]	Oligomeric state	T_{opt} [°C]	pH_{opt}	Activators	Inhibitors	Thermal inactivation	References
Alkalilactibacillus ikkense	NR (119.1[a])	NR	20–30	8.0	NR	NR	5 min at 50 °C	Schmidt and Stougaard (2010)
Arthrobacter sp. B7 (gene 15)	NR (111.0[a])	NR	40	7.2	Mg^{2+}, Mn^{2+}, 2-ME	EDTA, Cu^{2+}, Ca^{2+}	10 min at 50 °C	Trimbur et al. (1994)
Arthrobacter sp. C2-2 izosyme 1	550 (110.8[a])	4	40	7.5	DTT, Mg^{2+}, Mn^{2+}	EDTA, Cu^{2+},Al^{3+}, Tris	10 min at 50 °C	Karasová-Lipovová et al. (2003)
Arthrobacter sp. 20B	460 (113.7[a])	4	25	6.0–8.0	DTT, 2-ME, Na^{+}, K^{+}, Mn^{2+}	Pb^{2+}, Zn^{2+}, Cu^{2+}, pCMB	1 min at 60 °C	Białkowska et al. (2009)
Arthrobacter sp. SB	463 (114.0[a])	4	18	7.0	Mg^{2+}, Mn^{2+}, K^{+}	EDTA, D-galactose	10 min at 37 °C	Coker and Brenchley (2006), Coker et al. (2003)
Arthrobacter psychrolactophilus F2	548 (111.7[a])	4	10	8.0	NR	NR	5 min at 50 °C	Nakagawa et al. (2006a)
Arthrobacter sp. ON14 (*gal A*)	NR (111.4[a])	NR	15	8.0	Na^{+}, K^{+}, Mg^{2+}, Mn^{2+}	Zn^{2+}, Cu^{2+}	20 min at 50 °C	Xu et al. (2011)
Arthrobacter sp. 32 cB	257 (109.6[a])	2	28	8.0	Mg^{2+}, DTT	Mn^{2+}, Ni^{2+}, Co^{2+}, Ca^{2+}, EDTA, glutathione ox., cysteine, TCEP, D-galactose	5 min at 44 °C	Pawlak-Szukalska et al. (2014)
Flavobacterium sp. 4214	<66 (114.3[a])	1	42	7.5	NR	NR	NR	Sørensen et al. (2006)
Pseudoalteromonas haloplanktis TAE 79	>300 (118.1[a])	4	45	8.5	Mg^{2+}, Mn^{2+}, Ca^{2+}, Li^{+}, 2-ME	EDTA, Zn^{2+},Cu^{2+},Ni^{2+}	60 min at 45 °C	Hoyoux et al. (2001)

(continued)

Table 15.1 (continued)

Organism	Molecular mass [kDa]	Oligomeric state	T_{opt} [°C]	pH_{opt}	Activators	Inhibitors	Thermal inactivation	References
Paracoccus sp. 32d	161 (81.7[a])	2	40	7.5	–	Ca^{2+}, Mn^{2+}, Ni^{2+}, Co^{2+}, DTT, glutathione ox., D-glucose, D-galactose	15 min at 50 °C	Wierzbicka-Woś et al. (2011)
Pseudoalteromonas sp. 22b	490 (117.1[a])	4	40	6.0–8.0	Na^{+}, K^{+}, Mg^{2+}, Mn^{2+}, DTT, glutathione, 2-ME	Zn^{2+},Cu^{2+},Ni^{2+}, Pb^{2+}, pCMB, EDTA, D-glucose	2 min at 50 °C	Turkiewicz et al. (2003), Cieśliński et al. (2005), Makowski et al. (2007)

NR not reported, *2-ME* 2-mercaptoethanol, *pCMB* 4-chloromercuribenzoic acid, *TCEP* tris(2-carboxyethyl)phosphine, *DTT* dithiothreitol, *EDTA* ethylenediaminetetraacetic acid, *Tris* tris(hydroxymethyl)aminomethane

[a]Molecular mass of the monomer, calculated from the amino acid sequence

4.2 *Cold-Active β-D-Galactosidases Belonging to the GH42 Family*

Several cold-adapted β-D-galactosidases belonging to the GH42 family have been isolated and biochemically characterized from psychrotolerant bacteria belonging to the genus *Arthrobacter* (Hildebrandt et al. 2009; Gutshall et al. 1995) *Planococcus* (Hu et al. 2007; Sheridan and Brenchley 2000) and *Carnobacterium* (Coombs and Brenchley 1999), as well as metagenome DNA (Wang et al. 2010).

Selected properties of this group of cold-active β-D-galactosidases are presented in Table 15.2. The molecular mass of subunit of these cold-active β-D-galactosidases is around 70 kDa, and these enzymes generally exist as dimers or trimers in their native forms. The apparent optimal temperature and pH values for these enzymes range from 20 °C to 50 °C and pH 6.5–7.0, respectively. Moreover, cold-active enzymes belonging to the GH42 family showed thermal inactivation at temperatures in the range of 40–50 °C. Interestingly, Mg^{2+}, Mn^{2+}, Ca^{2+}, Co^{2+} ions behave as enzymatic activators to the β-D-galactosidase enzyme from *Arthrobacter* sp. B7 (gene 12) (Gutshall et al. 1995), but act as enzymatic inhibitors of the β-D-galactosidase enzyme from *Arthrobacter* sp. 32c (Hildebrandt et al. 2009). Moreover, Mn^{2+} ions inhibited the activities of the β-D-galactosidase enzyme derived from *Planococcus* sp. SOS Orange (Sheridan and Brenchley 2000) and the enzyme encoded by the *zd410* gene derived from a metagenomic library. For comparison, Co^{2+} ions inhibit the activity of enzymes derivate from *Planococcus* sp. L4 (Hu et al. 2007) and *Planococcus* sp. SOS Orange (Sheridan and Brenchley 2000). All of the cold active enzymes belonging to the GH42 family are inhibited by Cu^{2+} ions.

4.3 *Cold-Active β-D-Galactosidases Belonging to the GH35 Family*

The β-D-galactosidase enzymes derived from *Arthrobacter* sp. B7 (gene 14) (Gutshall et al. 1997) and *Carnobacterium piscicola* BA (gene *bgaC*) (Coombs and Brenchley 1999) are the only reported members of the GH35 family to have been isolated from cold-adapted microorganisms.

The β-D-galactosidase from *Arthrobacter* sp. B7 (gene 14) exists as a dimer with a molecular mass of 110 kDa. The thermal inactivation of this enzyme occurs at 65 °C after 10 min, although it remains stable at a temperature of 20 °C for 2 h. The optimal activity of this enzyme was detected at pH 6.5, which could be potentially useful in terms of the removal of lactose from milk. The substrate specificity of the GH35 enzyme from *Arthrobacter* sp. B7 was determined using *p*-nitrophenyl (*p*-NP)-linked substrates, which showed that this enzyme exhibited activity towards *p*-NP-β-galactopyranoside and *p*-NP-β-fucopyranoside. The product of gene 14 expression also cleaved β-1,3 linkages, which were assayed using Galβ1-3GlcNAcb1-3Galβ1-4Glc-7-amino methyl coumarin as a substrate.

Table 15.2 Biochemical characterization of cold-active β-D-galactosidases belonging to family GH42

Organism	Molecular mass [kDa]	Oligomeric state	T_{opt} [°C]	pH_{opt}	Thermal Inactivation	Substrate specificity	Activators	Inhibitors	Reference
Arthrobacter sp. 32c	195(76.1[a])	3	50	6.5	NR	Lactose, ONPG, PNPG	Ethanol	Cu^{2+}, Zn^{2+}, Mn^{2+}, Ni^{2+}, Fe^{2+}, Co^{2+}, Ca^{2+}, Mg^{2+}, red glutathione, D-glucose	Hildebrandt et al. (2009)
Arthrobacter sp. B7 (Gene *12*)	NR(71[b])	NR	45–50	6.6	Less than 15 min at 50 °C	Lactose, ONPG, PNPG	Mg^{2+} Mn^{2+}, Ca^{2+}, Co^{2+}, 2-ME	Cu^{2+}, EDTA	Gutshall et al. (1995)
Planococcus sp. L4	156(77.3[a])	2	20	6.8	10 min at 45 °C	Lactose, ONPG, PNPG	Na^{+}, K^{+} at low concentration	Zn^{2+},Cu^{2+},Ni^{2+}, Co^{2+},	Hu et al. (2007)
Planococcus sp. SOS Orange	155(77.5[a])	2	42	6.5	10 min at 55 °C	ONPG, PNPG, PNPF, ONPF	None	Zn^{2+},Cu^{2+}, Ni^{2+},Co^{2+}, Mn^{2+}	Sheridan and Brenchley (2000)
Metagenome DNA (Gene *zd410*)	NR(78.6[a])	NR	38	7.0	1 h at 50 °C	Lactose, ONPG, PNPG, pNP-β-D-arabinoside, pNP-β-D-glucuronide, pNP-β-D-mannoside	Na^{+}, K^{+}, Ca^{2+}	Cu^{2+}, Fe^{2+}, Zn^{2+}, Mn^{2+}	Wang et al. (2010)
Carnobacterium piscicola BA (Gene *bgaB*)	NR(76.8[a])	NR	30	NR	10 min at 40 °C	ONPG, PNPG, pNP β-D-fucoside, pNP β-D-galacturonide	NR	NR	Coombs and Brenchley (1999)

NR not reported, *ONPF* *o*-nitrophenyl-β-D-fucopyranoside, *PNPF* *p*-nitrophenyl-β-D-fucopyranoside, *ONPG* *o*-nitrophenyl-β-D-galactopyranoside, *PNPG* *p*-nitrophenyl-β-D-galactopyranoside, *EDTA* ethylenediaminetetraacetic acid, *2-ME* 2-mercaptoethanol

[a]Molecular mass of the monomer, calculated from the amino acid sequence

[b]Molecular mass of the monomer, estimated by SDS-PAGE

The second β-D-galactosidase belonging to the GH35 family is encoded by the *bgaC* gene of *Carnobacterium piscicola* BA. In contrast to the other GH35 family enzyme described above, this enzyme has only been preliminary characterized, and its apparent optimal temperature was found to be at 40 °C. Furthermore, the half life of this enzyme was found to be 15 min at 45 °C. It has also been reported that the BgaC enzyme can be inhibited by Zn^{2+}, Ni^{2+} and Cu^{2+} ions (Coombs and Brenchley 2001).

4.4 Cold-Active β-D-Galactosidases Belonging to Any GH Family

There is a general lack of literature data pertaining to some cold-active β-D-galactosidases, which has prevented them from being assigned into a specific GH family. The enzymatic properties of these enzymes are shown in Table 15.3.

A comparison of the data shown in Tables 15.1, 15.2 and 15.3 (columns: "MW of monomer" and "Oligomeric state of enzymes") suggests that the cold-active β-D-galactosidases derived from *Halomonas* sp. S62 and *Pseudoalteromonas* sp. TAE 79b could be members of the GH42 and GH2 families, respectively. Moreover, the significant differences in the optimum pH for enzymatic activities of the enzymes derived from *Guehomyces pullulans* 17–1, *Guehomyces pullulans* R1 (Table 15.3) and several other cold-active β-D-galactosidases (Tables 15.1, 15.2 and 15.3) are undoubtedly a consequence of their origin. In contrast to the other enzymes presented in the tables, which were isolated from different bacterial species, these two cold-active β-D-galactosidases were derived from psychrotolerant yeast-like fungi. As mentioned above, a low pH optimum is also characteristic of β-D-galactosidases isolated from mesophilic fungi (e.g., *Aspergillus oryzae*).

5 Cold-Active β-D-Galactosidases: Biotechnological Potential

5.1 Cold-Active β-D-Galactosidases: The Hydrolysis of Lactose in Milk

As far as we know, 26 cold-active β-D-galactosidases have been isolated and characterized to date. Consideration of the current state of the art for these enzymes revealed that all of the scientific reports that have been published in this field have focused on the detailed characterization of the hydrolytic activity of these enzymes using *o*-nitrophenyl-β-D-galactopyranoside (ONPG) as a substrate. Furthermore, all of these reports confirmed that these enzymes effectively catalyzed the hydrolysis of this substrate at low temperatures. However, we have found that studies pertaining to the hydrolysis of lactose in milk have only been carried out for 12 of the characterized and publicized cold-active β-D-galactosidases (Table 15.4)

Table 15.3 Biochemical characterization of cold-active β-D-galactosidases not assigned to any of the known GH families

Organism	Molecular mass [kDa]	Oligomeric state	T_{opt} [°C]	pH_{opt}	Thermal inactivation	Activators	Inhibitors	Substrate specificity	References
Guehomyces pullulans 17-1	335(170[a])	2	50	4.0	2.5 h at 60 °C	Li^+	Ca^{2+}, Mg^{2+}, Co^{2+}, Ag^+, Cu^{2+},Fe^{3+}, Hg^{2+}, Fe^{2+}, iodoacetic acid, SDS, 1,10-phenanthroline, DTT	ONPG, lactose	Song et al. (2010)
Guehomyces pullulans R1	NR	NR	50	4.0	1 h at 50 °C	NR	NR	Lactose, ONPG	Nakagawa et al. (2006b)
Halomonas sp. S62	130[b](63[a])	2	45	7.0	5 min at 60 °C	Fe^{2+}, Mn^{2+}, Na^+, Ca^{2+}	Fe^{3+}, Cu^{2+}	ONPG, lactose	Wang et al. (2013)
Bacillus subtilis KL88	90[b]	NR	50	6.0	NR	Na^+, K^+ , Li^+, 2-ME ethanol, DTT	Cu^{2+}, Fe^{2+}, Fe^{3+}, Zn^{2+}, Ca^{2+} , Mg^{2+}, Mn^{2+}, D-glucose, D-galactose	Lactose, glucose-β-D-galactopirano side, methyl-β-D-galactopyranosideside, methylotio-β-galactopyranoside, ONPG, PNPG	Rahim and Lee (1991), Torres and Lee (1995)
Pseudoalteromonas sp. TAE 79b	513 (110[a])	4	26	9.0	1 h at 40 °C	Na+, K+, Mg^{2+}, Mn^{2+}	Ca^{2+}, Zn^{2+} Hg^{2+}, Cu^{2+}, H_2O_2, D-galactose	ONPG, lactose	Fernandes et al. (2002)
Rahnella aquatilis 14-1	NR(60[a])	NR	35	6.5–7.0	1.5 h at 45 °C	Mn^{2+}	Cu^{2+} , Zn^{2+} , Fe^{3+}, Al^{3+}, Ca^{2+}, Mg^{2+}, Na^+	ONPG, lactose	Liu et al. (2008)

ONPG *o*-nitrophenyl- β-D-galactopyranoside, *PNPG* *p*-nitrophenyl- β-D-galactopyranoside, *DTT* dithiothreitol, *2-ME* 2-mercaptoethanol, *SDS* sodium dodecyl sulfate, *NR* not reported

[a]Molecular mass of the monomer, estimated by SDS-PAGE

[b]Relative molecular mass determinated by native PAGE

Table 15.4 Cold active β-D-galactosidases – results of research on the hydrolysis of lactose in milk

Enzyme origin	Specific activity [U/mg]	Enzymatic activity (0–20 °C)	Amount of enzyme	Hydrolysis of lactose in milk reaction conditions			References
				T [°C]	Time [h]	Lactose hydrolysis rate [%]	
Arthrobacter psychrolactophilus F2	33.3	80–100 % (0–10 °C)	1 U/ml of milk	10	24	80	Nakagawa et al. (2006a)
Arthrobacter sp. ON14	25.4	30–100 % (0–15 °C)	200 μg/ml of milk (5.08U/ml of milk)	4	8	100	Xu et al. (2011)
Guehomyces pullulans R1	1.25	20 % (0 °C)	10 U/ml of milk	10	96	80	Nakagawa et al. (2006b)
Halomonas sp. S62	118.45	50–70 %	0.15 U/10 μl of milk	7	24	60	Wang et al. (2013)
Paracoccus sp. 32d	40.98	ca. 0–30 %	1 U/ml of milk	10	11 24	91 97	Wierzbicka-Woś et al. (2011)
Planococcus sp. L4	NR	27–100 %	2.5 μg of enzyme	5	1	36	Hu et al. (2007)
Pseudoalteromonas haloplanktis TAE79	138.2	NR	5 μg/500 μl of milk (0.691 U/500 μl of milk)	25 4	0.5 50 min	26 33	Hoyoux et al. (2001)
Pseudoalteromonas sp. TAE 79b	256.0	28 % (5 °C)	1 U/ml of milk	10	4 8	30 40	Fernandes et al. (2002)
Rahnella aquatilis 14-1	321.7	40 % (15 °C)	0.5 ml/ml of milk (6.37 U/ml of milk)	30	2	80	Liu et al. (2008)
metagenome-derived β-D-galactosidase gene *zd410*	NR	11–54 %	20 U/g of lactose in milk (1 U/1 ml of milk)	4 20 25	1 1 1	4.2 22.8 36.2	Wang et al. (2010)
Arthrobacter sp. SB	25.4	50–100 %	98 μg/ml of milk	2.5	7.5	80	Coker and Brenchley (2006)
Arthrobacter sp. 32cB	213.3	31–65 % (5–20 °C)	1 U/ml of milk 2 U/ml of milk	10	24	70 90	Pawlak-Szukalska et al. (2014)

belonging to the GH2 and GH42 families. Notably, there is lack of literature data concerning the hydrolysis of lactose using cold-active β-D-galactosidases belonging to the GH35 family.

Furthermore, studies pertaining to the effects of the Ca^{2+} ions in milk, as well as the D-galactose and D-glucose (products of the hydrolysis of lactose in milk) on the hydrolytic activity of these enzymes have only been conducted using ONPG as a substrate. It is noteworthy that studies of this type could also be conducted using lactose as a substrate. For example, we found that Ca^{2+}, D-galactose and D-glucose are inhibitors of the cold-active β-D-galactosidase from *Paracoccus* sp. 32d, but that the same enzyme was also highly effective for the hydrolysis of lactose in milk (Wierzbicka-Woś et al. 2011). The results of research concerning the inhibition of ONPG by Ca^{2+}, D-glucose or D-galactose do not therefore provide a sufficient basis for assessing the appropriateness of an enzyme for the hydrolysis of lactose.

Moreover, it is noteworthy that the K_M and k_{cat} values are crucial factors for estimating the amount of enzyme that is needed to achieve the effective hydrolysis of a substrate (e.g., the lactose in milk in industrial processes). This is an important parameter that is generally taken into account when assessing the economic viability of any industrial process involving the use of an enzyme as a biocatalyst. Analysis of K_M values presented in Table 15.5 revealed that ONPG was the preferred substrate of all of the enzymes listed in the table. Moreover, the K_M values for lactose as a substrate were high for most of the enzymes reported in the table. These data therefore show that these enzymes have a low affinity to lactose. Furthermore, the k_{cat}/K_M parameter, which is a measure of how efficiently an enzyme converts a substrate to a particular product, has only been determined for ten of the enzymes shown in Table 15.5.

Based on the data shown in the table, we concluded that the current state of the art for research pertaining to cold-active β-D-galactosidases was insufficient for making reliable comparisons about the suitability of known enzymes for the hydrolysis of lactose in milk. However, it has been reported that the enzymatic efficiencies of some of the cold-active enzymes shown in Table 15.4 towards the hydrolysis of lactose in milk at low temperatures are comparable with those of the β-D-galactosidase from *K. lactis* towards the hydrolysis of lactose in milk at moderate temperatures. To the best of our knowledge, there have been only two patents published in the literature pertaining to the use of cold-active β-D-galactosidases from *Alkalilaclibacillus ikkense* (Stougaard and Shmidt, Patent No.: US 8,288,143 B2) and *Pseudoalteromonas haloplanktis* (Hoyoux et al. Patent No.: US 6,727,084 B1) for the hydrolysis of lactose in dairy products and cow's milk in processes conducted at low temperatures.

The use of cold-active β-D-galactosidases was also evaluated as immobilized enzymes. The *Pseudoalteromonas* sp. 22b LacZ enzyme was immobilized on glutaraldehyde-treated chitosan beads. The apparent optimum temperature for activity of the immobilized enzyme was higher, optimum pH range was wider and its thermostability was slightly enhanced. Moreover, the immobilized β-D-galactosidase was inhibited neither by D-glucose nor by D-galactose. The activity of immobilized enzyme was maintained for at least 40 days of continuous lactose

Table 15.5 Kinetic parameters of a cold-active β-D-galactosidases

Enzyme origin	T [°C]	K_m [mM]		V_{max} [U/mg]		k_{cat} [s^{-1}]		k_{cat}/K_m [$s^{-1}mM^{-1}$]		References
		ONPG	lactose	ONPG	lactose	ONPG	lactose	ONPG	lactose	
Arthrobacter sp. B7 gene 15	30	0.4	16	1.182	117	NR	NR	NR	NR	Trimbur et al. (1994)
Arthrobacter sp. B7 gene 12	30	0.57	4.81	254	3.97	NR	NR	NR	NR	Gutshall et al. (1995)
Arthrobacter sp. C2-2 izosyme 1	10	NR	344.2	NR	NR	NR	324	NR	0.9	Karasová-Lipovová et al. (2003)
Arthrobacter sp. SB	15	0.8	11.5	NR	NR	100	5.3	125	0.5	Coker and Brenchley (2006)
Arthrobacter sp. 32c	10	5.75	77.54	NR	NR	52.4	1.76	9.12	0.023	Hildebrandt et al. (2009)
Arthrobacter sp. 32cB	10	1.52	16.56	16.74	17.44	30.55	31.84	20.16	1.92	Pawlak-Szukalska et al. (2014)
Arthrobacter psychrolactophilus F2	10	2.8	50	NR	NR	93.5	18	33.39	0.36	Nakagawa et al. (2006a)
Carnobacterium piscicola BA gene *bgaB*	30	1.7	NR	450	NR	588	NR	NR	NR	Coombs and Brenchley (1999)
Flavobacterium sp. 4214	25	0.65	NR	62	NR	NR	NR	NR	NR	Sørensen et al. (2006)
Guehomyces pullulans 17-1	50	3.3	NR	9.2	NR	NR	NR	NR	NR	Song et al. (2010)
Guehomyces pullulans R1	10	NR	50.5	NR	NR	NR	NR	NR	NR	Nakagawa et al. (2006b)
Halomonas sp. S62	35	2.9	NR	NR	NR	390.3	NR	NR	NR	Wang et al. (2013)
	45	NR	32.06	NR	NR	NR	269.5	NR	NR	
Paracoccus sp. 32d	10	1.17	2.94	NR	NR	71.81	43.23	61.38	15.06	Wierzbicka-Woś et al. (2011)
Planococcus sp. L4	10	3.8	11.2	NR	NR	195	62	51.3	5.5	Hu et al. (2007)
Planoccocus sp. SOS Orange	10	4.5	NR	80	NR	104	NR	23	NR	Sheridan and Brenchley (2000)
Pseudoalteromonas haloplanktis TAE 79	25	NR	2.4	NR	NR	NR	33	NR	13.7	Hoyoux et al. (2001)
Pseudoalteromonas sp. 22b	20	0.28	3.3	NR	NR	312	157	1114	47.5	Turkiewicz et al. (2003)
Pseudoalteromonas sp. TAE 79b	25	0.16	NR	11.7	NR	NR	NR	NR	NR	Fernandes et al. (2002)
metagenome-derived β-galactosidase gene *zd410*	10	4.7	27.1	77	56	66.7	33	14.2	1.21	Wang et al. (2010)

NR not reported

hydrolysis in milk, at 15 °C in a column reactor. The degree of lactose hydrolysis reached 93 % at an enzyme/substrate ratio of 30 U per 1 g of lactose within 18 h at 15 °C. In addition, the immobilized *Pseudoalteromonas* sp. 22b β-D-galactosidase was stable for at least 12 months at 4 °C (Makowski et al. 2007). The β-D-galactosidase from *Pseudoalteromonas* sp. TAE 79b was immobilized on Sepharose by adsorption or covalent coupling. The immobilized preparations were more stable than the soluble enzyme, but their ability to hydrolyze lactose was similar to that of the soluble β-D-galactosidase (Fernandes et al. 2002). In summary, immobilized preparations of cold-active β-D-galactosidases could be important in view of industrial applications.

5.2 *Cold-Active β-D-Galactosidases: Synthesis of β-Galactopyranosides*

Very few studies have been reported to date concerning the synthesis of GOSs and other β-galactopyranosides using cold-active β-D-galactosidases. In 1991, Rahim and Lee published the results of their research towards the synthesis of oligosaccharides using a β-D-galactosidase from psychrotolerant *Bacillus subtilis* KL88. In this particular study, the authors showed that the cold-active enzyme could effectively produce oligosaccharides using lactose as substrate (Rahim and Lee 1991). The highest concentration of GOSs was reached using a reaction mixture containing 20 % lactose, which was incubated at 10 °C for 4 h. Under these conditions, the reaction involving the hydrolysis of lactose was markedly decreased, with the reaction involving the formation of GOSs being the more favorable reaction. In 2003, Karasová-Lipovová et al. characterized another cold-active β-D-galactosidase from the psychrotolerant Antarctic bacterium *Arthrobacter* sp. C2-2, which could not only catalyze the hydrolysis of lactose but could also catalyze transglycosylation reactions. In this article, it was reported that the concentration of trisaccharides in the reaction mixture reached a plateau (34 mM) after a reaction time of approximately 10 h at 15 °C and 0.68 M lactose (beginning lactose concentration at reaction mixture). Moreover, the concentration of tetrasaccharides was much lower (6 mM) and it was achieved after a reaction time of 50 h. According to the authors of this report, these results showed that the β-D-galactosidase from *Arthrobacter* sp. C2-2 could be used (probably also in an immobilized form) for the synthesis of GOSs, either directly in milk at low temperatures or for the production of galactooligosaccharides from whey or whey permeate as an functional additive in dairy products. For comparison, the commercially available enzyme from *K. lactis* produces 40 mM trisaccharides at 40 °C. In a separate report, Nakagawa et al. (2006a) demonstrated that the efficiency of the cold-active rBglAp β-D-galactosidase from *Arthrobacter psychrolactophilus* F2 for the synthesis of GOS trisaccharides at 10 °C was similar to that reported for the corresponding enzyme from *Arthrobacter* sp. C2-2. However, there was a lack of information in this article concerning the production of tetrasaccharides under these reaction conditions. For comparison,

Pawlak-Szukalska et al. (2014) reported the efficient production of tri- and tetrasaccharides in lactose solutions (292–584 mM) containing the cold-active β-D-galactosidase from *Arthrobacter* sp. 32cB after an incubation time of 24 h at 10 and 20 °C or 6 h at 30 °C. The reactions incubated at 20 and 30 °C were also found to contain small quantities of pentasaccharides. Moreover, in contrast to the enzymes mentioned above, the β-D-galactosidase from *Arthrobacter* sp. 32cB was also used to catalyze the synthesis of heterooligosaccharides such as lactulose (i.e., galactosyl-fructose), galactosyl-xylose and galactosyl-arabinose from lactose and the appropriate monosaccharide. The highest quantities of lactulose were obtained after an incubation period of 8 h at 30 °C. Similar results were also obtained during galactosylation of D-xylose and L-arabinose. It was also found that the enzyme from *Arthrobacter* sp. 32cB effectively catalyzed the galactosylation of 1-butanol whereas the yields of other glycosylated alcohols, namely 2-propanol, 1-hexanol and cyclohexanol were lower. Furthermore, the aromatic glycoside salicin (2-(hydroxymethyl)phenyl-β-D-glucopyranoside) was also galactosylated using lactose as a galactosyl donor and the β-D-galactosidase from *Arthrobacter* sp. 32cB as a biocatalyst. The ability to synthesize of alkyl galactosides has also been demonstrated for the cold-active β-D-galactosidase from Antarctic *Pseudoalteromonas* sp. 22b (Makowski et al. 2009). This enzyme efficiently catalyzed the synthesis of galactosylated derivatives of C3–C6 alcohols. It is noteworthy that the cold-active β-D-galactosidase from *Pseudoalteromonas* sp. 22b performed more efficiently in terms of its ability to synthesize alkyl galactosides (i.e., the yield almost doubled in some cases) when the reaction mixtures contained organic solvents (below 50 %, v/v). This result is therefore similar to that observed for the mesophilic counterparts of these enzymes.

In summary, based on the data presented above, it could be concluded that cold-active β-D-galactosidases have the potential to be used as interesting biocatalysts for the production of functional food ingredients, cosmetics and pharmaceuticals, and could be especially useful for the glycosylation of chemicals which are thermolabile. It is noteworthy that a national application was recently filed at the Polish Patent Office concerning the use of the cold-active β-D-galactosidase from *Arthrobacter* sp. 32cB for the synthesis of GOSs (Pawlak et al. Patent application No.: PL 407,475).

5.3 *Cold-Active β-D-Galactosidases: Production of D-Tagatose*

D-Tagatose is a rare natural ketohexose and an isomer of the aldohexose D-galactose. This monosaccharide is similar in sweetness to sucrose, but its calorific value is much lower. In 2001, D-tagatose was designated as a Generally Recognized as Safe (GRAS) product by the U.S. Food and Drug Administration, which means that it can be used in beverages, dietetic candies and other dietary products as a low-calorie and low-glycaemic sweetener. D-Tagatose has also been reported to exhibit numerous health benefits such as prebiotic, anti-cariogenic and antioxidant properties

(Levin 2002; Oh 2007). Furthermore, this sugar is tested as a new drug for the treatment of type 2 diabetes and obesity. The results of a recent phase 3 clinical trial demonstrated that D-tagatose could effectively lower glycosylated hemoglobin A1c levels in patients with type 2 diabetes. D-Tagatose treatment also reduces fasting blood glucose levels and total LDL-cholesterol concentrations (Lu et al. 2008; Ensor et al. 2015).

An economically feasible method of D-tagatose manufacturing can be realized through isomerization of D-galactose with calcium hydroxide or using an L-arabinose isomerase. D-Galactose can be obtained from the hydrolysis of pure lactose or lactose contained in whey or whey permeate using β-D-galactosidase. The resulting mixture of D-glucose and D-galactose could then be separated and the pure D-galactose subsequently isomerized to give D-tagatose (Beadle et al. Patent No.: US 5,002,612,26; Oh 2007).

In this context, the cold-active β-D-galactosidase from the marine bacterium *Pseudoalteromonas haloplanktis* was studied in terms of its ability to affect the hydrolysis of lactose in whey permeate. A lactose conversion above 96 % was obtained in 24 h at 23 °C and pH 7.0 in whey permeate diluted three times (corresponding to dry matter content of 10 % w/w and 122 g L^{-1} lactose), using 0.1 mL of crude cell extract per 1 mL of whey permeate (corresponding to 0.05 mg mL^{-1} of the enzyme). The β-D-galactosidase from *P. haloplanktis* was inhibited by both D-glucose and D-galactose, although the effect was stronger in the presence of D-glucose (Van de Voorde et al. 2014). D-Galactose can also be obtained using a recombinant strain of *Pichia pastoris* secreting β-D-galactosidase from the psychrotolerant bacterium *Arthrobacter chlorophenolicus*. Growing the recombinant yeast strain in the whey permeate results in the hydrolysis of lactose and the utilization of the resulting D-glucose, while the D-galactose product remains in the medium. Hydrolysis efficiencies of around 90 % were achieved after 168 h at 30 °C in whey permeate containing up to 120 g L^{-1} lactose. In this case, the inhibitory effect of D-glucose was omitted (Wanarska and Kur 2012, Wanarska et al. Patent No.: PL 217,153).

D-Tagatose can also be produced directly from whey permeate using the *P. pastoris* strain secreting β-D-galactosidase from *A. chlorophenolicus* and a recombinant L-arabinose isomerase from the psychrotolerant bacterium *Arthrobacter* sp. 22c. Cultivation of the recombinant yeast strain in a whey permeate with a lactose content of 110 g L^{-1} supplemented with L-arabinose isomerase (0.2 mg mL^{-1}) resulted in a 90 % efficiency of lactose conversion, a 30 % isomerization of D-galactose to D-tagatose and the complete utilization of D-glucose within 144 h (Wanarska and Kur 2012, Wanarska et al. Patent No.: PL 216,683).

6 Concluding Remarks and Future Trends

Cold-active enzymes are already used in many industrial and biotechnological processes, and many other processes could also benefit from their effects, because reduced temperatures can be beneficial in many different ways. Processes of this type could save energy and production costs, improve hygiene, maintain taste and several other organoleptic properties, and reduce the risk of contamination. Cold-active enzymes are already used in the synthesis of fine chemicals, environmental biotechnology, energy and biofuels production, and in the pharmaceutical, medical, detergent, textile and food industries (Marx et al. 2007; Cavicchioli et al. 2011). Specific examples include the use of cold-active β-D-galactosidases. In recent times there has been a growing trend towards the isolation and characterization of novel β-D-galactosidase enzymes.

It is noteworthy that there are two enormous reservoirs available for the bioprospecting of cold-adapted activities of interest. The large currently uncultivable microorganisms could potentially be used by improved cultivation conditions or by using metagenomic approaches, and each of these technologies has its own unique advantages and disadvantages (Vester et al. 2014, 2015). Culture-based methods result in organisms that are recognized to produce potentially novel, active enzymes, but the rediscovery rate can be very high. Sequence-based metagenomics can be used to identify large numbers of genes encoding putative enzyme activities, however there is no guarantee that the genes can be expressed as active enzymes in available heterologous hosts. Function-based metagenomics may result in novel, functionally active enzymes, but the hit rate can be extremely low.

At present, recombinant DNA technology may be used to express and optimize the production of different β-D-galactosidases from the most diverse sources in microorganisms that are known for their highly efficient heterologous protein production (Oliveira et al. 2011). To date, almost all cold-active β-D-galactosidases presented in this review are recombinant enzymes, which were effectively produced in different *E. coli* expression systems. This prospect greatly expands the scope of potential applications for β-D-galactosidase and their economically effective exploitation in industrial processes. Contemporary molecular biology techniques combined with bioprocess engineering procedures can be used to optimize protein production, resulting in technically and economically effective enzyme production systems.

Two ways can be observed: one related to the basic hydrolytic activity of β-D-galactosidases, where novel cold-active and thermophilic enzymes have been described, and the other related to the characterization of innovative proteins with activities besides their hydrolytic activities. Even though the hydrolytic activity of β-D-galactosidase has classically been used in the milk industry to obtain lactose-free products, there is still plenty of space for new developments. It is envisaged that use of cold active β-D-galactosidases will be progressively incorporated into the dairy industry. Based on the key role played by GOSs in the field of functional foods, β-D-galactosidases with additional activities (e.g., transglycosylation activity) will undoubtedly attract considerable attention from researchers. Taking into account that the GOS products presently available on the market were not designed in rational way, there is still a need to discover good microbial sources of β-D-galactosidases with high transgalactosylation activities, and to characterize the individual components of the GOS mixtures obtained with these biocatalysts. In actual fact, this is currently a very active research field that will most probably continue to grow. Overall, there is undoubtedly a trend that β-D-galactosidase enzymes will work up into promising synthetic tools.

The cold-active β-D-galactosidases, despite of their numerous advantages which make them attractive biocatalysts for industrial processes carried out at low temperatures, are not used presently in the processes mentioned above. It is not possible to specify the one major reason for this state of affairs. The literature study revealed that the most of cold active β-D-galactosidases are effectively produced in active forms in different *E. coli* expression systems. However, to the best of our knowledge, any data about the production of cold-active β-D-galactosidases in the scale larger than lab scale were publicized so far. Because of this, it is difficult to predict the real cost of cold-active β-D-galactosidases production on a scale sufficient to cover today's needs of the industry on this group of hydrolytic enzymes.

Acknowledgments This work was supported by grant from the National Centre for Research and Development of Poland (PBS1/A9/7/2012).

References

Beadle JR, Saunders JP, Wajda TJ Jr. Process for manufacturing tagatose. US Patent 5,002,612, 26

Białkowska AM, Cieśliński H, Nowakowska KM, Kur J, Turkiewicz M (2009) A new β-galactosidase with a low temperature optimum isolated from the Antarctic *Arthrobacter* sp. 20B: gene cloning, purification and characterization. Arch Microbiol 191:825–835

Boon MA, Janssen AEM, van't Riet K (2000) Effect of temperature and enzyme origin on the enzymatic synthesis of oligosaccharides. Enzyme Microb Techol 26:271–281

Cavicchioli R, Charlton T, Ertan H, Mohd OS, Siddiqui KS, Williams TJ (2011) Biotechnological uses of enzymes from psychrophiles. Microb Biotechnol 4:449

Cieśliński H, Kur J, Białkowska A, Baran I, Makowski K, Turkiewicz M (2005) Cloning, expression, and purification of a recombinant cold-adapted β-galactosidase from antarctic bacterium *Pseudoalteromonas* sp. 22B. Protein Expr Purif 39:27–34

Coker JA, Brenchley JE (2006) Protein engineering of a cold-active beta-galactosidase from *Arthrobacter* sp. SB to increase lactose hydrolysis reveals new sites affecting low temperature activity. Extremophiles 10:515–524

Coker JA, Sheridan PP, Loveland-Curtze J, Gutshall KR, Auman AJ, Brenchley JE (2003) Biochemical characterization of a β-galactosidase with a low temperature optimum obtained from an Antarctic *Arthrobacter* isolate. J Bacteriol 185:5473–5482

Coombs JM, Brenchley JE (1999) Biochemical and phylogenetic analyses of cold-active β-galactosidase from the lactic acid bacterium *Carnobacterium piscicola* BA. Appl Environ Microbiol 65:5443–5450

Coombs JM, Brenchley JE (2001) Characterization of two new glycosyl hydrolases from the lactic acid bacterium *Carnobacterium piscicola* strain BA. Appl Environ Microbol 67:5094–5099

Coté A, Brown WA, Cameron D, van Walsum GP (2004) Hydrolysis of lactose in whey permeate for subsequent fermentation to ethanol. J Dairy Sci 87:1608–1620

Cruz R, Cruz VD, Belote JG, Khenayfes MD, Dorta C, Oliveira LHD (1999) Properties of a new fungal beta-galactosidase with potential application in the dairy industry. Rev Microbiol 30:265–271

Di Lauro B, Strazzulli A, Perugino G, La Cara F, Bedini E, Corsaro MM, Rossi M, Moracci M (2008) Isolation and characterization of a new family 42 β-galactosidase from the thermoacidophilic bacterium *Alicyclobacillus acidocaldarius*: identification of the active site residues. Biochim Biophys Acta 1784:292–301

Ensor M, Banfield AB, Smith RR, Wiliams J, Lodder RA (2015) Safety and efficacy of D-tagatose in glycemic control in subjects with type 2 diabetes. J Endocrinol Diab Obes 3:1065
Fernandes S, Geueke B, Delgado O, Coleman J, Hatti-Kaul R (2002) β-Galactosidase from a cold-adapted bacterium: purification, characterization and application for lactose hydrolysis. Appl Microbiol Biotechnol 58:313–321
Guimaraes WV, Dudey GL, Ingram LO (1992) Fermentation of sweet whey by ethanologenic *Escherichia coli*. Biotechnol Bioeng 40:41–45
Gutshall K, Trimbur D, Kasmir J (1995) Analysis of a novel gene and beta-galactosidase isozyme from a psychrotrophic *Arthrobacter* isolate. J Bacteriol 177:1981–1988
Gutshall K, Wang K, Brenchley JE (1997) A novel *Arthrobacter* β-galactosidase with homology to eucaryotic β-galactosidases. J Bacteriol 179:3064–3067
Harju M (1987) Lactose hydrolysis. Bull Int Dairy Fed 212:50–54
Harju M, Kallioinen H, Tossavainen O (2012) Lactose hydrolysis and other conversions in dairy products: technological aspects. Int Dairy J 22:104–109
Hildebrandt P, Wanarska M, Kur J (2009) A new cold-adapted beta-D-galactosidase from the Antarctic *Arthrobacter* sp. 32c – gene cloning, overexpression, purification and properties. BMC Microbiol 9:151
Hoyoux A, Francois JM, Dubois P, Baise E, Jennes I, Genicot S, Gerday C. Cold-active beta-galactosidase, the process for its preparation and the use thereof. US Patent 6,727,084 B1
Hoyoux A, Jennes I, Dubois P, Genicot S, Dubail F, Franc JM, Baise E, Feller G, Gerday C (2001) Cold-adapted β-galactosidase from the Antarctic psychrophile *Pseudoalteromonas haloplanktis*. Appl Environ Microbiol 67:1529–1535
Hu JM, Li H, Cao LX, Wu PC, Zhang CT, Sang SL, Zhang XY, Chen MJ, Lu JQ, Liu YH (2007) Molecular cloning and characterization of the gene encoding cold-active beta-galactosidase from a psychrotrophic and halotolerant *Planococcus* sp. L4. J Agric Food Chem 55:2217–2224
Johansen AG, Vegarud GE, Skeie S (2002) Seasonal and regional variation in the composition of whey from Norwegian Cheddar-type and Dutch-type cheeses. Int Dairy J 12:621–629
Juers DH, Matthews BW, Huber RE (2012) LacZ beta-galactosidase: structure and function of an enzyme of historical and molecular biological importance. Protein Sci 21:1792–1807
Jurado E, Camacho F, Luzón G, Vicaria JM (2002) A new kinetic model proposed for enzymatic hydrolysis of lactose by a β-galactosidase from *Kluyveromyces fragilis*. Enzyme Microb Technol 31:300–309
Karasová-Lipovová P, Strnad H, Spiwok V, Malá S, Králová B, Russell NJ (2003) The cloning, purification and characterisation of a cold-active b-galactosidase from the psychrotolerant Antarctic bacterium *Arthrobacter* sp. C2-2. Enzyme Microb Technol 33:836–844
Kosugi A, Murashima K, Doi RH (2002) Characterization of two noncellulosomal subunits, ArfA and BgaA, from *Clostridium cellulovorans* that cooperate with the cellulosome in plant cell wall degradation. J Bacteriol 184:6859–6865
Kren V, Sedmera P, Havlicek V, Fiserova A (1992) Enzymatic galactosylation of ergotalkaloids. Tetrahedron Lett 47:7233–7236
Ladero M, Perez MT, Santos A, Garcia-Ochoa F (2003) Hydrolysis of lactose by free and immobilized beta-galactosidase from *Thermus* sp. strain T2. Biotechnol Bioeng 81:241–252
Lee JH, Kim YS, Yeom SJ, Oh DK (2011) Characterization of a glycoside hydrolase family 42 β-galactosidase from *Deinococcus geothermalis*. Biotechnol Lett 33:577–583
Levin GV (2002) Tagatose, a new GRAS sweetener and health product. J Med Food 5:23–36
Liu WY, Shi YW, Wang XQ, Wang Y, Wei CQ, Lou K (2008) Isolation and identification of a strain producing cold-adapted β-galactosidase, and purification and characterisation of the enzyme. Czech J Food Sci 26:284–290
Lu Y, Levin GV, Donner TW (2008) Tagatose, a new antidiabetic and obesity control drug. Diabetes Obes Metab 10:109–134
Makowski K, Białkowska M, Szczesna-Antczak M, Kalinowska H, Kur J, Cieslinski H, Turkiewicz M (2007) Immobilized preparation of cold-adaptedand halotolerant antarctic β-galactosidase as a highlystable catalyst in lactose hydrolysis. FEMS Microbiol Ecol 59:535–542

Makowski K, Białkowska A, Olczak J, Kur J, Turkiewicz M (2009) Antarctic, cold-adapted β-galactosidase of *Pseudoalteromonas* sp. 22b as an effective tool for alkyl galactopyranosides synthesis. Enzyme Microbial Technol 44:59–64

Marx JC, Collins T, D'Amico S, Feller G, Gerday C (2007) Cold-adapted enzymes from marine Antarctic microorganisms. Marine Biotechnol 9:293

Mehaia MA, Alverez J, Cheryan M (1993) Hydrolysis of whey permeate lactose in a continuous stirred tank membrane reactor. Int Dairy J 3:179–192

Mlichová Z, Rosenberg M (2006) Current trends of β-galactosidase application in food technology. J Food Nutr Res 45:47–54

Nakagawa T, Fujimoto Y, Ikehata R, Miyaji T, Tomizuka N (2006a) Purification and molecular characterization of cold-active beta-galactosidase from *Arthrobacter psychrolactophilus* strain F2. Appl Microbiol Biotechnol 72:720–725

Nakagawa T, Ikehata R, Uchino M, Miyaji T, Takano K, Tomizuka N (2006b) Cold-active acid β-galactosidase activity of isolated psychrophilic-basidiomycetous yeast *Guehomyces pullulans*. Microbiol Res 161:75–79

Nath A, Mondal S, Chakraborty S, Bhattacharjee C, Chowdhury R (2014) Production, purification, characterization, immobilization, and application of β-galactosidase: a review. Asia-Pac J Chem Eng 9:330–348

Oh DK (2007) Tagatose: properties, applications, and biotechnological processes. Appl Microbiol Biotechnol 76:1–8

Oliveira C, Guimarães PM, Domingues L (2011) Recombinant microbial systems for improved β-galactosidase production and biotechnological applications. Biotechnol Adv 29:600

Onishi N, Tanaka T (1996) Purification and properties of a galacto- and gluco-oligosaccharide-producing beta-glycosidase from *Rhodotorula minuta* IFO879. J Ferment Bioeng 82:439–443

Pawlak A, Wanarska M, Popinigis A, Kur J. β-D-galaktozydaza *Arthrobacter* sp. S3*, β-D-galaktozydaza *Arthrobacter* sp. 32cB, sekwencje nukleotydowe szczepu *Arthrobacter* sp. S3* i *Arthrobacter* sp. 32cB kodujące β-D-galaktozydazę oraz sposób wytwarzania mleka o obniżonej zawartości laktozy, galaktooligosacharydów, heterooligosacharydów i glikozylowanych związków chemicznych z wykorzystaniem tych enzymów. Patent application no: PL 407,475

Pawlak-Szukalska A, Wanarska M, Popinigis AT, Kur J (2014) A novel cold-active β-D-galactosidase with transglycosylation activity from the Antarctic *Arthrobacter* sp. 32cB – gene cloning, purification and characterization. Process Biochem 49:2122–2133

Poltorak OM, Chukhrai ES, Pilipenko OS, Atyaksheva LF, Beregalov AE (2007) A comparative study of the structure and properties of β-galactosidases. Russ J Phys Chem A 81:808–812

Rahim KA, Lee BH (1991) Specificity, inhibitory studies, and oligosaccharide formation by beta-galactosidase from psychrotrophic *Bacillus subtilis* KL88. J Dairy Sci 74:1773–1778

Saishin N, Ueta M, Wada A, Yamamoto I (2010) Properties of β-galactosidase purified from *Bifidobacterium longum* subsp. *longum* JCM 7052 grown on gum arabic. Int J Biol Macromol 10:23–31

Santos A, Ladero M, García-Ochoa F (1998) Kinetic modeling of lactose hydrolysis by a β-galactosidase from *Kluyveromyces fragilis*. Enzyme Microb Technol 22:558–567

Scheckermann C, Wagner F, Fischer L (1997) Galactosylation of antibiotics using the beta-galactosidase from *Aspergillus oryzae*. Enzyme Microb Technol 20:629–634

Schmidt M, Stougaard P (2010) Identification, cloning an expression of a cold-active β-galactosidase from a novel arctic bacterium, *Alkalilactibacillus ikkense*. Environ Technol 3:1107–1114

Sheridan PP, Brenchley JE (2000) Characterization of a salt-tolerant family 42 β-galactosidase from a psychrophilic antarctic *Planococcus* isolate. Appl Environ Microbiol 66:2438–2444

Shimizu R, Shimabayashi H, Moriwaki M (2006) Enzymatic production of highly soluble myricitrin glycosides using beta-galactosidase. Biosci Biotechnol Biochem 70:940–948

Song C, Liu G-L, Xu J-L, Chi Z-M (2010) Purification and characterization of extracellular β-galactosidase from the psychrotolerant yeast *Guehomyces pullulans* 17–1 isolated from sea sediment in Antarctica. Process Biochem 45:954–960

Sørensen HP, Porsgaard TK, Kahn RA, Stougaard P, Mortensen KK, Johnsen MG (2006) Secreted β-galactosidase from *Flavobacterium* sp. isolated from low-temperature environment. Appl Microbiol Biotechnol 70:548–557

Stevenson DE, Stanley RA, Furneaux RH (1993) Optimization of alkyl β-D-galactopyranoside synthesis from lactose using commercially available β-galactosidases. Biotechnol Bioeng 42:657–666

Stougaard P, Schmidt M. Cold-active beta-galactosidase, a method of producing same and use of such enzyme. US 8,288,143 B2

Torres MJ, Lee BH (1995) Cloning and expression of β-galactosidase from psychotrophic *Bacillus subtilis* KL88 into *Escherichia coli*. Biotechnol Lett 17:123–128

Trimbur D, Gutshall K, Prema P, Brenchley JE (1994) Characterization of a psychrotrophic *Arthrobacter* gene and its cold-active-galactosidase. Appl Environ Microbiol 60:4544–4552

Turkiewicz M, Kur J, Białkowska A, Cieśliński H, Kalinowska H, Bielecki S (2003) Antarctic marine bacterium *Pseudoalteromonas* sp. 22b as a source of cold-adapted β-galactosidase. Biomol Eng 20:317–324

Van de Voorde I, Goiris K, Syryn E, Van den Bussche C, Aerts G (2014) Evaluation of the cold-active *Pseudoalteromonas haloplanktis* β-galactosidase enzyme for lactose hydrolysis in whey permeate as primary step of D-tagatose production. Process Biochem 49:2134–2140

Van Laere KMJ, Abee T, Schols HA, Beldman G, Voragen AGJ (2000) Characterization of a novel beta-galactosidase from *Bifidobacterium adolescentis* DSM 20083 active towards transgalactooligosaccharides. Appl Environ Microbiol 66:1379–1384

Vester JK, Glaring MA, Stougaard P (2014) Discovery of novel enzymes with industrial potential from a cold and alkaline environment by a combination of functional metagenomics and culturing. Microb Cell Fact 13:72

Vester JK, Glaring MA, Stougaard P (2015) Improved cultivation and metagenomics as new tools for bioprospecting in cold environments. Extremophiles 19:17–29

Vetere A, Paoletti S (1998) Separation and characterization of three beta-galactosidases from *Bacillus circulans*. BBA Gen Subj 1380:223–231

Wanarska M, Hildebrandt PŁ, Kur JW. Sposób otrzymywania D-galaktozy z laktozy. Patent no: PL217,153

Wanarska M, Hildebrandt PŁ, Kur JW. Sposób wytwarzania D-tagatozy z laktozy z wykorzystaniem rekombinantowego szczepu drożdży *Pichia pastoris* wytwarzającego β-D-galaktozydazę *Arthrobacter chlorophenolicus* i izomerazy arabinozowej *Arthrobacter* sp.22c. Patent no: PL216,683

Wanarska M, Kur J (2005) β-D-Galactosidases – sources, properties and applications. BioTechnologia 4:46–62

Wanarska M, Kur J (2012) A method for the production of D-tagatose using a recombinant *Pichia pastoris* strain secreting β-D-galactosidase from *Arthrobacter chlorophenolicus* and a recombinant L-arabinose isomerase from *Arthrobacter* sp. 22c. Microb Cell Fact 11:113

Wang K, Li G, Qin Y, Znang CT, Liu YH (2010) A novel metagenome-derived β-galactosidase: gene cloning, overexpression, purification and characterization. Appl Microbiol Biotechnol 88:155–165

Wang GX, Gao Y, Hu B, Lu XL, Liu XY, Jiao BH (2013) A novel cold-adapted β-galactosidase isolated from *Halomonas* sp. S62: gene cloning, purification and enzymatic characterization. World J Microbiol Biotechnol 29:1473–1480

White PL, Wynn-Williams DD, Russell NJ (2000) Diversity of thermal responses of lipid composition in the membranes of the dominant culturable members of an Antarctic fellfield soil bacterial community. Antarct Sci 72:386–393

Wierzbicka-Woś A, Cieśliński H, Wanarska M, Kozłowska-Tylingo K, Hildebrandt P, Kur J (2011) A novel cold-active β-D-galactosidase from the *Paracoccus* sp. 32d – gene cloning, purification and characterization. Microb Cell Fact 10:108

Xu K, Tang X, Gai Y, Mehmood M, Xiao X, Wang F (2011) Molecular characterization of cold-inducible beta-galactosidase from *Arthrobacter* sp. ON14 isolated from Antarctica. J Microbiol Biotechnol 21:236–242

Chapter 16
Engineering of Extremophilic Phosphotriesterase-Like Lactonases for Biotechnological Applications

Elena Porzio, Immacolata Del Giudice, and Giuseppe Manco

1 Introduction

Organophosphate compounds (OPs) constitute the core of pesticides (i.e. paraoxon, parathion, coumaphos, diazinon, dimethoate and chlorpyrifos) and many nerve-based warfare agents (i.e. sarin, tabun, soman and VX) (Table 16.1).

Pesticides are commonly used in agriculture to obtain high productivity, thanks to the control of plants or animals life (pests). However, although they certainly improve the quality and quantity of the agricultural production, their extensive application has caused pollution of aquatic systems, currently posing risks also to human health.

OPs were introduced only in the last century but their use has increased rapidly and now they represent about 38 % of the total pesticides applied worldwide (Singh 2009). In 2008, China produced about 1.74 million tons of 300 different types of pesticides, which has made this country the largest producer and user of such compounds worldwide (Jin et al. 2010).

The latest estimate of the Environmental Protection Agency (EPA) on the annual application of pesticides is more than five million pounds worldwide and one million pounds in the USA (Gavrilescu 2005; Grube et al. 2011). Currently, it is estimated that 200,000 t of OPs are stocked in the world (Singh 2009) and that self-poisoning of workers is very frequent in agriculture settings (Aardema et al. 2008; Calvert et al. 2008; Eddleston et al. 2005).

The fate of the pesticides in the environment depends on several aspects, including soil physico-chemical properties, topography, weather, agricultural management practices and, ultimately, chemical properties of each pesticide (water solubility, tendency to adsorb to the soil, and pesticide persistence) (Tiryaki and Temur 2010).

E. Porzio (✉) • I. Del Giudice • G. Manco
Institute of Protein Biochemistry, CNR, Via P. Castellino 111, 80131 Naples, Italy
e-mail: e.porzio@ibp.cnr.it; g.manco@ibp.cnr.it

P.H. Rampelotto (ed.), *Biotechnology of Extremophiles*, Grand Challenges in Biology and Biotechnology 1, DOI 10.1007/978-3-319-13521-2_16

Table 16.1 PLL substrates

	Name	Structure	Use
OP pesticides	Paraoxon	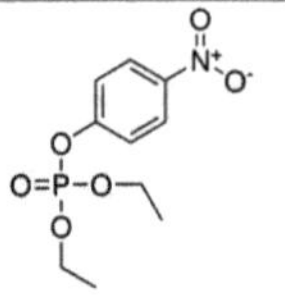	Insecticide
	Parathion	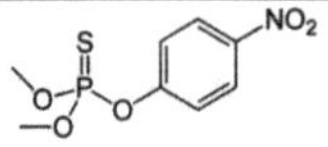	Insecticide
	Coumaphos	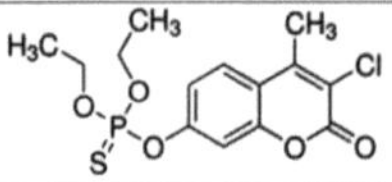	Acaricide
	Chlorpyrifos	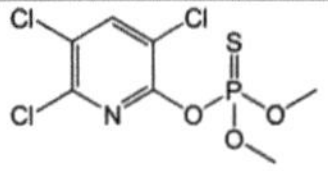	Insecticide
	Diazinon	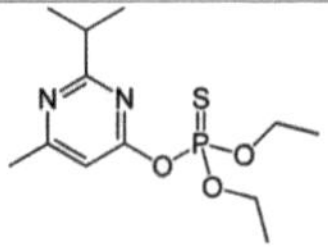	Insecticide
	Dimethoate	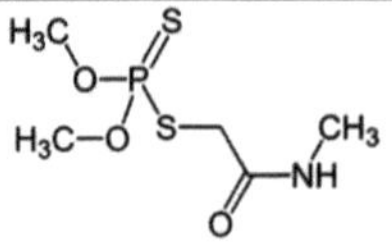	Insecticide
OP nerve gases	Tabun	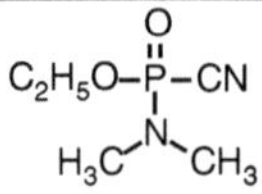	CWA[a]
	Sarin	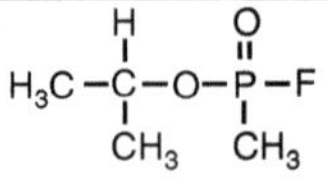	CWA[a]
	Soman	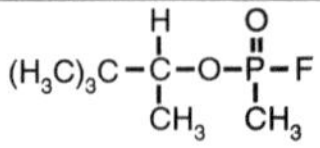	CWA[a]
	VX	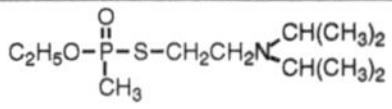	CWA[a]
Lactones	AHL[b]	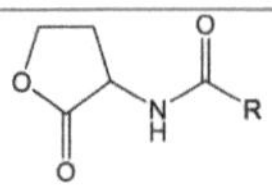	–
	3-oxo-AHL[b]	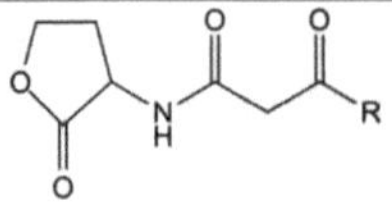	–

Commonly employed organophosphorus compounds and general structure of acyl-homoserine lactone (AHL)

[a]Chemical warfare agents

[b]R corresponds to different size of acyl chain

Usually, the OPs have low persistence in soil because sunlight exposure, water and microbial hydrolysis cause their rapid degradation (Caceres et al. 2010; Ragnarsdottir 2000). However, their over-use, together with the storage of more than half a million tons of obsolete, prohibited or outdated pesticides (Ortiz-Hernàndez et al. 2013), as well as the limitation or absence in many developing countries of exposure-control programs, led to their anomalous accumulation in soil and water runoff and their magnification through the food chain. The high toxicity of these compounds led the Food and Agriculture Organization of the United Nations (FAO) and World Health Organization (WHO) to establish international food standards and guidelines codes for food safety (Codex Alimentarius) (FAO 2014).

In addition to their use as pesticides, in the early twentieth century extremely poisonous OPs were synthesized and used as chemical warfare agents in World War II (Raushel 2002) and, recently, in terroristic attacks (Tokio subway 1995) and poisoning of civilians (Syria 2013).

OPs are esters of phosphorus with various combinations of oxygen, carbon, sulphur and nitrogen attached, resulting in six different subclasses: phosphates, phosphonates, phosphorothioates, phosphorodithioates, phosphorothiolates and phosphoramidates (Can 2014). They have a general structure in which the phosphorus is linked by a double bond to either an oxygen atom (oxon-OPs), or to a sulphur atom (thion-OPs), and by ester linkages to alkoxy or amino groups, and to another group (halogen, aliphatic, aromatic, or heterocyclic groups).

Based on how the subgroups are oriented around the central phosphorus atom, they have either an R_P or S_P chiral configuration, where the S_P one usually represents the more toxic isomer.

The most studied toxic effect of OPs is the irreversible inhibition of a key enzyme in cholinergic transmission, acetylcholinesterase (AChE), by the covalent phosphorylation of the serine residue within the active site. Such inhibition causes the accumulation of the neurotransmitter acetylcholine in the neuron-neuron or neuron-muscle junctions, determining lacrimation, hypersalivation, bronchial hypersecretion and bronchoconstriction, skeletal muscle fasciculation and twitching, ataxia, respiratory failure, convulsions, hypothermia and, ultimately, death (Carey et al. 2013).

Recent studies on the toxicological properties of OPs also reported gene mutations, chromosomal aberrations, DNA damage (Ojha and Gupta 2014), alteration of semen quality and sperm chromatin (Salazar-Arredondo et al. 2008), and insulin resistance (Lasram et al. 2014). Moreover, the involvement of OPs in cancerogenesis and endocrine disorders has been reported (Gupta 2005).

Since OPs pose hazards to human health and wildlife, it is necessary to realize strategies for the assessment of the respective risk, for their environmental monitoring, and for the treatment of waste and remediation of contaminated sites.

The traditional chromatography and spectroscopy analytical techniques, usually employed in environmental monitoring, show high sensitivity and selectivity, but require a lot of expertise and time, and are very expensive. Detection via biosensors represents a good complement or alternative to the classical methods by simplifying or eliminating some limiting steps, such as sample preparation and making field-testing easier and faster with significant decrease in cost per analysis (Marrazza 2014).

Many of the classical methods used in the OP decontamination field, including chemical treatment, electrochemical oxidation and reduction, volatilization, incineration or photodecomposition (Kiss and Viràg 2009), are expensive, not eco-friendly, and inefficient to remove low contaminant concentrations (LeJeune et al. 1998). The presence of a robust complement of metabolic pathways and enzymes necessary to use different xenobiotic compounds makes the microbial organisms possible candidate in several promising biotechnology platforms (biomonitoring and bioremediation, destruction of nerve agent stockpiles, use as bioscavengers in medical prophylaxis). However, different drawbacks, including strict dependence of growth on temperature, oxygen, and supply of nutrients, stability under harsh environmental conditions, inhibition of growth in the presence of different chemicals, have limited their application (Singh and Walker 2006).

Therefore enzymes, which hydrolyze phosphoester or phosphothiol bonds in organophosphates, are ideal candidate to be used in the above-discussed applications; indeed, they show stable half-lives, broad substrate ranges, high specificity, work very rapidly and are effective against racemic mixtures and individual stereoisomers (Tsai et al. 2010a, b). These enzymes, generally named phosphotriesterases (PTEs; EC 3.1.8.1), were isolated and characterized from different microorganisms (Islam et al. 2010; Singh 2009; Zhang et al. 2005).

To improve upon desirable traits (stability, activity, stereoselectivity and substrate specificity) of several well-known organophosphatases, many researchers have combined different mutational strategies (directed evolution and rational design) (Goldsmith et al. 2012; Iyer and Iken 2015; Tsai et al. 2012). In particular, rational design methods have provided numerous successful solutions but, over the last years, directed evolution, especially when combined with rational approaches (protein structure analysis and bioinformatics tools) has become the best method to design focused libraries (Cobb et al. 2013a, b; Denard et al. 2015). Directed evolution consists of iterative cycles of random mutagenesis and/or DNA recombination to generate diversity in a target gene sequence and allows the identification of the desired protein variants by high-throughput screening (Cobb et al. 2012). Over the years, directed evolution has enabled the engineering and improvement of known protein features, leading to the generation of optimized biocatalysts to employ in several industrial applications. Moreover, it has also helped the design and optimization of biocatalysts with activities not previously encountered in Nature (Denard et al. 2015).

This review will describe some of the recent achievements in the engineering of OP-hydrolyzing biocatalysts, with a special focus on the evolution of the promiscuous activities of Phosphotriesterase-like Lactonases, which represent the new appealing candidates in the OP remediation.

Promiscuous functions are of great importance because they represent a repertoire of enzymatic activities that can be recruited when the environment changes and a new activity becomes important for fitness. Moreover, they provide excellent opportunities for in vitro evolution of new functions to employ in biotechnology and drug development.

2 Organophosphorus Hydrolases

To date, many enzymes able to hydrolyze OP compounds have been identified, such as organophosphorus hydrolases (OPHs), methyl parathion hydrolases (MPHs), serum paraoxonases (PONs) and microbial prolidases/OP acid anhydrolases.

2.1 OPH Hydrolases

A variety of genes different in terms of form, location or host organism involved in OP degradation has been identified so far. The first and best characterized *opd* ('OP degradation') genes and their associated protein products (PTEs) were identified in *Sphingobium fuliginis* ATCC 27551 (previously, *Flavobacterium* sp.) (Kawahara et al. 2010; Singh and Walker 2006) and *Brevundimonas diminuta* GM (previously, *Pseudomonas diminuta* GM) (Harper et al. 1988; Segers et al. 1994); these enzymes showed a strong hydrolytic activity towards OP compounds, in particular, paraoxon (Dumas et al. 1989a).

The protein products of the first identified *opd* genes were initially designated as parathion hydrolases and then classified as OPHs.

OPH from *B. diminuta* (named *bd*PTE in this review) belongs to the amidohydrolase superfamily (subtype I) (Seibert and Raushel 2005) and derives from a gene included in a transposable element of a large plasmid. Structural data showed that *bd*PTE is a homodimer with $(\beta/\alpha)_8$-barrel structural fold including a catalytic binuclear centre at the C-terminal end (Benning et al. 1994, 1995, 2001; Vanhooke et al. 1996), containing two metal ions with zinc or cobalt as the favorite cofactors (Dumas et al. 1989a, b; Benning et al. 2001). The residues involved in the catalytic site are: four histidines (H55, H57, H201, and H230) that directly interact with the two divalent cations, an aspartate (D301) and a carboxylated lysine (K169) (Benning et al. 2000).

In *bd*PTE structure, as well as in the structure of other members of the amidohydrolase superfamily, two loops, one (loop 7) involved in the substrate binding and hydrolytic mechanism and the other (loop 8) responsible for the substrate specificity, have been identified (Seibert and Raushel 2005).

*bd*PTE hydrolyzes different organophosphate nerve agents, such as sarin, soman and VX that were used in recent years as chemical warfare agents (Caldwell and Raushel 1991). Moreover, to date the enzyme is the most efficient in the paraoxon hydrolysis ($k_{cat}/K_M = 4 \times 10^7\ M^{-1}s^{-1}$) (Caldwell et al. 1991) among the characterized OP degrading genes. Interestingly, different promiscuous activities (carboxylesterase and lactonase activities) have been detected for *bd*PTE (Roodveldt and Tawfik 2005), but the natural substrate is not yet identified; this suggests that its catalytic activity could have recently evolved from a preexisting enzyme, due to the widespread use of high levels of organophosphate insecticides throughout the world (Raushel and Holden 2000).

In a recent review of Jackson et al., the structural determinants of the high catalytic efficiency towards paraoxon of *bd*PTE were deeply investigated. Two dominant conformational sub-states ('closed' and 'open') with a low-energy structural transition were described for *bd*PTE. The 'closed' state seems to facilitate the substrate recognition and the lowering of the reaction activation energy but not the fast diffusion of the substrate/product; conversely, the 'open' state seems to facilitate mainly the enzyme/product complex dissociation rather than the rate of hydrolysis reaction (Jackson et al. 2008).

This study demonstrated the importance of the conformational changes in the optimization of enzyme catalysis, a concept that has also been revealed for GkaP in the work of Zhang et al. (2015).

A close homolog of *bd*PTE showing organophosphatase activity was isolated from *Agrobacterium radiobacter* P230 and designated as OpdA (Horne et al. 2002, 2003). OpdA was able to hydrolyze diethyl OPs (paraoxon, coumaphos, parathion, diazinon, coroxon) and, at a higher rate than *bd*PTE, dimethyl substrates (methyl-parathion, phosmet, fenthion, dMUP) (Horne et al. 2002).

The presence of *opd*-like genes has also been demonstrated in several bacterial genome sequencing projects, although these sequences are much more distant (30–40 % identity at the amino acid level) from those found in *Flavobacterium*, *B. diminuta* and *Agrobacterium radiobacter*.

In particular, an OPD-like protein, designated as PHP (phosphotriesterase homology protein), has been isolated from *Escherichia coli*, purified and characterized (Buchbinder et al. 1998). Biochemical analysis showed that PHP of *E. coli* (ePHP) is a monomer with two zinc ions; its structure is similar to that of *bd*PTE and the residues that coordinate the metal ions are conserved (excluding K169) (Buchbinder et al. 1998). The native function of this OPD homolog is not known. As for an ePHP mutant, only weak esterase and phosphotriesterase activities have been reported (Roodveldt and Tawfik 2005).

A protein with features similar to PHP is the product of gene *mll7664* from *Mesorhizobium loti*. Similarly to ePHP, this enzyme has a glutamate bridging the two metals (by similarity) and the main activity observed is the carboxylesterase activity, with promiscous phosphotriesterase and phosphodiesterase activity. Accordingly this enzyme has been dubbed PLC, phosphotriesterase-like carboxylesterase. By mutation of the glutamate to a lysine, the carboxylesterase activity was completely abolished whereas the phosphodiesterase activity became the main activity (Mandrich and Manco 2009).

In these years many efforts have focused on the improvement of the catalytic efficiency of the known OPH enzymes. For example, several directed evolution strategies have been employed to improve *bd*PTE hydrolytic activity against phosphorothioate substrates, such as methyl parathion and chlorpyrifos, which are hydrolyzed at rates 30 and 1000-fold slower respect to paraoxon (Cho et al. 2002, 2004; Dumas et al. 1989a). By a mutagenesis strategy based on DNA shuffling, a variant of *bd*PTE (22A11) showing an improvement of the methyl parathion hydrolytic activity (25-fold) was identified (Cho et al. 2002).

By using the evolved variant (22A11) as template, a new mutagenesis approach was combined with a chlorpyrifos screening and led to the identification of a variant (B3561) able to hydrolyze chlorpyrifos at a rate 725-fold higher than the wild-type (k_{cat}/K_M value of 2.2×10^8 $M^{-1}s^{-1}$) (Cho et al. 2004). Some substitutions (A80V, I274N, and K185R), found in both variants (B3561 and 22A11), cause conformational changes resulting in a widening of the active site and in a stabilization of the enzymatic metal-free state (Cho et al. 2004, 2006).

Many reports have focused on modifying the stereoselectivity of *bd*PTE, in order to make the enzyme more active on the toxic S_P isomers of OP compounds. Structural studies identified three binding pockets (small, large, and leaving group) of *bd*PTE as responsible for the interaction with the substituents of the phosphorus center of the substrates (Chen-Goodspeed et al. 2001a, b). A deep study on the determinants of the *bd*PTE stereoselectivity has demonstrated that the preference for chiral substrates is determined by the small sub-site size. Indeed, the substitution of a glycine residue with an alanine causes an increase of the S_P enantiomer preference by reducing the small subsite size (Chen-Goodspeed et al. 2001b). On the contrary, mutations (I106A, F132A and S308A) that cause a widening of the small sub-site, decrease the S_P preference (Chen-Goodspeed et al. 2001b).

Preferential mutations collected from several studies (K185R, H254Q, H254G, H257F, H257W, I274N and L303T) were chosen to design mutagenic libraries and to create highly effective *bd*PTE variants. From the screening of these libraries, GWT-F5 variant has emerged, which showed catalytic efficiencies for the S_P analogs of nerve agents (soman, tabun, sarin, cyclosarin, VX) ranging between 10^2 M^{-1} s^{-1} and 10^6 M^{-1} s^{-1} (Benning et al. 1995; Madej et al. 2012; Tsai et al. 2012).

The power of the in vitro evolution was also highlighted by a recent work in which nine substitutions in *Flavobacterium* OPH (A80V, I106V, F132D, K185R, D208G, H257W, I274N, S308L, and R319S) were used as scaffold for the creation of a synthetic construct (M9), which showed an increase of the rate of catalysis on VX of 35-fold respect to wild type (Jeong et al. 2014). A prediction of M9 structure was obtained and used in a docking simulation with VX substrate. From this analysis two additional critical residues (L271 and Y309) were identified, which, if substituted with alanines, showed an activity fivefold higher than M9 one (Jeong et al. 2014).

In vitro evolution efforts have been also focused on other OP-degrading enzymes. For example, in order to increase the catalytic activity of *Agrobacterium* OpdA on malathion, a combinatorial active site saturation testing (CASTing) was used (Naqvi et al. 2014). In this mutagenesis approach the protein tridimensional structure is analyzed in order to choose groups of few amino acids in the binding pocket which can be randomized simultaneously to generate small mutagenic libraries easy to screen. The CASTing strategy, used for the evolution of the malathion hydrolytic activity of OpdA, led to the identification of a double mutant (S308L/Y309A) which showed a widening of the active site and an increase of 5000-fold in catalytic efficiency towards malathion, reaching the highest value reported for this substrate (Naqvi et al. 2014).

2.2 MPH Hydrolases

A distinct OP-degrading pathway is represented by *mpd* (methyl parathion degradation) genes which confer hydrolytic activity towards methyl paraoxon, methyl parathion, and chlorpyrifos. Based on their similar metallo-β-lactamase domains, the protein products of *mpd* genes were designated as methyl parathion hydrolases (MPHs). The *mpd* genes were found in *Achromobacter*, *Ochrobactrum*, *Stenotrophomonas* and *Pseudomonas* but none of these shows a significant homology to *opd* or other OP-degrading gene (Cui et al. 2001; Yang et al. 2006; Zhang et al. 2005).

Recently, thanks to whole genome sequencing projects, many putative *mpd* homologues were discovered and cloned; a phylogenetic analysis revealed a separated evolution of these genes from *opd* genes.

The isolation of many *mpd* genes from soil bacteria of China (Liu et al. 2005; Zhang et al. 2005, 2006; Zhongli et al. 2001), suggested there is an environmental influence on *mpd* evolution. An AHL lactonase from *Bacillus thuringiensis*, belonging to the β-lactamase superfamily, showed some promiscuous organophosphatase activities, suggesting that it is possible an evolution of OPH and MPH from different lactonase enzymes (Afriat et al. 2006).

MPH enzyme from *Pseudomonas sp.* WBC-3 is a dimer containing a mixed binuclear zinc centre per subunit, where one of the zinc ions is usually replaced by cadmium (Dong et al. 2005). It has been demonstrated that three aromatic residues at the entry of the active site have a crucial role in the determination of affinity towards methyl parathion; indeed, any substitution in these positions results in a significant loss of catalytic activity on the tested substrate (Dong et al. 2005).

Moreover, a newly identified MPH from *Pseudomonas pseudoalcaligenes* (named OPHC2) showed an unexpected thermal resistance and a broad substrate activity spectrum. OPHC2 catalyzes the hydrolysis of a lactone and different phosphotriesters and esters (Gotthard et al. 2013). Its high T_m (97.8 °C) is probably due to the presence of an extended dimerization surface and an intramolecular disulfide bridge (typical of thermostable proteins) and makes OPHC2 a good candidate for OP decontamination (Gotthard et al. 2013).

2.3 Serum Paraoxonases (PONs)

Another interesting group that showed a critical role in OP metabolism is represented by mammalian lactonase/arylesterase enzymes, usually named as serum paraoxonases (Draganov 2010). These enzymes are components of the high-density lipoproteins, which seem to be implicated in the inactivation of toxic by-products of lipid oxidation (Blum et al. 2006; Mackness et al. 2000). The structure of the PON proteins is similar to two lactonases exhibiting promiscuous organophosphatase activity: squid diisopropyl fluorophosphatase (DFPase) and the human senescence marker protein-30 (Belinskaya et al. 2012).

In humans the following PON variants were found: PON1, PON2 and PON3; extensive reports have demonstrated that only PON1 is able to hydrolyze the P–O bond in insecticides, such as paraoxon or chlorpyrifos oxon, and G-series nerve agents (Draganov 2010; Josse et al. 2001; Rochu et al. 2007).

Given its importance as bioscavenger in pharmacological applications, PON1 enzyme was used as template of many evolution attempts. PON proteins are not soluble when produced in recombinant expression systems, therefore, human, mouse and rabbit PON1 genes were mutagenized by DNA shuffling to obtain highly soluble, recombinant enzymes (Aharoni et al. 2004). An interesting variant, named rePON1, was identified and employed as start point for protein engineering studies (Aharoni et al. 2004; Harel et al. 2004; Madej et al. 2012). Functional analysis on rePON1 identified L69, V346, and H115 in the active site as key residues in the catalysis (Harel et al. 2004). Substitutions of these residues were used as target to change the enzymatic stereoselectivity. L69V and V346A substitutions enhance S_P catalysis of soman and diisopropylfluorophosphate (DFP), and H115W substitution increases the specificity for the P–S bond (Amitai et al. 2006; Rochu et al. 2007).

In another study a combinatorial saturation mutagenesis of well-characterized PON1 catalytic sites was used to obtain variants with higher catalytic efficiency on cyclosarin (10^7 M^{-1} $min^{-1)}$. This value is consistent with an appropriate prophylactic protection in vivo; indeed, protection assays in mice demonstrated that, in the presence of these variants, the survivability of mice increases up to 6 h after acute exposure to cyclosarin (Gupta et al. 2011). A chimeric PON1 mutant with high catalytic efficiency towards soman and cyclosarin was also obtained by protein engineering but, however, its activity is still very low on the phosphoramidate tabun and absent on VX (Worek et al. 2014). Other achievements in the evolution of the interesting properties of PON1 are well-described in a recent review (Iyer and Iken 2015).

2.4 *Microbial Prolidases/OP Acid Anhydrolases*

Despite their main function, many enzymes able to hydrolyze P–O and P–F bonds in G-type nerve agents, were identified and designated as OP acid anhydrases (OPAAs) (EC 3.1.8.2). In particular, prolidase/OPAA enzymes with hydrolytic activity towards G-series nerve agents were isolated from *Alteromonas* species (Theriot et al. 2011; Vyas et al. 2010).

These enzymes, belonging to the prolidase family (Theriot et al. 2011), were found in all domains of life (from bacteria to humans) and represent novel thermostable templates for directed evolution studies. Despite the success achieved in *bd*PTE and PON1 evolution, mutagenesis strategies addressed to engineer efficient prolidases for organophosphorus degradation were underutilized.

OPAA proteins are similar to prolidase enzymes; indeed, they are able to hydrolyze dipeptides that have a proline at the carboxyl-terminal position but are also active on many substrates of OPH and MPH enzymes. They show an high similarity of their structure and their catalytic and stereoselective mechanisms with OPH or

MPH (Cheng and DeFrank 2000) but not significant sequence homology; it is now suggested that OPAA could have evolved from an ancestral prolidase (Theriot and Grunden 2011; Vyas et al. 2010).

Two prolidases were identified in hyperthermophilic archaeons *Pyrococcus furiosus* (Pfprol) and *Pyrococcus horikoshii* (Ph1prol); they were characterized and used as templates for directed evolution (Theriot et al. 2010a, b, 2011; Theriot and Grunden 2011).

In order to obtain thermophilic enzymes with high activity at lower temperatures, many mutagenesis efforts were employed; indeed, the first studies on prolidases from *P. furiosus* (*Pf*prol) have focused on the increasing of its activity at 35 °C, 50 °C, and 70 °C (Theriot et al. 2010a). By random mutagenesis of *Ph1*prol from *P. horikoshii*, four mutants (A195T/G306S, Y301C/K342N, E127G/E252D, and E36V) with high thermostability and increased prolidase and phosphotriesterase activity in a wider range of temperatures were obtained (Theriot et al. 2011). The mutations responsible for the catalytic improvement were located mainly in the loops and linker regions of the enzyme and seem to cause structural changes in regions away from the active site (Madej et al. 2012; Theriot et al. 2010b, 2011).

To date, *bd*PTE from *Pseudomonas/Brevundiminas*, OpdA from *Agrobacterium* and PON1 from mammalian are the best studied organophosphate-degrading enzymes. These enzymes and their evolved mutants showed significant activity against OPs and nerve gases but their low stability in solution limits their application in OP biosensing and remediation.

At this purpose, in recent years more researches were focused on a new family of OP-degrading enzymes (Phosphotriesterase-Like Lactonases) of which many extremophilic members were identified.

3 Phosphotriesterase-Like Lactonases (PLLs)

Phosphotriesterase-Like Lactonase (PLL) family includes a group of enzymes that have main lactonase activity on lactones and acyl-homoserin lactones (AHLs) and, in addition, low promiscuous phosphotriesterase activity towards organophosphate compounds (OPs) (Table 16.1).

PLLs have first been identified in the hyperthermophilic crenearchaeon *S. solfataricus* (*Sso*Pox) and *S. acidocaldarius* (*Sac*Pox) (Afriat et al. 2006; Merone et al. 2005; Porzio et al. 2007) and also in mesophilic organisms such as *Mycobacterium tuberculosis* (PPH), *Rhodococcus erythropolis* (AhlA), *Deinococcus radiodurans* (*Dr*OPH/*Dr*0930) (Afriat et al. 2006; Hawwa et al. 2009b) and *Mycobacterium avium* subsp. *Paratuberculosis* K-10 (MCP) (Chow et al. 2009). Then new thermostable PLLs from *Geobacillus stearothermophilus* (GsP) (Hawwa et al. 2009a) and from *Geobacillus kaustophilus* (GKL/*Gka*P) (Chow et al. 2010) have been reported. Very recently other PLLs have been characterized from *Sulfolobus islandicus* (*Sis*Lac) (Hiblot et al. 2012b), and from *Vulcanisaeta moutnovskia* (*Vmo*Lac/VmutPLL) (Kallnik et al. 2014) (Table 16.2).

Table 16.2 Classification of PLLs and main structural and kinetic information

Organism	Enzyme name	Structure (PDB code)	Sequence (UniProtKB code)	Stoichiometry and metal content	Catalytic efficiency for *ethyl*-paraoxon[a] k_{cat}/K_M (s^{-1} M^{-1})	Main references[b]
Brevundimonas diminuta	*bd*PTE[c]	1HZY	P0A434	Homodimer Zn^{2+}–Zn^{2+}	4×10^7	Benning et al. (2001)
Hyperthermophilic archaea						
Sulfolobus solfataricus	*Sso*Pox	2VC5	Q97VT7	Homodimer Fe^{3+}–Co^{2+}	3.75×10^3	Afriat et al. (2006), Merone et al. (2005)
Sulfolobus acidocaldarius	*Sac*Pox	Not solved	Q4J6Z8	–	2.66×10^4	Afriat et al. (2006), Porzio et al. (2007), (2013), Bzdrenga et al. (2014)
Sulfolobus islandicus	*Sis*Lac	4G2D	C4KKZ9	Homodimer Fe^{2+}–Co^{2+}	6.98×10^2	Hiblot et al. (2012b)
Vulcanisaeta moutnovskia	*Vmo*Lac	4RDY	F0QXN6	Homodimer Co^{2+}–Co^{2+}	1.86	Kallnik et al. (2014), Hiblot et al. (2015)
Thermophilic/extremophilic bacteria						
Deinococcus radiodurans	*Dr*0930/ *Dr*OPH	3FDK	Q9RVU2	Homodimer Zn^{2+}–Zn^{2+}	1.39	Afriat et al. (2006), Hawwa et al. (2009b), Xiang et al. (2009)
Geobacillus stearothermophilus	GsP	3F4D	D0VX06	Homodimer Co^{2+}–Co^{2+}	3.28×10^3	Hawwa et al. (2009a)
Geobacillus kaustophilus	GKL/*Gka*P	3ORW	Q5KZU5	Homodimer Fe^{3+}–Zn^{2+}	4.5	Chow et al. (2010), Zeng et al. (2011) to be published
Mesophilic bacteria						
Mycobacterium tuberculosis	PPH	4IF2	P9WHN9	Monomer Zn^{2+}–Zn^{2+}	8.65	Afriat et al. (2006), Zhang et al. to be published

(continued)

Table 16.2 (continued)

Organism	Enzyme name	Structure (PDB code)	Sequence (UniProtKB code)	Stoichiometry and metal content	Catalytic efficiency for *ethyl*-paraoxon[a] k_{cat}/K_M (s^{-1} M^{-1})	Main references[b]
Rhodococcus erythropolis	AhlA	2R1N	Q93LD7	Monomer Fe^{2+}–Co^{2+}	0.50	Afriat et al. (2006), Jackson et al. (2008)
Mycobacterium avium subsp. paratuberculosis K-10	MCP	Not solved	GI:41409766[d]	–	4.1	Chow et al. (2009)

[a]The reported values of specificity constant on ethyl-paraoxon have been done at the optimal temperature for each group of enzymes: *bd*PTE at 25 °C; PLLs from hyperthermophilic archaea at 70 °C; PLLs from extremophilic bacteria at 35 °C; PLLs from mesophilic bacteria at 25 °C

[b]Main references for each enzyme are reported, where it is possible to find main information about structure and catalytic properties summarized in the table

[c]*bd*PTE in this table is only to compare its catalytic activity with that of PLL enzymes

[d]NCBI code

The PLL family is related to bacterial phosphotriesterases (PTEs) from a structural and biochemical point of view (Afriat et al. 2006; Elias et al. 2008). Most of PLL members shows a low sequence identity (about 30 %) with the best characterized phosphotriesterase *bd*PTE. At the beginning most of them has been identified as putative phosphotriesterase and were called “Paraoxonases” (Pox) because able to degrade pesticides such as paraoxon (Merone et al. 2005; Porzio et al. 2007).

However, further structural, phylogenetic, and biochemical studies have revealed that these enzymes have a proficient lactonase activity, beside the weak phosphotriesterase activity (Afriat et al. 2006). Therefore these new enzymes were designed as PLLs and they share the same $(\beta/\alpha)_8$ barrel fold of *bd*PTE with a conserved binuclear metal center, essential for the catalysis (Seibert and Raushel 2005). Both PTE and PLL families belong to the amidohydrolase superfamily, but some extremophilic PLLs are structurally different from the mesophilic PTEs, in particular for loop 7, that is practically absent in the extremophilic enzymes, and for loop 8 where the few sequence differences are more important from a structural point of view. We performed a structural alignment by using the known 3D structure of these PLLs, and it confirms that, except for AhlA from *R. erythropolis*, all the PLLs have major differences respect to *bd*PTE in the loop 7 (Fig. 16.1).

Fig. 16.1 Structured-based sequence alignment among representatives of PLLs and *bd*PTE from Fig. 16.2. *Sso*Pox, *Vmo*Lac (from hyperthermophilic archaea); GsP, *Gka*P (from extremophilic bacteria); PPH, Ahla (from mesophilic bacteria). The structured-based alignment was performed by using Swiss-PdbViewer 4.1.0. Conserved residues are labeled by asterisks. The loop 7 and loop 8 are indicated by *black square frames*. Conserved residues that constitute the binuclear metal center are labeled in *gray*

In fact, thanks to the first solved 3D structure of a PLL (*Sso*Pox), very important structural differences emerged (Elias et al. 2008). The structure of the archaeal PLL shows an atypical active site topology and a single hydrophobic channel that perfectly accommodates lactone substrates such as AHLs (Elias et al. 2008). In Fig. 16.2, the superimposition of the structure of *bd*PTE with those of six PLLs (two from hyperthermophilic archaea, *Sso*Pox and *Vmo*Lac two from extremophilic bacteria, GsP and *Gka*P; two from mesophilic bacteria, PPH and Ahla) indicates that the proteins are homologous. It's worth noting that the length of loop 7 in all PLLs, except for AhlA (pink), is shorter than *bd*PTE (gray). Loop 8 of all PLLs has almost the same length as in *bd*PTE, but its topology is different for *Gka*P (green) and GsP (blue), where it is more detached from the protein core than that of *bd*PTE, and for *Vmo*PLL; in the latter case loop 8 is rigid and structured to form an α-helix, differently in comparison with the thermostable PLL *Sso*Pox (Fig. 16.2). Previously, some studies have supposed that the elongation of loop 7 in *bd*PTE forms a short α-helix, which may provide an active

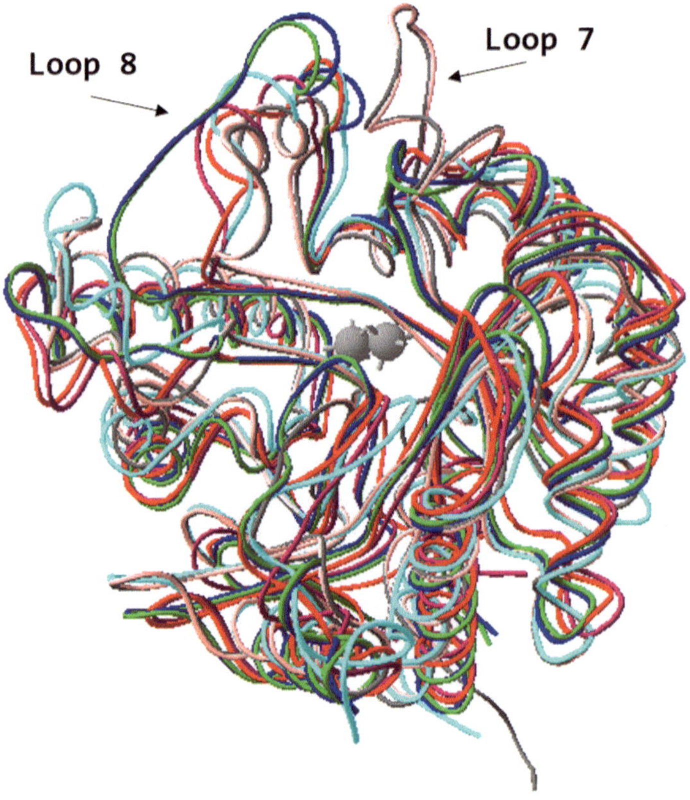

Fig. 16.2 Structural superposition of *bd*PTE with six PLLs. The structural superposition was performed by using Swiss-PdbViewer 4.1.0. *bd*PTE from *B. diminuta* (*gray*), *Sso*Pox from *S. solfataricus* (*red*); *Vmo*Lac from *V. moutnovskia* (*magenta*), GsP from *G. stearothermophilus* (*blue*), *Gka*P from *G. kaustophilus* (*green*), PPH from *M. tuberculosis* (*cyan*), Ahla from *R. erythropolis* (*pink*); metal cations are represented as *gray spheres*

"cap" that narrows the active site mouth and may thereby increase phosphotriesterase activity toward the substrate paraoxon (Afriat et al. 2006).

All the PLLs possesses a homo- or hetero-binuclear metal center which generates a catalytic hydroxide ion involved in the hydrolysis of both OPs and lactones. The main difference between the two hydrolytic activities is the transition state geometry: sp^3 coordination for phosphotriesters and pentacoordinated state for lactones (Elias et al. 2008). The coexistence of the two activities in the same active site suggests the use of the same catalytic machinery with the optimization for substrate specificity linked to environmental adaptation.

The active site of PLLs contains three sub-sites that are very well adapted for the lactone binding: a small sub-site, a large sub-site and a hydrophobic channel (Elias et al. 2008). The aliphatic chain of the lactones binds within the hydrophobic channel, the large sub-site accommodates the amide group of the N-acyl chain, and the small sub-site positions the lactone ring. Moreover, biochemical analyses confirmed that PLL enzymes are natural lactonases because they have a high activity on acyl-homoserine lactones. Thus it was supposed that PTE has evolved from a yet unknown PLL, whose primary activity was the hydrolysis of quorum sensing homoserine lactones (HSLs) (Afriat et al. (2006). Recently it has been reported that a nine-amino-acid deletion alongside an adjacent point mutation of *bd*PTE gave an intermediate (PTEΔ7-2/254R) that exhibits both the ancestral acyl homoserine lactonase activity (k_{cat}/K_M values of up to 2×10^4), (undetectable in *bp*PTE), and low paraoxonase activity (Afriat et al. 2012). Bifunctional intermediates with changed specificity have a powerful role in the divergence of new enzymatic functions that are also due to a very important structural changes of loop remodeling.

The bifunctional activity profile of PTEΔ7-2/254R is compatible with the expected properties of an evolutionary intermediate along a trajectory leading from an HSL with a promiscuous phosphotriesterase activity to a highly efficient phosphotriesterase with a weak, promiscuous lactonase activity (Afriat et al. 2012). The *Sso*Pox lactonase activity was tested with synthetic substrates (thiolactones) (Fig. 16.3) but also with real acyl-homoserine lactones (Afriat et al. 2006). However, docking studies and structural analyses demonstrated that the mode of binding is likely to be different (Merone et al. 2010). Starting from this observation, we found a trend of activity on thiolactones quite different from that on AHL lactones, because the specificity constant decreases with the increase of the lateral acyl chain (Fig. 16.3).

AHLs, also named as autoinducer 1 (AI-1), are small signal molecules, produced mainly by gram-negative bacteria, and mediate cell to cell communication known as quorum sensing (QS). When these molecules accumulate in the extracellular environment until a critical threshold, the transcriptional profile of the bacteria is altered (Hentzer et al. 2003).

It is reported that the virulence and the biofilm formation of some pathogens are regulated by QS (Costerton et al. 1999; Dickschat 2010; Jones et al. 2010; Popat et al. 2008), suggesting that the quenching of this mechanism (quorum quenching) could be an interesting strategy against multi-resistant pathogen bacteria which use AHL based QS like *P. aeruginosa* (Amara et al. 2011; Dong et al. 2000, 2001; Ma et al. 2009; Reimmann et al. 2002). Therefore, as lactonases, PLLs can hydrolyze AHLs and quench the QS mechanism, as seen for AiiA, human paraoxonases (Dong et al. 2001;

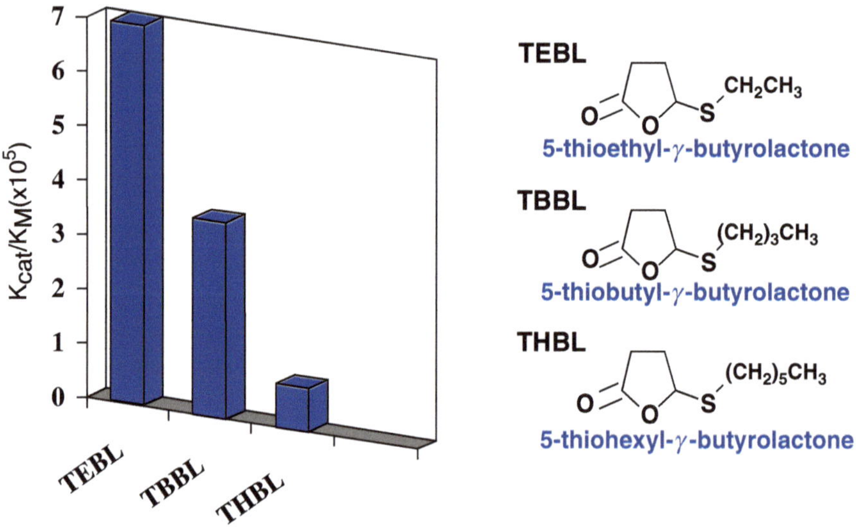

Fig. 16.3 Lactonase activity of *Sso*Pox on thiolactones with different acyl chain length. The assays were performed at 70 °C as reported in Afriat et al. (2006). *Sso*Pox shows a k_{cat}/K_M of 7×10^5, 3×10^5 and 0.8×10^5 M^{-1} s^{-1} on TEBL, TBBL and THBL respectively

Ma et al. 2009), and *Sso*Pox, *Sac*Pox and *Dr*0930 that, as reported, are able to reduce swarming PAO motility (Mandrich et al. 2011). Moreover *Sso*Pox has also recently been shown to decrease virulence factors expression and biofilm formation in vitro and to reduce rat mortality in a rat pulmonary infection model (Hraiech et al. 2014).

By considering they have both lactonase and phosphotriesterase activity, PLLs represent a very interesting candidate for biotechnological application as quorum quenching agent (Dong et al. 2000) or OPs biodecontaminant (Singh 2009).

However, at present it is still unclear which is (are) their biological function(s), also because in many cases PLLs were found in microorganisms that do not produce or are regulated by AHLs; therefore, probably they could act on the mechanism of communication of other microrganisms. Furthermore, some members of the PLL family efficiently hydrolyze gamma and/or delta oxo-lactones, but not AHLs (Chow et al. 2010; Xiang et al. 2009). Anyway the interest in these enzymes, in particular for biotechnological applications in decontamination field, lies in exploring phosphotriesterase activity. Most of the extremophilic PLLs are able to hydrolyze different OP pesticides, even if at different extent. As representative compound it is reported the catalytic efficiency on ethyl-paraoxon of the described PLLs (Table 16.2). Among the archaeal PLLs, the *Sso*Pox promiscuous catalytic activity against paraoxon is 3.75×10^3 $s^{-1}M^{-1}$, tenfold lower than that of *Sac*Pox (2.66×10^4 $s^{-1}M^{-1}$) at 70 °C (Merone et al. 2010; Porzio et al. 2013). These values are higher than *Sis*Lac catalytic efficiency (6.98×10^2 $s^{-1}M^{-1}$) at 70 °C (Hiblot et al. 2012b).

*Sso*Pox and *Sac*Pox are both in dimeric form, which is consistent with the crystal structures of *Sis*Lac, *Dr*0930, GsP and *Gka*P (Table 16.2).

The recent discovered hypertermophilic PLL from *V. mountnovski* (*Vmo*Lac) shows a low catalytic efficiency against paraoxon of 1.86 s^{-1} M^{-1} (Hiblot et al. 2015), comparable to those of *Dr*0930 (1.39 s^{-1} M^{-1}) (Hawwa et al. 2009b; Mandrich et al. 2013) and *Gka*P

(4.5 s^{-1} M^{-1} at 35 °C; 1.1×10^2 at 75 °C), even if all of them are highly stable enzymes. Among the others extremophilic bacterial PLLs, GsP from *G. stearothermophilus* has the highest value of k_{cat}/K_M on paraoxon that is 3.28×10^3 s^{-1} M^{-1} at the same comparable temperature (35 °C) (Hawwa et al. 2009a). All the mesophilic PLLs, PPH, AhlA and MCP, exhibit a weak phosphotriesterase activity. For hydrolysis of paraoxon by AhlA, only the k_{cat}/K_M ratio was estimated (0.5 s^{-1} M^{-1}), while MCP and PPH exhibited a higher value of 4.1 and 8.6 M^{-1} s^{-1}, respectively (Afriat et al. 2006; Chow et al. 2009).

4 In Vitro Evolution of Extremophilic PLLs

The promiscuous phosphotriesterase activity of PLL enzymes on OP compounds suggests that they may constitute a "generalist" intermediate from which PTE may have arisen. However, the possibility that *M. loti* PLC (more similar to ePHP) represents such starting point could not be discarded (Fig. 16.4).

Recently, various mutagenesis strategies, such as site-directed mutagenesis and directed evolution, have been used to enhance the promiscuous OP-degrading activ-

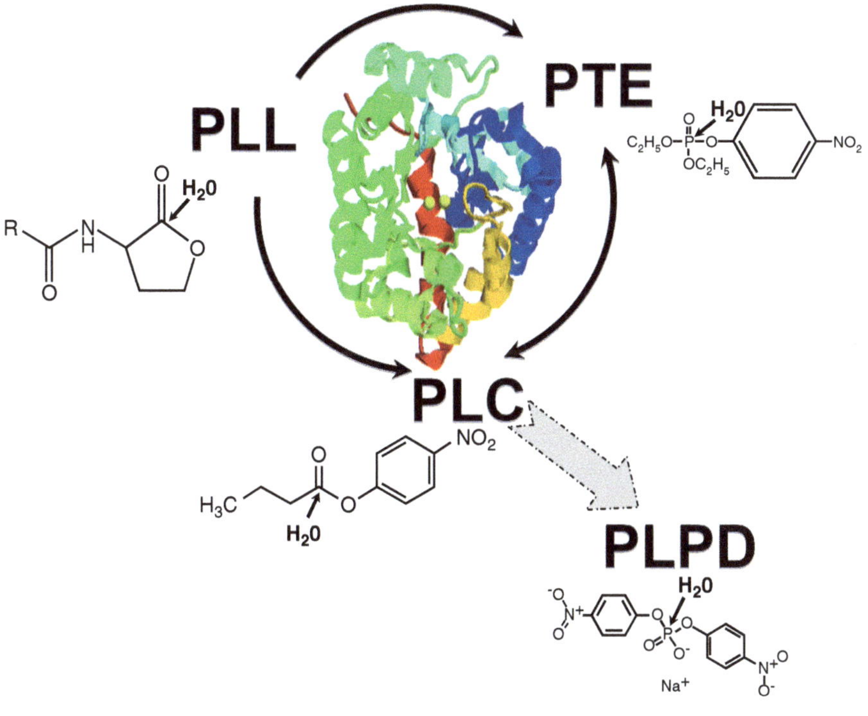

Fig. 16.4 Representation of likely evolution paths among members of the Amidohydrolase Superfamily. Schematic representation of the possible evolution of three different members of the amidohydrolase superfamily: *PLL* Phosphotriesterase-Like Lactonase, *PTE* PhosphoTriestErase, *PLC* Phosphotriesterase-Like Carboxylesterase, enzymes. *PLPD* Phosphotriesterase-Like PhosphoDiesterase was generated from PLC by a single mutation

ity of several PLLs; this approach efficiently reshapes enzymatic proficiency and specificity, and provides a powerful evidence of the evolvability of PLL enzymes for OP decontamination.

An important difference between PLL and PTE enzymes is that loop 7 and loop 8 have a different length and topology. As reported above, both the loops in the amidohydrolase superfamily are often found to interact with the substrates and to play important role in determination of substrate specificity (Seibert and Raushel 2005). The association of protein and solvent engineering for *Sso*Pox has led to kinetics parameters (k_{cat} and specificity constant) more similar to those of the PTE, but it was exceptionally more stable and so it represent an interesting detoxification device (Merone et al. 2010). The active site structure of *Sso*Pox shows a low elasticity to changes afforded by temperature because various mutations have led to different activity/temperature profiles (Merone et al. 2010).

This means that in the active site a compromise is required between the rigidity necessary to withstand high temperatures assuring at the same time specificity, and the flexibility which is a prerequisite for catalytic activity (Tehei et al. 2005). In detail, variants Y97W, I98F, Y97W/I98F showed profiles very similar to wild type even though with different degrees of activation. As already reported in Elias et al. (2008), single mutant Y97W suggests an involvement of position 97 in the correct orientation of the substrate for catalysis. The single mutant W263F emerged as the best variant, but activating effects were observed for all mutants (Table 16.3).

Table 16.3 Mutants of PLLs obtained by in vitro evolution of paraoxonase activity

PLL	Important identified residues and relevant location/role in the structure	Relevant evolved variants	Effect on catalytic efficiency for *ethyl*-paraoxon (k_{cat}/K_M)	References
***Sso*Pox**	M79 Y97 *(interaction with the substrate)* I98 I261 (*Loop 8*) W263 (*Loop 8*)	M79V Y97W I98F W263Q W263F I261F M79V/W263Q I98F/I261F Y97W/I98F Y97W/I261F Y97W/W263Q Y97W/W263F Y97W/I98F/I261F	Increase from 1.7- to 6-fold the wt activity	Merone et al. (2010)
		W263L,M, F, D, I, P, Y, V, A, C, E, G, Q, N, S, H, T, R, K	Increase from 1.7- to 23-fold the wt activity	Hiblot et al. (2013)

(continued)

Table 16.3 (continued)

PLL	Important identified residues and relevant location/role in the structure	Relevant evolved variants	Effect on catalytic efficiency for *ethyl*-paraoxon (k_{cat}/K_M)	References
***Sis*Lac**	E14 Y34	E14K Y34H E14K/Y34Q	Increase from 3- to 17-fold the wt activity	Hiblot et al. (2012b)
***Dr*0930**	F26 *(small subsite)* T68 *(small subsite)* Y97 *(metal binding)* Y98 *(leaving group)* G207 R228 *(large subsite)* W269 *(small subsite)* W287 *(large subsite)* D71 *(leaving group)* A85 E101 *(leaving group)* V35 A43 A44 V64 V65 N75 *(large subsite)* F26 *(small subsite)* M234*(large subsite)* Y28A *(small subsite)* S48 V235 *(large subsite)* R272 A103 L231 *(large subsite)*	F26G T68S Y97W Y98F G207D R228A W269A W287G D71G/E101G D71G/A85G/V165M V35L/A43T/A44G/V64G/V65G/D71G/N75T/E101G F26G/M234R Y28A/Y97W, Y28C/Y97W, Y28G Y28W/S48N/Y97C D71G/E101G/M234I D71G/E101G/M234L D71G/E101G/V235L D71G/E101G/R272C/W287R Y97G Y97I A103V L231D L231R/W287G	Increase from 1.5- to 559-fold the wt activity	Hawwa et al. (2009b)
		Y28L/D71N/E101G/E179D/V235L/L270M	Increase until 10^2 fold the wt activity	Meier et al. (2013)

(continued)

Table 16.3 (continued)

PLL	Important identified residues and relevant location/role in the structure	Relevant evolved variants	Effect on catalytic efficiency for *ethyl*-paraoxon (k_{cat}/K_M)	References
***Gka*P**	E15 F28 *(small binding pocket)* D73 Y99 *(in the leaving group subsite)* T171 F228 *(at the base of loop 7)* N269 *(inside loop 8)* V270 *(inside loop 8)* W271 *(inside loop 8)* G273 *(inside loop 8)* F276 I299	Y99L D73Y/Y99L E15V/Y99L/G273D/F276V/I299S Y99L/T171S/V270G/G273D Y99L/W271C F28I/Y99L/T171S/F228L/N269S/ V270G/G273D F28I/D73Y/Y99L/T171S/F228L/N269S/V270G/G273D F28I/Y99L/T171S/F228L/N269S/ V270G/W271C/G273D F28I/D73Y/Y99L/T171S/F228L/ N269S/V270G/W271C/G273D	Increase from 3- to 230-fold the wt activity	Zhang et al. (2012)

W263F is sixfold more efficient towards ethyl-paraoxon and it also shows a little improvement of rate of hydrolysis towards the warfare agent cyclosarin (Merone et al. 2010).

It has also been reported that the residue W263 plays an important role in substrate binding in the *Sso*Pox active site, at the dimer interface in loop 8, and influence the protein stability, enzymatic activity and promiscuity (Hiblot et al. 2013). Recent structural analyses of *Sso*Pox show that all the 19 different substitutions of the W263 residue increase the *Sso*Pox paraoxonase activity, by inducing very subtle changes in the loop 8 positioning with no effect on the very short loop 7 (Hiblot et al. 2013) (Table 16.3).

Hiblot et al. (2012b) reported that, on the basis of *Sis*Lac and *Sso*Pox structural analysis, variations at the positions 14 and 34 seem to have a big impact between the two structures, with a possible effect on dimerization changes (position 34 of *Sis*Lac) or stability (positions 14, 34 of *Sis*Lac). The analysis of T_m of the mutants E14K and Y34Q shows that single substitution destabilize *Sis*Lac, but the combination of both of them helps to restore partially the stability (Hiblot et al. 2012b). Additionally, the rational-designed mutants activity have been tested on different substrates, and in particular it was observed an increase of until 17-fold (E14K) the wild type *Sis*Lac (Hiblot et al. 2012b) (Table 16.3).

To improve the organophosphatase activity of *Dr*0930, in particular on ethyl and methyl paraoxon, site-directed mutagenesis, random mutagenesis, and site-saturation mutagenesis were also used. More than 30,000 potential mutants were screened, and a total of 26 mutated enzymes were purified and characterized kinetically (Hawwa et al. 2009a). In total, all the mutagenesis efforts raised the specificity for paraoxon until 559-fold respect to wt (Table 16.3).

This study underline that by using an iterative approach to mutagenesis, it's possible to achieve a large rate enhancements when mutations are made in already active mutants. Moreover, the effect of new mutations on kinetic parameters allows to have important structural and functional information of the enzyme (Hawwa et al. 2009b). In addition other authors described rational protein design and random mutagenesis methodologies that were combined with an efficient in vitro screening to identify highly efficient hydrolase variants with enhanced degradation ability (Meier et al. 2013). The best mutant displayed a catalytic activity more than two orders of magnitude higher than the wild-type *Dr*0930 (Table 16.3).

As regards the *Geobacillus* sp, by combining rational and random mutagenesis strategies, new variants were successfully obtained after four rounds of screening for *Gka*P from *G. kaustophilus*. The catalytic efficiency (k_{cat}/K_M) of the best variant against ethyl-paraoxon was about 40-fold higher than that of the D71G/E101G/V235L *Dr*0930 variant, and comparable to that of the W263F*Sso*Pox variant (Zhang et al. 2012) (Table 16.3).

Eight amino acid changes in *Gka*P have enhanced the catalytic efficiency for ethyl-paraoxon by nearly two orders of magnitude. Among the eight mutations (F28I, Y99L, T171S, F228L, N269S, V270G, W271C, and G273D), some of these substitutions were located far from the catalytic β-metal, whereas others were within the vicinity of the catalytic site. Tyrosine 99, which is located at the entrance of the binuclear metal center and is thought to reside in the leaving group sub-site, is a conserved residue in the PLL family that appears to play a critical role in stabilizing the lactone ring in a favorable state (Elias et al. 2008). The *Gka*P variants showed a strong inverse relationship between thermostability and catalytic activity, even if they were still remarkably thermostable (Zhang et al. 2012).

Beside the in vitro evolution approach, it is worth noting that metals in the active site have an important role in influencing the kinetic parameters, as reported in *bd*PTE (Rochu et al. 2004). Recently we reported that promiscuous paraoxonase activity of *Sac*Pox seems to be more sensitive than the main lactonase activity to the metal species in the binuclear metal centre (Porzio et al. 2013). In fact the enzyme prepared in presence on Mn^{2+} cations (*Sac*Pox-Mn^{2+}) shows about 30- and 19-fold increase in paraoxon or me-paraoxon efficiency respectively, with respect to the enzyme prepared with Cd^{2+} (*Sac*Pox-Cd^{2+}). This finding suggest a new interesting approach to be used to increase phosphotriesterase promiscuous activity for biotechnological application.

Finally, the functional and structural homologies suggest that PTE evolved from an as yet unknown PLL, using its promiscuous paraoxonase activity as an essential starting point. As demonstrated by the properties of *Sso*Pox, the paraoxonase activity of PLLs can be considerably high, even without compromising the lactonase activity, and perhaps while acquiring other activities such as aryl esterase. In that respect, *Sso*Pox resembles a "generalist" intermediate from which PTE may have emerged (Aharoni et al. 2005; Afriat et al. 2006).

The recent specialization as a phosphotriesterase, as seen in *bd*PTE, probably through an insertion into loop 7, did not completely eliminate the ancestor's lactonase activity (Afriat et al. 2006). The promiscuous carboxylesterase and phosphodiesterase activities observed in *bd*PTE and PLLs in turn paved the way for the prediction of the existence in Nature of PTE-like enzymes with main carboxylesterase or phosphodiesterase activities. It has been discovered that an homologue of ePHP, from *Mesorhizobium loti*, is an efficient carboxylesterase with promiscuous phosphotriesterase activity, and therefore it was called *Mlo*PLC (Phosphotriesterase-Like Carboxylesterase) (Mandrich and Manco 2009). It shows also a low promiscuous phosphodiesterase and lactonase activity, and we suppose it may be originated from either an ancestral PLL-like or PTE-like enzyme, considering that it maintains both these promiscuous activities (Fig. 16.4). Moreover the discovery of *Mlo*PLC supports the theory on the enzyme evolution starting from promiscuous activities in the amidohydrolase superfamily (Afriat et al. 2006; Aharoni et al. 2005; Khersonsky et al. 2006). However, the conversion of *Mlo*PLC into an excellent phosphodiesterase by means of a single mutation, in parallel with the complete loss of carboxylesterase activity, is a rare case of substrate assisted gain-of-function via stabilization of the binuclear metal center (Mandrich and Manco 2009). This scenery let us to speculate the existence of a new intermediate, PLC-like, between PTE and PLL, from which, during evolution by sequence/structural rearrangements, phosphodiesterase activity could have been arising (Fig. 16.4). The mutant obtained shows a complete loss of carboxylesterase and lactonase activities (Mandrich and Manco 2009), suggesting that a Phosphotriesterase-like Phosphodiesterase enzyme (PLPD) may represent a final point in this evolutionary route, supposing probably other interesting intermediates after PLC to be discovered and explored.

5 Potentiality of Evolved OP-Degrading Enzymes in Biotechnological Applications

The rapid accumulation of data on evolved OP-degrading enzymes led to renewed interest in developing new strategies for several biotechnological applications, such as therapy and prophylaxis of OP poisoning, OP biomonitoring and OP remediation.

5.1 Therapy and Prophylaxis of OP Poisoning

The human PON1 provides the organism with a natural protection from low-dose intake of some organophosphates, such as paraoxon, but is not efficient in the case of assumption of high doses of deadly organophosphates, such as chemical warfare agents. Therefore, mutants of *bd*PTE, OpdA and PON1 have been obtained and assayed for in vivo prophylaxis of OP pesticides and nerve agents (Masson and Rochu 2009). Despite the achievements of these studies, their promising therapeutic application requires a deep study of efficient transport and delivery methods in order to reduce the decrease or the full loss of the enzymatic activity. Furthermore, others important issues regard their expression, pharmacokinetics and humanization.

An improvement of the thermal stability and an extension of the mean residence time in mice (from minutes to days) was reached by PEGylation of these enzymes (Trovaslet-Leroy et al. 2011) (for a recent review, see Nachon et al. 2013).

5.2 OP Biomonitoring

While ELISA kits are commercially available for the monitoring of few OPs, in the recent years, enzymatic sensors were deeply investigated for the rapid OP detection (Carullo et al. 2015).

OP biosensors used today are inhibition-based rather than catalysis-based. Enzyme inhibition requires regular replacement of the enzyme part and is measured via amperometric detection of thiocholine, hydrogen peroxide, or p-aminophenol (Wanekaya et al. 2008). These compounds are produced by hydrolysis of acetylcholine (thiocoline) or aminophenyl acetate (p-aminophenol) by AChE, or choline oxidation (hydrogen peroxide) by choline oxidase.

Other OP biosensors use a potentiometric transduction based on the pH variation caused by acetic acid reduction.

AChE-based biosensors are sensitive but suffer of many drawbacks, such as the limited selectivity since AChE is inhibited by several neurotoxins (carbamates, heavy metals), the impossibility to be reused without regeneration with reactivators such as pyridine 2-aldoxime, and the unsuitability for real-time monitoring. Also stability over time is a problem. Recently we investigated the opportunity to use the thermostable EST2 from *Alicyclobacillus acidocaldarius* as substitute of AChE (Febbraio et al. 2011).

The use of catalysis-based systems (typically using OPHs) is extremely attractive for OP biosensing because they have broad substrate recognition, and the hydrolysis product (p-nitrophenol) can be also quantitatively detected by electrochemical and optical methods (Wanekaya et al. 2008).

The use of higher ionic strength buffers (in which the enzymes show the maximum activity for all time course of the measures) represents the main advantage of the use of optical biosensors rather than the potentiometric ones (Wanekaya et al. 2008).

An example of the discussed application is represented by the combination of a mutant of *bd*PTE (H254R/H257L), showing an increased catalytic rate for P–S bond hydrolysis, with carbon nanotubes for the sensitive, rapid and selective amperometric detection of V-type nerve agents and demeton-S insecticide (Joshi et al. 2006).

5.3 OP Remediation

In recent years, a strong public opposition to the use of polluting chemical methods has led to the research and development of alternative eco-friendly devices to remediate OPs. The use of microorganisms has limited applicability because poses many issues (dependence of growth on temperature, presence of growth inhibitors such as toxic compounds and some metabolites, and so forth). Therefore enzymes represent a feasible and good alternative.

Many examples of the application of OP-degrading enzymes in the remediation were reported.

The presence of a permeability barrier represented by the outer membrane hinders the OP/OP degrading enzymes interaction in the cell (Richins et al. 1997); to overcame this problem, different surface anchoring motifs, such as Lpp-OmpA chimera, ice nucleation protein (INP), or autotransporter have been used to address these enzymes to the cell surface (Li et al. 2008; Richins et al. 1997; Shimazu et al. 2001).

By using an INP anchoring motif, Yang et al. (2013) reported the functional display of the PTE-GFP fusion on the cell surface of *Sphingobium japonicum* UT26. The strain containing the fusion showed the ability to hydrolyze OPs and allowed to track the fluorescence during the time course of bioremediation.

The same anchoring motif was successfully used to display an improved mutant of OPH from *B. diminuta* on the cell surface of *E. coli* (Tanga et al. 2014). This engineered biocatalysts showed the highest paraoxonase activity among the already reported OPH-displayed strains, an optimal temperature of 55 °C and an high stability (100 % activity over 1 month at room temperature) (Tanga et al. 2014).

Pinjari et al. (2013) have heterologously expressed *bd*PTE on the membrane of *Pseudomonas* sp. Ind01, a strain isolated from the activated sludge of a pesticide-manufacturing sites (Pinjari et al. 2012). The engineered strain was very efficient in the remediation of different commercially-available OP pesticides (Pinjari et al. 2013).

Many strategies have been used to immobilize, encapsulate or entrap OP-degrading enzymes to create materials with preserved organophosphatase activity. In a recent paper, mesoporous thin films carried over glass slides were used as host matrices for enzymatic remediators. These bio-catalysts were active and very sensitive (down to 15 μM concentration for paraoxon) and, moreover, they were easly prepared, and reused several times with no significant loss in catalytic activity (Francic et al. 2014).

Gao et al. (2014) reported the covalent immobilization of OpdA enzyme on highly porous nonwoven polyester fabrics for OP degradation. First, these materials were activated with ethylenediamine; the enzyme was, then, immobilized by glutaraldehyde, a bifunctional crosslinker. Following the enzymatic immobilization, the K_M for methyl parathion was increased, the pH profile was widened, and the enzymatic stability was enhanced. In batch mode, this remediation system could hydrolyze 20 μM methyl parathion in un-buffered water (Gao et al. 2014).

Despite these applications, many data report a gradual inactivation of *bd*PTE at temperatures up to 35 °C until a full loss of catalytic activity at 60 °C (Caldwell and Raushel 1991) and, moreover, all attempts to immobilize the enzyme were linked to short-term stability, hi gh costs and possible changes of the kinetic properties (Braatz 1994; Caldwell et al. 1991; Gill and Ballesteros 2000; Grimsley et al. 2001; LeJeune et al. 1997).

Therefore it is crucial to find alternative highly resistant and stable enzymes. As already described, phosphotriesterase activity has been found as promiscuous component in several hydrolases such as prolidase, PLL and MBL enzymes isolated from different extremophilic hosts. The exceptional thermal stability of the extremophilic bacterial and archaeal enzymes offers many biotechnological advantages in industrial processes such as high resistance to harsh conditions (presence of solvents or detergents), minimized contamination potential, extreme stability and increased enzymatic longevity. Furthermore, the purification costs of the enzymes expressed in mesophilic hosts are affordable, indeed, the contaminating proteins deriving from the host can be precipitated by simple thermal fractioning.

Thanks to the link between protein stability and evolvability (Bloom et al. 2006), many reports used highly stable scaffold for next-generation engineering technologies.

For this purpose, many efforts have been performed to use extremophilic PLL enzymes as remediation devices. To date, only two examples of practical applications of PLLs have been reported, even if one of them was not properly applied in OP remediation. In fact, the first report deals with the employment of the lactonase activity of *Sso*Pox on acyl-homoserine lactones (Ng et al. 2011). The enzyme was absorbed onto nanoalumina membranes; in order to interfere with quorum sensing signaling, bacterial cultures were treated with the immobilized enzyme. The results open up new scenarios of biotechnological applications, for example, it could be possible to use lactonases immobilized on filtration membranes for the control of undesirable microbial activities in water purification systems (Ng et al. 2011).

In the second report, the advantage of a mutant of *Sso*Pox (W263F) as detoxification tool was assessed. The enzyme was dissolved in different buffered aqueous solvents (30 % ethanol, 30 % or 50 % methanol and 0.1 % sodium-dodecyl-sulphate) to analyze its activity under stressing denaturing environment, typical, for example, of the conditions usually employed in the toxin extraction from contaminated soils (Merone et al. 2010; Hiblot et al. 2012a). In one case the results were compared with those obtained with *bd*PTE from *B. diminuta*. W263F outperforms *bd*PTE under most of the tested conditions; in 15 min at room temperature about 99.5 % of paraoxon was hydrolized in 30 % methanol and ethanol and 0.1 % sodium-dodecyl-sulphate (Merone et al. 2010).

6 Concluding Remarks and Future Perspectives

It is generally accepted the theory that thermostable enzymes are the best candidate for the in vitro evolution since they well-accept destabilizing mutations, which can lead to new or improved enzymatic functions. The investigation reported here outlines the in vitro evolution potential of PLL enzymes and their successful employment in OP detoxification. It is well-known that also *p*-nitrophenol produced by paraoxon/parathion hydrolysis is toxic (120-fold less than the original OP) (Munnecke 1979), but, in the last years, different soil bacteria able to use *p*-nitrophenol as carbon source were studied offering the opportunity to completely remove the toxicity of such compounds (Spain and Gibson 1991).

We have described many examples of OP-degrading enzymes with considerably improved activities that can be used in combination with PNP-utilizing bacteria for a full degradation of the OP compounds (Shimazu et al. 2001).

Acknowledgments This research was supported from the Italian Ministry for University and Research (MIUR) (project PON01_01585 to G.M.).

References

Aardema H, Meertens JH, Ligtenberg JJ, Peters-Polman OM, Tulleken JE, Zijlstra JG (2008) Organophosphorus pesticide poisoning: cases and developments. Neth J Med 66(4): 149–153

Afriat L, Roodveldt C, Manco G, Tawfik DS (2006) The latent promiscuity of newly identified microbial lactonase is linked to a recently diverged phosphotriesterase. Biochemistry 45: 13677–13686

Afriat-Jurnou L, Jackson CJ, Tawfik DS (2012) Reconstructing a missing link in the evolution of a recently diverged phosphotriesterase by active-site loop remodeling. Biochemistry 51: 6047–6055

Aharoni A, Gaidukov L, Yagur S, Toker L, Silman I, Tawfik DS (2004) Directed evolution of mammalian paraoxonases PON1 and PON3 for bacterial expression and catalytic specialization. Proc Natl Acad Sci USA 101(2):482–487

Aharoni A, Gaidukov L, Khersonsky O, Mc QGS, Roodveldt C, Tawfik DS (2005) The 'evolvability' of promiscuous protein functions. Nat Genet 37:73–76

Amara N, Krom BP, Kaufmann GF, Meijler MM (2011) Macromolecular inhibition of quorum sensing: enzymes, antibodies, and beyond. Chem Rev 111:195–208

Amitai G, Gaidukov L, Adani R, Yishay S, Yacov G, Kushnir M et al (2006) Enhanced stereoselective hydrolysis of toxic organophosphates by directly evolved variants of mammalian serum paraoxonase. FEBS J 273(9):1906–1919

Belinskaya T, Pattabiraman N, diTargiani R, Choi M, Saxena A (2012) Differences in amino acid residues in the binding pockets dictate substrate specificities of mouse senescence marker protein-30, human paraoxonase1, and squid diisopropylfluorophosphatase. Biochim Biophys Acta 1824(5):701–710

Benning MM, Kuo JM, Raushel FM, Holden HM (1994) Three-dimensional structure of phosphotriesterase: an enzyme capable of detoxifying organophosphate nerve agents. Biochemistry 33:15001–15007

Benning MM, Kuo JM, Raushel FM, Holden HM (1995) Three-dimensional structure of the binuclear metal center of phosphotriesterase. Biochemistry 34:7973–7983

Benning MM, Hong SB, Raushel FM, Holden HM (2000) The binding of substrate analogs to phosphotriesterase. J Biol Chem 275:30556–30560

Benning MM, Shim H, Raushel FM, Holden HM (2001) High resolution X-ray structures of different metal-substituted forms of phosphotriesterase from *Pseudomonas diminuta*. Biochemistry 40:2712–2722

Bloom JD, Labthavikul ST, Otey CR, Arnold FH (2006) Protein stability promotes evolvability. Proc Natl Acad Sci USA 103(15):5869–5874

Blum MM, Löhr F, Richardt A, Rüterjans H, Chen JC (2006) Binding of a designed substrate analogue to diisopropyl fluorophosphatase: implication for the phosphotriesterase mechanism. J Am Chem Soc 128:12750–12757

Braatz JA (1994) Biocompatible polyurethane-based hydrogels. J Biomater Appl 9:71–96

Buchbinder JL, Stephenson RC, Dresser MJ, Pitera JW, Scanlan TS, Fletterick RJ (1998) Biochemical characterization and crystallographic structure of an *Escherichia coli* protein from the phosphotriesterase gene family. Biochemistry 37:5096–5106

Bzdrenga J, Hiblot J, Gotthard G, Champion C, Elias M, Chabriere E (2014) SacPox from the thermoacidophilic crenarchaeon Sulfolobus acidocaldarius is a proficient lactonase. BMC Res Notes 7:333

Cáceres T, Megharaj M, Venkateswarlu K, Sethunathan N, Naidu R (2010) Fenamiphos and related organophosphorus pesticides: environmental fate and toxicology. Rev Environ Contam Toxicol 205:117–162

Caldwell SR, Raushel FM (1991) Detoxification of organophosphate pesticides using a nylon based immobilized phosphotriesterase from *Pseudomonas diminuta*. Appl Biochem Biotechnol 31:59–63

Caldwell SR, Newcomb JR, Schlecht KA, Raushel FM (1991) Limit of diffusion in the hydrolysis of substrate by the phosphotriesterase from *Pseudomonas diminuta*. Biochemistry 30:7438–7444

Calvert GM, Karnik J, Mehler L, Beckman J, Morrissey B, Sievert J, Barrett R, Lackovic M, Mabee L, Schwartz A, Mitchell Y, Moraga-McHaley S (2008) Acute pesticide poisoning among agricultural workers in the United States, 1998–2005. Am J Ind Med 51(12):883–898

Can A (2014) Quantitative structure-toxicity relationship (QSTR) studies on the organophosphate insecticides. Toxicol Lett 230(3):434–443

Carey JL, Dunn C, Gaspari RJ (2013) Central respiratory failure during acute organophosphate poisoning. Respir Physiol Neurobiol 189(2):403–410

Carullo P, Cetrangolo GP, Mandrich L, Manco G, Febbraio F (2015) Fluorescence spectroscopy approaches for the development of a real-time organophosphate detection system using an enzymatic sensor. Sensors (Basel) 15(2):3932–3951

Cheng TC, DeFrank JJ (2000) Hydrolysis of organophosphorus compounds by bacterial prolidases. In: Zwanenburg B, Mikolajczyk M, Kielbasinski P (eds) Enzymes in action: green solutions for chemical problems, vol 33. Kluwer, Dordrecht, pp 243–262

Chen-Goodspeed M, Sogorb MA, Wu F, Hong SB, Raushel FM (2001a) Structural determinants of the substrate and stereochemical specificity of phosphotriesterase. Biochemistry 40(5): 1325–1331

Chen-Goodspeed M, Sogorb MA, Wu F, Raushel FM (2001b) Enhancement, relaxation, and reversal of the stereoselectivity for phosphotriesterase by rational evolution of active site residues. Biochemistry 40(5):1332–1339

Cho CMH, Mulchandani A, Chen W (2002) Bacterial cell surface display of organophosphorus hydrolase for selective screening of improved hydrolysis of organophosphate nerve agents. Appl Environ Microbiol 68(4):2026–2030

Cho CMH, Mulchandani A, Chen W (2004) Altering the substrate specificity of organophosphorus hydrolase for enhanced hydrolysis of chlorpyrifos. Appl Environ Microbiol 70(8):4681–4685

Cho CMH, Mulchandani A, Chen W (2006) Functional analysis of organophosphorus hydrolase variants with high degradation activity towards organophosphate pesticides. Protein Eng Des Sel 19(3):99–105

Chow JY, Wu L, Yew WS (2009) Directed evolution of a quorum-quenching lactonase from *Mycobacterium avium* subsp. *paratuberculosis* K-10 in the amidohydrolase superfamily. Biochemistry 48:4344–4353

Chow JY, Xue B, Lee KH, Tung A, Wu L, Robinson RC, Yew WS (2010) Directed evolution of a thermostable quorum-quenching lactonase from the amidohydrolase superfamily. J Biol Chem 285(52):40911–40920

Cobb RE, Si T, Zhao H (2012) Directed evolution: an evolving and enabling synthetic biology tool. Curr Opin Chem Bio 16:285–291

Cobb RE, Chao R, Zhao H (2013a) Directed evolution: past, present, and future. AIChE J 59:1432–1440

Cobb RE, Sun N, Zhao H (2013b) Directed evolution as a powerful synthetic biology tool. Methods 60:81–90

Costerton JW, Stewart PS, Greenberg EP (1999) Bacterial biofilms: a common cause of persistent infections. Science 284:1318–1322

Cui ZL, Li SP, Fu GP (2001) Isolation of methyl parathion-degrading strain M6 and cloning of the methyl parathion hydrolase gene. Appl Environ Microbiol 67:4922–4925

Denard CA, Ren H, Zhao H (2015) Improving and repurposing biocatalysts via directed evolution. Curr Opin Chem Biol 25:55–64

Dickschat JS (2010) Quorum sensing and bacterial biofilms. Nat Prod Rep 27(3):343–69

Dong YH, Xu JL, Li XZ, Zhang LH (2000) AiiA, an enzyme that inactivates the acylhomoserine lactone quorum-sensing signal and attenuates the virulence of *Erwinia carotovora*. Proc Natl Acad Sci USA 97:3526–3531

Dong YH, Wang LH, Xu JL, Zhang HB, Zhang XF et al (2001) Quenching quorum-sensing-dependent bacterial infection by an N-acyl homoserine lactonase. Nature 411:813–817

Dong YJ, Bartlam M, Sun L, Zhou YF, Zhang ZP, Zhang CG et al (2005) Crystal structure of methyl parathion hydrolase from *Pseudomonas* sp. WBC-3. J Mol Biol 353:655–663

Draganov DI (2010) Lactonases with organophosphatase activity: structural and evolutionary perspectives. Chem Biol Interact 187(1):370–372

Dumas DP, Caldwell SR, Wild JR, Raushel FM (1989a) Purification and properties of the phosphotriesterase from *Pseudomonas diminuta*. J Biol Chem 264(33):19659–19665

Dumas DP, Wild JR, Raushel FM (1989b) Diisopropylfluorophosphate hydrolysis by a phosphotriesterase from *Pseudomonas diminuta*. Biotechnol Appl Biochem 11:235–243

Eddleston M, Eyer P, Worek F, Mohamed F, Senarathna L, von Meyer L, Juszczak E, Hittarage A, Azhar S, Dissanayake W, Sheriff MH, Szinicz L, Dawson AH, Buckley NA (2005) Differences between organophosphorus insecticides in human self-poisoning: a prospective cohort study. Lancet 366(9495):1452–1459

Elias M, Dupuy J, Merone L, Mandrich L, Porzio E, Moniot S, Rochu D, Lecomte C, Rossi M, Masson P, Manco G, Chabriere E (2008) Structural basis for natural lactonase and promiscuous phosphotriesterase activities. J Mol Biol 379:1017–1028

FAO (2014) Tools for the implementation of the international code of conduct on pesticide management. http://www.fao.org/agriculture/crops/thematic-sitemap/theme/pests/code/toolkits/en/. Accessed 12 May 2014

Febbraio F, Merone L, Cetrangolo GP, Rossi M, Nucci R, Manco G (2011) Thermostable esterase 2 from *Alicyclobacillus acidocaldarius* as biosensor for the detection of organophosphate pesticides. Anal Chem 83(5):1530–1536. doi:10.1021/ac102025z

Francic N, Bellino MG, Soler-Illia GJAA, Lobnik A (2014) Mesoporous titania thin films as efficient enzyme carriers for paraoxon determination/detoxification: effects of enzyme binding and pore hierarchy on the biocatalyst activity and reusability. Analyst 139:3127–3136

Gao Y, Truong YB, Cacioli P, Butler P, Kyratzis IL (2014) Bioremediation of pesticide contaminated water using an organophosphate degrading enzyme immobilized on nonwoven polyester textiles. Enzyme Microb Technol 54:38–44

Gavrilescu M (2005) Fate of pesticides in the environment and its bioremediation. Eng Life Sci 6:497–526

Gill I, Ballesteros A (2000) Bioencapsulation within synthetic polymers (part 2): non-sol–gel protein-polymer composites. Trends Biotechnol 18:469–479

Goldsmith M, Ashani Y, Simo Y, Ben-David M, Leader H, Silman I et al (2012) Evolved stereoselective hydrolases for broad-spectrum G-type nerve agent detoxification. Chem Biol 19(4): 456–466

Gotthard G, Hiblot J, Gonzalez D, Elias M, Chabriere E (2013) Structural and enzymatic characterization of the phosphotriesterase OPHC2 from *Pseudomonas pseudoalcaligenes*. PLoS One 8, e77995. doi:10.1371/journal.pone.0077995

Grimsley JK, Singh WP, Wild JR, Giletto A (2001) A novel, enzyme-based method for the wound-surface removal and decontamination of organophosphorous nerve agents. Bioact Fibers Polymers 742:35–49

Grube A, Donaldson D, Kiely T, Wu L (2011) Pesticides industry sales and usage 2006–2007 market estimates. US Environmental Protection Agency, Washington, DC. http://www.epa.gov/opp00001/pestsales/07pestsales/market_estimates2007.pdf

Gupta RC (2005) Toxicology of organophosphate and carbamate compounds. Academic, London

Gupta RD, Goldsmith M, Ashani Y, Simo Y, Mullokandov G, Bar H et al (2011) Directed evolution of hydrolases for prevention of G-type nerve agent intoxication. Nat Chem Biol 7(2):120–125

Harel M, Aharoni A, Gaidukov L, Brumshtein B, Khersonsky O, Meged R et al (2004) Structure and evolution of the serum paraoxonase family of detoxifying and anti-atherosclerotic enzymes. Nat Struct Mol Biol 11(5):412–419

Harper LL, McDaniel CS, Miller CE, Wild JR (1988) Dissimilar plasmids isolated from *Pseudomonas diminuta* MG and a *Flavobacterium* sp. (ATCC 27551) contain identical *opd* genes. Appl Environ Microbiol 54(10):2586–2589

Hawwa R, Aikens J, Turner RJ, Santarsiero BD, Mesecar AD (2009a) Structural basis for thermostability revealed through the identification and characterization of a highly thermostable phosphotriesterase-like lactonase from *Geobacillus stearothermophilus*. Arch Biochem Biophys 488:109–120

Hawwa R, Larsen SD, Ratia K, Mesecar AD (2009b) Structure-based and random mutagenesis approaches increase the organophosphate-degrading activity of a phosphotriesterase homologue from *Deinococcus radiodurans*. J Mol Biol 393:36–57

Hentzer M, Wu H, Andersen JB, Riedel K, Rasmussen TB et al (2003) Attenuation of *Pseudomonas aeruginosa* virulence by quorum sensing inhibitors. EMBO J 22:3803–3815

Hiblot J, Gotthard G, Chabriere E, Elias M (2012a) Characterisation of the organophosphate hydrolase catalytic activity *of Sso*Pox. Sci Rep 2:779

Hiblot J, Gotthard G, Chabriere E, Elias M (2012b) Structural and enzymatic characterization of the lactonase SisLac from *Sulfolobus islandicus*. PLoS ONE 7(10), e47028

Hiblot J, Gotthard G, Elias M, Chabriere E (2013) Differential active site loop conformations mediate promiscuous activities in the lactonase *Sso*Pox. PLoS ONE 8(9), e75272

Hiblot J, Bzdrenga J, Champion C, Chabriere E, Elias M (2015) Crystal structure of VmoLac, a tentative quorum quenching lactonase from the extremophilic crenarchaeon *Vulcanisaeta moutnovskia*. J Sci Rep 5:8372

Horne I, Sutherland TD, Harcourt RL, Russell RJ, Oakeshott JG (2002) Identification of an *opd* (organophosphate degradation) gene in an *Agrobacterium* isolate. Appl Environ Microbiol 68(7):3371–3376

Horne I, Qiu X, Russell RJ, Oakeshott JG (2003) The phosphotriesterase gene *opdA* in *Agrobacterium radiobacter* P230 is transposable. FEMS Microbiol Lett 222(1):1–8

Hraiech S, Hiblot J, Lafleur J, Lepidi H, Papazian L, Rolain JM, Raoult D, Elias M, Silby MW, Bzdrenga J, Bregeon F, Chabriere E (2014) Inhaled lactonase reduces Pseudomonas aeruginosa quorum sensing and mortality in rat pneumonia. PLoS ONE 9(10), e107125

Islam SM, Math RK, Cho KM, Lim WJ, Hong SY, Kim JM, Yun MG, Cho JJ, Yun HD (2010) Organophosphorus hydrolase (OpdB) of *Lactobacillus brevis* WCP902 from kimchi is able to degrade organophosphorus pesticides. J Agric Food Chem 58(9):5380–5386

Iyer R, Iken B (2015) Protein engineering of representative hydrolytic enzymes for remediation of organophosphates. Biochem Eng J 94:134–144

Jackson CJ, Foo JL, Kim HK, Carr PD, Liu JW, Salem G, Ollis DL (2008) In crystallo capture of a Michaelis complex and product-binding modes of a bacterial phosphotriesterase. J Mol Biol 375(5):1189–1196

Jeong YS, Choi JM, Kyeong HH, Choi JY, Kim EJ, Kim HS (2014) Rational design of organophosphorus hydrolase with high catalytic efficiency for detoxifying a V-type nerve agent. Biochem Biophys Res Commun 449(3):263–267

Jin F, Wang J, Shao H, Jin M (2010) Pesticide use and residue control in China. J Pestic Sci 35:138–142

Jones MB, Peterson SN, Benn R, Braisted JC, Jarrahi B et al (2010) Role of luxS in *Bacillus anthracis* growth and virulence factor expression. Virulence 1:72–83

Joshi KA, Prouza M, Kum M, Wang J, Tang J, Haddon R, Chen W, Mulchandani A (2006) V-type nerve agent detection using a carbon nanotube-based amperometric enzyme electrode. Anal Chem 78:331

Josse D, Lockridge O, Xie W, Bartels CF, Schopfer LM, Masson P (2001) The active site of human paraoxonase (PON1). J Appl Toxicol 21(S1):S7–S11

Kallnik V, Bunescu A, Sayer C, Bräsen C, Wohlgemuth R, Littlechild J, Siebers B (2014) Characterization of a phosphotriesterase-like lactonase from the hyperthermoacidophilic crenarchaeon *Vulcanisaeta moutnovskia*. J Biotechnol 190:11–17

Kawahara K, Tanaka A, Yoon J, Yokota A (2010) Reclassification of a parathion degrading *Flavobacterium* sp. ATCC 27551 as *Sphingobium fuliginis*. J Gen Appl Microbiol 56(3): 249–255

Khersonsky O, Roodveldt C, Tawfik DS (2006) Enzyme promiscuity: evolutionary and mechanistic aspects. Curr Opin Chem Biol 10:498–508

Kiss A, Virág D (2009) Photostability and photodegradation pathways of distinctive pesticides. J Environ Qual 38(1):157–163

Lasram MM, Dhouib IB, Annabi A, El Fazaa S, Gharbi N (2014) A review on the molecular mechanisms involved in insulin resistance induced by organophosphorus pesticides. Toxicology 322:1–13

LeJeune KE, Mesiano AJ, Bower SB, Grimsley JK, Wild JR, Russell AJ (1997) Dramatically stabilised phosphotriesterase polymers for nerve agent degradation. Biotechnol Bioeng 54:105–114

LeJeune KE, Wild JR, Russell AJ (1998) Nerve agents degraded by enzymatic foams. Nature 395(6697):27–28

Li C, Zhu Y, Benz I, Schmidt MA, Chen W, Mulchandani A, Qiao C (2008) Presentation of functional organophosphorus hydrolase fusions on the surface of *Escherichia coli* by the AIDA-I autotransporter pathway. Biotechnol Bioeng 99:485–490

Liu H, Zhang JJ, Wang SJ, Zhang XE, Zhou NY (2005) Plasmid-borne catabolism of methyl parathion and p-nitrophenol in *Pseudomonas* sp. strain WBC-3. Biochem Biophys Res Commun 334(4):1107–1114

Ma F, Wang Y, Zhang Y, Xiong N, Yang B et al (2009) Heterologous expression of human paraoxonases in *Pseudomonas aeruginosa* inhibits biofilm formation and decreases antibiotic resistance. Appl Microbiol Biotechnol 83:135–141

Mackness B, Durrington PN, Mackness MI (2000) The paraoxonase gene family and coronary heart disease. Curr Opin Lipidol 13:357–363

Madej T, Addess KJ, Fong JH, Geer LY, Geer RC, Lanczycki CJ et al (2012) MMDB: 3D structures and macromolecular interactions. Nucleic Acids Res 40(Database issue): D461–D464

Mandrich L, Manco G (2009) Evolution in the amidohydrolase superfamily: substrate-assisted gain of function in the E183K mutant of a phosphotriesterase-like metal-carboxylesterase. Biochemistry 48(24):5602–5612

Mandrich L, Porzio E, Merone L, Febbraio F, Nucci R, Manco G (2011) Exploring thermostable quorum quenching lactonases to counteract bacterial infections in cystic fibrosis. In: Mendez-Vilas A (ed) Science and technology against microbial pathogens. Research, development and evaluation. Proceedings of the international conference on antimicrobial research (ICAR2010). World Scientific, pp 150–154

Mandrich L, Di Gennaro S, Palma A, Manco G (2013) A further biochemical characterization of DrPLL the thermophilic lactonase from *Deinococcus radiodurans*. Protein Pept Lett 20(1): 36–44

Marrazza G (2014) Piezoelectric biosensors for organophosphate and carbamate pesticides: a review. Biosensors 4:301–317

Masson P, Rochu D (2009) Catalytic bioscavengers against toxic esters, an alternative approach for prophylaxis and treatments of poisonings. Acta Nat 1(1):68–79

Meier MM, Rajendran C, Malisi C, Fox NG, Xu C, Schlee S, Barondeau DP, Hocker B, Sterner R, Raushel FM (2013) Molecular engineering of organophosphate hydrolysis activity from a weak promiscuous lactonase template. J Am Chem Soc 135:11670–11677

Merone L, Mandrich L, Rossi M, Manco G (2005) A thermostable phosphotriesterase from the archaeon *Sulfolobus solfataricus*: cloning, overexpression and properties. Extremophiles 9(4):297–305

Merone L, Mandrich L, Porzio E, Rossi M, Müller S, Reiter G, Worek F, Manco G (2010) Improving the promiscuous nerve agent hydrolase activity of a thermostable archaeal lactonase. Bioresour Technol 101(23):9204–9212

Munnecke DM (1979) Hydrolysis of organophosphate insecticides by an immobilized-enzyme system. Biotechnol Bioeng 21:2247–2261

Nachon F, Brazzolotto X, Trovaslet M, Masson P (2013) Progress in the development of enzyme-based nerve agent bioscavengers. Chem Biol Interact 206:536–544

Naqvi T, Warden AC, French N, Sugrue E, Carr PD, Jackson CJ et al (2014) A 5000-fold increase in the specificity of a bacterial phosphotriesterase for malathion through combinatorial active site mutagenesis. PLoS ONE 9(4), e94177

Ng FSW, Wright DM, Seah SYK (2011) Characterization of a phosphotriesterase-like lactonase from *Sulfolobus solfataricus* and its immobilization for disruption of quorum sensing. Appl Environ Microbiol 77(4):1181–1186

Ojha A, Gupta Y (2014) Evaluation of genotoxic potential of commonly used organophosphate pesticides in peripheral blood lymphocytes of rats. Hum Exp Toxicol pii, 0960327114537534. [Epub ahead of print]

Ortiz-Hernández ML, Sánchez-Salinas E, Castrejón Godínez ML, Dantan González E, Popoca Ursino EC (2013) Mechanisms and strategies for pesticide biodegradation: opportunity for waste, soils and water cleaning. Rev Int Contam Ambie 29:85–104

Pinjari AB, Novikov B, Rezenom YH, Russell DH, Wales ME, Siddavattam D (2012) Mineralization of acephate, a recalcitrant organophosphate insecticide is initiated by a *Pseudomonad* in environmental samples. PLoS One 7, e31963

Pinjari AB, Pandey JP, Kamireddy S, Siddavattam D (2013) Expression and subcellular localization of organophosphate hydrolase in acephate-degrading *Pseudomonas* sp. strain Ind01 and its use as a potential biocatalyst for elimination of organophosphate insecticides A.B. Lett Appl Microbiol 57:63–68

Popat R, Crusz SA, Diggle SP (2008) The social behaviours of bacterial pathogens. Br Med Bull 87:63–75

Porzio E, Merone L, Mandrich L, Rossi M, Manco G (2007) A new phosphotriesterase from *Sulfolobus acidocaldarius* and its comparison with the homologue from *Sulfolobus solfataricus*. Biochimie 89(5):625–636

Porzio E, Di Gennaro S, Palma A, Manco G (2013) Mn^{2+} modulates the kinetic properties of an archaeal member of the PLL family. Chem Biol Interact 203(1):251–256

Ragnarsdottir KV (2000) Environmental fate and toxicology of organophosphate pesticides. J Geol Soc 157:859–876

Raushel FM (2002) Bacterial detoxification of organophosphate nerve agents. Curr Opin Microbiol 5(3):288–295

Raushel FM, Holden HM (2000) Phosphotriesterase: an enzyme in search of its natural substrate. Adv Enzymol Relat Areas Mol Biol 74:51–93

Reimmann C, Ginet N, Michel L, Keel C, Michaux P et al (2002) Genetically programmed autoinducer destruction reduces virulence gene expression and swarming motility in *Pseudomonas aeruginosa* PAO1. Microbiology 148:923–932

Richins RD, Kaneva I, Mulchandani A, Chen W (1997) Biodegradation of organophosphorus pesticides by surface-expressed organophosphorus hydrolase. Nat Biotechnol 15:984–987

Rochu D, Viguie N, Renault F, Crouzier D, Froment MT, Masson P (2004) Contribution of the active site metal cation to catalytic activity and to conformational stability of phosphotriesterase: a thermo- and pH-dependence study. Biochem J 380:627–633
Rochu D, Chabriere E, Masson P (2007) Human paraoxonase: a promising approach for pre-treatment and therapy of organophosphorus poisoning. Toxicology 233(1):47–59
Roodveldt C, Tawfik DS (2005) Shared promiscuous activities and evolutionary features in various members of amidohydrolase superfamily. Biochemistry 44:12728–12736
Salazar-Arredondo E, Solis-Heredia MJ, Rojas-Garcia E, Ochoa IH, Quintanilla Vega B (2008) Sperm chromatin alteration and DNA damage by methyl-parathion, chlorpyrifos and diazinon and their oxon metabolites in human spermatozoa. Reprod Toxicol 25:455–460
Segers P, Vancanneyt M, Pot B, Torck U, Hoste B, Dewettinck D et al (1994) Classification of *Pseudomonas diminuta* Leifson and Hugh 1954 and Pseudomonas vesicularis Büsing, Döll, and Freytag 1953 in *Brevundimonas* gen. nov. as *Brevundimonas diminuta* comb. nov. and *Brevundimonas vesicularis* comb. nov., respectively. Int J Syst Bacteriol 44(3):499–510
Seibert CM, Raushel FM (2005) Structural and catalytic diversity within the amidohydrolase superfamily. Biochemistry 44(17):6383–6391
Shimazu M, Mulchandani A, Chen W (2001) Simultaneous degradation of organophosphorus pesticides and p-nitrophenol by a genetically engineered *Moraxella* sp. with surface-expressed organophosphorus hydrolase. Biotechnol Bioeng 76:318–324
Singh BK (2009) Organophosphorus-degrading bacteria: ecology and industrial applications. Nat Rev Microbiol 7(2):156–164
Singh BK, Walker A (2006) Microbial degradation of organophosphorus compounds. FEMS Microbiol Rev 30(3):428–471
Spain JC, Gibson DT (1991) Pathway for biodegradation of p-nitrophenol in a *Moraxella* sp. Appl Environ Microbiol 57:812–819
Tanga X, Lianga B, Yi T, Manco G, Palchetti I, Liua A (2014) Cell surface display of organophosphorus hydrolase for sensitive spectrophotometric detection of p-nitrophenol substituted organophosphates. Enzyme Microb Technol 55:107–112
Tehei M, Madern D, Franzetti B, Zaccai G (2005) Neutron scattering reveals the dynamic basis of protein adaptation to extreme temperature. J Biol Chem 280:40974–40979
Theriot CM, Grunden AM (2011) Hydrolysis of organophosphorus compounds by microbial enzymes. Appl Microbiol Biotechnol 89(1):35–43
Theriot CM, Du X, Tove SR, Grunden AM (2010a) Improving the catalytic activity of hyperthermophilic *Pyrococcus* prolidases for detoxification of organophosphorus nerve agents over a broad range of temperatures. Appl Microbiol Biotechnol 87(5):1715–1726
Theriot CM, Tove SR, Grunden AM (2010b) Characterization of two proline dipeptidases (prolidases) from the hyperthermophilic archaeon *Pyrococcus horikoshii*. Appl Microbiol Biotechnol 86(1):177–188
Theriot CM, Semcer RL, Shah SS, Grunden AM (2011) Improving the catalytic activity of hyperthermophilic Pyrococcus horikoshii prolidase for detoxification of organophosphorus nerve agents over a broad range of temperatures. Archaea 2011:5651271–5651279
Tiryaki O, Temur C (2010) The fate of pesticide in the environment. J Biol Environ Sci 4(10):29–38
Trovaslet-Leroy M, Musilova L, Renault F, Brazzolotto X, Misik J, Novotny L, Froment MT, Gillon E, Loiodice M, Verdier L, Masson P, Rochu D, Jun D, Nachon F (2011) Organophosphate hydrolases as catalytic bioscavengers of organophosphorus nerve agents. Toxicol Lett 206:14–23
Tsai PC, Bigley A, Li Y, Ghanem E, Cadieux CL, Kasten SA et al (2010a) Stereoselective hydrolysis of organophosphate nerve agents by the bacterial phosphotriesterase. Biochemistry 49(37):7978–7987
Tsai PC, Fan Y, Kim J, Yang L, Almo SC, Gao YQ, Raushel FM (2010b) Structural determinants for the stereoselective hydrolysis of chiral substrates by phosphotriesterase. Biochemistry 49(37):7988–7997
Tsai PC, Fox N, Bigley AN, Harvey SP, Barondeau DP, Raushel FM (2012) Enzymes for the homeland defense: optimizing phosphotriesterase for the hydrolysis of organophosphate nerve agents. Biochemistry 51(32):6463–6475

Vanhooke JL, Benning MM, Raushel FM, Holden HM (1996) Three dimensional structure of the zinc containing phosphotriesterase with the bound substrate analog diethyl 4-methylbenzylphosphonate. Biochemistry 35:6020–6025

Vyas NK, Nickitenko A, Rastogi VK, Shah SS, Quiocho FA (2010) Structural insights into the dual activities of the nerve agent degrading organophosphate anhydrolase/prolidase. Biochemistry 49(3):547–559

Wanekaya AK, Chenb W, Mulchandani A (2008) Recent biosensing developments in environmental security. J Environ Monit 10:703–712

Worek F, Seeger T, Goldsmith M, Ashani Y, Leader H, Sussman JS, Tawfik D, Thiermann H, Wille T (2014) Efficacy of the rePON1 mutant IIG1 to prevent cyclosarin toxicity in vivo and to detoxify structurally different nerve agents in vitro. Arch Toxicol 88(6):1257–1266

Xiang DF, Kolb P, Fedorov AA, Meier MM, Fedorov LV et al (2009) Functional annotation and three-dimensional structure of Dr0930 from *Deinococcus radiodurans*, a close relative of phosphotriesterase in the amidohydrolase superfamily. Biochemistry 48:2237–2247

Yang C, Liu N, Guo X, Qiao C (2006) Cloning of *mpd* gene from a chlorpyrifos-degrading bacterium and use of this strain in bioremediation of contaminated soil. FEMS Microbiol Lett 265:118–125

Yang C, Liu R, Yuan Y, Liu J, Cao X, Qiao C, Song C (2013) Construction of a green fluorescent protein (GFP)-marked multifunctional pesticide-degrading bacterium for simultaneous degradation of organophosphates and γ-hexachlorocyclohexane. J Agric Food Chem 61(6): 1328–1334

Zhang R, Cui ZL, Jiang J, He J, Gu X, Li S (2005) Diversity of organophosphorus pesticide-degrading bacteria in a polluted soil and conservation of their organophosphorus hydrolase genes. Can J Microbiol 51(4):337–343

Zhang R, Cui ZL, Zhang X, Jiang J, Gu JD, Li SP (2006) Cloning of the organophosphorus pesticide hydrolase gene clusters of seven degradative bacteria isolated from a methyl parathion contaminated site and evidence of their horizontal gene transfer. Biodegradation 17(5): 465–472

Zhang Y, An J, Ye W, Yang G, Qian ZG, Chen HF, Cui L, Feng Y (2012) Enhancing the promiscuous phosphotriesterase activity of a thermostable lactonase (GkaP) for the efficient degradation of organophosphate pesticides. Appl Environ Microbiol 78(18):6647–6655

Zhang Y, An J, Yang GY, Bai A, Zheng B, Lou Z, Wu G, Ye W, Chen HF, Feng Y, Manco G (2015) Active site loop conformation regulates promiscuous activity in a lactonase from *Geobacillus kaustophilus* HTA426. PLoS ONE 10(2), e0115130

Zhongli C, Shunpeng L, Guoping F (2001) Isolation of methyl parathion-degrading strain M6 and cloning of the methyl parathion hydrolase gene. Appl Environ Microbiol 67(10):4922–4925

Chapter 17
α-Amylases from *Archaea*: Sequences, Structures and Evolution

Štefan Janeček

1 Introduction

α-Amylase (EC 3.2.1.1) together with β-amylase and glucoamylase belong to best known amylolytic enzymes. The α-amylase catalyses in an endo-fashion the hydrolysis of α-1,4-glucosidic linkages in starch and related α-glucans (Janecek et al. 2014). In contrast to both exo-acting β-amylase and glucoamylase catalysing predominantly the liberation of β-maltose and β-glucose, respectively, from the non-reducing end of the same substrates (Ray and Nanda 1996; Sauer et al. 2000; Kumar and Satyanarayana 2009; Marin-Navarro and Polaina 2011), the products of the α-amylase cleavage ranging from longer maltooligosaccharides to α-limit dextrins are more variable (Vihinen and Mantsala 1989; MacGregor et al. 2001). Unlike β-amylases and glucoamylases that employ the inverting mechanism of the α-glucosidic bonds hydrolysis, the result of the α-amylase activity is the retention of the α-anomeric configuration of the liberated hydroxyl group in all formed products (McCarter and Withers 1994; Davies and Henrissat 1995). These three amylolytic enzymes are thus different from the evolutionary point of view (Janecek 1994a) since the obviously related function of α-glucosidic bond cleavage is brought about by completely different amino acid sequences of α-amylases, β-amylases and glucoamylases that use their own catalytic machineries (Pujadas et al. 1996; Coutinho and Reilly 1997; Janecek 1997).

Depending on the source of the enzyme, an α-amylase may be able, in addition to hydrolysis, to catalyse to a very less extent also a transglycosylation reaction, as

Š. Janeček (✉)
Laboratory of Protein Evolution, Institute of Molecular Biology, Slovak Academy of Sciences, Dubravska cesta 2, SK-84551 Bratislava, Slovakia

Faculty of Natural Sciences, Department of Biology, University of SS Cyril and Methodius, Nam. J. Herdu 2, SK-91701 Trnava, Slovakia
e-mail: Stefan.Janecek@savba.sk; Stefan.Janecek@ucm.sk

P.H. Rampelotto (ed.), *Biotechnology of Extremophiles*, Grand Challenges in Biology and Biotechnology 1, DOI 10.1007/978-3-319-13521-2_17

for example some bacterial, fungal and even mammalian α-amylases do (Brzozowski and Davies 1997; Aghajari et al. 2002; Ramasubbu et al. 2003; Li et al. 2005). It is therefore very important to pay a relevant attention to a serious biochemical characterisation supported by a precise bioinformatics knowledge when assigning the α-amylase specificity to an amylolytic enzyme displaying a starch-hydrolysing activity (Janecek et al. 2014). The detailed *in silico* analysis may sometimes be crucial (Janecek et al. 1995) since, for example, cyclodextrin glucanotransferase with its main transglycosylating activity and only a side ability to catalyse hydrolysis (i.e. to a less extent) is very similar to typical α-amylases (Leemhuis et al. 2003a, b).

With regard to their taxonomic origin, there is also a substantial difference between both β-amylases and glucoamylases in comparison with the α-amylases. β-Amylases are best known to originate from plants, much less from *Bacteria* and eventually also from fungi (Ray and Nanda 1996; Kim et al. 2000, 2001a). Glucoamylases are well-known from *Bacteria* and fungi and also from *Archaea* (Marin-Navarro and Polaina 2011). This means that there are neither β-amylases nor glucoamylases of animal origin, the former being not found in *Archaea* and the latter in plants, too. On the other hand, the α-amylases have been identified throughout all the three domains of life, i.e. *Bacteria*, *Archaea* and *Eucarya*, so that for a long time these enzymes have been well-known from microbes, plant and animals (Janecek 1994b; Janecek et al. 1999). It is worth mentioning, however, that there seem to be some prokaryotes, especially among the *Archaea*, that obviously do not contain any glycoside hydrolase in their genome, i.e. they seem to exist without any α-amylase, too (Coutinho and Henrissat 1999).

Nevertheless the huge number and a great variety of possible producers together with a high industrial demand (Leveque et al. 2000b; Pandey et al. 2000; Bertoldo and Antranikian 2002; van der Maarel et al. 2002; Gupta et al. 2003; Sharma and Satyanarayana 2013) really make the α-amylases one of the best studied and most utilised enzyme. The *Archaea* with their remarkable environmental niche they often occupy may offer in this respect α-amylases with unique and highly desired properties.

1.1 CAZy Classification System

The above-mentioned functional similarities, but clear sequence-structural, mechanistic and thus also evolutionary differences between α-amylases, β-amylases and glucoamylases are unambiguously reflected in the Carbohydrate-Active EnZyme (CAZy) database (Lombard et al. 2014) where these enzymes define their own families. The CAZy classification system, originally developed as a sequence-based classification of glycoside hydrolases (GH) (Henrissat 1991), currently represents a well-accepted and reputable database of all enzymes that catalyse the formation, degradation and modification of glycosidic linkages (http://www.cazy.org/). Whereas β-amylases and glucoamylases formed their monospecific families GH14 and GH15, respectively, each counted only five sequences in 1991, the α-amylase was known from the beginning as the main representative of a larger polyspecific, the so-called α-amylase family GH13 with 40 members covering the specificities of isoamylase, pullulanase, amylopullulanase, neopullulanase, cyclodextrin glucanotransferase, α-glucosidase, dextran glucosidase and some exo-acting amylases, in addition to α-amylase (Henrissat 1991). The existence of such a polyspecific α-amylase family was supported by both experimental and *in silico* studies (Svensson 1988; MacGregor and Svensson 1989; Jespersen et al. 1991; Takata et al. 1992). Today (April 2015), the α-amylase family GH13 with about 24,000 sequences and more than 30 different enzyme specificities belongs to the largest GH families among the all 133 ones established until now (Janecek et al. 2014; Lombard et al. 2014).

Since the CAZy system has been primarily based on comparison of amino acid sequences, it should not be much surprised that the same enzyme specificity can be found in more than one family. In other words, the same enzyme activity can be executed by different proteins that do not share mutual sequence similarities. And this is also the case of α-amylase, that within the CAZy database, in addition to the originally defined family GH13, has been classified also in the families GH57 and GH119 (Lombard et al. 2014). The α-amylases from *Archaea* are present only in the main α-amylase family GH13 and the second smaller α-amylase family GH57 (Janecek et al. 2014).

The main α-amylase family is the family GH13 (Janecek et al. 2014). It forms, together with circularly permuted glucan sucrases from the family GH70 (MacGregor et al. 1996; Vujicic-Zagar et al. 2010) and 4-α-glucanotransferases from the family GH77 (Przylas et al. 2000; Godany et al. 2008), the clan GH-H (MacGregor et al. 2001). Currently the specificity of α-amylase is assigned to GH13 subfamilies GH13_1, 5, 6, 7, 15, 24, 27, 28, 36, 37 (e.g., Da Lage et al. 2004, 2013; van der Kaaij et al. 2007; Hostinova et al. 2010; Lei et al. 2012; Majzlova et al. 2013; Puspasari et al. 2013) and, possibly, a few more not yet defined (Janecek et al. 2014). The members of the α-amylase family GH13 employ a retaining reaction mechanism (MacGregor et al. 2001), share 7 conserved sequence regions (CSRs) (Janecek 2002) and adopt a $(\beta/\alpha)_8$-barrel (i.e. the so-called TIM-barrel) domain with the GH13 catalytic machinery (Matsuura et al. 1984; Uitdehaag et al. 1999): aspartic acid as catalytic nucleophile, glutamic acid as proton donor and aspartic acid as transition state stabilizer at strands β4, β5 and β7, respectively.

The α-amylases from the family GH57 (Janecek 2005), which at present counts about 1100 members (Lombard et al. 2014) and less than 10 enzyme specificities (Janecek et al. 2014), use the same retaining mechanism (Palomo et al. 2011), but they have their own 5 CSRs (Zona et al. 2004) and different catalytic machinery (Imamura et al. 2003): glutamic acid as catalytic nucleophile and aspartic acid as proton donor at strands β4 and β7, respectively, within a $(\beta/\alpha)_7$-barrel (i.e. the so-called incomplete TIM-barrel) fold. Interestingly, based on an in silico analysis and structure prediction (Janecek and Kuchtova 2012) the family GH119 α-amylases (only 10 members) may share all above-mentioned features with the family GH57, but it currently does not contain any archaeal sequences (Lombard et al. 2014).

2 Archaeal α-Amylases from the Family GH13

First α-amylases and/or amylolytic enzymes from *Archaea* were isolated from hyperthermophilic strains of *Pyrococcus* and described at the beginning of 1990s (Koch et al. 1990, 1991; Laderman et al. 1993b). Although the first primary structure of an archaeal "amylase" was announced in 1993 from *Pyrococcus furiosus* (Laderman et al. 1993a), its sequence was obviously unrelated to typical α-amylases known at that time that were assigned to the family GH13 (Janecek 1998). Indeed, it was used as a fundamental member of the second α-amylase family GH57 (Henrissat and Bairoch 1996) and currently is rather considered to be a 4-α-glucanotransferase (Laderman et al. 1993a; Blesak and Janecek 2012). The first determined amino acid sequence of a real family GH13 archaeal α-amylase that has also been biochemically characterized was therefore the one from *Pyrococcus* sp. KOD1 (Tachibana et al. 1996); the strain was later re-classified as *Thermococcus kodakaraensis* (Atomi et al. 2004).

When more α-amylases from *Archaea* became available, their remarkable close evolutionary relatedness with plant counterparts was independently revealed (Janecek et al. 1999; Jones et al. 1999). Prior to that, the α-amylases from plants were known to occupy in the evolutionary tree their own branch adjacent to the branch leading to liquefying α-amylases from bacilli (Janecek 1994b). The three groups of mutually closely related α-amylases from *Archaea*, plants and bacteria (these should be either of a liquefying type or produced intracellularly) were assigned their individual α-amylase subfamilies GH13_7, GH13_6 and GH13_5, respectively, when the entire α-amylase family GH13 was officially divided into subfamilies (Stam et al. 2006). At present, there is another group of α-amylases, distinct from the one of the subfamily GH13_7, produced by halophilic *Archaea* that, however, has not been assigned a GH13 subfamily as yet (Lombard et al. 2014). Unfortunately, only a few proteins of that group have already been expressed and partly characterised as potential α-amylases (Kobayashi et al. 1994; Hutcheon et al. 2005).

In addition to α-amylases, there have been sequenced and experimentally described also other archaeal amylolytic enzymes in the family GH13 (Lombard et al. 2014), for example, cyclodextrin glucanotransferase (Rashid et al. 2002; Lee et al. 2006; Bautista et al. 2012), pullulan hydrolase type III (Niehaus et al. 2000; Ahmad et al. 2014), amylopullulanase (Li et al. 2013), cyclomaltodextrinase/maltogenic amylase (Hashimoto et al. 2001; Li et al. 2010; Jung et al. 2012; Park et al. 2013; Sun et al. 2015), maltooligosyltrehalose trehalohydrolase (Kobayashi et al. 1996; Feese et al. 2000), maltooligosyltrehalose synthase (Maruta et al. 1996; Kobayashi et al. 2003) and trehalose synthase (Chen et al. 2006). For some of them also the GH13 subfamily number has already been assigned, such as for cyclodextrin glucanotransferase, cyclomaltodextrinase/maltogenic amylase, maltooligosyltrehalose trehalohydrolase, maltooligosyltrehalose synthase and trehalose synthase – the subfamilies GH13_2, GH13_20, GH13_10, GH13_26 and GH13_16, respectively, but for some others, this information has still not been released (Lombard et al. 2014).

2.1 The Archaeal α-Amylase Subfamily GH13_7

The archaeal α-amylases that belong to the subfamily GH13_7 are produced purely just by thermococci of Euryarchaeota (Tachibana et al. 1996; Dong et al. 1997; Jorgensen et al. 1997; Frillingos et al. 2000; Leveque et al. 2000a; Lim et al. 2007; Wang et al. 2008), namely the genera of *Pyrococcus*, *Thermococcus* and *Palaeococcus* (Lombard et al. 2014). They all should obey the three-domain structural organization typical for the family GH13 (Janecek et al. 2014), i.e. the catalytic TIM-barrel domain A with domain B usually of irregular secondary structure protruding as a long loop out of the barrel between the third β-strand and the third α-helix, and with the domain C adopting typically structure of an antiparallel β-sandwich that succeeds the catalytic barrel (Linden et al. 2003). Other family GH13 enzyme specificities and even the α-amylases from sources other than *Archaea* may contain additional domains (MacGregor et al. 2001; Janecek et al. 2014).

The α-amylases produced by hyperthermophilic archaeons, classified into the subfamily GH13_7, contain several exclusive sequence features that were originally identified by Janecek et al. (1999). They are present almost only in their plant counterparts from the subfamily GH13_6 (Fig. 17.1). These features are as follows

No.	Subfamily	Source	CSR VI β2	CSR I β3	CSR V loop3	CSR II β4	CSR III β5	CSR IV β7	CSR VII β8
1	GH13_7	*Palaeococcus_pacificus*_A0A075LR97	GITSIWLPP	GLEVYADIVINH	YPDIC	GWRFDYVKG	YAVGEYWD	FVANHD	GQPTIFYRD
2		*Pyrococcus_furiosus*_O08452	GISAIWLPP	GIKVIADVVINH	FPDIC	GWRFDYVKG	WAVGEYWD	FVANHD	GQPVIFYRD
3		*Pyrococcus_woesei*_Q7LYT7	GISAIWLPP	GIKVIADVVINH	FPDIC	GWRFDYVKG	WAVGEYWD	FVANHD	GQPVIFYRD
4		*Thermococcus_hydrothermalis*_O93647	GISAIWIPP	NMKVIADIVINH	YPDIC	AWRFDYVKG	WAVGEYWD	FVANHD	GQPAIFYRD
5		*Thermococcus_kodakaraensis*_O33476	GISAIWIPP	GIKVIADIVINH	FPDIA	AWRFDYVKG	WAVGEYWD	FVANHD	GQPVIFYRD
6		*Thermococcus_onnurineus*_B6YUG5	GIAAIWLPP	NMKVVADIVINH	FPDID	AWRFDYVKG	WAVGEYWD	FVANHD	GQPMIFYRD
7		*Thermococcus_sp._Rt3*_O50200	GISAIWIPP	GIKVIADIVINH	FPDIA	AWRFDYVKG	LAVGEYWD	FVANHD	GQPVIFYRD
8	GH13_7	*Capnocytophaga_ochracea*_C7M7T3	GVDRIWLPP	GLQVIADIVIGH	EQDLS	GWRFDYVKS	FAVGELWD	FTANHD	GYPCIFYSD
9		*Croceibacter_atlanticus*_A3U4D8	GIDAIWLPP	GISVIADIVINH	FPDLC	GWRFDYVKG	FSVGEYWD	FVANHD	GYPTLFYKD
10		*Flavobacterium_johnsoniae*_A5FII4	GIGSIWLPP	NIKVYADIVINH	FPDLC	GWRFDYVKG	FSVGELWD	FVTNHD	GYPTIFYRD
11		*Gramella_forsetii*_A0M3B1	GVDRIWLPV	DLEVIADIVINH	ETNLD	GWRFDYVKG	FSVGENFD	FTANHD	GYPTIFYSD
12		*Kordia_algicida*_A9E1U8	GIDAIWLPP	RLSVIADIVLNH	FADLS	GWRFDYVKG	FSVGEYWD	FVTNHD	GYPTIFYRD
13		*Sinomicrobium_sp._5DNS001*_L7Y1I6	GVDRIWLPV	GLEVIADIVLNH	EQDLC	GWRFDYVKG	FSVGEYWD	FVANHD	GYPTVFYGD
14	GH13_6	*Micromonas_pusilla*_C1MR05	GFTQIWLPP	NILPVLDAVLNH	SPNID	GIRFDFSKG	FAVGEFWD	FIDNHD	GTPSVFWDH
15		Apple_Q5BLY0	GFTSAWLPP	KVRAMADIVINH	VPNID	DFRFDFARG	FSVGEYWD	FLDNHD	GIPTVFYDH
16		Barley_low_pI_isozyme_P00693	GVTHVWLPP	GVQAIADIVINH	APDID	AWRLDFARG	LAVAEVWD	FVDNHD	GIPCIFYDH
17		Bean_Q9ZP43	GITHVWLPP	GIKCLADIVINH	APDID	GWRFDFVKG	FAVGEKWD	FIDNHD	GTPSIFYDH
18		Kiwi_Q5BLY1	GFTVVWLPP	GIRVLGDVVLNH	APNID	GWRLDFVRG	FAVGEYWD	FIENHD	GTPAVFYDH
19		Maize_Q41770	GATHVWLPP	GVKCVADVVINH	APDID	GWRLDFAKG	FVVAEIWS	FVDNHD	GTPCIFYDH
20		Potato_Q41442	GFTTAWLPP	KVRAMADIVINH	VPNID	DFRFDFAKG	FAVGEYWD	FIDNHD	GIPSVFFDH
21	GH13_6	*Microscilla_marina*_A1ZMR5	GFSMVWLPP	NIKPIADIVVNH	GRDLD	GWRYDYVKG	FSVGELWP	FVDNHD	GIPCVYWPH
22		*Saccharophagus_degradans*_Q21NA2	GATHVWFAP	GIDSVADIVINH	ARDID	GLRYDYSKG	FCVGEVWT	FVDNHD	GIPTVYWAH
23	GH13_1	*Aspergillus_oryzae*_P0C1B3	GFTAIWITP	GMYLMVDVVANH	LPDLD	GLRIDTVKH	YCIGEVLD	FVENHD	GIPIIYAGQ
24		*Saccharomycopsis_fibuligera*_D4P4Y7	GFTAIWISP	DMLLMVDIVTNH	LPDLR	GLRIDSAKH	YSVGEVFQ	FVENHD	GIPVIYYGQ
25	GH13_5	*Bacillus_licheniformis*_P06278	GITAVWIPP	DINVYGDVVINH	YADID	GFRLDAVKH	FTVAEYWQ	FVDNHD	GYPQVFYGD
26		*Histoplasma_capsulatum*_A0T074	GVTSILLPP	EIRIIWDTVLNH	FSNLD	GLRLDAAKH	LLVAEYWK	FVMNHD	GYPCLFYGD
27	GH13_15	Fruit_fly_P08144	GYAGVQVSP	GVRTYVDVVFNH	LRDLN	GFRVDAAKH	YIVQEVID	FVDNHD	GTPRVMSSF
28		Meal_worm_P56634	GFGGVQISP	GVRIYVDAVINH	LRDLN	GFRVDAAKH	FIYQEVID	FVDNHD	GTTRIMSSF
29	GH13_24	Shrimp_Q26193	GFAGVQVSP	GVRIYVDAVINH	LNDLN	GFRIDASKH	FIFQEVID	FIDNHD	GYTRVMSSY
30		Human_pancreas_P04746	GFGGVQVSP	GVRIYVDAVINH	LLDLA	GFRLDASKH	FIYQEVID	FVDNHD	GFTRVMSSY
31	GH13_27	*Aeromonas_hydrophila*_P22630	GYKQVLISP	GIAVYADVVLNH	LPDLD	GFRVDAVKH	HVFGEVIT	FAITHD	GSPLVYSDH
32		*Xanthomonas_campestris*_Q56791	GYRKVLVAP	GVETYADVVFNH	LPDLL	GFRVDAAKH	YVFGEVIT	FAVTHD	GVPMVYTDN
33	GH13_28	*Bacillus_subtilis*_P00691	GYTAIQTSP	GIKVIVDAVINH	LYDWN	GFRFDAAKH	FQYGEILQ	WVESHD	STPLFFSRP
34		*Lactobacillus_amylovorus*_Q48502	GYTAVQTSP	NIRIIVDATLND	FYDWN	GFRYDAATH	FQYGEVLQ	WVESHD	SVPLFFDRP
35	GH13_32	*Streptomyces_limosus*_P09794	GYGYVQVSP	GVKVVADSVINH	LADLD	GFRIDAAKH	YWKQEAIH	FVDNHD	GSPDVHSGY
36		*Thermomonospora_curvata*_P29750	GFGAVQVSP	GVKIYVDAVINH	LADLK	GFRIDAAKH	YIFQEVIA	FVVNHD	GTPKVMSSY
37	GH13_36	*Anaerobranca_gottschalki*_Q5I942	GVNGIWLTP	GIKVIIDLVINH	MPDLN	GFRLDAAKH	YLVGEIWD	FLSNHD	GDPYIFAGE
38		*Halothermothrix_orenii*_Q8GPL8	GVNGIWLMP	GIKVIIDLPINH	MPDLN	GFRLDGAMH	YLVGEVWD	FLTNHD	GNPFIYYGE
39	GH13_37	*Photobacterium_profundum*_Q6LIA8	GMNAVWLTP	GLYVFFDGVFGH	-----	GWRLDQAYQ	YMVAEIWN	MLGNHD	GPITLYYGD
40		Uncultured_bacterium_D9MZ14	GMNAIWLTP	GLYVFFDGVFGH	-----	GWRLDQAYQ	YMVAEIWN	MLGNHD	GPITLYYGE
41	GH13_??	*Halalkalicoccus_jeotgali*_D8J7H2	GIDVVWLTP	DIRLIMDLVANH	QPDLN	GFRLDAIAH	LTVGEVGT	FLSNHD	GTPFVYQGE
42	GH13_??	*Haloarcula_hispanica*_Q4A3E0	GYDAIQVPP	DLDVVADAVINH	LKDLK	GIRWDAAKH	WTVGEVLD	FVSNHD	GYPRVYSNR
43	GH13_??	*Haloarcula_japonica*_L8B068	GVDAIWLTP	GFKVLFDLVCNH	IANFN	GFRCDMAWA	LLLDETIP	YAENHD	GAPLLYAGQ
44	GH13_??	*Haloarcula_marismortui*_Q5UZY3	GVDVVWLTP	GIRVVFDLVINH	IPNLN	GFRCDVAWG	LLLDETVP	YVDNHD	GTPMVYAGQ
45	GH13_??	*Natronococcus_amylolyticus*_Q60224	GVSAIWIPQ	DIDVIVDIVLNH	LPSMD	GLRIDAAAH	WRVGEVWD	FVQNHD	GMPMLYRGG

Fig. 17.1 Conserved sequence regions of the family GH13 α-amylases. The enzymes represent the individual taxonomic sources belonging to various GH13 subfamilies with emphasis on *Archaea*. The currently unassigned archaeal members are marked with a *double question mark*. The sequence features characteristic of archaeal α-amylases from the subfamily GH13_7 are highlighted in *yellow*. The catalytic triad is signified by *black-and-white* inversion. These conserved sequence regions (CSR-I to CSR-VII) cover mostly the individual β-strands of the catalytic TIM-barrel domain (Janecek 2002). The individual sequences are identifiable by their UniProt accession numbers (UniProt Consortium 2015) succeeding the source name

(*Thermococcus hydrothermalis* α-amylase numbering): (i) the CSR-VI (strand β2) – Ile42 (right after the conserved glycine) and two adjacent prolines Pro48-Pro49 at the start and at the end, respectively; (ii) the CSR-I (strand β3) – Ile107 succeeding the conserved aspartate; (iii) the CSR-V (domain B) – Ile196; (iv) the CSR-II (strand β4) – (Ala194)-Trp195, Tyr199 and Gly202 at the beginning, in the middle and at the end, respectively; (v) the CSR-III (strand β5) – Ala219 and Tyr223-Trp224 succeeding the conserved tryptophan and catalytic proton donor (Glu222), respectively; (vi) the CSR-IV (strand β7) – Ala286; and (vii) the CSR-VII (strand β8) – Gln309 (right after the conserved glycine), tripeptide Ile312-Phe313-Tyr314 and Asp316 at the beginning, in the middle and at the end, respectively.

Some of the residues described above have been recognized as playing important roles in α-amylases from both plants and *Archaea* (Kadziola et al. 1998; Linden et al. 2003). Thus, e.g., as seen in the complex structure of barley α-amylase with acarbose (Kadziola et al. 1998), glycine located at the end of the CSR-II is a calcium-ion ligand and the tryptophan positioned two residues after the catalytic proton donor in the CSR-III interacts in the active site via stacking with one of the acarbose rings. Both glycine and tryptophan play probably analogical roles also in the α-amylases from *Archaea*, as documented by the structure of *Pyrococcus woesei* α-amylase (Linden et al. 2003). It is of note, however, that the tryptophan in the position $i+2$ with respect to catalytic proton donor (i.e. the glutamic acid in the CSR-III in the strand β5) is not unique for the archaeal α-amylases from the subfamily GH13_7, since, for example, it is frequently present in the "intermediary" α-amylases from the subfamily GH13_36 (Sivakumar et al. 2006; Majzlova et al. 2013), and/or even the members from both the oligo-1,6-glucosidase and neopullulanase subfamilies (Oslancova and Janecek 2002) playing a similar binding role (Hondoh et al. 2003, 2008) being substantially flexible (Kobayashi et al. 2015).

Figure 17.2 illustrates evolutionary relationships among the selected α-amylases from various subfamilies of the family GH13, emphasizing the close relatedness of archaeal and plant counterparts. Concerning the most closely related α-amylases to those from *Archaea*, it has become evident that it is a group of α-amylases from *Bacteria* classified in the same subfamily GH13_7 (Fig. 17.2). Interestingly, all of them exclusively originate from various genera of flavobacteria (Lombard et al. 2014). The fact that one of the flavobacterial GH13_7 α-amylases has already been expressed and biochemically characterised, the one from *Sinomicrobium* sp. 5DNS001 is of special importance, since despite the close homology with hyperthermophilic archaeal α-amylases the enzymes from flavobacteria are only weakly thermostable (Li et al. 2014).

Concerning the plant α-amylases, a few bacterial enzymes exhibit their features, too, but those are only putative proteins (Lombard et al. 2014). In this context, the eventuality of a horizontal gene transfer within the *Saccharophagus degradans* genome (Da Lage et al. 2004) possessing the α-amylase plant-like copy (Fig. 17.2) should also be taken into account.

The α-amylases from *Archaea* and plants (i.e. the subfamilies GH13_7 and GH13_6) occupy the adjacent branches in the evolutionary tree (Fig. 17.2) only if the tree is based on the best conserved parts of their sequences (Janecek et al. 1999),

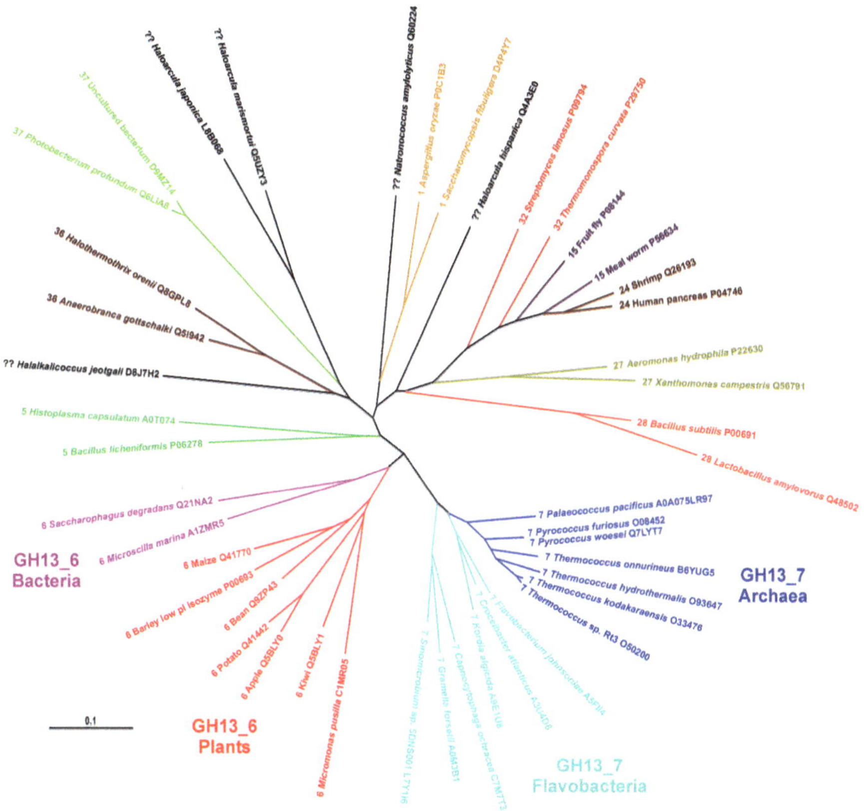

Fig. 17.2 Evolutionary tree of the family GH13 α-amylases. The tree is based on the alignment of seven conserved sequence regions. The tree was calculated and displayed using the programs Clustal-X (Jeanmougin et al. 1998) and TreeView (Page 1996), respectively

e.g., a tree based on CSRs (Fig. 17.1). Otherwise the exclusive relatedness of archaeal and plant α-amylases may be impaired by the enzymes from the subfamily GH13_5 (Da Lage et al. 2004; van der Kaaij et al. 2007; Godany et al. 2010) that covers liquefying bacterial α-amylases, their intracellular counterparts from virulent fungi and even from a few potential representatives of *Archaea* from Methanosarcinales of as yet unknown specificity (Janecek et al. 2014; Lombard et al. 2014).

It is worth mentioning that all archaeal, flavobacterial and even plant α-amylases share sequence features that discriminate them from the remaining sources (Fig. 17.1), although at the tertiary structure level no obvious differences are seen (Fig. 17.3), except for the longer domain B present in bacterial α-amylases from the closely related subfamily GH13_5 (Fig. 17.3d). Simultaneously the α-amylases from *Archaea* have to possess the additional sequence features that have to be unique only to them. It is very reasonable to trace the sequence features that discriminate from each other the representatives of archaeal α-amylases from those

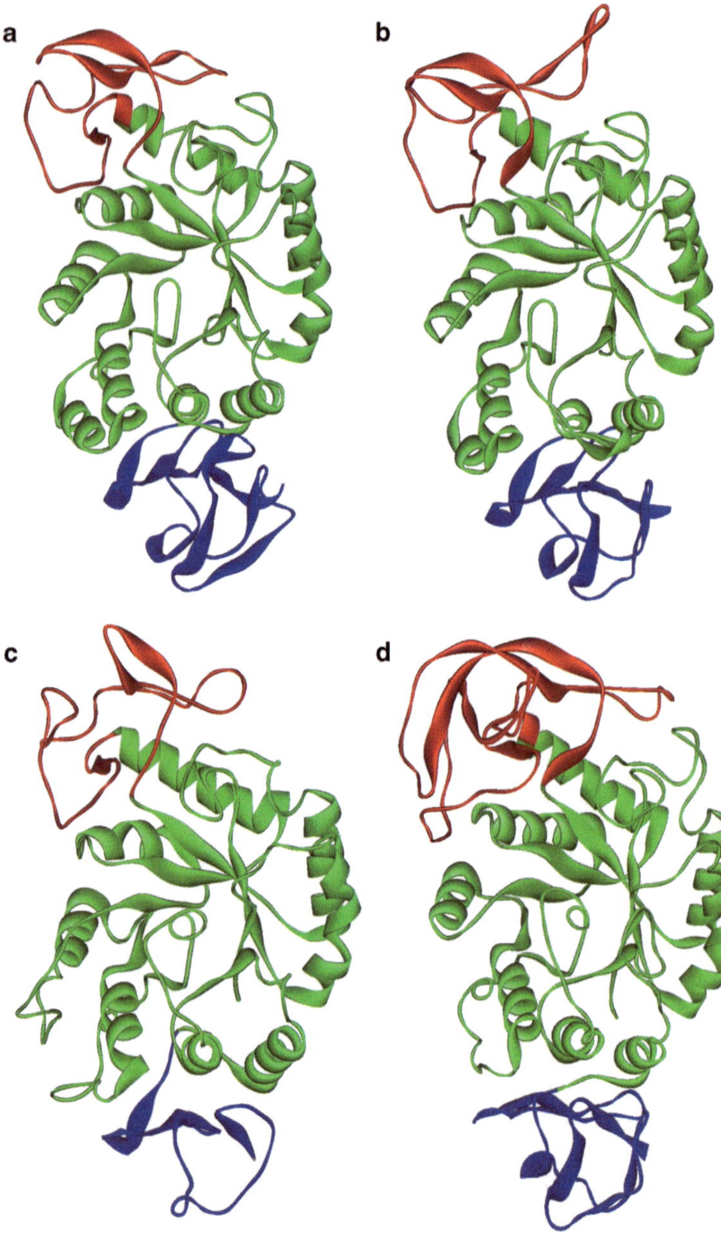

Fig. 17.3 Tertiary structures of archaeal α-amylase and its close homologues. (**a**) GH13_7 – *Pyrococcus woesei* (PDB code: 1MWO; Linden et al. 2003); (**b**) GH13_7 – *Sinomicrobium* sp. 5DNS001 (UniProt accession No.: L7Y1I6; Li et al. 2014); (**c**) GH13_6 – *Hordeum vulgare* – barley isozyme AMY-1 (PDB code: 1P6W; Robert et al. 2003); (**d**) GH13_5 – *Bacillus licheniformis* (PDB code: 1BLI; Machius et al. 1998). The individual domains are colored as follows: catalytic $(\beta/\alpha)_8$-barrel – *green*, domain B – *red*, domain C – *blue*. The flavobacterial α-amylase structure (b) was modeled using the Phyre-2 server (Kelley and Sternberg 2009) based on the *P. woesei* α-amylase structure (1MWO) as template (The structures were retrieved from the Protein Data Bank (PDB; Deshpande et al. 2005) and visualized with the program WebLabViewerLite (Molecular Simulations, Inc.))

of both flavobacterial and plant enzymes. This means to reveal the features that might be responsible for the extra thermostability of the α-amylases from *Archaea* that exhibit the temperature optima around and above 80 °C (Leveque et al. 2000b; Linden and Wilmanns 2004). One of such features could be the alanine from the CSR-IV (strand β7) that has no correspondence in the plant counterparts, and, more importantly, the second position in the CSR-VII (strand β-8) where archaeal α-amylases possess invariant glutamine instead of an invariantly conserved tyrosine present in the flavobacterial homologues (Fig. 17.1).

For example, the α-amylase from *P. furiosus*, exhibiting optimal activity at 100 °C (Dong et al. 1997) and undergoing irreversible denaturation above 106 °C (Brown et al. 2013), is stabilized by Zn^{2+} coordinated by Cys165 (Savchenko et al. 2002; Linden et al. 2003), which is the last residue of the CSR-V (Fig. 17.1). Interestingly, that cysteine is not invariantly conserved in all archaeal GH13_7 α-amylases, but it is also found in flavobacterial counterparts from the same subfamily (Fig. 17.1) that should, however, be less thermostable in general (Li et al. 2014). The only characterized representative of GH13_7 α-amylases from flavobacteria, the one from *Sinomicrobium* sp. 5DNS001, showed optimal activity at 50 °C, but its thermostability was significantly improved after introducing a disulfide bond within the domain C by site-directed mutagenesis (Li et al. 2014). Two archaeal α-amylases from thermococci – *T. hydrothermalis* (Leveque et al. 2000a; Horvathova et al. 2006) and *T. onnurineus* (Lim et al. 2007) – were studied by protein engineering approaches. Whereas the Y39I mutant of the former α-amylase, mimicking the *Thermotoga maritima* counterpart (Liebl et al. 1997), was slightly destabilized indicating the role of Tyr39 in the enzyme thermostability (Godany et al. 2010), the latter enzyme achieved a higher thermostability by recovery zinc-binding residues, especially the above-mentioned cysteine from the end of the CSR-V (Lim et al. 2007). That cysteine seems to be more characteristic for the archaeal α-amylases from pyrococci (Fig. 17.1).

2.2 Archaeal GH13 "α-Amylases" from Halophiles

In recent few years, a group of potential α-amylases from halophilic *Archaea* has been added into the family GH13 (Lombard et al. 2014). Originally, there was only one such "α-amylase" from *Natronococcus amylolyticus* (Kobayashi et al. 1994) that was, however, shown to be rather a maltotriose-producing amylase (Kobayashi et al. 1992). Being evidently different from sequences of typical hyperthermophilic archaeal α-amylases from the subfamily GH13_7 (Fig. 17.1), the problem is that most of those halophilic *Archaea*-originated amylolytic enzymes are only putative proteins from genome sequencing projects (Zorgani et al. 2014), or their amino acid sequences are not available (Perez-Pomares et al. 2003; Fukushima et al. 2005; Moshfegh et al. 2013). A few of them have already been assigned to the subfamily GH13_31 (Lombard et al. 2014) to be probable α-glucosidases (Stam et al. 2006). This can also be supported by the analysis of soluble starch hydrolysis products by the enzyme from *Haloarcula japonica* yielding mainly maltose (Onodera et al. 2013).

The family GH13 characteristic CSRs of five selected potential α-amylases from halophilic *Archaea* are shown in Fig. 17.1. These enzymes obviously do not contain the features characteristic of archaeal α-amylases from the subfamily GH13_7, the fact that is reflected also in the evolutionary tree (Fig. 17.2). The putative amylolytic enzyme from *Halalkalicoccus jeogtali* does exhibit the clear QPDLN signature in the CSR-V typical for the members of the oligo-1,6-glucosidase subfamily (Oslancova and Janecek 2002). Its position in the evolutionary tree on the branch leading to α-amylases from the subfamily GH13_36 (Fig. 17.2) that are closely related to both oligo-1,6-glucosidase and neopullulanase subfamilies (Majzlova et al. 2013) is in agreement with that. Whereas the two amylolytic enzymes from *Haloarcula japonica* (yielding mainly maltose; Onodera et al. 2013) and *Haloarcula marismortui* go well with the members of the subfamily GH13_37 (Lei et al. 2012), which are usually the α-amylases from marine bacteria lacking domain B (Janecek et al. 2014), the one from *Haloarcula hispanica* is on a solitary branch adjacent to a large cluster of α-amylases from subfamilies GH13_15, 24, 27, 28 and 32 covering bacterial and animal sources (Fig. 17.2). The enzyme from *Natronococcus amylolyticus* that possesses, however, the specificity of a maltotriose-forming amylase (Kobayashi et al. 1992, 1994) shares the branch with the extracellular fungal α-amylases from the subfamily GH13_1 (Fig. 17.2), represented by the well-known Taka-amylase A from *Aspergillus oryzae* (Matsuura et al. 1984). Interestingly, when the five sequences of potential halophilic archaeal α-amylases were sent to a fold recognition server Phyre-2 (Kelley and Sternberg 2009), the tertiary structures of α-amylases as the best templates were used only for the enzymes from *Haloarcula japonica* (GH13_5) and *Natronococcus amylolyticus* (GH13_24, GH13_5 and GH13_15), whereas those of oligo-1,6-glucosidase (GH13_31) and neopullulanase (GH13_20) subfamilies were identified for the enzymes from *Halalkalicoccus jeogtali* and both *Haloarcula japonica* and *Haloarcula marismortui*, respectively. More insightful biochemical characterization studies are thus necessary in order to reach a conclusion concerning the exact specificity of these amylolytic enzymes from halophilic *Archaea*.

3 Archaeal α-Amylases from the Family GH57

The family GH57 was created (Henrissat and Bairoch 1996) based on the existence of sequences of two "α-amylases" from extremely thermophilic prokaryotes that hardly contain the attributes characteristic of the main α-amylase family GH13 (Janecek 1998). They were originally considered as α-amylases – one from bacterium *Dictyoglomus thermophilum* (Fukusumi et al. 1988) and the other one from archaeon *Pyrococcus furiosus* (Laderman et al. 1993a) – but they both represent 4-α-glucanotransferases (Janecek et al. 2014) due to their biochemical re-evaluation (Laderman et al. 1993b; Nakajima et al. 2004) supported by detailed *in silico* studies (Blesak and Janecek 2012). It should be taken into account, however, that although recent bioinformatics analyses revealed 50–60 true potential α-amylase sequences in the family GH57 (Janecek and Blesak 2011; Blesak and Janecek 2012), the only evidence concerning the presence of the α-amylase specificity comes from biochemical characterization of the α-amylase from methanogenic archaeon *Methanococcus jannaschii* shown to be able to degrade soluble starch (Kim et al. 2001b). This α-amylase was first identified as a putative protein in the genome sequencing project (Bult et al. 1996) and the uncertainty about its true α-amylase specificity is based on its ability to degrade also pullulan with more than 80 % relative rate in comparison with starch (Kim et al. 2001b). In addition to other amylolytic specificities (Janecek et al. 2014), there are currently two more closely related amylases in the family GH57 (Blesak and Janecek 2013), but one is a maltogenic amylase (or maltose-forming exo-amylase) (Jeon et al. 2014; Jung et al. 2014; Park et al. 2014) and the other one represents a non-specified amylase (Wang et al. 2011).

The family GH57 α-amylases represented by the *Methanococcus jannaschii* α-amylase should be approximately 500-residue long proteins consisting of two structural parts (Janecek and Blesak 2011) that are indispensable for any family GH57 member (Imamura et al. 2003; Palomo et al. 2011): the catalytic incomplete TIM-barrel succeeded by a bundle of 3–4 α-helices. Although the catalytic machinery is conserved within the barrel at the strands β4 (CSR-3) and β7 (CSR-4) (Fig. 17.4), the helical bundle is also considered as belonging to the catalytic area of the family GH57 structure, since it contains functionally essential residues in the CSR-5 (Palomo et al. 2011). Other GH57 enzyme specificities may contain additional domains (Blesak and Janecek 2012). Since the CSR-1 is located very close to the N-terminus (Fig. 17.4), there is probably no signal peptide in α-amylases.

Most of the sequence-structural features mentioned above apply also for another group of the family GH57 members, which is closely related to α-amylases, the so-called group of α-amylase-like homologues (Janecek et al. 2014). The most important difference is that the α-amylase-like homologues lack one or both catalytic residue, i.e. they either possess their catalytic machinery incomplete or lack it at all (Janecek and Blesak 2011). Thus the catalytic nucleophile and the proton donor, Glu145 and Asp237, respectively, in the *Methanococcus jannaschii* α-amylase, are most often replaced by a serine and a glutamic acid, respectively (Fig. 17.4).

Despite the eventual loss of enzymatic activity in the α-amylase-like homologues, they obviously exhibit a close evolutionary relatedness to their enzymatically active counterparts, i.e. to family GH57 α-amylases (Blesak and Janecek 2013). This is clearly seen also in the sequence logos (Fig. 17.4) that – based on the originally identified CSRs in this family (Zona et al. 2004) – were suggested to define the so-called sequence fingerprints of a given enzyme specificity within the family (Blesak and Janecek 2012). For both the α-amylases and their α-amylase-like homologues, the positions 1 (CSR-1) and 12 (CSR-2) with mostly glutamic acid and arginine (or glutamic acid), respectively, may be considered as unique since all remaining specificities contain a histidine and tryptophan, respectively, in corresponding positions (Fig. 17.4). Position 13 (CSR-2) is moreover occupied

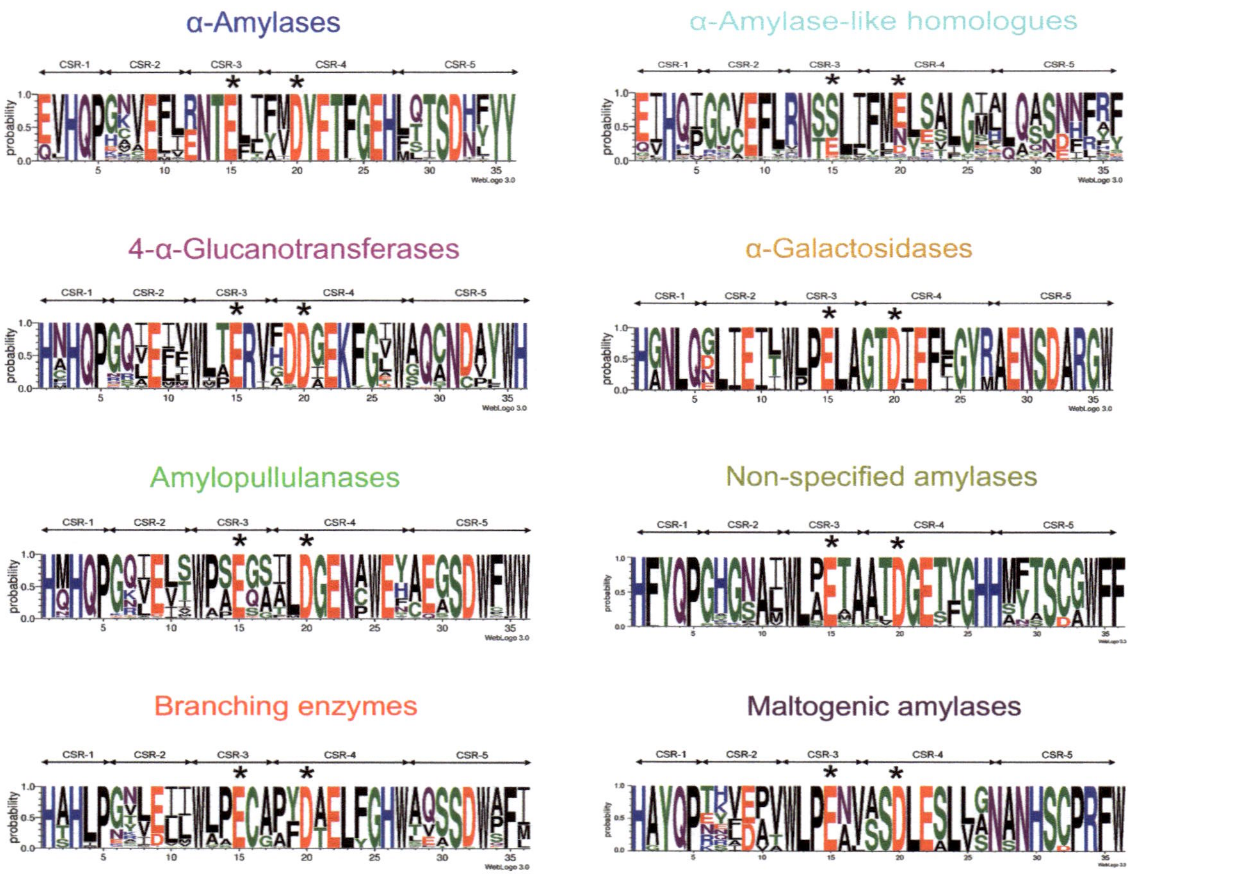

Fig. 17.4 Sequence logos of the family GH57 enzyme specificities. α-Amylases, 56 sequences; α-amylase-like homologues; 48 sequences; 4-α-glucanotransferases, 46 sequences; amylopullulanases, 99 sequences; branching enzymes, 158 sequences; α-galactosidases, 8 sequences; non-specified amylases, 127 sequences; maltogenic amylases, 12 sequences. CSR-1, residues 1–5; CSR-2, residues 6–11; CSR-3, residues 12–17; CSR-4, residues 18–27; CSR-5, residues 28–36. The catalytic nucleophile (No. 15, glutamic acid) and the proton donor (No. 20, aspartic acid) are indicated by *asterisks* (Adapted according to Janecek and Blesak (2011) and Blesak and Janecek (2012, 2013))

by almost invariant asparagine. On the other hand the positions 21 (CSR-4) and 35–36 (CSR-5) with invariant tyrosines should be a feature that can be used to discriminate the family GH57 α-amylase from its α-amylase-like homologue (Janecek and Blesak 2011).

Of note may also be the occurrence of the two proteins throughout the taxonomy spectrum: while the α-amylases originate mostly from *Archaea*, the α-amylase-like homologues are found mainly among *Bacteria* (Janecek and Blesak 2011). Since a similar phenomenon has been observed also in the main α-amylase family GH13 (Janecek et al. 1997; Fort et al. 2007; Gabrisko and Janecek 2009), it cannot be excluded that the family GH57 α-amylase-like homologues may play a role different from an amylolytic (even enzymatic) activity.

4 Conclusions

The world of amylolytic enzymes seems to be very diverse and reflects different evolutionary histories for α-amylases, β-amylases and glucoamylases (Janecek 1994a). The α-amylase itself is an amazing enzyme which, although it had to obey a definition criterion on employing the retaining mechanism of α-1,4-glucosidic bond cleavage, may have evolved under more sequence-structural templates currently known as the families GH13 and GH57 (and eventually GH119). The α-amylases from *Archaea*, present in both major families GH13 and GH57, with their unique features and properties (Janecek et al. 2014) represent, together with other glycoside hydrolases and transglycosylases (Bissaro et al. 2015), a relevant offer for expanding our own knowledge as well as for their practical utilization.

α-Amylases from *Archaea*, applied either alone or in combination with other related amylolytic enzymes, such as pullulanases and amylopullulanases, are extremely attractive for their potential use in various industrial branches (for reviews, see Leveque et al. 2000b; Hii et al. 2012; Nisha and Satyanarayana 2013), where they could replace, e.g., their less thermostable bacterial and fungal counterparts. Importantly, archaeal amylopullulanases, similar to α-amylases, are also found in both families GH13 and GH57 (Zona et al. 2004; Blesak and Janecek 2012; Jiao et al. 2013). As mentioned above, the family GH57 contains additional amylolytic specificities very closely related to α-amylase, namely the maltogenic amylase (or maltose-forming exo-amylase) (Jeon et al. 2014; Jung et al. 2014; Park et al. 2014) and the so-called non-specified amylase (Wang et al. 2011) that both are produced also by *Archaea* (Blesak and Janecek 2013). With regard to industrial applications of enzymes from extremophiles in general, reviewed recently by Elleuche et al. (2014, 2015), marine extremophiles may deserve a special attention (Dalmaso et al. 2015).

Acknowledgement Financial support from the Slovak Grant Agency VEGA (grant No. 2/0150/14) and the Slovak Research and Development Agency APVV (contract No. LPP-0417-09) is gratefully acknowledged.

References

Aghajari N, Roth M, Haser R (2002) Crystallographic evidence of a transglycosylation reaction: ternary complexes of a psychrophilic α-amylase. Biochemistry 41(13):4273–4280

Ahmad N, Rashid N, Haider MS, Akram M, Akhtar M (2014) Novel maltotriose-hydrolyzing thermoacidophilic type III pullulan hydrolase from *Thermococcus kodakarensis*. Appl Environ Microbiol 80(3):1108–1115

Atomi H, Fukui T, Kanai T, Morikawa M, Imanaka T (2004) Description of *Thermococcus kodakaraensis* sp. nov., a well studied hyperthermophilic archaeon previously reported as *Pyrococcus* sp. KOD1. Archaea 1(4):263–267

Bautista V, Esclapez J, Perez-Pomares F, Martinez-Espinosa RM, Camacho M, Bonete MJ (2012) Cyclodextrin glycosyltransferase: a key enzyme in the assimilation of starch by the halophilic archaeon *Haloferax mediterranei*. Extremophiles 16(1):147–159

Bertoldo C, Antranikian G (2002) Starch-hydrolyzing enzymes from thermophilic archaea and bacteria. Curr Opin Chem Biol 6(2):151–160

Bissaro B, Monsan P, Fauré R, O'Donohue MJ (2015) Glycosynthesis in a waterworld: new insight into the molecular basis of transglycosylation in retaining glycoside hydrolases. Biochem J 467(1):17–35

Blesak K, Janecek S (2012) Sequence fingerprints of enzyme specificities from the glycoside hydrolase family GH57. Extremophiles 16(3):497–506

Blesak K, Janecek S (2013) Two potentially novel amylolytic enzyme specificities in the prokaryotic glycoside hydrolase α-amylase family GH57. Microbiology 159(12):2584–2593

Brown I, Dafforn TR, Fryer PJ, Cox PW (2013) Kinetic study of the thermal denaturation of a hyperthermostable extracellular α-amylase from *Pyrococcus furiosus*. Biochim Biophys Acta 1834(12):2600–2605

Brzozowski AM, Davies GJ (1997) Structure of the *Aspergillus oryzae* α-amylase complexed with the inhibitor acarbose at 2.0 Å resolution. Biochemistry 36(36):10837–10845

Bult CJ, White O, Olsen GJ, Zhou L, Fleischmann RD, Sutton GG, Blake JA, FitzGerald LM, Clayton RA, Gocayne JD, Kerlavage AR, Dougherty BA, Tomb JF, Adams MD, Reich CI, Overbeek R, Kirkness EF, Weinstock KG, Merrick JM, Glodek A, Scott JL, Geoghagen NS, Venter JC (1996) Complete genome sequence of the methanogenic archaeon, *Methanococcus jannaschii*. Science 273(5278):1058–1073

Chen YS, Lee GC, Shaw JF (2006) Gene cloning, expression, and biochemical characterization of a recombinant trehalose synthase from *Picrophilus torridus* in *Escherichia coli*. J Agric Food Chem 54(19):7098–7104

Coutiho PM, Henrissat B (1999) Life with no sugars? J Mol Microbiol Biotechnol 1(2):307–308

Coutinho PM, Reilly PJ (1997) Glucoamylase structural, functional, and evolutionary relationships. Proteins 29(3):334–347

Da Lage JL, Feller G, Janecek S (2004) Horizontal gene transfer from Eukarya to bacteria and domain shuffling: the α-amylase model. Cell Mol Life Sci 61(1):97–109

Da Lage JL, Binder M, Hua-Van A, Janecek S, Casane D (2013) Gene make-up: rapid and massive intron gains after horizontal transfer of a bacterial α-amylase gene to Basidiomycetes. BMC Evol Biol 13:40

Dalmaso GZ, Ferreira D, Vermelho AB (2015) Marine extremophiles: a source of hydrolases for biotechnological applications. Mar Drugs 13(4):1925–1965

Davies G, Henrissat B (1995) Structures and mechanisms of glycosyl hydrolases. Structure 3(9):853–859

Deshpande N, Addess KJ, Bluhm WF, Merino-Ott JC, Townsend-Merino W, Zhang Q, Knezevich C, Xie L, Chen L, Feng Z, Green RK, Flippen-Anderson JL, Westbrook J, Berman HM, Bourne PE (2005) The RCSB protein data bank: a redesigned query system and relational database based on the mmCIF schema. Nucleic Acids Res 3(1):D233–D237

Dong G, Vieille C, Savchenko A, Zeikus JG (1997) Cloning, sequencing, and expression of the gene encoding extracellular α-amylase from *Pyrococcus furiosus* and biochemical characterization of the recombinant enzyme. Appl Environ Microbiol 63(9):3569–3576

Elleuche S, Schröder C, Sahm K, Antranikian G (2014) Extremozymes – biocatalysts with unique properties from extremophilic microorganisms. Curr Opin Biotechnol 29:116–123
Elleuche S, Schäfers C, Blank S, Schröder C, Antranikian G (2015) Exploration of extremophiles for high temperature biotechnological processes. Curr Opin Microbiol 25:113–119
Feese MD, Kato Y, Tamada T, Kato M, Komeda T, Miura Y, Hirose M, Hondo K, Kobayashi K, Kuroki R (2000) Crystal structure of glycosyltrehalose trehalohydrolase from the hyperthermophilic archaeum *Sulfolobus solfataricus*. J Mol Biol 301(2):451–464
Fort J, de la Ballina LR, Burghardt HE, Ferrer-Costa C, Turnay J, Ferrer-Orta C, Uson I, Zorzano A, Fernandez-Recio J, Orozco M, Lizarbe MA, Fita I, Palacin M (2007) The structure of human 4F2hc ectodomain provides a model for homodimerization and electrostatic interaction with plasma membrane. J Biol Chem 282(43):31444–31452
Frillingos S, Linden A, Niehaus F, Vargas C, Nieto JJ, Ventosa A, Antranikian G, Drainas C (2000) Cloning and expression of α-amylase from the hyperthermophilic archaeon *Pyrococcus woesei* in the moderately halophilic bacterium *Halomonas elongata*. J Appl Microbiol 88(3):495–503
Fukushima T, Mizuki T, Echigo A, Inoue A, Usami R (2005) Organic solvent tolerance of halophilic alpha-amylase from a Haloarchaeon, *Haloarcula* sp. strain S-1. Extremophiles 9(1):85–89
Fukusumi S, Kamizono A, Horinouchi S, Beppu T (1988) Cloning and nucleotide sequence of a heat-stable amylase gene from an anaerobic thermophile, *Dictyoglomus thermophilum*. Eur J Biochem 174(1):15–21
Gabrisko M, Janecek S (2009) Looking for the ancestry of the heavy-chain subunits of heteromeric amino acid transporters rBAT and 4F2hc within the GH13 α-amylase family. FEBS J 276(24):7265–7278
Godany A, Vidova B, Janecek S (2008) The unique glycoside hydrolase family 77 amylomaltase from *Borrelia burgdorferi* with only catalytic triad conserved. FEMS Microbiol Lett 284(1):84–91
Godany A, Majzlova K, Horvathova V, Vidova B, Janecek S (2010) Tyrosine 39 of GH13 α-amylase from *Thermococcus hydrothermalis* contributes to its thermostability. Biologia 65(3):408–415
Gupta R, Gigras P, Mohapatra H, Kumar Goswami V, Chauhan B (2003) Microbial α-amylases: a biotechnological perspective. Process Biochem 38(11):1599–1616
Hashimoto Y, Yamamoto T, Fujiwara S, Takagi M, Imanaka T (2001) Extracellular synthesis, specific recognition, and intracellular degradation of cyclomaltodextrins by the hyperthermophilic archaeon *Thermococcus* sp. strain B1001. J Bacteriol 183(17):5050–5057
Henrissat B (1991) A classification of glycosyl hydrolases based on amino acid sequence similarities. Biochem J 208(2):309–316
Henrissat B, Bairoch A (1996) Updating the sequence-based classification of glycosyl hydrolases. Biochem J 316(2):695–696
Hii SL, Tan JS, Ling TC, Ariff AB (2012) Pullulanase: role in starch hydrolysis and potential industrial applications. Enzyme Res 2012:921362
Hondoh H, Kuriki T, Matsuura Y (2003) Three-dimensional structure and substrate binding of *Bacillus stearothermophilus* neopullulanase. J Mol Biol 326(1):177–188
Hondoh H, Saburi W, Mori H, Okuyama M, Nakada T, Matsuura Y, Kimura A (2008) Substrate recognition mechanism of α-1,6-glucosidic linkage hydrolyzing enzyme, dextran glucosidase from *Streptococcus mutans*. J Mol Biol 378(4):913–922
Horvathova V, Godany A, Sturdik E, Janecek S (2006) α-Amylase from *Thermococcus hydrothermalis*: re-cloning aimed at the improved expression and hydrolysis of corn starch. Enzyme Microb Technol 39(6):1300–1305
Hostinova E, Janecek S, Gasperik J (2010) Gene sequence, bioinformatics and enzymatic characterization of α-amylase from *Saccharomycopsis fibuligera* KZ. Protein J 29(5):355–364
Hutcheon GW, Vasisht N, Bolhuis A (2005) Characterisation of a highly stable α-amylase from the halophilic archaeon *Haloarcula hispanica*. Extremophiles 9(6):487–495
Imamura H, Fushinobu S, Yamamoto M, Kumasaka T, Jeon BS, Wakagi T, Matsuzawa H (2003) Crystal structures of 4-α-glucanotransferase from *Thermococcus litoralis* and its complex with an inhibitor. J Biol Chem 278(21):19378–19386

Janecek S (1994a) Parallel β/α-barrels of α-amylase, cyclodextrin glycosyltransferase and oligo-1,6-glucosidase versus the barrel of β-amylase: evolutionary distance is a reflection of unrelated sequences. FEBS Lett 353(2):119–123
Janecek S (1994b) Sequence similarities and evolutionary relationships of microbial, plant and animal α-amylases. Eur J Biochem 224(2):519–524
Janecek S (1997) α-Amylase family: molecular biology and evolution. Prog Biophys Mol Biol 67(1):67–97
Janecek S (1998) Sequence of archaeal *Methanococcus jannaschii* α-amylase contains features of families 13 and 57 of glycosyl hydrolases: a trace of their common ancestor? Folia Microbiol 43(2):123–128
Janecek S (2002) How many conserved sequence regions are there in the α-amylase family? Biologia 57(11):29–41
Janecek S (2005) Amylolytic families of glycoside hydrolases: focus on the family GH-57. Biologia 60(16):177–184
Janecek S, Blesak K (2011) Sequence-structural features and evolutionary relationships of family GH57 α-amylases and their putative α-amylase-like homologues. Protein J 30(6):429–435
Janecek S, Kuchtova A (2012) In silico identification of catalytic residues and domain fold of the family GH119 sharing the catalytic machinery with the α-amylase family GH57. FEBS Lett 586(19):3360–3366
Janecek S, Svensson B, MacGregor EA (1995) Characteristic differences in the primary structure allow discrimination of cyclodextrin glucanotransferases from α-amylases. Biochem J 305(2):685–686
Janecek S, Svensson B, Henrissat B (1997) Domain evolution in the α-amylase family. J Mol Evol 45(3):322–331
Janecek S, Leveque E, Belarbi A, Haye B (1999) Close evolutionary relatedness of α-amylases from Archaea and plants. J Mol Evol 48(4):421–426
Janecek S, Svensson B, MacGregor EA (2014) α-Amylase – an enzyme specificity found in various families of glycoside hydrolases. Cell Mol Life Sci 71(7):1149–1170
Jeanmougin F, Thompson JD, Gouy M, Higgins DG, Gibson TJ (1998) Multiple sequence alignment with Clustal X. Trends Biochem Sci 23(10):403–405
Jeon EJ, Jung JH, Seo DH, Jung DH, Holden JF, Park CS (2014) Bioinformatic and biochemical analysis of a novel maltose-forming α-amylase of the GH57 family in the hyperthermophilic archaeon *Thermococcus* sp. CL1. Enzyme Microb Technol 60:9–15
Jespersen HM, MacGregor EA, Sierks MR, Svensson B (1991) Comparison of the domain-level organization of starch hydrolases and related enzymes. Biochem J 280(1):51–55
Jiao YL, Wang SJ, Lv MS, Fang YW, Liu S (2013) An evolutionary analysis of the GH57 amylopullulanases based on the DOMON_glucodextranase_like domains. J Basic Microbiol 53(3):231–239
Jones RA, Jermiin LS, Easteal S, Patel BK, Beacham IR (1999) Amylase and 16S rRNA genes from a hyperthermophilic archaebacterium. J Appl Microbiol 86(1):93–107
Jorgensen S, Vorgias CE, Antranikian G (1997) Cloning, sequencing, characterization, and expression of an extracellular α-amylase from the hyperthermophilic archaeon *Pyrococcus furiosus* in *Escherichia coli* and *Bacillus subtilis*. J Biol Chem 272(26):16335–16342
Jung TY, Li D, Park JT, Yoon SM, Tran PL, Oh BH, Janecek S, Park SG, Woo EJ, Park KH (2012) Association of novel domain in active site of archaic hyperthermophilic maltogenic amylase from *Staphylothermus marinus*. J Biol Chem 287(11):7979–7989
Jung JH, Seo DH, Holden JF, Park CS (2014) Maltose-forming α-amylase from the hyperthermophilic archaeon *Pyrococcus* sp. ST04. Appl Microbiol Biotechnol 98(5):2121–2131
Kadziola A, Søgaard M, Svensson B, Haser R (1998) Molecular structure of a barley α-amylase-inhibitor complex: implications for starch binding and catalysis. J Mol Biol 278(1):205–217
Kelley LA, Sternberg MJ (2009) Protein structure prediction on the Web: a case study using the Phyre server. Nat Protoc 4(3):363–371
Kim HO, Park JN, Sohn HJ, Shin DJ, Choi C, Im SY, Lee HB, Chun SB, Bai S (2000) Cloning and expression in *Saccharomyces cerevisiae* of a β-amylase gene from the oomycete *Saprolegnia ferax*. Biotechnol Lett 22(18):1493–1498

Kim HO, Park JN, Shin DJ, Lee HB, Chun SB, Bai S (2001a) A gene encoding *Achlya bisexualis* β-amylase and its expression in *Saccharomyces cerevisiae*. Biotechnol Lett 23(14):1101–1107

Kim JW, Flowers LO, Whiteley M, Peeples TL (2001b) Biochemical confirmation and characterization of the family-57-like α-amylase of *Methanococcus jannaschii*. Folia Microbiol 46(6):467–473

Kobayashi T, Kanai H, Hayashi T, Akiba T, Akaboshi R, Horikoshi K (1992) Haloalkaliphilic maltotriose-forming α-amylase from the archaebacterium *Natronococcus* sp. strain Ah-36. J Bacteriol 174(11):3439–3444

Kobayashi T, Kanai H, Aono R, Horikoshi K, Kudo T (1994) Cloning, expression, and nucleotide sequence of the α-amylase gene from the haloalkaliphilic archaeon *Natronococcus* sp. strain Ah-36. J Bacteriol 176(16):5131–5134

Kobayashi K, Kato M, Miura Y, Kettoku M, Komeda T, Iwamatsu A (1996) Gene cloning and expression of new trehalose-producing enzymes from the hyperthermophilic archaeum *Sulfolobus solfataricus* KM1. Biosci Biotechnol Biochem 60(11):1882–1885

Kobayashi M, Kubota M, Matsuura Y (2003) Refined structure and functional implications of trehalose synthase from *Sulfolobus acidocaldarius*. J Appl Glycosci 50(1):1–8

Kobayashi M, Saburi W, Nakatsuka D, Hondoh H, Kato K, Okuyama M, Mori H, Kimura A, Yao M (2015) Structural insights into the catalytic reaction that is involved in the reorientation of Trp238 at the substrate-binding site in GH13 dextran glucosidase. FEBS Lett 589(4):484–489

Koch R, Zablowski P, Spreinat A, Antranikian G (1990) Extremely thermostable amylolytic enzyme from the archaebacterium *Pyrococcus furiosus*. FEMS Microbiol Lett 71(1–2):21–26

Koch R, Spreinat A, Lemke K, Antranikian G (1991) Purification and properties of a hyperthermoactive α-amylase from the archaeobacterium *Pyrococcus woesei*. Arch Microbiol 155(6):572–578

Kumar P, Satyanarayana T (2009) Microbial glucoamylases: characteristics and applications. Crit Rev Biotechnol 29(3):225–255

Laderman KA, Asada K, Uemori T, Mukai H, Taguchi Y, Kato I, Anfinsen CB (1993a) α-Amylase from the hyperthermophilic archaebacterium *Pyrococcus furiosus*. Cloning and sequencing of the gene and expression in *Escherichia coli*. J Biol Chem 268(32):24402–24407

Laderman KA, Davis BR, Krutzsch HC, Lewis MS, Griko YV, Privalov PL, Anfinsen CB (1993b) The purification and characterization of an extremely thermostable α-amylase from the hyperthermophilic archaebacterium *Pyrococcus furiosus*. J Biol Chem 268(32):24394–24401

Lee HS, Shockley KR, Schut GJ, Conners SB, Montero CI, Johnson MR, Chou CJ, Bridger SL, Wigner N, Brehm SD, Jenney FE Jr, Comfort DA, Kelly RM, Adams MW (2006) Transcriptional and biochemical analysis of starch metabolism in the hyperthermophilic archaeon *Pyrococcus furiosus*. J Bacteriol 188(6):2115–2125

Leemhuis H, Rozeboom HJ, Dijkstra BW, Dijkhuizen L (2003a) The fully conserved Asp residue in conserved sequence region I of the α-amylase family is crucial for the catalytic site architecture and activity. FEBS Lett 541(1–3):47–51

Leemhuis H, Rozeboom HJ, Wilbrink M, Euverink GJ, Dijkstra BW, Dijkhuizen L (2003b) Conversion of cyclodextrin glycosyltransferase into a starch hydrolase by directed evolution: the role of alanine 230 in acceptor subsite +1. Biochemistry 42(24):7518–7526

Lei Y, Peng H, Wang Y, Liu Y, Han F, Xiao Y, Gao Y (2012) Preferential and rapid degradation of raw rice starch by an α-amylase of glycoside hydrolase subfamily GH13_37. Appl Microbiol Biotechnol 94(6):1577–1584

Leveque E, Haye B, Belarbi A (2000a) Cloning and expression of an α-amylase encoding gene from the hyperthermophilic archaebacterium *Thermococcus hydrothermalis* and biochemical characterisation of the recombinant enzyme. FEMS Microbiol Lett 186(1):67–71

Leveque E, Janecek S, Belarbi A, Haye B (2000b) Thermophilic archaeal amylolytic enzymes. Enzyme Microb Technol 26(1):2–13

Li C, Begum A, Numao S, Park KH, Withers SG, Brayer GD (2005) Acarbose rearrangement mechanism implied by the kinetic and structural analysis of human pancreatic α-amylase in complex with analogues and their elongated counterparts. Biochemistry 44(9):3347–3357

Li D, Park JT, Li X, Kim S, Lee S, Shim JH, Park SH, Cha J, Lee BH, Kim JW, Park KH (2010) Overexpression and characterization of an extremely thermostable maltogenic amylase, with an optimal temperature of 100 degrees C, from the hyperthermophilic archaeon *Staphylothermus marinus*. N Biotechnol 27(4):300–307

Li X, Li D, Park KH (2013) An extremely thermostable amylopullulanase from *Staphylothermus marinus* displays both pullulan- and cyclodextrin-degrading activities. Appl Microbiol Biotechnol 97(12):5359–5369

Li C, Du M, Cheng B, Wang L, Liu X, Ma C, Yang C, Xu P (2014) Close relationship of a novel Flavobacteriaceae α-amylase with archaeal α-amylases and good potentials for industrial applications. Biotechnol Biofuels 7(1):18

Liebl W, Stemplinger I, Ruile P (1997) Properties and gene structure of the *Thermotoga maritima* α-amylase AmyA, a putative lipoprotein of a hyperthermophilic bacterium. J Bacteriol 179(3):941–948

Lim JK, Lee HS, Kim YJ, Bae SS, Jeon JH, Kang SG, Lee JH (2007) Critical factors to high thermostability of an α-amylase from hyperthermophilic archaeon *Thermococcus onnurineus* NA1. J Microbiol Biotechnol 17(8):1242–1248

Linden A, Wilmanns M (2004) Adaptation of class-13 α-amylases to diverse living conditions. Chembiochem 5(2):231–239. doi:10.1002/cbic.200300734

Linden A, Mayans O, Meyer-Klaucke W, Antranikian G, Wilmanns M (2003) Differential regulation of a hyperthermophilic α-amylase with a novel (Ca, Zn) two-metal center by zinc. J Biol Chem 278(11):9875–9884

Lombard V, Golaconda Ramulu H, Drula E, Coutinho PM, Henrissat B (2014) The carbohydrate-active enzymes database (CAZy) in 2013. Nucleic Acids Res 42(Database Issue):D490–D495

MacGregor EA, Svensson B (1989) A super-secondary structure predicted to be common to several α-1,4-D-glucan-cleaving enzymes. Biochem J 259(1):145–152

MacGregor EA, Jespersen HM, Svensson B (1996) A circularly permuted α-amylase-type α/β--barrel structure in glucan-synthesizing glucosyltransferases. FEBS Lett 378(3):263–266

MacGregor EA, Janecek S, Svensson B (2001) Relationship of sequence and structure to specificity in the α-amylase family of enzymes. Biochim Biophys Acta 1546(1):1–20

Machius M, Declerck N, Huber R, Wiegand G (1998) Activation of *Bacillus licheniformis* α-amylase through a disorder→order transition of the substrate-binding site mediated by a calcium-sodium-calcium metal triad. Structure 6(3):281–292

Majzlova K, Pukajova Z, Janecek S (2013) Tracing the evolution of the α-amylase subfamily GH13_36 covering the amylolytic enzymes intermediate between oligo-1,6-glucosidases and neopullulanases. Carbohydr Res 367:48–57

Marin-Navarro J, Polaina J (2011) Glucoamylases: structural and biotechnological aspects. Appl Microbiol Biotechnol 89(5):1267–1273

Maruta K, Mitsuzumi H, Nakada T, Kubota M, Chaen H, Fukuda S, Sugimoto T, Kurimoto M (1996) Cloning and sequencing of a cluster of genes encoding novel enzymes of trehalose biosynthesis from thermophilic archaebacterium *Sulfolobus acidocaldarius*. Biochim Biophys Acta 1291(3):177–181

Matsuura Y, Kusunoki M, Harada W, Kakudo M (1984) Structure and possible catalytic residues of Taka-amylase A. J Biochem 95(3):697–702

McCarter JD, Withers SG (1994) Mechanisms of enzymatic glycoside hydrolysis. Curr Opin Struct Biol 4(6):885–892

Moshfegh M, Shahverdi AR, Zarrini G, Faramarzi MA (2013) Biochemical characterization of an extracellular polyextremophilic α-amylase from the halophilic archaeon *Halorubrum xinjiangense*. Extremophiles 17(4):677–687

Nakajima M, Imamura H, Shoun H, Horinouchi S, Wakagi T (2004) Transglycosylation activity of *Dictyoglomus thermophilum* amylase A. Biosci Biotechnol Biochem 68(11):2369–2373

Niehaus F, Peters A, Groudieva T, Antranikian G (2000) Cloning, expression and biochemical characterisation of a unique thermostable pullulan-hydrolysing enzyme from the hyperthermophilic archaeon *Thermococcus aggregans*. FEMS Microbiol Lett 190(2):223–229

Nisha M, Satyanarayana T (2013) Recombinant bacterial amylopullulanases: developments and perspectives. Bioengineered 4(6):388–400

Onodera M, Yatsunami R, Tsukimura W, Fukui T, Nakasone K, Takashina T, Nakamura S (2013) Gene analysis, expression, and characterization of an intracellular α-amylase from the extremely halophilic archaeon *Haloarcula japonica*. Biosci Biotechnol Biochem 77(2):281–288

Oslancova A, Janecek S (2002) Oligo-1,6-glucosidase and neopullulanase enzyme subfamilies from the α-amylase family defined by the fifth conserved sequence region. Cell Mol Life Sci 59(11):1945–1959

Page RD (1996) TreeView: an application to display phylogenetic trees on personal computers. Comput Appl Biosci 12(4):357–358

Palomo M, Pijning T, Booiman T, Dobruchowska JM, van der Vlist J, Kralj S, Planas A, Loos K, Kamerling JP, Dijkstra BW, van der Maarel MJ, Dijkhuizen L, Leemhuis H (2011) *Thermus thermophilus* glycoside hydrolase family 57 branching enzyme: crystal structure, mechanism of action, and products formed. J Biol Chem 286(5):3520–3530

Pandey A, Nigam P, Soccol CR, Soccol VT, Singh D, Mohan R (2000) Advances in microbial amylases. Biotechnol Appl Biochem 31(2):135–152

Park JT, Song HN, Jung TY, Lee MH, Park SG, Woo EJ, Park KH (2013) A novel domain arrangement in a monomeric cyclodextrin-hydrolyzing enzyme from the hyperthermophile *Pyrococcus furiosus*. Biochim Biophys Acta 1834(1):380–386

Park KH, Jung JH, Park SG, Lee ME, Holden JF, Park CS, Woo EJ (2014) Structural features underlying the selective cleavage of a novel exo-type maltose-forming amylase from *Pyrococcus* sp. ST04. Acta Crystallogr D Biol Crystallogr 70(6):1659–1668. doi:10.1107/S1399004714006567

Perez-Pomares F, Bautista V, Ferrer J, Pire C, Marhuenda-Egea FC, Bonete MJ (2003) α-Amylase activity from the halophilic archaeon *Haloferax mediterranei*. Extremophiles 7(4):299–306

Przylas I, Tomoo K, Terada Y, Takaha T, Fujii K, Saenger W, Straeter N (2000) Crystal structure of amylomaltase from *Thermus aquaticus*, a glycosyltransferase catalysing the production of large cyclic glucans. J Mol Biol 296(3):873–886

Pujadas G, Ramirez FM, Valero R, Palau J (1996) Evolution of β-amylase: patterns of variation and conservation in subfamily sequences in relation to parsimony mechanisms. Proteins 25(4):456–472

Puspasari F, Radjasa O, Noer A, Nurachman Z, Syah Y, van der Maarel M, Dijkhuizen L, Janecek S, Natalia D (2013) Raw starch degrading α-amylase from *Bacillus aquimaris* MKSC 6.2: isolation and expression of the gene, bioinformatics and biochemical characterization of the recombinant enzyme. J Appl Microbiol 114(1):108–120

Ramasubbu N, Ragunath C, Mishra PJ (2003) Probing the role of a mobile loop in substrate binding and enzyme activity of human salivary amylase. J Mol Biol 325(5):1061–1076

Rashid N, Cornista J, Ezaki S, Fukui T, Atomi H, Imanaka T (2002) Characterization of an archaeal cyclodextrin glucanotransferase with a novel C-terminal domain. J Bacteriol 184(3):777–784

Ray RR, Nanda G (1996) Microbial β-amylases: biosynthesis, characteristics, and industrial applications. Crit Rev Microbiol 22(3):181–199

Robert X, Haser R, Gottschalk TE, Ratajczak F, Driguez H, Svensson B, Aghajari N (2003) The structure of barley α-amylase isozyme 1 reveals a novel role of domain C in substrate recognition and binding: a pair of sugar tongs. Structure 11(8):973–984

Sauer J, Sigurskjold BW, Christensen U, Frandsen TP, Mirgorodskaya E, Harrison M, Roepstorff P, Svensson B (2000) Glucoamylase: structure/function relationships, and protein engineering. Biochim Biophys Acta 1543(2):275–293

Savchenko A, Vieille C, Kang S, Zeikus JG (2002) *Pyrococcus furiosus* α-amylase is stabilized by calcium and zinc. Biochemistry 41(19):6193–6201

Sharma A, Satyanarayana T (2013) Microbial acid-stable α-amylases: characteristics, genetic engineering and applications. Process Biochem 48(2):201–211

Sivakumar N, Li N, Tang JW, Patel BK, Swaminathan K (2006) Crystal structure of AmyA lacks acidic surface and provide insights into protein stability at poly-extreme condition. FEBS Lett 580(11):2646–2652

Stam MR, Danchin EG, Rancurel C, Coutinho PM, Henrissat B (2006) Dividing the large glycoside hydrolase family 13 into subfamilies: towards improved functional annotations of α-amylase-related proteins. Protein Eng Des Sel 19(12):555–562

Sun Y, Lv X, Li Z, Wang J, Jia B, Liu J (2015) Recombinant cyclodextrinase from *Thermococcus kodakarensis* KOD1: expression, purification, and enzymatic characterization. Archaea 2015:397924

Svensson B (1988) Regional distant sequence homology between amylases, α-glucosidases and transglucanosylases. FEBS Lett 230(1–2):72–76

Tachibana Y, Mendez Leclere M, Fujiwara S, Takagi M, Imanaka T (1996) Cloning and expression of the α-amylase gene from the hyperthermophilic archaeon Pyrococcus sp. KOD1, and characterization of the enzyme. J Ferment Bioeng 82(3): 224–232

Takata H, Kuriki T, Okada S, Takesada Y, Iizuka M, Minamiura N, Imanaka T (1992) Action of neopullulanase. Neopullulanase catalyzes both hydrolysis and transglycosylation at α-(1,4)- and α-(1,6)-glucosidic linkages. J Biol Chem 267(26):18447–18452

Uitdehaag JC, Mosi R, Kalk KH, van der Veen BA, Dijkhuizen L, Withers SG, Dijkstra BW (1999) X-ray structures along the reaction pathway of cyclodextrin glycosyltransferase elucidate catalysis in the α-amylase family. Nat Struct Biol 6(5):432–436

UniProt Consortium (2015) UniProt: a hub for protein information. Nucleic Acids Res 43(Database Issue):D204–D212

van der Kaaij RM, Janecek S, van der Maarel MJEC, Dijkhuizen L (2007) Phylogenetic and biochemical characterization of a novel cluster of intracellular fungal α-amylase enzymes. Microbiology 153(12):4003–4015

van der Maarel MJEC, van der Veen B, Uitdehaag JC, Leemhuis H, Dijkhuizen L (2002) Properties and applications of starch-converting enzymes of the α-amylase family. J Biotechnol 94(2):137–155

Vihinen M, Mantsala P (1989) Microbial amylolytic enzymes. Crit Rev Biochem Mol Biol 24(4):329–418

Vujicic-Zagar A, Pijning T, Kralj S, Lopez CA, Eeuwema W, Dijkhuizen L, Dijkstra BW (2010) Crystal structure of a 117 kDa glucansucrase fragment provides insight into evolution and product specificity of GH70 enzymes. Proc Natl Acad Sci U S A 107(50):21406–21411

Wang S, Lu Z, Lu M, Qin S, Liu H, Deng X, Lin Q, Chen J (2008) Identification of archaeon-producing hyperthermophilic α-amylase and characterization of the α-amylase. Appl Microbiol Biotechnol 80(4):605–614

Wang H, Gong Y, Xie W, Xiao W, Wang J, Zheng Y, Hu J, Liu Z (2011) Identification and characterization of a novel thermostable *gh-57* gene from metagenomic fosmid library of the Juan de Fuca Ridge hydrothemal vent. Appl Biochem Biotechnol 164(8):1323–1338

Zona R, Chang-Pi-Hin F, O'Donohue MJ, Janecek S (2004) Bioinformatics of the family 57 glycoside hydrolases and identification of catalytic residues in amylopullulanase from *Thermococcus hydrothermalis*. Eur J Biochem 271(14):2863–2872

Zorgani MA, Patron K, Desvaux M (2014) New insight in the structural features of haloadaptation in α-amylases from halophilic *Archaea* following homology modeling strategy: folded and stable conformation maintained through low hydrophobicity and highly negative charged surface. J Comput Aided Mol Des 28(7):721–734

Chapter 18
Proteins of DNA Replication from Extreme Thermophiles: PCR and Beyond

Dennis W. Grogan

1 Overview

The intrinsically thermostable proteins of extreme- and hyper-thermophiles have been used to develop an impressive, and perhaps bewildering, array of techniques for experimental molecular biology and environmental, forensic, and medically related diagnostics. This chapter provides a cursory and selective overview of the principles underlying some of these diverse techniques, using natural DNA replication as a frame of reference. It first summarizes the basic features of DNA replication in living cells and then examines conventional PCR. This comparison provides a context for summarizing some of the enhancements of PCR, as well as the development of distinct methods, all of which exploit intrinsically stable proteins of thermophilic archaea and bacteria (Table 18.1). Although a single-subunit DNA polymerase remains the centerpiece of most of the techniques described in this chapter, the innovations typically involve supplementing or modifying this enzyme with the functional diversity of proteins produced by other thermophiles, including some that have never been cultured.

2 Biological DNA Replication

Cells normally exert strict control over the timing of DNA synthesis. The biochemical strategies for this controlled initiation differ among organisms, but all involve proteins that are poised to unwind the duplex at one or more specific sites

D.W. Grogan (✉)
Department of Biological Sciences, University of Cincinnati,
614 Rieveschl Hall, Cincinnati, OH 45221-0006, USA
e-mail: grogandw@ucmail.uc.edu

P.H. Rampelotto (ed.), *Biotechnology of Extremophiles*, Grand Challenges in Biology and Biotechnology 1, DOI 10.1007/978-3-319-13521-2_18

Table 18.1 Intrinsically thermostable proteins of DNA synthesis

Type	Technological uses	Relevant properties
DNA polymerases		
Bacterial, A-family	Conventional PCR	Robust elongation
Bacterial, phage	Whole-genome amplification	Strand-displacing polymerization
Archaeal, B-family	High-fidelity PCR	Proofreading (3′ exonuclease); bind template dU
Archaeal, Y-family	PCR of damaged DNA	Bypass lesions, non-processive
DNA-binding proteins	Fused to DNA polymerase	Increased processivity
DNA ligases	single-nucleotide genotyping	Highly specific for correct base-pairing
Accuracy/repair proteins		
MutS	SNP detection	Binds mismatches in ds DNA
dUTPase	High-fidelity PCR	Prevents dUTP incorporation
Uracil DNA glycosylase	High-fidelity PCR	Excises dU from DNA

(the origin(s) of replication) in response to some form of cellular signal. The subsequent steps are also very similar in all cells, although archaea, bacteria, and eukaryotes may use different proteins to catalyze them. Details regarding the molecular features of DNA replication and how they differ among the three domains have been summarized by O'Donnell et al. (2013).

Once triggered to initiate replication, proteins associated with the origin unwind it to create a bubble large enough to allow assembly of the replicative helicase. This helicase is a protein complex that encircles one DNA strand and moves along it, driven by ATP hydrolysis, thereby forcing the paired strands apart. Two helicase complexes, one on each strand of the duplex, move out from the origin, enlarging the DNA bubble as they go. Each helicase now becomes the center of a corresponding replication fork; the following descriptions focus on one such fork, which has two 'arms' created by the helicase.

Polymerization reactions proceed simultaneously on both arms of the replication fork, but in opposite directions, as dictated by the anti-parallel orientation of the two strands in dsDNA. On one arm, the "leading strand" polymerizes in the same direction as the fork progresses, and this process can, in principle, continue indefinitely. In contrast, the "lagging strand" elongates in the opposite direction as overall fork progression, demanding frequent repetition of the following synthesis cycle as new ssDNA is created by the helicase: (i) DNA primase uses the ssDNA as a template to synthesize a short RNA; (ii) this RNA primer is recognized as substrate by the replicative DNA polymerase, which binds to its 3′ end and carries out its basic function, i.e., primer extension. A toroidal protein complex (the 'sliding clamp') associated with the polymerase encircles the DNA and holds the polymerase on its template. (iii) The DNA polymerase encounters the previous RNA primer and dissociates. (iv) The primer is removed and replaced by DNA to leave only a nick, which is then (v) sealed by DNA ligase (O'Donnell et al. 2013).

On the two completed arms emerging from the replication fork, a process occurs which can be considered a form of DNA repair (DNA mismatch repair) as well as the last stage of synthesis. Specialized proteins (homologs of the *E. coli* MutS and MutL proteins) scan the duplex DNA for mismatches. Where mismatches are found, these proteins direct the destruction of the newly synthesized strand, creating a gap approximately 1 kbp long. The resulting gaps are then filled in by the replicative DNA polymerase, thereby correcting the original synthesis error (Li 2008).

3 Polymerase Chain Reaction (PCR)

Although both processes necessarily have similarities, conventional PCR is strikingly simple compared to biological DNA synthesis, involving only three steps and one protein. In PCR, as in living cells, duplex DNA is not a substrate for DNA synthesis; thus PCR does not begin when the ingredients (template DNA, primers, dNTPs, and thermostable DNA polymerase) are combined, but only after the complete reaction mixture is heated to near 95 °C, denaturing all duplex DNA. As a unimolecular reaction, denaturation is complete within a few seconds; it is followed by cooling to a temperature that allows the primers present in the mixture (usually two synthetic oligonucleotides) to anneal specifically to their complements in the ssDNA Because the primer concentration is usually relatively high, also this second step is rapid, following "pseudo-first-order" kinetics.

Primers that have annealed (base-paired) to the high-MW template strands are then recognized as substrate by the monomeric DNA polymerase and extended. The configuration of primer annealing is critical for successful amplification of an interval of the template DNA. Oligonucleotides must bind to sites within a short distance (usually a few kb) of each other, to allow successful extension from one site to the other, and both 3′ ends must be oriented to the inside of this interval. In the third step, the temperature is increased to approximately 72°; this accelerates the extension of correctly annealed primers, a process which must continue until the 3′ end which originated from one primer passes the binding site of the opposing primer. The three-step cycle (denaturation, primer annealing, primer extension) is then repeated, roughly doubling with each cycle the amount of dsDNA in the mixture that corresponds to the interval between the two priming sites.

The simplicity of PCR distinguishes it from genome replication in several respects, with corresponding functional consequences. In PCR, for example, high temperature accomplishes the unwinding performed in vivo by helicase, and exogenous primers confine replication to a fixed interval. This interval typically corresponds to only a small fraction of the template DNA sequence, but its rate of replication is very high. The performance of PCR can be enhanced by certain proteins from extreme- and hyper-thermophiles (Table 18.1), whereas other thermostable proteins provide for other techniques of DNA analysis distinct from PCR, with their own characteristic uses. These proteins, and their technological uses, are discussed below, with reference to the aspect of natural DNA replication that they affect or mimic.

4 Temporal Control and 'Hot-Start' Techniques

Although precise timing of initiation does not present a challenge for most PCR applications, the residual polymerase activity exhibited at low temperature (i.e., during set-up) can compromise primer specificity in some situations. For example, if the 3′ end of a primer anneals by only a few base pairs to a non-target-site gap in the DNA sample, or another primer, and is extended, a new primer may be created which can yield aberrant product. Such problems have been addressed by various "hot-start" PCR techniques. Most of these methods involve adding the DNA polymerase to the reaction in a reversibly inhibited form, which becomes catalytically active only when heated during the denaturation phase of PCR. The use of high temperature to reverse the inhibition is made possible by the intrinsic thermostability of the DNA polymerases used in PCR, but the method of inhibition itself can take a variety of forms, including antibody binding, reagent sequestering, and chemical modification (Paul et al. 2010; Barnes and Rowlyk 2002; Kermekchiev et al. 2003; Kaboev et al. 2000; Louwrier and van de Valk 2005; Liu and Liu 2011).

5 Sequence Specificity and Its Implications for Whole-Genome Amplification Versus Genotyping

The primary advantage of conventional PCR is the specificity with which it amplifies a short interval of a much longer DNA sequence. However, if the goal is to replicate DNA indiscriminately, i.e., to carry out whole-genome amplification, this regional specificity must be overcome. One of the most generally used approaches for whole-genome synthesis in vitro is the "multiple-displacement amplification" (MDA) technique (Dean et al. 2002) which synthesizes a dsDNA indiscriminately at constant temperature. MDA uses thermophilic DNA polymerases with strand-displacement (SD) activity; this feature solves the strand-separation problem in a way distinct from both the strategy of biological replication (ATP-dependent helicases) and that of PCR (thermal denaturation). An SD polymerase acts, in a sense, as its own helicase; when it extends a primer, it displaces any strand it encounters, thereby converting it to ssDNA. The displaced strand provides a potential new template, creating the possibility of sustained isothermal DNA synthesis. The reaction is typically performed at a constant, but moderately high temperature, and suitable polymerases have been identified and cloned from moderately thermophilic bacteria, extremely thermophilic bacteria, and phage harvested from thermal environments (Hafner et al. 2001; Chander et al. 2014).

Realizing this potential of sustained isothermal synthesis requires a way of providing and replenishing DNA synthesis primers. Nicked duplex DNA can serve as a substrate for one round of synthesis, and new rounds can be triggered by adding endonucleases or similar activities that create new nicks as the old ones are extended. Under these conditions, the SD polymerase will catalyze continuous production of single-stranded DNA (with a heterogeneous length distribution) at a constant rate from both strands of the original duplex template. If, in addition to the nicking activity, low-specificity ("random") primers are added, the process produces duplex DNA exponentially, as the ssDNA displaced by synthesis provides priming sites for a new round of synthesis (Dean et al. 2002). Unlike biological DNA replication, however, the continual, spatially random re-initiation of new synthesis generates DNA with a complex, highly branched structure, which complicates some uses of the product (For a succinct but comprehensive overview of SD and related isothermal DNA synthesis methods, see Yan et al. 2014).

At the other extreme of specificity, certain genetic analyses, such as scoring single nucleotide polymorphisms (SNPs), require a narrower focus (i.e., a single base-pair) and a higher stringency than PCR can provide. In principle, Taq will not extend a mismatched 3′ end, and this has been used to design various forms of allele-specific PCR, but in practice, the desired goal of 'all vs. nothing' discrimination between one base and another at a specific location is often elusive. DNA ligases provide more-reliable discrimination among bases because they have negligible activity with a 3′ end that is not correctly base-paired. Thus, two adjacent oligonucleotide probes annealed to a template are ligated successfully to each other only if a particular base occurs in the template DNA opposite the 3′ end of the 'upstream' probe. Running multiple ligation reactions, each with a different base at the 3′ end of the 'upstream' probe, thus allows unambiguous identification of which base occurs at that site (or which two, in the case of heterozygous diploids).

The ligases of hyperthermophilic archaea provide special advantages for these techniques (Wiedmann et al. 1994). The thermostability of these enzymes allows multiple rounds of thermal denaturation and ligation, so that ligation product can accumulate, making it detectable. In addition, the archaeal ligases use ATP to activate phosphodiester bond formation, which is a technical advantage over thermostable bacterial ligases, which use the significantly more expensive and chemically labile NAD. Reaction mixtures contain template DNA, hybridization probes, ATP, and ligase; multiple rounds of annealing and ligation are performed by thermal cycling, analogous to PCR. If this cycling is performed using only two probes that anneal next to each other, the result is the linear increase of ligated product with each subsequent cycle, and the technique is called ligation detection reaction (LDR). If, in addition, a corresponding pair of probes for the complementary strand of the query site is added, ligated product accumulates exponentially, as in PCR, because the ligated product made in one round serves as additional template for the next round. This exponential mode is termed ligase chain reaction (LCR) (Wiedmann et al. 1994).

Because the ligated product is short, detecting it generally involves capillary or polyacrylamide gel electrophoresis. Alternatively, product (or its absence) can be detected in melting curves performed after a sufficient number of ligation cycles have let the product (if any) accumulate. In this method, it is important that the ligated product paired to its complement has a T_m separated by several degrees from those of both the template and unligated probes.

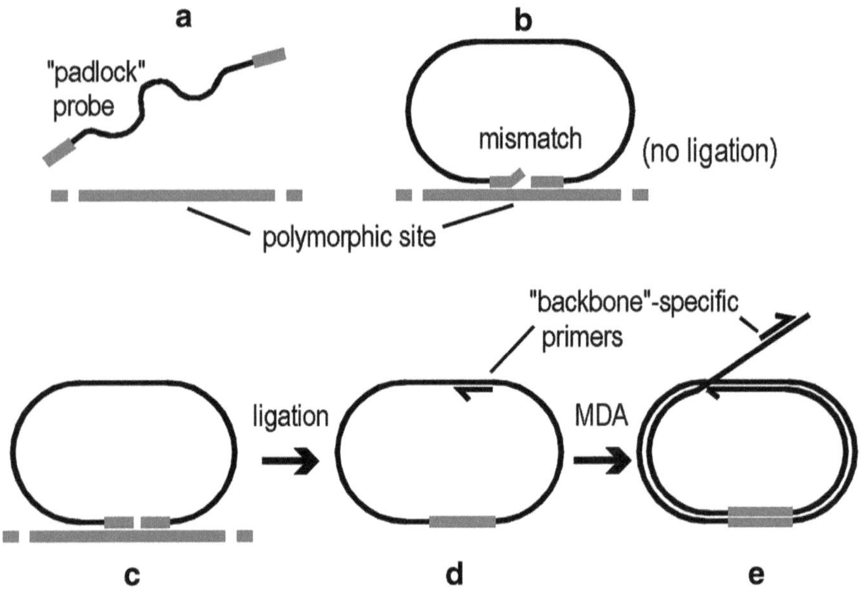

Fig. 18.1 The rolling-circle amplification (RCA) method of scoring single-nucleotide polymorphisms

Another approach to detection uses a different probe topology, in which both sides of the query site are represented by the two ends of one relatively long, ssDNA, called a 'padlock probe'. This rolling-circle amplification (RCA) technique (Lizardi et al. 1998; Nilsson et al. 2006) is illustrated in Fig. 18.1. If the probe ends correspond to the polymorphic site, the base-pairing of both ends of this probe to the template DNA generates a small single-stranded circle by ligation. Although a thermostable ligase and thermal cycling provides a way to build up levels of the circular product, the amplification needed for detection is usually achieved by using the ssDNA circles as template. In this case, a conventional primer specific for the central region, or 'backbone', of the padlock probe initiates rolling-circle DNA synthesis by a thermostable SD DNA polymerase (Fig. 18.1d, e). This isothermal process produces long concatemers of the probe sequence in those reactions where the probe has circularized through ligation. Variations on this theme of circularization/ rolling-circle amplification have been devised and applied to high-throughput genotyping (Alsmadi et al. 2003; Yan et al. 2014; Chang et al. 2015).

6 Replication Accuracy and Its Technological Importance

Taq remains one of the most effective DNA polymerases for conventional PCR, due, in part, to its lack of 3′ exonuclease activity, which allows fast and robust chain elongation. This results in a relatively high error rate, however, which creates a problem if the amplification product contains a functional gene that is to be cloned. For example, with a mis-incorporation rate of 3×10^{-4} per nucleotide inserted (Tindall and Kunkel 1988), Taq will leave an error in up to 30 % of the strands it makes from a 1 kbp target. Because the product of each cycle serves as a template for the next cycle, a reaction that gives 25 doublings of DNA will yield error-free DNA molecules at an expected fraction of about $(0.7)^{25} = 10^{-4}$ of the total. Thus, gene cloning and other uses that require preservation of the sequence in the amplified product demand polymerization accuracies much higher than that of native Taq polymerase.

Some of the earliest solutions to this problem replaced Taq with thermostable proofreading DNA polymerases, generally B-family DNA polymerases from extreme- or hyper-thermophilic archaea. In contrast to Taq, these DNA polymerases have strong 3′ exonuclease (proofreading) activity. This results in much lower mis-incorporation rates, although it also creates the danger of probe shortening during PCR setup. Since hyperthermophilic archaea (HA) grow at higher temperatures than *Thermus* does, their polymerases are generally more thermostable, thus decreasing the loss of activity at later cycles in PCR programs, when the demand for polymerase is also the greatest.

With these advantages of archaeal DNA polymerases comes a property that complicates PCR and related techniques, however, and remains somewhat mysterious in terms of its biological significance. Archaeal DNA polymerases bind tightly to dU that they encounter in the template strand, which impedes nascent strand synthesis (Lasken et al. 1996). The technical obstacle has been overcome in various ways which improve the performance of PCR and illustrate some of the biological implications of dU-induced stalling.

One modification that has improved PCR performance by the DNA polymerases of HA (such as *Pfu*) involves supplementing the reaction mixture with a thermostable dUTPase (Hogrefe et al. 2002). The success of this technique suggests that much of the template dU encountered during PCR results from cytosine deamination before, not after, incorporation. This is consistent with in vitro studies demonstrating relatively poor discrimination against dUTP by DNA polymerases of HA (Gruz et al. 2003). Another effective technique involves supplementing the archaeal polymerase with a thermostable uracil DNA glycosylase (UDG) (Liu and Liu 2011). Since the effect of the UDG will be to destroy the dU-containing strand as a usable template in any subsequent cycle, the fact that this supplementation improves PCR performance implies that one of the problems caused by dU binding is sequestration of otherwise active DNA polymerase molecules. This, in turn, suggests that the unproductive polymerase:dU complex can be very stable, and raises questions about what impact such a complex would have in vivo. A third solution to the problem has been made possible by identifying a specific dU-binding site in the DNA polymerases of HA that mediates the stalling (Fogg et al. 2002). Introducing mutations into HA polymerases to remove this pocket has resulted in extremely thermostable and accurate DNA polymerases insensitive to the occurrence of dU in its template (Gaidamaviciute et al. 2010).

The extremely high level of accuracy achieved by DNA replication in living cells reflects a significant contribution from post-replicational mismatch repair (MMR), which decreases the error rate by an additional factor of 100–1000 over that achieved by proofreading polymerases alone. The critical steps of MMR are mediated by homologs of the *E. coli* MutS and MutL proteins, although the overall process remains difficult to reproduce with high efficiency in vitro (Li 2008). The first step of MMR is binding of the MutS homolog to a mismatch, and this property can be used to detect mismatches in vitro. The intrinsically thermostable MutS protein of a *Thermus* species has been fluorescently labelled, and its binding has shown to detect mismatches, although the efficiency of detection varies with the identity of the mismatched bases (Cho et al. 2007). No hyperthermophilic archaea have MutS and MutL proteins, however, and the biological significance of this absence remains unclear (White and Grogan 2008).

At the other extreme of the accuracy spectrum, certain situations, such as directed evolution experiments, call for many changes to be introduced into a gene, and conventional PCR using Taq is often too accurate for these methods to be efficient. PCR using Taq can be made highly erroneous, however, by adding Mn^{++} to the reaction and skewing the concentration of individual dNTPs. The density of mutations can be controlled by appropriate adjustment of reaction conditions, and is reported to be higher than purely chemical or purely biological mutagenesis (Rasila et al. 2009).

In principle, enzymes of extreme- and hyper-thermophiles that generate mutations in vivo could also be harnessed to provide an alternative route to mutagenic PCR. Error-prone thermostable DNA polymerases are well-known, as exemplified by the Y-family DNA polymerases of hyperthermophilic archaea, which have been crystallized with a variety of DNA substrates and analyzed extensively in biochemical terms. Large catalytic sites and absence of proofreading activity allow these enzymes to insert nucleotides opposite diverse DNA lesions; this supports their biological function, which is to mediate trans-lesion DNA synthesis (TLS). The catalytic features of these TLS polymerases also limit the accuracy of replicating intact template, making some of them (Dbh of *Sulfolobus acidocaldarius*, for example) among the most error-prone DNA polymerases known (Yang and Woodgate 2007; Pata 2010). However, consistent with the biological imperative to limit the genetic costs of lesion bypass, all TLS polymerases, including the extremely thermostable ones, have low intrinsic processivity, and thus synthesize only a few base pairs at a time (Pata 2010; Waters et al. 2009). In principle, modifications to increase the processivity of these enzymes, or simply adding high concentrations to synthesis reactions, could support an alternative form of error-prone PCR. However, the properties observed for the *Sulfolobus* enzymes (Potapova et al. 2002; Wilson and Pata 2008) argue that they would probably create a high proportion of frameshift mutations, which would be unacceptable for generating diversity in functional genes. It remains to be determined whether this limitation applies to other intrinsically thermostable TLS polymerases, such as those from extremely thermophilic bacteria.

7 Lesion By-Pass and Amplification of Damaged DNA

Although thermostable Y-family polymerases of hyperthermophilic archaea have not been exploited as mutagens, they do show promise in the distinct but related context where relaxed polymerase specificity for the template improves extension and enables otherwise difficult samples to be analyzed. Consistent with the biological function of TLS polymerases in supporting replication past DNA damage, adding these polymerases to a standard PCR reaction improves the amplification of damaged DNA (McDonald et al. 2006). Although reported for UV damage, this strategy should, in principle, also apply to template DNA damaged by depurination, oxidation, and related degradation processes. The problem of amplifying damaged DNA has also been approached by engineering of the conventional (replicative) polymerases (Ghadessy et al. 2004; d'Abbadie et al. 2007; Jozwiakowski et al. 2014), using methods such as compartmentalized self-replication, described below (Making PCR resistant to inhibitors).

8 Incorporation of Unnatural Bases

A challenge related to the amplification of damaged DNA is incorporation of non-native nucleotides opposite a normal template. This is essential to processes such as fluorescent labeling of nucleic acids as primers or hybridization probes, DNA sequencing, post-synthetic covalent attachment to surfaces, etc. These applications require a polymerase to accept and insert chemically modified dNTPs which are larger than natural dNTPs and which normally would be excluded from the polymerase active site. Much of the progress in this area has come from the engineering of Taq polymerase to relax its specificity for its monomeric substrate (dNTPs). Molecular-evolution approaches (see below), have produced variants of Taq with greatly increased extension of terminally mismatched primers or incorporation of fluorescently labeled dNTPs (Laos et al. 2013; Gardner and Jack 2002). The variants have not exhibited a corresponding increase in error rate of synthesis, thus indicating that they retain accurate selection of natural dNTPs.

"Relaxed" forms of thermostable DNA polymerases produced by these efforts exhibit a range of properties, exemplified by a modified *Thermococcus* polymerase that synthesizes RNA from a DNA template (Cozens et al. 2012). Similarly, a variety of enhanced or new techniques are made possible by such polymerases. Examples include improved incorporation of reversible dye terminators for next-generation sequencing, amplification of ancient DNA (see below), and the use of artificial bases to create new base-pairs that can be faithfully replicated along with the natural A:T and G:C base-pairs (Laos et al. 2014).

9 Making PCR Resistant to Inhibitors

Biological DNA replication occurs in the stable, favorable biochemical milieu within living cells, whereas many uses of PCR require DNA synthesis in the presence of substances that interfere with the process. The versatility of PCR obviously increases with increased robustness, which, in turn, generally depends on the processivity of the extension reaction. Processivity of the DNA polymerase itself has been enhanced by forming certain chimeras (Yamagami et al. 2014), while other groups have explored ways to reproduce in vitro the biological strategy for processivity by including a sliding clamp (Motz et al. 2002; Ribble et al. 2015). An analogous, and commercially successful, method uses a generalized dsDNA-binding protein of Sulfolobus, which mimics the sliding clamp when fused to the DNA polymerases (Wang et al. 2004). This, in turn, greatly enhances the overall performance of PCR, making it feasible for unpurified specimens such as whole blood, for example (Trombley-Hall et al. 2013).

The performance of PCR in matrices containing various inhibitors has also been improved dramatically by applying generalized protein-evolution techniques to the polymerase (Kermekchiev et al. 2009). A particularly effective technique, compartmentalized self-replication (CSR), is uniquely relevant to thermostable polymerases because it enables these enzymes to amplify their own genes. As depicted in Fig. 18.2, the CSR technique involves cloning a mutagenized DNA polymerase gene in a suitable expression vector and transforming *E. coli* with the resulting plasmid library (Ghadessy et al. 2001). Individual cells, each expressing a different variant from the library, are encapsulated separately with dNTPs, primer, and buffer in droplets of a water-in-oil emulsion. Thermal cycling of the emulsion lyses the cell in each compartment, enabling each polymerase to amplify its own gene, according to its catalytic properties. The resulting products are pooled and re-cloned to produce a second-generation library suitable for a second round of selection (Ghadessy et al. 2001), and the process can be repeated for several cycles.

The method not only selects improved polymerase genes on the basis of the performance of the encoded enzyme, but it can do so under particular conditions of interest, such as in the presence of inhibitors. In this way, Taq polymerases modified with multiple amino acid substitutions have been selected to give greatly enhanced performance under unfavorable conditions. The inherent plasticity of enzyme properties with respect to amino acid sequence, and the ability of CSR to recover combinations of amino acid changes that improve performance under PCR conditions of choice, suggests that variant polymerases will continue to be tailored individually with specific properties addressing particular applications. Examples include amplification of ancient DNA (d'Abbadie et al. 2007; Baar et al. 2011), tolerance of fluorescently labelled dNTPs (Ramsay et al. 2010), and shortened PCR cycles (Arezi et al. 2014).

CSR as a protein engineering method appears to be confined to thermostable DNA polymerases or SD polymerases, as these represent the few enzymes capable of replicating their own genes in *E. coli* lysates. In principle, however, CSR should be capable of selecting any protein or protein variant that improves the PCR performance of a DNA polymerase "in trans". This reflects the feasibility of including the gene for a potentially enhancing protein on an expression vector along with the DNA polymerase gene. Thus, the ancillary protein can be mutated to generate a library of variants, and each variant can be inserted into a construct also expressing the polymerase. The population of constructs is then subjected to selection for amplification of the accessory protein gene by appropriate primers, analogous to the conventional CSR procedure. Alternatively, if the mutant library can be made as sufficiently long inserts, it also may prove feasible to mutagenize and apply selection to both the polymerase and ancillary protein genes concurrently.

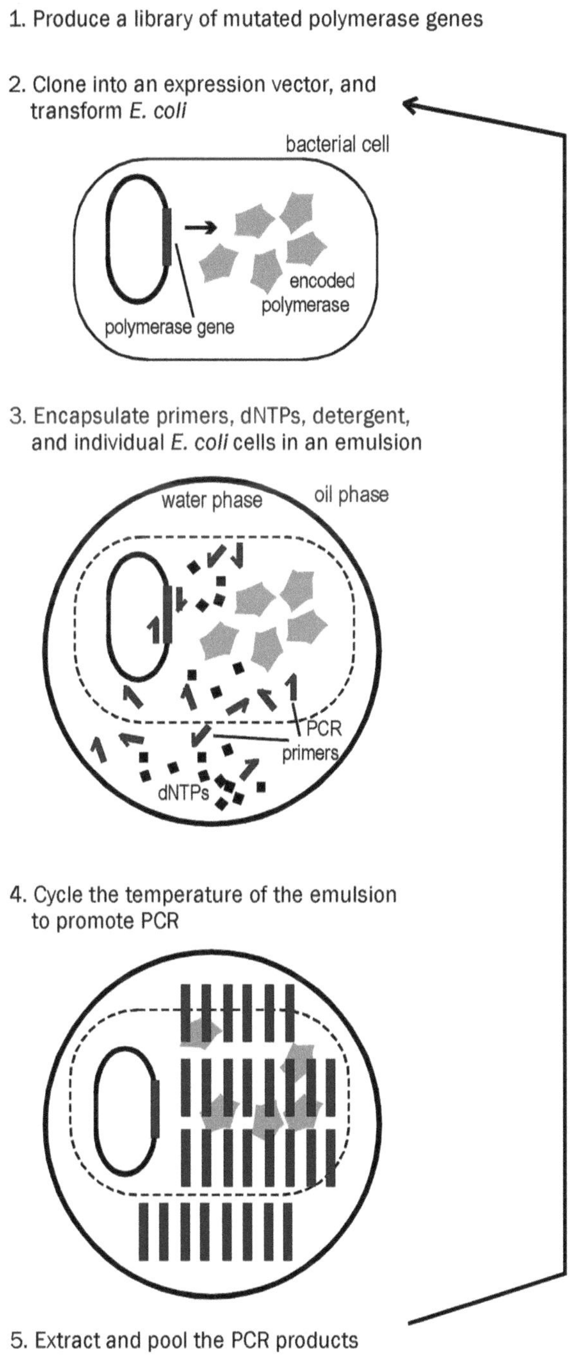

Fig. 18.2 Compartmentalized self-replication (CSR) of DNA polymerase genes

10 Conclusion

The intrinsically thermostable proteins which package, replicate and repair the genomes of extreme- and hyper-thermophilic archaea and bacteria provide the enzymatic basis for a spectrum of increasingly sophisticated molecular-genetic techniques. Diverse forms of PCR continue to dominate the landscape of these techniques, which underscores the flexible utility of DNA synthesis in vitro and the robustness which proteins from thermophiles lend to this and related enzymatic processes. Many of these methods use ancillary proteins to enhance the minimal mode of DNA replication provided by conventional PCR, and in certain respects these enhancements represent closer approximations of genome replication in living cells. Although some research has pursued the goal of complete and accurate genome replication in an explicitly biomimetic way, often the most effective techniques result from unnatural combinations of proteins and create processes without any obvious biological precedent. The pace of innovation shows no sign of slowing, as new approaches, new enzymes, and new protein-engineering techniques are combined in new permutations to address a widening scope of molecular-genetic challenges and possibilities.

References

Alsmadi OA, Bornarth CJ, Song W, Wisniewski M, Du J, Brockman JP, Faruqi AF, Hosono S, Sun Z, Du Y, Wu X, Egholm M, Abarzua P, Lasken RS, Driscoll MD (2003) High accuracy genotyping directly from genomic DNA using a rolling circle amplification based assay. BMC Genomics 4:21

Arezi B, McKinney N, Hansen C, Cayouette M, Fox J, Chen K, Lapira J, Hamilton S, Hogrefe H (2014) Compartmentalized self-replication under fast PCR cycling conditions yields Taq DNA polymerase mutants with increased DNA-binding affinity and blood resistance. Front Microbiol 5:408

Baar C, d'Abbadie M, Vaisman A, Arana ME, Hofreiter M, Woodgate R, Kunkel TA, Holliger P (2011) Molecular breeding of polymerases for resistance to environmental inhibitors. Nucleic Acids Res 39, e51

Barnes WM, Rowlyk KR (2002) Magnesium precipitate hot start method for PCR. Mol Cell Probes 16:167–171

Chander Y, Koelbl J, Puckett J, Moser MJ, Klingele AJ, Liles MR, Carrias A, Mead DA, Schoenfeld TW (2014) A novel thermostable polymerase for RNA and DNA loop-mediated isothermal amplification (LAMP). Front Microbiol 5:395

Chang K, Deng S, Chen M (2015) Novel biosensing methodologies for improving the detection of single nucleotide polymorphism. Biosens Bioelectron 66:297–307

Cho M, Chung S, Heo SD, Ku J, Ban C (2007) A simple fluorescent method for detecting mismatched DNAs using a MutS-fluorophore conjugate. Biosens Bioelectron 22:1376–1381

Cozens C, Pinheiro VB, Vaisman A, Woodgate R, Holliger P (2012) A short adaptive path from DNA to RNA polymerases. Proc Natl Acad Sci USA 109:8067–8072

d'Abbadie M, Hofreiter M, Vaisman A, Loakes D, Gasparutto D, Cadet J, Woodgate R, Paabo S, Holliger P (2007) Molecular breeding of polymerases for amplification of ancient DNA. Nat Biotechnol 25:939–943

Dean FB, Hosono S, Fang L, Wu X, Faruqi AF, Bray-Ward P, Sun Z, Zong Q, Du Y, Du J, Driscoll M, Song W, Kingsmore SF, Egholm M, Lasken RS (2002) Comprehensive human genome amplification using multiple displacement amplification. Proc Natl Acad Sci USA 99:5261–5266

Fogg MJ, Pearl LH, Connolly BA (2002) Structural basis for uracil recognition by archaeal family-B DNA polymerases. Nat Struct Biol 9:922–927

Gaidamaviciute E, Tauraite D, Gagilas J, Lagunavicius A (2010) Site-directed chemical modification of archaeal *Thermococcus litoralis* Sh1B DNA polymerase: acquired ability to read through template-strand uracils. Biochim Biophys Acta 1804:1385–1393

Gardner AF, Jack WE (2002) Acyclic and dideoxy terminator preferences denote divergent sugar recognition by archaeon and Taq DNA polymerases. Nucleic Acids Res 30:605–613

Ghadessy FJ, Ong JL, Holliger P (2001) Directed evolution of polymerase function by compartmentalized self-replication. Proc Natl Acad Sci USA 98:4552–4557

Ghadessy FJ, Ramsay N, Boudsocq F, Loakes D, Brown A, Iwai S, Vaisman A, Woodgate R, Holliger P (2004) Generic expansion of the substrate spectrum of a DNA polymerase by directed evolution. Nat Biotechnol 22:755–759

Gruz P, Shimizu M, Pisani FM, De Felice M, Kanke Y, Nohmi T (2003) Processing of DNA lesions by archaeal DNA polymerases from *Sulfolobus solfataricus*. Nucleic Acids Res 31:4024–4030

Hafner GJ, Yang IC, Wolter LC, Stafford MR, Giffard PM (2001) Isothermal amplification and multimerization of DNA by Bst DNA polymerase. Biotechniques 30:852–856

Hogrefe HH, Hansen CJ, Scott BR, Nielson KB (2002) Archaeal dUTPase enhances PCR amplifications with archaeal DNA polymerases by preventing dUTP incorporation. Proc Natl Acad Sci USA 99:596–601

Jozwiakowski SK, Keith BJ, Gilroy L, Doherty AJ, Connolly BA (2014) An archaeal family-B DNA polymerase variant able to replicate past DNA damage: occurrence of replicative and translesion synthesis polymerases within the B family. Nucleic Acids Res 42:9949–9963

Kaboev OK, Luchkina LA, Tretiakov AN, Bahrmand AR (2000) PCR hot start using primers with the structure of molecular beacons (hairpin-like structure). Nucleic Acids Res 28, e94

Kermekchiev MB, Tzekov A, Barnes WM (2003) Cold-sensitive mutants of Taq DNA polymerase provide a hot start for PCR. Nucleic Acids Res 31:6139–6147

Kermekchiev MB, Kirilova LI, Vail EE, Barnes WM (2009) Mutants of Taq DNA polymerase resistant to PCR inhibitors allow DNA amplification from whole blood and crude soil samples. Nucleic Acids Res 37, e40

Laos R, Shaw R, Leal NA, Gaucher E, Benner S (2013) Directed evolution of polymerases to accept nucleotides with nonstandard hydrogen bond patterns. Biochemistry 52:5288–5294

Laos R, Thomson JM, Benner SA (2014) DNA polymerases engineered by directed evolution to incorporate non-standard nucleotides. Front Microbiol 5:565

Lasken RS, Schuster DM, Rashtchian A (1996) Archaebacterial DNA polymerases tightly bind uracil-containing DNA. J Biol Chem 271:17692–17696

Li GM (2008) Mechanisms and functions of DNA mismatch repair. Cell Res 18:85–98

Liu XP, Liu JH (2011) Characterization of family IV UDG from *Aeropyrum pernix* and its application in hot-start PCR by family B DNA polymerase. PLoS ONE 6, e27248

Lizardi PM, Huang X, Zhu Z, Bray-Ward P, Thomas DC, Ward DC (1998) Mutation detection and single-molecule counting using isothermal rolling-circle amplification. Nat Genet 19:225–232

Louwrier A, van der Valk A (2005) Thermally reversible inactivation of Taq polymerase in an organic solvent for application in hot start PCR. Enzyme Microb Technol 36:947–952

McDonald JP, Hall A, Gasparutto D, Cadet J, Ballantyne J, Woodgate R (2006) Novel thermostable Y-family polymerases: applications for the PCR amplification of damaged or ancient DNAs. Nucleic Acids Res 34:1102–1111

Motz M, Kober I, Girardot C, Loeser E, Bauer U, Albers M, Moeckel G, Minch E, Voss H, Kilger C, Koegl M (2002) Elucidation of an archaeal replication protein network to generate enhanced PCR enzymes. J Biol Chem 277:16179–16188

Nilsson M, Dahl F, Larsson C, Gullberg M, Stenberg J (2006) Analyzing genes using closing and replicating circles. Trends Biotechnol 24:83–88
O'Donnell M, Langston L, Stillman B (2013) Principles and concepts of DNA replication in bacteria, Erchaea, and Eukarya. Cold Spring Harb Perspect Biol 5:pii: a010108. doi:10.1101/cshperspect.a010108
Pata JD (2010) Structural diversity of the Y-family DNA polymerases. Biochim Biophys Acta 1804:1124–1135
Paul N, Shum J, Le T (2010) Hot start PCR. Methods Mol Biol 630:301–318
Potapova O, Grindley ND, Joyce CM (2002) The mutational specificity of the Dbh lesion bypass polymerase and its implications. J Biol Chem 277:28157–28166
Ramsay N, Jemth AS, Brown A, Crampton N, Dear P, Holliger P (2010) CyDNA: synthesis and replication of highly Cy-dye substituted DNA by an evolved polymerase. J Am Chem Soc 132:5096–5104
Rasila TS, Pajunen MI, Savilahti H (2009) Critical evaluation of random mutagenesis by error-prone polymerase chain reaction protocols, *Escherichia coli* mutator strain, and hydroxylamine treatment. Anal Biochem 388:71–80
Ribble W, Kane SD, Bullard JM (2015) Long-range PCR amplification of DNA by DNA polymerase III holoenzyme from *Thermus thermophilus*. Enzyme Res 2015:837–842
Tindall KR, Kunkel TA (1988) Fidelity of DNA synthesis by the *Thermus aquaticus* DNA polymerase. Biochemistry 27:6008–6013
Trombley-Hall A, McKay Zovanyi A, Christensen DR, Koehler JW, Devins Minogue T (2013) Evaluation of inhibitor-resistant real-time PCR methods for diagnostics in clinical and environmental samples. PLoS One 8, e73845
Wang Y, Prosen DE, Mei L, Sullivan JC, Finney M, Vander Horn PB (2004) A novel strategy to engineer DNA polymerases for enhanced processivity and improved performance in vitro. Nucleic Acids Res 32:1197–1207
Waters LS, Minesinger BK, Wiltrout ME, D'Souza S, Woodruff RV, Walker GC (2009) Eukaryotic translesion polymerases and their roles and regulation in DNA damage tolerance, Microbiol Mol Biol Rev 73:134–154
White MF, Grogan DW (2008) DNA stability and repair. In: Robb FT, Antranikian G, Grogan DW, Driessen AJ (eds) Thermophiles: biology and technology at high temperatures. CRC Press, Boca Raton, pp 179–188
Wiedmann M, Wilson WJ, Czajka J, Luo J, Barany F, Batt CA (1994) Ligase chain reaction (LCR) – overview and applications. PCR Methods Appl 3:S51–S64
Wilson RC, Pata JD (2008) Structural insights into the generation of single-base deletions by the Y-family DNA polymerase Dbh. Mol Cell 29:767–779
Yamagami T, Ishino S, Kawarabayasi Y, Ishino Y (2014) Mutant Taq DNA polymerases with improved elongation ability as a useful reagent for genetic engineering. Front Microbiol 5:461
Yan L, Zhou J, Zheng Y, Gamson AS, Roembke BT, Nakayama S, Sintim HO (2014) Isothermal amplified detection of DNA and RNA. Mol Biosyst 10:970–1003
Yang W, Woodgate R (2007) What a difference a decade makes: insights into translesion DNA synthesis. Proc Natl Acad Sci USA 104:15591–15598

Chapter 19
Technical Developments for Vegetable Waste Biomass Degradation by Thermophiles

Annarita Poli, Ilaria Finore, Annabella Tramice, Paola Di Donato, Barbara Nicolaus, and Licia Lama

1 World Vegetable Waste Biomass Production: Overview of Main Lignocellulosic Residues Production

Biomass is the main renewable feedstock that can be used for the sustainable generation of energy and useful molecules. Biomass exploitation for the production of bioenergy and platform chemicals could afford great reduction of greenhouse gas emissions, thus it represents a valuable alternative to the present fossil fuels-based economy. In this line, exploitation of biomass could represent a sustainable environmental practice for a biobased economy in which no resource (arable land, crops) is diverted from food chain to biofuels production. In this frame green chemistry is presently the most promising approach to convert biomass into bioenergy (biofuels and biopower) and biochemicals, according to the integrated biorefinery approach. Indeed, by means of a combination of chemical, physical and biochemical technologies, different kinds of biomass are used as starting materials to produce chemicals, biopolymers, biotechnologically useful enzymes/microorganisms and energy. The generation of energy and chemicals typically based on non-renewable fossil resources, shifts to a sustainable and low environmentally impacting system.

A big portion of biomass is represented by vegetable wastes and residues that are generated in huge amounts from plants and crops, during all the phases of cultivation

A. Poli • I. Finore • A. Tramice • B. Nicolaus (✉) • L. Lama
Institute of Biomolecular Chemistry ICB, CNR,
via Campi Flegrei 34, 80078 Pozzuoli, NA, Italy
e-mail: barbara.nicolaus@icb.cnr.it

P. Di Donato
Institute of Biomolecular Chemistry ICB, CNR,
via Campi Flegrei 34, 80078 Pozzuoli, NA, Italy

Department of Sciences and Technologies, Centro Direzionale,
University of Naples 'Parthenope', Naples, Italy

P.H. Rampelotto (ed.), *Biotechnology of Extremophiles*, Grand Challenges in Biology and Biotechnology 1, DOI 10.1007/978-3-319-13521-2_19

and industrial processing. These residues can be defined as: primary wastes, i.e. harvesting residues like for example crop residues (stover, stalks, bagasse) and forestry residues; secondary wastes, that are generated by agro-industries i.e. industries that use agriculture products (husks, exhausted pulps, peels and seeds) or forestry products (wood chips).

A main fraction of these wastes is represented by lignocellulosic residues, that are currently used for several purposes like for example biofuel production or extraction of value added compounds (natural antioxidants, lipids, oligosaccharides, fibres and proteins).

In particular, lignocellulose residues are constituted of cellulose, hemicellulose and lignin that, in natural sources, are found as interconnected macromolecular structures, i.e. fibrils. Their heterogeneous structure due to the variable proportion of cellulose, hemicellulose and lignin in diverse plant sources, is perhaps one of the major hindrances in developing as much as possible universal multi-step procedures, which include mechanical and chemical pre-treatments and also enzyme-based bioconversion processes. Their focus is to ferment the sugars, obtained from biomass feedstocks, to generate ethanol besides other useful molecules (FitzPatrick et al. 2010).

Cellulose, as important renewable source of bioenergy, is involved in several treatments useful for its conversion to fermentable glucose (Bhalla et al. 2013). Differently, the hemicellulose for its intrinsic structure, can furnish also oligosaccharides extremely interesting in the field of nutrition as prebiotics and then used for several purposes, like for example nutrition research and applications (Azevedo Carvalho et al. 2013).

Nevertheless these residues are still an under-utilised feedstock that, by means of bioprocesses, could be used for the production of several bioproducts like for example biotechnologically useful extremophilic microorganisms and enzymes. Indeed, thanks to their complex chemical composition, these residues could be exploited as sole carbon source to promote and sustain microbial growth. Therefore, waste biomass could afford a cheap and environmentally friendly way to produce useful microorganisms (like for example extremophiles) and their related enzymes. The latter are the object of increasing interest since they find manifold applications in the production of second generation biofuels or value added compounds like for example biologically active oligosaccharides that could be used for the production of other useful chemicals.

Every year, million tonnes of these kinds of wastes are produced and therefore high amounts of cellulose and hemicellulose are lost. As reported in Table 19.1, significantly huge amounts of this cheap and renewable feedstock are available. Food wastes that are rich sources of cellulose and hemicellulose comprise fruits, vegetables and cereal residues that are produced by agro-industries like canning or packing industries.

Cellulose and hemicellulose are the main constituents of pomace, the waste of processing of fruits like apples, black currant, cherry, chokeberry and pears that are usually employed for juice or jams production. Apple pomace is one of the most abundant pomace residue and is a very rich source of cellulose and hemicellulose that account for 43.6 % and 24.4 % of the dry waste weight, respectively. About 0.5

Table 19.1 Annual production of main vegetable waste biomass that are renewable sources of cellulose and hemicellulose

Category	Waste type	Quantity/year	Reference
Food wastes	Apple pomace	3.42 Mton	Nawirska and Kwasniewska (2005)
	Black currant, Cherry, Chokeberry, Pear pomace	0.5 Mton	Nawirska and Kwasniewska (2005)
	Orange peels	7.4 Mton	Sanchez-Vazquez et al. (2013)
	Mandarin peels	3.1 Mton	
	Carrot residues	≈11 Mton	Nawirska and Kwasniewska (2005), Sanchez-Vazquez et al. (2013)
	Corn bran	7.8 Mton	Rose et al. (2010)
	Tomato	11 Mton	Tommonaro et al. (2008), Das and Sing (2004)
Crop residues	Cassava bagasse	≈200 Mton	Das and Sing (2004)
	Sugar cane bagasse	300 Mton	Cardona et al. (2010)
	Corn stover	696 Mton	Sanchez-Vazquez et al. (2013)
	Barley straw	≈5.6–9.8 Mton	Das and Sing (2004)
	Oat straw	≈10.5–15.7 Mton	Das and Sing (2004)
	Rice straw	731 Mton	Das and Sing (2004), Binod et al. (2010)
	Sorghum straw	≈28–43 Mton	Das and Sing (2004)
	Wheat straw	550 Mton	Kuan and Liong (2008), Das and Sing (2004)
Forestry/wood residues	Wood residues	210 Mm^3	FAO data (http://faostat3.fao.org/home/E)
	Wood pellets	22 Mton	
	Chips and particles	250 Mm^3	

million tonnes of pomace are produced from the other above listed fruits: complex residues from processing of black currant, cherry, chokeberry and pears are on average made up on a weight basis of 34.6 % cellulose and 33.5 % hemicellulose (Nawirska and Kwasniewska 2005). Citrus fruits also generate abundant wastes that are cheap sources of cellulose and hemicellulose: on a weight basis in the case of orange pomace they represent 37.1 % and 11.0 % respectively of the dry weight, while for mandarin they account for 22.6 % and 6.0 % respectively (Sanchez-Vazquez et al. 2013). High cellulose contents can be found also in carrot and bran wastes: residues of carrot selection and processing for juice production are composed for 51.6 % w/w and 12.3 % w/w respectively by cellulose and hemicellulose; corn bran, that is generated by dry-milling of grains, accounts for about 7.8 Mtons that are composed by 70.0 % w/w of cellulose and by 28.0 % w/w of hemicellulose. Finally, canning residues from tomato processing can also be a valuable source of hemicellulose and cellulose that represent 7.5–11 % and 9.1 % respectively of the weight of these residues, mainly composed by peels and seeds.

Crop residues comprise the vegetable material left on the ground after harvesting i.e. stalks, leaves, straw (that on average represent more than 50 % of the crop), stover or bagasse. These wastes contain high amounts of cellulose and hemicellulose, therefore they are at world level the main renewable source lignocellulose (Sanchez-Vazquez et al. 2013). The main waste sources of lignocellulose are: corn stover (comprising leaves, shells, and stalks) that contains about 33.0 % w/w cellulose and 26.0 % w/w hemicellulose (Sanchez-Vazquez et al. 2013); rice straw that contains about 27.8 % w/w cellulose and 32.0 % w/w hemicellulose (Das and Sing 2004; Binod et al. 2010); wheat straw that is composed by both cellulose and hemicellulose that respectively account for 39.0 % and 36.0 % of dry biomass weight (Kuan and Liong 2008; Das and Sing 2004). Barley, oat and sorghum straw are produced in lesser amounts, nevertheless they also are very rich in cellulose and hemicellulose that on average respectively represent 38.6 % and 24.3 % of dry waste biomass weight (Das and Sing 2004).

Forestry residues are almost exclusively lignocellulosic materials and in general they comprise the biomass remaining in forests after harvesting and all the kinds of residues produced by wood processing industries (pellets; chips and particles). According to FAOSTAT database, the total world solid volume of forest wood residues produced in 2013 was about 210 million cubic meters: such biomass includes logging residues, excess small trees, rough or rotten dead wood. Usually these residues are either left in the forest or burned. With regard to wood processing wastes, about 22 million tonnes of wood pellets were produced in 2013: usually they are exploited as pellet fuel and are produced by using sawdust, residues of lumber's milling or wastes of manufacture of wood, furniture and construction. Finally the other main woody residues that could represent a cheap lignocellulose reserve is represented by chips and particles i.e. wood that has been reduced to small pieces and is suitable for pulping, fibreboard production. These wastes too, thank to their high cellulose content, could be used as cheap fermentation media for microorganisms and enzyme's production.

1.1 Main Current Technologies for Pre-treatment of Lignocellulose Waste Biomass

Food wastes, crop and forestry/wood residues are all polysaccharide rich materials that, as previously mentioned, could serve for several biotechnological applications. Indeed, thanks to their rich chemical composition, they can be used as growth media to produce useful microorganisms (like for example extremophiles) and their related enzymes: with this regard, several examples of wastes to be used as fermentation media are available in literature (Das and Sing 2004; Di Donato et al. 2011). However lignocellulosic residues are a recalcitrant biomass since plant cell wall's cellulose is deeply inset in a network composed of hemicellulose and lignin that hamper microorganisms or enzyme's access to the polysaccharide matrix. Therefore a kind of pre-treatment is necessary to open the lignin sheath in order to make

Table 19.2 Main techniques for waste lignocellulosic biomass pre-treatments

Pre-treatment type	Method	Biomass	Reference
Physical	Comminution	Crop residues (wheat straw, rice straw, bagasse); forestry residues	Zheng et al. (2014)
	Extrusion	Crop residues (barley straw, maize)	
	Liquid hot water	Crop residues (wheat and rice straw; sugarcane bagasse)	
Chemical	Acid hydrolysis	Crop residues (straws; sugarcane bagasse)	Sun and Cheng (2002)
	Alkaline hydrolysis	Food wastes (pomace); crop residues (wheat and rice straw, corn stover, sugarcane bagasse); forestry residues (leaves, wood residues)	Chang and Holtzapple (2000), Zhang et al. (2007), Kim et al. (2003)
	Organosolv	Crop residues (wheat and rice straw, sugarcane bagasse); forestry residues (wood residues)	Sun and Cheng (2002)
Combined physico-chemical	Afex	Crop residues (wheat and rice straw, corn stover); forestry residues (wood chips)	Sun and Cheng (2002)
	Steam explosion	Crop residues (wheat straw, corn stover); forestry residues (wood chips)	
Biological	Fungi digestion	Crop residues (wheat straw); forestry residues (wood chips)	Sun and Cheng (2002)

cellulose accessible for further processes. Different pre-treatment techniques are described in literature and they include physical, chemical, biological or combined treatments (Table 19.2): depending on the process applied, partial hydrolysis of hemicellulose as well as degradation of the lignin matrix may occur. The choice of the most appropriate pre-treatment is determined by the nature of lignocellulosic biomass and by the wanted products, therefore it is not possible to identify the best technique. In Table 19.2 are listed some of the most representative technologies that are currently used for different biomass pre-treatment, either at laboratory or at industrial scale.

Comminution is a physical pre-treatment that is implemented by means of milling or grinding machines and is used to reduce biomass particle size. Although this technique can affect the ultrastructure of cellulose, it is useful to increase the accessible surface area of the polymer; moreover it allows to reduce both the extent of crystallinity and the polymerization of cellulose thus improving its enzyme digestion. Extruders are schematically constituted by a barrel inside which a driving screw is allowed to move along the barrel: by means of this technique different mechanical treatments are applied to the biomass i.e. friction, mixing and shearing forces. Extrusion allows more accessibility of cellulose and in some cases it can also cause partial depolymerization of cellulose, besides degradation of both hemicellulose and lignin matrices. In liquid hot water (LHW) technique, biomass is treated at high pressures that allow to hot liquid water to penetrate plant cell walls thus hydrating cellulose: in such a way hemicellulose can be solubilised while lignin is partially removed. Therefore by increasing the accessible surface area of cellulose, LHW can improve subsequent cellulose degradation by microorganisms and enzymes (Zheng et al. 2014).

Acid hydrolysis is a chemical method that can be implemented by using either diluted or concentrated acids. In diluted acids pre-treatment, biomass is sprayed with sulphuric, maleic or fumaric acid and then it is heated up to 160–220 °C for a small number of minutes. Such a treatment usually causes hemicellulose removal and increase in porosity: in this way enzyme hydrolysis affords oligosaccharides or monomer sugars for further applications (for example fermentation to ethanol). In concentrated acids pre- treatment, sulphuric or hydrochloric are the most widely used acids: although this technique affords high yields for cellulose depolymerisation (that doesn't require further enzyme hydrolysis) it is more costly due to the corrosion of apparatuses and to the need of recycling of reagents (Sun and Cheng 2002).

Alkaline hydrolysis can be carried out by using calcium or sodium hydroxides or aqueous ammonia either at low or at higher temperatures (Kim et al. 2003): the main effect of these treatments, that require long reaction times, is the solubilisation of lignin (up to 70–80 %) and the very limited degradation of sugars (Chang and Holtzapple 2000). Finally, organosolv processes allow to remove lignin by treating biomass with pure organic solvents (ethanol, methanol, acetone, ethylene glycol) or their mixtures: the treatment is carried out either at low or at high temperatures depending on the biomass type. The main advantage is represented by solubilisation of lignin and hemicellulose depolymerisation, that consequently improve cellulose purity and accessibility to enzymes or microorganisms (Sun and Cheng 2002).

The most known combined physico-chemical treatments are ammonia fiber explosion (AFEX) and catalysed steam explosion. AFEX technique is based on liquid ammonia treatment that is conducted at high values of temperature and pressure (Bals et al. 2010) for about 30 min. This method affords minor hemicellulose depolymerization but efficient removal of lignin, although the cost of ammonia strongly limits its use. On the other hand steam explosion is more convenient for both economical and environmental aspects: indeed it requires minor quantities of chemicals and a lower energy input, therefore being a less environmentally impacting technique. Typically, biomass is treated with saturated steam at high-pressures and in the presence of 1 % H_2SO_4 as catalyst: the process lasts for few minutes at 160–260 °C and 0.69–4.83 MPa, then the material is returned to atmospheric pressure. By means of this process, hemicellulose degradation and lignin solubilisation are quite complete, and the potential of cellulose enzymatic hydrolysis is significantly increased (Sun and Cheng 2002).

Biological pre-treatments exploit white or brown or soft rot-fungi, that act by degrading hemicellulose and lignin matrices. These methods are very low-impacting since they require mild treatment conditions and very low energy inputs, nevertheless their application is still limited to lab-scale processes due to the low rate of biological degradation of lignin polymer and of hemicellulose whose sugars are also assimilated by fungi (Sun and Cheng 2002). Presently the only cost-effective methods are steam explosion besides liquid hot water or acid hydrolysis pre-treatments.

1.2 Extraction Procedures of Xylan from Vegetable Biomass

One of the conditions of an efficient use of biomass such as cereal straws and grasses is the separation of the main components, polysaccharides and lignin, by a relatively mild processes, able to guarantee their minimum physical and chemical modifications as well as an efficient extraction procedure. Various processes have been described for the exhaustive extraction of hemicellulose to obtain xylan useful for enzymatic applications (Aachary and Prapulla 2009).

However, before starting any pre-treatment, the raw material should be prepared by washing with organic solvents (ethanol or ethyl acetate) to separate impurities and other molecules (e.g., waxes and pectin) and make the following processes (like xylo-oligosaccharides steps) more simple (Brienzo et al. 2009). In some cases, alcohols and ketones could be exploited in the collection of soluble xylan or to concentrate the released XOS (xylo-oligosaccharides) after enzymatic or chemical reactions (Akpinar et al. 2009).

Dilute acid treatment, dilute alkali extraction, autohydrolysis, enzymatic hydrolyses represent the current pre-treatments employed to isolate hemicellulosic polymers from agro-residues (Fig. 19.1) (Azevedo Carvalho et al. 2013; Chapla et al. 2012; Aachary and Prapulla 2009; Akpinar et al. 2010). In some cases, organic solvent are used as extraction co-solvents, but these procedures, although efficient, are not totally eco-friendly and economical. Each of these procedures has its advantages and its drawbacks.

Autohydrolysis is the deacetylation of xylan by the thermal hydrolysis of hemicellulose to generate acetic acid. It is essential to operate with specialized equipment for reaching high temperatures and pressures (Kabel et al. 2002).

In recent cases, hydrothermal methods consider the combined use of microwave heating or steam explosion procedures. According to the temperature and time of working, it is possible to modulate the degree of de-polymerization and then the molecular weight interval of xylan/xylo-oligosaccharides obtained (Aguedo et al. 2014). Hemicellulose material is usually recovered in a liquid phase, whereas the precipitate solids are rich of cellulose, which is available for other purposes (Azevedo Carvalho et al. 2013). Unfortunately, the products obtained from lignocellulosic materials by this procedure were contaminated with several undesirable

Fig. 19.1 Pre-treatments for xylan extraction and procedures for XOS production

components such as lignin, furfural, and others, thereby demanding further purification (Zhu et al. 2006).

Acid pretreatments consist of the exposure of hemicelluloses to acid solutions. Regardless of the vegetable initial biomass, these extractions are carried out in acidic mild conditions (generally from 0.01 to 0.5 M of H_2SO_4) to limit the amount of xylose released and avoid or decrease the resulting corrosion. However, acid methods are always coupled with the steaming of lignocellulosic material (>120 °C) allowing the removal of lignin: in this case the time of process is important for modulating the degradation of xylan to xylose and for avoiding its complete de-polymerization and the undesirable formation of toxic furfural and hydroxymethylfurfural (HMF) (Akpinar et al. 2009). Indeed, the production of monosaccharides and these toxic components represent the major impediment of the acid treatment that requires a greater degree of purification also for these extraction procedures.

Alkali treatments have been widely accepted because they destroy the cell wall structure and cleave the hydrogen bonds, the ester linkages with acetyl units and hydroxycinnamic acids and covalent bonds with lignin (mainly α-benzyl ether linkages) in the cell wall matrix, liberating hemicelluloses polymers in the aqueous media. The limitation is that the alkaline pre-hydrolysis does not preserve the acetyl groups on the xylan chains (Lama et al. 2014; Bian et al. 2012). Aqueous solutions of potassium, sodium, barium, calcium and lithium hydroxide have been widely used to recover hemicelluloses at different temperatures, in particular from 4 °C to 40 °C (Bian et al. 2012). Nevertheless, the favorite alkali is the potassium hydroxide because the potassium acetate produced during the neutralization of the alkali extract is more soluble in alcohol used for precipitation than other acetate (Lawther et al. 1996). According to nature of lignocelluosic material, the percentage of alkali used could be different. Moreover, sometimes the presence of co-solvents (e.g. ethanol) could be employed (Li et al. 2015). A good procedure of xylan recovery involves both alkali at higher concentration and thermal hydrolysis: it was described for corncob biomass (Samanta et al. 2012).

Akaline pretreatments with hydrogen peroxide (2–4 %) at lower temperature (20–60 °C) furnish good results in delignification and hemicellulose solubilization from straws and grasses, preserving its chemical structure, as reported for sugarcane bagasse hemicellulose extraction (Brienzo et al. 2009).

The microwave irradiation coupled with alkali solutions may be an alternative pretreatment approach for lignocellulosic materials at lower temperature to overcome the problems related to the production at high temperature (autohydrolysis) of undesired and harmful compounds for the following enzymatic procedures (Zhu et al. 2006).

Pre-treatments with selected enzymes could be more attractive because it does not generate undesirable by-products or high amounts of monosaccharides and does not require special instruments, which work at high temperatures, for example. Unfortunately, they are useful only for susceptible materials such as citrus peels to produce XOS with direct enzymatic treatment of xylan-containing materials (Aachary and Prapulla 2011).

Microwave-assisted enzymatic hydrolysis has recently considered as a possible method to produce xylo-oligosaccharides by reducing the reaction time and avoiding undesirable products; it was described for the production of prebiotics from wheat bran by using commercial xylanases (Wang and Lu 2013).

The extracted xylan, obtained from the pretreatment of the lignocellulosic material, could be converted to monosaccharides by strong acidic reactions or treated by enzymatic hydrolysis (Azevedo Carvalho et al. 2013) if the aim is the preparation of oligosaccharides. In fact enzymatic hydrolysis is evaluated as the best option to produce xylo-oligosaccharides (XOS) for the food industry.

Preparation of XOS is performed by an endo-xylanase, which hydrolyses β-1,4 bonds in the main chain of xylan. If enzymatic complex are used for this aim, they must have a low activity of exoxylanases to avoid or reduce the production of xylose, which inhibits the XOS formation (Fig. 19.1) (Vázquez et al. 2005).

2 Waste Biomass Development as Alternative Carbon Source for Extremophilic Microbial Growth

A still not completely explored opportunity in waste biomass improvement is its use as alternative source for supporting microbial growth. The microbes can be considered as an inestimable factory of molecules with numerous uses in several fields and for different industrial processes. For this reason, the reduction of costs due to the needed chemicals for microbial fermentation even more so at industrial scale, represents an attracting prospective. On the other hand, this change permits to restore a commercial value to the waste so extending the duration life cycle of starting material. In addition, this could offer a manner to amortise the disposal waste costs (Di Donato et al. 2011). The above mentioned concepts are more amplified when the factories are represented by the extremophilic microorganisms, so called because they prefer environmental niches characterized by one or more stress-factors, such as high or low temperature, pressure and pH, absence of oxygen, presence of radiations, high saline and metal concentration. Therefore, they are able to thrive under stressful conditions and so their use result more advantageous than the mesophilic microorganisms.

Knowing of these organisms allows us to better understand how life originated and developed on Earth but also allows us to understand how we can improve some processes or obtain better results from already known processes (Mastascusa et al. 2014). These particular microorganisms, in fact, to survive and thrive in such adverse conditions, have developed special properties such as chemical structures and components specific for cell membranes, incredible metabolic processes, mechanisms of energy transformation and regulation of the intracellular environment (Antranikian and Egorova 2007). Thermophilic bacteria are currently used in the production of alcohols and other biologically active compounds (carotenoids, amino acids, antibiotics), in the removal of metal ions and organic compounds from waste solids or aqueous. Thermozymes find application in food industry for the syrups production with a high content of sugar (amylase, xylose isomerase, pullulanase) and to improve the organoleptic properties (pectinase) or the digestibility of certain foods (beta galactosidase); several thermoactive protease and lipase are used in the detergent industry and in the process of baking; the protease thermolysin from *Bacillus thermoproteolyticus* is used in the synthesis of the dipeptide aspartame, a low calorie sweetener (Ogino et al. 1995). Thermophilic enzymes are also used in the medical and biological precursors of such drugs, or, as the DNA polymerase and DNA ligase thermophilic, used in the PCR (Polymerase Chain Reaction) used in diagnostic medicine, molecular biology and taxonomy.

The adapting strategies of extreme microbes allowed them to carry out all metabolic processes even if in the presence of chemico-physical stress factors. These strategies come in handy in biotechnology researches, by means of pharmaceutical, food, genetic, feed, environmental, etc. Indeed, various industrial production processes are based on the exploitation of the extremophile potentialities together with their bio-molecules (enzymes, polymers, lipids, etc.) (Elleuche et al. 2014). For all those motivations, it's clear how convenient could be to search for a cheaper manner to growth extremophilic microorganisms. Serious attention is being paid in the research to this subject matter and several efforts are already reported into the literature.

An example of interesting molecules are the polyhydroxyalcanoates, PHAs; they are biodegradable polymers produced essentially by halophilic bacteria, microorganisms growing in presence of high salt concentrations. The PHAs are accumulated into the microbial cells in granules as reserve energy and as strategy to contrast the unbalance growing conditions. The bio-compatibility of PHAs has allowed its utilization up to industrial level in several fields (packaging, medicine, agriculture, drug delivery etc.) (Shrivastav et al. 2013).

Danis et al. (2015) studied the production of polyhydroxyalcanoates by the halophilic archaea *Natrinema* strain 1KYS1 in a cheaper manner using wastes as sole carbon and energy sources such as corn starch, sucrose, whey, melon, apple and tomato wastes. Among the waste investigated, the media containing corn starch resulted the most suitable for the PHA cell accumulation, with a yield of 53.14 % of dry cell weight (Danis et al. 2015). Also the haloarchaea *Haloterigena hispanica* was tested for its capability to utilize tomato and carrot wastes for growth and polyhydroxybutyric acid (PHB) production. The best substrate was the carrot that gave a PHB yield comparable to that obtained with standard complex broth, 1.25 ± 0.05 mg/g^{-1} of dry cell and 1.35 ± 0.06 mg/g^{-1} of dry cell, respectively (Di Donato et al. 2014). Huang et al. (2006) investigated the chance to increase the (PHA)s accumulation into the cells of another extremely halophilic archaea *Haloferax mediterranei* by using as carbon source pre-treated rice bran and starch, coupled to a repeated batch fermentation for growth. The highest PHA yield reached was of 55.6 % of dry cell weight.

In field of polymers, many attempts are doing continuously mainly for microbial exopolysaccharides (EPS), of which the biotechnological applications are already known and they extend from food to medicine, pharmacology, nutraceutical, cosmetic, herbicides and insecticides (Nwodo et al. 2012). There are several kinds of EPS produced by alternative carbon sources, but in the field of extremophilic microorganisms the levan can be cited; its an homopolymer of fructose, usable in food, medical, pharmaceutical and agricultural applications (Rhee et al. 2005). Speaking of which, an example of fermentation on substrates deriving from exhausted productive processes for extracellular levan production is the study of the halophilic microorganism *Halomonas smyrniensis* AAD6. The microbial growths were set up by using starch molasses and sugar beet molasses, after diverse pretreatments, as alternative sources to sucrose; beet molasses resulted to be better utilized from the microorganism for the EPS release (12.4 g/L^{-1}) (Küçükaşik et al. 2011).

Among the possible molecules of microbial origin, the enzymes represent probably the most exploited ones in productive industrial processes. Especially the thermophilic enzymes, for their thermostability, tolerance to organic solvents and metals, always find purpose in several fields. Deeply investigated are the amylase, cellulase and xylanase activities, so called for their ability to hydrolyse the starch, cellulose and xylan polymers, respectively. The polysaccharides are the most common in nature and represent an inestimable source of by-products (mono and oligosaccharides) with high economical potential. The amylase commercial applications are ascribable to pharmaceutical, paper, food, textile, fuel, detergent and starch industries (de Souza and Magalhães 2010). Therefore, the thermophilic *Anoxybacillus amylolyticus* was grown in submerged and solid state fermentation on diverse carbon sources, comprised agro-wastes, to investigate the microorganism capability to utilize them for growing and producing α-amylase. The waste sources tested and added to minimal medium (1 %, w/v) were rhizomes from *Arundo donax* L., steam and leaf from *Cynara cardunculus* and potato peel wastes. The growth and enzymatic yields were compared to those obtained from conventional chemicals such as yeast extract and soluble starch. The rhizomes from *Arundo donax* L., in submerged fermentation conditions, showed to be the most suitable inducer for the recovery of amylase from *A. amylolyticus*, exactly 2.2-fold higher respect to that observed when the microorganism was grown on yeast extract (Finore et al. 2014). Another thermophilic microorganisms tested for its ability to utilize waste sources was *Geobacillus thermoleovorans* subsp. *stromboliensis*; it produced an extracellular α-amylase able to hydrolyse raw starches also in presence of high ethanol concentration; it makes this enzyme a suitable candidate for a cloning procedure in heterologous host such as *Saccharomyces cerevisiae* for a simultaneous saccharification and fermentation in order to obtain ethanol directly from un-treated starch (Finore et al. 2011; Kasavi et al. 2012). Therefore, *G. thermoleovorans* was grown both in batch and dialysis fermentation in presence of the following waste biomass: lemon, tomato, fennel, carrot and rhizomes from *Arundo donax* L. The investigated amylase activity reached the highest value in presence of the rhizomes from *Arundo donax* L., its specific activity was 110 % higher than that collected under standard complex medium; while lemon wastes gave only 39 % of specific activity compared to the complex medium (Di Donato et al. 2014). The industrial processes involving xylanase enzymes are numerous and concern the animal feed, paper industry, cellulose bleaching, etc. (Polizeli et al. 2005). *Thermobacillus xylanilyticus* is a thermophilic bacterium with complete stable hemicellulotic enzymes. Its growth was tested on wheat bran and straw and compared to glucose and xylans. The experiments showed that the microorganism utilized all tested carbon sources and in any case, the most abundant activity produced was the xylanase, while the medium composition affected the relative presence of the debranching enzymes; more in detail, the esterase activity was more produced in presence of wheat straw and the arabinofuranosidase activity resulted more abundant when the bacterium grew on straw (Rakotoarivonina et al. 2012). Also *Geobacillus thermantarcticus* is a thermophilic microorganism producing an extracellular thermophilic xylanase and a β-xylosidase (Lama et al. 2004). For this bacterium the cheaper medium containing steam and leafs of *Cynara cardunculus* increased the xylanase release up to the 160 % respect to that measured in standard growth conditions (Di Donato et al. 2014).

Fennel, carrot and tomato wastes were evaluated as possible substrates for the growth of the halophilic bacterium *Halobacillus alkaliphilus* and the intracellular α-glucosidase production was investigated. Waste biomass was added at 1 % (w/v) in minimal medium both in dialysis and in batch fermentation. All vegetable wastes supported the growth of microorganism and in the presence of fennel waste the enzyme yield was found comparable with the respect to the activity measured in standard growth conditions (Di Donato et al. 2014).

3 Extremophilic Enzymes Useful in Biomass Conversion to Obtain Biofuel

Through acid or enzymatic treatments, the cellulose and hemicelluloses are converted into hexose and pentose sugars. Enzymatic hydrolysis by (hemi)cellulases is the better method, indeed it allows to reach higher conversions and, at the same time, results more eco-friendly because less toxic respect to the acid hydrolysis. Fermentation of all free sugars into ethanol is obtained by yeasts or bacteria. Nowadays, the use of commercial enzymes for hemicelluloses and celluloses conversion, during the ethanol production process, results the most expensive aspect in the production of bioethanol (Elleuche et al. 2014). In fact the available enzymes have been produced by mesophilic organisms and the enzymatic reactions are performed at ≤50 °C. This caused slow and, sometimes, incomplete enzymatic hydrolysis with low yields of bioconversion of lignocellulose biomass; furthermore high amount of enzyme is necessary and there is also an high contamination risk (Bhalla et al. 2013).

To solve these problems, many efforts have been made by the research for improving the hydrolysis procedures, increasing the cellulase activities, optimizing the reaction parameters, such as appropriate enzyme and substrate combination, cascade-reactions, enzyme reusing.

During last studies aiming to optimize the lignocellulosic biomass hydrolysis, statistical approach coupled with factorial design was applied; therefore enzymes of various origin were combined and mixed in an appropriate amounts (Zhou et al. 2009). A considerable decrease in the proteins content used has been reached (twofold) to obtain glucose from glucan and xylose from xylan, 99 % and 88 % conversion yields, respectively, so validating the statistical approach (Berlin et al. 2005). Several efforts are underway to reduce the cost and maximize enzyme production.

Some of the strategies include enhancing the enzymatic catalytic abilities by increasing the specific activity (by directed evolution and site directed mutagenesis) and thereby minimizing enzyme dosage or reduce the cost of enzyme production by improving cellulase titers during fermentation (through process engineering approaches by using cheap substrates including biomass, producing enzymes near biorefinery, or expression of enzyme in plants). An optimal enzymatic system in biorefinery should be engendered in situ and usable in continuous culture, having elevate activity, being stable under process conditions such as high temperature and inhibitory compounds presence (aldehydes), and possessing a considerable half-life.

3.1 Thermophilic Bacteria and Thermostable Enzymes

Several problems connected to the conversion of biomass into biofuels could be solved by utilizing extremophilic microorganisms and their thermostable enzymes.

Extremophiles have always found applications in the frame of bioenergies; furthermore recently, new attention has been pointed out on to these particular forms of life. Indeed extremophilic microorganisms are able to survive under unique environmental conditions, by means of low and high pH, temperature, elevate salinity or pressure (Bhalla et al. 2013). In addition, these microorganisms thrive in similar niches thanks to metabolic and physiological strategic adaptations. Extremozymes show kinetics that enable integration into processes carried out under stressful conditions.

In going over the knowledge concerning the applications of extremophilic microorganisms in biofuel process production processes, it is evident that most of microbial sources are essentially thermophilic. This has not astound, because thermophiles are able to put up with pH, temperature and environmental variations, proprieties which make them interesting for several applications into commercially valuable fuel production processes (Bhalla et al. 2013).

The employ of extremophilic, in particularly of their resistant thermoenzymes, represents a great potential in the bioconversion of biomass, crucial step for the generation of a commercial process. It's widely reported in the literature that extremozymes in biofuel production are mainly collocate in the hydrolysis of polymers from the various feedstocks.

Biofuel industries need enzymatic activities highly substrate-specific, not inhibited by the end-products accumulation and stable even during the process parameters variation. Enzymes produced by acidophilic and thermophilic microorganisms are obtaining always more interest because of their characteristic properties.

Indeed, to utilize thermostable enzymes implies numerous advantages during the processes of lignocellulosic biomass hydrolysis; first of all, the high temperature guarantees a major solubility of both reactants and products, so resulting in a more efficient process, mainly in terms of reaction velocity and therefore in quantity of request enzyme (Zhang et al. 2011); a lower contamination risk and, in this manner an improved productivity; an easier recovery of volatile released compounds (e.g. ethanol); reduced energy costs associated to the cooling after the thermal treatment; an higher stability that permits a longer time of hydrolysis and a potentially more flexible production process.

It is commonly known that the bacteria producing thermostable cellulases and xylanases are the most able microbes in the depolymerization of lignocellulose (Liang et al. 2011).

Research is also interested on activities able to put up with acid and heat which both may improve the lignocellulose processing. These enzymatic activities could be derived by iper-thermophilic microbes/extremophiles (Bhalla et al. 2013). The applicative potential of such microorganisms together with their enzymes could increase, mainly, in the biofuel industry.

3.2 Cellulose Deconstruction

Cellulose represent the most abundant polymer of all plants and the most copious organic molecule on globe. Cellulose is a glucan homopolysaccharide comprised of β-D-glucopyranose units linked together by β-1,4-glycosidic bonds. Cellulases hydrolyse the glycosidic bonds of both crystalline and amorphous cellulose (Mischnick and Momcilovic 2010), and according to the cellulase, different portions and chains of cellulose are targeted.

The conversion of cellulose to monomer of glucose entails the interventation of endocellulases (Enzyme Commission EC 3.2.1.4), exocellulases (cellobiohydrolases, CBH, EC 3.2.1.91; glucanohydrolases, EC 3.2.1.74), and beta-glucosidases (EC 3.2.1.21). The endocellulases act randomly on internal glycosidic bonds; it leads to a fast reduction of polymer extension and a progressive enhancement of reducing sugars. Exocellulase activities cut cellulose by removing essentially cellobiose starting from both the reducing and the not-reducing ends. In this manner, it's obtained a rapid increase in reducing sugars concentration but small variation in terms of polymer length. The hydrolytic action of endocellulase and exocellulase enzymes is synergic and results in cello-oligosaccharides and cellobiose release, which are then converted into glucose by beta-glucosidase (Kumar et al. 2008).

However, the conversion procedures of lignocellulose biomass are crucial; it's necessary to search for new and more efficient enzymes. The most problematic and slow step in biofuel production from cellulose is represented by the initial enzymatic attack on the ordered and unsoluble structure of crystalline cellulose.

A diverse combination of hydrolytic activities is necessary for a total degradation of hemicellulose of various origin. Even though cellulases hydrolyze one type of bond, crystalline substrates because of their of wide pattern of bonding require the participation of a set of enzymes or multi-component systems named cellulosomes (Kumar et al. 2008). They consist of a multi-domain scaffoldin that carries at least one Carbohydrate Binding Module (CBM) and various cohesion modules.

Essentially the carbohydrate-binding module CMB are non-catalytic modules, that facilitates the targeting of the enzymes to the insoluble polymers, and the dockerin module that mediates the binding of the catalytic module via cohesion-dockerin interactions, improving the degradation efficiency, mainly on raw lignocellulose biomass. A more efficient degradation of cellulose, in terms of speed and costs would produce both environmental and economic advantages, motivating trials using enzyme mixtures, as well as engineered cells, and is still a stimulus for the researchers.

The main properties of thermostable endoglucanases from many thermophilic and hyperthermophilic bacteria are reported in Table 19.3. It's reported that a great number of bacteria and fungi are source of thermostable cellulases. Thermophilic and mesophilic fungal genera belonging to the *Rhizopus, Trichoderma, Aspergillus* and *Sclerotium, Thermoascus thermophile* var. *coprophile, Chaetomium thermophile, Sporotrichum thermophile, Coniochaeta ligniaria* (Barnard et al. 2010) presented cellulases, but the hydrolysis of cellulose was often not complete. As an instance, one of the major bad aspect of *Trichoderma* is that it possesses a low amount of beta-glucosidase enzyme (Rahman et al. 2009). The union of cellulases, cellobiose dehydrogenase and glycoside hydrolase 61 (GH61) family of proteins, found in many species of thermophilic fungi (Dimarogona et al. 2012) have been reported to drive to an interesting increase in to lignocellulose hydrolysis (Horn et al. 2012).

Pyrococcus (Kim and Ishikawa 2010), *Sulfolobus* (Girfoglio et al. 2012) *Thermotoga* (Hong et al. 2007), *Geobacillus* (Rastogi et al. 2011) and *Thermus* (Antranikian and Egorova 2007) represent some Archaea and Bacteria examples in which thermostable endoglucanases have been found. Moreover, in *Pyrococcus* and *Thermus* spp. (Chang et al. 2001; Xiangyuan et al. 2001) exocellulases and glucosidases have been described. The improvements in terms of enzyme specificity and activity have been achieved by molecular biology (Table 19.3). Since the degradation for example of lignocellulosic materials requires a huge amount of enzymes (Lynd et al. 2005), the cellulases employed came from recombinant strains of aerobic fungi, such as *Trichoderma reesei* (syn. *Hypocrea jecorina*) and *Humicola insolens* (Karlsson et al. 2002).

The absence of CBM is responsible for unable of several hyperthermophilic microorganisms to decompose crystalline cellulose efficiently at temperatures above 75 °C, even if the employment of a multidomain hyperthermophilic cellulase could be responsible of a good degradation of lignocellulose at temperature above 90 °C (Graham et al. 2011).

Thermostable endoglucanases could operate at different optimum pH values. *Acidothermus cellulolyticus* presents an endoglucanase that works optimally at pH of 5.0 and a temperature of 80 °C (Lindenmuth and McDonald 2011). The thermoacidophilic *Alicyclobacillus* sp. A4 presents an extremely acidic β-1,4-glucanase (optimum pH 2.6 and 65 °C) (Bai et al. 2010a). *Bacillus* KSM-S237 shows a thermostable alkaline endoglucanase (optimum pH 8.6–9.0), and stored more than 30 % of the original activity after exposure at temperature of 100 °C, at pH 9.0 for 0.17 h (Hakamada et al. 1997). The use of these thermoacidophilic and thermoalkaliphilic enzymes are very widespread for several reason. For example the ligno-

Table 19.3 Characteristics of thermostable endoglucanases from various thermophilic and hyperthermophilic bacteria

Organisms	Enzymes	pH optimum	Temperature optimum (°C)	Stability	References
Thermotoga sp.	Exoglucanase	6.8–7.8	100–105	Half-life 70 min at 108 °C	
T. neapolitana	Endoglucanase Cel B	6.0	95	Half-life of 2.16 h at 106	Bok et al. (1998)
T. neapolitana	Endoglucanase Cel A	6.0–6.6	100	Half-life of 0.43 h at 110	Bok et al. (1998)
Rhodothermus marinus	Endoglucanase	7.0	95	Retained 50 % of its activity after 3.5 h at 100 °C and 80 % after 16 h at 90 °C	Hreggvidsson et al. (1996)
Alicyclobacillus acidocaldarius ATCC27009	Endoglucanase	4.0	80	Retained 60 % of activity after incubation for 1 h at 80 °C	Eckert and Schneider (2003)
Caldibacillus cellulovorans	Endoglucanase	6.5–7.0	80	Half-lifeof 0.53 h at 80 °C, and 0.03 h at 85 °C, and retained 83 % activity after 3 h at 70 °C	Huang and Monk (2004)
Geobacillus thermoleovorans T4	Endoglucanase	7.0	70	Retained more than 10 % of the original activity for 1 h at 90 °C and 100 °C	Tai et al. (2004)
Geobacillus sp. DUSEL R7 NA	Endoglucanase	5.0	75	Retained 26 % activity at 60 °C up to a period of 300 h	Rastogi et al. (2009)
Alicyclobacillus sp. A4	Endoglucanase	2.6	65	Retained 90 % activity after 1 h at 60 °C	Bai et al. (2010a)
Clostridium thermocellum	Endoglucanase	7.0	70	50 % of activity remained after 48 h at 60 °C	Romaniec et al. (1992)
Clostridium stercorarium	Endoglucanase	6.0–6.5	90	Stable for several days	Bronnenmeier and Staudenbauer (1990)
Rhodothermus marinus	Endoglucanase	7.0	95	50 % of activity remained after 3.5 h at 100 °C, 80 % after 16 h at 90 °C	Hreggvidsson et al. (1996)
Acidothermus cellulolyticus	Endoglucanase	5.0	83	Inactivated at 110 °C	Himmel et al. (1994)
Thermosipho sp. strain 3	Endoglucanase	5.6	70	Retained 50 % after 90 h at 70 °C in the presence of Ca2	Dipasquale et al. (2014)
E. coli expressing gene from *Thermoanaerobacter tengcongensis* MB4	Endoglucanase	6.0–6.5	75–80	Half-life of ½h at 82 °C	Liang et al. (2011)
E. coli expressing gene from *Clostridium thermocellum*	Endoglucanase	6.4	80	Cumulative activity after 1/2 h at 90 °C was 15 % of that at 80 °C	Zverlov et al. (2005)

cellulosic biomass have to be treated by acids or alkali, followed by neutralization step, before the action of degrading enzymes (Zambare et al. 2011). On the other hand when thermoacidophilic and thermoalkaliphilic enzymes are used for the hydrolysis of lignocellolosic biomass, the neutralization step could be eliminated (Zambare et al. 2011).

3.3 Hemicellulose Deconstruction

Hemicellulose is a complex polymer that could be homo- and hetero-polymer and contains units of xylo- manno-, gluco- and galacto-pyranose that constitute the main chain and is responsible for 25–35 % of lignocellulosic biomass (Jain et al. 2014). Pentoses such as D-xylose, D-arabinose, hexoses such as D-mannose, D-glucose, D-galactose and sugar acids, in addition to other substituents in the branching, make this heteropolymer very complex and unique for each plant source. Xylan represents the major component of hemicelluloses and also the second polymer for its abundance representing about one-third of the renewable biomass available on our planet (Dhiman et al. 2008). From structural point of view xylan is a heteropolymer in which the repeating unit is formed by beta-1,4-linked xylose (main chain) that appears decorated with 4-O-methylglucuronopyranosyl, alpha-L arabinofuranosyl, alpha-D-glucuronyl residues, acetyl, feruloyl, and/or *p*-coumaroyl units (Sun et al. 2005).

Hardwoods are the natural source of xylan as O-acetyl-4-O-methylglucuronoxylan, instead softwoods represented the source of arabino-4-O-methylglucuronoxylans: they have next to the 4-O-methylglucuronic acid a substitution with α-(1,3)-L-arabinofuranosyl residues. In the endosperm with high content of starch and the external parts of cereals, it is possible to recover arabinoxylans with a substitution of the β-(1,4)-D-xylopyranose backbone at position C2 or C3 with α-L-arabinofuranose, which can be link by esterification with phenolic acids, and/or 4-O-methyl-D-glucuronic acid. In woody part of grasses and cereals are isolated several polymers such as (glucurono)arabinoxylans linked by acetylation and esterification with ferulic acid. Differently, in the case of cereal stalks, seeds and gums, these heteropolymers are more complex thank to many substitutions of monosaccharides or oligosaccharides (Sedlmeyer 2011). Hardwood and softwood beside the presence of xylan and glucomannan respectively, are also the source of xyloglucan, glucomannan, galactoglucomannan and arabinogalactan.

Several enzymes are necessary in order to hydrolyse completely the complex structure of hemicellulose into fermentable sugars such as pentoses (D-xylose and D-arabinose), hexoses (D-glucose, D-galactose and D-mannose) and sugar acids (Subramaniyan and Prema 2000, 2002). Glycoside hydrolases, polysaccharide lyases, carbohydrate esterases are some example of digesting enzyme for hemicellulose hydrolysis: these enzyme act together to break glycosidic and ester bonds, and in the removal substituents from chains (Sweeney and Xu 2012). The enzymes involved are mainly endo-beta-1,4 xylanases ([EC 3.2.1.8]), xylan 1,4-beta-xylosidases

([EC 3.2.1.37]), alpha-L-arabinofuranosidases ([EC 3.2.1.55]), alpha-glucuronidases ([EC 3.2.1.139]), acetylxylan esterases ([EC 3.1.1.72]), feruloyl esterases ([EC 3.1.1.73]), mannan endo-1,4-beta-mannanases ([EC 3.2.1.78]), beta-1,4-mannosidases ([EC 3.2.1.25]), and arabinan endo-1,5-alpha-L-arabinosidases ([EC 3.2.1.99]) (Collins et al. 2005).

In the hydrolysis of hemicelluloses, endoxylanases and exoxylanases are useful to start the break the cross-linked polymers, then β-xylosidases convert xylo-oligosaccharides to xylose with xylo-oligomers of various lengths; on the other hand α-arabinofuranosidase breaks arabinose units in both furanose and pyranose forms; methyl glucuronic acid substitutes are hydrolised by α-glucoronisidase while acetylxylan esterase and ferulic acid esterase hydrolyses acetyl substitutes, arbinose and ferulic acids. To be noted that several of these enzymes are able to hydrolise also other compounds: this aspect make hard to know their enzymatic activities when a lignocellulosic material is digested. Nowadays there are many examples of protein combinations with good activity and high resistance to inhibition that contain hydrolitic enzymes that came from fungi or bacteria.

In Table 19.4 are listed examples of endoxylanases isolated from thermophilic and hyperthermophilic microorganisms. Thermostable xylanases are mainly produced by bacteria and fungi (Collins et al. 2005). Examples of fungi producing thermostable xylanase are *Thermoascus aurantiacus* (Zhang et al. 2011), *Rhizomucor miehei* (Fawzi 2011), *Thermomyces lanuginosus* (Singh et al. 2003), *Nonomuraea flexuosa* (Zhang et al. 2011), *Laetiporus sulphureus* (Lee et al. 2009), *Talaromyces thermophiles* (Maalej et al. 2009). Since xylanases produced by bacteria possess a higher optimum temperature and higher thermostability are usually employed for lignocellulosic materials deconstruction with respect to xylanases from fungi (Bhalla et al. 2013). Bacteria belong to *Alicyclobacillus, Anoxybacillus, Paenibacillus, Thermoanaerobacterium, Actinomadura, Nesterenkonia, Enterobacter, Acidothermus, Cellulomonas, Bacillus, Geobacillus, Thermotoga* genera have been found to possess thermostable xylanases.

Geobacillus thermantarcticus, a thermophiles microorganism collected from Antarctica, produced extracellular xylanase and β-xylosidase. These activities show interesting properties for biotechnological applications, such as optimal pH and temperature activity, thermostability, the absence of cellulolytic activities and high portions of low-member xylo-oligomers. *G. thermantarcticus,* when use xylan as unique organic source, presents a characteristic xylan digestion system in two steps that is advantageous for recovery of hydrolysis products by modulating growth conditions and physico-chemical parameters (Lama et al. 2004).

Other sources of thermostable xylanases are represented by thermoalkaliphiles, thermoacidophiles, and thermohalophiles. *Thermoanaerobacterium saccharolyticum* NTOU1 is a marine halophilic bacterium whose xylanase stored 71 % activity for 24 h when incubated in the presence of 2 M NaCl (Hung et al. 2011). *Alicyclobacillus* sp. A4 possesses an interesting xylanase stables in a wide range of pH from 3.8 to 9.4: this enzyme, that has been cloned in *Escherichia coli,* stored 90 % enzyme activity for 1 h after incubation at 60 °C (Bai et al. 2010b). Other examples of thermostable xylanase has been found in *Enterobacter* sp. MTCC 5112 (Khandeparkar and Bhosle 2006).

Table 19.4 Characteristics of thermostable xylanase from various thermophilic and hyperthermophilic bacteria

Organisms	Enzymes	pH optimum	Temperature optimum (°C)	Stability	References
Thermotoga sp. strain FjSS3-B. 1	β-1,4-xylanase	5.5	80	Half-lives of 1.5, 0.13 and <0.03 h at 95 °C, 100 °C, and 105 °C, respectively	Simpson et al. (1991)
Bacillus stearothermophilus T-6	β -1,4-xylanase	6.5	75	Half-lives of about 14.5 and 0.33 h at 70 °C and 75 °C, respectively	Khasin et al. (1993)
B. flavothermus strain LB3A	β -1,4-xylanase	7.0	70	Half-life of 0.16 h at 80 °C	Sunna et al. (1997)
B. thermoleovorans strain K-3d	β -1,4-xylanase	7.0	70–80	Half-life of 0.3 h at 80 °C	Sunna et al. (1997)
Bacillus sp. strain SPS-0	β -1,4-xylanase	6.0	75	Retained 80 % activity for 4 h at 70 °C in presence of xylan, and 20 % without xylan	Bataillon et al. (2000)
B. licheniformis	β -1,4-xylanase I	7.0	70	Retained more than 90 % of activity for 1 h at 50 °C and 60 °C for X-I and X-II, respectively	Damiano et al. (2006)
	β -1,4-xylanase II	8.0–10.0	75		
B. subtilis	β -1,4-xylanase	8.0	60	Half-lives at 60 °C, 70 °C and 80 °C were 16.2, 9.6 and 2.8 h, respectively	Saleem et al. (2011)
Clostridium sp. TCW1	β -1,4-xylanase	6.0	75	NA	Lo et al. (2011)
Bacillus sp.	β -1,4-xylanase	6.5, 8.5 and 10.5	50	Retained full activity at 50 °C for more than 23 h	Sapre et al. (2005)
Enterobacter sp. MTCC	β -1,4-xylanase	9.0	100	Retained 85 % and 64 % of its activity for 18 h at 60 °C and 70 °C, respectively	Khandeparkar and Bhosle (2006)
Paenibacillus macerans IIPSP3	β -1,4-xylanase	4.5	60	Half-life of 6 h at 60 °C and 2 h at 90 °C	Dheeran et al. (2012)
E. coli expressing *Alicyclobacillus* sp. A4	β -1,4-xylanase	7.0	55	Half-life of 6.5, 0.28, and 0.05 h at 60 °C, 65 °C, and 70 °C, respectively	Bai et al. (2010b)
Geobacillus thermantarcticus	β -1,4-xylanase	5.6	80	Half life at 70 °C of 24 h and at 80 °C for 50 min	Lama et al. (2004)
	β-xylosidase	6.0	70	Retained full activity at 1 h at 60 °C	

3.4 Lignin Deconstruction

Lignin structure is a complex network originated by the oxidative coupling of three phenolic precursors i.e. coniferyl alcohol, sinapyl and *p*-coumaryl. Precursors form, respectively, guiacyl, syringyl and hydroxyphenyl phenylpropanoid subunits. Since lignin is a poor source of fermentable carbon, it has to be removed to allow efficient biomass processing as it effectively hinder the access to the cellulose and hemicellulose polymers. Several microorganisms have been studied that are able to depolymerize lignin by means of enzymes or chemical oxidative mechanisms. Fungi (e.g. basidiomycetes) could hydrolyse lignin by means of different kind of peroxidases including manganese (EC 1.11.1.13), lignin (EC 1.11.1.14), versatile (EC 1.11.1.16) and phenol oxidases (laccases) (EC 1.10.3.2). Also other enzymes are required for the oxidative degradation of lignin such as cellobiose dehydrogenase (EC 1.1.99.18), glyoxal oxidase (EC 1.2.3.5), aryl alcohol oxidase (EC 1.1.3.7) and cellobiose/quinone oxidoreductase (EC 1.1.5.1). These enzymes are able of producing H_2O_2 that destroy the lignin polymer in a non-specifical manner (Wong 2009).

3.5 Thermostable Enzymes: Their Overexpression for Lignocellulose Degradation

The majority of thermophilic bacteria, even under optimal growth conditions, do not generate significant amounts of enzymes: as an example at 60° *Geobacillus* spp. produce endoglucanase activities from 0.0113 U/ml C (Tai et al. 2004) up to 0.058 U/ml (Rastogi et al. 2009). Similarly at 55 °C *Geobacillus* sp. has been reported to produce 0.064 U/ml endoglucanase (Abdel-Fattah et al. 2007). Therefore in order to use such thermostable enzymes at industrial scale, it is necessary to over produce them by overexpression in suitable hosts, usually *E. coli*, but also species that belong to *Bacillus* genus such as *B. subtilis* and *B. megaterium*. In addition, *Pichia pastoris* has been described as good producer of recombinant cellulase and xylanase (Sriyapai et al. 2011; Lindenmuth and McDonald 2011). An endoglucanase produced by the thermophile *Geobacillus* sp. 70PC53 showing optimum activity at 65 °C, was efficiently expressed in *E. coli* (Ng et al. 2009). *E. coli* was also exploited for the expression of a xylanase gene from a novel thermophilic strain *Geobacillus* sp. MT-1: both the recombinant and wild-type xylanases showed their optimal activity at 70 °C, besides similar activity profiles in the temperature range from 20 °C to 90 °C (Wu et al. 2006). Interestingly also other endoglucanases like those from *Bacillus subtilis* strain I15 (Yang et al. 2010), *Bacillus* sp. (Afzal et al. 2010), *B. subtilis* (Li et al. 2000, 2009), *Thermoanaerobacter tengcongensis* MB4 (Liang et al. 2011) and finally *Fervidobacterium nodosum* Rt17-B1 (Wang et al. 2010) were successfully expressed in *E. coli*. The production of recombinant cellulases and xylanases has been implemented also by means of other hosts like *Bacillus megaterium*, *B. subtilis* and *Pichia pastoris* (Lindenmuth and McDonald 2011). The last host showed to be more effective in the case of the thermostable xylanase from *Actinomadura* sp. S14: the recombinant enzyme obtained by expression in *P. pastoris* was more thermostable (50 % activity retained after 2 h of incubation at 80 °C) that obtained by expression in *E. coli* (30 % activity retained after 2 h of incubation at 80 °C) (Sriyapai et al. 2011). Such results suggested that expression in mesophilic systems could result in different post-translational modifications such as probably protein folding or glycosylation (Gao et al. 2012). Today, few data are still available regarding the expression of lignocellulose depolymerizing enzymes in thermophilic hosts.

3.6 Starch Hydrolysis by Extremophiles

Starch is the main carbohydrate energy reserve in many cereals like maize, wheat, rice, oat, potato, cassava, etc. that represent a major potential feedstock for the sustainable generation of energy (as gas or liquid biofuels) and chemicals (Kumar et al. 2007).

Starch degrading enzymes belong to the alpha-amylase superfamily: such group include numerous enzymes that show high similarity in primary sequence and act by means of a retaining catalytic mechanism (Sinnott 1990) thus releasing sugars in the alpha-configuration. The alpha-amylase superfamily is grouped in the glycoside hydrolase clan GH-H, and encompasses three sequence-related families of the GH13 family. The specificity variability of such enzymes is due to the specific consensus sequences, and to a variable number of domains, that in turn result in different hydrolytic or transferase activity, as well as in diverse substrate specificity.

The industrial process of starch's conversion to glucose is a two-step process, namely it requires a first energy-intensive step of liquefaction followed by the saccharification (Sivaramakrishnan et al. 2006). Liquefaction is carried out at high temperatures (above 100 °C) thus it requires the employ of highly thermostable enzymes. Thermophilic α-amylases produced by the hyperthermophilic archaea belonging to the genera *Methanocaldococcus*, *Pyrococcus*, *Sulfolobus* and *Thermococcus* (Kim and Peeples 2006; Van et al. 2007; Yang et al. 2004) have been widely investigated. These enzymes showed optimal activity typically around 90 °C besides possessing an impressive thermostability with activity being retained also after 4 h treatment at 120 °C. Most remarkable examples are the amylase activity produced by *Methanocaldococcus jannaschii* (T_{opt} 120 °C; T1/2 = 50 h at 100 °C), by *Pyrococcus furiosus* (T_{opt} 100 °C; T1/2 = 13 h at 98 °C) and by *Thermococcus kadakaraensis* (T_{opt} 90 °C; T1/2 = 24 h at 70 °C) (Antranikian and Egorova 2007). Also the moderately thermostable amylase produced by *Bacillus licheniformis* is extensively exploited by starch industry (Bravo Rodriguez et al. 2006). In addition, also glycoside hydrolases from halophilic species have been studied because of their tolerance to high salt and solvent concentrations (Antranikian and Egorova 2007). In particular also some haloarchaea's amylases have been described since they display significant activity in several solvents (like for example toluene, benzene or chloroform), and in the presence of high salt concentrations (up to 4.5 M NaCl) or at high pH values (upto 10) (Antranikian and Egorova 2007). Also other hydrolytic enzymes are required to complete starch's degradation like for example glucoamylase (EC 3.2.1.3) and α-glucosidase (EC 3.2.1.20). Glucoamylase enzymes release α-1,4-linked D-glucose units from the non-reducing ends of the polymer: interesting thermostable and acidophilic glucoamylases are those produced by some archaeal species belonging to the genus *Sulfolobus* (Kim et al. 2004), to the genus *Picrophilus* (Topt 90 °C; T1/2=24 h at 90 °C; $pH_{opt}=2$) and to the genus *Thermoplasma* (T_{opt} 75–90 °C; T1/2=24–40 h at 60–90 °C; $pH_{opt}=2–5$) (Serour and Antranikian 2002). α-Glucosidase acts on α-1,4-bonds in dimers, trimers and tetramers of D-glucose: remarkable examples of such enzymes are those found in by species of the genera *Sulfolobus, Ferroplasma, Pyrococcus* and *Thermococcus* that produce highly thermotolerant enzymes (Chang et al. 2001; Piller et al. 1996; Schiraldi et al. 2000).

4 Advance Procedures for Vegetable Biomass Degradation

In the range of second-generation biofuels, also named advanced biofuels, the chemical characterization of de-starched polysaccharidic and oligosaccharidic materials, which are recovered from vegetable biomass feedstock, is essential for their exploitation in the production of value added-compounds. In particular, it is matter of lignocellulosic materials (as agro-residues and forestry biomass), which represent a precious low cost and abundant renewable biomass resource with considerable potential for the bioconversion to special bio-products (Kamm and Kamm 2004).

In this section, an overview on the chemical procedures for the characterization of compounds obtained from lignocellulosic materials will be presented. Particular attention will be devoted to the oligosaccharides and glyco-conjugated prepared by using extremophilic enzymes.

4.1 Monosaccharidic Composition of Hemicellulosic Extract

In order to know the nature of hemicelluloses extracted from vegetable biomass, it is essential to analyze their monomeric composition. Arabionose/xylose or glucuronic acids/xylose ratios are considered indicative of the degree of linearity o branching of hemicelluloses (Verbruggen et al. 1995).

The monomeric composition analysis of hemicelluloses extracted from vegetable biomass is essential to know their nature.

Usually, they are exhaustively hydrolyzed with sulfuric acid, in different experimental conditions. In the case of wheat bran extract, for example, hydrolysis was carried out in 1 M H_2SO_4 for 3 h at 100 °C; for eucalyptus cell wall extract and hemicellulose from barley straw for 2.5 h at 105 °C.

After neutralization, total monosaccharides (xylose, rhamnose, arabinose, mannose, glucose, galactose, glucuronic and galacturonic acids) were analyzed by HPAEC-PAD (High-performance anion-exchange chromatography with amperometric detection) systems in proper elution conditions (Aguedo et al. 2014; Lina et al. 2006).

In the case of sugarcane bagasse hemicelluloses, neutral monosaccharides were hydrolysed by 2 M trifluoroacetic acid (TFA) in 2 h at 120 °C and were converted in alditol peracetylated and analyzed by GC (Sun et al. 2004).

Uronic acids were analyzed in arabinoxylans from wheat bran by hydrolyzing with TFA 2 N for 4 h at 105 °C (Aguedo et al. 2014). Following the classical procedure for detecting the primary structure of saccharidic chains, monosaccharide composition is determined by GC-MS as paracetylated methyl glycosides; in the case of hemicellulose from rhizome of *Arundo donax*, methanolysis was performed at 1.25 M HCl/MeOH, 80 °C and overnight (Lama et al. 2014).

It is worth to note that sometimes, monosaccharidic composition of totally hydrolyzed xylan extracts could be qualitatively and quickly detected by Thin-layer chromatography (TLC) analyses in proper eluting conditions (*n*-BuOH/AcOH/H_2O 6:2:2 by vol., or EtOAc/ H_2O/ AcOH/2-propanol/HCOOH/, 25:15:10:5:1: by vol) (Lama et al. 2014).

4.2 *Chromatographic Characterization of Oligosaccharides Produced from Hemicellulosic Agro-residues*

High-Performance Anion-Exchange Chromatography (HPAEC) is the typical procedure employed for the chromatographic analyses of xylo-oligosaccharides obtained by enzymatic or chemical hydrolyses of xylan extracts. The identification of oligosaccharides and their concentrations values are recovered by using proper standards.

HPAEC-PAD analyses of arabinoxylo-oligosaccharides (AXOS) produced from wheat arabinoxylan (AX) by using Shearzyme (GH10 endo-1,4-β-D-xylanase) and two α-L-arabinofuranosidases (AXH-m and AXHd3).

HPAEC-PAD analysis of these reactions furnished chromatographic profiles in which not only the shortest xylo-oligosaccharides are eluted from the column in advance, but the corresponding elution order is followed within the collections of singly and doubly α-L-Arabinofuranosyl (Ara*f*) substituted arabinoxylan oligosaccharides (AXOS); furthermore the α-1-Ara*f* linkage position on the β-D-xylopyranosyl (Xyl*p*) unit also influences the elution order (Pastell et al. 2008).

Recently, an extracellular endoxylanase from *Bacillus halodurans* TSEV1, which resulted stable to heat and alkaline pHs, was cloned and expressed in *E. coli* and then exploited in the xylo-oligosaccharides production from several agro-residues. The monitoring of the saccharifications was performed by high pressure liquid chromatography (HPLC) equipped with a differential refractive index detector (Kumar and Satyanarayana 2014).

However, sometimes a qualitative monitoring by TLC analyses can result sufficient for the identification of xylo-oligosaccharides produced from agro-residues; the degradation study by TLC of the wheat bran hemicellulose (TLC system solvent: *n*-butanol:ethanol:water 5:3:2 by vol) for the oligosaccharides production by using a thermostable endoxylanase from the thermophilic bacterium *Geobacillus thermodenitrificans* TSAA1 is a useful example (Anand et al. 2013). Furthermore, a TLC investigation of enzymatic digestions using *A. donax* hemicelluloses extract and different thermopilic enzymatic preparations (cell-free extract of *T. neapolitana* and *Thermoanaeorbacterium thermostercoris*, extracellular suspension of *Geobacillus thermantarcticus*, commercial xylanases from *Thermomyces lanuginosus* and *Thermotoga maritima*) resulted fundamental for the individuation of unknown oligosaccharides into mixture reactions, which were later spectroscopically characterized (Lama et al. 2014).

4.3 Spectroscopic Investigation of Hemicellulosic Fractions Before and After Enzymatic Digestion: NMR, MS, FT-IR Analyses

1D and 2D NMR investigations of xylan polysaccharides from lignocellulosic sources are essential before any procedure designed to produce xylo-oligosaccharides or their value-added derivatives.

Nuclear magnetic resonance (NMR) analyses gives structural information about the nature, the configuration and relative content of monosaccharide essential to identifying the structure of several hemicellulosic extract. In most of cases, the homogeneity of polysaccharidic extracts corresponds to relatively simple spectra with well resolved signals.

In general, in protonic and carbon NMR spectra of arabino(glucurono)xylans and their oligosaccharidic derivatives it is possible to detect some diagnostic signals, of which intensity is variable, according to the branching of saccharidic chains (Jin et al. 2009).

In Table 19.5 we summarise peculiar values of chemical shifts belonging to hemicellulose structures.

In the case of hemicellulose isolated from rhizomes of *A. donax*, ^{13}C NMR spectrum in *d*-DMSO at 40 °C of hemicellulosic substrate showed intense and diagnostic signals attributed to xylan main chain as reported in Table 19.5. These signals represent, in general, the major signals of (1→4)-linked-β-xylan (Bendahou et al. 2007). Furthermore, in the ^{13}C NMR of *A. donax* hemicellulose extract, small signals were present corresponding to arabinose residues. These values confirmed the arabinoxylan structure of polysaccharidic extract and their intensity suggested a partially de-branched skeleton.

The analysis of ^{13}C and ^{1}H NMR spectra (in D_2O) of hemicellulose extracts at different temperature from sugarcane bagasse (Table 19.5) presented together with the signals of β-xylose chain and arabinose residues, also other signals of smaller intensity belonging to 4-O-methyl-α-D-glucuronic acid units (Bian et al. 2012).

Table 19.5 Diagnostic chemical shifts in ^{1}H and ^{13}C spectra of xylan from lignocellulosic sources

Compounds	Structural elements	NMR ^{1}H (ppm)	^{13}C (ppm)
(acetyl)arabinoglucoronoxylans (Hoffmann et al. 1992a, b)			
	Anomeric sites of β-D-xylose residues which are substituted at C-2 and C-3 (disubstituted), C-3 (monosubstituted), or unsubstituted	4.5:4.8	
	Anomeric sites of Ara*f* belonging to short side chains	~5.4	
	Sites 2:5 belonging to xylose and arabinose residues	3.1:4.3	
	Anomeric site of methyl glucuronic acid	5.3	
	Methyl signals of methyl glucuronic acid	1.8	
	Acetyl groups of xylan chain	2.6	
Arabinoxylan isolated from rhizomes of *A. donax* (Lama et al. 2014)			
	Anomeric site of xylan main chain		101.7
	Site 4 of 1,4-β-linked –xyloses		75.3
	Site 2,3,5 of xylose in 1,4 β-linked xyloses		73.9, 72.5, 63.2
	Anomeric site of arabinofuranose residues		107.1
	Site 2: 5 of α-L-Ara*f* linked to position 3 of xylose residue		87.3, 86.0, 80.2, 77.8, 61.8
Hemicellulose from sugarcane bagasse (Bian et al. 2012)			
	Site 2:5 of 1,4-β-linked xylose residues of main chain	4.34 (H-1), 3.96 (H-5 eq), 3.66 (H-4), 3.40 (H-3), 3.25 (H-5ax), 3.18 (H-2)	102.32, 75.89, 74.91, 73.28, 63.27
	Site 2:5 of α-L-Ara*f*		109.50, 86.46, 80.23, 78.37, 61.72
	Sites1: 6 and OCH_3-4 substitution of 4-O-methy-α-D-glucuronic acid residues		177.01 (C-6), 97.46 (C-1), 75.3 (C-2) 72.12 (C-3), 73.86 (C-5), 79.26 (C-4), 59.53 (OCH_3-4)
Hemicellulose from *Eucalyptus* cell walls (Li et al. 2015)	Sites 1,2,5 6 and OCH_3-4 substitution of 4-O-methy-α-D-glucuronic acid residues	5.14 (H-1), 4.15 (H-5), 3.49 (H-2), 3.35 (–OCH3)	

Analogously, in the protonic spectra of hemicellulose extract of *Eucalyptus* cell walls, recently investigated, although the signals of β-xylan backbone were predominant, minor signals were detected and attributed to 4-O-methyl-α-glucuronic acid residues (Li et al. 2015).

2D-NMR experiments, such as COSY (COrrelated SpectroscopY), TOCSY (TOtal Correlation SpectroscopY), HSQC (Heteronuclear Single Quantum Correlation), HSQC-EDITED (multiplicity-edited HSQC), HMBC (Heteronuclear Multiple-Bond Correlation Spectroscopy), NOESY (Nuclear Overhauser Spectroscopy) are necessary for the assignment of the protonic and carbon value signals of each position within each monosaccharidic residue; it allows to establish the primary structure of polysaccharidic chains and/or also the sequence of monosaccharide units into an oligosaccharide skeleton.

COSY, TOCSY and HSQC experiments are essential for the identification of spin systems; positioning of acidic residues or arabinoses, for example, along the xyloses backbone is possible by evaluating the long-range correlations C–H in HMBC experiments or dipolar coupling H–H in NOE experiments.

In fact, an exhaustive 2D-NMR spectroscopic investigation resulted essential to characterize the structure of a pentasaccharide (Fig. 19.2, compound 1) and a tetrasaccharide (Fig. 19.2, compound 2) obtained by enzymatic digestion of hemicellulose extract of *Arundo donax* rhizome and by using the commercial *Thermomyces lanuginosus* xylanase and the xylanase from cell free extract of *Thermoanaeorbacterium thermostercoris* (Lama et al. 2014).

The analyses of oligosaccharides released from xylans by mass spectrometry parallel and support the NMR investigations. Usually, Matrix-Assisted Laser Desorption/Ionisation Time-Of-Flight Mass Spectrometry (MALDI-TOF-MS) and

1

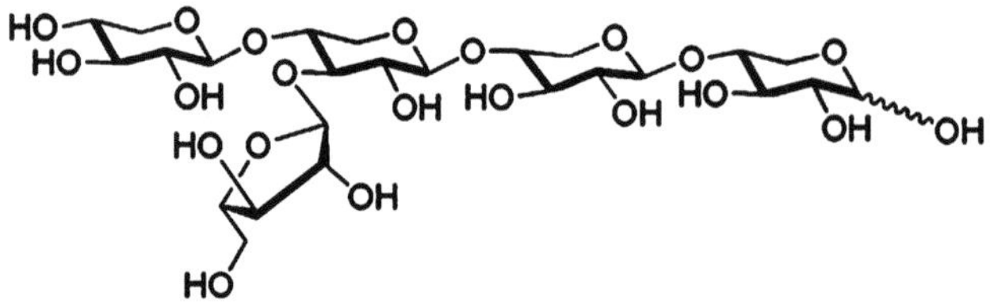

2

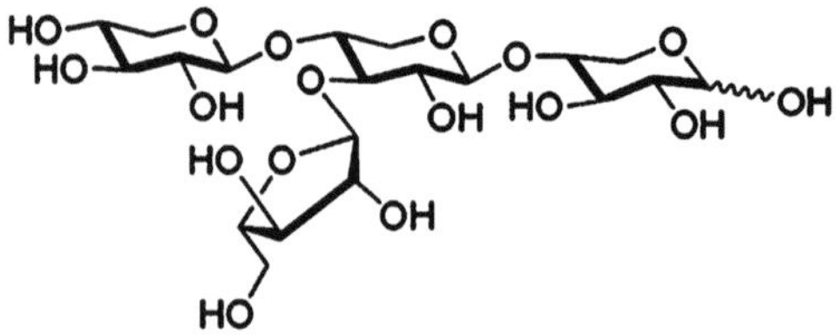

Fig. 19.2 The pentasaccharide β-D-Xylp-(1–4)-[α-L-Araf-(1–3)]-β- D-Xylp-(1–4)-β-D-Xylp-(1–4)-β-D-Xylp (1) from hydrolysis mixture using T. lanuginosus enzyme and β-D-Xylp-(1–4)-[α-L-Araf-(1–3)]- β-D-Xylp-(1–4)-D-Xylp (2) the tetrasaccharide, from the hydrolysis reaction mixture using crude extract of *T. thermostercoris*

electrospray mass spectrometry (ESI-MS) are the most common methods of analysis (Aachary and Prapulla 2011; Pastell et al. 2008).

In some cases, these analyses are performed in association with separation procedures such as size exclusion chromatography (SEC), high-performance anion-exchange chromatography (HPAEC) and reversed phase high-performance liquid chromatography (RPHPLC).

Lately, positive tandem mass spectrometry using ESI has been proposed for the characterization of underivatized or acetylated neutral and acidic XOS (Reis et al. 2005).

Furthermore, in the case of a mixture of xylo-oligosacchrides produced by enzymatic digestion for example, the MS studies furnish the xylo-oligosaccharide molecular weights, which in combination with information about the primary structure of initial xylan, allow possible structures to be proposed. On the other hand, certainty about the structures supposed can only be obtained by tandem mass spectrometry investigations (ESI-MS/MS) (Reis et al. 2005).

Infrared spectroscopy has been widely used for investigating the functional groups of polysaccharides (Li et al. 2015).

In IR spectra of hemicelluloses, stretching signal of C=O belonging to acetyl, uronic, and ferulic ester groups is recorded at about 1745 cm^{-1} together with a bending signal at 1249 cm^{-1} corresponding to –C–O– in ester groups.

Furthermore, in the carbonyl group stretching region, the signals at 1463, 1426, 1382, and 1318 cm^{-1} are attributed to –CH2 symmetric bending, CH and OH bending, OH in-plane bending, and –CH wagging, respectively (Sun et al. 1996). However, the presence of a band at 903 cm^{-1} is indicative of the dominant β-glycosidic linkages between the sugar units in the hemicellulosic extracts (Robert et al. 2005).

In the 1120–1000 cm^{-1} region, signals with multiple peaks is a distinctive characteristic of highly substituted arabinoxylans, and the arabinose substitution at C-3 site of xylose residues furnishes a band at 1173 cm^{-1} (Subba and Muralikrishna 2004). In general, the intensity of bands between 990 and 900 cm^{-1} is strongly dependent on the amount of Ara*f* units along the xylan backbone (Robert et al. 2005).

4.4 Transglycosylation Processes with Hemicelluloses Extracts

An alternative to exploit lignocellulosic biomass for the biofuel recovery, could be the syntheses of high value-added molecules. In this context, xylanases which naturally hydrolyze xylans into oligoxylosides and xylose monomers and xylosidases which hydrolyse oligoxylosides into xylose monosaccharides, result extremely useful in transglycosylation processes for the production xylo-conjugated.

As we know, alkyl substituted saccharides represent nonionic surfactants that are currently added in liquid and powder detergents, in pharmaceuticals preparations and in personal care commodities. Alkylxylosides, as example of the important group of the pentose-based surfactants, have very stimulating surfactant properties (Xu et al. 2012), and these molecules can be produced from lignocellulosic biomass in order to decrease the cost of production by enzymatic synthesis in environment-friendly conditions.

In this context, a recombinant β-xylosidase from *Geobacillus thermodenidrificans* TSAA1 has been recently used for the production of alkyl xylosides, starting from aliphatic alcohols with short chains (C1-C5) as acceptors and a mixture of XOs, which were enzymatically produced from a de-starched wheat bran extract, as donors.

In proper reaction conditions and after 16 h of reaction between methanol and wheat bran hydrolysates, the corresponding band of methyl xyloside was well detected on TLC plate; furthermore in HPLC chromatogram, the presence of peaks corresponding to methyl xyololigosaccharides was recorded other than that of methyl xyloside (Jain et al. 2014).

More efficient transglycosylation reactions were carried out in presence of a xylanase from *Thermobacillus xylanilyticus* (Tx-xylanase) and a commercial xylanase (Novozymes NS-50030). In this case, birchwood or oat spelt xylans were used as donor or also alternatively xylo-oligosaccharides generated from hydrothermally pretreated and destarched wheat bran; the acceptors were aliphatic alcohols with growing chain length (from methanol to decanol) (Ochs et al. 2011).

Experiments with Tx-xylanase and NS-50030 xylanase were focused on the analysis of reaction parameters with the aim to induce their best transglycosylation capacity in presence of a partially water miscible alcohol (pentan-1-ol) and a non water-miscible alcohol (octan-1-ol) as acceptors.

Destarched wheat bran was subjected to autohydrolysis; after 1 h of reaction at 135 °C, the arabinoxylan present in the bran was solubilized. The oligosaccharidic solution after filtration and lyophilization was used as donor at a concentration of 2 % (w/v) of arabinoxylan equivalent. This donor was used in a 1 h reaction with octan-1-ol (20 %, v/v), tert-butanol (20 %, v/v) as co-solvent and Tx-xylanase (20 IU mL^{-1}).

Surprisingly, the highest total yield was recorded (222.2 mg g^{-1} arabinoxylan equivalent) when using the supernatant obtained from pretreated wheat bran, with respect to the process in which 2 % (w/v) birchwood xylan was used as substrate (146.6 mg g^{-1} arabinoxylan equivalent).

It was previously reported that 9-fluorenylmethyl glycosides showed interesting antiviral and antiproliferative on ovarian cancer cells activities (Tramice et al. 2008, 2009).

Recently, hemicellulolitic extract from *Arundo donax* rhizomes was employed as donor in transglycosylation processes with the aim to produce 9-fluorene methanol xylosides (Lama et al. 2014) by using *T. maritima* xylanase and xylanase/β-xylosidase activities from *T. neapolitana*, *T. thermostercoris* and *G. thermantarcticus* crude enzymatic homogenates. Methanol extracts of reactions performed with each enzymatic system were analyzed by MS spectroscopy, revealing the production of mono- and di-xylosylated derivatives of 9-fluorenyl methanol. Signals at *m/z* 351 and 483 $[M+Na]^+$ were recorded. Furthermore, in the reaction with *G. thermantarcticus* crude homogenate, the presence into the reaction medium of trixylosylated 9-fluorenyl methanol was secured by a signal in the reaction MS spectrum at m/z 651 $[M+Na]^+$.

5 Biotechnological Application of Extremophiles in the Valorization of Waste Biomass: Biofuels and Biohydrogen Production

One of the main applications of extremophilic microorganisms is their use for the conversion of waste derived from agriculture and forestry in order to obtain biofuels. In particular, lignocellulosic agricultural and forestry wastes, that represent the starting materials for second-generation biofuels, are no-food materials that aren't in competition with food, resulting irrelevant to the increase of the food prices. In view of this, the search of new thermophilic bacteria with interesting thermostable enzymes is a critical point to overlap the best utilization of these wastes biomass. In order to access to cellulose and hemicellulose, lignocellulosic start material needs of several pretreatments, usually a thermomechanically pretreatment to open the lignin sheath, following by chemical treatment as described in Table 19.2. After that, enzymatic hydrolysis are requested, usually cellulases, hemicellulases and xylanase in order to release fermentable sugars. These enzymatic hydrolysis are usually performed at <50 °C and in that condition the rate of hydrolysis appears slow, the enzyme concentration have to be high as well as the risks related to microbial contamination and in general the yields of fermentable sugars appear low. In this scenario the use of thermophilic microorganisms and their thermostable enzymes is very useful to overcome the limitations of lignocellulosic biomass conversion and to improve the whole process (Bhalla et al. 2013). In fact, the employ of these thermozymes, since they usually require working temperature ranging from 50 °C to 80 °C, implicates a shorter hydrolysis times, an increase of reagent and product solubility, a higher rate of hydrolysis, a lower microbial contamination and a facilitation in volatile product recovery (Zhang et al. 2011; Liang et al. 2011).

Since thermozymes possess a great adaptability to work in different pH values, this feature makes these enzymes ideal actors for lignocellulose conversion by extremophilic microorganisms, overall in the case in which acid or alkaline chemical pretreatments are required. The lignocellulosic biomass conversion could be achieved by using the thermozymes able to make an efficient lignocellulose deconstruction: the use of a mix of thermostable enzymes in combination with ethanologenic thermophiles, increase the yield of hydrolysis and in this condition the fermentation could be carried out at higher temperatures using just one fermentator (Thermophilic Simultaneous Saccharification and Fermentation -SSF) (Podkaminer et al. 2011; Shaw et al. 2008). This bioprocess was used for example by Shaw et al. (2008) that employed the thermophile *Thermoanaerobacterium saccharolyticum* strain ALK2 obtaining a 2.5-fold reduction in terms of quantity of enzymes required to achieve equivalent hydrolysis for SSF process at 50 °C compared to SSF performed by *S. cerevisiae* that worked at 37 °C. In literature there are several examples of fermentative thermophilic bacteria (fermentation temperature ranging from 45 °C to 70 °C) that are able to use xylose, glucose, cellobiose, galactose, mannose but also mixed sugars that derive from biological decomposition of lignocellulosic biomass such as corn stover, in order to obtain ethanol; in this case the ethanol production lies in the interval between 0.16 and 0.47 g of ethanol/g of saccharides or cellulose reduced even if other products could be achieved such as acetate, lactate, pyruvate and succinate (Cai et al. 2011; Cripps et al. 2009; Georgieva et al. 2008). However, the best results have been obtained using Consolidated Bioprocessing (CBP) that could involve saccharolytic fermentative thermophiles using the pretreated lignocellulose biomass (Olson et al. 2012). The most challenging task associated with CBP is the selection of appropriate microorganisms or a microbial community (*consortium*) that can digest the lignocellulosic materials producing ethanol. For this aim the raw material should not required any pre-treatments such as chemical, physical or enzymatic actions, and just a reduction of particle size should be enough (Paulová et al. 2014). In CBP the production of hydrolytic enzymes, the degradation of lignocellulosic waste and the use of the monosaccharides released by fermentation occurred efficiently in a single reactor, led up to a cost reduction of a fourfold in comparison with SSCF (simultaneous saccharification and co-fermentation) (Lynd et al. 2005). There are two categories of microorganisms used in the CBP: category I or cellulase producers and category II or ethanol producers. At first category belong for example *Clostridium thermocellum, Geobacillus thermoglucosidans, Thermoanaerobacter mathranii* and cellulolytic fungi (*Tricoderma reesei, Paecilomyces variotii*) (Paulová et al. 2014). Category II CBP producers include ethanol producers microorganisms conveniently engineered such as *Saccharomyces cerevisiae, Kluyveromyces marxianus* and *Zymomonas mobilis* (Paulová et al. 2014). These saccharolytic thermophiles have been improved by genetic engineering providing constructed microorganisms containing a set of interesting hydrolytic enzymes. This is the example of *Clostridium thermocellum*, able to convert plant biomass into ethanol by CBP. In order to increase ethanol tolerance and to obtain a high yield in ethanol production, it has been mutagenised at level of alcohol dehydrogenase as reported by Brown et al. (2011). The possibilities to use a microbial consortium combining both category CBP producers (I and II) have been also reported. Ethanol production has been also studied using a microbial community consisting of *S. cerevisiae*, *T. reesei* and *Scheffersomyces stipitis*. It was observed a maximum ethanol concentration of 9.8 g/l^{-1} using an acid pretreated wheat straw with a theoretical ethanol yield of 69 % (the theoretical ethanol yield of 0.51 g/g^{-1} is taken as 100 % and represent the yield achieved by *S. cerevisiae* during ethanol fermentation of glucose). Other approaches to decrease the costs linked to lignocellulose conversion to biofuels were the use of immobilized biocatalysts, such as cellulolytic enzymes and microorganisms. Apparently, the microbial cell immobilization gave better results compared with cellulase immobilization in that a decrease in enzymatic activities occurred after enzyme immobilization (Paulová et al. 2014).

Moreover, several efficient hydrogen cell factories have been developed using anaerobic thermophiles able also to produce thermostable cellulolytic and xylanolytic enzymes. *Caldicellulosiruptor saccharolyticus* represented a suitable candidate for hydrogen production using lignocellulosic biomass such as sugarcane bagasse and sweet sorghum (VanFossen et al. 2009). The efficiently of the whole process of bioconversion of lignocellulosic biomass into biofuels reached the maximum expression when *Caldicellulosiruptor bescii* DSM 6725 was used in a single reactor. In fact, this microorganism, with an optimum growth temperature of 80 °C, was shown to convert untreated biomass such as bermuda grass and switch grass into hydrogen (Yang et al. 2009). In this case, the high growth temperature of *C. bescii* helped the recovery of volatile products (ethanol) from bioreactor (Chang and Yao 2011).

6 Conclusion

The active research of new thermophilic bacteria with interesting thermostable enzymes able to convert lignocellulosic materials could improve the productivity of this process and the energy consumption. In addition to classic approaches, the knowledge of draft genome sequence of thermophilic microorganisms able to degrade lignocellulosic substrates, revealing genes encoding cellulose or xylan-degrading enzymes, could help into exploration for the biofuel production processes. In fact, even using all the information concerning both the waste composition and the microorganism pathways, not all attempts in waste utilization for microbial growth result be successfully. Indeed, the last tendencies recognize the helpful contribute coming from the information contained into microbial genome. The knowledge of whole genome could permit a preliminary screening of the suitable extremophilic microorganisms to be used for the waste biomass utilization (Studholme 2015).

In the world energy scenario, the absolute certainty consists of the quickly fossil fuel depletion that, coupled with the boost population, has imposed the search, at every level, for an efficient and valuable alternative. Different renewable energy are currently being examined. In this chapter, the vegetable waste biomass are consider a source of added value products. As extensively argued, they are produced in huge amount and need to be correctly disposed of during expansive appropriate practices. The development of procedures aiming to reuse and valorize useful waste materials. Therefore, the innovative idea for all industrial and agricultural residues is to tend to a zero-waste emission. In the Fig. 19.3 is reported the degradation of waste biomass (rhizomes of *Arundo donax* L.), utilizing thermophilic microorganisms, to obtain polymeric components (starch, lignin, cellulose, xylan) and their monosaccharides and oligosaccharides useful for biofuel production. In addition, vegetable biomass could be exploited as organic and energy supporting materials for thermophiles growths and their enzyme hyper-production (Finore et al. 2014; Lama et al. 2014). In this integrated system the residues of each treatment represent the starting materials for a further transformation with the aim of zero-emission processes.

In this contest the emergent Synthetic Biology offers great potential to overcome the challenges associated with lignocellulose conversion. Synthetic biology is considered a recent discipline with the interesting properties to construct new biological systems (Keasling 2008) and in the field of renewable sources, it represents a cheaper way to get biofuels and/or chemicals (Nieves et al. 2015). This discipline is based on the design and the build of new biological components ranging from enzymes, metabolic pathways, genetic circuits up to the whole cells, in order to assembly an inte-

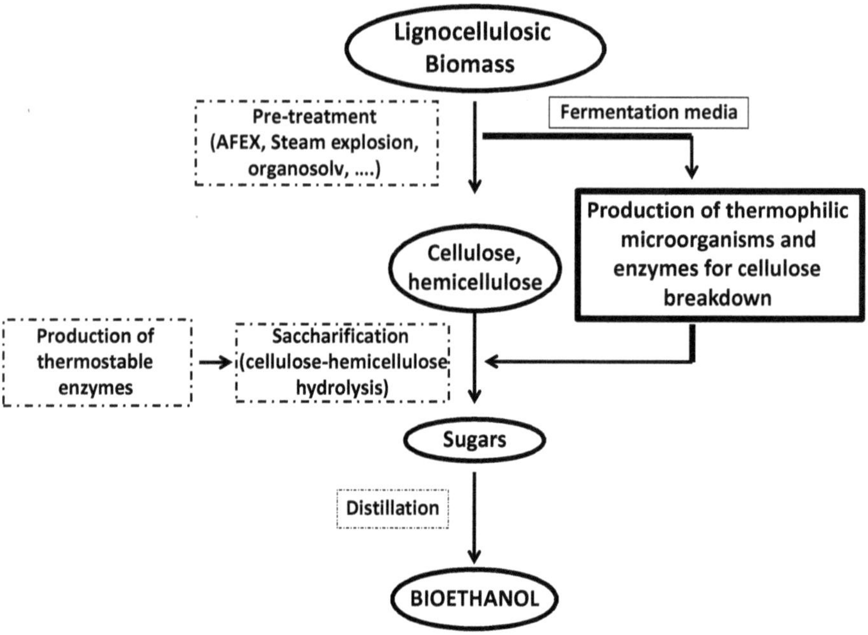

Fig. 19.3 Scheme of integrated treatment of lignocellulosic biomass from rhizomes of *Arundo donax* L

grated systems that can solve specific problem. In the case of lignocellulose conversion for example, the main goals for microbial catalysts are represented by a good utilization of xylose and resistance to furan aldehydes (Sandoval et al. 2012; Wang et al. 2013). Next step will be the construction of a cell able to provide a lignocellulosic conversion thank to the new developed genetic techniques taking advantages from the genome sequence of lignocellulose-degrading thermophiles.

Conflict of Interest Annarita Poli, Ilaria Finore, Annabella Tramice, Paola Di Donato, Barbara Nicolaus, and Licia Lama declare that they have no conflict of interest.

Acknowledgment This work was partially supported by the project *PON03PE_00107_1 BioPoliS* "Sviluppo di tecnologie verdi per la produzione di BIOchemicals per la sintesi e l'applicazione industriale di materiali POLImerici a partire da biomasse agricole ottenute da sistemi colturali Sostenibili nella Regione Campania".

References

Aachary AA, Prapulla SG (2009) Value addition to corncob: production and characterization of xylo-oligosaccharides from alkali pretreated lignin-saccharide complex using *Aspergillus oryzae* MTCC 5154. Bioresour Technol 100:991–995

Aachary AA, Prapulla SG (2011) Xylo-oligosaccharides (XOS) as an emerging prebiotic: microbial synthesis, utilization, structural characterisation, bioactive properties, and applications. Compr Rev Food Sci Food Saf 10(1):2–16
Abdel-Fattah YR, El-Helow ER, Ghanem KM, Lotfy WA (2007) Application of factorial designs for optimization of avicelase production by a thermophilic *Geobacillus* isolate. Res J Microbiol 2:13–23
Afzal S, Saleem M, Yasmin M, Naz M, Imran M (2010) Pre and post cloning characterization of a β-1,4-endoglucanase from *Bacillus* sp. Mol Biol Rep 37:1717–1723
Aguedo M, Fougnies C, Dermiencec M, Richel A (2014) Extraction by three processes of arabinoxylans from wheat bran and characterization of the fractions obtained. Carbohydr Polym 105:317–324
Akpinar O, Erdogan K, Bostanci S (2009) Production of xylo-oligosaccharides by controlled acid hydrolysis of lignocellulosic materials. Carbohydr Polym 344:660–666
Akpinar O, Erdogan K, Bakir U, Yilmaz L (2010) Comparison of acid and enzymatic hydrolysis of tobacco stalk xylan for preparation of xylo-oligosaccharides. LWT-Food Sci Technol Int 43(1):119–125
Anand A, Kumar V, Satyanarayana T (2013) Characteristics of thermostable endoxylanase and β-xylosidase of the extremely thermophilic bacterium *Geobacillus thermodenitrificans* TSAA1 and its applicability in generating xylooligosaccharides and xylose from agro-residues. Extremophiles 17:357–366
Antranikian G, Egorova K (2007) Extremophiles, a unique resource of biocatalysts for industrial biotechnology. In: Gerday C, Glansdorff N (eds) Physiology and biochemistry of extremophiles. ASM Press, Washington, DC, pp 361–406
Azevedo Carvalho AF, de Oliva Neto P, da Silva Fernandes D, Pastore GM (2013) Xylo-oligosaccharides from lignocellulosic materials: chemical structure, health benefits and production by chemical and enzymatic hydrolysis. Food Res Int 51:75–85
Bai Y, Wang J, Zhang Z, Pengjun Shi P, Luo H, Huang H, Luo C, Yao B (2010a) Expression of an extremely acidic beta-1,4-glucanase from thermoacidophilic Alicyclobacillus sp. A4 in Pichia pastoris is improved by truncating the gene sequence. Microb Cell Fact 9:33
Bai Y, Wang J, Zhang Z, Yang P, Shi P, Luo H, Meng K, Huang H, Yao B (2010b) A new xylanase from thermoacidophilic *Alicyclobacillus* sp. A4 with broad-range pH activity and pH stability. J Ind Microbiol Biotechnol 37:187–194
Bals B, Rogers C, Jin M, Balan V, Dale B (2010) Evaluation of ammonia fibre expansion (AFEX) pretreatment for enzymatic hydrolysis of switchgrass harvested in different seasons and locations. Biotechnol Biofuels 3:1–11
Barnard D, Casanueva A, Tuffin M, Cowan D (2010) Extremophiles in biofuel synthesis. Environ Technol 31:871–888
Bataillon M, Cardinali APN, Castillon N, Duchiron F (2000) Purification and characterization of a moderately thermostable xylanase from *Bacillus* sp. strain SPS-0. Enzyme Microb Technol 26:187–192
Bendahou A, Dufresne A, Kaddami H, Habibi Y (2007) Isolation and structural characterization of hemicelluloses from palm of Phoenix dactylifera L. Carbohydr Polym 68(3):601–608
Berlin A, Gilkes N, Kurabi A, Bura R, Tu MB, Kilburn D, Saddler J (2005) Weak lignin binding enzymes. A novel approach to improve the activity of cellulases for hydrolysis of lignocellulosics. Appl Biochem Biotechnol 121:163–170
Bhalla A, Bansal N, Kumar S, Bischoff KM, Sani RK (2013) Improved lignocellulose conversion to biofuels with thermophilic bacteria and thermostable enzymes. Bioresour Technol 128:751–759
Bian J, Peng F, Peng XP, Xu F, Sun RC, Kennedy JF (2012) Isolation of hemicelluloses from sugarcane bagasse at different temperatures: structure and properties. Carbohydr Polym 88:638–645
Binod P, Sindhu R, Singhania RR, Vikram S, Devi L, Nagalakshmi S, Kurien N, Sukumaran RK, Pandey A (2010) Bioethanol production from rice straw: an overview. Bioresour Technol 101:4767–4774

Bok J, Dienesh A, Yernool D, Eveleigh D (1998) Purification, characterization and molecular analysis of thermostable cellulases CelA and CelB from *Thermotoga neapolitana*. Appl Environ Microbiol 64:4774–4781

Bravo Rodriguez V, Jurado Alameda E, Martinez Gallegos JF, Reyes Requena A, Garcia Lopez AI (2006) Enzymatic hydrolysis of soluble starch with an alpha-amylase from *Bacillus licheniformis*. Biotechnol Prog 22(3):718–722

Brienzo M, Siqueira AF, Milagres AMF (2009) Search for optimum conditions of sugarcane bagasse hemicellulose extraction. Biochem Eng J 46:199–204

Bronnenmeier K, Staudenbauer W (1990) Cellulose hydrolysis by a highly thermostable endo-1,4-glucanase (Avicelase I) from Clostridium stercorarium. Enzyme Microb Technol 12:431–436

Brown SD, Guss AM, Karpinets TV, Parks JM, Smolin N, Yang S, Land ML, Klingeman DM, Bhandiwad A, Rodriguez M Jr, Raman B, Shao X, Mielenz JR, Smith JC, Keller M, Lynd LR (2011) Mutant alcohol dehydrogenase leads to improved ethanol tolerance in *Clostridium thermocellum*. Proc Natl Acad Sci U S A 108:13752–13757

Cai Y, Lai C, Li S, Liang Z, Zhu M, Liang S, Wang J (2011) Disruption of lactate dehydrogenase through homologous recombination to improve bioethanol production in *Thermoanaerobacterium aotearoense*. Enzyme Microb Technol 48:155–161

Cardona CA, Quintero JA, Paz IC (2010) Production of bioethanol from sugarcane bagasse: status and perspectives. Bioresour Technol 101:4754–4766

Chang VS, Holtzapple MT (2000) Fundamental factors affecting biomass enzymatic reactivity. Appl Biochem Biotechnol 84–86:5–37

Chang T, Yao S (2011) Thermophilic, lignocellulolytic bacteria for ethanol production: current state and perspectives. Appl Microbiol Biotechnol 92:13–27

Chang ST, Parker KN, Bauer MW, Kelly RM (2001) Alpha-glucosidase from *Pyrococcus furiosus*. Methods Enzymol 330:260–269

Chapla D, Pandit P, Shah A (2012) Production of xylooligosaccharides from corncob xylan by fungal xylanase and their utilization by probiotics. Bioresour Technol 115:215–221

Collins T, Gerday C, Feller G (2005) Xylanases, xylanase families and extremophilic xylanases. FEMS Microbiol Rev 29:3–23

Cripps RE, Eley K, Leak DJ, Rudd B, Taylor M, Todd M, Boakes S, Martin S, Atkinson T (2009) Metabolic engineering of *Geobacillus thermoglucosidasius* for high yield ethanol production. Metab Eng 11:398–408

Damiano VB, Ward R, Gomes E, Alves-Prado HF, Da Silva R (2006) Purification and characterization of two xylanases from alkalophilic and thermophilic *Bacillus licheniformis* 77–2. Appl Biochem Biotechnol 129–132:289–302

Danis O, Ogan A, Tatlican P, Attar A, Cakmakci E, Mertoglu B, Birbir M (2015) Preparation of poly(3-hydroxybutyrate-co-hydroxyvalerate) films from halophilic archaea and their potential use in drug delivery. Extremophiles. doi:10.1007/s00792-015-0735-4

Das H, Sing SK (2004) Useful byproducts from cellulosic wastes of agriculture and food industry – a critical appraisal. CRC Cr Rev Food Sci 44:77–89

de Souza PM, Magalhães PDO (2010) Application of microbial α-amylase in industry – a review. Braz J Microbiol 41(4):850–861

Dheeran P, Nandhagopal N, Kuma S, Jaiswal YK, Adhikari DK (2012) A novel thermostable xylanase of *Paenibacillus macerans* IIPSP3 isolated from the termite gut. J Ind Microbiol Biotechnol 39(6):851–860

Dhiman SS, Sharma J, Battan B (2008) Industrial aspects and future prospects of microbial xylanases: a review. BioResources 3:1377–1402

Di Donato P, Fiorentino G, Anzelmo G, Tommonaro G, Nicolaus B, Poli A (2011) Re-use of vegetable wastes as cheap substrates for extremophile biomass production. Waste Biomass Valoriz 2:103–111

Di Donato P, Finore I, Anzelmo G, Lama L, Nicolaus B, Poli A (2014) Biomass and biopolymer production using vegetable wastes as cheap substrates for extremophiles. Chem Eng Trans 38:163–168

Dimarogona M, Topakas E, Olsson L, Christakopoulos P (2012) Lignin boosts the cellulase performance of a GH-61 enzyme from *Sporotrichum thermophile*. Bioresour Technol 110:480–487

Dipasquale L, Romano I, Picariello G, Calandrelli V, Lama L (2014) Characterization of a native cellulase activity from an anaerobic thermophilic hydrogen-producing bacterium *Thermosipho* sp. strain 3. Ann Microbiol 64:1493–1503

Eckert K, Schneider E (2003) A thermoacidophilic endoglucanase (CelB) from *Alicyclobacillus acidocaldarius* displays high sequence similarity to arabinofuranosidases belonging to family 51 of glycoside hydrolases. Eur J Biochem 270:3593–3602

Elleuche S, Schröder C, Sahm K, Antranikian G (2014) Extremozymes – biocatalysts with unique properties from extremophilic microorganisms. Curr Opin Biotechnol 29:116–123

Fawzi EM (2011) Highly thermostable xylanase purified from *Rhizomucor miehei* NRL 3169. Acta Biol Hung 62:85–94

Finore I, Kasavi C, Poli A, Romano I, Toksoy Oner E, Kirdar B, Dipasquale L, Nicolaus B, Lama L (2011) Purification, biochemical characterization and gene sequencing of a thermostable raw starch digesting alpha-amylase from *Geobacillus thermoleovorans* subsp. stromboliensis subsp. nov. World J Microbiol Biotechnol 27:2425–2433

Finore I, Di Donato P, Poli A, Kirdar B, Kasavi C, Toksoy EO, Nicolaus B, Lama L (2014) Use of agro waste biomass for α-amylase production by *Anoxybacillus amylolyticus*: purification and properties. J Microb Biochem Technol 6:320–326

FitzPatrick M, Champagne P, Cunningham MF, Whitney RA (2010) A biorefinery processing perspective: treatment of lignocellulosic materials for the production of value-added products. Bioresour Technol 101(23):8915–8922

Gao L, Gao F, Wang L, Geng C, Chi L, Zhao J, Qu Y (2012) N-glycoform diversity of cellobiohydrolase I from *Penicillium decumbens* and synergism of nonhydrolytic glycoform in cellulose degradation. J Biol Chem 287:15906–15915

Georgieva TI, Mikkelsen MJ, Ahring BK (2008) Ethanol production from wetexploded wheat straw hydrolysate by thermophilic anaerobic bacterium *Thermoanaerobacter* BG1L1 in a continuous immobilized reactor. Appl Biochem Biotechnol 145:99–110

Girfoglio M, Rossi M, Cannio R (2012) Cellulose degradation by *Sulfolobus solfataricus* requires a cell-anchored endo-β-1-4-Glucanase. J Bacteriol 194(18):5091–5100

Graham JE, Clark ME, Nadler DC, Huffer S, Chokhawala HA, Rowland SE, Blanch HW, Clark DS, Robb FT (2011) Identification and characterization of a multidomain hyperthermophilic cellulase from an archaeal enrichment. Nat Commun 2:375

Hakamada Y, Koike K, Yoshimatsu T, Mori H, Kobayashi T, Ito S (1997) Thermostable alkaline cellulase from an alkaliphilic isolate, *Bacillus* sp. KSMS237. Extremophiles 1:151–156

Himmel M, Adney W, Tucker M, Grohmann K (1994) Thermostable purified endoglucanase from *Acidothermus cellulolyticus* ATCC 43068. US Patent 5,275,944

Hoffmann RA, Geijtenbeek T, Kamerling JP, Vliegenthart JF (1992a) 1H-N.M.R. study of enzymically generated wheat-endosperm arabinoxylan oligosaccharides: structures of hepta- to tetradeca-saccharides containing two or three branched xylose residues. Carbohydr Res 223:19–44

Hoffmann RA, Kamerling JP, Vliegenthart JFG (1992b) Structural features of a water-soluble arabinoxylan from the endosperm of wheat. Carbohydr Res 226(2):303–311

Hong SY, Lee JS, Cho KM, Math RK, Kim YH, Hong SJ, Cho YU, Cho SJ, Kim H, Yun HD (2007) Construction of the bifunctional enzyme cellulase-beta-glucosidase from the hyperthermophilic bacterium *Thermotoga maritima*. Biotechnol Lett 29(6):931–936

Horn SJ, Vaaje-Kolstad G, Westereng B, Eijsink VG (2012) Novel enzymes for the degradation of cellulose. Biotechnol Biofuels 5:45

Hreggvidsson GO, Kaiste E, Holst O, Eggertsson G, Palsdottir A, Kristjansson JK (1996) An extremely thermostable cellulase from the thermophilic eubacterium *Rhodothermus marinus*. Appl Environ Microbiol 62:3047–3049

Huang XP, Monk C (2004) Purification and characterization of a cellulose (CMCase) from a newly isolated thermophilic aerobic bacterium *Caldibacillus cellulovorans* gen. nov., sp. nov. World J Microbiol Biotechnol 20:85–92

Huang TY, Duan KJ, Huang SY, Chen CW (2006) Production of polyhydroxyalkanoates from inexpensive extruded rice bran and starch by *Haloferax mediterranei*. J Ind Microbiol Biotechnol 33:701–706

Hung KS, Liu SM, Tzou WS, Lin FP, Pan CL, Fang TY, Sun KH, Tang SJ (2011) Characterization of a novel GH10 thermostable, halophilic xylanase from the marine bacterium *Thermoanaerobacterium saccharolyticum* NTOU1. Process Biochem 46:1257–1263

Jain I, Kumar V, Satyanarayana T (2014) Applicability of recombinant beta-xylosidase from the extremely thermophilic bacterium *Geobacillus thermodenitrificans* in synthesizing alkylxylosides. Bioresour Technol 170:462–469

Jin AX, Ren JL, Peng F, Xu F, Zhou GY, Sun RC, Kennedy JF (2009) Comparative characterization of degraded and non-degradative hemicelluloses from barley straw and maize stems: composition, structure, and thermal properties. Carbohydr Polym 78:609–619

Kabel MA, Carvalheiro F, Garrote G, Avgerinos E, Koukios E, Parajò JC, Girio FM, Schols HA, Voragen AGJ (2002) Hydrothermally treated xylan rich by-products yield different classes of xylo-oligosaccharides. Carbohydr Polym 50(1):47–56

Kamm B, Kamm M (2004) Biorefinery-systems. Chem Biochem Eng Q 18(1):1–6

Karlsson J, Momcilovic D, Wittgren B, Schulein M, Tjerneld F, Brinkmalm G (2002) Enzymatic degradation of carboxymethyl cellulose hydrolyzed by the endoglucanases Cel5A, Cel7B, and Cel45A from *Humicola insolens* and Cel7B, Cel12A and Cel45Acore from Trichoderma reesei. Biopolymers 63(1):32–40

Kasavi C, Finore I, Lama L, Nicolaus B, Oliver SG, Toksoy EO, Kirdar B (2012) Evaluation of industrial *Saccharomyces cerevisiae* strains for ethanol production from biomass. Biomass Bioenergy 45:230–238

Keasling JD (2008) Synthetic biology for synthetic chemistry. ACS Chem Biol 3:64–76

Khandeparkar R, Bhosle NB (2006) Purification and characterization of thermoalkalophilic xylanase isolated from the *Enterobacter* sp. MTCC 5112. Res Microbiol 157:315–325

Khasin A, Alchanati I, Shoham Y (1993) Purification and characterization of a thermostable xylanase from *Bacillus stearothermophilus* T-6. Appl Environ Microbiol 59:1725–1730

Kim HW, Ishikawa K (2010) Complete saccharification of cellulose at high temperature using endocellulase and beta-glucosidase from *Pyrococcus* sp. J Microbiol Biotechnol 20(5):889–892

Kim JW, Peeples TL (2006) Screening extremophiles for bioconversion potentials. Biotechnol Prog 22(6):1720–1724

Kim TH, Kim JS, Sunwoo C, Lee YY (2003) Pretreatment of corn stover by aqueous ammonia. Bioresour Technol 90(1):39–47

Kim MS, Park JT, Kim YW, Lee HS, Nyawira R, Shin HS, Park CS, Yoo SH, Kim YR, Moon TW, Park KH (2004) Properties of a novel thermostable glucoamylase from the hyperthermophilic archaeon Sulfolobus solfataricus in relation to starch processing. Appl Environ Microbiol 70(7):3933–3940

Kuan YH, Liong MT (2008) Chemical and physicochemical characterization of agrowaste fibrous materials and residues. J Agric Food Chem 56:9252–9257

Küçükaşik F, Kazak H, Güney D, Finore I, Poli A, Yenigün O, Nicolaus B, Toksoy EO (2011) Molasses as fermentation substrate for levan production by *Halomonas* sp. Appl Microbiol Biotechnol 89:1729–1740

Kumar V, Satyanarayana T (2014) Secretion of recombinant thermo-alkali-stable endoxylanase of polyextremophilic *Bacillus halodurans* TSEV1 and its utility in generating xylooligosaccharides from renewable agro-residues. Process Biochem 49(11):1875–1883

Kumar S, Kumar P, Satyanarayana T (2007) Production of raw starch-saccharifying thermostable and neutral glucoamylase by the thermophilic mold thermomucor indicae-seudaticae in submerged fermentation. Appl Biochem Biotechnol 142:221–230

Kumar R, Singh S, Singh OV (2008) Bioconversion of lignocellulosic biomass: biochemical and molecular perspective. J Ind Microbiol Biotechnol 35:377–391

Lama L, Calandrelli V, Gambacorta A, Nicolaus B (2004) Purification and characterization of thermostable xylanase and β-xylosidase by the thermophilic bacterium *Bacillus thermantarcticus*. Res Microbiol 155(4):283–289

Lama L, Tramice A, Finore I, Anzelmo G, Calandrelli V, Pagnotta E, Tommonaro G, Poli A, Di Donato P, Nicolaus B, Fagnano M, Mori M, Impagliazzo A, Trincone A (2014) Degradative actions of microbial xylanolytic activities on hemicelluloses from rhizome of *Arundo donax*. AMB Expr 4:55–64

Lawther JM, Sun RC, Banks WB (1996) Effects of extraction conditions and alkali type on yield and composition of wheat straw hemicelluloses. J Appl Polym Sci 60(11):1827–1837

Lee JW, Park JY, Kwon M, Choi IG (2009) Purification and characterization of a thermostable xylanase from the brown-rot fungus *Laetiporus sulphureus*. J Biosci Bioeng 107(1):33–37

Li W, Zhang WW, Yang MM, Chen YL (2000) Cloning of the thermostable cellulase gene from newly isolated *Bacillus subtilis* and its expression in *Escherichia coli*. Mol Biotechnol 40:195–201

Li W, Huan X, Zhou Y, Ma Q, Chen Y (2009) Simultaneous cloning and expression of two cellulase genes from *Bacillus subtilis* newly isolated from Golden Takin (Budorcas taxicolor Bedfordi). Biochem Biophys Res Commun 383:397–400

Li HY, Sun SN, Zhou X, Peng F, Sun RC (2015) Structural characterization of hemicelluloses and topochemical changes in Eucalyptus cell wall during alkali ethanol treatment. Carbohydr Polym 123:17–26

Liang C, Xue Y, Fioroni M, Rodriguez-Ropero F, Zhou C, Schwaneberg U, Ma Y (2011) Cloning and characterization of a thermostable and halo-tolerant endoglucanase from Thermoanaerobacter tengcongensis MB4. Appl Microbiol Biotechnol 89:315–326

Lina L, Ping Z, Yaqi C, Mou S (2006) High-performance anion exchange chromatography with pulsed amperometric detection for simultaneous determination of monosaccharides and uronic acids. Chin J Anal Chem 34(10):1371–1374

Lindenmuth BE, McDonald KA (2011) Production and characterization of Acidothermus cellulolyticus endoglucanase in *Pichia pastoris*. Protein Expr Purif 77:153–158

Lo YC, Huang CY, Cheng CL, Lin CY, Chang JS (2011) Characterization of cellulolytic enzymes and bioH_2 production from anaerobic thermophilic Clostridium sp. TCW1. Bioresour Technol 102:8384–8392

Lynd LR, van Zyl WH, McBride JE, Laser M (2005) Consolidated bioprocessing of cellulosic biomass: an update. Curr Opin Biotechnol 16:577–583

Maalej I, Belhaj I, Masmoudi NF, Belghith H (2009) Highly thermostable xylanase of the thermophilic fungus *Talaromyces thermophilus*: purification and characterization. Appl Biochem Biotechnol 158:200–212

Mastascusa V, Romano I, Di Donato P, Poli A, Della Corte V, Rotundi A, Bussoletti E, Quarto M, Pugliese MG, Nicolaus B (2014) Extremophiles survival to simulated space conditions: an astrobiology model study. Orig Life Evol Biosph 44(3):231–237

Mischnick P, Momcilovic D (2010) Chemical structure analysis of starch and cellulose derivatives. Adv Carbohydr Chem Biochem 64:117–210

Nawirska A, Kwasniewska M (2005) Dietary fibre fractions from fruit and vegetable processing waste. Food Chem 91:221–225

Ng IS, Li CW, Yeh YF, Chen PT, Chir JL, Ma CH, Yu SM, Ho TH, Tong CG (2009) A novel endoglucanase from the thermophilic bacterium *Geobacillus* sp. 70PC53 with high activity and stability over a broad range of temperatures. Extremophiles 13:425–435

Nieves LM, Panyon LA, Wang X (2015) Engineering sugar utilization and microbial tolerance toward lignocellulose conversion. Front Bioeng Biotechnol 3:17

Nwodo UU, Green E, Okoh AI (2012) Review bacterial exopolysaccharides: functionality and prospects. Int J Mol Sci 13(11):14002–14015

Ochs M, Muzard M, Plantier-Royon R, Boris E, Rémond C (2011) Enzymatic synthesis of alkyl β-D-xylosides and oligoxylosides from xylans and from hydrothermally pretreated wheat bran. Green Chem 13(9):2380–2388

Ogino H, Yasui K, Shiotani T, Ishihara T, Ishikawa H (1995) Organic solvent-tolerant bacterium which secretes an organic solvent-stableproteolytic enzyme. Appl Environ Microbiol 61:4258–4262

Olson DG, McBride JE, Shaw AJ, Lynd LR (2012) Recent progress in consolidated bioprocessing. Curr Opin Biotechnol 23(3):396–405

Pastell H, Tuomainen P, Virkki L, Tenkanen M (2008) Step-wise enzymatic preparation and structural characterization of singly and doubly substituted arabinoxylo-oligosaccharides with non-reducing end terminal branches. Carbohydr Res 343(18):3049–3057

Paulová L, Patáková P, Branská B, Rychtera M, Melzoch K (2014) Lignocellulosic ethanol: technology design and its impact on process efficiency. Biotechnol Adv. doi:10.1016/j.biotechadv.2014.12.002

Piller K, Daniel RM, Petach HH (1996) Properties and stabilization of an extracellular alpha-glucosidase from the extremely thermophilic archaebacteria *Thermococcus* strain AN1: enzyme activity at 130 degrees C. Biochim Biophys Acta 1292(1):197–205

Podkaminer KK, Shao X, Hogsett DA, Lynd LR (2011) Enzyme inactivation by ethanol and development of a kinetic model for thermophilic simultaneous saccharification and fermentation at 50 C with Thermoanaerobacterium saccharolyticum ALK2. Biotechnol Bioeng 108:1268–1278

Polizeli ML, Rizzatti R, Monti HF, Terenzi JA, Jorge JA, Amorim DSJ (2005) Xylanases from fungi: properties and industrial applications. Appl Microbiol Biotechnol 67:577–591

Rahman Z, Shida Y, Furukawa T, Suzuki Y, Okada H, Ogasawara W, Morikawa Y (2009) Application of Trichoderma reesei cellulase and xylanase promoters through homologous recombination for enhanced production of extracellular beta-glucosidase I. Biosci Biotechnol Biochem 73:1083–1089

Rakotoarivonina H, Hermant B, Monthe N, Rémond C (2012) The hemicellulolytic enzyme arsenal of *Thermobacillus xylanilyticus* depends on the composition of biomass used for growth. Microb Cell Fact 11:159

Rastogi G, Muppidi GL, Gurram RN, Adhikari A, Bischoff KM, Hughes SR, Apel WA, Bang SS, Dixon DJ, Sani RK (2009) Isolation and characterization of cellulose-degrading bacteria from the deep subsurface of the Homestake gold mine, Lead, South Dakota, USA. J Ind Microbiol Biotechnol 36:585–598

Rastogi G, Bhalla A, Adhikari A, Bischoff KM, Hughes SR, Christopher LP, Sani RK (2011) Characterization of thermostable cellulases produced by *Bacillus* and *Geobacillus* strains. Bioresour Technol 101(22):8798–8806

Reis A, Pinto P, Evtuguin DV, Neto CP, Domingues P, Ferrer-Correia AJ, Domingues MRM (2005) Electrospray tandem mass spectrometry of underivatised acetylated xylo-oligosaccharides. Rapid Commun Mass Spectrom 19(23):3589–3599

Rhee SK, Song KB, Kim CH, Park BS, Jang EK, Jang KH (2005) Levan. Biopolymers 5. doi:10.1002/3527600035.bpol5014

Robert P, Marquis M, Barron C, Guillon F, Saulnier L (2005) FT-IR investigation of cell wall polysaccharides from cereal grains. Arabinoxylan infrared assignment. J Agric Food Chem 53(18):7014–7018

Romaniec M, Fauth U, Kobayashi T, Huskisson N, Barker P, Demain A (1992) Purification and characterization of a new endoglucanase from Clostridium thermocellum. Biochem J 283:69–73

Rose DJ, Inglett GE, Liu SX (2010) Utilisation of corn (*Zea mays*) bran and corn fiber in the production of food components. J Sci Food Agric 90:915–924

Saleem M, Aslam F, Akhtar MS, Tariq M, Rajoka MI (2011) Characterization of a thermostable and alkaline xylanase from *Bacillus* sp. and its bleaching impact on wheat straw pulp. World J Microbiol Biotechnol 28(2):513–522

Samanta AK, Jayapal N, Kolte AP, Senani S, Suresh KP, Sampath KT (2012) Enzymatic production of xylooligosaccharides from alkali solubilized xylan of natural grass (*Sehima nervosum*). Bioresour Technol 112:199–205

Sanchez-Vazquez SA, Hailes HC, Evans JRG (2013) Hydrophobic polymers from food waste: resources and synthesis. Polym Rev 53:627–694

Sandoval NR, Kim JY, Glebes TY, Reeder PJ, Aucoin HR, Warner JR et al (2012) Strategy for directing combinatorial genome engineering in *Escherichia coli*. Proc Natl Acad Sci U S A 109:10540–10545

Sapre MP, Jha H, Patil MB (2005) Purification and characterization of a thermoalkalophilic xylanase from *Bacillus* sp. World J Microbiol Biotechnol 21:649–654

Schiraldi C, Martino A, Acone M, Di Lernia I, Di Lazzaro A, Marulli F, Generoso M, Carteni M, De Rosa M (2000) Effective production of a thermostable alpha-glucosidase from *Sulfolobus solfataricus* in *Escherichia coli* exploiting a microfiltration bioreactor. Biotechnol Bioeng 70(6):670–676

Sedlmeyer FB (2011) Xylan as by-product of biorefineries: characteristics and potential use for food applications. Food Hydrocoll 25(8):1891–1898

Serour E, Antranikian G (2002) Novel thermoactive glucoamylases from the thermoacidophilic Archaea *Thermoplasma acidophilum*, *Picrophilus torridus* and *Picrophilus oshimae*. Anton van Leeuw 81(1–4):73–83

Shaw AJ, Podkaminer KK, Desai SG, Bardsley JS, Rogers SR, Thorne PG, Hogsett DA, Lynd LR (2008) Metabolic engineering of a thermophilic bacterium to produce ethanol at high yield. Proc Natl Acad Sci U S A 105:13769–13774

Shrivastav A, Kim HY, Kim YR (2013) Advances in the applications of polyhydroxyalkanoate nanoparticles for novel drug delivery system. BioMed Res Int 581684. doi:10.1155/2013/581684

Simpson HD, Haufler UR, Daniel RM (1991) An extremely thermostable xylanase from the thermophilic eubacterium *Thermotoga*. Biochem J 277:413–417

Singh S, Madlala AM, Prior BA (2003) *Thermomyces lanuginosus*: properties of strains and their hemicellulases. FEMS Microbiol Rev 27:3–16

Sinnott ML (1990) Catalytic mechanisms of enzymic glycosyl transfer. Chem Rev 90:1171–1202

Sivaramakrishnan S, Gangadharan D, Nampoothiri KM, Soccol CR, Pandey A (2006) α-amylases from microbial sources – an overview on recent developments. Food Technol Biotechnol 44:173–184

Sriyapai T, Somyoonsap P, Matsui K, Kawai F, Chansiri K (2011) Cloning of a thermostable xylanase from *Actinomadura* sp. S14 and its expression in *Escherichia coli* and *Pichia pastoris*. J Biosci Bioeng 111:528–536

Studholme DJ (2015) Some (bacilli) like it hot: genomics of *Geobacillus* species. Microb Biotechnol 8(1):40–48

Subba MVSST, Muralikrishna G (2004) Structural analysis of arabinoxylans isolated from native and malted finger millet (*Eleusine coracana*, ragi). Carbohydr Res 339(14):2457–2463

Subramaniyan S, Prema P (2000) Cellulase-free xylanases from *Bacillus* and other microorganisms. FEMS Microbiol Lett 183(1):1–7

Sun Y, Cheng J (2002) Hydrolysis of lignocellulosic materials for ethanol production: a review. Bioresour Technol 83(1):1–11

Sun RC, Lawther JM, Banks WB (1996) Fractional and structural characterization of wheat straw hemicelluloses. Carbohydr Polym 29(4):325–331

Sun JX, Sun XF, Sun RC, Su YQ (2004) Fractional extraction and structural characterization of sugarcane bagasse hemicelluloses. Carbohydr Polym 56(2):195–204

Sun XF, Sun RC, Fowler P, Baird MS (2005) Extraction and characterization of original lignin and hemicelluloses from wheat straw. J Agric Food Chem 53(4):860–870

Sunna A, Moracci M, Rossi M, Antranikian G (1997) Glycosyl hydrolases from hyperthermophiles. Extremophiles 1:2–13

Sweeney MD, Xu F (2012) Biomass converting enzymes as industrial biocatalysts for fuels and chemicals: recent developments. Catalysts 2:244–263

Tai SK, Lin HP, Kuo J, Liu JK (2004) Isolation and characterization of a cellulolytic *Geobacillus thermoleovorans* T4 strain from sugar refinery wastewater. Extremophiles 8:345–349

Tommonaro G, Poli A, De Rosa S, Nicolaus B (2008) Tomato derived polysaccharides for biotechnological applications: chemical and biological approaches. Molecules 13(6):1384–1398

Tramice A, Arena A, De Gregorio A, Ottanà R, Maccari R, Pavone B, Arena N, Innello D, Vigorita MG, Trincone A (2008) Facile biocatalytic access to 9-fluorenyl methyl polyglycosides: evaluation of antiviral activity on immunocompetent cells. ChemMedChem 3(9):1419–1426

Tramice A, Melck D, Virno A, Randazzo A, Motta A, Trincone A (2009) Enzymatic synthesis and 3-D structure of anti-proliferative acidic (MeGlcA) xylotetrasaccharide. J Mol Catal B Enzym 61(3–4):129–135

Van TT, Ryu SI, Lee KJ, Kim EJ, Lee SB (2007) Cloning and characterization of glycogen-debranching enzyme from hyperthermophilic archaeon *Sulfolobus shibatae*. J Microbiol Biotechnol 17(5):792–799

VanFossen AL, Verhaart MR, Kengen SM, Kelly RM (2009) Carbohydrate utilization patterns for the extremely thermophilic bacterium *Caldicellulosiruptor saccharolyticus* reveal broad growth substrate preferences. Appl Environ Microbiol 75:7718–7724

Vázquez MJ, Garrote G, Alonso JL, Domiguez H, Parajò JC (2005) Refining of autohydrolysis liquors for manufacturing xylo-oligosaccharides: evaluation of operational strategies. Bioresour Technol 96(8):889–896

Verbruggen MA, Beldman G, Voragen AGJ (1995) The selective extraction of glucuronoarabinoxylans from Sorghum endosperm cell walls using barium and potassium hydroxide solutions. J Cereal Sci 21(3):271–282

Wang TH, Lu S (2013) Production of xylooligosaccharide from wheat bran by microwave assisted enzymatic hydrolysis. Food Chem 138(2–3):1531–1535

Wang J, Bai Y, Yang P, Shi P, Luo H, Meng K, Huang H, Yin J, Yao B (2010) A new xylanase from thermoalkaline *Anoxybacillus* sp. E2 with high activity and stability over a broad pH range. World J Microbiol Biotechnol 26:917–924

Wang X, Yomano LP, Lee JY, York SW, Zheng H, Mullinnix MT, Shanmgam KT, Ingram LO (2013) Engineering furfural tolerance in *Escherichia coli* improves the fermentation of lignocellulosic sugars into renewable chemicals. Proc Natl Acad Sci U S A 110:4021–4026

Wong DW (2009) Structure and action mechanism of ligninolytic enzymes. Appl Biochem Biotechnol 157(2):174–209

Wu S, Liu B, Zhang X (2006) Characterization of a recombinant thermostable xylanase from deep-sea thermophilic *Geobacillus* sp. MT-1 in East Pacific. Appl Microbiol Biotechnol 72:1210–1216

Xiangyuan H, Shuzheng Z, Shoujun Y (2001) Cloning and expression of thermostable beta-glycosidase gene from *Thermus nonproteolyticus* HG102 and characterization of recombinant enzyme. Appl Biochem Biotechnol 94:243–255

Xu W, Osei-Prempeh G, Lema C, Devis Oldham E, Aguilera RJ, Parkin S, Rankin SE, Knutson BL, Lehmler HJ (2012) Synthesis, thermal properties and cytotoxicity evaluation of hydrocarbon and fluorocarbon alkyl-β-D xylopyranoside surfactants. Carbohydr Res 349:12–23

Yang SJ, Lee HS, Park CS, Kim YR, Moon TW, Park KH (2004) Enzymatic analysis of an amylolytic enzyme from the hyperthermophilic archaeon Pyrococcus furiosus reveals its novel catalytic properties as both an alpha-amylase and a cyclodextrin-hydrolyzing enzyme. Appl Environ Microbiol 70(10):5988–5995

Yang SJ, Kataeva I, Hamilton-Brehm SD, Engle NL, Tschaplinski TJ, Doeppke C, Davis M, Westpheling J, Adams MW (2009) Efficient degradation of lignocellulosic plant biomass, without pretreatment, by the thermophilic anaerobe *Anaerocellum thermophilum* DSM 6725. Appl Environ Microbiol 75:4762–4769

Yang D, Weng H, Wang M, Xu W, Li Y, Yang H (2010) Cloning and expression of a novel thermostable cellulase from newly isolated *Bacillus subtilis* strain I15. Mol Biol Rep 37:1923–1929

Zambare VP, Bhalla A, Muthukumarappan K, Sani RK, Christopher LP (2011) Bioprocessing of agricultural residues to ethanol utilizing a cellulolytic extremophile. Extremophiles 15(6):11–618

Zhang Y-HP, Ding SY, Mielenz JR, Cui J-B, Elander RT, Laser M, Himmel ME, McMillan JR (2007) Fractionating recalcitrant lignocellulose at modest reaction conditions. Biotechnol Bioeng 97(2):214–223

Zhang J, Siika-Aho M, Puranen T, Tang M, Tenkanen M, Viikari L (2011) Thermostable recombinant xylanases from *Nonomuraea flexuosa* and *Thermoascus aurantiacus* show distinct properties in the hydrolysis of xylans and pretreated wheat straw. Biotechnol Biofuels 4:12

Zheng Y, Zhao J, Xu F, Li Y (2014) Pretreatment of lignocellulosic biomass for enhanced biogas production. Prog Energy Combust Sci 42:35–53

Zhou J, Wang Y-H, Chua J, Luoa L-Z, Zhuanga Y-P, Zhanga S-L (2009) Optimization of cellulase mixture for efficient hydrolysis of steam-exploded corn stover by statistically designed experiments. Bioresour Technol 100:819–825
Zhu S, Wu Y, Yu Z, Zhang X, Li H, Gao M (2006) The effect of microwave irradiation on enzymatic hydrolysis of rice straw. Bioresour Technol 97(15):1964–1968
Zverlov VV, Schantz N, Schmitt-Kopplin P, Schwarz WH (2005) Two new major subunits in the cellulosome of Clostridium thermocellum: xyloglucanase Xgh74A and endoxylanase Xyn10D. Microbiology 151:3395–3401

Chapter 20
A Strategy for Designing Thermostable Enzymes by Reconstructing Ancestral Sequences Possessed by Ancient Life

Satoshi Akanuma and Akihiko Yamagishi

1 Introduction

The question what kind of life our oldest ancestor was has been a long-running controversy because the answer will help us understand the characteristics and environment of our universal ancestor and, importantly, how it evolved to exist in ever changing environments on Earth. The recent expansion of genome data of bacteria and archaea have made us possible to conduct phylogenetic analyses that will unveil the evolution of life on Earth. In addition, it is now possible to infer an ancestral gene and resurrect the ancestral protein from the gene (Thornton 2004). Resurrecting ancestral protein sequences and characterizing their properties are currently a powerful means to predict ancestor's characteristics and its environment because a protein's properties should reflect its host's characteristics and environment (Gaucher et al. 2010; Boussau and Gouy 2012).

In this chapter, we introduce our recent study (Akanuma et al. 2013a), in which we experimentally addressed the question: What was the growth temperature of the last common ancestor of extant life? The approach used to answer this question involved phylogenetic tree building to infer the amino acid sequences of archaeal and bacterial ancestral nucleoside diphosphate kinases (NDK), expressing the genes encoding the inferred sequences, and characterizing the unfolding temperatures of the gene products. The study eventually provided the first experimental support for the concept that

S. Akanuma
Department of Applied Life Sciences, Tokyo University of Pharmacy and Life Sciences, 1432-1, Horinouchi, Hachioji, Tokyo 192-0392, Japan

Faculty of Human Sciences, Waseda University, Shinjuku, Japan

A. Yamagishi (✉)
Department of Applied Life Sciences, Tokyo University of Pharmacy and Life Sciences, 1432-1, Horinouchi, Hachioji, Tokyo 192-0392, Japan
e-mail: yamagish@toyaku.ac.jp

P.H. Rampelotto (ed.), *Biotechnology of Extremophiles*, Grand Challenges in Biology and Biotechnology 1, DOI 10.1007/978-3-319-13521-2_20

the universal common ancestor of all extant life was a (hyper)thermophile that lived at a very high temperatures. Moreover, the procedure for ancestral sequence reconstruction serve as a reliable means of creating extremely thermally stable proteins (Gaucher et al. 2003, 2008; Akanuma et al. 2011, 2013a; Butzin et al. 2013).

2 The Last Universal Common Ancestor "Commonote"

All extant life on earth can be divided into three domains of life: Archaea, Bacteria and Eukarya (Woese 1987; Woese et al. 1990). A phylogenetic tree built by comparing the small subunit ribosomal RNA (rRNA) sequences indicates that all of the extant life are the descendants of the single common ancestor, the last universal common ancestor (Fig. 20.1). Although the common ancestor has been referred to as LUCA, LCA, or senancestor, we herein refer to as "Commonote" (Yamagishi et al. 1998). What kind of organism the Commonote was is the common interest for all of humanity because its answer will help us understand the coevolution of our universal ancestor and the environment of early earth where the ancestor lived.

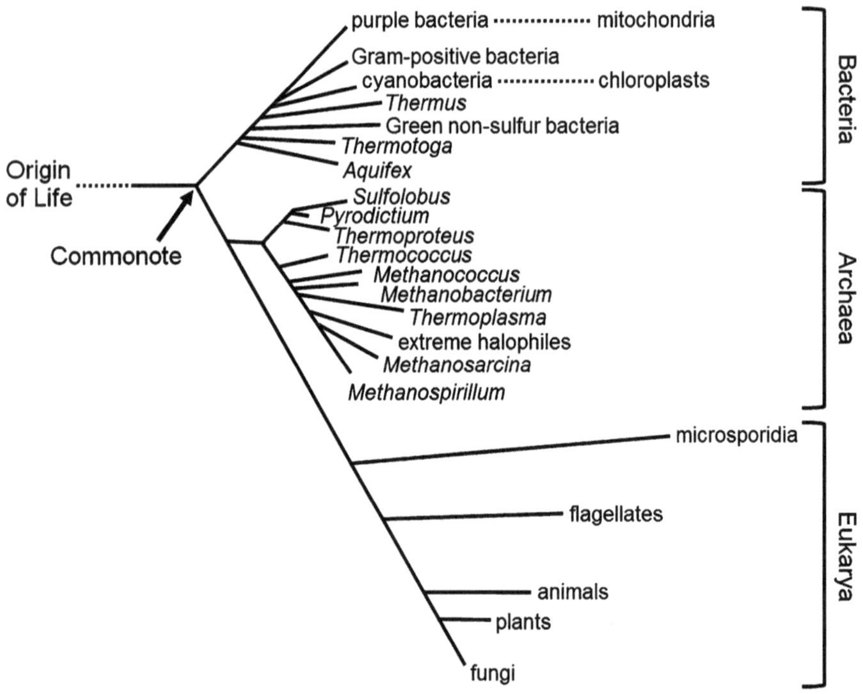

Fig. 20.1 The most-referenced phylogenetic tree constructed from SSU-rRNA sequences (Woese et al. 1990; Yamagishi et al. 1998). The position of the Commonote is indicated

3 Early Studies Focusing on the Environment Temperature of the Commonote

The environment temperature of ancient life has long been an interesting topic. Many researchers have addressed the environment temperature of the Commonote. In early, the common ancestors of Archaea and of Bacteria have been proposed to be hyperthermophilic (Pace 1991; Stetter 1996) because hyperthermophilic archaea and bacteria are located near the basal positions in the most-referenced phylogenetic tree constructed from rRNA sequences (Woese 1987; Achenbach-Richter et al. 1988). Therefore, the Commonote was also parsimoniously predicted to be hyperthermophilic. However, it might be possible that the oldest ancestor lived in a cooler environment and afterwards adapted to higher temperatures (Greaves and Warwicker 2007). Moreover, Brochier and Philippe (2002) analyzed the bacterial phylogeny and proposed that the first phyla that emerged in the tree was not hyperthermophilic organisms. They also suggested that the emergency of hyperthermophilic bacteria within the bacterial domain was a result from the second adaptation to the high temperature.

The evolution of reverse gyrase also suggests that the primitive organisms could not be hyperthermophilic. Reverse gyrase is an ATP-dependent type I DNA topoisomerase and only found in thermophilic organisms. This protein is thought to play an essential role in the adaptation to high temperatures (Forterre 2002; Atomi et al. 2004) because all known hyperthermophilic organisms contain a gene for reverse gyrase (Heine and Chandra 2009). Reverse gyrase consists of non-related two domains, a topoisomerase domain and a helicase domain (Declais et al. 2000). Assuming that these domains originated and evolved independently, and later fused to generate reverse gyrase, the first organism could not be a hyperthermophile (Forterre 1996). However, this argument does not rule out the possibility that reverse gyrase had emerged prior to the appearance of the Commonote.

4 Computational Analysis of Ancestral Sequences

Ancestral states of nucleotide sequences suggested the non-thermophilic ancestry of life. The underlying concept of this approach is that the G+C content of the stem region of an extant prokaryotic ribosomal RNA correlates well with the environmental temperature of its host. Galtier et al. (1999) estimated the ancestral G+C content in ribosomal RNA sequence, which was not compatible with high temperature-living organisms (Galtier et al. 1999). However, use of different computational algorisms led to conflicting conclusions (Di Giulio 2000, 2003a, b).

Ancestral states of the amino acid compositions of proteins were also used to estimate the environmental temperatures of ancient organisms. Brooks et al. (2004) used an expectation-maximization method to compute amino acid compositions of a set of proteins of the universal ancestor. The ancestral amino acid composition of this protein set was similar to those of extant thermophilic organisms rather than to those of extant mesophiles, suggesting that the universal ancestor was thermophilic. Groussin and Gouy (2011) performed a comprehensive analysis of the archaeal phylogeny. They estimate ancestral G+C contents of rRNAs and amino acid compositions of proteins, and then inferred the optimal growth temperature of the ancestral archaea corresponding to each internal node of the archaeal phylogeny. Eventually, they predicted that the oldest archaeal ancestor was hyperthermophilic and extant mesophilic species have adapted to cooler environments. Another study computing both G+C content in ancestral rRNA sequence and the amino acid composition in ancestral protein sequences suggested that both the last common ancestors of Archaea and of Bacteria were thermophilic, but the universal ancestor was mesophilic (Boussau et al. 2008). Groussin et al. (2013) also calculated the ancestral amino acid compositions of proteins, and therefore the optimal environmental temperatures of early organisms. Although the estimated environmental temperatures (33–68 °C) for the universal ancestor was higher than that (1–37 °C) calculated by Boussau et al. (2008), the universal ancestor was still estimated to have lived at lower temperature than the common ancestors of Archaea and of Bacteria. Thus, a number of computational works have focused on the environmental temperatures at which the ancient organisms lived, but conclusive evidence characterizing those temperatures has not been obtained from those studies. Moreover, the assumption that an ancestral G+C content directly correlated with the environmental temperature of an ancient organism might be too simplistic. It is not apparent that the correlation observed between the optimal environmental temperature of a modern organism and the G+C content of its ribosomal RNA can be used to estimate the environmental temperature of an ancient organism living in primitive earth. In addition, we know well that the thermal stabilities of proteins is often very sensitive to a few amino acid substitutions. Therefore, the conclusion obtained from the amino acid composition analysis in ancestral protein sequences remained inferential if there is no empirical testing. In the following sections, we introduce our recent ancestral sequence reconstruction study that provide the empirical evidence for the existence of a (hyper) thermophilic universal common ancestor (Akanuma et al. 2013a).

5 Ancestral Sequence Reconstruction of Nucleoside Diphosphate Kinase

We reconstructed ancestral NDK sequences that might be possessed by the last common archaeal and bacterial ancestors. The ancestor of NDK family is thought to have emerged early in evolution because most extant organisms, from bacteria to human, contain the gene(s) that encode a member of this family of proteins. Modern NDK seeuences are relatively well conserved, allowing residues in ancient proteins to be predicted with a high degree of confidence. A more important reason to choose NDK is the fact that, as shown in Fig. 20.2, the unfolding temperature of a NDK correlates strongly with the optimal environmental temperature of its host organism (Akanuma et al. 2013a). In other words, the thermal stability of each NDK works as a molecular thermometer for its host's natural environmental temperature.

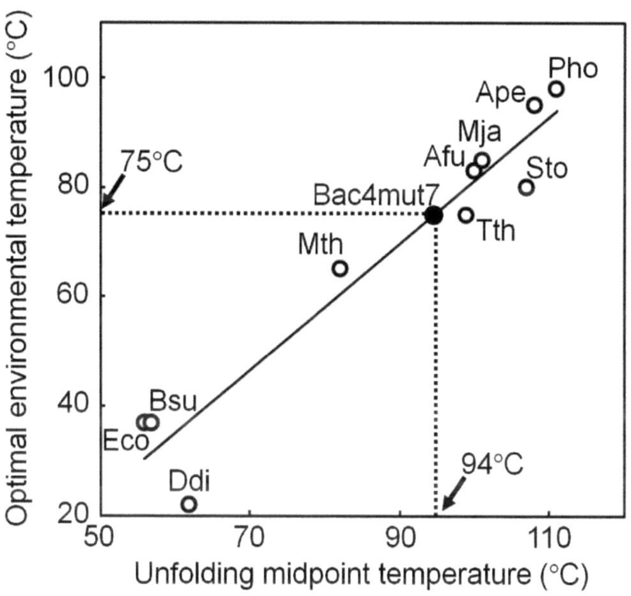

Fig. 20.2 Relationship between the unfolding temperature of microbial NDKs and the optimum environmental temperatures of their hosts. The loweest estimate for the environmental temperature (75 °C) of the Commonote was found using the calibration curve and the unfolding temperature (94 °C) of Bac4mut7. Ddi, *Dictyostelium discoideum*; Eco, *E. coli*; Bsu, *Bacillus subtilis*; Mth, *Methanothermobacter thermoautotrophicus*; Tth, *T. thermophilus*; Afu, *Archaeoglobus fulgidus*; Mja, *Methanocaldococcus jannaschii*; Sto, *S. tokodaii*; Ape, *Aeropyrum pernix*; Pho, *Pyrococcus horikoshii*

Figure 20.3 describes a procedure for reconstructing the ancestral amino acid sequences. First, extant homologous amino acid sequences were retrieved from public databases. Then, the amino acid sequences were aligned to generate a multiple sequence alignment. Insertions and gaps within the sequences contained in a multiple alignment often interfere with the process of inferring ancestral sequences. Fortunately, only a few insertions/gaps (indels) are found in a multiple alignment of NDK sequences. Nevertheless, we manually adjusted to correct for the indels positions. A phylogenetic tree was built using the resulting alignment. Although two methods, the maximum–likelihood (ML) method (Yang et al. 1995) and the Bayesian method (Yang and Rannala 1997), have ubiquitously been employed for calculating the tree topology and the ancestral sequence, we used only ML method because the ML phylogenetic algorithm probably reproduces accurate ancestral sequences (Hanson-Smith et al. 2010). Using the multiple sequence alignment of NDKs, we built two maximum likelihood phylogenetic trees, whose topologies partly differed from each other. Eukaryotic sequences were omitted from our analysis because “Eukarya” appeared as parts of “Bacteria” and was not grouped into an own monophyletic group in a preliminary tree building including eukaryotic sequences. The eukaryotic NDKs used in the analysis might have transferred from bacterial ancestors. Next, based on the sequences and the topologies of the trees,

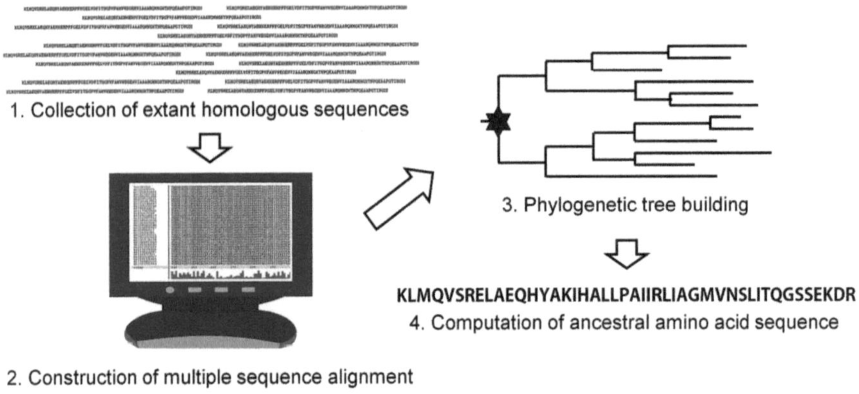

Fig. 20.3 Schematic representation of the procedure to reconstruct an ancestral protein sequence

ancestral NDK sequences corresponding to the last common ancestors of Archaea and of Bacteria were inferred using two program, CODEML in PAML (Yang 1997) and GASP (Edwards and Shields 2004). The former model was used for giving ancestral sequences and the latter model for giving the position of gaps. The archaeal ancestors were named Arc3 and Arc4, and the bacteria ancestors were named Bac3 and Bac4.

Groussin et al. (2015) suggested that accounting for the duplication, horizontal transfer, and loss of gene through evolution improves the accuracy of ancestral sequences inferred. Therefore, the physical properties of a reconstructed protein may be affected by the tree topology used to infer the ancestral sequence. The topologies of the two NDK trees differ somewhat from the species tree that was built by comparing rRNA sequences (Woese 1987). We therefore constructed another phylogenetic tree that contained the small-subunit rRNA sequences of the species whose NDK sequences were used to calculate the sequences of Arc3/4 and Bac3/4. Using the topology of the phylogenetic tree constructed from the rRNA sequences and replacing each rRNA sequence by the NDK sequence of its corresponding species, the sequences for the nodes corresponding to the last common ancestors of Archaea and Bacteria were inferred and named Arc5 and Bac5, respectively.

6 Resurrection of Ancestral NDKs to Estimate the Environment Temperatures of Ancient Life

To estimate the environmental temperature of the Commonote, we synthesized the genes encoding the inferred ancestral amino acid sequences by a PCR-mediated whole gene synthesis technique. The resulting ancestral genes were expressed in *Escherichia coli* and the ancestral NDKs were purified. The purified ancestral NDKs are catalytically active and significantly stable around 100 °C or even above

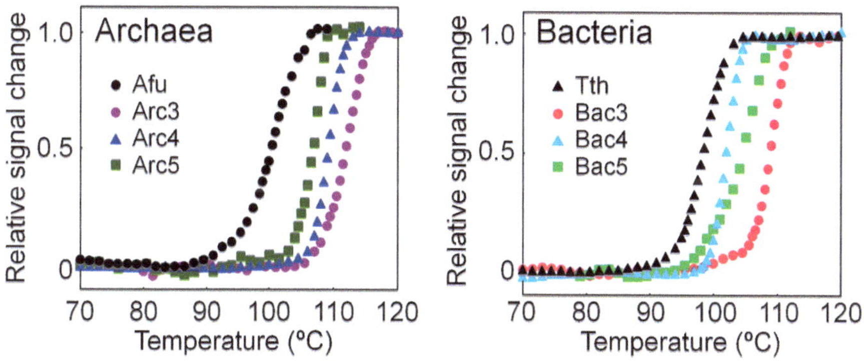

Fig. 20.4 Temperature-induced unfolding of extant thermophilic and ancestral NDKs monitored by change in ellipticity at 222 nm. The plots were normalized with respect to the baselines of the native and denatured states. *Afu* hyperthermophilic archaeon *Archaeoglobus fulgidus* NDK, *Tth* extreme thermophilic bacterium *Thermus thermophilus* NDK

(Fig. 20.4). The results are robust to the topologies of the trees that were used to reconstruct the ancestral sequences. According to the unfolding temperatures of the reconstructed NDKs and the calibration curve (Fig. 20.2), the optimum environmental temperatures of the last common bacterial and archaeal ancestors are ~80–93 °C and ~81–97 °C, respectively (Akanuma et al. 2013a).

7 Estimation of the Environment Temperature of the Commonote

The precise sequence of the Commonote's NDK could not be determined in our study, because the trees used to infer the ancestral sequences were unrooted. To determine the position of the Commonote on a phylogenetic tree, we should have constructed a composite tree of two or more paralogous proteins that diverged from their common ancestral protein prior to the age of the Commonote (Akanuma et al. 2013b). However, we expect that the position of the Commonote would be placed between the root of Archaea and the root of Bacteria. In addition, of the 139 reconstructed residues in the ancestral NDK sequences, 115 are identical among the sequences of the six reconstructed NDKs (Arc3/4/5 and Bac3/4/5). Therefore, the 115 amino acids are also likely to present at the same positions in the Commonte's NDK sequence. Moreover, the Commonote's sequence is parsimoniously expected to have had residues present in either the sequence of the last common archaeal or bacreial ancestor. Therefore, for the 24 non-conserved positions, the Commonte's NDK sequence would have residues contained in at least one of the inferred NDK sequences for the archaeal and bacterial ancestors. We therefore tried to identify the sequence for the lowest stability among the all possible Commonote's NDK sequences.

Because Bac4 shows the lowest unfolding temperature among the reconstructed ancestral NDKs, we individually replaced the residues at the 24 positions of Bac4 by residues found in the other ancestral NDKs at the same positions. We also tested the effect of the combinations of amino acid substitutions that are located within 5 Å of each other on the stability of Bac4. As the result, we found that seven amino acid substitutions decreased the thermal stability of Bac4. We then simultaneously introduced all of the destabilizing amino acid substitutions into Bac4, thus producing Bac4mut7. The unfolding temperature 94 °C of the Bac4mut7 is the lowest estimate of the unfolding temperature of the Commonote's NDK. Accordingly, the Commonote's environment temperature is estimated to be 75 °C or higher as calculated using the calibration curve shown in Fig. 20.2. Thus, the Commonote was likely a (hyper)thermophile that lived at 75 °C or higher temperature (Akanuma et al. 2013a).

As mentioned above, a composite phylogenetic tree of paralogous proteins that might be diverged prior to the age of the Commonote must be built to determine the root of the tree. Two composite trees, one for elongation factor Tu/1 (Iwabe et al. 1989; Baldauf et al. 1996) and the other for H^+ ATPase (Gogarten et al. 1989), suggested that the Commonote was located within the branch connecting the common ancestors of Archaea and Bacteria. The composite trees of 3-isopropylmalte dehydrogenase and isocitrate dehydrogenase (Miyazaki et al. 2001), and of aliphatic aminoacyl tRNA synthases (Brown and Doolitle 1995; Fournier et al. 2011) also supports the position of the root. However, other studies suggested that the Commonote was a member of Bacteria (Cavalier-Smith 2002, 2006a, b, 2010; Lake et al. 2008, 2009). We assumed that the correct Commonote's node would be placed on the branch that connects the Archaea and Bacteria clades. Because the accuracy of the Commonote's sequence relies largely on the position of the root in the tree, we should note here that our conclusion is based on the hypothesis that the Commonote would exist within the branch that connects the roots of Bacteria and Archaea.

Our protein resurrection study relied on a simplistic approach to infer ancestral amino acid sequences. We used a homologous substitution model that exploited the approximation of constant global amino acid compositions in proteins over evolution and across lineages. Boussau et al. (2008) pointed out that the use of a non-homogeneous substitution model is crucial to estimate ancestral states of genes and proteins accurately. In the analysis with a non-homogeneous model, amino acid composition and substitution matrix are constant at all branches, which is not the case for the true evolution because extant proteins have different amino acid compositions. In contrast, in the analysis with a non-homogeneous model, amino acid composition and substitution probabilities among amino acids are allowed to change during evolution. The evidences supporting the thermophilicity of the universal ancestor have been obtained from the ancestral sequences inferred using homogeneous substitution models (Di Giulio 2000, 2003a, b; Brooks et al. 2004; Akanuma et al. 2013a). Very recently, we re-inferred the ancestral NDK sequences using a non-homogeneous substitution model. The thermal stabilities of the newly reconstructed NDKs showed that not only the archaeal and bacterial ancestors but also the Commonote's NDKs have extreme thermal stability. Therefore we can hold our conclusion even if we used the substitution model that relaxes this constraint by allowing different global amino acid compositions in different lineages of the trees. The results will be published elsewhere.

8 Other Studies to Synthesize Ancestral Proteins

Other studies also synthesized ancestral proteins experimentally after the ancestral amino acid sequences was reconstructed computationally. The empirical technique provided experimental evidence for the environmental temperatures experienced by ancient bacteria. Gaucher et al. (2003) reproduced the ancestral elongation factor Tu and reported that the bacterial ancestor was thermophilic, rather than hyperthermophilic or mesophilic. They also performed a comprehensive analysis of the bacterial phylogeny and estimated the environmental temperatures corresponding to the internal nodes of the entire bacterial domain. Then, they observed that the thermophilic ancestor adapted to cooler environment progressively as the temperature of the ancient ocean decreased (Gaucher et al. 2008). In addition, the reconstructed elongation factor Tu works correctly in an extant thermophilic organisms (Zhou et al. 2012). Similarly, Hart et al. (2014) investigated how modern mesophilic and thermophilic ribonuclease H1 (RNH) have adapted to low and high temperatures, respectively. The thermal stability of the ancestral RNH was between those of the mesophilic and thermophilic RNH. Moreover, the unfolding temperatures of intermediate ancestors found in the thermophilic lineage gradually increased over evolution, whereas a steep drop and subsequent slow decrease in unfolding temperature was observed for evolutionary intermediates in the mesophilic lineage. Another ancestral sequence reconstruction experiment provided the evidence that the recent common ancestor of Thermotogales lived at higher temperatures than its descendants (Butzin et al. 2013).

Ancestral sequence reconstruction was also used for studying the evolution of the universal genetic code table. Fournier and Alm (2015) reconstructed the amino acid sequence of the common ancestor of tyrosyl and tryptophanyl aminoacyl-tRNA synthetases. The ancestral synthetase that might existed prior to the last universal common ancestor contained tyrosines within its sequences but tryptophan was absent, suggesting that tryptophan added later to the amino acid repertoire used for protein synthesis.

9 Ancestral Sequence Reconstruction for Designing Thermally Stable Proteins

Designing of proteins with great thermal stability is still challenging. The conventional methods based on rational design rely on a high-resolution three-dimensional structure for the design of mutations to improve the thermal stability of the protein (Ulmer 1983). Several rational design approaches have been proposed to improve the thermal stability of a protein. Those methods include improving hydrophobic interactions and interior core packing (Clark et al. 2004; Dong et al. 2008), increasing the number of ion-pairs and ion-pair networks on the protein surface (Christodoulou et al. 2003; Tanaka et al. 2006), introducing a disulfide bond (Ivens et al. 2002), and increasing the number of ion-pairs or improving hydrophobic interactions between subunits (Kirino et al. 1994; Cheung et al. 2005). However, the effect of each modification is usually small and often strongly depends on its structural context. Improvement of a protein's thermal stability was also performed by a computational protein design method (Korkegian et al. Science 2005). The ancestral sequence reconstruction described here provides a new method for producing mutated proteins with enhanced thermal stability. Assuming that the ancestral organisms were (hyper)thermophilic, ancestral residues would be responsible for protein's thermostability to a much greater extent than non-ancestral residues. We have validated the reliability of the method by producing and characterizing mutants in which one or a small number of the residue(s) found in the phylogenetically predicted "ancestral" sequence had substituted for the original one(s). Two homologous enzymes from the extremely thermophiles were used in such studies. A leucine biosynthetic enzyme, 3-isopropylmalate dehydrogenase (IPMDH), and an enzyme in TCA cycle, isocitrate dehydrogenase (ICDH), are thought to have been duplicated from a common ancestor of the enzymes before the age of the Commonote (Hurley and Dean 1994; Zhang and Koshland 1995; Suzuki et al. 1997). We constructed a composite tree of the two paralogous enzymes and then inferred ancestral IPMDH and ICDH amino acid sequences that might be hosted by the Commonote. We then individually introduced inferred ancestral residues into the intrinsically thermally stable IPMDHs from the hyperthermophilic archaeon, *Sulfolobus tokodaii*, and from the extreme thermophilic bacterium, *Thermus thermophilus*, and into ICDH from the extreme thermophilic archaeon, *Caldococcus noboribetus*. In those studies, as summarized in Table 20.1, 45–67 % of the mutants constructed were more thermally stable than the respective wild-type enzymes (Miyazaki et al. 2001; Iwabata et al. 2005; Watanabe et al. 2006).

A similar trend was also observed in the study using an enzyme involved in the translation system of *T. thermophilus*. We inferred ancestral amino acid sequences of α_2 type glycyl-tRNA synthetase (GlyRS) and then introduced each or combination of inferred ancestral amino acids into GlyRS from *T. thermophiles*. Thermal stability measurement showed that, as listed in Table 20.1, six of eight mutants tested showed enhanced thermal stability compared to wild-type GlyRS (Shimizu et al. 2007). Thus, these mutagenesis studies clearly demonstrated that introduction of ancestral amino acid residues found in reconstructed ancestral amino acid sequences further improved the thermal stabilities of the intrinsically thermally stable enzymes from (hyper)thermophiles with high probability. Recently, it was demonstrated that introduction of ancestral amino acids inferred using a dataset that contains only eukaryotic sequences also improves the stability of a fungal enzyme (Semba et al. 2015). In addition, the catalytic efficiencies of the ancestral mutants are often same as or higher than those of the respective wild-type enzymes.

Table 20.1 Unfolding temperatures of designed proteins produced by a site-directed mutagenesis or a whole gene synthesis in combination with ancestral sequence reconstruction

Target protein	Wild-type/mutants/ancestors	T_m (°C)	Citation
S. tokodaii IPMDH	Wild-type	96	Miyazaki et al. (2001)
	Met91Leu+Ile95Leu	99[a]	
	Lys152Arg+Gly154Ala	97[a]	
	Lys152Arg	96	
	Gly154Ala	98[a]	
	Ala259Ser+Phe261Pro	97[a]	
	Tyr282Leu	95	
T. thermophiles IPMDH	Wild-type	87	Watanabe et al. (2006)
	Phe53Leu	86	
	Pro56Glu	89[a]	
	Arg58Leu	84	
	Val61Ile	87	
	Leu134Asn	91[a]	
	His179Lys	87	
	Val181Thr	89[a]	
	Asp184His	88[a]	
	Ser261Asn	86	
	Pro324Thr	87	
	Ala35Glu	88[a]	
C. noboribetus ICDH	Wild-type	88	Iwabata et al. (2005)
	Tyr309Ile+Ile310Leu	88	
	Ile321Leu	89[a]	
	Ala325Pro+Gly326Ser	91[a]	
	Ala336Phe	74	
T. thermophiles GlyRS	Wild-type	80	Shimizu et al. (2007)
	Gln29Ala	82[a]	
	Met167Leu+Val173Ile	84[a]	
	Asp202Glu+Ser205Arg	78	
	Val232Thr	82[a]	
	Tyr242Phe	81[a]	
	Phe314Tyr+Gly317Ser	77	
	Ala455Cys	81[a]	
	Val479Ile	83[a]	
Elongation factor Tu	*T. thermophilus* elongation factor Tu	77	Gaucher et al. (2008)
	Bacterial common ancestor	65	
	Bacterial common ancestor	73	
DNA gyrase ATPase domain	*T. thermophilus* ATPase domain	90	Akanuma et al. (2011)
	Ancestral ATPase domain	90	
NDK	*Pyrococcus horikoshii* NDK	111	Akanuma et al. (2013a)
	Archaeal common ancestor Arc1	114	
	Archaeal common ancestor Arc2	109	
	Archaeal common ancestor Arc3	112	
	Archaeal common ancestor Arc4	109	
	Archaeal common ancestor Arc5	108	
	Bacterial common ancestor Bac1	99	
	Bacterial common ancestor Bac2	98	
	Bacterial common ancestor Bac3	109	
	Bacterial common ancestor Bac4	102	
	Bacterial common ancestor Bac5	107	

(continued)

Table 20.1 (continued)

Target protein	Wild-type/mutants/ancestors	T_m (°C)	Citation
Myo-inositol-3-phospate	*Thermococcus sibiricus* MM 739	81	Butzin et al.
synthase (MIPS)	MIPS	85	(2013)
	Thermotoga sp. str. RQ2 MIPS	81	
	T. maritima MSB8 MIPS	89	
	Ancestor of *Thermotoga* ATM_T1	89	
	Ancestor of *Thermotoga* ATM_T1	89	
	Ancestor of *Thermotoga* ATM_T1	89	
	Ancestor of *Thermotoga* ATM_T1	>99	
	Ancestor of *Thermococcus* ACM_C1	>99	
	Ancestor of *Thermococcus* ACM_C1	>99	
	Common ancestor of *Thermotoga*	>99	
	and *Thermococcus* AAM_A1		
	Common ancestor of *Thermotoga*		
	and *Thermococcus* AAM_A2		
RNH	*T. thermophiles* RNH	89	Hart et al.
	E. coli RNH	68	(2014)
	Common ancestor of *T. thermophilus*	77	
	and *E. coli* RNHs Anc1	77	
	Ancestor of thermophilic lineage	83	
	Anc2	70	
	Ancestor of thermophilic lineage	68	
	Anc3	67	
	Ancestor of mesophilic lineage AncA	68	
	Ancestor of mesophilic lineage AncB		
	Ancestor of mesophilic lineage AncC		
	Ancestor of mesophilic lineage AncD		

[a]Mutants with elevated T_m compared with their original enzyme

We also used the ancestral sequence reconstruction technique to design a sequence for the deepest nodal positions of phylogenic trees composed of ATPase domains of 16 bacterial DNA gyrases (Akanuma et al. 2011). The designed ancestral sequence was then synthesized and characterized. The thermal stability of the ancestral ATPase domain is comparable to that of the corresponding domain of *T. thermophilus* DNA gyrase (Table 20.1). Moreover, the ancestral ATPase domain is catalytically more active than is the *T. thermophiles* ATPase domain when the protein's concentration is greater than 15 uM. Therefore, the reconstruction method is an effective way for creating the complete sequence of a catalitically active protein with great thermostability even when a small set of homologous sequences were available.

The method for designing thermally stable proteins using the ancestral sequence reconstruction does not require any knowledge of the structure of an enzyme under study and only relies on the amino acid sequences of homologous enzymes, which can be, in most cases, found in the growing public databases. The recent expansion of genome projects of bacteria and archaea will provide the necessary resources for the construction of a phylogenetic tree that is essential for computing ancestral sequences. Designing of thermally stable proteins by the ancestral sequence reconstruction method is often more beneficial than the use of extant thermophilic counterparts as it can often produce proteins that are more thermally stable than hyperthermophilic proteins whose sequences are contained in the phylogenetic tree used to infer the ancestral sequences (Akanuma et al. 2013a; Butzin et al. 2013). Especially, the method has merit when thermophilic homologues are unavailable. Thus, resurrection of ancient sequences provides a generic and effective tool for creating thermally stable proteins.

Directed evolution is an alternative way to modify protein's thermostability or other properties (Arnold et al. 2001). Directed evolution is a powerful method because it does not require any information about the structure and function of the targeted protein. However, it relies on the construction of a large library size containing millions of independent variants, which may require the consumption of costs and time. However, a phylogeny-based approach can take advantage of the sequence space that has already been limited through the natural evolution process (Gaucher 2007). The approach termed Reconstructing Evolutionary Adaptive Paths (REAP), which uses phylogenetic information of a protein family, can help to find sites for mutation that may alter a protein's function. The use of REAP together with a directed molecular evolution technique reduced the required sequence space to identify a DNA polymerase variant capable of accepting non-standard nucleoside (Chen et al. 2010; Cole and Gaucher 2011).

10 Conclusion

Investigation of the early life is important for our comprehensive understanding of the coevolution of life and earth's environment. To address this issue, we reconstructed NDK sequences that might be possessded by the anicient organisms and characterized their properties, which must have optimized to their environments. The thermal stabilities of the resurrected NDKs that might exist 3,500–3,800 million years ago are comparable to those of extant (hyper)thermophilic proteins. Therefore, the common ancestors were likely to thrive at very high temperatures. Given the (hyper)thermophilicity of ancient life, the ancestral sequence resurrection will provide a reliable means of creating thermally stable enzymes, which only relies on the homologous amino acid sequences.

Conflict of Interest Satoshi Akanuma, and Akihiko Yamagishi declare that they have no conflict of interest.

References

Achenbach-Richter L, Gupta R, Zillig W, Woese CR (1988) Rooting the archaebacterial tree: the pivotal role of *Thermococcus celer* in archaebacterial evolution. Syst Appl Microbiol 10:231–240

Akanuma S, Iwami S, Yokoi T, Nakamura N, Watanabe H, Yokobori S, Yamagishi A (2011) Phylogeny-based design of a B-subunit of DNA gyrase and its ATPase domain using a small set of homologous amino acid sequences. J Mol Biol 412:212–225

Akanuma S, Nakajima Y, Yokobori S, Kimura M, Nemoto N, Mase T, Miyazono K, Tanokura M, Yamagishi A (2013a) Experimental evidence for the thermophilicity of ancestral life. Proc Natl Acad Sci U S A 110:11067–11072
Akanuma S, Yokobori S, Yamagishi A (2013b) Thermophilic microbes in environmental and industrial biotechnology. In: Satyanarayana T, Litterchild J, Kawarabayasi Y (eds) Comparative genomics of thermophilic bacteria and archaea. Springer, Berlin, pp 331–349
Arnold FH, Wintrode PL, Miyazaki K, Gershenson A (2001) How enzymes adapt: lessons from directed evolution. Trends Biochem Sci 26:100–106
Atomi H, Matsumi R, Imanaka T (2004) Reverse gyrase is not a prerequisite for hyperthermophilic life. J Bacteriol 186:4829–4833
Baldauf SL, Palmer JD, Doolittle WF (1996) The root of the universal tree and the origin of eukaryotes based on elongation factor phylogeny. Proc Natl Acad Sci U S A 93:7749–7754
Boussau B, Gouy M (2012) What genomes have to say about the evolution of the Earth. Gondwana Res 21:483–494
Boussau B, Blanquart S, Necsulea A, Lartillot N, Gouy M (2008) Parallel adaptations to high temperatures in the Archaean eon. Nature 456:942–945
Brochier C, Philippe H (2002) Phylogeny: a non-hyperthermophilic ancestor for bacteria. Nature 417:244
Brooks DJ, Fresco JR, Singh M (2004) A novel method for estimating ancestral amino acid composition and its application to proteins of the Last Universal Ancestor. Bioinformatics 20:2251–2257
Brown JR, Doolittle WF (1995) Root of the universal tree of life based on ancient aminoacyl-tRNA synthetase gene duplications. Proc Natl Acad Sci U S A 92:2441–2445
Butzin NC, Lapierre P, Green AG, Swithers KS, Gogarten JP, Noll KM (2013) Reconstructed ancestral Myo-inositol-3-phosphate synthases indicate that ancestors of the Thermococcales and Thermotoga species were more thermophilic than their descendants. PLoS One 8:e84300
Cavalier-Smith T (2002) The neomuran origin of archaebacteria, the negibacterial root of the universal tree and bacterial megaclassification. Int J Syst Evol Microbiol 52:7–76
Cavalier-Smith T (2006a) Cell evolution and Earth history: stasis and revolution. Philos Trans R Soc Lond B Biol Sci 361:969–1006
Cavalier-Smith T (2006b) Rooting the tree of life by transition analyses. Biol Direct 1:19
Cavalier-Smith T (2010) Deep phylogeny, ancestral groups and the four ages of life. Philos Trans R Soc Lond B Biol Sci 365:111–132
Chen F, Gaucher EA, Leal NA, Hutter D, Havemann SA, Govindarajan S, Ortlund EA, Benner SA (2010) Reconstructed evolutionary adaptive paths give polymerases accepting reversible terminators for sequencing and SNP detection. Proc Natl Acad Sci U S A 107:1948–1953
Cheung YY, Lam SY, Chu WK, Allen MD, Bycroft M, Wong KB (2005) Crystal structure of a hyperthermophilic archaeal acylphosphatase from Pyrococcus horikoshii – structural insights into enzymatic catalysis, thermostability, and dimerization. Biochemistry 44:4601–4611
Christodoulou E, Rypniewski WR, Vorgias CR (2003) High-resolution X-ray structure of the DNA-binding protein HU from the hyper-thermophilic Thermotoga maritima and the determinants of its thermostability. Extremophiles 7:111–122
Clark AT, McCrary BS, Edmondson SP, Shriver JW (2004) Thermodynamics of core hydrophobicity and packing in the hyperthermophile proteins Sac7d and Sso7d. Biochemistry 43:2840–2853
Cole MF, Gaucher EA (2011) Exploiting models of molecular evolution to efficiently direct protein engineering. J Mol Evol 72:193–203
Declais AC, Marsault J, Confalonieri F, de La Tour CB, Duguet M (2000) Reverse gyrase, the two domains intimately cooperate to promote positive supercoiling. J Biol Chem 275:19498–19504
Di Giulio M (2000) The universal ancestor lived in a thermophilic or hyperthermophilic environment. J Theor Biol 203:203–213
Di Giulio M (2003a) The universal ancestor and the ancestor of bacteria were hyperthermophiles. J Mol Evol 57:721–730
Di Giulio M (2003b) The universal ancestor was a thermophile or a hyperthermophile: tests and further evidence. J Theor Biol 221:425–436

Dong H, Mukaiyama A, Tadokoro T, Koga Y, Takano K, Kanaya S (2008) Hydrophobic effect on the stability and folding of a hyperthermophilic protein. J Mol Biol 378:264–272

Edwards RJ, Shields DC (2004) GASP: Gapped Ancestral Sequence Prediction for proteins. BMC Bioinf 5:123

Forterre P (1996) A hot topic: the origin of hyperthermophiles. Cell 85:789–792

Forterre P (2002) A hot story from comparative genomics: reverse gyrase is the only hyperthermophile-specific protein. Trends Genet 18:236–237

Fournier GP, Alm EJ (2015) Ancestral reconstruction of a pre-LUCA aminoacyl-tRNA synthetase ancestor supports the late addition of Trp to the genetic code. J Mol Evol 80:171–185

Fournier GP, Andam CP, Alm EJ, Gogarten JP (2011) Molecular evolution of aminoacyl tRNA synthetase proteins in the early history of life. Orig Life Evol Biosph 41:621–632

Galtier N, Tourasse N, Gouy M (1999) A nonhyperthermophilic common ancestor to extant life forms. Science 283:220–221

Gaucher EA (2007) Ancestral sequence reconstruction as a tool to understand natural history and guide synthetic biology: realizing (and extending) the vision of Zukerkandl and Pauling. Oxford University Press, Oxford, pp 20–33

Gaucher EA, Thomson JM, Burgan MF, Benner SA (2003) Inferring the palaeoenvironment of ancient bacteria on the basis of resurrected proteins. Nature 425:285–288

Gaucher EA, Govindarajan S, Ganesh OK (2008) Palaeotemperature trend for Precambrian life inferred from resurrected proteins. Nature 451:704–707

Gaucher EA, Kratzer JT, Randall RN (2010) Deep phylogeny-how a tree can help characterize early life on Earth. Cold Spring Harb Perspect Biol 2:a002238

Gogarten JP, Kibak H, Ditrrich P, Taiz L, Bowman EJ, Bowman BJ, Manolsono MF, Poole RJ, Date T, Oshima T, Konishi J, Dendai K, Yoshida M (1989) Evolution of the vacuolar H^+-ATPase: implications for the origin of eukaryotes. Proc Natl Acad Sci U S A 86:6661–6665

Greaves RB, Warwicker J (2007) Mechanisms for stabilisation and the maintenance of solubility in proteins from thermophiles. BMC Struct Biol 7:18

Groussin M, Gouy M (2011) Adaptation to environmental temperature is a major determinant of molecular evolutionary rates in archaea. Mol Biol Evol 28:2661–2674

Groussin M, Boussau B, Charles S, Blanquart S, Gouy M (2013) The molecular signal for the adaptation to cold temperature during early life on Earth. Biol Lett 9:20130608

Groussin M, Hobbs JK, Szollosi GJ, Gribaldo S, Arcus VL, Gouy M (2015) Toward more accurate ancestral protein genotype-phenotype reconstructions with the use of species tree-aware gene trees. Mol Biol Evol 32:13–22

Hanson-Smith V, Kolaczkowski B, Thornton JW (2010) Robustness of ancestral sequence reconstruction to phylogenetic uncertainty. Mol Biol Evol 27:1988–1999

Hart KM, Harms MJ, Schmidt BH, Elya C, Thornton JW, Marqusee S (2014) Thermodynamic system drift in protein evolution. PLoS Biol 12:e1001994

Heine M, Chandra SB (2009) The linkage between reverse gyrase and hyperthermophiles: a review of their invariable association. J Microbiol 47:229–234

Hurley JH, Dean AM (1994) Structure of 3-isopropylmalate dehydrogenase in complex with NAD+: ligand-induced loop closing and mechanism for cofactor specificity. Structure 2:1007–1016

Ivens A, Mayans O, Szadkowski H, Jurgens C, Wilmanns M, Kirschner K (2002) Stabilization of a (betaalpha)8-barrel protein by an engineered disulfide bridge. Eur J Biochem 269:1145–1153

Iwabata H, Watanabe K, Ohkuri T, Yokobori S, Yamagishi A (2005) Thermostability of ancestral mutants of *Caldococcus noboribetus* isocitrate dehydrogenase. FEMS Microbiol Lett 243:393–398

Iwabe N, Kuma K, Hasegawa M, Osawa S, Miyata T (1989) Evolutionary relationship of archaebacteria, eubacteria, and eukaryotes inferred from phylogenetic trees of duplicated genes. Proc Natl Acad Sci U S A 86:9355–9359

Kirino H, Aoki M, Aoshima M, Hayashi Y, Ohba M, Yamagishi A, Wakagi T, Oshima T (1994) Hydrophobic interaction at the subunit interface contributes to the thermostability of 3-isopropylmalate dehydrogenase from an extreme thermophile *Thermus thermophilus*. Eur J Biochem 220:275–281

Korkegian A, Black ME, Baker D, Stoddard BL (2005) Computational thermostabilization of an enzyme. Science 308:857–860
Lake JA, Servin JA, Herbold CW, Skophammer RG (2008) Evidence for a new root of the tree of life. Syst Biol 57:835–843
Lake JA, Skophammer RG, Herbold CW, Servin JA (2009) Genome beginnings: rooting the tree of life. Philos Trans R Soc Lond B Biol Sci 364:2177–2185
Miyazaki J, Nakaya S, Suzuki T, Tamakoshi M, Oshima T, Yamagishi A (2001) Ancestral residues stabilizing 3-isopropylmalate dehydrogenase of an extreme thermophile: experimental evidence supporting the thermophilic common ancestor hypothesis. J Biochem 129:777–782
Pace NR (1991) Origin of life – facing up to the physical setting. Cell 65:531–533
Semba Y, Ishida M, Yokobori SI, Yamagishi A (2015) Ancestral amino acid substitution improves the thermal stability of recombinant lignin-peroxidase from white-rot fungi, Phanerochaete chrysosporium strain UAMH 3641. Protein Eng Des Sel. doi:10.1093/protein/gzv023
Shimizu H, Yokobori S, Ohkuri T, Yokogawa T, Nishikawa K, Yamagishi A (2007) Extremely thermophilic translation system in the common ancestor commonote: ancestral mutants of Glycyl-tRNA synthetase from the extreme thermophile *Thermus thermophilus*. J Mol Biol 369:1060–1069
Stetter KO (1996) Hyperthermophilic procaryotes. FEMS Microbiol Rev 18:149–158
Suzuki T, Inoki Y, Yamagishi A, Iwasaki T, Wakagi T, Oshima T (1997) Molecular and phylogenetic characterization of isopropylmalate dehydrogenase of a thermoacidophilic archaeon, Sulfolobus sp. strain 7. J Bacteriol 179:1174–1179
Tanaka T, Sawano M, Ogasahara K, Sakaguchi Y, Bagautdinov B, Katoh E, Kuroishi C, Shinkai A, Yokoyama S, Yutani K (2006) Hyper-thermostability of CutA1 protein, with a denaturation temperature of nearly 150 °C. FEBS Lett 580:4224–4230
Thornton JW (2004) Resurrecting ancient genes: experimental analysis of extinct molecules. Nat Rev Genet 5:366–375
Ulmer KM (1983) Protein engineering. Science 219:666–671
Watanabe K, Ohkuri T, Yokobori S, Yamagishi A (2006) Designing thermostable proteins: ancestral mutants of 3-isopropylmalate dehydrogenase designed by using a phylogenetic tree. J Mol Biol 355:664–674
Woese CR (1987) Bacterial evolution. Microbiol Rev 51:221–271
Woese CR, Winker S, Gutell RR (1990) Architecture of ribosomal RNA: constraints on the sequence of "tetra-loops". Proc Natl Acad Sci U S A 87:8467–8471
Yamagishi A, Kon T, Takahashi G, Oshima T (1998) From the common ancestor of living organisms to protoeuk-aryotic cell. In: Wiegel J, Adams M (eds) Thermophiles: the keys to molecular evolution and the origin of life? Taylor & Francis, London, pp 287–295
Yang Z (1997) PAML: a program package for phylogenetic analysis by maximum likelihood. Comput Appl Biosci 13:555–556
Yang Z, Rannala B (1997) Bayesian phylogenetic inference using DNA sequences: a Markov Chain Monte Carlo method. Mol Biol Evol 14:717–724
Yang Z, Kumar S, Nei M (1995) A new method of inference of ancestral nucleotide and amino acid sequences. Genetics 141:1641–1650
Zhang T, Koshland DE Jr (1995) Modeling substrate binding in *Thermus thermophilus* isopropylmalate dehydrogenase. Protein Sci 4:84–92
Zhou Y, Asahara H, Gaucher EA, Chong S (2012) Reconstitution of translation from *Thermus thermophilus* reveals a minimal set of components sufficient for protein synthesis at high temperatures and functional conservation of modern and ancient translation components. Nucleic Acids Res 40:932–7945

Chapter 21
A Systems Biology View on Bacterial Response to Temperature Shift

Marco Fondi, Emanuele Bosi, Angelina Lo Giudice, and Renato Fani

1 Introduction

The term "*extreme environments*" is used to define ecological niches characterized by harsh chemical/physical conditions that challenge most of the life forms, at the point they are characterized by a restricted species diversity. Even though extremophiles can be found across all the domains of life, as well as among viruses, the most prevalent ones are the microbes, mainly due to their ability to adapt to sharp environmental changes, to their metabolic versatility and to their capability of surviving in oligotrophic environments.

Living organisms inhabiting these environments have developed peculiar mechanisms to cope with these extreme conditions, in such a way that they mark the chemical-physical boundaries of life on Earth.

In general, a distinction can be made between adaptation (i.e. the process of genetic change that accumulates over many generations in response to an organism's specific environmental niche) and acclimation (short-term physiological adjustments in response to transitory changes in environmental conditions) (Morgan-Kiss et al. 2006).

The study of the mechanisms adopted by extremophilic organisms to overcome the selective pressure acting in the ecological niches they occupy is interesting from

M. Fondi • E. Bosi • R. Fani (✉)
Department of Biology, University of Florence,
Via Madonna del Piano 6, 50019 Sesto Fiorentino, Firenze, Italy
e-mail: renato.fani@unifi.it

A.L. Giudice
Institute for Coastal Marine Environment, National Research Council (IAMC-CNR),
Spianata San Raineri 86, I-98124 Messina, Italy

Department of Biological and Environmental Sciences, University of Messina,
Viale F. Stagnod'Alcontrès 31, I-98166 Messina, Italy

P.H. Rampelotto (ed.), *Biotechnology of Extremophiles*, Grand Challenges in Biology and Biotechnology 1, DOI 10.1007/978-3-319-13521-2_21

both applied and basic biology viewpoint. Cold-adapted microorganisms, for example, can be exploited in the construction of low-temperature protein expression, facilitating the overproduction of thermo-labile proteins (Papa et al. 2007; Miyake et al. 2007). Furthermore, cold-active enzymes from these microorganisms have proven to be useful for many areas of biotechnology (e.g. food-processing enzymes) and molecular biology (e.g. the use of alkaline phosphatase for dephos-phorylating DNA vectors before cloning) (Cavicchioli et al. 2011).

1.1 Mechanisms of Cold Adaptations

Cold temperatures represent the most common of the extreme conditions. Indeed, the marine waters account for an important portion of the Earth's biosphere and the majority of them (~90 %) is at a temperature not higher than 5 °C (Russell 1990). Moreover, polar regions account for 15 % of the Earth's surface. The microbial biodiversity of these environments is somehow constrained by such a challenging climate, resulting in a lack of competition in the extreme cold for the organisms that manage to adapt (Pearce 2012). Nonetheless, their metabolism is hindered by a number of factors, like a reduced catalytic activity, lower substrate affinity (Gerday et al. 1997), lower thermal energy and reaction rate (Collins et al. 2008). These factors, together with an increase in water viscosity and a decrease in membrane fluidity (Graumann and Marahiel 1996), concur to slower cellular growths in respect to those found in mesophilic niches. Additionally, a lower temperature range allowing for ice formation can lead to additional stress effects such as mechanical disruption, oxidative damage and osmotic imbalance (Tanghe et al. 2003).

To cope with these conditions, organisms adapted to the cold have evolved a number of different strategies. Overall, at molecular level these adaptations include the production of enzymes specifically evolved to perform better at lower temperatures, the synthesis of cryoprotectants, the production of specific lipids to change the membrane fluidity, and an increased expression of response-related genes. The next paragraphs will illustrate some examples of how microorganisms can exploit such mechanisms for responding to temperature perturbations.

1.1.1 Cold-Active Enzymes

Enzymes evolved to perform better in the cold have high turnover rates and catalytic efficiency at low temperatures, despite having a reduction in stability at higher temperatures. These proteins are characterized by an increased structural flexibility, either at the active site or at whole protein level, causing a reduction in the activation energy required for the enzymatic catalysis. The downside of the increased flexibility is the loss of stability at moderate and high temperatures (Feller and Gerday 2003).

This feature is caused by multiple factors: indeed, the comparison of the amino acid sequence of these enzymes with their mesophilic homologs has revealed a reduced number of Pro and Arg residues and increased number of Gly residue stretches. Mutagenesis experiments on these clusters, for example, have shown that they play a key role in the establishment of the psychrophilic properties of the TAB5-like alkaline phosphatase (Mavromatis et al. 2002). Consequently, the biosynthetic pathways leading to the production of such amino acids have been shown to be more active in response to a temperature down-shift (see Sect. 2.3 for an example).

1.1.2 Cryoprotection Molecules

Among the detrimental effects of low temperature, the water freezing and the production of ice crystals lead to lower concentrations of active water, which may cause cell lysis due to the osmotic stress. Indeed, while the intracellular ice can act as a solute, thus drawing water into the cell, the extra-cellular formation of ice can cause membrane fractures and osmotic pressure shift (Mavromatis et al. 2002).

To counteract these effects, cold-adapted organisms produce different molecules acting as cryoprotectants (Kawahara 2008). These can be either (i) small molecules, such as sugars, glycols and amines, that act lowering the water freezing temperature (Bouvet and Ben 2003), (ii) anti-freeze proteins, which bind to the surface of embryonic ice crystals thus inhibiting their expansion, or (iii) cryoprotectant proteins, which can either prevent protein cold denaturation, by binding to the surface of labile proteins (Kawahara 2008), or assist the folding of other essential proteins.

1.1.3 Lipid Production

As previously mentioned, one of the major effects of the cold-shock is the low membrane flexibility, making them more prone to rupturing and mechanical lysis. To avoid this and to maintain the membrane in a liquid-crystalline state, the cell can change the membrane's fatty acid composition by modulating the expression of different specific enzymes (i.e. fatty acid reductases and desaturases). The fatty acids modifications include introduction of double bonds by desaturases, a reduction of the fatty acid chain length and the introduction of branching by methyl groups. Microorganisms possess different number and types of desaturases and, usually, they have one housekeeping enzyme and other cold-inducible ones (Klein et al. 1999).

1.2 Mechanisms of Heat Adaptations

High temperatures can be critical for a cell, causing the disruption of weak bonds, which, in turn, is responsible for the denaturation of macromolecules, including proteins and nucleic acids. To survive in such conditions, some organisms (thermophiles) have developed enzymes with peculiar properties, such as thermostability and/or optimality at temperatures >70 °C (or 110 °C in some cases). Also, to counteract the effects of a heat-shock, the cell can activate the so-called heat-shock response. The principal effectors of this stress response are a group of conserved proteins, namely the heat-shock proteins (HSPs), which act as chaperones, helping the other proteins to fold and maintain the correct structure, or as proteases, degrading the unfolded proteins. The regulation of this response in bacteria has been well characterized using the *Escherichia coli* model, in which it has been observed that an alternative sigma factor, referred to as σ32, acts as a positive regulator of the synthesis of HSPs, increasing the affinity of RNA polymerase for the HSP promoters (Morimoto et al. 1997). A complex molecular mechanism is responsible for the heat-shock response and is based on the regulation of σ32 factor activity, which is controlled at two different levels, namely a translational level, and a post-translational one. The translational regulation is based on the sequence of the mRNA of the gene encoding the σ32 factor, *rpoH*. At mesophilic temperature this RNA is folded in a secondary structure that partially covers the translation starting site, leading to a low translation rate, whereas at higher temperatures the *rpoH* mRNA changes its three-dimensional structure and it is no more able to cover the translation starting site, thus allowing the efficient expression of *rpoH* (Morita et al. 1999). The other level of regulation is based on a feedback loop, in which the constitutively expressed HSPs (DnaK, DnaJ and GrpE) direct σ32 to the metalloprotease FtsH. Following a temperature upraise, the number of denatured proteins increases, which sequesters the HSPs, thus increasing the σ32 levels (Arsene et al. 2000). Hyperthermophilic organisms can survive at temperatures close to 100 °C. For this reason, their macromolecules need to have a greater thermostability than those of mesophiles (Sterner and Liebl 2001). Given such feature, these enzymes may be used as model systems in several scientific disciplines (including biology, chemistry and physics) with the aim of understanding a key point in enzyme evolution, that is the upper temperature limit for enzyme functionality (Vieille and Zeikus 2001).

As shown above, several studies aimed at deciphering the molecular strategies involved in the short term response to temperature shifts (acclimation) have been performed using mesophilic organisms (e.g. *E. coli*) as model organisms. It should be pointed, however, that these acclimation mechanisms may be different (or absent at all) in cold- and heat-adapted microbes. Interestingly, in at least a few examples, some general trends (phenotypes) appear to be shared by mesophiles and extremophiles as in the case of metabolic adjustments following a temperature downshift (see Sect. 4.3). Moreover, for what concerns cold-adaptation mechanisms, it has been shown that different cold-adapted microorganisms may express distinct sets of cold-inducible proteins, thus revealing the use of different strategies to cope with cold environments. Even in this case, general trends (at least among psychrophiles) can be depicted; these include the induction of the RNA chaperone CspA and (different forms of) peptidylprolyl isomerase (PPIase) and the modulation of modulation of RNA polymerase (Kawamoto et al. 2007).

2 -Omics Approaches for Studying Response to Temperature Changes

Since the advent of techniques allowing the massive production of biological data (e.g. Next Generation Sequencing, NGS), the biological focus has shifted from single target (characterization of a single gene, operon, protein etc.) to a collective characterization and quantification of the biological molecules, which are informally referred to as "omics". From a more general viewpoint, the approach to biology has changed from a prevalent *reductionist* approach, which successfully identified the single components of complex biological structures, to a *holistic* approach, in which the main topic is the study of the relationships between different biological components. The latter approach, also known as *systems biology*, allows modelling and discovering emergent properties of complex biological systems, which might not be identified through a reductionism-oriented approach.

Here some of the advances on the understanding of adaptations to temperature change achieved using the -omics approach are presented, which have been divided according to the -omics technology exploited (genomics, transcriptomics, proteomics, and phenomics).

2.1 Genomics

The advent of massive sequencing technologies, also known as NGS, allowed for the rapid sequencing of complete genomes with ever-decreasing low costs, thus making sequencing projects feasible for most of research laboratories. In time, the repositories of biological data have been saturated with genomic sequences from a large number of organisms. These databases contain thousands of annotated gene sequences, which can be used to annotate novel uncharacterized genes. For instance, the *Heat Shock Protein Information Resource* (HSPIR) database (Ratheesh et al. 2012) contains 9902 protein records from 227 complete genomes of prokaryotic and eukaryotic species, while the Cold Shock domain DataBase (CSDBase) (Weber et al. 2002) is a resource providing information on proteins containing cold-shock domains and bacterial cold-shock response proteins.

Taking advantage of the presence of annotated sequences and databases, the analysis of genomic sequences allows identifying putative genes related to the specific stress response and to compare these with literature. These putative genes can be further investigated, either by using different *in silico* techniques or *in vitro* analyses, to strengthen these evidences. Many examples in which genomics have been fruitfully used to investigate the genomic basis of temperature shift adaptation have been reported, two of which are represented by the analysis of the genome sequence of *Pseudoalteromonas haloplanktis* TAC125, a marine bacterium isolated from Antarctic sea (Medigue et al. 2005), and of *Colwellia psychrerythraea*

34H (Methe et al. 2005). In both cases genomic signatures characteristic of a cold-adapted lifestyle were identified, including the deletion of whole pathways producing reactive oxygen species to cope with the increased solubility of oxygen at low temperature and expansions of gene families related to cell membrane synthesis and to the uptake or synthesis of compounds that in part may confer cryotolerance.

In general, the availability of the genomic sequence of a microorganism is required for the integration of other phenotypic data, which allows (i) the determination of functional implications of the genes, (ii) the identification of functional modules, and (iii) to have insights into the interaction network of genes and gene products. For these reasons, the determination of the genomic sequence of a microorganism is usually the first step of a workflow including analyses based on different -omics techniques (and possibly their integration, see Sect. 3.2).

2.2 Transcriptomics

The analysis of the whole body of transcripts (the *transcriptome*) produced by an organism in different conditions can be used to correlate the differential expression of gene sets with the different experimental conditions tested. In the case of stress response, the pattern of differential gene expression can provide insights into the role of genes in the cellular physiological reaction to stressful events (i.e. temperature change). When analysing the transcriptome change, some considerations should be made before inferring meaningful conclusions. First, it has been shown that, even in the case of persistent stress, gene expression responses are transient, meaning that the new steady-state mRNA levels will be similar to those relative to normal condition (Lopez-Maury et al. 2008). The physiological shift of gene expression undergoing during stress is characterized by the activation of some stress-specific modules (gene clusters) and the down regulation of growth-related genes. Second, the change in the transcription level of a given gene might be not directly related to effective changes in the gene expression level (Griffin et al. 2002), since regulatory events may still occur at translational and post-translational levels (i.e. mRNA and/or protein degradation). This means that, in ultimate analysis, the quantification of the protein level can be used, in addition with the quantification of transcript levels, to measure the real effective gene expression.

Transcriptomics has been used extensively to study the effects of rapid temperature shift (in both directions) on various (non extremophilic) microorganisms, such as *E. coli* (Jozefczuk et al. 2010), *Bacillus subtilis* (Kaan et al. 2002) *Shewanella oneidensis* (Gao et al. 2004, 2006), often in conjunction with other -omics procedures (proteomics, metabolomics). The results of these works highlighted some general trends, such as the down-regulation of sets of metabolic genes associated with energy production and biosynthesis (growth-genes), with the up-regulation of stress-specific genes, encoding the effectors of the shock response, as well as the differential expression of regulatory genes such as *hlyU* and *narQ*.

2.3 Proteomics

As already mentioned, it has been experimentally proven that in *Saccharomyces cerevisiae*, the changes in the mRNA levels are not linearly related to actual changes in protein expression levels (Griffin et al. 2002). From a mathematical viewpoint, it has been demonstrated that, even when the simplest gene networks are taken into account, a detailed understanding of the control of these networks requires information at both mRNA and protein level (Hatzimanikatis and Lee 1999). This points out to the need of a quantification of the expressed proteome and the integration with other omics data. This kind of approach has been used to provide deep insights into the cold/heat adaptation/acclimation. For example, the quantitative analysis of the *Sphingopyxis alaskensis* (Ting et al. 2010) proteome recovered the general features of low temperature physiology, starting from an increased abundance of enzymes involved in the synthesis of amino acids such as tryptophan, histidine and proline, whose importance in the response to cold stress has been reported in different organisms (Liu et al. 2002; Kawamura et al. 1994), to the over-expression of lipid desaturase and specific cold-shock response proteins. Notably, this analysis revealed the over-expression of poorly characterized proteins, which prompts for further analysis on these molecules.

Two works have recently described the growth dynamics and the proteomes of *P. haloplanktis* TAC125 cells grown at 4 °C and 18 °C (Piette et al. 2010, 2011). These authors showed that this bacterium has a doubling time of 1 h 40 min at 18 °C and that this value drops to 4 h when the cells are grown at 4 °C. These studies also led to the identification of 123 differentially expressed genes among the two temperatures (83 down- and 40 up-regulated). Overall, this proteomic analysis revealed that the acclimation to low temperature was characterized by a depression of the metabolism, i.e. the under-expression of the proteins related to compound degradation, biosynthesis and energy production, and an activation of the cold-shock response proteins. The other effects of response to cold-stress are the down-regulation of iron metabolism and the up-regulation of cold-specific oxidative stress response proteins. These effects can be explained, since this down-regulation mostly shuts down the Krebs cycle and respiratory chain, there is lower need for ferric ions, while during the conditions of low metabolism, one of the oxidative stress effectors, the glutathione, is poorly regenerated since there are low levels of NADPH (Piette et al. 2010, 2011).

2.4 Phenomics

Phenomics (high-throughput phenotyping) allows exploring the phenotypic space of a given organism and deriving phenotype capabilities (e.g. the capability of metabolizing certain carbon sources in respect to others) in an automated and large-scale fashion. This is typically performed through Phenotype MicroArrays (PMs)

(Bochner et al. 2001), which uses cellular respiration (i.e. NADH reduction) and consequent production of a purple colour as a reporter system for overall metabolic activity (Fondi and Lio 2015).

To the best of our knowledge, no work concerning the phenomics of cold-adapted bacteria (or other extremophiles) grown at different temperatures has been carried on up to now. However, preliminary data obtained on two Antarctic *P. haloplanktis* strains, that is TAC125 and TB41, isolated from different environmental niches, i.e. water column and an Antarctic sponge, respectively, revealed that the phenomic pattern of the two strains are quite different between them and between different growth temperatures (4 °C and 15 °C). Indeed, the two strains, even sharing some phenotypic properties at both growth temperatures, exhibited different growth rates on different substrates, suggesting the existence of a temperature-related effect on the range of compounds that can be metabolized. Also, this analysis points towards a certain metabolic plasticity even between strains belonging to the same species.

3 Systems Biology for Studying Response to Temperature Changes

Despite their stand-alone importance, no single -omics approach can, on its own, unravel the complexities of fundamental microbiology (Zhang et al. 2010). Multi-omics approaches have thus spread among diverse research areas, including bio-based fuel (Zhu et al. 2013) and biopharmaceutical production (Schaub et al. 2012), medical research (Wiench et al. 2013) and host-pathogen interactions (Ansong et al. 2013). The integration and interpretation of such diverse data-types may be considered one of the key challenges of present-day bioinformatics, due to different data formats, high data dimensionality and need for data normalization (Fondi and Lio 2015).

Recently, the integration of such data-types with genomic derived information has proven to represent one of the most efficient ways for interpreting high throughput data in a biologically consistent way. In particular, given the link existing between phenotypic features of microorganisms and their underlying metabolism, genome-scale metabolic reconstructions are a highly exploited in this context.

In the next sections of this chapter we will provide a general overview of metabolic network reconstruction and modelling, procedures integration of -omics data with metabolic modelling and a recent study case in which such methodologies have been exploited for studying, for example, acclimation to low temperatures.

3.1 Genome Scale Metabolic Reconstruction and Metabolic Modelling

Strictly speaking, the metabolic model of a given microorganism (also called metabolic network) is the list of all the reactions which is supposed to be able to carry out. Using sequence similarity as a proxy, the metabolic repertoire of a microbe can be automatically computed by means of several annotation servers (e.g. RAST (Overbeek et al. 2014a), KBASE (http://kbase.us), Metacyc (Caspi et al. 2014)). Nowadays, such task is relatively easy and quick to achieve, once an accurate and reliable genome annotation has been obtained (Fig. 21.1a).

Unfortunately, these *de-novo* reconstructions rarely succeed in representing all the possible functional and physiological states of a given microorganism and this incompleteness is usually due to the fact that (i) they often do not include essential metabolic steps for sustaining *in silico*-predicted rates of cell division (metabolic gaps most often linked to wrong or incomplete genome functional annotation), (ii) key issues of embedded reactions such as stoichiometry, directionality, and charge are sometimes missing or erroneous and, finally, (iii) since draft models are mainly reconstructed on the basis of homology with respect to other (closely related) microorganisms, they will not include organism-specific metabolic pathways (often responsible for key phenotypic features). For these reasons, extensive manual refinement is usually necessary to produce a metabolic reconstruction that will represent the real metabolic space of the microorganism under study.

A comprehensive protocol that describes the main steps required to reconstruct metabolic networks to be used for computational modelling was assembled by Thiele and Palsson (Thiele and Palsson 2010), comprising (at least) 96 different steps, most of which consist in extensive revision of available literature and experimental data (Fig. 21.1b). Omics-derived data can be of great help in this stage, guiding gap-filling and model expansion procedures. Genomics, proteomics, and transcriptomics, for example, have been shown to be of valuable help in refining draft-metabolic models and allowing them more closely to represent the real phenotype of an organism (see (Fondi and Lio 2015) for a review).

Once a metabolic reconstruction has been gap-filled and missing information has been added, the simulation process can start (Fig. 21.1c). Roughly speaking, two possible (mathematical) approaches exist for simulating the distribution of cellular metabolic fluxes within the cell: stoichiometric and kinetic modelling. A kinetic model consists of a set of reactions that can be expressed by means of ordinary or partial differential equations (ODEs and PDEs, respectively) (Tomar and De 2013). Given the large amount of parameters required (available only for a very limited number of model organisms/reactions), the use of kinetic models is restricted to small metabolic systems. Conversely, stoichiometric (constraints-based) modelling can be applied to larger (genome scale) systems since this approach only requires the information on the stoichiometry of the metabolic reactions stoichiometry (Oberhardt et al. 2009). Flux Balance Analysis (FBA) is currently the most widely adopted constraint-based modelling technique to compute the resulting balance of all the chemical reactions predicted to be active in the cell. In other words, once boundary conditions have been defined (i.e. the list the available nutrient sources and their uptake fluxes), FBA is used for deriving a feasible set of steady-state fluxes that optimizes a defined cellular objective, usually the maximization of biomass production within a metabolic network. Overall, FBA predictions can be used to infer key physiological features such as growth rate (moles of biomass produced in a given amount of time), the list of compounds that can be metabolized by the cell, the changes in fluxes distribution at the system level following a perturbation

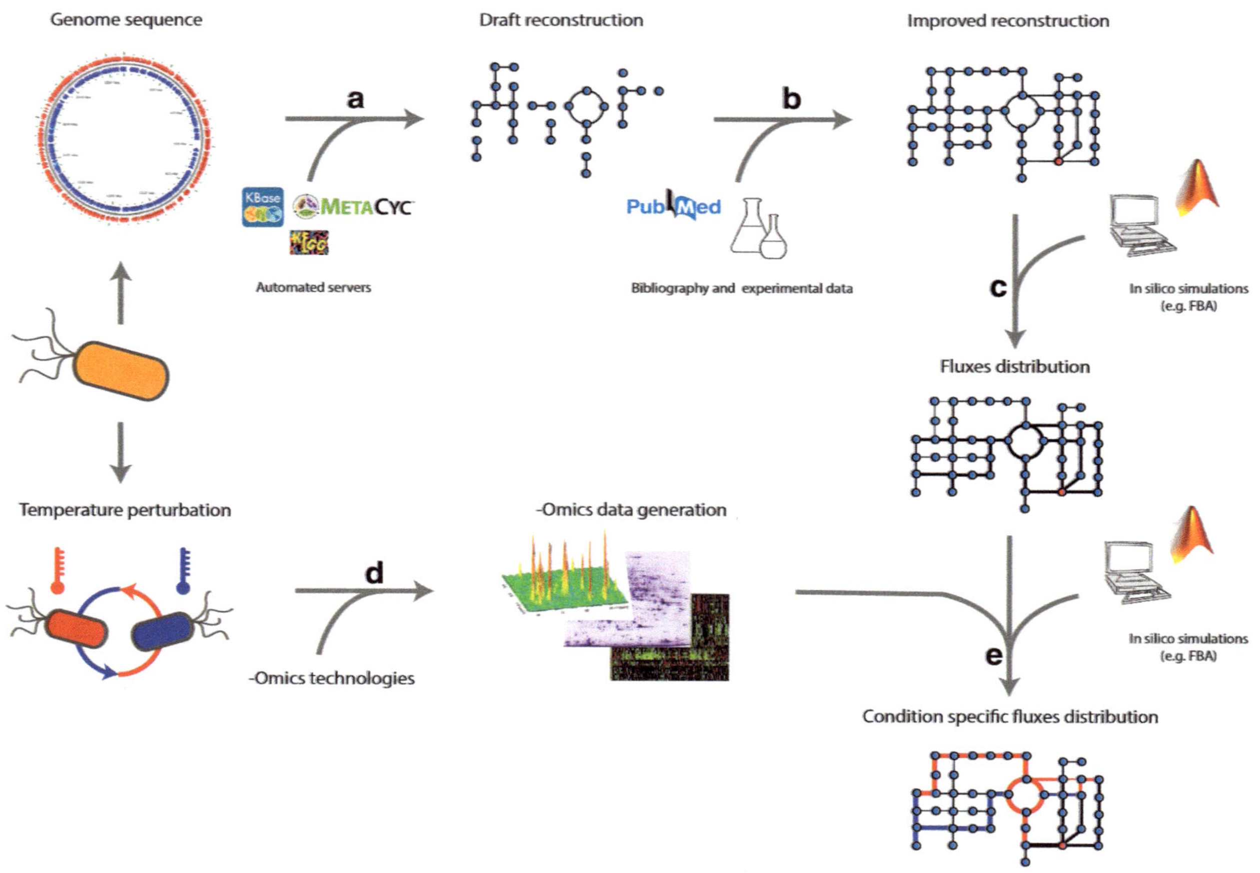

Fig. 21.1 A hypothetical pipeline for metabolic modelling and -omics integration following a temperature shock experiment

(e.g. temperature shift or growth medium composition). In general, for these modelling outcomes to be considered robust, a thorough comparison with experimental data should be carried out. The reliability of metabolic models in predicting growth phenotypes is usually assessed by comparing them against large-scale growth tests (e.g. Biolog Phenotype Microarray) or experimentally calculated growth rates.

Despite the purpose of a metabolic reconstruction defines itself the level of accuracy required at this stage, an overall agreement between experimental phenotypic data and simulation outcomes is nowadays required.

3.2 Data Integration

Omics-derived data is increasingly more often used to assist and improve metabolic model predictions and to provide a system-level understanding of the cellular behaviour. In a hypothetical scenario of a temperature shock study, -omics data can be collected after the cells have undergone heat or cold shock (Fig. 21.1d). In the last decade, such approach has been extensively exploited for studying both cold shock response (Gao et al. 2006; Kaan et al. 2002; Garnier et al. 2010) and adaptation to warming/heat shock (Garcia-Descalzo et al. 2014; Li et al. 2011; Wang et al. 2014). Despite such studies have provided useful insights in the identification and understanding of genome-scale level trends of temperature-shock adaptation/acclimation, it has become clear that no single -omics analysis can fully unravel the complexities of fundamental microbiology (Zhang et al. 2010). In this context, genome scale metabolic models (together with modelling techniques such as FBA) provide a robust and meaningful platform for predicting the metabolic adjustments that may follow a temperature shock. Indeed, many diverse experimental data can be mixed with metabolic modelling (Fig. 21.1e) to get to a more realistic and comprehensive view of cellular circuits and their alterations following environmental perturbations; this can be achieved, for example, by integrating measures on the quantification of gene products and metabolites, including their interaction (Zhang et al. 2010; Fondi and Lio 2015).

3.3 Integrating Expression Data and Metabolic Modelling for Studying Temperature Shift Response

In at least four examples, such approach has been exploited for studying the metabolic consequences of temperature shock in microbes (Tong et al. 2013; Fondi et al. 2014; Topfer et al. 2012; Navid and Almaas 2012). By integrating time-resolved transcriptomics data with flux-based methods, Topfer et al. (2012) revealed cold and heat shocks-induced metabolic acclimation in *Escherichia coli*. These authors were able to characterize the transitional behaviour of the overall network following a temperature shift and, at the same time, of single specific reactions. In particular, these authors found two types of reactions that exhibited significantly different activity patterns across the analysed time-points (both after either cold or heat shock). The first type included those reactions inferred to be active across all time-points and predicted to be functionally related to indispensable metabolic processes for sustaining growth (e.g. amino acids biosynthesis, nucleotide, nucleoside and nucleobase interconversion and tricarboxylic acids cycle). The second embedded those reactions whose usage was not constant during the progression of stress application. Functionally speaking these reactions were shown to be involved in processes like catabolism of (some) amino acids, organic acids and glucose, conversion of pyruvate to acetyl-CoA and biosynthesis of amino acids belonging to the aspartate family. Moreover, this analysis showed a partial overlap between reactions activity patterns in response to cold and heat shocks. However, the authors point out how, although the same biological processes are involved in the response to both temperature stresses, the temporal usage in terms of (in)activation may slightly differ and this may further amplify the effect of genes specific for reacting to cold/heat stress (Topfer et al. 2012).

With the specific focus of exploring the metabolic consequences of a downshift in culture temperature in the anaerobic thermophilic bacterium *Thermoanaerobacter tengcongensis*, metabolic network modelling and proteome data were combined by Tong et al. (2013). By perturbing the culture temperature from 75 °C to 55 °C the authors collected differential proteomes and combined such data with network modelling and, in particular, robustness analysis (the prediction of variations in the trend of cell growth rate following a perturbation of the reaction flux in a metabolic network). This study demonstrated how robustness analysis can correctly predict in most cases (73 %) the effect of perturbation-related changes in protein abundance on the overall cellular growth and physiology. For those cases in which no agreement was identified between FBA predictions and protein expression data, independence between reactions activity and enzyme abundance was postulated by the authors. Most of these reactions were found to be involved in amino acids biosynthesis (i.e. valine, isoleucine, leucine and threonine) and, in these cases a regulation at different levels (including temperature, buffers, modifications and/or allosteric structure) was inferred.

Gene expression levels of *Yersinia pestis* were exploited to study the metabolic changes following variations in temperature (Navid and Almaas 2012). Interestingly, for this strain it had been demonstrated that a transition from a non-virulent to a virulent phenotype can occur after an increase of growth temperature from 26 °C to 37 °C (Chromy et al. 2005; Konkel and Tilly 2000). Gene expression data for two strains (namely *Y. pestis* 201 and KIM5) were collected and used for analysing the metabolic underpinnings for the observed phenotypic behaviours. In the case of *Y. pestis*, the cell's primary metabolic response involves saving energy by lowering the flux through non- essential pathways. Production of purines is one such of processes since it can be compensated by a slower rate of nucleotide degradation. Concomitant to decreased ROS (Reactive Oxidative Species) production, the cell increases the flux through those reactions (e.g., catalase-peroxidase) that protect cellular macromolecules from interactions with Using the catabolism of some amino acids as sources of carbon is another strategy predicted to be active in such conditions. Even in this case, computational simulations integrated with information on gene expression were found to be in good agreement with experimental observations concerning the physiology of these strains in response to temperature perturbation.

In a recent work (Fondi et al. 2014), we have integrated different -omics data (including phenomics, proteomics and genomics) with metabolic modelling to globally investigate possible metabolic adjustments of *P. haloplanktis* TAC125 during growth at low temperature. The next paragraph will describe such an approach.

As shown in this section, to date and for what concerns their response to temperature shifts, only two extremophiles have been investigated by means of -omics integration with metabolic modelling. Given the good predictive accuracy, however, this approach is promising when trying to elucidate the system level metabolic consequences of adaptation/acclimation to changes in growth temperature.

4 The *P. haloplanktis* TAC125 Case Study

One of the model organisms of cold-adapted bacteria is represented by *P. haloplanktis* TAC125 (*Ph*TAC125). This bacterium has been isolated from a sea water sample of Antarctica. Given its ability to rapidly multiply at low temperatures it has been proposed as an alternative host for the soluble overproduction of heterologous proteins, (Duilio et al. 2004; Wilmes et al. 2010; Rippa et al. 2012; Corchero et al. 2013). Also, this bacterium is capable of thriving in a wide temperature range (from 4 °C up to close to 30 °C). For these reasons, a detailed understanding of its metabolic landscape is crucial for an effective exploitation of its biotechnological potential and for a systems-level understanding of its cellular physiology.

In this context, a genome-scale reconstruction of *Ph*TAC125 metabolism may be valuable, allowing the identification of system level properties of its phenotypic space, including the response to environmental perturbations (e.g. changes in growth medium nutrient composition and temperature shock).

4.1 Model Overview

Starting from an initial draft metabolic reconstruction of *Ph*TAC125 obtained using RAST and post-processed using ModelSeed (Overbeek et al. 2014b), a thorough manual refinement was carried out following the main steps listed in (Thiele and Palsson 2010). Also, this reconstruction was completed by integrating information from *Ph*TAC125 original genome annotation (Medigue et al. 2005) and from different functional databases, including KEGG (Kanehisa 2002), BRENDA (Scheer et al. 2011) and MetCyc (Caspi et al. 2006). The main features of the reconstructed

Table 21.1 Overall features of the reconstructed *P. haloplanktis* TAC125 metabolic model

P. haloplanktis **TAC125 genome**	
Genome size in bp (Chr1 + Chr2)	3850272
N. of protein encoding genes	3484
P. haloplanktis **TAC125 model**	
N. of genes (% of coding genes)	721 (20.7)
N. of reactions	1322
Gene-associated	1146
Non gene-associated (exchange reactions)	176 (85)
N. of metabolites	1133

*Ph*TAC125 metabolic network are reported in Table 21.1. The genome-scale metabolic model of *Ph*TAC125 (named iMF721, following the standard naming scheme (Reed et al. 2003) includes information on 721 genes (ORFs; 20.7 % of the *Ph*TAC125 protein-encoding genes), 1133 metabolites and 1322 reactions.

Noteworthy, this model includes 97 % of the metabolic genes expressed at 16 °C on a complex medium (Wilmes et al. 2011) and thus, in principle, it almost entirely represents the metabolic potential of this microorganism in such conditions.

4.2 Testing the Metabolic Model

The reconstructed model was then checked against a diverse and large set of reference data. In particular, model-predicted growth rates were compared with those obtained experimentally (Wilmes et al. 2010; Giuliani et al. 2011).

In Fig. 21.2 the comparison between model-predicted growth rates and experimentally determined ones is reported, revealing an overall agreement between the two sets. Furthermore, Biolog Phenotype Microarray (PM) data was used to evaluate and correct iMF721 growth predictions. This can be done by comparing the *in silico* flux value across biomass assembly reaction and the cellular respiratory activity measured during phenotype microarray experiments. *In silico* growth on 64 carbon sources was simulated by imposing each of them as the sole carbon source. To do so, its uptake rate was set to the arbitrary value of 1 mmol/g*h^{-1} (under aerobic conditions). In most cases, discrepancies between the outcome of the model and PM data allowed to "gapfill" the model, adding a metabolic reaction in the network and/or missing transport reactions. After this was repeated for each of the carbon sources, in 84 % of the cases (54 out of 64) the outcomes of *in silico* simulations matched results from *in vivo* experiments.

Results from both quantitative and qualitative testing of the model fall in the range of those from most of the metabolic reconstructions available (see, for example (Schatschneider et al. 2013; Bartell et al. 2014; Fang et al. 2011; Durot et al. 2008)), suggesting that iMF721 can be used for reliable modelling of the central metabolism of *Ph*TAC125.

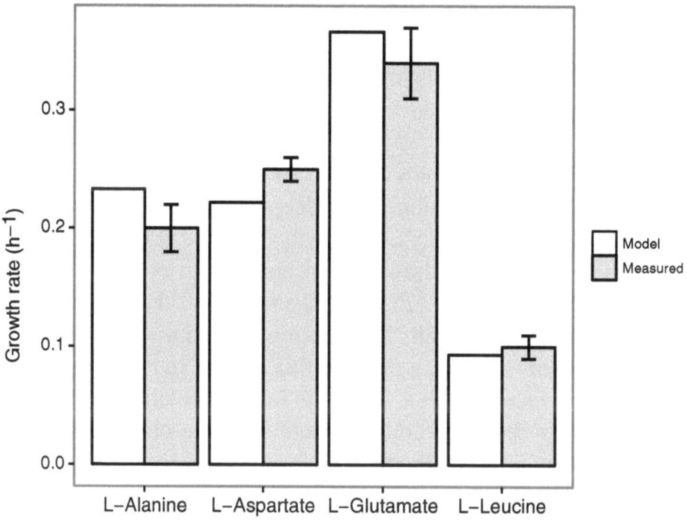

Fig. 21.2 *Ph*TAC125 predicted (*white bars*) and experimentally (*grey bars*) determined growth rates

4.3 Studying the Response to Cold shock

The overall agreement between experimental data and modelling outcomes suggests that, in principle, the iMF721 model might be exploited for understanding the metabolic response of *Ph*TAC125 in changing conditions (e.g. following a temperature switch) at the system level. Accordingly, we integrated protein abundance data with constraints-based modelling of the iMF721 model. Proteomes and growth rates of *Ph*TAC125 at 4 °C and 18 °C have been obtained (Piette et al. 2010, 2011), revealing the presence of 123 differentially expressed genes following a temperature downshift (83 down- and 40 up-regulated). Importantly, the iMF721 model embeds gene-protein rules (GPR) for a large fraction of the (metabolic) genes being differentially expressed in two conditions (54 out of 60.90 %).

As previously mentioned (see Sect. 3.2), protein abundance data (and, similarly, gene expression data) can be combined with metabolic modelling by modulating the admissible flux through each reaction in relation to the expression level of the corresponding enzyme. In the present study-case, up- and down-regulation ratios were combined with the iMF721 metabolic model with MADE (Metabolic Adjustment by Differential Expression) (Jensen and Papin 2011). Using gene/protein expression values, this tool creates a sequence of binary expression states ("1" and "0" for "on" and "off" reactions, respectively) that matches the expression variation across all the measurements and that can be incorporated in a metabolic reconstruction. By doing so MADE produces models simulating the real metabolic functional state of the cell, given the initial gene expression values. Following such approach two distinct functional metabolic states can be depicted, i.e. the original iMF721 model and the one derived from turning “off” the reactions of down-regulated genes and thus simulating growth at lower temperature. By doing so, 33 genes (involved in 54 reactions in the model) were switched off (since they were down-regulated at 4 °C), whereas 12 of them (involved in 35 reactions) were switched on in the 4 °C model.

Such approach allows the study the metabolic consequences of temperature-related changes in gene expression and a deep analysis of the changes in fluxes distribution across the *Ph*TAC125’s. The analysis of the altered fluxes distribution revealed that the most evident drawback of changes in protein abundance is the adjustment of the overall *Ph*TAC125 metabolism towards the reduction of the activity of its central pathways. Overall, 209 reactions varied their flux between the two conditions, 141 of them showing a reduced flux at 4 °C in respect to 18 °C, and 68 reactions showing an increased flux. This is in line both with the number of induced *vs*. repressed genes in the two conditions and with the observed decrease of the *Ph*TAC125 growth rate at 4 °C (Piette et al. 2010, 2011).

The pathways involved in the biosynthesis of compounds and the energy production (Fig. 21.3) were those that mostly embedded those reactions displaying a reduced switch following the temperature switch. These included most of the reactions involved in the biosynthesis of purine and pyrimidine precursors, glycolysis and pentose phosphate pathway. Most likely, lower fluxes in these pathways hamper both energy production and biosynthesis of important intermediates for nucleic and amino acids assembly.

Remarkably, reactions involved in amino acids degradation (e.g. cystathionine beta-lyase, threonine dehydratase) displayed an increased flux following the temperature downshift. Also 23 reactions devoted to fatty acids biosynthesis/metabolism (7 from fatty acids biosynthesis and 16 from fatty acids elongation) increased their flux at 4 °C. Among them we found, for example, (i) acetyl-CoA carbon-dioxide ligase, providing the malonyl-CoA substrate for biosynthesis of fatty acids and (ii) malonyl CoA-acyl carrier protein transacylase, representing a critical enzyme responsible for the transfer of the malonyl moiety to holo-acyl carrier protein (ACP) forming the malonyl-ACP intermediates in the initiation step of type II fatty acid synthesis (FAS II) in bacteria (Ruch and Vagelos 1973). According to our modelling outcomes it seems that one of the possible metabolic reprogramming following changes in gene expression due to a temperature downshift involves the reduction of the activity the main central metabolic pathways. Its overall metabolism appears to be reprogrammed towards lowering the costs associated with amino acids and nucleotide biosynthesis whereas amino acids degradation and fatty acid-related metabolism seem to acquire specific importance in this particular physiological condition.

The adjustments predicted by functional modelling of iMF721 find several matches in the current -omics derived knowledge of cold perturbation responses and adaptation in other microbes, including non-cold-adapted ones. It was found, for example, that *Bacillus subtilis* down regulate amino acids and nucleotides biosynthetic genes as a consequence of a temperature downshift. Conversely, in the same condition, amino acids degradation and fatty acids metabolism related genes

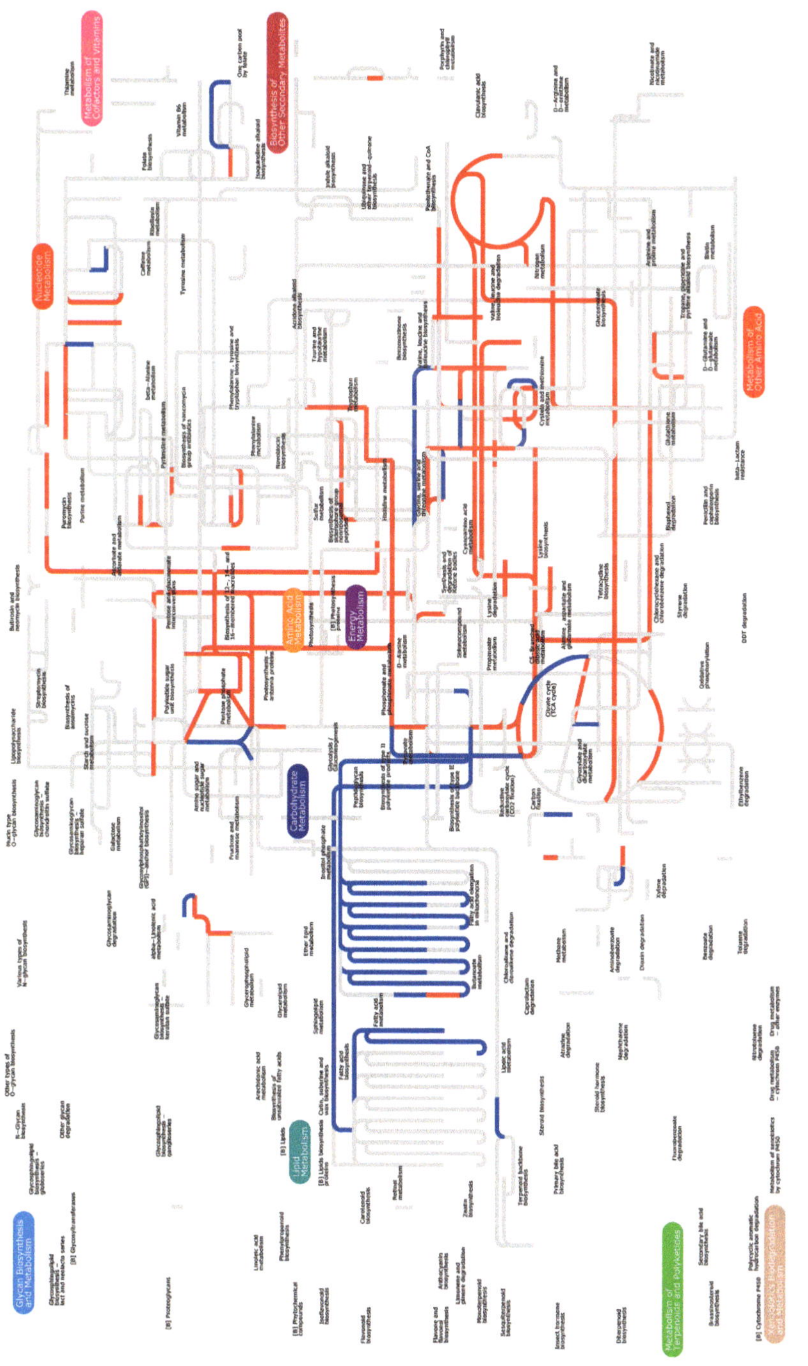

Fig. 21.3 *Ph*TAC125 metabolic map showing the changes in the distribution of metabolic fluxes following a temperature downshift. *Red* and *blue lines* represent a decrease and an increase (of at least a factor 2) in reaction fluxes, respectively. *Grey lines* represent reactions with no significant changes in fluxes values

increased their expression (Kaan et al. 2002). Also, many evidences support a link between fatty acids metabolism and acclimation/adaptation to cold temperature in bacteria. In a proteomics study of *S. alaskensis*, for example, the *de novo* synthesis of fatty acids was connected to adaptation to growth at low temperature (Ting et al. 2010). Also, a relatively large set of genes predicted be involved in fatty acid metabolism was retrieved in psychrophilic microorganisms by means of comparative genomics (Methe et al. 2005), as well as in cold environments in general by metagenomics approaches (Simon et al. 2009; Varin et al. 2012).

5 Conclusions

In this chapter we have provided an overview of the possible approaches which can be exploited for gaining a systems-level understanding of adaptation and acclimation to temperature perturbations. In this context, -omics technologies (and, more importantly, their integration) are increasingly more often used to pursue a holistic understanding of the cellular behaviour. Indeed, we would like to stress the importance of integrating large, multi-scale, -omics datasets for a deep and system-level view of microbial cell organization and physiology in response to environmental stresses as cold/heat shock. This approach is promising and may be used in the future to depict a unifying scheme (if any) of the molecular response to temperature shift in both mesophilic and extremophilic microorganisms.

Finally, we have shown the predictive potential of integrating expression data with functional modelling of the Antarctic bacterium *P. haloplanktis* TAC125 for identifying possible, biologically consistent, metabolic changes related to the remodulation of gene expression in response to temperature perturbations. Remarkably, given the relative ease at which -omics experiments can be performed nowadays, this general framework may be used, in the future, to characterize and predict the metabolic response of microbes to many other environmental perturbations, including pH, pressure and salinity. This, in turn, may have important drawbacks when trying to elucidate the consequences on the whole microbial community of large-scale climate changes and ocean dynamics over the next years.

Conflict of Interest Marco Fondi, Emanuele Bosi, Angelina Lo Giudice, and Renato Fani declare that they have no conflict of interest.

References

Ansong C, Deatherage BL, Hyduke D, Schmidt B, McDermott JE, Jones MB, Chauhan S, Charusanti P, Kim YM, Nakayasu ES, Li J, Kidwai A, Niemann G, Brown RN, Metz TO, McAteer K, Heffron F, Peterson SN, Motin V, Palsson BO, Smith RD, Adkins JN (2013) Studying *Salmonellae* and *Yersiniae* host-pathogen interactions using integrated 'omics and modeling. Curr Top Microbiol Immunol 363:21–41

Arsene F, Tomoyasu T, Bukau B (2000) The heat shock response of *Escherichia coli*. Int J Food Microbiol 55(1–3):3–9

Bartell JA, Yen P, Varga JJ, Goldberg JB, Papin JA (2014) Comparative metabolic systems analysis of pathogenic *Burkholderia*. J Bacteriol 196(2):210–226

Bochner BR, Gadzinski P, Panomitros E (2001) Phenotype microarrays for high-throughput phenotypic testing and assay of gene function. Genome Res 11(7):1246–1255

Bouvet V, Ben RN (2003) Antifreeze glycoproteins: structure, conformation, and biological applications. Cell Biochem Biophys 39(2):133–144

Caspi R, Foerster H, Fulcher CA, Hopkinson R, Ingraham J, Kaipa P, Krummenacker M, Paley S, Pick J, Rhee SY, Tissier C, Zhang P, Karp PD (2006) MetaCyc: a multiorganism database of metabolic pathways and enzymes. Nucleic Acids Res 34(Database issue):D511–D516

Caspi R, Altman T, Billington R, Dreher K, Foerster H, Fulcher CA, Holland TA, Keseler IM, Kothari A, Kubo A, Krummenacker M, Latendresse M, Mueller LA, Ong Q, Paley S, Subhraveti P, Weaver DS, Weerasinghe D, Zhang P, Karp PD (2014) The MetaCyc database of metabolic pathways and enzymes and the BioCyc collection of Pathway/Genome Databases. Nucleic Acids Res 42(Database issue):D459–D471

Cavicchioli R, Charlton T, Ertan H, Mohd Omar S, Siddiqui KS, Williams TJ (2011) Biotechnological uses of enzymes from psychrophiles. Microbial Biotechnol 4(4):449–460

Chromy BA, Choi MW, Murphy GA, Gonzales AD, Corzett CH, Chang BC, Fitch JP, McCutchen-Maloney SL (2005) Proteomic characterization of *Yersinia pestis* virulence. J Bacteriol 187(23):8172–8180

Collins T, Roulling F, Piette F, Marx J-C, Feller G, Gerday C, D'Amico S (2008) Fundamentals of cold-adapted enzymes. In: Marx J-C, Gerday C (eds) Psychrophiles: from biodiversity to biotechnology. Springer, Berlin, pp 211–227

Corchero JL, Gasser B, Resina D, Smith W, Parrilli E, Vazquez F, Abasolo I, Giuliani M, Jantti J, Ferrer P, Saloheimo M, Mattanovich D, Schwartz S Jr, Tutino ML, Villaverde A (2013) Unconventional microbial systems for the cost-efficient production of high-quality protein therapeutics. Biotechnol Adv 31(2):140–153

Duilio A, Tutino ML, Marino G (2004) Recombinant protein production in Antarctic Gram-negative bacteria. Methods Mol Biol 267:225–237

Durot M, Le Fevre F, de Berardinis V, Kreimeyer A, Vallenet D, Combe C, Smidtas S, Salanoubat M, Weissenbach J, Schachter V (2008) Iterative reconstruction of a global metabolic model of *Acinetobacter baylyi* ADP1 using high-throughput growth phenotype and gene essentiality data. BMC Syst Biol 2:85

Fang K, Zhao H, Sun C, Lam CM, Chang S, Zhang K, Panda G, Godinho M, Martins dos Santos VA, Wang J (2011) Exploring the metabolic network of the epidemic pathogen *Burkholderia cenocepacia* J2315 via genome-scale reconstruction. BMC Syst Biol 5:83

Feller G, Gerday C (2003) Psychrophilic enzymes: hot topics in cold adaptation. Nat Rev Microbiol 1(3):200–208

Fondi M, Lio P (2015) Multi -omics and metabolic modelling pipelines: challenges and tools for systems microbiology. Microbiol Res 171C:52–64

Fondi M, Maida I, Perrin E, Mellera A, Mocali S, Parrilli E, Tutino ML, Lio P, Fani R (2014) Genome-scale metabolic reconstruction and constraint-based modelling of the Antarctic bacterium *Pseudoalteromonas haloplanktis* TAC125. Environ Microbiol. doi:10.1111/1462-2920.12513

Gao H, Wang Y, Liu X, Yan T, Wu L, Alm E, Arkin A, Thompson DK, Zhou J (2004) Global transcriptome analysis of the heat shock response of *Shewanella oneidensis*. J Bacteriol 186(22):7796–7803

Gao H, Yang ZK, Wu L, Thompson DK, Zhou J (2006) Global transcriptome analysis of the cold shock response of *Shewanella oneidensis* MR-1 and mutational analysis of its classical cold shock proteins. J Bacteriol 188(12):4560–4569

Garcia-Descalzo L, Garcia-Lopez E, Alcazar A, Baquero F, Cid C (2014) Proteomic analysis of the adaptation to warming in the Antarctic bacteria *Shewanella frigidimarina*. Biochim Biophys Acta 1844(12):2229–2240

Garnier M, Matamoros S, Chevret D, Pilet MF, Leroi F, Tresse O (2010) Adaptation to cold and proteomic responses of the psychrotrophic biopreservative *Lactococcus piscium* strain CNCM I-4031. Appl Environ Microbiol 76(24):8011–8018

Gerday C, Aittaleb M, Arpigny JL, Baise E, Chessa JP, Garsoux G, Petrescu I, Feller G (1997) Psychrophilic enzymes: a thermodynamic challenge. Biochim Biophys Acta 1342(2):119–131

Giuliani M, Parrilli E, Ferrer P, Baumann K, Marino G, Tutino ML (2011) Process optimization for recombinant protein production in the psychrophilic bacterium Pseudoalteromonas haloplanktis. Process Biochem J 46:953–959

Graumann P, Marahiel MA (1996) Some like it cold: response of microorganisms to cold shock. Arch Microbiol 166(5):293–300

Griffin TJ, Gygi SP, Ideker T, Rist B, Eng J, Hood L, Aebersold R (2002) Complementary profiling of gene expression at the transcriptome and proteome levels in *Saccharomyces cerevisiae*. Mol Cell Proteomics 1(4):323–333

Hatzimanikatis V, Lee KH (1999) Dynamical analysis of gene networks requires both mRNA and protein expression information. Metab Eng 1(4):275–281

Jensen PA, Papin JA (2011) Functional integration of a metabolic network model and expression data without arbitrary thresholding. Bioinformatics 27(4):541–547

Jozefczuk S, Klie S, Catchpole G, Szymanski J, Cuadros-Inostroza A, Steinhauser D, Selbig J, Willmitzer L (2010) Metabolomic and transcriptomic stress response of *Escherichia coli*. Mol Syst Biol 6:364

Kaan T, Homuth G, Mader U, Bandow J, Schweder T (2002) Genome-wide transcriptional profiling of the *Bacillus subtilis* cold-shock response. Microbiology 148(Pt 11):3441–3455

Kanehisa M (2002) The KEGG database. Novartis Found Symp 247:91–101; discussion 101–103, 119–128, 244–152

Kawahara H (2008) Cryoprotectants and ice-binding proteins. In: Margesin R, Schinner F, Marx J-C, Gerday C (eds) Psychrophiles: from biodiversity to biotechnology. Springer, Heidelberg, pp 229–246

Kawamoto J, Kurihara T, Kitagawa M, Kato I, Esaki N (2007) Proteomic studies of an Antarctic cold-adapted bacterium, *Shewanella livingstonensis* Ac10, for global identification of cold-inducible proteins. Extremophiles 11(6):819–826

Kawamura D, Yamashita I, Nimi O, Toh-e A (1994) Cloning and nucleotide sequence of a gene conferring ability to grow at a low temperature on *Saccharomyces cerevisiae* tryptophan auxotrophs. J Biosci Bioeng 77(1):1–9

Klein W, Weber MH, Marahiel MA (1999) Cold shock response of Bacillus subtilis: isoleucine-dependent switch in the fatty acid branching pattern for membrane adaptation to low temperatures. J Bacteriol 181(17):5341–5349

Konkel ME, Tilly K (2000) Temperature-regulated expression of bacterial virulence genes. Microbes Infect 2(2):157–166

Li JS, Bi YT, Dong C, Yang JF, Liang WD (2011) Transcriptome analysis of adaptive heat shock response of *Streptococcus thermophilus*. PLoS One 6(10):e25777

Liu S, Graham JE, Bigelow L, Morse PD, Wilkinson BJ (2002) Identification of Listeria monocytogenes genes expressed in response to growth at low temperature. Appl Environ Microbiol 68(4):1697–1705

Lopez-Maury L, Marguerat S, Bahler J (2008) Tuning gene expression to changing environments: from rapid responses to evolutionary adaptation. Nat Rev Genet 9(8):583–593

Mavromatis K, Tsigos I, Tzanodaskalaki M, Kokkinidis M, Bouriotis V (2002) Exploring the role of a glycine cluster in cold adaptation of an alkaline phosphatase. Eur J Biochem 269(9):2330–2335

Medigue C, Krin E, Pascal G, Barbe V, Bernsel A, Bertin PN, Cheung F, Cruveiller S, D'Amico S, Duilio A, Fang G, Feller G, Ho C, Mangenot S, Marino G, Nilsson J, Parrilli E, Rocha EP, Rouy Z, Sekowska A, Tutino ML, Vallenet D, von Heijne G, Danchin A (2005) Coping with cold: the genome of the versatile marine Antarctica bacterium *Pseudoalteromonas haloplanktis* TAC125. Genome Res 15(10):1325–1335

Methe BA, Nelson KE, Deming JW, Momen B, Melamud E, Zhang X, Moult J, Madupu R, Nelson WC, Dodson RJ, Brinkac LM, Daugherty SC, Durkin AS, DeBoy RT, Kolonay JF, Sullivan SA,

Zhou L, Davidsen TM, Wu M, Huston AL, Lewis M, Weaver B, Weidman JF, Khouri H, Utterback TR, Feldblyum TV, Fraser CM (2005) The psychrophilic lifestyle as revealed by the genome sequence of *Colwellia psychrerythraea* 34H through genomic and proteomic analyses. Proc Natl Acad Sci U S A 102(31):10913–10918

Miyake R, Kawamoto J, Wei YL, Kitagawa M, Kato I, Kurihara T, Esaki N (2007) Construction of a low-temperature protein expression system using a cold-adapted bacterium, *Shewanella* sp. strain Ac10, as the host. Appl Environ Microbiol 73(15):4849–4856

Morgan-Kiss RM, Priscu JC, Pocock T, Gudynaite-Savitch L, Huner NP (2006) Adaptation and acclimation of photosynthetic microorganisms to permanently cold environments. Microbiol Mol Biol Rev 70(1):222–252

Morimoto RI, Kline MP, Bimston DN, Cotto JJ (1997) The heat-shock response: regulation and function of heat-shock proteins and molecular chaperones. Essays Biochem 32:17–29

Morita MT, Tanaka Y, Kodama TS, Kyogoku Y, Yanagi H, Yura T (1999) Translational induction of heat shock transcription factor sigma32: evidence for a built-in RNA thermosensor. Genes Dev 13(6):655–665

Navid A, Almaas E (2012) Genome-level transcription data of *Yersinia pestis* analyzed with a new metabolic constraint-based approach. BMC Syst Biol 6:150

Oberhardt MA, Chavali AK, Papin JA (2009) Flux balance analysis: interrogating genome-scale metabolic networks. Methods Mol Biol 500:61–80

Overbeek R, Olson R, Pusch GD, Olsen GJ, Davis JJ, Disz T, Edwards RA, Gerdes S, Parrello B, Shukla M, Vonstein V, Wattam AR, Xia F, Stevens R (2014a) The SEED and the rapid annotation of microbial genomes using Subsystems Technology (RAST). Nucleic Acids Res 42(Database issue):D206–D214

Overbeek R, Olson R, Pusch GD, Olsen GJ, Davis JJ, Disz T, Edwards RA, Gerdes S, Parrello B, Shukla M, Vonstein V, Wattam AR, Xia F, Stevens R (2014b) The SEED and the rapid annotation of microbial genomes using subsystems technology (RAST). Nucleic Acids Res 42(Database issue):D206–D214

Papa R, Rippa V, Sannia G, Marino G, Duilio A (2007) An effective cold inducible expression system developed in Pseudoalteromonas haloplanktis TAC125. J Biotechnol 127(2): 199–210

Pearce DA (2012) Extremophiles in Antarctica: life at low temperatures. Springer, Dordrecht

Piette F, D'Amico S, Struvay C, Mazzucchelli G, Renaut J, Tutino ML, Danchin A, Leprince P, Feller G (2010) Proteomics of life at low temperatures: trigger factor is the primary chaperone in the Antarctic bacterium Pseudoalteromonas haloplanktis TAC125. Mol Microbiol 76(1):120–132

Piette F, D'Amico S, Mazzucchelli G, Danchin A, Leprince P, Feller G (2011) Life in the cold: a proteomic study of cold-repressed proteins in the antarctic bacterium *Pseudoalteromonas haloplanktis* TAC125. Appl Environ Microbiol 77(11):3881–3883

Ratheesh RK, Nagarajan SN, Arunraj PA, Sinha D, Veedin Rajan VB, Esthaki VK, D'Silva P (2012) HSPIR: a manually annotated heat shock protein information resource. Bioinformatics 28(21):2853–2855

Reed JL, Vo TD, Schilling CH, Palsson BO (2003) An expanded genome-scale model of *Escherichia coli* K-12 (iJR904 GSM/GPR). Genome Biol 4(9):R54

Rippa V, Papa R, Giuliani M, Pezzella C, Parrilli E, Tutino ML, Marino G, Duilio A (2012) Regulated recombinant protein production in the Antarctic bacterium *Pseudoalteromonas haloplanktis* TAC125. Methods Mol Biol 824:203–218

Ruch FE, Vagelos PR (1973) The isolation and general properties of *Escherichia coli* malonyl coenzyme A-acyl carrier protein transacylase. J Biol Chem 248(23):8086–8094

Russell NJ (1990) Cold adaptation of microorganisms. Philos Trans R Soc Lond B Biol Sci 326(1237):595–608

Schatschneider S, Persicke M, Watt SA, Hublik G, Puhler A, Niehaus K, Vorholter FJ (2013) Establishment, in silico analysis, and experimental verification of a large-scale metabolic network of the xanthan producing *Xanthomonas campestris* pv. campestris strain B100. J Biotechnol 167(2):123–134

Schaub J, Clemens C, Kaufmann H, Schulz TW (2012) Advancing biopharmaceutical process development by system-level data analysis and integration of omics data. Adv Biochem Eng Biotechnol 127:133–163

Scheer M, Grote A, Chang A, Schomburg I, Munaretto C, Rother M, Sohngen C, Stelzer M, Thiele J, Schomburg D (2011) BRENDA, the enzyme information system in 2011. Nucleic Acids Res 39(Database issue):D670–D676

Simon C, Wiezer A, Strittmatter AW, Daniel R (2009) Phylogenetic diversity and metabolic potential revealed in a glacier ice metagenome. Appl Environ Microbiol 75(23):7519–7526

Sterner R, Liebl W (2001) Thermophilic adaptation of proteins. Crit Rev Biochem Mol Biol 36(1):39–106

Tanghe A, Van Dijck P, Thevelein JM (2003) Determinants of freeze tolerance in microorganisms, physiological importance, and biotechnological applications. Adv Appl Microbiol 53:129–176

Thiele I, Palsson BO (2010) A protocol for generating a high-quality genome-scale metabolic reconstruction. Nat Protoc 5(1):93–121

Ting L, Williams TJ, Cowley MJ, Lauro FM, Guilhaus M, Raftery MJ, Cavicchioli R (2010) Cold adaptation in the marine bacterium, *Sphingopyxis alaskensis*, assessed using quantitative proteomics. Environ Microbiol 12(10):2658–2676

Tomar N, De RK (2013) Comparing methods for metabolic network analysis and an application to metabolic engineering. Gene 521(1):1–14

Tong W, Chen Z, Cao Z, Wang Q, Zhang J, Bai X, Wang R, Liu S (2013) Robustness analysis of a constraint-based metabolic model links cell growth and proteomics of *Thermoanaerobacter tengcongensis* under temperature perturbation. Mol Biosyst 9(4):713–722

Topfer N, Jozefczuk S, Nikoloski Z (2012) Integration of time-resolved transcriptomics data with flux-based methods reveals stress-induced metabolic adaptation in *Escherichia coli*. BMC Syst Biol 6:148

Varin T, Lovejoy C, Jungblut AD, Vincent WF, Corbeil J (2012) Metagenomic analysis of stress genes in microbial mat communities from Antarctica and the High Arctic. Appl Environ Microbiol 78(2):549–559

Vieille C, Zeikus GJ (2001) Hyperthermophilic enzymes: sources, uses, and molecular mechanisms for thermostability. Microbiol Mol Biol Rev 65(1):1–43

Wang ZX, Zhou XZ, Meng HM, Liu YJ, Zhou Q, Huang B (2014) Comparative transcriptomic analysis of the heat stress response in the filamentous fungus *Metarhizium anisopliae* using RNA-Seq. Appl Microbiol Biotechnol 98(12):5589–5597

Weber MH, Fricke I, Doll N, Marahiel MA (2002) CSDBase: an interactive database for cold shock domain-containing proteins and the bacterial cold shock response. Nucleic Acids Res 30(1):375–378

Wiench B, Chen YR, Paulsen M, Hamm R, Schroder S, Yang NS, Efferth T (2013) Integration of different "-omics" technologies identifies inhibition of the IGF1R-Akt-mTOR signaling cascade involved in the cytotoxic effect of shikonin against leukemia cells. Evid Based Complement Alternat Med 2013:818709

Wilmes B, Hartung A, Lalk M, Liebeke M, Schweder T, Neubauer P (2010) Fed-batch process for the psychrotolerant marine bacterium *Pseudoalteromonas haloplanktis*. Microb Cell Fact 9:72

Wilmes B, Kock H, Glagla S, Albrecht D, Voigt B, Markert S, Gardebrecht A, Bode R, Danchin A, Feller G, Hecker M, Schweder T (2011) Cytoplasmic and periplasmic proteomic signatures of exponentially growing cells of the psychrophilic bacterium *Pseudoalteromonas haloplanktis* TAC125. Appl Environ Microbiol 77(4):1276–1283

Zhang W, Li F, Nie L (2010) Integrating multiple 'omics' analysis for microbial biology: application and methodologies. Microbiology 156(Pt 2):287–301

Zhu H, Ren X, Wang J, Song Z, Shi M, Qiao J, Tian X, Liu J, Chen L, Zhang W (2013) Integrated OMICS guided engineering of biofuel butanol-tolerance in photosynthetic *Synechocystis* sp. PCC 6803. Biotechnol Biofuels 6(1):106

Chapter 22
Experimental Microbial Evolution of Extremophiles

Paul Blum, Deepak Rudrappa, Raghuveer Singh, Samuel McCarthy, and Benjamin Pavlik

1 Experimental Microbial Evolution Theory and Applications in Model Organisms

Experimental microbial evolutions (EME) involves studying closely a microbial population after it has been through a large number of generations under controlled conditions (Kussell 2013). Adaptive laboratory evolution (ALE) selects for fitness under experimentally imposed conditions (Bennett and Hughes 2009; Dragosits and Mattanovich 2013). However, experimental evolution studies focusing on the contributions of genetic drift and natural mutation rates to evolution are conducted under non-selective conditions to avoid changes imposed by selection (Hindré et al. 2012).

To understand the application of experimental evolutionary methods to extremophiles it is essential to consider the recent growth in this field over the last decade using model non-extremophilic microorganisms. This growth reflects both a greater appreciation of the power of experimental evolution for testing evolutionary hypotheses and, especially recently, the new power of genomic methods for analyzing changes in experimentally evolved lineages. Since many crucial processes are driven by microorganisms in nature, it is essential to understand and appreciate how microbial communities function, particularly with relevance to selection. However, many theories developed to understand microbial ecological patterns focus on the distribution and the structure of diversity within a microbial population comprised of single species (Prosser et al. 2007). Therefore an understanding of the concept of species is needed. A common definition of species using a genetic concept is a

P. Blum (✉) • D. Rudrappa • R. Singh • S. McCarthy
School of Biological Science, University of Nebraska, Lincoln, NE, USA
e-mail: pblum1@unl.edu

B. Pavlik
Department of Chemical and Biomolecular Engineering, University of Nebraska, Lincoln, NE, USA

P.H. Rampelotto (ed.), *Biotechnology of Extremophiles*, Grand Challenges in Biology and Biotechnology 1, DOI 10.1007/978-3-319-13521-2_22

group of interbreeding individuals that is isolated from other such groups by barriers of recombination (Prosser et al. 2007). An alternative ecological species concept defines a species as set of individuals that can be considered identical in all relevant ecological traits (Cohan 2001). This is particularly important because of the abundance and deep phylogenetic complexity of microbial communities. Cohan postulated that "bacteria occupy discrete niches and that periodic selection will purge genetic variation within each niche without preventing divergence between the inhabitants of different niches". The importance of gene exchange mechanisms likely in bacteria and archaea and therefore extremophiles, arises from the fact that their genomes are divided into two distinct parts, the core genome and the accessory genome (Cohan 2001). The core genome consists of genes that are crucial for the functioning of an organism and the accessory genome consists of genes that are capable of adapting to the changing ecosystem through gain and loss of function. Strains that belong to the same species can differ in the composition of accessory genes and therefore their capability to adapt to changing ecosystems (Cohan 2001; Tettelin et al. 2005; Gill et al. 2005). Additional ecological diversity exists in plasmids, transposons and pathogenicity islands as they can be easily shared in a favorable environment but still be absent in the same species found elsewhere (Wertz et al. 2003). This poses a major challenge for studying ALE and community microbial ecology indicating a continued need to develop a fitting theory that connects the fluid nature of microbial communities to their ecology (Wertz et al. 2003; Coleman et al. 2006). Understanding the nature and contribution of different processes that determine the frequencies of genes in any population is the biggest concern in population and evolutionary genetics (Prosser et al. 2007) and it is critical for an understanding of experimental evolution. Tatum and Lederberg (Tatum and Lederberg 1947) discovered laterally transferred genes in *E. coli*. However sequencing of the genomes of two isolates of *Helicobacter pylori* revealed that >6 % of the genes are unique (Alm et al. 1999). When recombination is rare and limited to few genes, almost all other genes will be transmitted through vertical inheritance. Mutations will accumulate slowly over the period and will result in irreversible divergence of lineages. In contrast, when recombinational events are common, genes will release themselves from the rest of the genome and diversity in genes linked to adaptive alleles will be purged through selection (Polz et al. 2006). This type of pan-mictic population structure has been reported in *Neisseria gonorrhoeae* and *Rhizobium meliloti* lineages (Smith et al. 1993).

Comparisons with founder or starting strains are made to quantify genotypic and phenotypic changes for example, DNA sequencing of specific genes or entire genomes is used to determine the mutational differences among evolved strains (Kussell 2013). Adaptive laboratory evolution (ALE) is a common method in EME studies to gain insights into the basic mechanisms of molecular evolution and adaptive landscapes that accumulate in microbial populations during long term selection under predefined growth conditions (Bennett et al. 1990). Over the past two to three decades, there have been an increasing number of experiments to understand the adaptive landscapes during ALE especially with *E. coli* and *Saccharomyces cerevisiae* (Bennett et al. 1990; Paquin and Adams 1983). Microbial adaptation to new environmental conditions mainly occurs via two different mechanisms: alteration of gene regulation without any heritable genetic change, or, selection of novel adaptive phenotypes conferred by stable mutations. Experimental evolution allows phenotypic changes to be associated with growth conditions that eventually select for traits (Hardison 2003). EME studies have led to significant insights and experimental proof for evolutionary biology (Bennett et al. 1990; Paquin and Adams 1983). The long term EME study of *E. coli* by Lenski's group remains at the forefront of these studies where a single parallel *E. coli* adaptation experiment has exceeded 50,000 generations (Sniegowski et al. 1997; Lenski et al. 1998; Cooper and Lenski 2000). These studies along with similar experiments conducted by others have provided insights into the genetic basis of increased microbial fitness (Barrick et al. 2009), their implications during evolution (Woods et al. 2011), understanding of population size, evolvability (Bloom et al. 2007; Elena et al. 2007; Draghi et al. 2010) and clonal interference (Kao and Sherlock 2008).

Experimental evolution allows the microorganisms to evolve through propagation thus relying on the inherent capacity of the organism to introduce mutations (Sauer 2001). Fast generation time, ease of maintaining large population sizes and storing them make microorganisms well suited to the study of ALE in a laboratory setting (Elena and Lenski 2003). However extensive cultivation periods are required for selecting a desired phenotype with a limited natural mutation rate. Adding to this, identifying mutations necessary for conferring a particular phenotype is difficult since neutral mutations end-accumulate thus making an "omic" based approach necessary to understand the observed phenotype (Bro and Nielsen 2004). Furthermore, experimental evolution provides an advantage over reverse genetics-based approaches that employ targeted activation or inactivation of genes in that ALE can result in occurrence of mutations of unexpected composition that provide gain of function for the organism (Conrad et al. 2011). Growing knowledge about the biochemical and physiological natures of some of the model organisms along with the use of genetic and mapping techniques developed in the 1980s and 1990s, has opened up new avenues (Helling et al. 1987; Rosenzweig et al. 1994; Treves et al. 1998; Kinnersley et al. 2014; Ferenci 2007; Adams et al. 1992). These same approaches have been used to test both experimentally-evolved and naturally-occurring microorganisms for genome amplification, deletion, insertion and rearrangement of genes or sequences (Cooper et al. 2003; Philippe et al. 2007; Kadam et al. 2008; Bachmann et al. 2012; Gresham et al. 2008; Wenger et al. 2011).

2 Technologies Relevant to Performing Experimental Microbial Evolution

For both extremophilic and non-extremophilic microbes, methods of cultivation are fundamental, as they dictate physiologic status and therefore mutation formation. For this reason, they are critical to the field of experimental evolution. Extended laboratory timescales of microbial growth in selective environments are also crucial. Under these conditions, microbes are advantageous as evolutionary models due to their large population size, short generation time, relatively small genome and generally high supply of mutations (Gresham and Dunham 2014). Extremophiles may exhibit elongated generation times and but maintain normal mutation rates (Grogan et al. 2001), necessitating longer culture durations to achieve equivalent degrees of mutation formation. The understanding and application of continuous cultivation and serial batch culture passage has expanded the use of microbial systems to address evolutionary questions. Although paramount to experimental evolution, these culturing technologies have remained relatively unused in the ALE of extremophiles and future research is warranted.

Continuous long term culturing techniques for the purpose of mutation accumulation and strain evolution was popularized in the mid-twentieth century. The chemostat is a continuous culturing device first described by (Monod 1942). The modern chemostat was introduced by Novick and Szilard and applied to the study of genetic changes of microorganisms (Novick and Szilard 1950). The basic concept of this culturing device is the continuous removal of medium with concurrent replacement of fresh medium at a defined rate. Microbial populations will grow in proportion to this rate. Contemporary versions of this device have been scaled downwards to achieve larger numbers of experimental replicates or to the point of single cell analysis (Dénervaud et al. 2013). Discoveries made using chemostats are rooted in physiological responses to environmental stresses. The nature of some of these studies include: variation between and within species, mutation rates, mutational takeovers, population changes, quorum sensing, genomic rearrangements, emerging diversity, metabolism and energetics, and membrane transport (Ferenci 2008). Chemostats present a research opportunity for extremophilic experimental evolution in that the current use is mainly in bioprocess applications (Lorantfy et al. 2014).

Similar to the chemostat is the turbidostat, popularized for experimental microbial evolution by (Bryson and Szybalski 1952). The main difference is that nutrients are not limited, and that fresh medium introduction is proportional to maximum culture turbidity, resulting in maximum growth rates of microorganisms in mid-exponential phase (Gresham and Dunham 2014). Both the chemostat and turbidostat maintain a steady-state environment, but represent minimal or optimal resource conditions, respectively. These conditions result in different types of mutations based on nutrient stress or growth enhancement and overall strain fitness improvement. Turbidostats have been employed in the study of evolutionary responses to anti-microbial compounds (Avrahami-Moyal et al. 2012; Toprak et al. 2012). As with chemostats, turbidostats have seen limited use in the experimental evolution of extremophiles.

Long term batch culture intended for the evolution of microbes requires serial dilution or passaging of cells at a known dilution and duration of culture (Atwood et al. 1951). This technique provides dynamic environment for microbial growth that is less conditionally consistent than continuous culture. Under this regimen, cells may be selected during all phases of cell growth. Moreover, this introduces a complex environment for non-continuous selection as in the case of continuous culturing devices. This approach has recently been introduced in the experimental evolution of the extremophile *Sulfolobus solfataricus* and is a prudent approach to the ALE of extremophiles without knowing the physiological responses required to design feeding regimes in continuous cultures (McCarthy 2015).

2.1 High-Throughput Methods Evaluate Genotypic and Phenotypic Evolution in Extremophiles

The method of analysis of evolutionary changes in microbial genomes has increased in throughput and resolution through the recent development of "omic" technologies, and must be considered regardless of the species of microorganism. High-throughput technologies such as whole genome sequencing, transcriptomics, microarrays and chromatin immunoprecipitation (ChIP-Seq) have enhanced resolution of the relationships between observable strain characteristics (traits) and the underlying molecular mechanisms of evolution. Understandably, high-throughput technologies have become indispensable to the study of EME. Advances in DNA sequencing throughput has expanded the research potential of experimental microbial evolution in extremophiles (Araya et al. 2010; Barrick et al. 2009; Hong and Gresham 2014; Kvitek and Sherlock 2013; Lang et al. 2013; Herring et al. 2006). Since the first archaeal genome was sequenced for *Methanocaldococcus jannaschii* (Bult et al. 1996), many more genomes have become available for extremophiles (Allers and Mevarech 2005). Gene expression data can also be obtained in high throughput formats using RNA sequencing (RNA-seq) or microarrays (DeRisi et al. 1997; Lashkari et al. 1997; Dunham et al. 2002; Gresham et al. 2006; Wurtzel et al. 2010), or proteomics (Rigaut et al. 1999). In addition protein-protein interactions (Schwikowski et al. 2000) and protein-DNA interactions (ChIP-seq) have been developed for EME model organisms and extended to extremophiles (Gresham et al. 2006; Dunham et al. 2002; Wilbanks et al. 2012)

3 Extending Experimental Evolution to Extremophiles

3.1 Specific Challenges Associated with Extremophiles

The application of experimental evolution to extremophiles must overcome specific difficulties associated with their cultivation and biophysical issues pertaining to their macromolecules arising from their extremophilic environments. Working with extremophiles in the laboratory requires special techniques and equipment to maintain their extreme environments and complex nutritional requirements. Many extremophiles have relatively slow growth rates compared to other microbial model systems. This means that ALE studies require longer time periods to achieve equivalent number of generations. For example, this is especially true for extremophiles cultivated under lithoautotrophic conditions where energy limitation constrains replication rate (McCarthy et al. 2014; Maezato et al. 2012).

In the context of macromolecules, where EME impacts their identities, proteins constitute a specific barrier while lipids and various classes of small molecules can also present challenges. Proteins derived from hyperthermophilic extremophiles require high temperature for analysis and therefore the use of specialized instrumentation. Proteins derived from halophilic extremophiles are particularly difficult to purify and characterize because they are unstable in low salt concentrations and their fractionation is impeded by the incompatibility of methods such as electrophoresis and ion-exchange with high salt concentrations (Anfinsen et al. 1995). Recombinant expression of extremophile proteins in non-extremophile hosts such as *E. coli* often face a similar problem: outside of the extreme conditions of their native environments halophilic proteins have a tendency to misfold and aggregate (Allers 2010). For these reasons, examining the performance of evolved extremophilic macromolecules within the native extremophile host becomes critical and depends on the availability of genetic systems (Maezato et al. 2012).

3.2 Cultivation and Preservation of Extremophiles

One example of an extremophilic life style is a hyperthermophilic anaerobic bacterium, *Thermotoga maritima* that was isolated from geothermal heated marine sediment at Vulcano, Italy (Huber and Stetter 1998). While most of the members of Thermotogales were isolated from hot springs and deep-sea hydrothermal vents, few of the members are able to cope with partially oxidative condition due to the partially-oxygenated hot sediments and fluids in hydrothermal vent ecosystems (Rusch et al. 2005). *T. maritima*, despite it description as a strict anaerobe, has been reported to grow in the presence of 0.5 % v/v oxygen (Le Fourn et al. 2008). Lack of abundant carbon in its marine thermal vent environment likely led to the ability of this organism to metabolize a broad diversity of sugars without apparent selectivity. For example this organism possess a significantly higher number of ABC-type substrate transporters than most other organisms (Nelson et al. 1999; Ren et al. 2007) *T. maritima* and its closely related relatives, *T. neapolitana*, *Thermotoga sp. strain RQ2*, *T. naphthophila* and *T. petrophila* thrive between 55 °C and 90 °C and pH range 5.5–9.5 using simple and complex carbohydrates (Chhabra et al. 2003; Conners et al. 2005). *T. maritima* can be cultivated in batch culture in biological replicates using Hungate tubes or serum bottles supplemented with diverse carbon sources. Tubes can be sealed with butyl rubber stoppers (Bellco Biotechnology), crimped with metal collars and the headspace can be exchanged with nitrogen gas. Medium inoculation can use sterile 1 cc syringes attached to 20½G needles and cultures incubated anaerobically at 80 °C typically overnight.

T. maritima is also of interest due to an apparent extensive degree of lateral gene transfer from an archaeal donor and a tendency to undergo continued genome size reduction (Nelson et al. 1999; Singh et al. 2015). Moreover, it is of applied interest due to its ability to produce molecular hydrogen at rates surpassing those of other microorganisms (Schröder et al. 1994). Ongoing experimental evolution studies concerned with expanding this capacity through the use of transient gene inactivation by targeted chromosomal recombination combined with purifying selection have established novel cell lines with unique properties (Singh et al. 2015). The term transient gene inactivation refers to the use of temporary gene disruption of a chromosomal locus resulting from consecutive single crossover events. The term purifying selection is a classic genetic method that involves maintaining selective growth conditions while isolating clonal populations.

Another example of an extremophile is the thermoacidophile *Sulfolobus solfataricus* which was isolated from volcanic hot springs in both Italy and the United States (Brock et al. 1972). This organism grows from a temperature range of 65–90 °C and a pH range of 2.5–5.0 with optimum growth conditions of 80 °C and pH 3.0 (Brock et al. 1972; Grogan 1989). In the laboratory this organism is cultivated in a modified basal salts medium (Brock et al. 1972): complex media is supplemented with 0.2 % tryptone and minimal media is supplemented with 0.2 % sugars such as glucose. Cultures are incubated at 80 °C in glass screw-capped flasks with aeration in orbital baths (Rolfsmeier and Blum 1995; Bini et al. 2002; Worthington et al. 2003) and growth is monitored by light absorption at a wavelength of 540 nm. *S. solfataricus* is a model organism in the archaea and it has an established genetic system (Maezato et al. 2011). This makes it ideal for ALE studies as genetic changes seen in experimental evolution experiments with this organism can be reconstructed in wild-type cell lines to verify their effects. *S. solfataricus* is of applied interest as a source of heat and acid stable enzymes for various industrial processes such as trehalose production and cellulose degradation (Antranikian et al. 2005). Ongoing experimental evolution studies in this organism have generated cell lines with increased acid stability (McCarthy 2015).

Extremophiles can be stored for long periods through the preparation of a frozen permanent. This becomes essential during ALE to preserve intermediate stages of evolutionary changes resident within distinct microbial populations. To achieve this end, mid-exponential phase cultures are collected by centrifugation, washed with fresh medium, amended with 7 % (v/v) dimethyl sulfoxide (DMSO). The sample is then mixed and flash frozen using an ethanol-dry ice bath. Storage at −80 °C provides a long term method to preserve culturability (Maezato et al. 2012).

4 Examples of Experimental Microbial Evolution Using Extremophiles

A broad range of genome changes underlie evolutionary changes in non extremophilic microbes. Gene duplication, deletion (Kunin and Ouzounis 2003), translocation, inversion (Suyama and Bork 2001), lateral gene transfer (Garcia-Vallvé et al. 2000; Nelson et al. 1999) and transposition (Kidwell and Lisch 2000) are some of the phenomenon that result in genome evolution (Fraser-Liggett 2005). Genome evolution in extremophiles can occur in a similar fashion because various genetic elements such as direct repeats (DR), inverted repeats (IR) and transposable elements exist in extremophiles. Since DRs and IRs have the potential to manipulate the dynamics of a genome (Ussery et al. 2004), extremophiles should be capable of undergoing processes involving these sequences to result in genome evolution. The origins of DRs in non-thermophiles have been proposed to arise from lateral gene transfer, slip strand pairing and genome hopping (Achaz et al. 2002; Romero and Palacios 1997). As lateral gene transfer in Thermotogales has been described, these extremophiles can be used to study genome evolution. Close repeats or tandem repeats with smaller spacer regions are removed by RecA independent recombination (Bi and Liu 1994; Lovett et al. 1993) whereas direct repeats with a longer spacer depending on selection pressure may be stably maintained in the genome and are less likely to be deleted by illegitimate recombination (Lovett et al. 1994; Chédin et al. 1994). The biological relevance of repetitive sequence is to provide phenotypic variation in two ways. Repeats located in the regulatory region of a gene can modulate the expression of the gene (van Ham et al. 1993) and location inside a coding region could result in protein truncation when gene rearrangement results in loss of the repeat. Repetitive sequence retention also depends on selective pressure and this aspect of genomic instability can be experimentally tested.

Transposable elements are hypothesized to be both harmful and occasionally beneficial to their hosts (Schneider and Lenski 2004) by providing a source of genetic diversity through mutations, duplications, and genome rearrangements (Kidwell and Lisch 2000). Since transposable elements are broadly distributed in extremophiles their activity is again relevant to evolutionary change. In Lenski's long-term *E. coli* evolution experiments several pivotal changes in the different lineages of evolved cell lines were linked to transpositions, however the transposition rates did not correlate with the rates of adaptation: fitness showed a rapid increase in early generations and then sharply decelerated over time, while the rate of transposition remained relatively constant (Barrick et al. 2009; Schneider and Lenski 2004). In another EME study on thermal adaptation in *E. coli* the phenotypes of half of the derived heat-adapted cell lines could be explained by duplication events in a similar genomic region and a corresponding up-regulation of the duplicated genes. However, the other half of the adapted cell lines showed no duplications or transposition events and no change in expression of those same genes, indicating that these adapted through a different but unknown pathway (Riehle et al. 2001). These studies may be predictive of analogous efforts using extremophiles.

4.1 *The Hyperthermophilic Bacterial Anaerobe* **Thermotoga maritima** *and Deletion Formation*

T. maritima is one of the most well characterized members of the bacterial phylum Thermotogae. This phylum contains a broad range of bacterial extremophiles noted for their thermophilicity and anaerobiosis (Huber and Stetter 1998). *T. maritima* has been reported to undergone lateral gene transfer from an archaeon (Nelson et al. 1999), making it a promising model system to investigate experimental genome

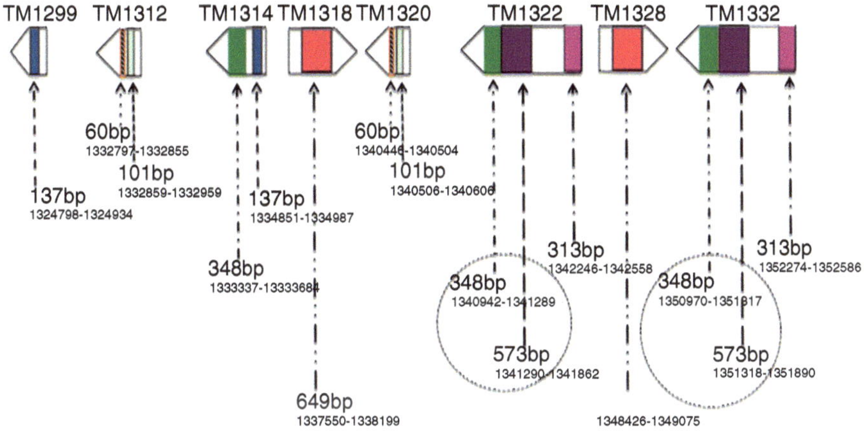

Fig. 22.1 A potential hot spot of genome evolution in *T. maritima*. A specific colored block represents direct repeats of conserved length. Genomic coordinates and gene locus tags are shown according to the genome of *T. maritima* reported (Nelson et al. 1999)

evolution, and an 8 kb deletion in the type strain supports the occurrence of genome evolution in this organism (Zhaxybayeva et al. 2009). Bioinformatic analysis to find direct repeats (DRs) confirms the presence of various small and larger DRs in this organism that could contribute to genome evolution (Fig. 22.1). To identify the location of DRs that might contribute to genome evolution, the *T. maritima* genome was scanned to find those DRs ranging from 50 to 1500 bp using REPuter (Kurtz et al. 2001). The *T. maritima* genome reported by Nelson (Nelson et al. 1999) was used to describe the coordinates and gene locus tags. In *T. maritima*, repetitive sequences surround a variable spacer region and clustering of various DRs has generated a larger DR. This was evident when a region between TM1299 and TM1332 where a bigger DR generated via clustering, was identified as a potential hot spot of genome evolution. Various arrangements of *T. maritima* DRs are presented (Fig. 22.2). The biggest DRs of 921 bp exist in TM1322 (coordinates; 1340942–1341862) and TM1332 (1350970–1351890) surrounding a 10 kb spacer region. Considering the occurrence of large DRs (921 bp) in the genome, their location could constitute a hot spot for deletion of the intervening region. Such events would continue genome evolution in this bacterium. While the organism has been shown to evolve by deleting an 8 kb gene naturally, an experimental evolutionary approach used to manipulate metabolite formation resulted in deletion of genes (Singh et al. 2015). A transient gene inactivation by targeted chromosomal recombination combined with purifying selection established novel cell lines with unique properties (Singh). Genome resequencing identified a 10 kb deletion between TM1322 and TM1332 that resulted from crossover between flanking 921 bp DRs that deleted the intervening region (1341863–1351890). This is the first report of experimental microbial genome evolution in *T. maritima*. The 10 kb deletion strain of *T. maritima* was named Tma200 and due to its altered phenotype of hydrogen production may provide a strategy for further evolution of its extremophilic traits (Singh et al. 2015).

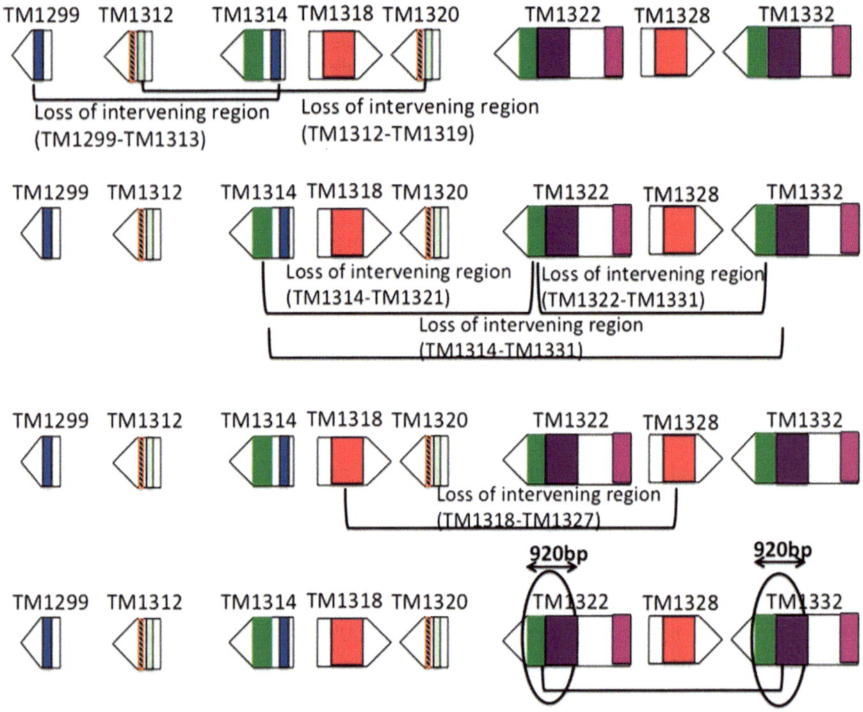

Fig. 22.2 A proposed schematic of genome deletions mediated via cross-over between various direct repeats. Specific colored block represents direct repeats of conserved length. Identical direct repeats that could undergo homologous recombination are presented as a bracketed regions

4.2 The Archaeon **Sulfolobus solfataricus** *and the Role if Insertion Sequence Elements*

Sulfolobus solfataricus is an extremely thermoacidophilic member of the Crenarchaeotal phylum of the Archaea. Ongoing studies concerned with evolving new cell lines with increased thermoacidophily have generated a series of novel lineages (McCarthy). Aspects of these lineages pertaining to ALE are presented here. Unlike other species in this genus, *S. solfataricus* has a genome that is rich in insertion sequence (IS) elements with over 200 IS elements (10 % of genome) and its genome is predicted to be continually changing due rearrangements caused by these transposable elements (Redder and Garrett 2006; Brügger et al. 2004). It undergoes a high frequency of transposition and its gene order (synteny) is very different from other members of the *Sulfolobus* genus, indicating that it has undergone multiple rearrangements (Brügger et al. 2004).

Several evolved *S. solfataricus* cell lines were isolated by extensive serial passage to select for increased acid resistance in an experimental evolution experiment (Fig. 22.3) (McCarthy). Intermediate isolates were purified to clonality using a solid complex medium and the clonal isolates were re-screened for the acid resistance

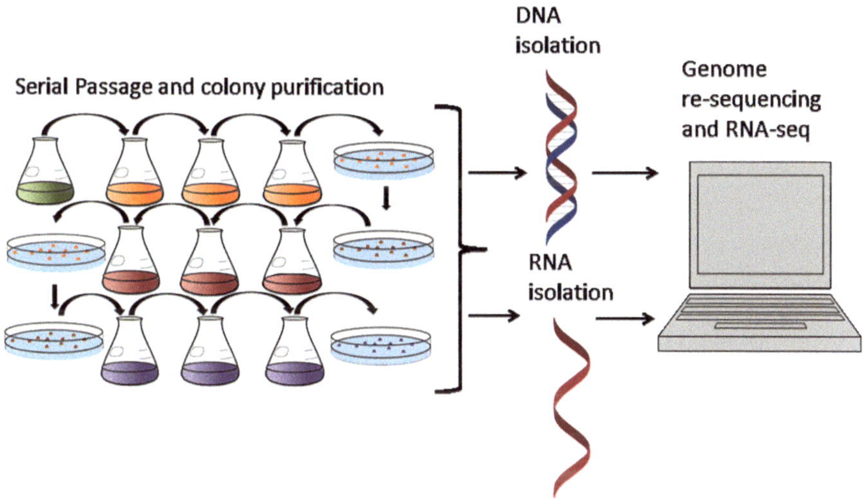

Fig. 22.3 Experimental microbial evolution by serial passage. Cells are sub-cultured into a slightly more extreme condition (indicated by changing coloration of the medium), passaged several times until they adapt to this condition, and then the condition is adjusted and the process is repeated. Colonies are isolated from adapted cultures and screened for the unique phenotype. The genomes and transcriptomes of experimentally evolved colonies are then examined to determine its cause

phenotype. These cell lines maintained acid resistance after passaging at pH 3.0, indicating that their phenotype was stable and heritable rather than a transient response or stress-induced trait (McCarthy). The acid adapted *S. solfataricus* cell lines were resistant to a 150-fold greater acid concentration than the wild-type and was predicted to have a significant number of adaptive mutations based on transcriptomic studies. RNA-seq analysis showed high (+5-fold or more) up-regulation of 27 transposons and transposases throughout the entire genome in the adapted strains at low pH compared to the transcriptomes of wild-type cell lines growing at optimal pH. Eighteen of these elements were up-regulated tenfold or more and eight were up-regulated over 30-fold. The most up-regulated IS elements were all members of the same families including transposases ISC1217, 1234/ST1916, and IS1 (Fig. 22.4). Curiously however, genome re-sequencing of these strains showed very few transposition events. Two of the three adapted cell lines had a single event each and the third showed no transposition. The frequency of transposition in the adapted cell lines appeared to decrease, even though the expression of many transposase genes was increased at low pH. The lack of transpositions in the evolved cell lines indicates that these were not a source of adaptation and this was surprising given the high frequency of IS elements in the genome.

The overall transcriptomes of the acid-adapted cell lines gave additional clues to their mechanism of acid resistance. Many changes were observed in expression of putative transporters and genes predicted to be involved in signal transduction as well as overall metabolism (Fig. 22.5). Examining transcriptomics using RNAseq is relatively new in prokaryotes and combining this approach with genomics of ALE represents a powerful approach for looking at the relative contributions of mutations and changes in gene expression to adaptation and the interaction between them.

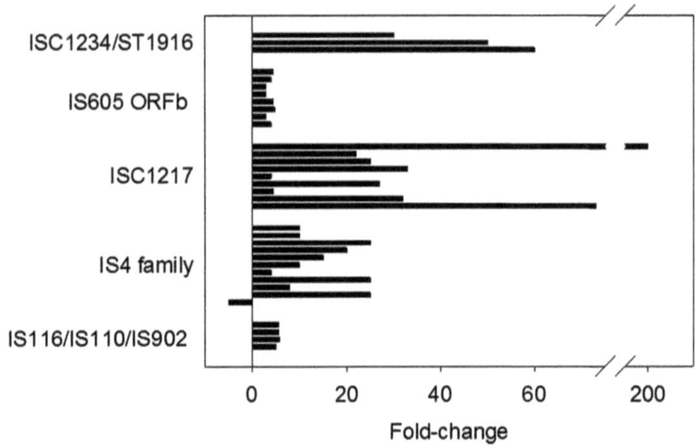

Fig. 22.4 Transcriptomic profile of insertion elements in *S. solfataricus* evolved to strong acid resistance by EME. Bars represent the expression changes of individual transposase genes relative to the expression in pH 3.00-grown wild-type *S. solfataricus*. Most of these elements were up-regulated in the evolved cell line, with the greatest changes seen in ISC1217 and ISC1234/ST1916 family transposable elements

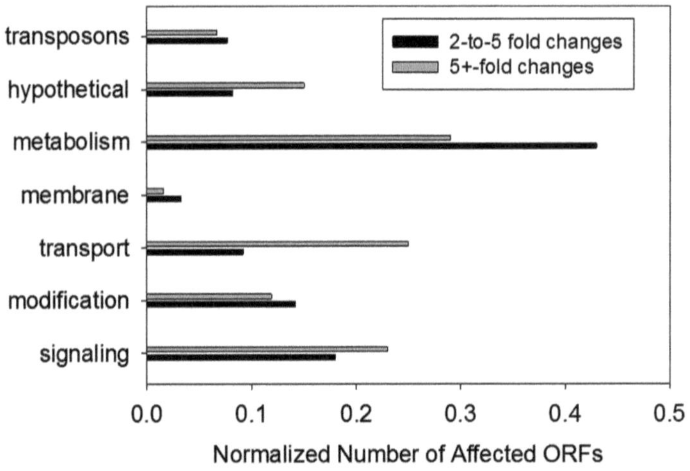

Fig. 22.5 Classes of ORFs with altered expression in experimentally evolved *S. solfataricus*. The number of ORFs in several functional categories that had small (two-to-fivefold) and large (5+-fold) expression changes in a pH 1.00-grown evolved cell line normalized to the total number of ORFs in each category in the genome are shown. The fold-changes are relative to pH 3.00-grown wild-type *S. solfataricus*. Biological replicates of all RNA-seq samples had Pearson correlation coefficients greater than 0.96

Acknowledgments This article was supported by funds from the University of Nebraska Cell Development Facility.

Conflict of Interest Paul Blum, Deepak Rudrappa, Raghuveer Singh, Samuel McCarthy and Benjamin Pavlik declare that they have no conflict of interest.

References

Achaz G, Rocha EP, Netter P, Coissac E (2002) Origin and fate of repeats in bacteria. Nucleic Acids Res 30:2987–2994

Adams J, Puskas-Rozsa S, Simlar J, Wilke CM (1992) Adaptation and major chromosomal changes in populations of *Saccharomyces cerevisiae*. Curr Genet 22:13–19

Allers T (2010) Overexpression and purification of halophilic proteins in *Haloferax volcanii*. Bioeng Bugs 1:288–290

Allers T, Mevarech M (2005) Archaeal genetics-the third way. Nat Rev Genet 6:58–73

Alm RA, Ling LS, Moir DT, King BL, Brown ED, Doig PC, Smith DR, Noonan B, Guild BC, deJonge BL, Carmel G, Tummino PJ, Caruso A, Uria-Nickelsen M, Mills DM, Ives C, Gibson R, Merberg D, Mills SD, Jiang Q, Taylor DE, Vovis GF, Trust TJ (1999) Genomic-sequence comparison of two unrelated isolates of the human gastric pathogen *Helicobacter pylori*. Nature 397:176–180

Anfinsen CB, Edsall JT, Richards FM, Eisenberg DS (1995) Advances in protein chemistry. Academic Press, San Diego

Antranikian G, Vorgias CE, Bertoldo C (2005) Extreme environments as a resource for microorganisms and novel biocatalysts. Adv Biochem Eng Biotechnol 96:219–262

Araya CL, Payen C, Dunham MJ, Fields S (2010) Whole-genome sequencing of a laboratory-evolved yeast strain. BMC Genomics 11:88

Atwood KC, Schneider LK, Ryan FJ (1951) Selective mechanisms in bacteria. Cold Spring Harb Symp Quant Biol 16:345–355

Avrahami-Moyal L, Engelberg D, Wenger JW, Sherlock G, Braun S (2012) Turbidostat culture of *Saccharomyces cerevisiae* W303-1A under selective pressure elicited by ethanol selects for mutations in SSD1 and UTH1. FEMS Yeast Res 12:521–533

Bachmann H, Starrenburg MJ, Molenaar D, Kleerebezem M, van Hylckama Vlieg JE (2012) Microbial domestication signatures of *Lactococcus lactis* can be reproduced by experimental evolution. Genome Res 22:115–124

Barrick JE, Yu DS, Yoon SH, Jeong H, Oh TK, Schneider D, Lenski RE, Kim JF (2009) Genome evolution and adaptation in a long-term experiment with *Escherichia coli*. Nature 461:1243–1247

Bennett AF, Hughes BS (2009) Microbial experimental evolution. Am J Physiol Regul Integr Comp Physiol 297:29

Bennett AF, Dao KM, Lenski RE (1990) Rapid evolution in response to high-temperature selection. Nature 346:79–81

Bi X, Liu LF (1994) recA-independent and recA-dependent intramolecular plasmid recombination. Differential homology requirement and distance effect. J Mol Biol 235:414–423

Bini E, Dikshit V, Dirksen K, Drozda M, Blum P (2002) Stability of mRNA in the hyperthermophilic archaeon *Sulfolobus solfataricus*. RNA 8:1129–1136

Bloom JD, Lu Z, Chen D, Raval A, Venturelli OS, Arnold FH (2007) Evolution favors protein mutational robustness in sufficiently large populations. BMC Biol 5:29

Bro C, Nielsen J (2004) Impact of 'ome' analyses on inverse metabolic engineering. Metab Eng 6:204–211

Brock TD, Brock KM, Belly RT, Weiss RL (1972) *Sulfolobus*: a new genus of sulfur-oxidizing bacteria living at low pH and high temperature. Arch Mikrobiol 84:54–68

Brügger K, Torarinsson E, Redder P, Chen L, Garrett RA (2004) Shuffling of *Sulfolobus* genomes by autonomous and non-autonomous mobile elements. Biochem Soc Trans 32:179–183

Bryson V, Szybalski W (1952) Microbial selection. Science 116:45–46

Bult CJ, White O, Olsen GJ, Zhou L, Fleischmann RD, Sutton GG, Blake JA, FitzGerald LM, Clayton RA, Gocayne JD, Kerlavage AR, Dougherty BA, Tomb JF, Adams MD, Reich CI, Overbeek R, Kirkness EF, Weinstock KG, Merrick JM, Glodek A, Scott JL, Geoghagen NS, Venter JC (1996) Complete genome sequence of the methanogenic archaeon, *Methanococcus jannaschii*. Science 273:1058–1073

Chédin F, Dervyn R, Ehrlich SD, Noirot P (1994) Frequency of deletion formation decreases exponentially with distance between short direct repeats. Mol Microbiol 12:561–569

Chhabra SR, Shockley KR, Conners SB, Scott KL, Wolfinger RD, Kelly RM (2003) Carbohydrate-induced differential gene expression patterns in the hyperthermophilic bacterium *Thermotoga maritima*. J Biol Chem 278:7540–7552

Cohan FM (2001) Bacterial species and speciation. Syst Biol 50:513–524

Coleman ML, Sullivan MB, Martiny AC, Steglich C, Barry K, Delong EF, Chisholm SW (2006) Genomic islands and the ecology and evolution of *Prochlorococcus*. Science 311:1768–1770

Conners SB, Montero CI, Comfort DA, Shockley KR, Johnson MR, Chhabra SR, Kelly RM (2005) An expression-driven approach to the prediction of carbohydrate transport and utilization regulons in the hyperthermophilic bacterium *Thermotoga maritima*. J Bacteriol 187:7267–7282

Conrad TM, Lewis NE, Palsson BØ (2011) Microbial laboratory evolution in the era of genome-scale science. Mol Syst Biol 7:42

Cooper VS, Lenski RE (2000) The population genetics of ecological specialization in evolving *Escherichia coli* populations. Nature 407:736–739

Cooper TF, Rozen DE, Lenski RE (2003) Parallel changes in gene expression after 20,000 generations of evolution in *Escherichia coli*. Proc Natl Acad Sci USA 100:1072–1077

Dénervaud N, Becker J, Delgado-Gonzalo R, Damay P, Rajkumar AS, Unser M, Shore D, Naef F, Maerkl SJ (2013) A chemostat array enables the spatio-temporal analysis of the yeast proteome. Proc Natl Acad Sci USA 110:15842–15847

DeRisi JL, Iyer VR, Brown PO (1997) Exploring the metabolic and genetic control of gene expression on a genomic scale. Science 278:680–686

Draghi JA, Parsons TL, Wagner GP, Plotkin JB (2010) Mutational robustness can facilitate adaptation. Nature 463:353–355

Dragosits M, Mattanovich D (2013) Adaptive laboratory evolution – principles and applications for biotechnology. Microb Cell Fact 12:1475–2859

Dunham MJ, Badrane H, Ferea T, Adams J, Brown PO, Rosenzweig F, Botstein D (2002) Characteristic genome rearrangements in experimental evolution of *Saccharomyces cerevisiae*. Proc Natl Acad Sci USA 99:16144–16149

Elena SF, Lenski RE (2003) Evolution experiments with microorganisms: the dynamics and genetic bases of adaptation. Nat Rev Genet 4:457–469

Elena SF, Wilke CO, Ofria C, Lenski RE (2007) Effects of population size and mutation rate on the evolution of mutational robustness. Evolution 61:666–674

Ferenci T (2007) Bacterial physiology, regulation and mutational adaptation in a chemostat environment. In: Poole RK (ed) Advances in microbial physiology, vol 52. Elsevier Academic Press, San Diego

Ferenci T (2008) Bacterial physiology, regulation and mutational adaptation in a chemostat environment. In: Poole RK (ed) Advances in microbial physiology, vol 53. Elsevier Academic Press, San Diego

Fraser-Liggett CM (2005) Insights on biology and evolution from microbial genome sequencing. Genome Res 15:1603–1610

Garcia-Vallvé S, Palau J, Romeu A (2000) Horizontal gene transfer in bacterial and archaeal complete genomes. Genome Res 10:1719–1725

Gill SR, Fouts DE, Archer GL, Mongodin EF, Deboy RT, Ravel J, Paulsen IT, Kolonay JF, Brinkac L, Beanan M, Dodson RJ, Daugherty SC, Madupu R, Angiuoli SV, Durkin AS, Haft DH, Vamathevan J, Khouri H, Utterback T, Lee C, Dimitrov G, Jiang L, Qin H, Weidman J, Tran K, Kang K, Hance IR, Nelson KE, Fraser CM (2005) Insights on evolution of virulence and resistance from the complete genome analysis of an early methicillin-resistant *Staphylococcus aureus* strain and a biofilm-producing methicillin-resistant *Staphylococcus epidermidis* strain. J Bacteriol 187:2426–2438

Gresham D, Dunham MJ (2014) The enduring utility of continuous culturing in experimental evolution. Genomics 104:399–405

Gresham D, Ruderfer DM, Pratt SC, Schacherer J, Dunham MJ, Botstein D, Kruglyak L (2006) Genome-wide detection of polymorphisms at nucleotide resolution with a single DNA microarray. Science 311:1932–1936

Gresham D, Desai MM, Tucker CM, Jenq HT, Pai DA, Ward A, DeSevo CG, Botstein D, Dunham MJ (2008) The repertoire and dynamics of evolutionary adaptations to controlled nutrient-limited environments in yeast. PLoS Genet 4:e1000303

Grogan DW (1989) Phenotypic characterization of the archaebacterial genus *Sulfolobus*: comparison of five wild-type strains. J Bacteriol 171:6710–6719

Grogan DW, Carver GT, Drake JW (2001) Genetic fidelity under harsh conditions: analysis of spontaneous mutation in the thermoacidophilic archaeon *Sulfolobus acidocaldarius*. Proc Natl Acad Sci USA 98:7928–7933

Hardison RC (2003) Comparative genomics. PLoS Biol 1:e58

Helling RB, Vargas CN, Adams J (1987) Evolution of *Escherichia coli* during growth in a constant environment. Genetics 116:349–358

Herring CD, Raghunathan A, Honisch C, Patel T, Applebee MK, Joyce AR, Albert TJ, Blattner FR, van den Boom D, Cantor CR, Palsson BØ (2006) Comparative genome sequencing of *Escherichia coli* allows observation of bacterial evolution on a laboratory timescale. Nat Genet 38:1406–1412

Hindré T, Knibbe C, Beslon G, Schneider D (2012) New insights into bacterial adaptation through in vivo and in silico experimental evolution. Nat Rev Microbiol 10:352–365

Hong J, Gresham D (2014) Molecular specificity, convergence and constraint shape adaptive evolution in nutrient-poor environments. PLoS Genet 10:e1004041

Huber H, Stetter KO (1998) Hyperthermophiles and their possible potential in biotechnology. J Biotechnol 64:39–52

Kadam SV, Wegener-Feldbrügge S, Søgaard-Andersen L, Velicer GJ (2008) Novel transcriptome patterns accompany evolutionary restoration of defective social development in the bacterium *Myxococcus xanthus*. Mol Biol Evol 25:1274–1281

Kao KC, Sherlock G (2008) Molecular characterization of clonal interference during adaptive evolution in asexual populations of *Saccharomyces cerevisiae*. Nat Genet 40:1499–1504

Kidwell MG, Lisch DR (2000) Transposable elements and host genome evolution. Trends Ecol Evol 15:95–99

Kinnersley M, Wenger J, Kroll E, Adams J, Sherlock G, Rosenzweig F (2014) Clonal reinforcement drives evolution of a simple microbial community. PLoS Genet 10:e1004430

Kunin V, Ouzounis CA (2003) The balance of driving forces during genome evolution in prokaryotes. Genome Res 13:1589–1594

Kurtz S, Choudhuri JV, Ohlebusch E, Schleiermacher C, Stoye J, Giegerich R (2001) REPuter: the manifold applications of repeat analysis on a genomic scale. Nucleic Acids Res 29:4633–4642

Kussell E (2013) Evolution in microbes. Annu Rev Biophys 42:493–514

Kvitek DJ, Sherlock G (2013) Whole genome, whole population sequencing reveals that loss of signaling networks is the major adaptive strategy in a constant environment. PLoS Genet 9:e1003972

Lang GI, Rice DP, Hickman MJ, Sodergren E, Weinstock GM, Botstein D, Desai MM (2013) Pervasive genetic hitchhiking and clonal interference in forty evolving yeast populations. Nature 500:571–574

Lashkari DA, DeRisi JL, McCusker JH, Namath AF, Gentile C, Hwang SY, Brown PO, Davis RW (1997) Yeast microarrays for genome wide parallel genetic and gene expression analysis. Proc Natl Acad Sci USA 94:13057–13062

Le Fourn C, Fardeau ML, Ollivier B, Lojou E, Dolla A (2008) The hyperthermophilic anaerobe *Thermotoga maritima* is able to cope with limited amount of oxygen: insights into its defence strategies. Environ Microbiol 10:1877–1887

Lenski RE, Mongold JA, Sniegowski PD, Travisano M, Vasi F, Gerrish PJ, Schmidt TM (1998) Evolution of competitive fitness in experimental populations of *E. coli*: what makes one genotype a better competitor than another? Antonie van Leeuwenhoek 73:35–47

Lorantfy B, Seyer B, Herwig C (2014) Stoichiometric and kinetic analysis of extreme halophilic Archaea on various substrates in a corrosion resistant bioreactor. New Biotechnol 31:80–89

Lovett ST, Drapkin PT, Sutera VA Jr, Gluckman-Peskind TJ (1993) A sister-strand exchange mechanism for recA-independent deletion of repeated DNA sequences in *Escherichia coli*. Genetics 135:631–642

Lovett ST, Gluckman TJ, Simon PJ, Sutera VA Jr, Drapkin PT (1994) Recombination between repeats in *Escherichia coli* by a recA-independent, proximity-sensitive mechanism. Mol Gen Genet 245:294–300

Maezato Y, Dana K, Blum P (2011) Engineering thermoacidophilic archaea using linear DNA recombination. Methods Mol Biol 765:435–445

Maezato Y, Johnson T, McCarthy S, Dana K, Blum P (2012) Metal resistance and lithoautotrophy in the extreme thermoacidophile *Metallosphaera sedula*. J Bacteriol 194:6856–6863

McCarthy S, Ai C, Wheaton G, Tevatia R, Eckrich V, Kelly R, Blum P (2014) Role of an archaeal PitA transporter in the copper and arsenic resistance of *Metallosphaera sedula*, an extreme thermoacidophile. J Bacteriol 196:3562–3570

McCarthy S, Johnson T, Pavlik B, Payne S, Schackwitz W, Martin J, Lipzen A, Keffler E, Blum P (2015) Expanding the limits of thermoacidophily in the archaeon Sulfolobus solfataricus by adaptive evolution. Appl Environ Microbiol. PII: AEM.03225-15 [Epub ahead of print], PMID: 26590281

Monod J (1942) Recherche sur la croissance des cultures bactériennes. Hermann and Cie, Paris

Nelson KE, Clayton RA, Gill SR, Gwinn ML, Dodson RJ, Haft DH, Hickey EK, Peterson JD, Nelson WC, Ketchum KA, McDonald L, Utterback TR, Malek JA, Linher KD, Garrett MM, Stewart AM, Cotton MD, Pratt MS, Phillips CA, Richardson D, Heidelberg J, Sutton GG, Fleischmann RD, Eisen JA, White O, Salzberg SL, Smith HO, Venter JC, Fraser CM (1999) Evidence for lateral gene transfer between Archaea and bacteria from genome sequence of *Thermotoga maritima*. Nature 399:323–329

Novick A, Szilard L (1950) Description of the chemostat. Science 112:715–716

Paquin C, Adams J (1983) Frequency of fixation of adaptive mutations is higher in evolving diploid than haploid yeast populations. Nature 302:495–500

Philippe N, Crozat E, Lenski RE, Schneider D (2007) Evolution of global regulatory networks during a long-term experiment with *Escherichia coli*. Bioessays 29:846–860

Polz MF, Hunt DE, Preheim SP, Weinreich DM (2006) Patterns and mechanisms of genetic and phenotypic differentiation in marine microbes. Philos Trans R Soc Lond B Biol Sci 361:2009–2021

Prosser JI, Bohannan BJM, Curtis TP, Ellis RJ, Firestone MK, Freckleton RP, Green JL, Green LE, Killham K, Lennon JJ, Osborn AM, Solan M, van der Gast CJ, Young JPW (2007) The role of ecological theory in microbial ecology. Nat Rev Microbiol 5:384–392

Redder P, Garrett RA (2006) Mutations and rearrangements in the genome of *Sulfolobus solfataricus* P2. J Bacteriol 188:4198–4206

Ren Q, Chen K, Paulsen IT (2007) TransportDB: a comprehensive database resource for cytoplasmic membrane transport systems and outer membrane channels. Nucleic Acids Res 35:D274–D279

Riehle MM, Bennett AF, Long AD (2001) Genetic architecture of thermal adaptation in *Escherichia coli*. Proc Natl Acad Sci USA 98:525–530

Rigaut G, Shevchenko A, Rutz B, Wilm M, Mann M, Seraphin B (1999) A generic protein purification method for protein complex characterization and proteome exploration. Nat Biotechnol 17:1030–1032

Rolfsmeier M, Blum P (1995) Purification and characterization of a maltase from the extremely thermophilic crenarchaeote *Sulfolobus solfataricus*. J Bacteriol 177:482–485

Romero D, Palacios R (1997) Gene amplification and genomic plasticity in prokaryotes. Annu Rev Genet 31:91–111

Rosenzweig RF, Sharp RR, Treves DS, Adams J (1994) Microbial evolution in a simple unstructured environment: genetic differentiation in *Escherichia coli*. Genetics 137:903–917

Rusch A, Walpersdorf E, de Beer D, Gurrieri S, Amend JP (2005) Microbial communities near the oxic/anoxic interface in the hydrothermal system of Vulcano Island, Italy. Chem Geol 224:169–182

Sauer U (2001) Evolutionary engineering of industrially important microbial phenotypes. In: Nielsen J, Eggeling L, Dynesen J, Gárdonyi M, Gill RT, Graaf AA, Hahn-Hägerdal B, Jönsson LJ, Khosla C, Licari R, McDaniel R, McIntyre M, Miiller C, Nielsen J, Cordero Otero RR, Sahm H, Sauer U, Stafford DE, Stephanopoulos G, Wahlbom CE, Yanagimachi KS, Zyl WH (eds) Metabolic engineering. Springer, Berlin/Heidelberg

Schneider D, Lenski RE (2004) Dynamics of insertion sequence elements during experimental evolution of bacteria. Res Microbiol 155:319–327

Schröder C, Selig M, Schönheit P (1994) Glucose fermentation to acetate, CO_2 and H_2 in the anaerobic hyperthermophilic eubacterium *Thermotoga maritima*: involvement of the Embden-Meyerhof pathway. Arch Microbiol 161:460–470

Schwikowski B, Uetz P, Fields S (2000) A network of protein-protein interactions in yeast. Nat Biotechnol 18:1257–1261

Singh R, Gradnigo J, White D, Lipzen A, Martin J, Schackwitz W, Moriyama E, Blum P (2015) Complete genome sequence of an evolved *Thermotoga maritima* isolate. Genome Announc 3:e00557–15

Singh R, White D, Kelly R, Noll K, Blum P. Uncoupling fermentative synthesis of molecular hydrogen from biomass formation in *Thermotoga maritima*. (in preparation)

Smith JM, Smith NH, O'Rourke M, Spratt BG (1993) How clonal are bacteria? Proc Natl Acad Sci USA 90:4384–4388

Sniegowski PD, Gerrish PJ, Lenski RE (1997) Evolution of high mutation rates in experimental populations of *E. coli*. Nature 387:703–705

Suyama M, Bork P (2001) Evolution of prokaryotic gene order: genome rearrangements in closely related species. Trends Genet 17:10–13

Tatum EL, Lederberg J (1947) Gene recombination in the bacterium *Escherichia coli*. J Bacteriol 53:673–684

Tettelin H, Masignani V, Cieslewicz MJ, Donati C, Medini D, Ward NL, Angiuoli SV, Crabtree J, Jones AL, Durkin AS, Deboy RT, Davidsen TM, Mora M, Scarselli M, Margarit Rosy I, Peterson JD, Hauser CR, Sundaram JP, Nelson WC, Madupu R, Brinkac LM, Dodson RJ, Rosovitz MJ, Sullivan SA, Daugherty SC, Haft DH, Selengut J, Gwinn ML, Zhou L, Zafar N, Khouri H, Radune D, Dimitrov G, Watkins K, O'Connor KJ, Smith S, Utterback TR, White O, Rubens CE, Grandi G, Madoff LC, Kasper DL, Telford JL, Wessels MR, Rappuoli R, Fraser CM (2005) Genome analysis of multiple pathogenic isolates of Streptococcus agalactiae: implications for the microbial "pan-genome". Proc Natl Acad Sci USA 102:13950–13955

Toprak E, Veres A, Michel JB, Chait R, Hartl DL, Kishony R (2012) Evolutionary paths to antibiotic resistance under dynamically sustained drug selection. Nat Genet 44:101–105

Treves DS, Manning S, Adams J (1998) Repeated evolution of an acetate-crossfeeding polymorphism in long-term populations of *Escherichia coli*. Mol Biol Evol 15:789–797

Ussery DW, Binnewies TT, Gouveia-Oliveira R, Jarmer H, Hallin PF (2004) Genome update: DNA repeats in bacterial genomes. Microbiology 150:3519–3521

van Ham SM, van Alphen L, Mooi FR, van Putten JP (1993) Phase variation of *H. influenzae* fimbriae: transcriptional control of two divergent genes through a variable combined promoter region. Cell 73:1187–1196

Wenger JW, Piotrowski J, Nagarajan S, Chiotti K, Sherlock G, Rosenzweig F (2011) Hunger artists: yeast adapted to carbon limitation show trade-offs under carbon sufficiency. PLoS Genet 7:e1002202

Wertz JE, Goldstone C, Gordon DM, Riley MA (2003) A molecular phylogeny of enteric bacteria and implications for a bacterial species concept. J Evol Biol 16:1236–1248

Wilbanks EG, Larsen DJ, Neches RY, Yao AI, Wu CY, Kjolby RA, Facciotti MT (2012) A workflow for genome-wide mapping of archaeal transcription factors with ChIP-seq. Nucleic Acids Res 40(10):e74

Woods RJ, Barrick JE, Cooper TF, Shrestha U, Kauth MR, Lenski RE (2011) Second-order selection for evolvability in a large *Escherichia coli* population. Science 331:1433–1436

Worthington P, Blum P, Perez-Pomares F, Elthon T (2003) Large-scale cultivation of acidophilic hyperthermophiles for recovery of secreted proteins. Appl Environ Microbiol 69:252–257

Wurtzel O, Sapra R, Chen F, Zhu Y, Simmons BA, Sorek R (2010) A single-base resolution map of an archaeal transcriptome. Genome Res 20:133–141

Zhaxybayeva O, Swithers KS, Lapierre P, Fournier GP, Bickhart DM, DeBoy RT, Nelson KE, Nesbø CL, Doolittle WF, Gogarten JP, Noll KM (2009) On the chimeric nature, thermophilic origin, and phylogenetic placement of the Thermotogales. Proc Natl Acad Sci USA 106:5865–5870

Chapter 23
Solid-Binding Peptides: Immobilisation Strategies for Extremophile Biocatalysis in Biotechnology

Andrew Care, Peter L. Bergquist, and Anwar Sunna

1 Introduction

Biocatalysis is a widely used tool for chemical syntheses that are complementary to purely chemical methods. Enzymes are efficient and versatile biological catalysts that operate under ambient conditions while displaying high regio-, and enantio-, chemo-, and stereo- selectivity. In contrast to chemical catalysts, no protection or de-protection of functional groups is required and shorter routes to an end product are achieved with reduced waste products from side reactions. Accordingly, enzyme-catalysed processes have become increasingly significant in several large-scale industrial areas as the interest and emphasis on 'green chemistry' has developed.

In the past, enzymes generally have been studied either in solution or immobilised on a solid support. Significant progress has been made on a range of immobilisation

A. Care
ARC Centre of Excellence for Nanoscale BioPhotonics (CNBP), Macquarie University, Sydney, NSW, Australia

P.L. Bergquist
Department of Molecular Medicine and Pathology, Medical School, University of Auckland, Auckland, New Zealand

Department of Chemistry and Biomolecular Sciences, Macquarie University, Sydney, NSW, Australia

Biomolecular Frontiers Research Centre, Macquarie University, Sydney, NSW, Australia

A. Sunna (✉)
Department of Chemistry and Biomolecular Sciences, Macquarie University, Sydney, NSW, Australia

Biomolecular Frontiers Research Centre, Macquarie University, Sydney, NSW, Australia
e-mail: anwar.sunna@mq.edu.au

P.H. Rampelotto (ed.), *Biotechnology of Extremophiles*, Grand Challenges in Biology and Biotechnology 1, DOI 10.1007/978-3-319-13521-2_23

supports (Brady and Jordaan 2009; Hanefeld et al. 2009). The widespread industrial use of free enzymes has been limited by a number of factors such as stability problems at high process temperatures, poor performance in non-aqueous media, difficulties in recovery and reuse, all contributing to the high cost of the catalyst. Immobilised enzymes on the other hand provide several advantages over free enzymes including repetitive use in single batch and continuous bioprocesses, possible reuse over multiple cycles, improved enzyme stabilisation over the soluble free form, rapid termination of reactions and easy enzyme removal, as well as elimination of enzyme–product contamination (Raghuvanshi and Gupta 2010). However the immobilisation of proteins remains a particularly challenging task, mainly due to their heterogeneous nature and the marginal stability of the active tertiary structure over the denatured (and inactive) random coil structure. Recent reviews summarise progress and the status of the field (Camarero 2008; Hanefeld et al. 2009).

1.1 *Immobilisation of Proteins*

The literature is replete with descriptions of immobilised proteins (usually but not exclusively enzymes) and the methods used can be divided into (a) adsorption (b) covalent binding (c) entrapment/encapsulation; and (d) cross-linking. Cowan and Fernandez-Lafuente (2011) published a comprehensive account of thermophilic enzymes that have been immobilised, largely emphasising chemical immobilisation on solid supports. A summary of selected immobilised thermophilic enzymes for each method discussed is presented in Table 23.1.

1.1.1 Adsorption

Most of the methods available for immobilising proteins onto solid supports traditionally have relied on non-specific physical adsorption. Different types of interactions including, H-bonding, van der Waals forces and ionic and hydrophobic interactions mediate the adsorption of proteins onto the surface of water-insoluble carriers (Hanefeld et al. 2009). The main advantage of this technique is that is simple, performed under mild conditions and enzymes do not require pre-treatment or chemical modifications. However, a main disadvantage of this method is that the adsorbed enzyme may leach from the carrier in aqueous media due to the weak binding interactions involved in the immobilisation process.

A number of thermostable lipases have been immobilised successfully on hydrophobic supports by physical adsorption (Palomo et al. 2004; Wilson et al. 2006). For example, the adsorption of lipases from *Thermus thermophilus* and *Thermus aquaticus* onto octadecyl-Sepabeads was shown to enhance catalytic activity 10-fold at mesophilic temperatures and increase functional stability (Palomo et al. 2004). Conversely, immobilisation of these lipases via ionic adsorption and covalent attachment resulted in a 20–30 % decrease in enzyme activity.

Table 23.1 Methods used in the immobilisation of thermophilic enzymes

Enzyme	Organism	Immobilisation method	Reference
Alcohol dehydrogenase	*Thermoanaerobacter brockii*	Adsorption	Miroliaei (2007)
Alcohol dehydrogenase	*Thermus thermophilus* HB27	Covalent binding Adsorption	Rocha-Martin et al. (2009)
Alcohol oxidoreductase	*Sulfolobus solfataricus*	Covalent binding	Raia et al. (1995)
Aminoacylase	*Thermococcus litoralis*	Entrapment/encapsulation Adsorption	Ngamsom et al. (2010)
L-arabinose isomerase	*Thermoanaerobacter mathranii*	CLEAs	Jørgensen et al. (2004)
β-glucosidase	*Caldicellulosiruptor saccharolyticus*	Covalent binding	Tran et al. (2013)
Caldolysin	*Thermus aquaticus* T351	Covalent binding Adsorption	Cowan et al. (1987)
Catalase	*Thermus thermophilus*	Covalent binding CLEAs	Hidalgo et al. (2003)
CelB	*Caldicellulosiruptor saccharolyticus*	Covalent binding	Tran et al. (2013)
Dehalogenase	*Sulfolobus tokodaii*	Adsorption	Bachas-Daunert et al. (2009)
Diguanylate cyclase	*Thermotoga maritima*	Entrapment/encapsulation	Pasunooti et al. (2010)
α-galactosidase	*Thermus sp. strain* T2	Covalent binding Adsorption	Filho et al. (2008)
β-galactosidase	*Sulfolobus solfataricus*	Covalent binding	Jørgensen et al. (2004)
	Thermus sp. strain T2	Covalent binding Adsorption	Pessela et al. (2004)
α-glucosidase	*Thermococcus* AN1	Chemical modification	Piller et al. (1996)
β-glucosidase	*Pyrococcus furiosus*	Covalent binding	Fischer et al. (1996)
Glutamate dehydrogenase	*Thermus thermophilus*	Covalent binding	Bolivar et al. (2009)
Hydrogenase	*Pyrococcus furiosus*	Adsorption	Greiner et al. (2003)
(+)-λ-lactamase	*Sulfolobus solfataricus* MT4	CLEAs	Hickey et al. (2009)
L-lactate dehydrogenase	*Clostridium thermohydrosulfuricum*	Entrapment/encapsulation	Lehn and Schmidt (1997)
	Thermus caldophilus GK24	Chemical modification	Taguchi et al. (1984)
Lipase	*Thermus thermophilus*	Adsorption	Palomo et al. (2004)
	Thermus aquaticus	Adsorption	Palomo et al. (2004)
Proteinase	*Thermus* Rt41A	Covalent binding	Wilson et al. (1994)
Pyrophosphatase	*Pyrococcus furiosus*	Covalent binding	Dong et al. (2014)
Xylanase	*Thermotoga* FjSS3-B.1	Covalent binding	Simpson et al. (1991)
D-xylose isomerase	*Thermus aquaticus*	Covalent binding CLEAs	Fernandez-Lafuente (2009)
	Thermotoga maritima	Covalent binding	Bandlish et al. (2002)
	Thermotoga neapolitana 5068	Covalent binding	Bandlish et al. (2002)

1.1.2 Covalent Binding

This immobilisation technique involves chemical modification of the enzyme to facilitate the formation of covalent bonds between the enzyme and the carrier. Covalent binding is mediated mainly by functional groups (e.g., amino, hydroxyl, thiol, carboxyl, sulfhydryl) on the enzyme. As a pre-requisite for this technique, the modification must be performed under conditions that do not affect the enzyme's catalytic function. In particular, the active site of the enzyme should not be affected by the chemical modification or reagents used. Covalent attachment will prevent the enzymes leaching from the support, but this method has the disadvantage that the enzyme molecules may be partially or fully denatured by the chemical process, rendering the immobilised enzyme less effective than the protein in solution. The organisation and orientation of the immobilised enzyme on a solid support is very difficult to control by covalent bonding-based immobilisation techniques. The implication of this lack of control is that enzyme activity might be reduced if essential amino acid residues located near to the active site are involved in immobilisation of the enzyme. Furthermore, lack of orientation control may result in limited access of the substrate to the active site of the enzyme and accordingly, immobilised enzymes generally may display lower activity and higher apparent Michaelis constants when compared to free enzymes (Kress et al. 2002; Cao 2006).

A thermophilic inorganic pyrophosphatase from the archaeon *Pyrococcus furiosus* has been attached covalently to amino-functionalised silica beads using glutaraldehyde as a coupling agent (Dong et al. 2014). In contrast to the free enzyme, the immobilised enzyme exhibited higher activity, and increased pH and thermal stability. The immobilised enzyme also retained over 50 % of its initial activity after the hydrolysis of inorganic pyrophosphate in 12 sequential reactions at 95 °C.

One alternative approach to covalent coupling reactions is plasma immersion ion implantation (PIII), a physical process that creates active radicals on the surfaces of materials that form covalent bonds with biomolecules without the need for chemical linkers. This technique was used recently to immobilise the cellulase CelB and β-glucosidase BglA from the thermophilic bacterium *Caldicellulosiruptor saccharolyticus* onto polystyrene sheets (Tran et al. 2013). The immobilised enzymes retained their functionality and were combined to hydrolyse carboxymethyl cellulose to glucose. However, it should be noted that PIII treatments are time-consuming and impractical on a large scale.

1.1.3 Entrapment/Encapsulation

Entrapment of an enzyme involves retention in a polymer network such as gels (e.g., polyacrylamide, calcium alginate, and gelatine), a silica sol-gel, a silica cage or a hollow fibre and often involves the synthesis of the polymer matrix in the presence of the enzyme. The polymeric network must allow for the reaction to occur by diffusion of reactants and products while avoiding leaching of the enzyme into the medium. Frequently, the physical conditions are too weak to prevent leaching and covalent attachment may be required. The main advantages of this method are decreased leaching, improved stability and free diffusion of low molecular weight substrates. However, mass transfer resistance to substrates, products and inhibitors can limit entrapment while restricted access of macromolecular substrates represents another disadvantage. This immobilisation technique is treated in more detail by others (Betancor and Luckarift 2008; Hudson et al. 2008; Hartmann and Kostrov 2013; Magner 2013). It was used to immobilise a thermoalkalophilic esterase enzyme isolated from *Geobacillus* sp. in calcium alginate beads, which were then coated with a layer of silicate (Gülay and Şanlı-Mohamed 2012). This coating prevented enzyme leaching, resulting in improved enzyme loading, immobilisation yields and reusability. Moreover, the entrapped esterase exhibited better operational, thermal and storage stabilities than the free enzyme.

1.1.4 Cross-Linking

Crosslinking with a bifunctional reagent (e.g. glutaraldehyde) has been used to prepare cross-linked enzyme crystals (CLECs) but more recently cross-linked enzyme aggregates (CLEAs) as carrier-free entities have been developed to simplify the procedure and simultaneously effect purification of the enzyme (Sheldon 2011; Sheldon and van Pelt 2013). CLEAs are prepared by aggregating the enzymes by addition of common protein precipitants (e.g., ammonium sulphate, polyethylene glycol, ethanol). These physical aggregates are held together by non-covalent interactions and can redissolve when dispersed in an aqueous medium. A subsequent cross-linking step assures that the insolubility and pre-organised structure of the aggregate is maintained through covalent bond formation and thus the catalytic activity of the aggregated enzyme is preserved. The main advantages of this technique are the production of strong systems with availability of highly concentrated enzyme activity, low production costs and relatively high stability of the system. However, the ideal conditions for CLEAs formation vary from one enzyme to another and must be tested to avoid loss or reduction of catalytic activity.

Thermostable CLEAs of l-arabinose isomerase from *Thermoanaerobacter mathranii* and a β-glycosidase from *Sulfolobus solfataricus* were immobilised by cross-linking with glutaraldehyde and polyethylenimine (Jørgensen et al. 2004). In combination, these CLEAs were shown to mediate the two-step conversion of D-galactose to D-tagatose at 65 °C in a single reaction with a 38 % conversion.

2 Immobilisation Using Solid-Binding Peptides

Solid-binding peptides (SBPs) are short amino acid sequences that selectively recognise and bind strongly to the surfaces of a wide range of solid materials, including metals (Naik et al. 2002b; Hnilova et al. 2008; Heinz et al. 2009; Li et al. 2009; Chiu et al. 2010; Forbes et al. 2010), carbon materials (Wang et al. 2003; Kulp et al. 2005; Pender et al. 2006; Cui et al. 2010), semiconductors (Whaley et al. 2000; Lee et al. 2002; Estephan et al. 2011), polymers and minerals (Li et al. 2002; Serizawa et al. 2007; Gungormus et al. 2008; Roy et al. 2008). These unique peptides can act as molecular linkers to direct the orientated immobilisation of biomolecules (e.g., proteins and enzymes) onto solid matrices without compromising their biological activity. Also, they have the capacity to regulate the synthesis and assembly of nanostructures with varying complexities and functions. Several recent reviews on SBPs give comprehensive accounts of the status of the field (Sarikaya et al. 2003; Baneyx and Schwartz 2007; Tamerler et al. 2010; Care et al. 2015). There has been a recent interest in utilising these peptides for inorganic materials synthesis, assembly and formation under mild ambient conditions (Ball 2001; Seeman and Belcher 2002; Sarikaya et al. 2003). Consequently, SBPs have been used as molecular tools in a number of nanobiotechnological applications, such as the functionalisation of nanomaterials (Lu et al. 2014); generation of inorganic nanocomposites (Inoue et al. 2014) and hybrid biomaterials (Chen et al. 2014); and nanobioelectronics (Dang et al. 2011; Sano et al. 2013).

SBPs can exhibit high affinity and selectivity for their respective solids with binding affinities (K_d) in the nM to sub-μM range. They display also large negative binding energies (Tamerler et al. 2006; Seker et al. 2007; Hnilova et al. 2008; Kacar et al. 2009a; So et al. 2009; Tang et al. 2013). This strong binding is the combined result of multiple non-covalent interactions, including electrostatic, hydrophobic, polar and hydrogen bonds. However, many of the mechanisms underlying SBP recognition, selectivity and affinity are still not well understood due to the dynamic and complex nature of the peptide-material interface. Thus far, the diverse properties of peptides (e.g., composition, sequence, conformation, structure, chemistry, and charge) in conjunction with the surfaces of solid materials (e.g., oxidation state, charge, crystallographic orientation, and defects) and the surrounding solution (e.g., water or solvent) have all been shown to contribute to binding interactions (Naik et al. 2002b; Goede et al. 2004; Nel et al. 2009; Sapsford et al. 2013; Tang et al. 2013; Slocik and Naik 2014; Care et al. 2015). Further complexity arises from the dynamic nature of the assembled peptide–material surface interface. Peptides are in constant motion in this interface, tending to diffuse, reorient and adapt themselves to the lowest energy structures.

The majority of SBPs have been isolated using combinatorial display technologies (Brown 1997; Whaley et al. 2000; Naik et al. 2002a; Sarikaya et al. 2003; Thai et al. 2004), specifically phage display (Smith 1985; Hoess 2001) and cell-surface display (Charbit et al. 1986; Wittrup 2001, see Fig. 23.1). These methods are relatively simple, robust and allow large libraries of peptides to be screened for material binding activity. Once selected, whole phage particles (Whaley et al. 2000; Huang et al. 2005) or bacterial cells (Park et al. 2009) that display the peptide can be used as solid-binding agents. The peptide can be synthesised using solid-state synthetic methods (Nochomovitz et al. 2010); or the peptide as part of a fusion protein can be used to facilitate the immobilisation of biomolecules onto solid surfaces (Ko et al. 2009). Recently, further advances have been made in selection of new SBPs using computational approaches (e.g., knowledge-based design, molecular modelling and bioinformatics). However, the success of this approach requires bioinformatics tools and computational approaches that are combined with the availability of more

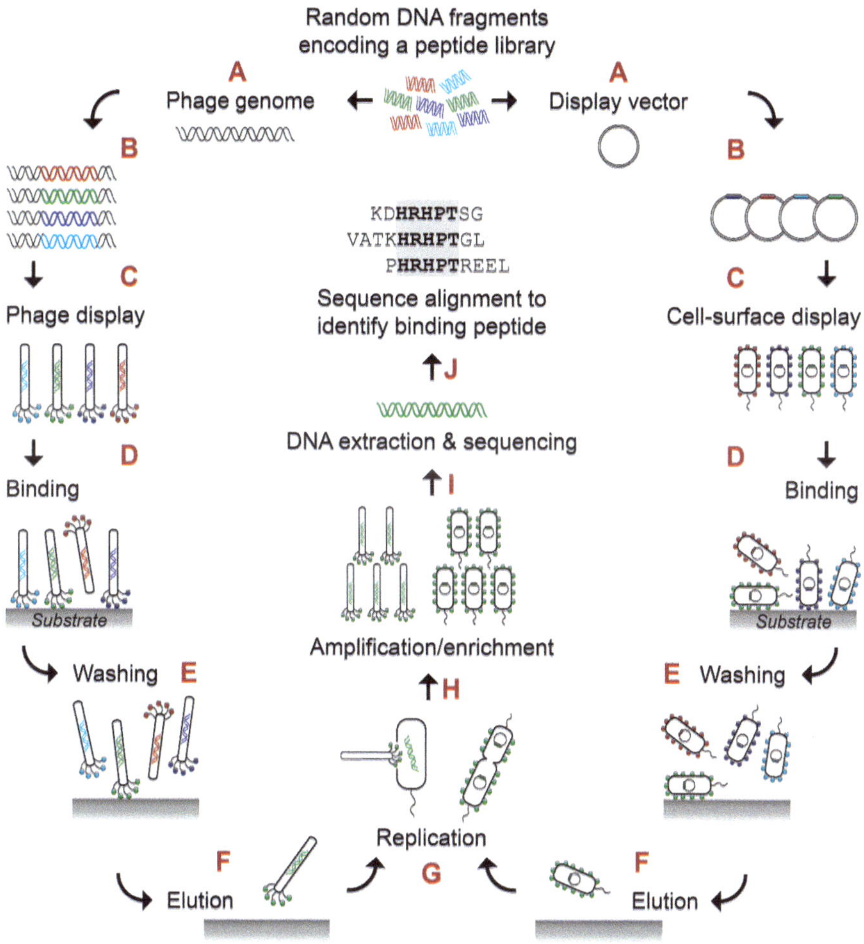

Fig. 23.1 Illustration showing the phage (*left-hand side*) and cell (*right-hand side*) surface display technologies used to isolate peptides that bind to solid substrates. To screen for SBPs, large and diverse peptide libraries (10^8–10^9) encoded by randomly generated DNA fragments are fused to proteins expressed on the surface of phage particles or bacterial cells (*A–C*). Thus, each phage or cell displays multiple copies of a unique peptide that is firmly linked to its encoding gene packaged within. Peptides that exhibit high affinity towards a solid substrate can be isolated from these libraries using a selection process known as biopanning where a library of phage or cell-displayed peptides is incubated with a solid substrate (*D*), phage or bacteria that do not bind are removed by washing (*E*), and those bound tightly to the substrate are eluted (*F*). The eluted phage or bacteria are then amplified (*G–H*) and subjected to additional binding and amplification steps (*C–H*) to enrich the population of peptides that bind. After three to four rounds of biopanning, the DNA of individual colonies or plaques are sequenced to allow identification of the amino acid sequence of peptides that bind tightly to the solid substrate (*I–J*) (Adapted from Sarikaya et al. 2003)

experimentally-determined binding data to allow the generation of peptides with improved properties (Oren et al. 2007, 2010; Evans et al. 2008; Notman et al. 2010; Masica et al. 2010; Walsh 2014). However, up until now the practical application of SBPs has not been extended to large-scale industrial biotechnology.

SBPs have been shown to control the orientation at which an enzyme attaches to a solid surface without compromising its native conformation and catalytic function (Yang et al. 2011 and Table 23.2). For example, Kacar et al. (2009b) investigated the immobilisation of alkaline phosphatase (AP) on gold surfaces via a gold-binding peptide (GBP1). In contrast to unmodified AP, GBP1-AP exhibited specific binding affinity to gold and assembled into densely packed uniform monolayers upon immobilisation that displayed better orientation and higher enzymatic activity per unit of area than the enzyme without GBP1. GBP1 also has been shown to facilitate the immobilisation of enzymes without affecting their native conformation (Yang et al. 2011). SBPs with binding affinity to either silica (Naik et al. 2002a) or iron oxide (Brown 1992) have been used to facilitate the immobilisation of a bioremediation-related enzyme (haloalkane dehalogenase) onto silica-coated or uncoated iron oxide magnetic nanoparticles (Johnson et al. 2008). In both cases, the enzyme retained its activity. The attachment of enzymes to magnetic nanomaterials allows their straightforward recovery and reuse in multiple reactions (Puri et al. 2013).

2.1 Application of SBPs in Biocatalysis

Many insoluble materials can be used as carriers for immobilisation of enzymes. However, the mode of interaction between the enzyme and the carrier can have a critical effect on the stability and catalytic efficiency of the immobilised enzyme. SBPs bind their corresponding solids under mild conditions and without the need for any chemical modifications or physical treatments, which not only ensures enzyme integrity but also supports their introduction into environmentally-friendly biocatalytic processes. Industrial-scale biocatalysis generally requires materials that are inexpensive and stable at high temperatures, pressures or extreme pH's. Inorganic materials are excellent carriers due to their structural and operational stability and their lack of susceptibility to microbial degradation. However, laborious trial-and-error methods are required to identify the ideal immobilisation strategy for specific enzyme/carrier combinations (Garcia-Galan et al. 2011; Sassolas et al. 2012). SBPs, on the other hand, can be selected for virtually any material using combinatorial display technologies, enabling the attachment of enzymes onto any preselected matrix. Despite these promising characteristics, the practical application of SBPs currently is limited to the immobilisation of enzymes in chemo/biosensors and bioassays that function under ambient conditions (Yang et al. 2011; Cetinel et al. 2013, see Table 23.2). These accounts of immobilisation generally employ exotic and expensive laboratory-based matrices and may not be realistic economically for large-scale processes. There are few reports regarding inexpensive matrices and little is known about the use of SBPs for the immobilisation of thermophilic enzymes and their feasibility in industrial-scale biocatalysis.

Throughout this chapter we will outline our views on the potential use of SBPs as parts of fusion proteins in the isolation, purification and reuse of enzymes from one class of extremophiles, thermophiles, using readily-available and inexpensive

Table 23.2 Enzymes immobilised using SBPs

Binding matrix	SBP sequence	Enzyme	Source	Thermophilicity[a]	References
Diamond-like carbon	HFYPGANRSTTQGGGS ANLHQTAASAKNSAPQK-SENRKVPFYSHSRTRENNRSIYTA	Alkaline phosphatase	*Escherichia coli*	Mesophilic	Gabryelczyk et al. (2013), Gabryelczyk et al. (2014)
Gold	MHGKTQATSGTIQS	Alkaline phosphatase	*Escherichia coli*	Mesophilic	Brown (1997), Kacar et al. (2009b)
Silica-coated iron oxide nanoparticles	MSPHPHPRHHHT	Haloalkane dehalogenase	*Xanthobacter autotrophicus*	Mesophilic	Johnson et al. (2008)
Iron oxide nanoparticles	RRTVKHHVN	Haloalkane dehalogenase	*Xanthobacter autotrophicus*	Mesophilic	Johnson et al. (2008)
Gold	WALRRSIRRQSY	L-lactate dehydrogenase	*Bacillus stearothermophilus*	Thermophilic[b]	Cetinel et al. (2013)
Synthetic zeolite	VKTQATSREEPPRLPSKHRP	Alkaline phosphatase	*Escherichia coli*	Mesophilic	Nygaard et al. (2002)
Gold nanoparticles	MHGKTQATSGTIQS	Organophosphorus hydrolase	*Flavobacterium* sp	Mesophilic	Yang et al. (2011)

[a]Thermophilicity of the source organism
[b]This enzyme is thermostable up to 60 °C, however all enzymatic reactions were performed at room temperature only

silica-based matrices as examples of low-cost immobilised biocatalysts for industrial biotechnological processes. We will explore also the concept of a molecular toolbox of suitable parts that are compatible with extremophile enzymes that could be employed for the design of biocatalytic modules in a number of areas of applied biotechnology.

2.2 *Silica-Binding SBPs*

Silica-based solid matrices like silica, zeolite and mesoporous silica are considered the most suitable inorganic matrices for enzyme immobilisation in industrial processes (Blanco et al. 2004; Ho et al. 2004; Pierre 2004). These inorganic supports offer unique properties including high mechanical strength and stability, are chemically inert and can be used under a wide range of operating pressures and conditions. They also possess high surface areas, which can be modified chemically and functionalised for ligand attachment.

Silica adsorption is one of the simplest and most economical pathways to immobilise enzymes but the weak binding interactions involved may result in enzyme leaching in an aqueous medium. Furthermore, as with the case of other non-specific protein attachment methods, the protein may not end up immobilised in a correctly orientated manner. Recently, a selective and more specific technique based on silica-specific SBPs has been developed for immobilising proteins and enzymes onto silica-based inorganic matrices (Brown 1992; Taniguchi et al. 2007; Sunna et al. 2013a; Abdelhamid et al. 2014; Coyle and Baneyx 2014). For example, Sunna et al. (2013a) reported on an SBP that displayed high binding affinity towards natural zeolite and commercial-grade synthetic zeolites, silica and silica-containing materials. They genetically fused this SBP sequence [(VKTQATSREEPPRLPSKHRPG)$_4$ VKTQTAS (Nygaard et al. 2002)], to the C-terminus of Green Fluorescent Protein (GFP, Table 23.5) and the resulting fusion protein was shown to bind selectively to natural zeolite (Fig. 23.2). Similarly, they fused the SBP to an antibody-binding protein and the resulting fusion protein was shown to mediate the orientated immobilisation of antibodies onto silica-containing materials within minutes and without the need for complex surface chemical modification (Sunna et al. 2013b; Care et al. 2014a; Lu et al. 2014). This system was used to functionalise silica-coated lanthanide-doped upconversion nanocrystals with antibodies to facilitate the binding and luminescent labelling of the waterborne protozoan pathogen, *Giardia lambia* (Lu et al. 2014).

Silica-binding SBPs have been used also for silica-based protein affinity purification (Fig. 23.3). Abdelhamid et al. (2014) incorporated the C-terminal 14-aa peptide CotB1p (SGRARAQRQSSRGR) from the *Bacillus cereus* spore coat protein CotB1 to mCherry as a model fusion protein to demonstrate the efficiency of this affinity purification method. The purity and yield obtained using the CotB1p silica-binding SBP was comparable to values reported for conventional affinity tags (Lichty et al. 2005). Similarly, Coyle and Baneyx (2014) reported a new dodecapeptide

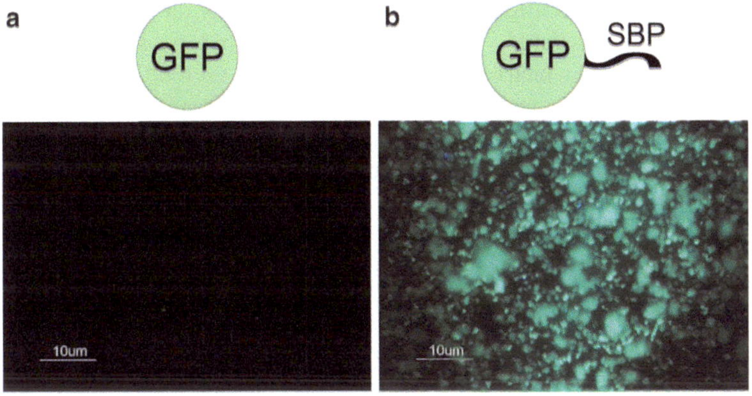

Fig. 23.2 Binding and fluorescence micrographs of recombinant Green Fluorescent Protein (GFP) and GFP-SBP (Table 23.5) to natural zeolite. (**a**) GFP without SBP does not bind to natural zeolite and the zeolite fraction displayed no fluorescence. (**b**) Only when the GFP was expressed recombinantly with the fused SBP (GFP-SBP) does the protein bind to natural zeolite and display fluorescence

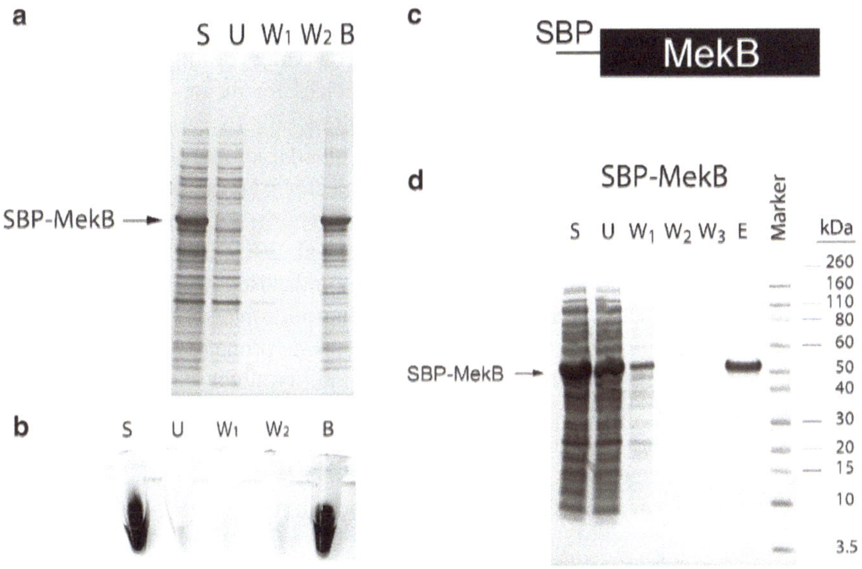

Fig. 23.3 Zeolite binding assay and partial purification of recombinant SBP-MekB esterase (Table 23.5) from *E. coli* crude extract. (**a**) SDS-PAGE gel of binding assay of SBP-MekB to synthetic zeolite. (**b**) Esterase activity of binding assay fractions. (**c**) Diagrammatic representation of SBP-MekB construction with the SBP at the N-terminus. (**d**) SDS-PAGE gel of partial purification of SBP-MekB from *E. coli* crude extracts using zeolite binding technology. *S* starting *E. coli* soluble proteins crude extract, *U* unbound fraction, *W1–3* wash fractions 1, 2 and 3, *B* zeolite-bound fraction, *E* SBP-MekB final eluted fraction from zeolite

purification tag (DSARGFKKPGKR) that confers affinity for unmodified silica on proteins to which it is fused. They developed an inexpensive silica-based protein purification scheme in which recombinant proteins (i.e., GFP, mCherry, and Maltose Binding Protein) were tagged with the silica-binding SBP and selectively recovered from clarified lysates by specific adsorption on silica gel columns. The tagged proteins were then released from the silica by incubation with L-lysine to achieve purities of 80–90 %.

2.3 Matrices for Silica-Binding SBPs

The use of unmodified silica in protein purification has been used relatively rarely (Ghose et al. 2004). Methods in which a biological recognition molecule is immobilised onto a silica or silica-based sensing substrate are used more frequently. Methods involving covalent immobilisation onto a surface are generally achieved in three steps. First, there is the modification of the surface to add specific functional groups; followed by covalent attachment of a crosslinker through one of its reactive groups; and finally, covalent linking of the recognition element to the remaining reactive sites of the crosslinker. Some methods have been described in biosensor-binding assays that allow the attachment of the biological recognition molecules onto sensing surfaces with retention of their full functionality, for example, the orientated immobilisation of antibodies onto silicon wafers (Taniguchi et al. 2007; Ikeda et al. 2009) or nanowires as electronic devices for the sensitive and direct detection of biological analytes (Patolsky et al. 2006).

Most early enzyme immobilisation studies involved microporous silica and zeolites. Recent studies have focussed on the use of mesoporous matrices on which the enzyme is bound on the external surface because the pore size is too small to allow entry of the protein (Hudson et al. 2008; Popat et al. 2011; Zhou and Hartmann 2012; Magner 2013). These compounds can be synthesised readily with ordered and defined pore structures, a limited pore size distribution and can be chemically modified with functional groups. Although they may be prone to enzyme leaching, they can be covalently immobilised to increase stability albeit with the drawbacks mentioned already (Gaffney et al. 2012; Puddu and Perry 2012; Hartmann and Kostrov 2013). Puddu and Perry (2012) have emphasised that peptide adsorption on silica nanoparticles involves hydrophobic interactions and are governed by the peptide identity. Their work showed that it was possible to change peptide binding on silica by altering the binding environment or changing the surface properties of the silica.

Zeolites are naturally-occurring inorganic materials that have a highly ordered structure. Also they can be synthesised as nanocrystals with pores and channel systems in the molecular size range of 0.3–3.0 nm. They have desirable characteristics as inorganic support materials for the immobilisation of proteins that include mechanical and chemical resistance and high surface area (Breck 1973). Zeolites can be modified with respect to their basic/acidic nature by varying the Si/Al ratio or by introducing different metals into the crystalline framework and are stable under both wet and dry conditions. The properties of the silica-binding SBP [(VKTQATSREEPPRLPSKHRPG)$_4$VKTQTAS] fused to a truncated form of *Streptococcus* Protein G' (Goward et al. 1990) and the resulting recombinant fusion protein (LPG) has been tested extensively for its binding affinity toward several industrially-relevant silica-containing matrices (Sunna et al. 2013a), including zeolites. LPG binding was tested against 13 commercial synthetic zeolites belonging to 8 different zeolite families. LPG displayed low affinity (40–60 %) to Valfor 100 (Linde Type A, LTA) and Molecular Sieve 13X (Faujasite-X, FAU-X). LPG exhibited no affinity to CBV400 (Faujasite-Y, FAU-Y), CP914C (Ferrierite, FER), CP814E (Beta Polymorph A, BEA) and CBV2314 (ZSM-5, MFI). The highest binding affinity (100 % binding) was displayed against Mordenite (CBV10A and CBV21A) and Faujasite-Y (CBV100 and CBV300) types of zeolites. Table 23.3 summarises the properties of the synthetic zeolites tested and the degree of binding as shown stylistically by the proportions of bound and unbound LPG.

Table 23.3 Properties of some of the synthetic zeolites used to study the silica-binding SBP specificity[a]

Zeolites	SiO_2/Al_2O_3 (mole ratio)	Nominal cation	Na_2O weight (%)	Unit cell size (Å)	Surface area (m^2/g)	Binding[b] U B
Faujasite (FAU)						
(FAU-Y)						
CBV 100	5.1	Sodium	13.0	24.65	900	
CBV 300	5.1	Ammonium	2.8	24.68	925	
CBV 400	5.1	Hydrogen	2.8	24.50	730	
(FAU-X)						
Molecular sieve 13X	5.1	Sodium	14.6	24.94	700	
Mordenite (MOR)						
CBV 10A	13.0	Sodium	65.0	N.A[c]	425	
CBV 21A	20.0	Ammonium	0.08	N.A	500	
Ferrierite (FER)						
CP 914c	20.0	Ammonium	0.05	N.A	400	
Beta Polymorph A (BEA)						
CP 814E	25.0	Ammonium	0.05	N.A	680	
ZSM-5 (MFI)						
CBV 2314	23.0	Ammonium	0.05	N.A	425	
Linden type A (LTA)						
Valfor 100	1.1	Sodium	15.0[c]	N.A	71[d]	

[a]Unless otherwise specified all data provided by the manufacturer except for the SDS-PAGE of LPG binding to material

[b]SDS-PAGE of LPG binding to material. *U* unbound fraction, *B* bound fraction. Wash fractions not shown

[c]*NA* data not available

[d]Hui and Chao (2008)

Table 23.4 Properties of some of the commercial-grade silica used to study the silica-binding SBP specificity[a]

Silica	SiO_2 (%)	Particle size (μm)	Pore size (Å)	Surface area (m^2/g)	Binding[b] U B
Silica (precipitated)	99.0[c]	N.A[d]	Non-porous[e]	62	
Silica gel grade 646	99.4	250–500	107–189	275–375	
Silica Davisil LC 60A	99.4	40–63	60	500–600	
Silica gel 60 Scharlau	99.4	60–200	60	500	
Silica Rhodoline HP 34 M	94.0	2.6	NA	157	
Silica Tixosil 38	94.0	15	NA	190	
Silica Tixosil 38A	94.0	60	NA	230	
Silica Tixosil 68	94.0	250	NA	160	
Silica Perkasil SM 660	98.0	17	NA	200	
Silica Perkasil KS 300-PD	98.0	14.5	NA	125	
Silica Elfadent SM 514	98.0	12	NA	125	
Silica Durafill 200[f]	82.0	6.0	NA	80	

[a]Unless otherwise specified all data provided by the manufacturer except for the SDS-PAGE of LPG binding to material
[b]SDS-PAGE of LPG binding to material. *U* unbound fraction, *B* bound fraction. Wash fractions not shown
[c]Çaykara and Güven (1998)
[d]*NA* data not available
[e]Furlong (1982)
[f]Sodium magnesium aluminium silicate (8.5 % Al_2O_3, 2 % MgO)

Twelve different types of commercial silica materials were tested for their ability to immobilise LPG. LPG exhibited high binding affinity (90–100 %) to 10 samples, but did not bind to Silica LC 60A (50 % binding affinity) and Silica Gel 60 (30 % binding affinity). Table 23.4 summarises the main properties of the silica samples tested.

There are economic advantages to the use of appropriate zeolites and silica powders as matrices for enzyme immobilisation. Zeolites are used frequently in industrial processes requiring bulk supplies such as water purification gas separation, reprocessing of nuclear wastes, laundry detergents, in the construction industry, for soil treatment and other agricultural procedures and for cat litter. Chemically-synthesised zeolites are used in catalytic cracking and hydrochemical cracking of petroleum and as molecular sieves. A more complete review of zeolite applications can be found in Naber et al. (1994) and Rhodes (2010). In a similar fashion, industry is familiar with silica powders for steel refining, aluminium casting, in the chemical and semi-conductor and electronic industries and as an anticaking agent for foodstuffs and a thickener for paint as well as a constituent of the construction industry in several forms of concrete. Prices quoted for the less pure forms of silica appear to be between US$10–US$500/metric ton (www.alibaba.com). Accordingly, the cost of the matrix for immobilisation of industrial enzymes should not be a major factor. The interested reader can find further details on applications in the text of Kogel et al. (2006) and at www.monographs.iarc.fr/ENG/Monographs/vol100C-14.pdf

2.4 *Silica-Binding SBP as Part of Fusion Proteins with Thermophilic Enzymes*

There are very few reports on thermostable enzymes immobilised to solid materials using SBPs. Up until now, most research has involved the silica-binding SBP described above and partly has been reported recently by Care et al. (2014b) and otherwise is summarised in this section in the form of unpublished results.

2.4.1 Immobilisation

So far, the silica-binding SBP reported by Sunna et al. (2013a) has been fused successfully to more than 20 proteins or enzymes (Care et al. 2015). The effect of this SBP was tested by introducing it into plasmids containing the genes for different thermophilc enzymes with optimal temperatures for activity ranging between 50 °C and 100 °C. A range of fusion proteins were constructed comprising a selection of either N- or C-terminal SBP genetically fused to the industrially-relevant thermostable enzymes (Table 23.5). All the recombinant SBP fusion proteins were expressed as soluble proteins in *E. coli* and retained their catalytic function (Fig. 23.4). In all cases the silica-binding SBP mediated the specific binding of the fusion protein to the zeolite matrix. Zeolite-binding assays using the corresponding version of the enzyme without the fused SBP displayed either no specific binding or just non-specific adsorption to the zeolite (Fig. 23.4b). Although the molecular mass of the proteins is not directly related to their adsorption efficiency, their adsorption to zeolites has been shown to be pH-dependent with maximum adsorption reached when the pH was just at or below the pI of the protein (Klint and Eriksson 1997; Tavolaro et al. 2006). At pH's below their pI, proteins are positively charged and will adsorb onto the external surfaces of zeolites (overall negative charge) via electrostatic attraction. At pH's above their pI, proteins are negatively charged and there is minimal adsorption to zeolite due to repulsion between the negatively charged protein and zeolite. Any adsorption under these conditions will be likely the result of other attractive forces (e.g., hydrophobic interactions). The SBP has been reported to mediate the specific binding of proteins to zeolite across a wide pH range (5–9), indicative of a true-pH independent binding affinity (Sunna et al. 2013a).

SBP sequences commonly are positively charged due to their high proportion of basic amino acids residues. SBPs also show tendencies toward amino acid residues that promote a less stable ordered structure (Sarikaya et al. 2003; Ikeda and Kuroda 2011; Sunna et al. 2013a). This intrinsically disordered state of the SBP supports

Table 23.5 SBPs recombinant fusion proteins produced in *E. coli*

Construction	Function	Source	SBP position	Function retained	Zeolite binding	References
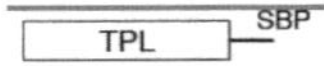	Lipase	Metagenomic enrichment	C[a]	Y	Y	Bell et al. (2002)
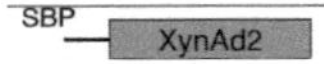	Xylanase	*Caldibacillus cellulovorans*	N[b]	Y	Y	Sunna et al. (2000)
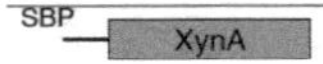	Xylanase	Environmental DNA	N	Y	Y	Sunna and Bergquist (2003)
	Xylanase	*Dictyoglomus thermophilum*	N	Y	Y	Morris et al. (1998), Sunna et al. (2013a)
	Mannanase	*Caldicellulosiruptor saccharolyticus*	N	Y	Y	Sunna (2010)
	Mannanase	*Dictyoglomus thermophilum*	C	Y	Y	Gibbs et al. (1999), Care et al. (2014b)
	β-glucosidase	*Caldicellulosiruptor saccharolyticus*	N	Y	Y	Love et al. (1988)
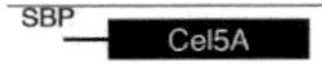	Endoglucanase	*Thermotoga maritima*	N	Y	Y	Chhabra et al. (2002)
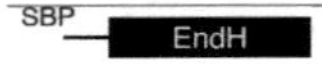	Endoglucanase	*Dictyoglomus thermophilum*	N	Y	Y	Coil et al. (2014)

The ability to retain the enzyme activity function after SBP incorporation and the zeolite binding affinity of the fusion proteins is shown
[a]C, SBP is at C-terminus of fusion protein
[b]N, SBP is at N-terminus of fusion protein

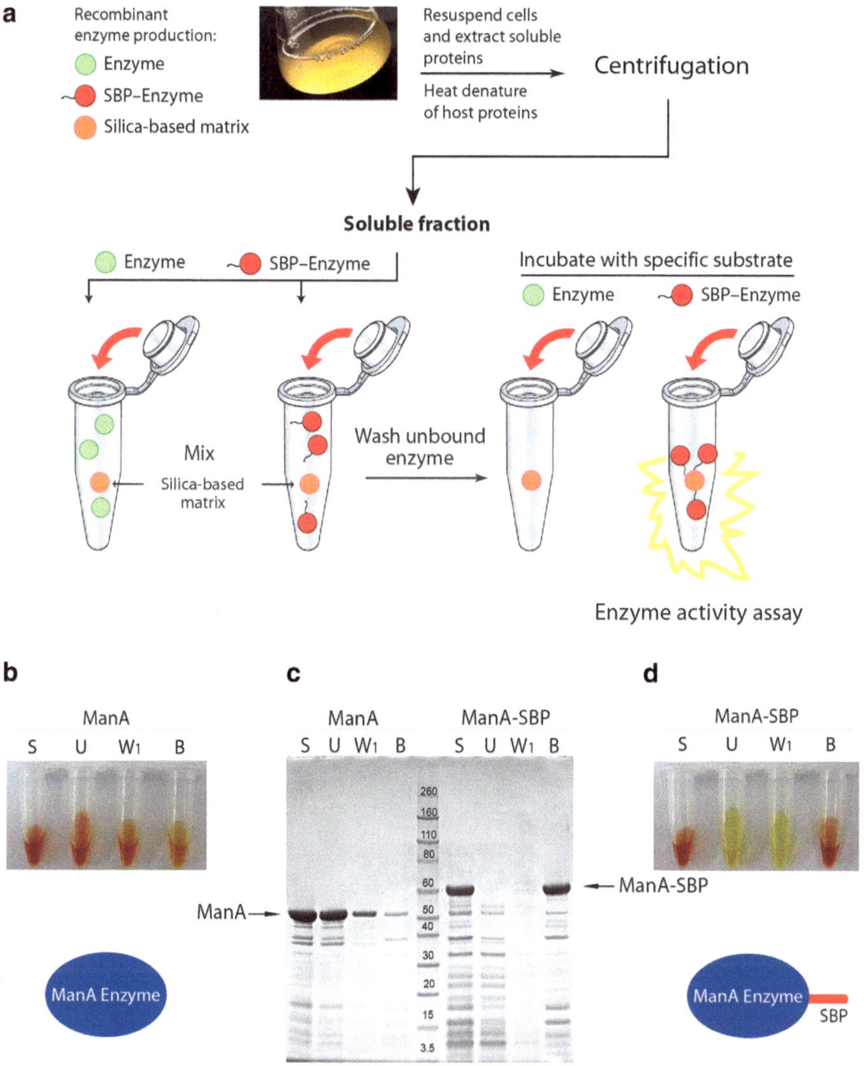

Fig. 23.4 Silica-based matrix binding and activity assay. (**a**) Schematic diagram of a standard extraction of recombinant soluble proteins from *E. coli* cells and subsequent silica-based matrix binding assay. (**b**) *Dictyoglomus thermophilum* thermostable ManA (Table 23.5) mannanase activity of binding assay fractions. (**c**) SDS-PAGE gel of zeolite-binding assays performed with partially purified ManA and ManA-SBP. (**d**) Mannanase activity of ManA-SBP binding assay fractions. Mannanase activity assay was performed with locust bean gum (LBG) as substrate at 80 °C for 10 min. Reducing sugars released from LBG were determined by the 3,5-dinitrosalicylic acid (DNS) assay. *S* starting *E. coli* soluble proteins crude extract, *U* unbound fraction, *W1* first wash fraction (wash fractions 2 and 3, were not loaded onto gel), *B* zeolite-bound protein fraction

flexibility and plasticity, which facilitates ionic interactions between the SBP's positively charged residues and the negatively charged surfaces of silica-containing materials (Collino and Evans 2008). Accordingly, it is expected that SBPs are in a constantly unfolded state that facilitates their adsorption and conformational adaptation to the substrate (e.g., zeolite, silica) at the binding interface (Ikeda and Kuroda 2011).

Polyester, polyamide or cotton textile fabrics have been introduced recently as new alternative matrices for immobilisation of biomolecules (Opwis et al. 2014). These fabrics represent relatively inexpensive support materials with flexible and open structures, mechanical strength and high surface area. Their open structures allow for high substrate turn-over. Various chemical (cross-linking) and non-chemical (photo-active cross-linking) strategies for the immobilisation of enzymes on textiles have shown the potential of this technique, resulting in fabrics with high enzyme load, high activity and enzyme reusability (Opwis et al. 2005, 2014).

An alternative to the above strategies for enzyme immobilisation on a textile carrier is the thermal deposition of the carrier directly onto the fabric. The potential of this approach has been demonstrated recently by directly immobilising a GFP fusion protein carrying a silica-binding SBP (Fig. 23.5). The fusion protein was immobilised successfully onto the silica-loaded fabric via the SBP's strong affinity to silica (Sunna unpublished results).

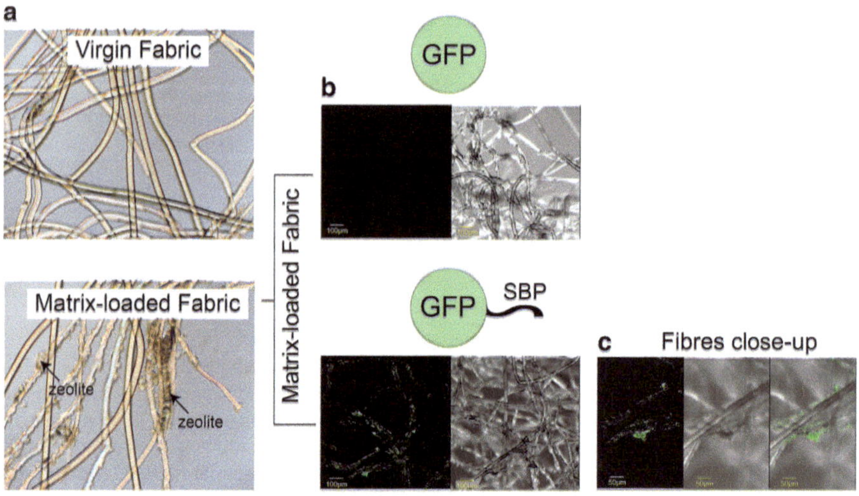

Fig. 23.5 Binding of GFP and GFP-SBP to virgin and zeolite-loaded fabrics. (**a**) Phase contrast microscope images of virgin and zeolite-loaded fabric fibres. (**b**) Confocal microscope images of virgin and zeolite-loaded fabric after incubation with GFP (*top*) and GFP-SBP (*bottom*). *Left-black panels* show GFP fluorescence; *right panels*, corresponding DIC images. (**c**) Close up of confocal microscope image of zeolite-loaded fabric treated with GFP-SBP. *Left panel* shows GFP fluorescence; *middle panel* shows corresponding DIC image and *right panel* shows GFP fluorescence image overlayed on the DIC image

2.4.2 Reuse/Recycling

Despite the advances in biotechnology and protein engineering, the widespread use of enzymes in specific industrial applications is often hindered by their lack of long-term operational stability, cumbersome recovery and poor reuse characteristics (Sheldon and van Pelt 2013). Immobilisation of enzymes onto an insoluble matrix or support offers the advantage of handling the enzyme in a solid form rather than as a free enzyme in a liquid formulation. A more convenient handling of enzyme preparations and easy separation from reaction products reduces the costs of downstream processing and simplifies enzyme applications. Furthermore, immobilisation frequently has been shown to improve enzyme stability and result in increased resistance to changes in pH and temperatures (Cherry and Fidantsef 2003). Compared to free enzymes, immobilised enzymes can be easily recovered and reintroduced into the reaction, thus allowing for cost-effective multiple reuse of the biocatalyst. Reuse through immobilisation provides cost advantages and especially facilitates operations where there is a high biocatalyst cost and allows support of continuous processes. Enzyme cost represents one of the biggest contributions to the overall expense of industrial bioprocesses and from a manufacturing cost perspective, the yield of immobilised enzyme activity is determined by the immobilisation method and the amount of soluble enzyme used. Improved operational performance and reuse of immobilised enzymes results in higher catalytic productivities (kg product/kg enzyme), which has a directly positive effect on enzyme cost per kg product (Sheldon and van Pelt 2013).

Ideally, enzyme immobilisation for large-scale industrial applications should be a simple procedure that does not require highly purified enzyme preparations or costly support matrices that would result in a biocatalyst that is commercially non-viable. Accordingly, SBPs-mediated immobilisation using low cost supports like silica and zeolite show significant potential for the immobilisation of enzymes for industrial applications. Thermophilic enzymes immobilised onto silica-containing materials via fusion to SBPs have been shown to remain functional at high operational temperatures and are fully reusable for several cycles (Fig. 23.6, Care et al. 2015).

2.4.3 Challenges

SBP's normally are identified by their binding activity in dilute laboratory buffers under ambient conditions. As a result, many SBPs are unstable under harsh industrial conditions, resulting in dissociation from their corresponding solids or complete degradation. Filamentous M13 bacteriophages have been shown to remain viable at pH's from 3 to 11 and temperatures below 80 °C (Branston et al. 2013). Thus, phage display performed with commercially available M13 bacteriophage display libraries (e.g. from New England Biolabs) represents a potential method to isolate SBPs that are compatible with such industrial processes. To achieve a robust response, we propose the selection of SBPs using combinatorial display libraries

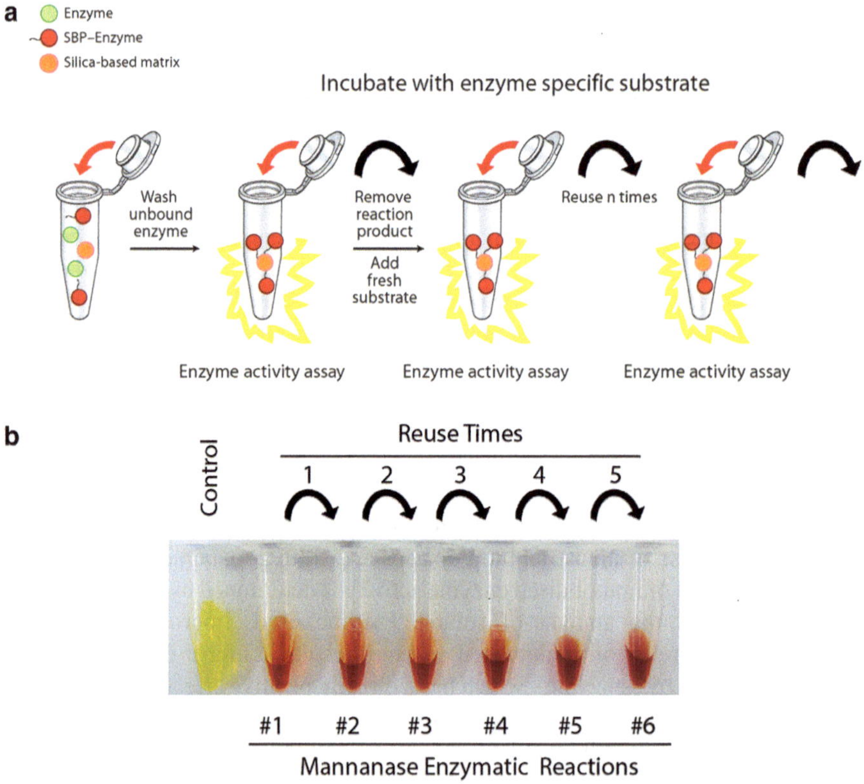

Fig. 23.6 Silica-based matrix binding and enzyme recycling. (**a**) Schematic diagram of a standard binding assay using partially purified recombinant soluble proteins from *E. coli* cells and subsequent silica-based matrix binding assay and enzyme recycling. (**b**) *D. thermophilum* thermostable ManA-SBP (Table 23.5) mannanase was immobilised on synthetic zeolite and reused (recycled) in enzymatic reactions six times over a 60 min assay period at 80 °C (10 min/cycle). Mannanase activity assay was performed with locust bean gum (*LBG*) as substrate and reducing sugars released were determined by the 3,5-dinitrosalicylic acid (*DNS*) assay. *C* control

that bind solid materials in the presence of conditions that they will come into contact with before or during their industrial application (e.g. extreme pH and high temperatures). This action would ensure that the stability and binding function of the isolated SBPs were retained under industrial conditions, enabling their fusion to thermophilic enzymes and consequently, improving their applicability in industrial biocatalysis.

In general, an SBP is fused genetically to a protein of interest and the resulting fusion protein produced in *E. coli*. Although *E. coli* is a common host for recombinant protein production, it does not secrete large amounts of protein naturally. Thus, partial or complete cell disruption is required to recover any *E. coli*-produced protein, a major obstacle in its application as a production host for industry. Filamentous fungi such as *Aspergillus* and *Trichoderma* are the usual work-horses for industrial protein production because of their high yields and ability to secrete enzymes into their media. Recently, the filamentous fungus *Trichoderma reesei* was tested as an alternative host to achieve higher yields and extracellular production of a thermostable β-mannanase-SBP fusion protein, ManA-SBP (Table 23.5 and Fig. 23.4, Care et al. 2014b). It was shown that the majority of extracellular proteins secreted by *T. reesei* do not bind to zeolite and that the *E. coli*-produced SBP was degraded by extracellular host proteases present in *T. reesei* supernatant. It was proposed that the eukaryotic expression system of *T. reesei* may produce a glycosylated form of the SBP that is resistant to proteolysis. However, introduction into *T. reesei* of a codon-optimised version of the ManA-SBP fusion protein resulted in detection in the growth medium of only the thermostable ManA enzyme without its fused SBP. Subsequently, a rational approach was used to produce multiple linker derivatives expected to show resistance to *T. reesei* proteases. They were produced in *E. coli* and assayed for zeolite binding specificity and protease stability. Unfortunately, all the derivatives were demonstrated to show no specific affinity towards zeolite and only one exhibited *T. reesei* protease resistance.

As a result, in order to ameliorate the degradation of the SBP sequence by *T. reesei* proteases, we have suggested that combinatorial display technologies could be utilised to isolate new peptides that bind to zeolite in the presence of *T. reesei* proteases during the biopanning steps (i.e., in the presence of *T. reesei* extracellular supernatant). This action should allow the protease-resistant and zeolite-binding functions of the selected SBPs to be retained within the *T. reesei* growth medium following their secretion. However, the effective expression of these SBPs in *T. reesei* is not guaranteed and obviously would need to be assessed experimentally.

It should be noted also that combinatorial display technologies have their disadvantages. All peptide libraries exhibit some degree of compositional, positional, and expression bias, which may cause an under-representation of peptides that have favourable properties and functions (Umlauf et al. 2014). Furthermore, it has been shown that biopanning procedures have an intrinsic bias towards positively charged peptides that bind through electrostatic interactions, and discriminate against peptides that bind through non-electrostatic interactions and thus many strongly-binding SBPs may not be selected (Puddu and Perry 2012).

3 Prospects for Thermostable Enzymes Immobilised Using SBPs

3.1 Synthetic Biology

With advances in the field of synthetic biology, most projects have focused in combining metabolic engineering and design with systems biology. This in vivo (top-down) approach to synthetic biology remains extremely complex, difficult and laborious (Kwok 2010; Hodgman and Jewett 2012). In vitro (bottom-up) cell-free synthetic biology on the other hand provides more engineering flexibility and eliminates cell viability constraints and physical barriers (cell wall and membranes) encountered with living cells in the in vivo approach.

Production cost and yields of thermostable enzymes remain a fundamental obstacle for their application in large-scale biomanufacturing (Turner et al. 2007). One strategy to overcome these factors is the use of recombinant technology to express thermophilic enzymes in mesophilic hosts like *E. coli*, which allows for easy denaturation of host proteins at high temperatures and represents a simple way to obtain relatively pure proteins without costly and cumbersome procedures. This strategy involves automated codon usage optimisation, which is a standard option currently offered by commercial gene synthesis providers, to achieve improved expression and yield of heterologous recombinant proteins. Furthermore, advances in protein engineering, especially computational protein design, have resulted in new ways to design enzymes that exhibit the desired properties of specificity, activity and stability (for a review see Khare and Fleishman 2013). Recently, Ye et al. (2012) demonstrated the suitability of thermostable enzymes in cell-free synthetic biology. They constructed a simple cell-free pathway by synthetic metabolic engineering which included four main steps: (a) selection of thermostable enzymes; (b) expression in *E. coli*; (c) inactivation of host indigenous proteins at high temperature and partial purification of thermostable enzymes and (d) rational combination of enzymes to achieve stoichiometrical substrate conversion. Using these principles they constructed a synthetic chimeric Embden-Meyerhof (EM) pathway using seven glycolytic enzymes from *Thermus thermophilus*, *Pyrococcus horikoshii* and *Thermococcus kodakarensis* that had been expressed recombinantly in *E. coli*. The synthetic pathway produced a stoichiometric amount of lactate from glucose with an overall ATP turnover number of 31.

Despite all these advances and the exploitation of thermophilic enzymes in cell-free synthetic biology, the widespread introduction of these enzymes in biotechnological processes is often hampered by problems of enzyme yields and the operational stability and difficulty of recovering and re-using the enzymes. Zhang et al. (2011) and You and Zhang (2013) have recently postulated the use of thermostable enzymes immobilised on solid matrices as an alternative cell-free synthetic biology approach to produce biocatalytic modules for low-cost biomanufacturing. Combining enzyme immobilisation on solid matrices and enzyme reuse would facilitate the introduction of thermostable enzymes into large-scale operations and has significant potential to reduce final product costs and improve the yield of a process. SBPs have been fused genetically to thermostable enzymes (Table 23.5) and the resulting fusion proteins (SBP–Enzyme) have been immobilised successfully onto solid inorganic matrices including silica and zeolite (Care et al. 2014b; Care et al. 2015). SBP–Enzymes immobilised on the solid matrix were shown to retain their intrinsic thermostability and catalytic functions.

In the same way as inorganic and organic compounds can act as nanoscale building blocks which can be combined to assemble functional nanomaterials, SBPs can be introduced into a molecular toolbox for the rational designs of stable basic building blocks (individual immobilised enzymes) with predefined functions and catalytic properties. The molecular toolbox (Fig. 23.7) would have three main

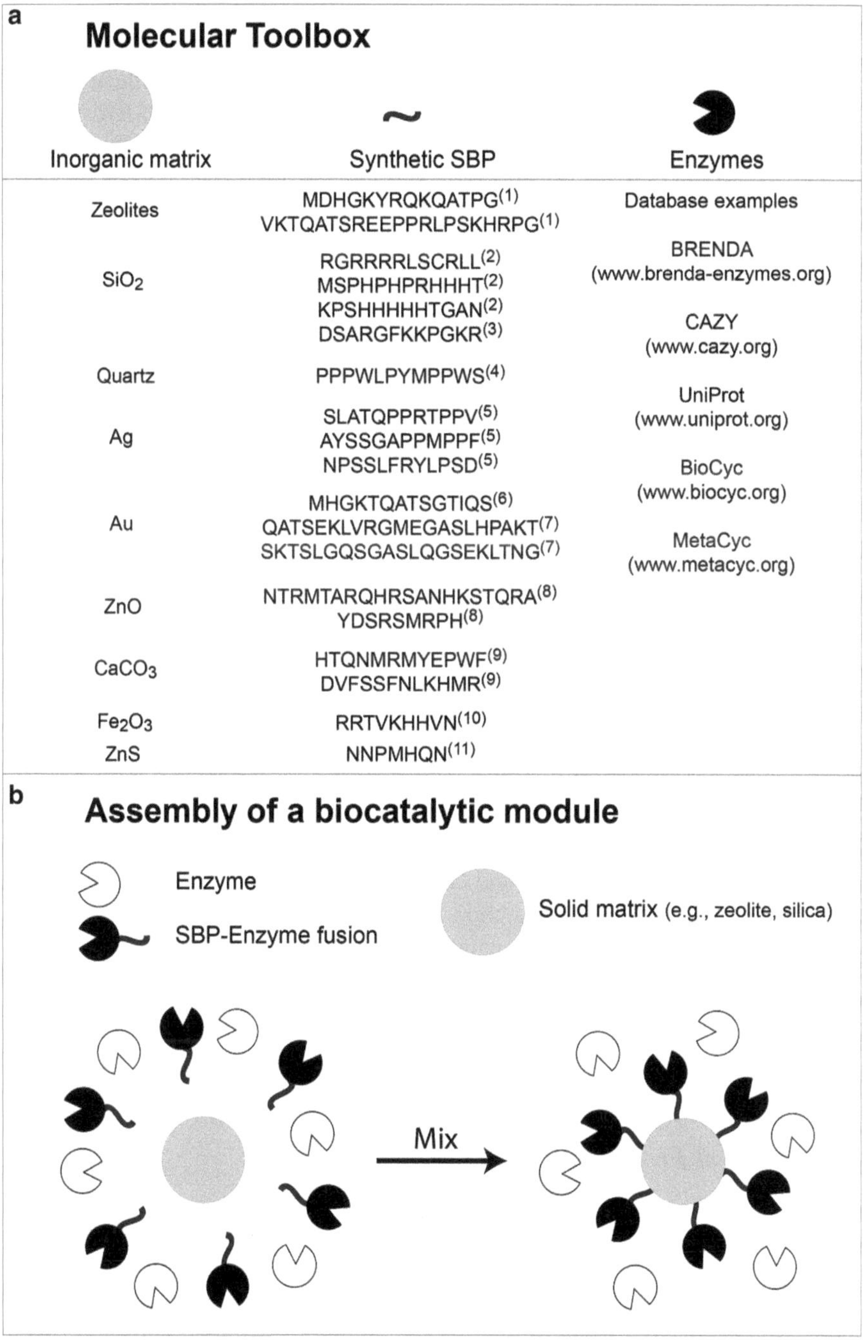

Fig. 23.7 Molecular toolbox (**a**) The three main interchangeable building blocks of the molecular toolbox include; low-cost inorganic matrices, matrix specific SBPs and thermostable enzymes. References for matrix-specific SBPs: (*1*) Nygaard et al. 2002; (*2*) Naik et al. 2002a; (*3*) Coyle and Baneyx 2014; (*4*) Oren et al. 2007; (*5*) Naik et al. 2002b; (*6*) Brown 1997; (*7*) Brown et al. 2000; (*8*) Kjaergaard et al. 2000; (*9*) Gaskin et al. 2000; (*10*) Brown 1992; (*11*) Lee et al. 2002. (**b**) Assembly of biocatalytic modules with predefined functions and catalytic properties

interchangeable building blocks; low-cost inorganic matrices (e.g., silica, zeolite), matrix specific SBPs and thermostable enzymes (e.g., glycoside hydrolases). By rationally combining SBPs, solid matrices and functional thermostable enzymes, researchers would have the flexibility to 'pick and mix' multiple building blocks for the assembly of natural and non-natural pathways and/or biocatalytic modules could be designed for efficient biopolymer degradation.

Such a molecular toolbox would rely on continued improvements in bioinformatics software, large-scale data-sharing initiatives (databases) and the ability to convert digital sequences into genetic material (*de novo* synthesis) without a natural template in short periods of time (Mueller et al. 2009). This strategy is driven further by the continuous decrease in the cost of gene synthesis. Currently the cost of gene synthesis is similar to the cost of column-synthesis of the oligonucleotides required for their assembly. However, transitioning gene synthesis to array-based oligonucleotides is expected to result in cost reduction to approximately \$1 per 10^3–10^5 bp (Kosuri and Church 2014).

Several biological databases (biochemical, genome, proteins or enzyme, pathways) that are relevant to this approach are available in the public domain. BRENDA (Braunschweig Enzyme Database, www.brenda-enzymes.org) has functional and property data for over 6500 enzyme entries. Another important protein database is UniProt (Universal Protein Resource, www.uniprot.org), which is a collection of sequences and annotation data of proteins. CAZy (Carbohydrate-Active enZYmes, www.cazy.org) is a database that provides structural and biochemical information about enzymes involved in the synthesis, metabolism and hydrolysis of carbohydrates. CAZy currently has a collection of over 3895 entries. Pathway databases describe the biochemistry of specific metabolic processes and contain data primarily on chemical pathways, reactions and components. Pathway databases like BioCyc (www.biocyc.org) contain information of 5500 pathways/genome databases plus software tools for graphical representation in pathways maps. The metabolic pathway database MetaCyc (www.metacyc.org), is a collection of 2260 pathways from 2600 different organisms. It stores a representative sample of each experimentally-elucidated pathway.

3.2 *Biofuel Production*

The reliance on fossil fuels and their consequent release of greenhouse gases has emerged as a global problem that is forecast to have serious consequences for the environment. The instability of oil prices has been unsettling commercially and these two factors have encouraged the search for alternative fuels and energy from renewable biological resources. An early source has been grain-based ethanol but the examination of energy in versus energy out in the process has been the subject of dispute but at best is marginal. Such first generation biofuels are a mature industry that is based on traditional yeast fermentation. Second generation biofuels have made use of biomass that is not a food source or does not grow on otherwise arable land and farms, and from plantation and forestry wastes, all of which are included by the term 'cellulosic ethanol'. The associated technologies utilise enzymes and whole cell catalysts to produce biodegradable products that result in less waste and lower carbon emissions than those produced from fossil resources as well as being sustainable in the long term. Recent advances in metabolic engineering, sequencing, systems biology and synthetic biology have opened new opportunities for tailoring the genetics and physiology of microorganisms involved in alcohol fermentations. Microbial cell factories that are designed for practical applications have been assembled using natural and *de novo* engineered pathways to redirect carbon flows to fuels and other desirable biologically-sourced chemicals (Dellomonaco et al. 2010; Hollinshead et al. 2014).

A major issue in biofuel production are enzyme costs for the breakdown of biomass polymers to produce sugars for fermentation into liquid fuels. With second generation biofuels, particularly lignocellulosic ethanol, the 5- (pentoses) and 6-carbon sugars (hexoses) together with lignin are the major constituents of the biomass and enzyme costs have been a major economic factor that has retarded the development of biofuels. Solutions proposed have included Consolidated Bioprocessing (CBP), where there is simultaneous saccharification and fermentation (SSF) of the biomass using anaerobic thermophilic bacteria (Lynd et al. 2002) and the use of genetically engineered thermophilic ethanologenic bacteria such as *Bacillus thermoglucosidasius* (now *Geobacillus thermoglucosidans*) where the acetogenic and lactogenic pathways have been inactivated genetically (Cripps et al. 2009).

Others have reported the deconstruction of switchgrass lignocellulose biomass that had not been pretreated by the hyperthermophilic anaerobic bacterium *Caldicellulosiruptor bescii* at 78 °C with the solubilisation of cellulose, hemicellulose and lignin and the production of small amounts of ethanol besides the major products comprising organic acids (Kataeva et al. 2013; Basen et al. 2014). This area has been reviewed comprehensively by Blumer-Schuette et al. (2014), who examined the enzyme repertoire of the *Caldicellulosiruptor* family as participants in CBP. There are some notable deficiencies in enzymes that hydrolyse cellulose and it should be possible to use individual thermophilic strains as 'genetic reservoirs' for the supplementation of strains with these deficiencies. As the genetic systems for most thermophilic strains are rudimentary it may be advantageous to complement CBP strains with appropriate amounts of the deficient enzyme added externally rather than the gene(s) being engineered into the ethanologenic strain. These genes could be cloned and expressed in a genetically-tractable host such as *E. coli* with the silica-binding SBP described here attached to allow binding to inexpensive natural zeolite for enzyme recovery and recycling.

The possibility of using enzymes on nanoparticles for biofuel production has been reviewed briefly by Puri et al. (2013) but the matrices suggested for immobilisation appear to be expensive to produce. Recently, Cho et al. (2012) described the co-immobilisation of three cellulases on Au-doped magnetic silica nanoparticles for the degradation of cellulose. An endo-, and an exo-cellulase apparently from *Trichoderma* and a β-glucosidase from *Thermotoga maritima* were cysteine-tagged and adsorbed onto gold nanoparticles and gold-doped magnetic nanoparticles and their ability to hydrolyse model soluble and insoluble substrates measured. Gold-doped magnetic nanoparticles appeared to be more suitable for immobilisation but the activity decreased over time, presumably as a result of enzyme leaching and inactivation of the endo- and exo-cellulase components that are not thermostable as the assay was performed at 80 °C.

In general, using enzymes immobilised on natural silica or zeolite via the silica-binding SBP will facilitate the simple removal and recovery of the immobilised enzyme after performing its action and allow for it reuse. Such a system would allow the simple transfer and action of individual genes for enzymes producing sugars such as arabinose and mannose from the appropriate biomass and would allow tailoring of hydrolysis conditions by mixing and matching enzyme proportions according to origin of biomass. For example, cellulosic ethanol production from *Pinus* might need increased amounts of mannanase and mannosidase as compared to hardwood feedstocks. Equally, it should be feasible to construct hydrolytic pathways on the same zeolite matrix or to make an artificial blend with individual immobilised enzymes in the most desirable proportions depending on the substrate in a 'pick and mix' fashion.

The classical scheme for the enzymatic degradation of cellulose involves three major enzyme groups for hydrolysis; endoglucanases, cellobiohydrolases (exoglucanases) and β-glucosidases. All these enzymes are hydrolases acting synergistically on the β-1,4-glycosidic bonds of the polysaccharide (Wood and Garcia-Campayo 1990; Teeri 1997). Enzymatic degradation at high temperatures is desirable to facilitate the conversion of the polysaccharide into fermentable sugars for biotechnological applications like biofuels but enzyme cost is a critical factor in industrial biotransformation and processing. Accordingly, enzyme immobilisation via SBPs represents an enabling technology that provides functional and stable biocatalysts. Figure 23.8 shows the main enzymes involved in the classical degradation of cellulose to glucose and a simplified example of how biocatalytic modules described in the previous section could be introduced into a simple degradation pathway for polysaccharides. Product inhibition (cellobiose) during the process may have a detrimental effect on the action of the enzymes involved. The ability to combine SBPs, inexpensive solid matrices (e.g., zeolite, silica) and thermostable cellulose-degrading enzymes into functional biocatalytic modules would allow for a much simpler process and better operational stability and flexibility, especially for the prospect of sequential hydrolysis and enzyme removal and recycling. This sequence not only would result in better control of product inhibition but also increase the economic value of the enzymes.

3.3 Gas Phase Catalysis and CO_2 Capture

Enzymes have wide applications in industry as catalytic proteins. However, a major limitation is the instability of many enzymes to heat, leading to the loss of enzyme activity through denaturation of the proteins. It has been shown previously that

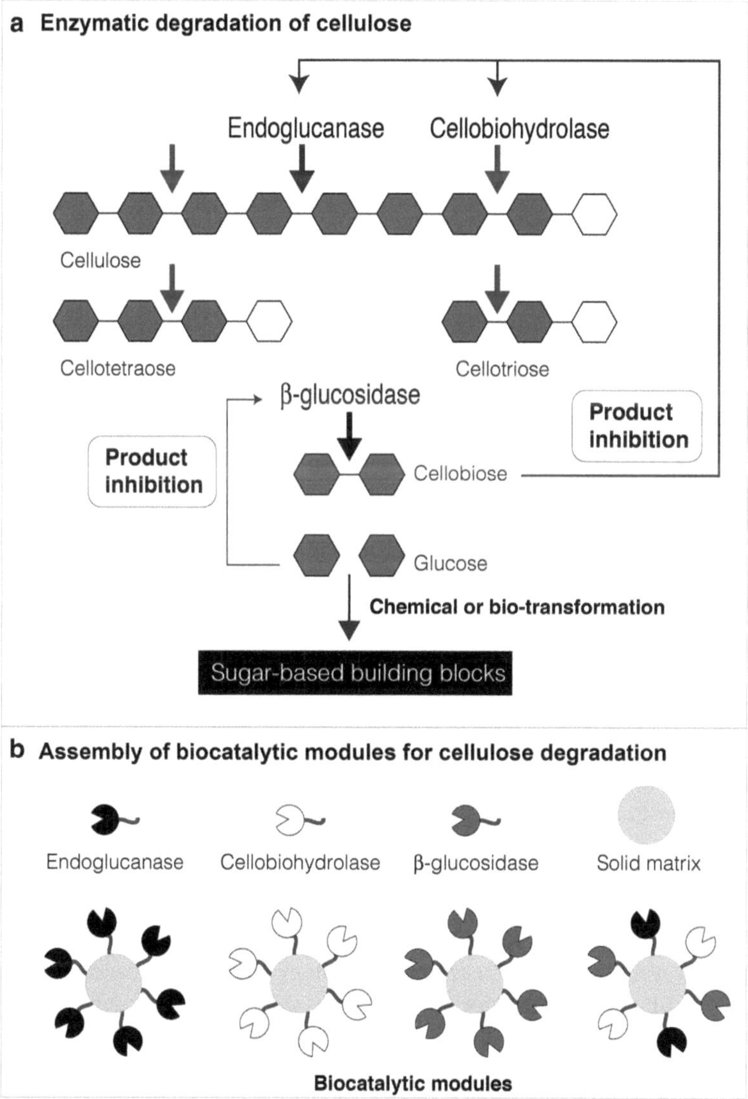

Fig. 23.8 Cellulose conversion to glucose. (**a**) Classical scheme for the enzymatic degradation of cellulose showing the main three enzymes involved; endoglucanases, cellobiohydrolases and β-glucosidases. (**b**) Assembly of functional biocatalytic modules for cellulose degradation

when enzymes are dry (containing less than 3 % water), they are remarkably stable to heat. For example, dry trypsin has been shown to exhibit long-term stability well above 100 °C, whereas the same protein rapidly denatures at 60 °C when fully hydrated (>20 % water by weight) (Mullaney 1966). It is generally accepted that protein hydration is essential for enzyme catalysis to occur. However, when the hydration of enzymes is sufficient to make them active, they are also unstable.

When enzymes are dry enough to be stable, they are not active. While there are varying estimates of the degree of protein hydration necessary for activity, a threshold value of about 0.2 g H_2O g^{-1} protein, corresponding to a hydration of 40–45 %, is generally accepted (Dunn and Daniel 2004).

Conducting biocatalysis with reactants present in the gas phase is a relatively new development and much research has focused on the hydrolysis of halogenated compounds. Some advantages of gas phase catalysis are as follows. The addition of solvents is avoided and the production of by-products is much more limited than aqueous systems, simplifying purification and processing. Mass transfer in the gas phase is more efficient than in solution and the reactions are unlikely to be limited by diffusion, while inhibitory, volatile or toxic substances are removed continuously. Further, an enzyme in a dehydrated form is more resistant to thermal inactivation (Trivedi et al. 2006) thus allowing the use of higher temperatures compared to aqueous systems. Also, unconventional reactions are possible such as transfer reactions or syntheses using hydrolases (Lamare and Legoy 1993).

There is little recent information on basic mechanisms of catalysis with substrates in the gas phase apart from some reports by Dunn and Daniel (2004) on the degree of hydration necessary for enzyme activity. Lind et al. (2004) have investigated the activity of hydrated lipases as esterases with vapour phase ethyl butyrate as substrate. These experiments allowed a direct comparison of the activities of enzymes of different hydrations without diffusional constraints and showed that activity could be demonstrated at much lower hydrations than previously reported.

There is a substantial body of evidence that the combustion of fossil fuels is a major contributor to global climate change. Coal accounts for approximately 25 % of the world energy supply and 40 % of the carbon emissions. The scale of carbon capture and sequestration to effect a significant reduction of world-wide carbon emissions is staggeringly large. The enzyme carbonic anhydrase has been identified as having the potential to accelerate CO_2 capture from major emitters such as coal-powered generators of electricity in a biomimetic process (Fisher et al. 2012). The focus on enzyme-assisted carbon capture has been reviewed recently (Pierre 2012; Boone et al. 2013). However, the current enzyme technology platform enhances rather than replaces existing post-combustion solvent processes for carbon capture.

Carbonic anhydrase catalyses the hydration of CO_2 to bicarbonate and a proton at a turnover number approaching 10^6/s (Ho et al. 2008). Unfortunately, the poor stability and activity of most naturally derived carbonic anhydrases under the harsh conditions of the post-combustion capture process (temperatures from 50 °C to >125 °C; high concentrations of organic amines, heavy metals, nitrogen and sulphur oxides) have limited the use of these enzymes. Further limitations of current enzymatic-aided CO_2 capture and storage process include poor mass transfer and the quick build-up of inhibitors. An improved biomimetic CO_2 capture system will require (a) better or ideally non-limiting interface mass transfer, (b) immobilised enzymes without conformational activity loss, and (c) in-situ removal of inhibitors. Immobilisation of carbonic anhydrase is the most common method used to stabilise the enzyme and to limit exposure to denaturing conditions in carbon capture. Immobilisation matrices include polyamide (Belzil and Parent 2005) chitosan

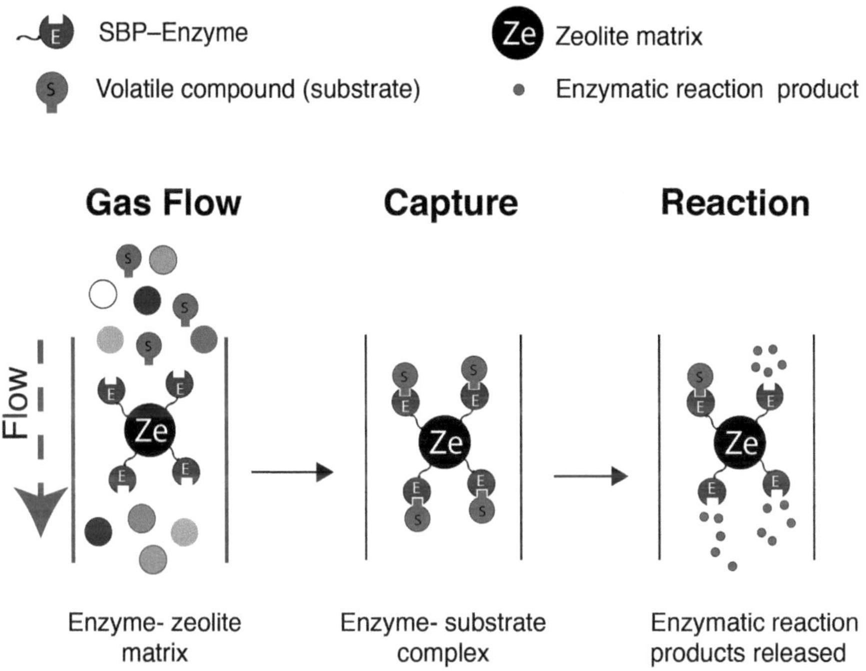

Fig. 23.9 Application concept of SBP-Enzyme immobilisation technology for a general gas phase catalytic reaction using a zeolite and/or silica solid matrix

(Sharma et al. 2011) and alkyl sepharose (Azari and Nemat-Gorgani 1999), microparticles and nanoparticles (Yadav et al. 2011) and foams (Kanbar and Ozdemir 2010). The activity lost on immobilisation can be substantial but this fact is seldom evident because of the nature of the assays employed, particularly using the *p*-nitrophenol esterase activity that is only suitable for α-class CAs (Savile and Lalonde 2011).

An alternative approach that would contribute to the resolution of these concerns involves examining the feasibility of operating carbon capture at a high temperature with enzymes specifically designed to be stable and reusable in a continuous gas phase process using a low-cost natural matrix (Fig. 23.9). The carbonic anhydrase from the hyperthermophile, *Sulphurhydrogenibium azorense* is claimed to be the fastest enzyme reported for CO_2 capture (Capasso et al. 2012) and it could be modified by fusion to SBP sequences, thus allowing its specific immobilisation to solid inorganic materials. We have described the new generation of SBPs that specifically mediate the immobilisation of enzymes onto synthetic and natural zeolites as reported earlier (Sunna et al. 2013a; Care et al. 2014b). Furthermore, several studies (Kim et al. 2012; Hasan et al. 2013) suggest that adsorption processes using porous materials such as zeolites show promising alternatives for carbon capture because of their selective CO_2 adsorption and large capacities. Zeolites have a high affinity for polar molecules such as H_2O and CO_2 (Breck 1973) and are especially important in partitioning an analyte gas from a mixture of gases. This feature is due to its uniform molecular pore size, polarity, reversible and selective adsorption, and adsorption capacity. The combination of a low-cost natural matrix and its specific SBP would lead to a new class of optimized and robust carbonic anhydrases for use in all classes of biomimetic carbon capture, from flue gas to fermentation broths. In addition it would advance the under-developed area of gas phase catalysis, a technique that has been largely ignored but has major advantages such as speed, lack of end-product inhibition, more efficient mass transfer than in solution and reactions that are not limited by diffusion.

4 Conclusions

SBPs selectively recognise and bind to of a wide range of solid materials. Recently, there has been considerable interest regarding SBPs that display binding affinity to silica-based solid matrices like silica, zeolite and mesoporous silica, which are considered to be the most suitable inorganic matrices for enzyme immobilisation in industrial processes. There is a substantial amount of information in the literature on the use of SBPs in conjunction with silica-based synthetic materials mostly in the area of nanobiotechnology. However, most studies involving SBP-mediated protein/enzyme immobilisation have focussed on biosensors, construction of nanomaterials and functionalisation. Silica-based support matrices like zeolite and silica are used commonly in industry in a number of applications and the provision for bulk supply is well established. Natural zeolites, which are mined and available already at very low costs, represent an important bulk support matrix for several applied processes. However, there is little information regarding their employment in added-value applications except for a few published studies on the use of natural zeolites as immobilisation matrices.

In the context of the immobilisation of thermostable enzymes, silica-binding SBPs have been described that are stable at high temperatures and over a wide pH range. Genetic constructions incorporating the SBP sequence as a fusion with the sequences of thermostable enzymes have demonstrated successfully the retention of the enzyme catalytic activity and the solid-binding property of the SBP in the recombinant fusion protein (SBP-enzyme) that can be used for the recovery and biocatalytic reuse of the proteins.

A further application of SBPs is envisaged as part of a molecular toolbox for the design of stable basic building blocks (individual immobilised enzymes) with predefined functions and catalytic properties for the construction biocatalytic modules. These biocatalytic modules can be introduced to construct simple hydrolysis pathways or be combined to suit the composition of the substrate, for example, in the production of sugars for fermentation into alcohols. There is the possibility of using silica-based matrices to immobilise hyperthermophilic carbonic anhydrase using SBPs for carbon capture. This could lead to reduced greenhouse gases and at least the development of a process to purify CO_2 as a 'merchant gas' by-product and would appear to be a superior solution compared to underground sequestration or mineralisation. However, considerably more research needs to be done on the suitability of gas phase catalysis in the process of carbon capture. While the exotic and biomedically-relevant aspects of SBP's have dominated research publications, there are many opportunities in environmentally-friendly 'green chemistry' that could benefit from suitably-focussed research projects.

References

Abdelhamid MA, Motomura K, Ikeda T et al (2014) Affinity purification of recombinant proteins using a novel silica-binding peptide as a fusion tag. Appl Microbiol Biotechnol 98(12): 5677–5684

Azari F, Nemat-Gorgani M (1999) Reversible denaturation of carbonic anhydrase provides a method for its adsorptive immobilization. Biotechnol Bioeng 62(2):193–199

Bachas-Daunert P, Sellers Z, Wei Y (2009) Detection of halogenated organic compounds using immobilized thermophilic dehalogenase. Anal Bioanal Chem 395(4):1173–1178

Ball P (2001) Life's lessons in design. Nature 409(6818):413–416

Bandlish RK, Michael Hess J, Epting KL et al (2002) Glucose-to-fructose conversion at high temperatures with xylose (glucose) isomerases from *Streptomyces murinus* and two hyperthermophilic *Thermotoga* species. Biotechnol Bioeng 80(2):185–194

Baneyx F, Schwartz DT (2007) Selection and analysis of solid-binding peptides. Curr Opin Biotechnol 18(4):312–317

Basen M, Rhaesa AM, Kataeva I et al (2014) Degradation of high loads of crystalline cellulose and of unpretreated plant biomass by the thermophilic bacterium *Caldicellulosiruptor bescii*. Bioresour Technol 152:384–392

Bell PJL, Sunna A, Gibbs MD et al (2002) Prospecting for novel lipase genes using PCR. Microbiology 148(8):2283–2291

Belzil A, Parent C (2005) Methods of chemical immobilization of an enzyme on a solid support. Biochem Cell Biol 83(1):70–77

Betancor L, Luckarift HR (2008) Bioinspired enzyme encapsulation for biocatalysis. Trends Biotechnol 26(10):566–572

Blanco RM, Terreros P, Fernández-Pérez M et al (2004) Functionalization of mesoporous silica for lipase immobilization: characterization of the support and the catalysts. J Mol Catal B: Enzym 30(2):83–93

Blumer-Schuette SE, Brown SD, Sander KB et al (2014) Thermophilic lignocellulose deconstruction. FEMS Microbiol Rev 38(3):393–448

Bolivar JM, Rocha-Martin J, Mateo C et al (2009) Coating of soluble and immobilized enzymes with ionic polymers: full stabilization of the quaternary structure of multimeric enzymes. Biomacromolecules 10(4):742–747

Boone CD, Gill S, Habibzadegan A et al (2013) Carbonic anhydrase: an efficient enzyme with possible global implications. Int J Chem Eng 2013:6

Brady D, Jordaan J (2009) Advances in enzyme immobilisation. Biotechnol Lett 31(11): 1639–1650

Branston S, Stanley E, Ward J et al (2013) Determination of the survival of bacteriophage M13 from chemical and physical challenges to assist in its sustainable bioprocessing. Biotechnol Bioprocess Eng 18(3):560–566

Breck DW (1973) Zeolite molecular sieves: structure, chemistry, and use. Wiley, New York

Brown S (1992) Engineered iron oxide-adhesion mutants of the *Escherichia coli* phage lambda receptor. Proc Natl Acad Sci U S A 89(18):8651–8655

Brown S (1997) Metal-recognition by repeating polypeptides. Nat Biotechnol 15(3):269–272

Brown S, Sarikaya M, Johnson E (2000) A genetic analysis of crystal growth. J Mol Biol 299(3):725–735
Camarero JA (2008) Recent developments in the site-specific immobilization of proteins onto solid supports. Pept Sci 90(3):450–458
Cao L (2006) Carrier-bound immobilized enzymes: principles, application and design. Wiley-VCH, Weinheim
Capasso C, De Luca V, Carginale V et al (2012) Biochemical properties of a novel and highly thermostable bacterial alpha-carbonic anhydrase from *Sulfurihydrogenibium yellowstonense* YO3AOP1. J Enzyme Inhib Med Chem 27(6):892–897
Care A, Chi F, Bergquist P et al (2014a) Biofunctionalization of silica-coated magnetic particles mediated by a peptide. J Nanopart Res 16(8):1–9
Care A, Nevalainen H, Bergquist P et al (2014b) Effect of *Trichoderma reesei* proteinases on the affinity of an inorganic-binding peptide. Appl Biochem Biotechnol 173(8):2225–2240
Care A, Bergquist PL, Sunna A (2015) Solid-binding peptides: smart tools for nanobiotechnology. Trends Biotechnol 33(5):259–268
Çaykara T, Güven O (1998) Effect of preparation methods on thermal properties of poly(acrylic acid)/silica composites. J Appl Polym Sci 70(5):891–895
Cetinel S, Caliskan HB, Yucesoy DT et al (2013) Addressable self-immobilization of lactate dehydrogenase across multiple length scales. Biotechnol J 8(2):262–272
Charbit A, Boulain JC, Ryter A et al (1986) Probing the topology of a bacterial membrane protein by genetic insertion of a foreign epitope; expression at the cell surface. EMBO J 5(11):3029–3037
Chen AY, Deng Z, Billings AN et al (2014) Synthesis and patterning of tunable multiscale materials with engineered cells. Nat Mater 13(5):515–523
Cherry JR, Fidantsef AL (2003) Directed evolution of industrial enzymes: an update. Curr Opin Biotechnol 14(4):438–443
Chhabra SR, Shockley KR, Ward DE et al (2002) Regulation of endo-acting glycosyl hydrolases in the hyperthermophilic bacterium *Thermotoga maritima* grown on glucan- and mannan-based polysaccharides. Appl Environ Microbiol 68(2):545–554
Chiu CY, Li Y, Huang Y (2010) Size-controlled synthesis of Pd nanocrystals using a specific multifunctional peptide. Nanoscale 2(6):927–930
Cho EJ, Jung S, Kim HJ et al (2012) Co-immobilization of three cellulases on Au-doped magnetic silica nanoparticles for the degradation of cellulose. Chem Commun 48(6):886–888
Coil DA, Badger JH, Forberger HC et al (2014) Complete genome sequence of the extreme thermophile *Dictyoglomus thermophilum* H-6–12. Genome Announc 2(1):e00109–e00114
Collino S, Evans JS (2008) Molecular specifications of a mineral modulation sequence derived from the aragonite-promoting protein n16. Biomacromolecules 9(7):1909–1918
Cowan DA, Fernandez-Lafuente R (2011) Enhancing the functional properties of thermophilic enzymes by chemical modification and immobilization. Enzyme Microb Technol 49(4):326–346
Cowan DA, Daniel RM, Morgan HW (1987) Some observations on the inhibition and activation of a thermophilic protease. Int J Biochem 19(5):483–486
Coyle BL, Baneyx F (2014) A cleavable silica-binding affinity tag for rapid and inexpensive protein purification. Biotechnol Bioeng 111(10):2019–2026
Cripps RE, Eley K, Leak DJ et al (2009) Metabolic engineering of *Geobacillus thermoglucosidasius* for high yield ethanol production. Metab Eng 11(6):398–408
Cui Y, Kim SN, Jones SE et al (2010) Chemical functionalization of graphene enabled by phage displayed peptides. Nano Lett 10(11):4559–4565
Dang X, Yi H, Ham M-H et al (2011) Virus-templated self-assembled single-walled carbon nanotubes for highly efficient electron collection in photovoltaic devices. Nat Nanotechnol 6(6):377–384
Dellomonaco C, Fava F, Gonzalez R (2010) The path to next generation biofuels: successes and challenges in the era of synthetic biology. Microb Cell Fact 9(1):3
Dong Q, Yan X, Zheng M et al (2014) Immobilization of a thermostable inorganic pyrophosphatase from the archaeon *Pyrococcus furiosus* onto amino-functionalized silica beads. J Appl Polym Sci 131(17):8669–8675

Dunn RV, Daniel RM (2004) The use of gas–phase substrates to study enzyme catalysis at low hydration. Phil Trans R Soc Lond 359(1448):1309–1320
Estephan E, Saab MB, Martin M et al (2011) Phages recognizing the indium nitride semiconductor surface via their peptides. J Pept Sci 17(2):143147
Evans JS, Samudrala R, Walsh TR et al (2008) Molecular design of inorganic binding polypeptides. MRS Bull 33:514–518
Fernandez-Lafuente R (2009) Stabilization of multimeric enzymes: strategies to prevent subunit dissociation. Enzyme Microb Technol 45(6–7):405–418
Filho M, Pessela BC, Mateo C et al (2008) Immobilization–stabilization of an α-galactosidase from *Thermus sp.* strain T2 by covalent immobilization on highly activated supports: selection of the optimal immobilization strategy. Enzyme Microb Technol 42(3):265–271
Fischer L, Bromann R, Kengen SW et al (1996) Catalytical potency of beta-glucosidase from the extremophile *Pyrococcus furiosus* in glucoconjugate synthesis. Biotechnology (NY) 14(1):88–91
Fisher Z, Boone CD, Biswas SM et al (2012) Kinetic and structural characterization of thermostabilized mutants of human carbonic anhydrase II. Protein Eng Des Sel 25(7):347–355
Forbes LM, Goodwin AP, Cha JN (2010) Tunable size and shape control of platinum nanocrystals from a single peptide sequence. Chem Mater 22(24):6524–6528
Furlong D (1982) Adsorption of Tris(2,2′-bipyridine)ruthenium(II) cations at silica/aqueous solution interfaces. Aust J Chem 35(5):911–917
Gabryelczyk B, Szilvay GR, Salomaki M et al (2013) Selection and characterization of peptides binding to diamond-like carbon. Colloids Surf B Biointerfaces 110:66–73
Gabryelczyk B, Szilvay GR, Linder MB (2014) The structural basis for function in diamond-like carbon binding peptides. Langmuir 30(29):8798–8802
Gaffney D, Cooney J, Magner E (2012) Modification of mesoporous silicates for immobilization of enzymes. Top Catal 55(16–18):1101–1106
Garcia-Galan C, Berenguer-Murcia Á, Fernandez-Lafuente R et al (2011) Potential of different enzyme immobilization strategies to improve enzyme performance. Adv Synth Catal 353(16):2885–2904
Gaskin DJH, Starck K, Vulfson EN (2000) Identification of inorganic crystal-specific sequences using phage display combinatorial library of short peptides: a feasibility study. Biotechnol Lett 22(15):1211–1216
Ghose S, McNerney TM, Hubbard B (2004) Preparative protein purification on underivatized silica. Biotechnol Bioeng 87(3):413–423
Gibbs MD, Reeves RA, Sunna A, Bergquist PL (1999) Sequencing and expression of a β-mannanase gene from the extreme thermophile *Dictyoglomus thermophilum* Rt46B.1, and characteristics of the recombinant enzyme. Curr Microbiol 39(6):351–357
Goede K, Busch P, Grundmann M (2004) Binding specificity of a peptide on semiconductor surfaces. Nano Lett 4(11):2115–2120
Goward CR, Murphy JP, Atkinson T et al (1990) Expression and purification of a truncated recombinant streptococcal protein G. Biochem J 267(1):171–177
Greiner L, Schroder I, Muller DH et al (2003) Utilization of adsorption effects for the continuous reduction of NADP+ with molecular hydrogen by *Pyrococcus furiosus* hydrogenase. Green Chem 5(6):697–700
Gülay S, Şanlı-Mohamed G (2012) Immobilization of thermoalkalophilic recombinant esterase enzyme by entrapment in silicate coated Ca-alginate beads and its hydrolytic properties. Int J Biol Macromol 50(3):545–551
Gungormus M, Fong H, Kim IW et al (2008) Regulation of in vitro calcium phosphate mineralization by combinatorially selected hydroxyapatite-binding peptides. Biomacromolecules 9(3):966–973
Hanefeld U, Gardossi L, Magner E (2009) Understanding enzyme immobilisation. Chem Soc Rev 38(2):453–468
Hartmann M, Kostrov X (2013) Immobilization of enzymes on porous silicas: benefits and challenges. Chem Soc Rev 42(15):6277–6289

Hasan MMF, First EL, Floudas CA (2013) Cost-effective CO2 capture based on in silico screening of zeolites and process optimization. Phys Chem Chem Phys 15(40):17601–17618
Heinz H, Farmer BL, Pandey RB et al (2009) Nature of molecular interactions of peptides with gold, palladium and pd-au bimetal surfaces in aqueous solution. J Am Chem Soc 131(28):9704–9714
Hickey AM, Ngamsom B, Wiles C et al (2009) A microreactor for the study of biotransformations by a cross-linked γ-lactamase enzyme. Biotechnol J 4(4):510–516
Hidalgo A, Betancor L, Lopez-Gallego F et al (2003) Design of an immobilized preparation of catalase from *Thermus thermophilus* to be used in a wide range of conditions: structural stabilization of a multimeric enzyme. Enzyme Microb Technol 33(2–3):278–285
Hnilova M, Oren EE, Seker UOS et al (2008) Effect of molecular conformations on the adsorption behavior of gold-binding peptides. Langmuir 24(21):12440–12445
Ho L-F, Li S-Y, Lin S-C et al (2004) Integrated enzyme purification and immobilization processes with immobilized metal affinity adsorbents. Process Biochem 39(11):1573–1581
Ho MT, Allinson GW, Wiley DE (2008) Reducing the cost of CO2 capture from flue gases using membrane technology. Ind Eng Chem Res 47(5):1562–1568
Hodgman CE, Jewett MC (2012) Cell-free synthetic biology: thinking outside the cell. Metab Eng 14(3):261–269
Hoess RH (2001) Protein design and phage display. Chem Rev 101(10):3205–3218
Hollinshead W, He L, Tang YJ (2014) Biofuel production: an odyssey from metabolic engineering to fermentation scale-up. Front Microbiol 5:344
Huang Y, Chiang C-Y, Lee SK et al (2005) Programmable assembly of nanoarchitectures using genetically engineered viruses. Nano Lett 5(7):1429–1434
Hudson S, Cooney J, Magner E (2008) Proteins in mesoporous silicates. Angew Chem Int Ed 47(45):8582–8594
Hui KS, Chao CYH (2008) Methane emissions abatement by multi-ion-exchanged zeolite A prepared from both commercial-grade zeolite and coal fly ash. Environ Sci Technol 42(19):7392–7397
Ikeda T, Kuroda A (2011) Why does the silica-binding protein "Si-tag" bind strongly to silica surfaces? Implications of conformational adaptation of the intrinsically disordered polypeptide to solid surfaces. Colloids Surf B Biointerfaces 86(2):359–363
Ikeda T, Hata Y, Ninomiya K-i et al (2009) Oriented immobilization of antibodies on a silicon wafer using Si-tagged protein A. Anal Biochem 385(1):132–137
Inoue I, Watanabe K, Yamauchi H et al (2014) Biological construction of single-walled carbon nanotube electron transfer pathways in dye-sensitized solar cells. ChemSusChem 7(10):2805–2810
Johnson A, Zawadzka A, Deobald L et al (2008) Novel method for immobilization of enzymes to magnetic nanoparticles. J Nanopart Res 10(6):1009–1025
Jørgensen F, Hansen OC, Stougaard P (2004) Enzymatic conversion of d-galactose to d-tagatose: heterologous expression and characterisation of a thermostable l-arabinose isomerase from *Thermoanaerobacter mathranii*. Appl Microbiol Biotechnol 64(6):816–822
Kacar T, Ray J, Gungormus M et al (2009a) Quartz binding peptides as molecular linkers towards fabricating multifunctional micropatterned substrates. Adv Mater 21(3):295–299
Kacar T, Zin MT, So C et al (2009b) Directed self-immobilization of alkaline phosphatase on micro-patterned substrates via genetically fused metal-binding peptide. Biotechnol Bioeng 103(4):696–705
Kanbar B, Ozdemir E (2010) Thermal stability of carbonic anhydrase immobilized within polyurethane foam. Biotechnol Prog 26(5):1474–1480
Kataeva I, Foston MB, Yang S-J et al (2013) Carbohydrate and lignin are simultaneously solubilized from unpretreated switchgrass by microbial action at high temperature. Energy Environ Sci 6(7):2186–2195
Khare SD, Fleishman SJ (2013) Emerging themes in the computational design of novel enzymes and protein–protein interfaces. FEBS Lett 587(8):1147–1154
Kim J, Lin L-C, Swisher JA et al (2012) Predicting large CO_2 adsorption in aluminosilicate zeolites for postcombustion carbon dioxide capture. J Am Chem Soc 134(46):18940–18943

Kjaergaard K, Sørensen JK, Schembri MA et al (2000) Sequestration of zinc oxide by fimbrial designer chelators. Appl Environ Microbiol 66(1):10–14
Klint D, Eriksson H (1997) Conditions for the adsorption of proteins on ultrastable zeolite Y and its use in protein purification. Protein Expr Purif 10(2):247–255
Ko S, Park TJ, Kim H-S et al (2009) Directed self-assembly of gold binding polypeptide-protein A fusion proteins for development of gold nanoparticle-based SPR immunosensors. Biosens Bioelectron 24(8):2592–2597
Kogel JE, Trivedi NC, Barker JM et al (2006) Industrial minerals & rocks: commodities, markets, and uses, 7th edn. Society for Mining, Metallurgy, and Exploration, Littleton
Kosuri S, Church GM (2014) Large-scale de novo DNA synthesis: technologies and applications. Nat Methods 11(5):499–507
Kress J, Zanaletti R, Amour A et al (2002) Enzyme accessibility and solid supports: which molecular weight enzymes can be used on solid supports? An investigation using confocal Raman microscopy. Chem Eur J 8(16):3769–3772
Kulp JL, Shiba K, Evans JS (2005) Probing the conformational features of a phage display polypeptide sequence directed against single-walled carbon nanohorn surfaces. Langmuir 21(25):11907–11914
Kwok R (2010) Five hard truths for synthetic biology. Nature 463(7279):288–290
Lamare S, Legoy M-D (1993) Biocatalysis in the gas phase. Trends Biotechnol 11(10):413–418
Lee SW, Mao C, Flynn CE et al (2002) Ordering of quantum dots using genetically engineered viruses. Science 296(5569):892–895
Lehn C, Schmidt H-L (1997) Stability and stabilization of enzymes from mesophilic and thermophilic organisms in respect to their use in FIA-systems for the determination of L-lactate and acetate. J Chem Technol Biotechnol 69(2):161–166
Li CM, Botsaris GD, Kaplan DL (2002) Selective in vitro effect of peptides on calcium carbonate crystallization. Cryst Growth Des 2(5):387–393
Li YJ, Whyburn GB, Huang Y (2009) Specific peptide regulated synthesis of ultrasmall platinum nanocrystals. J Am Chem Soc 131(44):15998–15999
Lichty JJ, Malecki JL, Agnew HD et al (2005) Comparison of affinity tags for protein purification. Protein Expr Purif 41(1):98–105
Lind PA, Daniel RM, Monk C et al (2004) Esterase catalysis of substrate vapour: enzyme activity occurs at very low hydration. Biochim Biophys Acta 1702(1):103–110
Love DR, Fisher R, Bergquist PL (1988) Sequence structure and expression of a cloned β-glucosidase gene from an extreme thermophile. Mol Gen Genet 213(1):84–92
Lu Y, Zhao J, Zhang R et al (2014) Tunable lifetime multiplexing using luminescent nanocrystals. Nat Photon 8(1):32–36
Lynd LR, Weimer PJ, van Zyl WH et al (2002) Microbial cellulose utilization: fundamentals and biotechnology. Microbiol Mol Biol Rev 66(3):506–577
Magner E (2013) Immobilisation of enzymes on mesoporous silicate materials. Chem Soc Rev 42(15):6213–6222
Masica DL, Schrier SB, Specht EA et al (2010) De novo design of peptide-calcite biomineralization systems. J Am Chem Soc 132:12252–12262
Miroliaei M (2007) Studies on the activity and stability of immobilized thermophilic alcohol dehydrogenase. Sci Iranica 14(2):112–117
Morris DD, Gibbs MD, Chin CW et al (1998) Cloning of the xynB gene from *Dictyoglomus thermophilum* Rt46B.1 and action of the gene product on kraft pulp. Appl Environ Microbiol 64(5):1759–1765
Mueller S, Coleman JR, Wimmer E (2009) Putting synthesis into biology: a viral view of genetic engineering through de novo gene and genome synthesis. Chem Biol 16(3):337–347
Mullaney PF (1966) Dry thermal inactivation of trypsin and ribonuclease. Nature 210(5039):953
Naber JE, de Jong KP, Stork WHJ et al (1994) Industrial applications of zeolite catalysis. In: Weitkamp J, Karge HG, Pfeifer H, Hölderich W (eds) Studies in surface science and catalysis, vol 84. Elsevier, Amsterdam, pp 2197–2219

Naik RR, Brott LL, Clarson SJ et al (2002a) Silica-precipitating peptides isolated from a combinatorial phage display peptide library. J Nanosci Nanotechnol 2(1):95–100
Naik RR, Stringer SJ, Agarwal G et al (2002b) Biomimetic synthesis and patterning of silver nanoparticles. Nat Mater 1(3):169–172
Nel AE, Madler L, Velegol D et al (2009) Understanding biophysicochemical interactions at the nano-bio interface. Nat Mater 8(7):543–557
Ngamsom B, Hickey AM, Greenway GM et al (2010) Development of a high throughput screening tool for biotransformations utilising a thermophilic l-aminoacylase enzyme. J Mol Catal B: Enzym 63(1–2):81–86
Nochomovitz R, Amit M, Matmor M et al (2010) Bioassisted multi-nanoparticle patterning using single-layer peptide templates. Nanotechnology 21(14):145305
Notman R, Oren EE, Tamerler C et al (2010) Solution study of engineered quartz binding peptides using replica exchange molecular dynamics. Biomacromolecules 11(12):3266–3274
Nygaard S, Wendelbo R, Brown S (2002) Surface-specific zeolite-binding proteins. Adv Mater 14(24):1853–1856
Opwis K, Knittel D, Bahners T et al (2005) Photochemical enzyme immobilization on textile carrier materials. Eng Life Sci 5(1):63–67
Opwis K, Straube T, Kiehl K et al (2014) Various strategies for the immobilization of biocatalysts on textile carrier materials. Chem Eng Trans 38:223–238
Oren EE, Tamerler C, Sahin D et al (2007) A novel knowledge-based approach to design inorganic binding peptides. Bioinformatics 23(21):2816–2822
Oren EE, Notman R, Kim IW et al (2010) Probing the molecular mechanisms of quartz-binding peptides. Langmuir 26:11003–11009
Palomo JM, Segura RL, Mateo C et al (2004) Improving the activity of lipases from thermophilic organisms at mesophilic temperatures for biotechnology applications. Biomacromolecules 5(1):249–254
Park TJ, Zheng S, Kang YJ et al (2009) Development of a whole-cell biosensor by cell surface display of a gold-binding polypeptide on the gold surface. FEMS Microbiol Lett 293(1):141–147
Pasunooti S, Surya W, Tan SN et al (2010) Sol–gel immobilization of a thermophilic diguanylate cyclase for enzymatic production of cyclic-di-GMP. J Mol Catal B: Enzym 67(1–2):98–103
Patolsky F, Zheng G, Lieber CM (2006) Fabrication of silicon nanowire devices for ultrasensitive, label-free, real-time detection of biological and chemical species. Nat Protoc 1(4):1711–1724
Pender MJ, Sowards LA, Hartgerink JD et al (2006) Peptide-mediated formation of single-wall carbon nanotube composites. Nano Lett 6(1):40–44
Pessela BCC, Mateo C, Fuentes M et al (2004) Stabilization of a multimeric β-galactosidase from *Thermus* sp. strain T2 by immobilization on novel heterofunctional epoxy supports plus aldehyde-dextran cross-linking. Biotechnol Prog 20(1):388–392
Pierre AC (2004) The sol-gel encapsulation of enzymes. Biocatal Biotransfor 22(3):145–170
Pierre AC (2012) Enzymatic carbon dioxide capture. ISRN Chem Eng 2012:22
Piller K, Daniel RM, Petach HH (1996) Properties and stabilization of an extracellular alpha-glucosidase from the extremely thermophilic archaebacteria *Thermococcus* strain AN1: enzyme activity at 130 degrees C. Biochim Biophys Acta 1292(1):197–205
Popat A, Hartono SB, Stahr F et al (2011) Mesoporous silica nanoparticles for bioadsorption, enzyme immobilisation, and delivery carriers. Nanoscale 3(7):2801–2818
Puddu V, Perry CC (2012) Peptide adsorption on silica nanoparticles: evidence of hydrophobic interactions. ACS Nano 6(7):6356–6363
Puri M, Barrow CJ, Verma ML (2013) Enzyme immobilization on nanomaterials for biofuel production. Trends Biotechnol 31(4):215–216
Raghuvanshi S, Gupta R (2010) Advantages of the immobilization of lipase on porous supports over free enzyme. Protein Pept Lett 17(11):1412–1416
Raia CA, D'Auria S, Guagliardi A et al (1995) Characterization of redox proteins from extreme thermophilic archaebacteria: studies on alcohol dehydrogenase and thioredoxins. Biosens Bioelectron 10(1–2):135–140
Rhodes CJ (2010) Properties and applications of zeolites. Sci Prog 93(Pt 3):223–284

Rocha-Martin J, Vega DE, Cabrera Z et al (2009) Purification, immobilization and stabilization of a highly enantioselective alcohol dehydrogenase from *Thermus thermophilus* HB27 cloned in *E. coli*. Process Biochem 44(9):1004–1012

Roy MD, Stanley SK, Amis EJ et al (2008) Identification of a highly specific hydroxyapatite-binding peptide using phage display. Adv Mater 20(10):1830–1836

Sano K-I, Miura A, Yoshii S et al (2013) Nonvolatile flash memory based on biologically integrated hierarchical nanostructures. Langmuir 29(40):12483–12489

Sapsford KE, Algar WR, Berti L et al (2013) Functionalizing nanoparticles with biological molecules: developing chemistries that facilitate nanotechnology. Chem Rev 113(3):1904–2074

Sarikaya M, Tamerler C, Jen AKY et al (2003) Molecular biomimetics: nanotechnology through biology. Nat Mater 2(9):577–585

Sassolas A, Blum LJ, Leca-Bouvier BD (2012) Immobilization strategies to develop enzymatic biosensors. Biotechnol Adv 30(3):489–511

Savile CK, Lalonde JJ (2011) Biotechnology for the acceleration of carbon dioxide capture and sequestration. Curr Opin Biotechnol 22(6):818–823

Seeman NC, Belcher AM (2002) Emulating biology: building nanostructures from the bottom up. Proc Natl Acad Sci U S A 99(2):6451–6455

Seker UOS, Wilson B, Dincer S et al (2007) Adsorption behavior of linear and cyclic genetically engineered platinum binding peptides. Langmuir 23(15):7895–7900

Serizawa T, Techawanitchai P, Matsuno H (2007) Isolation of peptides that can recognize syndiotactic polystyrene. ChemBioChem 8(9):989–993

Sharma A, Bhattacharya A, Shrivastava A (2011) Biomimetic CO2 sequestration using purified carbonic anhydrase from indigenous bacterial strains immobilized on biopolymeric materials. Enzyme Microb Technol 48(4–5):416–426

Sheldon RA (2011) Characteristic features and biotechnological applications of cross-linked enzyme aggregates (CLEAs). Appl Microbiol Biotechnol 92(3):467–477

Sheldon RA, van Pelt S (2013) Enzyme immobilisation in biocatalysis: why, what and how. Chem Soc Rev 42(15):6223–6235

Simpson HD, Haufler UR, Daniel RM (1991) An extremely thermostable xylanase from the thermophilic eubacterium *Thermotoga*. Biochem J 277(Pt 2):413–417

Slocik JM, Naik RR (2014) Peptide-nanoparticle strategies, interactions and challenges. In: Walsh TR, Knecht MR (eds) Bio-inspired nanotechnology. Springer, New York, pp 1–16

Smith GP (1985) Filamentous fusion phage: novel expression vectors that display cloned antigens on the virion surface. Science 228(4705):1315–1317

So CR, Kulp JL, Oren EE et al (2009) Molecular recognition and supramolecular self-assembly of a genetically engineered gold binding peptide on Au{111}. ACS Nano 3(6):1525–1531

Sunna A (2010) Modular organisation and functional analysis of dissected modular β-mannanase *Cs*Man26 from *Caldicellulosiruptor* Rt8B.4. Appl Microbiol Biotechnol 86(1):189–200

Sunna A, Bergquist PL (2003) A gene encoding a novel extremely thermostable 1,4-β-xylanase isolated directly from an environmental DNA sample. Extremophiles 7(1):63–70

Sunna A, Gibbs MD, Bergquist PL (2000) A novel thermostable multidomain 1,4-β-xylanase from '*Caldibacillus cellulovorans*' and effect of its xylan-binding domain on enzyme activity. Microbiology 146(11):2947–2955

Sunna A, Chi F, Bergquist PL (2013a) A linker peptide with high affinity towards silica-containing materials. N Biotechnol 30(5):485–492

Sunna A, Chi F, Bergquist PL (2013b) Efficient capture of pathogens with a zeolite matrix. Parasitol Res 112(7):2441–2452

Taguchi H, Matsuzawa H, Ohta T (1984) l-Lactate dehydrogenase from *Thermus caldophilus* GK24, an extremely thermophilic bacterium. Eur J Biochem 145(2):283–290

Tamerler C, Oren EE, Duman M et al (2006) Adsorption kinetics of an engineered gold binding peptide by surface plasmon resonance spectroscopy and a quartz crystal microbalance. Langmuir 22(18):7712–7718

Tamerler C, Khatayevich D, Gungormus M et al (2010) Molecular biomimetics: GEPI-based biological routes to technology. Biopolymers 94(1):78–94

Tang Z, Palafox-Hernandez JP, Law WC et al (2013) Biomolecular recognition principles for bionanocombinatorics: an integrated approach to elucidate enthalpic and entropic factors. ACS Nano 7(11):9632–9646
Taniguchi K, Nomura K, Hata Y et al (2007) The Si-tag for immobilizing proteins on a silica surface. Biotechnol Bioeng 96(6):1023–1029
Tavolaro A, Tavolaro P, Drioli E (2006) Influence of synthesis parameters on vanadium-silicalite-1 crystal growth prepared with fluoride-containing media. J Cryst Growth 289(2):609–616
Teeri TT (1997) Crystalline cellulose degradation: new insight into the function of cellobiohydrolases. Trends Biotechnol 15(5):160–167
Thai CK, Dai H, Sastry MSR et al (2004) Identification and characterization of Cu_2O- and ZnO-binding polypeptides by *Escherichia coli* cell surface display. J Biotech Bioeng 87(2):129–137
Tran CTH, Nosworthy NJ, Kondyurin A et al (2013) CelB and β-glucosidase immobilization for carboxymethyl cellulose hydrolysis. RSC Adv 3(45):23604–23611
Trivedi AH, Spiess AC, Daussmann T et al (2006) Study on mesophilic and thermophilic alcohol dehydrogenases in gas-phase reaction. Biotechnol Prog 22(2):454–458
Turner P, Mamo G, Karlsson E (2007) Potential and utilization of thermophiles and thermostable enzymes in biorefining. Microb Cell Fact 6(1):9
Umlauf BJ, McGuire MJ, Brown KC (2014) Introduction of plasmid encoding for rare tRNAs reduces amplification bias in phage display biopanning. Biotechniques 58(2):81–84
Walsh TR (2014) Fundamentals of peptide-materials interfaces. In: Walsh TR, Knecht MR (eds) Bio-inspired nanotechnology. Springer, New York, pp 17–36
Wang SQ, Humphreys ES, Chung SY et al (2003) Peptides with selective affinity for carbon nanotubes. Nat Mater 2(3):196–200
Whaley SR, English DS, Hu EL et al (2000) Selection of peptides with semiconductor binding specificity for directed nanocrystal assembly. Nature 405(6787):665–668
Wilson SA, Peek K, Daniel RM (1994) Immobilization of a proteinase from the extremely thermophilic organism *Thermus* Rt41A. Biotechnol Bioeng 43(3):225–231
Wilson L, Palomo JM, Fernández-Lorente G et al (2006) Improvement of the functional properties of a thermostable lipase from *Alcaligenes* sp. via strong adsorption on hydrophobic supports. Enzyme Microb Technol 38(7):975–980
Wittrup KD (2001) Protein engineering by cell-surface display. Curr Opin Biotechnol 12(4):395–399
Wood T, Garcia-Campayo V (1990) Enzymology of cellulose degradation. Biodegradation 1(2–3):147–161
Yadav R, Satyanarayanan T, Kotwal S et al (2011) Enhanced carbonation reaction using chitosan-based carbonic anhydrase nanoparticles. Curr Sci 100(4):520–524
Yang M, Choi BG, Park TJ et al (2011) Site-specific immobilization of gold binding polypeptide on gold nanoparticle-coated graphene sheet for biosensor application. Nanoscale 3(7):2950–2956
Ye X, Honda K, Sakai T et al (2012) Synthetic metabolic engineering-a novel, simple technology for designing a chimeric metabolic pathway. Microb Cell Fact 11(1):120
You C, Zhang YHP (2013) Cell-free biosystems for biomanufacturing. In: Zhong J-J (ed) Future trends in biotechnology, vol 131. Springer, Berlin, pp 89–119
Zhang YHP, Myung S, You C et al (2011) Toward low-cost biomanufacturing through in vitro synthetic biology: bottom-up design. J Mater Chem 21(47):18877–18886
Zhou Z, Hartmann M (2012) Recent progress in biocatalysis with enzymes immobilized on mesoporous hosts. Top Catal 55(16–18):1081–1100

Chapter 24
Molecular Dynamics Simulations to Study Structure-Function Relationship in Psychrophilic Enzymes

Elena Papaleo, Matteo Tiberti, and Gaetano Invernizzi

1 Activity-Stability-Flexibility in Psychrophilic Enzymes

It is crucial to unravel the details of the structural mechanisms that rule the relationship between thermal stability, activity and dynamics in psychrophilic enzymes both for fundamental research and industrial applications (Gerday et al. 2000). Indeed, enzymes isolated from cold-adapted organisms are often of interest for their features in terms of high activity at very low temperatures, thermolability and unusual specificity, making them suitable for a large spectrum of industrial applications. The study of enzyme cold adaptation has also a broader and more general relevance. We have to consider that cold-adapted organisms are able to successfully grow and proliferate in very challenging and restrictive habitats for life. It is thus surprising to observe how those organisms evolved to feature metabolic fluxes comparable to those exhibit by the mesophilic counterparts at their own optimal growth temperature (van den Burg 2003; Struvay and Feller 2012).

Different adaptation mechanisms have been exploited by cold-adapted organisms. Among them, the optimization of their enzymatic repertory is one of the most exciting. Indeed, if we think that temperature is one of the fundamental environmental factors for life and that reaction rates is dramatically reduced when the medium temperature decreases from 37 °C to 10 °C (Feller and Gerday 2003; Siddiqui and Cavicchioli 2006), it is really surprising that biological activities can be recorded even at temperatures as low as −20 °C (Cary et al. 2010). Thus,

E. Papaleo (✉) • G. Invernizzi
Structural Biology and NMR Laboratory, Department of Biology,
University of Copenhagen, Copenhagen, Denmark
e-mail: elena.papaleo@bio.ku.dk

M. Tiberti
Department of Biotechnology and Biosciences, University of Milano-Bicocca,
Piazza della Scienza 2, 20126 Milan, Italy

P.H. Rampelotto (ed.), *Biotechnology of Extremophiles*, Grand Challenges in Biology and Biotechnology 1, DOI 10.1007/978-3-319-13521-2_24

psychrophilic organisms had to evolve and optimize their enzymatic repertory to survive to extremely cold environments.

In particular, psychrophilic enzymes are generally characterized by higher thermal lability and higher catalytic efficiency (k_{cat}/K_m) at low temperatures with respect to their warm-adapted counterparts (Siddiqui and Cavicchioli 2006; Struvay and Feller 2012). Their maximal activity is shifted to lower temperatures with respect to the mesophilic enzymes, reflecting the weak stability of psychrophilic enzymes, which are generally prone to inactivation and unfolding at moderate temperatures. It has also been shown that enzyme cold adaptation is usually "incomplete", since the activity of most of the psychrophilic enzymes around 0 °C, although high, is generally lower than the one of the mesophilic homologs at 37 °C (Siddiqui and Cavicchioli 2006).

k_{cat}/K_m is generally optimized by both increasing k_{cat} at the expenses of K_m or optimizing both the kinetic parameters in different cold-adapted enzymes (Siddiqui and Cavicchioli 2006; Struvay and Feller 2012). The increase in k_{cat} is also related to a decrease in activation free energy of the catalysed reaction, and in particular to a decrease of activation enthalpy, which was speculated to be structurally achieved by a decrease in the number of enthalpy-driven interactions that need to be broken to reach the reaction transition state. These aspects have been also considered as an indication of low stability of cold-adapted enzymes and of enhanced structural flexibility, at least in the proximity of the catalytic site. This was the first suggestion of a relationship between activity and stability in thermal adaptation and they structurally reflect in a higher flexibility of the three-dimensional (3D) architecture of the protein (Fields 2001; Somero 2004).

Nevertheless, the link between activity, flexibility and thermal stability in cold-adapted enzymes is still debated. The intrinsic thermal lability of cold-adapted enzymes, together with their enhanced low temperature activity, suggests a direct connection between them. Indeed, activity at low temperatures may require a weakening of intramolecular interactions that, in turn, results in reduced stability. Otherwise, the relationship seems not be so clear and straightforward, since low thermal stability could be also related to random genetic drift. It might be indeed just a consequence of lack of evolutionary pressure for stable enzyme in low temperature habitats (Wintrode and Arnold 2000; Leiros et al. 2007; Fedøy et al. 2007). Moreover, the existence of unusual cold-adapted enzymes, which feature both unusual thermostability and high catalytic efficiency, have been reported (Leiros et al. 2007; Fedøy et al. 2007), along with proofs of the ability to decouple stability and activity in in vitro studies (Wintrode and Arnold 2000; Jónsdóttir et al. 2014). All these observations make the definition of activity-stability-flexibility relationship even more challenging.

The hypotheses of an intimate connection between structural rearrangements, protein flexibility, catalytic activity and thermal stability in cold-adapted enzymes have stimulated the scientific community to define the mechanisms related to enzyme cold-adaptation in atomic details with particular attention to the effects mediated by residue substitutions using, for instance, mutational studies.

2 Localized Flexibility and Enzyme Cold Adaptation

The enhanced flexibility of cold-adapted enzymes has been shown not be necessarily spread over the whole 3D structure (global flexibility) but it can be localized in specific regions that affect even from distal site the surroundings of the catalytic site (Olufsen et al. 2005; Papaleo et al. 2006, 2008, 2011a, b; Chiuri et al. 2009; Pasi et al. 2009; Mereghetti et al. 2010; Tiberti and Papaleo 2011; Martinez et al. 2011; Isaksen et al. 2014).

The low thermal stability of psychrophilic enzymes supports the flexibility hypothesis, but its verification is not straightforward due to the intrinsic difficulties in defining and measuring flexibility. Indeed, we can describe flexibility in terms of dynamic motions and thus relate this to a specific timescale in which certain dynamics occur or we can consider it as related to the degree of the deformation of the structure at a certain temperature.

The study at the atom level of protein flexibility in solution is a challenging task and only few experimental techniques can really win the challenge, such as NMR, EPR, or neutron scattering and very recently the sampling of electron density from X-ray crystallography (Fraser and Jackson 2011). We will not treat them in this chapter since our focus is the usage of molecular dynamics simulations, i.e. a computational technique, to tackle the study of cold adapted enzymes. Nevertheless, the aforementioned experimental techniques (Tehei et al. 2005; Heidarsson et al. 2009; van den Bedem and Fraser 2015), and NMR especially, are the more promising approaches to integrate with simulations, on which we cannot completely rely as we will discover in the next pages.

3 An Ensemble Description of Protein Structures and the Importance of Protein Dynamics

In the last decades, while researchers in cold-adaptation field started to take advantage from experimental and computational structural studies to solve the enigma of enzyme cold adaptation, an increasing amount of evidence supported the strict relationship not only between protein structure and function, but also between enzyme dynamics and activity (Henzler-Wildman et al. 2007; Henzler-Wildman and Kern 2007; Nashine et al. 2010). Moreover, the notion that proteins are dynamic rather than static entities and are thus better described as an ensemble of conformations in solution starts to be widely accepted by the community in protein science (Hilser et al. 2006; Acuner Ozbabacan et al. 2010; Woldeyes et al. 2014).

Indeed, proteins are not static molecules and they experience conformational changes across a number of different sub-states over different timescales. Some of those motions are also called 'breathing motions' of the protein structure. The transitions between the different states often depend on concerted motions of groups of residues involving hinge and rocking motions in timescales from 10^{-8} to 10^{3} s. It has been also suggested that many fluctuations, involving side-chain or main-chain motions and that originate from rotations, stretching or torsional motions (10^{-12} s) can underlie the large structural motions (Eisenmesser et al. 2005).

It thus becomes critical to take into account the timescales that are accessible to the technique that one wants to employ for the structural and computational studies of cold-adapted enzymes. Conventional MD simulations, thanks to the progresses in the field, can routinely reach the microsecond (10^{-6} s) or in exceptional cases the millisecond timescale (10^{-3} s) but still suffer of limitations when it comes to compare the simulated timescale to the dynamics that the same protein would experience in solution in the experiments. Solutions are available to overcome these problems. Indeed, MD simulations often encounter the risk to entrap the protein for a too long time in local minima. Enhanced sampling methods coupled to atomistic MD physical models have been also proposed (Spiwok et al. 2014; Barducci et al. 2015) to overcome limitations of conventional MD simulations, such as temperature or Hamiltonian replica exchange, metadynamics approaches or restraining simulations using ensemble averaged NMR chemical shifts (Camilloni et al. 2012, 2013; Camilloni and Vendruscolo 2014). The two latters especially hold promising in the capability to describe with high accuracy with respect to the experimental data protein dynamics occurring on the millisecond timescale and even beyond (Camilloni et al. 2012; Sutto and Gervasio 2013; Palazzesi et al. 2013; Papaleo et al. 2014b).

As stated above, enzyme catalysis involves the 'breathing' of particular sites of the enzyme structure, enabling for example the accommodation of the substrate or assuming states that resemble the transition state for the enzymatic reaction. The ease of such conformational changes might be an important determinant of catalytic efficiency. This observation is even more relevant if we consider that there are now many examples of proteins in which, even in the free (substrate-unbound) state, we can observe conformations that resemble functionally important states, such as substrate- or ligand-bound or the transition state conformations (Boehr et al. 2009; Ma and Nussinov 2010; Kosugi and Hayashi 2012). These states are very often just a minor population (even lower than 10 %) of the conformational ensemble of the free enzyme in solution and are difficult to characterize in details (Baldwin and Kay 2009; Mittermaier and Kay 2009).

The discoveries in the field of protein dynamics also impact on the way in which we study the structural determinants of cold-adaptation. An ensemble view of protein structures has to be applied to the investigation of cold-adapted enzymes more widely and as a complementary technique to the biochemical characterization of these enzymes. MD simulations of psychrophilic enzymes have been pioneers in this.

4 Molecular Dynamics Simulations of Proteins: An Overview, Limitations and Advantages

If it is clear that we need to rely on a ensemble description of protein structures and that dynamics over different timescales are important for protein function, and thus also for the understanding of cold-adapted enzymes at the atomic level, it should also be clear that suitable approaches for these studies, at the computational level, are molecular simulations. Among the different techniques available in this context, atomistic explicit solvent molecular dynamics is one of the most promising ones. Indeed, it allows us not to renounce to an atomistic description of either the protein and the solvent and to sample dynamics over different timescales (from the picosecond to the millisecond timescale, as stated in Paragraph 3). It is beyond the efforts of this chapter to summarize the principles beyond this methodology that has been applied to proteins for the first time many years ago (McCammon et al. 1977) and recently three eminent scientists in the field of protein simulations have been recognized by a Nobel Prize (Smith and Roux 2013; Nussinov 2014). Several review articles, and book chapters are already available to become familiar with this technique (Rapaport 1998; Dror et al. 2012; Leach 2001; van Gunsteren and Berendsen 1990). Nevertheless, we want to recall here some important practical aspects that can be crucial for a researcher that wants to approach to these techniques to study cold-adapted enzymes.

A brief scheme of how MD simulations work is reported and discussed in Fig. 24.1.

Simulations are based on the model of the system under study i.e., on a representation as close as possible to nature of the system behaviour. In principle, we could recur to first-principle physics to model atomic motions with the computer. Nonetheless, it is not easily applicable to use quantum mechanics (QM) theory for the thousands of atoms that make up a protein and to study long timescales. Therefore, other methods need to be evoked. Indeed, the widely accepted Born-Oppenheimer approximation allows us to decouple the nuclear and electronic behaviour of the system under investigation since the electron cloud equilibrates quickly for each instantaneous configuration of the heavy nuclei. This allows to avoid to take the electrons explicitly into account and the motion of the nuclei can then be expressed as a nuclear potential energy surface (PES). Given the PES, we can use classical mechanics to follow the dynamics of the nuclei and, in turn, the dynamics of the protein or biomolecule of interest. Since we have discarded the first-principle representation, however, we need to somehow define how the nuclei interact with each other. We thus need a potential energy function $V(\mathbf{r;p})$ that determines the energy of the system starting from some parameters (p) and the 3D coordinates of the atoms (r). Such potential functions are called force fields. Force fields are designed to adequately represent the physics of the system of interest, and this is usually attained by building the force field so that it approximates the relevant regions of the ab initio Born-Oppenheimer surface or correctly reproduce the experimental data. The functional form of a typical force field used in MD simulations of biomolecules is the sum of many contributions, which are meant to represent the molecule behaviour. Bond and angle vibrations around the equilibrium values are represented using Hook's law and a periodic version of it is used for torsions. Nonbonding electrostatic interactions are taken into account through the Coulomb potential, while Van der Waals interactions between atoms are usually modeled through the so-called Lennard-Jones potential. In a simulation one of the goal is to calculate emerging average properties from the motions of protein atoms, given certain selected thermodynamic conditions. This is possible because of the ergodic

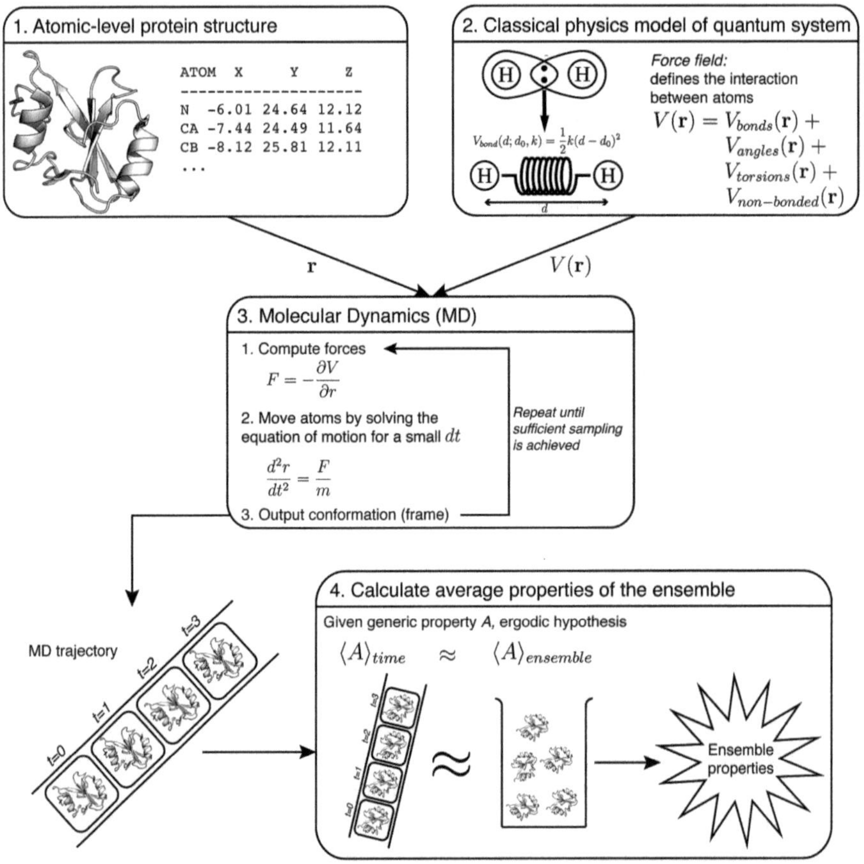

Fig. 24.1 Schematic representation of the workflow for MD simulations. Every simulation starts from two essential pieces of information. The first one is the detailed structure of the protein that needs to be simulated (1), in which the three-dimensional (3D) coordinates of all the atoms or all the heavy atoms have to be specified. Once this starting structure has been selected, it is further processed by adding solvent (usually water), counter-ions (usually Na^+ and Cl^-) to make the simulation as close to reality as possible. The second requirement is a physical model that specifies the interaction between the atoms in the simulation (2). In the molecular mechanics framework, interactions between atoms are approximated by means of classical physics, so that electrons do not need to be explicitly taken into account. For instance, the potential energy that defines the covalent bond has the form of Hooke's law, which is used in the description of classical springs. Several potential terms build up the total potential energy function V, which depends exclusively on a set of parameters (which collectively make up the so-called "force field") and the position of atoms. Once both these requirements are satisfied, several preparation steps can be carried out to bring the system to the desired thermodynamic conditions (not shown here). Finally, the productive MD simulation can run (3), and a number of time-consecutive conformations are written as output (the MD trajectory) (4). Once this is done, average properties of the system can be computed from the trajectory, as if it was a collection of independent structures sampled according to the Boltzmann distribution (5). This is possible under the assumption of the ergodic hypothesis, which states that, if sampling is carried out for enough time, the average properties extracted from the trajectory will converge to the properties of the system, in the given thermodynamic conditions (ensemble)

hypothesis, which states that simulations a protein for a time long enough will eventually allow to calculate properties that are representative of the average behaviour of the protein in the thermodynamic context. In MD simulations the time evolution of the particles, whose interactions are described by a force field, is studied by iteratively solving the Newton's equations of motions. This involves first the calculations of the forces acting on the atoms from the potential energy function. Then, a new conformation of the protein is calculated by numerically integrating the Newton's equation of motion for a small time step, usually of few femtoseconds at most. This process is repeated a number of times, until a sufficient timescale has been sampled in the simulation to study the properties of interest. We should notice that before collecting the productive MD simulations several steps are needed to ensure the proper physical behaviour of the system. Indeed, the protein under study has to be soaked in a box of water molecules, ions might need to be added, and several preparatory steps (which are short MD simulations themselves) are needed to bring the system at the selected thermodynamic conditions.

The main outcome of a MD simulation is the MD trajectory, which is the collection of structures generated over the simulation time. From this ensemble of structures, following the ergodic hypothesis, the properties of interest can be calculated.

In brief, it should be clear that two main aspects are important in approaching to MD simulations i.e., the physical model selected to describe the protein and the solvent (i.e. the force field) and the coverage of the conformational space accessible to the molecule with the correct Boltzmann distribution achieved in the simulation (i.e. the sampling).

The efforts done by the community in the last years pointed out a major issue in MD simulations, i.e. the sensitivity of protein dynamics to the force fields used to carry out the simulations. Current MD force fields have been extensively evaluated and validated by comparison with experimental data that are probes of dynamics over different timescales, such as parameters that can be measured by NMR or other biophysical spectroscopies (Lindorff-Larsen et al. 2012; Best et al. 2012). It turned out that the description of the same protein with different force field can provide different information on protein dynamics and some of the new generation force fields (such as CHARMM22* or AMBER99SB*-ILDN), in which backbone or side chain corrections have been added, seem to perform better in describing dynamics that is comparable to the NMR data than some older force fields (such as CHARMM22 and old OLPS versions). The selection of the force fields is thus crucial and, when possible, a cross-validation against experimental biophysical data on the system under investigation is preferable before trusting the simulation results too much.

On one hand, the high accuracy that some of those force fields have achieved is very encouraging in the direction of successfully using MD simulations to study protein dynamics in details and to compare, for example, different variants of an enzyme or different homologs adapted to different conditions, such as extremophilic enzymes. Indeed, the most accurate force fields have been shown capable to unfold and refold small fast folding domains to their native structure (Lindorff-Larsen et al. 2011; Piana et al. 2013), to sample the conformational space around the protein native state of folded proteins (Lindorff-Larsen et al. 2012; Martín-García et al. 2015) and to quantitatively estimate the population of minor states in enzymes and the effects induced upon mutations on these states (Papaleo et al. 2014b). On the other hand, atomistic classical force fields are not perfect and there are still many limitations, such as for example the risk to overestimate salt bridges in solution (Debiec et al. 2014; Jónsdóttir et al. 2014), especially for charged residues not involved in electrostatic networks or solvent-exposed (Jónsdóttir et al. 2014), or even in treating metal ions bound to metallo-proteins or -enzymes (Calimet and Simonson 2006; Zhu et al. 2013; Li et al. 2013) and explicitly account for polarization effects (Halgren and Damm 2001). Improvements in these directions are thus needed, especially in the study of cold-adapted enzymes where many targets for the study are metal-binding enzymes or where the role of salt bridges can (or cannot) be important for temperature adaptation.

5 The Folding Funnel Model of Cold-Adapted Enzymes

D'Amico and coworkers (2003) have proposed the integration of biochemical and biophysical data on psychrophilic enzymes in a folding funnel model which describes the folding-unfolding reactions of cold-adapted enzymes. They proposed that differentially temperature-adapted enzymes also possess differently shaped funnels (Fig. 24.2). The height of the funnel represents the folding free energy, which corresponds to the conformational stability, whereas the unfolded state occupies the upper region of the funnel. In their model, the edge of the funnel for the cold-adapted proteins is larger respect to the one of warm-adapted counterparts, corresponding to a broader distribution of the unfolded states. During the folding process, the free energy levels decrease, as well as the broadness of the conformational ensemble accessible to protein structures, in agreement with the general funnel theory applied to protein folding (Onuchic and Wolynes 2004). The walls of the funnel have been suggested to present a different amount of roughness in psychrophilic and thermophilic enzymes (Fig. 24.2). Thermophilic proteins have overall a quite corrugate funnel. Since structures of cold-adapted enzymes generally unfold cooperatively without intermediates, due to few intramolecular interactions, the funnel slopes are steeper and smoother. The bottom of the folding funnel of a warm-adapted protein is represented as a single global minimum divided by large energy barriers from other minor populated minima and overall little conformational freedom. The bottom of a psychrophilic protein is wider and rugged, as it represents a collection of many conformers separated by low energy barriers, resulting in a more labile and flexible protein.

According to a conformational selection scenario (Ma and Nussinov 2010), the substrate will bind to the enzyme populations competent for the interaction with it and, as a consequence, a population shift toward these binding-prone conformations is observed, leading to an 'active' structural ensemble. In the case of cold-adapted enzymes where the bottom of the folding funnel is rugged, the aforementioned population shift requires only modest free energy changes for the interconversion of the different conformational states, thus explaining the role of increased flexibility, for example, in facilitating the binding of the substrate.

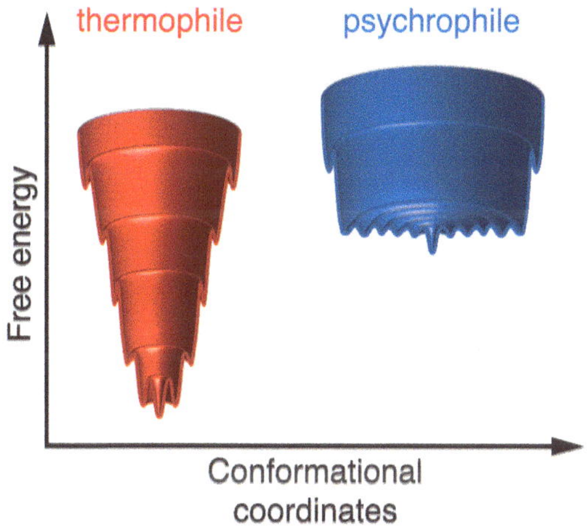

Fig. 24.2 Folding funnel model for thermophilic and psychrophilic enzymes. The folding free energy is shown as a function of reaction coordinates that account for the conformation of thermophilic (*red*) and psychrophilic (*blue*) enzyme. Here the funnels have been cut vertically to expose the internal structure. The top of the funnels is occupied by unfolded conformations, while native conformations are found at the bottom. The different heights of the funnels and ruggedness of the bottom exemplify the energy barriers for the interconversion between different substates, as well as the different degree of structural fluctuations between the two extremophiles

The computational study of the cold-adapted funnel model is a challenging task since it requires the proper sampling and representation of the free energy surface accessible to the protein under investigation in both its folded and unfolded state. Moreover, the free energy landscape (FEL) of a protein is a complex multidimensional landscape and in simulations we need thus to reduce it to two or three coordinates (reaction coordinates). The reaction coordinates to describe the FEL have to be collective variables, i.e. they need to capture the main features of the conformational ensemble under study. The FEL description achieved is strictly dependent on the choice of the reaction coordinates. Suitable descriptors to be employed as reaction coordinates can be, for example, the root mean square deviation (rmsd) of specific groups of residues (i.e, the ones in a loop that is expected to have different conformational preferences in cold- and warm-adapted homologs), the gyration radius, the first principal components from Principal Component Analysis (PCA) or contact maps. In a equilibrated thermodynamic system, the free energy can be then estimated from the probability density function of one or more of these reaction coordinates (Poland 2001).

In the case of cold-adapted enzymes, we have provided a qualitative representation of the FEL from multi-replicate all-atom MD simulations (Papaleo et al. 2009) of serine proteases and uracil-DNA glycosylases (Mereghetti et al. 2010). This study was limited to the description of the bottom of the folding funnel of cold- and warm-adapted counterparts, i.e. the region related to conformational fluctuations around the native state. We hope that this study can stimulate further investigation

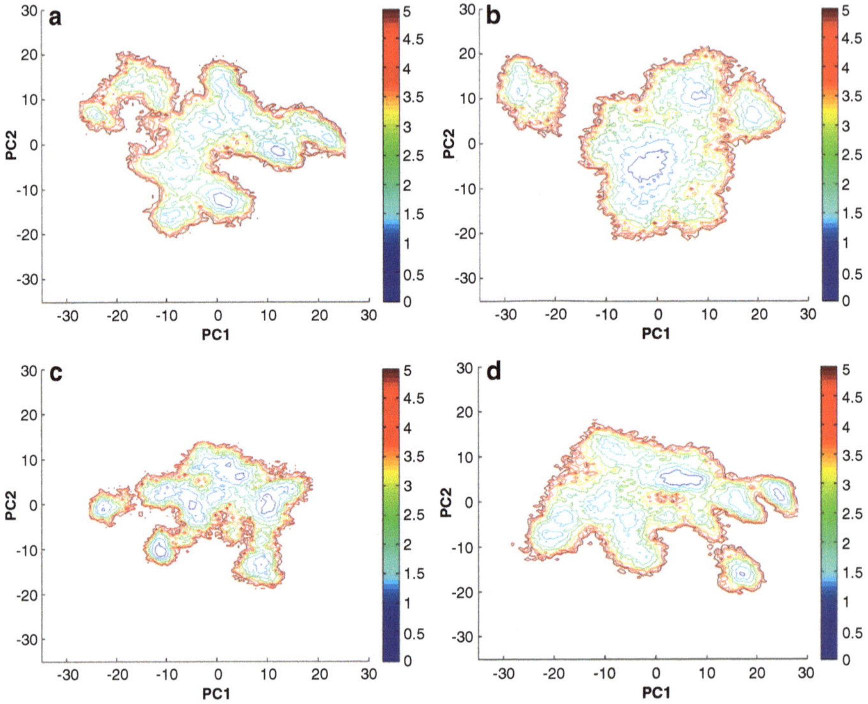

Fig. 24.3 Two-dimensional FEL profiles of cold- and warm-adapted enzymes using as reaction coordinates the two first principal components from PCA analysis of MD trajectories of a mesophilic (**a**, **b**) and a psychrophilic (**c**, **d**) serine protease at 283 (**a**, **c**) and 310 K (**b**, **d**) (Reprinted adapted with permission from (Mereghetti et al. 2010). Copyright 2010 American Chemical Society)

aiming at describing also the unfolded state of differently temperature-adapted enzymes. In our study we employed different properties, such as principal components from PCA of the MD ensemble (Amadei et al. 1993; Garcia et al. 1992), rmsd of loops in the surrounding of the catalytic site, and the protein radius of gyration (Fig. 24.3) as reaction coordinates. These computational studies allowed to test the hypothesis of the folding funnel model of cold-adapted enzymes mentioned above with regard to the bottom of the funnel. Indeed, the FEL from the MD ensemble of cold- and warm-adapted enzymes showed an intrinsic tendency of cold-adapted variants to explore more structural basins and a more rugged flat bottom landscape, which favours the interconversion among several metastable states.

Recently, many methods for enhanced sampling of the conformational space accessible to proteins have been proposed (Spiwok et al. 2014; Barducci et al. 2015) and can be combined to atomistic MD force fields. Thus, one intriguing direction for the future would be to extend the MD investigation mentioned above to the whole FEL, including the study of folding and unfolding in cold- and warm-adapted homologs, if we will also be able to improve the solvent models for MD simulations at temperatures higher than the ones at which the current solvent models have been parametrized. Until good solvent models along many range of temperatures will be developed, we cannot rely on MD simulations alone in studying extremophilic enzymes at different temperatures and a good strategy would be to always integrate this kind of calculations with experiments (Invernizzi et al. 2009; Ganjalikhany et al. 2012).

6 The Need for Structures

Due to the relatively low availability of structural data on cold-adapted enzymes, the limited availability of cold-adapted proteins for comparative purposes is a huge constraint and limitation. Several homologous enzymes that are adapted to different temperature with high-resolution atomic structures must exist to allow a suitable comparison. Most of the MD studies carried out so far only compared mesophilic and psychrophilic enzymes, since few examples are known for which also the 3D structure of the thermophilic counterpart is available (Tiberti and Papaleo 2011; Sigtryggsdóttir et al. 2014).

To overcome the lack of 3D structures of psychrophilic enzymes and their homologs, homology modelling or other modelling techniques (Eswar et al. 2008) are in principle useful to increase the subset of available structures for comparison. Nevertheless, a lot of caution is needed in this case. Indeed, homology models are not as accurate as experimental structures and their accuracy dramatically decrease with the decrease in sequence identity and similarity between targets and templates. If we consider that fine details often make the difference between cold- and heat-adapted enzymes, it becomes clear that homology models by themselves and in absence of an experimental validation are not the best of the strategies. This issue is even more critical for MD applications, since it has been shown that simulations started from models suffer of many limitations and often encounter the risk to sample portions of the conformational space with limited functional and structural relevance (Fan and Mark 2004; Raval et al. 2012). Therefore, a tight cross-talk with experiments becomes a necessary requirement when we employ models to describe the structure and dynamics of cold-adapted enzymes (Parravicini et al. 2013; Papaleo et al. 2014a).

We should also remember that models are just a way to overcome the lack of experimental structures but that more and more efforts are also needed experimentally to cover the gap of unknown structures from psychrophilic organisms.

7 A Recap on MD Simulations of Cold-Adapted Enzymes (1999–2015)

Table 24.1 reports a summary of all the classical MD simulations studies published so far for cold-adapted enzymes alone or in comparison with their warm-adapted counterparts, in which simulation length, force field used, number of replicates, temperature and methods for analyses are briefly summarized along with the reference to the original publication.

Table 24.1 Summary of MD simulations studies of cold-adapted enzymes published since 1999

Protein	Length and replicates	Force field	Temperature (K)	Analysis	Reference
Trypsin (psychrophile and mesophile)	0.5–1.2 ns	CHARMM22	300	rmsf 2D structures	Brandsdal et al. (1999)
Trypsin (psychrophile and mesophile)	6 ns (4 replicates)	GROMOS87	300	rmsf 2D structures SAS Hydrogen bonds	Papaleo et al. (2005)
Uracil-DNA-glycosylases (psychrophile, mesophile and mutant variants)	1–2 ns	AMBER95	300	Rmsf	Olufsen et al. (2005)
Thermolysin (psychrophile, mesophile and thermophile)	3 ns	AMBER94	273 and 300	rmsf Salt bridges Cation-PI Hydrogen bonds Cavities	Adekoya et al. (2006)
Elastases (psychrophile and mesophile)	12 ns (4 replicates)	GROMOS87	283 and 310	Clustering 2D structures Hydrogen bonds rmsf SAS S^2 order parameters Salt bridges	Papaleo et al. (2006, 2007)
Uracil-DNA glycosylase (psychrophile and mesophile)	6–9 ns	Parm99 (implicit solvent)	375, 400 and 425	SAS 2D structure rmsf Hydrogen bonds	Olufsen et al. (2007)

Table 24.1 (continued)

Protein	Length and replicates	Force field	Temperature (K)	Analysis	Reference
α-amylase, citrate synthase, malate dehydrogenase, alkaline protease, xylanase (psychrophile, mesophile, thermophile)	1–1.4 ns	GROMOS87	300 280–330	rmsf PCA	Spiwok et al. (2007)
Trypsin (psychrophila, mesophile)	10 ns (4 replicates)	GROMOS96	300	rmsf PCA 2D structure S^2 order parameters	Papaleo et al. (2008)
Termolysin (psychrophilic, mesophilic, thermophilic)	30 ns	CHARMM22	280 and 310	rmsf Hydrogen bonds Salt bridges	Xie et al. (2009)
Esterase (psychrophila)	10 ns	GROMOS96	277, 293 and 318	rmsf SAS 2D structure Hydrogen bonds	D'Auria et al. 2009 Aurilia et al. (2009)
α-amylase (psychrophile and mesophile)	6–11 ns (6 replicates)	GROMOS87	298	Hydrogen bonds 2D structure rmsf Anisotropic temperature factors Correlated motions PCA	Pasi et al. (2009)
Elastase and Uracil-DNA glycosylase (psychrophile and mesophile)	12 ns (10–12 replicates)	GROMOS96	283 and 310	PCA FEL Clustering Configurational entropy	Mereghetti et al. (2010)

(continued)

Table 24.1 (continued)

Protein	Length and replicates	Force field	Temperature (K)	Analysis	Reference
Subtilisin (psychrophile, mesophile and thermophile)	20 ns (4–8 replicates)	GROMOS96 43a1	283, 300 and 343	PCA 2D structure Hydrogen bonds SAS rmsf Salt bridge networks	Tiberti (2009)
Subtilisin (psychrophile and thermostable mutants)	50 ns	Not indicated	283 and 363	PCA rmsf 2D structure Clustering	Martinez et al. (2011)
α-amylase (psychrophila, mesophile and mutants)	6–10 ns (6 replicates)	GROMOS96	300	2D structure Hydrophobicity index PCA Hydrogen bonds Correlated motions Salt bridge networks rmsf Anisotropic temperature factor Self-organizing maps	Papaleo et al. (2011), Fraccalvieri et al. (2012)
GH5 cellulase (psychrophile, mesophile, thermophile and thermostable mutants)	10 ns	AMBER ff03	300, 400 and 500	2D structure rmsf SAS	Badieyan et al. (2012)
Chitinase (psychrophile)	10 ns	GROMOS96	273, 288 and 300	Clustering rmsf SAS Salt bridges Hydrogen bonds	Ramli et al. (2012)

Table 24.1 (continued)

Protein	Length and replicates	Force field	Temperature (K)	Analysis	Reference
Lipase (psychrophile)	30 ns (3 replicates)	AMBER ff99SB	278, 308 and 323	rmsf Radial distribution function of water NMA PCA Distance fluctuation map	Ganjalikhany et al. (2012)
Alkaline phosphatase (psychrophile)	40–60 ns (4 replicates)	GROMOS96 43a1	303	2D structure PCA rmsf Correlated motions Salt bridge networks PSN	Papaleo et al. (2013)
Acyl aminoacyl peptidase (psychrophile)	60 ns	GROMOS96 43a1	290	2D structure Salt bridges	Parravicini et al. (2013)
Acyl aminoacyl peptidase (psychrophile)	90 ns (7 replicates)	CHARMM22-CMAP	290	2D structure Intramolecular networks PSN	Papaleo et al. (2014)
Subtilisin (psychrophile and thermophile)	100 ns (2 replicates)	AMBER ff99SB-ILDN	300	rmsf SAS	Sigtryggsdóttir et al. (2014)
Trypsin (psychrophilic and mesophilic)	100 ns	OPLS2005	300	rmsf	Isaksen et al. (2014)

We report here on atomistic simulations of cold-adapted enzymes using classical force fields so that other kinds of computational techniques (such as quantum-mechanics approaches or coarse grained methods) are not in this list, along with studies done on cold-adapted proteins without an enzymatic activity

Pioneering MD studies of cold-adapted enzymes track back to 1999 and account for approximately 1–2 ns timescale using single MD runs for cold- and warm-adapted enzymes.

In many of more recent MD works, the usage of multi-replicate MD simulations guarantees a wider conformational sampling and the possibility to identify differences in flexibility between cold- and warm-adapted enzymes. The reproducibility of the results is a critical point in MD simulations, and the usage of multiple replicates can not only compensate for the lack of sampling but also allows to verify if the differences that we see between two different systems are statistically significant or just occurs in one unique simulation.

MD simulations of cold-adapted enzymes published so far have been analyzed with many different tools to characterize not only flexibility patterns but also long-range communication, changes in the networks of intra- and intermolecular interactions, structural changes, different conformational substates.

Among the techniques to analyze MD data and applied so far to the study of cold-adapted enzymes we have already mentioned PCA, also known as Essential Dynamics (ED), which aims at extracting informative directions of motions in a multidimensional space by reducing the overall complexity of the trajectories and isolating the larger amplitude motions.

Commonly employed metrics to evaluate protein flexibility in MD simulations of extremophilic enzymes are the per-residue root mean square fluctuations (rmsf) (Fig. 24.4) or anisotropic temperature factors from the MD trajectories using as a reference the average structure from the simulation or even back-calculated S2 generalized order parameters. Caution has to be taken when we employ rmsf or B-factors to estimate protein flexibility from a MD trajectory. They are indeed strongly dependent on the average structure used as a reference and the better way to proceed should be to calculate these metrics on shorter time-windows with respect to the full simulation length to see if there are any statistically significant differences in flexibility patterns of two different proteins. Indeed, since cold- and warm-adapted homologs share very often the same 3D fold and few amino acidic substitutions, the differences in rmsf or B-factor are more likely differences in the intensity of the fluctuations than differences in the presence or absence of certain peaks. It thus becomes crucial to address if the differences observed in the rmsf intensity are not ascribable to noise in the analysis but are truly genuine.

Other analyses go in the direction to define quantitative indexes of similarity within the ensemble of conformations of the cold- and warm-adapted variants, i.e. defining the overlap between the population of the two ensembles. This has been done for a long time for example evaluating the root mean square inner product (rmsip) or other overlap metrics upon PCA analysis. Rmsip is often calculated for the first 20 principal components of the PCA matrix and can range from 0 to 1, where a rmsip value equal to 1 is achieved if the two MD simulations sample identical essential subspaces and 0 when the two subspaces are completely orthogonal.

Other methods employ metrics to estimate cross-correlated motions and can range from Pearson Correlation to methods based on mutual information. Once a map of residues featuring correlated motions is achieved, it can be also represented

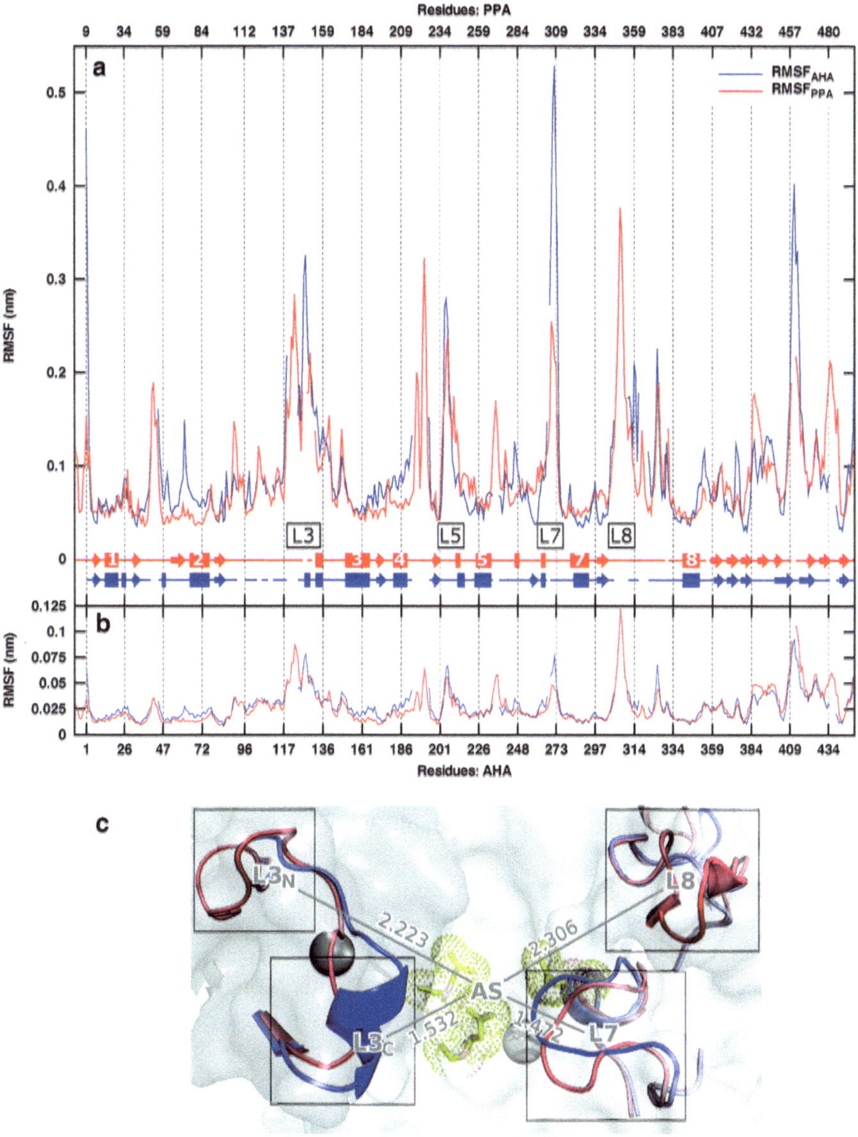

Fig. 24.4 Flexibility profiles from MD simulations calculated as rmsf of Cα atoms in cold- (AHA) and warm- (PPA) adapted counterparts. The rmsf profiles for the two proteins have been aligned according to a structural alignment between the two experimental structures and they are shown as thick lines that are broken in correspondence of alignment gaps. (**a**) we show here the rmsf profile of the whole concatenated trajectories (i.e., merging all the replicates together). (**b**) the rmsf profile over 100 ps intervals is shown. (**c**) The most flexible loops are highlighted on the 3D structure. The distances are ensemble averages and are calculated between the centers of mass of the groups and expressed in nm (Reprinted adapted with permission from (Pasi et al. 2009). Copyright 2009 American Chemical Society)

on the 3D structure to better highlight different dynamical patterns in different proteins. Also these methods suffer of convergence issues and it is thus very important to calculate and then average them on much shorter timescales than the one covered by the whole simulation length.

Another very recent approach is inspired to graph theory and describe the intra- and intermolecular interactions as networks where each protein residue is a node and a link is included in the network if two residues interact each other. This allows for example to identify hubs (i.e., highly interconnected residues within the networks that might have a structural role) and paths of long-range communication between distal residues.

It is although time to also move to other state-of-the-art approaches integrated to MD simulations, such as the enhanced sampling methods mentioned above, which can provide more accurate and comprehensive results.

8 A Family-Centred Point of View in Comparative Structural and Dynamics Studies of Cold- and Warm-Adapted Enzymes

It is well known that psychrophilic enzymes adopt diverse structural strategies to increase structural flexibility, such as the weakening of intramolecular hydrogen bonds, optimizing protein-solvent interactions, decreasing the packing of the hydrophobic core, enhancing the solvent accessibility of hydrophobic side chains and reducing the number of ion-pair networks (Russell 2000; Gianese et al. 2002; Feller and Gerday 2003; Siddiqui and Cavicchioli 2006; Adekoya et al. 2006; Tronelli et al. 2007).

Some general features can be still identified, such as for example the tendency of cold-adapted enzymes to have enhanced localized flexibility in specific regions that can locally (Papaleo et al. 2006, 2008) or long-range (Papaleo et al. 2011a, 2012, 2013; Fraccalvieri et al. 2012) affect the catalytic site. Moreover, in three different studies (Pasi et al. 2009; Papaleo et al. 2011a; Isaksen et al. 2014), carried out on two different enzymes, it has been suggested by simulations that one of the main differences between cold- and warm-adapted enzymes is located on the protein surface so that cold-adapted enzymes have evolved toward 'softer', i.e. more flexible and with less intramolecular interactions, regions with respect to their warm-adapted counterparts.

Even if it can be tempting, however, a general theory on enzyme cold-adaptation cannot be formulated yet. Similarities in the strategies employed at the molecular level can be identified in proteins sharing the same 3D and same functional residues, i.e. proteins belonging to the same family and for some extend superfamily (Papaleo et al. 2011b). Moreover, the networks of intra and intermolecular interactions and correlated motions related to these mechanisms can change in the different enzyme families and are strictly connected to specific features of the 3D fold (Papaleo et al. 2011a, b).

The analysis of many proteomes also confirmed the lack of a unique theory of cold-adaptation mechanisms. Indeed, Gu and Hilser (Gu and Hilser 2009) reported a homogeneous modulation of structural flexibility and stability across the components of the proteome of organisms adapted to different habits, but also that the molecular mechanisms of temperature adaptation can still significantly vary in different proteins.

The investigation so far collected suggests that each enzyme exploits diverse structural strategies to adapt to low temperatures, strategies that are often difficult to precisely identify.

In this context, we should keep in mind for comparative studies, as the ones achieved with MD simulations, that the sequence similarity between psychrophilic and mesophilic counterparts should be relatively high to make the comparison effectively meaningful. Subtle structural effects are hard to estimate if pronounced differences in the protein architecture are found. This is even more important if we consider that very often we are searching for very subtle modifications that are sufficient to adapt a protein to cold temperatures. To overcome these problems, comparative MD studies among homologous enzymes with high sequence identities and similarities (higher than 60–70 %) of differently temperature-adapted enzymes are strongly encouraged and have already been applied in some cases (Papaleo et al. 2008; Sigtryggsdóttir et al. 2014). We would encourage to make this a more routinely and general approach in the study of cold-adaptation at the atomic level.

For example, MD investigations on cold-adapted serine-proteases, clarify how distinct members of this superfamily have addressed the detrimental effects of low temperature on protein activity and stability (Papaleo et al. 2006, 2007, 2008; Isaksen et al. 2014). As an example, the finding that the separation between psychrophilic and mesophilic trypsins or elastases, is subsequent to the separation of trypsins from elastases, is an indication that cold-adapted trypsins and elastases independently evolved identical strategies to optimize flexibility at low temperatures, and is a striking case of molecular evolutionary convergence (Papaleo et al. 2008).

The many structural and computational studies mentioned in this paragraph suggest, thus, that even if a general theory for enzyme cold adaptation cannot be postulated, enzymes sharing function and 3D architecture are expected to adopt similar solutions to tune their flexibility and stability in a way that they can carry out their function under extreme low-temperature conditions, pointing out an evolutionary convergence on structural and dynamical properties of homologue cold-adapted enzymes. These evidences suggest that even if common structural strategies in cold adaptation cannot be formulated, a family-centered point of view will be extremely useful in the comparative analyses of cold- and warm-adapted enzymes.

9 Future Perspectives

MD simulations demonstrated their potential in the study of structural and dynamical properties of cold-adapted enzymes in the last two decades. It is now time to make more progresses in the field taking advantage of the usage of new and more accurate force fields and new methods to enhance the sampling in the simulations or to account for experimental restraints derived for example from NMR data, such as chemical shifts which are probe of dynamics over different timescales. In the application of conventional MD, a step forward is needed both in terms of sampling beyond the microsecond timescale but also in order to assess the reproducibility of the results. New methods can also be applied to the analyses of the MD data to identify for examples differences in conformational transitions that the proteins undergoes in solution and differences in the population of major and minor states of the conformational ensemble. MD simulations can also have a big potential in the study of cold adapted enzymes for applicative purposes. They indeed provide a computationally affordable technique to screen and to design mutant variants with increased local flexibility or enhanced protein stability.

Conflict of Interest Elena Papaleo, Matteo Tiberti, and Gaetano Invernizzi declare that they have no conflict of interest.

References

Acuner Ozbabacan SE, Gursoy A, Keskin O, Nussinov R (2010) Conformational ensembles, signal transduction and residue hot spots: application to drug discovery. Curr Opin Drug Discov Devel 13:527–537

Adekoya OA, Helland R, Willassen N-P, Sylte I (2006) Comparative sequence and structure analysis reveal features of cold adaptation of an enzyme in the thermolysin family. Proteins 62:435–449

Amadei A, Linssen AB, Berendsen HJ (1993) Essential dynamics of proteins. Proteins 17:412–425

Aurilia V, Rioux-Dube JF, Marabotti A, Pezolet M, D'Auria S (2009) Structure and dynamics of cold-adapted enzymes as investigated by FT-IR spectroscopy and MD. The case on an esterase from Pseudoalteromonas haloplanktis. J Phys Chem B 113:7753–7761

Badieyan S, Bevan DR, Zhang C (2012) Study and design of stability in GH5 cellulases. Biotechnol Bioeng 109:31–44

Baldwin AJ, Kay LE (2009) NMR spectroscopy brings invisible protein states into focus. Nat Chem Biol 5:808–814

Barducci A, Pfaendtner J, Bonomi M (2015) Tackling sampling challenges in biomolecular simulations. Methods Mol Biol 1215:151–171

Best RB, Zhu X, Shim J et al (2012) Optimization of the additive CHARMM all-atom protein force field targeting improved sampling of the backbone φ, ψ and side-chain $\chi(1)$ and $\chi(2)$ dihedral angles. J Chem Theory Comput 8:3257–3273

Boehr DD, Nussinov R, Wright PE (2009) The role of dynamic conformational ensembles in biomolecular recognition. Nat Chem Biol 5:789–796

Brandsdal BO, Heimstad ES, Sylte I, Smalås AO (1999) Comparative molecular dynamics of mesophilic and psychrophilic protein homologues studied by 1.2 ns simulations. J Biomol Struct Dyn 17:493–506

Calimet N, Simonson T (2006) Cys(x)His(y)-Zn2+ interactions: possibilities and limitations of a simple pairwise force field. J Mol Graph Model 24:404–411

Camilloni C, Vendruscolo M (2014) Statistical mechanics of the denatured state of a protein using replica-averaged metadynamics. J Am Chem Soc 136:8982–8991

Camilloni C, Robustelli P, De Simone A et al (2012) Characterization of the conformational equilibrium between the two major substates of RNase A using NMR chemical shifts. J Am Chem Soc 134:3968–3971
Camilloni C, Cavalli A, Vendruscolo M (2013) Replica-averaged metadynamics. J Chem Theory Comput 9:5610–5617
Cary SC, McDonald IR, Barrett JE, Cowan DA (2010) On the rocks: the microbiology of Antarctic Dry Valley soils. Nat Rev Microbiol 8:129–138
Chiuri R, Maiorano G, Rizzello A et al (2009) Exploring local flexibility/rigidity in psychrophilic and mesophilic carbonic anhydrases. Biophys J 96:1586–1596
D'Amico S, Marx J-C, Gerday C, Feller G (2003) Activity-stability relationships in extremophilic enzymes. J Biol Chem 278:7891–7896
D'Auria S, Aurilia V, Marabotti A, Gonnelli M, Strambini G (2009) Structure and dynamics of cold-adapted enzymes as investigated by phosphorescence spectroscopy. J Phys Chem B 113:13171–13178
Debiec KT, Gronenborn AM, Chong LT (2014) Evaluating the strength of salt bridges: a comparison of current biomolecular force fields. J Phys Chem B. doi:10.1021/jp500958r
Dror RO, Dirks RM, Grossman JP et al (2012) Biomolecular simulation: a computational microscope for molecular biology. Annu Rev Biophys 41:429–452
Eisenmesser EZ, Millet O, Labeikovsky W et al (2005) Intrinsic dynamics of an enzyme underlies catalysis. Nature 438:117–121
Eswar N, Eramian D, Webb B et al (2008) Protein structure modeling with MODELLER. Methods Mol Biol 426:145–159
Fan H, Mark AE (2004) Refinement of homology-based protein structures by molecular dynamics simulation techniques. Protein Sci 13:211–220
Fedøy A-E, Yang N, Martinez A et al (2007) Structural and functional properties of isocitrate dehydrogenase from the psychrophilic bacterium Desulfotalea psychrophila reveal a cold-active enzyme with an unusual high thermal stability. J Mol Biol 372:130–149
Feller G, Gerday C (2003) Psychrophilic enzymes: hot topics in cold adaptation. Nat Rev Microbiol 1:200–208
Fields PA (2001) Review: protein function at thermal extremes: balancing stability and flexibility. Comp Biochem Physiol A Mol Integr Physiol 129:417–431
Fraccalvieri D, Tiberti M, Pandini A (2012) Functional annotation of the mesophilic-like character of mutants in a cold-adapted enzyme by self-organising map analysis of their molecular dynamics. Mol Biosyst 8:2680–2691
Fraser JS, Jackson CJ (2011) Mining electron density for functionally relevant protein polysterism in crystal structures. Cell Mol Life Sci 68:1829–1841
Ganjalikhany MR, Ranjbar B, Taghavi AH, Tohidi Moghadam T (2012) Functional motions of Candida antarctica lipase B: a survey through open-close conformations. PLoS One 7:e40327
Garcia AE (1992) Large-amplitude nonlinear motions in proteins. Phys Rev Lett 68:2696–2699
Gerday C, Aittaleb M, Bentahir M et al (2000) Cold-adapted enzymes: from fundamentals to biotechnology. Trends Biotechnol 18:103–107
Gianese G, Bossa F, Pascarella S (2002) Comparative structural analysis of psychrophilic and meso- and thermophilic enzymes. Proteins 47:236–249
Gu J, Hilser VJ (2009) Sequence-based analysis of protein energy landscapes reveals nonuniform thermal adaptation within the proteome. Mol Biol Evol 26:2217–2227
Halgren TA, Damm W (2001) Polarizable force fields. Curr Opin Struct Biol 11:236–242
Heidarsson PO, Sigurdsson ST, Asgeirsson B (2009) Structural features and dynamics of a cold-adapted alkaline phosphatase studied by EPR spectroscopy. FEBS J 276:2725–2735
Henzler-Wildman K, Kern D (2007) Dynamic personalities of proteins. Nature 450:964–972
Henzler-Wildman KA, Lei M, Thai V et al (2007) A hierarchy of timescales in protein dynamics is linked to enzyme catalysis. Nature 450:913–916
Hilser VJ, García-Moreno EB, Oas TG et al (2006) A statistical thermodynamic model of the protein ensemble. Chem Rev 106:1545–1558

Invernizzi G, Papaleo E, Grandori R et al (2009) Relevance of metal ions for lipase stability: structural rearrangements induced in the Burkholderia glumae lipase by calcium depletion. J Struct Biol 168:562–570

Isaksen GV, Åqvist J, Brandsdal BO (2014) Protein surface softness is the origin of enzyme cold-adaptation of trypsin. PLoS Comput Biol 10:e1003813

Jónsdóttir LB, Ellertsson BÖ, Invernizzi G et al (2014) The role of salt bridges on the temperature adaptation of aqualysin I, a thermostable subtilisin-like proteinase. Biochim Biophys Acta. doi:10.1016/j.bbapap.2014.08.011

Kosugi T, Hayashi S (2012) Crucial role of protein flexibility in formation of a stable reaction transition state in an α-amylase catalysis. J Am Chem Soc 134:7045–7055

Leach AR (2001) Molecular modelling. Principles and applications, 2nd edn. Prentice Hall, Harlow/New York

Leiros H-KS, Pey AL, Innselset M et al (2007) Structure of phenylalanine hydroxylase from Colwellia psychrerythraea 34H, a monomeric cold active enzyme with local flexibility around the active site and high overall stability. J Biol Chem 282:21973–21986

Li P, Roberts BP, Chakravorty DK, Merz KM (2013) Rational design of particle mesh Ewald compatible Lennard-Jones parameters for +2 metal cations in explicit solvent. J Chem Theory Comput 9:2733–2748

Lindorff-Larsen K, Piana S, Dror RO, Shaw DE (2011) How fast-folding proteins fold. Science 334:517–520

Lindorff-Larsen K, Maragakis P, Piana S et al (2012) Systematic validation of protein force fields against experimental data. PLoS One 7:e32131

Ma B, Nussinov R (2010) Enzyme dynamics point to stepwise conformational selection in catalysis. Curr Opin Chem Biol 14:652–659

Martinez R, Schwaneberg U, Roccatano D (2011) Temperature effects on structure and dynamics of the psychrophilic protease subtilisin S41 and its thermostable mutants in solution. Protein Eng Des Sel 24:533–544

Martín-García F, Papaleo E, Gomez-Puertas P et al (2015) Comparing molecular dynamics force fields in the essential subspace. PLoS One 10:e0121114

McCammon J, Gelin B, Karplus M (1977) Dynamics of folded proteins. Nature 267:585–590

Mereghetti P, Riccardi L, Brandsdal BO et al (2010) Near native-state conformational landscape of psychrophilic and mesophilic enzymes: probing the folding funnel model. J Phys Chem B 114:7609–7619

Mittermaier AK, Kay LE (2009) Observing biological dynamics at atomic resolution using NMR. Trends Biochem Sci 34:601–611

Nashine VC, Hammes-Schiffer S, Benkovic SJ (2010) Coupled motions in enzyme catalysis. Curr Opin Chem Biol 14:644–651

Nussinov R (2014) The significance of the 2013 Nobel Prize in Chemistry and the challenges ahead. PLoS Comput Biol 10:e1003423

Olufsen M, Smalås AO, Moe E, Brandsdal BO (2005) Increased flexibility as a strategy for cold adaptation: a comparative molecular dynamics study of cold- and warm-active uracil DNA glycosylase. J Biol Chem 280:18042–18048

Olufsen M, Brandsdal BO, Smalås AO (2007) Comparative unfolding studies of psychrophilic and mesophilic uracil DNA glycosylase: MD simulations show reduced thermal stability of the cold-adapted enzyme. J Mol Graph Model 26:124–134

Onuchic JN, Wolynes PG (2004) Theory of protein folding. Curr Opin Struct Biol 14:70–75

Palazzesi F, Barducci A, Tollinger M, Parrinello M (2013) The allosteric communication pathways in KIX domain of CBP. Proc Natl Acad Sci USA 110:14237–14242

Papaleo E, Fantucci P, De Gioia L (2005) Effects of calcium binding on structure and autolysis regulation in trypsins. A molecular dynamics investigation. J Chem Theory Comput 1:1286–1297

Papaleo E, Riccardi L, Villa C et al (2006) Flexibility and enzymatic cold-adaptation: a comparative molecular dynamics investigation of the elastase family. Biochim Biophys Acta 1764:1397–1406

Papaleo E, Olufsen M, De Gioia L, Brandsdal BO (2007) Optimization of electrostatics as a strategy for cold-adaptation: a case study of cold- and warm-active elastases. J Mol Graph Model 26:93–103
Papaleo E, Pasi M, Riccardi L et al (2008) Protein flexibility in psychrophilic and mesophilic trypsins. Evidence of evolutionary conservation of protein dynamics in trypsin-like serine-proteases. FEBS Lett 582:1008–1018
Papaleo E, Mereghetti P, Fantucci P, Grandori R, De Gioia L (2009) Free-energy landscape, principal component analysis, and structural clustering to identify representative conformations from molecular dynamics simulations: the myoglobin case. J Mol Graph Model 27:889–899
Papaleo E, Pasi M, Tiberti M, De Gioia L (2011a) Molecular dynamics of mesophilic-like mutants of a cold-adapted enzyme: insights into distal effects induced by the mutations. PLoS One 6:e24214
Papaleo E, Tiberti M, Invernizzi G et al (2011b) Molecular determinants of enzyme cold adaptation: comparative structural and computational studies of cold- and warm-adapted enzymes. Curr Protein Pept Sci 12:657–683
Papaleo E, Renzetti G, Tiberti M (2012) Mechanisms of intramolecular communication in a hyperthermophilic acylaminoacyl peptidase: a molecular dynamics investigation. PLoS One 7:e35686
Papaleo E, Renzetti G, Invernizzi G, Asgeirsson B (2013) Dynamics fingerprint and inherent asymmetric flexibility of a cold-adapted homodimeric enzyme. A case study of the Vibrio alkaline phosphatase. Biochim Biophys Acta 1830:2970–2980
Papaleo E, Parravicini F, Grandori R et al (2014a) Structural investigation of the cold-adapted acylaminoacyl peptidase from Sporosarcina psychrophila by atomistic simulations and biophysical methods. Biochim Biophys Acta 1844:2203–2213
Papaleo E, Sutto L, Gervasio FL, Lindorff-Larsen K (2014b) Conformational changes and free energies in a proline isomerase. J Chem Theor Comput 10:4169–4174. doi:10.1021/ct500536r, 140820163733003
Parravicini F, Natalello A, Papaleo E et al (2013) Reciprocal influence of protein domains in the cold-adapted acyl aminoacyl peptidase from Sporosarcina psychrophila. PLoS One 8:e56254
Pasi M, Riccardi L, Fantucci P et al (2009) Dynamic properties of a psychrophilic alpha-amylase in comparison with a mesophilic homologue. J Phys Chem B 113:13585–13595
Piana S, Lindorff-Larsen K, Shaw DE (2013) Atomic-level description of ubiquitin folding. Proc Natl Acad Sci USA 110:5915–5920
Poland D (2001) Free energy distributions in proteins. Proteins 45:325–336
Ramli AN, Mahadi NM, Shamsir MS, Rabu A, Joyce-Tan KH, Murad AM, Illias RM (2012) Structural prediction of a novel chitinase from the psychrophilic glaciozyma antarctica PI12 and an analysis of its structural properties and function. J Comput Aided Mol Des 26:947–961
Rapaport D (1998) The art of molecular dynamics simulation. Cambridge University Press, Cambridge, UK
Raval A, Piana S, Eastwood MP et al (2012) Refinement of protein structure homology models via long, all-atom molecular dynamics simulations. Proteins 80:2071–2079
Russell NJ (2000) Toward a molecular understanding of cold activity of enzymes from psychrophiles. Extremophiles 4:83–90
Siddiqui KS, Cavicchioli R (2006) Cold-adapted enzymes. Annu Rev Biochem 75:403–433
Sigtryggsdóttir AR, Papaleo E, Thorbjarnardóttir SH, Kristjánsson MM (2014) Flexibility of cold- and heat-adapted subtilisin-like serine proteinases evaluated with fluorescence quenching and molecular dynamics. Biochim Biophys Acta 1844:705–712
Smith JC, Roux B (2013) Eppur si muove! The 2013 Nobel Prize in Chemistry. Structure 21:2102–2105
Somero GN (2004) Adaptation of enzymes to temperature: searching for basic "strategies". Comp Biochem Physiol B Biochem Mol Biol 139:321–333. doi:10.1016/j.cbpc.2004.05.003
Spiwok V, Lipovová P, Skálová T, Dusková J, Dohnálek J, Hasek J, Russell NJ, Králová B (2007) Cold-active enzymes studied by comparative molecular dynamics simulation. J Mol Model 13:485–497

Spiwok V, Sućur Z, Hošek P (2014) Enhanced sampling techniques in biomolecular simulations. Biotechnol Adv. doi:10.1016/j.biotechadv.2014.11.011
Struvay C, Feller G (2012) Optimization to low temperature activity in psychrophilic enzymes. Int J Mol Sci 13:11643–11665
Sutto L, Gervasio FL (2013) Effects of oncogenic mutations on the conformational free-energy landscape of EGFR kinase. Proc Natl Acad Sci 110:10616–10621. doi:10.1073/pnas.1221953110
Tehei M, Madern D, Franzetti B, Zaccai G (2005) Neutron scattering reveals the dynamic basis of protein adaptation to extreme temperature. J Biol Chem 280:40974–40979
Tiberti M, Papaleo E (2011) Dynamic properties of extremophilic subtilisin-like serine-proteases. J Struct Biol 174:69–83
Tronelli D, Maugini E, Bossa F, Pascarella S (2007) Structural adaptation to low temperatures—analysis of the subunit interface of oligomeric psychrophilic enzymes. FEBS J 274:4595–4608
Van den Bedem H, Fraser JS (2015) Integrative, dynamic structural biology at atomic resolution—it's about time. Nat Methods 12:307–318
Van den Burg B (2003) Extremophiles as a source for novel enzymes. Curr Opin Microbiol 6:213–218
Van Gunsteren WF, Berendsen HJC (1990) Computer simulation of molecular dynamics: methodology, applications and perspective. Angew Chem Int Eng Ed 29:992–1023
Wintrode PL, Arnold FH (2000) Temperature adaptation of enzymes: lessons from laboratory evolution. Adv Protein Chem 55:161–225
Woldeyes RA, Sivak DA, Fraser JS (2014) E pluribus unum, no more: from one crystal, many conformations. Curr Opin Struct Biol 28C:56–62
Xie BB, Bian F, Chen XL, He HL, Guo J, Gao X, Zeng YX, Chen B, Zhou BC, Zhang YZ (2009) Cold adaptation of zinc metalloproteases in the thermolysin family from deep sea and arctic sea ice bacteria revealed by catalytic and structure properties and molecular dynamics: new insights into relationship between conformational flexibility and hydrogen bonding. J Biol Chem 284:9257–9269
Zhu T, Xiao X, Ji C, Zhang JZH (2013) A new quantum calibrated force field for zinc—protein complex. J Chem Theory Comput 9:1788–1798

Chapter 25
Halobacterium Expression System for Production of Full-Length *Plasmodium falciparum* Circumsporozoite Protein

Wolf T. Pecher, Jong-Myoung Kim, Priya DasSarma, Ram Karan, Photini Sinnis, and Shiladitya DasSarma

1 Introduction

As an innovative platform for antigen display and vaccine development, we are employing an extremophilic microbe, *Halobacterium* sp. NRC-1, that provides a number of advantages for production of antigenic proteins and vaccines (DasSarma 2007; DasSarma et al. 2013). This member of the archaeal domain has been

W.T. Pecher
Department of Microbiology and Immunology, University of Maryland
School of Medicine, Baltimore, MD, USA

Institute of Marine and Environmental Technology, University System of Maryland, Baltimore, MD, USA

College of Arts and Sciences, University of Baltimore, Baltimore, MD, USA

J.-M. Kim
Department of Microbiology and Immunology, University of Maryland
School of Medicine, Baltimore, MD, USA

Institute of Marine and Environmental Technology, University System of Maryland, Baltimore, MD, USA

Department of Marine Bio-Materials and Aquaculture, PuKyong National University, Pusan, Republic of Korea

P. DasSarma • R. Karan • S. DasSarma (✉)
Department of Microbiology and Immunology, University of Maryland
School of Medicine, Baltimore, MD, USA

Institute of Marine and Environmental Technology, University System of Maryland, Baltimore, MD, USA
e-mail: sdassarma@som.umaryland.edu

P. Sinnis
Department of Molecular Microbiology and Immunology, Bloomberg
School of Public Health, Johns Hopkins University, Baltimore, MD, USA

P.H. Rampelotto (ed.), *Biotechnology of Extremophiles*, Grand Challenges in Biology and Biotechnology 1, DOI 10.1007/978-3-319-13521-2_25

determined to be non-toxic, lipopolysaccharide (LPS) free, and is biocompatible by virtue of its presence as a natural component of salt and salty condiments (http://www.els.net/WileyCDA/ElsArticle/refId-a0000394.html). The genome is completely sequenced and its biology has been studied in detail using genetic, transcriptomic, and proteomic methodologies (Berquist et al. 2006; DasSarma 2004; Ng et al. 1998, 2000). High-level regulated promoters have been used to develop expression vectors for the production of native and foreign proteins (Karan et al. 2013, 2014). Moreover, with *Halobacterium* sp., hypotonic conditions may be used to lyse cells and release expressed proteins in a stable form.

One of the most interesting applications of the *Halobacterium* sp. NRC-1 expression system is the display of antigenic protein arrays on nanoparticles called gas vesicles (GVNPs), that are about 300 nm long, lemon-shaped buoyant organelles used by cells for flotation and are both bioengineerable and highly adjuvanting (DasSarma and Arora 1997; DasSarma et al. 2013; Stuart et al. 2001, 2004). GVNPs are composed only of proteins, one of which (GvpC) can be used for fusion to foreign antigenic proteins for display on the external surface of the nanoparticle. GVNPs have been shown to successfully display SIV, *Chlamydia,* and *Salmonella* antigens, and are processed slowly by macrophages (Childs and Webley 2012; DasSarma and DasSarma 2015; DasSarma et al. 2014; Sremac and Stuart 2008, 2010; Stuart et al. 2001, 2004). For nanoparticles displaying *Salmonella* SopB antigens, immunized mice were found to exhibit reduced bacterial load in organs (DasSarma et al. 2014).

Our recent interest has been to apply the *Halobacterium* sp. NRC-1 expression system for production of the major surface protein of sporozoites, the pre-erythrocytic form of *Plasmodium falciparum* which is transmitted from mosquitoes to humans. Among diseases for which there is a critical need for vaccines, malaria is one of the most widespread, with 200 million infections and over half a million deaths per year, primarily from infection by *P. falciparum* (http://who.int/malaria/publications/world_malaria_report_2013/en/index.html). The circumsporozoite protein, or CSP, is known to be highly immunogenic and is encoded by a single-copy gene unique to *Plasmodium* (Nussenzweig and Nussenzweig 1989). CSP consists of an N-terminal region containing a signal sequence, a central region with multiple tetrapeptide repeats, and a C-terminal cell-adhesion domain that ends with a putative glycosylphosphatidylinositol (GPI) anchor addition sequence (Fig. 25.1) (Coppi et al. 2011; Dame et al. 1984; del Portillo et al. 1987). The number of tetrapeptide repeats may vary from 25 to 49 copies, depending on the source (Bowman et al. 2013). A subunit vaccine, RTS,S, based on *P. falciparum* CSP consisting of most of the C-terminal half of the protein, including 19 copies of the tetrapeptide repeat fused to the hepatitis B surface protein, has been produced in *Pichia pastoris*, tested in different formulations with various immunostimulatory compounds, and shown to provide partial protection. However, protection is limited in infants, and in long term immunity (Campo et al. 2014; Mo and Augustine 2014). The ultimate potential of any pre-erythrocytic vaccine is underscored by reports of sterile immunity obtained with irradiated sporozoites administered intravenously (Seder et al. 2013).

We sought to utilize the *Halobacterium* sp. NRC-1 expression system for production of full-length *P. falciparum* CSP. In order to maximize the production of parasite protein in *Halobacterium* sp., we designed a codon-optimized gene and cloned it into expression vectors for production of the CSP protein alone or as a fusion to the GVNPs, including the N- and C-terminal regions not present in the mature protein in *Plasmodium*, which may contribute to enhanced immunogenicity or immunostimulation (Kastenmüller et al. 2013; Sedegah et al. 2013). In this report, we provide evidence for the biosynthesis of full-length CSP.

2 Materials and Methods

2.1 *Design and Cloning of Codon-Optimized* P. falciparum *CSP Gene*

To design a codon-optimized *P. falciparum* CSP gene for expression in *Halobacterium* sp. NRC-1, we obtained the CSP gene sequence from the *P. falciparum* strain 3D7 (NCBI reference number XM_001351086.1). The codon usage table for predicted genes in the genome of *Halobacterium* sp. NRC-1 was used to replace rare and infrequent codons (Kennedy et al. 2001). A codon modification (AAC>GGC) was incorporated in the design to add a unique *KasI* restriction site, which resulted in the mutation of an asparagine to glycine residue at amino acid position 201 (Fig. 25.1). The codon-optimized gene, flanked by *AfeI* restriction sites, was synthesized commercially for cloning into *Halobacterium* sp. NRC-1 expression vectors (Life Technologies, Carlsbad, CA, USA).

2.2 *Construction and Culturing of Expression Strains*

The *AfeI* restriction fragment coding the engineered and codon-optimized full-length CSP gene was cloned into the *Halobacterium* expression plasmids, pDRK and pSD, after cleavage of each at the unique *AfeI* sites in these vectors (Karan et al. 2014). The constructed pDRKcsp6 plasmid (Fig. 25.2a) was transformed into *Halobacterium* sp. strain SD109 (DasSarma et al. 1988, 1995), a derivative of wild-type strain NRC-1 deleted for the active *gvp* gene cluster A in pNRC100 and pNRC200, to construct *Halobacterium* strain SD109 (pDRKcsp6). The pSDcsp20 plasmid construct (Fig. 25.2b) was also transformed into SD109 to construct *Halobacterium* strain SD109 (pSDcsp20). The CSP gene region of each plasmid was completely sequenced in both strands using primer walking. *Halobacterium* strains transformed with expression plasmids were grown in CM^+ media with 20 μg/ml mevinolin (Merck, Sharp, and Dohme, Rahway, NJ, USA). Cultures were grown at 42 °C at 220 rpm to stationary phase. Cell lysates and GVNPs were prepared as described previously (DasSarma et al. 1995, 2013).

```
       AfeI                 Signal peptide
       AGCGCTATGATGCGCAAGCTCGCCATCCTCAGCGTCAGCAGCTTCCTCTTCGTCGAGGCC
   1         M  M  R  K  L  A  I  L  S  V  S  S  F  L  F  V  E  A     18

       CTCTTCCAGGAGTACCAGTGCTACGGCAGCAGCAGCAACACCCGCGTCCTCAACGAGCTC
  19   L  F  Q  E  Y  Q  C  Y  G  S  S  S  N  T  R  V  L  N  E  L     38

       AACTACGACAACGCCGGCACCAACCTCTACAACGAGCTCGAGATGAACTACTACGGCAAG
  39   N  Y  D  N  A  G  T  N  L  Y  N  E  L  E  M  N  Y  Y  G  K     58

       CAGGAGAACTGGTACAGCCTCAAGAAGAACAGCCGCAGCCTCGGCGAGAACGACGACGGC
  59   Q  E  N  W  Y  S  L  K  K  N  S  R  S  L  G  E  N  D  D  G     78

       AACAACGAGGACAACGAGAAGCTCCGCAAGCCCAAGCACAAGAAGCTCAAGCAGCCCGCC
  79   N  N  E  D  N  E  K  L  R  K  P  K  H  K  K  L  K  Q  P  A     98
                                  Tetrapeptide repeats
       GACGGCAACCCCGACCCCAACGCCAACCCCAACGTCGACCCCAACGCCAACCCCAACGTC
  99   D  G  N  P  D  P  N  A  N  P  N  V  D  P  N  A  N  P  N  V    118

       GACCCCAACGCCAACCCCAACGTCGACCCCAACGCCAACCCCAACGCCAACCCCAACGCC
 119   D  P  N  A  N  P  N  V  D  P  N  A  N  P  N  A  N  P  N  A    138

       AACCCCAACGCCAACCCCAACGCCAACCCCAACGCCAACCCCAACGCCAACCCCAACGCC
 139   N  P  N  A  N  P  N  A  N  P  N  A  N  P  N  A  N  P  N  A    158

       AACCCCAACGCCAACCCCAACGCCAACCCCAACGCCAACCCCAACGCCAACCCCAACGCC
 159   N  P  N  A  N  P  N  A  N  P  N  A  N  P  N  A  N  P  N  A    178

       AACCCCAACGCCAACCCCAACGCCAACCCCAACGCCAACCCCAACGCCAACCCCAACGTC
 179   N  P  N  A  N  P  N  A  N  P  N  A  N  P  N  A  N  P  N  V    198
            KasI
       GACCCCGGCGCCAACCCCAACGCCAACCCCAACGCCAACCCCAACGCCAACCCCAACGCC
 199   D  P  G  A  N  P  N  A  N  P  N  A  N  P  N  A  N  P  N  A    218

       AACCCCAACGCCAACCCCAACGCCAACCCCAACGCCAACCCCAACGCCAACCCCAACGCC
 219   N  P  N  A  N  P  N  A  N  P  N  A  N  P  N  A  N  P  N  A    238

       AACCCCAACGCCAACCCCAACGCCAACCCCAACGCCAACCCCAACGCCAACCCCAACGCC
 239   N  P  N  A  N  P  N  A  N  P  N  A  N  P  N  A  N  P  N  A    258

       AACCCCAACGCCAACCCCAACGCCAACCCCAACGCCAACCCCAACAAGAACAACCAGGGC
 259   N  P  N  A  N  P  N  A  N  P  N  A  N  P  N  K  N  N  Q  G    278

       AACGGCCAGGGCCACAACATGCCCAACGACCCCAACCGCAACGTCGACGAGAACGCCAAC
 279   N  G  Q  G  H  N  M  P  N  D  P  N  R  N  V  D  E  N  A  N    298

       GCCAACAGCGCCGTCAAGAACAACAACAACGAGGAGCCCAGCGACAAGCACATCAAGGAG
 299   A  N  S  A  V  K  N  N  N  N  E  E  P  S  D  K  H  I  K  E    318

       TACCTCAACAAGATCCAGAACAGCCTCAGCACCGAGTGGAGCCCCTGCAGCGTCACCTGC
 319   Y  L  N  K  I  Q  N  S  L  S  T  E  W  S  P  C  S  V  T  C    338

       GGCAACGGCATCCAGGTCCGCATCAAGCCCGGCAGCGCCAACAAGCCCAAGGACGAGCTC
 339   G  N  G  I  Q  V  R  I  K  P  G  S  A  N  K  P  K  D  E  L    358

       GACTACGCCAACGACATCGAGAAGAAGATCTGCAAGATGGAGAAGTGCAGCAGCGTCTTC
 359   D  Y  A  N  D  I  E  K  K  I  C  K  M  E  K  C  S  S  V  F    378
                                GPI anchor
       AACGTCGTCAACAGCAGCATCGGCCTCATCATGGTCCTCAGCTTCCTCTTCCTCAACTAA
 379   N  V  V  N  S  S  I  G  L  I  M  V  L  S  F  L  F  L  N  *    398
       AfeI
       AGCGCT
```

Fig. 25.1 Sequence of the synthetic codon-optimized *P. falciparum* CSP gene for expression in *Halobacterium* sp. and translation product. *AfeI* and *KasI* restriction sites are shown with recognition sequences *double-underlined*. Tryptic peptides of CSP identified by LC-MS/MS analysis are *underlined*. The signal peptide, tetrapeptide repeats, and the putative GPI anchor are boxed

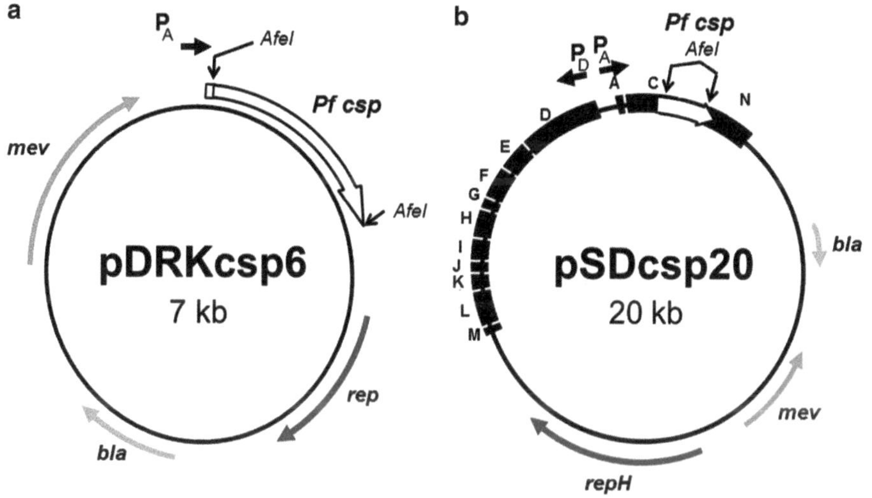

Fig. 25.2 *P. falciparum* CSP expression vectors pDRKcsp6 (**a**) and pSDcsp20 (**b**). The synthetic gene coding *P. falciparum* CSP (labeled *Pf csp*) is shown by *wide white arrows* between *AfeI* restriction sites used for cloning. Location of promoters are indicated by *narrow black arrows*. Locations of selection markers for mevinolin resistance (*mev*) and β-lactamase (*bla*) are shown as *narrow gray arrows*, as are locations of replicase genes (*rep*, *repH*). Location of the hexa-histidine tag is indicated by a small *white box* at the 5′-end of the CSP gene in pDRKcsp6 and locations of *gvpACN* and *gvpDEFGHIJKLM* genes are shown by *black boxes* in pSDcsp20

2.3 Western Blotting Analysis

Western blotting was carried out as previously described (Shukla and DasSarma 2004). Briefly, preparations of cell lysates were electrophoresed on 9 % or 12 % polyacrylamide-SDS gels. Proteins were transferred to 0.45 μm Immobilon-P polyvinylidene difluoride (PVDF) membranes (Millipore Corp., Boston, MA, USA). Membranes were blocked with PTM buffer (PBS, 0.1 % Tween 20, 5 % nonfat dry milk), followed by overnight incubation at 4 °C after addition of the CSP-specific monoclonal antibody 2A10 (1:1,000 dilution) (Wirtz et al. 1987). Membranes were washed with PTM buffer, and incubated with rabbit anti-mouse secondary antibodies (1:2,500 dilution) labeled with alkaline phosphatase (Sigma Aldrich, St. Louis, MO, USA). For detection of immunoreactive proteins, membranes were incubated in 1-Step NBT/BCIP substrate per manufacturer's specifications (Thermo Fisher Scientific, Rockford, IL, USA). Bands were quantified by densitometry using Image J (http://imagej.nih.gov/ij/).

2.4 Proteomic Analysis

To identify CSP by LC-MS/MS analysis, *Halobacterium* proteins were reduced with DTT, alkylated with iodoacetamide, and subsequently proteolyzed with trypsin (Shevchenko et al. 1996). Digested peptides were desalted by C18 stage-tip and eluted with 0.1 % trifluoroacetic acid in 60 % acetonitrile and resuspended in 0.1 % formic acid for LC–MS/MS analysis. Identification of peptides was performed on a Q-Exactive instrument (Thermo Scientific, Rockford, IL, USA) interfaced with a Proxionnano flow LC system. Peptide sequences were identified from isotopically resolved masses in MS and MS/MS spectra extracted with and without deconvolution using a Thermo Scientific MS2 processor and Xtract software. MS/MS spectra were searched against protein databases using the Sequest search engine interfaced with Proteome Discoverer 1.4 software.

3 Results

3.1 *Construction of* Halobacterium *sp. Expression Strains Containing the Codon-Optimized* P. falciparum *CSP Gene*

Due to the great difference in GC-composition between *P. falciparum* and *Halobacterium* sp., (19 % versus 66 % GC, respectively), we designed a synthetic CSP gene for expression in *Halobacterium* using the codon usage data from the genome of the wild-type NRC-1 strain (Fig. 25.1). The synthetic gene design contained 62.3 % GC, incorporating the full-length CSP coding region from *P. falciparum* strain 3D7, including the N-terminal signal and C-terminal anchor sequence, and was flanked by *AfeI* restriction sites for cloning.

For construction of expression plasmids, the synthetic CSP gene was inserted into plasmid pDRK at an *AfeI* site downstream of the *gvp*A promoter, resulting in plasmid pDRKcsp6 (Fig. 25.2a). The CSP encoded by pDRKcsp6 has an N-terminal 15 amino acid residue extension (MHHHHHHLKRLPRSA) containing a hexa-histidine tag, followed by the 397 amino acid residue full-length CSP (Fig. 25.1). The synthetic CSP gene was also cloned into expression plasmid pSD at the *AfeI* site in *gvp*C for expression of CSP and display on GVNPs in *Halobacterium* (Fig. 25.2b). The resulting pSDcsp20 encodes a fusion protein with 293 amino acids of GvpC followed by the full-length CSP.

3.2 *Expression of Full-Length* P. falciparum *CSP in* Halobacterium *sp.*

The expression plasmids, pDRKcsp6 and pSDcsp20, were transformed into *Halobacterium* sp. strain SD109, a gas vesicle-deficient mutant of the wild-type strain. Cultures were grown to stationary phase, collected by centrifugation, and lysed

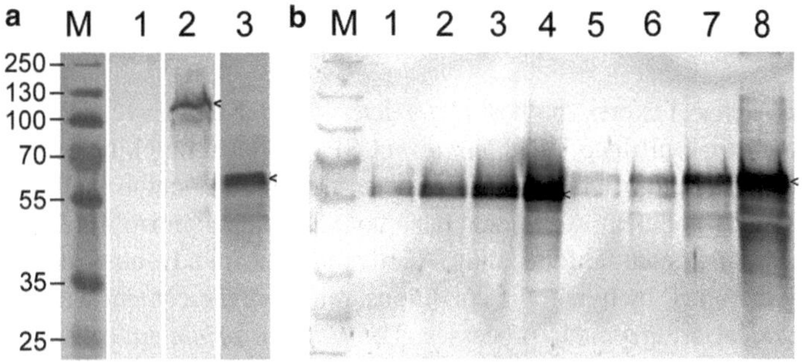

Fig. 25.3 Western blotting analysis of CSP expression strains. *Halobacterium* sp. SD109, SD109 (pSDcsp20), and SD109 (pDRKcsp6) (**a**, lanes 1–3, respectively) and *P. falciparum* sporozoite lysates (**b**, lanes 1–4, 1.9×10^2, 7.5×10^2, 3×10^3, and 1.2×10^4 cells, respectively) and *Halobacterium* sp. SD109 (pDRKcsp6) lysates (**b**, lanes 5–8, 8×10^4, 3×10^5, 1.3×10^6, and 5×10^6 cells, respectively), probed using the CSP monoclonal antibody 2A10. Lanes M contain molecular weight markers, with sizes indicated in kDa. *Arrowheads* indicate the prominent CSP band

using hypotonic conditions. Growth rates of the transformants were comparable to that of the parental strain (data not shown). Cell lysates were fractionated by polyacrylamide gel electrophoresis, transferred, and probed using the CSP monoclonal antibody 2A10. Protein bands corresponding to CSP at the expected positions, 120 kDa for strain SD109 (pSDcsp20) and 60 kDa for strain SD109 (pDRKcsp6), were found (Fig. 25.3a). Proteins of the *Halobacterium* sp. SD109 (pDRKcsp6) were also subjected to trypsin digestion followed by LC-MS/MS. Five unique peptides corresponding to *P. falciparum* CSP were observed, confirming the identity of the expressed protein (Fig. 25.1). The peptides identified covered 64/397 amino acids or 16 % of CSP; it should be noted that the entire tetrapeptide repeat region is devoid of trypsin sites and would not be expected to be detected by this method.

We determined the level of production of CSP in *Halobacterium* sp. by Western blotting analysis using *P. falciparum* sporozoite lysates as standards. Sporozoites were prepared from disrupted salivary glands of *Anopheles stephensi* adult females. Measured numbers of *P. falciparum* sporozoites and *Halobacterium* sp. SD109 (pDRKcsp6) cells were lysed, fractionated by SDS-PAGE, transferred, and probed with the CSP monoclonal antibody 2A10 (Fig. 25.3b). The CSP band in the sporozoites was observed at 55 kDa (lanes 1–4), while the free CSP band in *Halobacterium* was observed at 60 kDa (lanes 5–8), which is larger due to the presence of the signal sequence, the putative GPI anchor, and the hexa-histidine tag. Based on the intensity of these major bands apparent in the Western blot, we determined that 10^3 *P. falciparum* sporozoites produce similar quantities of CSP as 8.3×10^5 *Halobacterium* sp. SD109 (pDRKcsp6) cells. This level of expression represents 3.3 mg of free CSP per liter *Halobacterium* culture. The level of CSP fusion protein for display on GVNPs, observed as a 120 kDa protein, corresponds to about 15 % of this amount or 0.5 mg of CSP per liter of culture (Fig 25.3a and data not shown).

4 Discussion

Using the Archaeal expression host *Halobacterium* sp. NRC-1, we have documented the expression of full-length *P. falciparum* CSP. This extremophilic expression system provides a yield of CSP sufficient for immunological studies and the potential for vaccine development. Additional benefits include biocompatibility, non-toxicity, lack of LPS, and potential for scaling. Another significant advantage is that the cells are easily disrupted by hypotonic conditions, releasing the content without the need for mechanical or enzymatic processes. The *Halobacterium* sp. cells contain large quantities of intracellular GVNPs, which have been shown to be highly adjuvanting, eliciting strong and long-lived immune responses (Childs and Webley 2012; DasSarma and DasSarma 2015; DasSarma et al. 2013, 2014; Sremac and Stuart 2008, 2010; Stuart et al. 2001, 2004). The expression system permits CSP to be produced freely or potentially bound and displayed as an ordered array on the GVNPs. The CSP protein produced in *Halobacterium* is recognized by the 2A10 CSP monoclonal antibody, indicating that its antigenicity is retained in this host. These results may allow the development of multiple formulations for immunological testing and vaccine development in the future.

CSP sequences are available from diverse *Plasmodium* strains and several have been previously expressed in *Escherichia coli*, *P. pastoris*, yeast and other organisms (Dame et al. 1984; Kastenmüller et al. 2013; Kolodny et al. 2001; Plassmeyer et al. 2009; Young et al. 1985). However, in nearly all of these cases, successful expression has employed truncated forms of CSP lacking either the N-terminal signal peptide, the C-terminal putative GPI anchor regions, or both. Early attempts to generate full-length CSP in *E. coli* resulted in the expression of unstable products at low levels (Young et al. 1985). In our case, the entire CSP molecule, either free or fused to GvpC, is produced in *Halobacterium* sp. The presence of significant quantities of the entire CSP molecule offers a potentially attractive alternative to the other expression systems.

In a recent study, near full-length CSP (lacking the signal sequence and putative GPI anchor sequence) was shown to elicit a stronger antibody reaction in mice and non-human primates compared to the RTS,S vaccine that lacks the entire N-terminus of the CSP (Kastenmüller et al. 2013). Adjuvants were shown to differentially mediate effects on antibody and Th1 immunity as well as CD4+ T cell immunity and protection in mice. Haloarchaeal GVNPs may provide a significant boost to immune responses against displayed antigens, as suggested when *Salmonella* SopB antigen-GVNPs successfully reduced bacterial burden upon administration with live attenuated bacteria after challenge (DasSarma et al. 2014). Immunostimulatory effects of SopB-GVNPs were confirmed by significant increases in levels of IFN-γ, IL-2, and IL-9 in immunized mice.

The additional sequences present in full-length CSP produced in *Halobacterium* sp. that are lacking in the mature sporozoite CSP may potentially be either immunogenic or immunostimulatory in ways not observed with the truncated proteins. Consistent with this notion, the signal peptide of CSP contains epitopes recognized by CD8+ T cells of individuals from malaria endemic areas (Sedegah et al. 2013). Moreover, a combination of cell mediated (i.e. CD4+ Th1, CD8+ T cell) and humoral responses have been shown to be critical to provide protective immunity (Ménard et al. 2013; Schofield et al. 1987). Additional studies are necessary to determine the immunogenic effects of the full-length CSP produced in *Halobacterium* free or displayed on GVNPs.

The biocompatibility of the *Halobacterium* cells and GVNPs represents another significant distinguishing feature of this system (DasSarma et al. 2010). Furthermore, antigens fused to GNVPs were reported to be stable without refrigeration in *Halobacterium* cells for extended periods of time, suggesting that this platform is a valuable alternative for vaccines targeting diseases in developing countries (DasSarma et al. 2010, 2014). Although the ability to partially purify GVNPs by flotation may be valuable, the potential for direct use of CSP-containing cells or cell lysates can also be pursued and may offer advantages for immunogenicity. However, since in some cases, hexa-histidine tagged N-termini could facilitate purification of expressed proteins in other hosts, the inclusion of this affinity tag may be utilized, should purification be desired (Kolodny et al. 2001; Plassmeyer et al. 2009). Finally, the *Halobacterium* sp. CSP expression system may also be modified to express variant proteins, including shorter or truncated forms, should those become needed in future.

Conflict of Interest Wolf T. Pecher, Jong-Myoung Kim, Priya DasSarma, Ram Karan, Photini Sinnis, and Shiladitya DasSarma declare that they have no conflict of interest.

Acknowledgments This work was supported by Bill & Melinda Gates Foundation grant OPP1061509, and National Institutes of Health grant R03 AI107634 to SD and National Institutes of Health grant R01 AI056840 to PS. JK was supported by the PKNU Research Abroad Fund CD-2013-0914. We thank Susan Barnes and Folasade Ekulona for technical assistance, Stefanie Trop and Peter Dumoulin for valuable discussions, and Prof. F. Zavala for critical reading of the manuscript.

References

Berquist BR, Müller JA, DasSarma S (2006) Chapter 27. Genetic systems for halophilic archaea. In: Oren A, Rainey F (eds) Methods in microbiology, vol 35. Elsevier/Academic Press, Amsterdam, pp 649–680

Bowman NM, Congdon S, Mvalo T, Patel JC, Escamilla V, Emch M, Martinson F, Hoffman I, Meshnick SR, Juliano JJ (2013) Comparative population structure of *Plasmodium falciparum* circumsporozoite protein NANP repeat lengths in Lilongwe, Malawi. Sci Rep 3:1990

Campo JJ, Sacarlal J, Aponte JJ, Aide P, Nhabomba AJ, Dobaño C, Alonso PL (2014) Duration of vaccine efficacy against malaria: 5th year of follow-up in children vaccinated with RTS, S/AS02 in Mozambique. Vaccine 32(19):2209–2216. doi:10.1016/j.vaccine.2014.02.042

Childs TS, Webley WC (2012) In vitro assessment of halobacterial gas vesicles as a *Chlamydia* vaccine display and delivery system. Vaccine 30(41):5942–5948

Coppi A, Natarajan R, Pradel G, Bennett BL, James ER, Roggero MA, Corradin G, Persson C, Tewari R, Sinnis P (2011) The malaria circumsporozoite protein has two functional domains, each with distinct roles as sporozoites journey from mosquito to mammalian host. J Exp Med 208(2):341–356

Dame JB, Williams JL, McCutchan TF, Weber JL, Wirtz RA, Hockmeyer WT, Maloy WL, Haynes JD, Schneider I, Roberts D, Sanders GS, Reddy EP, Diggs CL, Miller LH (1984) Structure of the gene encoding the immunodominant surface antigen on the sporozoite of the human malaria parasite *Plasmodium falciparum*. Science 225(4662):593–599

DasSarma S (2004) Genome sequence of an extremely halophilic archaeon. In: Fraser CM, Read T, Nelson KE (eds) Microbial genomes. Humana Press, Totowa, pp 383–399

DasSarma S (2007) Extreme microbes: the salty side of life. Am Sci 95(3):224–231

DasSarma S, Arora P (1997) Genetic analysis of the gas vesicle gene cluster in haloarchaea. FEMS Microbiol Lett 153(1):1–10

DasSarma S, DasSarma P (2015) Gas vesicle nanoparticles for antigen display. Vaccines 3(3):686–702

DasSarma S, Halladay JT, Jones JG, Donovan JW, Giannasca PJ, Tandeau de Marsac N (1988) High-frequency mutations in a plasmid-encoded gas vesicle gene in *Halobacterium halobium*. Proc Natl Acad Sci U S A 85(18):6861–6865

DasSarma S, Robb FT, Place AR, Sowers KR, Schreier HJ, Fleischmann EM (eds) (1995) Archaea: a laboratory manual – halophiles. Cold Spring Harbor Laboratory Press, Plainview

DasSarma P, Coker JA, Huse V, DasSarma S (2010) Halophiles, industrial applications. In: Flickinger MC (ed) Encyclopedia of industrial biotechnology: bioprocess, bioseparation, and cell technology. Wiley, Hoboken, pp 1–43

DasSarma S, Karan R, DasSarma P, Barnes S, Ekulona F, Smith B (2013) An improved genetic system for bioengineering buoyant gas vesicle nanoparticles from haloarchaea. BMC Biotechnol 13:112. doi:10.1186/1472-6750-13-112

DasSarma P, Negi VD, Balakrishnan A, Karan R, Barnes S, Ekulona F, Chakravortty D, DasSarma S (2014) Haloarchaeal gas vesicle nanoparticles displaying *Salmonella* SopB antigen reduce bacterial burden when administered with live attenuated bacteria. Vaccine 32(35):4543–4549

del Portillo HA, Nussenzweig RS, Enea V (1987) Circumsporozoite gene of a *Plasmodium falciparum* strain from Thailand. Mol Biochem Parasitol 24(3):289–294

Karan R, Capes MD, DasSarma P, DasSarma S (2013) Cloning, overexpression, purification, and characterization of a polyextremophilic β-galactosidase from the Antarctic haloarchaeon *Halorubrum lacusprofundi*. BMC Biotechnol 13:3. doi:10.1186/1472-6750-13-3

Karan R, DasSarma P, Balcer-Kubiczek E, Weng RR, Liao C-C, Goodlett DR, Ng WV, DasSarma S (2014) Bioengineering radioresistance by overproduction of RPA, a mammalian-type single-stranded DNA-binding protein, in a halophilic archaeon. Appl Microbiol Biotechnol 98(4): 1737–1747

Kastenmüller K, Espinosa DA, Trager L, Stoyanov C, Salazar AM, Pokalwar S, Singh S, Dutta S, Ockenhouse CF, Zavala F, Seder RA (2013) Full-length *Plasmodium falciparum* circumsporozoite protein administered with long-chain poly(I·C) or the Toll-like receptor 4 agonist glucopyranosyl lipid adjuvant-stable emulsion elicits potent antibody and CD4+ T cell immunity and protection in mice. Infect Immun 81(3):789–800

Kennedy SP, Ng WV, Salzberg SL, Hood L, DasSarma S (2001) Understanding the adaptation of *Halobacterium* species NRC-1 to its extreme environment through computational analysis of its genome sequence. Genome Res 11:1641–1650

Kolodny N, Kitov S, Vassell MA, Miller VL, Ware LA, Fegeding K, De La Vega P, Sacci JB Jr, Lanar DE (2001) Two-step chromatographic purification of recombinant *Plasmodium falciparum* circumsporozoite protein from *Escherichia coli*. J Chromatogr B Biomed Sci Appl 762(1):77–86

Ménard R, Tavares J, Cockburn I, Markus M, Zavala F, Amino R (2013) Looking under the skin: the first steps in malarial infection and immunity. Nat Rev Microbiol 11(10):701–712

Mo AX, Augustine AD (2014) NIAID meeting report: improving malaria vaccine strategies through the application of immunological principles. Vaccine 32(10):1132–1138

Ng WV, Ciufo SA, Smith TM, Bumgarner RE, Baskin D, Faust J, Hall B, Loretza C, Seto J, Slagel J, Hood L, DasSarma S (1998) Snapshot of a large dynamic replicon in a halophilic archaeon: megaplasmid or minichromosome? Genome Res 8:1131–1141

Ng WV, Kennedy SP, Mahairas GG, Berquist B, Pan M, Shukla HD, Lasky SR, Baliga NS, Thorsson V, Sbrogna J, Swartzell S, Weir D, Hall J, Dahl TA, Welti R, Goo YA, Leithauser B, Keller K, Cruz R, Danson MJ, Hough DW, Maddocks DG, Jablonski PE, Krebs MP, Angevine CM, Dale H, Isenbarger TA, Peck RF, Pohlschroder M, Spudich JL, Jung K-H, Alam M, Freitas T, Hou S, Daniels CJ, Dennis PP, Omer AD, Ebhardt H, Lowe TM, Liang P, Riley M, Hood L, DasSarma S (2000) Genome sequence of *Halobacterium* species NRC-1. Proc Natl Acad Sci U S A 97(22):12176–12181

Nussenzweig V, Nussenzweig RS (1989) Rationale for the development of an engineered sporozoite malaria vaccine. Adv Immunol 45:283–334

Plassmeyer ML, Reiter K, Shimp RL Jr, Kotova S, Smith PD, Hurt DE, House B, Zou X, Zhang Y, Hickman M, Uchime O, Herrera R, Nguyen V, Glen J, Lebowitz J, Jin AJ, Miller LH, MacDonald NJ, Wu Y, Narum DL (2009) Structure of the *Plasmodium falciparum* circumsporozoite protein, a leading malaria vaccine candidate. J Biol Chem 284:26951–26963

Schofield L, Villaquiran J, Ferreira A, Schellekens H, Nussenzweig R, Nussenzweig V (1987) γ interferon, CD8+ T cells and antibodies required for immunity to malaria sporozoites. Nature 330(6149):664–666

Sedegah M, Kim Y, Ganeshan H, Huang J, Belmonte M, Abot E, Banania JG, Farooq F, McGrath S, Peters B, Sette A, Soisson L, Diggs C, Doolan D, Tamminga C, Villasante E, Hollingdale MR, Richie T (2013) Identification of minimal human MHC-restricted CD8+ T-cell epitopes within the *Plasmodium falciparum* circumsporozoite protein (CSP). Malar J 12:185

Seder RA, Chang LJ, Enama ME, Zephir KL, Sarwar UN, Gordon IJ, Holman LA, James ER, Billingsley PF, Gunasekera A, Richman A, Chakravarty S, Manoj A, Velmurugan S, Li ML, Ruben AJ, Li T, Eappen AG, Stafford RE, Plummer SH, Hendel CS, Novik L, Costner PJM, Mendoza FH, Saunders JG, Nason MC, Richardson JH, Murphy J, Davidson SA, Richie T, Sedegah M, Sutamihardja A, Fahle GA, Lyke KE, Laurens MB, Roeder M, Tewari K, Epstein JE, Sim BKL, Ledgerwood JE, Graham BS, Hoffman SL, the VRC 312 Study Team (2013) Protection against malaria by intravenous immunization with a nonreplicating sporozoite vaccine. Science 341(6152):1359–1365

Shevchenko A, Wilm M, Vorm O, Mann M (1996) Mass spectrometric sequencing of proteins silver-stained polyacrylamide gels. Anal Chem 68(5):850–858

Shukla HD, DasSarma S (2004) Complexity of gas vesicle biogenesis in *Halobacterium* sp. strain NRC-1: identification of five new proteins. J Bacteriol 186(10):3182–3186

Sremac M, Stuart ES (2008) Recombinant gas vesicles from *Halobacterium* sp. displaying SIV peptides demonstrate biotechnology potential as a pathogen peptide delivery vehicle. BMC Biotechnol 8:9

Sremac M, Stuart ES (2010) SIVsm Tat, Rev, and Nef1: functional characteristics of r-GV internalization on isotypes, cytokines, and intracellular degradation. BMC Biotechnol 10:54

Stuart ES, Sremac M, Morshed F, DasSarma S (2001) Antigen presentation using novel particulate organelles from halophilic archaea. J Biotechnol 88(2):119–128

Stuart ES, Morshed F, Sremac M, DasSarma S (2004) Cassette-based presentation of SIV epitopes with recombinant gas vesicles from halophilic archaea. J Biotechnol 114(3):225–237

Wirtz RA, Zavala F, Charoenvit Y, Campbell GH, Burkot TR, Schneider I, Esser KM, Beaudoin RL, Andre RG (1987) Comparative testing of monoclonal antibodies against *Plasmodium falciparum* sporozoites for ELISA development. Bull World Health Organ 65(1):39–45

Young JF, Hockmeyer WT, Gross M, Ballou WR, Wirtz RA, Trosper JH, Beaudoin RL, Hollingdale MR, Miller LH, Diggs CL, Rosenberg M (1985) Expression of *Plasmodium falciparum* circumsporozoite proteins in *Escherichia coli* for potential use in a human malaria vaccine. Science 228(4702):958–962

Index

P.H. Rampelotto (ed.), *Biotechnology of Extremophiles*, Grand Challenges in Biology and Biotechnology 1, DOI 10.1007/978-3-319-13521-2

MIX
Papier aus verantwortungsvollen Quellen
Paper from responsible sources
FSC® C105338

If you have any concerns about our products,
you can contact us on
ProductSafety@springernature.com

In case Publisher is established outside the EU,
the EU authorized representative is:
Springer Nature Customer Service Center GmbH
Europaplatz 3, 69115 Heidelberg, Germany

Printed by Libri Plureos GmbH
in Hamburg, Germany